Deep Water Canyons, Fans and Facies: Models for Stratigraphic Trap Exploration

Papers reprinted from the AAPG Bulletin,
Special Publications and Section Publications

Compiled by

Roderick W. Tillman
Cities Service Company
Tulsa, Oklahoma

Syed A. Ali
Gulf Oil Exploration & Production Company
New Orleans, Louisiana

Published September 1982 by
The American Association of Petroleum Geologists
Tulsa, Oklahoma, U.S.A.

Published September 1982
Library of Congress Catalog Card No: 82-70530
ISBN: 0-89181-547-3

Printed by
Edwards Brothers, Inc.
Ann Arbor, Michigan

Contents:

Introduction — Section 1

Hydrocarbons are and have been actively explored for in deep-water clastic sediments for over 40 years. Only in the last 15 years have a variety of models for deep-water exploration emerged. Walker (1978) and Normark (1970, 1978) have outlined several models useful for subsurface exploration and facies analysis. Walker's work is based almost entirely on data from ancient sediments, while Normark's work is based almost entirely on data from recent sediments. As can be seen in the reply by Nilsen (1980), there is not as yet unanimity of opinion on which models have the most universal application.

The papers included herein deal with hydrocarbon production from a variety of models and geometries of deep-water facies. Sullwold's (1961) paper was an early effort to suggest a methodology for hydrocarbon exploration. McPhearson (1978) and Weagant (1972) discuss specific prolific California deep water units, the Stevens (Miocene) and the Forbes (Upper Cretaceous) sandstones. Heretier et al (1979), and Thomas et al (1974), discuss the North Sea fields which produce from sediments deposited in deep water.

Cores of turbidites taken during deep sea drilling projects in the Atlantic, Pacific, and Indian oceans have been assembled at various locations. Descriptions of some backarc turbidite cores from the Tertiary of Alaska are made by Stewart (1977). Ericson et al (1952), described a number of Pleistocene and recent deep-sea turbidite cores from the North Atlantic. An upper Miocene fan deposit exposed onshore in southern California is described in the paper by Sullwold (1960).

Taken as a group, the papers included in this and the companion section are a cross section of the type of work done on deep water sediment during the last 30 years.

An extensive geographical listing of references on deep-water deposits is included at the back of this volume. The geographic areas are subdivided by continents and/or divided by basin or into other appropriate smaller geographic areas. A second bibliography consisting of an extensive list of references on deep-water deposits categorized alphabetically by author is included as well. Many papers from guidebooks which are not readily available are also included. References are believed to be nearly complete through 1980.

Roderick W. Tillman
Syed A. Ali

Reprinted for private circulation from
THE AMERICAN ASSOCIATION OF PETROLEUM GEOLOGISTS BULLETIN
Vol. 62, No. 6, June, 1978

Deep-Water Sandstone Facies and Ancient Submarine Fans: Models for Exploration for Stratigraphic Traps[1]

ROGER G. WALKER[2]

Abstract Five main facies of deep-water clastic rocks can be defined: classic turbidites, massive sandstones, pebbly sandstones, conglomerates, and debris flows (with slumps and slides). The classic turbidites consist of monotonously parallel-interbedded sandstones and shales without channeling; internal sedimentary structures include grading, parallel lamination, and cross-lamination. Massive sandstones are thicker, coarser, and commonly channelized. They lack the sedimentary structures of classic turbidites, but do contain evidence of dewatering during deposition. Pebbly sandstones tend to be well graded, and can contain parallel stratification and large-scale cross-stratification. Conglomerates are characterized by inverse and normal grading, parallel and cross-stratification, and commonly have a preferred clast fabric (imbrication). Both the pebbly sandstones and conglomerates commonly are channelized.

The facies can be fitted into a model of submarine-fan deposition. Modern fans are subdivided into an upper fan (suprafan), characterized by (1) a single deep channel with levees, (2) a middle fan, built up from suprafan lobes that periodically switch in position, and (3) a topographically smooth lower fan. The suprafan lobes have shallow, braided channels on their inner parts, but the outer suprafan lobes are smooth, and grade basinward into the smooth lower fan and basin plain.

The smooth suprafan lobes and lower fan are characterized by deposition of the classic turbidite facies, and the braided part of the suprafan lobes by massive and pebbly sandstones. When one lobe is abandoned and another starts to prograde elsewhere, the first lobe is blanketed by mud, forming a potential stratigraphic trap. The upper-fan channel is an area of coarse sediment deposition, or conglomerates where gravel and boulders are supplied to the basin. During fan progradation, thickening- and coarsening-upward facies sequences can be formed in a manner analogous to those of deltas. Fan channels also can be abandoned progressively, forming thinning- and fining-upward sequences similar to those of fluvial or distributary channels. These sequences can be identified on electric logs.

Where basin shales act as hydrocarbon-source areas, the classic turbidites can act as conduits, leading the hydrocarbons to the thicker, laterally coalesced massive and pebbly sandstones of the braided suprafan lobes. These bodies can be of the order of 25 km in diameter, and up to 100 m thick. The coarse deposits of the upper-fan channel also might form good reservoirs, being bounded by shales (levee deposits) on either side, and possibly by shales above if the fan-channel system is abandoned. Such channels can be tens of kilometers long, several kilometers wide, and a few hundred meters deep. Reservoirs may be present in all of these environments.

INTRODUCTION

Prolific reservoirs in the Los Angeles and Ventura basins, the Great Valley of California, and parts of the Texas and Louisiana Coastal Plain, among others, are producing from deep-water sandstones. The family of deep-water sandstones includes classic turbidites together with other coarser grained facies such as pebbly sandstones and conglomerates. The feature that they all share in common is that of depositional setting, because they all accumulated at one time as unstable piles of loose sediment in shallow (wave agitated) water, and all subsequently were resedimented by gravity into deeper water (consistently below storm wave base). The term "resedimented" implies no specific transportation process, but embraces everything from fully turbulent turbidity currents to debris flows moving as semirigid plugs.

This paper will review the various sandstones and conglomerates that geologists recognize as deep-water deposits. The entire suite of rocks belongs to the "resedimented coarse-clastic family," but excludes oceanic mudstones, oozes, and related fine-grained deposits. The various members of the family will be related to depositional environments known in modern submarine fans (Normark, this issue of AAPG *Bull.*), and near the end of this paper, I will suggest how the submarine-fan model may be used in petroleum exploration.

[1]Manuscript received, May 16, 1977; accepted, November 16, 1977.

[2]Department of Geology, McMaster University, Hamilton, Ontario, Canada.

This manuscript is an outgrowth of an SEPM turbidite short course held at Anaheim, California, in 1973, and a New Orleans Geological Society short course in 1976. The ideas also have been presented at AAPG Continuing Education courses. I therefore thank a very large number of colleagues for their comments and suggestions, particularly Bill Normark for his information on modern fans, Keith Skipper for his detailed comments on the manuscript, Frank E. Weagant for his comments on the Grimes gas field, and Emiliano Mutti and his Italian colleagues for initiating a formal classification scheme for turbidites and related deposits. I thank the National Research Council of Canada for continuing support of this work.

Article Identification Number
0149-1423/78/B006-0001$03.00/0

1

As well as the basins mentioned, turbidite reservoirs have been recognized in the southern fringes of the Hackberry wedge (Louisiana), and prospecting for more turbidite reservoirs is currently important in offshore southern California, and in the North Sea (Thomas et al, 1974; Fowler, 1975; Parker, 1975) where the giant Forties and Montrose fields are producing from turbidites with good reservoir qualities associated with upper and middle submarine-fan environments (Skipper, 1977).

With so much oil and gas in turbidites, it is unfortunate that so few producing fields have been described in detail. Important studies of areas involving turbidites and deep-water channels associated with hydrocarbon production include the Los Angeles basin (Barbat, 1958; Yerkes et al, 1965; Mayuga, 1970; Gardett, 1971), the Ventura basin (Nagle and Parker, 1971; Hsü, 1977), the Great Valley of California (Sullwold, 1961; Martin, 1963), Sacramento Valley (Edmondson, 1965; Dickas and Payne, 1967; Weagant, 1972), southern and offshore Louisiana (Paine, 1966; Sabate, 1968; Benson, 1971), Texas (Hoyt, 1959), and Pennsylvania (Dixon, 1972). Some of these basins will be examined in more detail later, after a discussion of the facies in the resedimented coarse-clastic family and their relation to submarine-fan depositional systems.

RESEDIMENTED COARSE-CLASTIC FAMILY

The purpose of this section of the paper is to introduce the members of the family, as presently conceived. There have been various classification schemes and jargon terms used (turbidites, fluxoturbidites, grain flow deposits, neptunites, etc.), but I now believe that quite a simple scheme will suffice as a framework for understanding the relations between members of the family. The scheme presented is a simplification of that given by Walker and Mutti (1973), which in turn was based on the excellent synthesis of a large amount of information by Mutti and his Italian colleagues (Mutti and Ricci Lucchi, 1972). The scheme also has the advantage of an increasingly sound experimental and theoretical basis (Middleton and Hampton, 1976).

The most important members of the family are: (1) classic turbidites, (2) massive sandstones, (3) pebbly sandstones, (4) clast-supported conglomerates, and (5) matrix-supported beds (debris flows, pebbly mudstones, slumps).

The first four members are believed now to have been deposited from flows in which fluid turbulence was important as a grain- and clast-supportive mechanism. During the final stages of deposition from the flow, other mechanisms may take over and develop a characteristic suite of sedimentary structures in each facies. The fifth facies includes all the matrix-supported beds (debris flow deposits, pebbly mudstones, and the like); during transport, fluid turbulence was a much less important mechanism.

Classic Turbidites

The features that distinguish classic turbidites (Figs. 1, 2) from other members of the resedimented family are: (1) very parallel bedding, with consistent alternations of sandstone and shale (normally without channeling or major changes in bed thickness laterally); and (2) a consistent set of internal sedimentary structures that can be described using the Bouma (1962) model (Fig. 3).

The term "classic" turbidite is applied because there has been very extensive study of such beds since the concept of turbidites and turbidity currents first was introduced (Kuenen and Migliorini, 1950; Natland and Kuenen, 1951). As a result, there is extensive general agreement among sedimentologists as to characteristics that define such beds—they have become "classic." These features include: (1) a suite of erosional markings associated with the sharp base of each sandstone bed (sole marks); (2) a suite of internal sedimentary structures within each sandstone bed (including overall graded bedding, with horizontal lamination and ripple cross-lamination); (3) a covering pelitic layer on top of each sandstone that, in outcrop, gives the characteristic monotonous alternation of sand-shale-sand-shale-sand-shale (Figs. 1, 2); (4) a bedding regularity such that individual beds can be traced for hundreds or thousands of meters laterally without appreciable thickness changes (Figs. 1, 2).

Sole marks are prominent in outcrops of classic turbidites, but rarely would be visible in cores. They are grouped into two types, tool marks, carved into the underlying substrate by tools (sticks, larger clasts, etc.) in the current, and scour marks, cut into the substrate by fluid scour alone. Both types of markings are vitally important in determining local and regional paleoflow patterns in basins. In oriented cores, grain orientation can be used equally successfully in determining flow directions (Hsü, 1977), and this technique will be discussed later. The sharp sandstone bases with scour and/or tool marks indicate the sudden and erosive appearance of a turbidity current in an area of former mud deposition in very quiet water.

Within the sandstone layer, classic turbidites contain a dazzling array of sedimentary structures (Dzulynski and Walton, 1965). The most important of these were grouped into a sequence

FIG. 1—Thin-bedded turbidite facies, Cretaceous rocks at Buellton, California.

FIG. 2—Proximal turbidites, Eocene Matilija Formation, north of Ojai, California. As compared with Figure 1, much higher sand/shale ratio, and much thicker individual sandstone beds. Stratigraphic top on left; figure circled for scale.

by Bouma (1962; Fig. 3), and they form the basis for the classic turbidite predictive model (Walker, 1976a, b). Bouma's division A consists of massive (structureless) or graded sandstone. Several studies have shown that within division A there is a preferred grain orientation (Spotts and Weser, 1964; Colburn, 1968; Onions and Middleton, 1968; Parkash and Middleton, 1970) with grains commonly oriented at some angle to the flow directions independantly determined from sole marks. Spotts and Weser (1964, p. 218) found a grain orientation that diverged from the sole-mark orientation by an average of about 45° counterclockwise, but Onions and Middleton (1968) found no consistent relation between grain orientation and sole marks; in fact, grain orientation varied as much as 90° on either side of the sole-mark directions. However, Colburn's (1968) data show a much closer agreement between grain orientations and sole marks, and Parkash and Middleton (1970) demonstrated that the grain orientations measured near the base of the bed deviated least from the sole-mark directions. They also showed increasing divergence upward through a bed. Onions and Middleton (1968) gave a useful summary of techniques, statistical procedures, and test for operator errors.

Grain orientation has been used as a paleoflow indicator in several producing fields. One of the best examples is from the Ventura field, California, where Hsü (1977, p. 149) demonstrated an east-west preferred grain fabric in oriented cores, with a presumed westerly flow for the Repetto turbidites. The flow direction agreed closely with that indicated by ripple cross-lamination in the same cores.

Bouma's division B is characterized by horizontal lamination, normally in medium- to fine-sand sizes. If the bed can be split along lamination planes, parting lineation commonly can be seen, reflecting the excellent grain orientation associated with this type of horizontal lamination (Allen, 1964). There is no extensive report in the literature of the relation between grain orientation in division B and sole-mark directions; where division B parting lineation can be compared with sole marks in the field, the directions commonly are very similar. In oriented cores, I tentatively would recommend a grain-orientation study of division B rather than division A, simply because the standard deviation about the vector mean-grain orientation is normally less in division B (i.e., the grain orientation is "better developed").

Division C is characterized by ripple cross-lamination, which in some beds may be convoluted. The cross-lamination can consist either of a single row of ripples on top of division A or B, or of a

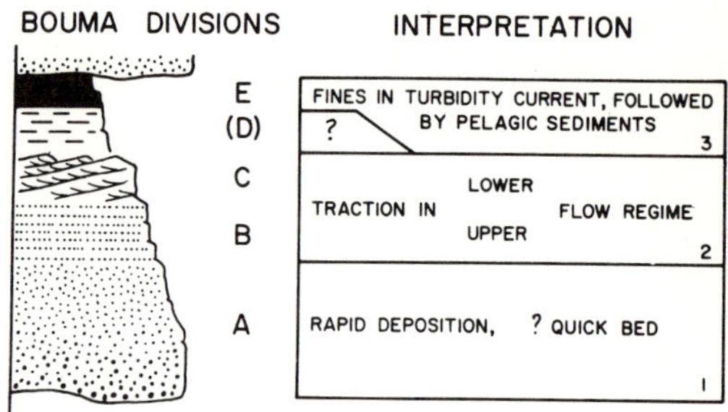

FIG. 3—Bouma model for classic turbidites. Division *A* is massive or graded, *B* is parallel laminated, *C* is rippled, *D* consists of faint laminations of silt and mud, and *E* is pelitic.

multiple set of climbing ripples. The single row probably indicates reworking of underlying sediment by the tail of the turbidity current (or possibly a semipermanent ocean current), whereas a set of climbing ripples indicates active deposition of sediment from the turbidity current during rippling (Walker, 1969). If the rate of deposition is too high, fluid is trapped between the grains, and the ripple cross-lamination becomes convoluted as the fluid escapes. Paleoflow directions measured from ripple cross-lamination are notoriously variable, and division C should be used as a paleocurrent indicator only if all else fails. It is possible to measure cross-lamination in oriented cores (see Hsü, 1977, p. 149-151), but it may be difficult to interpret the results in terms of paleoslope.

In Bouma's original definition, division C was overlain by division D, an upper division of horizontal lamination in silt and clay. This division cannot be seen in weathered or tectonized outcrops, and I prefer to describe it as (D) and include it with division E, the pelitic division. The pelitic material is normally a structureless gray silty clay, or darker gray clay. In many Tertiary basins where forams are abundant, fauna of most of division E consists of transported shallow-water benthonic forams. This implies that most of the sediment in division E was introduced into the basin by the turbidity current. However, the uppermost part of E may consist of a brown foraminiferal lutite with benthonic bathyal or abyssal forams—this represents the normal fine-grained deposition in the basin between turbidity currents (Natland, 1963).

The Bouma sequence now can be considered as a model for classic turbidites. It performs the four main functions of a model very well (a model should be a norm for purposes of comparison, a guide for future observations, a predictor in new situations, and a basis for interpretation; Walker, 1976a, b).

Hydrodynamic interpretation of classic turbidites—The Bouma sequence is an excellent basis for hydrodynamic interpretation (Walker, 1965; Harms and Fahnestock, 1965; Middleton and Hampton, 1976). Deposition can be envisaged in three phrases (Fig. 3).

The first stage involves the rapid deposition of grains from suspension. Continued shear of these grains by the flow, together with the escape of trapped pore water, tends to make the deposit (division A) massive and structureless, and it is perhaps surprising that there can be a preferred grain fabric. Details of this proposed mechanism were given by Middleton and Hampton (1976).

The second phase of deposition is characterized by traction of grains on the bed (Fig. 4). By comparison with experimental work, division B represents the plane bed with sediment movement of the upper flow regime. Divisions B and C can be formed either by reworking previously deposited sediment, or by continued deposition from the turbidity current at lower and lower flow velocities.

The third phase of deposition, divisions (D) and E, represents the quiet accumulation of fine sediments from the tail of the current, with possibly some hemipelagic deposition after the current has died away completely.

Predictive implications of Bouma model—In the preceding interpretation, each division of the Bouma sequence represents a progressively waning current, upward through the bed (Figs. 3, 5). Individual turbidity currents also wane progressively in their journey across the basin floor, and it follows from the model that near the point where deposition begins, turbidites will tend to begin with division A (Fig. 5). Progressively farther from this area, as the currents wane, beds will tend to begin with division B and, in the most distant areas, velocities will have fallen to the point where deposition begins with division C (Fig. 5).

Together with the change in sedimentary structures at the bases of beds, other systematic changes in turbidite characteristics take place from the point where deposition begins to distal depositional environments. Specifically, turbidites in the area of initial deposition tend to be thicker bedded (sandstones >about 20 to 30 cm) and coarser grained (division A typically medium sand or coarser). They also have a higher sand/shale ratio and individual beds tend to begin with Bouma's division A (Figs. 2, 3, 5).

These characteristics define a classic turbidite facies now termed "proximal" (the area of initial deposition). Until now, the term "distal" was used for the contrasting finer, thinner bedded, low sand/shale ratio, Bouma B and C type turbidites. It is now important to term this facies "thin-bedded turbidites" (Fig. 1) rather than distal, because the thin-bedded facies can be deposited in several depositional environments, not all of which are distal (Mutti, 1977). This problem is discussed later, and is illustrated in Figure 13.

The first three criteria—sandstone thickness, grain size, and sand/shale ratio—are particularly important because they may be interpreted from electric logs. The distinction between proximal and distal depositional environments is important in general basin analysis. In the case of hydrocarbon reservoirs, this distinction may be indicative of directions of hydrocarbon migration. For example, of the many reasons given by Barbat (1958) for the abundance of hydrocarbons in the Los Angeles basin, we may note specifically the interfingering of carrier and reservoir sands, and the lateral persistence of fine-grained rocks. Thus, the basin shales interbedded with the thin-bedded turbidite facies may make good source rocks. Oil and gas expelled from these source rocks may be stored in the thin-bedded turbidites, and if these sandstones have good permeability, it may migrate toward the basin margin. Here, the thicker and more extensive sands (proximal classic turbidites, massive and pebbly sandstones) may form good stratigraphic traps, as will be discussed later.

Massive Sandstones

The classic turbidite facies passes gradationally into the massive sandstone facies by a decrease in abundance of interbedded shales, by an increase in channeling and irregularity of bedding, by an increase in overall grain size, and by an increase in sandstone-bed thickness. Thus, the massive sandstone facies consists essentially of massive sandstones, without shaly interbeds (Fig. 6). In the Bouma terminology, a stratigraphic sequence of beds would be described as AAAAA, etc., because horizontal lamination (B), ripple cross-lamination (C), and interbedded fines (D and E) are typically absent. I suggest that the Bouma model is inappropriate with respect to this facies, particularly as it does not describe the one sedimentary structure that is present in some massive sandstones—*dish structure* (Fig. 7).

Typically, massive sandstones are 0.5 to 5 m thick, and may be composite (several flows welded or amalgamated together). Tool marks and scour marks are present on the bases of beds, but internally the beds are not only massive, but com-

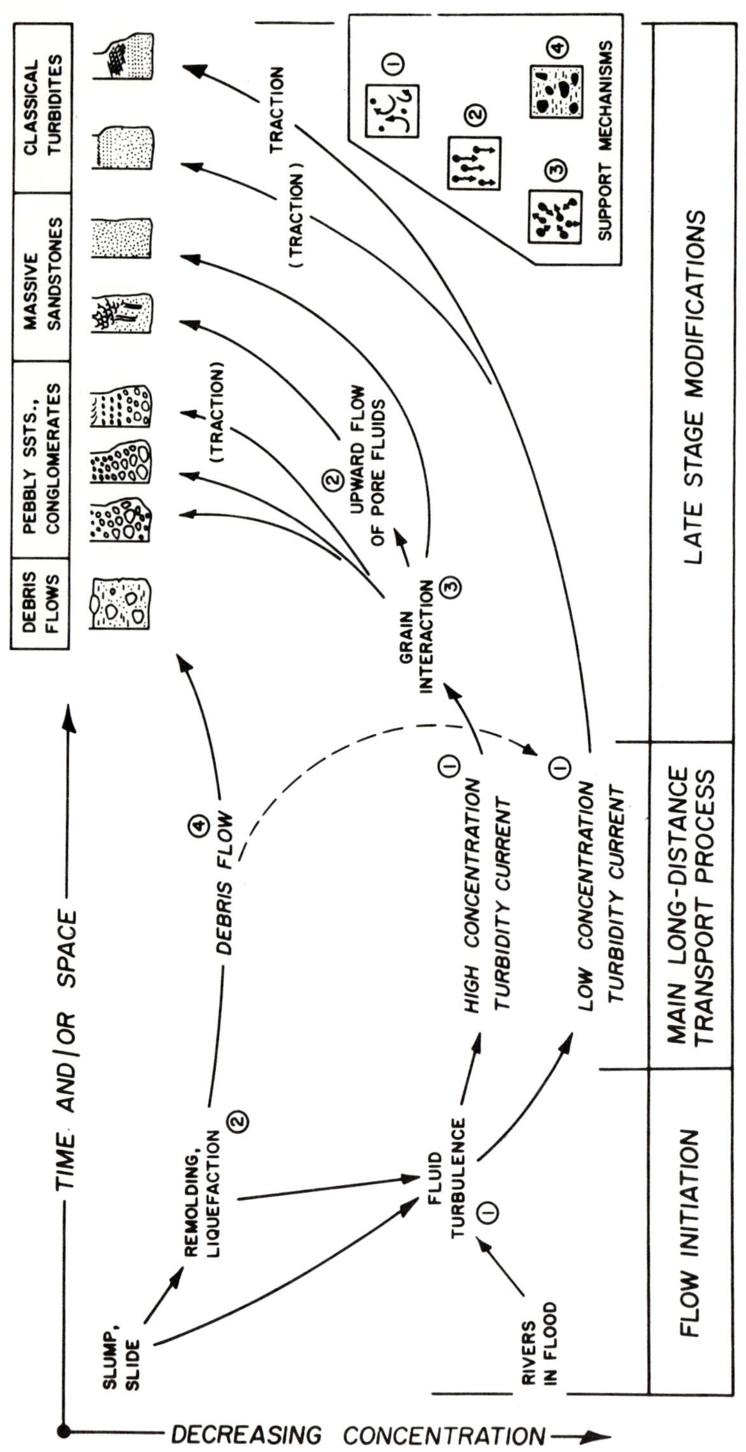

FIG. 4—Processes of initiation, long-distance transport, and deposition for currents transporting sediment into deep water. Framework is one of time and/or space, and concentration of flows. Grain-supporting mechanisms (insert, lower right) include: 1, fluid turbulence; 2, liquefaction; 3, collision between individual grains (dispersive pressure in grain flow); and 4, matrix strength (as in debris flow). Modified in discussion with Middleton (personal commun.) from Middleton and Hampton (1976), to show possibility of debris flows becoming turbulent, and to eliminate grain flows as long-distance transport processes.

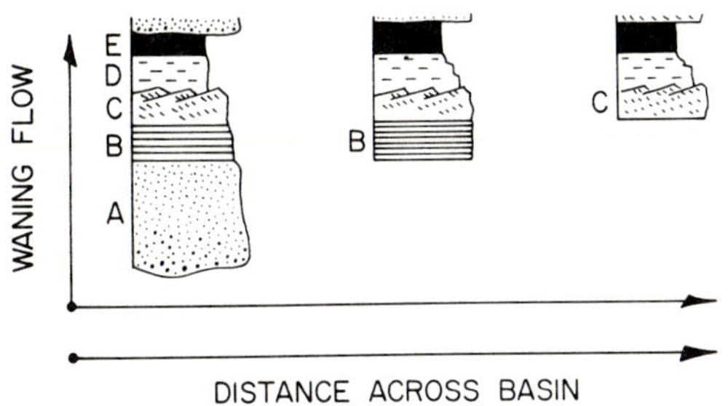

FIG. 5—Interpretation of *ABCDE* Bouma sequence in terms of waning flow suggests that groups of turbidites beginning with divisions *B* and *C* represent deposition from progressively slower flows. This can be related to increasing distance across basin, although it is emphasized in text that some *CDE* thin-bedded turbidites can be present on levees in proximal environments, and hence *CDE* turbidites are not necessarily distal.

FIG. 6—Massive sandstone facies, with no interbedded shales. Cambrian-Ordovician Cap Enragé Formation, Quebec Appalachians. Stratigraphic top on left. Compare with Figure 2.

FIG. 7—Dish structures and vertical fluid-escape pipes (pale). Within circle, fluid from below two adjacent dishes breaks through between dishes to form vertical pipe.

monly are ungraded. The only sedimentary structures reflect vertical intergranular fluid escape during deposition—vertical fluid-escape tubes, and dish structures (defined as thin, subhorizontal, flat-to-concave upward argillaceous laminations in sandstones; Fig. 7). Details of the origin of dish structures have been discussed by Lowe and LoPiccolo (1974) and Lowe (1975). The various fluid-escape features suggest deposition from a flow in which the final grain-support mechanism before deposition is the upward flow of pore fluid (fluidized flow; Lowe, 1976). It seems probable that the currents that transported the sediment most of the way into the basin were turbidity currents maintaining the sand in suspension. During late stages of transport and initial stages of deposition, grains became highly concentrated toward the base of the flow. The fluid trapped between those accumulating grains finally escaped upward, fluidizing the sand and developing the characteristic sedimentary structures (support mechanism 2 of Fig. 4). The massive sandstones without dish structure bear a final imprint that reflects grain collisions and a less forceful escape of pore water, resulting in massive beds without fluid-escape features (Fig. 6).

Pebbly Sandstones

The pebbly sandstone facies (Figs. 8, 9) is distinct from the massive sandstone facies, and it has not yet been established whether or not the facies grade into each other. In the Los Angeles and Ventura basins, pebbly sandstones are abundant

FIG. 8—Pebbly sandstone facies, Cambrian St. Damase Formation, Quebec Appalachians, shows irregular, probably loaded, base; excellent graded bedding; and crudely developed horizontal stratification in upper part of bed.

in outcrop, and form some of the major reservoirs.

Beds range from about 0.5 to over 5 m in thickness, and are characterized by sharp bases and absence of shaly interbeds. The Bouma model is not applicable. Sole marks tend to be large, with flute casts up to 1 m long on some beds. Internally, the beds commonly are well graded, from basal pebbly sandstone (clasts up to about 2 cm) up into medium (and rarely fine) sandstone. The most characteristic internal sedimentary structures are stratification and cross-stratification—dish structures and dewatering pipes are present but less common. The stratification commonly consists of alternating pebble-rich and pebble-poor layers, the layers having gradational bases and tops. Average layer thickness is in the range of 5 to 10 cm. The cross-stratification consists of medium-scale (sets 20 to 30 cm) trough or planar-tabular cross-beds of pebbly sandstone (Fig. 9). It must be emphasized that any form of cross-stratification larger than ripple cross-lamination (sets up to about 5 cm thick) is extremely rare in the resedimented coarse-clastic family, yet when medium-scale cross-stratification is present, it is normally in the pebbly-sandstone facies.

There is no model that attempts to organize the internal features of pebbly sandstones into a Bouma-like sequence, and it is not known whether there is a consistent relation between the various sedimentary structures (grading, horizontal stratification, cross-stratification, dish, and dewatering structures). Although imbrication of the coarser clasts commonly can be seen with the naked eye

FIG. 9—Pebbly sandstone facies, Cambrian-Ordovician Cap Enragé Formation, Quebec Appalachians. Two sets of cross-bedding below main pebbly horizon probably represent separate depositional event from main pebbly sandstone.

in the field, there has not been a comprehensive study of pebbly-sandstone fabrics. Field observations of many pebbly-sandstone formations in the Cambrian-Ordovician Appalachian flysch belt of Quebec, and in Cretaceous and Tertiary pebbly sandstones in California, suggest that the long axes of the grains parallel the flow (as in classic turbidites). In the Miocene pebbly sandstones at Dana Point, California, not only are the long axes mostly parallel with flow, but the long axes dip upstream to define the imbrication (Walker, 1975a). This is a very uncommon fabric, and has been discussed in detail by Davies and Walker (1974) and Walker (1975a, 1977).

In outcrop, the pebbly sandstones commonly are lenticular, and have irregular and scoured bases. Interbedded shales are uncommon. Many pebbly-sandstone formations are rather thick (tens or hundreds of meters) sheet sands, built up by lateral and vertical coalescing of a large num-

ber of individual graded beds. In places where this facies fingers out downcurrent into classic turbidites (as appears to happen in the Los Angeles basin), a perfect potential source-carrier-reservoir situation is established.

The transport mechanisms for the pebbly sandstones are probably similar to those for the massive sandstones, namely, a major phase of suspension by fluid turbulence as the sediment is swept into the basin, and late-stage modifications during deposition (mainly clast collisions) that give rise· to the structureless graded base (Figs. 4, 8). For reasons not yet understood, traction of clasts on the bed was important in many pebbly sandstones, giving rise to stratification and cross-stratification.

Clast-Supported Conglomerates

There is probably a gradation between the coarser grained pebbly-sandstone facies, and the finer grained and stratified conglomerates. Conglomerates are prominent members of the resedimented family in many areas (Appalachians of Quebec, California, Oregon), and may act as reservoir rocks in the Los Angeles and Ventura basins.

As a result of recent work on conglomerates there is some agreement on the important features to observe in the field. These include the type of grading (normal or inverse), the presence or absence of stratification (if present, its type, layer thickness, and fabric; Walker, 1975a), and the presence or absence of imbrication (Harms et al, 1975). The combination of these features has led to the proposal of three intergradational models for clast-supported conglomerates (Walker, 1975a, 1976b, 1977; Fig. 10).

Individual beds of conglomerate can range from a little under a meter to over 50 m (as in the Jurassic Otter Point conglomerates in southwestern Oregon). They have sharp, commonly channeled bases, and tend to be laterally unpersistent. Shale layers rarely are preserved between beds, if indeed they ever were deposited.

In the inverse to normally graded conglomerates (Figs. 10, 11), the inverse grading is rarely thicker than 20 to 30 cm, and passes up into normal grading or massive bedding. There tends to

be a well-developed preferred fabric in the form of an imbrication in which the long axes of the clasts are parallel with flow and dip upstream. The hydrodynamic implications of this form of imbrication have been discussed by Walker (1975a, 1977).

In the downcurrent direction, although not necessarily in the same beds, the inverse grading dies out, and graded-bed conglomerates are developed (Fig. 10). These lack inverse grading and stratification, but commonly have a well-developed imbrication. Even farther downstream, but not necessarily in the same beds, are the graded-stratified conglomerates (Fig. 10). Above the graded part of the bed, stratification can consist either of horizontal alternations of coarse and fine layers, or of cross-stratified gravels very similar to those shown in Figure 9. The cross-stratification is discussed in the following section.

In the field, it is very difficult to trace individual conglomerate beds in the downcurrent direction, and the downstream relations suggested in Figure 10 are based upon theoretical considerations of the formation of inverse grading and stratification (Davies and Walker, 1974; Walker, 1975a, 1977). It is not intended to imply in Figure 10 that any one current deposits a single bed that changes in character downstream. It *is* intended

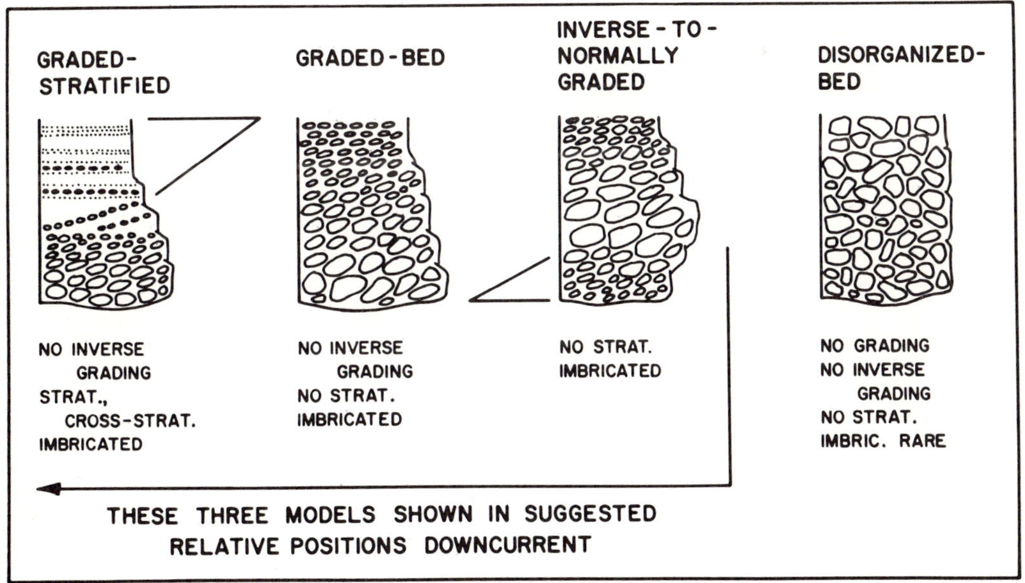

FIG. 10—Four models for resedimented conglomerates. Inversely to normally graded, graded-bed, and graded-stratified models are shown in relative downcurrent position, but this relation is suggested on theoretical grounds only.

FIG. 11—Inverse to normally graded conglomerate, also showing imbrication in both larger and smaller clasts. Clasts dip upstream, hence flow is to left. Cambrian-Ordovician Cap Enragé Formation, Quebec Appalachians.

to imply that currents which deposit sediment in the most proximal areas tend to produce inverse to normally graded beds, and that other currents might bypass the proximal areas, and begin deposition farther downstream, producing graded-stratified beds (Walker, 1975a).

Pebbly sandstones and conglomerates—fluvial or resedimented?—In some situations, there is a real possibility of confusion between pebbly sandstones and conglomerates in fluvial and deep-water environments. The basis for confusion lies in the features common to both environments—the abundance of channeling, abundance of massive and poorly stratified gravels, presence of large-scale cross-stratification, presence of graded (or "fining upward") gravels, and presence of imbrication.

In most situations, the distinction between the two environments can be based upon the associated facies. Fluvial sandstones and conglomerates may be associated with other flood-plain features, such as rootlets, calichelike concretions, and desiccation cracks, whereas in deep water there is an association with other members of the resedimented family, of which classic turbidites might be the most easily recognized.

The presence of a marine fauna also would favor a resedimented interpretation. Shallow (transported) or deep (in situ) faunal elements can be present in the resedimented conglomerates and associated fine-grained beds, but would not be present in fluvial conglomerates.

As a final criterion, the type of imbrication also may be used to distinguish fluvial and resedimented conglomerates. In fluvial situations, pebbles roll on the bed around their long axis, and the normal imbrication is long axis transverse to flow, with intermediate axis dipping upstream. In many resedimented conglomerates, the imbrication is long axis parallel with flow, with long axis dipping upstream. These different imbrications can be identified in unoriented cores, even if the regional paleocurrent directions were not known, and hence can be very important in areas where the only information is subsurface. Details of conglomerate fabrics were discussed by Davies and Walker (1974), Harms et al (1975, p. 136-137), and Walker (1975a, 1977), and an important discussion of very large-scale cross-stratification in resedimented conglomerates was given by Winn and Dott (1977).

Matrix-Supported Beds

This group of beds includes those that were transported into the basin in such a manner that the deposit consists of matrix-supported sand, pebbles, cobbles, and boulders (mainly subaqueous debris flows), and those that attained their texture by shorter distance movements within the depositional part of the basin (slumps).

Slumps (Fig. 12) can be on a small scale, involving a few beds that become broken and folded together, or can range up to thicknesses of tens of meters. In the latter case, tens or perhaps hundreds of beds can be involved, and the style of dislocation can range from immense open folds (Gregory, 1969) to complete disruption, mixing, and brecciation of the strata. In many published basin reconstructions, the orientation of the slump-fold axes has been used to establish the dip of the paleoslope. The assumptions behind this method can be very misleading, and the general

FIG. 12—Slumped facies, Cretaceous at Point Fermin, California. Laminated shales are slumped and contorted, and sandstone blocks (presumably torn from originally interbedded turbidites) are incorporated into slump.

problem of slump-fold orientation has been reviewed by Walker (1970, p. 223-226). A useful and illuminating case history has been documented by Lajoie (1972), and the method of plotting slump folds on stereonets was discussed by Hansen (1967, 1971).

The sedimentary features of subaqueous debris flows are relatively poorly documented, although the process has been discussed in detail by Hampton (1972). Bases of beds tend to be irregular, and they lack the normal suite of tool and scour marks. However, if some large blocks are in contact with the bed, broad "slide" marks can be produced. Internally, the beds are chaotic, and easily can be confused with tillites (again, the overall stratigraphic context and facies relations may be the determining factor in correct identification). Consistent preferred fabrics seem to be absent, although some debris flows locally may show some imbrication. Normal and inverse graded bedding are not developed consistently, although locally there may be some inverse grading at the base of some beds. It is well established that the clast-support mechanism in debris flows is matrix strength (Fig. 4). Because of this, large

clasts can "float" in the upper part of the flow, and upon deposition, these clasts can project upward above the top of the bed. In the field, upward-projecting clasts are the most characteristic and diagnostic features of debris flows. The Haymond boulder beds (Marathon basin, Texas; McBride, 1966) are a classic example.

GRADATIONS BETWEEN FACIES IN RESEDIMENTED FAMILY

The five facies described represent a simplification of a scheme first published by Mutti and Ricci Lucchi (1972) and Walker and Mutti (1973). We are concerned here not with subdivisions of the basic facies scheme, but the general extent to which the facies grade into each other. Most of the data concerning these gradations consist of casual observations rather than rigorously defined associations, and more information on the relations, particularly from an economic viewpoint, is urgently required.

It seems well established that within the classic turbidites there is a complete facies transition from thin bedded (Bouma C[D]E types, Figs. 1, 3,

5) to proximal (Bouma ABC[D]E or AE types, Figs. 2, 3, 5).

From limited published data, it also appears that there is a gradation in facies from classic proximal turbidites into massive sandstones. The gradation is characterized by a thickening of the sandstone beds, and loss of the monotonously regular sand/shale interbedded appearance of the outcrop (Fig. 2). The loss of bedding regularity is due to increasing amounts of scouring and channelling associated with the massive sandstone, resulting in composite sandstones without interbedded shales (Fig. 6).

The extent of a gradation between massive sandstones and pebbly sandstones, if any, is not known. It remains an important research topic, because it is difficult to construct a predictive basin model if the extent of facies transitions is not known. Although there are some formations in which both pebbly and massive sandstones are present, the differences (especially in the development of graded bedding, horizontal and cross-stratification in pebbly sandstones) between the facies suggest rather different depositional mechanisms. This may indicate in turn two distinct facies, rather than two types with a gradation between them.

However, there does appear to be a gradation between the pebbly sandstone and clast-supported conglomerate facies. In particular, the style of horizontal stratification in the pebbly sandstones is very similar to that of the graded-stratified conglomerates, and the gradation between the two facies is dominantly one of bed thickness and clast size. Similarly, the gradation among the various conglomerate facies is by loss of stratification (graded stratified → graded bed) and then appearance of inverse grading (graded bed → inverse to normally graded).

The suggested gradations between facies describe general relations among rock types, rather than changes that could be predicted within individual beds. A more detailed discussion of lateral and vertical facies relations—the basis of a predictive basin model—can be given only within the context of a submarine-fan model. Such a model, in a general way, expresses many of the detailed relations from a large number of individual studies.

SUBMARINE FANS

A detailed review of modern submarine fans has been given by Normark (this issue). There has been an important interplay between recent and ancient sediment studies in deriving the present fan model. The first studies directly applicable to geology were of the smaller California borderland fans (Gorsline and Emery, 1959). At about the same time, the first submarine-fan interpretation of ancient rocks was suggested, by Sullwold (1960, 1961), with reference to the late Miocene Tarzana fan in the Santa Monica Mountains. He emphasized in his interpretation the lenticularity of the sandstones (0 to 4,000 ft in 16 mi or 0 to 1,200 m in 26 km), and the radial paleocurrent directions fanning from a point source (? foot of canyon) north of the outcrop area.

As work continued in offshore southern California (Shepard and Einsele, 1962), definite sedimentary facies began to be associated with particular topographic parts of the fans and, in 1964, Hand and Emery recognized slope, canyon or channel, levee, and apron environments. They discussed the sediments from a geologic point of view, and even published an inferred map of current directions for the adjacent Newport, Oceanside, and Carlsbad fan systems. This work stimulated the first very detailed interpretation of an ancient fan, with descriptions of different turbidite facies, drawings and photographs of fan channels, and an overall fan stratigraphy (Walker, 1966a, b). This Late Carboniferous fan from northern England will be discussed in detail later.

Work on recent and ancient fans continued in the late 1960s, but increasing sophistication in seismic profiling led to important new insights into modern fan morphology and evolution (Normark, 1970). Shortly afterward important summaries of Italian fans were published (Mutti and Ghibaudo, 1972; Mutti and Ricci Lucchi, 1972) and these papers presented fan models very similar to that which Normark (1970) had suggested from recent-sediment studies. Since then, many recent and ancient-sediment studies have emphasized one fan model, which will be used in this paper as the basis for facies relations, fan stratigraphy, and sand-body prediction. Important recently published fan interpretations of ancient rocks include those of Nilsen and Simoni, 1973 (Butano Sandstone, Eocene, California); Mutti, 1974 (various Italian fans); Stanley, 1975 (guidebook to the Annot Sandstone, southern France); Kruit et al, 1975 (guidebook and discussion of the superbly exposed San Sebastian fan, northern coast of Spain); and Mutti, 1977 (Hecho Group, Eocene, Spain).

Relation of Facies to Fan Morphology

The proposed facies distributions suggested here (Fig. 13) are based on the morphologic subdivisions suggested by Normark (this issue). The upper fan is characterized by a single leveed channel that may have within it a sinuous, meandering thalweg channel flanked by relatively flat

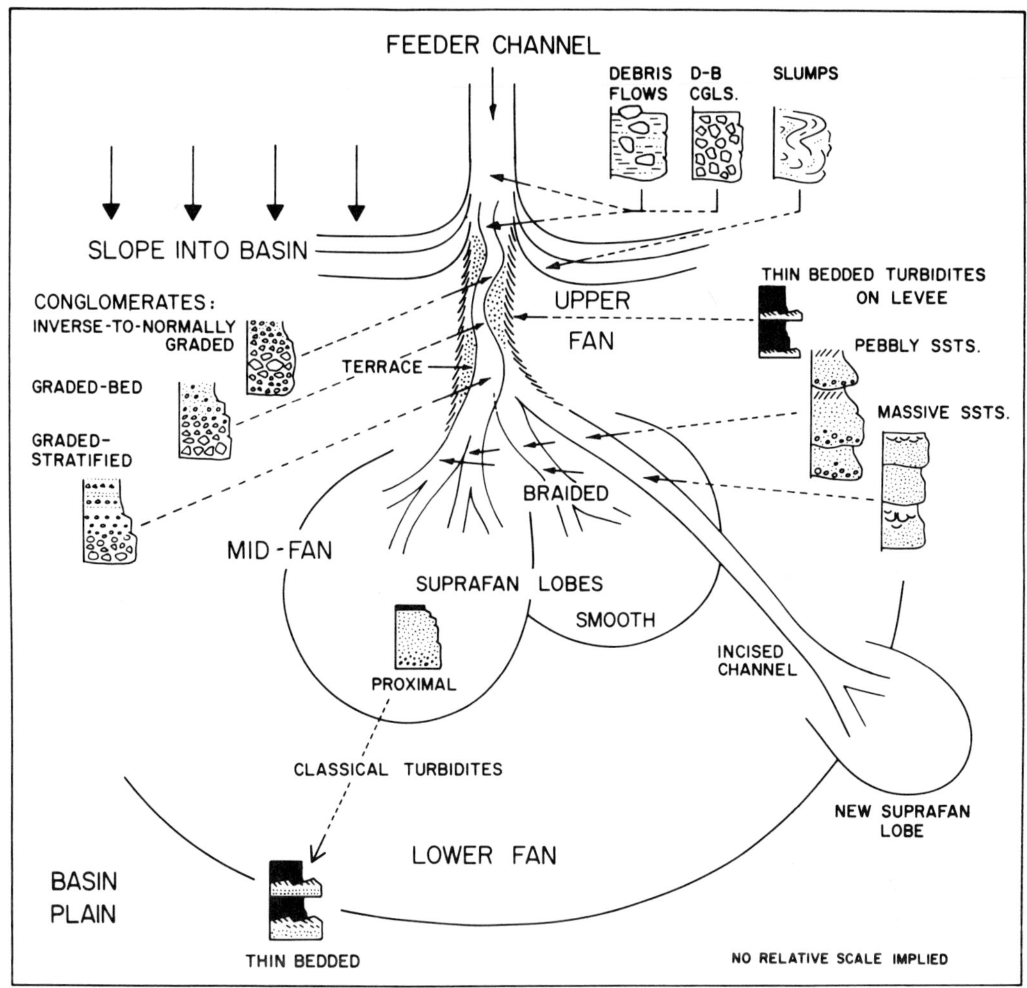

FIG. 13—Model of submarine-fan deposition, relating facies, fan morphology, and depositional environment. *D-B* indicates disorganized-bed conglomerates.

terraces (Buffington, 1964, p. 50-52; Shepard et al, 1969, Fig. 2). The middle fan is built up by deposition on suprafan lobes, which shift in position from time to time in the manner of switching delta lobes. The inner part of the suprafan lobes is characterized by shallow, nonleveed braided channels, whereas the outer part is smooth and merges imperceptibly with the smooth lower fan. This area is indistinguishable from the basin plain in most cases.

In Figure 13, the various facies of the resedimented family are shown in their interpreted positions. This interpretation is based on the detailed morphology described by Normark, the known relations among facies in ancient rocks, the abundance and depth of channeling associated with the various facies, and an unfortunately small number of recent sediment cores on modern fans.

Basin plain and lower fan—These topographically smooth, low-gradient areas (Fig. 13) are characterized by slow hemipelagic deposition, interrupted periodically by turbidity currents. Deposition on the smooth, featureless bed results in very regularly and parallel-bedded classic turbidites, which are thin bedded on the basin plain, but become thicker bedded toward the middle fan. Proximal turbidites, retaining the monotonous sand/shale interbedding of Figure 2, also suggest a smooth seafloor, probably the smooth outer part of the suprafan lobes (Fig. 13).

Suprafan lobes—Deposition in the braided-channel parts of the suprafan lobes will not result in such continuous parallel bedding as on the smooth lobe farther downslope. The most likely facies to be deposited in the channels are the massive and pebbly sandstones, both characterized in outcrop by lenticular bedding and shallow channels. As the channels braid and switch position, sand bodies will tend to coalesce, and any fine material that was deposited between channels will tend to be scoured away and not preserved. Relatively fine-grained, small-scale turbidity currents using these channels may deposit classic turbidites within them, and turbidites displaying Bouma sequences (Figs. 3, 5) are known from cores of several modern fan channels (Haner, 1971, on Redondo fan; Cleary and Conolly, 1974, on Hatteras fan). Conversely, unusually large and coarse-grained flows may transport gravel and boulders into the braided suprafan channels, dumping the coarse material as a conglomeratic lag on the channel floor. In general, however, the braided suprafan probably is dominated by massive and pebbly sandstones.

Upper fan—The main upper-fan channel (Fig. 13) is probably the area of deposition of the conglomeratic facies, if such coarse material is being supplied to the basin. The conglomerate first will be deposited in the thalweg channel, and perhaps also on the terraces when large flows spill out of the thalweg. Alternatively, conglomerates might be restricted to the thalweg channel, and sandstones could be deposited on the terraces (from finer material originally suspended higher in the flows). Relations in the upper-fan channel are conjectural because of extremely limited coring in modern examples, and because interpretations of ancient upper-fan channels have not, until very recently, been concerned with distinguishing thalweg and terrace deposits. The levees of the upper-fan channel tend to consist of fine-grained alternations of thin sandstone beds and mudstones. These belong to the thin-bedded turbidite facies, and in ancient examples easily could be confused with the similar facies on the basin plain. This problem will be considered again later.

Feeder channels—The feeder channels, or submarine canyons, act mainly as conduits for the sand and gravel moving out toward the fan. They may be plugged either by coarse materials (slumps, debris flows, conglomeratic or other coarse material as availalbe at source), or by very fine material (clays, mudstones). The latter commonly results from a relative rise of sea level, cut-ting the fan off from its original source of sediment; a good example of this is the abandoned, mud-filled, Mississippi feeder channel (Sabate, 1968).

STRATIGRAPHIC EVOLUTION OF FANS

The lateral facies relations already discussed and shown in Figure 13 now can serve to predict vertical stratigraphic sequences under conditions of active fan progradation. The concept of a specific stratigraphic sequence arising from fan progradation first was suggested by Walker (1966a), and a direct, explicit comparison of submarine fans with deltas was made by Mutti and Ghibaudo (1972) and Mutti and Ricci Lucchi (1972). By developing this comparison, they suggested that fan progradation would result in a coarsening-upward sequence very similar to that of a delta. They also compared the upper fan and suprafan channels to deltaic distributary channels.

It is now possible to relate specific facies to specific parts of prograding-fan systems. Progradation of the lower fan onto the basin plain should result in a sequence of classic turbidites in which the sandstone beds become slightly coarser grained and slightly thicker upsection. This is termed a thickening- and coarsening-upward sequence (C-U on Fig. 14). All of the turbidites in such a sequence would tend to be of the thin-bedded facies (sequence 1 of Fig. 14), similar to the "outer fan thin bedded turbidites" of Mutti (1977, p. 119).

Above the lower-fan sequence one would predict a sedimentary record of the middle fan. The smooth parts of the suprafan lobes would prograde in much the same way as the lower fan, and give rise to a thickening- and coarsening-upward sequence that would begin with classic turbidites, but would terminate upward in massive or pebbly sandstones (Fig. 14, sequences 3, 4; Fig. 15). Again, bed thickness, grain size, and sand/shale ratio would all increase upward. The middle fan area is gradually built up by lateral switching, coalescing, and superimposition of the suprafan lobes. Consequently, one lobe can become abandoned, and a different one become the main locus of deposition. The first lobe might receive a veneer of mudstone while the second lobe was being built, but eventual reestablishment of another lobe in the same position as the first would lead to superimposed thickening- and fining-upward sequences, perhaps separated by a veneer of mudstone.

Within the suprafan channels themselves, Mutti and Ghibaudo (1972, Table 2) suggested that a fining-upward sequence would be developed during channel filling and abandonment. The se-

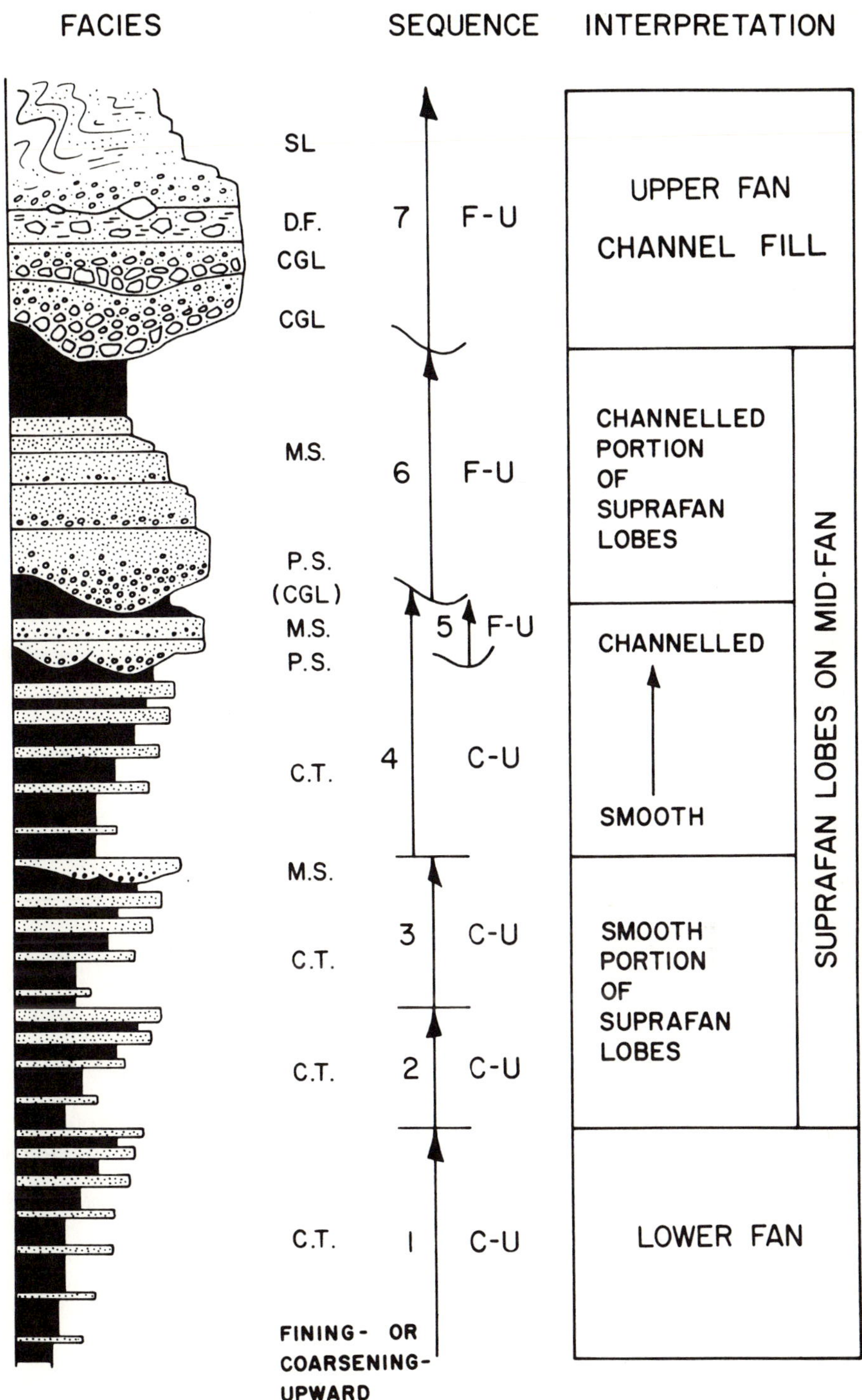

FIG. 14—Hypothetical stratigraphic sequence that could be developed during fan progradation. *C-U* represents thickening- and coarsening-upward sequence; *F-U* represents thinning- and fining-upward sequence; *C.T.*, classic turbidites; *M.S.*, massive sandstones; *P.S.*, pebbly sandstones; *CGL*, conglomerate; *D.F.*, debris flows; *SL*, slumps. Numbered sequences discussed in text.

FIG. 15—Thickening- and coarsening-upward sequence in Jurassic turbidites, southwestern Oregon. Beds are overturned, and base of sequence (right) is shown by white line. Offset by fault runs down cliff. Lower part of sequence consists dominantly of thin-bedded turbidites, center of sequence consists of proximal classic turbidites, and upper part of sequence (from circled figure to left edge of photo where beds disappear into beach) consists of massive sandstones. Similar to sequence 3 in Figure 14.

quence would be comparable in principle (but not in details of stratification) to the filling and abandonment of a fluvial or delta-top distributary channel. Thus, in Figure 14, sequence 5 represents a fining- and thinning-upward sequence (F-U represents fining upward) from pebbly to massive sandstones, and thence to mudstones after complete channel abandonment. Similarly, sequence 6 represents the abandonment of a deeper suprafan channel, and sequence 7 represents the filling and abandonment of the main upper-fan channel. Facies in this channel could include conglomerates, debris flows, and slumps.

Thus, the overall stratigraphy developed during continuous fan progradation would consist of one major coarsening-upward sequence (Fig. 14), on which would be superimposed smaller coarsening-upward sequences in the lower part (1-4 of Fig. 14; Fig. 15), and fining-upward sequences in the upper part (5-7 of Fig. 14). Under normal circumstances, these trends in bed thickness and grain size should be readily identified on electric logs but, without any other information, they would be extremely difficult or impossible to distinguish from trends within a prograding deltaic sequence.

Several modifications of this overall model are possible, and are based upon (1) fan abandonment and (2) accelerated progradation and channel entrenchment. Fans can be abandoned by cutting off their source of sediment, most commonly by a relative rise of sea level. The shoreline and shallow subtidal areas of maximum sand supply and longshore transport shift away from the basin center, thus cutting off supply from the canyon or feeder channel heads. The end result is that only fine suspended sediment can be deposited in the feeder channel, and as a veneer on the fan. Several of the large modern fans (Mississippi, Amazon, Astoria) have been starved since the post-Pleistocene rise of sea level; in the case of the Mississippi, the abundance of supply of fine sediments from the river has led to the filling of the old feeder channel (Sabate, 1968).

In cases of accelerated progradation (possibly related to a relative lowering of sea level, or accelerated supply of sediment, or both), the upper-fan channel begins to cut downward and outward across the fan. Much of the area of the old fan can be bypassed, and a new depositional lobe constructed seaward of the old fan. The channel cutting across the present La Jolla fan is a possible example of such a system. Normark and Piper (1969) related channel incision to the last rise in sea level but, importantly, the canyon head cut landward and remained in the surf zone, assuring a continuous supply of sediment during channel entrenchment.

This type of behavior in an ancient system perhaps could be detected by unusual facies relations, for example, a channel filled with conglomeratic facies cutting into a sequence of distal turbidites believed to have come from the same source. An incised channel of this type is sketched in Figure 13, and possible examples will be discussed in the following.

EXAMPLES OF ANCIENT SUBMARINE FANS

The discussion so far in this paper has centered on resedimented facies, their relation to a fan model, and facies sequences that would be predicted if the fan prograded. In this section, I propose to examine ancient submarine fans in the light of the model, to discuss details of facies relations and, where possible, give examples from hydrocarbon reservoirs. The various morphologic parts of the fan will be discussed first, and then one complete fan will be reviewed.

Feeder Channels

The term feeder channel is preferred over submarine canyon because of the immense scale suggested by the term canyon. Feeder channels have been reviewed recently by Whitaker (1974), who tabulated channel characteristics, and gave a full bibliography. There are only a few known cases of preservation of long feeder channels, but some of them are important hydrocarbon traps. These include the Meganos (Dickas and Payne, 1967), Yoakum (Hoyt, 1959), Mississippi (Sabate, 1968), Hackberry (Paine, 1966; Benson, 1971), Gevaram (Cohen, 1976), Rosedale (Martin, 1963), and Cook (Bloomer, 1977) channels; their major dimensions are shown in Table 1.

One surprising feature of the Meganos, Yoakum, Mississippi, and Gevaram channels is their monotonous, very fine-grained fill. In the case of the Mississippi, this is known to be due to a relative rise of sea level, trapping any coarse sediment at the new shoreline and supplying only fines to the abandoned channel. These fine sediments act as an updip seal to hydrocarbons in the sediments

outside the Meganos, Yoakum, and Mississippi channels. These channels all appear to have been cut by turbidity currents, and the deposits of these currents may well have formed fans spreading from the ends of the feeder channels. The coalesced suprafan sands of these and similar fans should make attractive drilling targets if they can be located, and other geologic conditions are suitable. The Meganos and Yoakum "fans," with their fine-grained channel fills, might be particularly attractive because the fill initially would have acted as an updip trap for any hydrocarbons that might have been present in the fan sediments. The survival of these stratigraphic traps obviously would depend on the subsequent structural histories of the areas.

The Gevaram Canyon, with its gray-black shaly fill (plus a few very porous sand lenses up to 10 m thick), is neither a reservoir itself, nor an updip seal. Cohen (1976) interpreted the canyon fill as the source of hydrocarbons that have accumulated in an immediately overlying reservoir (Helez formation). One would predict a fan lying at the end of the canyon, which now would be offshore, in the Mediterranean.

The Hackberry turbidite sandstones are prolific producers (Benson, 1971), and apparently were deposited in a structurally controlled trough rather than a feeder channel initially cut by turbidity currents. Under these conditions, the present submarine-fan model cannot predict whether a fan might be present downslope, or whether all turbidity-current deposition took place within the trough.

The Cook channel (lower Permian of west-central Texas, Bloomer, 1977, p. 355-357) is the smallest feeder channel listed in Table 1. Bloomer interpreted the sandstone fill of the channel as a bundle of turbidites, and the SP curves suggest a fining-upward sequence within the channel (Bloomer, 1977, Figs. 20, 21). The channel itself appears to be cut into the lower-slope environment. The North Bloodworth field is a stratigraphic trap producing from the channel-fill sandstones.

Upper Fan

The upper fan is characterized by one relatively deep, leveed channel that only rarely shifts in position. The coarse deposits within the channel tend to be surrounded by the finer levee deposits on either side. Because of the scale of these channels, tens or hundreds of meters deep, and hundreds of meters to kilometers wide, very few have been identified in outcrop. One possible example from the Quebec Appalachians is discussed here,

Table 1. Characteristics of some Feeder Channels

Channel	Age, Location	Length (km)	Width (km)	Depth (m)	Fill
Meganos	Late Paleocene, Sacramento Valley	80+	3.2 to 9.7	600	Silty shales
Yoakum	Mid-Eocene, Texas	80	16	900	Silty shales, more sand up channel
Mississippi	Pleistocene, Coastal La.	about 80	3.3	600	Clay
Hackberry	Oligocene, S.W. Louisiana	24	14	200+	Turbidites and shales
Rosedale	Late Miocene Bakersfield, Cal.	about 10	2.6	400	Turbidites and shales
Gevaram	Early Cretaceous, Israel	16+	up to 15	938	Gray-black silty shale
Cook	Early Permian, West-central Texas	5+	1.7	30+	Turbidites and shales, fining-upward sequence

and others will be discussed in the section on the Shale Grit (and see Walker, 1966a, b).

The Grosses Roches conglomerate in Quebec shows a channelized thinning- and fining-upward sequence; a diagrammatic section compiled from Hendry (1973) and from my own field notes is shown in Figure 16. The main channel contact itself (Fig. 17) is fairly steep, and cuts into bedded and massive silty mudstones without any evidence of interbedded turbidites. The observed depth of erosion is about 20 m. The lowest observed part of the fill consists of very lenticular sandstone and conglomerate beds that fit well with the disorganized-bed model (Fig. 10). A preferred fabric was observed in only one bed, and there was little evidence of grading or stratification. Above the channelized conglomerates and lenticular sandstones there is a unit of massive sandstones, followed by poorly exposed classic turbidites (Fig. 16). These in turn are cut by a second conglomerate-filled channel.

The Grosses Roches conglomerate is assigned tentatively to the upper fan (or possibly upstream part of the braided suprafan) because of the monotonous silty mudstone outside the channel, and the disorganized-bed conglomerates within the lower part of the channel fill. It is also a good example of a channelized thinning- and fining-upward sequence. The Grosses Roches conglom-

erates now outcrop only on one thrust slice, and no predictions as to the facies upstream and downstream can be checked.

The channel is relatively shallow compared with modern upper-fan channels (Normark, this issue). However, the 20-m (minimum) Grosses Roches channel and the 20 to 50-m channels in the Shale Grit (Walker, 1966b, discussed later) are among the deepest observed in outcrop. It is also possible that the observed channels are thalweg channels, incised into the deposits on the floor of a much deeper inner-fan channel. This suggestion has been made for deposits in the Cambrian-Ordovician Cap Enragé Formation (Quebec Appalachians; Johnson, 1974), where the larger channels themselves, if they exist, are not observed. In recent sediments, the best example of a meandering thalweg channel cutting through sediments on an essentially flat main-channel floor is in the La Jolla fan (Shepard et al, 1969, p. 392). The thalweg channel here is roughly 30 to 40 m deep and about 250 m wide (within a main valley 1.5 km wide and 100 to 120 m deep; these measurements are close to those of the Shale Grit channels discussed later; Fig. 22).

Upper-fan channel deposits, flanked by fine-grained levee sediments, are potential reservoirs. This is particularly so if the system is abandoned because of sea level rise, and a fine grained mud-

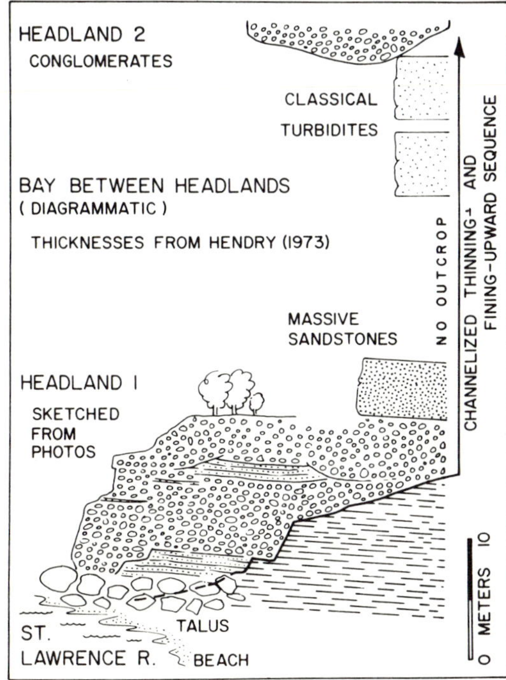

FIG. 16—Thinning- and fining-upward sequence within channel in Ordovician Grosses Roches conglomerate, Quebec Appalachians. *Headland 1* is sketched from photos (see Fig. 17); rest of sequence has been added in correct stratigraphic position from exposures of massive sandstones and classic turbidites in bay west of headland 1.

stone drape is laid down over the upper-fan channel deposits, and updip, in the feeder channel. The preservation of such a reservoir obviously depends on the subsequent structural history of the area.

Mid-Fan: Braided Channels on Suprafan Lobe

This part of the fan is potentially one of the best petroleum reservoirs; the sediments are fairly coarse grained, and lateral and vertical coalescing of channels gives rise to extensive reservoir sands that can be sealed by shale beds. The shale beds are not deposited one by one on top of each turbidite; rather, a thick shale bed (several meters) blankets one suprafan lobe when it is abandoned completely by channel switching (Fig. 18). Up-current, conglomerates may become more conspicuous if coarse material were available, and downcurrent, the massive and pebbly sandstones will tend to pass into classic proximal turbidites.

The first example of rocks interpreted as a braided suprafan-lobe deposit is the Capistrano Formation (upper Miocene) at San Clemente, California (Walker, 1975c). This classic area contains a series of eight channels nested within each other, and filled variously with pebbly and massive sandstones, some classic turbidites, and two conglomeratic beds. A map of the cliff face is shown in Figure 19. Only one wall of the channels (numbered in Fig. 19) can be seen, and the dip of the wall is between 5 and 20°. The estimated depths of the channels are 20 to 25 m (Walker, 1975c, p. 923). Channels 2 to 4 trend about 270 to 295°, and a second (and distinct) group of channels (5 to 8) trend about 235 to 240°. The entire channel complex is cut into bioturbated mudstones with rare, thin silty turbidites.

The channel fills are uncommon in that seven out of eight begin with a claystone or silty mudstone drape on the channel walls, 30 cm to 2.0 m thick. There is some evidence to suggest that these drapes were deposited by turbidity currents because, in one channel, sandy turbidites near the base can be seen to pass up the wall into silty mudstones. Presumably the sand load of the current moved close to the bed, and the finer suspended load was deposited higher on the channel wall. Most of the fill consists either of graded pebbly sandstones (lamination and cross-stratification very uncommon) or of classic turbidites. According to the model established earlier in this paper, the association of pebbly sandstones and classic turbidites would suggest deposition on the outer part of the braided part of the suprafan lobe, at the point where the channels are dying out and the topography is becoming smoother.

Thinning- and fining-upward sequences are not present in all the channels, but can be identified in the channel 5-6-7 complex (Walker, 1975c, p. 923).

The Capistrano Formation also crops out at Dana Point, a few miles north of San Clemente (Bartow, 1966; Normark and Piper, 1969; Piper and Normark, 1971). The dominant facies are conglomerates, graded pebbly sandstones and massive sandstones, with some slumping and extensive channeling (including the Doheny channel). In their reinterpretation, Piper and Normark (1971) specifically referred their "lower sands and conglomerates" (not part of the Doheny channel) to a suprafan depositional environment. Again, the main criterion for such an interpretation is the presence of channeling, in particular association with the massive and pebbly sandstone facies.

Because the fan model discussed in this paper dates from the work of Normark (1970), there are relatively few ancient fans that have been inter-

FIG. 17—Ordovician Grosses Roches conglomerate, Quebec Appalachians, in headland 1 (Fig. 16). Channel cuts into silty shales without turbidites (cliff and wave-cut platform), and is filled mostly by disorganized conglomerates and very lenticular sandstones. Lenticularity is caused mostly by erosion by overlying conglomerates. Channel base about 18 m above wave-cut platform at right edge of photo.

preted using Normark's terminology. Nevertheless, the conglomerate, pebbly sandstone, and massive sandstone facies of the braided suprafan are distributed abundantly in California, and I suggest that such units as the Cabrillo and La Jolla formations (San Diego–La Jolla area), the Cretaceous rocks of the Simi Hills (Colburn and Fritsche, 1973), the Matilija Sandstone (Link, 1975), and the Repetto Formation of the Los Angeles basin could all be reinterpreted, at least in part, with reference to the braided suprafan. In the Ventura basin, parts of the Repetto also may represent braided suprafans, an interpretation based upon isopach maps recently published by Hsü (1977). In Hsü's Figure 18, the producing AO_1 sand of the Ventura field is shown over an area 5 km long, 2.5 km wide. Isopachs show the sand in branching and rejoining elongate stringers, a pattern that essentially is braided. Individual sand stringers (? channels) are roughly 250 to 450 m wide, and up to 13 m thick. The sands pinch out into shales to the north and west, and the stringers are oriented ESE to WNW, suggest-

ing turbidity-current flow toward the west-northwest (Hsü, 1977, p. 143-150). There is little or no evidence in the electric logs published by Hsü (1977, Fig. 4) that the sands in these possible braided channels are in thinning- and fining-upward sequences.

Implications of Thinning- and Fining-Upward Sequences: Fan-Channel Deposits?

It has been suggested in the foregoing that thinning- and fining-upward sequences form by the progressive abandonment of channels in upperfan or braided-suprafan areas. The first interpretations along these lines were suggested by Mutti and Ghibaudo (1972) and Mutti and Ricci Lucchi (1972). The sequences recently have been emphasized by Ricci Lucchi (1975), but he did not discuss their origin in much detail.

Thinning- and fining-upward sequences, of the order of 10 to 50 m (and possibly to 100 m) thick, should be identified readily on electric logs. In a submarine-fan context, the problem is whether or not the sequences are always (or mostly) related

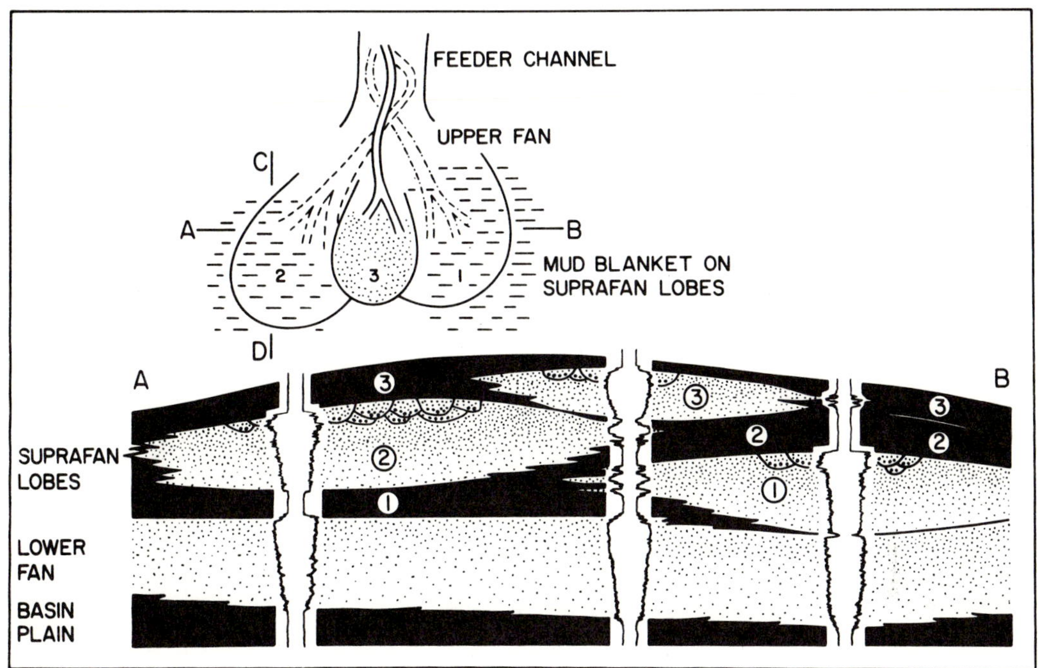

FIG. 18—Hypothetical vertical cross section across prograding lower- and middle-fan system. Lower-fan sequence is coarsening-upward, and each suprafan lobe has coarsening-upward sequence, from thin-bedded turbidites to channelized massive and pebbly sandstones. Each suprafan-lobe sequence also shales out laterally into mudstone drape that covers adjacent parts of fan. Sequence shown in cross section is result of suprafan-lobe switching, from position 1, to 2, to 3 (see plan view, upper part of diagram). Lobe switching may be related to changing meander patterns in upper-fan channel. Hypothetic electric logs illustrate coarsening-upward prograding-lobe sequences, fining-upward channel-fill sequences, and highlight problems of correlation in such a system. Compare with Figure 21, which is interpreted to represent cross section *C-D* of plan view of Figure 18.

to channel filling. The channel-fill interpretation suggests that a channel can be used as a funnel for turbidity currents, perhaps for a long time, until one particularly large, coarse flow partly plugs the base of the channel. During this time, the channel may have been nondepositional, or the channel floor may have been aggrading at about the same rate as the levee, thus maintaining the channel topography. However, after deposition of a large plug in the base, the channel may not be deep enough for subsequent flows which would overtop the banks and lead to eventual channel diversion. Thus the old channel gradually would be abandoned, most of the flows going elsewhere, and would be plugged up gradually with thinner and finer grained beds. These in fact may be overbank deposits from other more active channels.

This interpretation of thinning- and fining-upward sequences is perhaps the most plausible, unless one appeals to progressive changes of volume and grain size of material supplied at the basin margin (perhaps related to long-term tectonic controls).

The relation of thinning- and fining-upward sequences to channels has important implications in basin analysis and the prediction of facies distributions. The Cretaceous conglomerates at Wheeler gorge, California (Walker, 1975b), and the Jurassic Otter Point Formation of southwestern Oregon (Walker, 1977) both contain spectacular thinning- and fining-upward sequences which pass from conglomerates into massive sandstones, classic turbidites, and dark mudstones. In neither example is there any field evidence of channeling. The problem is easy to state but hard to answer. Do the sequences necessarily imply channeling? If so, one would predict a sandy fan system a little farther downcurrent (as was predicted earlier for the Meganos and Yoa-

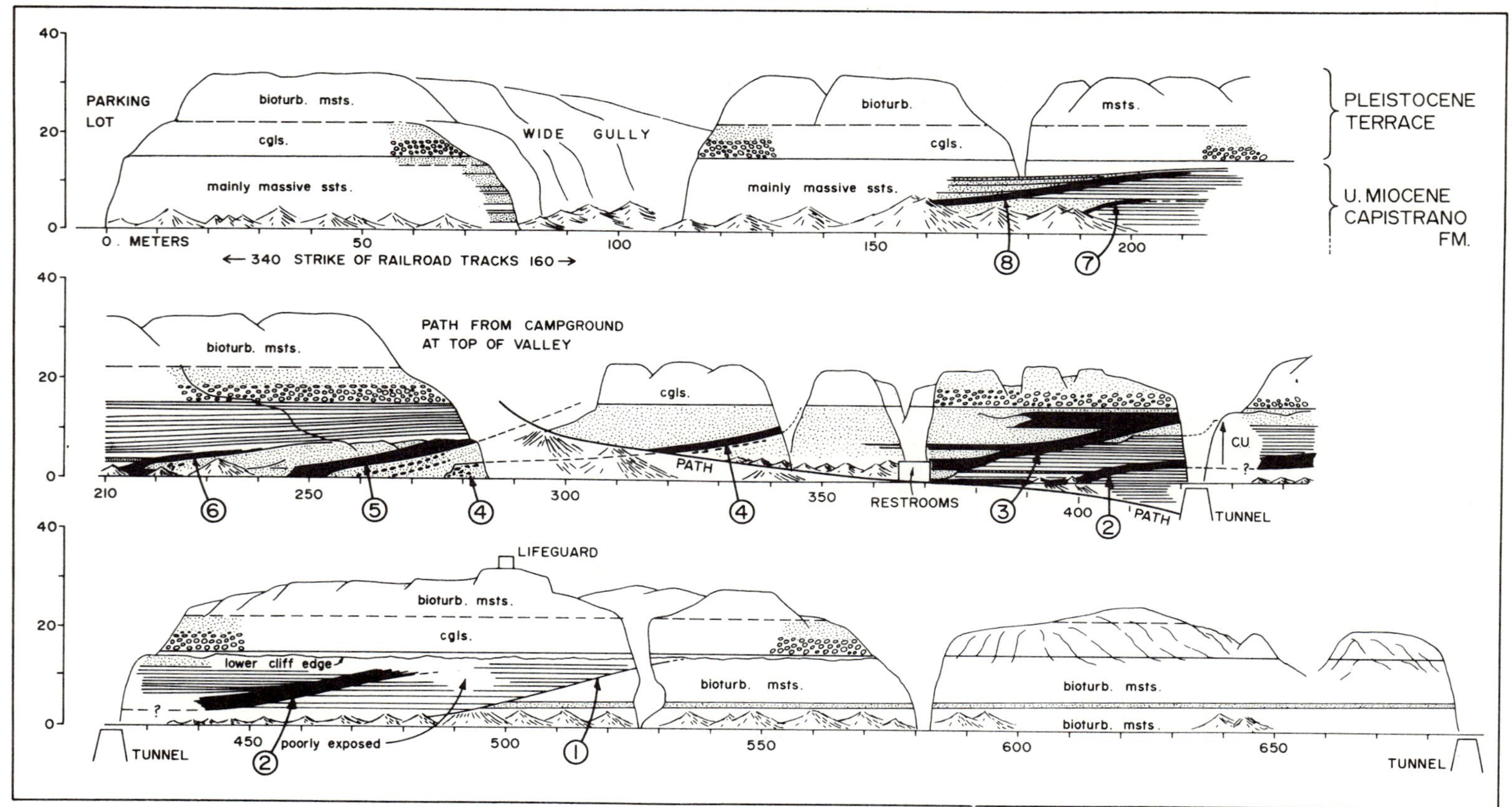

FIG. 19—Measured profile of cliff face at San Clemente, California. Individual channels are numbered, and black layers represent mudstone drapes on channel walls. Within Capistrano Formation, stippling represents massive and pebbly sandstones; horizontal ruling represents classic turbidites. Upper conglomerates and bioturbated mudstones form part of Pleistocene terrace. Vertical = horizontal scale, in meters.

kum channels). If the major conglomeratic thinning- and fining-upward sequences do not imply channeling, there is no model for predicting the facies downcurrent or for explaining the sequences themselves.

Incised-Fan Channels

It was emphasized earlier that the basic fan model describes the facies developed during steady progradation. However, under some circumstances, fan channels can become incised. If this happens, turbidity currents will bypass the fan, and the channels will tend to lengthen as they become more and more incised. As a result of lengthening, the channels will cut into previously deposited classic turbidites of the smooth suprafan lobe (Fig. 13). Such channels would not be predicted by the model to cut into classic turbidites unless they were to some extent incised.

A good example of this is the Cambrian St. Roch Formation at L'Islet Wharf, Quebec Appalachians (Hubert et al, 1970; Rocheleau and Lajoie, 1974). Here, a channel (Fig. 20) with depth of erosion exceeding 50 m cuts into a thickening- and coarsening-upward sequence of classic turbidites. The turbidites are neither truly proximal nor truly distal—they are somewhere in between. The flow direction for the turbidites averages 170° (my measurements, 19 readings); erosional markings on the channel base average 210° (flutes, grooves, scours; my measurements, 6 readings). It is therefore reasonable to assume

FIG. 20—Channel complex in Cambrian St. Roch Formation, Quebec Appalachians. Stratigraphic top on right. Channel (foreground and cliff in background) cuts into coarsening-upward sequence of classic turbidites (upper left). Channel fill consists of graded-bed and graded-stratified conglomerates, and massive sandstones. Arrow (foreground) shows basal, 7-m thick, graded-stratified conglomerate grading into massive sandstone.

that the turbidites were supplied from the same source as the channel fill, and that both are related parts of the same fan complex.

The channel itself has been filled in two or three stages. The lowermost 7 m (Fig. 20) consists of a graded-stratified conglomerate (3 m) that passes upward into massive sandstones (4 m). Higher in the channel, there are more alternations of conglomerate and massive sandstone that imply stages of filling and partial abandonment, followed by renewed scouring, and renewed filling. It is significant that the conglomerate facies is graded stratified (Fig. 20), because it was predicted in Figure 10 that this facies would be farthest downcurrent of any of the conglomerate facies. Hence of the various conglomerate facies, the graded-stratified would be the most likely in an extended, incised channel juxtaposed with classic turbidites. This is exactly the situation at L'Islet Wharf.

A second possible example of an incised channel is the Doheny channel at Dana Point, California. This channel has an exposed depth of about 20 m, and has been illustrated excellently by Bartow (1966) and Piper and Normark (1971). The basal channel fill consists of chaotic coarse massive sandstones. Individual beds are irregular, in places deeply channelled, and have gravelly or pebbly bases. The upper part of the channel fill is bedded, consisting of turbidites tens of centimeters thick, and extending laterally for over 100 m. Thus the fill is generally thinning- and fining-upward. The channel cuts into silts with interbedded fine sands 10 to 50 cm thick, interpreted as turbidites. The juxtaposition of a deep channel with coarse and pebbly sandstones at its base, cutting into a unit composed dominantly of siltstones, suggests that the channel is incised into the fan surface (Normark, personal commun., 1977).

Interpretation of channels incised into classic turbidites leads to the prediction of a new suprafan lobe beyond the incised channel (Fig. 13). Prediction of such lobes would be of possible importance in hydrocarbon exploration. A prediction of this type has been made by Normark (1974, p. 66) for the La Jolla and Monterey systems. The present La Jolla fan valley is incised into the fan as a result of headward erosion during the last rise in sea level. The head of the canyon remained in the shore zone, resulting in continuous sediment supply rather than abandonment and blanketing by mud. Beyond the foot of the present fan valley, about 15 km southeast and in water about 100 m deeper, there is an area of "hummocky suprafan morphology" (Normark, 1974, p. 66), but "further study, preferably with the deep tow, would be necessary to substantiate this observation."

Smooth Suprafan Depositional Lobes and Outer Fan

These two areas are very difficult to distinguish, both physiographically and sedimentologically. They lack two of the major features attributed to submarine fans, channels, and the massive and pebbly sandstone facies, and their recognition may depend upon an outward fanning of paleocurrent directions, and evidence of coarsening-upward facies sequences. Even this latter criterion may not be diagnostic, because thickening- and coarsening-upward sequences could develop in any prograding wedge of turbidites, not necessarily associated with submarine fans (Walker and Sutton, 1967). Nevertheless, a fan situation appears to be the most likely for the thickening- and coarsening-upward sequences, and exploration in an upcurrent direction may locate a coarse, coalesced sand body representing the braided suprafan.

There are probably a large number of thickening- and coarsening-upward sequences of classic turbidites in North America (Fig. 15), and particularly in California but, unfortunately, very few have been described. The best known examples are from the Mediterranean area, particularly the Messanagros Sandstone (Oligocene–lower Miocene, Island of Rhodes, Greece; Mutti, 1969) and the San Salvatore Sandstone (lower Miocene, Appennines, Italy; Mutti and Ghibaudo, 1972). The sequences vary from about 5 to 45 m (Ricci Lucchi, 1975, Tables 1, 2), and may include a few, to as many as 37 individual coarse layers.

Obviously it would be of predictive importance if the thickness of these sequences, and number of coarse beds within them, were related to features on the braided suprafan (for example, channel depth, bed thickness, and grain size). Perhaps a thickening- and coarsening-upward sequence that itself is rather thin (<5 m) implies only a minor progradation, possibly related to rather shallow suprafan channels and very rapid channel switching. Conversely, a thicker (20 m +) thickening- and coarsening-upward sequence would imply more stable, and probably deeper, suprafan channels. Unfortunately, there are no data to substantiate these hypotheses.

Suprafan Lobe Switching: Reservoir Implications

Normark (1970, 1974, this issue) has emphasized that suprafan lobes tend to switch in position periodically. The control may be either a switch in position of the upper-fan channel, or a switch in position of the meandering thalweg channel (Fig. 18). The result is the relatively abrupt abandonment of one lobe, and the beginning of progradation of another. After abandonment, the only sediment reaching the old lobe will consist of suspended fine sediments and gradually, during progradation of the new lobe, the old one will be smothered by a mud blanket. These mud blankets are potentially excellent sealing layers for hydrocarbons that may be trapped in marine and pebbly sandstones of the suprafan lobe.

Neither the thicknesses of the suprafan lobes, nor of the mud blankets, are known from recent sediment studies. Few ancient studies refer specifically to suprafan lobes and blanketing mudstones. Reinterpretation of some examples described in the literature, together with my own observations, suggests that average suprafan-lobe mudstone blankets range from 5 to 50 m (Cretaceous of Simi Hills, California; Upper Carboniferous of northern England, Walker, 1966a; Ventura basin, Hsü, 1977). This thickness of mudstone would form a very effective barrier to vertical hydrocarbon migration, and each individual suprafan lobe could be sealed off as an individual reservoir (Fig. 18). Similar shale layers overlie sands, some of them gas producers, in the Forbes Formation (Grimes gas field, California; Weagant, 1972, and personal commun., 1977). The sands are from about 10 to 60 m thick, but the blanketing shales are more variable, ranging "from a few tens to several hundred feet" (Weagant, personal commun., 1977).

In lateral extent, suprafan lobes on modern fans are roughly equidimensional and from 10 to 15 km across (up to 50 km on Monterey fan). Their thickness can be estimated only from interpretations of ancient rocks but, for individual lobes, the range may be of the order of 20 to 120 m.

The relations between suprafan-lobe sandstone bodies and blanketing shales, particularly the shaling out of the sands at the margin of the lobe, are well displayed in electric logs of the Ventura field recently published by Hsü (1977, Fig. 4). A simplification of some of these logs from the middle producing zone is shown in Figure 21. Between markers DA and DC in well 6 there appears to be a thinning- and fining-upward sequence about 40 m thick. It grades laterally into siltstones and mudstones about 30 m thick in well 1; these mudstones appear to blanket the underlying sand body (DC to DF, Fig. 21). The sand body between markers DC and DF appears to contain a thickening- and coarsening-upward sequence in well 1 about 65 m thick, which also partly shales out laterally toward well 6. The generalized paleoflow is approximately westerly (Hsü, 1977), roughly from well 6 toward well 1. The shaling out therefore is seen parallel with the paleoflow direction, and this is illustrated diagrammatically by the line CD in Figure 18. The shaling out takes place over a distance of about 8 km, but this does not necessarily define the size of the suggested suprafan lobes. If the submarine-fan model is used in more detail to interpret these

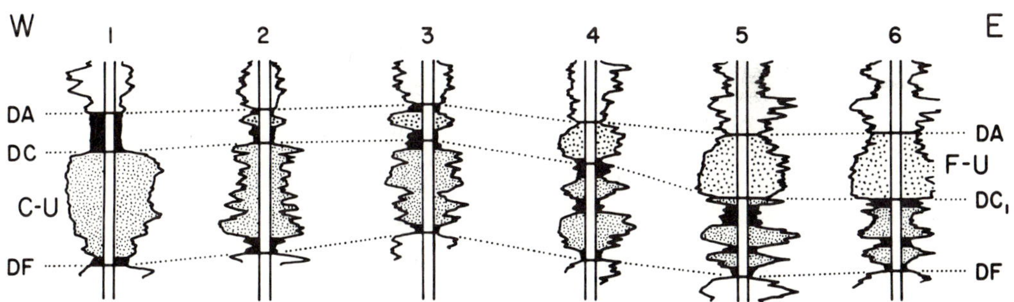

FIG. 21—Simplification of Hsü's (1977, Fig. 4) electric-log correlations of middle producing zones, lower Pliocene Repetto Formation, Ventura field, California. Correlations between E-log markers *DA*, *DC*, and *DF* are shown, and fining-upward (*F-U*) sequence in well 6, and coarsening-upward (*C-U*) sequence in well 1 are identified. Sandstones shale out, *DA* toward west (downcurrent) and *DC* toward east (upcurrent). This east to west downcurrent trend could be representative of cross section along *CD* of Figure 18. In order 1 through 6, the wells illustrated are Shell Taylor 505, Shell Taylor 349, TWA Hartman 44, TWA Lloyd 161, TWA Lloyd 165, and TWA V.L. & W. 84.

electric logs, it might be suggested that the thinning- and fining-upward DA-DC sands in wells 5 and 6 (Fig. 21) represent channel-fill sands. The thickening- and coarsening-upward DC-DF sand in well 1 is more suggestive of the unchanneled part of a prograding suprafan lobe. Thus the sandstone-mudstone packages between the electric-log markers appear to fit the switching suprafan-lobe model rather well.

THE SHALE GRIT: A COMPLETE ANCIENT SUBMARINE FAN

Several formations, or parts of formations, have been assigned to various depositional environments on submarine fans, and examples have been given earlier. However, there are very few descriptions of complete ancient submarine fans. By complete, I imply a prograding sequence from basin-plain turbidites, upward through fan deposits into silty mudstones deposited on a prograding slope. The fan deposits themselves must show evidence of channeling, and of the various facies sequences discussed earlier.

One of the best examples is the Shale Grit fan of northern England (Walker, 1966a, b). The term "Shale Grit" is a formal stratigraphic unit (dating from 1811) of formation status, and the fan is of Namurian (*Reticuloceras* R$_{1c}$ zone) age (late Carboniferous; Namurian is roughly equivalent to Morrowan plus Atokan).

The fan was deposited in a deep basin between two older limestone "highs" (Fig. 22). Supply of coarse-clastic material was consistently from the north or northeast, and the entire fan sequence, plus associated facies, represents the filling of the basin. The stratigraphic sequence is shown in Table 2. The original slope into the basin was created by block faulting at the southern margin on the northern limestone "high," and the stratigraphic sequence from the Edale Shales up into the Kinderscout Grit represents the southward progradation of facies down the basin, building up the observed stratigraphic sequence.

The initial clastic material being supplied to the southern (distal) end of the basin was black mud (up to 250 m thick), followed by classic turbidites of the Mam Tor Sandstones (100 to 130 m). According to Allen (1960, p. 194), sandstones increase in abundance upward and, by further thickening and coarsening, pass up into the Shale Grit. Thus the Mam Tor Sandstones would appear to constitute one overall thickening- and coarsening-upward sequence, and probably can be assigned to the basin-plain and lower-fan areas of the present submarine-fan model. The base of the Shale Grit was defined by the incoming of massive sandstones—beds consistently thicker than 60 cm, with many composite amalgamated sandstone beds without interbedded shales (Fig. 22). The sandstones are coarse (maximum grain size 3 to 4 mm), and commonly pebbly (up to 1 cm), and clearly belong to the massive and pebbly sandstone facies defined earlier in this paper (they originally were termed "facies C" by Walker, 1966a). As well as these facies, the Shale Grit also contains classic turbidites and thinly laminated dark mudstones. In the lower part of the Shale Grit (informally designated), classic turbidites are more abundant than the massive sandstones, and there are several thickening- and coarsening-upward sequences that begin with dark mudstones, and pass upward into classic turbidites and finally into massive sandstones (Fig. 22). I did not recognize these sequences as such in 1966, although several can be discerned in my original published stratigraphic sections (Walker, 1966a, Fig. 4, redrawn as Fig. 22 of this paper). Many more can be recognized in my unpublished sections (Walker, 1964), and they vary in thickness from a few meters to nearly 60 m. Only one major channel ($>$ 2 m deep) is known in the lower Shale Grit and hence, with the evidence of the facies, facies sequence, and general absence of channels, I now assign this part of the Shale Grit to the smooth outer parts of suprafan lobes. The thinly laminated dark mudstones in the Shale Grit probably represent quiet blanket deposition on top of an abandoned lobe, while a new lobe was forming elsewhere (Figs. 22, 23). They vary in thickness from 3 to 30 m, averaging about 10 m, and can be mapped continuously in the field for up to 10 km. The sequences of turbidites and massive sandstones between dark mudstones average about 30 m in thickness. It follows that an "average" prograding outer suprafan lobe is approximately 30 m thick and 4 to 10 km wide perpendicular to flow direction (Walker, 1966a, Table 3); this "average" lobe then was abandoned and draped with 10 m of dark mudstone.

The upper part of the Shale Grit differs from the lower part in two important ways; (1) the massive and pebbly sandstones are much more abundant than in the lower part and (2) there are nine major channels with observed depths up to 20 m (Fig. 22). The association of these particular facies with channels strongly suggests a braided-suprafan depositional environment. Thickening- and coarsening-upward sequences are present, up to 120 m thick, but contain coarser grained facies than the sequences in the lower part of the Shale Grit. The sequences contain nothing but massive and pebbly sandstones in their upper parts, and scouring and bed amalgamation is common. The major channels in the upper Shale Grit were de-

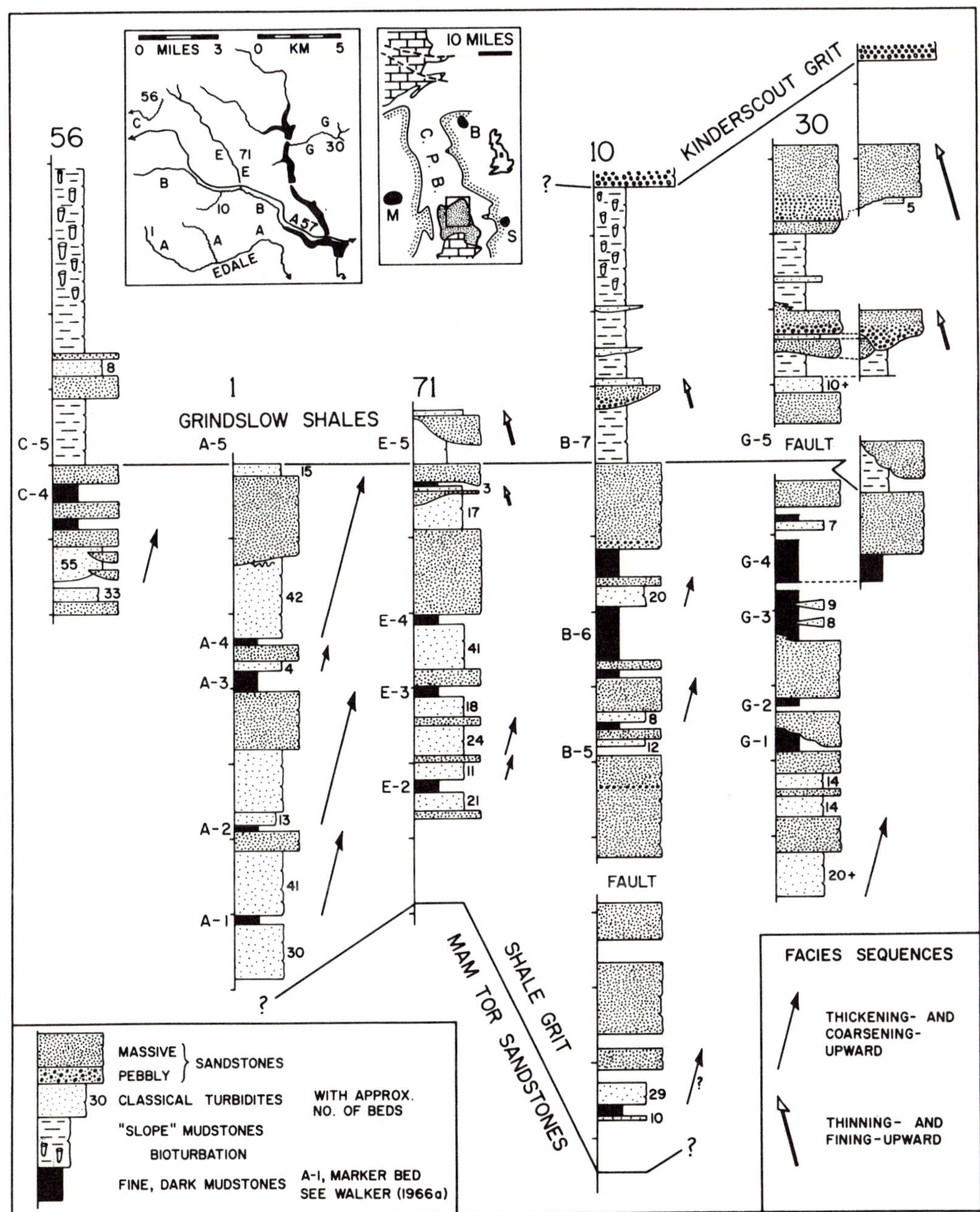

FIG. 22—Measured sections of Shale Grit and Grindslow Shales. Datum is at base of Grindslow Shales; thickness marks on left of columns are 25 m apart. Small inset map shows location of area at southern end of central Pennine basin, *C.P.B.*, between Sheffield, *S*, and Manchester, *M*. Larger inset map shows subdivision of area into smaller areas, *A, B* etc., in which mudstone marker beds can be mapped continuously. Numbers indicate location of measured sections. Highway A-57 links Sheffield and Manchester.

Table 2. Stratigraphy of Shale Grit Fan and Associated Rocks*

Formation	Approximate Thickness (m)	Lithology and Interpretation
Kinderscout Grit	150	Mainly coarse sandstones, some shales. Shallow water deltaic complex. (Collinson, 1969)
Grindslow Shales	100–120	Massive and laminated mudstones and shales. Mainly prograding slope deposits (Walker, 1966a, b; Collinson, 1969). Upper fan channels at base.
Shale Grit	130–240	Sandstones and shales. The upper part was mainly deposited on the braided suprafan; the lower part was deposited on smooth suprafan lobes.
Mam Tor Sandstones	100–130	Classical turbidites deposited on lower fan or basin plain (Allen, 1960).
Edale Shales	250	Black basinal mudstones.

*All formations are Late Carboniferous (Namurian) in age. Most Edale Shales belong to zones E, H, R_{1a} and R_{1b}. Uppermost Edale Shales, and all other formations, belong to zone R_{1c}.

scribed by Walker (1966b). Channel directions, interpreted from the strike of the channel walls, are toward south and southeast; this is in agreement with sole marks on the associated turbidites and massive sandstones. The fill is commonly fining upward (Fig. 22; Walker, 1966b, p. 1902), and there is evidence within the channel fills of slight channel shifting and minor periods of renewed scouring during channel filling. Again, this is very consistent with a braided-suprafan interpretation.

The upper fan, characterized by major channels stratigraphically surrounded by mudstones (Mutti and Ghibaudo, 1972), probably is represented by the lowermost part of the Grindslow Shales (Fig. 22). Here, there are seven major channels filled with fining-upward sequences of pebbly and massive sandstones. These channels probably merge imperceptably upslope into feeder channels, and the separate identification of the two is not possible. These channels have observed depths of up to 50 m, and mapping suggests widths of at least 1 km (Walker, 1966b). Many of them appear to be complex, multiple cuts. The silty mudstones outside the channels represent the bottom of the prograding slope and most of the sediment probably was deposited rapidly by spill-over from turbidity currents flowing down the feeder, and upper-fan channels.

The overall interpretation is shown in Figure 23. It is clear that the stratigraphic sequence (Table 2) represents an overall coarsening-upward prograding submarine fan. The system appears to have been fed by a very actively prograding, major delta (Pennine delta), and the slope and upper-fan environments, commonly not preserved, here have been buried rapidly by the delta.

The Shale Grit fan is particularly important because the mud blankets between individual suprafan lobes can be mapped in the field (Walker, 1964), and hence a quantitative reconstruction of the lobes can be attempted. It highlights the type of information that must be obtained from other ancient fans if a "general fan model" is to be used quantitatively in hydrocarbon exploration.

CLASSIC TURBIDITES WITHOUT SUBMARINE FANS

It is clear from the foregoing discussion that submarine fans are characterized by relatively coarse-grained facies associated with channels. In the absence of massive and pebbly sandstones, and in the absence of channels meters or tens of

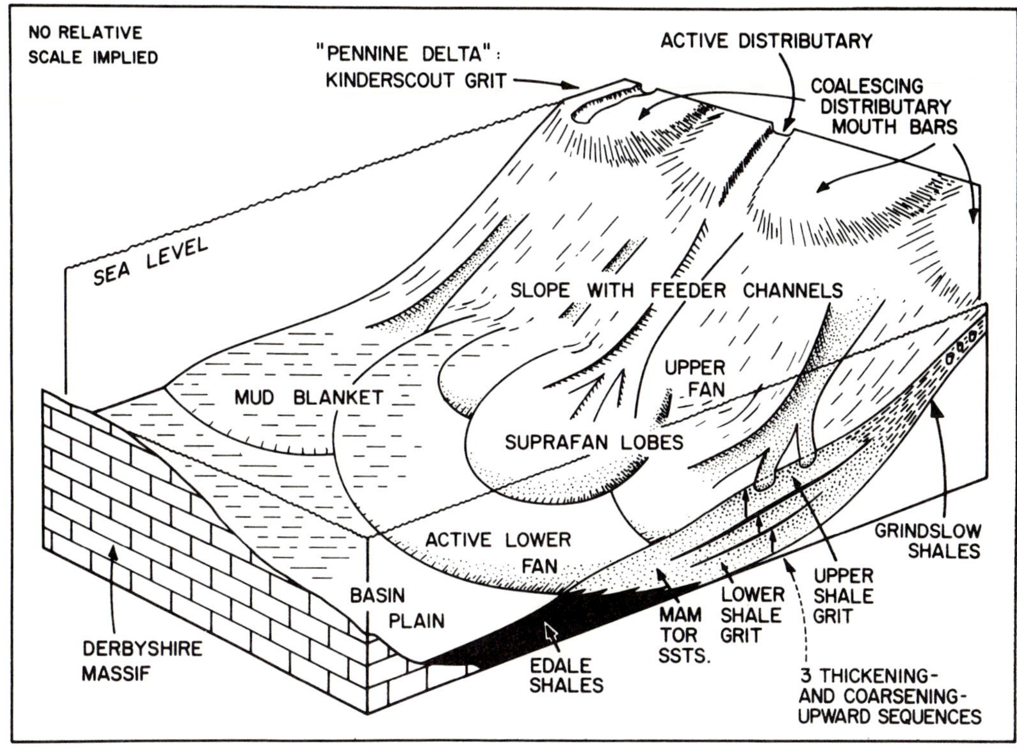

FIG. 23—Block diagram restoration of Shale Grit and associated formations. Modified from Walker (1966a). Compare with Figures 13 and 18.

meters deep, the fan model may be inappropriate or incorrect to use.

There are at least two situations in which classic turbidites are present in thick extensive sequences without evidence of associated coarse, channelized fans. The first is the exogeosynclinal situation, typified by the Middle Ordovician turbidites in Quebec (Enos, 1969) and the central Appalachians (McBride, 1962), and by the Mississippian and Pennsylvanian Stanley and Jackfork turbidites of the Ouachita Mountains (Morris, 1974). In these systems, paleocurrent flow is dominantly parallel with the basin axis, and hence parallel with tectonic strike. It generally is assumed that supply is from many points along the margins of these elongate troughs, and that turbidity currents flow downslope toward the basin axis, and then swing around to flow consistently parallel with the basin axis (see, for example, Morris, 1974, Figs. 14, 19). If fans of the type discussed in this paper are formed, they would be at the foot of the marginal slope into the basin. However, this environment normally is destroyed during continued uplift in the source area and thrusting of the turbidite sequences toward the craton, as typified in the Appalachian and Ouachita systems. Even within the axial turbidite systems prediction is difficult. It must be remembered that all of the prediction of facies relations implied by Figure 13 is based upon the spreading of turbidity currents from a single feeder channel. Within the long exogeosynclinal troughs, the turbidites in any one vertical stratigraphic sequence may have come from several different sources. Vertical trends (thickening or thinning upward) may be meaningless or absent, unless individual beds can be assigned to their own source, and it never has been demonstrated that this is possible.

A second situation of abundant classic turbidites without associated coarser facies (massive and pebbly sandstones), and without deep channels, is in prodeltaic areas on the craton. One of the best examples is the Upper Devonian turbidite sequence of the central Appalachians, in the lower part of the Catskill clastic wedge. In New York State, the only evidence of submarine fans consists of two superimposed thickening-upward sequences of classic turbidites (Walker and Sutton, 1967); these Upper Devonian units remain flat into northern Pennsylvania and form one of the reservoirs in the Bradford field (Dixon, 1972).

In the Susquehanna Valley area of Pennsylvania, there appear to be no bed thickness or grain-size trends within the 450-m thick sequences of classic turbidites (Walker, 1971). Where the Susquehanna Valley sequences are traced eastward for about 200 km toward the Lehigh Valley area (toward source), the classic turbidite sequences thicken upward into massive sandstones. However, apart from this facies change, there is no other indication that the submarine-fan model as discussed in this paper is applicable to the Upper Devonian "prodeltaic" turbidites.

Prodeltaic turbidites are present in other parts of the Upper Devonian and Mississippian clastic wedges of the Appalachians, and there is a thin turbidite sequence in the lower part of the Cretaceous Mesaverde clastic wedge in the Price, Utah, area (lower part of the Panther tongue). However, models have not been worked out for prodeltaic classic turbidites on the craton. Differences from the submarine-fan model discussed in this paper may arise from differences in grain size and turbidity-current generating mechanisms at the deltaic source, differences in length and angle of the prodelta slope on which the currents accelerate, the possible absence of a single main feeder channel on the delta slope, and relatively rapid switching of delta lobes resulting in changing points of supply of turbidity currents to their depositional areas.

EXPLORATION PREDICTIONS FROM FAN MODEL

As a conclusion to this review, I will summarize some important ways in which the fan model might be used in the prediction of stratigraphic traps. In emphasizing the general points, it will be necessary to ignore local problems of source and migration of the petroleum, diagenesis of the reservoir sands, and tectonic disruptions.

1. The study of the thickness, distribution, and grain size of the various facies suggests that a principal target would be the inner part of a suprafan lobe. Here, the massive and pebbly sandstones would be relatively coarse and thick, with fewer and thinner shales between individual beds. Also, each suprafan lobe may be sealed off by a mud blanket during lobe abandonment (Figs. 18, 21). Suprafan lobes may be identified from their coarsening-upward appearance on electric logs, together with associated fining-upward sequences due to channel filling on the suprafan lobes (Fig. 18). However, in the absence of any other information about the turbidite or deep-water aspect of the system, channelized suprafan lobes would be indistinguishable on electric logs from delta lobes with distributaries.

2. Suprafan lobes might be predicted at the downstream end of feeder channels and upper-fan channels. If they have not been displaced too much by subsequent tectonics, suprafan lobes might be found as downstream continuations of the Rosedale, Meganos, Yoakum, and Gevaram channels. If a positive identification of an upper-fan channel (as opposed to a slope feeder channel) could be made, the channel width or channel cross-sectional area may predict the fan radius, as suggested by Normark (this issue, Figs. 6A, B).

3. Unusual facies relations have important predictive value, particularly the discovery of incised channels in which the channel fill is uncommonly coarse, and cuts into facies that normally would be expected much farther downfan (Fig. 13). Such channels may well have new suprafan lobes at their downstream end. These would suggest potential drilling targets much farther out into the basin than would be expected without the evidence of channel incision. In this context, it must be remembered that the channels themselves might not be visible, but could be predicted by the presence of thinning- and fining-upward sequences. If such sequences are found beginning with massive and pebbly sandstones, or even conglomerates, and associated outside the sequences with dark mudstones and classic turbidites, the assemblage also might represent an incised-channel situation with a possible fan at the downstream end. The sandy Doheny channel (Piper and Normark, 1971) cutting into silts with classic turbidites, and the Wheeler gorge (California) conglomeratic thinning- and fining-upward sequences that rest on dark mudstones (Walker, 1975b) might be two examples where fans could be predicted downstream.

4. Several feeder channels are known from subsurface studies to be filled with fine-grained silts and clays. These can act as updip sealing layers for hydrocarbons in rocks outside the channels. The fine-grained fill appears to be related to channel abandonment, probably because of a relative rise of sea level (as is known for the Mississippi channel, Sabate, 1968). However, relative rise of sea level is a widespread phenomenon, and it is probable that adjacent canyons and channels also were abandoned and filled with fine-grained

materials. Thus the discovery of one major mud-filled feeder channel (Yoakum, Meganos, Mississippi, Gevaram) predicts that if adjacent feeders were present, they also probably will be mud filled, and perhaps act as updip sealing layers.

REFERENCES CITED

Allen, J. R. L., 1960, The Mam Tor Sandstones: a "turbidite" facies of the Namurian deltas of Derbyshire, England: Jour. Sed. Petrology, v. 30, p. 193-208.

———— 1964, Primary current lineation in the Old Red Sandstone (Devonian), Anglo-Welsh basin: Sedimentology, v. 3, p. 89-108.

Barbat, W. F., 1958, The Los Angeles basin area, California; in L. G. Weeks, ed., Habitat of oil: AAPG, p. 62-77.

Bartow, J. A., 1966, Deep submarine channel in upper Miocene, Orange County, California: Jour. Sed. Petrology, v. 36, p. 700-705.

Benson, P. H., 1971, Geology of the Oligocene Hackberry trend, Gillis English Bayou—Manchester area, Calcasieu Parish, Louisiana: Gulf Coast Assoc. Geol. Socs. Trans., v. 21, p. 1-14.

Bloomer, R. R., 1977, Depositional environments of a reservoir sandstone in west-central Texas: AAPG Bull., v. 61, p. 344-359.

Bouma, A. H., 1962, Sedimentology of some flysch deposits: Amsterdam, Elsevier, 168 p.

Buffington, E. C., 1964, Structural control and precision bathymetry of La Jolla submarine canyon, California: Marine Geology, v. 1, p. 44-58.

Cleary, W. J., and J. R. Connolly, 1974, Hatteras deep-sea fan: Jour. Sed. Petrology, v. 44, p. 1140-1154.

Cohen, Z., 1976, Early Cretaceous buried canyon: influence on accumulation of hydrocarbons in Helez oil field, Israel: AAPG Bull., v. 60, p. 108-114.

Colburn, I. P., 1968, Grain fabrics in turbidite sandstone beds and their relationship to sole mark trends on the same beds: Jour. Sed. Petrology, v. 38, p. 146-158.

———— and A. E. Fritsche, 1973, eds., Cretaceous stratigraphy of the Santa Monica Mountains and Simi Hills, southern California: SEPM Pacific Sec. Guidebook, 93 p.

Collinson, J. D., 1969, The sedimentology of the Grindslow Shales and the Kinderscout Grit: a deltaic complex in the Namurian of northern England: Jour. Sed. Petrology, v. 39, p. 194-221.

Davies, I. C., and R. G. Walker, 1974, Transport and deposition of resedimented conglomerates: the Cap Enragé Formation, Gaspé, Québec: Jour. Sed. Petrology, v. 44, p. 1200-1216.

Dickas, A. B., and J. L. Payne, 1967, Upper Paleocene buried channel in Sacramento Valley, California: AAPG Bull., v. 51, p. 873-882.

Dixon, W. H., 1972, Reservoir geology of the Sage Farm, Bradford oil field, McKean County, Pennsylvania, in Field trip guidebook: AAPG Eastern Sec., p. 111-1 - 111-22.

Dzulynski, S., and E. K. Walton, 1965, Sedimentary features of flysch and greywackes: New York, Elsevier, 274 p.

Edmondson, W. F., 1965, The Meganos Gorge of the southern Sacramento Valley: San Joaquin Geol. Soc. Selected Papers, no. 3, p. 36-51.

Enos, P., 1969, Anatomy of a flysch: Jour. Sed. Petrology, v. 39, p. 680-723.

Fowler, C., 1975, The geology of the Montrose field, in A. W. Woodland, ed., Petroleum and the continental shelf of northwest Europe, v. 1, Geology: New York, Wiley, p. 467-476.

Gardett, P. H., 1971, Petroleum potential of Los Angeles basin, California, in Future petroleum provinces of the United States—their geology and potential: AAPG Mem. 15, p. 298-308.

Gorsline, D. S., and K. O. Emery, 1959, Turbidity current deposits in San Pedro and Santa Monica basins off southern California: Geol. Soc. America Bull., v. 70, p. 279-290.

Gregory, M. R., 1969, Sedimentary features and penecontemporaneous slumping in the Waitemata Group, Whangaparaoa Peninsula, North Auckland, New Zealand: New Zealand Jour. Geology and Geophysics, v. 12, p. 248-282.

Hampton, M. A., 1972, The role of subaqueous debris flow in generating turbidity currents: Jour. Sed. Petrology, v. 42, p. 775-793.

Hand, B. M., and K. O. Emery, 1964, Turbidites and topography of north end of San Diego Trough, California: Jour. Geology, v. 72, p. 526-542.

Haner, B. E., 1971, Morphology and sediments of Redondo submarine fan, southern California: Geol. Soc. America Bull., v. 82, p. 2413-2432.

Hansen, E., 1967, Methods of deducing slip line orientations from the geometry of folds: Carnegie Inst. Washington Year Book 65, p. 387-405.

———— 1971, Strain facies: New York, Springer-Verlag, 207 p.

Harms, J. C., and R. K. Fahnestock, 1965, Stratification, bed forms and flow phenomena (with example from the Rio Grande), in Primary sedimentary structures and their hydrodynamic interpretation: SEPM Spec. Pub. 12, p. 84-115.

———— et al, 1975, Depositional environments as interpreted from primary sedimentary structures and stratification sequences: SEPM Short Course 2, Dallas, 161 p.

Hendry, H. E., 1973, Sedimentation of deep water conglomerates in lower Ordovician rocks of Quebec—composite bedding produced by progressive liquefaction of sediment?: Jour. Sed. Petrology, v. 43, p. 125-136.

Hoyt, W. V., 1959, Erosional channel in the middle Wilcox near Yoakum, Lavaca County, Texas: Gulf Coast Assoc. Geol. Socs. Trans., v. 9, p. 41-50.

Hsü, K. J., 1977, Studies of Ventura field, California, 1: Facies geometry and genesis of lower Pliocene turbidites: AAPG Bull., v. 61, p. 137-168.

Hubert, C., J. Lajoie, and M. A. Léonard, 1970, Deep sea sediments in the lower Paleozoic Québec Supergroup, in Flysch sedimentology in North America: Geol. Assoc. Canada Spec. Paper 7, p. 103-125.

Johnson, B. A., 1974, Deep sea fan-valley conglomerate: Cap Enragé Formation, Gaspe, Quebec: Master's thesis, McMaster Univ., 108 p.

Kruit, C., et al, 1975, Une excursion aux cones d'alluvions en eau profonde d'age Tertiaire près de San Sebastian (Province de Guipuzcoa, Espagne): 9th Internat. Cong. Sedimentologie, Nice, France, Excursion 23 Guidebook, 75 p. (in English).

Kuenen, P. H., and C. I. Migliorini, 1950, Turbidity currents as a cause of graded bedding: Jour. Geology, v. 58, p. 91-127.

Lajoie, J., 1972, Slump fold axis orientations—an indication of paleoslope?: Jour. Sed. Petrology, v. 42, p. 584-586.

Link, M. H., 1975, Matilija Sandstone—a transition from deep water turbidite to shallow marine deposition in the Eocene of California: Jour. Sed. Petrology, v. 45, p. 63-78.

Lowe, D. R., 1975, Water escape structure in coarse-grained sediments: Sedimentology, v. 22, p. 157-204.

———— 1976, Subaqueous liquefied and fluidized sediment flows and their deposits: Sedimentology, v. 23, p. 285-308.

———— and R. D. LoPiccolo, 1974, The characteristics and origins of dish and pillar structures: Jour. Sed. Petrology, v. 44, p. 484-501.

Martin, B. D., 1963, Rosedale channel—evidence for late Miocene submarine erosion in Great Valley of California: AAPG Bull., v. 47, p. 441-456.

Mayuga, M. N., 1970, Geology and development of California's giant—Wilmington oil field, in Geology of giant petroleum fields: AAPG Mem. 14, p. 158-184.

McBride, E. F., 1962, Flysch and associated beds of the Martinsburg Formation (Ordovician), central Appalachians: Jour. Sed. Petrology, v. 32, p. 39-91.

———— 1966, Sedimentary petrology and history of the Haymond Formation (Pennsylvanian), Marathon basin, Texas: Texas Bur. Econ. Geology Rept. Invest. 57, 101 p.

Middleton, G. V., and M. A. Hampton, 1976, Subaqueous sediment transport and deposition by sediment gravity flows, in D. J. Stanley, and D. J. P. Swift, eds., Marine sediment transport and environmental management: New York, Wiley Intersci., p. 197-218.

Morris, R. C., 1974, Carboniferous rocks of the Ouachita Mountains, Arkansas—a study of facies patterns along the unstable slope and axis of a flysch trough, in Symposium on the Carboniferous rocks of the southeastern United States: Geol. Soc. America Spec. Paper 148, p. 241-279.

Mutti, E., 1969, Sedimentologia delle Arenarie di Messanagros (Oligocene–Aquitaniano) nell' isola di Rodi: Soc. Geol. Italiana Mem., v. 8, p. 1027-1070.

———— 1974, Examples of ancient deep-sea fan deposits from Circum-Mediterranean geosynclines, in Modern and ancient geosynclinal sedimentation: SEPM Spec. Pub. 19, p. 92-105.

———— 1977, Distinctive thin-bedded turbidite facies and related depositional environments in the Eocene Hecho Group (south-central Pyrenees, Spain): Sedimentology, v. 24, p. 107-131.

———— and G. Ghibaudo, 1972, Un esempio di torbiditi di conoide sottomarina esterna: le Arenarie di San Salvatore (Formazione di Bobbio, Miocene) nell' Appennino di Piacenza: Mem. Acc. Sci. Torino, Classe Sci. Fis., Nat., ser. 4, n. 16, 40 p.

———— and F. Ricci Lucchi, 1972, Le torbiditi dell' Appennino settentrionale: introduzione all'analisi di facies: Soc. Geol. Italiana Mem., v. 11, p. 161-199.

Nagle, H. E., and E. S. Parker, 1971, Future oil and gas potential of onshore Ventura basin, California, in Future petroleum provinces of the United States—their geology and potential: AAPG Mem. 15, p. 254-297.

Natland, M. L., 1963, Paleoecology and turbidites: Jour. Paleontology, v. 37, p. 946-951.

———— and P. H. Kuenen, 1951, Sedimentary history of the Ventura basin, California, and the action of turbidity currents, in Turbidity currents and the transportation of coarse sediment into deep water: SEPM Spec. Pub. 2, p. 76-107.

Nilsen, T. H., and T. R. Simoni, 1973, Deep-sea fan paleocurrent patterns of the Eocene Butano Sandstone, Santa Cruz Mountains, California: U.S. Geol. Survey Jour. Research, v. 1, p. 439-452.

Normark, W. R., 1970, Growth patterns of deep-sea fans: AAPG Bull., v. 54, p. 2170-2195.

———— 1974, Submarine canyons and fan valleys: factors affecting growth patterns of deep-sea fans, in Modern and ancient geosynclinal sedimentation: SEPM Spec. Pub. 19, p. 56-68.

———— and D. J. W. Piper, 1969, Deep-sea fan valleys past and present: Geol. Soc. America Bull., v. 80, p. 1859-1866.

Onions, D., and G. V. Middleton, 1968, Dimensional grain orientation of Ordovician turbidite greywackes: Jour. Sed. Petrology, v. 38, p. 164-174.

Paine, W. R., 1966, Stratigraphy and sedimentation of subsurface Hackberry wedge and associated beds of southwestern Louisiana: Gulf Coast Assoc. Geol. Socs. Trans., v. 16, p. 261-274.

Parkash, B., and G. V. Middleton, 1970, Downcurrent textural changes in Ordovician turbidite greywackes: Sedimentology, v. 14, p. 259-293.

Parker, J. R., 1975, Lower Tertiary sand development in the central North Sea, in A. W. Woodland, ed., Petroleum and the continental shelf of northwest Europe. 1, Geology: London, Applied Sci. Pubs., p. 447-452.

Piper, D. J. W., and W. R. Normark, 1971, Re-examination of a Miocene deep-sea fan and fan valley, southern California: Geol. Soc. America Bull., v. 82, p. 1823-1830.

Ricci Lucchi, F., 1975, Depositional cycles in two turbidite formations of northern Apennines (Italy): Jour. Sed. Petrology, v. 45, p. 3-43.

Rocheleau, M., and J. Lajoie, 1974, Sedimentary structures in resedimented conglomerate of the Cambrian flysch, L'Islet, Quebec Appalachians: Jour. Sed. Petrology, v. 44, p. 826-836.

Sabate, R. W., 1968, Pleistocene oil and gas in coastal Louisiana: Gulf Coast Assoc. Geol. Socs. Trans., v. 18, p. 373-386.

Shepard, F. P., and G. Einsele, 1962, Sedimentation in San Diego Trough and contributing submarine canyons: Sedimentology, v. 1, p. 81-133.

———— R. F. Dill, and U. Von Rad, 1969, Physiography and sedimentary processes of La Jolla submarine fan and fan-valley, California: AAPG Bull., v. 53, p. 390-420.

Skipper, K., 1977, Offshore petroleum developments in northwest Europe—an update: Geosci. Canada, v. 4, p. 31-40.

Spotts, J. H., and O. E. Weser, 1964, Directional properties of a Miocene turbidite, California, *in* A. H. Bouma, and A. Brouwer, eds., Turbidites: New York, Elsevier, p. 198-221.

Stanley, D. J., 1975, Submarine canyon and slope sedimentation (Grès d'Annot) in the French Maritime Alps: 9th Internat. Cong. Sedimentologie, Nice, France, 129 p.

Sullwold, H. H., Jr., 1960, Tarzana fan, deep submarine fan of late Miocene age, Los Angeles County, California: AAPG Bull., v. 44, p. 433-457.

——— 1961, Turbidites in oil exploration, *in* J. A. Peterson, and J. C. Osmond, eds., Geometry of sandstone bodies: AAPG, p. 63-81.

Thomas, A. N., P. J. Walmsley, and D. A. L. Jenkins, 1974, Forties field, North Sea: AAPG Bull., v. 58, p. 396-406.

Walker, R. G., 1964, Some aspects of the sedimentology of the Shale Grit and Grindslow Shales (Namurian R_{1c}, Derbyshire) and the Westward Ho! and Northam Formations (Westphalian, north Devon): PhD thesis, Oxford Univ., 178 p.

——— 1965, The origin and significance of the internal sedimentary structures of turbidites: Yorkshire Geol. Soc. Proc., v. 35, p. 1-32.

——— 1966a, Shale Grit and Grindslow Shales: transition from turbidite to shallow water sediments in the Upper Carboniferous of northern England: Jour. Sed. Petrology, v. 36, p. 90-114.

——— 1966b, Deep channels in turbidite-bearing formations: AAPG Bull., v. 50, p. 1899-1917.

——— 1967, Turbidite sedimentary structures and their relationship to proximal and distal depositional environments: Jour. Sed. Petrology, v. 37, p. 25-43.

——— 1969, Geometrical analysis of ripple-drift cross-lamination: Canadian Jour. Earth Sci., v. 6, p. 383-391.

——— 1970, Review of the geometry and facies organization of turbidites and turbidite-bearing basins: *in* Flysch sedimentology in North America: Geol. Assoc. Canada Spec. Paper 7, 219-251.

——— 1971, Nondeltaic depositional environments in the Catskill clastic wedge (Devonian) of central Pennsylvania: Geol. Soc. America Bull., v. 82, p. 1305-1326.

——— 1975a, Generalized facies models for resedimented conglomerates of turbidite association: Geol. Soc. America Bull., v. 86, p. 737-748.

——— 1975b, Upper Cretaceous resedimented conglomerates at Wheeler Gorge, California: description and field guide: Jour. Sed. Petrology, v. 45, p. 105-112.

——— 1975c, Nested submarine-fan channels in the Capistrano Formation, San Clemente, California: Geol. Soc. America Bull., v. 86, p. 915-924.

——— 1976a, Facies models. 1, General introduction: Geosci. Canada, v. 3, p. 21-24.

——— 1976b, Facies models. 2, Turbidites and associated coarse clastic deposits: Geosci. Canada, v. 3, p. 25-36.

——— 1977, Deposition of upper Mesozoic resedimented conglomerates and associated turbidites in southwestern Oregon: Geol. Soc. America Bull., v. 88, p. 273-285.

——— and E. Mutti, 1973, Turbidite facies and facies associations, *in* G. V. Middleton, and A. H. Bouma, eds., Turbidites and deep-water sedimentation: SEPM Pacific Sec. Short Course, Anaheim, California, p. 119-157.

——— and R. G. Sutton, 1967, Quantitative analysis of turbidites in the upper Devonian Sonyea Group, New York: Jour. Sed. Petrology, v. 37, p. 1012-1022.

Weagant, F. E., 1972, Grimes gas field, Sacramento Valley, California, *in* Stratigraphic oil and gas fields: AAPG Mem. 16, p. 428-439.

Whitaker, J. H. McD., 1974, Ancient submarine canyons and fan valleys, *in* Modern and ancient geosynclinal sedimentation: SEPM Spec. Pub. 19, p. 106-125.

Winn, R. D., Jr., and R. H. Dott, Jr., 1977, Large-scale traction-produced structures in deep-water fan-channel conglomerates in southern Chile: Geology, v. 5, p. 41-44.

Yerkes, R. F., et al, 1965, Geology of the Los Angeles basin, California—an introduction: U.S. Geol. Survey Prof. Paper 420A, 57 p.

The American Association of Petroleum Geologists Bulletin
V. 54, No. 11 (November, 1970), P. 2170-2195, 25 Figs., 2 Tables

Growth Patterns of Deep-Sea Fans[1]

WILLIAM R. NORMARK[2]

La Jolla, California 92037

Abstract　The growth pattern of a deep-sea fan relates events in and around the fan-valleys to the structure and morphology of the open fan. The growth pattern cannot be determined without knowledge of the origin and recent history of the fan-valley system. The mapping of La Jolla and San Lucas deep-sea fans with the deep-towed instrument package developed at Marine Physical Laboratory of the Scripps Institution of Oceanography details the fine-scale morphology, structure, and internal fill of the fan-valleys and suggests the growth patterns of these fans.

The La Jolla fan, 20 km west of Scripps Institution, has one meandering fan-valley that extends across the entire fan. Except on the toe of the fan, the deeply incised valley has terraced walls with steeper walls on the outside of meanders. Very low-relief levees border the fan-valley in some localities. The present erosional valley bypasses the partly buried remnants of an older distributary system on the lower fan.

The San Lucas fan, off the southern tip of the peninsula of Baja California, shows a depositional lobe of sediment, or suprafan, below the short, leveed fan-valley extending from San Jose Canyon. The suprafan appears as a convex-upward bulge on a radial profile of the fan. The surface of the suprafan has a series of discontinuous depressions up to 55 m deep and 1 km wide. The depressions are generally asymmetric in cross section, commonly have terraced walls, and are underlain by coarse sand and gravel. They are interpreted to be channel remnants.

A model for deep-sea fan growth, based on this study, predicts that deposition on a fan will be localized in a suprafan at the end of large, leveed valleys commonly found on, and generally confined to, the upper reaches of deep-sea fans. The suprafan normally is on the midfan and is characterized by numerous smaller distributary channels. Rapid aggradation in the suprafan coupled with migration and meandering of the channels produces a surface marked by isolated depressions or channel remnants. Uniform deposition, producing a symmetrical half-cone morphology, results from the shifting through time of fan-valleys across the area of the fan.

[1] Manuscript received, January 8, 1970; accepted, April 27, 1970. Contribution of the Scripps Institution of Oceanography, new series.

[2] University of California, San Diego, Marine Physical Laboratory of the Scripps Institution of Oceanography. Present address: Department of Geology and Geophysics, University of Minnesota, Minneapolis, Minnesota 55455.

The success of the field work for this study is due in large part to J. R. Curray, F. N. Spiess, scientific leader for expeditions TOW MAS and TIPTOW, and the technicians and engineers of the deep-tow group, including M. S. McGehee, D. E. Boegeman, M. D. Benson, J. T. Donovan, F. W. Stone, and G. E. Forbes. The cooperation and help of the officers, crew, and scientific staff of the R/V *Thomas Washington* and ST-908 are greatly appreciated.

Fellow deep-tow enthusiasts, J. D. Mudie, R. L. Larson, B. P. Luyendyk, T. M. Atwater, and G. J. Miller, were of immense help in the manipulation and discussion of the geophysical records taken with the deep-tow instrument. E. L. Hamilton provided sound-velocity data from sediment cores collected during the surveys. F. J. Emmel discussed results of recent geophysical surveys on the Bengal fan, and G. B. Griggs, C. H. Nelson, and J. E. Andrews discussed their results from recent studies on deep-sea sedimentation.

D. J. W. Piper, a fellow student of deep-sea fans, and J. R. Curray provided many helpful suggestions during the early phases of this research. F. N. Spiess, J. R. Curray, P. Wilde, E. A. Silver, D. W. Scholl, and R. LeB. Hooke critically reviewed the manuscript.

The research was financed by the Office of Naval Research, the National Science Foundation, and the Deep Submergence Systems Project.

Introduction

Shepard suggested that "depositional fans" may represent turbidity-current material deposited as a result of a marked decrease in bed slope at the mouths of submarine canyons (Menard, 1955). Although most students of deep-sea fans (Menard, 1955) now accept this basic premise, there is little agreement as to the processes controlling fan morphology. Studies based on fan morphology and distribution of near-surface sediments are not in themselves adequate to determine the growth processes of the deep-sea fans. For this reason, the present study involves a comparative interpretation of the morphology, structure, and internal fill of deep-sea fan-valleys and their relation to the overall growth patterns of deep-sea fans. Major clues for constructing the growth patterns are gained from an understanding of (1) the origin of terraces, levees, and slump features associated with the fan-valleys, (2) the nature of the termination of the fan-valleys, and (3) the origin of fine-scale structures and the morphology of the fan surface outside the fan-valley. The deduced origin and recent history of the present fan-valleys on each fan then are related to the overall fan morphology to define the *growth pattern.*

A model for deep-sea fan growth combines the observations from the two fans described in this study with data on fans described in the literature. The model relates events in the fan-valleys to the development of the *shape* (lateral dimensions) and *profile* (radial surface profile) of deep-sea fans. If possible, the growth pattern for each fan is determined; otherwise, the reasons for the alteration of a growth pattern are evaluated. The model for fan growth is applied to the sublacustrine fan of the Rhone delta in Lake Geneva (Forel, 1885; Houbolt and Jonker, 1968). The processes governing deep-sea fan growth are shown to be somewhat analogous to those operating on subaerial alluvial fans.

EQUIPMENT

Although many investigators have concentrated on fan-valley systems of deep-sea fans, the morphology and internal structure of these features are not well known. With conventional echo-sounding devices it is possible only to determine the plan of fan-valleys: even this capability is handicapped in many surveys because of poor navigational systems. The only well-surveyed fan-valley to date is the La Jolla fan-valley (Shepard and Buffington, 1968); a precision shore-based radio navigation system (Buffington, 1960) provided accurate positioning for the survey.

Conventional wide-beam echo sounders provide no information from within narrow depressions on the sea floor (Krause, 1962), as shown in Figure 1. The upper trace of Figure 1A shows a typical profile over La Jolla fan-valley in 1,015 m of water taken with a conventional echo sounder operating on a 1-sec sweep. The lower trace (Fig. 1A) shows the same valley crossing as recorded by the narrow-beam sounding system on the deep-towed vehicle. All the terraces and other details in and near the fan-valley that are reflected by the narrow-beam echo sounder are obscured by the hyperbolic returns coming from near the upper edges of the valley on the conventional surface profile. The profiles of Figure 1A are from relatively shallow water. The problem of hyperbolic side echoes masking the true geometry becomes critical as the water depth increases, as illustrated in Figure 1B.

Study of the structure and morphology within fan-valleys obviously requires the use of a narrow-beam echo sounder. The deep-tow vehicle developed at Marine Physical Laboratory provides not only a precision narrow-beam echo sounder that operates near the sea floor, but also a 3.8-kHz reflection profiling system, side-looking sonars, stereo cameras, proton magnetometer, and a precision positioning system (Spiess *et al.*, 1967; Spiess and Mudie,

1970). The deep-tow instrument, or "fish," is maintained close to the sea floor on the end of an armored coaxial cable that provides power and commands to the fish and relays responses from the fish systems back to recorders on the towing ship (Spiess and Mudie, 1970).

FIELD WORK

The deep-tow data presented herein represent five separate field operations. The La Jolla fan, off southern California (Fig. 2), was surveyed during three separate short cruises: November 1967 and May 1968, test trips for the instrument, and May to June 1968, expedition TIPTOW TWO. During TIPTOW TWO, three bottom acoustic transponders (McGehee and Boegeman, 1966) provided accurate horizontal positioning for both the deep tow and the towing vessel, R/V *Thomas Washington*. The geographic positions of the transponders were determined by radar fixes on the San Diego coastline. The data from La Jolla fan are presented in two parts, the upper fan survey and the lower fan survey.

The deep-tow investigation of the San Lucas fan at the tip of Baja California (Fig. 2) included two separate surveys. Expedition TOW MAS, in November 1967, was terminated unexpectedly by loss of the deep-tow fish after six days of operation. In May 1968, expedition TIPTOW completed the work on the fan with a new instrument. Bottom acoustic transponders were used in both surveys although there was no common transponder in the two operations. Nevertheless, the relative positioning of the two transponder nets was accurate enough to allow recovery of the fish lost during the TOW MAS expedition by means of transponders dropped during expedition TIPTOW. Transponder navigation provides a very accurate relative positioning system with rms range errors (from three or more transponders to the fish) on the order of 4 m. (See Lowenstein and Mudie, 1967, and Spiess *et al.*, 1969, for more detailed discussions concerning positioning with bottom transponders.)

LA JOLLA FAN

The La Jolla deep-sea fan occupies the lower San Diego Trough west of La Jolla, California. The fan extends 35 km from the mouth of La Jolla submarine canyon to the base of Thirty Mile Bank, which forms the west scarp of the trough. The La Jolla fan-valley is continuous with the canyon and extends west and southwest across the entire fan (Fig. 3).

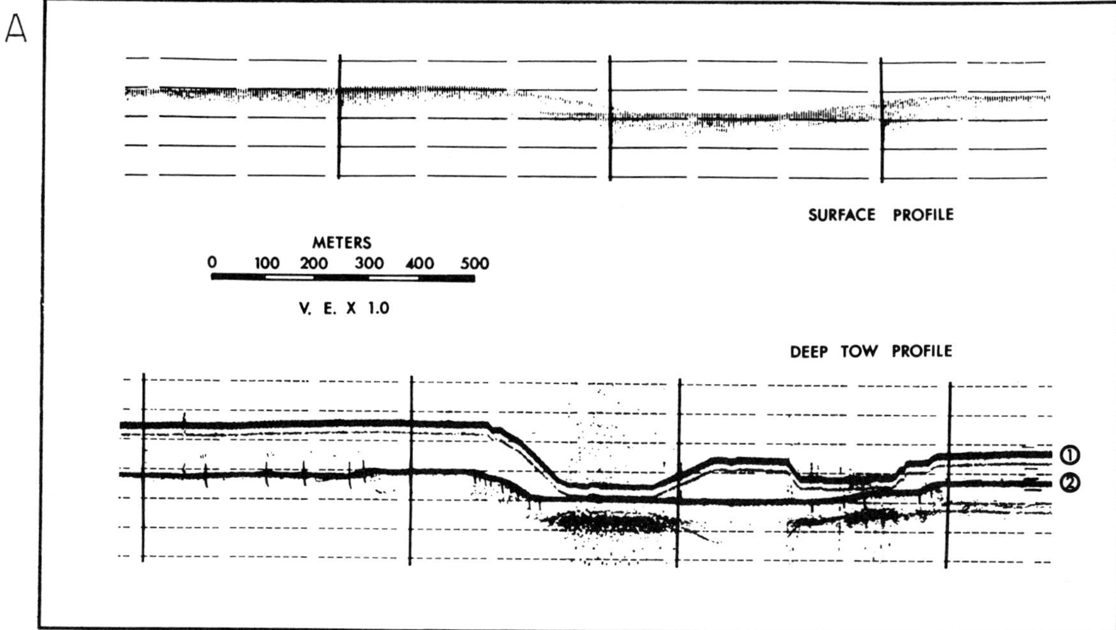

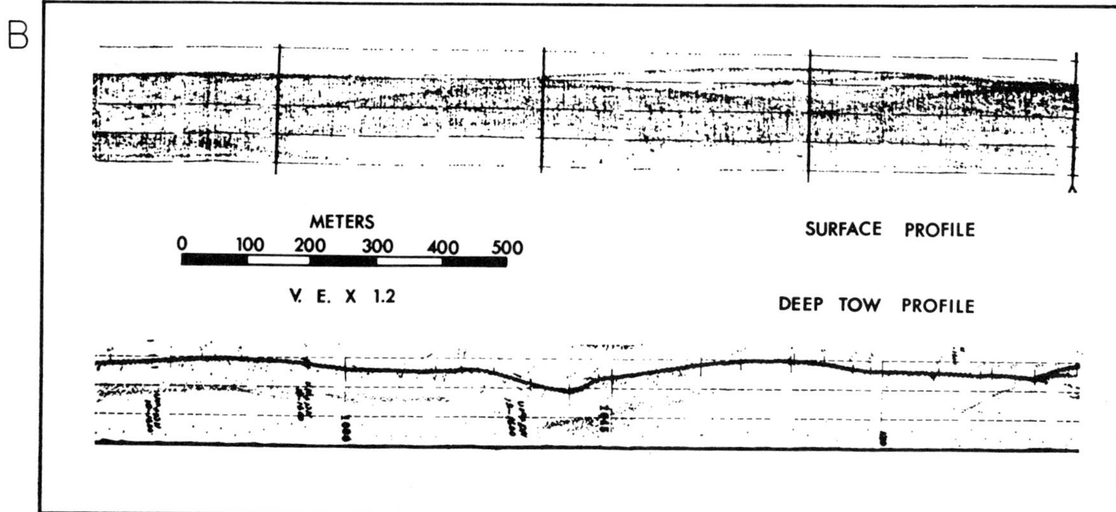

FIG. 1.—**A.** Comparison of bathymetric profiles taken with conventional wide-beam echo sounder on ship (surface profile) and with narrow-beam sounding system on deep-tow instrument (deep-tow profile labeled *2*). Upper trace, *1,* of deep-tow profile gives depth of water to fish. Difference between traces *2* and *1* gives fish height above sea floor. Lower trace, *2,* is true bathymetry obtained by addition of ranges for upward- and downward-looking echo sounder on deep tow. This profile is from La Jolla fan in 1,015 m of water.

B. Comparison of bathymetric profiles from San Lucas fan (Fig. 2) in 2,875 m of water taken with conventional wide-beam echo sounder on ship (surface profile) and with narrow-beam sounding system on deep-tow instrument (deep-tow profile). There is little resemblance between these profiles, and nearly 40 m of relief is obscured on surface profile.

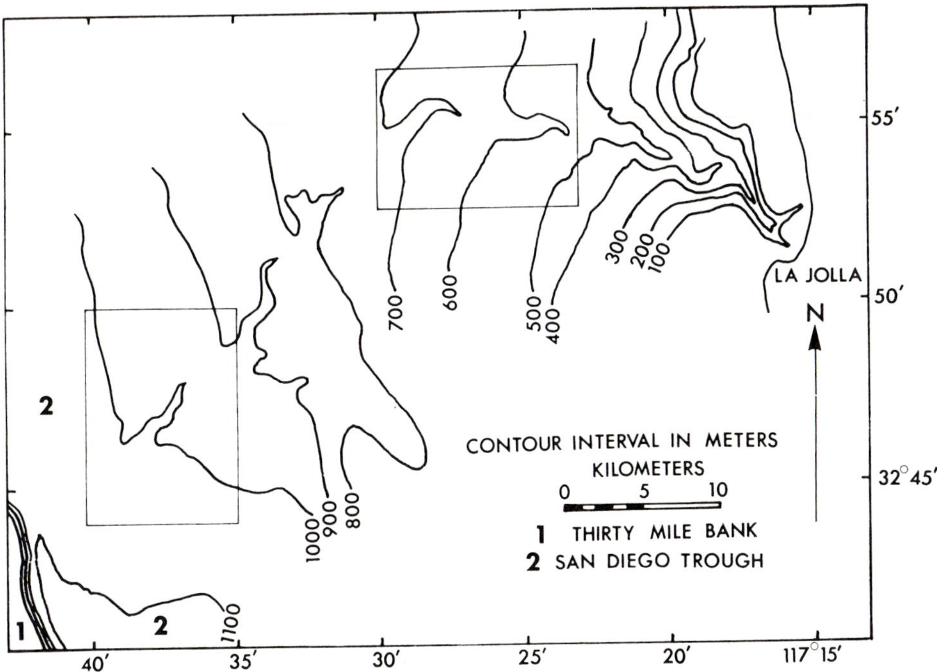

FIG. 3.—Location of survey areas on La Jolla fan. Bathymetry is modified after Shepard and Buffington (1968). CI = 100 m.

Many workers have studied the near-surface sediments of the fan, but the two important reviews are by Shepard and Einsele (1962) and Piper (1970). A comprehensive review combining sediment studies, deep submersible observations, and seismic reflection profiling was presented by Shepard *et al.* (1969). Shepard and Buffington (1968) prepared an excellent bathymetric map of the fan-valley at a 5-fm (9.15 m) contour interval. Moore (1965) and Normark *et al.* (1968) discussed fine-scale structures in and around the fan-valley.

Results

Upper fan survey.—The fan-valley on the upper fan is relatively straight and trends slightly north of west (Fig. 4). Near the western edge of the area surveyed, the valley bends sharply toward the southwest. The thalweg is

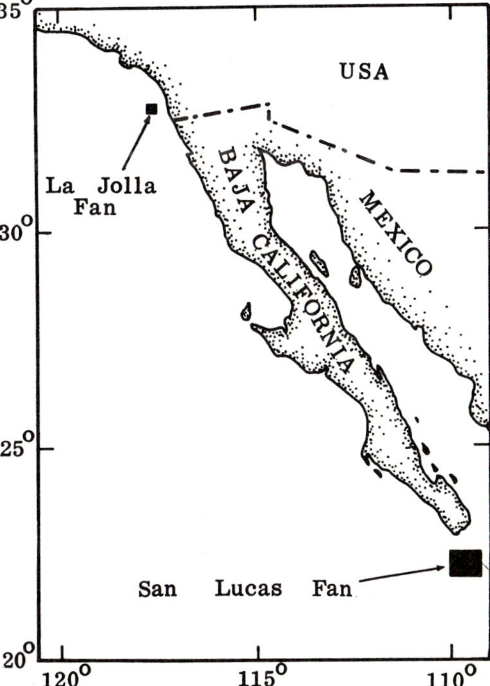

FIG. 2.—Location of La Jolla and San Lucas deep-sea fans.

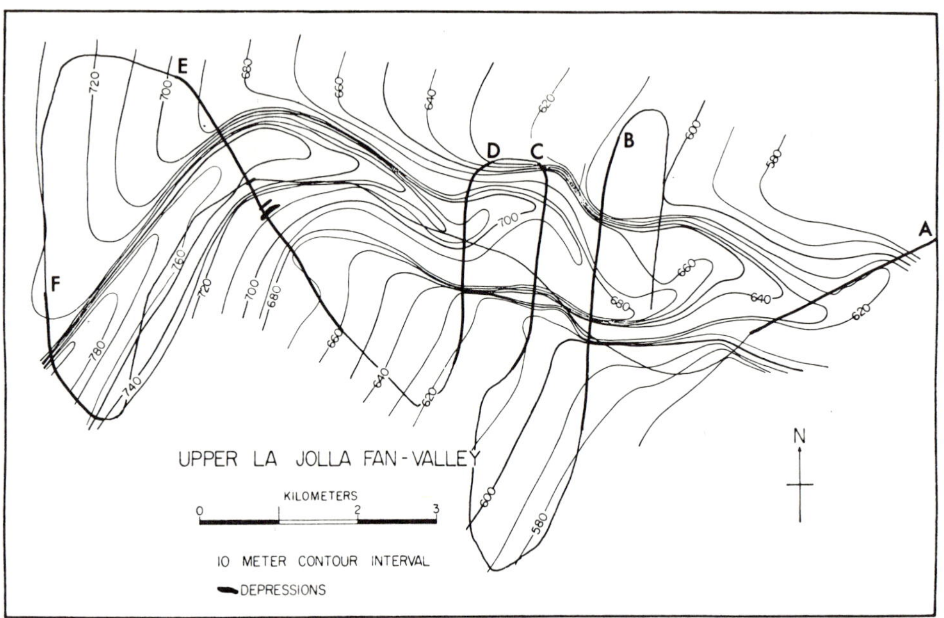

FIG. 4.—Bathymetry of upper La Jolla fan-valley from deep-tow data. Fish track and locations of profiles used in Figures 5 and 6 are shown. Cl = 10 m.

confined in a meandering channel within the straight trend of the fan-valley, and only locally is the valley form related to the thalweg's sinuous course. Broad, flat terraces are common within the fan-valley and are more prominent on the inside of meanders of the channel.

Line drawings of the fan-valley profiles having no vertical exaggeration are presented in Figure 5. The fan-valley is broad, flat floored, and flanked by wide, flat terrace surfaces. The narrow-beam soundings resolve several smaller terraces even on the steeper valley margins. Slopes between the terrace levels are commonly the steeper ones.

Little internal structure of the fan is discernible on seismic-reflection records from the deep tow. In most profiles, the lowest reflector is a smooth, flat surface generally parallel with the present sea floor. The material penetrated by the reflection system appears to thicken with increasing distance above the bottom of the fan-valley (Fig. 6). Little penetration was achieved into the sediments at the bottom of the fan-valley; however, short, discontinuous reflectors at 5 m below the valley floor were recorded on two crossings. Total penetration outside the fan-valley reached 24 m (assuming a sound velocity of 1.5 km/sec). Lower terraces (less than 10 m relief) within the fan-valley show 4–8 m of penetration and higher terraces show a range from 5 to 15 m.

The only exceptions to the smooth sea floor within the fan-valley are lines A and E, on the south side of the valley in Figure 5. These curious ridges or hummocks on the lower terrace levels apparently are unique to this area of the fan and are on the inside of the channel meanders. The reflection system did not define any internal structure of these features, and the side-looking sonars did not define their length or trend. These features may represent deposition on the inside of meanders similar to point-bar deposits in subaerial streams.

Lower fan survey.—The bathymetry of the lower fan-valley (Fig. 7) is similar to that of the upper fan except for the less relief. In general, the terraces are much less prominent on the lower fan, and the gross plan of the fan-valley shows much sharper meanders. The entire valley meanders and not just the thalweg. Steep slopes, commonly with narrow terraces, characterize the fan-valley walls on the outside of the meanders as on the upper fan (Figs. 8, 9). Slopes in excess of 60° have been measured, the steeper ones generally separating small, closely spaced terraces along the valley walls. One wall scarp 38 m high has a uniform slope of 54° and forms the entire relief of the valley at that point.

The lower fan also has depressions outside the main fan-valley (Fig. 7). Their general form suggests that they are in subsidiary chan-

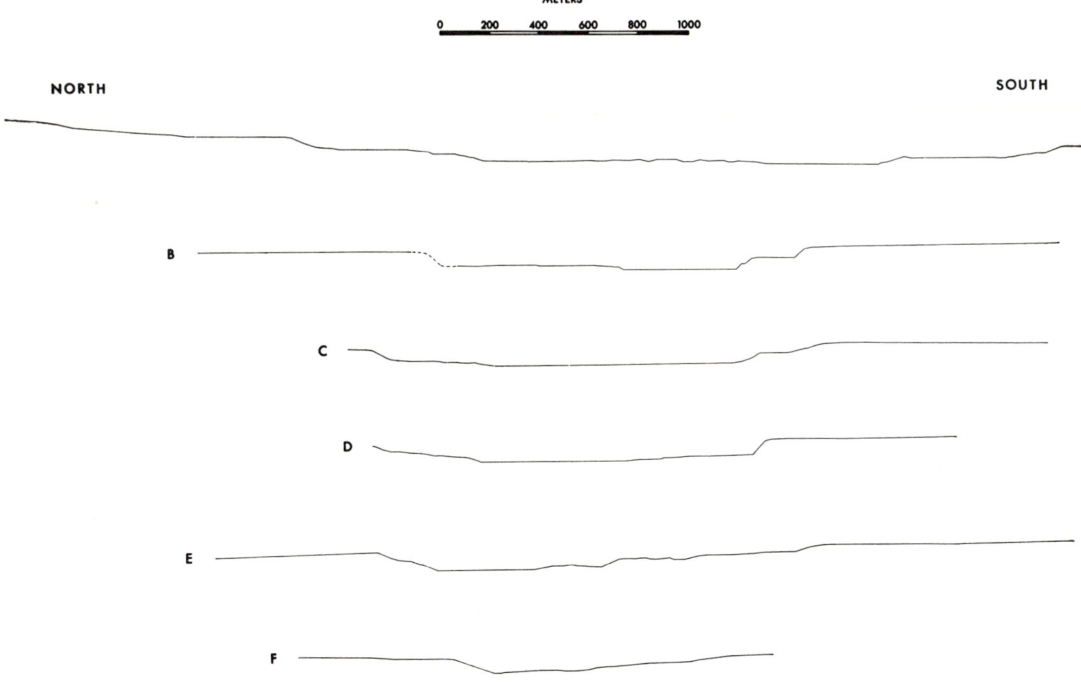

Fig. 5.—Line profile of upper La Jolla fan-valley with no vertical exaggeration taken with deep tow. Profile locations are shown in Figure 4.

nels, although they do not persist for any distance. The bathymetric map also suggests the presence of several smaller channels, but none of these is continuous between profiles (Figs. 8, 9).

The reflection profiles (Fig. 10) suggest that the fan-valley is eroded into approximately horizontally bedded sediments with reflecting horizons cropping out along the valley walls. Although most of the reflecting surfaces are discontinuous, there is evidence for contemporaneous erosion on the surface of the fan outside the fan-valley. Profile C (Fig. 10) shows a possible small channel east of the main valley that truncates some reflectors at the same level as reflectors truncated by the main valley, and there is little indication of filling of this feature.

The terraced morphology of the walls of the fan-valley is nearly ubiquitous on the lower fan except where the valley relief decreases to less than 10 m (Fig. 8). Some reflecting horizons crop out at or near the level of a terrace (Fig. 10, profiles B and C). The side-looking sonars on the deep tow were able to trace some of the

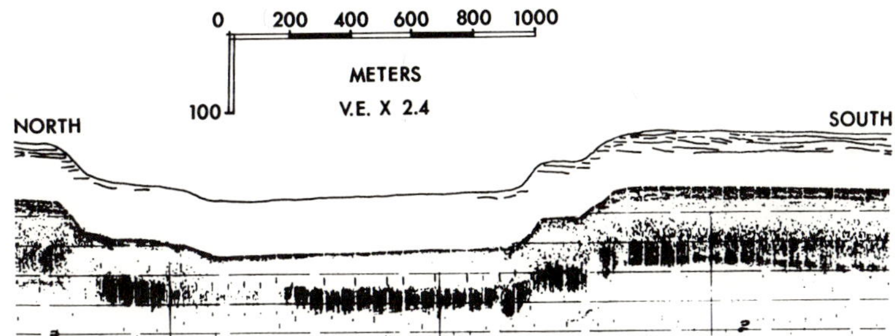

Fig. 6.—Line-drawing interpretation and photograph of 3.8-kHz reflection profile of upper La Jolla fan-valley. Note lack of acoustic penetration under floor of fan-valley as compared to outside valley.

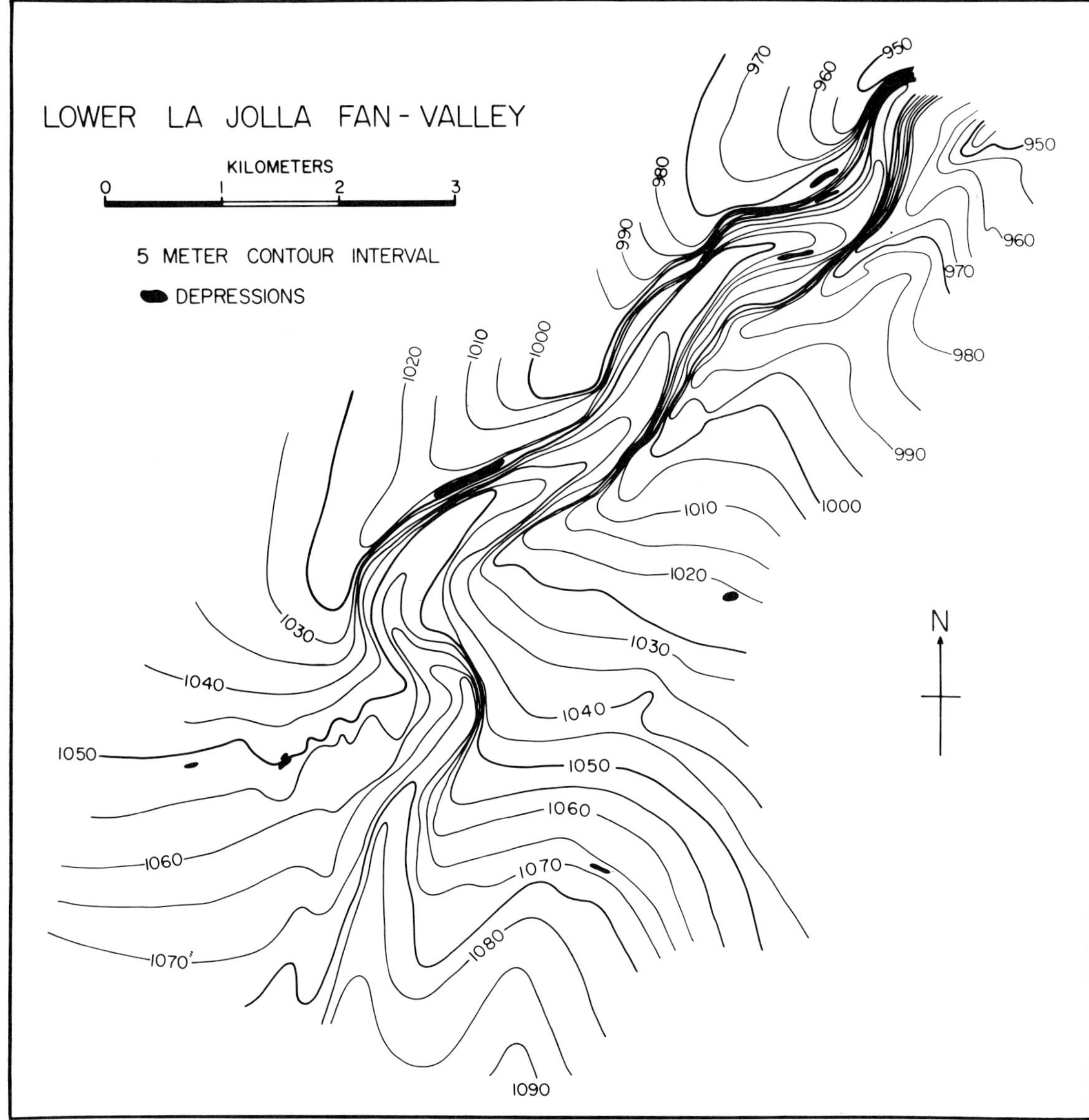

Fig. 7.—Bathymetry of lower La Jolla fan-valley. CI = 5 m.

narrower (< 100 m wide) terraces nearly 1 km along the valley walls (Fig. 11).

Discussion

Levees.—The deep-tow data and those of Shepard *et al.* (1969) suggest that most, if not all, of the relief of the fan-valley is due to erosion into the horizontally bedded sediments of the fan. The low-frequency sparker system used by Shepard and his coworkers does not show any of the features of levee growth documented by Hamilton (1967) on the Aleutian abyssal plain and by Andrews (1967) on a channel off the Bahama Islands using similar profiling equipment. The existence of the levees on La Jolla fan is based primarily on surface echo-sounding data. The levees generally are less than 1 km wide and have less than 10 m of relief. Low, broad, positive-relief features bordering the fan-valley are thought to indicate a levee

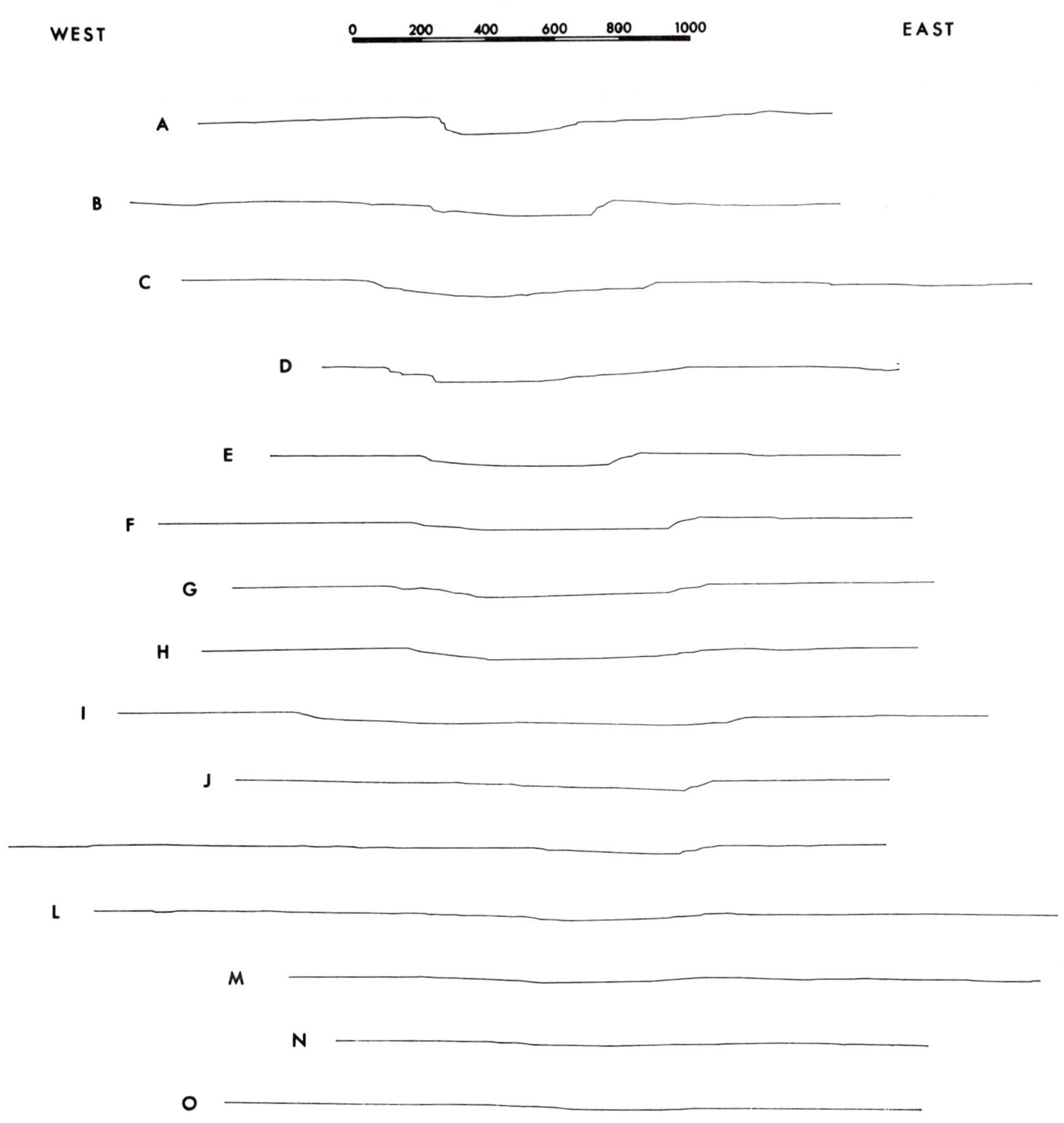

Fig. 8.—Line profiles of lower La Jolla fan-valley with no vertical exaggeration taken with deep tow. Profile locations are shown in Figure 9. Terraced morphology of fan-valley is maintained to line N.

type of deposit (Buffington, 1952; Shepard and Buffington, 1968). Shepard and Buffington (1968) believed that the levees are better developed on the northwest side of the fan-valley except on the lower fan between 915 and 1,050 m, where levees are present on both sides. They also described several areas where the levees appear as a series of *en echelon* ridges (Shepard and Buffington, 1968, p. 140).

Levee relief is not conspicuous within the deep-tow study areas except, perhaps, above 1,030 m east of the valley on the lower fan (Fig. 7). The internal structure of the fan-valley margins as recorded by the 3.8-kHz reflection system on the deep-tow fish does not clearly support levee development on La Jolla fan; leveelike features (profile B, west bank, and profile C, east bank; Fig. 10) are shown

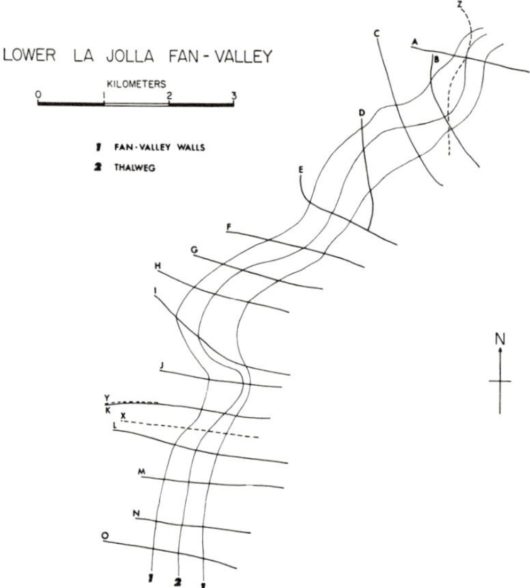

Fig. 9.—Location of profiles shown in Figures 8, 10, and 11 shown across outline of lower La Jolla fan-valley. Thalweg meanders inside valley walls and is found nearer outside wall on curves.

middle and upper fan-valley are the result of slumping of blocks from the sides of the valley. They argued that block slumping best explains (1) matched terrace levels on opposite sides of the valley, (2) the lack of persistence of a given terrace level along the valley axis, and (3) an observation that some of the terraces slope back toward the walls of the valley. The only mechanism of terrace formation they discussed that involved erosion was partial filling of the valley and subsequent rejuvenation; the resulting erosion would have left terraces that could be traced along the valley and matched across the valley. In light of the new data, these arguments can be revaluated.

Results of seismic reflection profiling with the deep tow show that little of the relief of the fan-valley is due to levee construction and that the entire valley appears to be an erosional feature. The steeper valley walls are on the outside of meanders where terraces are narrow and less more distinctly to be erosionally sculptured rather than depositionally controlled forms.[3]

Terraces and slumps.—Shepard and Buffington (1968) concluded that the terraces in the

[3] Normark *et al.* (1968) discussed the problems of identifying low-relief levees on deep-sea channels or fan-valleys by means of wide-beam surface echo sounding. Large levees with tens of meters of relief and tens of kilometers wide similar to those described by Hamilton (1967) and Shepard (1966) on the Monterey fan can be recognized with more confidence. In Figure 1 the apparent levee topography on the surface echo-sounding profile appears to be more probably of erosional origin on the deep-tow profile (see also profile G, Fig. 10).

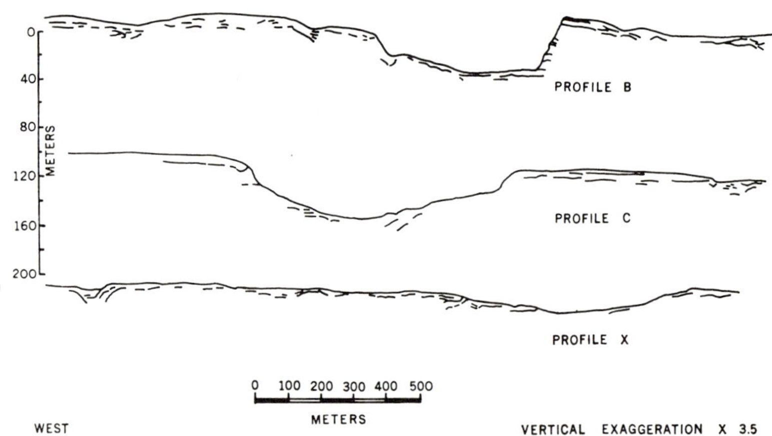

Fig. 10.—Line drawings of 3.8-kHz seismic reflection data on lower fan. Profile locations are given in Figure 9. Reflectors crop out on valley walls in all three profiles. Several possible partially filled older channels can be seen in western half of profile X. Two possible levee configurations are shown on east walls of profiles B and C. First, southeast wall of profile B suggests that levee form has been developed by rapid deposition within 200 m of top of valley wall, producing layers that dip away from fan-valley. Slope of sea floor away from fan-valley on this profile is steeper than on any other line (line B; Fig. 8). Second, southeast wall of profile C also appears to have levee morphology, but P-system shows that horizontally bedded sediments, truncated by fan-valley on south, crop out on back side of levee. This type of structure also could be considered erosional. Limited penetration by P-system prevents clearer picture.

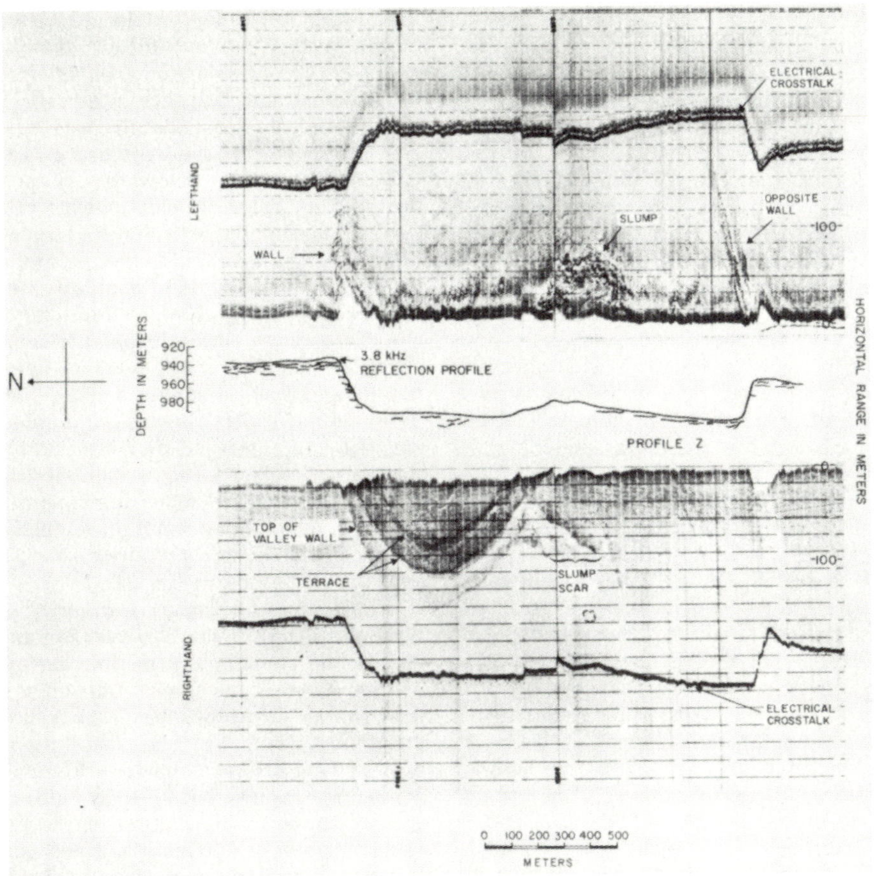

Fig. 11.—Side-looking sonars used to follow terraces along fan-valley walls with line drawing of 3.8-kHz reflection record in center (Fig. 9, profile Z). Upper side-looking sonar record is left-side record; lower one is right-side record. Dark returns in middle of both side-looking records that resemble bathymetric profile are electrical cross-talk not related to side-looking data. As fish moved downvalley (from left to right in this profile), right-side side-looking sonar recorded valley wall receding from path of fish. Two prominent terraces and upper edge of fan-valley were followed along small meander. As fish approached wall again (on right), fish moved above lowest terrace level and began to move farther up wall. Then returns from right side became very angular and discontinuous as fish rapidly receded from wall. At same time, left-side side-looking sonar recorded rounded elevation with very lumpy, lobate surface. This feature is interpreted as slump mass, and abrupt change in morphology of valley wall marks slump scar. A short distance farther, fish moved back over floor of fan-valley and 300 m later left-side sonar record showed fish approaching opposite wall of valley.

numerous. The inside walls on the meanders commonly have very wide terraces. It seems reasonable that the fan-valley morphology is the result of erosive current flow that incised the fan-valley into the nearly horizontal strata of the fan. Terraces formed on the more resistant layers as downcutting proceeded. The reflection profiling data (Fig. 10, profiles B and C) show reflecting horizons cropping out at the terrace level on the valley wall. The beds that uphold the terraces are probably thick sections of cohesive mud. More resistant beds of cohesive clay or silty clay are interbedded with lay- ers of cohesionless fine silt, and the more resistant clayey layers form small overhanging ledges (Moore, 1965). Pleistocene sediments are exposed within the fan-valley, probably as a result of downcutting of the valley (Shepard *et al.*, 1969).

Evidence of erosion of a deep-sea fan-valley, Doheny Channel, of the same scale as the La Jolla fan-valley, is recorded in turbidite deposits of Tertiary age in southern California. Doheny Channel (Bartow, 1966; Normark and Piper, 1969) is cut into interbedded silty shale and siltstone of the lower Capistrano Formation (upper Miocene). Several terraces 5–20 m wide are underlain by the thick shaly layers. These terraces are the same width as those commonly found in the La Jolla fan-valley (as shown by the deep-tow narrow-beam echo sounder).

Meandering of the flow within La Jolla fan-valley produced the steep walls on the outside of the curves and destroyed any large terraces that may have developed. Results from the precision bathymetry with the deep tow show that small terraces commonly are developed even on the steep walls on the outside of meanders. These terraces may be only 10–20 m wide and thus would not be detected by a wide-beam surface echo sounder. Even with the increased resolution with the deep-tow soundings, matching terraces across the valley is not possible.

Deposition on deep-sea fans occurs in lenses of interbedded sand, silt, and mud. The various lenses commonly wedge out across distances of hundreds of meters, and mapping of individual beds suggests that they measure only a few square kilometers (Bartow, 1964; Piper, 1970; Sullwold, 1960). It should be emphasized also that fan-valleys are very broad, shallow features, and if the dimensions of fan-valleys (1–5 km wide) are compared with the probable dimensions of the lenses of sediment that hold up the terraces, it is not surprising that terraces cannot be matched across the valleys.

The sand layers in the walls of La Jolla fan-valley are essentially unconsolidated, and the interbedded muds are only semi-indurated. It seems unlikely that these materials would slump so uniformly as to produce smooth, flat-topped terraces throughout the fan-valley as suggested by Shepard and Buffington (1968). Figure 11 illustrates a more probable slump deposit, with the irregular shape of the valley wall representing the slump scar.

The slump mass in Figure 11 obliterates two terraces that can be followed with the side-looking sonar for 600 m along the valley walls. These terraces produced distinctive returns on the side-looking record, yet they were barely noticeable on the bottom profile (plotted at a vertical exaggeration of 3:1) taken with the narrow-beam echo sounder on the deep-tow fish. The two terraces are on the outside of the meander, but retained their integrity throughout the traverse. This observation is further support for the formation of terraces by erosion of the valley into horizontally bedded sediments. It is unlikely that such long, narrow terraces can be slump blocks or that rejuvenation

within the fan-valley would preserve such a narrow strip of previous fill for great distances.

Lower fan distributary system.—Both the bathymetric map for the lower fan (Fig. 7) and the reflection data available (Fig. 10, profiles B, C) suggest the existence of a distributary-channel system below the 950-m contour on the fan. These profiles indicate no deposition within the channels, but the deepest profile (Fig. 10, profile X) shows several possible filled channels. These features are much smaller, both in relief and width, than the present fan-valley. The side-looking sonars also have detected elongate depressions on the open fan in areas of otherwise smooth topography south of the valley. Such discontinuous, small channels may have resulted from deepening of the present valley which eventually cut off an older distributary system; subsequent overflow from the main valley would have filled parts of the older channels and eventually smoothed over the topography (Fig. 12).

Distribution of surface sediments.—The reflection system on the fish used for the upper-fan survey could not penetrate the sediments on the floor of the upper fan-valley. Deepest penetration occurred outside the valley on the upper fan (Fig. 6). The distribution of the near-surface sediments on La Jolla fan may explain these results. Piper (1970) observed that the sediments in the bottom of the fan-valley consist mainly of fine massive sands, commonly containing mudlumps. Generally the thickness of the sand layers is unknown because of the poor penetration of coring devices. Lamination within the sands is uncommon and is present only at the tops of some beds. In general, there is little fine sediment, and the silty clay interbeds are mostly less than 5 cm thick.

Piper stated that on the open fan, ". . . sands become thinner and rarer with increasing distance from the main channel. The mud units become correspondingly thicker . . . " Most of the sand units show laminations, and massive sands with mudlumps are not found. Thick ($>$ 25 cm) mud layers are nearer the fan-valley on the upper fan than on the lower fan, and below 1,050 m no mud units thicker than 25 cm are found. It seems reasonable that the depth of acoustic penetration by the reflection system on the deep tow is related to the "percent sand" in the upper meter of the sediments on the fan (as most of Piper's observations were based on 60-cm-long box cores). The thinner the sand units (and the thicker the interbedded muds), the deeper is the penetration.

Summary

The available evidence suggests that La Jolla fan is not growing at present. Throughout most of the construction of the fan, the channels or fan-valleys were shorter and less entrenched than the present valley. Deposition was confined to the fan itself, and the growth of the low cone of sediment was aided by a system of shifting distributary channels on the middle and lower fan. Low-relief levees probably flanked many of the channels. The incision of the present fan-valley modified a system of smaller distributary channels that had formed on the lower fan (Fig. 12). This incision is the most recent development of the fan-valley system and may reflect an attempt to develop a new equilibrium profile (Normark and Piper, 1969). The valley meandered during incision, leaving a varied series of terraces along the walls of the valley. Remnants of the older distributary system are preserved because of slow and spotty deposition on the lower fan. Apparently, many currents moving down the fan-valley still have a high enough velocity to bypass the lower fan with little deposition of sediment (Piper, 1970).

SAN LUCAS FAN

The San Lucas deep-sea fan is a composite of four small coalescing fans in 3 km of water off the tip of the peninsula of Baja California, Mexico (Fig. 2). Sediment moves onto the fan through four submarine canyons—Cardonal, Vigia, San Lucas, and San Jose (Fig. 13). A deep-sea channel extending south from Tinaja Trough, a structural rift along the west side of the tip of the peninsula (western border of Fig. 13 lies along axis of trough), marks the western limit of the fan. Cabrillo Seamount marks the approximate eastern edge, but sediments of the lower fan extend onto low abyssal hills flanking the East Pacific Rise 60 km south and east of the mouth of San Jose Canyon.

The canyons feeding the fan include the "San Lucas Group" of submarine canyons that has been studied extensively by Shepard (1964; Shepard and Dill, 1966). The bathymetry of the fan itself is poorly known except for the apex of the cone of sediment below San Jose Canyon. The structure of the continental margin of the tip of Baja California, including parts of the San Lucas fan, was studied by means of continuous reflection profiling (Normark and Curray, 1968). At the base of the continental slope, the sediments of the fan are

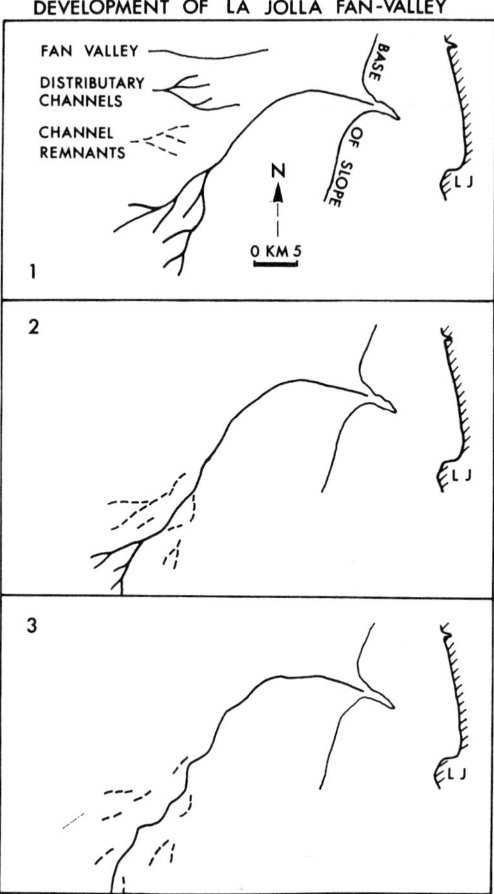

FIG. 12.—Recent history of La Jolla fan-valley. Incision of main valley distributary isolates upstream branches, and subsequent overbank deposition tends to smooth fan surface (**2**). Single valley cuts across entire fan as incision continues (**3**). Remnants of old distributaries are partly filled and smoothed, leaving discontinuous depressions on fan surface.

800–1,000 m thick and lie directly on the low-relief "basement," the seismic second layer of the oceanic crust (Phillips, 1964).

The San Lucas fan does not have a prominent fan-valley extending from any of the four submarine canyons. Short fan-valleys extend only 15–20 km below the mouths of the three eastern canyons. Shepard's bathymetric chart (1964, p. 180) shows a meandering valley south of San Jose Canyon between two irregular, hilly ridges. Bottom samples from the eastern ridge consist of sand. Seismic-reflection profiles (Fig. 14) show that these ridges are probably natural levees bordering the fan-valley. The section of sediment below the ridge

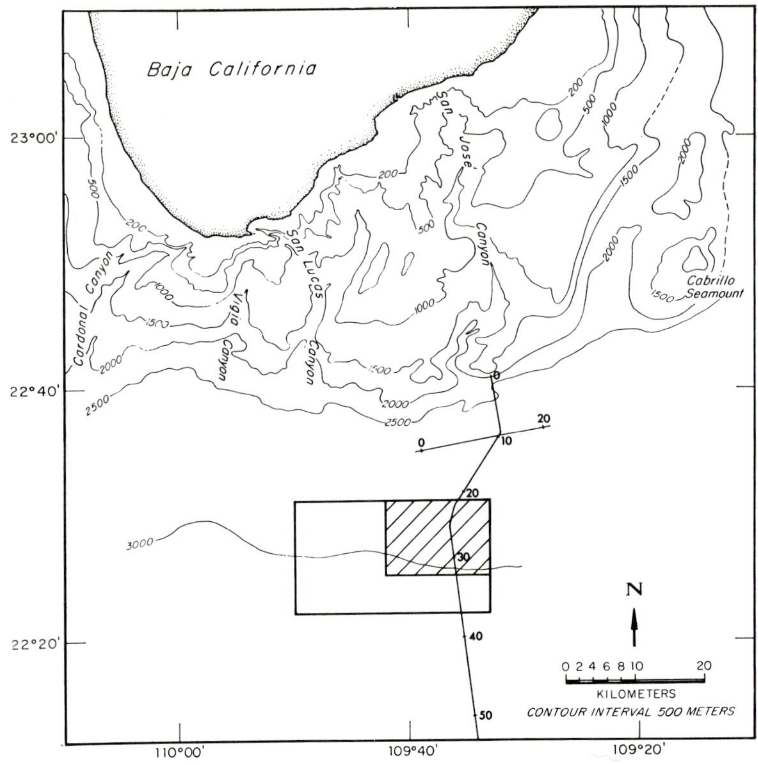

Fig. 13.—Bathymetry of tip of Baja California peninsula, after Shepard (1964). Location of deep-tow study areas on San Lucas fan are shown. Box encloses TOW MAS survey; hatched area denotes TIPTOW survey. Locations of seismic-reflection profiles of Figure 14 are shown with distance along profile in kilometers.

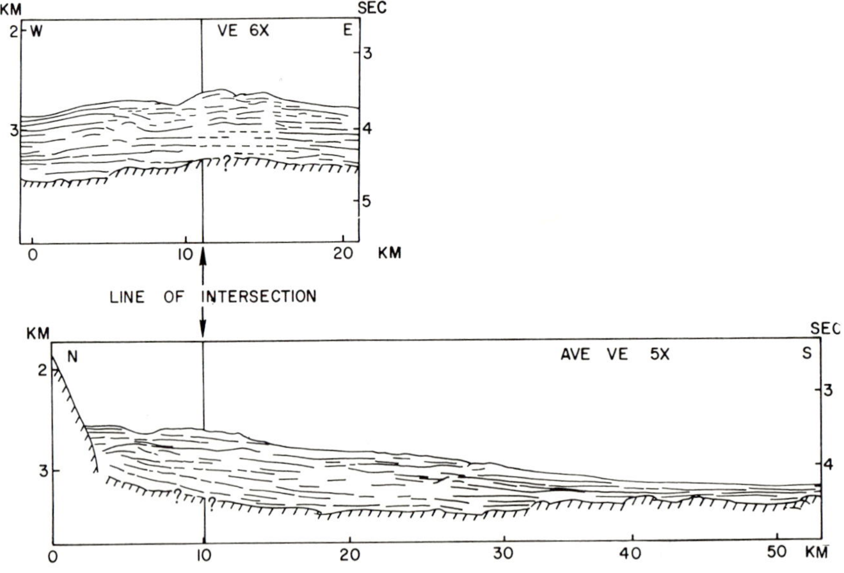

Fig. 14.—Seismic reflection profiles from San Lucas fan below San Jose Canyon (taken from Normark and Curray, 1968). Profile locations are shown in Figure 13. Study area lies between km 21 and 37 on north-south profile. See also Figure 23 for original record for north-south profile.

east of the valley exceeds 1 km and is the thickest anywhere on the San Lucas fan.

The San Lucas fan was selected for study of the area below the termination of the fan-valley associated with San Jose Canyon, and to relate events there to the growth of the fan. A north-south seismic-reflection profile across the fan (Fig. 14) shows irregular, hummocky topography with a generally convex-upward profile in the eastern part of the study area. The TOW MAS survey explored the western half of the fan (Figs. 13, 15) and the following TIPTOW survey (Fig. 16) concentrated on the area of hummocky relief. The steep gradient (22 m/km) through this region of hummocky relief is exceeded only at the apex of the fan, but the topography of the uppermost fan is modified by the hilly levees bordering the short fan-valley. The structure just beneath the hummocky topography is not decipherable from the reflection record. There is no clear fan-valley or channel morphology, and the reflection data suggest that the deeper sediments are horizontally bedded. The surface of the fan has a concave-upward profile on either side of the area of hummocky relief. The lower half of the fan has a smooth sea floor with very low gradients; near the toe of the fan, sediments lap onto low abyssal hills.

Results

TOW MAS survey.—Most of the area covered by TOW MAS survey (Fig. 17) has very low relief and was contoured at a 20-m interval because of relatively sparse data. Seen in profile, several shallow, channel-like trends suggested by the contours show none of the characteristic features of deep-sea fan valleys. Side-

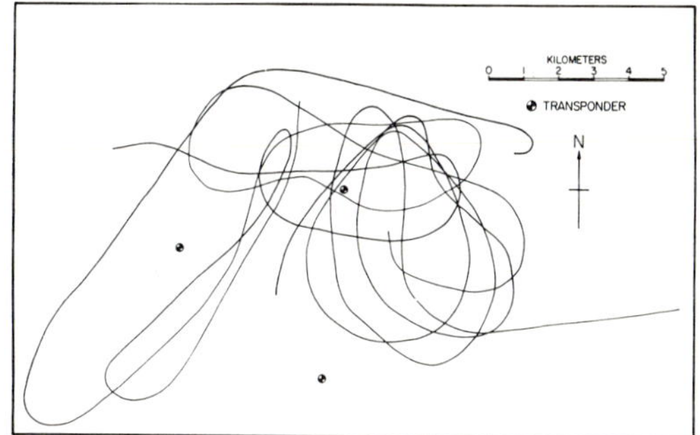

Fig. 16.—Fish track for expedition TIPTOW. Three transponders provided excellent positioning for deep-tow fish with rms error of 4 m for approximately 450 three-range fixes. TIPTOW and TOW MAS transponder nets were positioned with respect to each other by best fitting radar positions for surface, ship deep-tow topography, and deep-tow magnetics (Normark, 1969).

looking sonar records show that the bottom is smooth over these features with no scarps or irregularities indicative of channel walls. In the eastern end of the survey area, several steep-walled depressions were found. They could not be traced in any continuous system, and their lateral extent could not be determined.

TIPTOW survey.—The TIPTOW survey concentrated on the area of the depressions discovered during the earlier work. The closely spaced, accurately positioned traverses (Fig. 16) showed that the surface of the fan is indeed marked by a series of discontinuous depressions (Fig. 18). Although some of the depressions may appear to form a winding channel, good bathymetric control and side-looking sonar data confirm that many are unconnected features.

In the north-central section of the TIPTOW survey, contoured in Figure 18 at a 10-m interval, track coverage is extremely dense (Fig. 16). Side-looking sonar ranges overlap adjacent tracks across most of this area, and these data provided detailed morphology of several of the depressions. In this area, the depressions are somewhat irregular and trend nearly parallel with the gross contours of the fan. On the south, most of the depressions trend downfan and are arcuate in shape. Several are bordered by low leveelike relief. Profiles from five of them (Fig. 19) show terraced walls. The depressions generally are not symmetric either in relief or in development of the terraces.

Reflection profiles.—The information from the 3.8-kHz reflection profiling system is presented in Figure 20 as an isopach map of the

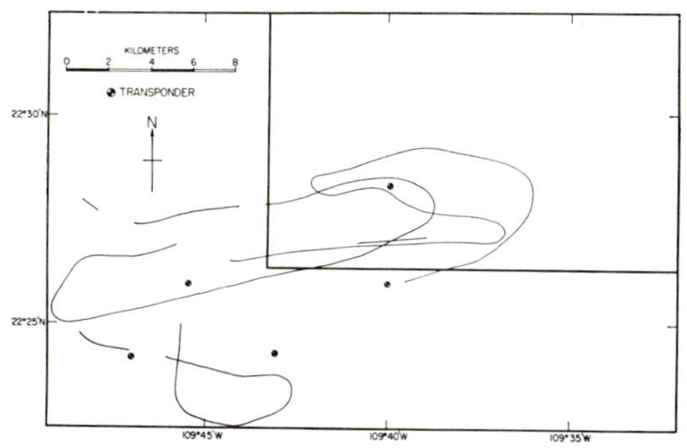

Fig. 15.—Fish track for TOW MAS survey. Rms position error for three or more range fixes was 6 m. Box outlines later TIPTOW survey area (Fig. 16). Two transponders along 109°40′W were dropped later in TOW MAS survey to extend study area eastward.

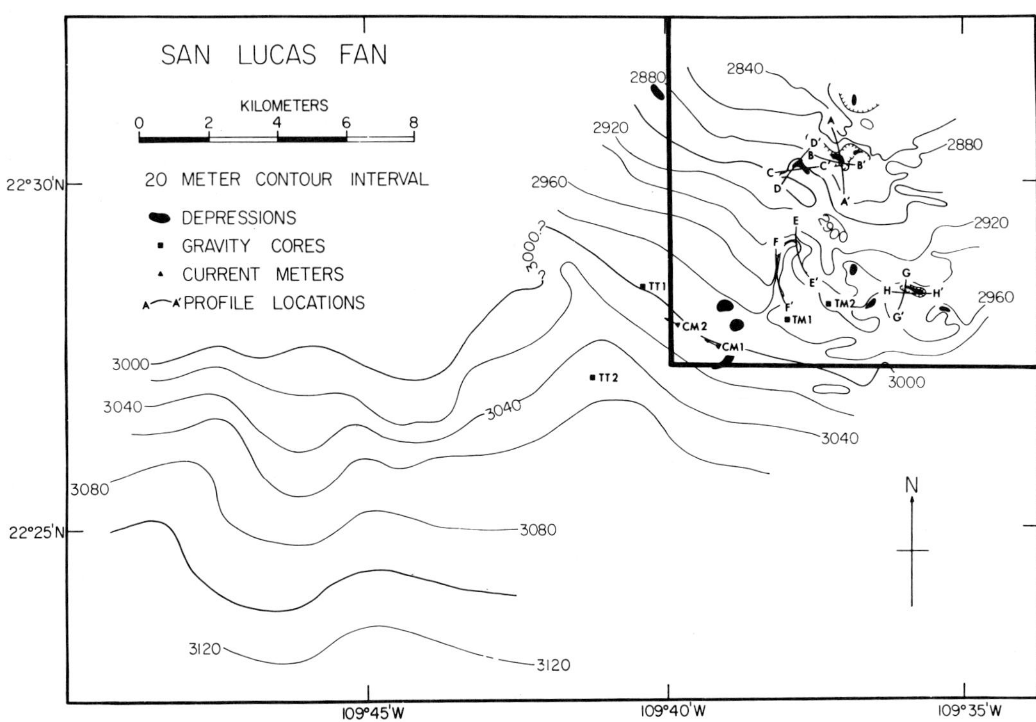

Fig. 17.—Generalized bathymetry of entire study area on San Lucas fan. Area enclosed in box is detailed in Figure 18. Location map for gravity core samples, current meters, and profiles is shown in Figure 19. Arrows on current-meter symbols indicate net current direction for bottom water (Normark, 1969).

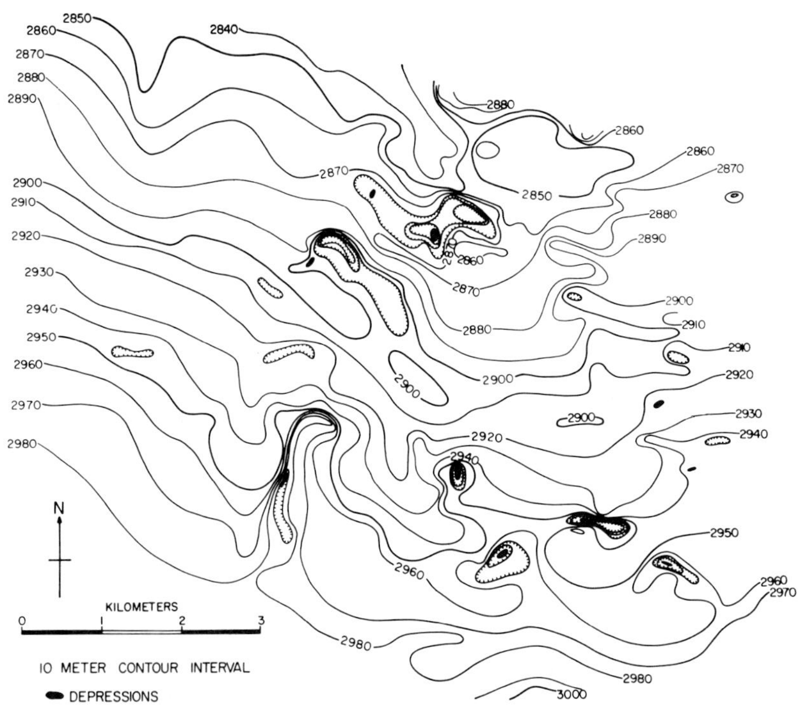

Fig. 18.—Detailed bathymetry of area of depressions on San Lucas fan. Closely spaced tracks across much of this area (Fig. 16) allowed many useful data on feature trends and additional soundings from side-looking sonar records to be used in constructing bathymetry (Normark, 1969, appendix I).

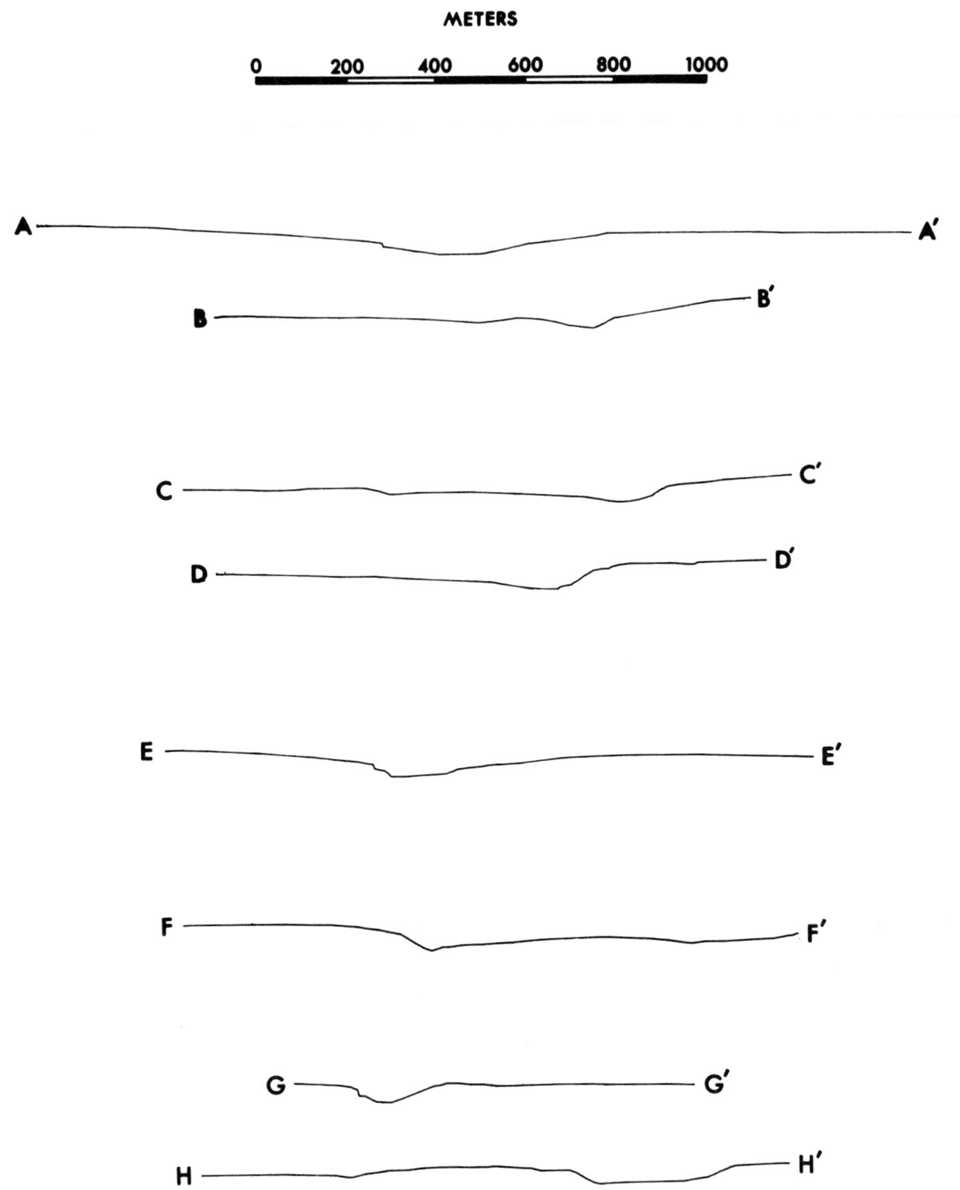

Fig. 19.—Line profiles over several depressions on San Lucas fan. Profile locations are given in Figure 17. No-vertical exaggeration. Profiles resemble in size and morphology, those from La Jolla fan valley (Figs. 5, 8).

depth of acoustic penetration, *i.e.,* the measured travel time to the deepest discernible reflector on the record. The fish height above the bottom ranged from 40 to 300 m, but had little effect on the depth of penetration. There was no penetration into the fan sediments at the bottoms of the depressions, but up to 13.4 m of penetration in the interdepression areas, with deeper penetration farther from the depressions. Northwest of the depressions the penetration exceeds 13.4 m with a maximum penetration of 40 m. This plot is not continued for the TOW MAS area because of the limited operation of the reflection system during the earlier survey (Normark, 1969).

Only the area of penetration greater than

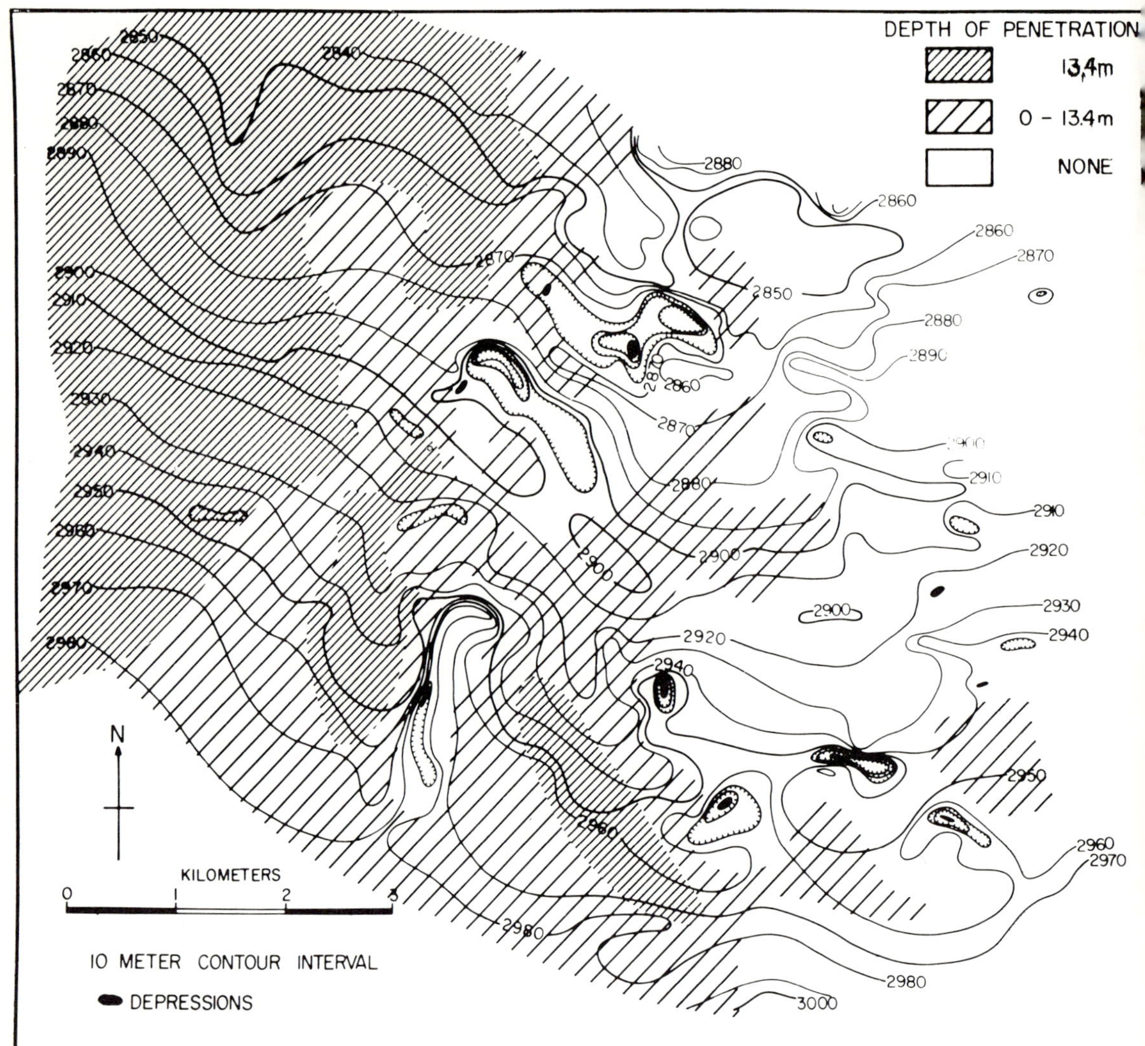

FIG. 20.—Sonar isopach map across area of depressions on San Lucas fan. Depth of penetration is plotted in meters; but arbitrary value of 13.4 m results from original plotting value in fathoms that was corrected to meters using sound velocity of sediments as measured in several gravity core samples (av. 1,550 m/sec) by E. L. Hamilton.

13.4 m shows reflecting horizons that can be traced for any distance. Little structure except for horizontally bedded sediments is suggested in this section of deepest penetration. The interdepression areas are characterized by indistinct, discontinuous returns that give no structural information. Records over the bottoms of the depressions typically show no subbottom reflectors of any kind. Where some penetration is recorded, it is similar to that of interdepression areas with very shallow ($<$ 7 m) returns.

Surface sediments.—The limited sampling of bottom sediments during the two deep-tow surveys (Normark, 1969) suggests a distribution of surface sediment similar to that of the La Jolla fan. Coarse sand and gravel are found within the depressions. Interbedded silty sands outside the depressions decrease in both thickness and grain size with increasing distance from the depressions. At greater distances, the predominant silty clays contain only a few thin (several centimeters) fine sand interbeds.

Discussion

Bathymetry.—The isolated depressions on the surface of the San Lucas fan are interpreted to be the remnants of a braiding and migrating channel system extending from the fan-valley near the apex of the fan. These depressions are negative-relief features and not part of a ridge and trough system. They are entrenchments in an area of otherwise smooth topography similar to the fan surface directly west. The hummocky topography described on the basis of the surface seismic-reflection profiles (Fig. 14) results from hyperbolic echoes from the edges of the depressions (Fig. 1B).

The cross-section profiles (Fig. 19) of the depressions resemble those of La Jolla fan-valley in size, relief, and morphology. Terraces are common on the walls of these depressions. Few profiles are symmetric. Steeper walls are common on the outside of arcuate depressions, and the terraces are wider on the opposite (or inside) wall. The fan surface away from the depressions is generally smooth, but low marginal ridges or mounds border several of the holes (Fig. 18).

Sediment distribution.—Little (< 7 m) or no subbottom penetration was achieved by the reflection system over the bottoms of the depressions (Fig. 20). The only available sample from a depression contained coarse sand and gravel similar to that described by Shepard (1964, p. 179) from the axis of San Jose Canyon. Intermediate penetration (< 15 m) characterized the interdepression areas and the two samples taken there consist of silty clay with up to 40 percent sand. West of the depressions, reflecting horizons as deep as 40 m below the sea floor were reached. One sample collected there consists of mostly mud with a few fine sand interbeds; the mud layers were 12–35 cm thick and the fine sand interbeds only 4–9 cm thick. The distribution of sediment types suggests a relation between the depth of penetration and the percentage of sand in the near-surface sediments. In general, the coarser the sediment, the poorer is the subbottom penetration.

Origin of Depressions

My observations suggest that the depressions are the result of erosion by channelized bottom currents into horizontally bedded fan sediments. This conclusion is supported by (1) the morphology of the depressions—arcuate, terraced walls, steep walls (narrower terraces) on the outside of curved depressions, and dimensions comparable to those of other fan-valleys, (2) the structure of the adjacent fan—truncation of flat-lying reflecting horizons near the depressions, and (3) the distribution of sediment—coarse sand and gravel inside the depression, becoming finer grained with distance from the channels. Initially a meandering fan-valley system formed, but a combination of circumstances transformed it into a series of disconnected depressions. Several mechanisms probably were involved. For example, the crescent-shaped depressions may be analogous to oxbows. Lateral migration of a fan-valley upstream might isolate short branches of the lower reaches of the system. A large slump mass generated by oversteepening could block the fan-valley and initiate deposition "upstream" from the slump, thus isolating the downstream segment of the valley. It is equally possible that not all turbidity currents would be large enough to traverse the entire length of the valley. Once sedimentation began within the valley, back-filling could have progressed (Andrews, 1967) until the upstream part was filled and the lower valley was abandoned. Depressions form in an area of rapid deposition; they cannot form by simple isolation of channel segments. In spite of uncertainties, I believe that the most reasonable interpretation of the data is that the depressions are the remnants of a former braiding and meandering fan-valley distributary system. A tentative conclusion from this reasoning is that small valleys, probably without levee development, are characteristic of the area of fan growth in the midfan (and perhaps upper fan) regions. This area of fan growth, which is characterized by numerous depressions, appears as a convex-upward bulge on the fan profile (Fig. 14).

MODEL FOR DEEP-SEA FAN GROWTH

A constructional model for deep-sea fan development utilizes the topologic and structural data obtained with the deep tow, as well as available published data on other deep-sea fans and on current concepts for abyssal-plain sedimentation. To combine the data from many deep-sea fans into one coherent model, a growth pattern must be established for each fan.

Growth Pattern

To unravel the growth pattern of a deep-sea fan it is necessary to determine the history of the fan-valleys that have developed. It is of utmost importance to determine whether the

TYPES OF FAN-VALLEYS

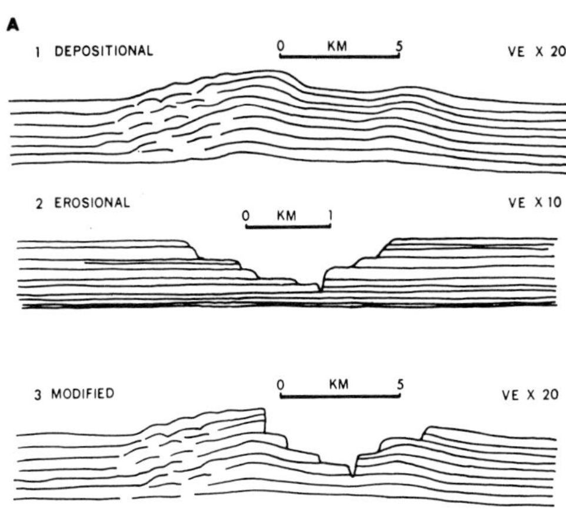

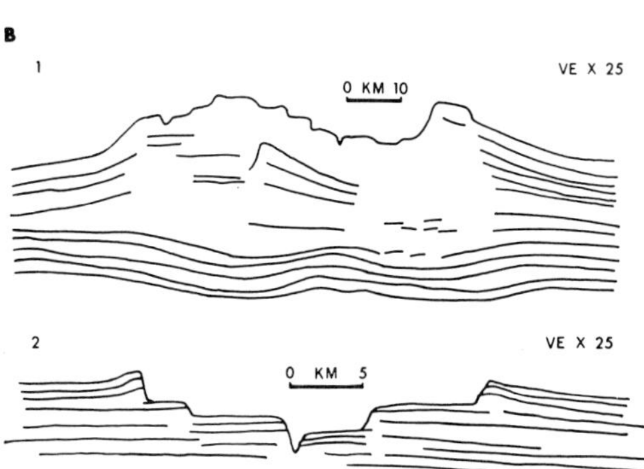

FIG. 21.—**A.** Types of fan-valleys; cross-sections presented are taken from seismic reflection profiles of real fan-valleys wherever possible: *1,* after Tufts abyssal plain channels (Hamilton, 1967); *2,* after La Jolla fan-valley; and *3,* after Monterey fan-valley (Normark, in press). Observations on distribution of the surface sediments around these should support this classification; however, only examples of erosional (La Jolla fan) and compound (Monterey fan) fan-valleys have been sampled sufficiently to recognize any systematic distribution of sediments with respect to the valleys.

B. Diagram illustrates two more configurations for leveed fan-valleys; both are based on reflection profiles from Bengal fan (Curray and Moore, 1968; F. Emmel, personal commun.). Large leveed valleys on fan apex (*1*) are clearly depositional features; weight of sediments in this feature has bowed down underlying layers. On the other hand, terraces are common on walls of valley, although this structure is more indicative of erosional valleys. Father south on Bengal fan, terraced wall morphology suggests erosional valley, but levees are present and levee structure is apparent from seismic reflection data (*2*). Fan-valley origin is not easy to determine. Samples shown in 1 and 2 are clearly only the extremes of a wide range of valley forms. A structural profile alone cannot uniquely identify a modified fan-valley (*3*).

present fan-valley system is in equilibrium with the present fan morphology. If the fan morphology represents deposition from a fan-valley pattern that predates the present fan-valley, the growth pattern of that fan cannot be deduced *a priori* by use of the present valley structure and morphology or the distribution of sediment within and around this valley. The growth patterns of deep-sea fans can be determined only if (1) the fans under consideration are growing at present or (2) the growth phase of the fan can be reconstructed after recognition of the alterations that have occurred since growth ceased.

The history of a fan-valley may be recorded in the internal structure of the fan sediments bordering the fan-valley. Accordingly, two fundamental types of fan-valleys are recognized (primarily by means of continuous reflection profiles) and are named on the basis of their inferred origins: (1) a *depositional* valley (Hamilton, 1967) commonly is bounded by levees with convex-upward bedding surfaces, and the valley floor may be built above the level of the surrounding sea floor; (2) an *erosional* valley (Laughton, 1968) cuts into previously deposited sediments. These definitions apply only on a large or generalized scale; for example, a depositional fan-valley may exhibit erosional sedimentary structures on the valley floor. Erosional and depositional fan-valleys represent the endpoints of a gradational series (Fig. 21). Changes from depositional to erosional conditions (or vice versa) can result in a *modified* valley form such as the present Monterey fan-valley (Normark, in press). None of these may be identified uniquely on the basis of a single structural profile, as emphasized in Figure 21B.

Constructional Model

The San Lucas fan appears to be the only fan described in this study that has preserved its depositional or growth pattern. This conclusion is based essentially on negative evidence in that there is no recognizable modification of the fan-valley pattern, *i.e.,* no evidence of fan erosion or extensive downcutting within the fan-valley. Whether the San Lucas fan is actively growing at present is a moot question, but certainly its formational morphology and structure have not been altered (Normark, 1969).

The morphology of the Astoria fan as described by Nelson (1968; personal commun.) compares very closely with that of San Lucas fan. The upper section of Astoria Canyon is filled, and the sediment load of the Columbia River no longer moves into and down the canyon. Instead, the sediment moves northward into Willipa Canyon, which feeds Cascadia Channel. The Astoria fan is essentially inactive (Nelson, 1968), compared to its development during times of lowered sea level, having lost its supply of sediment. Subsequent modification of the fan morphology is not recognized and the growth pattern of Astoria fan probably has been preserved.

Model description.—On the basis of my understanding of the San Lucas and Astoria fans, fan growth involves the formation of a leveed fan-valley (upper fan), a suprafan (usually on the mid-fan at the termination of the leveed valley), and a nearly flat-lying lower fan without channels. Under conditions of fan growth, the fan-valley (if one exists) extending from the mouth of the submarine canyon is depositional. The thalweg may follow an erosional channel within the valley, but the valley is characterized by natural levees and the valley floor may be built above the general level of the fan. The levees decrease in height and disappear down fan. Rapid radial deposition at the end of the leveed valley accelerates growth in this area, forming a depositional bulge or lobe on the fan surface. This depositional lobe is called a "suprafan" (Fig. 22). The suprafan is a small delta or fanlike deposit probably formed as the turbidity currents dissipate after leaving the confinement of the leveed valley. Deposition caused by spreading of turbidity currents was discussed by Buffington (1952), Johnson (1962), and Wilde (1965a, b).

On a radial profile of the fan, the suprafan appears as a low, convex-upward segment of an otherwise concave-upward profile. Channels are numerous on the suprafan. They are much smaller than the leveed fan-valley of the upper fan; their terracing is similar to that of erosional channels; they lack persistent levees; and they form a braiding and rapidly migrating system. Rapid shifting of these channels coupled with rapid aggradation of the suprafan may leave isolated channel remnants (San Lucas fan). If channel remnants are preserved as a series of depressions, deposition on the suprafan appears to be greatest in the vicinity of the channels. Deposition below the suprafan apparently results from broader, thinner flows, because not even small channels are developed on the lower fans. Figure 22 illustrates the morphology of this fan model; Figure 23 shows a longitudinal section of San Lucas fan for com-

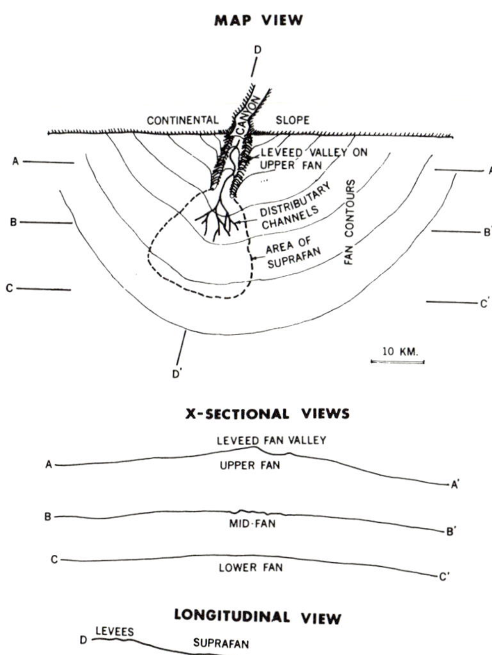

Fig. 22.—Model for deep-sea fan growth showing suprafan and simple morphology of fan related to growth area. Upper fan is characterized by large leveed fan-valley (*A-A'*), whereas midfan (or suprafan) shows many smaller distributary channels and some channel remnants or depressions (*B-B'*). Channels are absent from lower fan (*C-C'*). Suprafan appears as convex-upward bulge on longitudinal profile.

parison with the model profile, and representative surface echo-sounding profiles corresponding to the cross-sectional views of Figure 22.

Sands are found on the floor of the leveed fan-valley of the upper fan and on the suprafan. The coarsest fraction (coarse sand and gravel) is in the thalweg of the leveed valley and in the channels and depressions of the suprafan. Deposition of fine sand, silt, and silty clay occurs on the levees and the margin of the suprafan. Overflow deposition outside the levees and below the suprafan will be mostly silty clay with a few silt interbeds. Mud (silty clay) is found interbedded in all environments, but is thicker and more extensive with distance from the fan-valley floor and suprafan. This sediment distribution can be recognized on both San Lucas and Astoria fans (Nelson, 1968).

This model can be compared with other deep-sea fans only in a general way; published data on eight other fans (Tables 1, 2) are inade-

Table 1. Bathymetric Comparison of Deep-Sea Fans Ranked by Increasing Longitudinal Gradient

Name of Fan	Fan Dimensions (km)		Fan Gradient (m/km)	Depth (m)		Levees on Valleys	Apparent No. of Distributaries
	Length	Width		Apex	Base		
1. Bengal	2,600	1,100	1.1	1,500	4,400	yes	>20
2. Congo	>520	185	1.8	1,790	>2,700	yes	6
3. Rhone	300	250	3.0	1,900	2,800	yes	12
4. Astoria	165	100	4.6	2,080	2,840	yes	40
5. Monterey	320	280	4.8	3,050	4,600	yes	4–6
6. Hudson	150	150	6.7	2,100	3,100	yes	7
7. Delgada	300	330	8.0	2,200	4,000	—	1+?
8. Mississippi	220	150	8.3	1,280	2,950	yes	7
9. San Lucas	65	45	8.6	2,500	3,060	yes	at least 11 remnants
10. La Jolla	36	30	18.0	470	1,100	yes	1
11. Redondo	7.5	11	22.6	590	760	yes	4
Sublacustrine fan of Rhone delta, Lake Geneva	15	5.5	10.5	150	308	yes	7

Sources of data:

1. F. Emmel (personal commun.); Curray and Moore (1969).
2. Heezen *et al.* (1964).
3. Menard *et al.* (1965).
4. Taken from Nelson *et al.* (in press).
5. Wilde (1965a); Normark (in press).
6. Taken from Nelson *et al.* (in press).
7. Winterer *et al.* (1968); Wilde (1965a).
8. Taken from Nelson *et al.* (in press).
9. This study.
10. This study; Shepard and Buffington (1968).
11. Taken from Nelson *et al.* (in press).
Rhone delta fan, Lake Geneva, from Houbolt and Jonker (1968).

quate. The listings in Table 2 show that many deep-sea fans have undergone modification to the extent that no growth pattern can be discerned. It should be emphasized that features such as symmetry of outline, number of distributaries, and fan gradients, *per se,* are not defined rigidly by this model.

Control of fan shape.—A model for fan growth involving leveed fan-valleys and coalescing suprafans suggests that deposition may be localized to some extent. Deposition over the levees on the fan-valley, on the suprafan, and even on the lower fan where the turbidity currents spread out may include only a part of the entire fan area. A true fan morphology—a low, symmetric half-cone—will not form unless depositional sites through geologic time are random, leading to statistically uniform deposition on every sector of the fan. The position and length of the leveed fan-valley on the upper fan control the area of fan growth. To build a symmetrical fan, changes must occur in the leveed fan-valley on the upper fan; a new leveed valley must form near the apex and initiate growth on a new sector of the fan. Diversion to a new sector can occur if the existing fan-valley is blocked or if an extremely large turbidity current disrupts the existing levees. In

Table 2. Characteristics of Deep-Sea Fan Model Compared with Data Available for Deep-Sea Fans

Name of Fan	Growth Pattern Known	Leveed Valleys		Distributary Channel System	Suprafan	Outline of Fan
		Upper Fan	Lower Fan			
Bengal	no (insuff. data)	yes	yes?	yes	yes?	sym.
Monterey	no	yes	no	yes	no	asym.
Delgada	no	n.d.	n.d.	yes?	n.d.	asym.
Congo	no	yes	yes	yes	n.d.	sym.
Astoria[1]	yes	yes	no	yes	yes	asym.
Hudson	no	yes	n.d.	yes	n.d.	asym.
Rhone	no	yes	some	yes	yes?	asym.
Mississippi	no	yes	n.d.	yes	n.d.	n.d.
San Lucas[1]	yes	yes	no	yes	yes	sym.
La Jolla[1]	no	yes	no	yes (an older set)	no	asym.
Redondo	no	yes?	n.d.	yes	n.d.	n.d.
Hueneme-Mugu	no	yes	n.d.	yes	n.d.	n.d.
Lake Geneva, Rhone Delta Fan	yes	yes	no	yes	yes	asym.

[1] Model developed from studies in these fans.
n.d. = Not demonstrable with available data.

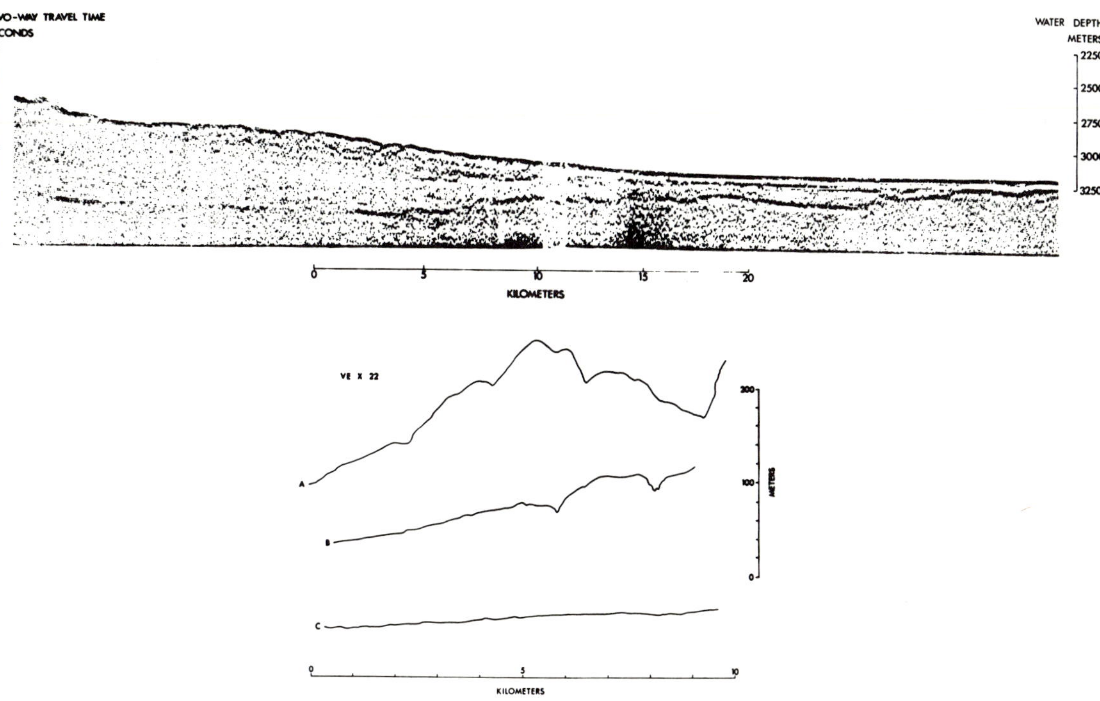

Fig. 23.—North-south (longitudinal) seismic-reflection profile and three line tracings of surface echo-sounding profiles from San Lucas fan illustrate relief and morphology envisioned in model for fan growth. Sections are comparable with those suggested in Figure 22. Suprafan appears on seismic reflection profile as convex-upward, hummocky area on upper midfan (see also Fig. 14). Only part of fan surface is shown in lower three echo-soundings tracings; these profiles emphasize type of relief comparable to sections of Figure 22.

either case, uniform deposition over the fan depends, more or less, on some catastrophic event to change the area of deposition.

In addition to building upward and laterally, levees tend to grow downstream because deposition of an unconfined (unchannelized) flow begins at the margins of the flow (Buffington, 1952, p. 477). Forward building (downstream) of the levees may proceed fast enough to prevent the formation of the suprafan morphology. Fans built from a source of well-sorted sediment may not develop the suprafan morphology because most of the deposition would be of the overflow type and levee growth (both upward and downfan) would be favored. If the levees did not build forward more rapidly than growth of the suprafan, the suprafan would build up and backfilling of the fan-valley could occur. However, as the valley floor and suprafan aggrade, the overflowing currents that formed the leveed valley would continue to increase levee height, resulting in elevation of the entire system. The longer growth continues, the more unstable the system

may become in terms of relief above the surrounding fan surface.

Menard *et al.* (1965) concluded that a typical deep-sea fan has an asymmetric outline. In some places, basement and/or basin morphology cause the asymmetry; my remarks are restricted to depositional processes that affect fan shape. Nelson (1968) recognized the asymmetry of the Astoria fan and tentatively accepted this in setting up his depositional model for deep-sea fans. The asymmetry is thought to reflect the tendency of turbidity currents in the northern hemisphere to hook left (looking downstream). Higher levees are formed on the right side of channels (looking downstream) in response to tilting of the surface of the channelized turbidity-current flow (Menard, 1955). Menard argued that through time the channels migrate leftward until confined at a point where leftward migration is impossible. The net result is an asymmetric fan.

Depositional channels on the Tufts abyssal plain have higher right-side levees, and leftward migration of these channels is common (Hamilton, 1967). The right-side levees on the Monterey fan are higher, but migration of the channel is not consistent in direction (Normark, 1969). In both cases, the lateral migration was only a few kilometers during deposition of the entire section of sediment, and thus this process is probably not important in developing fan morphology. Whether new channels preferentially break through the lower left-side levee and lead to asymmetric fan growth is not established. The old Monterey fan-valley broke through its right-side levee to capture the Ascension fan-valley (Normark, in press). On the Astoria fan, the main fan-valley is said to have shifted consistently to the left through time (Nelson *et al.*, in press); however, the present Astoria fan-valley is the youngest and is at the *right* of the next youngest fan-valley. Nelson *et al.* (in press) also observed that the left-side levee commonly is higher.

The asymmetry of outline of some deep-sea fans merely may reflect their youth. Many of these fans may be Pleistocene features (Menard, 1960; Menard *et al.*, 1965; Nelson *et al.*, in press) and, given more time, the outlines may develop more symmetry. In this sense, older and/or more rapidly growing fans should be more symmetric. In any case, it has not been established that leftward migration of the fan-valleys is the dominant factor determining fan shape.

Alternatively, the shape of deep-sea fans may reflect the importance of bottom currents in redistributing sediment brought to the fan by channelized turbidity currents. Asymmetric fans may be formed if bottom currents are strong enough to deflect sediment from its movement downslope (Wilde, 1965b). The bottom currents may deflect the turbidity currents that overflow or emerge from the fan-valleys. Erosion and redeposition of the fan sediments require more powerful bottom currents, and this effect should be less common. Heezen *et al.* (1966) concluded that deep geostrophic "contour" currents are capable of smoothing and shaping the continental rise along the western margin of the Atlantic Ocean.

Changes in Growth Patterns

La Jolla fan.—Piper (1970) suggested that 90 percent of the Holocene sediment contributed to La Jolla Canyon has bypassed the fan. It seems likely that sand-carrying turbidity currents now moving down the fan-valley are confined to the valley until they reach the lower fan (Normark, 1969). The fan is not being built up significantly at present. The small levees along the valley are a minor part of the relief in comparison to that caused by erosion.

The growth pattern of La Jolla fan cannot be reconstructed completely. The small levees bordering the valley may not have been deposited from currents moving through this fan-valley, although there is no evidence of an earlier depositional fan-valley. Small erosion channels on the lower fan below the 1,000-m contour suggest an earlier distributary system.

Monterey fan.—The Monterey fan has undergone a marked change since the growth of the large levees bordering the fan-valley (Normark, in press). The Monterey fan-valley has captured the more westerly Ascension fan-valley and subsequent erosion has deepened the valley —perhaps as much as 300 m. Deposition on Monterey fan now is confined to a deeply incised fan-valley and/or to the toe of the fan far south of the leveed valley. The growth pattern of Monterey fan cannot be reconstructed and the changes leading to the erosion of the present valley are unknown.

Bengal fan.—The tremendous Bengal fan (Curray and Moore, 1969; F. Emmel, personal commun.) is nearly 100 times larger than other known fans. The leveed fan-valleys on the upper fan are likewise much larger. The rate of deposition on the Bengal fan is relatively high, and a braiding system of leveed fan-valleys exists. Abandoned and buried fan-valleys also are common. Down fan the valleys become smaller and levee relief is much less, but no suprafan has been recognized definitely.

SUBLACUSTRINE FANS

A small sublacustrine fan near the eastern end of Lake Geneva has several characteristics described by my model for the growth of deep-sea fans. Leveed channels extending down the lacustrine Rhone delta foreslope terminate on a fan-shaped pile of sediments (Houbolt and Jonker, 1968). Deposition on the Rhone fan occurs from turbid, dense underflows generated by a difference in temperature between the river and lake waters as well as the sediment load of the river water (Houbolt and Jonker, 1968). The sedimentary structures are the same as those formed by turbidity currents, and the surface morphology, internal structure, and sediment distribution are essentially identical to those of the model for deep-sea fan growth. Channel width and levee–channel floor relief decrease markedly on the upper fan (Fig. 24), and a longitudinal profile of the fan shows a convex-upward bulge (Fig. 24). Channels on the middle fan are narrower and have small levees.

Houbolt and Jonker (1968) showed that the channel system controls the distribution of sand on the Rhone delta fan in Lake Geneva. Fine to medium sands are common on the central fan and channel bottom. On the lower fan, fine-medium sand is more commonly interbedded with silt and mud, and the reflection profiles show that the sediments on the toe of the fan interfinger with the predominantly muddy beds of the lake bottom. The sediment data suggest that the convex-upward relief in the longitudinal fan profile represents the area of most rapid deposition. Cores from the leveed channels on the delta foreslope are mostly silt and fine sand, indicating that the coarser material is kept effectively in the channel and is not deposited until it reaches the fan.

Alluvial Fans

The growth patterns of subaerial alluvial fans are somewhat analogous to those depicted in the model for deep-sea fan growth. Processes operating in a prominent channel, commonly incised on the fanhead (Bull, 1964; Eckis,

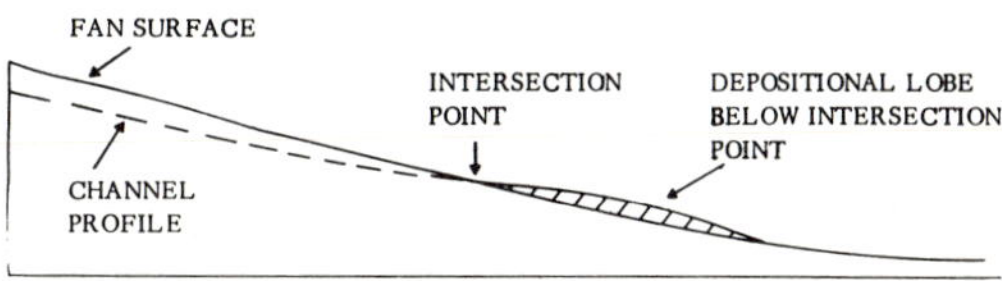

Fig. 25.—Longitudinal profile of subaerial alluvial fans (after Hooke, 1967). Growth profile, found locally on both laboratory and natural fans, shows convex-upward depositional lobe producing concave-convex-concave surface profile (after Hooke, 1967). Hachures indicate most recently deposited material. Dashed line indicates depth and slope of channel with respect to fan surface.

1928; Hooke, 1967; Tolman, 1909), control the shape and profile of the fan. Observations on models built in the laboratory showed that the main channel of laboratory fans intersects the fan surface near mid-fan (Hooke, 1967). Below this intersection point a "small secondary fan" may be built (Fig. 25). During Hooke's laboratory experiments the intersection point on the fan was controlled by the relative importance of debris flow and stream flow on any particular fan. The intersection point was farther down fan on fans built with more debris-flow deposits. On both laboratory and natural alluvial fans, the intersection point moves up fan with continued deposition until a new channel can form, shifting the locus of sedimentation (Bull, 1964; Eckis, 1928; Hooke, 1967).

Eckis (1928) described the blocking of flow in a fan channel by a debris dam formed during the flood stage of the flow; such damming and subsequent stream diversion also are considered important in diverting sediment to all segments of the fan. In both laboratory and field situations, once deposition begins within the channel, the intersection point shifts upstream and continued deposition decreases the relief of the channel. Subsequent flows then may be diverted more easily to other parts of the fan. Thus, deposition on alluvial fans is localized at any moment in time (Hooke, 1967), and the characteristic fan morphology develops only after there has been sufficient time for uniform deposition over the entire fan.

Both deep-sea and subaerial fans are built by channelized, sediment-laden currents; thus the comparable growth patterns reflect the similar functions of their channel distributary systems. In both environments, the development of the symmetric half-cone shape is a result of migra-

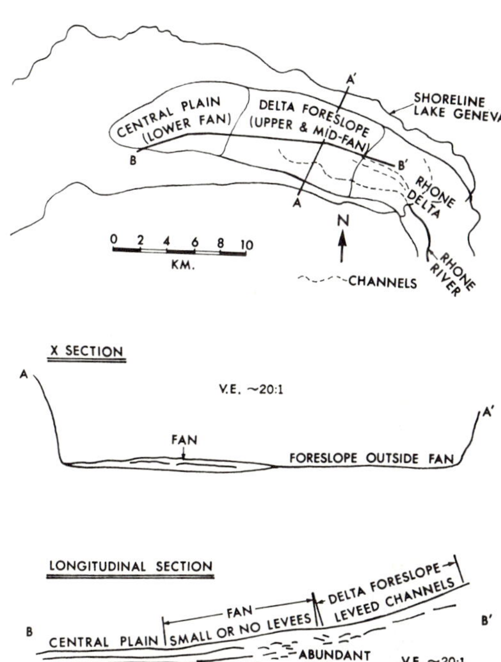

Fig. 24.—Sublacustrine fan of Rhone delta in Lake Geneva (after Houbolt and Jonker, 1968).

tion of the locus of deposition with time. The radial profiles of both deep-sea and subaerial fans show the segmented concave-convex-concave-upward form (Figs. 22, 23, 25) resulting from rapid deposition at the end of a prominent fan-valley or fan-head channel. This segmentation of the fan profile is the result of undisturbed fan growth and should not be confused with the segmentation caused by tectonic disturbance of the fan–source area system (Bull, 1964; Hooke, 1967). In both environments, events in a prominent channel on the upper fan control the fan shape and profile during growth on tectonically undisturbed fans.

Conclusion

A model for deep-sea fan growth based on available data on the morphology, structure, and sediment distribution of fan-valleys is proposed. Under conditions of fan growth, deposition is probably localized in time and space. A single leveed fan-valley on the upper fan feeds a distributary system of numerous smaller, ephemeral channels on the middle fan. Rapid deposition at the end of the leveed valley may build a small suprafan. Deposition on the lower fan occurs from sheetlike overflow (unchannelized currents). Aggradation within this system will continue until some catastrophic event leads to the development of a new leveed fan-valley on the upper fan. Those fans with strongly asymmetric shapes are probably morphologically young or are growing slowly; however, redistribution of sediment on the fan by bottom currents may lead to some asymmetry of shape. Given sufficient time, no tectonic disturbance, and a simple basement physiography, uniform deposition should occur within all sectors of the fan.

References Cited

Andrews, J. E., 1967, The Great Bahama Canyon and structure and sedimentary development of the outer channel of the Great Bahama Canyon: Ph.D. thesis, Miami Univ., 104 p.

Bartow, J. A., 1964, Stratigraphy and sedimentation of the Capistrano Formation, Dana Point area, Orange County, California: M.A. thesis, California Univ., Los Angeles, 102 p.

—— 1966, Deep submarine channel in upper Miocene, Orange County, California: Jour. Sed. Petrology, v. 36, p. 700–705.

Buffington, E. C., 1952, Submarine "natural levees": Jour. Geology, v. 60, p. 473–479.

—— 1960, Loran electronic surveying equipment (model 60)—operational appraisal: U.S. Navy Electron. Lab. Research Devel. Rept., v. 964, 32 p.

Bull, W. B., 1964, Geomorphology of segmented alluvial fans in western Fresno County, California: U.S. Geol. Survey Prof. Paper 352-E, p. 79–129.

Curray, J. R., and D. G. Moore, 1969, The Bengal deep-sea fan (abs.), in Abstracts for 1968: Geol. Soc. America Spec. Paper 121, p. 67.

Eckis, R., 1928, Alluvial fans of the Cucamonga district, southern California: Jour. Geology, v. 36, p. 224–247.

Forel, F. A., 1885, Les ravins sous-lacustres des fleuves glaciaires: Acad. Sci. Comptes Rendus, v. 101, p. 725–728.

Hamilton, E. L., 1967, Marine geology of abyssal plains in the Gulf of Alaska: Jour. Geophys. Research, v. 72, p. 4189–4213.

Heezen, B. C., C. D. Hollister, and W. F. Ruddiman, 1966, Shaping of the continental rise by deep geostrophic contour currents: Science, v. 152, p. 502–508.

—— R. J. Menzies, E. D. Schneider, W. M. Ewing, and N. C. L. Granelli, 1964, Congo submarine canyon: Am. Assoc. Petroleum Geologists Bull., v. 48, p. 1126–1149.

Hooke, R. L., 1967, Processes on arid region alluvial fans: Jour. Geology, v. 75, p. 438–460.

Houbolt, J. J. H. C., and J. B. M. Jonker, 1968, Recent sediments in the eastern part of the Lake of Geneva (Lac Leman): Geologie en Mijnbouw, v. 47, p. 131–148.

Johnson, M. A., 1962, Turbidity currents: Sci. Progress, v. 50, p. 257–273.

Krause, D. C., 1962, Interpretation of echo sounding profiles: Internat. Hydrog. Rev., v. 39, p. 65–122.

Laughton, A. S., 1968, New evidence of erosion on the deep ocean floor: Deep-Sea Research, v. 15, p. 21–29.

Lowenstein, C. D., and J. D. Mudie, 1967, On the optimization of transponder spacing for range-range navigation: Jour. Ocean Tech., v. 1, p. 29–31.

MeGehee, M. S., and D. E. Boegeman, 1966, MPL acoustic transponder: Rev. Sci. Instruments, v. 37, p. 1450–1455.

Menard, H. W., 1955, Deep-sea channels, topography, and sedimentation: Am. Assoc. Petroleum Geologists Bull., v. 39, p. 236–255.

—— 1960, Possible pre-Pleistocene deep-sea fans off central California: Geol. Soc. America Bull., v. 71, p. 1271–1278.

—— S. M. Smith, and R. M. Pratt, 1965, The Rhône deep-sea fan, in Submarine geology and geophysics: London, Butterworth, p. 271–286.

Moore, D. G., 1965, The erosional channel wall in La Jolla sea-fan valley seen from bathyscaph *Trieste II*: Geol. Soc. America Bull., v. 76, p. 385–392.

Nelson, C. H., 1968, Marine geology of Astoria deep-sea fan: Ph.D. thesis, Oregon State Univ., 287 p.

—— et al. (in press), Physiography of the Astoria canyon-fan system.

Normark, W. R., 1969, Growth patterns of deep-sea fans: Ph.D. thesis, California Univ., San Diego, 165 p.

—— (in press), Channel piracy on Monterey deep-sea: Deep-Sea Research.

—— and J. R. Curray, 1968, Geology and structure of the tip of Baja California, Mexico: Geol. Soc. America Bull., v. 79, p. 1589–1600.

—— and D. J. W. Piper, 1969, Deep-sea fan-valleys, past and present: Geol. Soc. America Bull., v. 80, p. 1859–1866.

—— et al., 1968, Detailed channel morphology of the La Jolla submarine fan-valley (abs.): Am. Geophys. Union Trans., v. 44, p. 212.

Phillips, R. P., 1964, Seismic refraction studies in Gulf

of California, *in* Tj. H. van Andel and G. G. Shor, Jr., eds., Marine geology of the Gulf of California: Am. Assoc. Petroleum Geologists Mem. 3, p. 90–121.

Piper, D. J. W., 1970, Transport and deposition of Holocene sediment on La Jolla deep-sea fan, California: Marine Geology, v. 8, p. 211–228.

Shepard, F. P., 1964, Sea-floor valleys of Gulf of California, *in* Tj. H. van Andel and G. G. Shor, Jr., eds., Marine geology of the Gulf of California: Am. Assoc. Petroleum Geologists Mem. 3, p. 157–192.

—— 1966, Meander in valley crossing a deep-ocean fan: Science, v. 154, p. 385–386.

—— and E. C. Buffington, 1968, La Jolla submarine fan-valley: Marine Geology, v. 6, p. 107–143.

—— and R. F. Dill, 1966, Submarine canyons and other sea valleys: Chicago, Rand McNally, 381 p.

—— —— and U. von Rad, 1969, Physiography and sedimentary process of La Jolla submarine fan and fan-valley, California: Am. Assoc. Petroleum Geologists Bull., v. 53, p. 390–420.

—— and G. Einsele, 1962, Sedimentation in San Diego Trough and contributing submarine canyons: Sedimentology, v. 1, p. 81–133.

Spiess, F. N., and J. D. Mudie, 1970, Small-scale topographic and magnetic features, *in* A. E. Maxwell, ed., The seas, V. IV: New York, Interscience.

—— *et al.*, 1967, Deeply-towed marine geophysical observation system (abs.): Am. Geophys. Union Trans., v. 48, p. 133.

—— *et al.*, 1969, Detailed geophysical studies on the northern Hawaiian arch using a deeply towed instrument package: Marine Geology, v. 7, p. 501–527.

Sullwold, H. H., Jr., 1960, Tarzana fan, deep submarine fan of late Miocene age, Los Angeles County, California: Am. Assoc. Petroleum Geologists Bull., v. 44, p. 433–457.

Tolman, C. F., 1909, Erosion and deposition in the southern Arizona bolson region: Jour. Geology, v. 17, p. 136–163.

Wilde, P., 1965a, Recent sediments of the Monterey deep-sea fan: Ph.D. thesis, Harvard Univ., 153 p.

—— 1965b, Estimates of bottom current velocities from grain size measurements for sediments from the Monterey deep-sea fan: Marine Tech. Soc. Am. Litt. Soc. Proc., v. 2, p. 718–722.

Winterer, E. L., J. R. Curray, and M. N. A. Peterson, 1968, Geologic history of the Pioneer fracture zone with the Delgada deep-sea fan northeast Pacific: Deep-Sea Research, v. 15, no. 5, p. 509–520.

Reprinted for private circulation from
THE AMERICAN ASSOCIATION OF PETROLEUM GEOLOGISTS BULLETIN
Vol. 62, No. 6, June, 1978

Fan Valleys, Channels, and Depositional Lobes on Modern Submarine Fans: Characters for Recognition of Sandy Turbidite Environments[1]

WILLIAM R. NORMARK[2]

Abstract The growth-pattern concept for modern submarine fans has been reviewed and broadened by additional data published or obtained in the last five years. The similarities in morphology, structure, and surficial-sedimentation patterns among modern fans from different geographic and geologic settings support a general growth-pattern model that can be applied to ancient turbidite deposits. Most submarine fans have three recognizable morphologic divisions that are related to distinct facies associations for sandy and coarser turbidites. (1) The large-leveed valley(s) of the upper fan produce wide (1 to 5 km) valley-floor deposits that are the coarsest on the fan and are deposited in meandering or braided, shallow channels within the general confines of the valley. These coarse deposits grade laterally into finer grained and more regularly bedded levee sands and silts. (2) The middle-fan region is recognized as a convex-upward depositional bulge on a radial profile and includes a depositional lobe or suprafan at the terminus of the leveed valley. The coarsening- and thickening-upward sequence of sandy turbidites on the upper suprafan are cut by numerous channels, channel remnants, and isolated depressions, whereas the lower suprafan is relatively free of such features. Suprafan channels are generally less than 1 km across and probably are filled by thinning- and fining-upward sequences. (3) The lower fan division is characteristically free of channel features (and coarse turbidites), is nearly flat-wing or ponded, and, therefore, is indistinguishable morphologically from basin-plain or abyssal-plain settings in many cases.

Basin shape and relief and the ultimate size of the fan appear less important than sediment-input parameters, such as the grain-size distribution and rate of sediment supply, in controlling development of the three morphologic divisions of the fan. Specifically, canyon-fed systems common along western North America tend to have a single-leveed valley terminating in a suprafan depositional lobe; some fans, such as the Monterey, have slightly more complex features where more than one canyon is involved in fan development. If the grain-size distribution is weighted toward the silt and clay fractions as in some delta-fed systems, the fans tend to have multiple-leveed valleys on the upper fan (although only one may be active at any given time), to have long valleys crossing much of the fan, and to lack (or have poorly developed) suprafan relief.

INTRODUCTION

Studies of modern marine turbidite sediments within the last 20 years resulted in a variety of depositional schemes or model submarine fans. Such models typically are derived from the overall surface morphology of a fan as determined by echo sounding, the gross structure of the turbidites determined through reflection-profiling techniques, and/or from the distribution and internal structures of the surficial sediments (Menard, 1955; Wilde, 1965; Shepard et al, 1969;

Piper, 1970; Normark, 1970a; Haner, 1971; Nelson and Kulm, 1973; Damuth and Kumar, 1975). The lack of resolution with conventional echosounding and reflection-profiling techniques, the relatively shallow penetration of fan sediments by standard sampling techniques, and the large size of most submarine fans (compared to exposures of ancient turbidites) may tend to distort the perspective of the exploration geologist if he is not careful how he attempts to use modern fan models as predictive tools. As a further complication, the fluctuation of sea level related to the Pleistocene glacial cycles has had marked effects on the sediment supply for modern submarine fans (Normark and Piper, 1969, 1972; Carson, 1971; Nelson and Kulm, 1973; Damuth and Kumar, 1975). The result is that for some fans the present distribution of surficial sediment largely may be unrelated to (not in equilibrium with) the gross morphology and structure of the fan and, without knowing the depositional history of the fan, i.e. its growth pattern, any derived models may be misleading (Normark, 1970a, 1974). As a result, submarine-fan depositional models based on ancient examples (but tempered by observations from the modern environment) may have more practical applications as exploration tools at this time (Mutti and Ricci Lucchi, 1972; Walker and Mutti, 1973; Mutti, 1974; Nelson and Nilsen, 1974; Walker, this issue AAPG *Bull.*).

This review of modern submarine fans will follow the growth-pattern concept of Normark (1970a, 1974) with emphasis on the similarities between modern fans despite orders-of-magnitude differences in size. Congruity between features from modern fans and the general depositional model developed by Walker (this issue) also will be stressed. The review focuses on the

[1]Manuscript received, August 19, 1977; accepted, November 8, 1977.

[2]U.S. Geological Survey, Menlo Park, California 94025.
This manuscript is the outgrowth of a New Orleans Geological Society short course in 1976. I thank F. N. Spiess and the engineers of the deep-tow group at the Marine Physical Laboratory, as their instrumentation provided the key data used in my interpretations. Discussions with D. J. W. Piper, G. R. Hess, J. E. Damuth, E. Mutti, C. H. Nelson, R. G. Walker, A. H. Bouma, P. R. Carlson, and J. Yount helped in the formulation and adjustment of my interpretations; the manuscript was reviewed critically by P. R. Carlson and R. G. Walker.

912

fine-scale geometry and structure of modern fans as derived primarily from geophysical data (conventional and deep-tow high-resolution reflection profiles, narrow-beam echo sounding, side-scanning sonar) and from a knowledge of the distribution of sand-size material within the surficial sediments (defined as the upper few meters, or at best, tens of meters of material that most commonly is sampled).

The fluctuation of sea level during the Pleistocene had marked effects on sedimentation patterns for some fans. In general, core samples of modern fans reflect Holocene processes adequately but often do not allow complete interpretation of Pleistocene events. Unfortunately, even the longer coring devices infrequently reach the shallowest subbottom reflectors observed with high-resolution reflection systems; thus, much of the available geophysical data from modern fans is, in a sense, lacking ground-truth samples. These hindrances are diminished by comparison of varied settings of modern fans.

It must be emphasized that the interpretations presented here reflect the writer's experience with deeply towed instrumentation (Spiess and Tyce, 1973) that allows much greater resolution of bathymetry and shallow structure than can be achieved with conventional surface-ship instrumentation. More comprehensive reviews on the nature and distribution of modern-fan turbidite sediments are available, e.g., Shepard et al (1969), Piper (1970), Haner (1971), Normark and Piper (1972), Nelson and Kulm (1973), Damuth and Kumar (1975), and Nelson (1976).

GROWTH PATTERNS OF DEEP-SEA FANS

The original growth pattern presented for deep-sea fans (Normark, 1970a) recognized three distinct morphologic divisions of the fan surface (Fig. 1): (1) the upper fan, which usually is characterized by leveed fan valleys; (2) the middle fan, where rapid deposition at the end of the leveed valleys builds a suprafan; and (3) the lower fan, which apparently is free of any major topographic relief and may correspond to a ponded, basin-plain environment where a fan is building into a restricted basin or to an abyssal plain for large fans.

The relative proportion of the fan surface comprising each division is quite variable and is a function of basin shape and of the type and rate of sediment supply (Figs. 2, 3). The shape of the middle and lower fan divisions for both the Navy and Monterey fans strongly reflects the influence of basin topography (Fig. 3A, B) whereas the lower fan division (as defined previously) for the Amazon fan (Fig. 3C) is relatively small and is

composed largely of the basin plain ponded on the flanks of the Mid-Atlantic Ridge. Despite the wide range in fan size, all the examples in Figure 3 exhibit leveed valley(s) on the upper fan, numerous distributary channels on the middle fan, and flat-lying, unchanneled deposits on the lower fan, including ponded-basin plain for Navy fan and abyssal plain for Amazon fan.

The growth-pattern concept was derived as a composite of features from several submarine fans, including La Jolla, San Lucas, Monterey, and Astoria (as described by Nelson, 1968) ranging in radial dimension from 60 to 300 km. Although three of these deep-sea fans are from continental-rise settings, the original growth pattern seems matched best to the smaller, ensialic fans of the southern California borderland; nevertheless, important similarities exist with many larger deep-sea fans that comprise continental-rise and abyssal-plain deposits. No one fan has been studied in sufficient detail to ascertain if it exhibits all characteristics of the general model. This study is not intended to be a comprehensive literature review of modern submarine fans. The following discussions avoid the incorporation of observations from fans for which the growth patterns are unknown or too little is known of the sedimentation history. The characteristics of each fan division are based on generalities drawn from many fans but usually are illustrated with specific examples from individual cases.

Upper Fan

If the sediment supply is restricted to a single submarine canyon, the upper fan generally has one active depositional valley continuous with the canyon. Depositional valleys (as opposed to incised, erosional channels) are characterized by prominent levees and the valley floors commonly are elevated above the surrounding level of the fan (Fig. 4A, B, C). The valley floor and associated levee deposits form a broad wedge of sediment that is especially easy to observe on reflection records across large fans (Curray and Moore, 1971; Damuth and Kumar, 1975). In the case of the Bengal fan (Fig. 4A), the leveed-valley complexes may reach 100 km in width and 500 m in thickness and overlie downwarped older fan sediments.

A conventional, low-frequency reflection profile across the Ascension fan valley on upper Monterey fan (Fig. 4B) illustrates Hamilton's (1967) observation that depositional valleys are formed by rather uniform deposition over a broad area, but that the rates of accumulation are slightly greater on the levee sites than elsewhere. Valley floors appear broad and smooth on con-

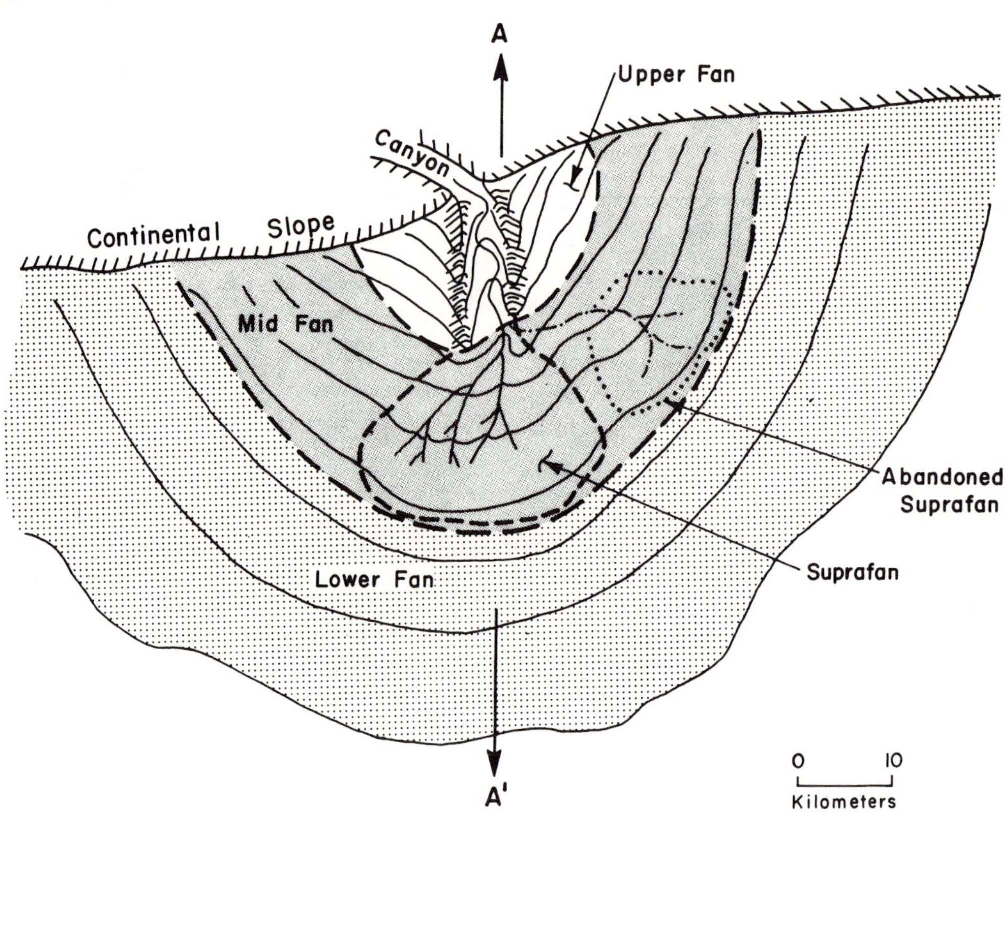

FIG. 1—Schematic representation of model for submarine-fan growth emphasizing active and abandoned depositional lobes termed suprafans (modified from Normark, 1970a).

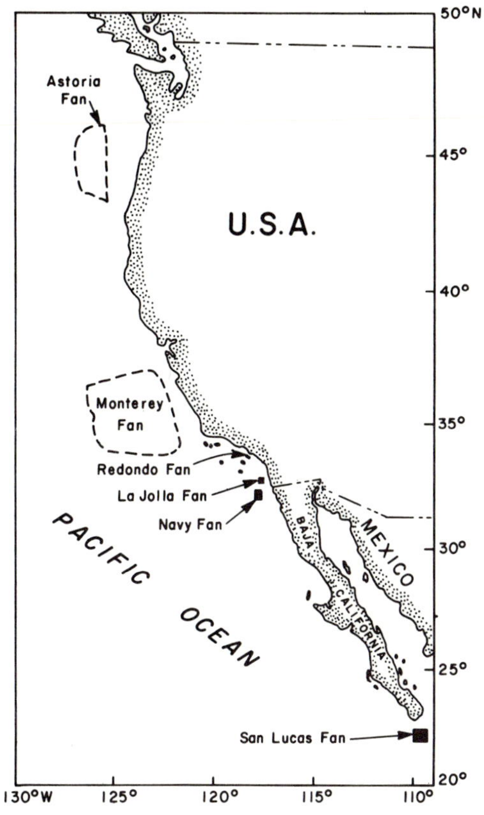

FIG. 2—Location map for fans from west coast of North America that are referred to frequently in text.

ventional reflection profiles. Although the floors of modern depositional fan valleys have aggraded above the level of the adjacent open fan, local relief on the floors is indicative of active channeling. Deep-tow bathymetric profiles show that shallow, irregular channels, to 8-m depth and 50 to 200 m wide, are common on the floor of the Navy fan valley (approx. 1 km wide) and slightly larger channel-like relief to 20-m depth is common on the 8-km wide floor of Ascension fan valley (Figs. 4B, 5). These valley-floor channels appear to form a braided system in some valleys whereas in others a single meandering thalweg channel is formed; both channel types are filled and abandoned with aggradation of the valley floor. The 100-m deep thalweg cutting Ascension valley (Figs. 4B, 5, IV) is uncharacteristic of a depositional regime within leveed valleys in that it represents an early stage of erosional downcutting (Hess and Normark, 1976). Even this channel

meanders sharply as it trends nearly normal to the levees of Ascension valley in the only deep-tow crossing of this feature.

Even the small sublacustrine Reserve fan has a well-developed leveed-valley system on the upper fan (Fig. 4D). The largest valley is bordered by a prominent wedge of sediment with a maximum levee crest to valley-floor relief of 18 m (Normark and Dickson, 1976). The floor of the Reserve fan valley has not aggraded appreciably although the levee section has filled in sizable areas of erosional channeling in prefan lake clays.

The comparative drawings of Figure 4 suggest a rough proportionality between the dimensions of the depositional valleys on the upper fan and the size of the fan. Using the fan radius for comparison, the width of the leveed valley complex, as determined by internal structure on reflection profiles and by physiography, and the cross-sectional area of the valley both increase with increasing fan size (Fig. 6A, B). Levee to valley-floor relief, fan-valley length, and the elevation of the valley floor above the surrounding fan also tend to be greater for the larger fans. Further, these relations can be shown to decrease rather uniformly in the downfan direction for a valley on any given fan. Fan valleys that have been modified extensively by erosional downcutting have not been included in these plots. The fan-valley dimensions were taken from bathymetric and seismic reflection profiles whenever possible and the innermost valley crossing on each fan was chosen while trying to avoid profiles that might include parts of the continental, basin, or delta slope. Two of the most variant points in Figure 6A and B (Mississippi and Laurentian fans) are from fans where published seismic reflection data are insufficient, and only bathymetric data were used.

These relations (Fig. 6) might be used for exploration purposes to estimate total basin size for an ancient fan deposit if a depositional valley is recognized. The size of the large leveed-valley complexes, however, which exceeds 10 km for the smallest borderland fan, suggests that it probably is difficult to recognize the actual width of the levees in ancient fan environments. Levee relief of several tens of meters may extend over 5 or 10 km, such that neither bed thickness nor geometry would be diagnostic for determining levee width. Erosional valleys are mapped more easily in ancient turbidites, but these may represent only braided or thalweg channels within the depositional valley floors. Even the relation using cross-sectional area may be difficult to apply to ancient turbidites unless full channel widths (levee to levee) are recognized.

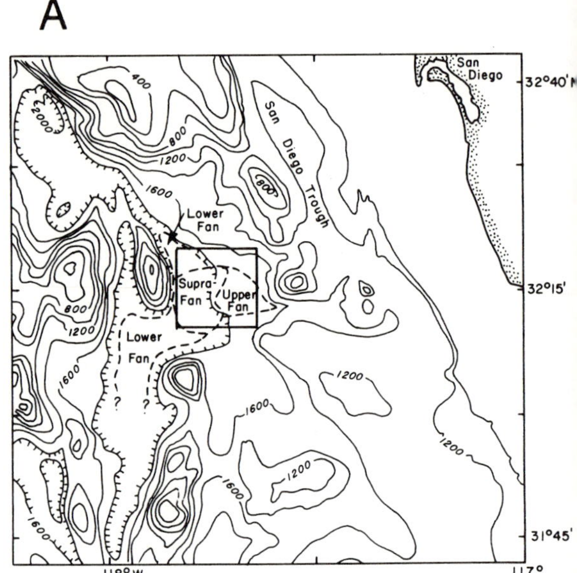

FIG. 3—Simplified bathymetric maps of **A**, Navy; **B**, Monterey; and **C**, Amazon submarine fans depicting upper-, middle-, and lower-fan divisions (modified from Normark and Piper, 1972; Chase et al, 1975; Damuth and Kumar, 1975, respectively). Boxed area in **A** shows location of map in Figure 9. Small tic marks in **C** show positions of channel crossings on echo-sounding lines as given by Damuth and Kumar (1975).

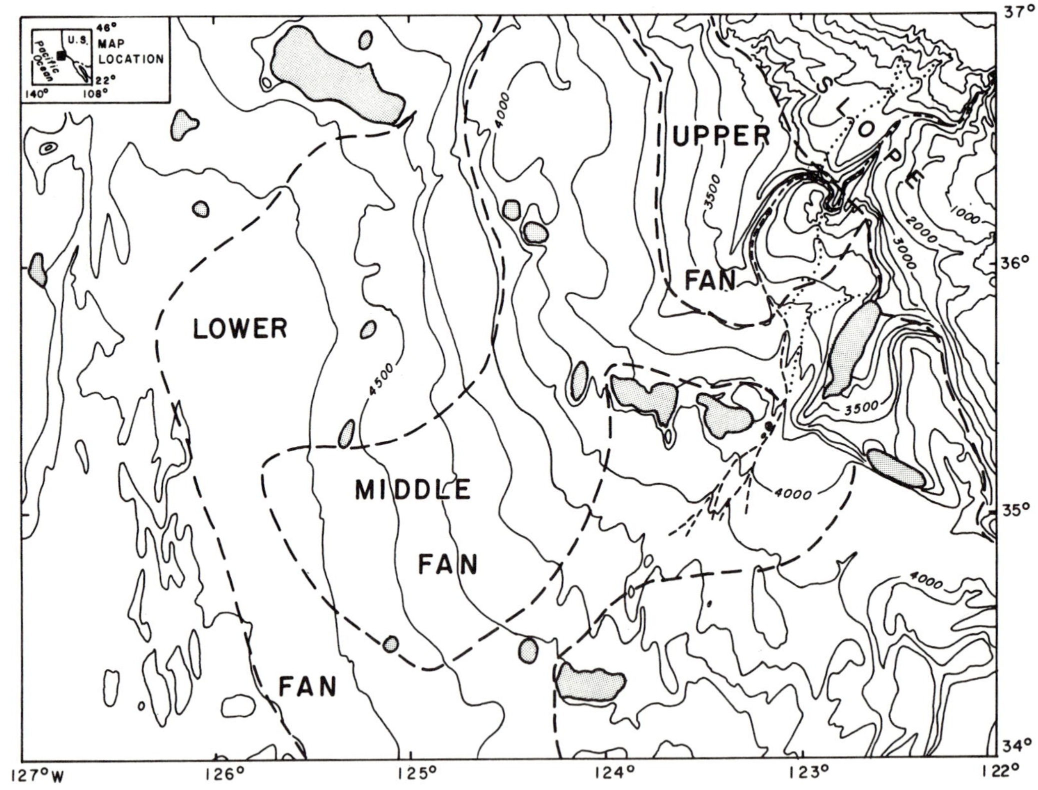

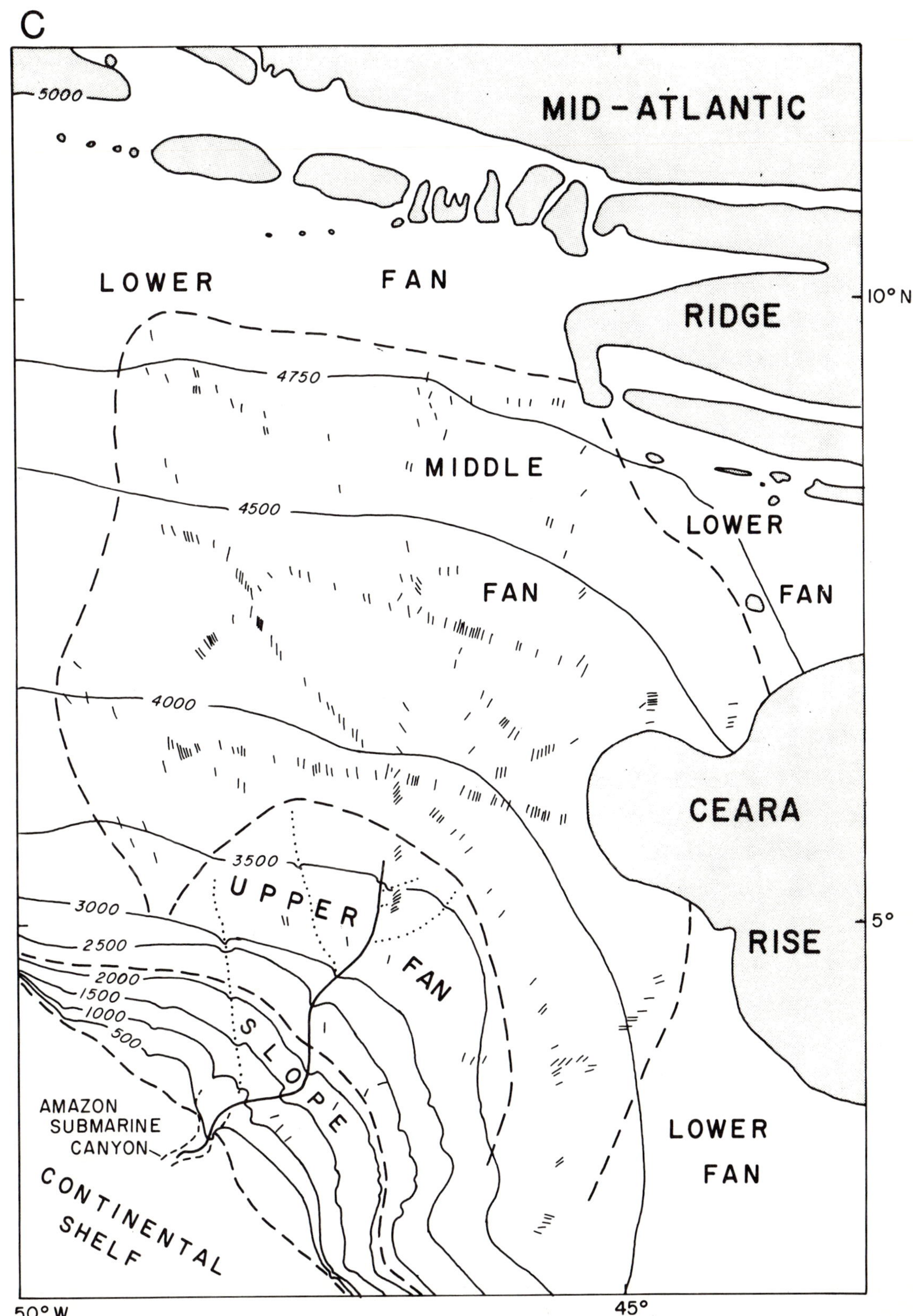

C
5000
MID-ATLANTIC
LOWER   FAN
RIDGE
10° N
4750
MIDDLE
4500
LOWER
FAN
FAN
4000
CEARA
3500
UPPER
3000
RISE
2500
5°
2000
FAN
1500
1000
SLOPE
500
AMAZON
SUBMARINE
CANYON
LOWER
FAN
CONTINENTAL
SHELF
50° W
45°

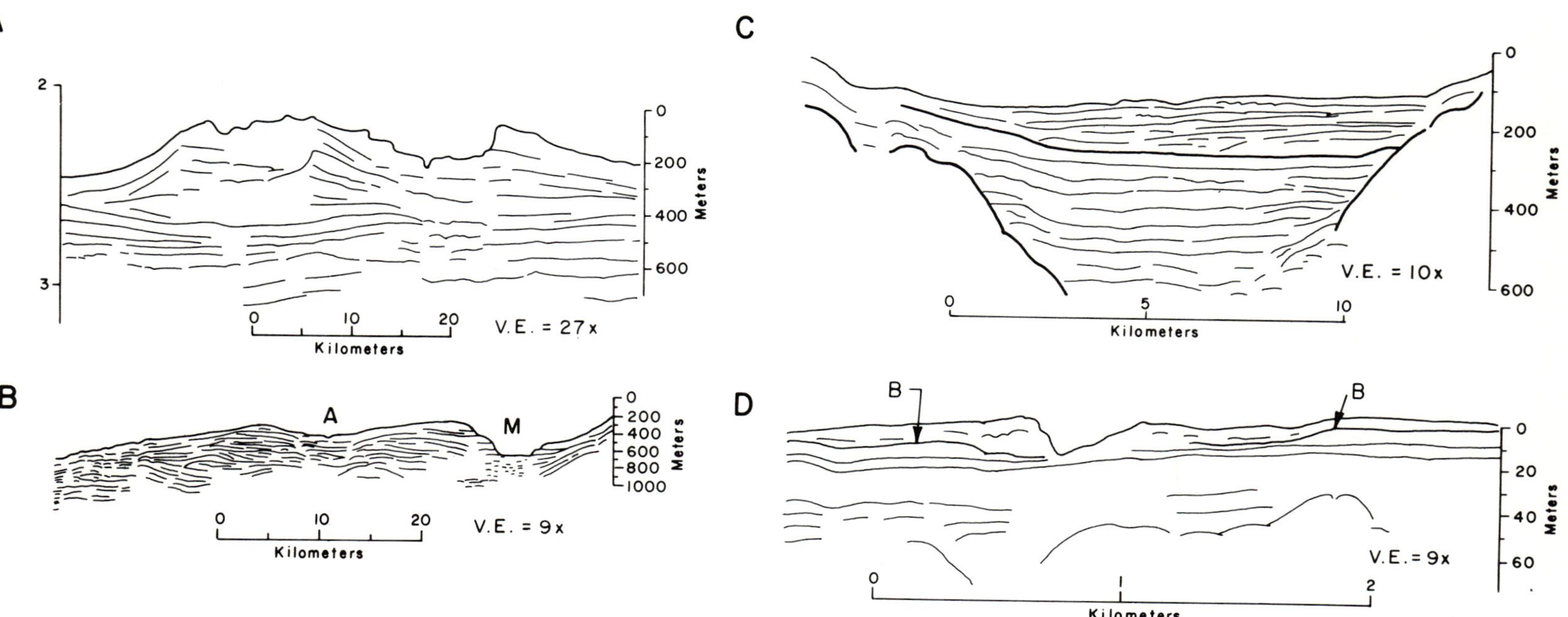

FIG. 4—Line drawings of reflection profiles across leveed valleys on upper fans: **A**, Bengal fan; **B**, Ascension Valley on Monterey fan; **C**, Navy fan; **D**, Reserve fan (modified respectively from Normark, 1970b; Curray and Moore, 1971; Normark, 1970b; Normark and Piper, 1972; Normark and Dickson, 1976). Vertical scales in meters, based on sound velocity in water; horizontal scales approximate. In **4B**, letters *A* and *M* denote axes of Ascension and Monterey fan valleys, respectively. In **4D**, letter *B* denotes base of fan (tailings) deposit.

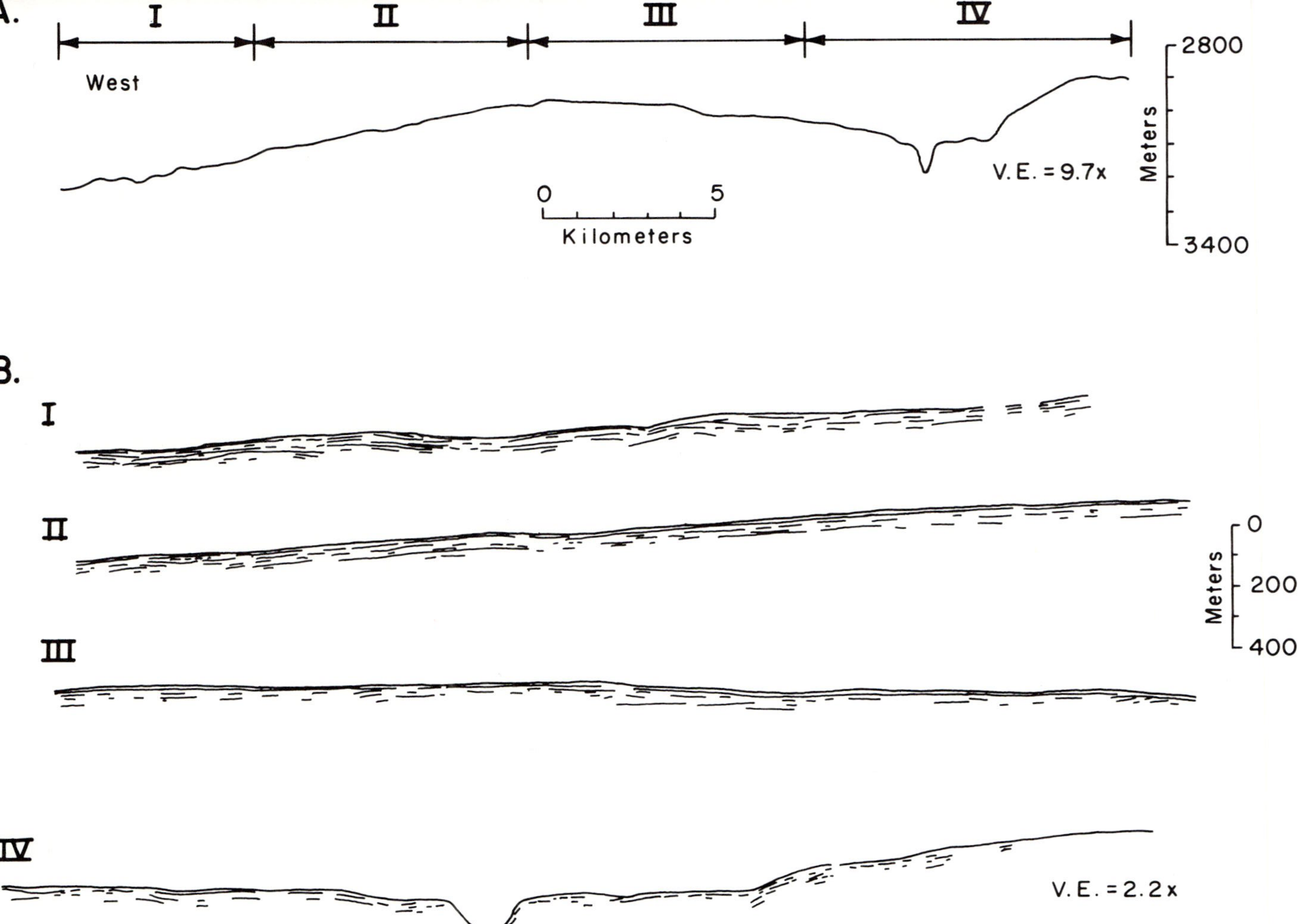

FIG. 5—A, Deep-tow bathymetric profile across Ascension fan valley. B, Line drawings of 4.0-kHz reflection profiles along profile shown in A. Adapted from Hess and Normark, 1976. Vertical scale in meters of water depth.

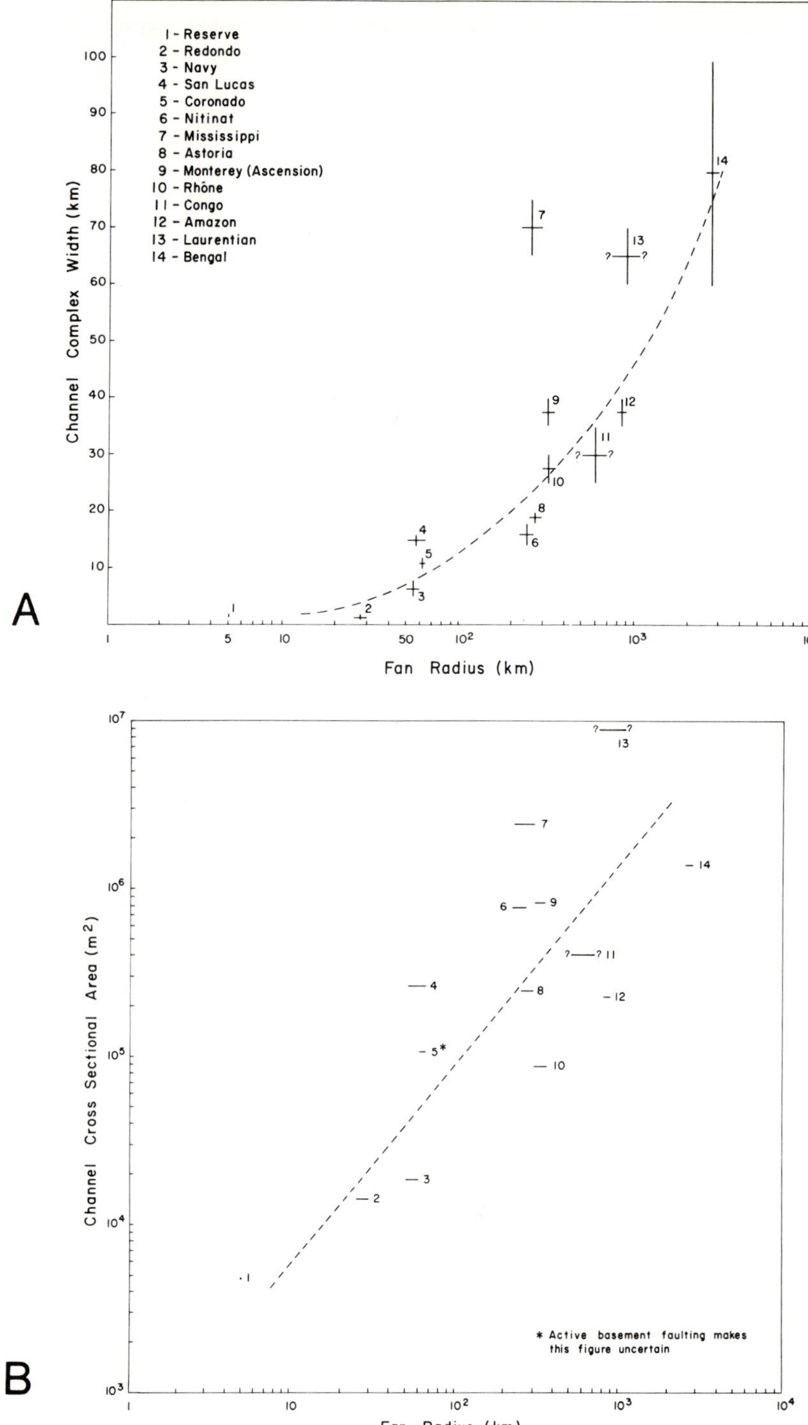

FIG. 6—**A**, Semilog plot of width of levee-valley complex of upper fan valley against fan size as represented by fan radius; numbers key to fan name. **B**, log-log relation between cross-sectional area of valley on upper fan and fan radius; numbers indicate same fan name as for 6A. For both plots, line length indicates estimate of uncertainty of values plotted.

During aggradational stages of depositional-valley development, thick, coarse sands grading to gravels are characteristic (Normark and Piper, 1972; Hess, 1974; Nelson, 1976) and, thus, in agreement with the depositional model presented by Walker (this issue). Lateral migration of valley-floor channels may cause undercutting and subsequent slumping of the more uniformly bedded and finer grained sediments of the valley walls. The best described sedimentary facies for the upper fan are the coarse-grained channel-fill sequences (Nelson and Kulm, 1973; Walker and Mutti, 1973; Mutti, 1974; Nelson, 1976; Walker, this issue).

Active depositional channels are floored by sandy sediments, and one would predict a decrease in sand content of valley-wall and levee sediments with increasing distance from the valley axis. Cores from the small Reserve fan in Lake Superior show a decrease in mean grain size (averaged for all cores within each geomorphic setting) from 0.053 mm in the channel axis, 0.031 mm on the levee crest, 0.022 mm on the middle levee to 0.017 mm at the edge of the levee (Normark and Dickson, 1976). Similar grain-size distributions are suggested for some modern submarine fans although no levee environment has been sampled adequately. On Astoria fan, a specific turbidite unit is present over much of the upper fan and thins from 250 mm on the valley floor to 20 mm about 50 km from the valley (Nelson, 1976). The Holocene green muds now blanketing many modern submarine fans (Carson, 1971; Normark and Piper, 1972; Nelson and Kulm, 1973; Damuth and Kumar, 1975) further complicate the sampling problem.

A similar decrease in grain size with distance away from the leveed valleys on the Bengal fan (Bay of Bengal) is reflected in a well-developed sorting of the biogenic component of sediments derived from turbidity currents passing through the valleys (Jim Yount, personal commun., 1976). The biogenic fraction with settling velocities equivalent to rounded quartz grains in the 0.03 to 0.06-mm size range settles out within 1 km of the levee crest whereas the 0.01 to 0.03-mm size-equivalent components are transported up to 50 km and the finer than 0.01-mm size-equivalent fraction may be several hundred kilometers from the valleys.

The depth of acoustic penetration observed with high resolution (3 to 4 kHz) reflection profiles can be related to the amount of sand in the near-surface sediments on a deep-sea fan (Normark, 1970a; Normark and Piper, 1972; Damuth and Kumar, 1975). The 4-kHz reflection profile taken with the deep-tow instrument of the Ascen-sion fan valley (Fig. 5B) shows that acoustic penetration increases rather uniformly across the backside of the levee (moving from the levee crest away from the valley floor). This profile (Fig. 5B, sections I-III) suggests, however, that bedding continuity is no greater over the smooth levee crest (in sandier sediments) than it is down the levee backside where an undulating or rolling topography has developed. This undulating topography is common over the backsides of the right-hand, higher levees of deep-sea channels in the northeast Pacific (Hamilton, 1967) and reflects a depositional process apparently related to turbidity currents overflowing their channels. The internal structure (Fig. 5B) suggests that these bed forms are elongate roughly parallel with the levee crests and during growth appear to migrate up-slope toward the levee crest (Hess and Normark, 1976). Such depositional features, resulting in onlap of lenticular beds reaching 1 km in length, may be common to smaller leveed valleys as well where the narrower levee width would make them less conspicuous on conventional seismic reflection profiles.

Not all major valleys on upper fans are depositional. Erosional valleys may be cut into low-relief, thin-bedded turbidite sediments commonly placing sands and other coarse clastic deposits of the valley floor into juxtaposition with interbedded silt and mud beds truncated at the valley walls. These erosional valleys dissect the fan surface and result in sediment bypassing much of the fan area (Normark and Piper, 1969; Normark, 1970b). The La Jolla fan valley, which cuts into fan sediments of low relief over much of its length and which does not exhibit distinct levee structures, extends almost to the lower limit of the fan itself (Shepard and Buffington, 1968). Only about 10% of the Holocene sediment entering the La Jolla submarine canyon system is deposited on the fan (Piper, 1970). On the Monterey fan, erosional downcutting of several hundred meters has occurred within a depositional valley whose floor previously was elevated well above the open-fan surface. This incised valley extends across much of the middle and lower fan before breaking into a series of distributary channels (Fig. 3B).

Middle Fan

The middle-fan area is distinguished by a depositional bulge that appears as a convex-upward segment on a radial profile (Fig. 1). This morphologic feature has been termed a suprafan (Normark, 1970a) and is present at the termination of the major active valley on the upper fan (Fig. 7). The suprafan is characterized by numerous, unle-

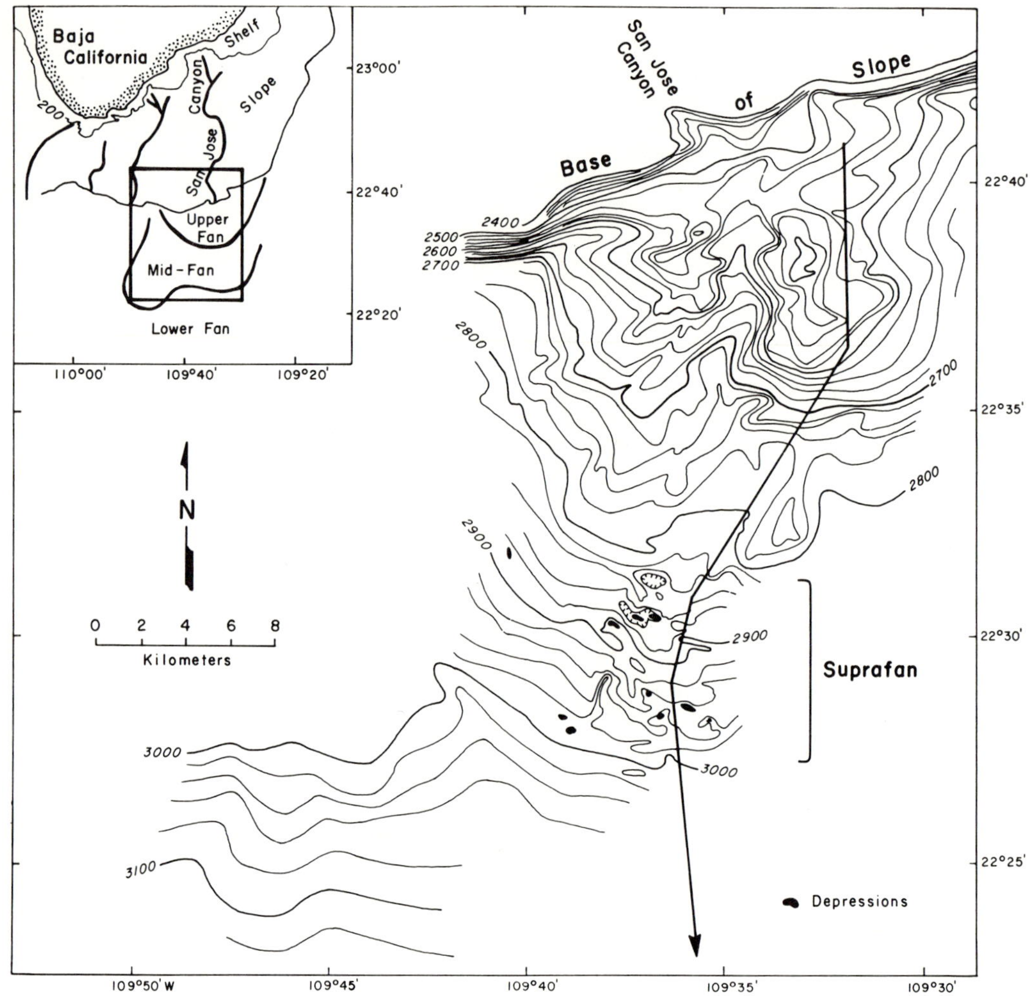

FIG. 7—Bathymetric map of part of upper and middle San Lucas fan from combining conventional echo-sounding data (shallower than 2,800 m) with deep-tow high-resolution sounding data (deeper than 2,800 m); modified from Normark (1970a) and Shepard and Dill (1966). Insert shows approximate fan division. Axes of San Jose and other canyons in slope area shown by heavy lines. Bold line pointing south locates reflection profile shown in Figure 8.

veed distributary channels or channel remnants that develop as a result of rapid migration and deposition within the distributary system. On conventional seismic reflection profiles, the suprafan surface appears hummocky with numerous hyperbolic "side echoes" and discontinuous subbottom reflectors (Fig. 8). Deep-tow profiles over the western part of the undissected suprafan on San Lucas fan at the tip of the peninsula of Baja California, Mexico, show many disconnected features comparable to erosional channels (Fig. 7). The deep-tow track coverage over the San Lucas suprafan was quite dense and analyses

of both narrow-beam echo-sounder and side-scan sonar records leave no doubt that the observed depressions do not connect into a simple, continuous channel network (Normark, 1970a). The distributary channel remnants, up to 2 km long and 50 m in relief, are arcuate and asymmetric, with an apparent remnant "thalweg" and terraced walls. Sands and even pebbles are present within the channel remnants and over the central part of the suprafan (Normark, 1970a). The depth of acoustic penetration with the deep-tow reflection system suggests that sands become less common toward the margins of the suprafan.

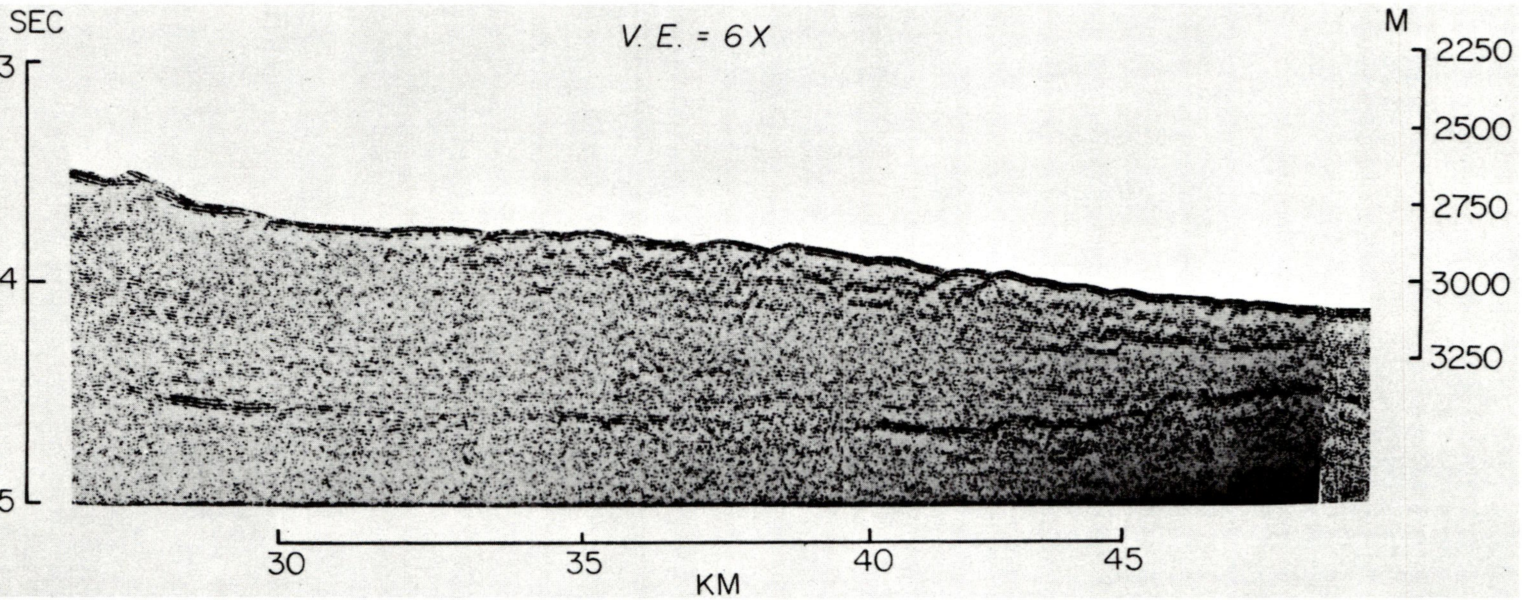

FIG. 8—Single-channel seismic reflection profile across upper fan and suprafan on San Lucas fan (from Normark and Curray, 1968). Horizontal scale in kilometers from apex of fan (approx.); see Figure 7 for location. Suprafan extends from 33 to 50 km on this scale. Vertical scales in seconds of two-way travel time and meters of water depth. Prominent reflector at 4.5-sec round trip travel time is top of oceanic crust (layer 2).

The suprafan on Navy fan is characterized by a lack of acoustic penetration on conventional seismic reflection records and by a hummocky surface topography (Fig. 9A, B). Toward the lower fan, the relief dies out and the depth of acoustic penetration increases (Fig. 9D). A comprehensive sampling of the surficial sediments (Normark and Piper, 1972) confirmed that the lack of acoustic penetration is related to the widespread presence of sands in the surficial sediments of the fan. Deep-tow data from the Navy suprafan show that the hummocky topography as seen on conventional echo-sounder profiles is the result of both the channel distributaries branching from the terminus of the leveed valley on the upper fan and numerous isolated depressions scattered across the middle and upper suprafan (Fig. 10). Unlike many of the San Lucas fan depressions (Fig. 7), many of the Navy fan features are roughly equidimensional, range in width from tens to hundreds of meters, and may represent a type of bed form for coarse turbidites. In addition, a rolling topography distinct from the depressions is present on the upper suprafan (Fig. 10, A-A'). Deep-tow reflection profiles show that surface relief of all kinds decreases down fan whereas the depth of penetration and continuity of reflectors increase. As noted previously, these acoustic signatures reflect a decrease in the proportion of sandy sediments down fan. The distributary channels, which do not have levees, terminate on the middle suprafan.

The suprafan concept, as originally defined, refers to a morphologic feature of modern, active fans. Application of this concept to other modern fans, in the absence of detailed deep-tow surveys or the equivalent, must be based on recognition of four characteristics: (1) the irregular, local bathymetric relief as seen on conventional reflection profiles (Fig. 8), (2) the overall convex-upward relief of this region of irregular relief (Fig. 8), (3) the association of these bathymetric characters with coarse-grained turbidites, and (4) the development of this morphology at, or immediately downslope of, the termination of the major valley on the upper fan (Figs. 7, 10).

Suprafan areas, exhibiting this association of features, have been identified on other deep-sea fans. Suprafan development can occur at the termination of erosional fan valleys as well. This has been incorporated into the schematic diagram by Walker (this issue, Fig. 3). Limited reflection profiles suggest that a suprafan may be developing in San Diego Trough below the termination of the incised La Jolla fan valley. In this case, the suprafan now is forming on the lowermost fan or adjacent basin plain (see following). The Monterey fan has three areas of possible suprafan morphology (Fig. 3B). The most active one is at the end of the incised Monterey fan valley on the physiographic lower fan; a second suprafan on the middle fan is at the termination of the beheaded Monterey East valley, which now receives sediment only from turbidity currents that can overflow from the incised Monterey fan valley on the upper fan (Hess and Normark, 1976). The third possible suprafan is abandoned and represents a previously active depositional area (the southwestward bulge in the middle-fan division [Fig. 3B]).

Several studies on modern submarine fans have noted the common appearance of a convex-upward segment on most radial profiles (Nelson, 1968, 1976; Haner, 1971; Damuth and Kumar, 1975). Shifting of the valley(s) on the upper fan will lead to growth of new suprafan lobes. Abandoned suprafans will be blanketed by muds, eventually smoothing over the characteristic local relief, but the convex-upward profile will remain (Fig. 11). The middle-fan division includes not only the active suprafan and/or distributary-channel system but also encompasses any fan areas with the convex-upward radial profile. Under this definition, the lower fan division is characteristically rather flat and smooth, being free of any recognizable channels in most cases (see below). On the Amazon fan (Fig. 11D), the middle fan bulge is subtle and the upper/middle-fan boundary is picked mostly on the recognition of numerous possible distributary channels (Fig. 2C and Damuth and Kumar, 1975).

Until further detailed deep-tow surveys together with comprehensive bottom sampling are available for additional submarine fans, the suprafan concept best may be expanded by comparison with ancient turbidites. The deltalike nature of deposition on a suprafan would result in coarsening- and thickening-upward sequences; channeling is expected to be most evident over the upper, aggraded part of the suprafan (Walker and Mutti, 1973; Walker, this issue). Suprafan development is thought to depend upon such variables as the local bottom slope and the rate and grain size of sediment supplied through the valley on the upper fan. In some cases, little more than a channel-mouth bar may form; in others, sand lobes comprised of coarsening- and thickening-upward sequences may be relatively free of channel development (Mutti, 1977; personal commun., 1974). For example, the southward bulge between the 3,000 and 3,120-m water-depth contours of San Lucas fan in the area immediately southwest of the present suprafan (Fig. 7) is similar in size to the adjacent suprafan, yet the deep-tow data

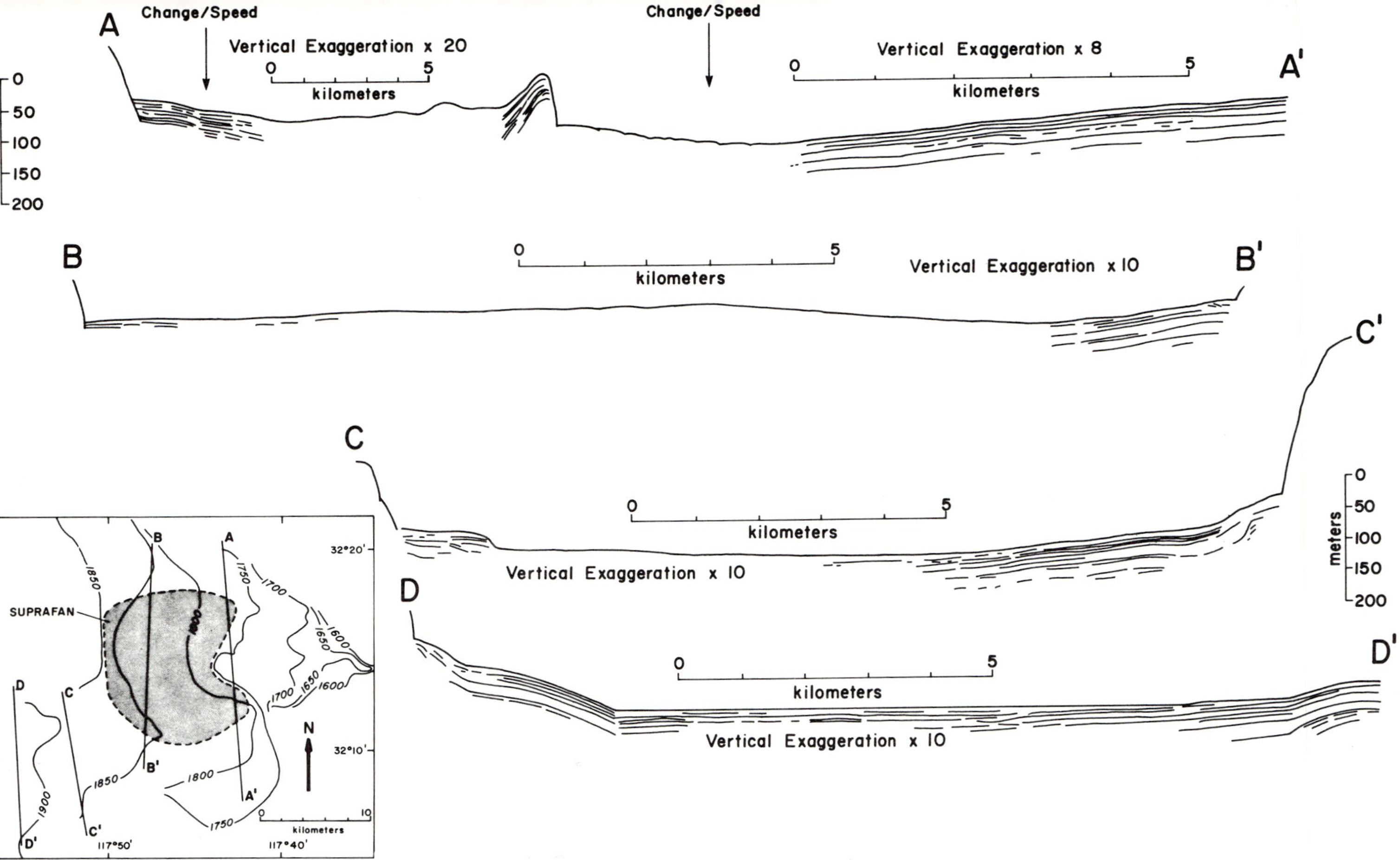

FIG. 9—Conventional 3.5-kHz high-resolution profiles from Navy fan (modified from Normark and Piper, 1972) illustrate lack of acoustic penetration over suprafan (lines A, B) grading to moderate penetration over lower fan (basin plain, line D). Suprafan on line A is near center of line around "change/speed" notation and on line B is region under scale bar in center. See inset for location and compare with Figure 3A.

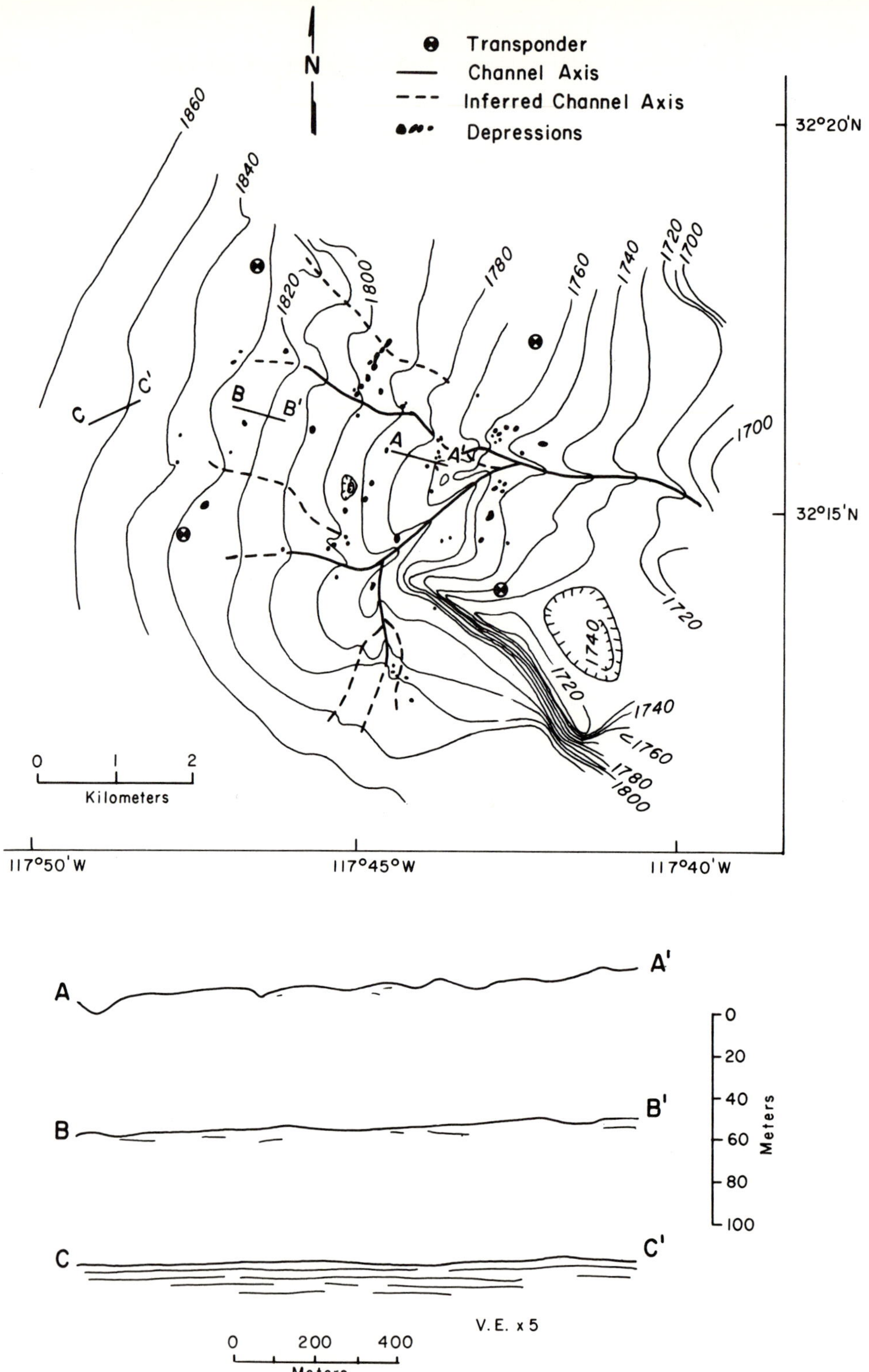

FIG. 10—Upper: Detailed map of central Navy fan from deep-tow soundings. Map extends from lower upper fan to lower suprafan (northwest corner). Side-scan data used to help map distributary channels and isolated depressions, which are drawn approximately to scale.

Lower: Line drawings of 4.0-kHz deep-tow seismic reflection profiles across suprafan.

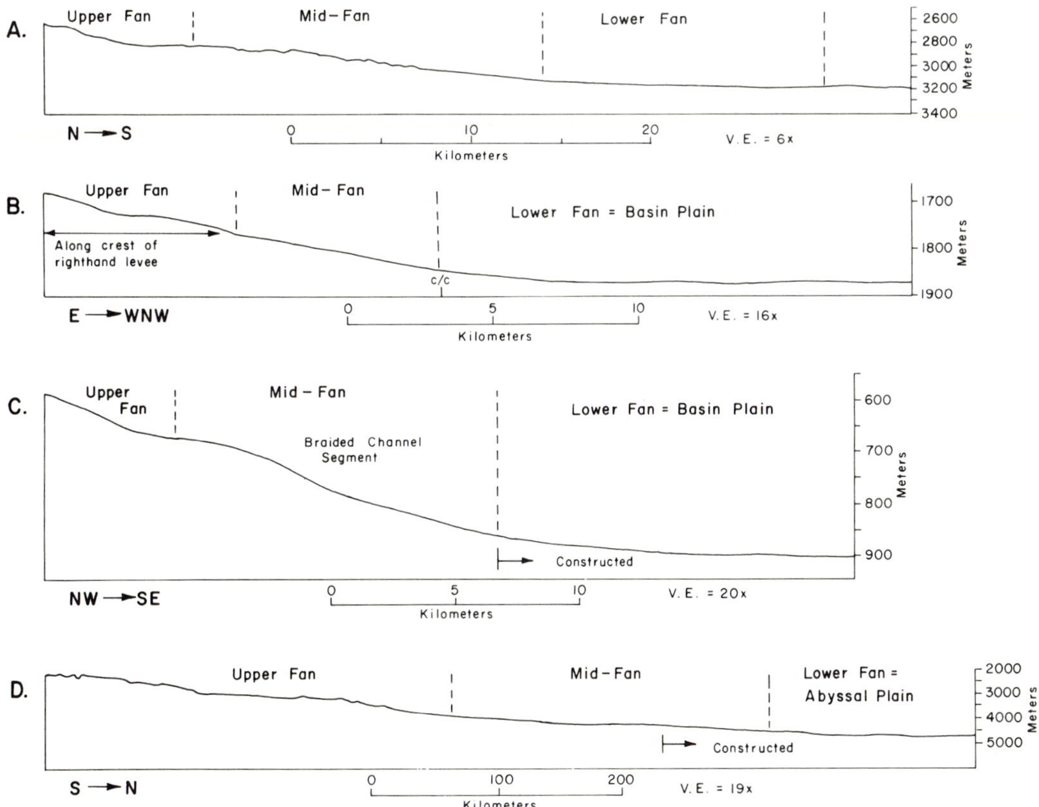

FIG. 11—Radial profiles of **A**, San Lucas; **B**, Navy; **C**, Redondo; and **D**, Amazon submarine fans. Middle-fan bulge is present in some form on all profiles and topographic similarities of lower fan, abyssal plain, or basin plain are obvious. Profile **A** traced from reflection profile of San Lucas fan (see Fig. 12); profile **B** constructed from Figure 4 of Normark and Piper (1972); profile **C** modified from Haner (1971) except for lower fan segment, which was constructed from additional sounding data; profile **D** traced and constructed from reflection profiles and maps in Damuth and Kumar (1975).

show no channels or other irregular surface relief. This bathymetric feature may be an older, sediment-smoothed suprafan or the topographic expression of a mud-covered "outer fan" sand lobe as described by Mutti (1974, 1977). The data from the best studied modern fans, especially Navy fan, suggest that unchanneled sand lobes comprised of thickening- and coarsening-upward sequences form the lower suprafan deposits. It is clear that the "outer fan" environment of Mutti and his coworkers is equivalent to the lower middle-fan division defined here.

Lower Fan (Basin Plain)

Only two of the submarine fans studied by the writer appear to have unaltered growth patterns—San Lucas and Navy fans. Both have well-developed lower fan divisions with nearly horizontal surfaces, uniform parallel bedding on reflection profiles (Fig. 12), and no evidence of channels. Distinct sand beds generally are fine grained, thin, and widespread and the grain size of the interbedded silts and silty clays decreases distally (Normark and Piper, 1972). The major difference between the two is that the lower Navy fan is ponded in two separate branches of South San Clemente basin whereas San Lucas fan has built over low-relief abyssal hills.

The similarities between these fans and others described in the literature suggest that depositional processes on basin plains (and, to a certain extent, on abyssal plains) are indistinguishable

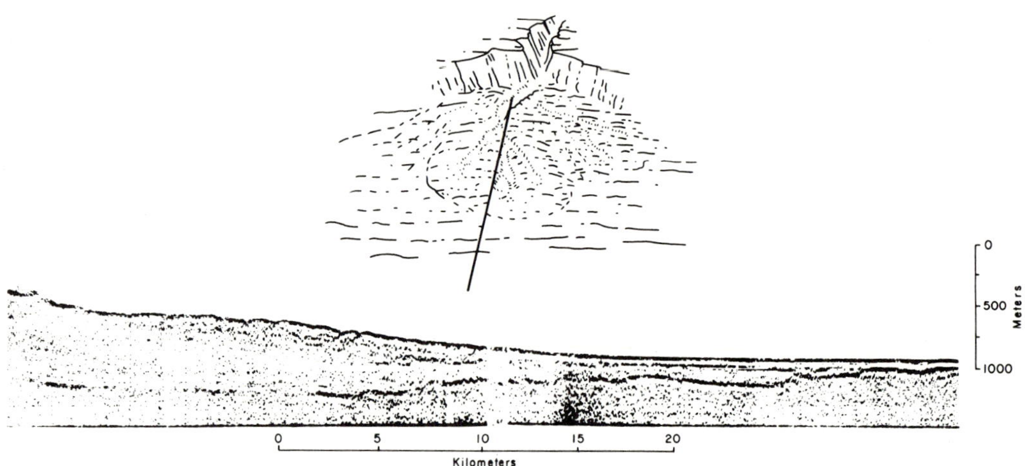

FIG. 12—Upper: Physiographic sketch of submarine-fan morphology (courtesy of Tau Rho Alpha) showing fan divisions. Lower: Seismic reflection record of north-south profile across San Lucas fan from flank of leveed valley on upper fan to low-relief abyssal hills at toe of lower fan (from Normark, 1970a). Horizontal scale approximate.

from those operating on the lower relief areas downslope of suprafans, especially on those fans not confined in topographic depressions. If the basin-plain or abyssal-plain deposits adjacent to Astoria and Amazon fans, for example, are recognized as the lower fan division, the comprehensive descriptions of these two large fans (Damuth and Kumar, 1975; Nelson, 1976) show strong parallels with the suprafan model. Further, the depositional model presented by Walker (this issue) recognizes the indistinguishable nature of outer-fan and basin-plain facies.

VARIATIONS IN GROWTH PATTERNS

Some modern submarine fans do not match the growth pattern developed in the foregoing. The rapid Holocene sea-level rise left many modern fans without a source of coarse clastic sediments, but this change alone in sediment regime should leave morphologic features unchanged. In these cases, the surface (Holocene) sediment blanket of pelagic and/or hemipelagic muds over much of the fans may not fit a sediment distribution that would be predicted from the morphology of the fan. The pre-Holocene channel shapes, distributary systems, and other major morphologic features, however, still would reflect the period of active growth. Good seismic reflection data are critical for the determination of the effect of sedimentologic changes on the growth pattern of a fan. For both the Astoria (Nelson and Kulm, 1973; Nelson, 1976) and Amazon (Damuth and Kumar, 1975) fans, the growth pattern for pre-

Holocene sedimentation appears to be slightly modified but still recognizable. The San Lucas and Navy fans (Normark, 1970a; Normark and Piper, 1972) probably have received little or no turbidite sediments in the late Holocene, but hemipelagic sedimentation also has been slow, resulting in an unaltered pre-Holocene growth pattern blanketed by a thin cover of Holocene hemipelagic mud.

For the Monterey and La Jolla fans, erosion within the main valley system has resulted in sediment bypassing the upper and middle fan areas (Piper, 1970; Hess and Normark, 1976). Occasional overbank muds represent the only turbidity-current deposition on the upper and middle fans in these cases. The incision of the valley systems may result from changes in the submarine-canyon gradient in response to sea-level fluctuation during the Pleistocene (Normark and Piper, 1969). There are some indications on both fans, however, of relict suprafan morphology that has since been cut through or bypassed by the incised channels.

The grain-size distribution and the rate of sediment supplied may determine whether a suprafan is formed. The large Bengal fan (Curray and Moore, 1971) does not exhibit the characteristic features of the middle-fan division. The upper fan has an extensive system of large leveed valleys, but because only one appears to be a continuation of the "Swatch-of-No-Ground" submarine canyon, the others are probably a succession of abandoned distributaries. Although these leveed valleys are comparable in size to those on the Monterey fan, they show little internal structure within the levee sequences. Many of the valleys cross almost the entire fan (2,500+ km) but none appear to terminate in a suprafan. Available core samples show little coarse sediment within the main valley, even on the upper fan, in contrast to the fans that are compatible with the suprafan model. The relative absence of internal levee structure (Fig. 4A) also may reflect a rather uniform grain size of the levee sediments. If there is little coarse sediment in suspension (or if the turbidity current has a limited bed load), rapid deposition at the end of the fan valley may not occur; instead, levee development by overbank deposition, enhanced near the valley terminus where the flow can spread laterally, results in relatively straight channels which will prograde rapidly down fan (Normark, 1974). In this case, no middle-fan convex-upward bulge would develop. A succession of channel-mouth bars (Mutti, 1977) may form if some coarse sediment is deposited where the turbidity currents leave the confines of the valley.

Deposition on the sublacustrine Reserve fan has produced thick (to 18 m) and wide (to 2.5 km) levee complexes (relative to the small size of the fan) that comprise most of the sediment in the fan area (Normark and Dickson, 1976). Like the Bengal fan, the leveed valleys have prograded rapidly down fan. Small suprafans (or channel mouth bars?) have developed but they are only a few meters thick; these suprafans are recognized on the basis of irregular local relief and slightly coarser sediments. Most of the sediments on Reserve fan are sandy silts, except for a few silty sands on the valley floor and suprafan. Because the grain-size distribution of sediment supplied to the fan has remained constant throughout deposition over 17 years, the grain-size distribution within the levee sediments should resemble closely the present surficial trends. Even very rapid sedimentation rates, to 1.1 m/year, have not favored suprafan development. It seems reasonable, therefore, that the enhanced levee growth is related to the rather uniform sandy-silt composition of the sediments transported by turbidity currents on Reserve fan.

The suprafan growth pattern pertains primarily to fans that receive sediments through submarine canyons whether they are small ensialic basin fans or whether they comprise large continental-rise prisms. All fans used for the original model, including all fans from the northeastern Pacific referred to in this text, fall into this category. On the other hand, submarine fans associated with deltas may lack the marked, convex-upward seg-

ment of the middle fan that results from suprafan development. The deposition of much of the coarse fraction on the delta platform would tend to produce muddy turbidity currents that favor levee growth over suprafan growth. Multiple-leveed valleys on the upper fan are more common on delta-fed fans than on canyon-fed systems, but exceptions to both cases are expected, e.g., Astoria canyon provides sediment to Astoria fan, but the upper fan has several leveed valleys (Nelson, 1968). Bengal fan and Reserve fan, at opposite ends of the spectrum of fan size, are the products of sediments derived from deltaic environments, whereas the Amazon fan (Damuth and Kumar, 1975) has some features compatible with both canyon-fed and delta-fed fans as discussed previously. The grain-size distribution of the sediment supplied to a fan probably is the key factor affecting the morphology of the fan. A variety of processes operating in coastal environments may affect the sediment supply for an adjacent fan, and the association of a delta with the absence of a suprafan may represent only an extreme case and, thus, should be used only with caution by the exploration geologist.

SEDIMENTARY ASSOCIATIONS ON MODERN FANS

Facies distinctions for submarine fans based on geophysical and surficial studies of modern turbidites may be inherently unsatisfying because of the lack of comparison with sedimentologic features at outcrop scale. Therefore, the review presented in the following should be used to help predict depositional trends in the application of the turbidite-fan depositional models presented by Mutti (1974) and Walker (this issue).

The upper fan may exhibit two sandy facies: the sediments of the major fan valley and the levee/overbank sequences. The coarsest sediments on the fan, sands and gravels, are found in interfingering channel-fill sequences across the floor of depositional fan valleys. These channel-fill complexes commonly are 1 to 5 km wide and may be 500 m or more thick, whereas the individual channel fills may exceed 20 m in thickness and several hundred meters in width. Paleocurrent indicators within the channel sands may be quite varied because of meanders and/or braided patterns within the overall confines of the main valley. The main depositional valley, especially near the head of the fan, may not meander significantly, and the sequences of interfingering channel sands may be several hundreds of meters (to several kilometers) thick and 5 or more km wide. The valley-floor sands are gradational with, and may interfinger with, sands of the adjacent levees.

The mean grain size of the sediment decreases across the levee, and the deep-tow profiles suggest that bedding may be continuous for hundreds of meters across the levee crest. Depositional bed forms across the backside of the levee may be expressed by relatively thin, lenticular (but not erosionally channeled) sand and silt beds.

Coarse sediments are also common within erosional valleys on upper fans (Wilde, 1965; Piper, 1970) as is the formation of a meandering thalweg (Shepard and Buffington, 1968; Komar, 1969). Overbank deposition generally is restricted and the channel-fill sands will be truncated abruptly at the valley margins with little interfingering with channel-margin sands. Erosional valleys on the upper fan cannot persist if fan aggradation is to occur. It seems reasonable, therefore, that erosional valleys will tend to be ephemeral and generally will not result in thick sequences of channel sands. Many ancient examples of erosional valleys contain sediment fill that becomes finer upward, having formed after the valley was abruptly abandoned (Piper and Normark, 1971; Walker, this issue).

Fans fed from submarine canyons tend to develop sandy suprafan lobes in the middle-fan area (Fig. 12). Unleveed channels, similar in size and shape to those in depositional-valley floors, form a braided or distributary system on the proximal part of the suprafan. Sediment within these channels is nearly as coarse as on the floor of the main valley on the upper fan, but the channels disappear and sands become finer toward the margins of the suprafan. The suprafan, which has a delta-like profile, forms a middle-fan lobe that contains sand beds that should become thicker and coarser upward. Channels are more common near the top of the sequence, also comparable with deltaic sedimentation. Suprafan channels and channel remnants generally lack levees, are less than 1 km wide, and may be filled with sand and silt beds that become finer and thinner upward as sedimentation shifts across the suprafan. A succession of suprafan lobes develops throughout the middle fan as the feeding valleys migrate or are cut off and new valleys develop. Each suprafan may be capped by pelagic muds. The sands of the suprafan in some cases may be contiguous with those in the feeding valley, and thus the whole sandy complex may have good reservoir potential for hydrocarbon accumulation.

The great width of modern levee-valley complexes (10 to 100 km) compared to their thickness of several hundred meters suggests that it will be difficult to identify the full extent of these sequences in ancient turbidite deposits. The relatively continuously bedded turbidites throughout the levee sequences probably are not even recognized as valley-related deposits in many cases. Only the braided or thalweg-channel deposits are easily recognized in outcrop. A more difficult identity problem, however, arising from the scale problems in comparing modern and ancient turbidites, concerns the proper interpretation of major distributary channels and depressions on the upper suprafan. The size of many suprafan relief features exceeds 500 m in width and 20 m depth and, if viewed in cross section in an outcrop, would be identified as a major channel sequence by the unsuspecting geologist, who then might place these features in an upper-fan depositional environment. The recognition of channel deposits, even if an individual channel is 1 or 2 km wide, is not, in itself sufficient to distinguish between upper fans (floor of leveed valley) and middle-fan (upper suprafan) environments.

Except for conditions that lead to incision of the major valley, the lower fan is relatively free of sands and channels. Bedding is laterally continuous and uniform in thickness over great distances; classical turbidite sequences dominate (Walker, this issue).

The suprafan depositional model as expanded by Walker (this issue) can be a useful exploration tool. However, submarine fans fed directly from deltaic systems or other coastal environments on the continental shelf that might cause restriction of the grain-size distribution, never may develop the thick clastic sequences representing suprafans and valley/channel complexes of the upper fan. Elongate (shoestring?) valley-floor sand bodies and associated fine-grained levee deposits or discontinuous channel-mouth bars may form the only clastic sections with potential for significant hydrocarbon accumulation. The depositional model will be less successful in predicting geometry and depositional trends in ancient fans of this type.

REFERENCES CITED

Carson, B., 1971, Holocene-Pleistocene sedimentation in Cascadia basin, northeast Pacific Ocean: 8th Internat. Sediment. Cong. Program with Abs., p. 16-17.

Chase, T. E., W. R. Normark, and P. Wilde, 1975, Oceanographic data of the Monterey deep-sea fan; IMR Tech. Rept. Ser. TR-58.

Curray, J. R., and D. G. Moore, 1971, Growth of the Bengal deep-sea fan and denudation in the Himalayas: Geol. Soc. America Bull., v. 82, p. 563-572.

Damuth, J. E., and N. Kumar, 1975, Amazon cone: morphology, sediments, age, and growth pattern: Geol. Soc. America Bull., v. 86, p. 863-878.

Hamilton, E. L., 1967, Marine geology of abyssal plains in the Gulf of Alaska: Jour. Geophys. Research, v. 72, p. 4189-4213.

Haner, B. E., 1971, Morphology and sediments of Redondo submarine fan: Geol. Soc. America Bull., v. 82, p. 2413-2432.

Hess, G. R., 1974, Submarine fanfare: comparison of modern and Miocene deep-sea fans: Master's thesis, Univ. Minnesota, 118 p.

———— and W. R. Normark, 1976, Holocene sedimentation history of the major fan valleys of Monterey fan: Marine Geology, v. 22, p. 233-251.

Komar, P. D., 1969, The channelized flow of turbidity currents with application to Monterey deep-sea fan channel: Jour. Geophys. Research, v. 74, p. 4544-4558.

Menard, H. W., 1955, Deep-sea channels, topography, and sedimentation: AAPG Bull., v. 39, p. 236-255.

Mutti, E., 1974, Examples of ancient deep-sea fan deposits from Circum-Mediterranean geosynclines, *in* Modern and ancient geosynclinal sedimentation: SEPM Spec. Pub. 19, p. 92-105.

———— 1977, Distinctive thin-bedded turbidite facies and related depositional environments in the Eocene Hecho Group (south-central Pyrenees, Spain): Sedimentology, v. 24, p. 107-131.

———— and F. Ricci Lucchi, 1972, Le torbiditi den "Appennino settentrionale:" introduzione all'analisi de facies: Soc. Geol. Italiana Mem., v. 11, p. 161-199.

Nelson, C. H., 1968, Marine geology of Astoria deep-sea fan: PhD thesis, Oregon State Univ., 287 p.

———— 1976, Late Pleistocene and Holocene depositional trends, processes, and history of Astoria deep-sea fan, northeast Pacific: Marine Geology, v. 20, p. 129-173.

———— and L. D. Kulm, 1973, Part II: Submarine fans and deep-sea channels, *in* Turbidites and deep-water sedimentation: SEPM Pacific Sec., Short Course Notes, Anaheim, California, p. 39-70.

———— and T. H. Nilsen, 1974, Depositional trends of modern and ancient deep-sea fans, *in* Modern and ancient geosynclinal sedimentation: SEPM Spec. Pub. 19, p. 69-91.

Normark, W. R., 1970a, Growth patterns of deep-sea fans: AAPG Bull., v. 54, p. 2170-2195.

———— 1970b, Channel piracy on Monterey deep-sea fan: Deep-Sea Research, v. 17, p. 837-846.

———— 1974, Submarine canyons and fan valleys: factors affecting growth patterns of deep-sea fans, *in* Modern and ancient geosynclinal sedimentation: SEPM Spec. Pub. 19, p. 56-68.

———— and J. R. Curray, 1968, Geology and structure of the tip of Baja California, Mexico: Geol. Soc. America Bull., v. 79, p. 1589-1600.

———— and F. H. Dickson, 1976, Sublacustrine fan morphology in Lake Superior: AAPG Bull., v. 60, p. 1021-1036.

———— and D. J. W. Piper, 1969, Deep-sea fan-valleys, past and present: Geol. Soc. America Bull., v. 80, p. 1859-1866.

———— 1972, Sediments and growth pattern of Navy deep-sea fan, San Clemente basin, California borderland: Jour. Geology, v. 80, p. 198-223.

Piper, D. J. W., 1970, Transport and deposition of Holocene sediment on La Jolla deep-sea fan, California: Marine Geology, v. 8, p. 211-227.

———— and W. R. Normark, 1971, Re-examination of a Miocene deep-sea fan and fan-valley, southern California: Geol. Soc. America Bull., v. 82, p. 1823-1830.

Shepard, F. P., and E. C. Buffington, 1968, La Jolla submarine fan-valley: Marine Geology, v. 6, p. 107-143.

———— and R. F. Dill, 1966, Submarine canyons and other sea valleys: Chicago, Illinois, Rand McNally, 381 p.

———— ———— and U. von Rad, 1969, Physiography and sedimentary processes of La Jolla submarine fan and fan-valley, California: AAPG Bull., v. 53, p. 390-420.

Spiess, F. N., and R. C. Tyce, 1973, Marine Physical Laboratory deep-tow instrumentation system: Scripps Inst. Oceanography Ref. 73-4, 37 p.

Walker, R. G., 1978, Deep-water sandstone facies and ancient submarine fans: models for exploration for stratigraphic traps: AAPG Bull., v. 62, p. 932-966.

———— and E. Mutti, 1973, Facies and facies associations, Part IV, *in* A. H. Bouma and G. V. Middleton, eds., Turbidites and deep-water sedimentation: SEPM Pacific Sec. Short Course, p. 119-158.

Wilde, P., 1965, Recent sediments of the Monterey deep-sea fan: PhD thesis, Harvard Univ., 153 p.

The American Association of Petroleum Geologists Bulletin
V. 64, No. 7 (July 1980) P. 1094-1112, 3 Figs.

DISCUSSIONS

Modern and Ancient Submarine Fans: Discussion of Papers by R. G. Walker and W. R. Normark[1]

TOR H. NILSEN[2]

INTRODUCTION

The AAPG *Bulletin* of June 1978 contained several papers that attempted to synthesize current concepts and models of deep-marine sedimentation for modern and ancient submarine fans. Walker (1978) presented a model for the development of submarine fan facies and its use in petroleum exploration. Normark (1978) presented an updated model of submarine fan development based primarily on studies of modern fans off the west coast of North America that also incorporated some concepts developed from studies of ancient fan deposits. Although I wish to direct most of my comments to Walker's paper, some of Normark's conclusions also require comment because of the stressed congruity of the two models. Several important aspects of these papers are misleading, in my opinion; perhaps the overall usefulness of the models should be thoroughly questioned in a more lengthy treatment.

I wish to address herein three major points: (1) the confusing misuse of the term "thick-bedded" for "proximal" by Walker; (2) problems introduced by application of the suprafan concept to ancient fan deposits, with resultant confusion regarding identification of middle- and outer-fan sequences; and (3) unfortunate overemphasis of channeled submarine fan deposits as exploration targets. The model presented by Walker is essentially a marriage of earlier models published by Mutti and Ricci Lucchi (1972) for ancient fans and Normark (1970) for modern fans. However, it has become clear to many workers that these two models may be applicable to two completely different types of fans fed by different types of sediment and characterized by different surface morphologies; thus, the attempt to combine these ancient and modern models into a single model with almost universal application may be misleading, particularly for petroleum exploration. These and other problems have resulted in difficulties for many workers in accepting and using both the earlier (Normark, 1970, 1974; Walker, 1970, 1976) and most recent versions of these models.

"PROXIMAL" VERSUS THICK-BEDDED SANDSTONE

In Walker's (1975) paper dealing with coarse-grained and thick-bedded deposits of Late Cretaceous age at Wheeler Gorge, California, the underlying, intercalated, and overlying thin-bedded turbidites were referred to as "distal" turbidites. Nelson et al (1977) questioned this use of the term "distal," interpreting the thin-bedded turbidites as channel-margin, levee, and interchannel deposits. By this interpretation, with which I agree, the thin-bedded turbidites are generically related to the thicker bedded and coarser deposits, which form thinning- and fining-upward megasequences, and are interpreted to be channel deposits.[3] In earlier papers, Nelson and Nilsen (1974, Table 4) and Nelson et al (1975) outlined the major distinguishing characteristics of thin-bedded turbidites deposited in proximal and distal settings, and Mutti (1977) clearly distinguished various thin-bedded turbidites deposited in different fan and basin-plain settings in a study of the Hecho

[1]Manuscript received, September 10, 1979; accepted, September 24, 1979.

[2]U.S. Geological Survey, Menlo Park, California 94025.

I thank A. H. Bouma, C. H. Campbell, G. Ghibaudo, M. H. Link, E. Mutti, C. H. Nelson, J. B. Southard, P. Stone, and J. C. Yount for helpful comments on a preliminary draft of this manuscript.

[3]Walker (1978, p. 954) states, with reference to the Wheeler Gorge outcrops and another sequence from Oregon, that ". . . in neither case is there any field evidence of channeling." However, there is considerable field evidence of channeling, best observed at (1) the base of the first or lowest thinning-upward megasequence, where the base of the lowest bed of conglomerate cuts gradually across the underlying sequence of thin-bedded turbidites and abundant rip-up clasts of the thin-bedded turbidites are incorporated in the lower part of the conglomerate, and (2) at the west flank of the third or highest thinning-upward megasequence, where Walker's sketch map shows the base of the lowest bed of conglomerate to be downcut several meters into the underlying thin-bedded turbidites at the top of the second megasequence, and where Fischer and Mattinson (1968) described injection of conglomerate and sandstone into the flanking thin-bedded turbidites and incorporation of large, partly deformed blocks of the latter into the conglomerate, forming a channel-margin breccia.

Group in northern Spain.

Walker (1978, p. 936) acknowledges some aspects of these differences, that is, that thin-bedded turbidites are deposited in various proximal and distal settings, and advocates use of the non-generic term "thin-bedded turbidite" rather than the previously used generic term "distal turbidite." However, in the same paper he misuses the term "proximal turbidite" for "thick-bedded turbidite" in a manner analogous to that of the previously misused and confusing "distal = thin-bedded" terminology. I question this use and will try to clarify the terms to avoid further confusion in the literature.

Walker (1978, Fig. 2) shows an outcrop photograph of the Eocene Matilija Sandstone from southern California as an example of a "proximal" turbidite, contrasting it with another photograph of thin-bedded turbidites. He refers to this outcrop (p. 953) as an example of proximal turbidites because of its "thicker individual sandstone beds" and "higher sandstone/shale ratio," inferring deposition on a braided suprafan. I would strongly argue that neither thickness of bedding nor sandstone to shale ratios indicate proximality, because thin-bedded turbidites with low sandstone to shale ratios are characteristic of slope and inner- and middle-fan interchannel and levee facies, and thick-bedded turbidites with high sandstone to shale ratios are characteristic of outer-fan lobe deposits, as defined by Mutti and Ricci Lucchi (1972). The beds shown in Walker's Figure 2 should be described as thick or (perhaps more appropriately) medium-bedded turbidites, until more information regarding source area, fan geometry, and paleocurrent patterns is known.

The problems of using the terms "proximal" and "distal" in relation to turbidite deposition, however, remain ambiguous. The strata shown in the Matilija outcrop are near the base of the lower part of the formation, which Link (1971, 1975) concluded to be a prograding deep-sea fan sequence. Acyclic basin-plain deposits are present below the Matilija, and the lowest part of the Matilija contains a thin section consisting of several thickening- and coarsening-upward megasequences considered by Link and me to be outer-fan lobe deposits as described by Mutti and Ricci Lucchi (1972). Above the outcrop shown in the photograph is a thick section of thicker bedded channeled beds of sandstone that form thinning- and fining-upward megasequences and grade upward into megafossil-bearing shallow-marine sandstone and shale of the upper middle part of the Matilija Sandstone. Thus, the whole sequence appears to represent a progradational event, with the strata in the photograph transitional between unchanneled thickening-upward megasequences

and channeled megasequences. I consider the outcrop in the photograph to be a very thick composite outer-fan lobe with several smaller megasequences and minor channeling within it; it formed by the intertonguing of several lobes or superposition of several lobes. The latter may have been possible because the Eocene Santa Ynez basin appears to have been narrow and restricted (Clarke et al, 1975; Nilsen and Clarke, 1975; Nilsen and McKee, 1979), thus limiting much lateral migration of distributary channels.

The Matilija strata crop out approximately in the central to proximal part of the Paleogene Santa Ynez basin. If basin-plain and outer-fan facies are considered distal and middle- and inner-fan facies are considered proximal, the strata would be distal if the facies interpretation is correct. The problem then becomes one of attempting usefully to define the terms proximal and distal and developing criteria for their use, if they are worth keeping as descriptive terms.

"Proximal" and "distal" have been used in two chief senses by turbidite workers: (1) relative distance from source of the depositional site, and (2) for submarine fans and associated deposits, basin-plain and outer-fan deposits have generally been considered "distal," and inner- and middle-fan as well as basin-margin slope deposits have generally been considered "proximal." The former use implies simple horizontal distance of transport of sediment and distance from source, whereas the latter requires facies interpretation.

Use of the terms in the second sense causes major problems, as shown by Figure 1, which shows one simple example that can cause confusion. Depositional site A, located on the basin plain, is clearly more proximal with regard to distance from source than the other lettered loca-

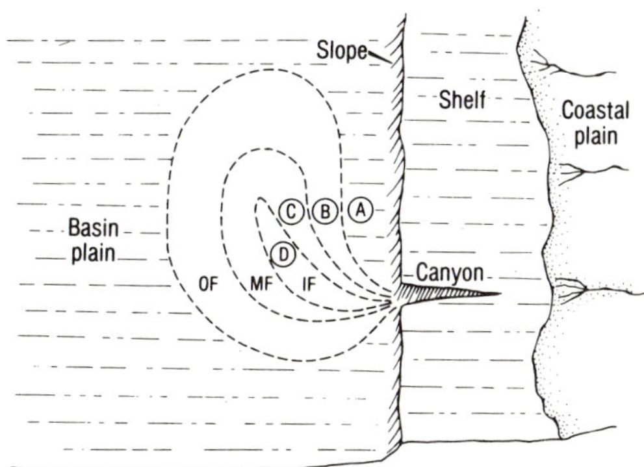

FIG. 1—Simplified diagram showing proximal and distal relations for submarine fan and associated depositional settings. *IF*, inner fan; *MF*, middle fan; *OF*, outer fan. *A, B, C, D* points discussed in text.

tions, which represent progressively more "proximal" facies according to the second usage of the terms. Furthermore, in prograding sequences, "proximal" inner- and middle-fan deposits will prograde out over and be deposited beyond or more distally than previously deposited "distal" basin-plain and outer-fan deposits, causing further confusion in applying the terms.

Because of these problems, I suggest abandoning the terms in the second sense and using them only in regard to relative distance from source, regardless of facies. They may be useful in some ways as loose field terms referring to facies but are probably best left out of the literature except when referring to distance from source. With channeled depositional systems that can hook, bend, prograde rapidly in local areas, and be supplied from several source areas, the terms "proximal" and "distal" must be used cautiously.

There are two possible exceptions to this suggested usage. The first involves basin-plain sequences where sediments are supplied from one source area and distributary channels and lobes are not present; in this example, "proximal" and "distal" can probably be used validly. This is the sense originally intended by Bouma (1962) and developed by Walker (1967) for what he now terms "classical" turbidites, the facies C and D basin-plain turbidites of Mutti and Ricci Lucchi (1972, 1975). The second example involves disrupted and faulted fan systems that can be reconstructed by sedimentary and structural studies. In this situation, it may be useful to refer to "proximal" and "distal" parts of the fan in a very general way, inferring that the parts originally formed a single submarine fan. Examples of single fans, to which the terms "proximal" and "distal" could be loosely applied to fragments broken by faulting, include the Mississippian Copper Basin Formation of south-central Idaho, broken by major thrust faults (Nilsen, 1977), and the Eocene Butano and Point of Rocks Sandstones, broken by the San Andreas strike-slip fault system (Clarke and Nilsen, 1973). Even for the last two formations, however, the fan facies should be used if possible.

SUBMARINE FAN MODELS AND SUPRAFAN CONCEPT

Mutti and Ricci Lucchi (1972) proposed a seven-facies, A to G, classification of individual beds of turbidite sequences, and suggested that the distribution and grouping of these seven facies defined five major facies associations: slope, inner fan, middle fan, outer fan, and basin plain. By their scheme, which was at the time thought to be applicable to most deep-sea fan deposits, the outer-fan facies association was distinguishable by the presence of thickening- and coarsening-up-

ward unchanneled lobe megasequences, the middle-fan facies association by the presence of thinning- and fining-upward channeled megasequences, and the inner-fan facies association by the presence of large bodies of conglomerate and sandstone.

This classification was modified (mainly by Walker) in the 1973 Walker and Mutti paper, in which thickening-upward lobe megasequences *and* thinning-upward channeled megasequences were grouped together as the middle-fan facies association, considered together to form "suprafan lobes." The outer-fan facies association by the Walker and Mutti (1973) scheme also contained thickening-upward megasequences, but these were thought not to have any morphologic expression (presumably by analogy to modern fans with suprafans) and to have been deposited downfan from the suprafan lobes.

The change in terminology and alteration of the original Mutti and Ricci Lucchi (1972) classification have caused innumerable problems in the literature, especially for English-speaking workers, because until recently the original Mutti and Ricci Lucchi paper was available only in Italian.[4] In addition, Mutti and Ricci Lucchi and coworkers have continued to use their original (1972) definitions of the middle- and outer-fan facies association in subsequent publications (Ghibaudo and Mutti, 1973; Mutti, 1974, 1977; Ricci Lucchi and Parea, 1974; Mutti and Ricci Lucchi, 1975; Ricci Lucchi, 1975a, b; Mutti et al, 1978).

The suprafan concept was proposed by Normark (1970) from studies of modern submarine fans off the west coast of North America and expanded in subsequent summary papers (Normark, 1974, 1978). The suprafan is a convex-upward depositional bulge present at the termination of the major inner-fan valley and is characterized by numerous unleveed distributary channels and channel remnants (Normark, 1978, p. 921-922). The suprafan is thought to be characteristic of the middle parts of submarine fans, whereas the outer fan is flat, smooth, and unchanneled. Normark (1978, p. 924-927) further stated that there are smooth, topographic bulges downfan from some suprafans; on San Lucas fan, the bulge is similar in size to the adjacent suprafan. He considers these smooth bulges as forming the lower suprafan or lower middle fan and as being equivalent to the outer-fan lobes of Mutti and Ricci Lucchi (1972). Normark (1978, p. 927) even states that these smooth bulges, especially

---

[4]An English translation is now available from the American Geological Institute, 5205 Leesburg Pike, Falls Church, Virginia 22041, and was published in International Geology Review, 1978, v. 20, p. 125-166.

on Navy fan, consist of thickening- and coarsening-upward sequences but does not state what type of data (presumably very long cores) he used nor does he present data to verify this conclusion.

Walker's (1978) model adapts this framework and also ingredients of the Mutti-Ricci Lucchi (1972) scheme into a single model that combines published modern and ancient models. However, I find the model confusing and difficult to use in the field and certainly not applicable to many ancient examples. The main problem is that ongoing work on ancient fan deposits appears to differentiate two major types of fans, those that seem to be describable by the Mutti and Ricci Lucchi system, and those that are characterized by suprafan lobes. The former seem to have well-defined channels and unchanneled lobes and to be composed of a variety of sediment types and, hence, were constructed by turbidity currents containing a mixture of sediment sizes. The latter appear to be composed mainly of sand and consist almost wholly of channeled deposits. Mutti (written commun., 1977) terms these fans "highly efficient" and "poorly efficient," respectively. Examples of the former are abundant in the literature (see summary in Ingersoll, 1978, for example) and perhaps form the great majority of ancient submarine fan deposits. The latter have been less well studied but include several examples from the lower Tertiary of California (Link and Nilsen, 1979; Nilsen, 1979a, b) and the Tertiary of northern Italy (Mutti, written commun., 1977). In "poorly efficient" fans, channeled sandy deposits grade abruptly downfan into basin-plain deposits, without the development of thickening- and coarsening-upward unchanneled lobes. These fans appear to be fed primarily by shelf sands transported down canyons by various types of mass flow, not necessarily high-energy turbidity currents, and to be abruptly dumped at the base of slope in a series of constantly shifting, generally small channels. An inner-fan channel extending outward from the submarine canyon may be present, but outer-fan facies in the sense of Mutti and Ricci Lucchi (1972) are generally not developed or are very poorly developed. These fans consist almost wholly of channeled suprafans (the type generally examined by Normark) and are fed primarily by littoral drift of shelf sand into submarine canyons.

Many other fans do not have suprafans because they are fed by river systems or other sources that supply a mixture of sediment sizes that trigger high-energy turbidity currents. In these fans, the Mutti-Ricci Lucchi (1972) scheme is applicable and, as demonstrated by many studies, is the most useful framework for analyzing most ancient fan deposits.

Thus, the suprafan concept appears applicable to only some fans. There is certainly no reason to expect that coarsening- and thickening-upward megasequences should be characteristic of the outer fan, the smooth outer parts of suprafan lobes, and also the transition from smooth to channeled parts of suprafan lobes, as shown by Walker (1978, Fig. 14). The thickening-upward sequences may not be present at all in submarine fans characterized by suprafan deposition, and the smooth outer fan of Normark (1970, 1978), which is morphologically almost indistinct from the basin plain, may be basin plain.

Because of the resulting differences in morphology, grain size, and mode of transport on the two types of fans, there will be many other differences than can be observed in outcrop, such as differences in the nature of thin-bedded turbidites deposited in channel margin, levee, and interchannel areas and in the types of surrounding basin-plain deposits. These will be summarized in forthcoming publications by several workers.

Of course, fans can change character through time, and the type of sediment supplied to them can change, so that both models may be applicable to the same fan for different intervals of fan growth, causing difficulties in unraveling the history of fans. Also, in some fans, such as the Macigno Formation of northern Italy, transitional sequences between the two models may develop (G. Ghibaudo, written commun., 1978).

In general, two other notes of caution should be stated with regard to models. First, models of modern fans developed substantially from morphologic studies and examinations of the upper few meters of mostly Holocene sediment should not be too hastily integrated with models of ancient fans developed from stratigraphic and paleogeographic studies of hundreds or thousands of meters of strata. The types of information provided by marine and onland studies are different, and, as I learned from an earlier attempt to integrate the two (Nelson and Nilsen, 1974), the difficulties are formidable. At this stage in the evolution of fan studies, a carefully planned seismic and drilling program for several modern fans is needed.

Second, studies of ancient fans in which only parts of fans are preserved are not very useful either for constructing models or for basing major interpretations of fan facies. Models and facies interpretations based on cross sections of one part of a fan, in which perhaps only channels are seen, are not so helpful as integrated studies of full systems. Thus, most of the ancient sequences that Walker (1978) cites as examples of his facies are easily interpretable in other ways. Rather than using two-dimensional examples of small parts of fans such as the Jurassic Otter Point Formation of Oregon, the Cretaceous sequence at Wheeler Gorge, the Capistrano Formation at Dana Point and San Clemente, the Cabrillo Formation in the San Diego area, and Paleozoic channels in Quebec, he would be far better using three-dimensional examples such as the Tyee Formation of Oregon; the Great Valley sequence, the Butano Sandstone, Point of Rocks Sandstone, Cantua Sandstone, The Rocks Sandstone, and other lower Tertiary sequences in California; the Hecho Group and San Sebastian area in northern Spain; and the Marnoso-Arenacea, Laga, and Macigno Formations in Italy.

APPLICATIONS TO PETROLEUM EXPLORATION

Walker (1978, p. 963) states that "a principal target would be the inner part of a suprafan lobe," and Wilde et al (1978, p. 981) state that "the best prospects for petroleum accumulation . . . are (1) near the apices of . . . fans within large leveed-valley systems and (2) on the middle fan . . ." Although it is certainly true that sand-filled channels of the inner and middle parts of both modern and ancient submarine fans are desirable exploration targets, explorationists should not think only in these terms. Outer-fan deposits, particularly in those fans termed "highly efficient" by Mutti and which can be described by the Mutti and Ricci Lucchi (1972) system, and which in my experience are the majority of ancient deep-sea fan deposits, have substantial and widespread accumulations of sand bodies that are not only suitable, but desirable exploration targets.

Outer-fan lobes can extend for lateral distances of at least 15 km, attain thicknesses of 10 to 50 m or more, and are characterized by high sandstone to shale ratios (see Mutti et al, 1978, for a description of the geometry of outer-fan lobes from the Laga Formation in east-central Italy). In many basins, because of restrictions imposed by basin geometry, outer-fan lobes may be even thicker and, like the sequence from the Matilija Sandstone shown by Walker (1978, Fig. 2), may form composite bodies, at least as interpreted by this writer. Work on some modern deep-sea fans and DSDP cores (e.g., the description of Astoria fan by Nelson, 1976) indicates substantial sand accumulations in outer-fan deposits beneath the Holocene accumulation of mud. Outer-fan sand deposits may also be better sorted and have more favorable porosities and permeabilities because of the progressive loss of lighter clay minerals, micas, and organic matter by preferential overbank spilling from channels and deposition of these components in channel-margin, levee, and inter-

channel areas of the inner and middle fan (see Nelson, 1976, for pertinent analyses from Astoria fan). Outer-fan lobes also contain sand that is medium to very coarse grained and individual beds several meters or more thick. Because of both their great lateral extent and position between shale-rich thin-bedded turbidites of the interlobe, lobe-fringe, and basin-plain facies, they are more likely to form good structural traps when folded.

There are good examples, especially from California, of hydrocarbon production from thickening- and coarsening-upward megasequences best interpreted as outer-fan lobes. Thus, Walker (1978) and Wilde et al (1978) are correct in suggesting that channel deposits are good exploration targets, but they are not so good that outer-fan deposits should not also receive equal emphasis. In many basins, outer-fan deposits may be better targets because of the discontinuous nature of many channel systems and the filling of many abandoned channels by mud rather than sand.

MISCELLANEOUS

Several smaller items also require comment. Walker (1978, p. 950) states that "the upper fan is characterized by one relatively deep, leveed channel that only rarely shifts in position." This may be true for many fans but certainly not all. Monterey fan (Wilde et al, 1978, p. 968) contains active inner-fan channels fed by several different submarine canyons; Astoria fan contains two active inner-fan channels (Nelson, 1976) and in the ancient record, the Butano Sandstone (Nilsen, 1979c) had at least two active inner-fan channels, and the Cretaceous Cabrillo Formation in the San Diego area had at least two major inner-fan channels (Nilsen and Abbott, 1979). Doubtless there are many additional examples, especially in fans fed by deltas that have constantly shifting distributary channels.

Walker (1978, p. 950-951) states that inner-fan channel deposits are characterized by thinning- and fining-upward sequences. Mutti and Ricci Lucchi (1972) did not characterize inner-fan channel deposits in this way, and it has been my experience that inner-fan deposits tend not to form well-defined thinning- and fining-upward megasequences; these features seem to be more characteristic of smaller middle-fan channels. Inner-fan channel sections are variable in character, depending upon whether the channel was straight or meandering, whether the thalweg channels were straight, meandering or braided, and whether sand, mud, or gravel constituted the main fill of the channel. Cycles certainly are common, but they are most commonly alternating beds of channeled conglomerate and sandstone without repetitive thinning- and fining-upward megasequences consisting of five or more beds. Examples of various types of inner-fan channel sequences that may be cyclic but do not form well-defined thinning- and fining-upward megasequences include the Butano Sandstone (Nilsen, 1979c), the Sitkinak Formation (Nilsen and Moore, 1979), the Cabrillo Formation (Nilsen and Abbott, 1979), the Copper Basin Formation (Nilsen, 1977), Mesozoic inner-fan deposits of southern Chile (Winn and Dott, 1977), and the Gottero Sandstone (Nilsen and Abbate, 1976).

Walker (1978, p. 962) states that "vertical trends (thickening or thinning upward) may be meaningless or absent" in elongate basins in which "paleocurrent flow is dominantly parallel with the basin axis." I do not believe that this is true, and there are numerous elongate basins with longitudinal flow in which vertical megasequences can be used to reconstruct the basin filling quite well. Good examples include the Hecho Group (Mutti, 1977), Marnoso-Arenacea and Laga Formations (Ricci Lucchi, 1975b), Macigno (G. Ghibaudo, written commun., 1978), San Sebastian area (Van Vliet, 1978), Great Valley sequence (Ingersoll, 1978), Chugach terrane (Nilsen and Bouma, 1977), Pico Formation of the Ventura basin (see Walker's use in his Figure 21 of the work of Hsü, 1977, for an example of interpretation of megasequences), and the Carboniferous turbidites of the Ouachita-Marathon basin system, which have a well developed axial-fan growth pattern defined by megasequences that has not yet been satisfactorily described in the literature.

Walker (1978, p. 952, and Fig. 19) describes the channel fills of the Capistrano Formation at San Clemente and points out that most fills do not form thinning- and fining-upward megasequences and that only the south sides of the channels are seen in outcrop. This is true for the part of the section shown in his Figure 19; however, farther north along the beach, the north flanks of the channels can be seen, and these margins are generally steeper than the south flanks. Thus, the size of the channels can be measured. Walker's point about the absence of thinning-upward megasequences is not valid because the bottom of the channels are not seen, only deposits on one of the channel walls, which begin with a mudstone drape on the wall. The outcrop on the right (south) of the parking lot, labeled "mainly massive sandstones" by Walker, contains two prominent thinning-upward megasequences, and all of the approximately eight channel fills north of the parking lot do form thinning- and fining-upward megasequences. Thus, the concept that thinning-

and fining-upward megasequences define channels does not necessarily apply to sequences that begin high on channel walls nor to abandoned channels filled with mudstone.

CONCLUSIONS

Walker (1978) and Normark (1978) have provided useful and generally accurate summaries of modern and ancient submarine fan deposits that will be helpful both now and in the future for petroleum exploration. However, both their models and exploration strategies based on them are certainly not universally accepted, and it seemed beneficial to the profession to raise some points that I know have also troubled many of my colleagues. I do this in the hope that better models and greater understanding of both modern and ancient turbidite systems may develop and to stimulate further discussion of these valuable papers.

REFERENCES CITED

Bouma, A. H., 1962, Sedimentology of some flysch deposits: Amsterdam, Elsevier, 168 p.

Clarke, S. H., Jr., and T. H. Nilsen, 1973, Displacement of Eocene strata and implications for the history of offset along the San Andreas fault, central and northern California, *in* Tectonic problems of the San Andreas fault system: Stanford Univ. Pubs. Geol. Sci., v. 13, p. 358-367.

———— D. G. Howell, and T. H. Nilsen, 1975, Paleogene geography of California, *in* D. W. Weaver et al, eds., Future energy horizons of the Pacific Coast—Paleogene symposium and selected technical papers: AAPG Pacific Sec., p. 121-154.

Fisher, R. V., and J. M. Mattinson, 1968, Wheeler Gorge turbidite-conglomerate series, California—inverse grading: Jour. Sed. Petrology, v. 38, p. 1013-1023.

Ghibaudo, G., and E. Mutti, 1973, Facies ed interpretazione paleoambientale delle Arenarie di Ranzano nei dintorni di Specchio (Val Pessola, Appennino Parmense): Soc. Geol. Italiana Mem., v. 12, p. 251-265.

Hsü, K. J., 1977, Studies of Ventura field, California, 1: Facies geometry and genesis of lower Pliocene turbidites: AAPG Bull., v. 61, p. 137-168.

Ingersoll, R. V., 1978, Submarine fan facies of the Upper Cretaceous Great Valley sequence, northern and central California: Sed. Geology, v. 21, p. 205-230.

Link, M. H., 1971, Sedimentology and environmental analysis of the Matilija Sandstone north of the Santa Ynez fault, Santa Barbara County, California: Master's thesis, California Univ., Santa Barbara, 106 p.

———— 1975, Matilija Sandstone—a transition from deep water turbidite to shallow marine deposition in the Eocene of California: Jour. Sed. Petrology, v. 45, p. 63-78.

———— and T. H. Nilsen, 1979, Sedimentology of The Rocks Sandstone and Eocene paleogeography of the northern Santa Lucia basin, California, *in* Tertiary and Quaternary geology of the Salinas Valley and Santa Lucia Range, Monterey County, California: SEPM Pacific Sec. Paleogeography Field Guide 4, p. 25-43.

Mutti, E., 1974, Examples of ancient deep-sea fan deposits from Circum-Mediterranean geosynclines, *in* Modern and ancient geosynclinal sedimentation: SEPM Spec. Pub. 19, p. 92-105.

——— 1977, Distinctive thin-bedded turbidite facies and related depositional environments in the Eocene Hecho Group (south-central Pyrenees, Spain): Sedimentology, v. 24, p. 107-131.

——— and F. Ricci Lucchi, 1972, Le torbiditi dell' Appennino settentrionale: introduzione all'analisi di facies: Soc. Geol. Italiana Mem., v. 11, p. 161-199.

——— ——— 1975, Turbidite facies and facies associations, *in* Examples of turbidite and facies associations from selected formations of the northern Apennines: 9th Internat. Cong. Sedimentology, Nice, Guidebook A11, p. 21-36.

——— T. H. Nilsen, and F. Ricci Lucchi, 1978, Outer fan depositional lobes of the Laga Formation (upper Miocene and lower Pliocene), east-central Italy, *in* D. J. Stanley and G. Kelling, eds., Sedimentation in submarine canyons, fans, and trenches: Stroudsburg, Pa., Dowden, Hutchinson & Ross, p. 210-223.

Nelson, C. H., 1976, Late Pleistocene and Holocene depositional trends, processes, and history of Astoria deep-sea fan, northeast Pacific: Marine Geology, v. 20, p. 129-173.

——— and T. H. Nilsen, 1974, Depositional trends of modern and ancient deep-sea fans, *in* Modern and ancient geosynclinal sedimentation: SEPM Spec. Pub. 19, p. 69-91.

——— E. Mutti, and F. Ricci Lucchi, 1975, Comparison of proximal and distal thin-bedded turbidites with current-winnowed deep sea sands: 10th Internat. Sed. Cong., Nice, Proc., v. 5, p. 317-325.

——— ——— ——— 1977, Discussion of paper by R. G. Walker—Upper Cretaceous resedimented conglomerates at Wheeler Gorge, California: description and field guide: Jour. Sed. Petrology, v. 47, p. 926-928.

Nilsen, T. H., 1977, Paleogeography of Mississippian turbidites in south-central Idaho, *in* J. H. Stewart et al, eds., Paleozoic paleogeography of the western United States: SEPM Pacific Sec., Pacific Coast Paleogeography Symp. 1, p. 275-300.

——— 1979a, Early Cenozoic stratigraphy, tectonics and sedimentation of the central Diablo Range between Hollister and New Idria, *in* T. H. Nilsen and T. W. Dibblee, Jr., eds., Geology of the central Diablo Range between Hollister and New Idria, California: Geol. Soc. America Cordilleran Sec. Guidebook, p. 31-55.

——— 1979b, Comparative sedimentology of the lower Tertiary Butano and Cantua deep-sea fan sequences, central California (abs.): Geol. Soc. America Abs. with Program, v. 11, p. 120.

——— 1979c, Sedimentology of the Butano Sandstone, Santa Cruz Mountains, California, *in* T. H. Nilsen and E. E. Brabb, eds., Geology of the Santa Cruz Mountains, California: Geol. Soc. America Cordilleran Guidebook, p. 30-39.

——— and E. Abbate, 1976, The Gottero Sandstone, a Late Cretaceous and Paleocene deep-sea fan complex in the Ligurian Apennines, northern Italy (abs.): Geol. Soc. America Program with Abs., v. 8, p. 1028-1029.

——— and P. L. Abbott, 1979, Turbidite sedimentology of the Upper Cretaceous Point Loma and Cabrillo Formations, San Diego, California, *in* P. L. Abbott, ed., Upper Cretaceous deep-sea fan deposits, San Diego, California: Geol. Soc. America Guidebook, p. 139-166.

——— and A. H. Bouma, 1977, Turbidite sedimentology and depositional framework of the Upper Cretaceous Kodiak Formation and related stratigraphic units, southern Alaska (abs.): Geol. Soc. America Abs. with Program, v. 9, p. 115.

——— and S. H. Clarke, Jr., 1975, Sedimentation and tectonics in the early Tertiary continental borderland of central California: U.S. Geol. Survey Prof. Paper 925, 64 p.

——— and E. H. McKee, 1979, Paleogene paleogeography of the western United States, *in* Cenozoic paleogeography of the western United States: SEPM Pacific Sec., Pacific Coast Paleogeography Symp. 3, p. 257-276.

——— and G. W. Moore, 1979, Reconnaissance study of Upper Cretaceous to Miocene stratigraphic units and sedimentary facies, Kodiak and adjacent islands, Alaska: U.S. Geol. Survey Prof. Paper 1093, 34 p.

Normark, W. R., 1970, Growth patterns of deep-sea fans: AAPG Bull., v. 54, p. 2170-2195.

——— 1974, Submarine canyons and fan valleys: factors affecting growth patterns of deep-sea fans, *in* Modern and ancient geosynclinal sedimentation: SEPM Spec. Pub. 19, p. 56-68.

——— 1978, Fan valleys, channels, and depositional lobes on modern submarine fans: characters for recognition of sandy turbidite environments: AAPG Bull., v. 62, p. 912-931.

Ricci Lucchi, F., 1975a, Depositional cycles in two turbidite formations of northern Apennines: Jour. Sed. Petrology, v. 45, p. 3-43.

——— 1975b, Miocene paleogeography and basin analysis in the Periadriatic Apennines, *in* C. H. Squyres, ed., Geology of Italy, v. 2: Libyan Arab Republic Earth Sci. Soc., p. 129-236.

Ricci Lucchi, F., and G. C. Parea, 1974, Cicli de deposizionali (megasequence) nelle torbiditi di conoide sottomarina—formazione della Laga (Appennino Marchigiano–Abruzzese): Atti. Soc. Nat. Mat. Modena, v. 104, p. 257-283.

Van Vliet, A., 1978, Early Tertiary deep-water fans of Guipuzcoa, northern Spain, *in* D. J. Stanley and G. Kelling, eds., Sedimentation in submarine canyons, fans and trenches: Stroudsburg, Pa., Dowden, Hutchinson & Ross, p. 190-209.

Walker, R. G., 1967, Turbidite sedimentary structures and their relationship to proximal and distal depositional environments: Jour. Sed. Petrology, v. 37, p. 25-43.

——— 1970, Review of the geometry and facies organization of turbidites and turbidite-bearing basins, *in* Flysch sedimentology in North America: Geol. Assoc. Canada Spec. Paper 7, p. 219-251.

———— 1975, Upper Cretaceous resedimented conglomerates at Wheeler Gorge, California: description and field guide: Jour. Sed. Petrology, v. 45, p. 105-112.

———— 1976, Facies models. 2, Turbidites and associated coarse clastic deposits: Geosci. Canada, v. 3, p. 25-36.

———— 1978, Deep-water sandstone facies and ancient submarine fans: models for exploration for stratigraphic traps: AAPG Bull., v. 62, p. 932-966.

———— and E. Mutti, 1973, Turbidite facies and facies associations, *in* G. V. Middleton and A. H. Bouma, eds., Turbidites and deep-water sedimentation: SEPM Pacific Sec. Short Course, p. 119-157.

Wilde, P., W. R. Normark, and T. E. Chase, 1978, Channel sands and petroleum potential of Monterey deep-sea fan, California: AAPG Bull., v. 62, p. 967-983.

Winn, R. D., Jr., and R. H. Dott, Jr., 1977, Large-scale traction-produced structures in deep-water fan-channel conglomerates in southern Chile: Geology, v. 5, p. 41-44.

Modern and Ancient Submarine Fans: Reply[1]

ROGER G. WALKER[2]

Nilsen has raised some important questions regarding the construction and use of fan models, and I am glad to be able to comment on some of his points. Models are improved by sharing experience, which is the main purpose of the discussion and reply.

I will comment on proximal-distal problems, on fan models, and on some of the other points raised. First, I agree that in a descriptive facies scheme, the term "thick-bedded turbidite" is to be preferred to "proximal turbidite." I have already incorporated the term thick bedded into the revised version of my latest paper (Walker, 1979). Historically, the proximal-distal idea stems from the assumed fanning out of a turbidity current on a smooth surface, with accompanying changes in bed thickness and sedimentary structures (Bouma, 1962; Parea, 1965; Walker, 1967). It is implicit (but unfortunately not explicit) in my proximal-distal interpretation of the Bouma sequence (Walker, 1967) that currents spread from a point source over an unchanneled surface, and Nilsen accepts this use in his discussion. Fortunately for turbidite studies, this simple scheme has been replaced by more complex interpretations of facies relations to submarine fan morphology (first discussed by Walker, 1966a, and emphasized by Mutti and Ricci Lucchi, 1972, and by many other authors since then).

I now agree with Nilsen that the terms proximal and distal should not be used in a formal facies terminology, but they retain an overall usefulness in describing changes from fan apex to basin plain, both in terms of geography and broad facies categories.

FAN MODELS AND SUPRAFAN LOBES

In his discussion, Nilsen comments on my modification of the original Mutti and Ricci Lucchi (1972) scheme (Fig. 1A), and the "innumerable problems in the literature" that it has caused. He continues to suggest that there may be two different fan types, one characterized by suprafan lobes, and one that fits the original Mutti and Ricci Lucchi scheme. I will reply to these points and will make some general comments on the evolution of fan models.

The stimulus to "alter" the original model came from an invitation to Mutti and me to participate in the SEPM Pacific Section short course held in conjunction with the Anaheim AAPG meeting in 1973. I prepared the notes (Walker and Mutti, 1973) with as much help from Mutti as the mail and phone services, and very tight deadlines, would allow. In 1972, it seemed imperative that some integration of ancient rock models (Mutti and Ricci Lucchi, 1972; Mutti and Ghibaudo, 1972; Fig. 1A, B), and recent sediment models (Normark, 1970; Fig. 2A), should be attempted, especially since the original Mutti and Ricci Lucchi work did not cite Normark (1970),

[1]Manuscript received, November 2, 1979; accepted, November 12, 1979.

[2]Department of Geology, McMaster University, Hamilton, Ontario L8S 4M1, Canada.

I am indebted to G. V. Middleton for his comments on an early draft of this manuscript.

Article Identification Number
0149-1423/80/B007-0006$03.00/0

Discussions

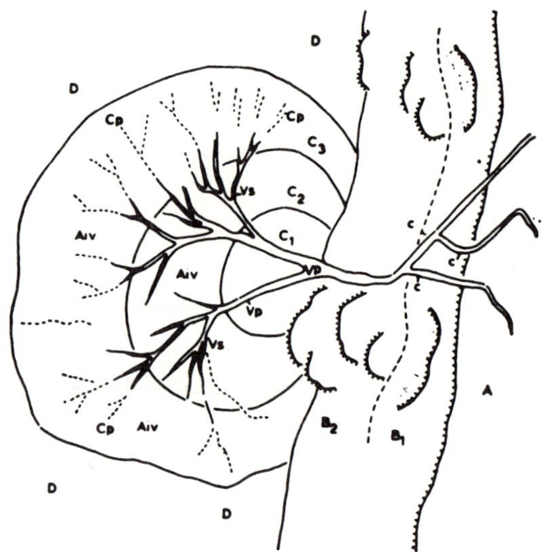

A. Mutti and Ricci Lucchi, 1972.

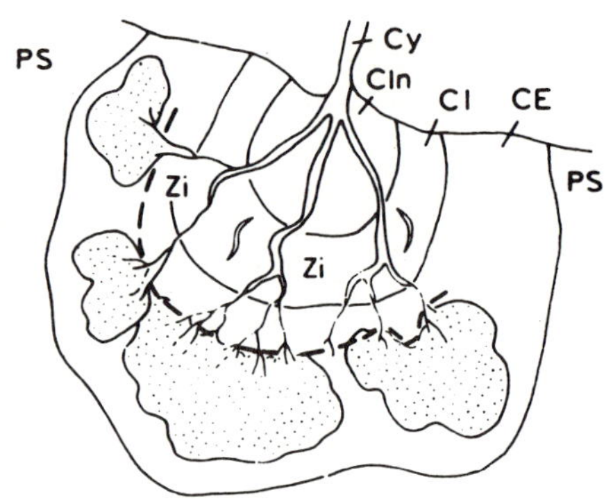

B. Mutti and Ghibaudo, 1972.

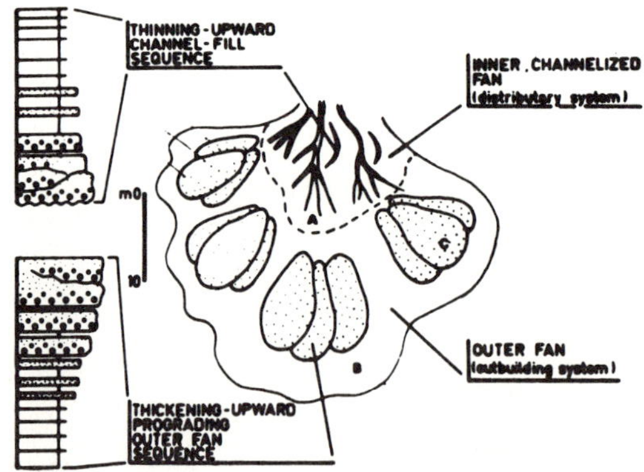

C. Mutti and Ricci Lucchi, 1974.

D. Mutti, 1977.

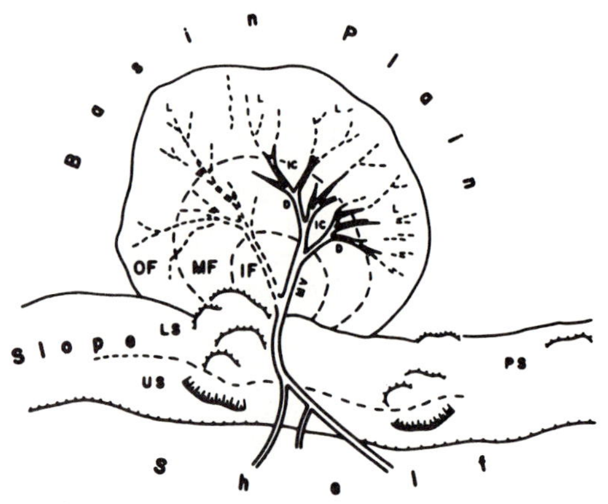

E. Ricci Lucchi, 1975a.

FIG. 1—Comparison of submarine fan models. In A: *A*, platform; B_1, B_2, upper and lower slope; C_1, C_2, C_3, inner, mid, and outer fans; *D*, submarine plain; *C*, canyon; *Vp*, main channels; *Vs*, secondary channels; *Cp*, peripheral channels; *Aiv*, interchannel area. In B: *Cy*, canyon; *CIn, CI, CE*, inner, mid, and outer fans; *Zi*, intermediate zone; *Ps*, submarine plain. In D: *CF*, channel fill; *IC*, interchannel area; *CMB*, channel-mouth bar; *SL*, outer-fan sandstone lobe; *LF*, lobe fringe; *BP*, basin plain. In E: *US, LS*, upper and lower slope; *PS*, passive slope; *IF, MF, OF*, inner, mid, and outer fan; *MV*, main valley; *D*, mid-fan distributary channels; *L*, lobes; *IC*, interchannel areas.

nor did it integrate much of the existing data on modern fans. Integrating recent *and* ancient sediment information into a model is essential for a thorough understanding of fluvial, deltaic, barrier-island, and shallow-marine systems.

The basic similarities of the original two models (Fig. 1A, B) and Normark's (1970) model with suprafan lobes (Fig. 2A) are obvious and their marriage led to the model in Figure 2B. The differences do not concern facies relations and facies sequences so much as fan terminology. In the models of Figure 1A and B, the depositional lobes were assigned to the outer fan, and the dominantly channeled system was assigned to the mid fan. This terminology is arbitrary. Normark (1970) suggested that the suprafan lobes consisted of an inner channeled part and an outer smooth part on a *topographically* defined (*not* arbitrarily defined) mid-fan area. It therefore seemed logical in late 1972 to integrate the Mutti-type models and the Normark model to produce the version shown in Figure 2B (Walker and Mutti, 1973).

I do not consider this change so fundamental that it should have caused "innumerable problems in the literature." Thickening-upward sequences were still assigned to lobe progradation and thinning-upward sequences to channel filling, as in the original Mutti models. The effect of the change was to alter the original arbitrary assignment of thickening-upward sequences to outer fan, and thinning-upward sequences to mid fan (Mutti and Ghibaudo, 1972), but the facies relations remained the same—lobes downfan from channels.

In reply to Nilsen's comments about two types of fan model, I would like to review briefly the evolutionary changes that have occurred in fan models since 1972. In the original model (Fig. 1A), the channeled part of the fan dies away into a smooth outer fan; depositional lobes were only briefly mentioned in the text (Mutti and Ricci Lucchi, 1972, p. 188; Nilsen's translation, 1978, p. 150). The same year, a similar model was published (Fig. 1B) showing depositional lobes on the outer fan (Mutti and Ghibaudo, 1972, Pl. II and p. 29-33). These were compared to deltaic "stream-mouth" or "channel-mouth bars." A much more significant variation was published by Mutti and Ricci Lucchi (1974), in which the lobes were shown detached from the channels (Fig. 1C). This modification appears to be due to Mutti's work in the Eocene Hecho Group of Spain. Subsequently, channel-mouth bars (Fig. 1D) were added to the model by Mutti et al (1975) and Mutti (1977).

The area between the ends of the channels and the beginnings of lobes is primarily a zone of bypassing, and this zone of bypassing is the fundamental difference between models such as those of Figure 2B and C, and those in Figure 1C and D. All through his discussion, Nilsen emphasizes the difference between my 1978 model (Fig. 2C) and the original Mutti and Ricci Lucchi model (Fig. 1A). The differences mostly involve terminology, not fundamental sedimentology; the real differences are in my 1978 model and the later versions of the Mutti model (Fig. 1C, D). However, the later versions of the Mutti models (Fig. 1C, D) have not yet been as widely accepted: Ricci Lucchi continued to use a 1972-type model (Fig. 1A) in his 1975 papers (Fig. 1E; Ricci Lucchi 1975a, b) and Ingersoll (1978) used a modification of Ricci Lucchi (1975b) to describe the Upper Cretaceous Great Valley sequence of California. I consider the Ricci Lucchi (1975a) model shown in Figure 1E very similar to my 1978 model (Fig. 2C) except in terminology, but I also consider it fundamentally different from the Mutti et al (1975) and Mutti (1977) model of Figure 1D.

FAN MODELS WITH DETACHED LOBES

By suggesting "two major types of fans" in his discussion, Nilsen contrasts "those that seem describable by the Mutti and Ricci Lucchi system [no citation], and those that are characterized by suprafan lobes." The problem is not that of contrasting Mutti and Ricci Lucchi (1972) with Walker (1978), where the problems are of terminology. The real problem is that of contrasting Mutti and Ricci Lucchi (1972) with Mutti and Ricci Lucchi (1974), where there are fundamental differences of facies distribution, bypassing zones, and attachment or detachment of lobes and channels. These differences are important in fan behavior, lobe development, and use of the model in prediction. Most workers appear to have assumed that when a turbidity current flows out of the end of a channel, it spreads laterally and hence decelerates rapidly, causing deposition of part of its load. In the Mutti models (either Fig. 1A or D), the channels gradually decrease in their topographic expression downfan and hence, even before the ends of the channels, the turbidity currents are probably spilling out of the channels and depositing interchannel thin-bedded turbidite facies. The tendency to spread laterally and deposit is established long before the flow spills from its few-meter-deep channel end. With a tendency to spread and deposit before reaching the channel terminus, it seems unlikely that the flow should then traverse a zone of bypassing. The zone of bypassing, it must be remembered, is based only upon interpretations of the Marnoso Arenacea and Hecho Group (Mutti and Ricci Lucchi, 1974, p. 579-580).

MAP VIEW

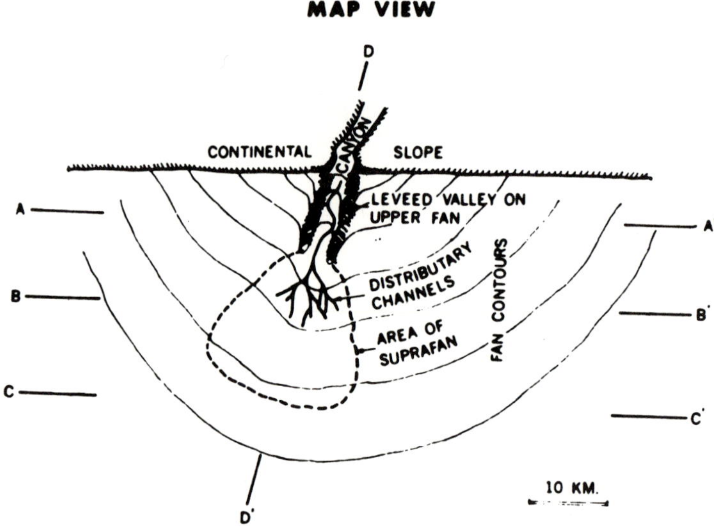

A. Normark, 1970.

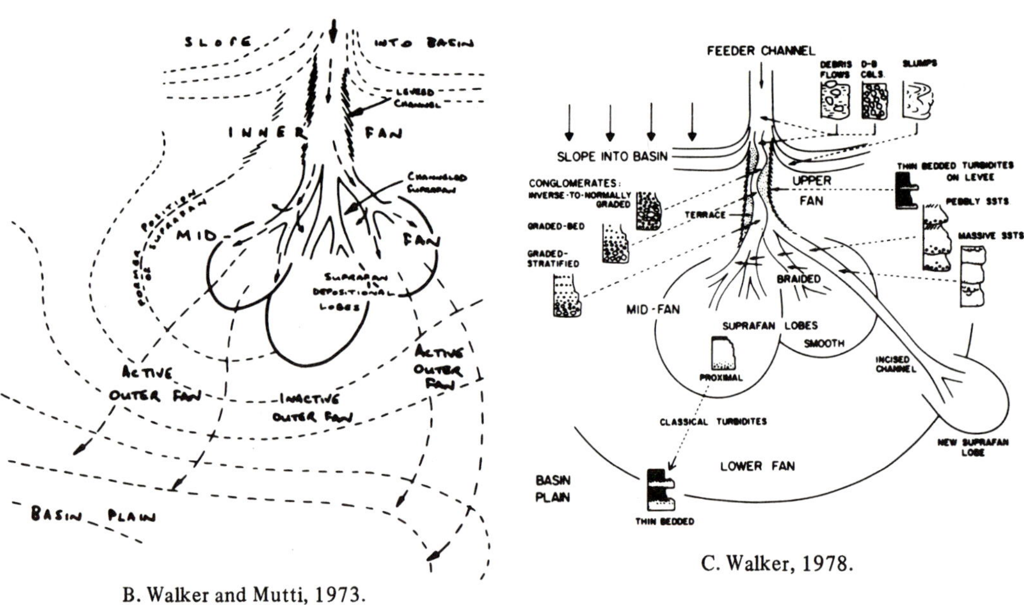

C. Walker, 1978.

B. Walker and Mutti, 1973.

FIG. 2—Comparison of fan model proposed by Normark (1978) from recent sediments with models combining recent and ancient sediment data.

The second problem of a model such as that shown in Figure 1D concerns the building of a prograding lobe, if the lobe is separated from its feeder channel. Mutti and coworkers have emphasized many times that such a lobe consists of thick-bedded coarse turbidites at the apex and thin-bedded, fine-grained turbidites at the lobe fringe. Progradation forms a thickening- and coarsening-upward sequence (Mutti and Ghibaudo, 1972; Mutti and Ricci Lucchi, 1972, 1974; Ricci Lucchi, 1975b; Mutti, 1977). Let us consider, then, a growing prograding lobe. By comparison with suprafans, I am assuming that such a lobe will have some topographic relief above the main fan surface. Thickening-upward sequences are commonly a few tens of meters in thickness. Even allowing for some differential compaction, an original relief of 10 to 20 m might be possible. First, the higher the lobe becomes, the greater the tendency for subsequent turbidity currents to flow around it rather than over it; thus the growth of a lobe inevitably would lead to its death. More significantly, I would suggest that even if subsequent turbidity currents did flow up over the growing lobe, the higher the lobe became the *thinner* the deposits of successive currents would be. The model in Figure 1D implies that turbidity currents cross the bypass zone and climb up on the lobe, and that the higher the lobe becomes, the more the currents climb up and deposit thicker and thicker beds. More probably the current would deposit thinner and thinner beds as they climb onto the higher and higher lobe, and hence the sequence would be *thinning-upward,* despite the standard interpretation of a thinning-upward sequence as a channel fill. Maybe there are such detached lobes in the geologic record but, if so, they have been called channels!

In a model such as that in Figure 2C, the channel is directly in contact with the lobe. Even without overall aggradation, mere progradation by channel extension across the suprafan will introduce more and more "proximal" (and presumably thicker) beds to what were previously lobe-fringe environments, forming the thickening-upward sequence. Net aggradation of channel system and lobe will emphasize the effect.

Thus the crux of the argument comes down to the zone of bypassing, because there are no fundamental differences in the models in Figures 1A, B, E, 2A, B, C.

MODELS WITH OR WITHOUT SUPRAFANS

In Nilsen's discussion, he envisages two types of fans: (A) fans with well-defined channels, unchanneled lobes, with a mixture of sediment types; (B) fans that are composed mostly of sand, and consist almost wholly of channeled deposits, where the sandy deposits grade abruptly downfan into basin-plain deposits without the development of thickening and coarsening-upward unchanneled lobes.

Nilsen has compared type A with a "Mutti and Ricci Lucchi (1972) system," and type B with a system "characterized by suprafan lobes." This comparison is incorrect. In the suprafan models (Fig. 2) the channels pass downfan into depositional lobes, not "abruptly . . . into basin plain deposits" as Nilsen states. I have emphasized that the only difference in models such as that in Figure 1B and the "suprafan" models is terminology. Thus Nilsen's fan type A embraces everything in Figures 1 and 2. His type B is not a suprafan type; by original definition (Normark, 1970, 1978) a suprafan has a smooth depositional lobe beyond the end of the channels. Nilsen admits that his type B fans "have been less well studied," and the examples he quotes have been described only in a guidebook, an abstract, in personal communications, or are in press. Thus I cannot yet comment on this distinction among fan types, nor on the suggested different flow mechanisms affecting fan growth.

With regard to different fan types, Nilsen also commented, "There is certainly no reason to expect that coarsening- and thickening-upward megasequences should be characteristic of the outer fan, the smooth outer parts of suprafan lobes, and also the transition from smooth to channeled parts of suprafan lobes, as shown by Walker (1978, Fig. 14)." I fail to see the reasoning behind this statement. Setting aside differences of terminology, the prograding smooth suprafan lobe is identical to the lobes in Figure 1B, yet this is the type example of the thickening-upward sequence (Mutti and Ghibaudo, 1972). Again, setting aside terminology, if one of the stippled lobes in Figure 1B is prograding, a similar record of progradation might reasonably be expected, albeit in overall thinner beds, on the nearby part of the basin plain. The third situation, of channeled suprafan building on top of smooth suprafan, is also likely to give a coarsening- and thickening-upward sequence. Mutti and Ghibaudo (1972, Pl. II) did not consider this situation, yet their diagram clearly suggests the channeled deposits to be coarser than the lobe deposits. Also, in the original Mutti and Ricci Lucchi scheme (Fig. 1A), the mid-fan channeled facies association is clearly described as coarser grained than the unchanneled outer fan association—progradation of a channeled system onto a smooth system (whether it is termed a suprafan or not) is therefore likely to give a thickening- and coarsening-upward sequence. For these reasons, I fail to understand Nilsen's comments cited at the beginning of this paragraph.

OTHER POINTS RAISED BY NILSEN

I agree with Nilsen that lobes of classical turbidites make good reservoirs. In the context of a model such as Figure 2C, these lobes would be connected to the channeled portion where one might expect thicker sands with fewer shaly permeability barriers. I also agree with Nilsen that lobes of the Figure 1D type would be good targets, and an example is possibly the Grimes gas field in California (Weagant, 1972). Here, production is from lobes about 10 km long, 5 km wide, and 10 to 60 m thick. The lobes do not appear to have channels directly attached to their upstream end and hence can be described by a Mutti-type model (Fig. 1D). Similarly, some of Hsü's (1977) data from the Ventura field (e.g., his Fig. 18) suggest channels that shale out in the downstream direction. I termed these suprafan channels but, in a Mutti-type model (Fig. 1D), they could represent the channeled mid-fan shaling out into a bypass zone. This is a superb chance to test the Mutti model, because it would predict depositional lobes in the downfan direction beyond the zone of bypassing, west of the area shown in Hsü's Figure 18. Maybe a reader familiar with the AO_1 sand west of the Shell Taylor lease can help.

With regard to Wheeler Gorge (Walker, 1975a; Nelson et al, 1977; Walker, 1977), I reexamined the section with Nilsen in January 1979 and agreed with him that the thin-bedded turbidites above the conglomerates are probably channel margin, levee, and interchannel. There is a much higher sand/shale ratio for the thin-bedded turbidites above the conglomerates than below, and beds are thicker with slump horizons, above the conglomerates. I continue to suggest that the much shalier sequence without slumps below the conglomerates is more probably a basin-plain deposit. My statement (Walker, 1978, p. 954) that there is "no field evidence of channeling" was sloppy. I intended to say that there is no field evidence that the whole conglomeratic sequence rests within a channel. There is indeed about 5 m of channeling at the base of the third conglomerate. The base of the first conglomerate is spectacularly planar, with large flute casts. It does not cut out more than 1 or 2 m of the underlying shales, but there are abundant rip-up clasts within the conglomerate. My point remains, that the thinning- and fining-upward sequences at Wheeler Gorge (and in Oregon) are not demonstrably in channels—by which I mean channels of a depth comparable to the thickness of the sequences.

My statement that "vertical trends (thickening or thinning upward) may be [*not* are] meaningless or absent" in elongate basins "in which paleoflow is dominantly parallel with the basin axis" remains true. Nilsen has directed his criticism out of the original context of the statement, which referred (Walker, 1978, p. 962) to an "exogeosynclinal situation," quoting examples from the Appalachian exogeosyncline in Quebec, the central Appalachians, and the Ouachitas. If supply is from many points along the side of the trough, there may be individual fans with distinct vertical facies sequences, but with paleoflow dominantly *toward* the axis. My comments applied in areas where flow was parallel with the axis, and where deposits from different sources may be stacked to produce random interbeds of geographically proximal and distal beds.

Finally, in replying to Nilsen's comments on San Clemente (Walker, 1975b), I regret that the channel margins north of the car park were very poorly exposed when I was working there in 1973 and 1974. However, because the northern margins cannot be matched with the southern margins, only the minimum width of the final channel can be measured, and Nilsen is a bit optimistic in his statement, "thus, the size of the channels can be measured." The bottoms of the channels cannot be seen, but by relating silty layers within the mud drapes on the channel walls to beds within the channels, I estimated minimum channel depths of about 20 to 25 m (Walker, 1975b, p. 923). About 15 m of channel fill can be seen in outcrop.

DATA BASE OF FAN MODELS

Nilsen has commented that I would be better off using "three-dimensional examples" and "integrated studies of full systems . . . rather than two-dimensional examples of small parts of fans." This raises the very important point: what actually is the data base for the various fan models? Nilsen gives the impression by his long list that there are many three-dimensional examples to choose from, but I believe that this list gives a false impression. For example, the Tyee has not been described in the literature in terms of a three-dimensional fan; the latest detailed work of which I am aware predates all the fan models (Lovell, 1969) and is a discussion of "proximality." The Great Valley sequence examined by Ingersoll (1978) interprets the Great Valley rocks in terms of fan facies, but it is not a three-dimensional example of a fan. Ingersoll (1976, Pl. 16) showed excellent detail in 16 measured sections, but these cover a northwest-southeast linear belt about 560 km long, and hence the average separation of the sections is about 32 km. Thus each section gives a view of vertical fan development at a point, but the sections cannot be correlated to show the development of a single fan, as Nilsen misleadingly implies. The Butano Sandstone, also cited by Nilsen, has a well-described paleocurrent pattern (Nilsen and Simoni, 1973), but there is no discussion of a three-dimensional picture of facies relations. The Point of Rocks Sandstone, the Cantua Sandstone, and The Rocks Sandstone are all mentioned without citation by Nilsen although none, to my knowledge, has been published and hence compilers of fan models cannot yet use these examples. The San Sebastian example is hardly three-dimensional; it embraces a narrow strip of shoreline about 25 km long. The Hecho Group example (Mutti, 1977) is also an *interpretation,* and does not present any evidence in the form of measured sections, facies sequences, or correlation diagrams showing lateral changes that compilers of fan models can use.

Thus I find Nilsen's list of "three-dimensional" examples very misleading, but it highlights the point that there are only a handful of well-studied three-dimensional full systems on which to base a model.

Consequently, the establishment of lateral and vertical facies relations in large single outcrops (San Clemente, Wheeler Gorge, etc) remains important and adds detail to the few larger scale three-dimensional studies.

The reason for reusing the Shale Grit–Grindslow Shales System (Pennsylvanian, northern England) as the type fan in my 1978 paper was precisely that it is a three-dimensional system, and that published correlation diagrams (Walker, 1966a, b; 1978) were based upon 88 measured sections (totaling 9,000 m) originally drawn at a scale of 0.4 cm to 15 m (in Walker, 1964) and mapped at a scale of 16 cm to the kilometer (Walker, 1964). Would that the California examples had been studied three dimensionally in this detail!

CONCLUSIONS

Nilsen's most substantive criticisms concern the development and evolution of fan models. The main change that I have made to the Mutti and Ricci Lucchi (1972) model is to take the channel and lobe systems (arbitrarily assigned to mid- and outer-fans, respectively) and place them both on the mid fan, following the topographic information published by Normark (1970). This change is one of terminology only, and does not affect lateral or vertical facies relations.

I have emphasized that a much more fundamental change has been made by Mutti and his colleagues, by detaching the depositional lobes from their channels and introducing a zone of by-passing.

Finally, Nilsen comments that "ongoing work on ancient fan deposits appears to differentiate two major types of fans." These are discussed as "Mutti and Ricci Lucchi (1972) systems" and systems "characterized by suprafan lobes." Nilsen is entangled in his own terminologic mess here, as the latter "suprafan-lobe type" consists "almost wholly of channeled deposits . . . without the development of . . . unchanneled lobes." The contrast between these two types of systems cannot be adequately judged, because most of the evidence for the "suprafan-lobe type" is unpublished. A more pressing problem is the contrast among models with lobes connected to channels (Figs. 1A, B, 2A, B, C) and models with zones of bypassing and detached lobes (Fig. 1C, D).

REFERENCES CITED

Bouma, A. H., 1962, Sedimentology of some flysch deposits: Amsterdam, Elsevier, 168 p.

Hsü, K. J., 1977, Studies of Ventura field, California, 1: Facies geometry and genesis of lower Pliocene turbidites: AAPG Bull., v. 61, p. 137-168.

Ingersoll, R. V., 1976, Evolution of the Late Cretaceous fore-arc basin of northern and central California: PhD thesis, Stanford Univ., 200 p.

———— 1978, Submarine fan facies of the Upper Cretaceous Great Valley sequence, northern and central California: Sed. Geology, v. 21, p. 205-230.

Lovell, J. P. B., 1969, Tyee Formation: a study of proximality in turbidites: Jour. Sed. Petrology, v. 39, p. 935-953.

Mutti, E., 1974, Examples of ancient deep-sea fan deposits from Circum-Mediterranean geosynclines, *in* Modern and ancient geosynclinal sedimentation: SEPM Spec. Pub. 19, p. 92-105.

———— 1977, Distinctive thin-bedded turbidite facies and related depositional environments in the Eocene Hecho Group (south-central Pyrenees, Spain): Sedimentology, v. 24, p. 107-131.

———— and G. Ghibaudo, 1972, Un esempio di torbiditi di conoide sottomarina esterna: le Arenarie di San Salvatore (formazione di Bobbio, Miocene) nell'appennino di Piacenza: Acad. Sci. Torino Mem., ser. 4, no. 16, 40 p.

———— and F. Ricci Lucchi, 1972, Le torbiditi dell'Appennino settentrionale: introduzione all'analisi di facies: Soc. Geol. Italiana Mem., v. 11, p. 161-199; English translation by T. H. Nilsen, 1978, Internat. Geology Rev., v. 20, p. 125-166; AGI Reprint Ser. 3.

———— 1974, La signification de certaines unités séquentielles dans les séries à turbidites: Soc. Géol. France Bull., v. 16, p. 577-582.

———— et al, 1975, Examples of turbidite facies and facies associations from selected formations of the northern Apennines: 9th Internat. Cong. Sedimentology, Nice. Guidebook A-11, 120 p.

Nelson, C. H., E. Mutti, and F. Ricci Lucchi, 1977, Upper Cretaceous resedimented conglomerates at Wheeler Gorge, California: description and field guide: Jour. Sed. Petrology, v. 47, p. 926-928.

Nilsen, T. H., and T. R. Simoni, Jr., 1973, Deep-sea fan paleocurrent patterns of the Eocene Butano Sandstone, Santa Cruz Mountains, California: U.S. Geol. Survey Jour. Research, v. 1, p. 439-452.

Normark, W. R., 1970, Growth patterns of deep-sea fans: AAPG Bull., v. 54, p. 2170-2195.

——— 1978, Fan valleys, channels, and depositional lobes on modern submarine fans: characters for recognition of sandy turbidite environments: AAPG Bull., v. 62, p. 912-931.

Parea, G. C., 1965, Evoluzione della parte settentrionale della geosynclinale appenninica dall'Albiano all'Eocene Superiore: Modena Acad. Naz. Sci. Lett. Art, v. 7, p. 3-97.

Ricci Lucchi, F., 1975a, Miocene paleogeography and basin analysis in the Periadriatic Apennines, in C. H. Squyers, ed., Geology of Italy, v. 2: Libyan Arab Republic Earth Sci. Soc., p. 129-236.

——— 1975b, Depositional cycles in two turbidite formations of northern Apennines: Jour. Sed. Petrology, v. 45, p. 3-43.

Walker, R. G., 1964, Some aspects of the sedimentology of the Shale Grit and Grindslow Shales (Namurian R1c, Derbyshire) and the Westward Ho! and Northam Formations (Westphalian, North Devon): PhD thesis, Oxford Univ., 3 v.

——— 1966a, Shale Grit and Grindslow Shales, transition from turbidite to shallow water sediments in the Upper Carboniferous of northern England: Jour. Sed. Petrology, v. 36, p. 90-114.

——— 1966b, Deep channels in turbidite-bearing formations: AAPG Bull., v. 50, p. 1899-1917.

——— 1967, Turbidite sedimentary structures and their relationship to proximal and distal depositional environments: Jour. Sed. Petrology, v. 37, p. 25-43.

——— 1975a, Upper Cretaceous resedimented conglomerates at Wheeler Gorge, California: description and field guide: Jour. Sed. Petrology, v. 45, p. 105-112.

——— 1975b, Nested submarine fan channels in the Capistrano Formation, San Clemente, California: Geol. Soc. America Bull., v. 86, p. 915-924.

——— 1977, Upper Cretaceous resedimented conglomerates at Wheeler Gorge, California: description and field guide: Reply: Jour. Sed. Petrology, v. 47, p. 928-930.

——— 1978, Deep-water sandstone facies and ancient submarine fans: models for exploration for stratigraphic traps: AAPG Bull., v. 62, p. 932-966.

——— 1979, Facies models 8. Turbidites and associated coarse clastic deposits, in R. G. Walker, ed., Facies models: Geol. Assoc. Canada, Reprint Ser. 1, p. 91-103.

——— and E. Mutti, 1973, Turbidite facies and facies associations, in G. V. Middleton and A. H. Bouma, eds., Turbidites and deep water sedimentation: SEPM Pacific Sec. Short Course, p. 119-157.

Weagant, F. E., 1972, Grimes gas field, Sacramento Valley, California, in Stratigraphic oil and gas fields: AAPG Mem. 16, p. 428-439.

Modern and Ancient Submarine Fans: Reply[1]

WILLIAM R. NORMARK[2]

INTRODUCTION

Nilsen's comments on my recent paper (Normark, 1978) and on others that appeared in the AAPG *Bulletin* for June 1978 provide a welcome opportunity to discuss problems that arise from the application of models based on modern fans to ancient rocks, and vice versa. The time is ideal for a brief review. My paper was an outgrowth of some notes originally prepared for a New Orleans Geological Society short course in 1976. In rewriting these notes for the *Bulletin,* I included some new data from a Navy fan deep-tow survey in 1976. Since then, the Navy fan study has been completed (Normark et al, 1979), and new geophysical and sample data have been obtained during four cruises to the Monterey and Delgada fans along the central California coast, and to the Laurentian fan in the North Atlantic. These new data, collected between May 1977 and December 1979, are still being processed. Without attempting to present details of this large data package, I believe that a critical review of Nilsen's comments and of my earlier ideas is, nevertheless, useful in light of this new information. Thus my remarks are based on personal studies of eight submarine or sublacustrine fans (Bengal, Delgada, La Jolla, Laurentian, Monterey, Navy, Reserve, and San Lucas), as well as on literature descriptions of others (Normark, 1978; Wilde et al, 1978).

Simply expanding our knowledge of modern and ancient fan systems will not eliminate discrepancies among the derived sedimentation models. Many of the problems under discussion

[1]Manuscript received, February 19, 1979; accepted, February 20, 1980.

[2]U.S. Geological Survey, Menlo Park, California 94025.

Discussions with D. J. W. Piper, P. Wilde, G. R. Hess, and D. G. Howell on subjects related to my submarine-fan studies in the last few years were helpful in constructing this reply. I thank M. E. Field and G. R. Hess for constructive reviews of the manuscript.

result directly , as Nilsen's discussion states, from disparity in the types of data used to construct these sedimentation models (Normark et al, 1978, p. 593-594). Studies of modern fans rely on surface morphology, sediment samples of the uppermost sedimentary layers (cores longer than 10 m are uncommon), and seismic-reflection profiles that provide only an approximate idea of the internal structure of the deposit. Core samples, however, rarely penetrate to even the shallowest reflector observed on high-resolution reflection profiles. Without visual continuity along the seafloor or the ability to trace individual bedding surfaces acoustically within the fan sequence, it is difficult to follow a given depositional unit across a fan using core samples unless a distinctive mineralogy is present in the bed (as for the Astoria fan; cf. Nelson, 1976). Without deep-tow sonar soundings or sophisticated multibeam echosounding systems, our knowledge of fan morphology is highly approximate as well.

Models based on ancient fan studies, however, rely on sedimentologic details from isolated outcrops. Details of individual beds commonly can be traced across and sometimes between the outcrops. The critical missing aspect is the shape of the fan surface that corresponds to any given time or event. Structural disruption and differential compaction of the deposits further complicate attempts to reconstruct the fan morphology of ancient systems. It is not surprising, therefore, that certain morphologic features recognized on modern fans are not recognized in ancient turbidite sequences, and that coarse-grained sequences seen in ancient fans are difficult to define on modern fans.

Nilsen's discussion addresses a variety of subjects drawn from three different papers in the June 1978 *Bulletin*. Rather than follow his subject order, I have restructured the material for brevity. The key subjects I address include: (1) clarification of the suprafan concept for modern fans, (2) problems that result from use of morphologic terms in describing ancient systems, (3) applicability of the "original" Mutti–Ricci Lucchi (1972) model to modern fans, (4) requirements for more than two models (suprafan type, Normark, 1970; Mutti–Ricci Lucchi type), (5) potential petroleum reservoirs on lower fans, and (6) specific goals for future work.

SUPRAFAN CONCEPT

The applicability of the suprafan concept to ancient fan deposits is one of three major problems envisioned by Nilsen; he finds that the model which includes the suprafan concept, as presented by Walker (1978), is confusing and difficult to use in the field. Although there are few published descriptions of suprafan deposits for ancient rocks, as Nilsen points out, they can be found for smaller modern fans. Because the suprafan is a *morphologic* feature of some modern fans, it is not surprising that a suprafan is not easily recognizable in ancient deposits. The best examples of modern suprafan development are from the San Lucas fan in deep water at the tip of the Baja California Peninsula and the Navy fan in South San Clemente basin off southern California (Normark, 1970; Normark and Piper, 1972; Normark et al, 1979). Until the Navy fan was resurveyed in detail, the hummocky surface of the upper suprafan appeared to result from hyperbolic echoes from the edges of numerous depressions that were thought to be distributary channels or channel remnants (Normark, 1970). The suprafan concept, therefore, was extended to include any channeled lobate area of active sand deposition, no matter how small (Normark and Dickson, 1976, Fig. 3) or large (Wilde et al, 1978, Fig. 6). In retrospect, this extension of the concept probably has caused some confusion; the stricter definition given in the following draws heavily on extensive deep-tow geophysical surveying and sedimentary sampling on the Navy fan (Normark et al, 1979).

A suprafan is defined as the area of active sand deposition on a fan; it forms downfan from the termination of a leveed valley on the upper fan and is recognized morphologically by: (1) a convex-upward bulge in the radial profile of the fan, (2) approximately lobate contours, and (3) numerous smaller scale forms, including distributary channels, channel segments, isolated depressions, and positive-relief features (perhaps even large bed forms). The varied local relief (item 3), referred to as mesotopography (Normark et al, 1979), is generally restricted to the upper part of the suprafan, as delineated by lobate contours and a convex-upward bulge. Detailed examination of the Navy fan shows that only one distributary channel (generally unleveed) may be active at any time and that small (1 km wide) smooth lobes form at the ends of the distributary channels (Normark et al, 1979). These smooth lobes do not have channels, although they are commonly bounded around the outer edges by small channellike segments which, however, do not connect with the feeding distributary channel. The mesotopography common on the upper suprafan decreases in size downfan and dies out on the lower suprafan. The smooth lobes also terminate on the lower suprafan. Generally, conventional surface-ship echo sounding or reflection profiling can detect (as numerous overlapping hyperbolic echoes) only the larger relief features of the upper suprafan.

Nilsen questions the conclusion that the lower suprafan may consist of thickening- or coarsening-upward sequences. Although it is true that no cores have sampled through the lower suprafan deposits on Navy or San Lucas fans, available shallow cores from both fans show a general thickening and coarsening of sand beds from the fringe to the apex of the suprafan (Normark, 1970, p. 2186; Normark et al, 1979). That this lateral change in bed thickness and grain size is expected to extend vertically in a prograding depositional system is merely a statement of Walther's law of correlation of facies (namely, that the same succession of facies observed laterally is also present in vertical succession; cf. discussions by Krumbein and Sloss, 1963, p. 318, and Blatt et al, 1972, p. 187-188).

PROBLEMS WITH USE OF MORPHOLOGIC TERMS

The suprafan is not the only morphologic feature of modern fans that is awkward to reconcile with interpretations of ancient deposits. The definition of inner-, middle-, and outer-fan settings for ancient systems is based on interpretation of depositional environments according to facies characteristics, whereas the upper-, middle-, and lower-fan distinction commonly used for modern fans is based on morphologic features including slope. Thus sedimentologic or facies distinctions for depositional settings on ancient fans may not be equivalent to the morphologic counterparts on modern fans. One geologist's "outer fan" may be another's "middle fan," so that combining studies of different fans to produce a single sedimentation model is not straightforward.

In general, the use of morphologic terms derived from modern analogs to describe the features of ancient sedimentary deposits can be misleading. For example, levees are a distinctive relief feature bordering valleys on the upper fan in modern systems. Such levee relief above the surrounding fan surface may not be demonstrable without extensive outcrops in ancient fan systems, and thus the term "overbank deposit" is more appropriate than "levee deposit." Large areas of the upper fan receive overbank deposits, but levee construction occurs only along the valley. Although overbank deposition on modern fans also occurs along distributary channels on the suprafan, levee relief rarely develops.

OTHER MODELS

Both Normark (1978, p. 928) and Nilsen agree that no single fan-sedimentation model is universally applicable. Certainly not all modern submarine fans have a suprafan depositional bulge! At the same time, the Mutti–Ricci Lucchi (1972)

model does not appear to apply directly to any of the submarine fans that I have studied. The thickening- and coarsening-upward unchanneled lobe sequences of their outer-fan facies associations have not been recognized. All areas of active sand deposition on modern fans that I've studied are either channeled or are associated with suprafans. The apparent absence of unchanneled depositional sand lobes on modern fans made the suggestion attractive that the lower unchanneled suprafan may be equivalent to the outer-fan sand lobes in the Mutti–Ricci Lucchi model. If Nilsen is correct in his suggestion that the morphologically smooth outer fan of Normark (1970, 1978) is the basin-plain equivalent of outer-fan sand lobes in the facies-association model, then the lower suprafan (if one exists) would correspond to the outer fan of the Mutti–Ricci Lucchi scheme.

Unchanneled thickening- and coarsening-upward megasequences need not be restricted to the Mutti–Ricci Lucchi type of fan, as intimated by Nilsen. As noted earlier, the lower smoother part of the prograding suprafan is probably a thickening- and coarsening-upward deposit, as are (probably) the small lobes present on the suprafan below the individual distributary channels. Thus, although I agree with Nilsen that the suprafan concept does not apply to all fans, I find that the Mutti–Ricci Lucchi (1972) scheme is not necessarily the most probable alternative model.

Some submarine fans have depositional lobes intermediate between a suprafan and outer-fan sand lobe. The Monterey fan has a large area of active sand deposition associated with numerous channels that appear to form a distributary system covering a lobate area 50 by 100 km (Wilde et al, 1978, Fig. 6). Little evidence exists, however, of the mesotopographic features and convex-upward bulge that are the other major characteristics of a suprafan lobe. Although it may, therefore, be misleading to refer to that area of sand deposition as a suprafan, the system certainly does not resemble the Mutti–Ricci Lucchi type of lobe because of the numerous channels present.

Fans with prominent leveed valleys fed by rivers or deltaic systems appear to lack the local relief and convex-upward profile in the midfan region used to characterize a suprafan (Normark, 1978, p. 928-929). I have further suggested (p. 924) that the grain-size distribution of sediment supplied to the fan determines the extent or type of morphologic features that develop on the midfan. Suprafans develop on fans that receive a relatively high proportion of coarse sediment (mostly sand) at the site where turbidity currents rapidly deposit coarse material at the termination of the leveed valley of the upper fan. Fans with a dominantly fine-grained sediment source tend to have pronounced leveed-valley systems (Normark, 1978, p. 928).

The Noyo system on the Delgada fan (Normark and Hess, 1980) has no recognizable distributary-channel system. Cores from the Noyo channel system contain little sand, in contrast to cores from similar-size valleys on the Monterey fan. The Delgada fan consists of a leveed valley that extends across half the fan radius and whose levees form an elongate lobe as large as 75 km across. The valley system is the primary morphologic feature on an otherwise gently sloping ramp of continental-rise deposits. The large leveed valleys are characteristic of upper fans, so this type of fan would appear to have only upper- and lower-fan morphologic divisions. I have suggested (Normark, 1978) that both the Bengal and Reserve fans may be examples of this type of fan.

I feel that a full spectrum of submarine fans eventually will be recognized. A levee-dominated system forms where little coarse sediment is supplied, whereas a suprafan system is of the sand-rich type, in which most of the sand is deposited at the termination of the upper-fan valley. The Mutti–Ricci Lucchi model would apply to the intermediate type, in which the available sand may be transported longer distances ("highly efficient" fans; cf. E. Mutti written communication cited by Nilsen).

CHANNELS AS HYDROCARBON RESERVOIRS

Nilsen feels there is an ". . . unfortunate overemphasis of channeled submarine fan deposits as exploration targets" in both Walker (1978) and Wilde et al (1978), and that more consideration should be given to outer-fan sand deposits. Wilde et al, using the Monterey fan as an example, did not rule out unchanneled sand bodies as reservoirs; they emphasized that the areas of active sand deposition on the middle and lower fan could be collecting units for the migration and concentration of petroleum (Wilde et al, 1978, p. 979). Because all known sand lobes on the Monterey fan, however, are associated with distributary-channel systems, Wilde et al further suggested that the petroleum would probably migrate updip through the coarser fill of these distributary channels. The upper-fan channel-fill deposits would thus be the best potential reservoirs, unless the broad sand lobes were sealed off from the main channels.

Nilsen further suggests that outer-fan sand lobes may be better sorted than upper-fan sand beds and have more favorable porosities and permeabilities. The opposite, however, is seen in our samples from the Monterey fan: the cleanest sand was found in the Ascension Valley floor on the upper fan and sand beds on the main depositional lobe contain more clay than clean sand beds in Monterey and Ascension Valleys (Hess and Normark, 1976, p. 245, Fig. 6).

Wilde et al (1978) stressed the importance of channeled deep-water submarine-fan sand bodies as possible reservoirs for recoverable hydrocarbons today. Their main thesis was to determine whether petroleum may be generating and accumulating in a modern continental-rise setting. If organic material is being converted to hydrocarbons on the Monterey fan, this conversion would appear to be limited to the thick upper fan and to a few deep, filled basins on the middle fan (Wilde et al, 1978, p. 973). For this reason, also, they stressed that upper-fan channel-fill sand deposits would be the most likely sites for hydrocarbon accumulation today. Nilsen is right in suggesting that outer-fan sand lobes, where they exist, could be excellent petroleum traps in ancient fan systems.

MISCELLANEOUS

Nilsen disagrees with Walker's (1978, p. 950) statement that "the upper fan is characterized by one relatively deep, leveed channel . . ." Although several exceptions are known, two of those mentioned by Nilsen require further comment. Although the Bengal fan has been shown to have several large valleys on the upper fan, recent mapping by Emmel et al (in prep.) shows ". . . that at any one time only one active channel is connected to the canyon." On the Monterey fan are two major canyon (and associated fan valley) systems that supply most of the sediment to the fan; only one canyon/valley system, however, may be active at any time (Hess and Normark, 1976). During interglacial high sea levels, little sand reaches the fan through the relatively inactive Ascension Canyon; the Monterey Submarine Canyon cuts across the shelf and receives most of the sediment reaching the fan today. During low stands of sea level, the Ascension Canyon received much of the littoral-drift sediment moving along the coast and thus may dominate fan growth during those periods. Although the Delgada fan off central California does have two major canyon systems, it appears to consist of two separate coalescing fan units (Normark and Hess, 1980).

These observations stress the importance of understanding the history of sedimentation on a modern fan before using morphologic concepts to interpret sedimentation patterns. The formation of suprafans or sand lobes, or even the absence of any such distinctive depositional features, must be related to the appropriate valley and to the time of active growth.

FUTURE WORK

As more fan systems, both ancient and modern, are documented, I suspect that the number of sedimentation models will proliferate. A full spectrum of morphologic features and patterns of coarse-sediment distribution may be expected. Nevertheless, many of the problems I review here in light of Nilsen's discussion might be alleviated by improved sampling and surveying systems on modern fans. The new hydraulic piston corer developed by the Deep Sea Drilling Project has promise for coring up to 100 m of uppermost fan deposits; use of this device could demonstrate if and where thickening- and coarsening-upward sequences are on modern fans. Long-range side-scanning sonar and multibeam echo-sounding systems can provide much improved images of fan morphology, especially in areas of distributary channels, suprafan relief (mesotopography), and other depositional-lobe features; and this morphology can be closely tied to sedimentary relations to construct a more comprehensive model for interpreting ancient fans.

More work on ancient fan systems is needed as well. I agree with Nilsen that studies of only fragments of ancient fans may be misleading for constructing models of sedimentation. I further caution that the recognition of components of the Mutti–Ricci Lucchi facies associations in scattered outcrops does not in itself demonstrate the existence of a submarine fan.

REFERENCES CITED

Blatt, H., G. Middleton, and R. Murray, 1972, Origin of sedimentary rocks: Englewood Cliffs, N.J., Prentice-Hall, 634 p.

Emmel, F. J., J. R. Curray, and D. G. Moore, Migration of fan valleys on the Bengal fan, northeastern Indian Ocean, in prep.

Hess, G. R., and W. R. Normark, 1976, Holocene sedimentation history of the major fan valleys of Monterey fan: Marine Geology, v. 22, p. 233-251.

Krumbein, W. C., and L. L. Sloss, 1963, Stratigraphy and sedimentation: San Francisco, Calif., W. H. Freeman, 660 p.

Mutti, E., and F. Ricci Lucchi, 1972, Le torbiditi dell'Appennino settentrionale: introduzione all'analisi di facies: Mem. Soc. Geol. Ital., v. 11, p. 161-199.

Nelson, C. H., 1976, Late Pleistocene and Holocene depositional trends, processes, and history of Astoria deep-sea fan, northeast Pacific: Marine Geology, v. 20, p. 129-173.

Normark, W. R., 1970, Growth patterns of deep-sea fans: AAPG Bull., v. 54, p. 2170-2195.

——— 1978, Fan valleys, channels, and depositional lobes on modern submarine fans: characters for recognition of sandy turbidite environments: AAPG Bull., v. 62, p. 912-931.

——— and F. H. Dickson, 1976, Sublacustrine fan morphology in Lake Superior: AAPG Bull., v. 60, p. 1021-1036.

——— and G. R. Hess, 1980, Quaternary growth patterns of California submarine fans in Quaternary depositional environments of Pacific coast, Pacific coast Paleogeography Symposium 4: SEPM Pacific Sec., p. 201-210.

——— ——— and F. N. Spiess, 1978, Mapping of small scale (outcrop-size) sedimentological features on modern submarine fans: Offshore Technology Conf. Proc., p. 593-598.

——— and D. J. W. Piper, 1972, Sediments and growth patterns of Navy deep-sea fan, San Clemente basin, California borderland: Jour. Geology, v. 80, p. 198-223.

——— ——— and G. R. Hess, 1979, Distributary channels, sand lobes, and mesotopography of Navy submarine fan, California borderland, with applications to ancient fan sediments: Sedimentology, v. 26, p. 749-774.

Walker, R. G., 1978, Deep-water sandstone facies and ancient submarine fans: models for exploration for stratigraphic traps: AAPG Bull., v. 62, p. 932-966.

Wilde, P. W., W. R. Normark, and T. E. Chase, 1978, Channel sands and petroleum potential of Monterey deep-sea fan, California: AAPG Bull., v. 62, p. 967-983.

TURBIDITES IN OIL EXPLORATION[1]

HAROLD H. SULLWOLD, Jr.[2]
North Hollywood, California

ABSTRACT

The rapid accumulation of data, mainly from field observations and oceanographic research in the past few years, has forced the conclusion that turbidites are not freaks of nature but are very commonplace, especially in environments of deep-water sedimentation. Billions of barrels of oil have been produced from turbidites in the Los Angeles basin alone. A thorough knowledge of their characteristic features and mode of origin would be of great use in petroleum exploration and development in areas where turbidites exist. By virtue of their mode of deposition, turbidites have peculiar syngenetic structures which serve as useful recognition clues and as indicators of current direction and sea-bottom topography. Much more data collecting is necessary before these potential clues can be fully utilized in predicting shape, size, and trend of turbidites from isolated well data, but the possibilities are enormous. Turbidites are deposited in low places on the sea floor, and their over-all geometry is controlled by the shape of these low places. Channel, fan, and blanket-like shapes have been observed. Descriptions in the literature of entire turbidites, showing their complete geometry, are lacking, but four incomplete or generalized examples are given from the California Tertiary.

INTRODUCTION

A *turbidity current* is a muddy current which is able to move, even in still water, by virtue of its own muddiness, and the sediment which it deposits is a *turbidite*. Turbidity currents as a common mechanism for the deposition of sandstones have received recognition only in the last decade. In this time much has been written describing the internal and external features of turbidites, both as recognition clues and as evidence for mode of deposition. The increasing tempo of study of turbidites on the modern sea floor has provided valuable data as to their present habitat and over-all shape. Directions of ancient turbidity current flow have been established in many areas from oriented structures observed on outcrops. As yet, however, there is a severe lack of descriptions in the literature of the over-all shape of individual turbidites. It is hoped this situation will soon be remedied.

Because turbidites are potential oil and gas reservoirs, the importance of predicting their shape and size in the prospecting or early field development stage is obvious. Internal distribution of porosity and permeability in turbidites also appears to have some peculiarities of importance. In relatively deep water of subsiding basins, deposi-

tion of sands by turbidity currents is now believed to be the rule rather than the exception. This dumping is in contrast to the winnowing and re-working which takes place on stable shelves. Most turbidites have the property of displaying their directions of transport by virtue of their contained syngenetic, oriented structures. This is an extremely useful tool in working out paleogeography—an obvious asset in the search for oil.

It now appears that virtually all the oil sands in the prolific Los Angeles basin are turbidites; this basin has produced nearly 5 billion barrels of oil. Turbidite oil sands are common in other areas in California also. More and more clastic formations are taking their places on the list of recorded turbidites, both in and out of the oil country. Examples include sandstones of the Permian Delaware Mountain group of Texas and New Mexico (Hull, 1957), Pennsylvanian Atoka formation of Oklahoma (E. V. Winterer, personal communication), Ordovician Martinsburg and Reedsville formations of the Appalachian Mountains (McBride, 1960; Van Houten, 1954), and Devonian Portage facies in the Appalachian Mountains (McIver, 1960). The Gulf Coast Tertiary becomes progressively deeper water in character toward the Gulf (Lowman, 1949) and should contain turbidites in its deeper water facies, for such are now being deposited in the modern gulf (Bates, 1953). Caution is advised in accepting all "turbidite" labels, however, for some descriptions have ap-

---

[1] Manuscript received December 1, 1960.
[2] George H. Roth and Associates, consulting geologists.

peared in the literature which do not fit the turbidity-current concept, especially those reported from shallow water on stable shelf areas.

This paper is a brief review of turbidites with emphasis on their geometry and their significance in oil exploration. The examples were hurriedly assembled, mostly from unpublished work of others, and represent the best that could be obtained in the time available and in the present state of knowledge of the subject. The responsibility for all statements herein rests entirely with the writer.

PRINCIPLES OF TURBIDITE DEPOSITION

Turbidity currents have been recognized for years under such names as density currents and suspension currents, but it was not until 1950, when Kuenen and Migliorini demonstrated that graded bedding was a natural consequence of turbidity currents, that their importance as a mechanism for sediment transport in the sea came to be recognized. Since then, many studies have been made of marine sedimentary rocks, modern sediments on the sea floor, and deposits in experimental tanks, which marshal evidence that these currents are common now, and were common in the past. Indeed, they constitute a very important sedimentation mechanism. Under certain conditions, transport of materials by these currents is probably the dominant process in operation.

A turbidity current is opaque and muddy (by definition) because of its load of suspended material (hence, *suspension current*) which gives the current greater density than the surrounding water (hence, *density current*). The mechanism for getting the material into suspension could be heavily loaded flood waters or a submarine slump in unconsolidated sediments at the edge of the continental shelf or delta or at the head of a submarine canyon. The trigger action causing the slump could be an earthquake, unusually severe storm, excess local deposition by other sorts of currents, or submarine seeps. Once the sediments begin to slump they mix with the surrounding water, and the resulting turbid mixture, because of its increased density, can move as a unit down the slope. Thus, even in bodies of standing water, a current is born. One can create miniature turbidity currents by poking one's finger into a

water-filled footprint in muddy soil.

A turbidity current is capable of transporting rock and mineral particles in some proportion to its density, velocity, and viscosity. Velocity depends on the slope of the sea bottom for a given current. The current moves downward along the bottom. On smooth slopes it should spread widely; on irregular floors it should follow low areas, probably meandering much like streams on land. Where the balance between density and velocity of the current and texture of the bottom sediment permits, the current may erode. It is becoming increasingly evident that the great submarine canyons, thousands of feet deep and tens of miles long, have been cut by these turbidity currents. All turbidity currents must eventually lose velocity as the slope lessens. Where this occurs, deposition begins. Deposition is generally accompanied by sorting as the heavier particles settle out first along with entrapped portions of the finer material (matrix?), resulting in a graded bed or lamina. Sorting may also take place while the current is moving, with the heavier particles advancing toward the toe of the current. As the material within the current is generally a mixture of many particle sizes, the resulting turbidite will, on the whole, be poorly sorted; yet, because of the particle selection during deposition, sorting at any particular level within the deposit will differ from the average for the turbidite, and sorting will improve upward in the deposit because of the decrease in coarse particles.

Turbidites differ from most other deposits in that they are nearly instantaneous. These sandstones do not accumulate at a steady rate, but a single deposit, possibly tens of feet thick, may require only a few hours or a few days, and it may be followed by a thin pelagic deposit representing many years or centuries of accumulation. A single turbidite, therefore, will not grade laterally into an equivalent thickness of lutite but will sharply disappear into a bedding plane. In areas of rapid subsidence and active upland erosion, turbidites may follow one another rather rapidly with intervening pelagic deposits thin or missing. In at least one area, repetition of thin turbidites has been related to annual cycles of organic growth and rainfall (Riveroll and Jones, 1954).

Turbidites should thus be expected in geosynclinal sequences in fairly deep to very deep water, on or below slopes. They would not be expected in shallow water within wave-base depths, for here they would be destroyed by reworking and winnowing if, indeed, there were sufficient slope for turbidity currents to get started.

In recent years the quickening pace of oceanographic research has resulted in a growing body of evidence which undeniably demonstrates the great importance of turbidity currents as a mechanism for erosion and transportation on the modern sea floor. Especially active along these lines have been the Lamont Geological Observatory, U. S. Naval Electronic Laboratory, Scripps Institution of Oceanography, and Allan Hancock Foundation. Their findings include sand deposits with shallow-water organic remains on the abyssal plains hundreds of miles from land; submarine canyons with well developed sandy deltas in water more than 10,000 feet deep; sudden evacuation of large volumes of sediment from heads of submarine canyons; breakage of a series of submarine cables by something that was travelling at the rate of 50 knots along a front 210 miles wide, and left a deposit of silt behind; natural levees bordering some of the more prominent channels; and topographic control of deposition wherein nearshore basins trap the sediment and fill up before more distant low places receive sediment. These phenomena are no longer surprising to the oceanographers, but are now expected as a matter of course. Heezen (1959) has very ably summarized early and recent thinking on deep-sea sedimentation.

RECOGNITION FEATURES

Because turbidites are deposited in a manner markedly different from the traditionally accepted sedimentary mechanisms, they have characteristic properties and structures which might be considered their "signature." These structures and textures are not all unique to turbidites, but, considered as a group and in the light of the rapidly accumulating experimental and sea-floor observations, they now appear to fit into a logical and well documented genetic pattern. These features have been discussed by Kuenen (1953), Kuenen and Carozzi (1953), and Crowell (1955), and there is now a large body of literature on the genesis of individual features, features in certain stratigraphic units, and features in modern sediments. European geologists have been especially active in this respect. There is no need to describe these features here again in detail, but they must be discussed briefly inasmuch as they form part of the geometry of these sandstone bodies.

GENERAL FEATURES

Perhaps most significant of all (though difficult to prove in older formations) is the presence of these sands in deep water. The great abyssal plains of the oceans are now thought to be built by turbidity currents (Heezen, Tharp, and Ewing, 1959). Miocene and Pliocene sandy and pebbly turbidites in California are interbedded with shales containing Foraminifera indicating water depths of 2,000–8,000 feet. Multiple repetitions of sands and deep-water shales rule out a shallow-water origin for the sands. Care must be taken to avoid associating coarse sands and conglomerates, crossbedding, scour and fill, and even shallow-water fossils, wholly with a shallow-water environment, for these features may be imparted by turbidity currents in deep water. Turbidites need not be confined to deep water, however.

Poor sorting is an expected feature in turbidites. This, of course, depends upon the type of material included in the original slump which started the current. It would be an unusual situation that would provide well-sorted sand without a significant proportion of clay and silt becoming involved in the slumping or transportation process. A problem arises in that sorting of sands is commonly expressed in terms of Trask's sorting coefficient which compares the grain sizes of the first and third quartiles of the cumulative grain-size curve. These quartiles do not take into account the tails which may consist of coarse sand grains or pebbles at one end and clay at the other; these portions constitute half of the rock. Thus, a wellsorted sand in the Trask system may be classified as poorly sorted by some other system such as that of Payne's (1942, p. 1707). The large percentage of silt and clay in these sands has caused them to be termed wackes and graywackes by most workers. This poor sorting strongly suggests rapid dumping and lack of winnowing and reworking—a suggestion generally supported by grain angularity and high feldspar content. The turbidity-current origin of any well-sorted sand should be viewed with suspicion. Sorting is directly related to porosity, and turbidites must therefore have less original porosity than shallow-water sands which have had the benefit of winnowing. That this decrease in porosity is not a serious handicap to the oil industry is illustrated by the Los Angeles basin.

Repetition of similar sandstone beds separated by normal pelagic finer grained clastics is common. It stands to reason that once the conditions on the sea floor become favorable to the development of turbidity currents, these conditions will persist long enough to permit many repetitions. A slight change in conditions may be reflected in the structural and textural characteristics of the turbidites. These conditions include sea-bottom slope, sea-bottom cohesiveness, type of material in the source area, and type and frequency of triggering mechanisms. The quantitative relating of certain structures and textures to certain conditions must await additional fact gathering.

A turbidity current is clearly not an ideal habitat for any kind of life. Fossils in growth position are lacking in turbidites. Fossils are rare, and those found can commonly be shown to have been transported. The writer spent 110 days in the field examining outcrops of Miocene deep-water turbidites near Los Angeles and found only one megafossil, a thoroughly worn oyster shell; Foraminifera were abundant in the enclosing shales, but those in the sands were clearly transported, as evidenced by mixed faunas and their presence in cross-bedding (Sullwold, 1960a, p. 441). Modern turbidites containing shallow-water faunas have been found at a depth of nearly 12,000 feet, 120 miles from the nearest appropriate biotope (Phleger, 1951). Others have been found with green grass at a depth of 4,500 feet (Heezen, 1956). Algae, requiring life-giving sunlight, have been found in turbidites at 26,000 feet in the bottom of the Puerto Rico Trough (Ericson, Ewing, and Heezen, 1952). Turbidity currents can not only transport living or recently deceased creatures from their natural habitat into deeper waters in wholesale lots, but can also erode fossil-bearing rocks, thus introducing older, perhaps extinct, forms into the fauna. The implications of this

willy-nilly mixing of ages and environments are enormous in working out fossil time ranges and paleoecology, and, especially, in detailed paleontologic correlation. The writer has the utmost admiration for the ability of those paleontologists who have endeavored, and largely succeeded, in bringing some sort of order out of this chaos.

INTERNAL FEATURES

Graded bedding is the most characteristic internal feature of turbidites. A single graded bed must have been deposited as a single depositional unit reflecting a definite beginning and end of the supply or the transporting agent. A series of graded beds could not have been deposited by normal oceanic currents with a continuous supply of material. Turbidity currents are the most logical explanation for the great bulk of marine graded beds. Reservoir characteristics clearly differ vertically within a single graded bed. A single core analysis would not necessarily be representative, and might be the source of considerable error in calculating fluid volume, even in a relatively thin bed. Studies of the relationship of porosity and permeability to graded bedding in turbidites are needed.

Other internal features include convolute bedding, current bedding, shale inclusions, charcoal fragments, oriented grains, and imbricate arrangement of larger clasts. These are less significant than graded bedding in reservoir calculations, but have varying degrees of utility in recognition of turbidites and orientation of flow directions.

EXTERNAL FEATURES

The most characteristic external feature of a turbidite is the sharp basal contact—a natural consequence of dumping an exotic mixture of sediment upon the sea floor. No gradational contact can be envisioned here, nor do such exist, in the writer's experience. Conversely, the upper contact is commonly seen to grade imperceptably upward into normal pelagic lutites—a consequence of the similarity in grain size of last-deposited particles in the turbidite and later lutites. This gradation may be locally destroyed by erosion by subsequent currents, hence it is not as universal as the sharp base.

Sole markings of various sorts are very distinctive and have received much attention partly because of speculation as to their origin and partly because of their great usefulness as current indicators. Birkenmajer (1959) has classified bed surface structures into 18 categories and provided an excellent bibliography. Sole markings or hicroglyphs may be subdivided into a few major groups. (1) One group is a result of load deformation wherein the sudden application of the load (turbidite) on a muddy bottom causes the mud to deform. The deformation takes the form of load folds in the mud, load casts on the under surface of the sandstone, load pockets within the sandstone, and load waves of mud squirting up into the sandstone (Sullwold, 1959 and 1960b). (2) Another group consists of tracks or grooves cut into the sea floor by objects dragged by the current. They were observed, illustrated, and correctly interpreted by James Hall (1843) nearly 120 years ago, and more recently dealt with in detail by Dzulynski and Radomski (1955). (3) A third group, also erosional, includes those markings made by the current itself in the sea floor without the benefit of "tools," including flow markings and flutes.

Where these markings are seen on the base of the sandstone, as is commonly the case, rather than at the top of the underlying shale, they are raised casts of the original impression of the mud on the sea floor, and the word "cast" is consequently combined with the structural term, as *flute-cast* and *load-cast*. These markings are nearly all linear features providing excellent bearings of the current, and some, by pointing upstream, even provide absolute direction of the current. A careful examination of bedding planes in cores could provide valuable data on current directions when related to known attitude of the strata.

The top surface of a turbidite is generally planar; the upper part of the turbidite is commonly laminated and gradational with the overlying lutites. Perhaps the most significant departure from this situation is the common presence of ripple marks at the top. As the sand is deposited from a current traveling along the sea floor, it is not surprising that occasionally the relationship of grain size to velocity is such that ripple marks (and associated current bedding) form. These too, are excellent current indicators. In the

California Tertiary ripple marks are more common at the top of thin beds, from a fraction of an inch to 6 inches, than in thicker beds.

All these relatively small elements in the geometry of turbidite bodies have their significance in the science of petroleum geology. They serve as valuable tools, first, in recognizing turbidites as such, then in determining the direction of travel. With this latter knowledge paleogeography can be worked out in much finer detail—an obvious asset in the search for oil. Furthermore, much promise is offered that by these studies, the trend of a turbidite sand body may be predicted from examination of cores in a single well. It is too soon yet, but studies are under way with this goal in mind. These structures are also useful in differentiating top and bottom of beds in complex structural areas such as in California. Once the current directions have been established in a given area at a given horizon, a core in a wildcat well could be oriented, provided it contained a recognizable, oriented feature. Prior to that, the data for fixing current directions would have to come from outcrops or from cores where the attitude was known by structural position or by oriented dips.

Calculation of volume of recoverable oil in place in turbidites should be tempered with considerations of the sharp basal and lateral contacts of each turbidite, the change in porosity and permeability within each turbidite corresponding to the grading, and the probable presence of thin impermeable pelagic lutite layers between the individual turbidites.

GROSS GEOMETRY

The geometry of special interest in this volume is the over-all shape of sandstone bodies. Unfortunately, published detailed descriptions of complete turbidites are lacking. However, sufficient observations have been made in the field and on the sea floor to permit reasonable statements as to the expected shapes of complete ancient turbidites—a necessary and significant approach in cases where the ancient ones are oil bearing.

The typical situation on the modern sea floor is for a turbidity current to flow down a submarine canyon, debouch over a fan at the mouth of the canyon, then spread out over a wide front on the basin floor, seeking out the lowest path, avoiding the topographic highs. Variations of this pattern may be caused by levee topping, deflections due to oceanic currents, centrifugal forces at curves, or momentum at slope reversals. The current may deposit all or part of its load at any part of its course, depending on the density, velocity, and load of the current. Several tentative conclusions may be reached from this hypothetical situation.

It is clear that a single turbidite will _not_ have a gradual change-of-facies relationship, laterally, with the enclosing normal pelagic sediments (Fig. 1). It will either fill a low area on the sea floor whose mud surface forms a continuous bed enclosing the bottom and sides of the turbidite, or it will fill a channel eroded into the sea floor by the forward part of the same current or by a previous current. Upward, a turbidite may grade into finer and finer clastics eventually inseparable from subsequent pelagic sediments. These single turbidites may be from less than an inch to many feet in thickness. The writer has seen single sedimentation units up to 10 feet thick in the California Tertiary, and turbidites more than 20 feet thick are reported from 15,000 feet in the Atlantic (Ericson and others, 1951).

As it is the habit of turbidites to follow one another in relatively rapid succession once the proper conditions prevail, outcrop sections com-

Fig. 1.—Bundle of sandstone turbidites (light) overlain by a shale unit (dark). Prominent sandstone in the right center of the shale unit is a single graded bed containing shale flakes. At featheredge it passes between the enclosing shale laminae without any evident erosion; no facies change is present. Individual turbidites in sandstone bundle were either deposited in sufficiently rapid succession to preclude deposition of normal pelagic shales, or any lutites that were deposited were stripped off by intervening turbidity currents. Shale slabs may be seen in sandstone near shovel handle. Shovel (lower right) is 3½ feet long. Horizontal streaks are man made. Modelo formation, upper Miocene, Santa Monica Mountains, Los Angeles, California.

monly show a series or "bundle" of sandstone beds with the intervening normal pelagic strata (shale) very thin, or missing altogether (Fig. 1). Where not well exposed, the bundle of sands would appear gradually to shale out at the edges. This would be particularly true in the subsurface where well spacing would preclude a detailed study of the edges of the deposit. Too often such a boundary is interpreted as a single body of sand with either a sharp buttressing or truncated edge, or a gradational facies change to shale. The accompanying diagram (Fig. 2) shows the stratigraphic relationships in a hypothetical bundle of turbidites, and indicates the sort of correlation problems that might exist in an oil field. Obviously a clear understanding of turbidity-current deposition would greatly facilitate geologic interpretations of the geometry of turbidite bodies; treating them as offshore bars, blanket sands, deltas, or some other more traditional and familiar form would lead to erroneous results.

It is axiomatic that the bottom surface of any sedimentary body must coincide with the surface on which it is deposited. As turbidites fill hollows in the sea floor, their bottoms should ideally be concave upward. This shape will not always be discernable in widespread deposits because of the scale. Concavity may be absent in the case of deposition on an abyssal plain, engulfment of a low eminence on the sea floor, or levee topping.

The areal dimensions and shape in plan view of turbidites are perhaps of greatest interest to the petroleum industry. Unfortunately, it is here that information is weakest. The writer knows of no isopach map of an entire turbidite. Portions of turbidite deposits no doubt are included in the many isopach maps of sandstone stratigraphic traps in the literature, though positive accompanying evidence of their turbidity-current origin is lacking. Our understanding of the mode of deposition of turbidites gives us a basis upon which to postulate several regimes in which they might be deposited and in which characteristic geometry might develop. A tripartite grouping would include *channel* or submarine canyon deposits, *fan* deposits, and *basin-floor* deposits.

Channel deposits would be linear with bottoms concave upward in cross section and truncating the subjacent strata. They would contain the coarsest grains of the three groups; gravel is commonly found in modern canyon bottoms. The gross size of a channel deposit would be relatively limited compared to a basin floor deposit. Directional features within the deposit would indicate current direction parallel to the axis of the channel. A single core may be of great significance in determining the trend of suspected channel deposits by means of lithologic studies and careful observation of syngenetic structures. Slump structures might be associated with channel deposits because of the relatively steep adjacent slopes. During erosion of the channel or canyon its sedimentary fill would be local and temporary, but a change in conditions—such as a rise in sea level—would permit relatively permanent deposition.

In a *submarine fan* individual turbidites might be linear because of deposition in minor channels, or might be blanketlike. The relative volume of the turbidite compared to the microtopography of the fan surface would be important in this respect. Slopes and shapes of fan surfaces might vary widely and thus control the shapes of the sands therein. A radial pattern would be displayed by the current-direction indicators in the deposit as a whole, with the currents radiating from a point source—the mouth of the submarine canyon. Here, too, slumps might develop should the slopes be sufficiently steep. Average grain size would be smaller than in the canyons, but gravels still might exist. The entire deposit would be equidimensional rather than linear, conforming to the topography below and convex upward in cross section, thinning abruptly in the up-current direction and gradually in the down-current direction, perhaps coalescing with other fans, especially laterally. If a fan were relatively isolated rather than being sandwiched between other bundles of turbidites, it would constitute a bulge on the sea floor partly enclosed by much thinner normal pelagic strata, thus potentially forming the core of a compaction fold.

Menard (1960) has supplied us with a pictorial representation of a group of fans off the coast of central California (Fig. 3). Here a group of three submarine canyons heading near shore have huge fans at their mouths in water 9,000–12,000 feet

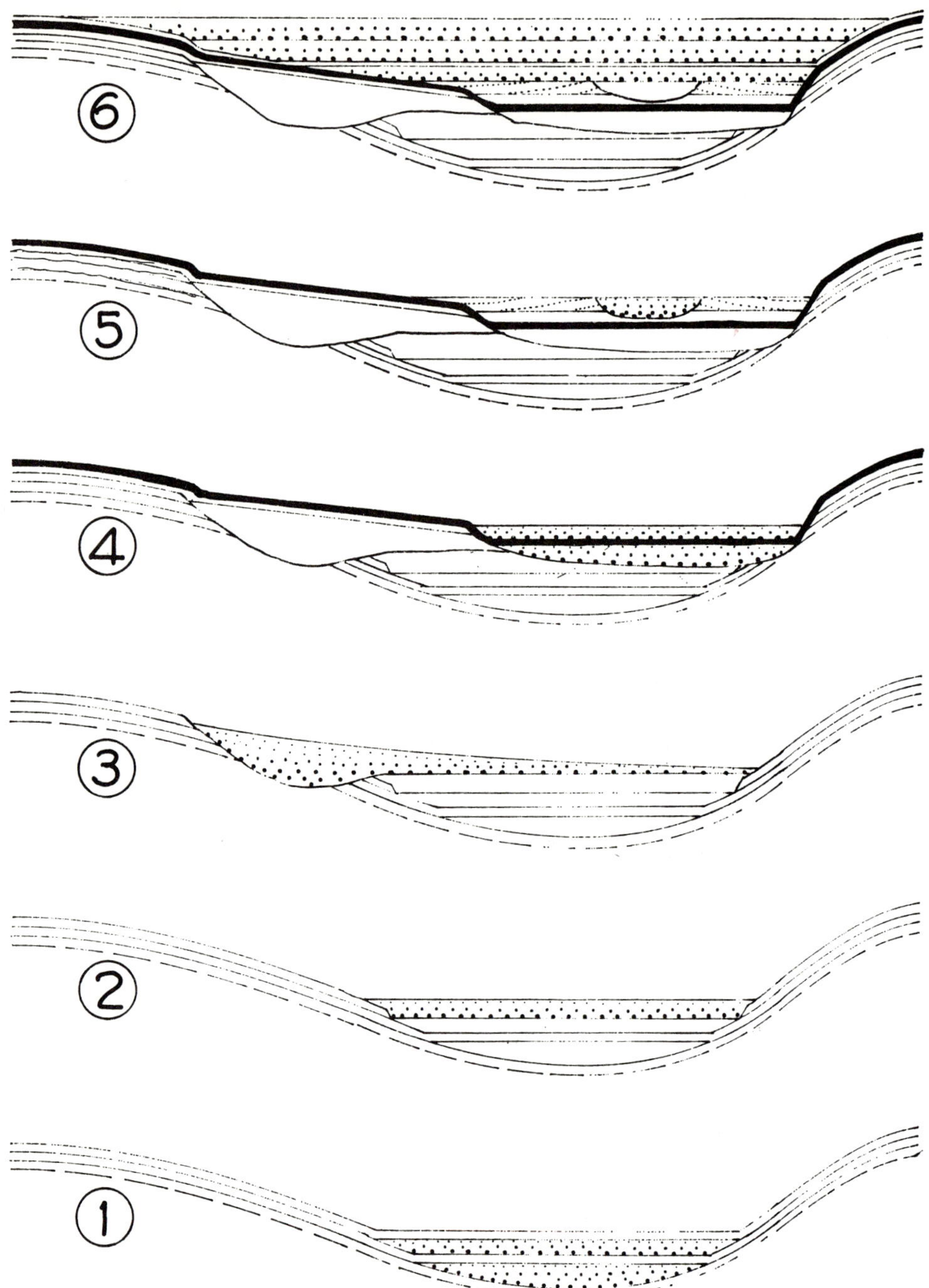

FIG. 2.—Hypothetical sequence in deposition of bundle of turbidites. No scale intended, but vertical is exaggerated probably by between 10 and 1,000. Assume view is in down-current direction and read the diagram from the bottom up. (1) A natural hollow, perhaps produced by tectonics or differential compaction (could also be erosional), receives two turbidites, each followed by a layer of pelagic lutite; no erosion involved. (2) The top pelagic layer is eroded off the floor by a turbidity current, and the third turbidite is deposited on the second. (3) A channel is eroded along the left side of the basin and filled to overflowing by a current under some sort of lateral stress permitting an asymmetric deposit. (4) The sloping surface is covered with pelagic sediment. A new channel is eroded in the new trough created by the sloping deposit. This channel is filled with two turbidites and an intervening pelagic layer (black). (5) A new channel is cut, and conditions are such that natural levees form, permitting the finer material at the top of the currents in the channel to spill over and dump their fine-grained loads (sloping dots) to the side of the main channel. Main channel then fills with a single deposit. (6) Three more graded beds are deposited, the first followed by a pelagic layer and the second followed by the third, without time for a pelagic layer; no erosion involved.

deep. The areal shape and bottom contour of these fans are clearly influenced by the pre-existing topography of the sea floor; for some ridges and fault scarps are buried, and others act as dams. The Monterey fan has a radius of about 200 miles. The two larger fans appear to have engulfed hills 1,000 feet high, and extrapolation of topography beneath the fans suggests a thickness at their apices of more than 4,000 feet.

Basin-floor turbidites differ from the first two types in being blanketlike, thinner, more extensive, and finer grained. Silt is the most common grain size on modern abyssal plains. Current-direction indicators would be more consistent than those on fans, and less consistent than those in channels. Variations in current directions would ensue from superposition of turbidites coming from different sides of the basin and also from meanderings of currents owing to slight tectonic tilts or bulges on the extensive floor. Relatively confined basins might act much like large canyons with the currents flowing down the axis of the trough. The possible huge areal extent of a turbidite on an abyssal plain is suggested by the turbidity current triggered by the Grand Banks earthquake of 1929, which travelled over an area in excess of 80,000 square miles and left a bed of graded silt one meter thick containing shallow-water microfossils.

Gorsline and Emery (1959) have described the turbidites on the sea floor off Los Angeles in the partially filled San Pedro and Santa Monica basins and their associated canyons and fans. Most of the basin-floor sediment is green clayey silt (in the top few feet studied), but contains interbeds of fine sand. Sand becomes coarser, thicker bedded, and more abundant toward the fans. Gravel is present on the fans and in the canyons. The basin

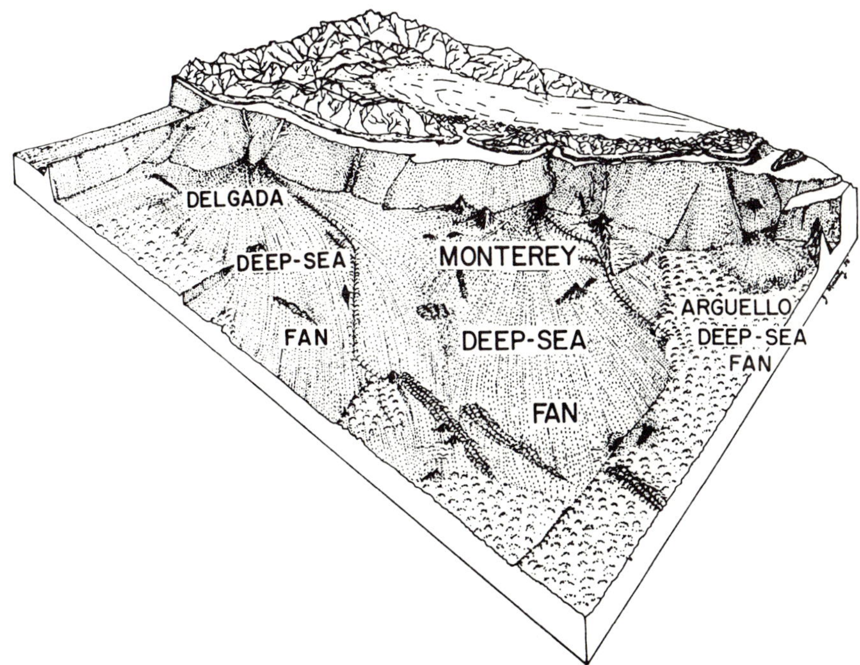

Fig. 3.—Deep-sea fans at mouths of Delgada, Monterey, and Arguello submarine canyons off coast of central California. The scarps from north to south (left to right) reflect the Mendocino, Pioneer, and Murray fracture zones. The Monterey fan has a radius of about 200 miles, and the sea floor lies at maximum depths of 14,000 feet. Sea-floor topography initially influences distribution, and is later engulfed by the vast quantities of sediment carried by turbidity currents. Canyons which do not head near shore and thus cannot trap sediment carried by long-shore currents on the shelf, have very small fans. From Menard, Geol. Soc. America Bull., Vol. 71, No. 8, 1960.

floors lie at a depth of about 3,000 feet and slope less than 0°11′. The areal shape of the entire sequence may be seen on the map (Fig. 4). Insufficient cores have been taken to map individual turbidites, nor is the total thickness known. The sediment is clearly filling a structural trough on the sea floor much like the already filled Los Angeles and Ventura basins on land. The shape of the basins is controlled in part by fault scarps.

EXAMPLES OF TURBIDITE GEOMETRY

The writer has not seen in the literature any isopach maps of entire sandstone bodies identified as turbidites. There is a great need for publication of studies of this sort. In lieu of ideal examples of complete single turbidites, the present paper includes a few incomplete or generalized examples of bundles of turbidites. These will serve the purpose of demonstrating some of the grosser aspects of size and shape of turbidity-current deposits. These illustrations are all from California, where the writer feels a certain amount of familiarity with the geology.

TARZANA FAN

The Tarzana fan (Sullwold, 1960a) is a clear-cut case of turbidity-current deposition. It is of upper Miocene age and is exposed on the north flank of the Santa Monica Mountains, in Southern California. Its thickness ranges from zero to 4,000 feet in a distance of 16 miles, with sandstone comprising about half, and siliceous shale and siltstone the other half. Foraminifera in the shales attest to water depths of about 3,000 feet, and the sands contain abundant syngenetic structures and textures characteristic of turbidites—such as graded bedding, poor sorting, convolute bedding, load deformation, slump structures, shale slabs, cross-bedding, ripple marks, sharp bottoms, gradational tops, charcoal fragments, transported faunas, and lack of indigenous faunas. A northern source is suggested by the radiation of abundant current indicators from a point source to the north; this is supported by the cross sectional lens shape of the deposit in outcrop, and by a tentative identification of its provenance area to the north. Figure 5 is a geologic map of the fan, redrawn to emphasize the shape and distribution of the mappable sandstone bodies or turbidite

bundles. As the rocks here are generally dipping about 20° northward, a down-dip view of the map provides one with a simulated cross section more detailed and accurate than could be constructed by conventional means. One finds it difficult to detect a pattern here which could be applied generally to other subsurface fans. Perhaps the very lack of pattern and the random distribution of sand offer hope that hitherto unsuspected sands in similar environments, elsewhere, may suddenly appear in the section at a favorable structural position.

STEVENS SAND, SAN JOAQUIN VALLEY

The Stevens sand, of late Miocene age, underlies an area 30 by 50 miles in the southern San Joaquin Valley, California, where it is a very important producer of oil in more than 20 oil fields. Its maximum thickness is more than 3,000 feet, of which two-thirds is sand. It is a series of turbidites alternating with normal marine organic shales whose fauna indicates a water depth of approximately 1,000 feet.[3] The deep-water Stevens grades eastward to shallow-water Santa Margarita sand (mostly non-turbid), and the latter to the nonmarine Chanac formation. Sand dominates in the central portion of the Stevens, but shale gradually increases to the north, west, and south until the sand disappears altogether. The sand is described as poorly sorted, silty to coarse to pebbly, locally containing siltstone pebbles, arkosic, subangular, unfossiliferous, lenticular, with silty to clayey matrix, and with rapid changes in permeability both vertically and horizontally.

Figure 6 is an isopach map of net sand in the combined Stevens-Santa Margarita facies, and the cross section (Fig. 7) further clarifies an understanding of its distribution and geometry. The writer is indebted to Ted L. Bear, consulting geologist, and Rollin Eckis, Richfield Oil Corporation, for access to unpublished manuscripts on the Stevens sand. Natland (1957, Pl. 6) has implied a turbidity current origin, and Eckis (unpublished manuscript) concluded as early as 1940 that deposition must have been by currents moving along the ocean floor (but without using the

---

[3] Water depth supplied by Manley L. Natland, Richfield Oil Corporation, Los Angeles.

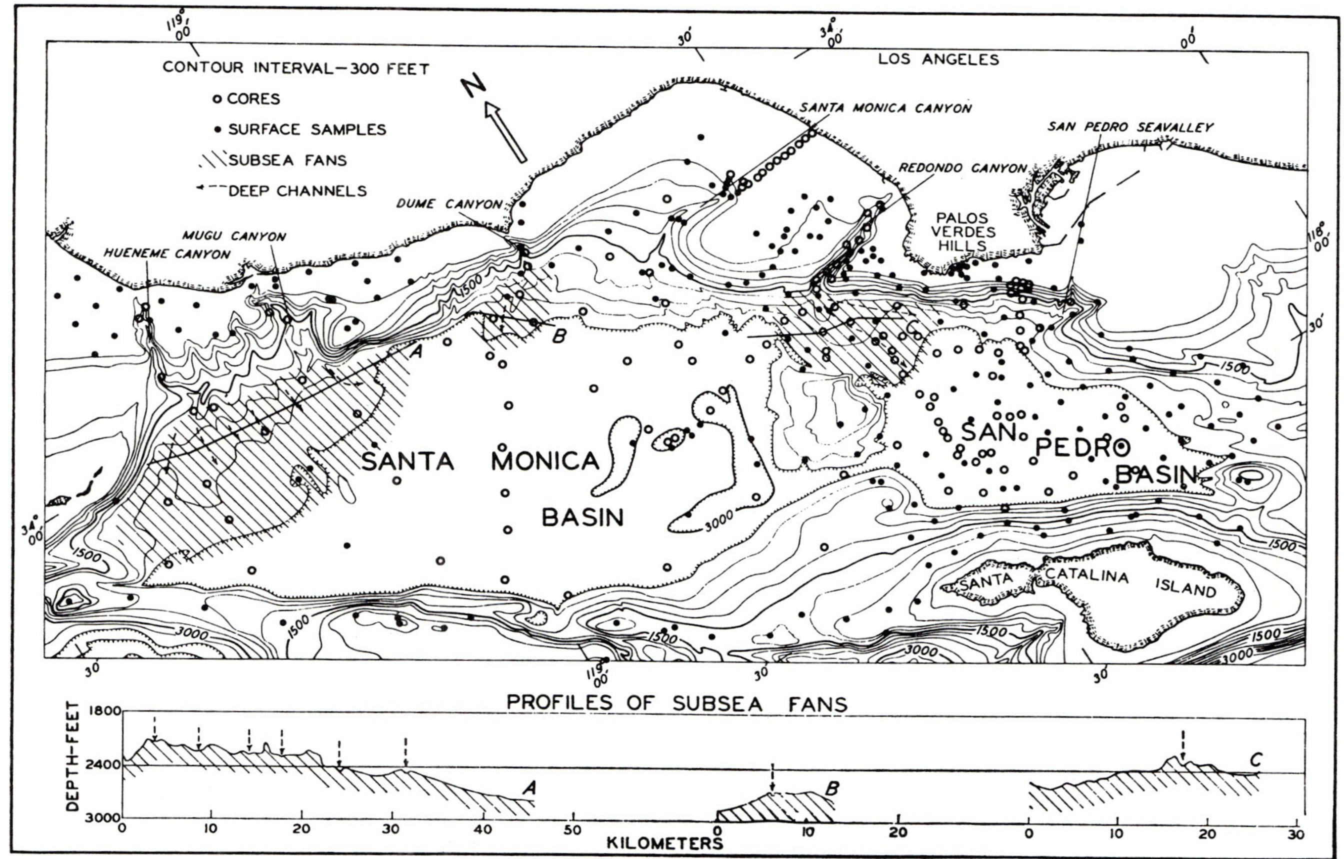

Fig. 4.—San Pedro and Santa Monica basins off coast of southern California. Basin floors are very flat and are underlain by sediment deposited in part by turbidity currents. The fans are reflected in topographic expression and in their coarser and thicker sands. Mineralogy of sands is very similar to sands being carried along shore on the shelf. From Gorsline and Emery, Geol. Soc. America Bull., Vol. 70, No. 3, 1959.

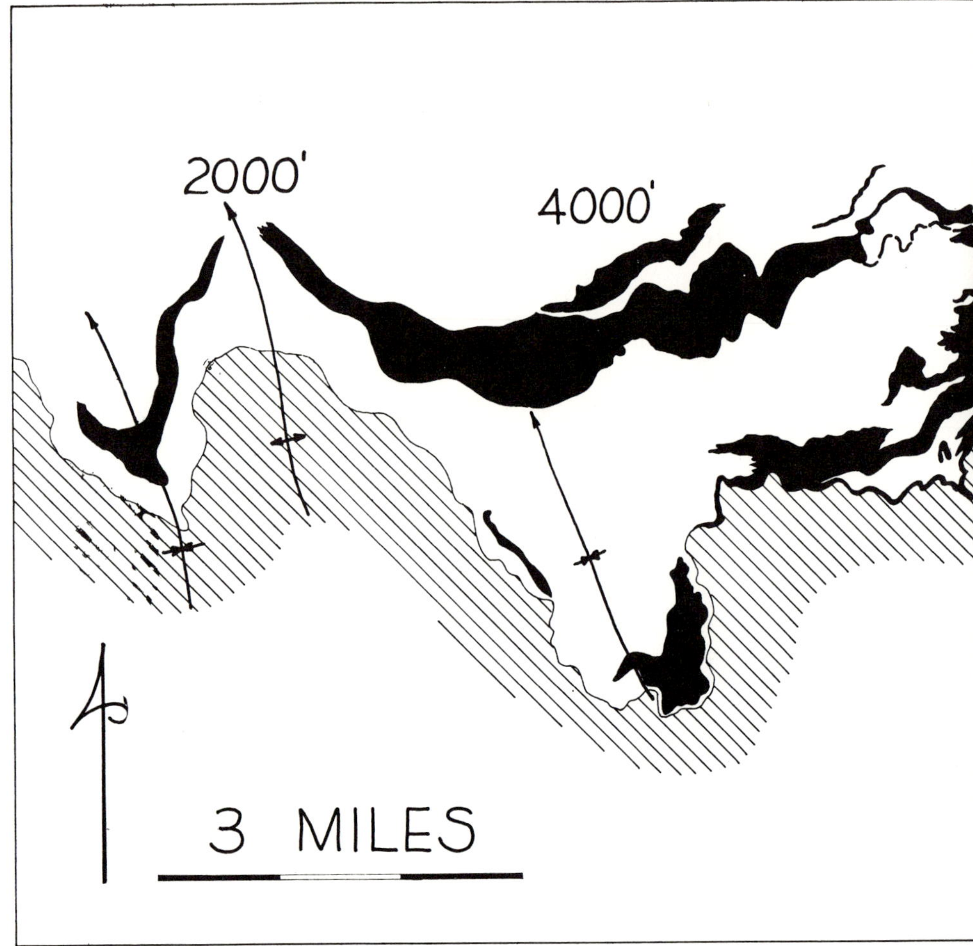

Fig. 5.—Simplified geologic map of Tarzana fan, bathyal upper Miocene, Los Angeles County, California. Mappable sandstone bodies (bundles of turbidites) are indicated in solid black. Older subjacent rocks are diagonally lined. Much of white area between sandstone beds is also sandstone, but finely laminated and interbedded with shale. Figures at top indicate total measured thickness of fan sediment at various places. Rocks are generally dipping about 20° north, so down-dip view simulates a cross section, providing some insight as to shape and size of turbidite bundles. Some minor irregularities are caused by topography, and some by folding, and one important fault interrupts the continuity, probably repeating the eastern portion of the fan. Base of diatomite marks approximate top of fan to east; top to west is less distinct. Adapted from Sullwold, 1960a.

magic words, turbidity current). The writer's only first-hand knowledge of the formation is an inspection of a single core that contained many typical syngenetic structures. No published work has been devoted to the Stevens sand as a turbidite, taking full advantage of the hitherto unused clues awaiting the worker who is willing to go a step farther than routine subsurface work. The evidence is sufficiently strong to assume that the Stevens sand is a bundle of turbidites, and its gross geometry is here presented as an example.

LOWER PLIOCENE, LOS ANGELES BASIN

Few areas in the world have produced so much oil from so small an area as the Los Angeles basin. Cumulative production is 4.9 billion bar-

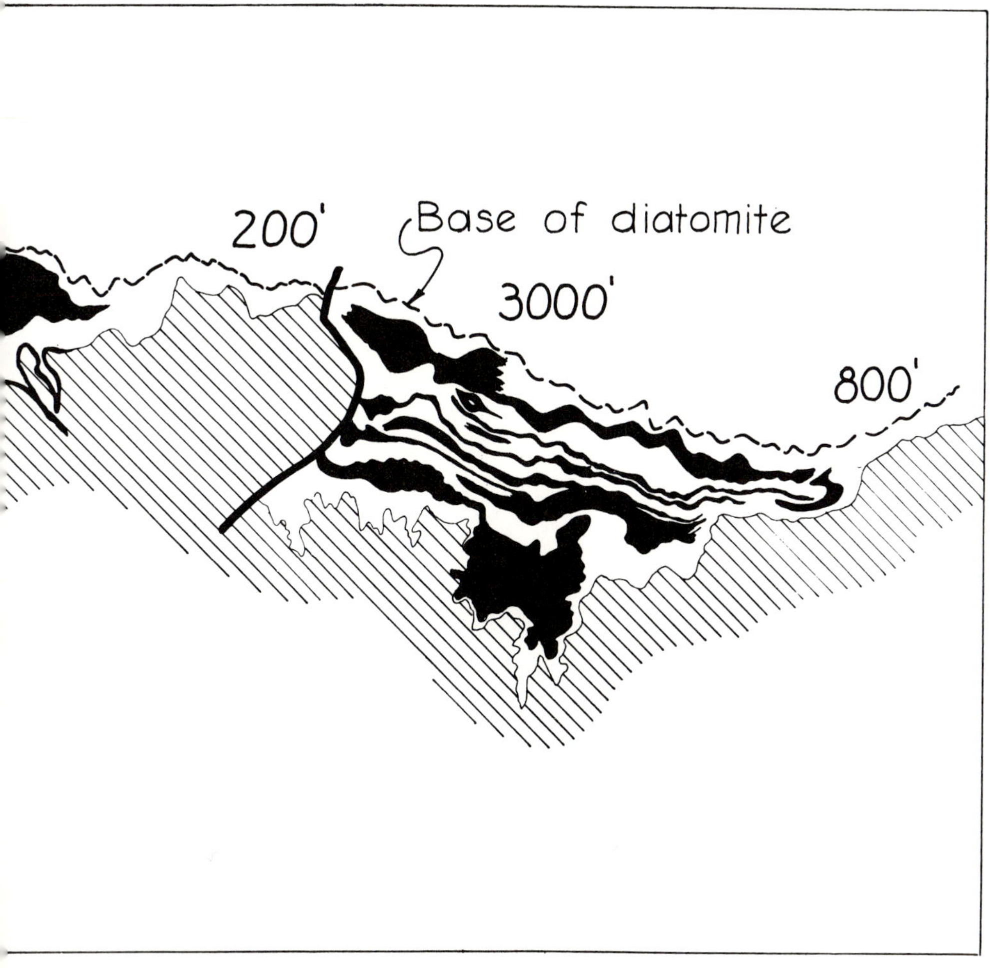

rels of oil from about 63 square miles of productive area, and nearly all of this oil comes from upper Miocene and lower Pliocene turbidites. About 20,000 feet of strata accumulated in this basin from middle Miocene to Recent time in a more or less continuous cycle of clastic sedimentation. Its general geology has been described by Wissler (1943), Driver (1948), and Barbat (1958), and countless papers have been written describing individual oil fields. Still, there are no published data on the detailed geometry of the turbidites.

More than half the oil produced from this area has come from the lower Pliocene Repetto "formation." This part of the section has recently been studied by Slosson (1958) and Conrey (1959) but the complete work has not yet been published. Repetto strata range from zero to 5,000 feet in thickness, and approximately half is sandstone. These sandstones were deposited in water depths of 6,000–8,000 feet, and are arkosic, silty to conglomeratic, laminated to massive, with graded bedding, sharp basal contacts, and local imbrication. It is generally agreed by all recent students of these deposits that the sandstones are turbidites. The gross geometry of the lower Pliocene Repetto sandstones is shown on the isopach map (Fig. 8) taken from Conrey. This map clearly shows a northeast source with a spread and

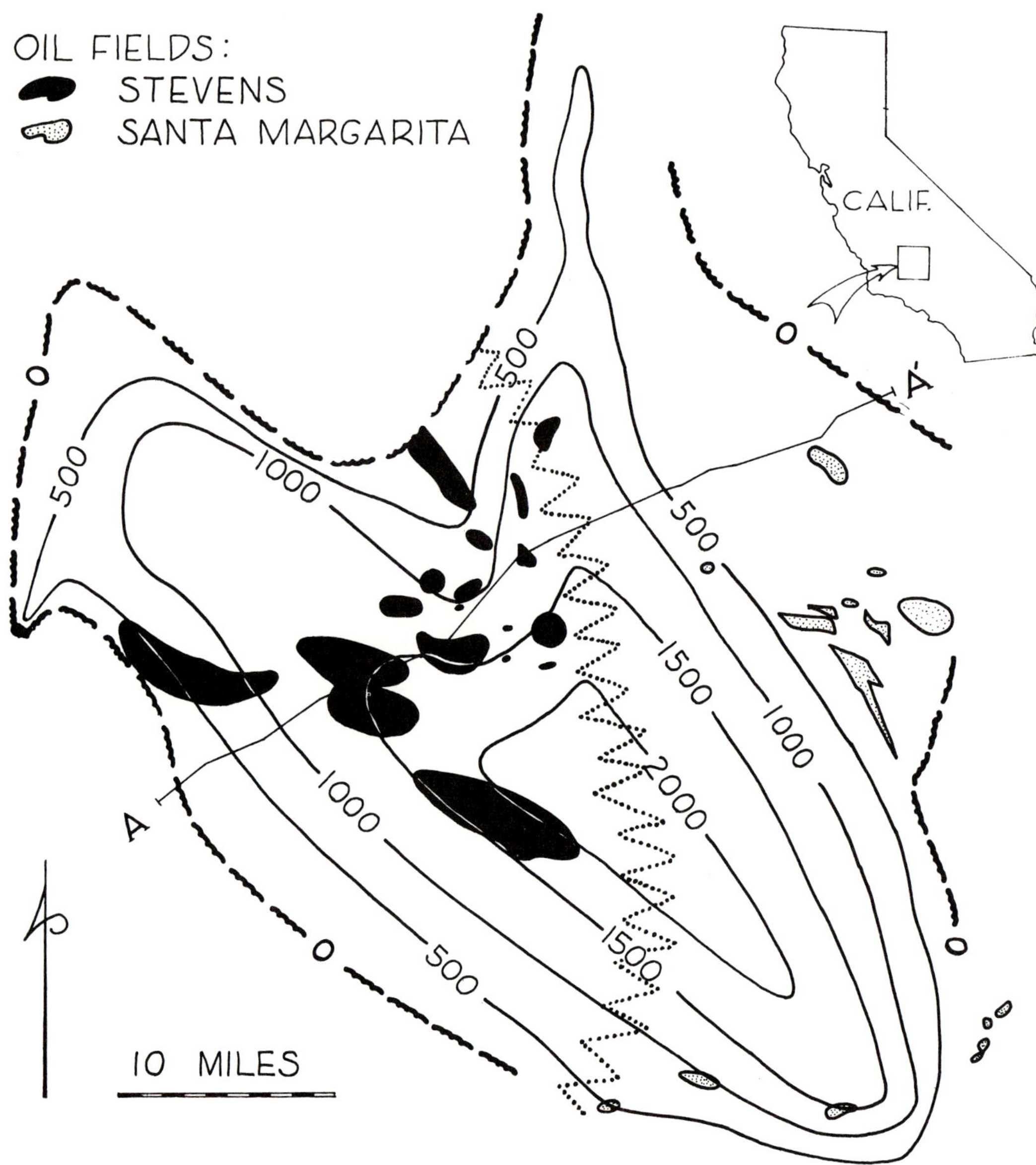

Fig. 6.—Net-sand isopachs, Stevens-Santa Margarita sand facies of the upper Miocene, San Joaquin Valley, California. Portion west of dotted line is the Stevens sand, a deep water turbidite facies; portion to east is shallow-water Santa Margarita sand. Nature of the facies change between the Stevens and Santa Margarita is not yet documented. See cross section A-A' (Fig. 7) for further elucidation. Adapted from unpublished reports of Rollin Eckis and Ted L. Bear.

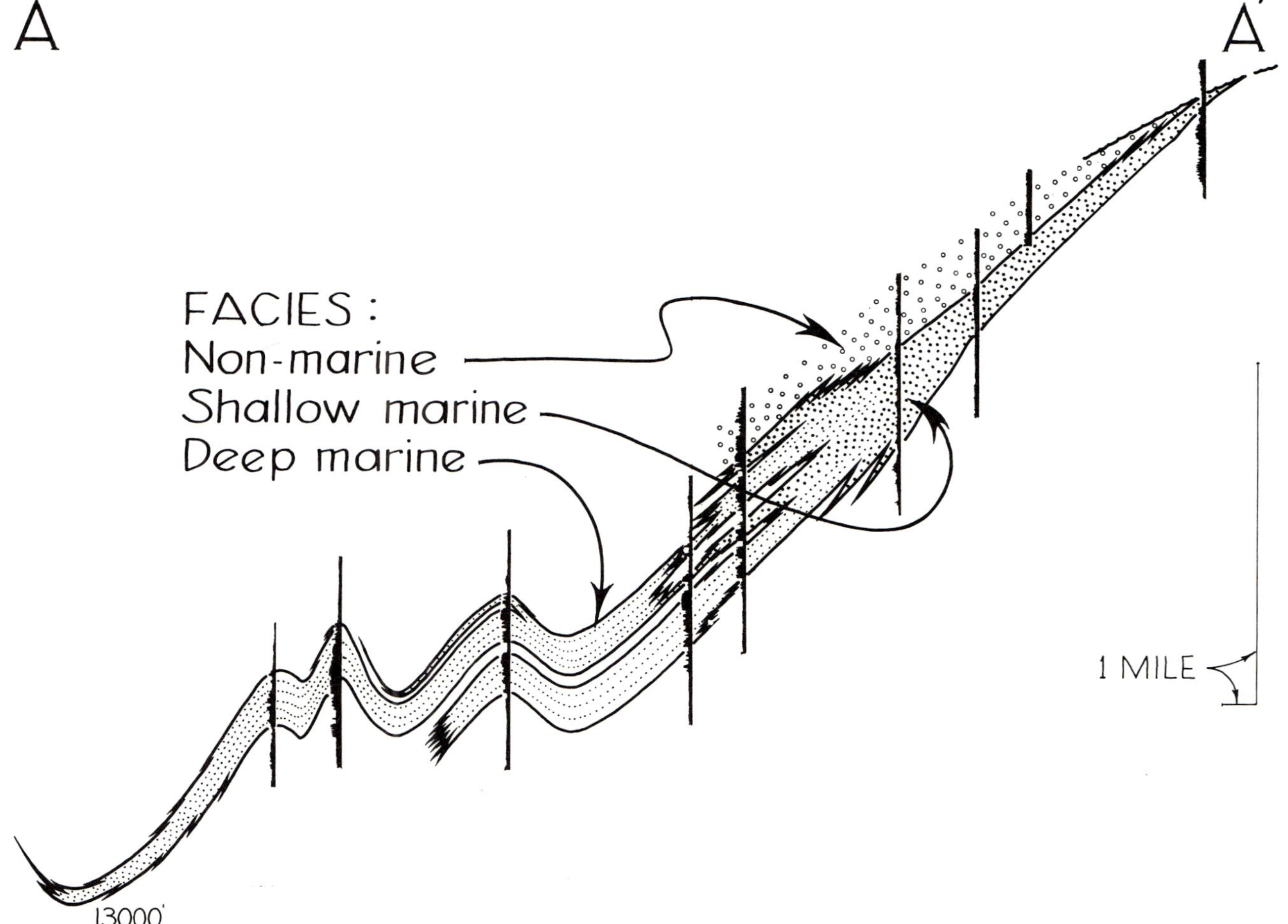

Fig. 7.—Section *A-A'*, Stevens and Santa Margarita sands, upper Miocene, San Joaquin Valley, California. The Stevens sand is the deep-water facies (fine parallel rows of dots), the Santa Margarita, the shallow-water facies (coarse dots), and the Chanac formation, the nonmarine facies (open circles). Self-potential curves indicate the sands by excursions to the left on the electric logs. Adapted from a cross section in an unpublished manuscript of Ted L. Bear. Trace is line *A-A'*, Fig. 6.

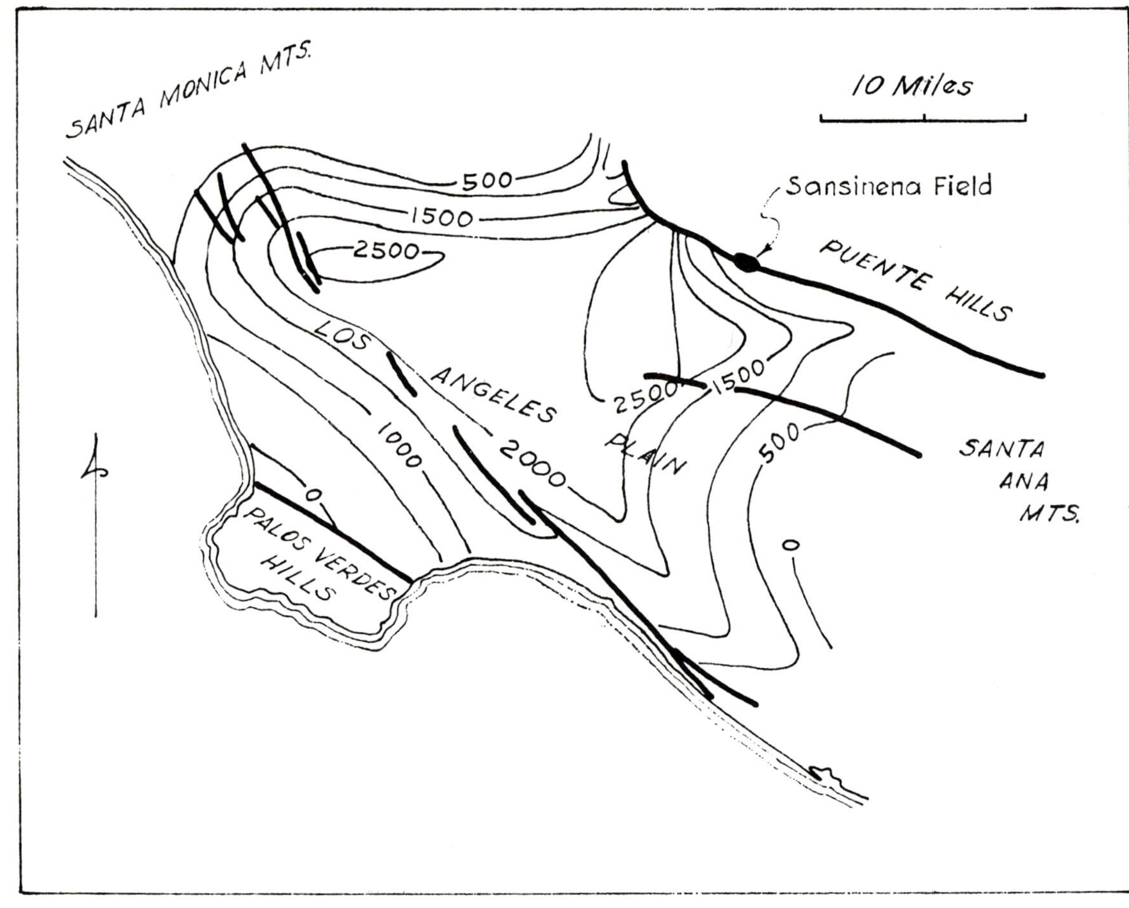

Fig. 8.—Isopach map of net sand in the Lower Pliocene Repetto "formation" in the Los Angeles basin, California. These sands are turbidites deposited in water 6,000–8,000 feet deep. Heavy lines are faults. From Conrey, 1959. Location of Sansinena oil field is shown as orientation for Fig. 9.

diminishing thickness away from that point. It is unfortunate that no detailed maps of individual turbidites or local bundles of turbidites are available.

EAST SANSINENA

The upper Miocene A-10 sand in the East Sansinena field is a more detailed example of a turbidite in the Los Angeles basin. The field has been described by A. F. Woodward (1958), and he and R. Miller kindly permitted the writer to study the Union Oil Company cross sections and to construct therefrom the accompanying isopach map (Fig. 9). In an effort to achieve detail, only the upper part of the A-10 sand is included. Inasmuch as the wells are directionally drilled from three drilling "islands," corrections for true thickness had to be made for the attitude of the boreholes as well as for the strata. The map is still a distortion, for although true thicknesses are shown, the plan view was not stretched out to undo the post-depositional structure, which is considerable. The field is small, 1 mile by $\frac{1}{2}$ mile, yet its structural relief is more than 2,000 feet.

The isopach map clearly delineates a channel or portion of a small basin. Its apparent trend agrees reasonably well with a northerly source area, generally considered valid for this part of the Los Angeles basin. The southern extension is unknown, for the sand plunges beyond the down-dip edge of the field into a deep syncline where well control is lacking. The sand was cored in five wells, but the core descriptions are not sufficiently detailed to provide proof of turbidite origin, nor have the cores been saved for further inspection. Descriptions show variations from sandy silt-stone to conglomerate, thinly laminated to massive, arkosic, gritty, poorly sorted—adequately fitting the qualifications for a turbidite. Furthermore, ecologic data (supplied by Manley Natland) indicate a depositional water depth here of 1,000–2,000 feet, and the depth increases southwestward in the general direction in which the isopachs also suggest deepening.

IMPORTANCE OF TURBIDITE STUDIES

The many billions of barrels of oil already found in turbidites testify to the importance of these rocks to the oil industry. Perhaps the significance of turbidites as oil reservoirs lies not in a superiority of reservoir characteristics, but in their environment of deposition. They are typically dumped into low places on the sea floor—areas low in oxygen—and are surrounded by muds rich in pelagic organic remains. The turbidity currents themselves may also enrich the organic content of the sea floor by introducing vast quantities of plant material. Thus, the reservoir is inserted directly into ideal source sediment, and the distance of primary migration of oil is greatly reduced.

Clearly, a full understanding of the genesis, lithology, and geometry of turbidites will greatly facilitate both oil development and oil exploration in relatively unstable parts of the earth's crust. The textural relationships within the sandstones should be taken into account in calculating reserves and reservoir mechanics. The geometry of the individual turbidites and bundles of turbidites constituting petroleum reservoirs is important for the same reasons. Oriented syngenetic structures serve to indicate direction of transport, and this, in turn, greatly enhances the understanding of paleogeography—an esteemed tool in oil exploration. For example, identification of a submarine fan in the subsurface would be an excel-

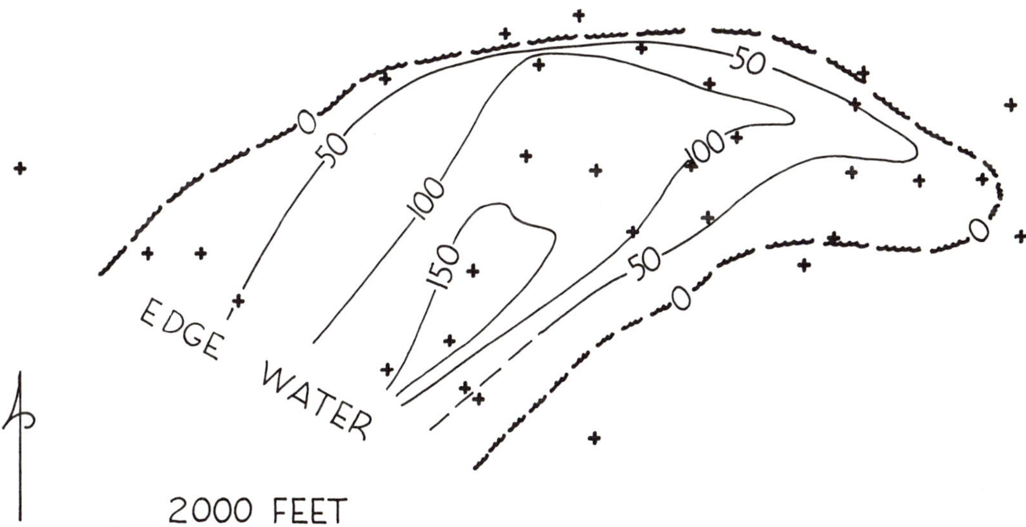

Fig. 9.—Isopach map of the upper part of the upper Miocene A-10 sand, principal producing zone in the East Sansinena field, Los Angeles County, California. Crosses indicate location of control points where slant-drilled wells penetrate sand. It appears that the turbidites were deposited in a southwest-sloping channel in water 1,000–2,000 feet deep. The geometry of this sandstone body makes its own trap. Constructed by the writer from data provided by A. F. Woodward and R. Miller, Union Oil Company of California. True sandstone thickness in feet. Location of East Sansinena field shown in Fig. 8.

lent guide toward finding the associated channel or canyon with its coarse-grained fill—an ideal reservoir. Furthermore, current indicators in the cores of an exploratory well may suggest the trend of a channel-type sand and reduce the risk on a follow-up well. These indicators may be used to orient cores and thus their contained dips. The mere routine correlations of electric logs, and construction of cross sections can be done with much more realism when the worker has a feeling for the genesis and geometry of the sands under study. And, too, a widespread turbidite constitutes a time marker fully as significant as an ash bed.

At the present time there appears to be more potential than actual use in the application of a knowledge of turbidites, for a large amount of observational data must be gathered before firm conclusions can be reached. The science is yet young. Even in California, where turbidites abound, there is a surprising lack of recognition of the utility of this new tool; its study is generally relegated to a few research workers.

Adequate coring is necessary fully to utilize the information locked up in subsurface turbidites, for it is visual inspection of the sandstones that provides the significant data. Various wireline devices are a poor substitute for geologic observation. Detailed outcrop studies are also of first-priority importance. Turbidite features have been completely overlooked in conventional geologic mapping, with a few recent exceptions. Old oil fields and cores should be reworked in the light of this new concept; perhaps unsuspected current trends will appear that will point to extensions and new pools. All of this will, of course, require more geologic labor, but the labor is largely geologic observation and footwork, and thus relatively inexpensive.

It is becoming increasingly obvious that turbidites are very common on the sea floor and in the geologic record—a fact little known and rarely taught when most of us were being educated. Surely the volume of turbidites in the modern sea must equal or exceed the volume of nearshore shallow-water sands. Although these volumes are not yet measurable, nor could they be projected quantitatively into the past, they certainly justify acceptance of ancient turbidites on a large scale, perhaps in a pre-eminent volumetric position.

REFERENCES

Barbat, W. F., 1958, The Los Angeles basin area, California, *in* Habitat of Oil, Am. Assoc. Petroleum Geologists, p. 62–77; also *in* A guide to the geology and oil fields of the Los Angeles and Ventura regions, Pacific Section, Am. Assoc. Petroleum Geologists, p. 37–49.

Bates, C. C., 1953, Rational theory of delta formation: Am. Assoc. Petroleum Geologists Bull., v. 37, no. 9, p. 2119–2162.

Birkenmajer, K., 1959, Classification of bedding in flysch and similar graded deposits: Studia Geologica Polonica, v. 3, 128 p.

Conrey, B. L., 1959, Sedimentary history of the early Pliocene in the Los Angeles basin, California: unpub. Ph.D. dissertation, Univ. Southern Calif., 273 p. Brief résumé published *in* A guide to the geology and oil fields of the Los Angeles and Ventura regions: Pacific Section, Am. Assoc. Petroleum Geologists, p. 51–54, 1958.

Crowell, J. C., 1955, Directional-current structures from the Prealpine Flysch, Switzerland: Geol. Soc. America Bull., v. 66, no. 11, p. 1351–1384.

Driver, H. L., 1948, Genesis and evolution of Los Angeles basin, California: Am. Assoc. Petroleum Geologists, v. 32, no. 1, p. 109–125.

Dzulynski, S., and Radomski, A., 1955, Origin of groove-casts in the light of turbidity current hypothesis: Acta Geologica Polonica, v. 5, p. 47–66.

Ericson, D. B., Ewing, M., and Heezen, B. C., 1951, Deep-sea sands and submarine canyons: Geol. Soc. America Bull., v. 62, no. 8, p. 961–966.

———— ———— and ————, 1952, Turbidity currents and sediments in the North Atlantic: Am. Assoc. Petroleum Geologists, v. 36, no. 3, p. 489–511.

Gorsline, D. S., and Emery, K. O., 1959, Turbidity-current deposits in San Pedro and Santa Monica basins off southern California: Geol. Soc. America Bull., v. 70, no. 3, p. 279–290.

Hall, James, 1843, Geology of New York; Part IV, Comprising the survey of the fourth geological district: Albany.

Heezen, B. C., 1956, Corrientes de turbidez del Rio Magdalena: Bol. Soc. Geog. Colombia, v. 51, 52, p. 135–143.

———— 1959, Dynamic processes of abyssal sedimentation—erosion, transportation, and redeposition on the deep-sea floor: Geophys. Jour., Royal Geog. Soc., v. 2, p. 142–163.

———— Tharp, M., and Ewing, M., 1959, The floors of the oceans; Part 1, The North Atlantic: Geol. Soc. America Spec. Paper 65, 122 p.

Hull, J. P. D., Jr., 1957, Petrogenesis of Permian Delaware Mountain sandstone, Texas and New Mexico: Am. Assoc. Petroleum Geologists Bull., v. 41, no. 2, p. 278–307.

Kuenen, Ph. H., 1953, Significant features of graded bedding: Am. Assoc. Petroleum Geologists, v. 37, no. 5, p. 1044–1066.

———— and Carozzi, A., 1953, Turbidity currents and sliding in geosynclinal basins in the Alps: Jour. Geology, v. 61, p. 363–373.

———— and Migliorini, C. I., 1950, Turbidity currents as a cause of graded bedding: Jour. Geology, v. 58, no. 2, p. 91–126.

Lowman, S. W., 1949, Sedimentary facies in Gulf Coast:

Am. Assoc. Petroleum Geologists Bull., v. 33, no. 12, p. 1939–1997.

McBride, E. F., 1950, Martinsburg-Reedsville paleo-currents (abstract): 45th Ann. Mtg. Am. Assoc. Petroleum Geologists and Soc. Econ. Paleontologists and Mineralogists, Atlantic City, program: p. 90–91.

McIver, N. L., 1960, Upper Devonian paleocurrents in the central Appalachian Mountains (abstract): Geol. Soc. America Bull., v. 71, no. 12, p. 1926.

Menard, H. W., 1960, Possible pre-Pleistocene deep-sea fans off central California: Geol. Soc. America Bull., v. 71, no. 8, p. 1271–1278.

Natland, M. L., 1957, Paleoecology of west coast Terti-ary sediments, *in* Treatise on marine ecology and paleocology, v. 2, edited by H. S. Ladd: Geol. Soc. America Mem. 67, Chap. 19, p. 543–572.

Payne, T. G., 1942, Stratigraphical analysis and envi-ronmental reconstruction: Am. Assoc. Petroleum Ge-ologists Bull., v. 26, no. 11, p. 1697–1770.

Phleger, F. B, 1951, Displaced Foraminifera faunas, *in* Turbidity currents and the transportation of coarse sediments to deep water—a symposium: Soc. Econ. Paleontologists and Mineralogists Special Pub. 21, p. 66–75.

Riveroll, D. D., and Jones, B. C., 1954, Varves and Foraminifera of the upper Puente formation (upper Miocene), Puente, California: Jour. Paleontology, v. 28, no. 2, p. 121–131.

Slosson, J. E., 1958, Lithofacies and sedimentary-paleogeographic analysis of the Los Angeles Repetto basin: unpub. Ph.D. dissertation, Univ. Southern Calif., 128 p.

Sullwold, H. H., Jr., 1959, Nomenclature of load de-formation in turbidites: Geol. Soc. America Bull., v. 70, no. 9, p. 1247–1248.

———— 1960a, Tarzana fan, deep submarine fan of late Miocene age, Los Angeles County, California: Am. Assoc. Petroleum Geologists Bull., v. 44, no. 4, p. 433–457.

———— 1960b, Load-cast terminology and origin of convolute bedding; further comments: Geol. Soc. America Bull., v. 71, no. 5, p. 635–636.

Van Houten, F. B., 1954, Sedimentary features of Mar-tinsburg slate, northwestern New Jersey: Geol. Soc. America Bull., v. 65, no. 8, p. 813–817.

Wissler, S. G., 1943, Stratigraphic relations of the pro-ducing zones of the Los Angeles basin oil fields: Calif. Div. Mines Bull. 118, p. 209–234.

Woodward, A. F., 1958, Sansinena oil field, *in* A guide to the geology and oil fields of the Los Angeles and Ventura regions: Pacific Section, Am. Assoc. Pe-troleum Geologists, p. 109–118.

The American Association of Petroleum Geologists Bulletin
V. 62, No. 11 (November 1978), P. 2243-2274, 23 Figs.

Sedimentation and Trapping Mechanism in Upper Miocene Stevens and Older Turbidite Fans of Southeastern San Joaquin Valley, California[1]

BRUCE A. MACPHERSON[2]

Abstract Most of the stratigraphic oil production in the southern San Joaquin Valley comes from the sandstones of the upper Miocene turbidites. Updip, time-equivalent rocks of the shallow-marine environment are second in productivity. Third in importance are the deep-marine, fractured, siliceous shales deposited beyond distal margins of the fan complex. Many underexplored areas remain in the San Joaquin Valley despite many years of field development and wildcat drilling.

The Miocene of the Bakersfield arch is representative of the classic arcuate type of turbidite deposits characteristic of modern submarine fans. Further paleogradient of the fans has been accentuated and not reversed through geologic time; stratigraphic oil fields are numerous. As in most California deep-marine troughs, the turbidite facies most easily reached by the drill, and in which most of the oil reserves have been developed, are in the mobile zone of basin-margin wedging. Two additional trapping aspects related primarily to depositional stratigraphy within the midfan facies are (1) intrabasin contemporaneous faults (analogous to Gulf Coast growth faults) resulting in expanded section and reverse drag on the downthrown block, and (2) compaction anticlines formed by post-sedimentary accommodation over submarine channel sands resulting in structural inversion with depth.

South of the Bakersfield arch is an earlier (pre-late Mohnian) turbidite basin, virtually unexplored and referred to informally as the "Maricopa subbasin," where drilling depths to potential Miocene pays range from 12,000 to 20,000 ft (3,600 to 6,000 m). Basic stratigraphic relations recognized in the Bakersfield arch may be applied in the subbasin to develop new hydrocarbon reserves.

INTRODUCTION

The Stevens sandstones have been the target of continued oil exploration for more than 40 years. Thus far, more than 450 million bbl of high-gravity crude has been proved with slightly more than 400 million bbl having been produced by 1973. These reserves are derived from only 15 fields on the Bakersfield arch. If the Stevens oil fields on the west side of the San Joaquin Valley (outside the scope of this study; Fig. 1) are included, the quoted reserves readily could be doubled for this most prolific pay.

Although the Bakersfield arch now might be considered a mature area of exploration, surprisingly few papers exist pertaining to Stevens stratigraphy. Yet the arch province, with its dense well control and numerous stratigraphic and structural traps, serves as an excellent pilot area for understanding the trapping mechanisms of the Stevens for future exploration for this zone as

well as for pre-Stevens turbidite sands in underexplored areas of the San Joaquin basin. All of the anatomical parts of fossil turbidites are preserved here and are available for study to those with access to the data. All depositional environments have been found to be productive, each with its unique trapping mechanisms for hydrocarbon accumulation.

In this study, the "Stevens sandstone" is an all-inclusive term encompassing more than 4,000 ft (1,200 m) of interbedded deep-marine sandstone and shale originating as two major coalescing submarine fans issuing from two source areas at the eastern end of the Bakersfield arch, one at Rosedale Ranch field and the other at Fruitvale field (Fig. 2). Each fan complex is named by the writer, Rosedale fan and Fruitvale fan, after its respective source.

Four discrete depositional cycles are recognized and informally named. In ascending order they are: "Rosedale," "Coulter," "Gosford," and "Bellevue."

The Stevens is of late Mohnian age but, herein, the early Mohnian Rosedale Sandstone of Martin (1963) also is included in the Stevens group (Fig.

AAPG grants permission for a *single* photocopy of this article for research purposes. Other photocopying not allowed by the 1978 Copyright Law is prohibited. For more than one photocopy of this article, users should send request, article identification number (see below), and $3.00 per copy to Copyright Clearance Center, Inc., One Park Ave., New York, NY 10006.

[1]Manuscript received, May 9, 1977; accepted, March 20, 1978.

[2]Continental Oil Co., Oklahoma City, Oklahoma 73112.

The writer is indebted to Continental Oil Co. for permission to publish this report. Special acknowledgment is due Richard Louie, formerly of Continental Oil Co., for his assistance in computer analysis of turbidite-sand wedging; R. Skalecke and Paul Hagood for preparation of the illustrations; and Sandra Medland for typing and editing the manuscript.

The writer thanks AAPG Pacific Section for permission to publish this modification of his earlier paper (MacPherson, 1977).

Article Identification Number
0149-1423/78/B011-0004$03.00/0

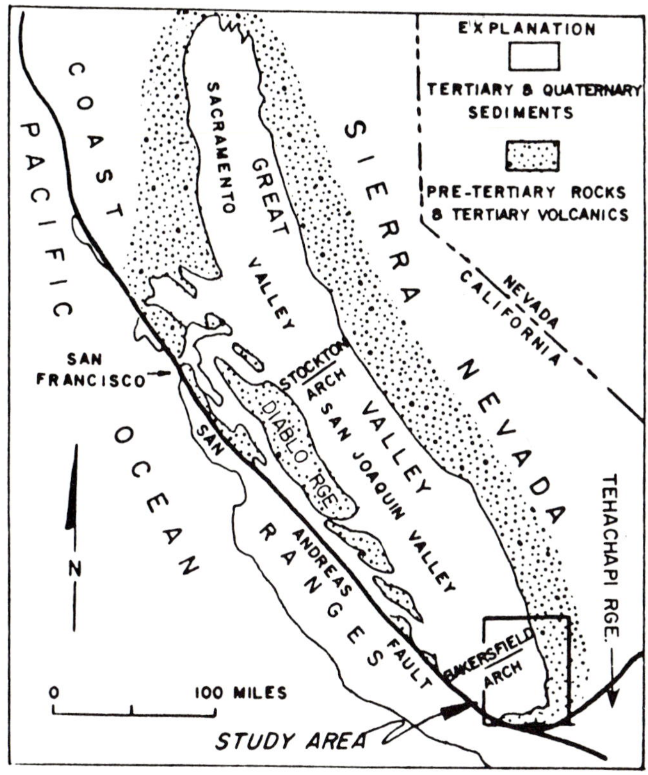

FIG. 1—Index map of Great Valley of California.

3). Other submarine fans ranging in age from early middle Miocene (Relizian) through early late Miocene (early Mohnian) are present south of the Bakersfield arch. To facilitate discussion these depositional features also have been named informally. The "Bena fan" and the "Valpredo fan" of the Maricopa subbasin, very sparsely drilled, should contain depositional facies similar to the Stevens and should offer similar opportunities for industry to develop new hydrocarbon reserves in the southern San Joaquin Valley.

PREVIOUS WORK

Sullwold (1961) first identified the Stevens sandstone as a turbidite deposit on the basis of his own observations, those of Natland (1957), and the unpublished works of Ted Bear, consultant, and of Rollin Eckis of the Richfield Oil Corp. Following Sullwold's pioneer report, Seiden (1965), Sanem and Stoddard (1965), and Callaway (1968) published detailed stratigraphic papers of local productive areas in the Stevens and related sandstones. Bazeley (1972) documented the exploratory tenacity required to find and develop new stratigraphic reserves in the upper Miocene sandstones. Maher et al (1975) recognized the submarine origin of the Stevens pay at Elk Hills.

Although the writer completed study on the Stevens sandstone in 1972, an attempt is made to correlate his conclusions with the later stratigraphic and sedimentologic analyses by Walker and Mutti (1973), Normark (1976), Walker (1976a, 1976b), and Le Blanc (1977). However, as the Stevens sandstone does not crop out and no intact conventional cores are available for study, it is not possible to categorize and compare the sedimentary structures, grain-size distribution, and bedding geometry which are so fundamental in precise classification of the many subfacies of turbidite fans recognized by these authors. Conclusions in this paper are based on observations made from the usual limited material available to the exploration geologist—electric logs, sample descriptions, and limited seismic data.

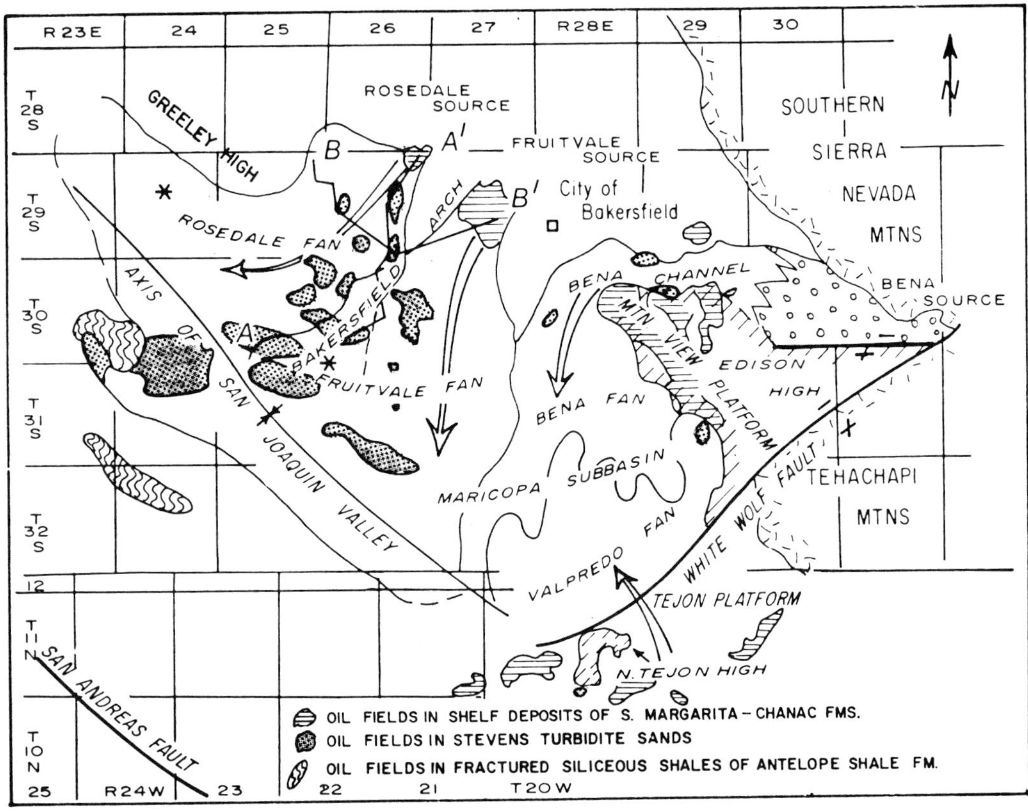

FIG. 2—Major turbidite fans of southern San Joaquin Valley. Rosedale-Fruitvale fans comprise Stevens sandstone.

REGIONAL STRATIGRAPHY

San Joaquin Valley

The San Joaquin Valley (Fig. 1) is the southernmost extension of the Great Valley of California, a narrow elongate marine basin between the Sierra Nevada Mountains on the east and the California Coast Ranges on the west. West-dipping Sierran granite floors the east half of the valley, where the structure is characterized by block faulting and broad, open folds. Franciscan metamorphic rocks underlie the western valley area where structural deformation is reflected in tight folding and thrust faulting.

Hoots et al (1954) recognized that the present form of the San Joaquin basin was embryonic during Late Cretaceous and/or Paleocene time. Throughout Cenozoic time the basin was the site of a marine seaway which was being filled by clastics derived from the ancestral Sierra, Tehachapi, and Coast Ranges. Ancient shorelines fluctuated near the present flanking mountain ranges.

Deep-marine environments lay slightly west of the present valley axis.

Marine sedimentation culminated in the Pliocene, after which the region was reduced to an intermontane basin characterized by internal drainage and filled by fluviolacustrine deposition in the form of coalescing bajadas interspersed with rather extensive playa lakes. Greatly reduced remnants of these lakes are present on the valley floor today.

Bakersfield Arch

A major feature of the southern San Joaquin Valley is the Bakersfield arch (also referred to as the "Kern arch," after Kern County in which it is located), a broad subsurface feature plunging west from the city of Bakersfield to the valley center. The arch separates the San Joaquin Valley into two subbasins. On the north lies the main San Joaquin Valley basin and on the south, the Maricopa subbasin. Uplift of the arch may have been initiated in late Paleocene or early Eocene

	CALIF. FORAM. STAGE	BAKERSFIELD ARCH	MARICOPA SUB BASIN	TEJON PLATFORM
UPPER	DELMONTIAN	STEVENS	STENDERUP	
	MOHNIAN		WICKER	RESERVE
MIDDLE	LUISIAN		PALLA O'NIELL / NOZU	VALV
	RELIZIAN			OLCESE
LOWER		UNKNOWN		

FIG. 3—Regional correlation chart for Miocene submarine-fan turbidites of southern San Joaquin Valley, California.

time, as Cretaceous and Paleocene deposits are absent over the arch and all the Cenozoic strata are thin except the upper Miocene and Pliocene to Pleistocene. Paleontologic studies by Bandy and Arnal (1960) showed that the arch formed a seafloor prominence in all but late Miocene time. During the late Miocene a heavy influx of flysch, the Stevens, overwhelmed this structural element and reduced its effect as a submergent high.

Maricopa Subbasin

The Maricopa subbasin is a deep structural depression oriented east-west across the southern end of the San Joaquin Valley and bounded on the north by the Bakersfield arch and on the south by the Tejon platform (Fig. 2). Deep marine conditions prevailed here until late Miocene time. Flysch deposition filled the basin throughout the middle and earliest late Miocene from sediment-dispersal centers in the east at the "Bena channel" and in the south on the Tejon platform. During this period of flysch deposition, the bathymetric high formed by the north-bounding arch was sediment starved. Only thin deposits of siliceous shale and siltstone—lateral equivalents of the Bena fans—were draped over the arch. By late Mohnian time the Bena channel ceased to function as a dispersal point for submarine fans, and coarse detritus backfilled the channel conduit. Silt deposition prevailed at the channel mouth and throughout the Maricopa subbasin. Subsequently, the source shifted northward to the Bakersfield arch. Henceforth, turbi-

dites entering the Maricopa deep were derived from excessive sand overflow from the north-flanking arch.

"Bena Channel" and "Edison High"

The terms, "Bena channel" and "Edison high," are introduced here to refer to a gravel-filled trough between the southern Sierra massif and a fault-bounded subsurface basement high at the east side of the Maricopa subbasin (Fig. 2). As stated previously, the Bena channel was an active source during Miocene time. From Relizian (early middle Miocene) through the early Mohnian (early late Miocene) all of the Bena fans issued from this source. The time-stratigraphic chart in Figure 3 shows the geographic distribution of turbidite fans of the southern San Joaquin Valley and their correlation.

Tejon Platform

The Tejon platform at the southern end of the San Joaquin Valley is bounded on the north by the White Wolf fault which separates the platform from the Maricopa subbasin. Benioff (1955) estimated about 12,000 ft (4,600 m) of vertical throw at basement level across the fault zone with the sense of movement down to the north. Epicenter plots of the 1952 earthquakes (Oakeshott, 1955) indicate a nearly vertical south-dipping fault plane. Most students of regional California tectonics hold that slight left-lateral wrench-fault motion occurred along the White Wolf fault at various times.

Miocene bathymetry maps by Bandy and Arnal (1960) show the Tejon platform as a narrow, persistently north-dipping shelf in existence since the Zemorrian or earliest Miocene. These maps do not date this fault, but I believe an incipient White Wolf fault existed in early middle Miocene time as part of the regional crustal breakup of California. The fault scarp may have controlled sedimentation in the southern end of the Maricopa subbasin in the manner of a contemporaneous fault. Thick wedges of sediment were deposited on the downthrown block on the north and relatively thin equivalents were retained on the platform on the south. In general, the platform area remained an active source area bounding the southern Maricopa deep throughout Miocene time.

STEVENS SANDSTONE OF BAKERSFIELD ARCH REGION

Regional Stratigraphy

Four major cycles of turbidite deposition for the late Miocene (late Mohnian) are recognized

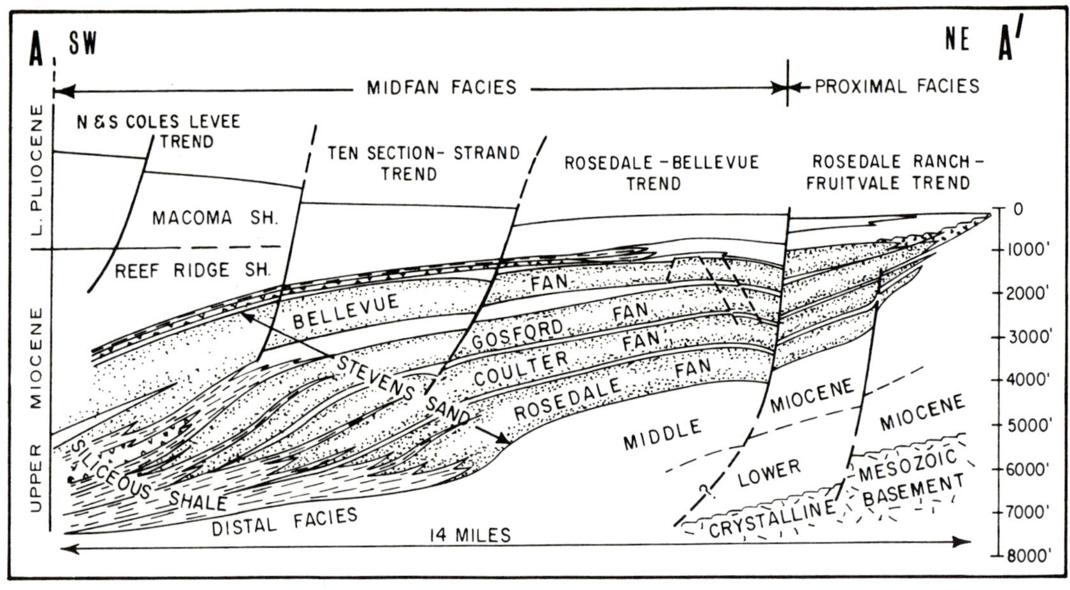

FIG. 4—Cross section *AA'* showing upper Miocene turbidite fans and related facies, Bakersfield arch, southern San Joaquin Valley, Kern County, California. Section parallels direction of transport. Location shown on Figure 8.

on the Bakersfield arch; collectively they comprise the Stevens sandstone (Figs. 2-4). A waning turbidite phase of the early Delmontian is assigned to the Reef Ridge sandstone, which may be represented better in the south-flanking Maricopa subbasin than on the arch proper.

Source areas are recognized in the subsurface at the eastern end of the arch at Rosedale Ranch and Fruitvale fields. These are thought to be submarine canyons cut into the Santa Margarita outer-shelf deposits and subsequently infilled by Chanac sediments to form the Rosedale and Fruitvale fans (Figs. 2, 3), named after the canyons from which they were derived.

Each fan is comprised of four discrete sandstone members; in ascending order they are Rosedale, Coulter, Gosford, and Bellevue. The Rosedale sandstone member should not be confused with the regional morphologic feature called "Rosedale fan," of which the Rosedale sandstone comprises only a single stratigraphic member. The Rosedale sandstone is also present in the Fruitvale fan.

In lithology the sequence is characterized by continued basinward offlap of successive turbidites. As each fan issued from its submarine canyon to occupy the space available on the ocean floor, the succeeding fan would then occupy successively deeper parts of the basin, and so on until the end of flysch sedimentation. Within each fan, widespread correlative shale members attest

to periods of local sand supply periodically diminishing to zero. These shales provide excellent markers for reconstruction of the sequential history in this otherwise monotonous section of lithologically identical sandstones. Siliceous shale characterizes the distal and far-lateral margins of each fan, and is retained to the very deepest parts of the Miocene basin beyond sand-transporting agents.

Fans issuing simultaneously from the Rosedale and Fruitvale canyons commingled along their lateral margins, introducing stratigraphic complexities in the central Bakersfield arch. At the north end of the arch, the Rosedale fan impinges the Greeley bathymetric high and, at the north and northeast, its distal limits have not been established. The Fruitvale fan (possibly in response to Coriolis force) traces an arc south-southeast where its upper sandstone members extend into the Maricopa subbasin.

Turbidite sedimentation terminated with the late Delmontian regional transgression, resulting in the deposition of the Reef Ridge Shale. The deep-marine siliceous shale facies shifted eastward to form the N-chert layer over the Bellevue turbidite. Weak development of lower Delmontian sandstones on the arch probably is indicative of diminished sand supply as the principal cause of the transgressive cycle.

Renewed uplift in the Sierras in Pliocene and Pleistocene time dispersed deltaic and nonmarine clastic wedges over areas formerly occupied by deep-marine turbidite fans. The Pleistocene to recent Kern bajada or piedmont fan extends like a blanket over both the Bakersfield arch and the Maricopa deep, both areas in which fan sedimentation prevailed for so long.

SHELF FACIES

The southeastern San Joaquin Valley is unique among California turbidite basins because shelf equivalents to the deep-sea fans still are preserved. In most basins, similar deposits have been removed by postdepositional erosion on the tectonically vulnerable basin edges.

Santa Margarita and Chanac sediments represent the shelf environment (Figs. 3, 4), fringing the ancestral Sierras and Tehachapis since early late Miocene time. Scattered outcrops of these formations have been mapped on the east and south basin areas. In outcrop the Chanac is totally nonmarine and unconformably overlies the marine Santa Margarita. Westward into the subsurface the unconformity appears to persist, but the Chanac grades laterally into marine sedimentary rocks. Chanac and Santa Margarita have not been differentiated in this subsurface study and only the term "Chanac" is used where shelf facies are recognized.

In the subsurface, Santa Margarita sediments originated as a prograding outer marine shelf during early Mohnian time. Shelf advance kept pace with the basinward progradation of the turbidite fans. Submarine canyons continually were cut into the leading edge of the advancing shelf at Rosedale Ranch and Fruitvale fields. These canyons served as source areas for flysch deposition.

During the Reef Ridge transgression, the submarine canyons were backfilled with Chanac-type sediments. The canyon fill formed an eastward (updip) pointing wedge of Chanac within the Santa Margarita Formation. It is suggested that the Chanac wedges are the origin of the stratigraphic oil at Fruitvale and Rosedale Ranch fields. No reasonable explanation for the origin of these fields has been advanced. I would suggest a similar origin for Kern River field through repetition of these events in Pliocene and Pleistocene time.

In some areas, the shelf (S$_1$, S$_2$ in Fig. 4) advanced over the lateral-margin equivalents of earlier turbidites and the basal contact with the subjacent Stevens is unconformable. The sequential history of this relation is rather complex and has resulted in shelf deposits overlying deep-marine sandstones near the eastern edge of the arch.

Near shore shelf deposits form bars, deltas, and other complex strandline features. The Chanac and equivalent facies are highly prolific oil producers on the periphery of the southeastern San Joaquin Valley. Because of its shallow depth the facies has been explored intensively and there probably is little unexplored area left for big future plays near the Bakersfield arch. However, Chanac–Santa Margarita production on basin flanks in underexplored areas is considered a favorable indication of deeper basin potential (Fig. 2).

TURBIDITE SEDIMENTATION

The "Stevens sandstone" is a group name applied to more than 5,000 ft (1,500 m) of aggregate thickness of deep-marine sandstone and shale reflecting several cycles of turbidite deposition. Nowhere do these sandstones crop out so it is impossible to say how closely lithogenetic units can be compared with those recognized as turbidites. In gross aspect based on subsurface control, the sandstones are fine to coarse grained, generally fining upward in some areas, especially near source; coarsening upward is more typical away from source. Interbedded shales are subsiliceous, sparsely foraminiferal, becoming highly siliceous and diatomaceous with increasing water depth. In this study, each turbidite fan has four distinct facies typified by unique lithogenetic units originating in response to discrete sedimentary processes (Figs. 4, 5). These facies are:

1. *Proximal*—Near or within submarine canyons; facies A and B of the inner fan of Walker and Mutti (1973); possibly correlative with the slope-channel association of Mutti and Ricci-Lucchi (1972).

2. *Midfan*—Delta-shaped apron on the seafloor plain; facies C through E of Walker and Mutti (1973), the suprafan of Normark (1970, 1976), and the middle-fan association of Mutti and Ricci-Lucchi (1972).

3. *Lateral margin (basin margin)*—Wedging peripheral edge of the midfan facies; not generally emphasized or recognized by turbidite authorities and highlighted here because of its extreme economic importance in California oil exploration.

4. *Distal margin*—Arcuate toe of the submarine fan; possibly equivalent to the F and G sequences of Walker and Mutti (1973); may incorporate the "classic turbidites" referred to by Walker and Mutti (1973) as the rhythmically bedded, laterally extensive laminae common in outcrops of deep-sea sediments.

PROXIMAL FACIES

Sedimentation

Submarine canyons represent dynamic environments of erosion and sediment transport. Only

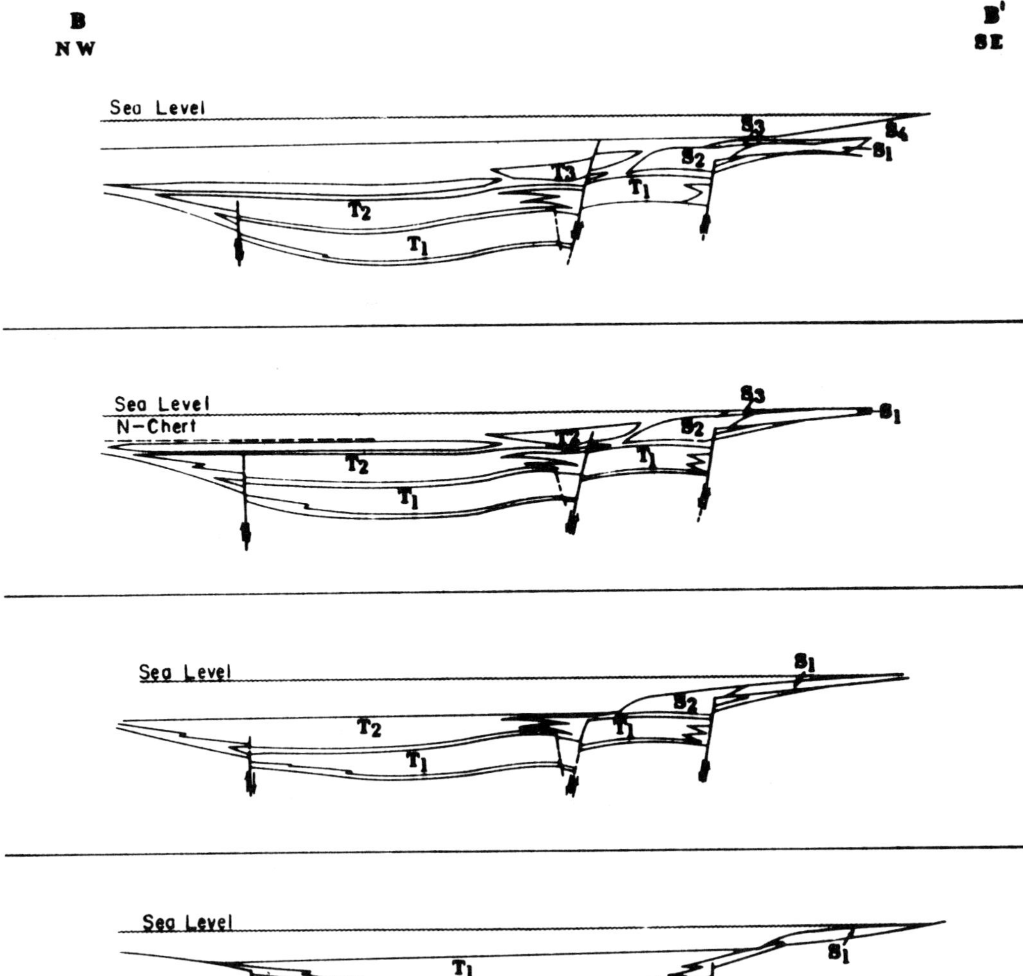

FIG. 5—Superposition of shelf sediments (*S*) over lateral margins of Stevens turbidite (*T*) fans. Sections are transverse to paleogradient. Location of *BB'* shown on Figure 8.

a small fraction of any turbidite is retained within the canyon proper. Most of the early sediment load bypasses the canyon, coming to rest beyond the mouth to construct a submarine fan or cone on the ocean floor. Locally a train of coarse detritus may be retained in the canyon similar to lag gravels in subaerial processes. In time the canyon is extended by headward erosion and the entrained coarse material, having a high angle of repose, gradually paves the canyon floor. Canyon-filling is completed when sediment supply diminishes and both coarse and fine materials are dumped in the canyon to fill the void. Mutti and Ricci-Lucchi (1972) recognized the final infill of submarine canyons in Italian fossil turbidites to consist of hemipelagic sediments usually found at the distal margin. In outcrop this would pose the paradoxical situation of channellike deposits and debris-flow material of the fluxoturbidites directly overlain by finely laminated shales containing deep-marine fauna.

In the Stevens sandstone, a proximal facies can be identified only for the apex of the Rosedale fan at Rosedale Ranch field. Coarse-grained con-

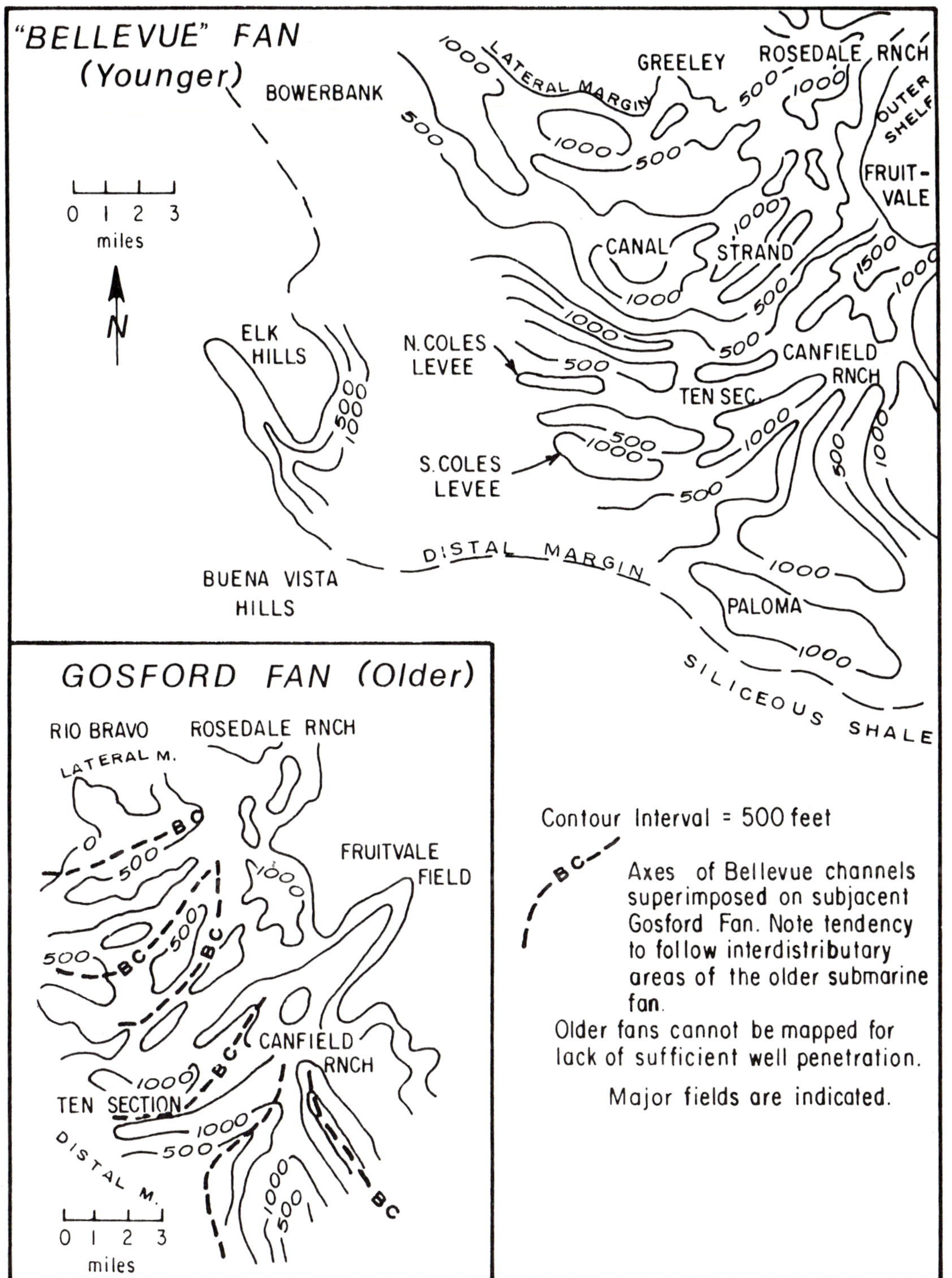

FIG. 6—Isopachs of Bellevue and Gosford turbidite fans on Bakersfield arch.

glomeratic sandstones, becoming finer upward, are present in the Gosford and Bellevue and lack of detailed stratigraphic correlation attests to the chaotic nature of the stratigraphic column. Presumably, the proximal facies of earlier sands (Coulter and Rosedale) have been erased by erosion or obscured by a continally prograding Santa Margarita shelf. Here, proximal sediments are overlain unconformably by finer grained lenticular Chanac deposits. Chanac sediments probably completely backfilled the canyon during the late Miocene–early Pliocene transgression.

Oil Occurrence

Submarine-canyon infill at Fruitvale field is composed entirely of Chanac debris. Apparently all of the Stevens sediment bypassed this site en route to the depocenter beyond the canyon mouth. Canyon-fill Chanac debris forms the principal oil reservoirs at Rosedale Ranch and Fruitvale fields. Vertical sealing of the reservoirs is accomplished by the overlying Reef Ridge and Macoma shales of the Miocene and Pliocene transgression.

MIDFAN FACIES

Sedimentation

The midfan turbidite facies includes most of the oil fields of the Bakersfield arch and provides the largest exploration target in underexplored basins where depositional systems can be compared to the Stevens sandstone. Normark (1976) described the midfan area as a "depositional bulge, convex-upward on a radial profile." Normark's midfan, also termed "suprafan," was defined as that part of the turbidite immediately downslope from the termination of the major active valley on the upper fan. The suprafan is characterized by numerous, unleveed distributary channels and channel remnants resulting from the rapid migration and deposition within the distributary system.

In the Stevens sandstone the midfan facies is comprised of distributary channels southwest of Rosedale Ranch and Fruitvale fields (Fig. 6). As stated previously, four major cycles of offlapping turbidite-sand deposition can be recognized on the arch (Fig. 4). Periodic cessation of sand influx was punctuated by widespread mud (shale) deposition. Regional correlation of these shales allows recognition of the several turbidite fans, not otherwise possible because the sand packets are lithologically identical. Informal names have been introduced for the separate fans for ease of reference (Figs. 3, 4). Lack of deep-well penetration in most areas makes feasible only the mapping of the upper two sandstones, Gosford and

Bellevue (Fig. 6). Most oil production has been found in these sandstones and unfortunately most operators elect not to drill to deeper sandstones if the upper two are dry.

Sections parallel with paleogradient as shown in Figure 4 illustrate each fan gradually thickening basinward (west) as more space became available on the ocean floor for sediment accumulation. Each fan had maximum sand accumulation near the medial line extending from its apex to its central distal margin, indicating that bottom conditions must have been fairly uniform to result in nearly symmetrical distribution. Sandstone content within a fan complex gradually diminishes toward the lateral and distal margins and is replaced by siliceous shale or subsiliceous claystone. Highly siliceous shales appear to be a function of water depth, being restricted to deeper basin areas.

Two of the most important depositional processes recognized within the midfan area are (1) contemporaneous or growth faulting and (2) distributary-channel transport.

Growth Faults

Sedimentation in midfan areas was accompanied by syndepositional down-to-the-basin faulting. A striking comparison can be made between the Gulf Coast growth faults and Stevens contemporaneous faulting. Like the Gulf Coast model, Stevens faults exhibit marked expansion of section on the downthrown side, reverse drag or rollover in beds predating the fault, and the presence of antithetic faults. However, an exception to the Gulf Coast model is that the master fault plane seems to be nearly vertical and only moderately asymptotic to bedding planes with increasing depth as indicated in Figures 4 and 5 (the section of Fig. 4 is constructed on a stratigraphic horizon, thus minimizing the effect of locally expanded sections caused by growth faulting).

Because fault planes rarely are cut by wells, they must be inferred chiefly by indirect evidence such as sudden thickening of discrete depositional units, anomalous fluid contacts, and local steep dip at the N-chert horizon directly overlying the Bellevue sandstone. Structure on the N-chert horizon (Fig. 7) only vaguely shows irregular, sublinear steep-dip alignments striking across the midfan facies approximately at right angles to the axis of the Bakersfield arch. When the N-chert structure is compared with the Macoma–Reef Ridge shale isopach (Fig. 8), zones of growth faulting become readily apparent. They are represented by abrupt linear "thicks" indicated by the isopachs. Evidently Stevens growth faults perpetuated themselves later to control deposition of

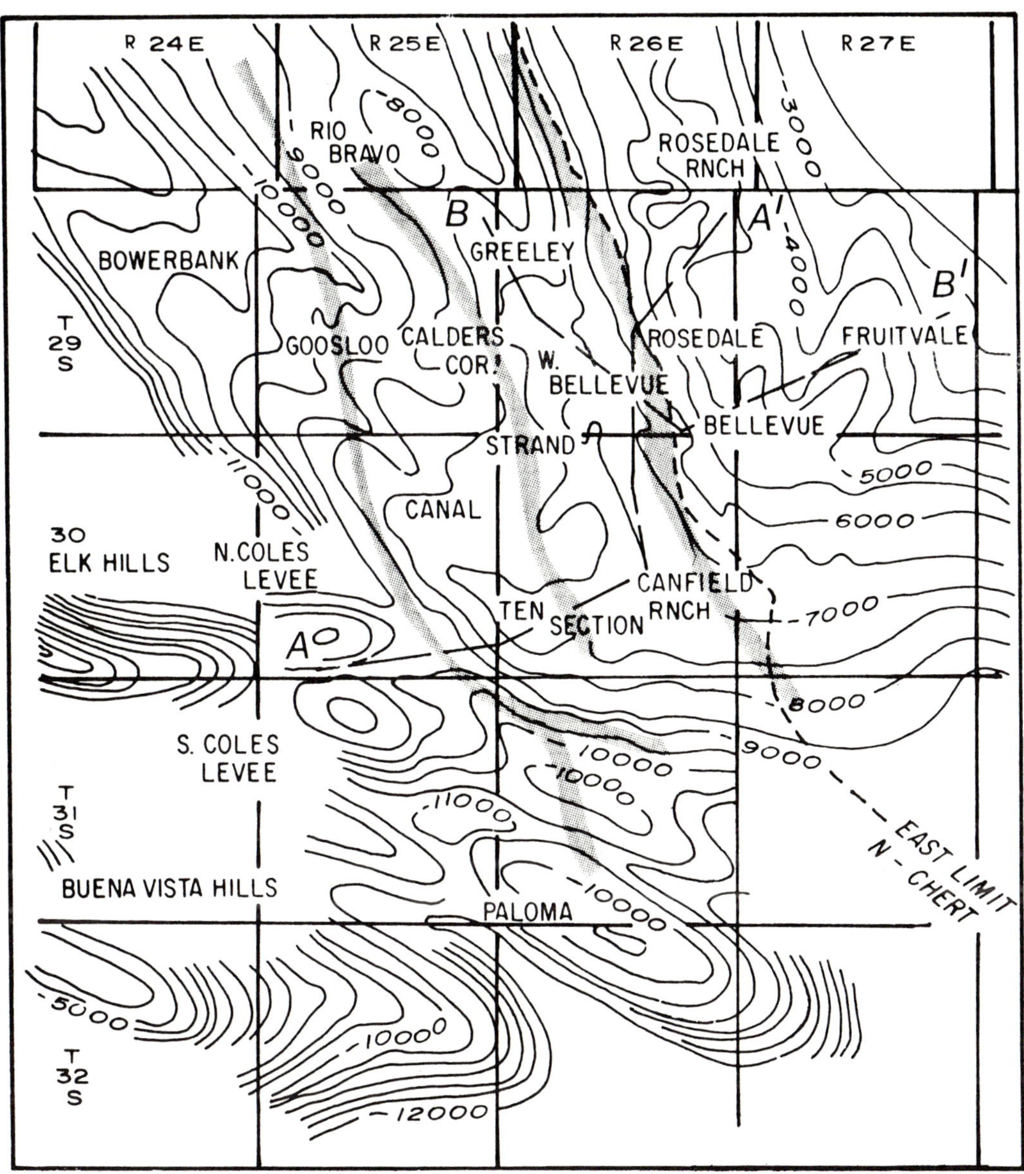

FIG. 7—Structure on N-chert overlying Stevens sandstone; east of N-chert limit, contours are on Santa Margarita. C.I. = 500 ft (150 m). *AA'*, Stratigraphic section (Fig. 4). Stippled areas are postulated growth-fault zones and generally coincide with steeper than average dips. Noses and terraces are due to drape over submarine channels. Compare with Figure 8 isopach.

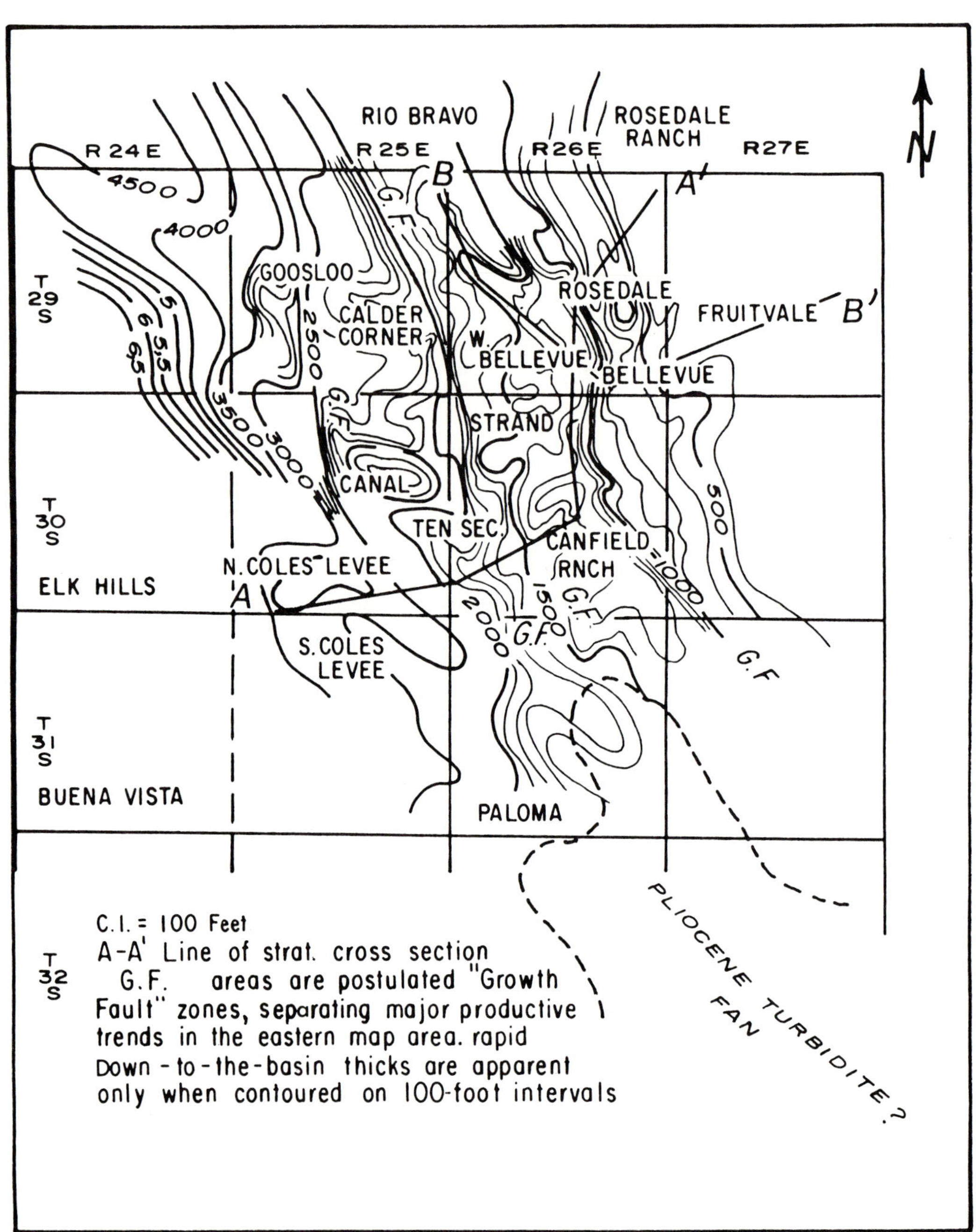

FIG. 8—Isopach of Miocene-Pliocene Reef Ridge and Macoma shale overlying youngest Stevens turbidite.

the overlying upper Miocene–lower Pliocene shales of the Reef Ridge and Etchegoin formations. Like Gulf Coast growth faults, those of the Bakersfield arch migrate basinward, becoming progressively younger as new clastic wedges are added to the section.

Origin—The most favorable environment for growth-fault tectogenesis is the outer marine-shelf environment where sand is deposited at a high depositional rate over the thick prisms of low-density (high water content) shales (Carver, 1968; Cloos, 1968; Bruce, 1973). Increased overburden pressure in time creates a glide plane much like a subaerial slump block, arcuate in plan view and concave upward in profile. Continued lubrication of the fault plane is provided by the gradually dewatering shales and results in an ongoing process of gravitational slumping and gliding which ceases only after sediment bypasses the downthrown fault block in search of a new depocenter.

All of the preceding conditions for growth-fault formation probably are met in the Stevens sandstone—thick flysch-derived sands rapidly deposited on several thousand feet of deep-marine lower and middle Miocene shales. Periodic earth tremors common to the area could further stimulate the process. Other turbidite sequences of California basins may exhibit midfan growth-fault systems but are difficult to document because of lack of well or seismic control. However, if present, such features might be detected by close-order seismic control in underexplored areas. A maximum 500 ft (150 m) of fault displacement has been measured, but faults having as little as 200 ft (60 m) of throw also trap oil. Structural closure in reverse-drag features usually amounts to less than a few hundred feet but is within the scope of seismic resolution.

Oil occurrence—Growth faults create hydrocarbon traps by effecting stratigraphic discontinuity between the sediment package on the downthrown block and thinner time-equivalent beds of the upthrown or stationary block. Oil accumulates in the downslumped basinward side of the fault, whereas only water is present in equivalent beds of the upthrown block. Thus successive linear trends of stratigraphic oil fields periodically traverse the midfan facies near the trace of the syndepositional fault zones.

A secondary trapping mechanism associated with growth faulting is the rollover-type structure (reverse drag) common at the toe of the fault on the downthrown block. As illustrated in Figures 4 and 5, the structural reversal usually occurs in sediments predating the particular fault in question. A closed structure is created by tensional stress normal to the fault plane resulting in rota-

tion of the downthrown block in the direction of the stratigraphic throw near the foot of the fault. Antithetic faulting generally accompanies reverse-drag structures and these faults are shallow displacements dipping in opposition (antithesis) to the dip of the plane of the parent growth fault.

Oil fields produced by growth-fault mechanics usually are hybrid composites of stratigraphic trapping in younger beds where the deposition was controlled by contemporaneous faulting whereas, at depth, the same field exhibits structural accumulation owing to reverse drag. The Bellevue growth-fault trend illustrated in Figure 9 has these characteristics. A stratigraphic section across Bellevue field (Fig. 10) demonstrates the effect of the parent fault on the deposition of the Bellevue sandstone and overlying Reef Ridge and Macoma shales. In this zone, trapping is entirely stratigraphic with regional dip the only structural control. In subjacent beds of the Gosford sandstone, a small structural accumulation of oil occurs at this fault zone. The Gosford sandstone predates the growth fault and trapping primarily is reliant on structural reversal against the plane of the Bellevue growth fault. Canfield Ranch, just on the south, in the deeper pay zones, has a sizable structural accumulation attributed to reverse drag.

Submarine Channels

Sediment transport within the midfan area was accomplished primarily by distributary-like subaqueous channels similar in morphology to subaerial bird's-foot or lobate deltas (Fig. 6). In all probability, each fan was built up gradually from the ocean floor by the oscillatory swinging of a major feeder channel emanating from a submarine canyon. In this manner, sediment layers gradually were built up until all bathymetric irregularities were smoothed out to form a low-relief bathyal plain with a slightly convex upper surface.

Oceanographic surveys of recent years have documented leveeing related to subaqueous-channel deposition in known subsea fans (Nelson and Kulm, 1973). It is reasonable to assume, therefore, that similar mechanics were operative during Stevens deposition. Detailed stratigraphic sections through the Stevens in select areas reveal periodic high concentrations of sandstone flanked by intervening areas of shale. High sandstone concentrations are correlated with sand-filled channel conduits and shaly areas with natural levees and areas of coarse-sediment bypass.

As sedimentation continues, differential compaction is continually in progress with substantially greater amounts of compaction in the shaly

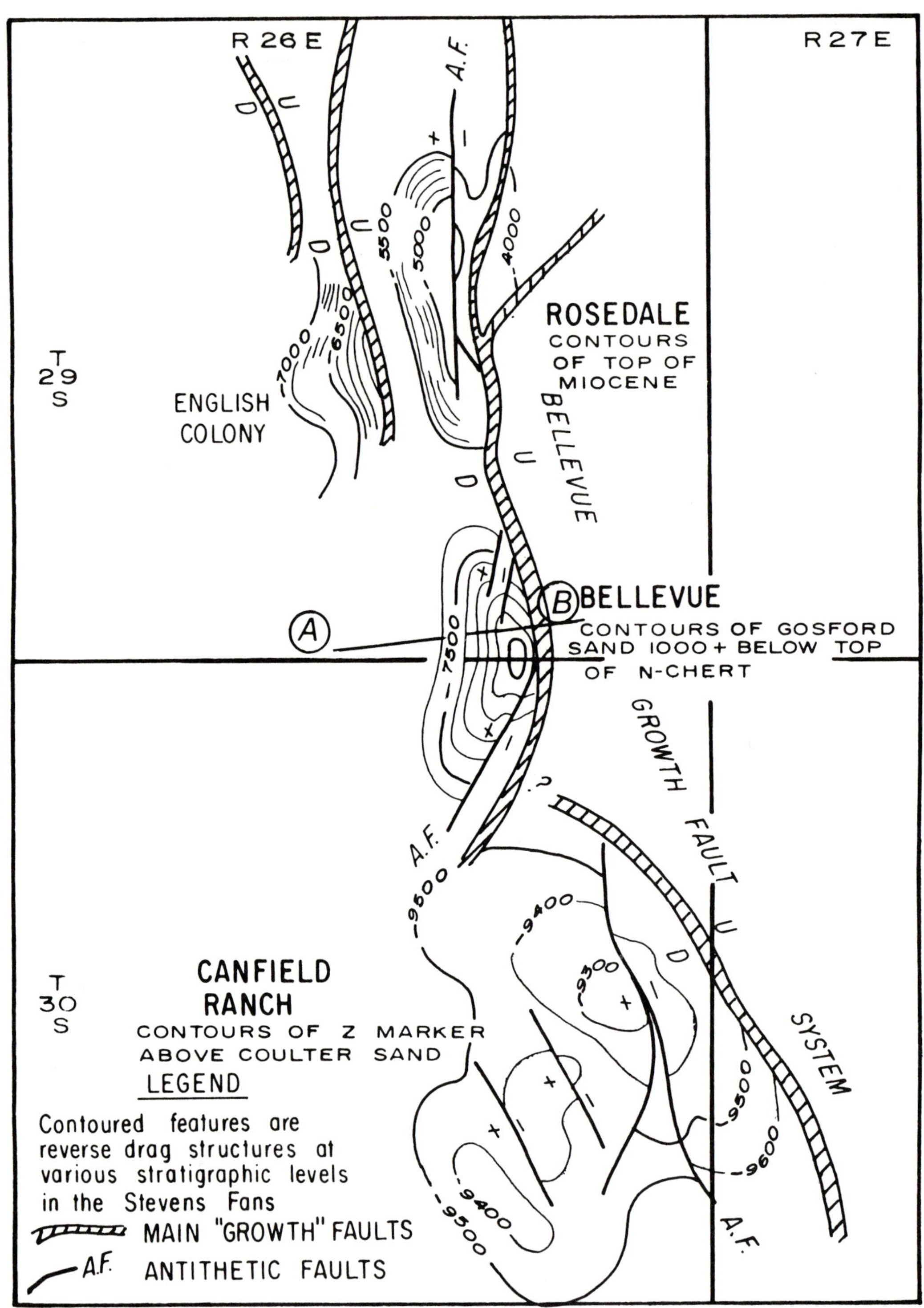

FIG. 9—Bellevue growth-fault system, Bakersfield arch. *AB*, Stratigraphic section (Fig. 10).

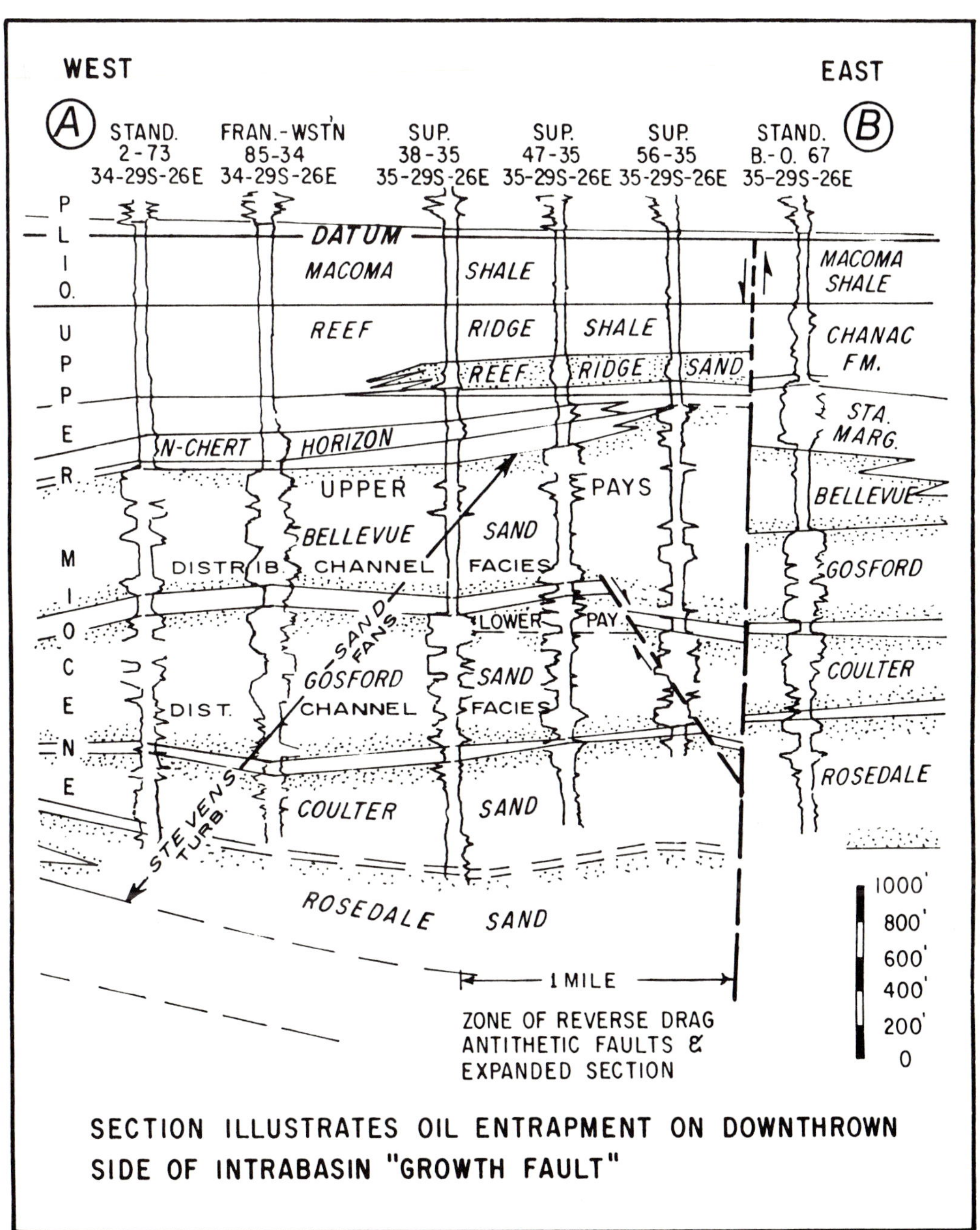

FIG. 10—Stratigraphic section across Bellevue field. See Figure 9 for location.

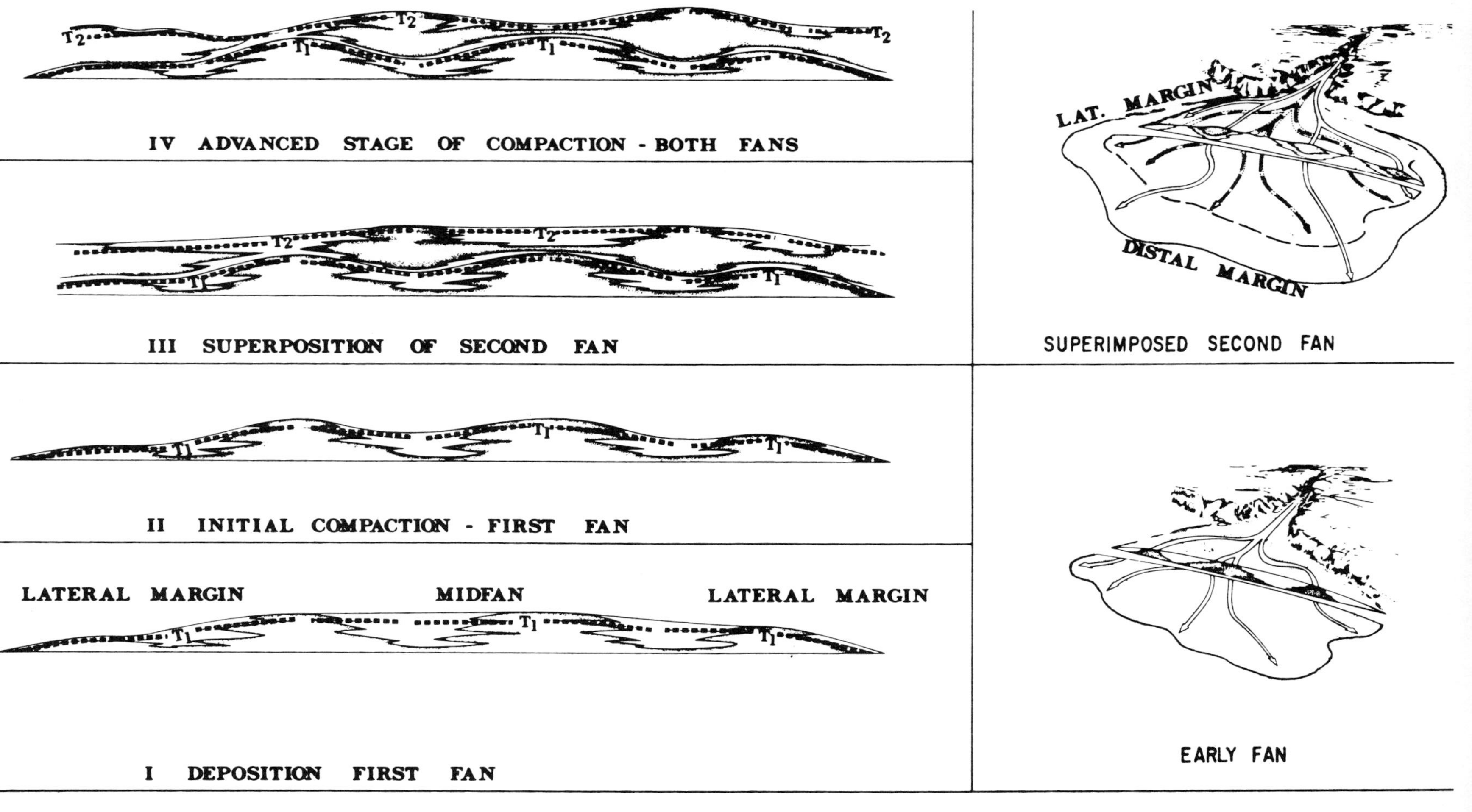

FIG. 11—Hypothetical reconstruction of depositional model for two Stevens turbidites illustrating sedimentary accommodation and alternate stacking with vertical offset of distributary channels on Bakersfield arch, Kern County, California. T_1, T_2 are time lines represented by thin shale zones.

area. Eventually sandstone-distribution patterns shift much as in modern deltas. Because turbidites are density flows, they seek the lowest topographic levels. Thus, the major patterns of turbidite flow respond to compacting shaly areas by abandoning the older distributary pattern and for a time occupying new ones. I refer to this mechanism as "sedimentary accommodation" and illustrate the concept in Figure 11. For simplicity, this illustration is limited to only two major depositional cycles. They may be likened to the vertical "tongue-in-groove" relation recognized in the mapped channel deposits of the Bellevue sandstone and the subjacent Gosford sandstone illustrated in Figure 6.

Repetition of this sequence through time results in alternate stacking of distributary channels through the vertical stratigraphic section. Compaction anticlines of relatively low structural relief form over the distributaries and shallow synclines within the shaly areas. Thus, because of alternate stacking, maps of different horizons reveal reversal of structural attitudes with depth. A broad, productive "anticline" on an upper marker is a nonproductive syncline on a lower marker. Inversely, broad productive anticlines on lower markers are nonproductive synclines on upper markers.

Although the writer arrived at the preceding observations independently, he now recognizes similar observations by other geologists. Ferm and Cavoroc (1968) described the alternate stacking of fluviodeltaic channels in the Allegheny Group of West Virginia. Brown (1969) noted a similar situation in the distributary systems of the Pennsylvanian to Permian deltas of north-central Texas. Walker (1976b) illustrated the vertical offset of channel axes for the Capistrano turbidites at San Clemente, California. This phenomenon would appear to be widespread geographically and throughout geologic time.

Oil occurrence—Sanem and Stoddard (1965) first recognized the marked areal offset of Stevens oil pools in the Greater Strand and Canal productive trend and attributed this relation to deflection of upper Stevens channels away from preexisting lower Stevens structural highs. The writer concurs with this explanation but also would suggest that alternate stacking of channel conduits as illustrated in Figures 12 to 14 is also a major contributing factor resulting in significant areal offset of productive zones through the several cycles of turbidite packets represented in the area.

Differential compaction over channel-sand "thicks" produces anticlinal noses like those shown in Figure 12 and on the N-chert structure (Fig. 7). Channel-sand reservoirs oriented down

regional dip, as the Strand, probably are sealed updip by the nearest growth fault, whereas channel reservoirs trending at right angles to regional dip are favorably oriented to trap of their own accord.

As recognized by Sanem and Stoddard (1965), the intricate depositional relations at Strand-Canal can be used in developing significant additional reserves in mature oil provinces but, as seismic techniques are improved, resolution of alternate channel systems may be possible in unexplored areas.

LATERAL-MARGIN FACIES
Sedimentation

Lateral margins are defined herein as that part of the turbidite fan extending from within the submarine canyon along the edge of the fan complex basinward to the highly arcuate toe or distal edge of the fan as shown in Figures 6 and 11. In comparison to the high-energy midfan environment, the lateral margin is a low-energy regime characterized by microchannels, thin sheetlike sandstones deposited by laminar flow, and a dominant lithology of deep-marine foraminiferal shales. Thin time-stratigraphic units (like the N-chert) overlying the fan complex can be observed to converge asymptotically with marker beds underlying the fans to form an attenuated sigmoidal prism. If, however, the lateral margin encroaches on a local bathymetric high, the wedge-shaped prism probably is fairly abrupt and loss of sand very rapid. Many California turbidite basins are bordered by steep fault-bounded topographic highs and, in such places, sand loss within the lateral-margin wedge may amount to several thousand feet within less than 1 mi (1.6 km). Examples of lateral-margin wedges are shown for several ancient turbidites of the San Joaquin Valley in Figures 15 and 18. The terms "lateral margin" and "basin margin" are used interchangeably in the following discussion.

"Basin-margin wedging," a term first coined by R. G. Hubbell of Continental Oil Co., is recognized as possibly the single most important hydrocarbon-trapping factor in California deep-marine basins. From sparker surveys on the outer continental shelf and subsurface studies in the onshore Los Angeles and Ventura basins, it can be determined that any topographic high on the seafloor will deflect density flows and create sand pinchouts toward the high. This deflection will continue until the basin has filled to a depth which will bury the high. Repeated many times, this depositional relation concentrates sands in basinal depressions and creates an overall facies change to shale on the basin margin and on intra-

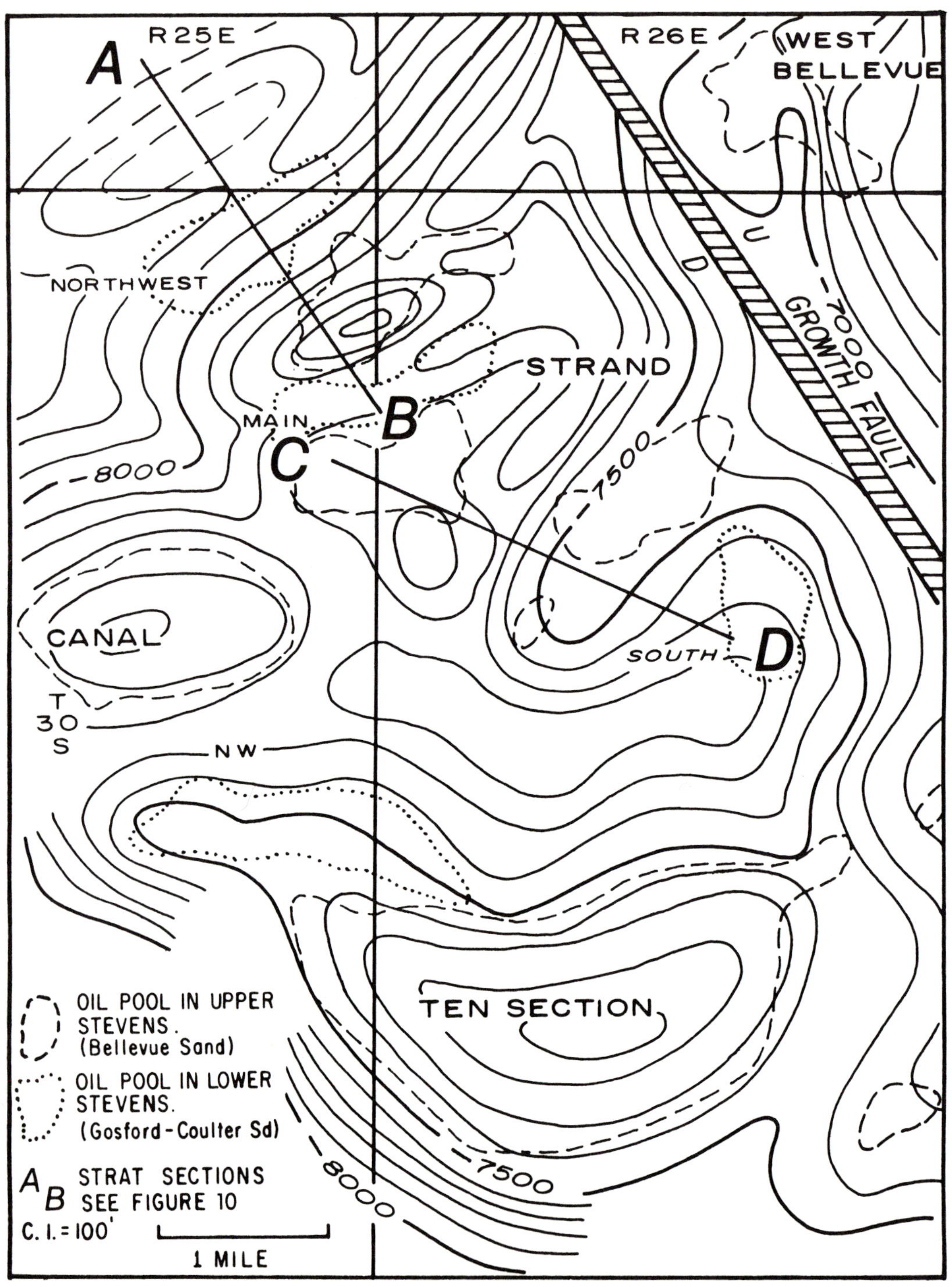

FIG. 12—Structure map of top of N-chert, Greater Strand area, illustrating vertical offset of oil pools. *AB, CD*, are cross sections of Figures 13 and 14.

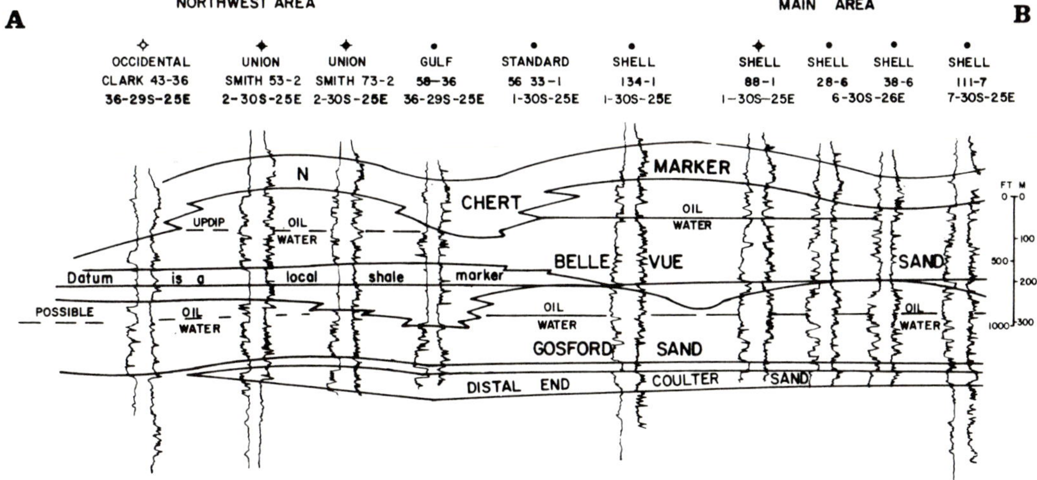

FIG. 13—Stratigraphic cross section through Strand oil field illustrating vertical offset of distributary turbidite channels. See Figure 12 for location.

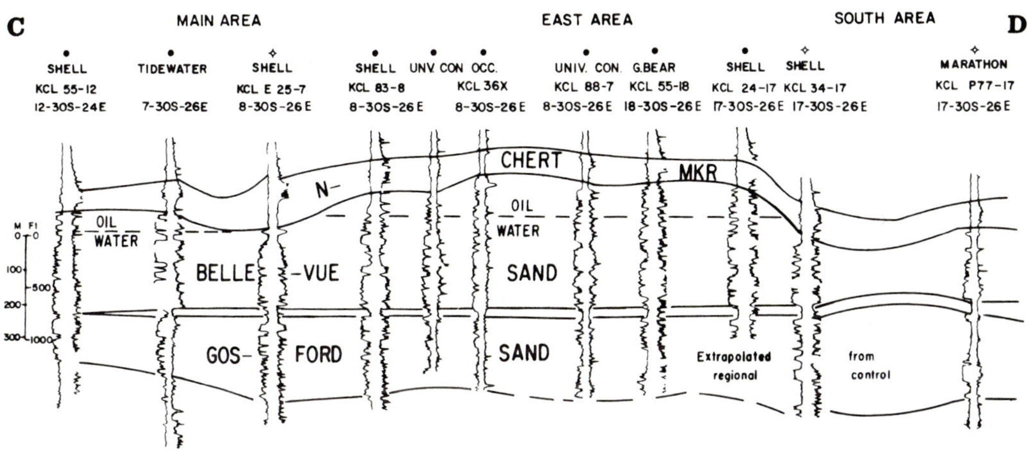

FIG. 14—Stratigraphic cross section through Strand oil field, illustrating vertical offset of distributary turbidite channels. See Figure 12 for location.

basin topographic highs. Basin-margin wedging in the Stevens sandstone is illustrated for the lateral margin of the Rosedale fan where it encroaches on the paleobathymetric high that existed at the present locality of Greeley oil field (Fig. 18) at the north flank of the Bakersfield arch (Fig. 2). Other subsurface fossil turbidites of the San Joaquin Valley, ranging in age from Paleocene to late Miocene (Figs. 15-18), characteristically show well-developed basin-margin wedging closely associated with oil or gas production attributed to stratigraphic trapping against paleotopographic highs and to porosity reduction as a result of facies change from sandstone to shale.

Sandstone loss within the basin-margin wedge (lateral margin) occurs at a constant rate. Thus, if the rate of wedging can be determined from a minimum of control points, a fairly accurate prediction can be made as to gross-sandstone content (equatable to potential pay). In the scatter plots for the several turbidites of the San Joaquin Valley given in Figures 15 through 18, the gross

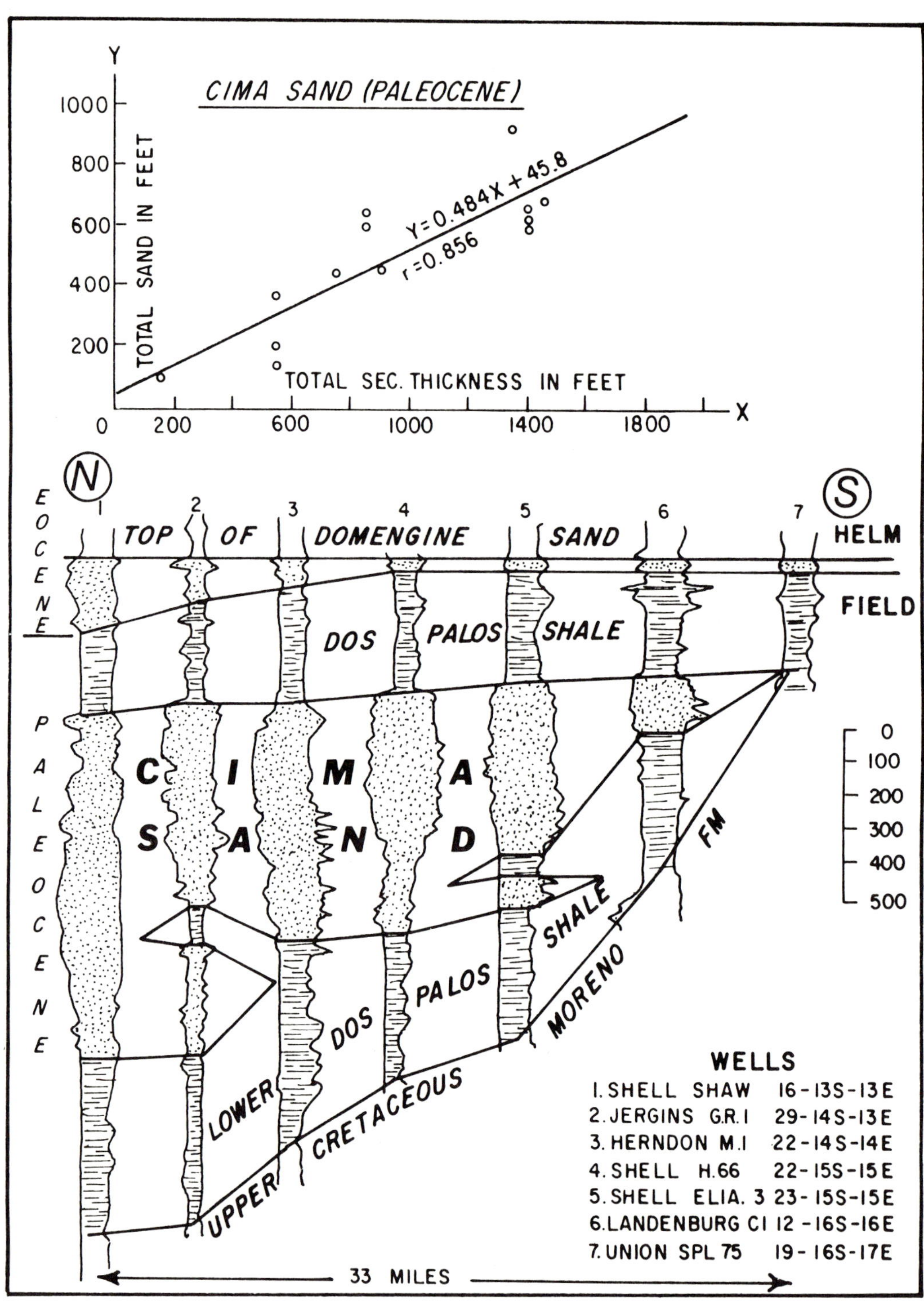

FIG. 15—Basin-margin wedging of Cima sandstone, north-central San Joaquin Valley.

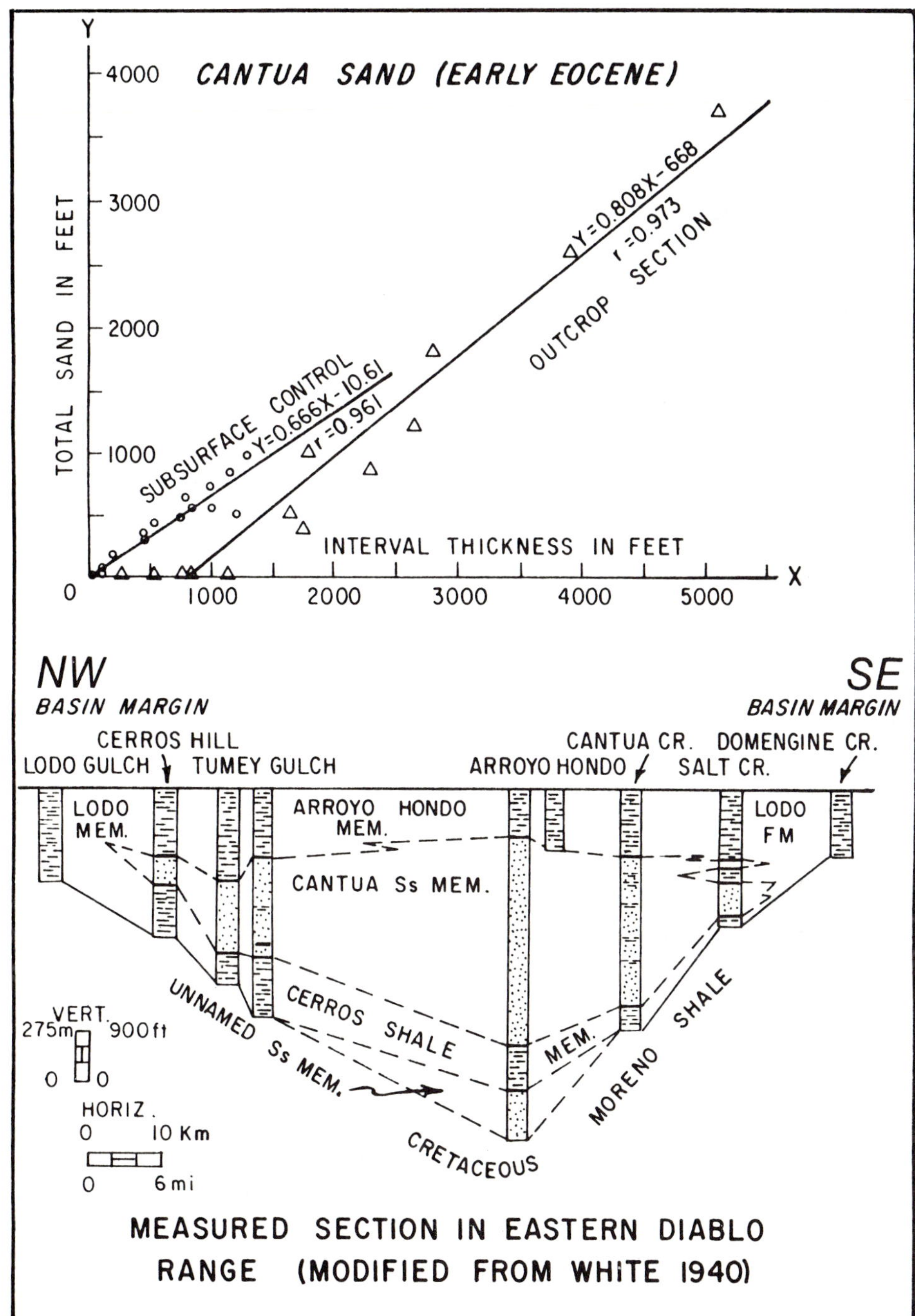

FIG. 16—Basin-margin wedging of Cantua sandstone from outcrop of Cantua turbidite, west side of San Joaquin Valley.

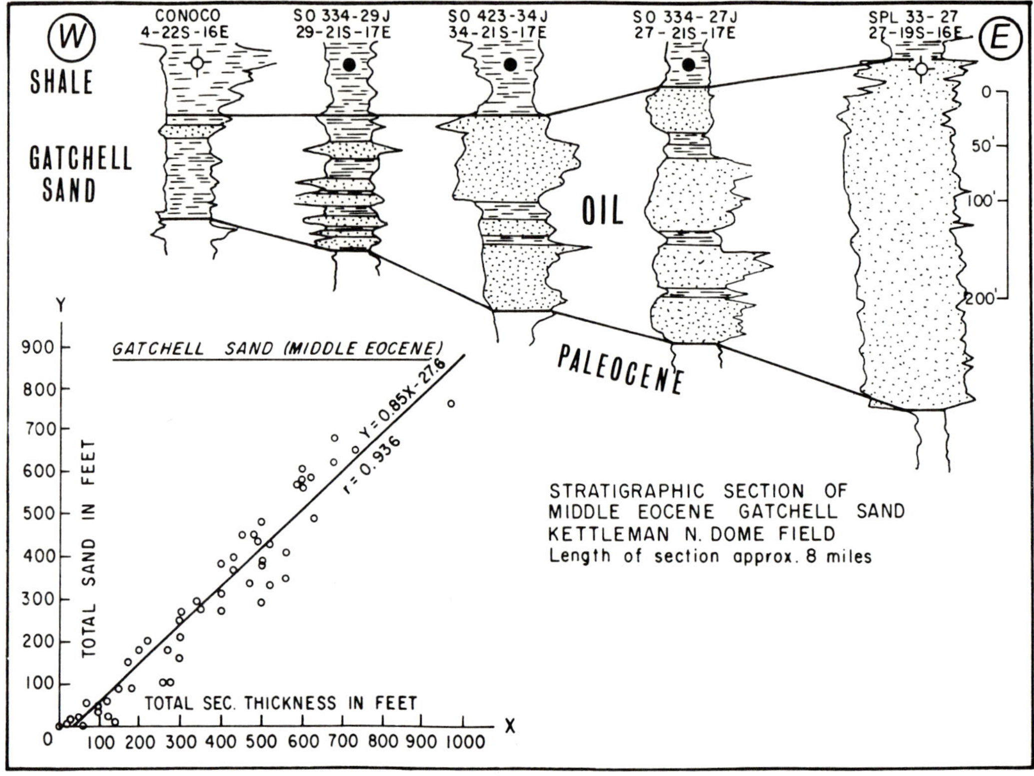

FIG. 17—Basin-margin wedging of Gatchell turbidite, central San Joaquin Valley.

interval is recorded on the X-axis and the gross sandstone on the Y-axis. In every sandstone, including the Stevens, the best least-squares curve of the form, $Y - aX + b$, is a straight line. Correlation coefficients (r), give values greater than 0.85 which is considered very good for most geologic analysis.

Oil Occurrence

Greeley field at the northern end of the Bakersfield arch (Figs. 6, 18) is the only oil field, within the scope of this study, which is attributed to stratigraphic trapping in the lateral-margin (basin margin) environment. However, it is suggested that the giant Spellacy, Monarch, and Leutholtz pools of Midway-Sunset field on the west side of the San Joaquin Valley are trapped by porosity pinchouts in the basin-margin environment. These are turbidite sandstones time-equivalent to the Stevens sandstone of the east side of the San Joaquin Valley.

Basin-margin or lateral-margin wedges should be detected readily by seismic surveys in underexplored areas such as the Maricopa subbasin in southeastern Kern County. On the basis of examples given in the preceding discussion, fairly accurate predictions of potential pays also might be made by methods similar to those used here.

DISTAL-MARGIN FACIES

Sedimentation

Le Blanc (1977) referred the distal part of the Jackfork turbidites of eastern Oklahoma to the fan-lobe sequence characterized by thin-bedded, laterally continuous interbedded sandstone and shale packets having a tendency to thicken upward in the section. In this environment, Le Blanc expected the vertical continuity of the reservoir sandstones to be extremely poor and indeed this appears to be borne out in the Stevens sandstone in similar depositional environments. Walker and Mutti (1973) referred to the distal portion of turbidites in terms of the D through G facies in their scheme of classification. They stated that shale dominates over sandstone, organic activity is high, and slump and slurried structures are common. All of these criteria would appear to be represented in the distal environments of the Stevens turbidites.

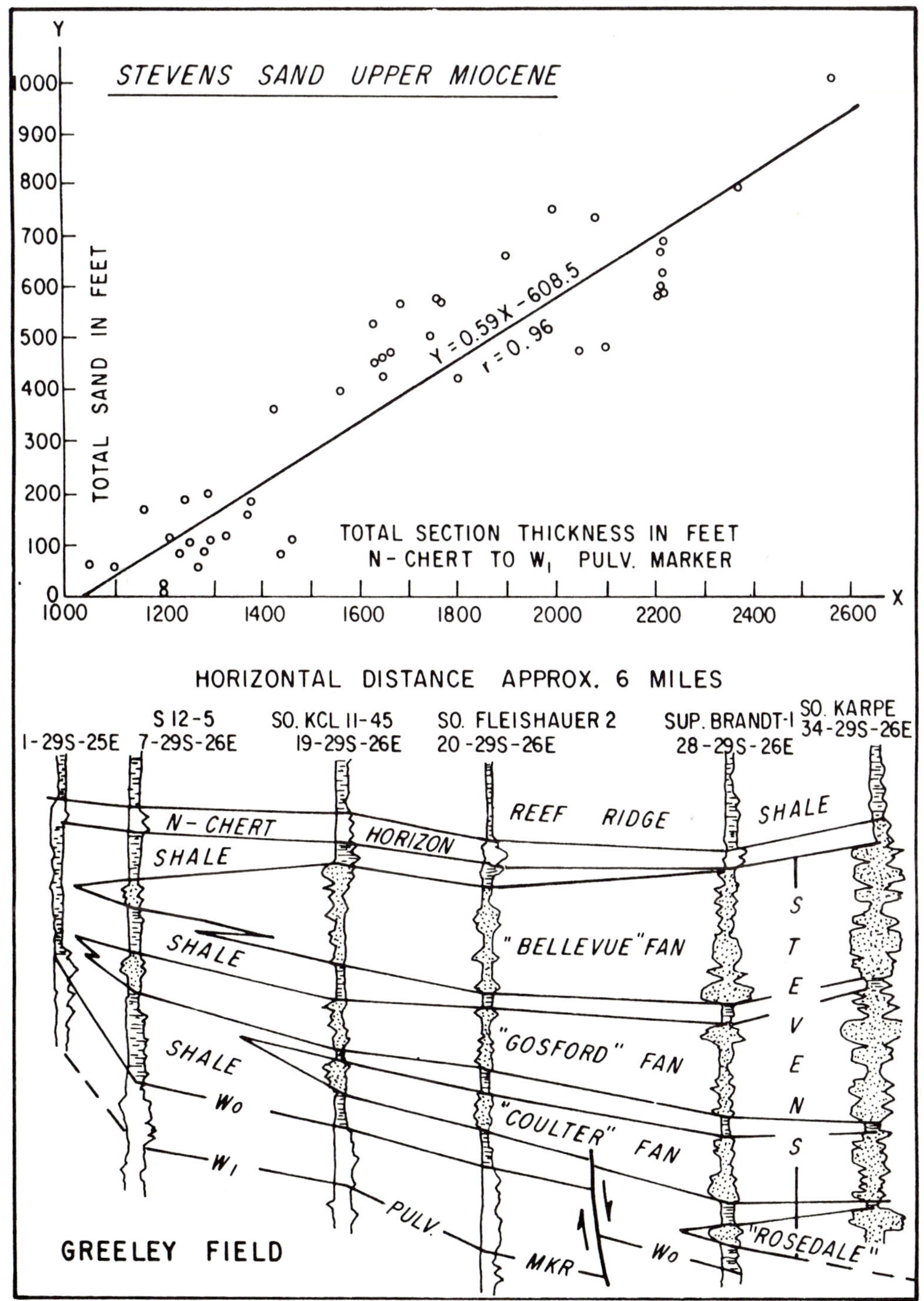

FIG. 18—Basin-margin wedging of Stevens turbidite fan at Greeley field, Bakersfield arch.

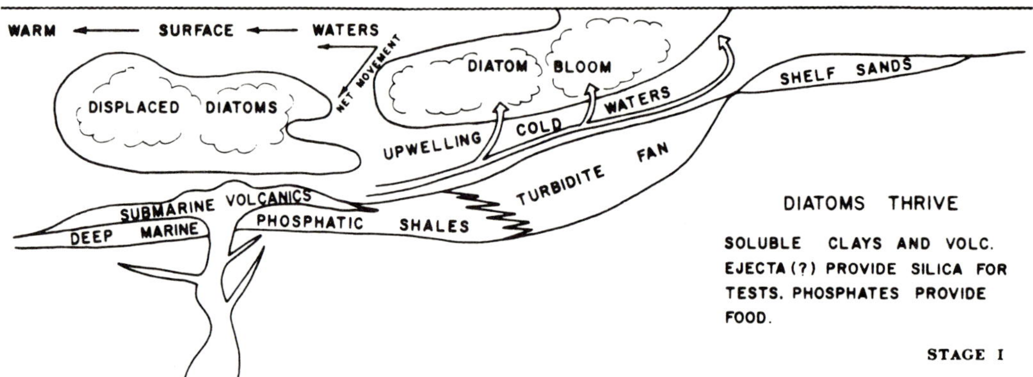

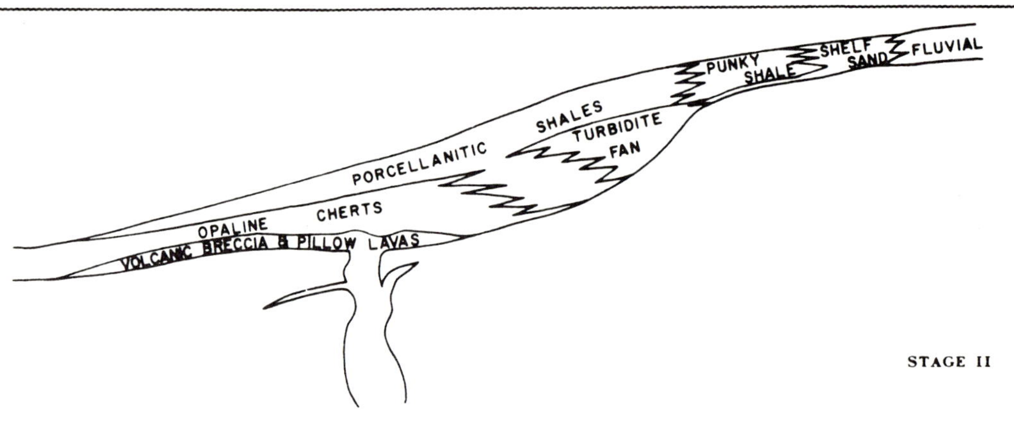

FIG. 19—Suggested origin of siliceous shales in distal turbidite fan.

Stevens turbidite sandstones grade basinward (Figs. 2, 4, 6) into a highly siliceous shale facies. The siliceous shales of the San Joaquin Valley were discussed by Regan (1953) and subdivided into three lithologic types: punky diatomite, porcelaneous brittle shale, and flinty opaline cherts. The opaline cherts occupy the deeper parts of the depositional basin. Regan was first to direct attention to the relation of these shale facies to the Stevens sandstones and their hydrocarbon potential.

Bramlette (1946) attributed the origin of the siliceous shales of the Monterey Formation of California to diagenetic processes affecting large concentrations of diatom tests characteristic of these rocks. Because the Antelope shale equivalents of the Stevens are part of the Monterey, Bramlette's conclusions would apply to the Antelope as well.

According to Bramlette, porcelaneous shales and opaline cherts are mutually intergradational and both are gradational with punky diatomite. Porcellanites commonly retain the original diatom skeletal structure and have measurable porosity and permeability. Opaline cherts on the contrary exhibit only shadowy organic structures, and most traces of the diatoms have been obliterated. Porosity and permeability are created in these rocks by shrinkage cracks in opaline gels and pervasive postdepositional shears and fractures. Bramlette, however, did not relate deep-water Miocene sandstones to the various siliceous-shale lithologies to combine the whole system into a dynamic framework.

A suggested relation between siliceous shales and deep-sea fans is given in the following discussion (see Fig. 19).

Diatom blooms would flourish in oceanic areas of cold, upwelling currents where phosphates provide nutrients to sustain a rich diatom biota. Areas of upwelling also would be preferred areas of turbidite deposition on the outer continental shelves where gravity flows would introduce abundant clay into the environment. Silica of clays is soluble in seawater, thus providing added silica for growth of diatom tests.

Temperature differential caused by the upwelling cold currents would force warm surface waters oceanward, propelling the diatom blooms out to sea. Eventually the diatom cloud must sink to the bottom. Submarine volcanism prevalent in Miocene time may have accelerated the settling process by lava flux, and additional silica enrichment of descending diatom blooms may have been provided by the volcanic ejecta.

Oil Occurrence

Siliceous shales form prolific reservoirs at Buena Vista Hills (Fig. 20) and Elk Hills where local flexures have created open oil-filled fracture systems directly updip from sandstone reservoirs. Low-gravity crude seems characteristic of the cherty shales whereas high-gravity oils have a preference for the porcellanites. The high-gravity oil typifies the San Joaquin Valley fractured shale reservoirs. The cherty facies is more prevalent in the Santa Maria Valley coastal basin.

Seismic mapping of facies change from sandstone to siliceous shale has been quite successful in the valley. Areas of siliceous shale are depicted by very high-amplitude reflections or strong energy bands correlating precisely with porcellanite and cherty areas mapped by well control. Any fault or fold should provide the necessary stress to fracture these shales, creating sufficient porosity and permeability for hydrocarbon accumulation.

TURBIDITE SEDIMENTATION IN MARICOPA SUBBASIN

The Maricopa subbasin is separated from the main part of the Great Valley geosyncline by the north-flanking Bakersfield arch (Fig. 2). It is bounded on the east by the Mountain View–Edison high and on the south by the Tejon platform. The western limits are unknown and probably originally lay west of the present trace of the San Andreas fault. Right-lateral wrench-fault movement may have displaced the western boundary at least as far north as Santa Cruz (about 170 mi or 272 km) since late Miocene time.

Actually, flysch sedimentation commenced earlier in the Maricopa subbasin than on the Bakersfield arch. As stated previously, the arch itself was a sediment-starved mildly submergent high through most of early and middle Miocene time. The gravel-filled Bena channel (Fig. 21) served as a source or dispersal center for submarine fans debouching into the northern Maricopa basin. The southern part of the basin was filled by sediment discharged across the Tejon platform over the rising lip of the White Wolf fault. Not until the early late Miocene (early Mohnian) did the Bakersfield arch receive any appreciable flysch sedimentation.

Explanation for Source Shift

It long has been a question as to why the Bakersfield arch was a depositional site for immense deposits of gravity-flow material like the Stevens sandstone late in Miocene time. Characteristically, bottom-seeking currents would have bypassed the arch and persisted in filling the Maricopa deep, keeping pace with local subsidence. The only sedimentation on the arch would result from progressive onlap as the Maricopa subbasin was filled.

The source shift is related to on-land stream capture of the ancestral Kern River and the following explanation is suggested.

Sierran drainage patterns probably were much the same in Miocene time as they are today. In the early middle Miocene (Fig. 21), the Kern River excavated a valley along the Kern Canyon fault, thus allowing southward flow down the spine of the Sierras, rather than down the shortest gradient west to the valley floor. Ancestral Kern River drainage emptied into the Maricopa subbasin at the foot of the White Wolf fault, east of the Edison high. About middle late Miocene time, uplift of the west block of the Kern Canyon fault caused rapid headward erosion of a relatively inconspicuous stream near Bakersfield—the embryonic west-flowing branch of the Kern River. This stream would capture the south-flowing Kern River because its headwaters were closer to the river than those of other west-flowing streams farther north. Two west-flowing former tribu-

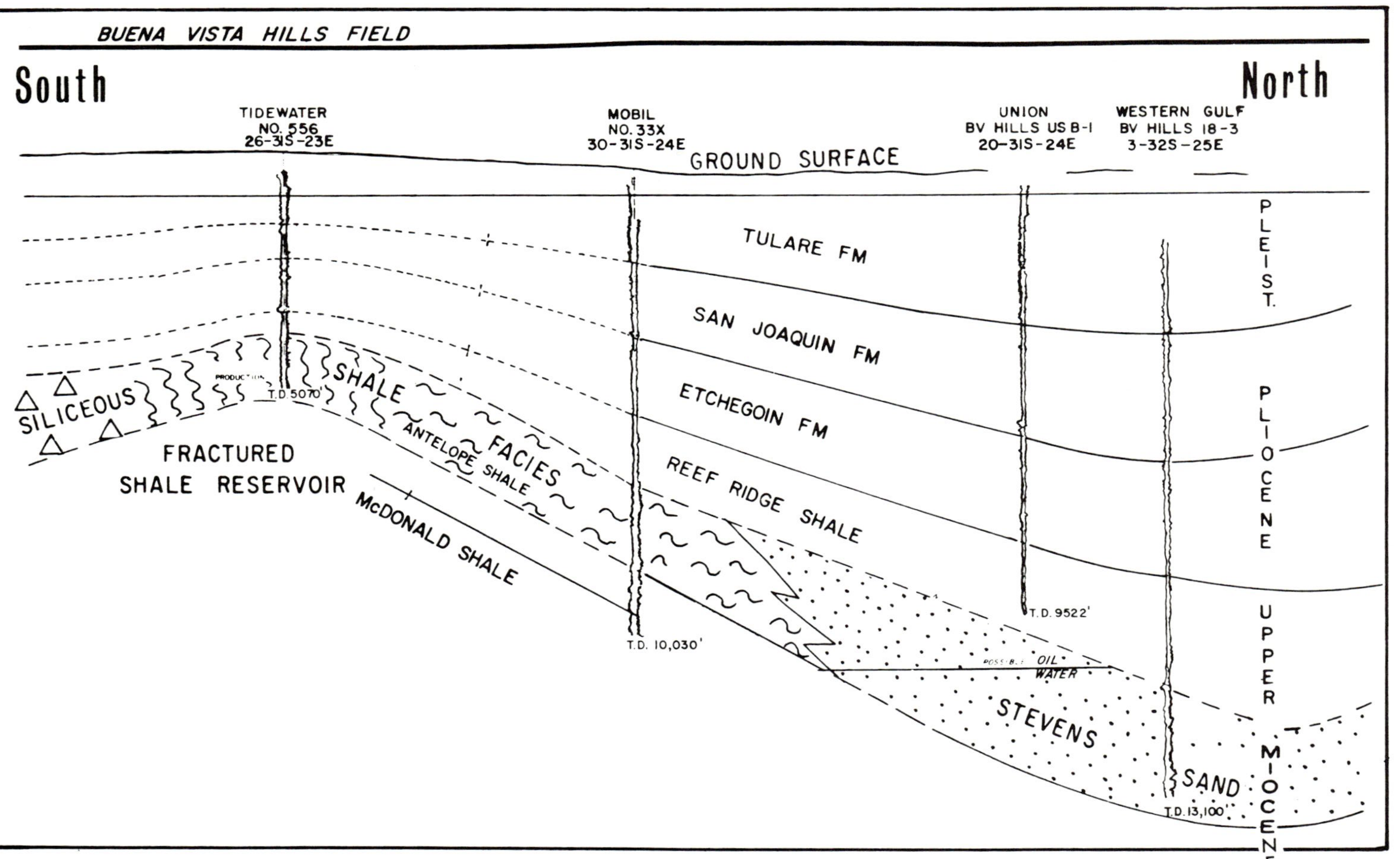

FIG. 20—Oil field in distal-shale facies of Stevens turbidite fan.

Bruce A. MacPherson

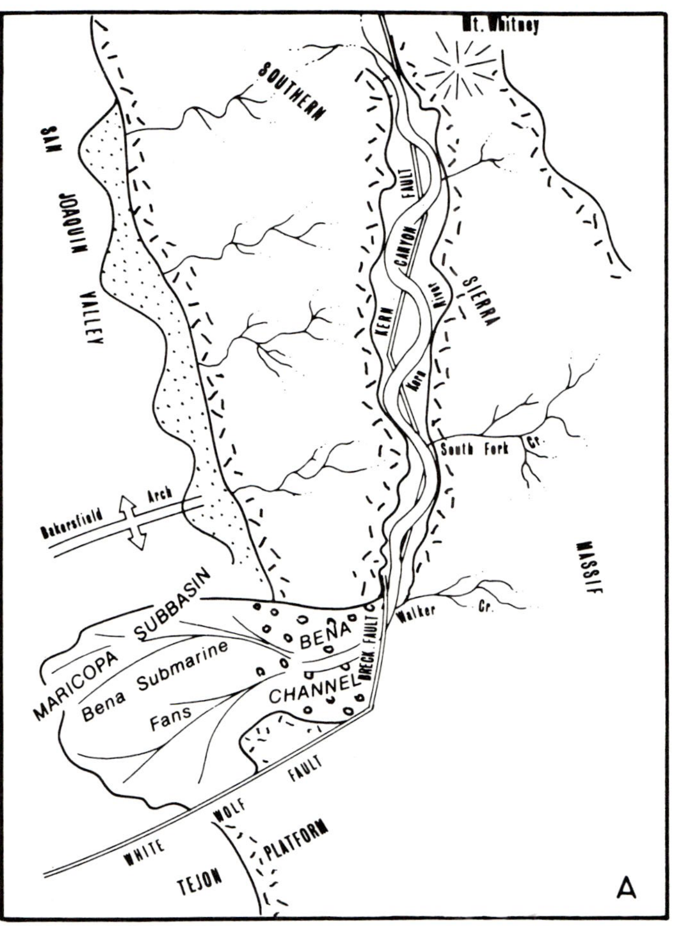

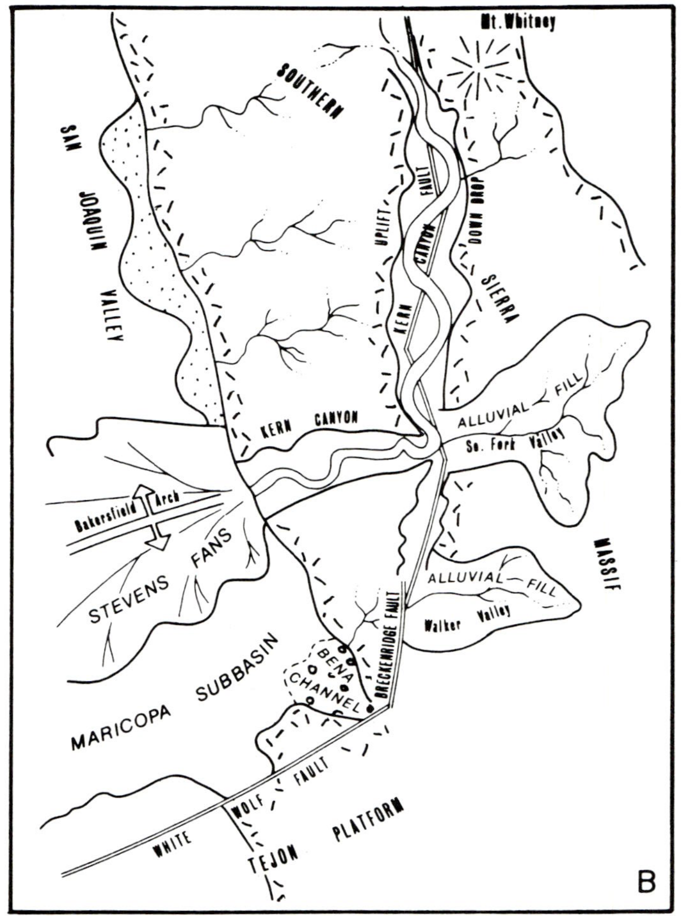

FIG. 21—Postulated history of ancestral Kern River during Miocene time. Paleogeography during **A**, middle Miocene and earliest late Miocene (Relizian–lower Mohnian); **B**, middle late Miocene (upper Mohnian to lower Delmontian).

taries to the ancestral Kern River were beheaded, resulting in alluviation of Walker and South Fork Valleys.

From the time of capture, the lower reach of the Kern River was abandoned and the Bena channel ceased to function as a source for the northern area of the Maricopa subbasin. The Bena channel filled its conduit with fanglomerate. Henceforth, the newly established lower course of the Kern River supplied abundant detritus to the Miocene sea with sediment influx and submarine currents capable of excavating canyons, at Rosedale Ranch and Fruitvale. These canyons became staging areas for the four cycles of flysch sedimentation represented here. Early turbidites were confined to the crest of the arch but, as available space was filled, later fans offlapped west, as well as off the flanks of the arch. Under these conditions the arch, although high, was overwhelmed by sediment. Coriolis force may have carried the last-deposited turbidites south-southeast to complete the filling of the Maricopa subbasin.

Submarine Fans

Only the periphery of the Maricopa subbasin is available for study through well control in the Mountain View–Edison field area and in the oil fields of the Tejon platform. Conditions in the basin center must remain a matter of geologic extrapolation and interpretation from seismic control, for too few wells have been drilled to penetrate the entire Miocene section.

The stratigraphic correlation chart in Figure 3 lists the prominent sandstone members thought to be of turbidite origin. Those originating from the Bena channel are referred to the informally named "Bena fan" whereas the sands derived from sources on the Tejon platform comprise the informally named "Valpredo fan" at the foot of the White Wolf fault.

Middle Miocene (Fig. 22)

During the early middle Miocene (Relizian) the Olcese sand was transported through the Bena channel; it was deposited near the eastern end of the Bakersfield arch, but mainly spread out south and southwest at least as far as the now-abandoned Kernsumner field. In was narrowly confined in the Bena channel north of the Mountain View–Edison high, eventually taking on the characteristics of an arcuate fan shape farther southwest in the basin.

Oil production has been established at Ant Hills (a faulted fold) and stratigraphically in a two-well pool along the lateral-margin facies on the downthrown block of the Mountain View fault. Olcese channels also are well represented in the south of the Tejon platform where they are

productive in stratigraphic traps flanking the bathymetric high of the North Tejon field.

Late middle Miocene (Luisian) is represented by the Nozu sandstone on the north and its southern counterpart, the Valv sandstone of the Tejon platform. Both Nozu and Valv sandstones are contributing stratigraphic oil in the periphery of the basin (Fig. 22).

Late Miocene (Fig. 23)

Lower upper Miocene (lower Mohnian) Wicker sandstone (also referred to as Palla-O'Niel in Kernsumner field) was deposited in the Bena channel and is productive at several localities in stratigraphic traps. Farther south and west, the sands spread out fanlike in a pattern similar to the subjacent Nozu and Olcese. In the south, on the Tejon platform, the time-equivalent Reserve-Pulv sandstones can be mapped to the trace of the White Wolf fault. They probably thicken on the downthrown block to form the upper member of the postulated Valpredo fan. These sandstones are highly productive in submarine channels and fractured-shale reservoirs at North Tejon and Tejon fields. At the same time, the Rosedale channel (Martin, 1963) was being excavated at the eastern extremity of the Bakersfield arch, possibly as a result of the northward shift of the ancestral Kern River and the eventual abandonment of the Bena channel as a significant dispersal area for the Maricopa depocenter.

Middle Late Miocene

In late Mohnian time, turbidite deposition from the Bena channel ceased. The channel became choked with Santa Margarita-type sediments which probably formed a sort of estuarine delta of considerable thickness. Santa Margarita barlike sands of the shallow-shelf environment were deposited fringing the Edison–Mountain View platform where they form highly productive stratigraphic traps. Similar conditions prevailed in the south on the Tejon platform. Turbidite sedimentation then shifted north to the site of the Bakersfield arch but, possibly because of the Coriolis force, the Stevens Fruitvale fan swept turbidite sands as far south into the Maricopa basin as the Stenderup pool of Mountain View field. Menard (1955) noted this sinistral migration in almost all westward-flowing turbidite fans in the recent sediments of the outer continental shelf. He attributed this phenomenon to the right-hand shift caused by the Coriolis force overbuilding the north side of the submarine fans beyond a certain critical depositional limit, thus causing the fan to migrate south to compensate for the oversteepening. It is possible that the Stevens fans, which

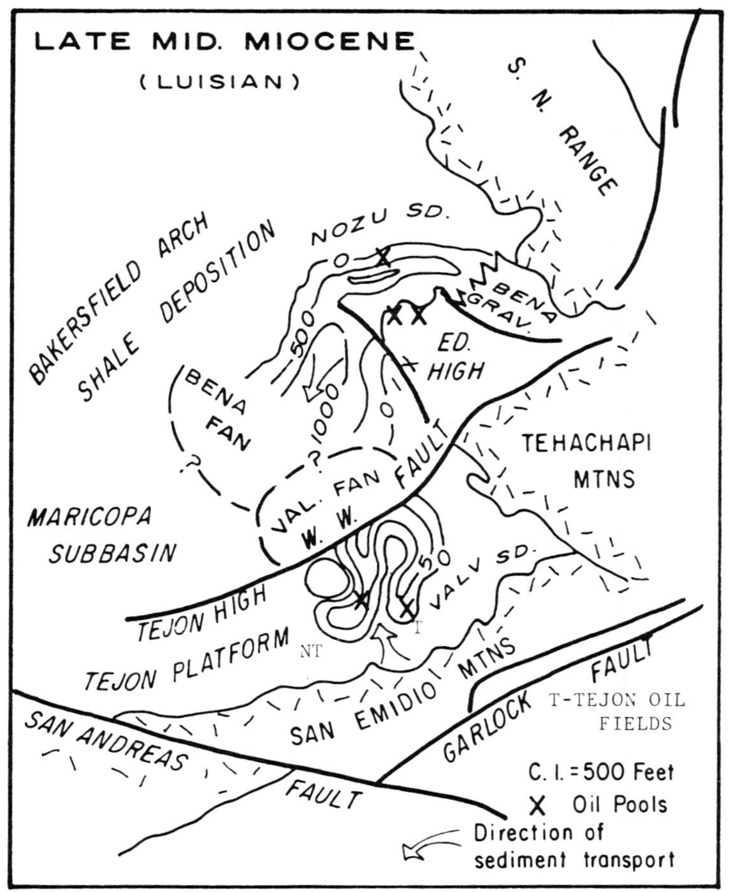

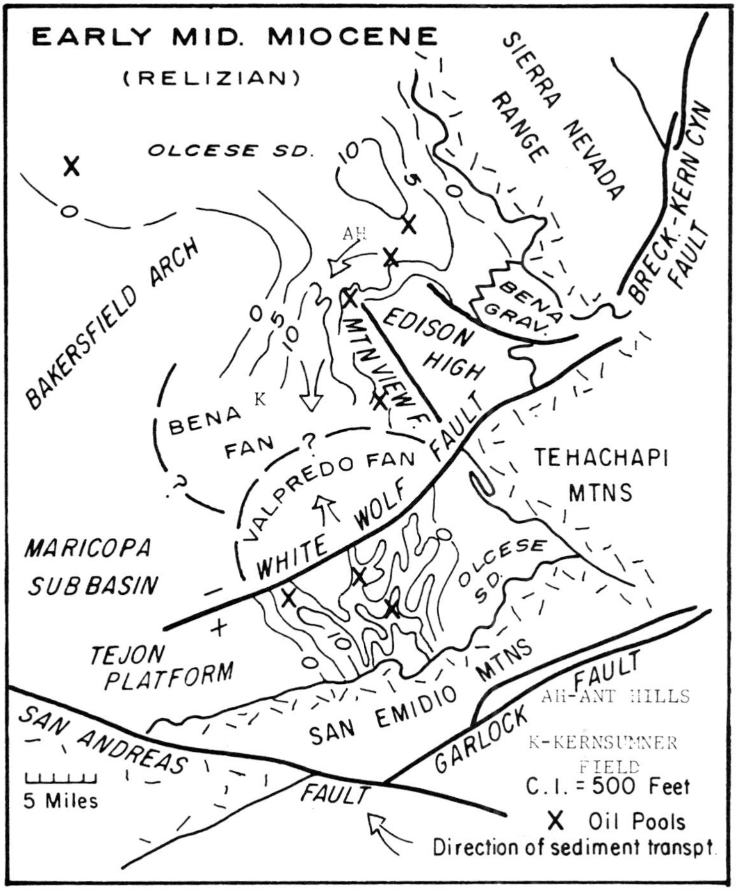

FIG. 22—Turbidite deposition in Maricopa subbasin during middle Miocene time.

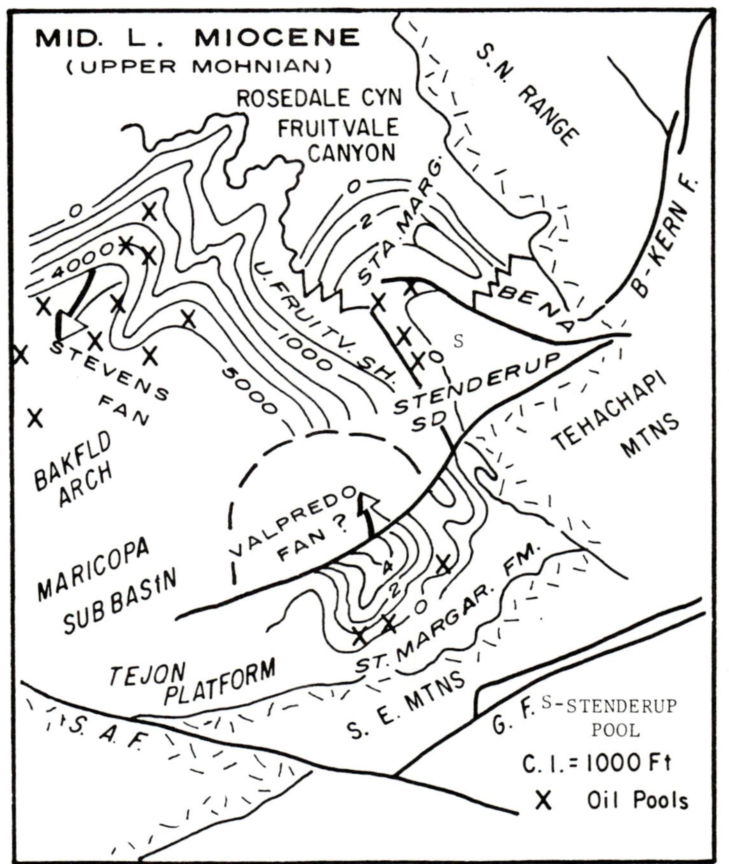

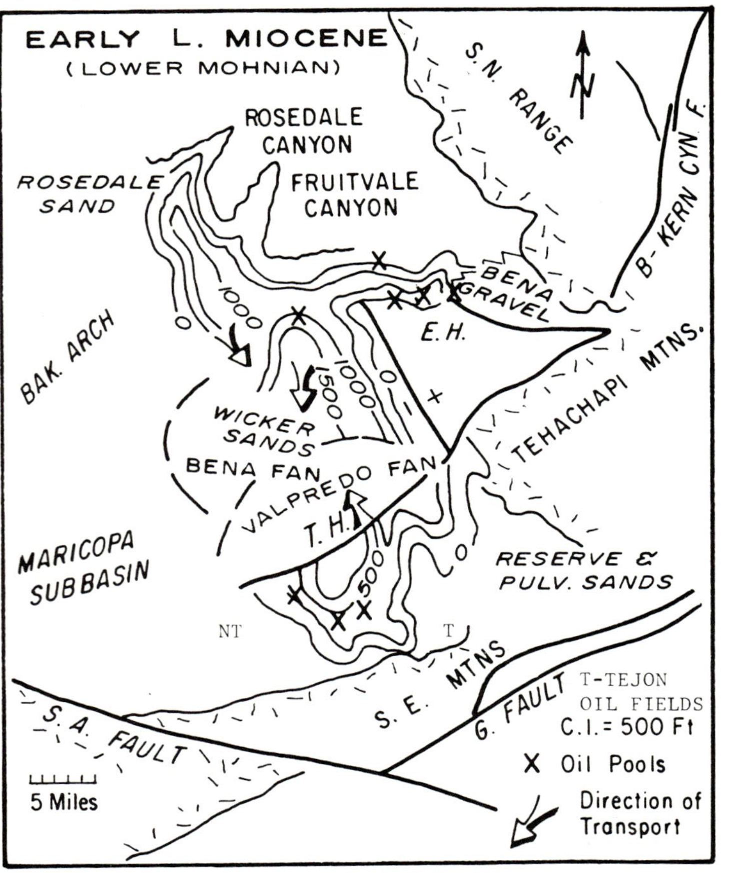

FIG. 23—Turbidite deposition in Maricopa subbasin during late Miocene time.

were westward flowing, responded to the same forces and thus eventually shifted southward (left) to complete the filling of the Maricopa basin.

Future Hydrocarbon Potential

According to Zieglar and Spotts (1976, p. 34), "In the Tejon depocenter [Maricopa subbasin], the lower part of the Pliocene and all the Miocene and older beds have subsided below 10,000 feet and this gross volume of about 1,800 cu mi may be considered to have entered or passed through the temperature zone for hycrocarbon generation." These authors also concluded that the Tertiary beds in the Maricopa basin have been in generation longer than any other depocenter in the San Joaquin Valley, and the 41 producing areas flanking the basin have produced more than 4.3 billion bbl of oil. Therefore, it is justifiable to conclude that an optimistic future awaits the explorer in the Maricopa basin who will evaluate the seismic control in view of the basic geologic relations readily observable in the Stevens and older turbidite fans of the southeastern San Joaquin Valley.

CONCLUSIONS

1. As viewed in this study, the Stevens sandstone is an all-inclusive term encompassing more than 4,000 ft (1,200 m) of interbedded deep-marine sandstone and shale that originated as two major coalescing submarine fans issuing from two source areas at the eastern end of the Bakersfield arch, one at Rosedale Ranch field and the other at Fruitvale field (Fig. 2). Each fan complex is named after its respective source. Four discrete depositional cycles are named informally in ascending order: "Rosedale," "Coulter," "Gosford," and "Bellevue." In formal nomenclature, the Stevens is assigned a late Mohnian age, but here the Rosedale sandstone of early Mohnian age also is included in the Stevens group.

The principal transporting agent for the sands comprising the Stevens submarine fans probably was the ancestral late Miocene Kern River. Upon entering the valley floor, a few miles east of Bakersfield, the stream (probably forming a bajada) bifurcated to form two major branches, one branch dispersing sediment through a shallow marine canyon at Rosedale Ranch field, the other through a similar canyon at Fruitvale field. From these sources, the fans were free to prograde westward on the ocean floor forming large arcuate deltalike features. This free pattern is in sharp contrast to most California fossil turbidites which usually were confined to elongate narrow fault-bounded troughs, where today only a narrow

band paralleling their basin margins is available for study, the axial parts being too deeply buried to be reached by the drill.

2. Submarine canyons at Rosedale Ranch field and Fruitvale field were areas of sediment bypass, and therefore contain little if any Stevens-type sandstone. They were backfilled with a chaotic, nonmarine appearing Chanac sand after the cessation of turbidite sedimentation during the early Pliocene (Macoma) transgression. Sediment prisms in the proximal areas form the present oil fields.

3. Growth faults downslope from the canyon mouths characterize the midfan facies. The submarine channels are less than 1 mi (1.6 km) wide but many hundreds of feet deep. Thick turbidite sand packets filled the channels, but the flanking levees and interchannel swales received only very thin sand—dominantly highly compactible silt and mud deposits. Less compactible channel-fill sands formed low-relief positive features on the ocean floor whereas the interchannel areas formed flanking depressions. Introduction of the next succeeding fan caused the new gravity-controlled channels to follow the depositionally created bathymetric lows of the interchannel areas on the subjacent fan. Thus, through time, thick channel deposits became "alternately stacked" vertically upward through more than 4,000 ft (1,200 m) of Stevens section. Structure contouring shows low-relief compactional anticlines over the old sand-filled channels and, paradoxically, synclines in the underlying highly compacted swale deposits of the subjacent fan—a phenomenon here referred to as "structural inversion due to differential compaction and sedimentary accommodation."

Most major oil fields of the Stevens sandstone are in the midfan facies within the submarine-channel regime. Alternate stacking of channel systems may satisfactorily account for the significant areal offset of oil pools recognized in most of the major fields having multiple pays. It may be a useful concept in finding new reserves in maturely developed areas but, most important, it serves as the geologic rationale for drilling every wildcat completely through the Stevens or other thick turbidites before they can be considered to have been tested.

4. Contemporaneous faulting also characterizes the midfan facies. Midfan aprons are periodically traversed by down-to-the-basin (west) gravity-slump or growth faults not dissimilar to the tectonogenetic larger scale features recognized in the Gulf Coast Tertiary. It is unknown if the fault planes of the Stevens growth faults become arcuate in profile with depth and eventually asymptot-

ic to the bedding planes, dying out in the thick shales of the underlying middle and lower Miocene, or if they continue to be nearly vertical with depth to the basement. In the writer's opinion, the Stevens growth faults may have been initiated by the extremely rapid flysch-type depositional loading of the Stevens turbidites on the underlying thick Miocene shales. Dewatering of the subjacent shales would contribute to the process of slumping through lubrication of incipient fault planes. Periodic movement along basement blocks or wrench faults coupled with occasional earthquakes common to the area would aid further in triggering the slumping process. Stevens growth faults affect trapping conditions in the following manner.

a. The faults form an updip seal for the channel sand reservoirs whose porosity trends ordinarily would be in unfavorable orientation on regional dip for trapping oil.

b. Reverse-drag features accompanied by antithetic faulting produce structural reversal on the downthrown rotated block at the foot of the growth fault. Usually this occurs in the older, deeper fan deposits. Thus prolific structural oil pools can be present at depth in fields whose stratigraphic oil at shallower depths is trapped only on homoclinal dip—another reason for deeper testing any Stevens field beyond the first shallow pay zone.

Recognition of such fault systems can be accomplished only by investigating several lines of evidence, because few wells penetrate a major growth-fault plane. Detailed field studies aided by seismic control are the best methods for fault mapping. Regional structure maps usually show sudden steep monoclinal dip along a narrow zone where the growth faults are suspected, but detailed isopach maps are more useful because they commonly record an abrupt explosive thickening across the trace of the suspect fault toward the downdropped block to reveal the syndepositional nature.

5. Other important stratigraphic traps within the Stevens turbidite facies occur along the lateral margins and within the distal siliceous shale environment of the abyssal plain. Lateral margins onlap bathymetric highs and are accompanied by increasing sandstone loss and lenticularity at a constant predictable rate. Stratigraphic time lines strongly converge over the seafloor high and the effect is one of overall wedging, readily mapped even where seismic data are the only available control. This relation is referred to as basin-margin wedging. Within the basin-margin wedge, porosity pinchouts are legion and multiple stratigraphic traps are to be expected flanking the

basin-bounding faults or bathymetric highs confining the lateral margins of the successive turbidite fans. Beyond the limits of the turbidite sand the environment is still one of dynamic sedimentation. Distal facies are characterized by thin sandstone grading into porcellanite and eventually to cherty shales. Silica of the shales is thought to have originated from the fossil tests of myriad diatom blooms feeding on the upwelling phosphates of the deeper marine environment. Much clay is introduced to the far reaches of the basin during flysch sedimentation. Silica of many clays, being very soluble in seawater, further enriches the environment and provides additional raw material for diatom tests. Ocean currents propel the diatom blooms seaward where they eventually settle in deeper water to become porcellanites and opaline cherts such as those of the central San Joaquin Valley. Fracturing of these extremely brittle rocks, usually over faults or folds, forms prolific oil reservoirs. High-gravity crude has an affinity for the porcellanites whereas low-gravity oils generally are concentrated in the cherty rocks. It is hoped that some or all of the described trapping mechanisms, primarily facies-controlled, will be useful in exploring the yet undrilled areas of the Stevens and pre-Stevens turbidite basins. New seismic techniques and improved seismic resolution should be a valuable aid in environmental detailing of target areas.

6. An earlier history of turbidite deposition is reflected in other submarine fans ranging in age from early middle Miocene (Relizian) through early late Miocene (early Mohnian). They are referred to in this report as the Bena and Valpredo fans, with depocenters in the Maricopa subbasin between the Bakersfield arch and the Tejon platform. Filling of the Maricopa subbasin is thought to have been accomplished when the ancestral Kern River emptied into the San Joaquin Valley south of its present east-west course, near the nexus of faults bounding the juncture of Tehachapi and Sierra Nevada Ranges. These fans are oil productive only on the basin margins at relatively shallow depths but should offer attractive future exploration targets in the underexplored Maricopa subbasin.

REFERENCES CITED

Bandy, O. L., and R. E. Arnal, 1960, Concepts of foraminiferal paleoecology: AAPG Bull., v. 44, p. 1921-1932.

Bazeley, W., 1972, San Emidio Nose oil field, *in* Stratigraphic oil and gas fields: AAPG Mem. 16, p. 297-316.

Benioff, H. V., 1955, Mechanism and strain characteristics of the White Wolf fault as indicated by the aftershock sequence, *in* Earthquakes in Kern County, California, during 1952: California Div. Mines Bull. 171, p. 199-202.

Bramlette, M. N., 1946, The Monterey Formation of California and origin of its siliceous rocks: U.S. Geol. Survey Prof. Paper 212, 57 p.

Brown, L. F., Jr., 1969, Geometry and distribution of fluvial and deltaic sandstone (Pennsylvanian and Permian), north-central Texas, *in* Geology of the American Mediterranean: Gulf Coast Assoc. Geol. Socs. Trans., v. 19, p. 23-47.

Bruce, C. H., 1973, Pressured shale and related sediment deformation: mechanism for development of regional contemporaneous faults: AAPG Bull., v. 57, p. 878-886.

Callaway, D. C., 1968, Habitat of oil on west side, San Joaquin Valley, *in* Geology and oil fields, west side southern San Joaquin Valley: AAPG, SEG, SEPM Pacific Secs. 43d Ann. Mtg. Guidebook, p. 21-25.

Carver, R. E., 1968, Differential compaction as a cause of regional contemporaneous faults: AAPG Bull., v. 52, p. 414-419.

Cloos, E., 1968, Experimental analysis of Gulf Coast fracturing patterns: AAPG Bull., v. 52, p. 420-444.

Ferm, J. C., and V. V. Cavaroc, Jr., 1968, A nonmarine sedimentary model for the Allegheny rocks of West Virginia, *in* Late Paleozoic and Mesozoic continental sedimentation, northeastern North America: Geol. Soc. America Spec. Paper 106, p. 1-19.

Hoots, H. W., T. L. Bear, and W. D. Kleinpell, 1954, Geological summary of the San Joaquin Valley, California, *in* Geology of southern California: California Div. Mines Bull. 170, p. 113-129.

Le Blanc, R. J., Sr., 1977, Distribution and continuity of sandstone reservoirs, pt. 1: Jour. Petroleum Technology, v. 29, p. 776-792.

MacPherson, B. A., 1977, Sedimentation and trapping mechanism in upper Miocene Stevens and older turbidite fans of the southeastern San Joaquin Valley, California, *in* Late Miocene geology and new oilfields of the southern San Joaquin Valley: AAPG Pacific Sec. Guidebook, p. 5-36.

Maher, J. C., R. D. Carter, and R. J. Lantz, 1975, Petroleum geology of the Naval Petroleum Reserve No. 1, Elk Hills, Kern County, California: U.S. Geol. Survey Prof. Paper 912, 109 p.

Martin, B., 1963, Rosedale channel—evidence for late Miocene submarine erosion in Great Valley of California: AAPG Bull., v. 47, p. 441-456.

Menard, H. W., Jr., 1955, Deep-sea channels, topo-graphy, and sedimentation: AAPG Bull., v. 39, p. 236-255.

Mutti, E., and F. Ricci-Lucchi, 1972, Le torbiditi dell' Appenino settentrionale: introdzione all'analisi di facies: Soc. Geol. Italiana Mem., v. 11, p. 161-199.

Natland, M. L., 1957, Paleoecology of West Coast Tertiary, *in* Paleoecology: Geol. Soc. America Mem. 67, v. 2, p. 543-571.

Nelson, C. H., and L. D. Kulm, 1973, Submarine fans and deep-sea channels, *in* Turbidites and deep-water sedimentation: SEPM Pacific Sec., p. 39-78.

Normark, W. R., 1970, Growth patterns of deep-sea fans: AAPG Bull., v. 54, p. 2170-2195.

——— 1976, Channels, suprafans, and depositional lobes: facies character for turbidite sand environments on modern fans, *in* R. S. Saxena, ed., Sedimentary environments and hydrocarbons, AAPG/NOGS short course: New Orleans Geol. Soc., p. 163-186.

Oakeshott, G. B., ed., 1955, Earthquakes in Kern County, California, during 1952: California Div. Mines Bull. 171, 283 p.

Regan, L. J., Jr., 1953, Fractured shale reservoirs of California: AAPG Bull., v. 37, p. 201-216.

Sanem, R. E., and R. R. Stoddard, 1965, Strati-structural traps in Stevens sand (abs.): AAPG Bull., v. 49, p. 1089.

Seiden, H., 1965, Asphalto—a sleeper among giants: AAPG Pacific Sec. 40th Ann. Mtg., Bakersfield, California, p. 4-15.

Sullwold, H. H., Jr., 1961, Turbidites in oil exploration, *in* Geometry of sandstone bodies: AAPG, p. 63-81.

Walker, R. G., 1976a, Resedimented coarse clastic family, facies, processes, and depositional models, *in* R. S. Saxena, ed., Sedimentary environments and hydrocarbons, AAPG/NOGS short course: New Orleans Geol. Soc., p. 129-162.

——— 1976b, Ancient submarine fans, *in* R. S. Saxena, ed., Sedimentary environments and hydrocarbons, AAPG/NOGS short course: New Orleans Geol. Soc., p. 187-217.

——— and E. Mutti, 1973, Turbidite facies and facies associations, *in* Turbidites and deep-water sedimentation: SEPM Pacific Sec., p. 119-157.

White, R. T., 1940, Eocene Yokut Sandstone north of Coalingua, California: AAPG Bull., v. 24, p. 1722-1751.

Zieglar, D. L., and J. H. Spotts, 1976, Reservoir and source bed history in the Great Valley of California, *in* Tomorrow's oil from today's provinces: AAPG Pacific Sec. Misc. Pub. 24, p. 19-38.

Grimes Gas Field, Sacramento Valley, California[1]

FRANK E. WEAGANT

Consulting Geologist, Bakersfield, California 93306

Abstract The Grimes gas field in the northern Sacramento Valley, California, produces from sandstones in the Upper Cretaceous Forbes Formation. These sandstones form a complex of lenticular stratigraphic traps. The Grimes field, with present annual production of 29 billion cu ft, is the third largest dry gas producer in the state.

The original Grimes prospect had as primary objectives the shallow Upper Cretaceous "Winters" sandstones; Forbes sandstones were secondary objectives. Subsurface data, supplemented by minor seismic information, were used to delineate the prospect. After the discovery in 1960, an extensive seismic survey resulted in modification of previous conceptions of the structure and helped greatly in development. Seismic information did not help delineate the lenticular sandstone traps. The Forbes sandstones were deposited in moderately deep water by deep-water currents, probably turbidity currents. Individual lenses are semielliptical, oriented generally in a northwesterly direction. They range in length from less than 1 to 8 or more miles (1.6–13 km).

A series of tectonic episodes, ending with the emplacement of igneous plugs at the Colusa high and Marysville Buttes in Pliocene time, has given the area its present structural configuration. The northern part of the area is dominated by the highs of the Marysville Buttes and Colusa areas, whereas the Grimes area proper is simply a southwest-dipping homocline. Faults of pre-Eocene origin are present in the Grimes area and are partly responsible for the Forbes accumulations there.

Fluid-pressure relations in the Grimes area are somewhat unusual. Formation pressures are well above hydrostatic and increase with depth at a gradient greater than the hydrostatic gradient; however, formation pressures decrease eastward from the west edge of the Grimes field. The pressure relations are probably important to the nature of presently existing gas accumulations.

Introduction

The Grimes gas field is one of a group of fields in northern California that represents a complex of semicontinuous stratigraphic gas accumulations in sandstones of the Forbes Formation. All these fields are shown on Figure 1 —(from north to south) West Butte, Butte Slough, Sutter Buttes, Sutter South, Sycamore, and Grimes. Although the entire area is discussed in this paper, the emphasis is on the Grimes field proper, for it is by far the largest of the group.

Location

The Grimes field is in the northern Sacra-

mento Valley gas basin. It is 40 mi (65 km) northwest of Sacramento. The complex of fields discussed herein is mainly in Sutter County but extends westward into Colusa County.

Discovery and Present Production

The Grimes field was discovered in January 1960 by the Cameron Franco Western 7–2 well, Sec. 7, T14N, R1E, near the town of Grimes. A drill-stem test of an Upper Cretaceous Forbes sandstone at 6,550–6,600 ft (1,995–2,030 m) yielded gas at a rate of more than 3,000 Mcf/day. Development and exploration spread rapidly to the northeast and northwest until the subject fields had been discovered and developed. The pioneer, however, was the old Marysville Buttes gas field near the eastern margin of what is now the Sutter Buttes gas field. This field was discovered in 1937 and, by 1943, seven wells had been completed as Forbes producers.

Many different reservoirs contain the Grimes and related accumulations; more than 20 different reservoirs produce in the Grimes field alone. These reservoirs occur within almost the entire Forbes section; within the Grimes area they range in depth from 5,500 to 8,500 ft (1,675 to 2,590 m).

The Grimes field at present is the third largest dry gas producer in the state; 1967 production was 29 billion cu ft. Total reserves, according to estimates made by area operators, exceed 200 billion cu ft. Total reserves from the other fields combined exceed 150 billion cu ft. Well spacing throughout the general Grimes area is 160 acres per well.

History of Subsurface and Seismic Exploration

The original geologic prospect in the Grimes area was based on the southeasterly projection of the Colusa high, which created a broad southeast-plunging nose in the prospect area. Shallow "Winters" sandstones supposedly were truncated to the north on this nose, thus forming the primary objectives for the prospect. Some sparse seismic information south of the Grimes area seemed to corroborate the presence of the nose.

The deeper Forbes sandstones were secon-

[1] Manuscript received, April 15, 1969.

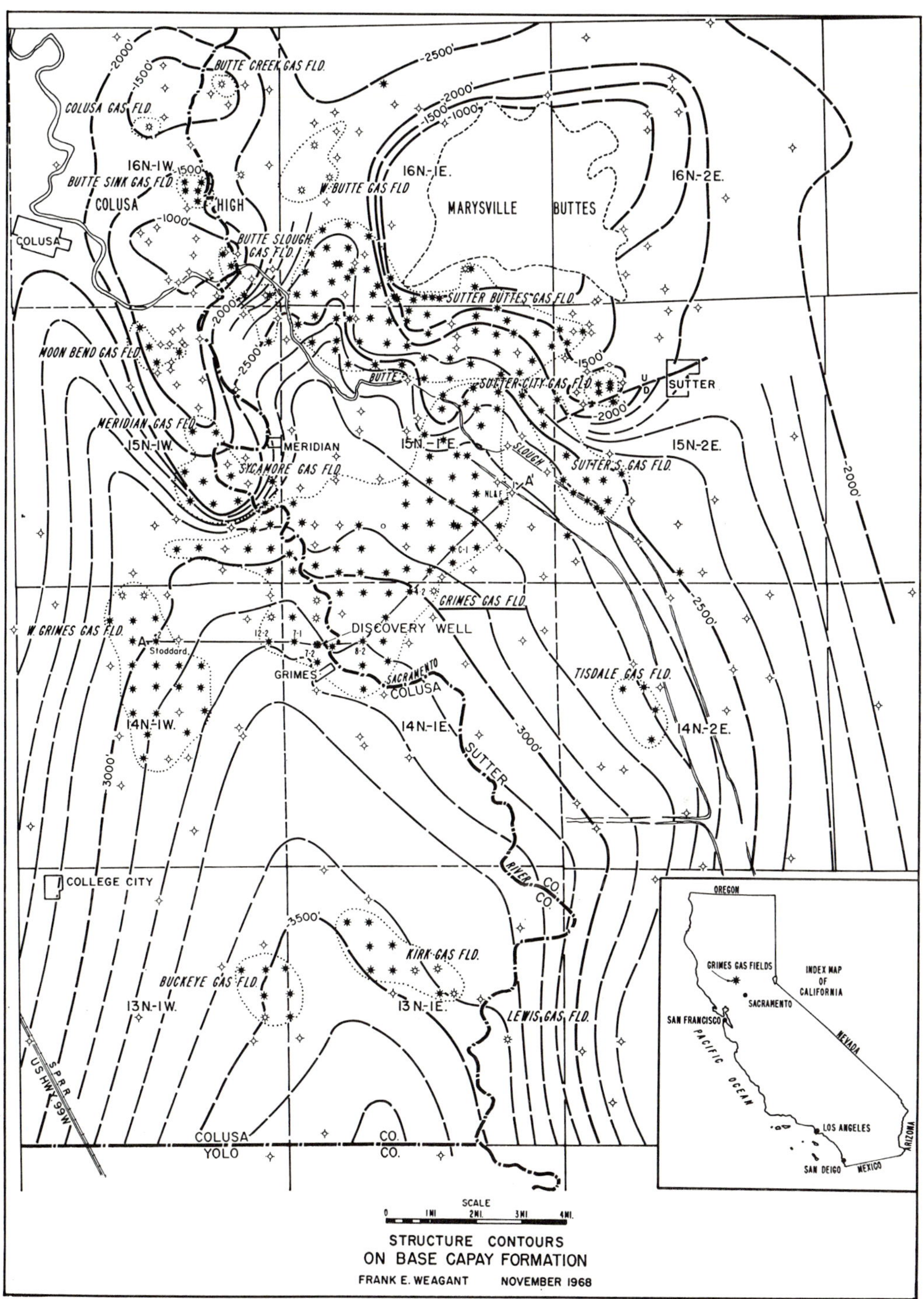

Fig. 1—Index map showing regional contours with base of Capay Formation as datum. CI = 100 ft. *A-A'* is trace of cross section in Figure 5.

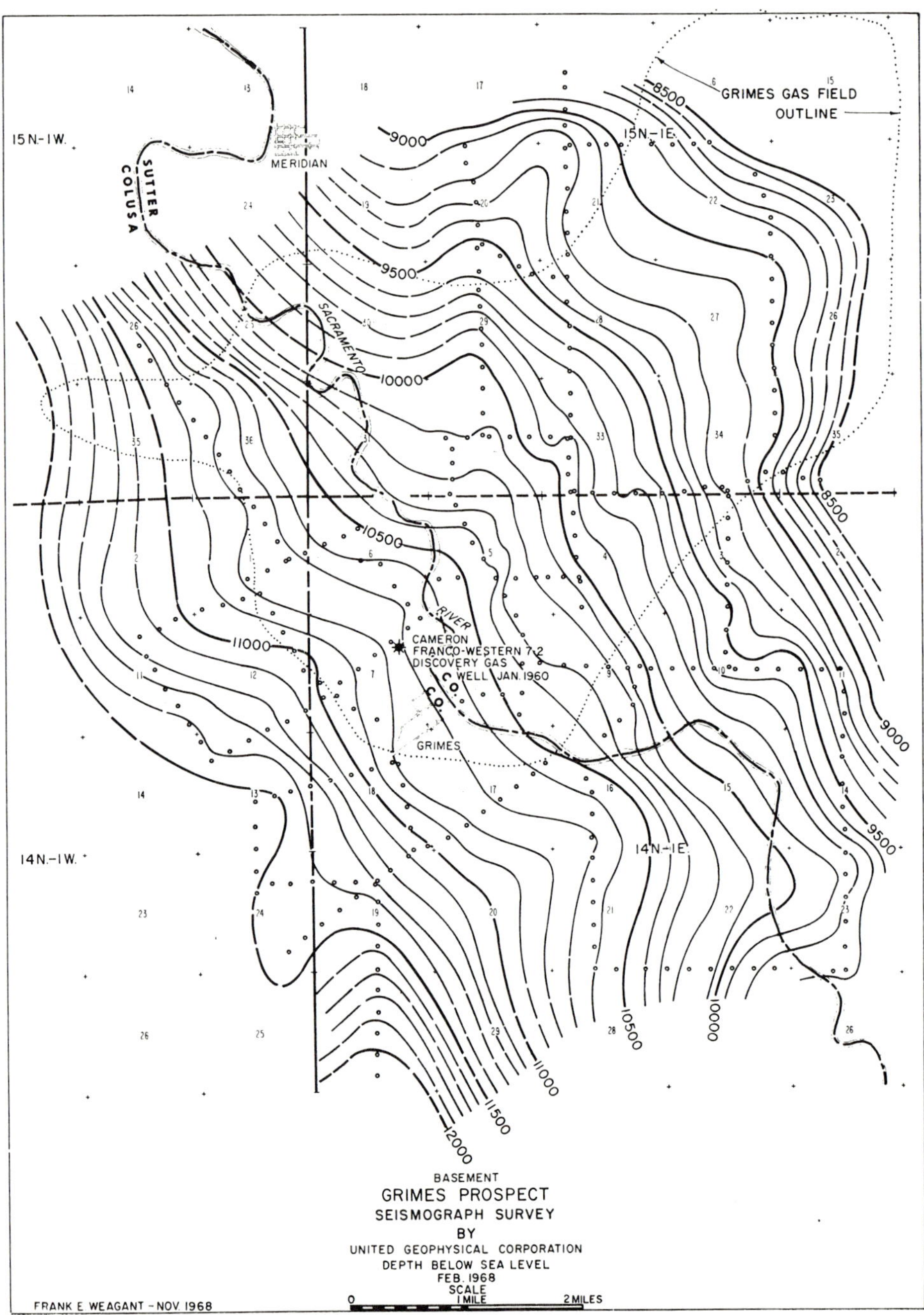

FIG. 2—Seismic contours of datum near basement. CI = 100 ft.

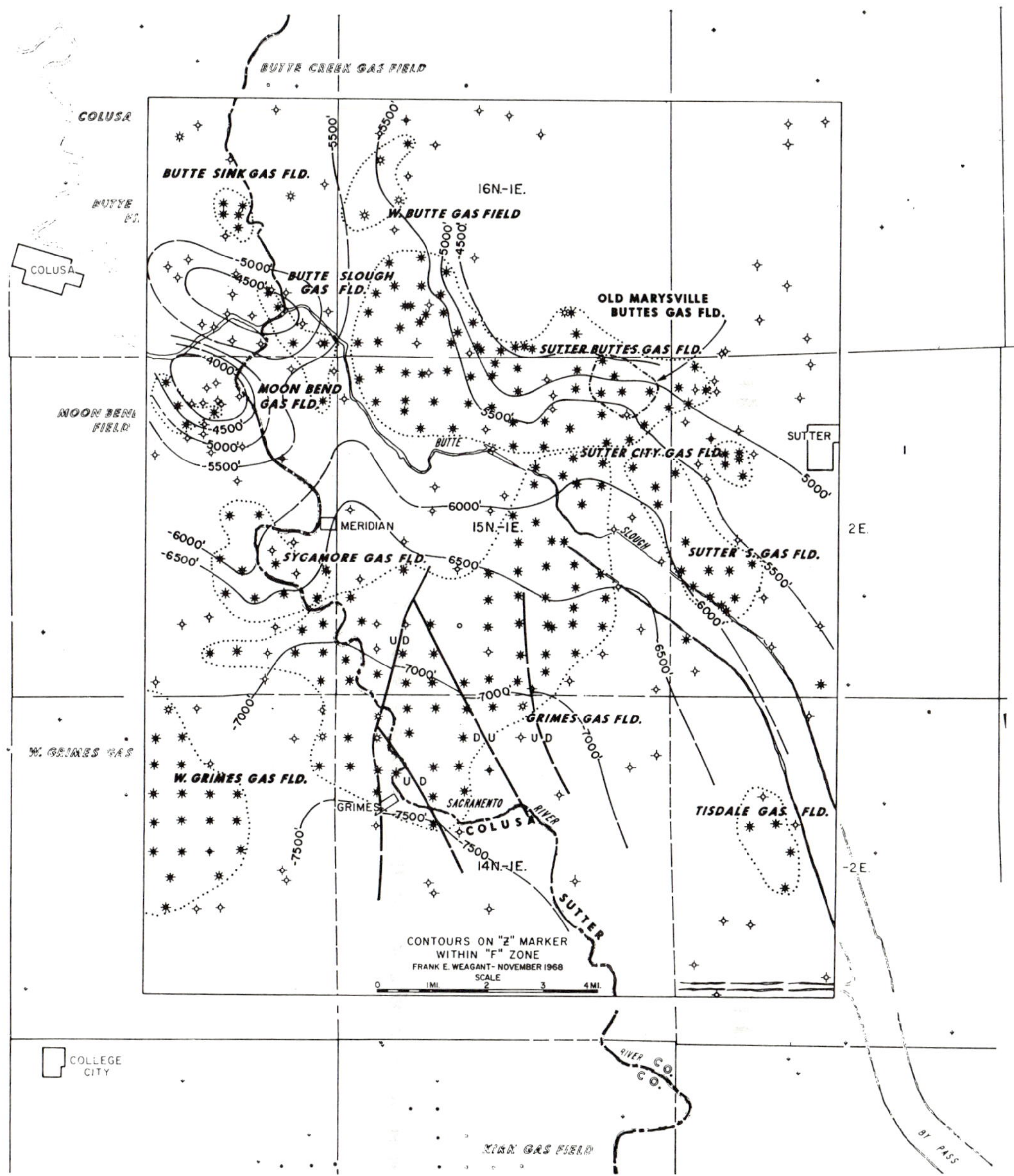

Fig. 3—Structure contours on Z marker horizon within F Zone. CI = 500 ft.

dary objectives. It was thought that the presence of well-developed Forbes sandstones at the Arbuckle and Marysville Buttes gas fields on either side (southwest and northeast, respectively) of the Grimes area enhanced the possibilities for similar sandstone development in the prospect area. However, there were no Forbes penetrations in the Grimes area to substantiate this idea.

After the discovery well was drilled, the companies involved ran an extensive seismic survey over the Grimes area. About 46 mi (74 km) of line was run over an approximately 48-sq mi (125 sq km) area. Shotholes were on 1,000-ft (305 m) spacing and light charges in shallow holes were used. The data were presented on plotted cross sections which were used to make the subsurface interpretations.

One variable-area cross section was made; however, it did not add to the interpretation, so the technique was not used. Reflections were obtained down to basement depth, but the quality of reflections was only fair.

Figure 2 is a seismic contour map at basement level prepared on the basis of this survey. The structure shown on this map is similar to the structure shown at a Forbes datum on a map prepared from subsurface well data (Fig. 3). The similarity of the two maps shows the general reliability of the seismic survey.

The seismic survey made it evident that the Grimes accumulation had to be primarily stratigraphic rather than structural. It showed the original concept of a southeast-plunging nose to be incorrect; instead, the basic structure of the area is a southwest-dipping homocline. This change in interpretation of the structure and the consequent realization that the accumulation is primarily stratigraphic were very important to the development plan for the Grimes field.

Unfortunately, the seismic data did not aid in the understanding or delineation of sandstone distribution.

STRATIGRAPHY OF FORBES FORMATION

Figure 4 shows rock units present in the Grimes area, but only the Forbes is described in detail herein.

The Forbes Formation consists of interbed-

Fig. 5—Cross-section *A-A'* through Grimes field. Trace is *A-A'* on Figure 1.

GEOLOGIC AGE	FORMATION	THICKNESS	LITHOLOGY
Pliocene to Recent	Tehama	1800-2000'	Interbedded sequence of continental sands, conglomerates and shales. Sands comprise about 80% of section.
	Unconformity		
Eocene	Undifferentiated but includes Domengine equivalent.	1000'	Conglomeratic, poorly sorted sands, and thin interbedded grey shales.
Eocene	Capay	250'	Light grey, glauconitic shale with a glauconitic gritstone at the base.
	Unconformity		
Upper Cretaceous D-1 Zone	Starkey	0-200'	Fine to medium grained sand.
Upper Cretaceous D-2 Zone	Winters	250-300'	Grey siltstones and silty shales.
Upper Cretaceous E-Zone	Sacramento Shale	200'	Dark grey siltstone and shale.
Upper Cretaceous E'-Zone	Kione	900-1000'	Grey, fine to medium grained sands, interbedded with grey siltstones.
Upper Cretaceous F-Zone (F-1 and F-2 Zones)	Forbes	3000-5500'	Dark grey siltstones and claystones with interbedded fine grained, friable, lenticular sands.
	Unconformity		
Upper Cretaceous G-Zone	Dobbins and Guinda		Grey soft shales and siltstones and a basal sand and conglomerate.
	Basement		Granite and other more basic igneous rocks

Fig. 4—Rock units present in Grimes area.

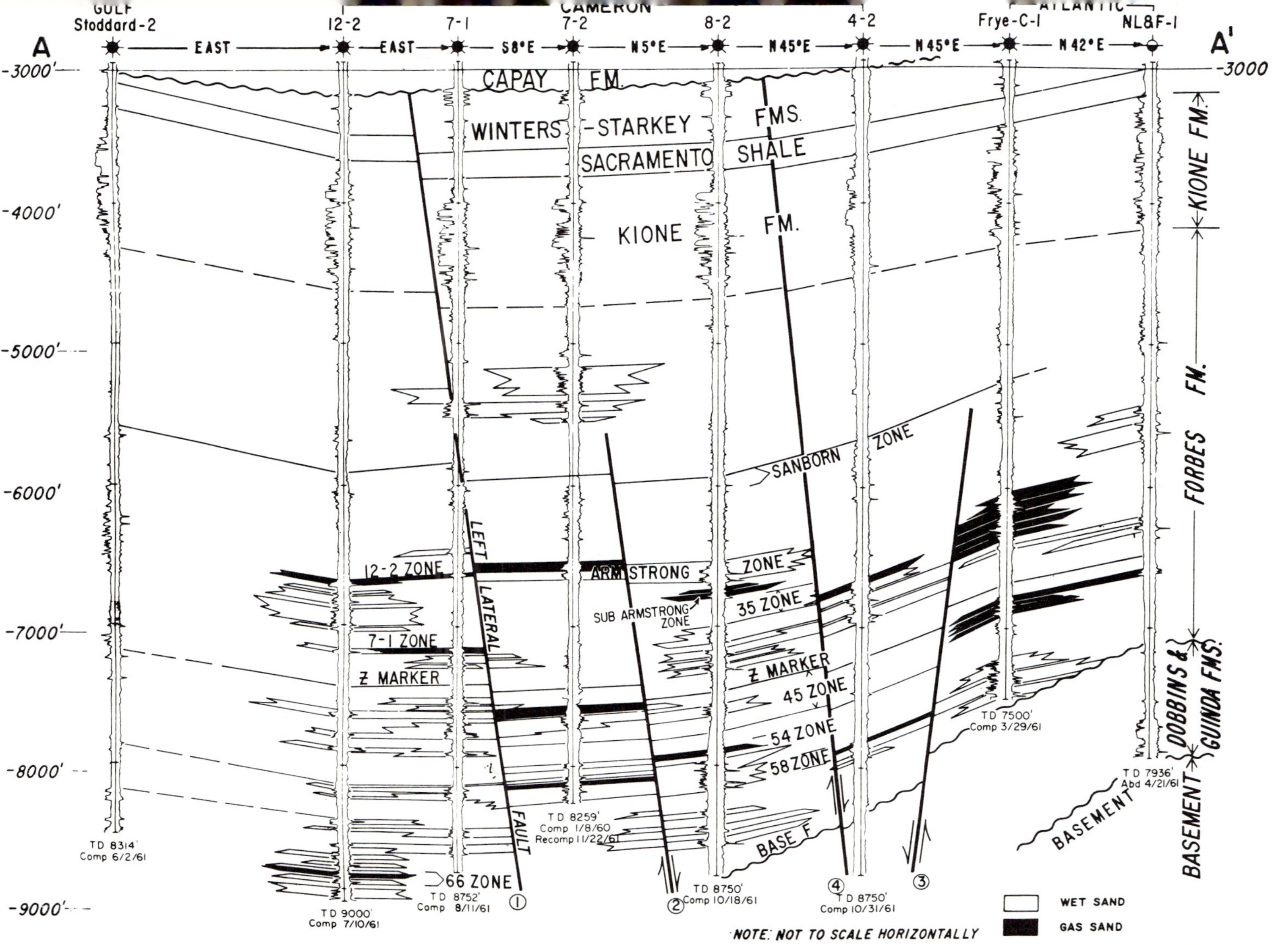

GULF   CAMERON   ATLANTIC
Stoddard-2   12-2   7-1   7-2   8-2   4-2   Frye-C-1   NL&F-1
A   A'
EAST   EAST   S8°E   N5°E   N45°E   N45°E   N42°E
-3000'   -3000
-4000'
-5000'
-6000'
-7000'
-8000'
-9000'
KIONE FM.
FORBES FM.
DOBBINS & GUINDA FMS.
BASEMENT
CAPAY FM.
WINTERS -STARKEY FMS.
SACRAMENTO SHALE
KIONE FM.
SANBORN ZONE
12-2 ZONE
ARMSTRONG ZONE
SUB ARMSTRONG ZONE
35 ZONE
7-1 ZONE
Z MARKER
Z MARKER
45 ZONE
54 ZONE
58 ZONE
BASE F
LEFT LATERAL FAULT
66 ZONE
TD 8314' Comp 6/2/61
TD 9000' Comp 7/10/61
TD 8752' Comp 8/11/61
TD 8259' Comp 1/8/60 Recomp 11/22/61
TD 8750' Comp 10/18/61
TD 8750' Comp 10/31/61
TD 7500' Comp 3/29/61
TD 7936' Abd 4/21/61
WET SAND
GAS SAND
NOTE: NOT TO SCALE HORIZONTALLY

ded dark gray claystones and siltstones and gray, fine-grained, friable, lenticular sandstones. These sediments have been assigned to the Upper Cretaceous F Zone (faunal zone of Goudkoff, 1945). Ditch and sidewall samples show that producing sandstones are fine grained; they range from very shaly to well-sorted and clean sandstones. Sidewall-sample analyses give porosities ranging from 25 to 30 percent and permeabilities ranging from 15 to 70 md. Relative thicknesses of sandstone and shale, as well as the range in thickness of the sandstones, can be seen on the cross section (Fig. 5).

The Forbes Formation, for the most part, was deposited in a moderately deep-water environment. A hiatus followed G Zone deposition (Dobbins and Guinda strata) and G Zone beds were tilted southwesterly. Forbes sediments then were deposited, onlapping G Zone sediments to the northeast. East of the Grimes area, the Forbes Formation fully overlaps the G Zone and lies directly on basement. The Forbes Formation thus thins markedly eastward across the area. Part of the thinning is truly stratigraphic, and part is due to onlap. The depositional strike is north-northwest, which is the general depositional strike of this part of the basin.

Forbes sands were deposited in a moderately deep-water environment, too far from shore to be affected by shoreline winnowing processes. The sandstones are the result of offshore basinal currents, probably turbidity currents. Geometry of some individual sandstone bodies of producing reservoirs is shown on Figures 6 and 7. In general, individual sandstone bodies that can be correlated are elongate, with the long axes parallel or subparallel with depositional strike. The individual sandstone lenses range in size from those present in only one or two wells to lenses from 2–3 mi (3–5 km) wide and more than 8 mi (13 km) long. The thickness of individual sandstones ranges from 8 to 60 ft (2–18 m).

STRUCTURE

General Setting

The Grimes and closely associated fields lie on the east flank of the Sacramento Valley gas basin. The regional southwest dip has been modified profoundly by intrusion of the Pliocene plugs at Marysville Buttes and the Colusa high. Figure 1, with the base of the lower Eocene Capay Formation as datum, shows the main elements of regional structure. These elements include (1) the main north-south-trending valley syncline, just west of the Grimes field and the Colusa high; (2) the Colusa high, a 10-mi-long (16 km) feature with subsidiary closures over the various plugs along the high; (3) the Marysville Buttes, a plug where the igneous rocks crop out over a large area; and (4) the regional southwest dip outside the area influenced by the plugs.

Upper Cretaceous

Within the Upper Cretaceous section, the structure is somewhat more complex than shown on Figure 1. Upper Cretaceous rocks are separated from lower Eocene rocks by a profound unconformity that affected the entire Sacramento Valley. In Late Cretaceous or early Eocene time, the basin was uplifted, tilted to the south, and folded into a broad syncline with a north-south axis. A fault pattern developed in the Grimes area that consisted mainly of small-displacement faults trending north to northwest. During the peneplanation preceding Eocene deposition, several thousand feet of Upper Cretaceous sediments was removed from the Grimes area. Figure 3 shows the structure on a marker within the Forbes Formation. The cross section (Fig. 5) shows the pre-Eocene faults and the structural discordance between the lower Eocene and the Upper Cretaceous.

Eocene

Eocene structure is similar to Upper Cretaceous structure except that the southwest component of dip is less and there is no evidence for the pre-Eocene faults (Fig. 1). All the major, and some of the minor, structural features are more evident on the contour map of the base of the Eocene than on the Forbes map, because more information is available at Eocene depths and correlation reliability is greater within the Eocene.

The structure shown on Figure 1 reflects a late Eocene–early Miocene tectonic episode and Pliocene intrusion of the plugs at Marysville Buttes and the Colusa high. In the late Eocene–early Miocene tectonic episode, the area was uplifted, folded, and beveled, and the nonmarine Miocene-Pliocene Tehama Formation was deposited.

Pliocene Volcanic Plugs

In the area of this study, emplacement of Pliocene plugs created the Colusa high and Marysville Buttes and greatly modified the previously existing structural pattern. At Marys-

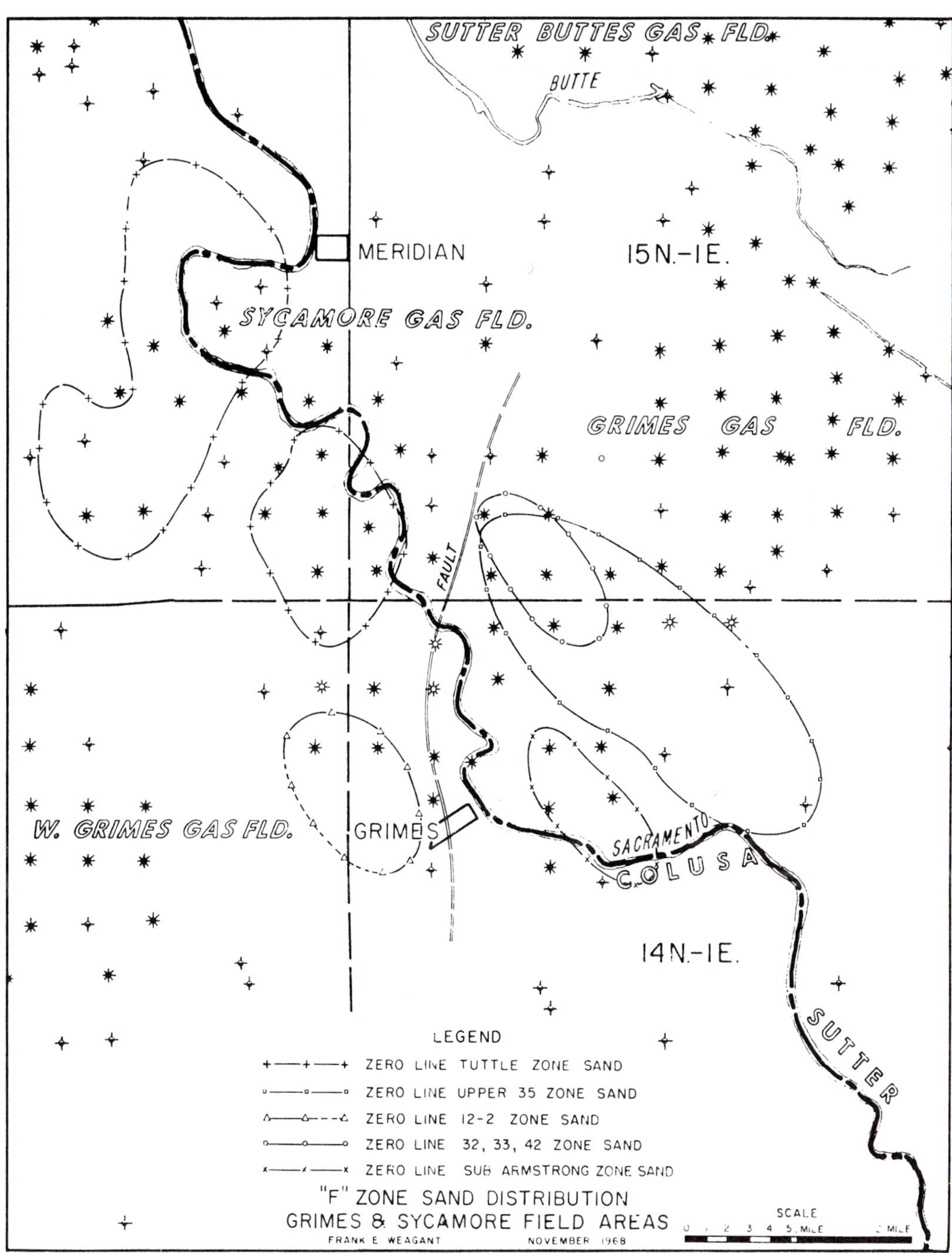

FIG. 6—Sandstone distribution of some F Zone reservoirs in Grimes and Sycamore areas.

ville Buttes the plug is a piercement intrusive that crops out over a large area. It is 4 mi (6 km) in diameter and has nearly vertical walls. The composition is mainly andesite porphyry with secondary intrusions of rhyolite porphyry. A series of plugs in the subsurface formed the Colusa high. These plugs and the associated sills and dikes are known from seismic surveys and deep drilling. The Colusa high is a 10-mi-long (16 km) closed anticline with a series of small closures along it.

These two centers of intrusion—the Colusa high and Marysville Buttes—have affected the structure of a broad area. In the area immediately surrounding the plugs, the dip is very steep—even vertical on the south flank of Marysville Buttes. Also around the plugs are structural irregularities probably caused by

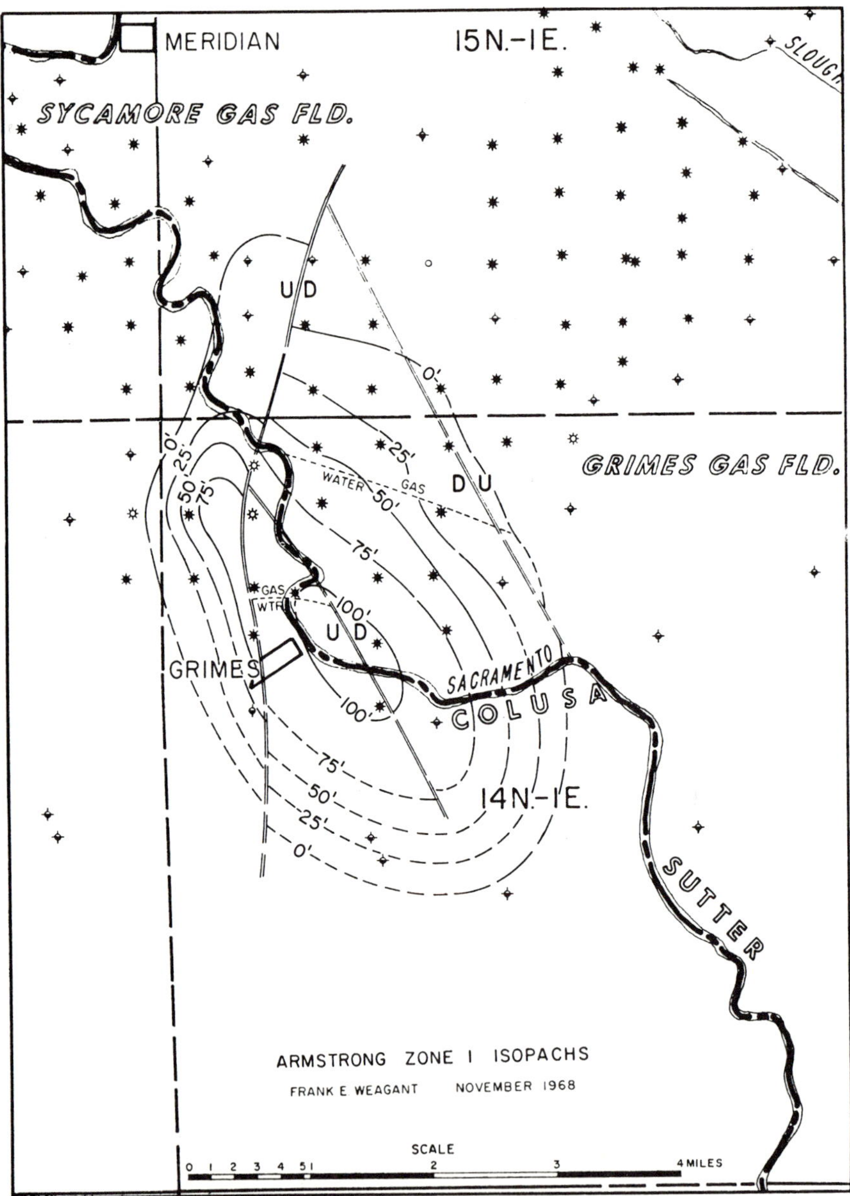

Fig. 7—Isopach map of "Armstrong Zone," Grimes field. CI = 25 ft.

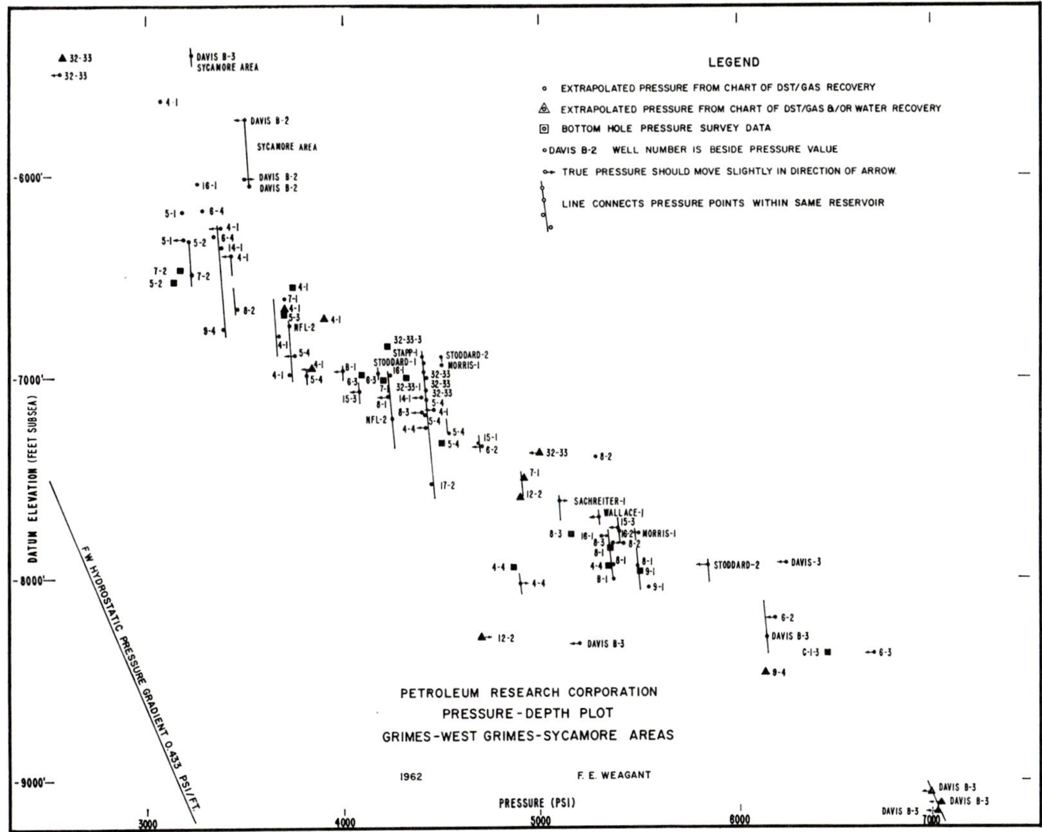

FIG. 8—A series of pressure-depth plots with data from many wells in area of Grimes, West Grimes, and Sycamore gas fields.

minor intrusions at depth. In a broad area around the Colusa high and Marysville Buttes the dip is much higher than is common for this part of the basin. Together, the Colusa high and Marysville Buttes form a broad intrabasin high with an east-west axis through the centers of the two features.

Gas Accumulation

Effect of Structure on Accumulation

The gas fields of the Grimes region occupy a diversity of structural positions, from anticlinal to synclinal to homoclinal. Because all these fields are stratigraphically very similar, it is evident that the accumulations are independent of the structural position. However, the broad regional high created by the intrusions of the Colusa high and Marysville Buttes may have caused a general migration of gas into the area. The only specific local structural features that help trap gas are the old pre-Eocene faults. It can be demonstrated that these faults are important to the present accumulation in the Grimes field. For example, the "Armstrong Zone" isopach map (Fig. 7) shows that zone to be cut by the principal and westernmost fault of the area. Gas accumulation in the "Armstrong Zone" lens is confined to that portion of the lens east of the fault. In addition, the old faults divide the area into different pressure blocks, some of which are more favorable than others to accumulation.

Time of Accumulation

The time that the present Forbes gas accumulated is a controversial subject. Geologists are divided into those favoring early and those favoring late migration. According to the early-stage migration hypothesis, gas generation and accumulation took place during F Zone time, and the high concentration of gas fields in the Grimes region is simply the result of the more favorable depositional environment.

The contrary argument holds that, if the gas were generated during F Zone time, accumulation in individual sandstone lenses would have taken place at shallow depths and low pressures. At present, the average pressure in the Grimes field is about 4,000 psi. Even if the sandstone lenses were originally completely filled with gas, fill-up now would be only a fraction of total reservoir volume; actually, fill-up in some of the lenses is more than half. Therefore, it seems improbable that the present accumulations occurred in F Zone time and have remained essentially unchanged since.

The Petroleum Research Corporation of Denver and the writer made a detailed fluid-pressure study of the Grimes field. The basic data were from multiple drill-stem tests run on most of the wells that had been drilled at the time of the study. This large amount of fluid-pressure information showed that (1) formation pressures increase markedly above hydrostatic starting at about 5,500 ft (1,675 m); (2) formation pressures differ considerably among the various fault blocks of the Grimes field for reservoirs at the same depths; and (3) sandstones in the fault blocks with lowest pressure contain more and larger accumulations. The abrupt increase in formation pressure at depths greater than 5,500 ft is shown by pressure-depth plots to occur in a series of discrete jumps across shale barriers. Formation pressures at the same depths decrease from fault block to fault block in an easterly direction; the amount of decrease is 200–300 psi. This easterly decrease in pressure shows that a hydrodynamic gradient with a strong easterly component exists in the area. The fact that good reservoirs in the fault blocks with lower pressure are more numerous than in the blocks with higher pressure is in accordance with hydrodynamic theory. According to this theory, hydrocarbons will tend to migrate from higher pressure reservoirs abutting a fault to lower pressure reservoirs on the opposite side. An example is the "Armstrong Zone" (Fig. 7), in which the portion of the sandstone lobe on the west side of the most westerly fault is barren of gas even though it is the structurally higher part of the lens. It is probable that the present unusual fluid-pressure system had its origin with the intrusions of the Colusa high and Marysville Buttes and the subsequent upfolding, erosion, and exposure of the gas-bearing section around Marysville Buttes.

Figure 8 illustrates some of the important fluid-pressure relations by use of a series of pressure-depth plots. Each plot consists of a steeply sloping line joining pressure points that usually fall in the gas phase of a single reservoir. A few plots include water-phase points.

The marked increase in pressure can be seen by the stairstep arrangement of the various pressure-depth plots with increase in depth. Formation-pressure differences among the various fault blocks are illustrated by the position of the plot at a 5,800-ft (1,770 m) average depth and 3,500 psi pressure relative to the plot at a 6,500-ft (1,980 m) average depth and 3,200 psi pressure. The first plot is at a higher pressure and lesser depth than the second and falls in the most westerly and most highly pressured fault block of the Grimes area. This series of plots also shows that many reservoirs are found in only one well, whereas some—such as that shown by the plot at a 7,200-ft (2,195 m) average depth and 4,400 psi pressure—are found in five or six wells.

The fluid-pressure study indicated that the existing Grimes. gas accumulations are very much dependent upon the present fluid-pressure environment. This fact suggests late-stage migration or late-stage modification of previously existing accumulations in the area. Following is a summary of the way in which generation, migration, and accumulation of gas in the Grimes region may have occurred.

1. *Late F Zone and E Zone time*—Gas accumulated in many Forbes sandstone lenses in the northern Sacramento Valley, where generative and sandstone conditions were favorable. This area probably included a large part of the south Grimes region. Depth of burial was fairly shallow and pressures were low.

2. *Late Cretaceous or early Eocene time*—Upper Cretaceous sediments were broadly folded and tilted southward in the tectonic episode separating the Late Cretaceous from the Eocene. The steeper dips increased the differential pressure across individual sandstone lenses, and the entry pressures of many enclosing shales and siltstones of higher permeability were exceeded. Some gas was then able to migrate in a northeasterly direction in the Grimes region.

3. *Late Eocene time*—The section again was folded and tilted, allowing further migration as discussed above.

4. *Pliocene time*—The plugs at Marysville Buttes and the Colusa high were emplaced. These plugs created a broad general uplift in addition to the sharp upturning adjacent to the plugs. The section around Marysville Buttes was eroded into the Upper Cretaceous Forbes Formation. At this time, the unusual fluid-pressure system present in the general Grimes area had its inception. Gas that had accumulated in the numerous sandstone lenses within and south of the Grimes area migrated northward under the influence of the increased dip and a different pressure environment. Gas then accumulated in those lenses with enclosing

shales and siltstones of sufficiently low permeability to trap gas under the new fluid-pressure system.

CONCLUSIONS

1. The Grimes gas field was discovered as the result of deeper drilling of a shallow prospect. Seismic methods were not responsible for the discovery.

2. Seismic mapping, done after the discovery, helped in development of the field because it gave good definition to the structure; however, this mapping did not help to define the lenticular sandstone traps.

3. Accumulations in the Grimes and related fields are in many north- to northwest-trending sandstone lenses. These accumulations are not dependent upon structural position within the Grimes region, but they may be related to the regional high created by igneous intrusions at Colusa and Marysville Buttes.

4. Gas probably accumulated first in Forbes sandstones of the northern Sacramento Valley in late F Zone or E Zone time. Much of this gas migrated northward as a result of the changed gradients resulting from Late Cretaceous and late Eocene tectonism. Later, in response to Pliocene intrusion, uplift, and erosion, and to inception of the present fluid-pressure environment, gas remigrated into the present reservoirs of the Grimes region.

SELECTED REFERENCES

Clark, E. W., chm., *et al.,* 1951, Cenozoic correlation section from north side Mt. Diablo to east side Sacramento Valley, California: Pacific Sec. Am. Assoc. Petroleum Geologists Correlation Sec. 1, scale 1 in. = 1,000 ft.

Cross, C. M., chm., *et al.,* 1954, Correlation section, northern Sacramento Valley, California: Pacific Sec. Am. Assoc. Petroleum Geologists Correlation Sec. 6, scale 1 in. = 500 ft.

Edmondson, W. F., 1962, Stratigraphy of the late Upper Cretaceous in the Sacramento Valley (California): San Joaquin Geol. Soc. Selected Papers, v. 1, p. 17–26.

Garrison, L. E., 1962, The Marysville (Sutter) Buttes, Sutter County, California: California Div. Mines and Geology Bull. 181, p. 69–72.

Goudkoff, P. P., 1945, Stratigraphic relations of Upper Cretaceous in Great Valley, California: Am. Assoc. Petroleum Geologists Bull., v. 29, no. 7, p. 956–1007.

Hackel, Otto, 1966, Summary of the geology of the Great Valley (California): California Div. Mines and Geology Bull. 190, p. 217–238.

Harding, T. P., chm., *et al.,* 1960, Correlation section, Sacramento Valley from Red Bluff to Rio Vista, California: Pacific Sec. Am. Assoc. Petroleum Geologists Correlation Sec. 13, scale 1 in. = 3 mi.

Hunter, G. W., 1955, Marysville Buttes gas field: California Div. Oil and Gas, California Oil Fields—Summ. Operations, v. 41, no. 1, p. 11–15.

Johnson, H. R., 1943, Marysville Buttes (Sutter Buttes) gas field: California Div. Mines Bull. 118, p. 610–615.

Sacramento Geological Society, 1961, Guidebook to the east-central Sacramento Valley field trip.

Safonov, Anatole, 1962, The challenge of the Sacramento Valley, California: California Div. Mines and Geology Bull. 181, p. 77–100.

Vaughn, R. H., 1962, The Arbuckle gas field, California: California Div. Mines and Geology Bull. 181, p. 112–119.

Weagant, F. E., 1962, Grimes gas field (California): San Joaquin Geol. Soc. Selected Papers, v. 1, p. 36–46.

The American Association of Petroleum Geologists Bulletin
V. 63, No. 11 (November 1979), P. 1999-2020, 18 Figs., 1 Table

Frigg Field—Large Submarine-Fan Trap in Lower Eocene Rocks of North Sea Viking Graben[1]

F. E. HERITIER,[2] P. LOSSEL,[3] and E. WATHNE[4]

Abstract In the deepest, axial part of the Viking sub-basin of the North Sea, the Frigg field, one of the world's largest offshore gas fields, straddles the border of the British and Norwegian continental shelf at lat. 60°N. The discovery well was drilled in 1971 on Norwegian block 25/1 in 100 m of water. Gas was discovered at a depth of 1,850 m in a lobate submarine fan representing the ultimate phase of a thick Paleocene deposit.

Sealed by middle Eocene open marine shales, the structure is mainly submarine-fan depositional topography enhanced by draping and differential compaction of sands. The area of structural closure is underlined by a typical "flat spot" on seismic sections and the gas column lies on a heavy oil disk. Chromatographic analysis shows that both oil and gas could be coming from underlying Jurassic source rocks.

Recoverable gas reserves are estimated to be about 200 billion cu m (7 Tcf). Production began September 15, 1977; the gas is brought ashore at St. Fergus in Scotland by a 360-km pipeline.

INTRODUCTION

The Frigg field straddles the line between the Norwegian and the United Kingdom sectors of the North Sea at lat. 59°50'N and long. 2°E. It is located approximately 190 km west-northwest of Haugesund, Norway, 180 km east of the Shetland Islands, and 390 km northeast of Aberdeen, Scotland (Fig. 1).

On the Norwegian side, the Frigg field lies within the Petronord group Block 25/1, with only a small part extending into Esso Block 30/10. On the United Kingdom side, the field lies mostly on Total Oil Marine and Elf Aquitaine License Blocks 10/1, 10/6, and 9/10a. A small part of the field extends into BP License Block 9/5.

Average water depth throughout the field is approximately 100 m.

Frigg was named after a goddess of the Norwegian pantheon.

HISTORY OF DISCOVERY

In 1965, the Petronord group, created by Elf Aquitaine, Total Oil Marine Norsk, and Norsk Hydro to search for oil on the Norwegian continental shelf south of 62°N lat., commenced an active program of seismic mapping which eventually led to the discovery of the Frigg field in July 1971.

In 1966, the first interpretation of a very loose seismic grid 15 × 20 km showed the Frigg structure on a mapping horizon which, at the time, was considered to be the top of the Upper Cretaceous chalk.

After the 1968 discovery by the Phillips Petronord group of the Cod field in Paleocene sands, more seismic lines were shot in the Tertiary basin along the border of the Norwegian and United Kingdom waters to make a selection of blocks in the second round of licensing.

In 1969, four of the eight blocks applied for were granted to the Petronord group under conditions whereby the Norwegian State could enter as a partner if a commercial discovery were found.[5] Block 25/1 and 25/2 were among the blocks granted. At that time a new 5 × 5 km seismic grid was shot. The interpretation of this survey permitted the definition of the Frigg structure, particularly at the top of the basal Tertiary sands.

On June 8, 1970, the United Kingdom Blocks 10/1 and 10/6 on the western flank of the structure were granted by the British authorities to the Total Oil Marine and Elf Aquitaine group.

In 1971, the official agreement between the Petronord group and the Norwegian State made possible the drilling of the first well on the Frigg

AAPG grants permission for a *single* photocopy of this article for research purposes. Other photocopying now allowed by the 1978 Copyright Law is prohibited. For more than one photocopy of this article, users should send request, article identification number (see below), and $3.00 per copy to Copyright Clearance Center, Inc., One Park Ave., New York, NY 10006.

[1]Manuscript received, April 26, 1978; accepted, May 29, 1979.

[2]Société Nationale Elf Aquitaine Norge (Production), Paris, France.

[3]Total Oil Marine Norsk, Paris, France.

[4]Norsk Hydro, Sandvika, Norway.

The writers thank many colleagues for valuable contributions and discussions, in particular, F. C. Duffaud for his help in preparing this paper. The writers also thank their respective companies, as well as Statoil, for permission to publish this paper.

[5]Den Norsk Stats Oljeselskap A/S (Statoil) became a partner of the Norwegian group in May 1973.

Article Identification Number
0149-1423/79/B011-0001$03.00/0

structure. The Petronord group spudded its first well in Block 25/1 using the *Neptune P 81* semi-submersible rig on March 30, 1971.

Well 25/1-1 was located at the crest of a very large but relatively low-relief structure with an area of closure on the seismic marker, then equated with the top of the Paleocene, of about 300 sq km.

At 1,812 m subsea, the well entered gas-bearing sands of early Eocene age. After penetrating a 135-m gas column, it entered an oil-bearing zone about 10 m thick. Gas-oil contact was established at 1,947 m subsea, and gas was subsequently tested at a rate of 24 MMCFGD (675,000 cu m/day) with a 7/8-in. (2.2 cm) choke. A major gas field had been discovered.

After that discovery, and to supplement the earlier work, two seismic surveys, totaling 550 line-km, were shot in both United Kingdom and Norwegian waters, and simultaneously a four-well appraisal program was started.

The first appraisal well, 25/1-2, was spudded in July 1971 about 5.5 km north of the discovery well and found a gas column of 49 m and an oil column of 10 m which confirmed the original gas discovery and proved an extension of the structure toward the north.

The second appraisal well, 25/1-3, was spudded in November 1971 about 5 km east-northeast of 25/1-1 and found a gas column of 17 m and an oil column of 10 m which proved a large extension of the structure toward the east.

The results of reinterpretation based on the 1971 seismic surveys, combined with data obtained from the first three wells, indicated a smaller and more complicated structure than had been envisaged originally.

The third appraisal well, 10/1-1, was spudded on the United Kingdom sector about 7 km southwest of 25/1-1, in a crestal position, and found a gas column of 92 m overlying 11 m of oil.

The fourth appraisal well, 10/1-2, was spudded about 4 km west of 25/1-1 to define the westward development and thickness of the Frigg sand facies. At a depth of 1,951 m, the well entered the oil zone in the lateral equivalent of the Frigg sand. This formation showed a facies of thin sand layers interbedded with shales, thus defining the western extension of the field.

At the same time, Esso drilled a stepout well in its Block 30/10, about 13 km north of 25/1-1, which found 2.5 m of gas-bearing sand and 7.5 m of oil-bearing sand that proved a small extension of the field into Esso 30/10 Block.

As the previous seismic grid was considered insufficient, a new detailed seismic survey consisting of 1,880 km of profiles was shot during the summer of 1973, along a regular grid of 1 × 1 km on the structure and 2 × 2 km off structure. The interpretation of this detailed survey and the results of the first six wells have thus confirmed a major gas field with a maximum gas column of 170 m in lower Eocene sands covering an area of about 115 sq km for the main accumulation. The decision to develop the find was made at that time.

REGIONAL SETTING

The Viking basin is a thick sedimentary domain which lies between the Norwegian shield on the east and the Shetland platform on the west, and between lat. 58 and 62°N. It is a collapsed area typical of rifts, which in terms of plate tectonics is the result of an aborted opening of the northern European continental shield (Fig. 2).

The basin's history involves two major geologic periods separated by the Cimmerian orogeny. The first period was a positive epicontinental sedimentary cycle which began with deposition of the red continental Triassic beds and ended with Oxfordian marine shales. The second was a negative open-marine sedimentary cycle, which began with deposition of deep-marine black shales of Kimmeridgian age and ended in the present shallow-marine conditions.

Pre-Cimmerian Period

The first cycle comprises successively the continental Triassic red clastics, the lower to middle Lias fluviodeltaic sandy deposits, middle to upper

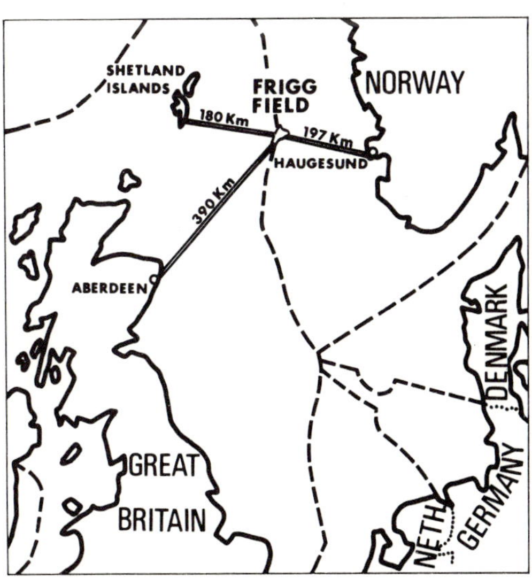

FIG. 1—Index map of central and northern North Sea.

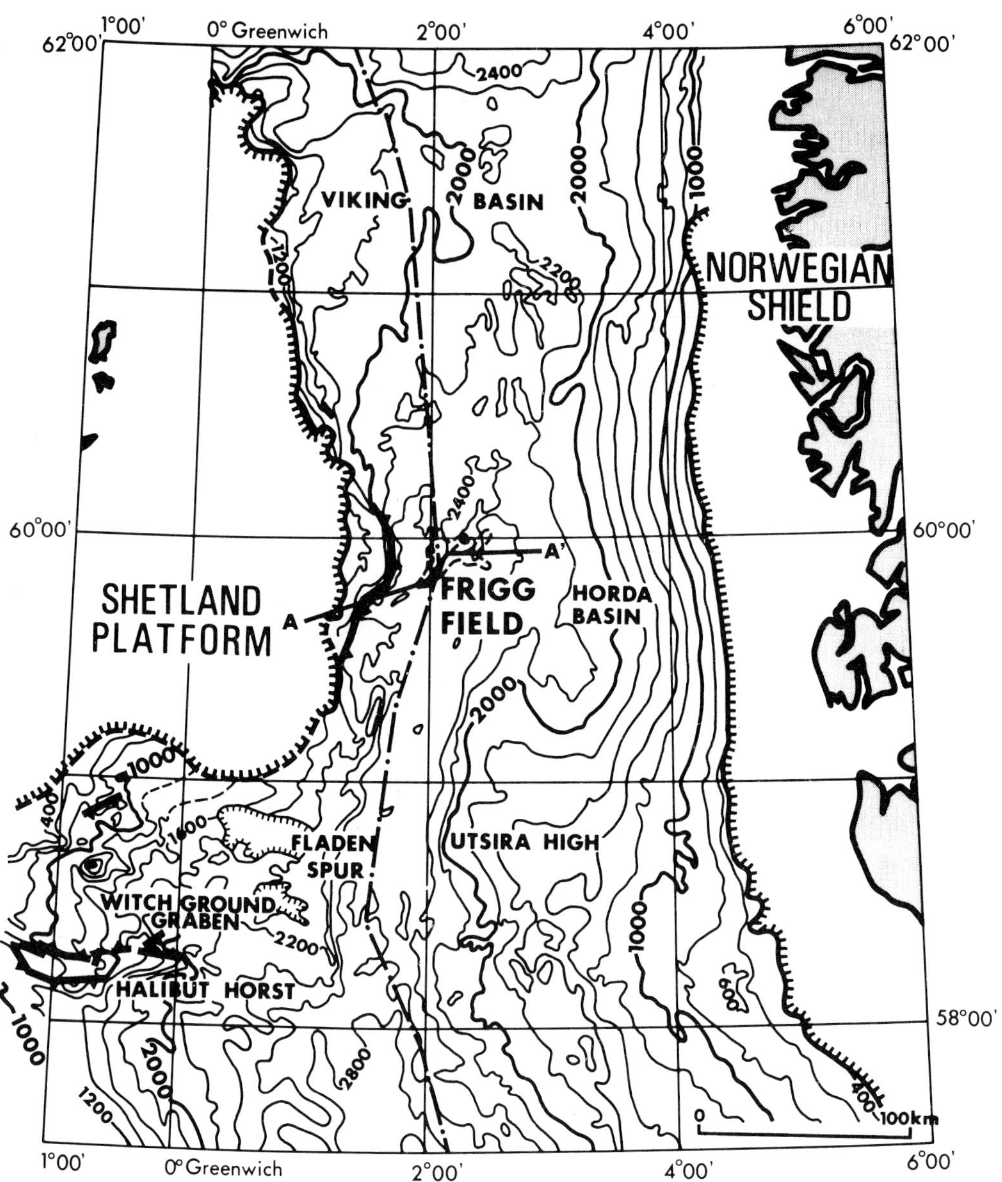

FIG. 2—Viking basin, structure map of top of Cretaceous. C.I. = 200 m. *AA'* is location of cross section shown in Figure 3.

Lias marine shales, Dogger regressive deltaic sandstones, and the Callovian-Oxfordian marine shales. The source of the clastics was mainly the Norwegian shield in the northeast. Oxfordian clastics in the southern part of the basin were derived from the Shetland platform.

The major facies and thickness changes of the second geologic period were related to a synsedimentary breakup of the basin into individual subbasins, or tectonic trends, bounded by large antithetic normal faults.

In the Frigg area, two of these subbasins (the Beryl embayment and the Frigg subbasin) are present. Figure 3, a geologic cross section, shows

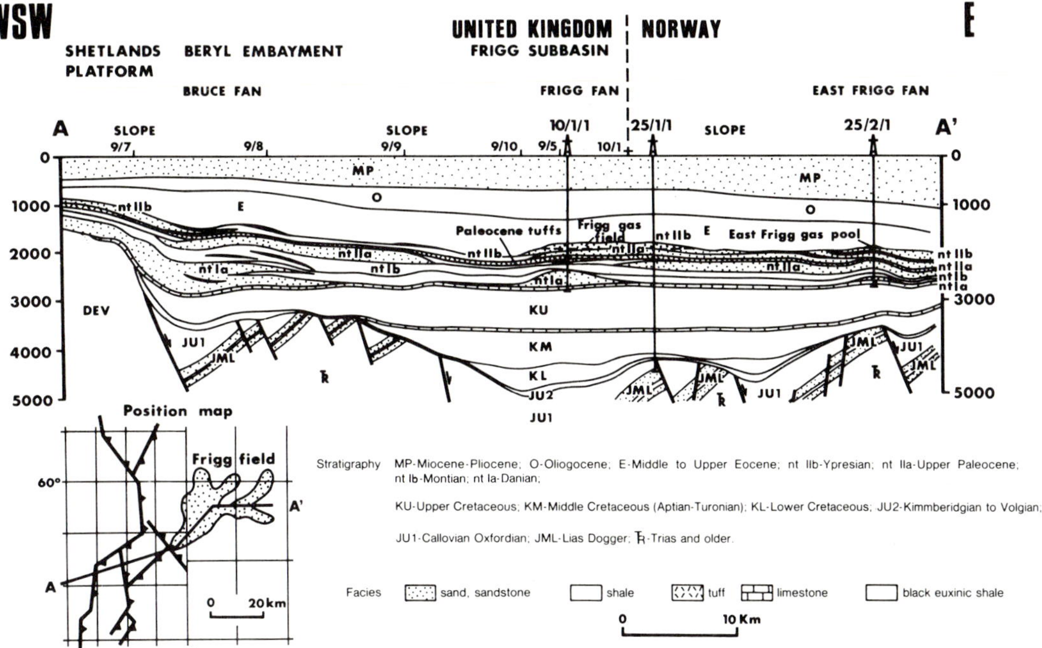

FIG. 3—Geologic cross section from Shetland platform to axis of Viking basin shows relation of pre-Cimmerian structures to successive Paleocene and Eocene fans. Depths are in meters. Location of section shown in Figure 2.

the presence beneath the Cimmerian unconformity of three major fault blocks—the Bruce, Frigg, and East Frigg structures.

Post-Cimmerian Period

After the Cimmerian orogeny, which was a phase of intense tectonic activity, erosion, and relief inversion at the end of the middle Oxfordian, the basin received much wider transgressive infilling. Sedimentation started in late Oxfordian–Kimmeridgian time with the deposition of richly organic, black radioactive shales characteristic of an euxinic deep-water marine environment.

During Early Cretaceous time a stronger connection into the Boreal sea became established; a Lower and middle Cretaceous shaly sequence onlaps the fault-block relief. The presence of a few limestone beds, of regional extent, emphasizes the eustatic low stand of the sea at the end of the Aptian, Cenomanian, and Turonian Stages.

The Upper Cretaceous in the Frigg area consists mostly of shale but includes chalky limestone beds of Campanian and Maestrichtian ages.

The Upper Cretaceous became mostly pure chalk in the southern North Sea.

In the Frigg area, inversion of structural relief related to the collapse of Utsira high occurred at the end of the Cretaceous. That event, together with a correlative rejuvenation of the Shetlands-Orcadian belt on the west, was the main cause of the strong offlap of sediments from west to east which characterizes the Tertiary of the Frigg area.

The Paleocene sediments include a large amount of clastic material that originated in the west and was brought into the basin by turbidity currents creating fan complexes at the foot of both the Shetland escarpment and the Fladen Group spur.

After the Paleocene regression, a new phase of marine shaly sedimentation started during the Eocene. It began with a very important clastic influx: the Ypresian Frigg sands.

The Oligocene is also represented by predominantly shaly marine sedimentation, but sandy deposition became more frequent at that time and finally predominated during the last regressive Miocene-Pliocene period.

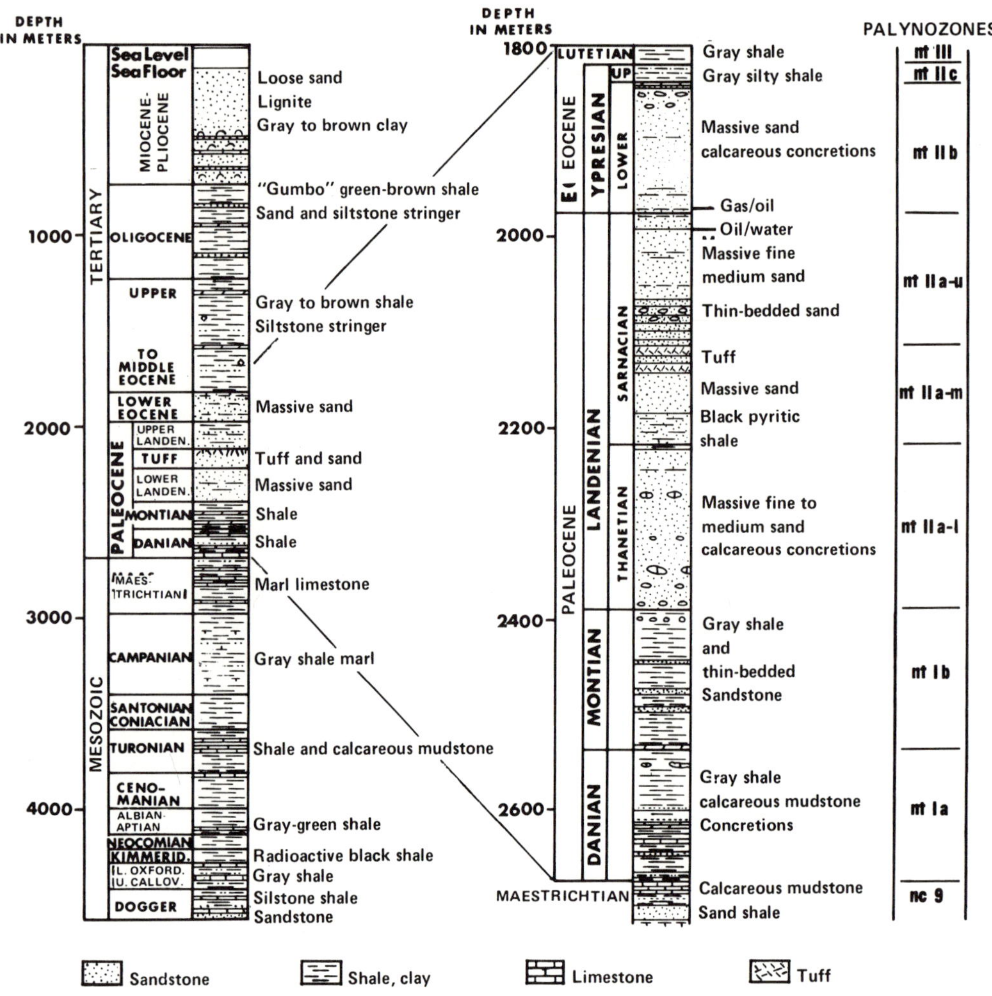

FIG. 4—Composite stratigraphic section of Frigg field showing palynologic zones of Paleocene and Eocene.

STRATIGRAPHY

Frigg 25/1-1

The Frigg discovery well 25/1-1, whose stratigraphic column is typical of the entire field, was drilled through a predominantly sandy Pliocene and Miocene section of pronounced continental character (Fig. 4). Under that section was a greenish-brown, soft, silty mudstone of late middle Oligocene age underlain by a monotonous section of brown, soft gumbo clay of early Oligocene to middle Eocene age.

Below a 58-m (190 ft) section of apple-green, soft, pyritic clay followed by brown-red silty shales, the drill reached the top of the Frigg sand at 1,836 m. This sand is of early Eocene age and constitutes the gas reservoir.

The top of the Paleocene, identified on the basis of palynologic evidence, was found at 1,976 m, about 175 m above the tuffaceous layers which are a prominent seismic marker over most of the North Sea. The total thickness of the Paleocene, which is predominantly sand, is 593 m.

Beneath the Paleocene sands are 104 m of greenish, silty shales with dolomite stringers of Danian age that overlie a very thick 1,322-m shaly section of Late Cretaceous age. The uppermost part of this sequence, consisting of alternations of chalky limestones and shales, is Maestrichtian; the more calcareous lower part is Turonian. Below lie 208 m of dark-gray undercompacted shales of Early Cretaceous age.

At this point the drill reached the strong deep seismic reflector which corresponds to the late Cimmerian unconformity and went through 77 m of highly radioactive (250 API) oil shale of Kimmeridgian age.

Below the Kimmeridgian shales are 282 m of dark-gray pyritic and calcareous shales of Callovian age, with numerous dolomite stringers in the upper part, then finally the water-bearing Dogger sandstone. Total depth is 4,570 m.

Paleocene–Lower Eocene—Frigg Field

The formations of the lower Tertiary are almost devoid of fauna and microfauna. Though the Maestrichtian is characterized by an abundant marine microfauna including both pelagic and benthonic forms, only some Globigerinidae are present in the Danian, and the later series provides only common arenaceous forms. Thus, the stratigraphy is mainly based on palynologic assemblages of both marine microplankton (dinoflagellates) and continental microflora (pollens and spores) which permit very accurate dating. These assemblages allow us to define eight zones, for which an age attribution is proposed as indicated on Table 1.

This correlation supports the idea that an initial period of marine sedimentation in the early Paleocene was followed first by a major regression in the late Paleocene and later by a new marine transgression in the early Eocene. These are regional events observed in other areas including the Paris basin. Nevertheless, an exact correlation with the classic stages of northwest Europe is difficult and, as indicated, the boundaries of some of the zones identified may not exactly coincide with the stage boundaries. The Frigg sequence probably represents a more complete and continuous stratotype than those of France, England, Belgium, or Denmark.

STRUCTURE

Structure of the Viking basin mapped on the top of the Cretaceous (Fig. 2) provides a regional model of the framework at the time of Paleocene deposition. The bordering platforms appear as shallow areas gradually deepening toward the central basin. A major feature, the Shetland escarpment, which is downthrown to the east from 500 to 1,000 m, represents an important structural discontinuity. This escarpment is considered to be a former shelf edge between the Shetland platform and the central deep-marine basin. In the Frigg area this escarpment has a maximum throw of 1,000 m.

The Frigg field is located on the western flank, not far from the deepest part of the Viking Tertiary embayment, in an area of low structural relief at the Tertiary base level where tectonism was of very mild intensity.

Table 1. Early Tertiary Faunal Assemblages, Frigg Field

Zones	Dinoflagellates	Associated Microflora	Proposed Age
nt III	*Areosphaeridium dictyoplokus*	Gymnosperms	Middle Eocene Lutetian
nt IIc	*Membranilarcia ursulae* *Wetzeliella edwardsii*	Gymnosperms Angiosperms	Early Eocene Ypresian
nt IIb	*Wetzeliella coleothrypta*	Angiosperms Gymnosperms	
nt IIa upper	*Detlandrea phosphoritica* *Wetzeliella pachyderma*	*Taxodiaceaepollenites hiatus* *Sequoiapollenites sp.*	Late Paleocene (Landenian)
nt IIa middle	*Wetzeliella meckenfeldensis* *Wetzeliella cf. lunaris*	*Taxodiaceaepollenites hiatus* *Sequoiapollenites sp.*	Sparnacian (to Ypresian)
nt IIa lower	*Wetzeliella hyperacantha* *Wetzeliella homomorpha*	*Caryapollenites sp.* *Platycaryapollenites sp.*	Thanentian (to Sparnacian)
nt Ib	*Areoligera senonensis* *Deflandrea speciosa*	Gymnosperms Pteridophytes, angiosperms	Early Paleocene Montian (to Thanetian)
nt Ia	*Eisenackia crassitabulata* *Paleoperidinium phrophorum*	Gymnosperms Pteridophytes	Danian

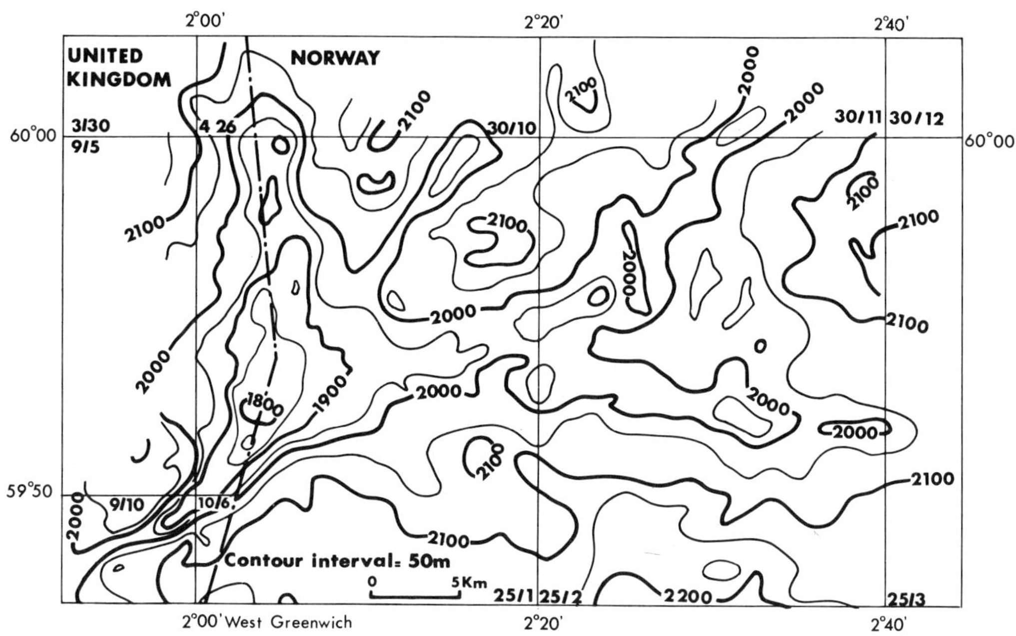

FIG. 5—Seismic structure of Frigg field at top of Frigg sand.

On seismic maps, the structure at the top of the Eocene appears as a low-amplitude, lobate, fan-shaped anticline, with a southwestern apex and three main lobes trending east, northeast, and north (Figs. 5, 6). Seismic maps of deeper horizons show that the Frigg field overlies a late Cimmerian faulted anticline which in turn overlies a complex Jurassic faulted block (Fig. 7). The migrated-depth contour map on the seismic marker very near the top of the reservoir indicates a closed area of 115 sq km with a vertical closure of about 170 m.

A well-defined "flat spot" can be recognized on most of the seismic sections which have been shot across the structure. This phenomenon, which is the seismic reflection due to the density contrast of the gas-liquid contact within a good reservoir, perfectly underlines the structural closed area (Fig. 8).

ORIGIN OF FRIGG STRUCTURE: DEEP-SEA-FAN HYPOTHESIS

The location of the Frigg structure, in the deepest part of the basin but not very far away from the Shetland platform escarpment, and its particular shape suggest a deep-sea fan. Enhancing this hypothesis is the monotonous character of the facies, which show no major variations through time.

The shales are gray to green, in places very pyritic. The sands are either massive or thin layered, generally fine to medium, coarser in the proximal areas, finer in the distal; they commonly show a bimodal distribution—a few coarser supported grains in a finer homogeneous sand. They generally contain shale clasts and are characterized by association of glauconite and carbonaceous detritus.

All these criteria, and particularly the association of glauconite and carbonaceous detritus, are typical of submarine sedimentation (Selley, 1976).

In such sedimentation, the differential compaction of sands and muds is an important factor in the depositional arrangement of the next clastic sediments, which have a tendency to be deposited above the shaly section on the flank of former thick sandy deposits (Figs. 9 to 12).

It can thus be inferred that the Frigg fan has been preserved to the present nearly in its final condition of deposition and that the structure is related mainly to the submarine-fan depositional topography enhanced by differential compaction of sands and muds (Fig. 13). In this respect the main structure represents the upper part of the fan and the apex its feeder channel. Lobes represent outer channels and levees in the middle part of the fan whereas the low areas between lobes are due to the compaction of the more shaly beds between the channels (Figs. 5, 6).

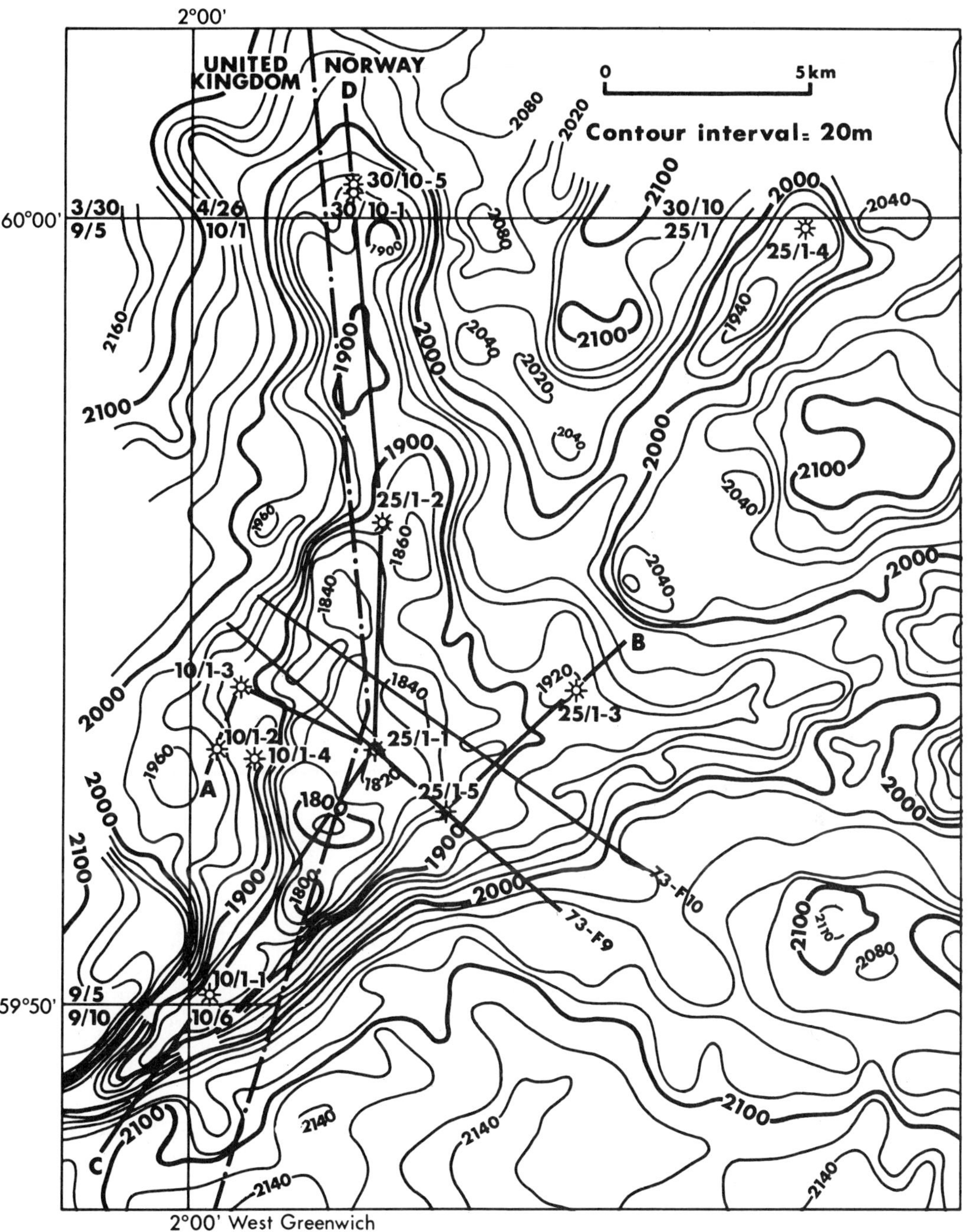

FIG. 6—Detailed structure map of top of Frigg sand in main Frigg field. *AB,CD* are locations of cross sections shown on Figures 9 and 10.

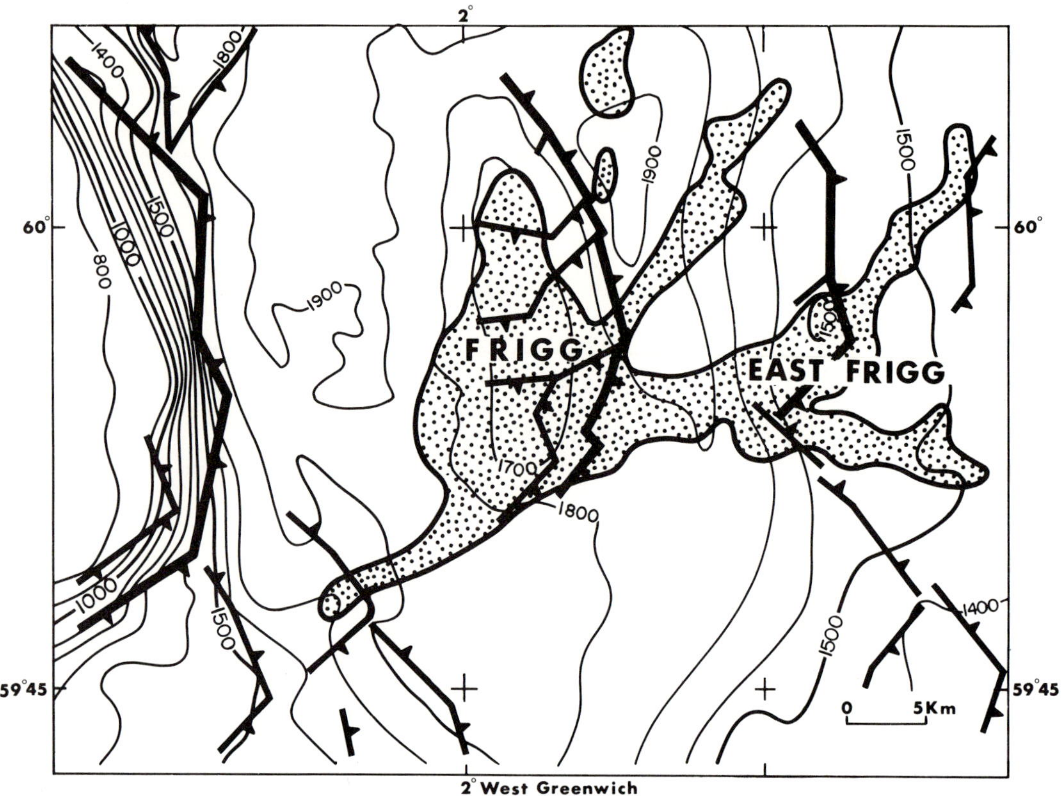

FIG. 7—Pre-Cimmerian structure (C.I. = 100 msec one-way time) and extent of Frigg sand (dotted). Axis of main Frigg field is parallel with axis of deep structure. Feeder channel follows deep fault zone which parallels old northeast-southwest Caledonian faults.

South of Frigg, the Beryl embayment, which is a zone of weakness of the Shetland escarpment, provides the fundamental control for the deposition of clastics in this part of the basin. In particular, in providing routes of transport, it is responsible for the position of several fan complexes. Detailed study of the thick Tertiary sediments deposited along the Shetland escarpment has shown a very important deep-sea-fan sedimentation.

Along the eastern Shetland escarpment, widely differing fan complexes are present. The type of slope, the amount of clastic material locally available, the rate of subsidence, and the variations of the sea level, together with the overriding structural influences already discussed, define the precise nature of each fan.

The general mode of deposition throughout the area is progradation, or offlap, of the deposits to the east or southeast. The maximum thickness of the series is generally in the proximal areas, such as the foot of the escarpment (Fig. 3). Farther east, differentiation is more important and results in the development of local fan complexes, which are much thicker and more sandy than the surrounding embayments. Such differentiation is typical of deep-sea-fan sedimentation where no important reworking of initial deposits occurs.

The fan complexes of the Viking basin form part of a regional system, which implicates on its eastern border a final progradation slope that is apparent on the seismic lines.

HISTORY OF TERTIARY SEDIMENTATION AND PALEOGEOGRAPHY

Early Paleocene (Fig. 14)

The first phase of sedimentation occurred in the early Paleocene and was characterized by rapid infilling of the basin under a general marine environment.

Danian—During the Danian stage (Table 1, nt Ia), in the southern area south of 59°, including the Witch Ground graben and the Viking basin,

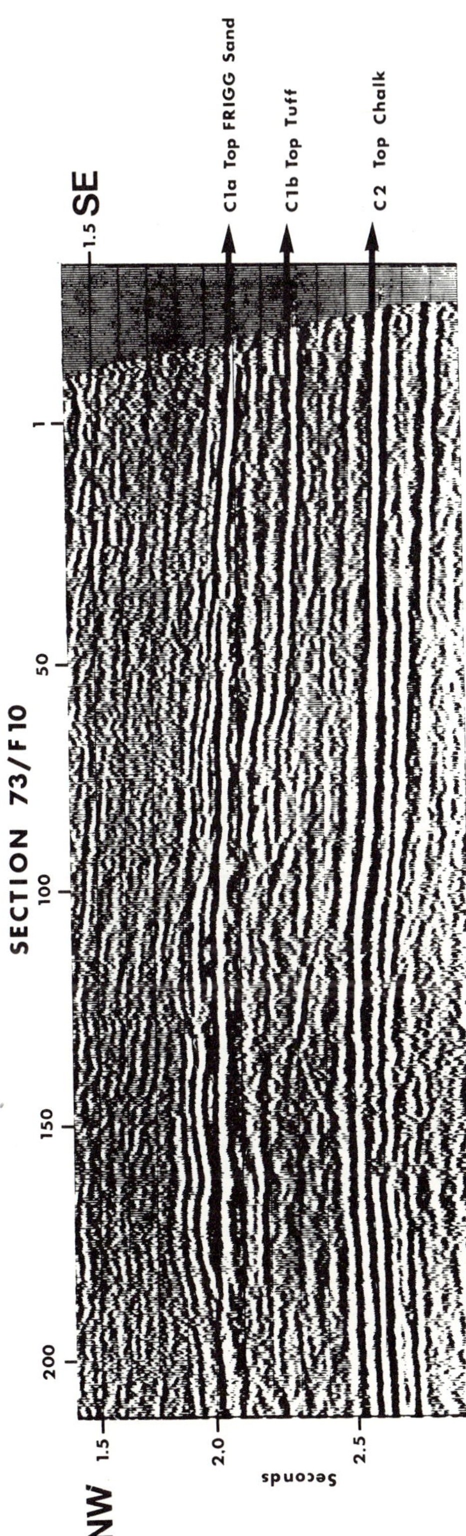

FIG. 8—Typical seismic section across Frigg field shows very clear "flat spot" at about 2 sec. This "flat spot" is real seismic reflection caused by density contrast of gas-liquid contact.

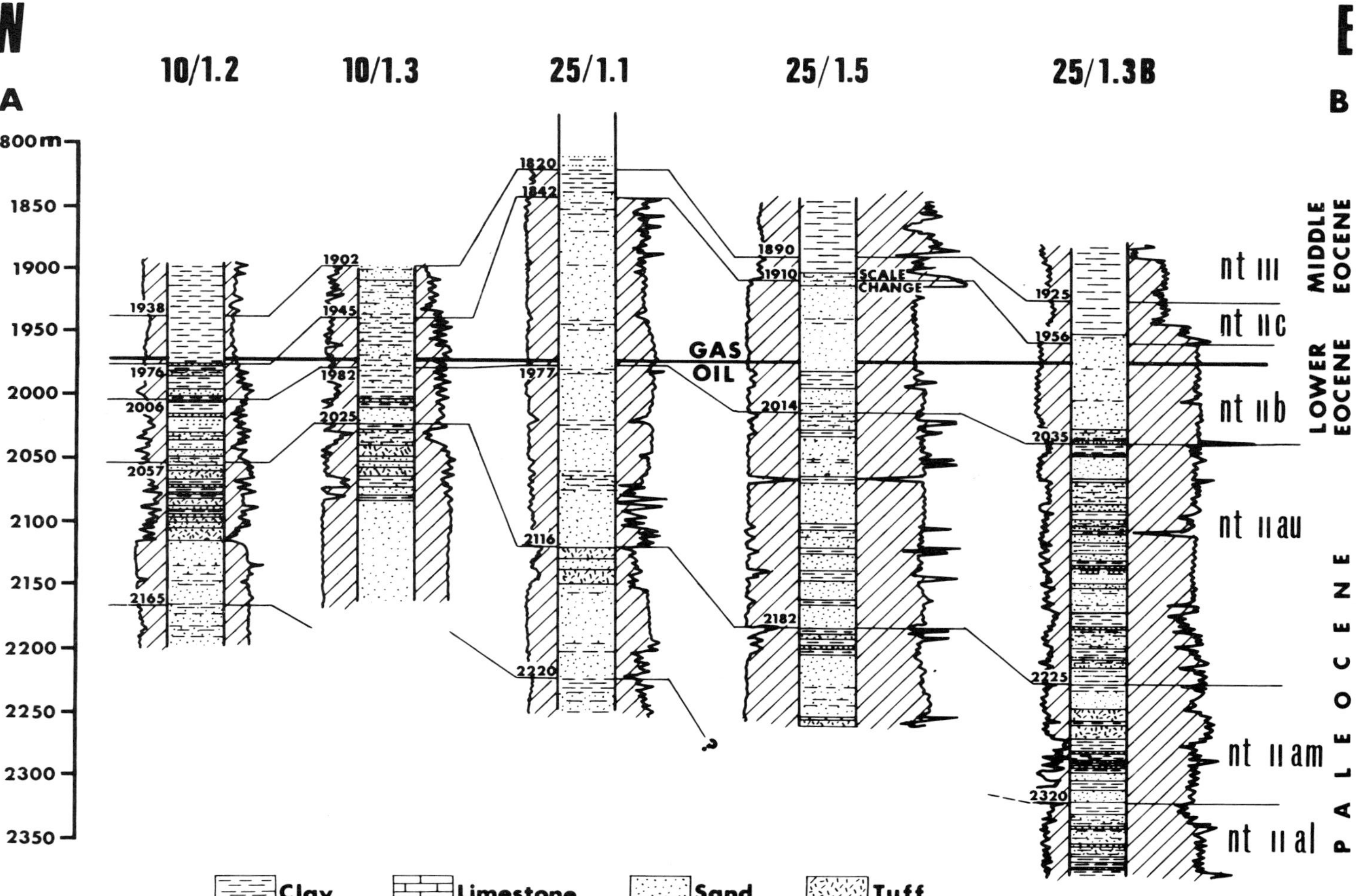

FIG. 9—West-east geologic section across Frigg structure shows eastward thickening of sand. For location see Figure 6.

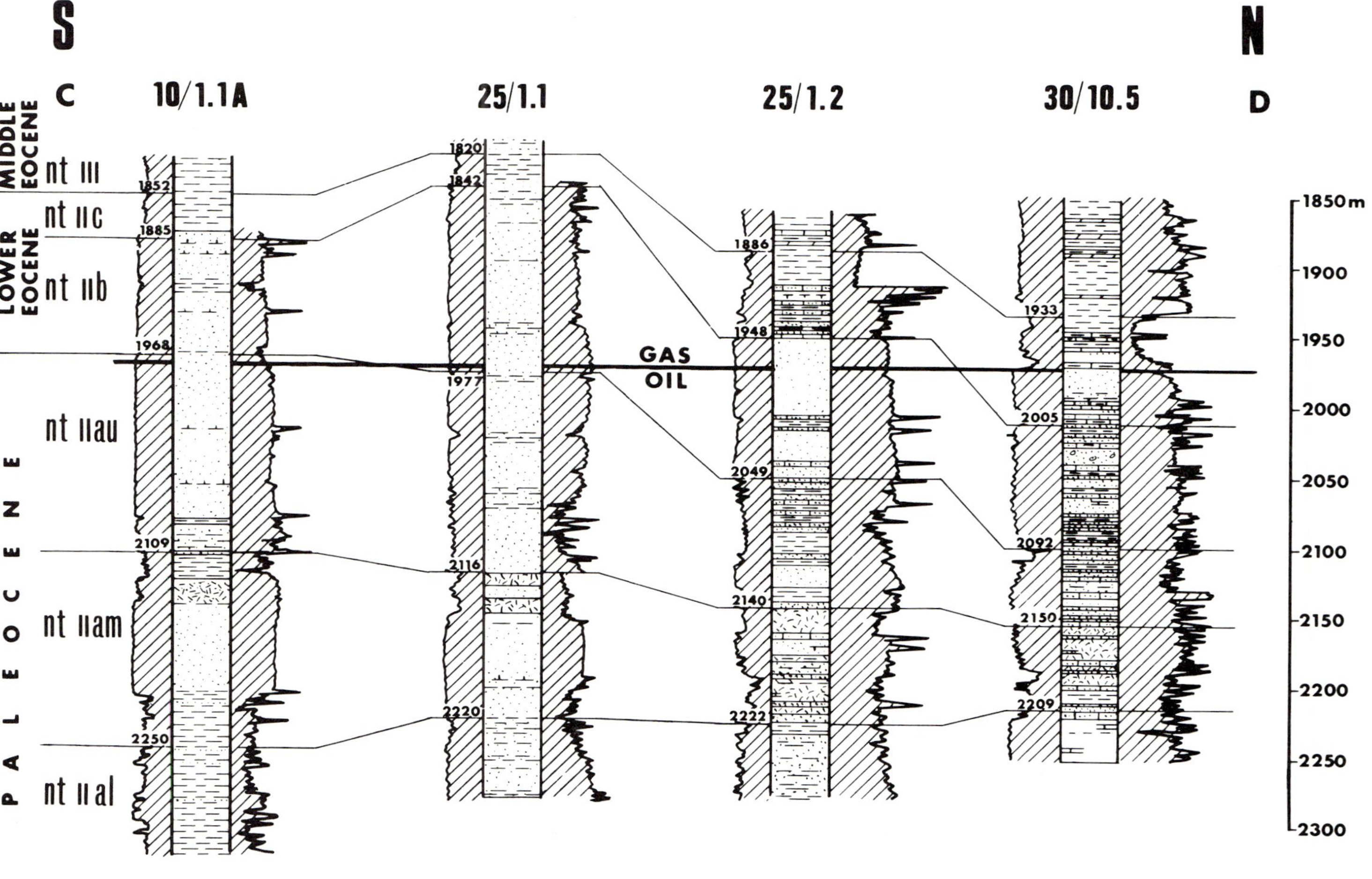

FIG. 10—North-south geologic section across Frigg structure shows northward thinning of sand. For location see Figure 6.

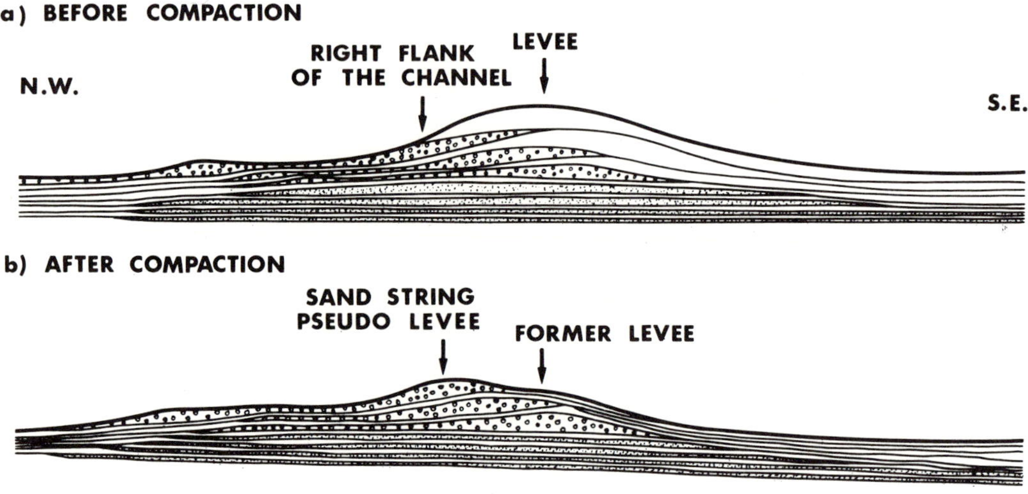

FIG. 11—Compaction of deep-sea-fan channel. **A**, levee at time of deposit. **B**, same levee after compaction; sand deposits on flanks of original levee are now topographic highs.

an initial phase of sedimentation 200 m thick is characterized by shale, marl, and limestone without detrital material (Fig. 14a).

The first important influx of clastic material occurred only in the late Danian and, at that time, two marine areas of thick sandy progradation were located in the Witch Ground graben and in the Beryl embayment. In the latter area, the Bruce fan shows a maximum thickness of 600 m of massive sands. After the first downslope infilling, progradation extended farther east and additional massive sand bodies were deposited farther into the basin in the Frigg area. These sand bodies (400 m thick at Frigg, 100 m thick at East Frigg) represent the first elements of the Frigg and East Frigg fans. They were deposited on the southeastern flank of both the Frigg and East Frigg Late Cretaceous anticlines. The relation is probably not random, but results from the antagonism between the turbidite flow and the structural obstacles. These low-amplitude structural obstacles created a velocity loss which allowed the coarser fraction of the flow to be deposited there. Apart from these events, thick distal shaly sedimentation occurred everywhere in the basin as well as on the slope areas between the Bruce, Frigg, and East Frigg fans.

At the end of the Danian, a new topography

developed as a result of differential compaction of the sand mounds and the surrounding shales. The new topography modified to a great extent the location of later sand bodies.

Montian—The Montian Stage (Table 1, nt Ib) was also a period of rapid basinal infilling (Fig. 14b). In the Witch Ground graben, offlap continued to the southeast, with deposition of thick downslope deposits. At the same time, new fan complexes appeared in the Viking basin (Sleipner, Heimdal, and Ninian fans), where only distal Danian shales had been previously deposited. Other fans continued to grow, including the Frigg and East Frigg fans where new sand bodies flank on the southeast the former Danian mounds. The Bruce fan complex, which had been infilled during the Danian, became a wide slope area with only shaly sedimentation.

Late Paleocene (Fig. 15a)

The late Paleocene regression brought important changes in the mode of deposition of the fan complexes, mainly because of a decrease in the amount of clastic material available and progressive establishment of a lower sea level.

Thanetian—The Thanetian was characterized

F. E. Heritier et al

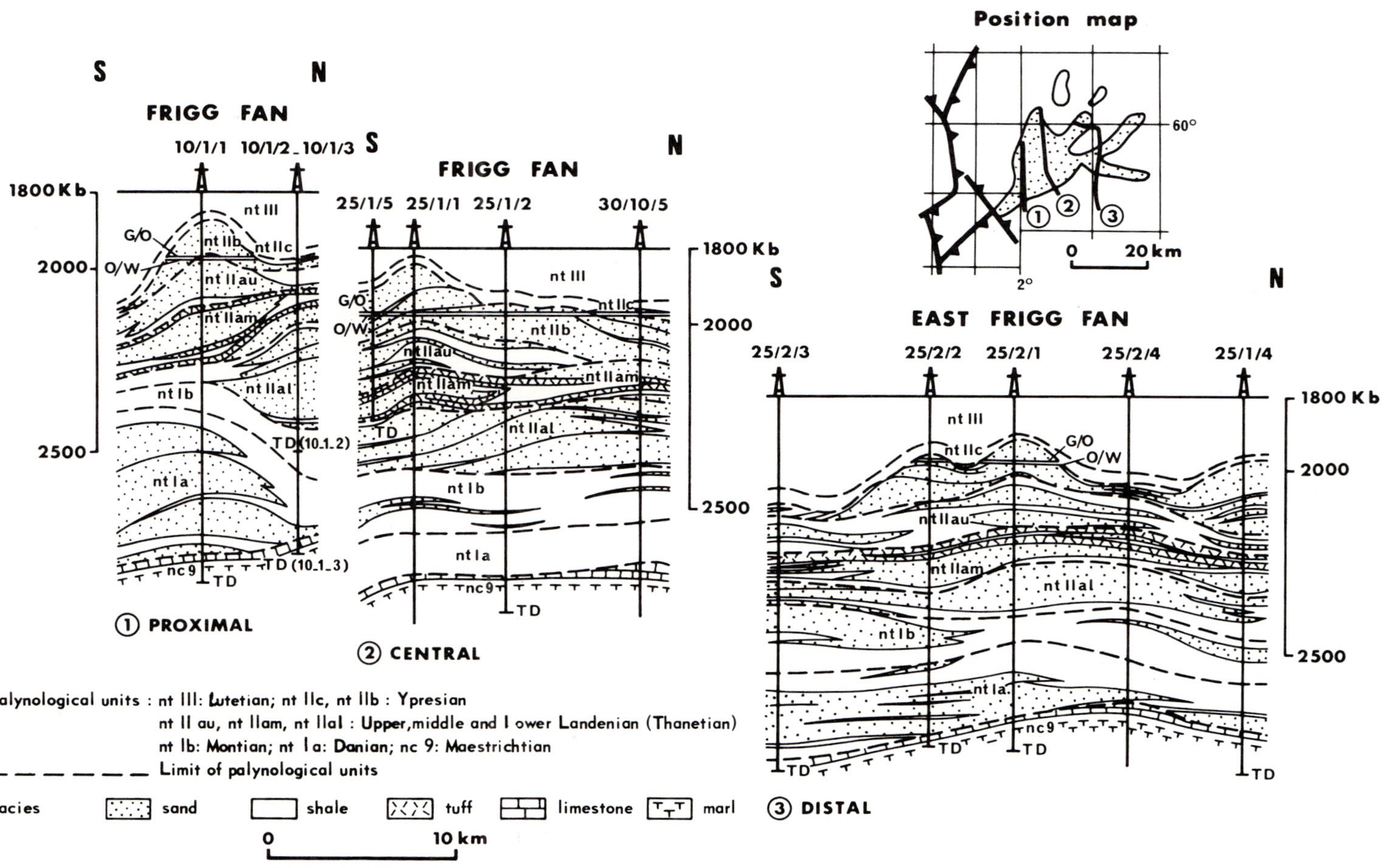

FIG. 12—Cross sections of Frigg field show deposition of thick sands (channels) on flanks of earlier structural features.

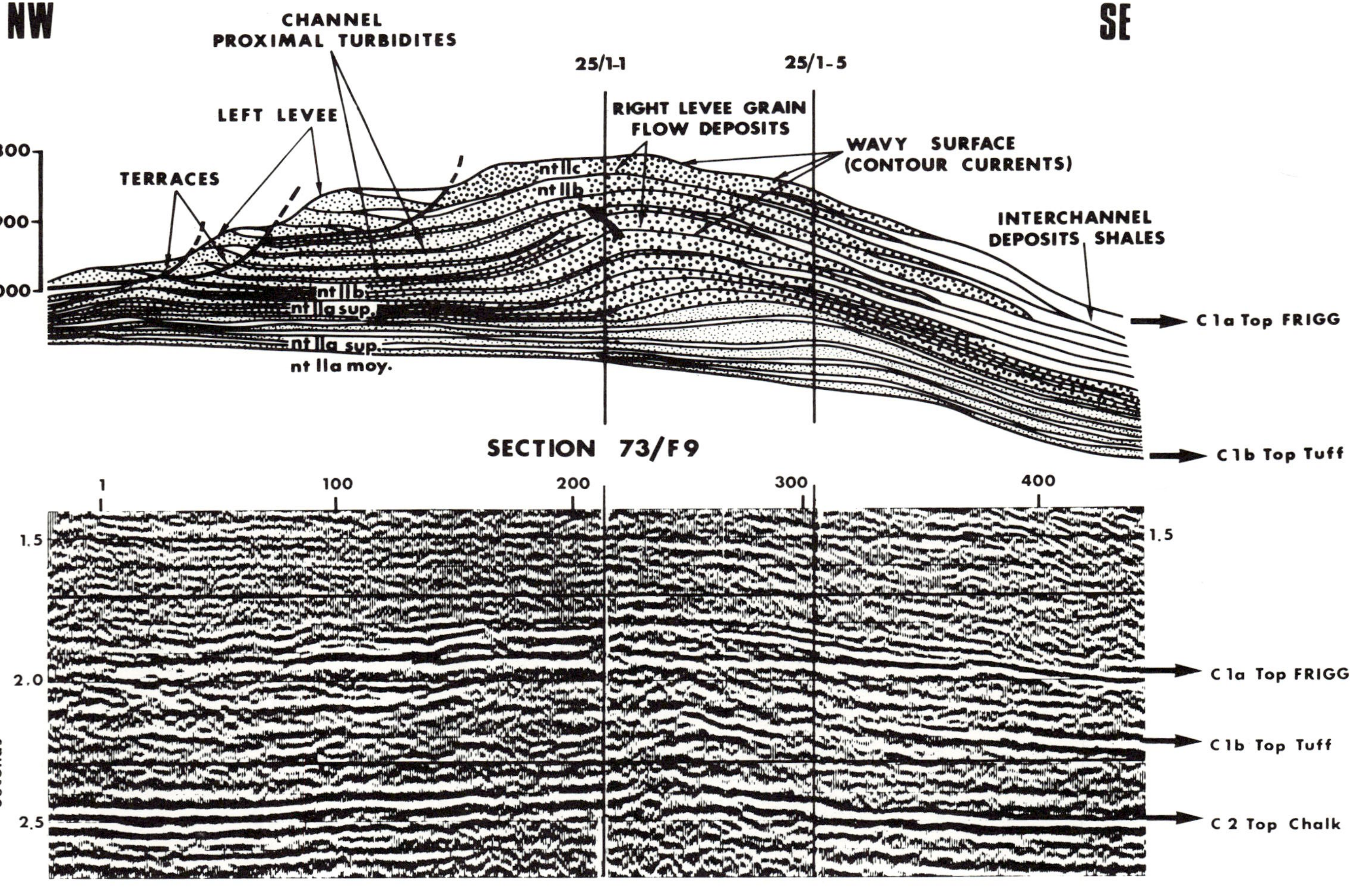

FIG. 13—Sedimentologic interpretation of typical Frigg seismic section crossing wells 25/1-1 and 25/1-5. Vertical exaggeration is 2×. Seismic "flat spot" is well defined.

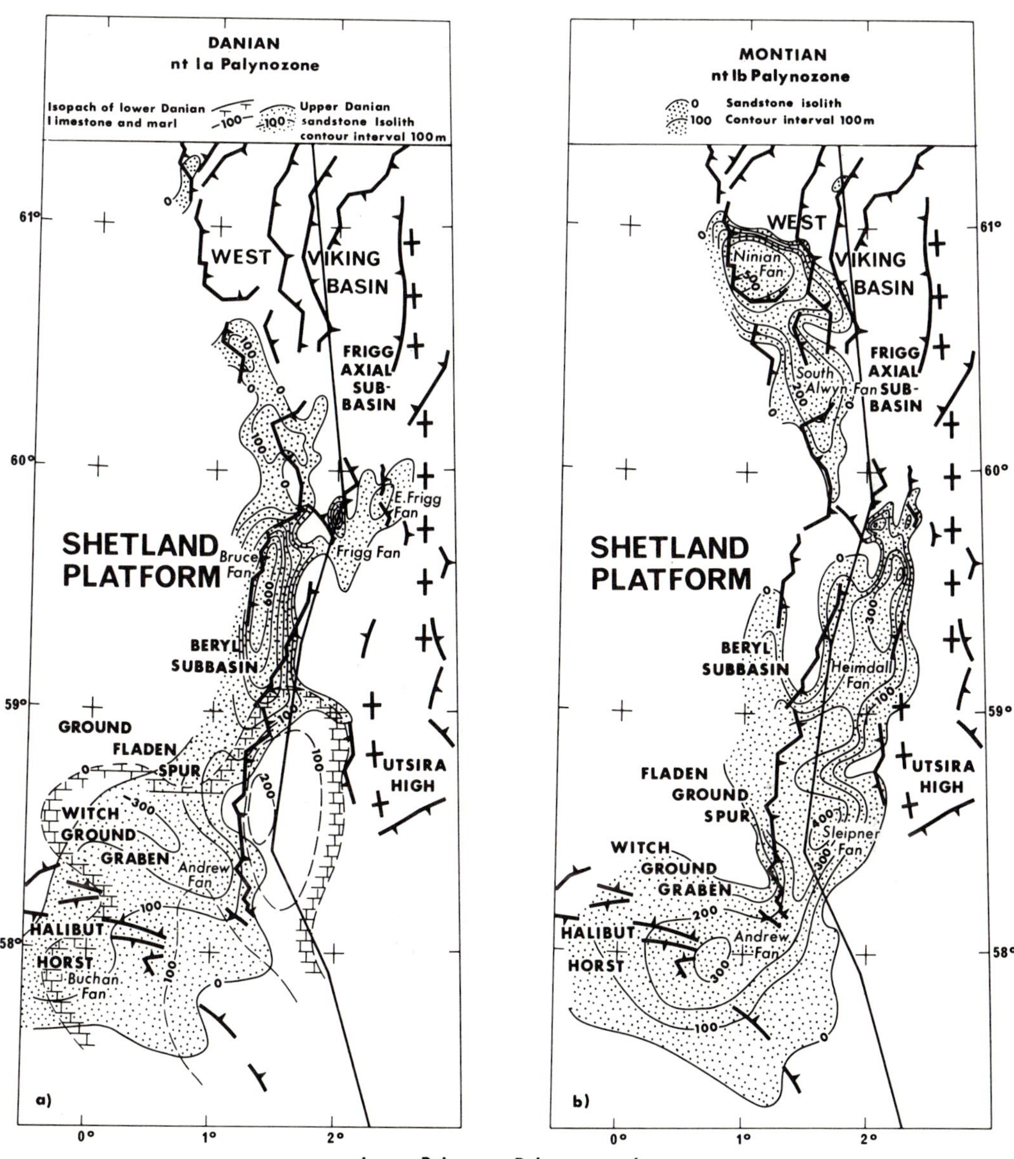

FIG. 14—Early Paleocene paleogeography. **A**, Danian. Structural framework of faults bordering Shetland platform and main pre-Cimmerian blocks of Viking basin, shows Danian episode of nonclastic sedimentation, south of 59°N lat., with maximum thickness of 200 m of limestone and marl. **b**, Montian sandstone isolith map shows spread of fan sedimentation over entire area.

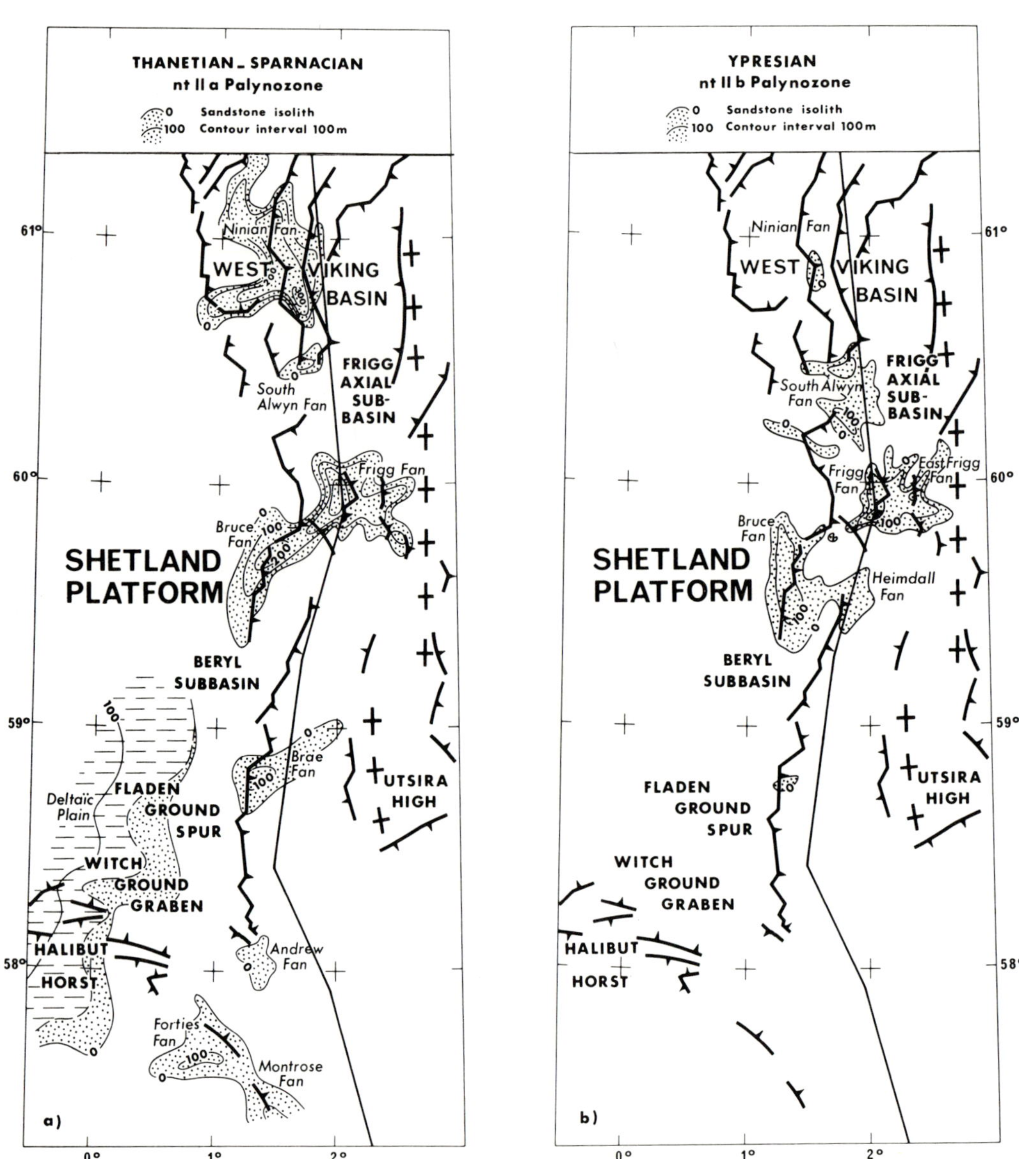

Upper Paleocene to Lower Eocene Paleogeography

FIG. 15—Late Paleocene and early Eocene paleogeography. **a,** Late Paleocene distal-plain complex is invading Witch Ground graben; all Thanetian sand bodies are very distal fans completely separated from deltaic belt by slope area of thin shaly marine sedimentation. **b,** Early Eocene marine transgression characterized by final influx of sands mainly localized in Frigg area, repeating initial distribution of Danian sands.

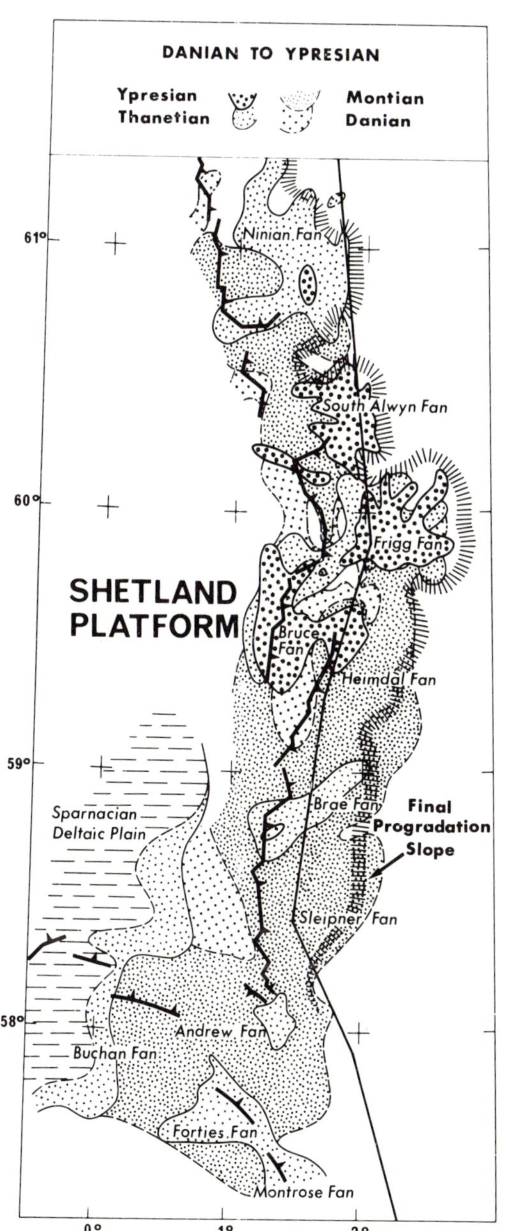

The Fan complex at the end of the Lower Eocene

FIG. 16—Fan complex at end of early Eocene. Cumulative map shows nearly 50 km wide area of thick clastic sediments at foot of Shetland platform. Eastern limit of clastic deposition is final progradation slope, missing on south where basin deepens. Perspective views are shown in Figures 17 and 18.

by deposition of a more distal fan complex. In Witch Ground graben, thick lower Paleocene sediments are overlain by thin shaly marine beds probably more representative of a prodeltaic environment. At the same time, the very distal fans of Forties, Montrose, and Cod fields were deposited in the Ekofisk basin. The same tendency is observed in the Viking basin, where new fans were deposited in the remaining low areas on the northern flanks of the original fans. The Frigg fan was developing in an area previously characterized by mainly shaly sedimentation between the Heimdal fan on the south and the South Alwyn fan on the north.

Sparnacian—The Sparnacian was the main period of regression. In the lower Sparnacian (Table 1, nt IIa middle) only local sand bodies are present. The sequence is widespread and relatively uniform in facies and thickness. Pyritic black shales are present at the base, indicating deposition in a more euxinic environment. The sequence also includes, near the top, a persistent layer of volcanic tuff.

The tuff very probably originated from the western Hebridean volcanic belt and is related to a major episode of volcanic activity resulting from the opening of the North Atlantic (Jacque and Thouvenin, 1975). At the same time, a thin deltaic complex prograding from the west developed in the Witch Ground graben and replaced the former sea, thus capping the very thick, early Paleocene fan system.

The early Paleocene fan complex, comprising a delta-front and a deltaic-plain facies, is well defined by the alternation of sand, shale, and coal layers. The present depth of the Sparnacian coals ranges from 900 to 1,200 m subsea, which may be interpreted as a regional paleo-sea level, representative of the late Paleocene. This paleo-sea level, which can be related in the Frigg area with the top of the escarpment, is probably the best direct evidence for the deep-sea character of the sedimentation, considering especially that it was a period of low stand of the sea.

The predominantly continental character of the "nt IIa middle" palynologic assemblage must be interpreted as due to the relative proximity of the deltaic plains.

The late Sparnacian (Table 1, nt IIa upper) was an episode of very local development represented by a persistent deltaic-plain facies in the northern Witch Ground graben. In the Frigg fan, the interval is represented by mainly sandy deposits, which constitute the base of the productive Frigg formation. The new sand lobes deposited at this time are on the southern flank of the former Thanetian fan.

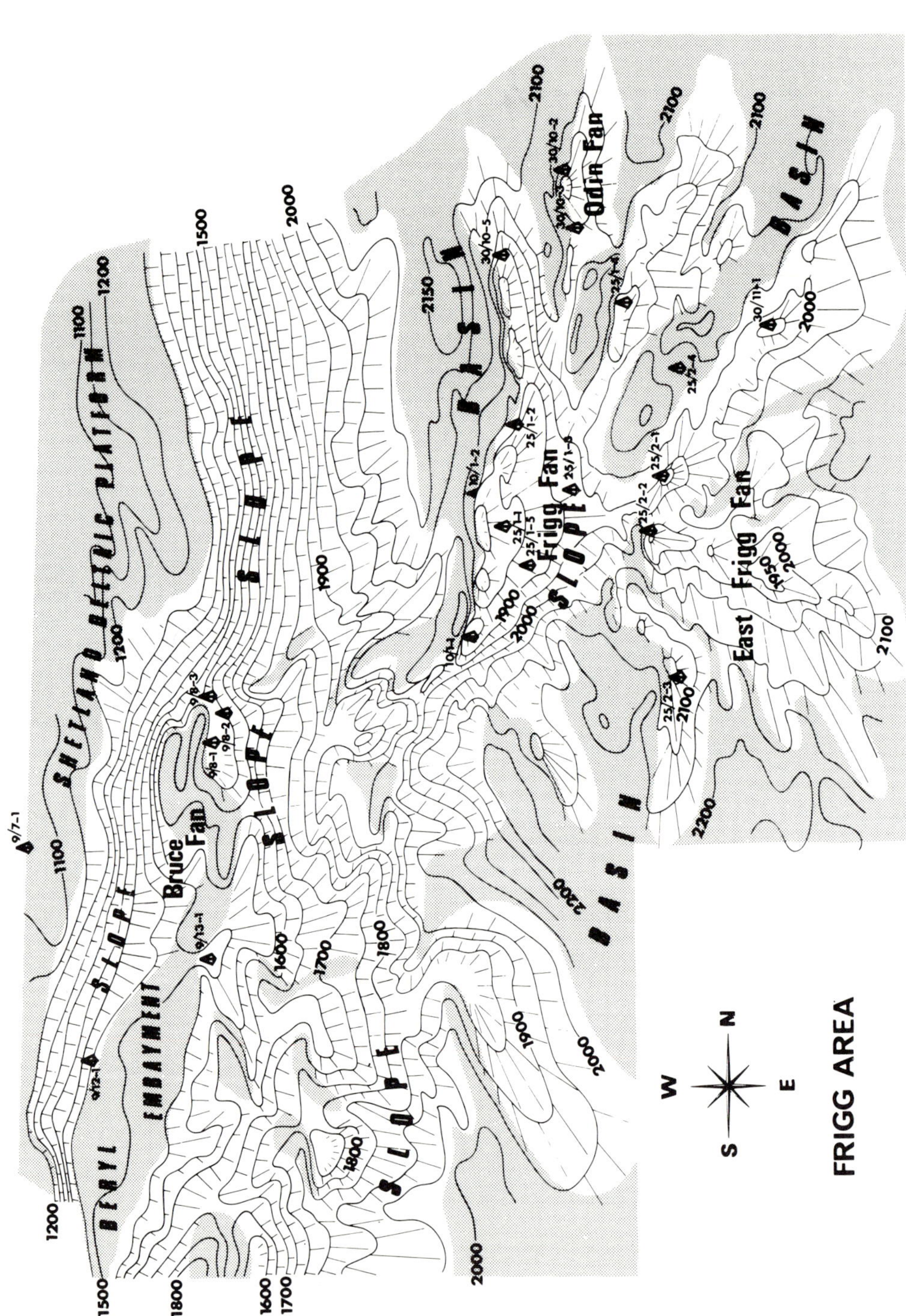

FIG. 17—Perspective view from east of Frigg field area shows landscape at end of lower Eocene which shows, in isobaths below sea bed, final shape of Frigg and East Frigg fan areas.

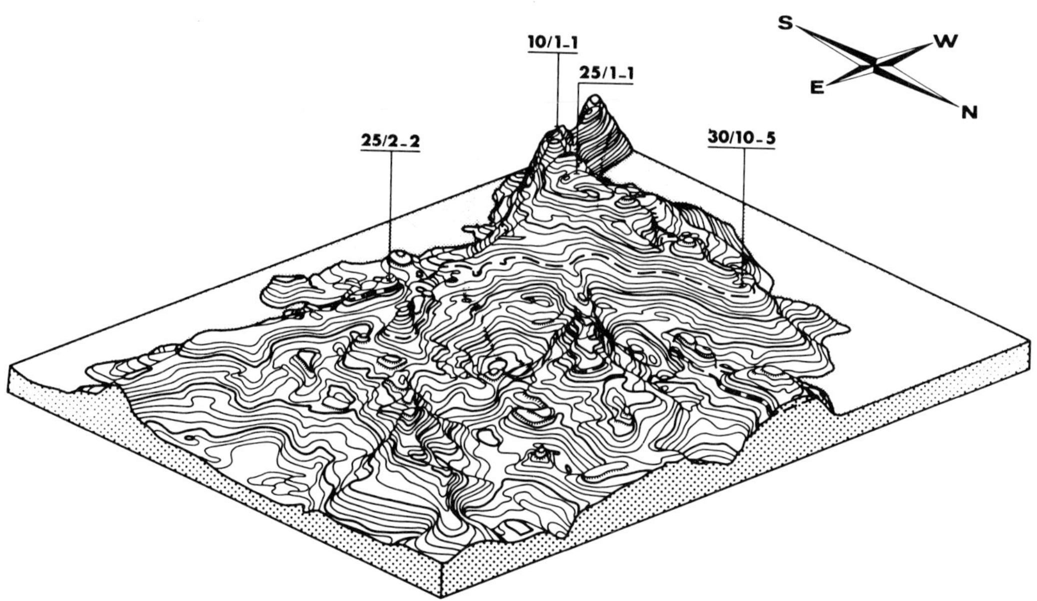

FIG. 18—Perspective view from northeast of Frigg field shows landscape of field and satellites. Drawn by computer from isobath map.

Early Eocene (Fig. 15b)

A new marine transgression of Ypresian age (Table 1, nt IIb) occurred in the basin at the beginning of the Eocene and was characterized by a final influx of clastic material into the basin. However, at that time thick sand was deposited only in the Bruce, Frigg, East Frigg, and South Alwyn fans. These final sand lobes were deposited on the southern flank of the previous structures because of differential compaction of the Thanetian and Sparnacian fan deposits (cf. Fig. 12).

The shaly lateral facies of the sands is either very thin or absent, which means that Frigg sands were originally deposited in mounds and that the relief observed today mainly reflects this original deposition rather than the differential compaction.

These sands were subsequently sealed by a thin upper Ypresian marine shale (nt IIc) which is replaced by red oxidized shales on the top of the structure. A very distal sand lobe of this phase has been identified on the northernmost part of the Frigg field in Esso 30/10-5 well. The extent of sand deposits and the sedimentologic landscape at the end of early Eocene time are shown in Figures 16 to 18.

HYDROCARBONS: COMPOSITION, DIAGENESIS, AND ORIGIN

Composition

The gas of the main Frigg field is a dry gas, containing only 3.7 g/cu m of condensate. Its percentage composition is as follows: N_2, 0.4; CO_2, 0.3; C^1, 95.5; C^2, 3.5; C^3, 0.04; C^4, 0.01; $>C^4$ . . . . The associated condensate is composed of 86.5% of $C^{11}+$.

The oil beneath the gas accumulation has a gravity of 23 to 24° API and is of a naphthenic composition. It is characterized by a saturated/aromatic hydrocarbon ratio of 3:76. The light compounds with a carbon number of $<C^{17}$ represent a very minor fraction of the oil, and the N alkanes are almost absent. Thus, the saturated hydrocarbons are mainly represented by cyclanes and isoalkanes.

Diagenesis

The type of oil characteristic of the main Frigg field may be considered as anomalous, suggesting biodegradation caused by bacteria. Other Paleocene or Eocene pools of the area do not show the same anomaly. For example, the East Frigg pool (well 25/2-1) produces gas with condensate and shows an underlying oil containing a significant fraction of N alkanes.

Origin

The less altered oil of East Frigg permitted a measurement of the ratio of pristane/n C^{17} to phytane/n C^{18}.

A value of 2 for that ratio corresponds with the readings obtained from Dogger and Lias source rocks in the area (values of 2 to 2.4), but differs significantly from those measured on Kimmeridgian radioactive shales (1.4). Such a correlation suggests that the Frigg oil probably was generated and subjected to an early phase of migration from the pre-Cimmerian, deeply buried Dogger and Lias source rocks. A similar correlation between hydrocarbons in lower Tertiary reservoirs and pre-Cimmerian source rocks has been clearly established by analysis of the condensate gas and oil of the Paleocene Heimdal field, only 35 km south of Frigg.

The Frigg gas probably results from a late phase of the diagenesis of the Jurassic source rocks, which show vitrinite-reflectance values ranging from 1.5 to 1.8. These values are characteristic of gas-zone diagenesis. By contrast, lower Tertiary and Cretaceous source rocks of the Frigg area show only a low degree of diagenesis. The vitrinite-reflectance value of 1.0 is reached only at a depth of 4,000 m in the deep Frigg wells. The deep origin of the Frigg gas is also mainly supported by its carbon-isotope composition, which shows a δC^{13} value of -43.3% which is characteristic of a deep gas.

The absence of propane and butane in the Frigg gas, which could be interpreted as being indicative of a shallow origin of the gas, probably results from the same biodegradation as that of the oil.

RESERVOIR CHARACTERISTICS, RESERVES, AND PRODUCTION PLANS

The gas-bearing sand in the discovery well 25/1-1 was tested after perforating the interval between 1,920 and 1,928 m. With a 7/8-in. (2.2 cm) choke, a gas flow of 673,000 cu m/day (24 MMcf/day) was measured at 60°F (16°C) and 14.7 psi (101 kPa). The buildup of the bottom-hole shut-in pressure was instantaneous and became steady at 2,835 psi (19,547 kPa) at 1,880 m. The maximum static wellhead pressure is 2,465 psi (17,000 kPa). These tests have confirmed the very good permeability of the reservoir which had been measured from cores.

The Frigg gas is sweet, dry, relatively free of noncombustible inert gases, and contains approximately 95% methane. The liquid condensate content is small and about 4 g/cu m.

At an initial absolute pressure of 199.25 bars (19,925 kPa; 2,890 psi), at a subsea depth of 1,912.5 m, the gas viscosity is 0.018 cp at a reservoir temperature of 60°C. The dimensionless gas compressibility factor (Z) is 0.865.

The reservoir is predominantly composed of clean, fine, and unconsolidated sand with good characteristics: porosities range from 25 to 32%, and average permeabilities from 1,200 to 1,600 md.

Gas-in-place and recoverable-gas-reserve calculations are based on the total accumulation in both Norway and the United Kingdom. The consulting firm of DeGolyer and MacNaughton estimated gas in place to be about 269 billion cu m (9.5 Tcf). Of this total, 60.82% is in the Norwegian sector and 39.18% is in the United Kingdom sector.

The probable recoverable gas depends on the effectiveness of the Frigg and Cod sand aquifers and could be 215 billion cu m (7.6 Tcf).

The development of this field is under way. The two drilling platforms (CDP1, DP2), the two production platforms (TP1, TCP2), the quarter platforms, the flare, and the two pipelines to St. Fergus in Scotland have been set up. On each of the two drilling platforms, 24 production wells will be drilled.

Drilling has already started on the two platforms. On CDP1, the 24 wells have been drilled, and have been completed with 65/8-in. screens to prevent the wells from producing sand, and with 75/8-in. tubing. On DP2, 16 wells have been drilled and completed.

Frigg production started in September 1977, and the first delivery from TP1 into the pipeline occurred on September 11, 1977. Gas sales at St. Fergus started on September 13, 1977, at a rate of 4 million cu m/day (140 MMcf), only 6½ years after discovery.

For the present, production from the already completed wells averaged 38 million cu m/day (1,350 MMcf/day).

The behavior of these wells is excellent, and no skin or turbulence effects have been noticed up to a rate of 2.3 million cu m/day (81 MMcf/day) per well.

The development of the field is scheduled to be completed by October 1, 1979, 8 years after the discovery, and by that time the average daily production should be 43 million cu m/day (1,520 MMcf).

The gas deposit is underlain by an oil zone 10 m thick. Extensive tests in well 25/1-3 have proved the impossibility of economically producing this heavy (24° API), naphthenic oil. Reserves in place amount to 125 million cu m (790 million bbl).

SELECTED REFERENCES

Blair, D., 1975, Structural styles in North Sea oil and gas fields, *in* A. W. Woodland, ed., Petroleum and the continental shelf of north-west Europe, v. 1, geology: New York, John Wiley & Sons, p. 327-337.

Fowler, C., 1975, The geology of the Montrose field, *in* A. W. Woodland, ed., Petroleum and the continental shelf of north-west Europe, v. 1, geology: New York, John Wiley & Sons, p. 467-476.

Jacque, M., and J. Thouvenin, 1975, Lower Tertiary tuffs and volcanic activity in the North Sea, *in* A. W. Woodland, ed., Petroleum and the continental shelf of north-west Europe, v. 1, geology: New York, John Wiley & Sons, p. 455-465.

Parker, J. R., 1975, Lower Tertiary sand development in the central North Sea, *in* A. W. Woodland, ed., Petroleum and the continental shelf of north-west Europe, v. 1, geology: New York, John Wiley & Sons, p. 447-453.

Selley, R. C., 1976, Subsurface environmental analysis of North Sea sediments: AAPG Bull., v. 60, p. 184-195.

Walker, R. G., 1978, Deep-water sandstone facies and submarine fans: models for exploration for stratigraphic traps: AAPG Bull., v. 62, p. 932-966.

Walmsley, P. J., 1975, The Forties field, *in* A. W. Woodland, ed., Petroleum and the continental shelf of north-west Europe, v. 1, geology: New York, John Wiley & Sons, p. 477-485.

Forties Field, North Sea[1]

A. N. THOMAS,[2] P. J. WALMSLEY,[2] and D. A. L. JENKINS[2]
London, England

Abstract The Forties field, a large oil pool discovered in 1970, is in the northern part of the British sector of the North Sea, 175 km (110 mi) east of Peterhead, Scotland, in water depths of 91 to 131 m (300–430 ft). The reservoir is a sandstone of Paleocene age at a depth of about 2,135 m (7,000 ft), at the base of a thick Cenozoic section consisting primarily of mudstone. The Paleocene sandstone/mudstone sequence is underlain by Danian and Maestrichtian chalk. The trap is a broad low-relief anticlinal feature with a closed area of 90 sq km (35 sq mi). Maximum gross oil column is 155 m (509 ft). Recoverable oil is estimated at 1.8 billion bbl from an in-place figure of 4.4 billion.

INTRODUCTION

The Forties field, in the British sector of the North Sea, is approximately 175 km (110 mi) due east of Peterhead, Scotland (Fig. 1). Most of the field falls within BP License Block 21/10, the eastern end extending into Shell/Esso Block 22/6. Approximate geographic coordinates are lat. 57°45′N, long. 0°45′E. The field is named "Forties" after the meteorological area in which it was discovered. Weather conditions are severe, with frequent gales, particularly in the winter months, and with waves exceeding 5 m (15 ft) for about one third of the time.

Water depths across the field range from 91 m (300 ft) in the southeast to 131 m (430 ft) in the northwest (see Fig. 8). The bottom generally consists of soft clay overlain by a variable thickness of mud.

HISTORY OF DISCOVERY

The Forties field was discovered in October 1970, when BP's well 21/10-1 found oil in Paleocene sands at a depth of about 2,135 m (7,000 ft). Block 21/10 formed part of a U.K. 2d Round license which had been awarded to BP in November, 1965. At that time little was known about the geology of the North Sea, particularly the northern part. Only five marine wells had been drilled in British waters at the time of application for the Forties license, and gas had yet to be discovered in the southern North Sea. It was realized that the North Sea covered a large Tertiary sedimentary basin, possibly overlying in part a thick development of older sediments, but the main interest was centered on the gas area in the south. Oil prospects in the north were thought at best to be highly speculative. Nevertheless, the possibility that oil could be present was recognized as early as 1965, although the subsequent success exceeded all expectations at that time.

During the mid 1960s the main exploration activity was in the southern North Sea. The first indication that the northern area might be an oil province was in July 1967 when the second well to be drilled in Norwegian waters, Esso 25/11-1, found oil shows. One or two small discoveries were made in Danish waters shortly afterward, and in June 1968 Phillips made a gas/condensate discovery, the Cod field, in Norwegian Block 7/11. However, none of these fields was large, and by the end of 1969, with over 50 exploration wells drilled in northern waters, hopes of making economic oil discoveries had begun to dwindle. In December 1969, Phillips made their important Ekofisk discovery in southwest Norwegian waters (Fig. 1). This, together with the simultaneous, but smaller, Montrose discovery by Amoco/Gas Council in UK Block 22/18, revived the interest of the petroleum industry. The tempo of exploration accelerated and further successes followed rapidly. In this climate BP spudded its exploration well 21/10-1 with Sea Quest in August 1970.

Early reconnaissance seismic work before 1965 had indicated a large structural nose plunging southeastward across Block 21/10 toward the deepest part of the North Sea Tertiary basin. A 5 × 5-km (3 × 3 mi) seismic grid shot in 1967 had defined this feature and had shown, at base of the Tertiary, 40 sq km (16 sq mi) of closure of low amplitude, cen-

[1] Manuscript received, July 19, 1973; accepted, September 6, 1973.

[2] The British Petroleum Company.

The writers are indebted to their colleagues in The British Petroleum Company Limited who have contributed directly or indirectly toward this account. We thank the chairman and directors of The British Petroleum Company Limited for permission to publish this paper.

186

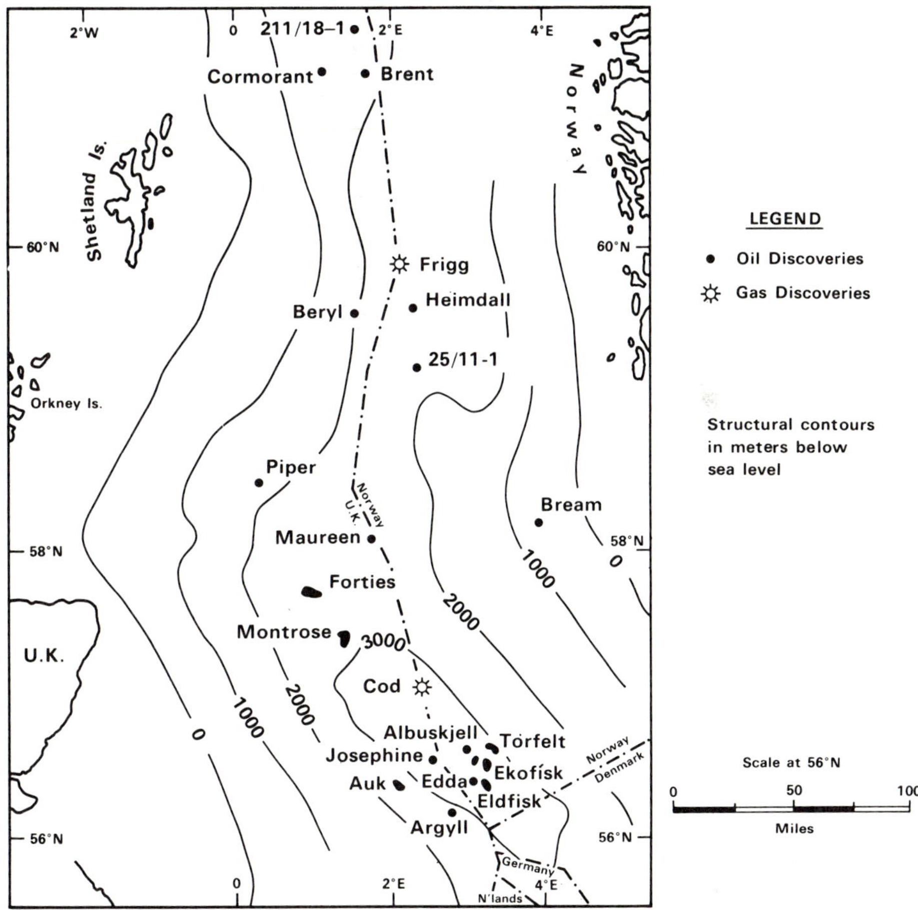

Fig. 1—North Sea hydrocarbon discoveries and base Tertiary contours.

tered in Block 21/10 on the axis of the nose. 21/10-1 was drilled on this feature.

At 2,132 m (6,994 ft) below RTE of 34 m (111 ft), the well entered Paleocene sands, which were indicated to be oil bearing from mud-gas and cuttings analysis. An oil-water contact was established at 2,251 m (7,385 ft) b.RTE, and subsequent testing produced 37° API low-sulphur oil at a rate of 4,730 bbl/day on a 54/64-in. surface choke. A field of major proportions had been discovered.

A detailed 1.5 × 1.5-km (1 × 1 mi) seismic survey was shot immediately to supplement the 1967 work, and the combined data were interpreted in the light of the information gained from the discovery well. The new interpretation indicated a larger closed area than originally had been envisaged.

The first appraisal well 21/10-2 was spudded in June 1971, 5.5 km (3.4 mi) northwest of the discovery well to delineate the field in that direction (Fig. 6). This well found an oil column of 33.5 m (110 ft) with an oil-water contact at the same depth as that found in 21/10-1. A second appraisal well, 21/10-3, then was spudded 7 km (4.3 mi) west of 21/10-1, but was junked at shallow depth. A replacement well, 21/20-3A, was drilled without incident and found an oil column of 126 m (413 ft) and once again the same oil-water contact.

At the same time Shell/Esso drilled a successful well in Block 22/6, although the sand development in the upper part of the Paleocene above the oil-water contact showed considerable deterioration compared with the other Forties wells. The same oil-water contact was present.

By this time plans for the construction of

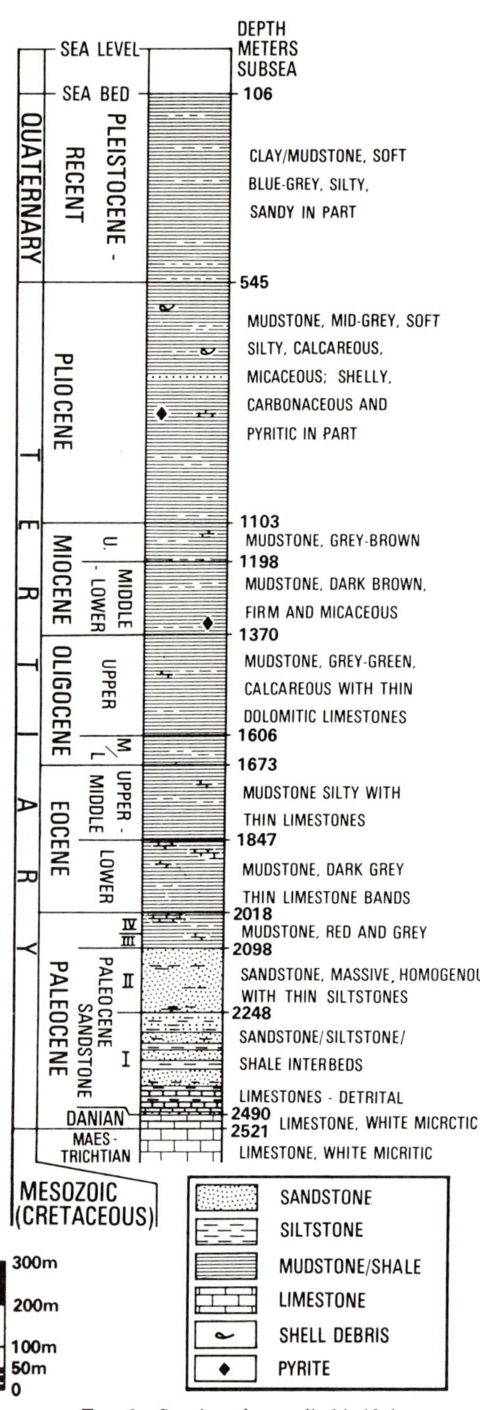

Fig. 2—Stratigraphy, well 21/10-1.

four platforms to develop Forties were well advanced, but some doubts remained as to the structure and sand distribution on the southern flank of the field. To select finally a site for the fourth drilling platform an additional well, 21/10-5, was drilled 5.5 km (3.5 mi) west-southwest of 21/10-1. This also proved successful and completed the appraisal of the field (Fig. 6). The first five wells therefore have confirmed a major oil field with an oil column of 155 m (509 ft) in Paleocene sandstone, and a closed area of about 90 sq km (35 sq mi).

STRATIGRAPHY

The Forties field is on the western flank of the North Sea Tertiary basin, the axis of the depositional trough trending approximately north-south along the median line of the continental shelf between the United Kingdom and Norway (Fig. 1). The Tertiary and Quaternary section within this trough can reach 3,500 m (12,000 ft) in thickness. Apart from fringe areas in the Netherlands and southern England, the basin is developed wholly offshore; Tertiary strata are absent from both eastern Scotland and Norway. The Tertiary section in the North Sea basin is almost entirely terrigenous, the predominant lithology being mudstone. Sandstone is present at many stratigraphic levels and in some areas can become the major lithotype within the Paleocene and Plio-Pleistocene. Producing sands are primarily of Paleocene age. The Tertiary sequence has been dated from microforaminiferal assemblages.

The central part of the Tertiary basin overlies a Mesozoic section, which also becomes attenuated near the Scottish coast. The generalized sequence comprises micritic limestones of Late Cretaceous age underlain by Lower Cretaceous mudstones. The Jurassic, where present, is a mudstone with subordinate sandstone, and the Triassic is a distinctive red-brown mudstone. A thin carbonate/anhydrite section, the Zechstein equivalent, caps the Paleozoic, with salt developed in the central part of the basin. The Zechstein is underlain by the Permian Rotliegendes sandstone.

The stratigraphic column for the Tertiary section of the discovery well in the Forties field, 21/10-1, is given in Figure 2. The oil accumulation is in sandstones of Paleocene age, which lie beneath a thick, monotonous section of gray to brown, variably calcareous and carbonaceous mudstones, ranging from upper Paleocene to Holocene. Sandstones are present in the Plio-Pleistocene and thin beds of limestone in the Eocene, but the post-Paleocene section is primarily argillaceous. The stratigraphic sub-divisions shown for the Tertiary in Figure 2 are based on micropaleontology. In well 21/10-1 the Paleocene is 502 m (1,647 ft) thick. The Eocene/Paleocene boundary is placed on paleontologic evidence at 2,018 m (6,621 ft) subsea, which is close to a prominent peak on the gamma-ray log and a thin limestone band which forms a useful lithologic marker. The basal 30 m (98 ft) of the Paleocene is white to gray micritic limestone of Danian age, but the overlying sequence is predominantly terrigenous. The exact boundary between the Danian and underlying Maestrichtian limestone is not well controlled paleontologically, but is taken at a gamma-ray sonic-log marker reflecting a change in limestone character from the clean compact micrites of the Maestrichtian to slightly argillaceous and sandy micrites of the Danian (Fig. 4).

The post-Danian section of the Paleocene has been divided into four distinctive litho-stratigraphic units (Fig. 3) by the late M. J. Wolfe and L. Aston of the BP Research Centre at Sunbury. The units are referred to by Roman numerals only and no formal terminology is proposed at this stage.

Unit I 2,248–2,490 M (7,375–8,169 Ft) Subsea

This is divisible into two "members," the lower, below 2,409 m (7,903 ft) subsea, including beds of detrital limestone, and the upper comprising argillaceous sandstone interbedded with siltstone and silty shale.

The lower member contains a basal sandstone 4.6 m (15 ft) thick overlain by a sequence of calcareous sandstone, limestone, and mudstone. The unit is characterized paleontologically by reworked Danian and Cretaceous faunas. The limestones are commonly detrital with disseminated quartz of sand grade and indicate contemporaneous erosion of Cretaceous and Danian carbonate rocks following uplift at the end of the Mesozoic. The thickness of the member increases northwestward across the field; the variations probably reflect partial infilling of an irregular depositional surface.

The upper member consists of interbedded sandstone, siltstone, and shale. Sandstone is the

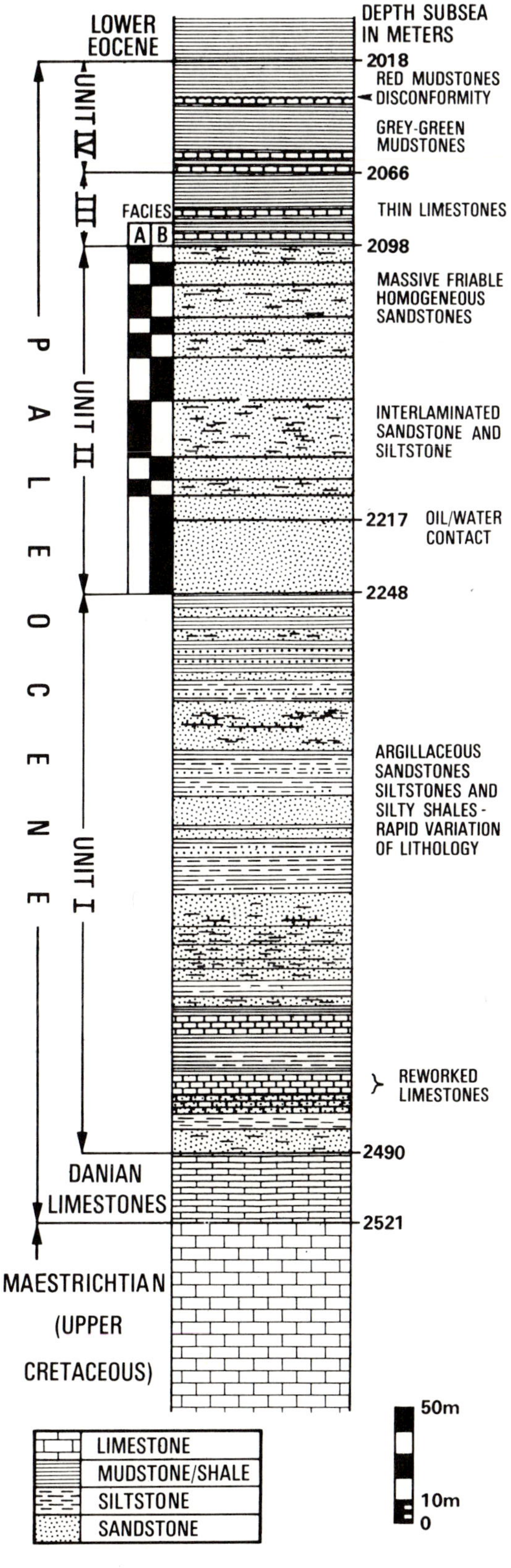

Fig. 3—Paleocene lithostratigraphy, well 21/20-1.

commonest lithology and is typically gray, fine grained, argillaceous, and poorly indurated. The sonic log is characterized by a rapidly fluctuating trace representing thinly interbedded lithologies. Thin bands of limestone and hard calcite-cemented sandstones are sparingly present throughout.

Unit II 2,098–2,248 M (6,883–7,375 Ft) Subsea

Unit II contains all the massive sandstones within the Paleocene and is the producing "formation" of the Forties field. It is also productive of oil and wet gas in other structures in the southern part of the northern North Sea. The dominant lithologies are sandstone and mudstone. Studies of core material from wells within the field have allowed the contrasting lithologies to be grouped into four facies, referred to as A, B, C, and D.

Facies A is characterized by fine-grained, locally silty sandstone, commonly with detrital mica and lignite, which is interbedded with laminated siltstone and shale as upward-fining, graded units commonly less than 1.5 m (5 ft) thick. The shales are typically kaolinitic and the fauna is sparse. The beds range from 2 cm to 1.7 m (5.5 ft) in thickness, and the sandstone percentage in the lithofacies varies between 20 and 55 percent, with the sands having moderate to good reservoir character. Flow and load structures are common, including ripple-drift lamination, load casting, deformed cross-bedding, contorted bedding, and slump folding and soft sediment faulting. At a few horizons are intraformational conglomerates with shale clasts up to several centimeters in size.

Facies B comprises the clean homogeneous sandstones. These vary in color from brown to almost white, in grain size from fine to coarse, and are typically clean and friable. Sorting is poor to moderate, but reservoir properties are excellent. Within thick sequences of sand there are clay laminae and impermeable calcite-cemented zones, and a few pebbly layers with 50 percent lithic fragments which are mainly quartzite, phyllite, and graphic granite. Fining upward sections are numerous, commonly with convolute and laminar bedding toward the top. Current bedding has not been observed. The sandstones of facies B in well 21/10-1 vary in thickness up to 35 m (115 ft), but elsewhere in the field can reach 80 m (260 ft). The thick sections are believed to be formed of superimposed sand units which are individually about 0.5–1.0 m (1.5–3 ft) thick. Fauna is again sparse, as in facies A.

Facies C is dominated by gray kaolinitic shales and graded siltstone-shale couplets. Occasional thin, dark-brown, fine-grained, sometimes graded sandstone beds are present. The base of these beds is unusually sharp, erosional, and some have sole marks, including sparse flute casts. The siltstones show wavy lamination, micro cross-lamination and lenticular bedding. Penecontemporaneous slump structures are common. The shales rarely are bioturbated and fauna is sparse.

Facies D is found only in the eastern part of the field. It is characterized by burrowed green waxy shales which have an abundant marine fauna and flora. Associated with this lithology are purple shales, black limestones, dolomitic mudstones, and sideritic concretions.

Over most of the field unit II is made up of facies A and B. Facies C is in the southern and eastern part of the field and facies D in the east. The distribution of the four facies appears to be related to thickness variations of unit II, facies C occurring within areas of thin development of the unit and facies B sands being well represented in areas of thick development. The central part of the field has the highest percentage of clean sands and will have the best production potential.

The change in facies across the field means that a correlation within unit II is not readily apparent from the logs of the widely spaced step-out wells. The variation in gross lithology is illustrated by the correlation diagram in Figure 4, which is drawn in a west to east direction from 21/10-3A, through 21/10-5 to 21/10-1. The argillaceous section in the upper part of unit II in 21/10-5 (facies C) is probably stratigraphically equivalent to the facies A and B sands in wells 21/10-1 and 21/10-3A, although an offlapping relation could be invoked.

Studies on this and related problems are continuing, but probably will not be resolved until further data from the production wells are obtained. The depositional environment of unit II is still under study, and it will be some time before it is elucidated completely. As additional data become available during the development-drilling program, it is expected that critical evidence of the provenance and mode of deposition will be found. At present we consider the most likely source of the sediment is from Scotland on the west and from locally eroded high areas composed of Upper Cretaceous and Dan-

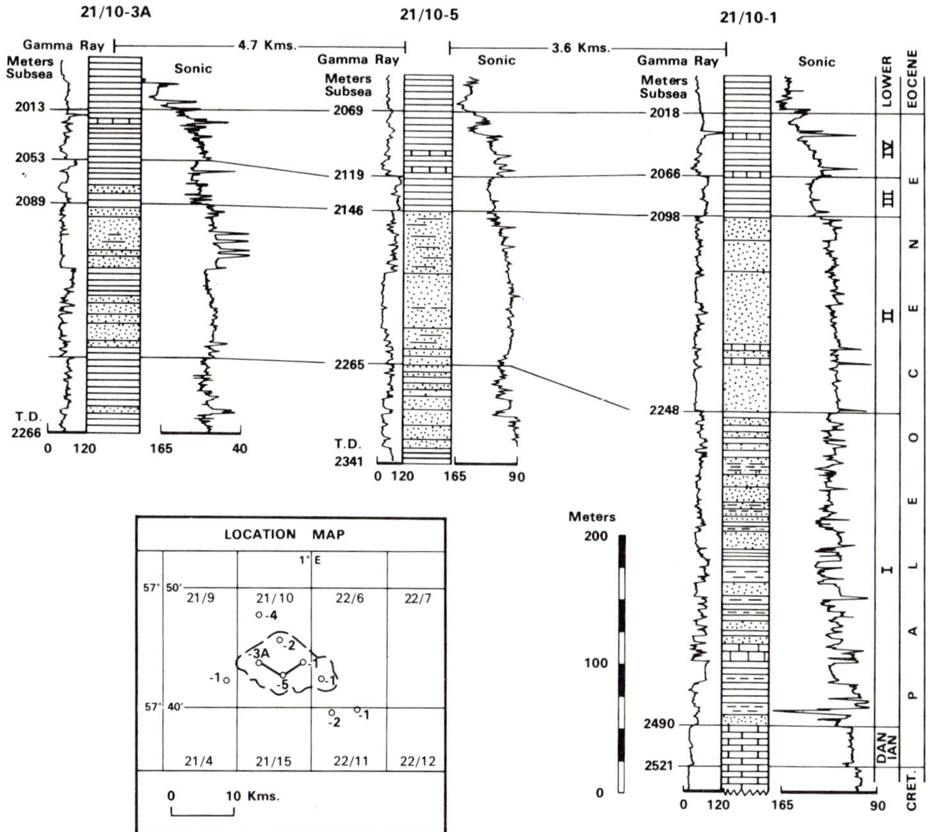

Fig. 4—Well correlation 21/10-3A, 21/10-5, 21/10-1.

ian chalky limestones. A Scottish provenance for the more arenaceous material is indicated by the predominance of quartzose grains, including quartzites and vein quartz, some phyllitic material, hornblende gneiss, pegmatite fragments and much graphic granite, and by the heavy-mineral suite predominantly of tourmaline, garnet, and zircon with small amounts of staurolite, rutile, sillimanite, hornblende, and pyroxene. These appear to indicate derivation from rocks akin to the Moinian and Lewisian of Scotland.

Unit III, 2,066–2,098 M (6,778–6,883 Ft) Subsea

Lithostratigraphic unit III forms the cap rock to the reservoir sands. The contact with unit II was cored in well 21/10-2 and indicates that the lithologic change is transitional. Unit III consists of dark gray, silty, lignitic shaley mudstones rich in montmorillonite. Thin beds of

sand are developed sparingly, with the proportion of sandstone increasing westward. Winnowed pockets of fish remains and several well-preserved skeletons have been found in cores. Planktonic Foraminifera are scarce, the fauna being dominated by siliceous diatoms and microplankton. The thickness and lithology of unit III are very constant across the field and the uniformity extends to nearby areas.

Unit IV 2,018–2,066 M (6,621–6,778 Ft) Subsea

This is essentially a mudstone unit, the mudstones typically being greenish-gray and slightly calcareous. However, the upper 10–20 m (33–66 ft) is red-brown in color and contains a characteristic red-stained calcareous foraminiferal assemblage in which *Globigerina* cf. *triloculinoides* predominates. These mudstones are separated from the underlying darker beds by a minor unconformity. Several thin pale clay beds with a distinctive mineralogy indicative of degraded volcanic ash are present, together with occasional thin limestone stringers. These lithologic features extend beyond the limits of the field and make logs useful for correlation. The thickness of the unit remains constant across the field, and it has a characteristic sonic-log pattern increasing in velocity with depth. This velocity increase serves as a good log-correlation feature, and also gives rise to a prominent seismic horizon which is used to map the configuration of the top of the reservoir.

STRUCTURE

The regional tilt at the base of the Tertiary sequence in the Forties area of the North Sea is down to the east, toward the axis of the Tertiary basin. In Block 21/10 this regional tilt is interrupted by a large east-southeast-directed nose which can be followed into Block 22/11. Reversal of dip on this nose provides closure for the hydrocarbon accumulation in the overlying Paleocene sands.

The isochrons on a seismic horizon 35 m (115 ft) above the top of the sands are given in Figure 5, and the depth contours on the top of the reservoir in Figure 6. A structural cross section east-west across the field is shown in Figure 7. Because of horizontal velocity gradients present in the late Tertiary section, the configuration of the structural contours on the top of the reservoir is different from that of the isochrons on the seismic horizon directly above. The velocity gradients are computed from the information from the wells within and adjacent to the field. The most significant change is an increase in closure at the western spill points of the structure, giving an approximate coincidence between structural spill point and the oil-water contact at 2,217 m (7,274 ft) subsea.

The depth contours on the top of the reservoir indicate that the structure is a broad dome elongated east-west with minor faulting affecting only the eastern extremity of the feature. The structure extends 16 km (10 mi) east-west by 8 km (5 mi) north-south, has a closed area of about 90 sq km (35 sq mi or 22,000 acres) and vertical closure of 155 m (509 ft).

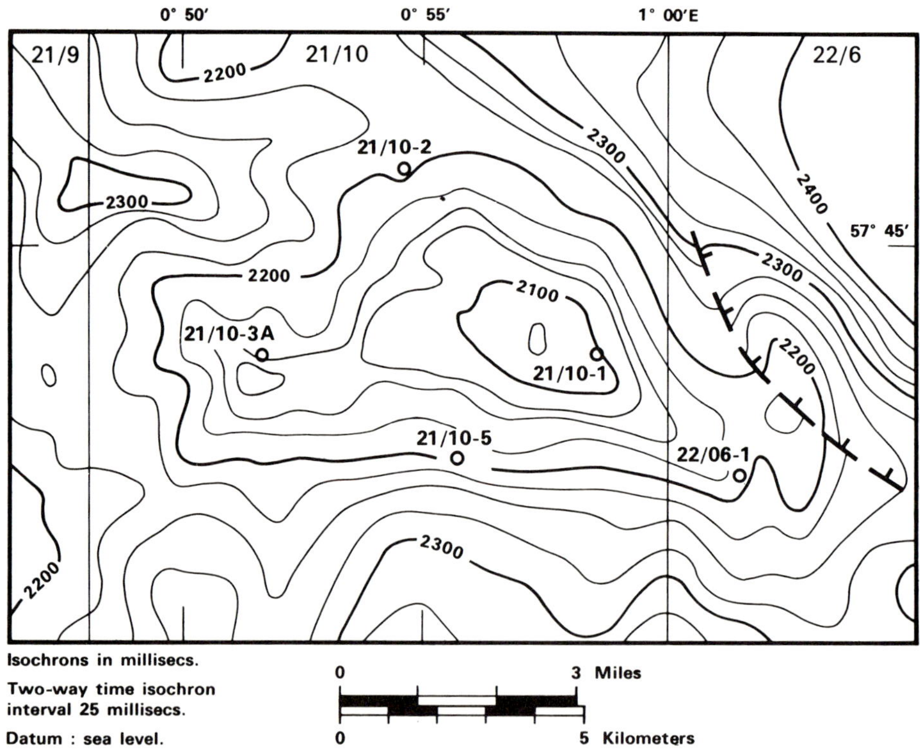

FIG. 5—Isochrons on seismic reflector overlying Paleocene reservoir.

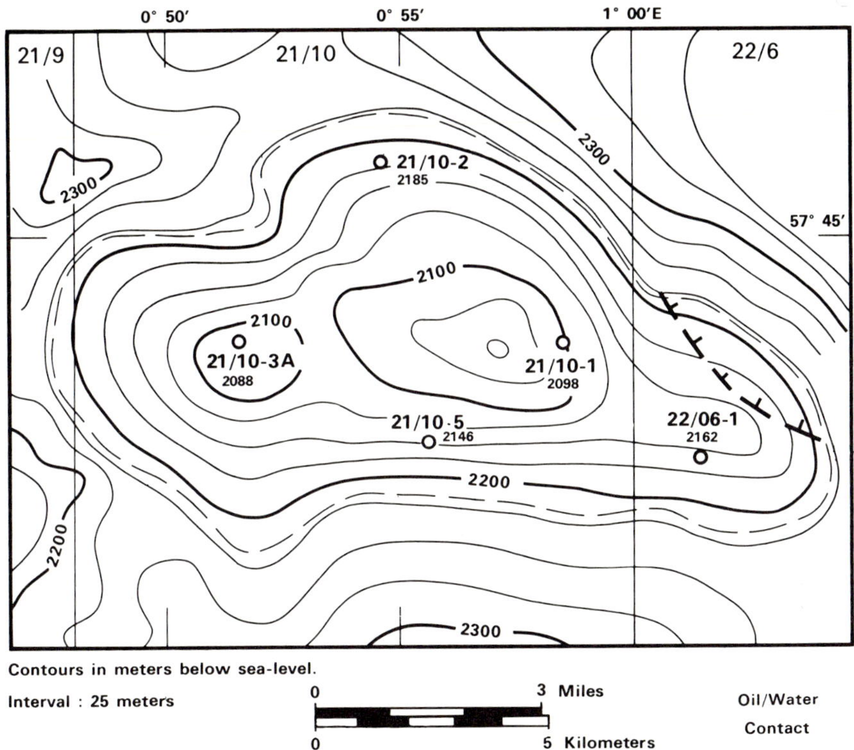

FIG. 6—Structural contours on top of Paleocene reservoir.

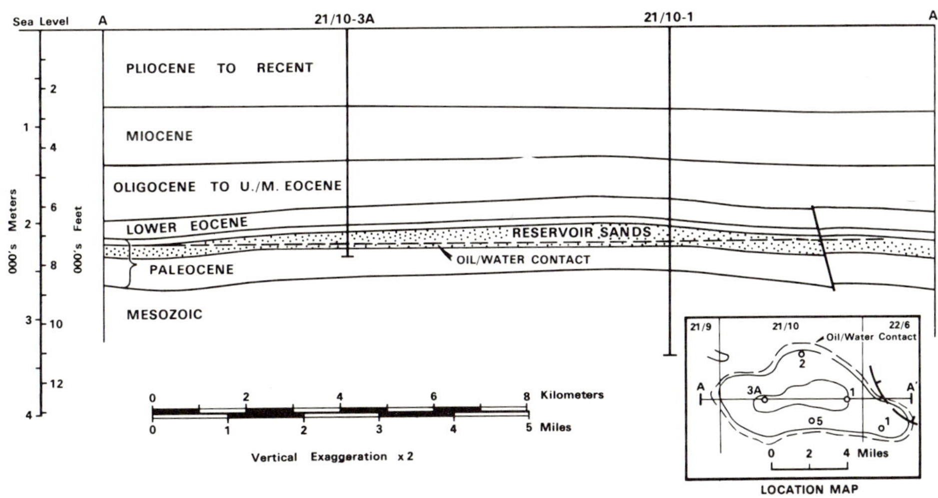

FIG. 7—Structural section east-west across field.

The Paleocene structure overlies a faulted high at the base-of-Cretaceous level. Seismic penetration below the Cretaceous is poor, but it does not appear likely that the structure in the deeper Mesozoic or Paleozoic is related directly to the base-of-Cretaceous uplift over which the Paleocene feature lies. The base-of-Cretaceous unconformity is a very pronounced, widespread, regional event affecting the whole of the North Sea basin. The Late Jurassic movements which preceded this unconformity were typified by block faulting, and a complex uplift of this type underlies the Forties field. Thickness variations in the Cretaceous and lower Tertiary indicate that the area of the field remained relatively positive during late Mesozoic and early Cenozoic times, a favorable factor for early migration of hydrocarbons, although at present insufficient data are available to assess the organic diagenesis of the Cretaceous and Paleocene shales. Structural expression above the Paleocene diminishes progressively. Turnover on the nose does not persist above the lower Miocene, and at the base-of-Pliocene level the beds dip uniformly southeastward across the field. Hence, structural growth was terminated in late Miocene time. Uplift on the Mesozoic fault block however may have ceased during the early Tertiary, with differential compaction allowing structural expression to persist through the Miocene section.

Reservoir Characteristics and Reserves

Well 21/10-1 was tested over the interval 2,108.5–2,113 m (6,918–6,933 ft) subsea and produced 37° API oil at a rate of 4,730 bbl/day on a 54/64-in. surface choke, with an estimated solution gas–oil ratio of 250 scf/bbl. The crude was low-sulphur (0.3 percent) and medium-wax at 8.5 percent. 21/10-3A was tested over the interval 2,092–2,137.5 m (6,865–7,013 ft) subsea at rates up to 3,260 bbl/day, the production being limited to this rate by equipment restrictions. Bottom-hole PVT samples during the test established the GOR at 330 scf/bbl.

Core analysis of the five wells, particularly 21/10-5 where extensive coring of the reservoir was carried out, indicates high porosities and permeabilities. Average well porosities range from 25 to 30 percent and permeabilities vary up to 3,900 md. With the exception of the tight calcite cemented layers, the permeability in facies B ranges from 1,000 md to 3,900 md. In facies A and C the sandy layers have a spread of permeabilities from 0.1 to 1,000 md with the most common permeability measurements lying in the range of 100–200 md.

Analysis of the flow test indicates that 21/10-1 and 21/10-3A are capable of producing rates in excess of 15,000 bbl/day. Both wells were, however, drilled in crestal position, where virtually the full oil column is present. An initial average-well rate of 8,000 bbl/day for development wells has been assumed. The initial reservoir presure is about 3,200 psi and with the oil being undersaturated there is no original gas cap.

Oil in place and recoverable oil calculations are based on the whole accumulation, including that part which falls in Shell/Esso Block 22/6. The calculations yield an average oil-in-place figure of about 1,400 bbl/acre ft, which in turn leads to a figure of about 4.4×10^9 bbl stock tank oil initially in place. A recovery of 40 percent would yield a figure of about 1.8×10^9 bbl recoverable oil.

Production Plans

Development drilling will take place from four fixed platforms, which will also serve as production platforms. The design of each platform allows for 36 drilling slots, although only 27 wells can be completed from each on the required spacing. It is planned to drill wells with a horizontal displacement of up to 2,195 m (7,200 ft), at a maximum of 55° from the vertical.

Spacing of 120 acres has been selected as the maximum required to ensure full recovery of the reserves over a 20–25 year period for an initial well rate of 8,000 bbls/day. Drilling is to take place over those parts of the field which are 100 ft above the oil-water contact; i.e., within an area of 15,500 acres. On this basis the field can be drilled by 106 wells deviated from the four platforms (Fig. 8).

The field will be developed as follows. The initial phase includes the emplacement of two drilling platforms with ancillary production equipment, a 32-in. submarine pipeline to Cruden Bay near Peterhead, a 36-in. land line from Cruden Bay to a gas separation plant at Kerse of Kinneil adjacent to BP's Grangemouth refinery where about half the production will be refined, and a marine terminal on the Firth of Forth for export of the balance. Fifty-four wells will be drilled from the first two platforms. It is estimated that this will lead to a peak production rate of 250,000 bbl/day. The

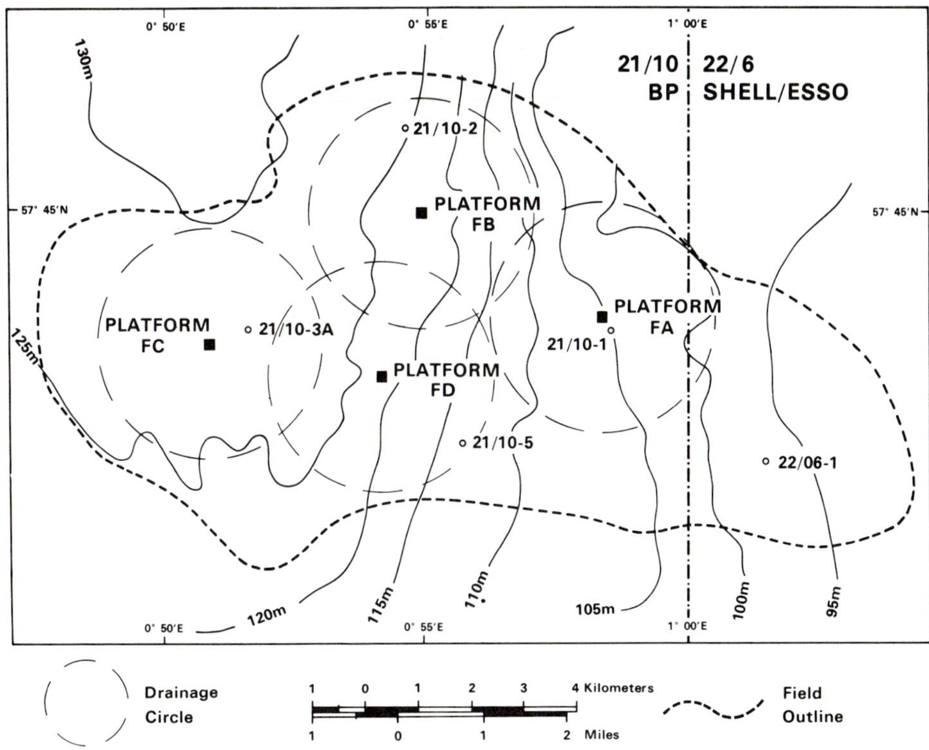

FIG. 8—Platform locations and bathymetry.

emplacement of two additional platforms and the extension of the storage capacity at the Firth of Forth will follow. Fifty-two wells will be drilled from the third and fourth platforms increasing the production from the field to an average rate of 400,000 bbl/day.

It is hoped that this ambitious program can be achieved by 1977 when the field should be on full production. On the basis of the well potentials assumed, this peak production of 400,000 bbl/day can be maintained for about three to four years before decline sets in.

Because of the undersaturated nature of the crude, pressure maintenance at about 2,500 psi will be effected to maintain well efficiency. This will be commenced early in the life of the field to delay the need for artificial lift as water cuts increase. Initially the oil and gas will be separated on the platforms at a pressure of 125 psig, and about a third of the solution gas will be piped ashore dissolved in the oil. About 20–25 percent of the remainder will be required as fuel on the platforms and thus, at peak throughputs about 15–20 MMscf/day will remain for disposal from each of the first two platforms.

During the drilling phase this of necessity must be flared, but subsequently it will be refrigerated for the recovery of natural gas liquids, which will be shipped to shore mixed with the crude. The gas balance at this stage indicates that about half of the total gas will be recovered, about 20 percent used as fuel, and the remaining 30 percent flared. Of the gas recovered about 50 percent will be in the form of dry gas, and the balance will comprise condensed liquids in the C_3–C_5 range.

The first two platforms are currently under construction, one at Nigg Bay on the Cromarty Firth, and the other at Middlesborough in Yorkshire. These mammoth structures will be larger and in greater water depths than any presently installed in the world. Conventional jacket templates will be floated to location, righted by controlled flooding, and pinned to the sea bed with piles. The largest, which is to be installed in 128 m (420 ft) of water, will be 146 m (480 ft) high, 61 × 76 m (200 × 250 ft) in plan, and weigh 17,000 tons. Each template will be topped by a three-level deck unit to accommodate drilling and production equipment. These

FIG. 9—Forties platform, as it would appear if standing in Edinburgh, Scotland.

will measure 43 × 52 m (140 × 170 ft) and each weigh 15,000 tons (Fig. 9).

The North Sea is notorious for its severe weather conditions, and design criteria have to cater for possible wind speeds of 114 knots and wave height of 29 m (94 ft).

The 32-in. submarine pipeline to Cruden Bay will be about 175 km (110 mi) long and, apart from the shoreward 24 km (15 mi), will be laid in water depths of 91–130 m (300–425 ft). The pipe will have a protective wrapping of fiberglass and coal tar enamel and an overall coating of 2.5 in. of reinforced concrete. It will be laid from a conventional lay barge during 1973–1974 and buried subsequently by removing the sea bed from under the pipe with high-pressure water jets.

At Cruden Bay the oil will enter a sealine receiver trap and then pass directly into the land line. An emergency flow tank will be provided in case of shutdowns.

The 36-in. land line will be 203 km (127 mi) long from Cruden Bay to the gas separation plant at Kerse of Kinneil. There the crude will be split, part going to BP Grangemouth Refinery and part going to the Firth of Forth terminal for export.

The foregoing account describes the discovery in 1970 by BP of the first major oil field in the British sector of the North Sea. Unlike the United States and Canada, the British regulations do not provide for the early release of well data, and it therefore is not possible to make public technical information without prejudice to one's competitive position in the industry. In this respect the writers apologize for the lack of discussion of certain aspects of the geology of the Forties field. Nevertheless, we hope that this general review will prove to be a useful contribution to the understanding of this new and exciting petroleum province and will stimulate others to release information for the benefit of all.

The American Association of Petroleum Geologists Bulletin
V. 62, No. 6 (June 1978), P. 967-983, 7 Figs., 3 Tables

Channel Sands and Petroleum Potential of Monterey Deep-Sea Fan, California[1]

P. WILDE,[2] W. R. NORMARK,[3] and T. E. CHASE[3]

Abstract The possibility of substantial petroleum resources in deep water adjacent to the continents in areas of hemipelagic sedimentation has been suggested by numerous authors. One such area adjacent to the United States is the continental rise off California between 33 and 40° N lat. The rise covers 200,000 sq km and consists of three major submarine fans, the Arguello, Monterey, and Delgada, with sediment thicknesses up to 3 km in water depths of 3,000 to 4,500 m. The fans are composed primarily of continental debris carried down submarine canyons and deposited on the fan through a system of branching and meandering submarine valleys and channels. During the development of the fan, the active channelway shifts periodically in response to locally increased gradients, producing abandoned channels. The coarser sediment naturally confined to the channel becomes covered, after abandonment, by finer grained material because of slumping, reworking, and subsequent fan deposition. Such channels, after sufficient burial, would make excellent stratigraphic traps for hydrocarbons, especially if abandonment occurs on the upper fan where the grain size of channel-fill sediments is coarser and the channel width and relief are greater. On the Monterey fan, potential stratigraphic traps would be seaward of the mouths of the Monterey-Carmel, Ascension, and Pioneer Canyons as shown by subbottom profiling. The proximity of such potential reserves to the United States and to existing refineries and markets in coastal areas makes buried channels on large deep-sea fans such as the Monterey fan particularly attractive prospects.

INTRODUCTION

The energy crises and the fuel shortages of the winter of 1973-74 demonstrated the dependence of the United States on foreign sources of petroleum at a time when land and near-shore domestic sources of petroleum appear insufficient to supply our needs. Emery (1973), Beck and Lehner (1974), and Thompson (1976) have suggested that vast new areas of petroleum resources may be in the sediments of the continental rise (and margin), sediments which consist of a ramp of terrigenous material lying in deep water at the base of the continental slope. In particular, Moore (1969) and Moore and Fullam (1973) emphasized the possibilities of substantial petroleum reserves from submarine fans; Caughey and Stuart (1976) reviewed the prospects for upper fan environments (as well as a buried Miocene fan) in the Gulf of Mexico, and Summerhayes et al (1977) described the petroleum potential of the Nile submarine fan.

The United States has extensive areas of continental-rise deposits off the Atlantic Coast (Heezen et al, 1959), in the Gulf of Mexico, in the Mis-

sissippi fan (Ewing et al, 1958; Caughey and Stuart, 1976), and off the Pacific Coast in the Monterey and Delgada fans (Menard, 1960, Fig. 1; Chase et al, 1975; Wilde et al, 1976a), and the Cascadia-Astoria fans (Hurley, 1964). These features are in water deeper than 3,000 m; accordingly, conventional shallow-water petroleum technology cannot be used to develop such deposits. However, the Deep-Sea Drilling Project using the D/V *Glomar Challenger* has demonstrated the technical feasibility of drilling in the deep sea, and recent developments (Stone, 1975; Snyder, 1976) suggest that completions at these depths are feasible. This paper evaluates the possibilities for economic petroleum deposits for the Monterey fan off the central California Coast (Fig. 1) on the basis of available seismic reflection profiles (Fig. 2a) and core samples (Fig. 2b). This is not to imply with certainty that there are commercial petroleum reserves in the Monterey fan. This can be determined only by drilling. The arguments and procedures used here, however, for assessing the potential of the Monterey fan, which is a relatively well-studied fan, may be helpful for the evaluation of other similar pro-

AAPG grants permission for a *single* photocopy of this article for research purposes. Other photocopying not allowed by the 1978 Copyright Law is prohibited. For more than one photocopy of this article, users should send request, article identification number (see below), and $3.00 per copy to Copyright Clearance Center, Inc., One Park Ave., New York, NY 10006.

[1]Presented in part under different title at the Circum-Pacific Energy and Mineral Resources Conference August 1974, Honolulu, Hawaii (Wilde et al, 1976b). Manuscript received, September 1, 1976; accepted, December 9, 1977.

[2]Lawrence Berkeley Laboratory, University of California, Berkeley, California 94720.

[3]U.S. Geological Survey, Menlo Park, California.

We wish to thank J. Hunt for his invaluable help in the complex problem of assessment of the potential source beds of petroleum in the deep ocean and for calling our attention to significant references on the subject. C. Summerhayes provided enthusiastic encouragement and the latest data on the Nile fan, and W. C. Luth helped strengthen our interpretative convictions. G. R. Hess provided unpublished data from his sediment studies on Monterey fan, and especial thanks go to F. P. Shepard and H. W. Menard, who stimulated our interests in the fan and whose earlier research efforts provided some of the samples and bathymetric data from selected areas.

Article Identification Number
0149-1423/78/B006-0002$03.00/0

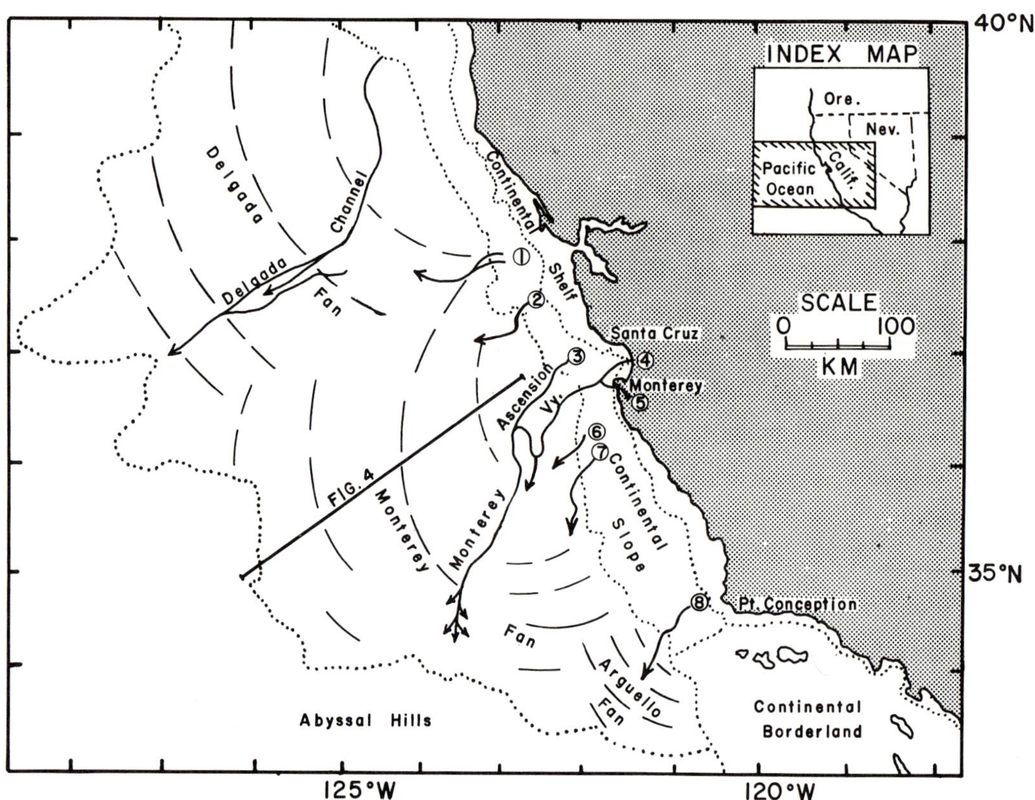

FIG. 1—Schematic index map of California continental rise showing extent of Delgada, Monterey, and Arguello submarine fans. Dotted lines mark outer edge of fans and limits of continental shelf and slope areas. Circled numbers indicate positions of canyons tributary to Monterey fan: *1*, Farallon; *2*, Pioneer; *3*, Ascension; *4*, Monterey; *5*, Carmel; *6*, Sur; *7*, Lucia-Partington; *8*, Arguello. Heavy southwest-trending line denotes radial cross section of fan shown in Figure 4.

spective areas on the continental margins of the United States.

MONTEREY FAN

The Monterey fan, extending from the latitudes of Point Conception to Santa Cruz, California, covers about 100,000 sq km of the seafloor (Fig. 1) heading in the Monterey Canyon with tributary canyons including Sur-Partington-Lucia, Carmel, Ascension, Pioneer, and Farallon (Shepard and Dill, 1966). The sediments of the fan are turbidites and hemipelagic muds with coarser deposits in submarine channels and fan valleys that extend from the mouths of the various canyons that debouch onto the fan (Wilde, 1965). The green color of the muds implies reducing conditions for most of the fan sediments. The provenance of the sand-size particles, the depositional

fabric and sedimentary structures, the location of coarse deposits in the axis of submarine channels (Wilde, 1965), and the presence of levees on the upper fan valleys (Shepard, 1966) all support the contention of Dill et al (1954) and Menard (1960) that most of the sediment on the fan was deposited and shaped by turbidity currents.

At present, however, there is no hard evidence that petroleum is present in economic amounts in the deposits of the continental rise. The three basic requirements for accumulation of petroleum hold for deep-sea deposits as well, that is: (1) source beds, fine-grained sediments rich in organic matter that have been heated sufficiently to generate hydrocarbons; (2) a reservoir, generally a porous and permeable sediment; and (3) a suitable trap to permit confined accumulation of the petroleum.

SOURCE BEDS

Organic Content

Philippi (1974) suggested that plankton-derived fatty acids in the marine muds off the California Coast could be converted to a paraffin-base oil. Thus, the organic component of hemipelagic muds of the Monterey fan that are the result of seasonal upwellings with consequent high organic productivity in the nearshore areas could be potential source material for petroleum, if present in sufficient quantities and transported to and preserved in the fan sediments. Organic material in fine-grained muds would tend to be preserved if transported to the site of deposition by turbidity currents; the relatively fast transport to the fan and the subsequent rapid burial would reduce the exposure time to any oxygenated bottom waters. Also, the lower part of any deposit more than 10 cm thick would be below the zone of active bioturbation (Goldhaber et al, 1977) and would be kept anoxic by the presence of sulfate-reducing bacteria (Berner, 1976), all of which would tend to preserve organic material in the sediment. Because of the episodic nature of turbidity flows, the apparent or average rates of deposition in hemipelagic areas are somewhat misleading, as they combine the very slow rates of pelagic deposition between turbidity flows with the rapid rates during deposition from turbidity currents. It is the rate during the actual deposition rather than the long-term average, which is computed from pelagic-fossil evidence, that determines whether the organic material has been exposed long enough to be oxidized or preserved. Thus the average rates of sedimentation for the toes of both the Monterey and Delgada fans are on the order of 30 m/m. y. since the late Miocene (Table 1A). The rate increases shoreward to 430 m/m.y. on the upper Monterey fan during the Holocene (Fig. 1; Table 1B). The Holocene rates on the upper fan are comparable to values from the continental borderland, particularly for the farther offshore basins (Emery and Bray, 1962). As Table 2A shows, the organic content in the borderland basins is much higher than that of the DSDP samples on Delgada fan, which suggests that the preserved organic content increases with the rate of deposition. Therefore, the average values for the organic content of the green muds on the upper fan might lie well within the range of values given for source beds by Trask and Patnode (1942). The only detailed analyses of the organic carbon of the sediments of the California deep-sea fans are those for samples from DSDP Sites 32, 33, and 34 on the toe of the Delgada fan (Table 2B). However,

Romankevich (1969) in his compilation for the whole Pacific shows that the organic content in the area of the fans is greater than 0.5% total organic carbon. These estimated values are at the low end of the range of organic content of source beds related to oil-producing zones in the marine sediments of the Los Angeles basin (Trask and Patnode, 1942, p. 129), but were considered adequate by Hunt (1974, p. 596) for petroleum generation. Hunt (1975) analyzed two samples of clayey mud from DSDP Site 34 on the toe of the Delgada fan and found unexpected amounts of C_4-C_7 hydrocarbons at shallow depths of about 100 m and low sediment temperatures of 9°C. Hunt attributed this occurrence to early diagenesis and a possible early indicator of the ability of sediments to generate hydrocarbons. Although no analyses of higher carbon number hydrocarbons have been done in the region, Hunt (1977, p. 102), using data from Philippi (1974) and Durand and Espitalie (1972), suggested "the ratio of C_4-C_{14} to C_{15}-C_{40} hydrocarbons is about 1:5." Hunt (1977, p. 101), summarizing previous work, stated that the depth profiles for hydrocarbon distribution "indicate that only about 10% of the C_{15}-C_{40} hydrocarbons in the entire sedimentary section are present prior to the catagenetic stage." This implies that the hemipelagic sediments on the Delgada fan could be potential source beds and that similar sediments on the Monterey fan on the south also may be potential source beds, as both fans head against the same coastal upwelling area of high organic productivity and have similar growth histories (Hein et al, 1974).

In addition to hemipelagic muds, there is indirect evidence that organic-rich sapropelic sediments that also could be potential source beds of petroleum may be found within the sediments of the Monterey fan. Organic-rich sapropelic sediments containing up to 10% organic carbon have been found in various DSDP sites, all in excess of 2,000-m depth of water (Thiede and Van Andel, 1977; Fischer and Arthur, 1977), and organic carbon levels reach 25% in thin beds from piston cores from the Levant platform of the Nile fan (Summerhayes et al, 1977). These sediments are thought to be deposited in an expanded oxygen-minimum zone during climatic optima and generally-poorer oceanic circulation. Fischer and Arthur (1977) proposed in their section on "Petroleum Source Beds" that the deposition of sapropelic sediments in such episodes greatly enhances the petroleum potential on the continental margins. In particular, they indicated that the Miocene was an interval of potential sapropel deposition, which would be during the early deposi-

P. Wilde, W. R. Normark, and T. E. Chase

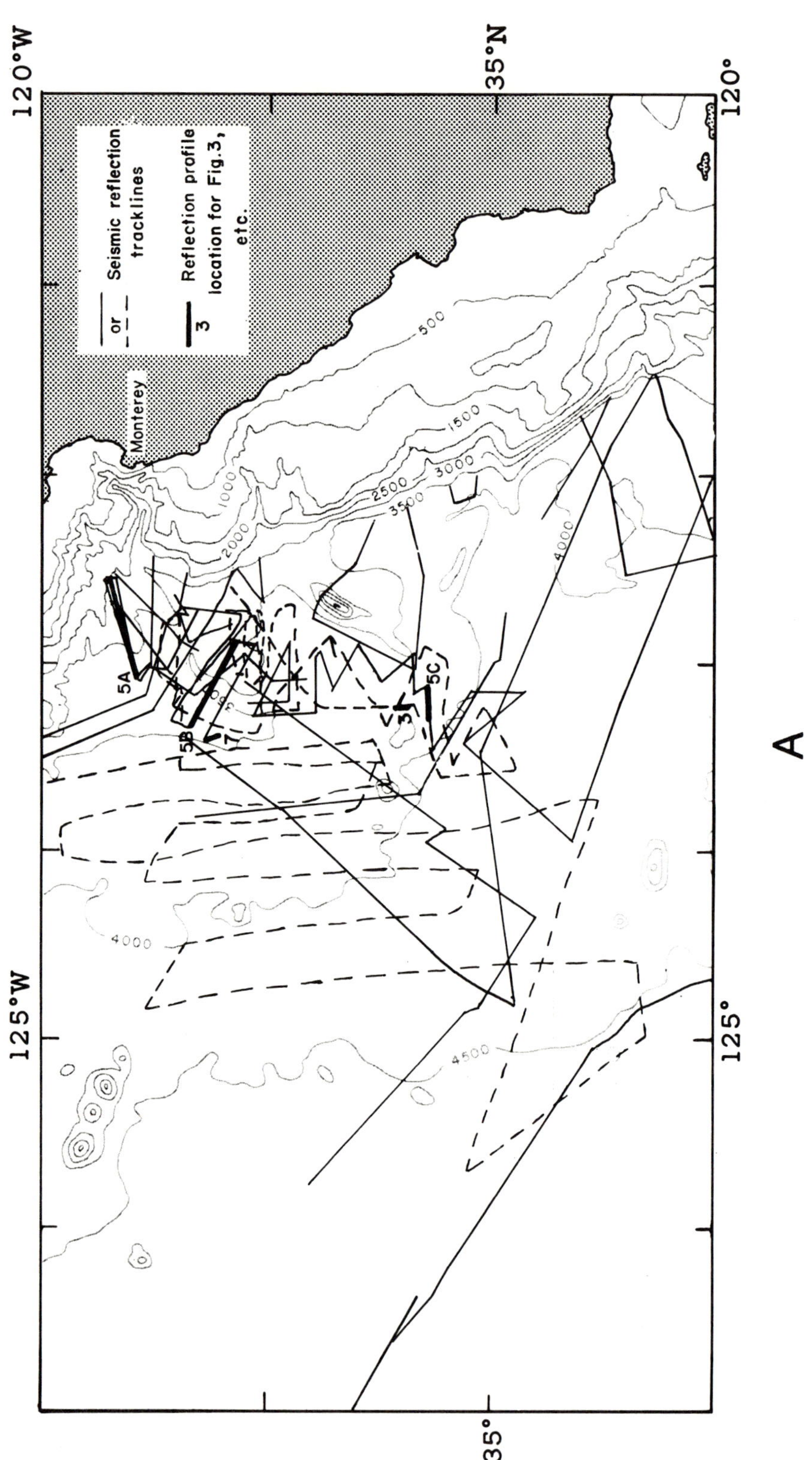

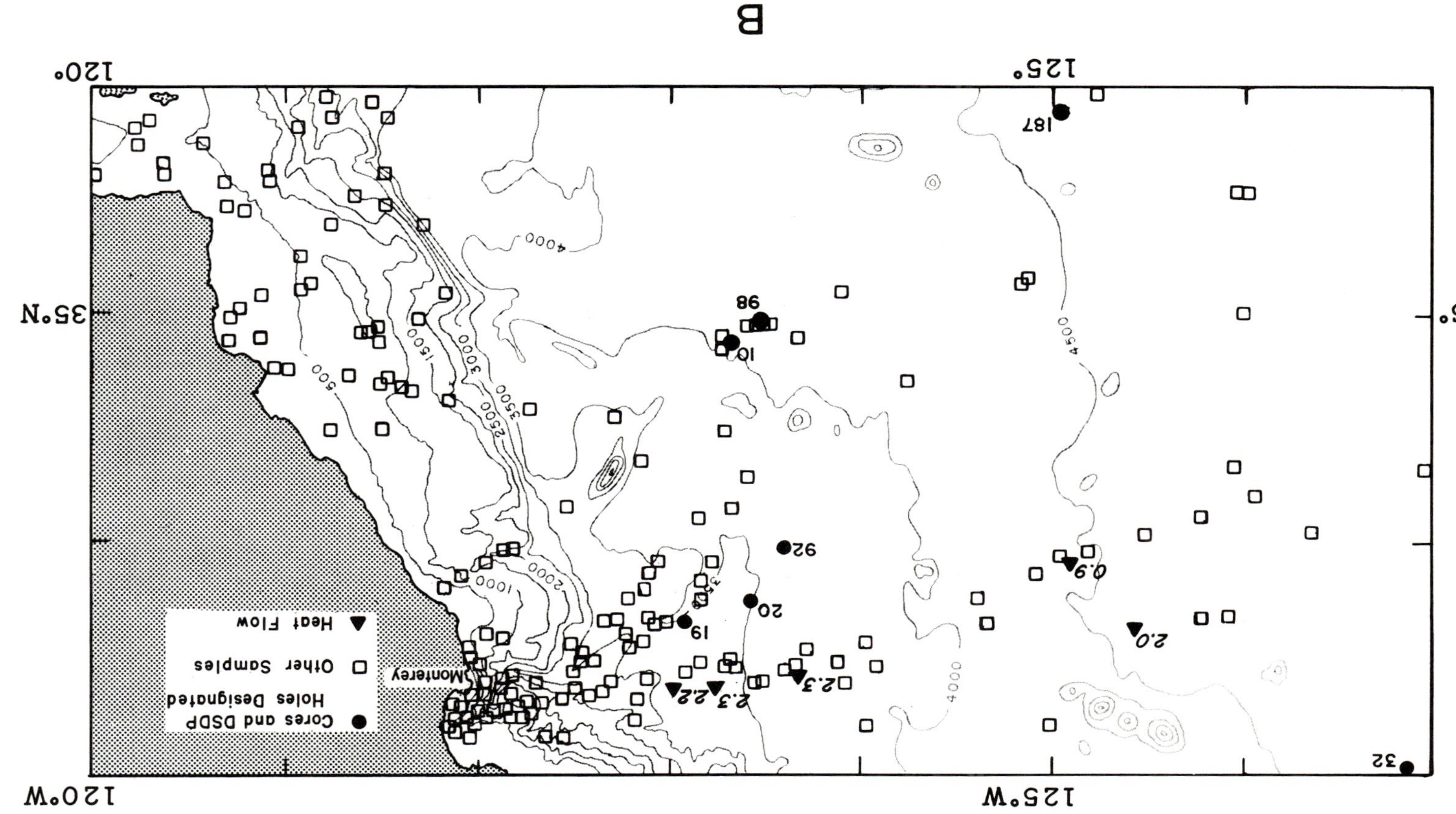

FIG. 2—a, Seismic reflection data for Monterey fan. Simplified 500-m contour bathymetry taken from Chase et al (1975) using map base from U.S. Naval Oceanographic Office 3000 series, Mercator projection. Dashed tracklines denote recent data obtained in April 1977; thick tracklines give location of profiles shown in Figures 3, 5, and 7. b, Sediment-core and heat-flow stations for Monterey fan. Same map base as in Figure 2a; core stations referred to in text or tables shown as solid dot with core number. Heat-flow values in standard heat-flow units (HFU) shown for each station.

TABLE 1. RATES OF SEDIMENTATION

A. Delgada fan
  After McManus and others (1970)

TIME	ABSOLUTE AGE (10^6 years)	RATE OF DEPOSITION (m/m.y.) SITE		
		32	33	34
Pleistocene	0.0 - 2.0	25	29	20
Late Pliocene	2.0 - 3.25	32	38	37
Early Pliocene	3.5 - 5.50	26	20	24
Late Miocene	5.50 - 11.5	10	16	19
Middle Miocene	11.5 - 14.0	4*	24	25
Early Miocene				
Early Oligocene		2*		2-3*

*pelagic sediments

B. Monterey fan
  After Hess (1974)

CORE	SCRIPPS CRUISE	HOLOCENE RATE OF DEPOSITION (m/m.y.)	DEPTH OF WATER (m)
19P	Antipode, Leg 1	432	3100
20P	Antipode, Leg 1	270	3400
92G	Montcanyon II	136	3500
98G	Montcanyon II	104	3800
187G	7 Tow, Leg 10	32	4300

tional history of the Monterey fan (minimum age of the basal fan sediments is upper Oligocene; Wilde, 1965). At present, in the waters overlying the Monterey fan, the oxygen-minimum zone (less than 1 ml/l O_2) is from about 500 to 1,200 m (CCOFI Rept., 1966), which would intersect the continental slope landward of the apex of the fan. Theide and Van Andel (1977, p. 305) gave the paleodepth of DSDP sapropels (greater than 5% organic carbon) as deep as 2,000 to 3,000 m or at the depth of the lower continental slope and the upper part of the present fan. Sediments from this zone in addition to muds from the shelf could be carried by slumps, debris flows, and turbidity currents in deeper waters of the fan and deposited rapidly enough to preserve their organic content.

The volume of sediment in the potential sapropelic zone in the oxygen minimum would vary directly with the thickness of the oxygen minimum and inversely with the bottom slope in the zone of intersection. During times of climatic optima, as in the Miocene, extensive sapropels that now are buried may have been deposited on the fan. This contention is supported by the work of Taylor (1976, p. 19-21) who found that the Miocene marine sediments in the area just southeast of the Monterey fan have substantially higher organic content than older or younger sediments.

Conversion to Hydrocarbons

There is a question whether the green muds or sapropels deposited on the fans have been buried

Table 2. Organic Content of Deep-Sea Sediments Off California

A. California Borderland
   After Emery (1960, p. 235)

	SANTA BARBARA BASIN (11 SAMPLES)	SANTA CATALINA BASIN (32)	CONT. SLOPE (15)	BLUE AND GREEN MUD (52)	SHALE (78)
PERCENT ORGANIC MATTER	5.8	8.6	5.4	7.66	1.25

B. Delgada fan

  (1) After Weser (1970, p. 586-588)

SITE	DEPTH OF WATER (m)	NO. SAMPLES	PERCENT ORGANIC CARBON	PERCENT TOTAL ORGANIC MATTER*	DISTANCE OFFSHORE (km)
32	4758	12	0.77	1.54	350
33	4284	30	0.43	0.86	295
34	4322	35	0.45	0.90	275

*Estimated using Trask and Patnode's (1942, p. 43) conclusion that total organic matter is approximately two times percent organic carbon.

  (2) From Hunt (1975)

SITE	SEDIMENT DEPTH (m)	AGE	TEMPERATURE °C	PERCENT ORGANIC CARBON	C_4-C_7 ALKANES PARTS PER 10^9
34	113	LOWER PLIOCENE	9°	0.56	5.5
	120	" "	9°	0.47	28

or heated sufficiently to convert the organic material into migrating hydrocarbons. Without analogs of large-scale, deep-water fan deposits on oceanic crust that are in production on land (Beck and Lehner, 1974), it is difficult to argue either way. As Connan (1974) has shown, petroleum genesis is a function of both time and temperature (Table 3) with catagenesis occurring initially in the range of 50 to 200°C (Hunt, 1977) and the range decreasing in temperature with age. Accordingly, higher than normal heat flow may proxy for deep burial in producing hydrocarbons for organic-rich sediments. A few heat-flow measurements on the upper part of the Monterey fan give values just over 2 HFU (Foster, 1962; Von Herzen, 1964; Lee and Uyeda, 1965; Grim, 1976) or about twice normal heat flow. Assuming reasonable thermal conductivities for the fan sediments (30 × 10⁻⁴ cal/cm °C sec; Bullard, 1963), the geothermal gradient would be conservatively 70°C/km.

Sedimentary sections filling local basement depressions on the Monterey fan are at least 2 km thick on the middle fan (Fig. 3) and about 3 km on the upper fan under the Ascension fan valley (Fig. 4), calculated by using an average sediment velocity of 2,100 m/sec as determined for the upper 2.5 sec of round trip travel time (D. McCul-

Table 3. Time-Temperature-Depth Relationships for Petroleum Generation

A	B	C	D
AGE 10^6 years	TEMPERATURE °C of CATAGENSIS [1]	DEPTH [2] IN KILOMETERS USING 70°C/km GRADIENT [3]	DEPTH [2] IN KILOMETERS USING 35°C/km GRADIENT [4]
2: Base of Pleistocene	182°C	2.6	5.2
5: Base of Pliocene	152°C	2.2	4.3
22: Base of Miocene	111°C	1.6	3.2
38: Base of Oligocene	97°C	1.4	2.8
55: Base of Eocene	89°C	1.3	2.5

[1] $\ln t = (5,879 \times \frac{1}{T}) - 12,224$

t = time in 10^6 years
T = temperature in degrees Kelvin
From Connan (1974, p. 2519)

[2] Depth in sediments below seafloor needed to attain temperature given in Column B; catagenic conditions exist below this depth for age given in Column A.

[3] Calculated from measured heat flow on upper fan of 2 HFU.

[4] "Normal" geothermal gradient in the oceans corresponding to approximately 1 HFU.

loch, 1977, personal commun.). Table 3 based on Connan's (1974) time-temperature regression curves for the threshold of petroleum generation suggests that Miocene sediments (5 to 25 m.y. old) must be buried from 1.5 to 2 km to attain the proper catagenetic temperature given a measured heat flow on the fan of 2 HFU. Thus early Miocene and older sediments in the middle fan and near the head of the fan would be above the threshold temperature. As the minimum age of the fan is Oligocene (Wilde, 1965) by sedimentary arguments and the age of the basement is Oligocene using seafloor magnetic-anomaly patterns (Pitman et al, 1974; LaBrecque et al, 1977), initial fan sediments in the deeper parts of the basement depressions would be above the threshold temperature even with a normal geothermal gradient of 35°C/km.

RESERVOIR BEDS AND TRAPS

If source beds are favorable, the next requirement for an economic petroleum deposit is a suitable reservoir. Cores from the Monterey fan indicate the presence of sands that may serve as reservoirs (Wilde, 1965; Hess, 1974). Pebble lenses and medium to fine-grained well-sorted sand layers tens of centimeters thick were found in a piston core taken in the Monterey channel at 4,000-m water depth and 280 km from the canyon head (Wilde, 1965, p. 138-139). McManus et al (1970, p. 16-17, 109) reported moderately well-sorted sands up to 40 cm thick at DSDP Site 32, and 200 cm of well-sorted friable sands in beds from 5 to 30 cm thick from core 6 of DSDP Site 34. In general, the sandy sediments deposited by turbidity currents on levees and on the fan surface are poorly sorted, with a muddy matrix that would make poor reservoir rocks. The sands deposited in channels, however, are coarser grained and better sorted, analogous to stream-channel deposits (Wilde, 1965; Nelson and Kulm, 1973).

Geometry of Sand Bodies

Valleys—The geometry of the sand bodies that could be potential reservoirs on a deep-sea fan (Moore, 1969; Moore and Fullam, 1973) depends upon the local depositional environments during channel development. As Wilde (1965), Nelson

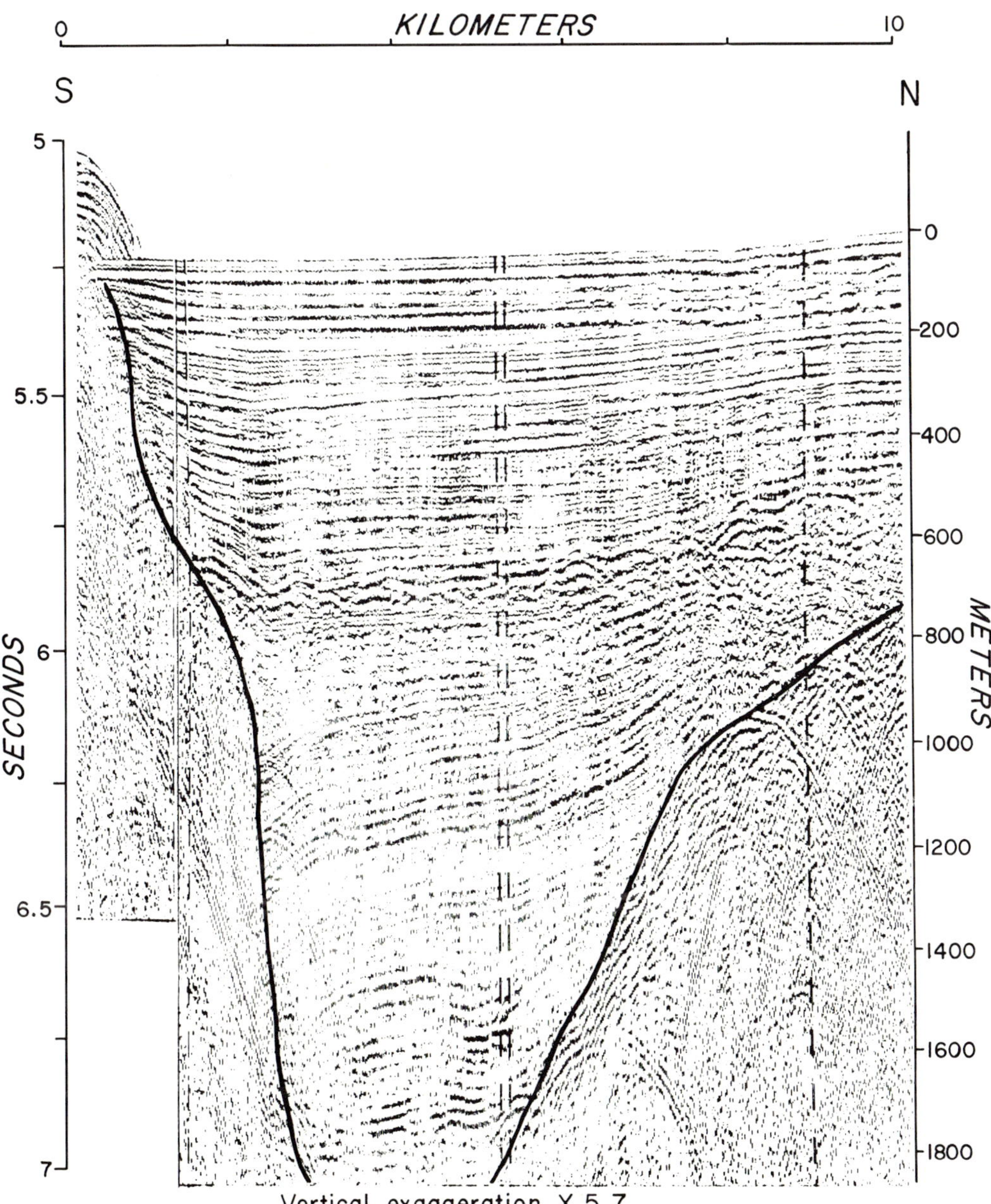

FIG. 3—Reflection profile across sediment-filled fracture zone trough on lower middle fan. Location given in Figure 2a. Vertical scales in seconds of round-trip travel time and in meters assuming a sound velocity of 2 km/sec. Heavy line denotes approximate shape of buried trough.

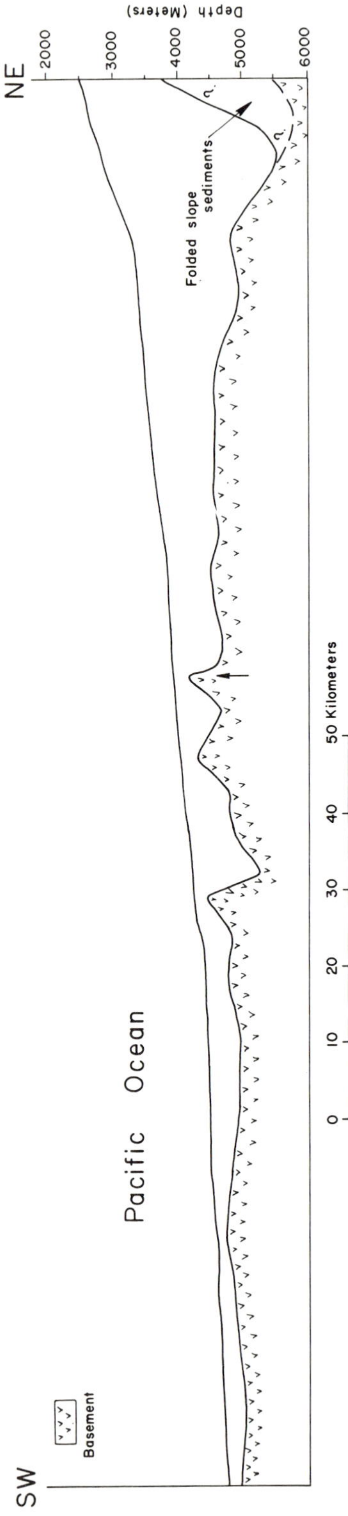

FIG. 4—Radial cross section of Monterey fan. Location is given in Figure 1. Vertical scale is depth below sea level in meters and thickness of fan sediments is calculated assuming sound velocity of 2 km/sec. At left of vertical pointing arrow, cross section is strike-normal projection of one reflection line; on right, section is based on depths and sediment thickness as interpolated between reflection profiles shown in dashed lines in Figure 2a. Total sediment thickness of fan apex from CDP reflection data (D. McCulloch, personal commun., 1977).

and Kulm (1973), and Normark (1970a, b; 1974) have shown, the development of a fan is not entirely constructional, and the sedimentary regime depends on the rate of supply and grain-size distribution of the material transported to the fan by turbidity currents.

Three valley-channel types, termed depositional, compound, and erosional (Normark, 1970b), are present on Monterey fan (Fig. 5). The geometry and lateral facies relations of sands associated with the different valley types in turn suggest major differences in their potential as reservoir sand bodies.

Where the fan valley is depositional, generally on the upper fan, prominent levees are constructed and both the valley floor and levee crests commonly build well above the level of the adjacent fan. The Ascension fan valley on the upper Monterey fan is such a well-developed constructional valley whose floor lies 300 m above the open fan (Fig. 5A). Sand units in the Ascension system are thicker, coarser, and cleaner than those on the adjacent levee crest (Hess, 1974). Deep-tow 3.8-kHz reflection profiles suggest that sands become less common down the back side of the levee (Normark, 1971). The gradient of depositional fan valleys is reduced as a result of continued sedimentation from turbidity currents. If the gradient becomes too low and the valley floor too elevated, the channel may abandon its old position and shift to a new area with a steeper gradient. Shepard (1966) has shown dramatic evidence for shifting of depositional channels on the upper fan where the Monterey fan valley goes through a tight meander. This shifting of channels beheaded the lower part of the Monterey East valley and left the upper part of the Ascension as a hanging tributary to the present Monterey canyon-valley system (Normark, 1970b). Similar forms of channel migration, piracy, and abandonment must occur frequently enough on all fans to produce their accurate shape.

In erosional (incised) fan valleys, levees are absent or poorly developed as in the case of the present lower Monterey fan valley (Fig. 5C). Erosional downcutting can modify significantly previously depositional valleys (Fig. 5B). Previously deposited turbidite and pelagic sediments on the fan are incorporated into the eroding flow and the local gradient may be steepened. Because overbank deposition of sediments coarser than muds is minimal in erosional channels, the sandy sediments in the valley do not interfinger with levee muds but generally have very abrupt lateral terminations against steep, terraced valley walls such as in the Miocene fan valleys of the Capistrano Formation of late Miocene and early Plio-

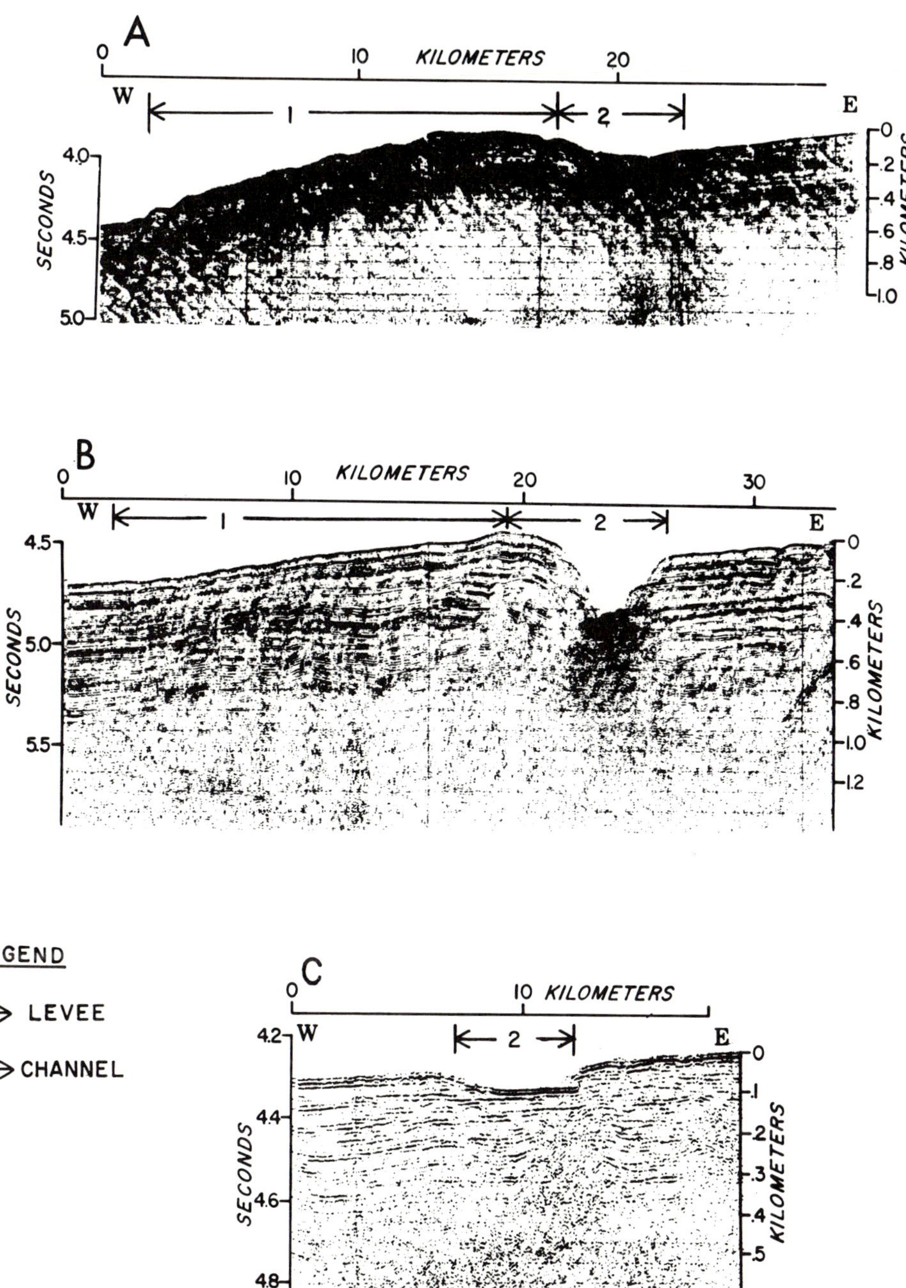

FIG. 5—Reflection profiles of different fan-valley types from Monterey fan. Vertical scales in seconds of round-trip travel time (left) and kilometers with an assumed sound velocity of 2 km/sec (right). Channel types: **A**, depositional; **B**, compound where erosion has deepened a leveed-fan valley; **C**, erosional, that is, no levees. Profile locations given in Figure 2a.

cene age in southern California (Bartow, 1966; Piper and Normark, 1971; Hess, 1974; Walker, 1975). These ancient fan valleys contain thick (several meters), massive sandstones and conglomerates that commonly become thinner and finer grained during later stages of channel filling. This change in grain size is probably a result of a gradual loss of sediment supply as the valley upstream cuts through its levee and begins to feed sediment to a new section of the fan.

The fan-valley classification emphasizes the (vertical) cross-sectional geometry of the valleys and the lateral facies relations with adjacent fan sediments. Recognition of these valley types is, of itself, insufficient for mapping potential reservoir sand bodies. Not all fan valleys develop as main distributary branches. The surface channel system on Monterey fan, for example, includes several types of overflow channels, which include recently abandoned major channels and channels receiving occasional overflows that have multiple origins.

Suprafans—On many fans where the turbidity currents carry appreciable amounts of sand-sized sediments, the valley on the upper fan terminates in a depositional lobe called a suprafan (Normark, 1970a). The suprafan area, generally present on the middle fan, appears as a bulge on a longitudinal fan profile and is characterized on conventional echo-sounding and seismic reflection profiles by an apparent hummocky local relief, which is the result of numerous, overlapping hyperbolic echoes. Detailed deep-tow studies of the suprafans on San Lucas and Navy submarine fans show that although some discrete, meandering distributary channels radiate from the terminus of the leveed valley on the upper fan, most of the suprafan relief consists of isolated depressions that are 2 to 50 m deep and 10 m to more than 1 km wide (Normark, 1970a; Normark et al, 1976). High-resolution reflection profiles and short core samples from these same fans suggest that the suprafan areas are composed primarily of sandy sediments.

The suprafan concept refers specifically to a physiographic feature present on modern submarine fans. The deltaic shape of well-developed suprafans and their location at the terminus of the main leveed valley (Fig. 6) suggest that they form broad (10 to 50 km) lenticular sandy complexes characterized by individual units that exhibit an upward increase in grain size and bed thickness. On the upper suprafan, depressions and distributary channels that presumably are floored by sands and coarser sediments are continuous with similar deposits of the adjacent leveed valley floor. The individual channel-fill sequences would exhibit thinning- and fining-upward beds.

Two active suprafan areas are inferred on Monterey fan. The largest and southernmost is associated with the present termination of the Monterey fan valley and has many distributary valleys. The smallest suprafan (36°N, 123°W) is probably the product of the beheaded Monterey East valley and is relatively inactive (Hess and Normark, 1976).

Development of Channels and Suprafans as Stratigraphic Traps

After channel abandonment, the former active channel probably is partly filled by slumps from the channel walls, overflow from nearby active distributaries, material reworked by contour currents, and eventually by pelagic sedimentation. As noted previously, the coarser and better sorted material is in the axis of the channel, whereas the poorly sorted, finer grained material is on the levees and the fan surface. Thus, the filling of the abandoned channel would produce a reasonably impermeable cap of silt and clay over the better sorted sands of the channel axis. For example, a buried channel from the Monterey fan (Fig. 7) is covered by an estimated 500 m of sediment, of which the uppermost 300 m probably is regularly bedded mud deposited by overflow from the nearby levee of the present Monterey fan valley.

After burial by green mud, the distributary channels and sandy sections within the suprafan would provide a network of entry points for the migration and concentration of petroleum upslope into the main channel. The gradient from the mouth of the submarine canyon to the suprafan areas maintained during fan growth results in the sands of the channels and the suprafans (the potential petroleum stratigraphic traps) being deposited with an upward tilt in the upchannel direction. On Monterey fan, the present slope or gradient decreases from 0°28′ at the apex of the fan to 0°07′ at the outer edge in a water depth of 4,700 m. Locally steeper slopes reaching 2° are present on the back sides of the levees (Fig. 5) and along the leading edges of suprafan lobes on some fans. Thus, if an updip seal is formed the abandoned channels would make excellent stratigraphic traps, with the petroleum migrating up the channel gradient to the coarse and more permeable sands of the channel near the apex of the fan (Moore and Fullam, 1973). Such seals could result from slumps of valley-wall sediments that would plug the valley, from channel abandonment that would lead to infilling the down fan-valley remnant with fine-grained sediments, and from tectonism. Differential compaction within the large and rapidly deposited leveed-valley

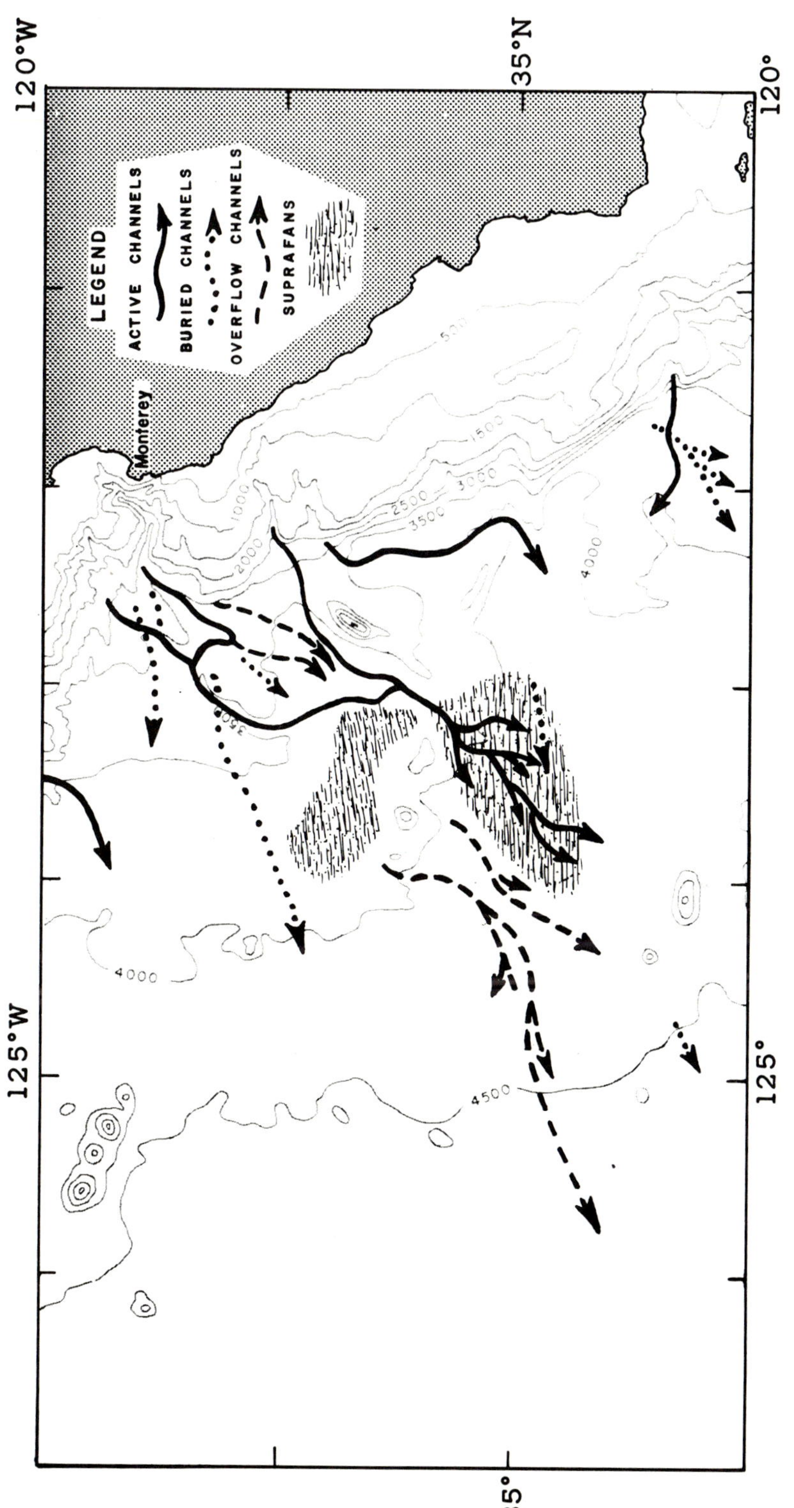

FIG. 6—Map of active and ancient channel systems on Monterey fan. Map base same as for Figure 2.

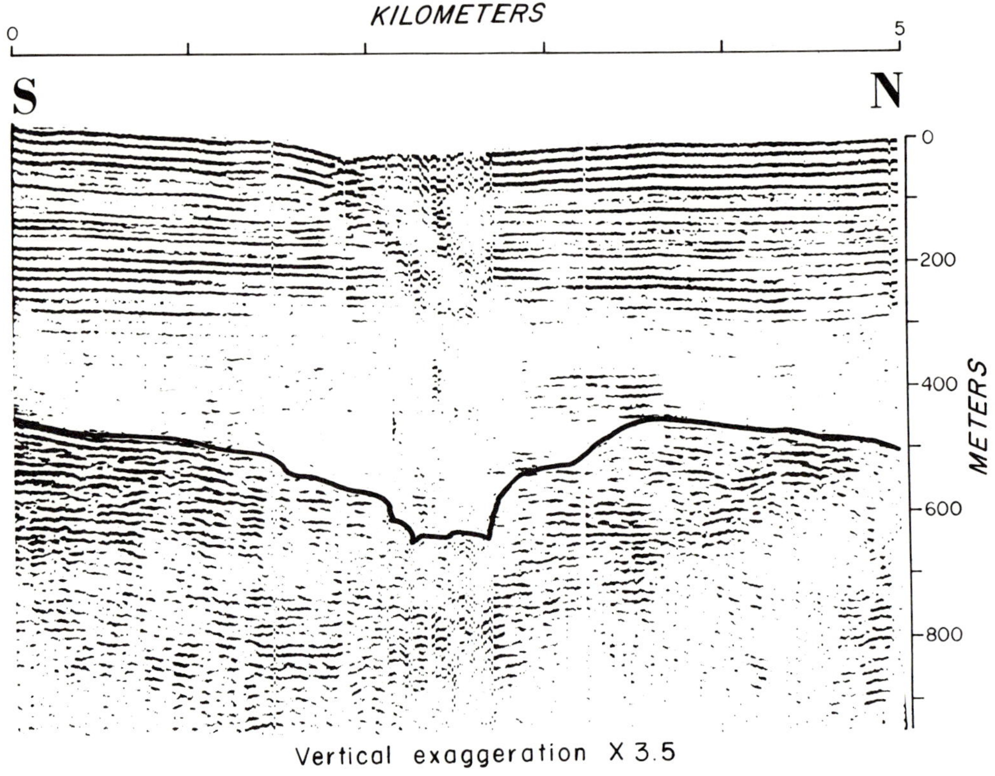

FIG. 7—Reflection profile of fan valley buried by about 500 m of sediment. Location given in Figure 2a; vertical scales same as for Figure 5.

complexes, where coarse sediments are common only on valley floors, could result in sufficient disruption of valley-floor sands to result in the formation of an updip seal.

Distribution of Abandoned Channels

The arcuate outline of most deep-sea fans is taken as evidence for the migration of valley distributary systems across the area of the fans during their growth (Menard, 1955, 1960; Wilde, 1965; Normark, 1970a; Nelson and Kulm, 1973). The present surface channels for both the Delgada and Monterey fans are confined to the southern sector of each fan. This suggests that a network of abandoned channels, now buried by subsequent deposition, extends in radiating patterns from the apex of the fans. Several westward-trending buried valleys (Fig. 6) have been identified tentatively in the Monterey fan sediments. One of the buried channels lies underneath the large levee of the present Monterey fan valley.

Dimensions of Channel Sands and Associated Sand Bodies

The channel widths on Monterey fan are on the order of 2 to 5 km with depths to 400 m (Wilde, 1965; Komar, 1969). The actual dimensions of the channels downstream are variable, with the present Monterey system traceable from the canyon mouth for about 250 km to the toe of the fan (Wilde, 1965). Any abandoned channel section would be shorter; for example, the beheaded section of the Monterey East channel is about 50 km long (Normark, 1970b). One of the buried channels trending west of the present Monterey system can be traced for at least 100 km. Because the rate of change of the gradients is greatest near the apex of the fan, one would expect that most of the abandonments would occur near the apex and that the abandoned segments would be rather long. The thickness of the potential reservoir part of the channel also would be variable. As noted previously from the DSDP samples, sandy sections to 2 m thick were found almost 300 km from shore. If the sediments filling the buried channel depicted in Figure 7 are similar to the Doheney Channel sands (Bartow, 1966; Piper and Normark, 1971) and the section further south at San Clemente (Hess, 1974; Walker, 1975), a sandy section over 100 m thick and 2 km wide would be implied. Thicker sections of sand could be expected near the apex of the fan, where the gradient decreases rapidly from more than 1° in the canyons to less than 1° on the fan surface and coarse sediment would tend to drop out of the flow.

The best potential traps for hydrocarbons may be the sands of the suprafan areas. For the small Navy and San Lucas fans, these areas are 5 to 10 km across and 100 m or more thick (Normark, 1970b; Normark et al, 1976). The lower parts of the suprafans are markedly less channelled, and geophysical and core data support Walker's (1976; Walker and Mutti, 1973) contention that the lower suprafan lobes correspond to the "outer-fan" sand lobes of Mutti and Ricci Lucchi (1972) where classic turbidite sequences are common. The lower (outer) fan environment on modern submarine fans (defined topographically) is equated with the basin-plain environment described for ancient turbidites (defined by interpretation of lithologic facies association). Thus, the sandy suprafan areas, which are channelled on their upper reaches, may provide relatively large, lenticular sand bodies that are excellent potential traps for hydrocarbons. On the Monterey fan, the largest (tentatively identified) suprafan may exceed 50 km in width.

The minimum volume of the fans off central California is on the order of 10^{14} cu m (Menard, 1960, p. 1274). The surface area of the fan now occupied by channels and receiving coarse sediment is about 1%. The suprafan areas may comprise another 3% of the fan surface. Taking the minimum volume of the fan as 10^{14} cu m, about 4 $\times$ 10^{12} cu m would be sandy complexes, assuming the ratio of channels and suprafans to total-fan area is constant throughout the development of the fans. Most of the sand would be concentrated on the suprafans and the major valleys of the upper fan. What percentage, if any, of these sands contains economic concentrations of petroleum, which would have migrated upward from muds exceeding the catagenetic limits (Connan, 1974), is certainly a speculation at this time.

PROSPECTS AND CONCLUSIONS

The best prospects for petroleum accumulation in stratigraphic traps in continental-rise environ-ments are (1) near the apices of the various fans within large leveed-valley systems and (2) on the middle fan both within deep basins and beneath the suprafan accumulations. In our discussion of the Monterey fan example, there has been no consideration of either tectonically or clathrate-produced traps that may be important reservoirs within particular continental-rise settings. The immense size of channel and suprafan systems, their proximity to refining facilities and markets in coastal areas, combined with our developing capacity to operate in the deep ocean make buried channels on deep-sea fans, as well as other areas on the continental rise with similar growth patterns, attractive targets for petroleum exploration in the near future.

REFERENCES CITED

Bartow, J. A., 1966, Deep submarine channel in upper Miocene, Orange County, California: Jour. Sed. Petrology, v. 36, p. 700-705.

Beck, R. H., and P. Lehner, 1974, Oceans, new frontier in exploration: AAPG Bull., v. 58, p. 376-395.

Berner, R. A., 1976, The benthic boundary layer from the viewpoint of a geochemist, in I. N. McCave, ed., The benthic boundary layer: New York, Plenum Press, p. 33-55.

Bullard, E. C., 1963, The flow of heat through the floor of the ocean, in M. N. Hill, ed., The sea, v. 3, The earth beneath the sea: New York, Wiley Intersci., p. 218-232.

California Co-operative Fisheries Investigations (CCO-FI) Report, 1966, Physical and chemical data, CCO-FI cruise 6404 and 6407: Scripps Inst. Oceanog. Ref. 66-20, 89 p.

Caughey, C. A., and C. J. Stuart, 1976, Where the potential is in the deep Gulf of Mexico: World Oil, v. 183, no. 1, p. 67-72.

Chase, T. E., W. R. Normark, and P. Wilde, 1975, Oceanographic data of the Monterey deep-sea fan, 34°-37°N, 120°-127°W: Univ. California Inst. Marine Resources Tech. Rept. TR-58, 2 p.

Connan, J., 1974, Time-temperature relation in oil genesis: AAPG Bull., v. 58, p. 2516-2521.

Dill, R. F., R. S. Dietz, and H. B. Stewart, 1954, Deep-sea channels and delta of the Monterey submarine canyon: Geol. Soc. America Bull., v. 65, p. 191-194.

Durand, B., and J. Espitalie, 1972, Formation et evolu-tion des hydrocarbures de C_1 a C_{15} et des gaz perma-nents dans les argiles du Toarcien du bassin de Paris, in Advances in organic geochemistry: Internat. Ser. Mon. Earth Sci., v. 33, p. 455-468.

Emery, K. O., 1970, The sea off southern California—a modern habitat of petroleum: New York, John Wiley, 366 p.

———— 1973, Oil on the shelf: Oceanus, v. 17, Spring, p. 11-17.

———— and E. E. Bray, 1962, Radiocarbon dating of California basin sediments: AAPG Bull., v. 46, p. 1839-1856.

Ewing, M., D. B. Erickson, and B. C. Heezen, 1958,

Sediments and topography of the Gulf of Mexico, *in* L. G. Weeks, ed., Habitat of oil: AAPG, p. 995-1053.

Fischer, A. G., and M. A. Arthur, 1977, Secular variations in the pelagic realm, *in* Deep marine carbonate environments: SEPM Spec. Pub. 25, p. 19-50.

Foster, T. D., 1962, Heat flow measurements in the northeast Pacific and in the Bering Sea: Jour. Geophys. Researth, v. 67, p. 2991-2993.

Goldhaber, M. B., et al, 1977, Sulfate reduction, diffusion, and bioturbation in Long Island Sound sediments: report of the Foam group: Am. Jour. Sci., v. 277, p. 193-237.

Grim, P. J., 1976, Terrestrial heat flow data: Boulder, Colorado, World Data Center A.

Heezen, B. C., M. Tharp, and M. Ewing, 1959, The floor of the oceans. I, The North Atlantic: Geol. Soc. America Spec. Paper 65, 122 p.

Hein, J. R., A. O. Allwardt, and G. B. Griggs, 1974, The occurrence of glauconite in Monterey Bay, California, diversity, origins, and sedimentary environment: Jour. Sed. Petrology, v. 44, p. 562-571.

Hess, G. R., 1974, Submarine fanfare; a comparison of modern and Miocene deep-sea fans: Master's thesis, Univ. Minnesota, 118 p.

———— and W. R. Normark, 1976, Holocene sedimentation history of the major fan valleys of Monterey fan: Marine Geology, v. 22, p. 233-251.

Hunt, J. M., 1974, Organic geochemistry of the marine environment, *in* Advances in organic geochemistry 1973: Paris, Editions Technit, p. 593-605.

———— 1975, Origin of gasoline range alkanes in the deep sea: Nature, v. 254, p. 411-413.

———— 1977, Distribution of carbon as hydrocarbons and asphaltic compounds in sedimentary rocks: AAPG Bull., v. 61, p. 100-104.

Hurley, R. J., 1964, Analysis of flow in Cascadia deep-sea channel, *in* Papers in marine geology: New York, Macmillan Co., p. 117-132.

Komar, P. D., 1969, The channelized flow of turbidity currents with application to Monterey deep-sea fan channel: Jour. Geophys. Research, v. 74, p. 4544-4558.

LaBrecque, J. L., D. V. Kent, and S. C. Cande, 1977, Revised magnetic polarity time scale for the Late Cretaceous and Cenozoic time: Geology, v. 5, p. 330-335.

Lee, W. H. K., and S. Uyeda, 1965, Review of heat flow data, *in* Terrestrial heat flow: Am. Geophys. Union Geophys. Mon. 8, p. 87-190.

McManus, D. A., et al, 1970, Initial reports of the Deep Sea Drilling Project, v. 5: Washington, D.C., U.S. Govt. Printing Office, 827 p.

Menard, H. W., 1955, Deep-sea channels, topography, and sedimentation: AAPG Bull., v. 39, p. 236-255.

———— 1960, Possible pre-Pleistocene deep-sea fans off central California: Geol. Soc. America Bull., v. 71, p. 1271-1278.

Moore, G. T., 1969, Interaction of rivers and oceans—Pleistocene petroleum potential: AAPG Bull., v. 53, p. 2421-2430.

———— and T. J. Fullam, 1973, Deep water channels and their potential value in petroleum localization: Gulf Coast Assoc. Geol. Soc. Trans., v. 23, p. 256-258.

Mutti, E., and F. Ricci Lucchi, 1972, Le torbiditi dell'Appenin. setten trionale: introduzione all'analis di facies: Soc. Geol. Italiana Mem., v. 11, p. 161-199.

Nelson, C. H., and L. D. Kulm, 1973, Submarine fans and channels, *in* Turbidites and deep water sedimentation: SEPM Pacific Sec. Short Course, p. 39-78.

Normark, W. R., 1970a, Growth patterns of deep-sea fans: AAPG Bull., v. 54, p. 2170-2195.

———— 1970b, Channel piracy on Monterey deep-sea fan: Deep-Sea Research, v. 17, p. 837-846.

———— 1971, Mini-topography of deep-sea fans: geometric considerations for facies interpretations in turbidites, *in* Geologic guidebook—Newport Lagoon to San Clemente, Orange County, California: SEPM Pacific Sec., p. 22-36.

———— 1974, Submarine canyons and fan valleys: factors affecting growth patterns of deep-sea fans, *in* Modern and ancient geosynclinal sedimentation: SEPM Spec. Pub. 19, p. 56-68.

———— D. J. W. Piper, and G. R. Hess, 1976, Distributary mesochannels, megaflutes, and microtopography of Navy submarine fan, California (abs.): Geol. Soc. America Abs. with Programs, v. 8, p. 1032.

Philippi, G. T., 1974, The influence of marine and terrestrial source material on the composition of petroleum: Geochim. et Cosmochim. Acta, v. 38, p. 947-966.

Piper, D. J. W., and W. R. Normark, 1971, Re-examination of a Miocene deep-sea fan and fan-valley, southern California: Geol. Soc. America Bull., v. 82, p. 1823-1830.

Pitman, W. C., R. L. Larson, and E. M. Herron, 1974, Age of the ocean basins determined from Magnetic anomaly lineations: Geol. Soc. America Misc. Map, scale approx. 1:34,500,000.

Romankevich, Ye. A., 1969, Organic carbon and nitrogen deposits in recent and Quaternary sediments of the Pacific Ocean: Oceanology, v. 8, p. 658-673.

Shepard, F. P., 1966, Meander in valley crossing a deep-ocean fan: Science, v. 154, p. 385-386.

———— and R. F. Dill, 1966, Submarine canyons and other sea valleys: Chicago, Rand McNally, 381 p.

Snyder, R. E., 1976, New concept unveiled for subsea completion, production: World Oil, v. 183, no. 4, p. 41-45.

Stone, G. R., 1975, Subsea completion systems: Offshore Services, v. 8, p. 34-48.

Summerhayes, C., D. A. Ross, and P. Stoffers, 1977, Nile submarine fan: sedimentation, deformation, and oil potential: 9th Offshore Tech. Conf. Proc., v. 1, OTC2731, p. 35-40.

Taylor, J. C., 1976, Geological appraisal of the petroleum potential of offshore southern California: the borderland compared to onshore coastal basins: U.S. Geol. Survey Circ. 730, 43 p.

Thiede, J., and T. H. Van Andel, 1977, The paleoenvironment of anaerobic sediments in the late Mesozoic South Atlantic Ocean: Earth and Planetary Sci. Letters, v. 33, p. 301-309.

Thompson, T. L., 1976, Plate tectonics in oil and gas exploration of continental margins: AAPG Bull., v. 60, p. 1463-1501.

Trask, P. D., and H. W. Patnode, 1942, Source beds of petroleum: AAPG, 561 p.

Von Herzen, R. R., 1964, Ocean floor heat flow measurements west of the United States and Baja California: Marine Geology, v. 1, p. 225-239.

Walker, R. G., 1975, Nested submarine-fan channels in the Capistrano Formation, San Clemente, California: Geol. Soc. America Bull., v. 86, p. 915-924.

———— 1976, The "resedimented coarse clastic family," facies processes and depositional models, in R. S. Saxena, ed., Sedimentary environments and hydrocarbons: New Orleans Geol. Soc. Short Course, p. 129-162.

———— and E. Mutti, 1973, Turbidite facies and facies associations, in G. V. Middleton, and A. H. Bouma, eds., Turbidites and deep water sedimentation: SEPM Pacific Sec. Short Course, Anaheim, California, p. 119-157.

Weser, O. E., 1970, Lithologic summary, in Initial reports of the Deep-Sea Drilling Project, v. 5: Washington, D.C., U.S. Govt. Printing Office, p. 569-620.

Wilde, P., 1965, Recent sediments of the Monterey deep-sea fan: California Univ., Berkeley, Hydrol. Eng. Lab. Rept. HEL 2-13, 153 p.

———— W. R. Normark, and T. E. Chase, 1976a, Oceanographic data off central California 37° to 40° North: Lawrence Berkeley Lab. Pub. 92, 2 p.

———— ———— ———— 1976b, Petroleum potential of continental rise off central California—summary, in Circum-Pacific energy and mineral resources: AAPG Mem. 25, p. 313-317.

Reprinted for private circulation from
THE AMERICAN ASSOCIATION OF PETROLEUM GEOLOGISTS BULLETIN
Vol. 62, No. 6, June, 1978

Neogene Turbidite Sedimentation in Komandorskiy Basin, Western Bering Sea[1]

RICHARD J. STEWART[2]

Abstract Sediments underlying the abyssal plain in Komandorskiy basin were recovered at Deep Sea Drilling Project Site 191 in the east-central part of the basin (168.1°E). Most sedimentary fill is diatomaceous silty clay. Interbedded in the silty clay are layers of turbidite sand and volcanic ash. The sand layers range in composition from volcanic lithic sand to chert-rich sand containing as much as 40% quartzose debris. The sand probably originates on the continental shelf in the westernmost Bering Sea, and is deposited by turbidity currents crossing the abyssal plain. Chert-rich sand may originate on the shelf between Mys Olyutorskiy and Ostrov Karaginskiy, and volcanic lithic sand may originate farther south toward Mys Kamchatskiy.

Below 500 m, sediments of the abyssal plain are altered extensively. The silty clay is compacted to mudstone, and sand layers are graywacke. Porosity in the sand layers is destroyed completely, and the deepest sand layer has textures that approach those of semischist. The graywacke layers contain considerable quantities of volcanic and sedimentary lithic debris, and may have originated on the continental shelf in the Komandorskiy Islands.

Neogene accumulation rates in Komandorskiy basin were about 60 m/m.y.$^{-1}$ in the late Miocene, increasing to 300 m/m.y.$^{-1}$ in late Pliocene and early Pleistocene time. These high rates were maintained by extensive continental-shelf erosion during late Cenozoic sea-level transits, and by exceedingly high diatom productivity in the overlying waters of the Bering Sea. A slight decrease in accumulation rates in the late Pleistocene may be the result of a decline in diatom productivity.

Unlithified Neogene deposits of Komandorskiy basin are distal representatives of contemporaneous deposits in continental-margin basins and have some characteristics of potential reservoir rocks. Their more proximal counterparts at shallower-water depths deserve increased attention as possible hydrocarbon sources.

INTRODUCTION

Komandorskiy basin is a major depression in the floor of the western Bering Sea and is separated from the main Aleutian basin by the Shirshov Ridge (Fig. 1). The sedimentary sequence underlying the floor of the basin was cored during Leg 19 of the Deep Sea Drilling Project at Site 191, 120 km west of Shirshov Ridge (Fig. 1). This paper describes the mineralogy and petrology of both unconsolidated and lithified sediment recovered from Site 191 with emphasis on the sand-sized material. The sediments are particularly significant because they may be equivalent to Neogene debris filling structural basins on the adjacent continental margin. If the sands in the deep Komandorskiy basin were derived from the

west and northwest, then these sands may be distal representatives of the nearshore facies in the continental-margin basins where most hydrocarbon exploration interest centers (Eremenko et al, 1973; Gnibidenko and Svarichevskiy, 1974). Also, Komandorskiy basin is one of many basins in the northern and western Pacific Ocean lying on the concave or "backarc" side of a major island arc, and its sediments may be a model for comparison with other backarc basins both modern and fossil.

GEOLOGIC SETTING AND SITE DESCRIPTION

Komandorskiy basin is isolated from the depths of the Pacific Ocean by the western Aleutian Ridge, crested in this region by the Komandorskiye Ostrova (Commander Islands, Fig. 1). Komandorskiy basin is a crudely triangular feature about 400 km long and 200 km wide, with a surface area of about 140,000 sq km. The floor of the basin is an almost horizontal abyssal plain lying at a water depth of 3,400 to 4,000 m (Udintsev et al, 1959; Chase et al, 1971; Scholl et al, 1974). The basin is bordered on the west by the volcanic terrane of the Kamchatskiy Poluostrov (Kamchatka Peninsula, Fig. 1), on the north by the Koryakskiy Khrebet (Koryak Mountains, Fig. 1), and on the east by Shirshov Ridge.

Exposures of Aleutian Ridge rocks on the Komandorskiye Ostrova are mainly Eocene through Miocene volcanic rocks, overlain unconformably by deeply eroded Pliocene stratovolcanoes (Zhegalov, 1964; Shmidt, 1973). In contrast, the Kamchatskiy Poluostrov is the site of extensive Pliocene-Pleistocene volcanism, including the towering, explosive volcanoes of the northern Kuril Island arc (Nalivkin, 1960; Avdeiko, 1971; Gnibidenko et al, 1972). The Koryakskiy coast consists of sedimentary, metamorphic, and vol-

[1]Manuscript received, November 17, 1975; accepted, July 7, 1976.

[2]Department of Geological Sciences, University of Washington, Seattle, Washington 98195.

I am grateful to the scientific staff of the Deep Sea Drilling Project, Leg 19, for discussions of the implications of sedimentary and tectonic events in the evolution of the western Bering Sea. Critical review of the manuscript by Joe Creager is acknowledged gratefully. This work was supported in part by National Science Foundation grant GA 36034.

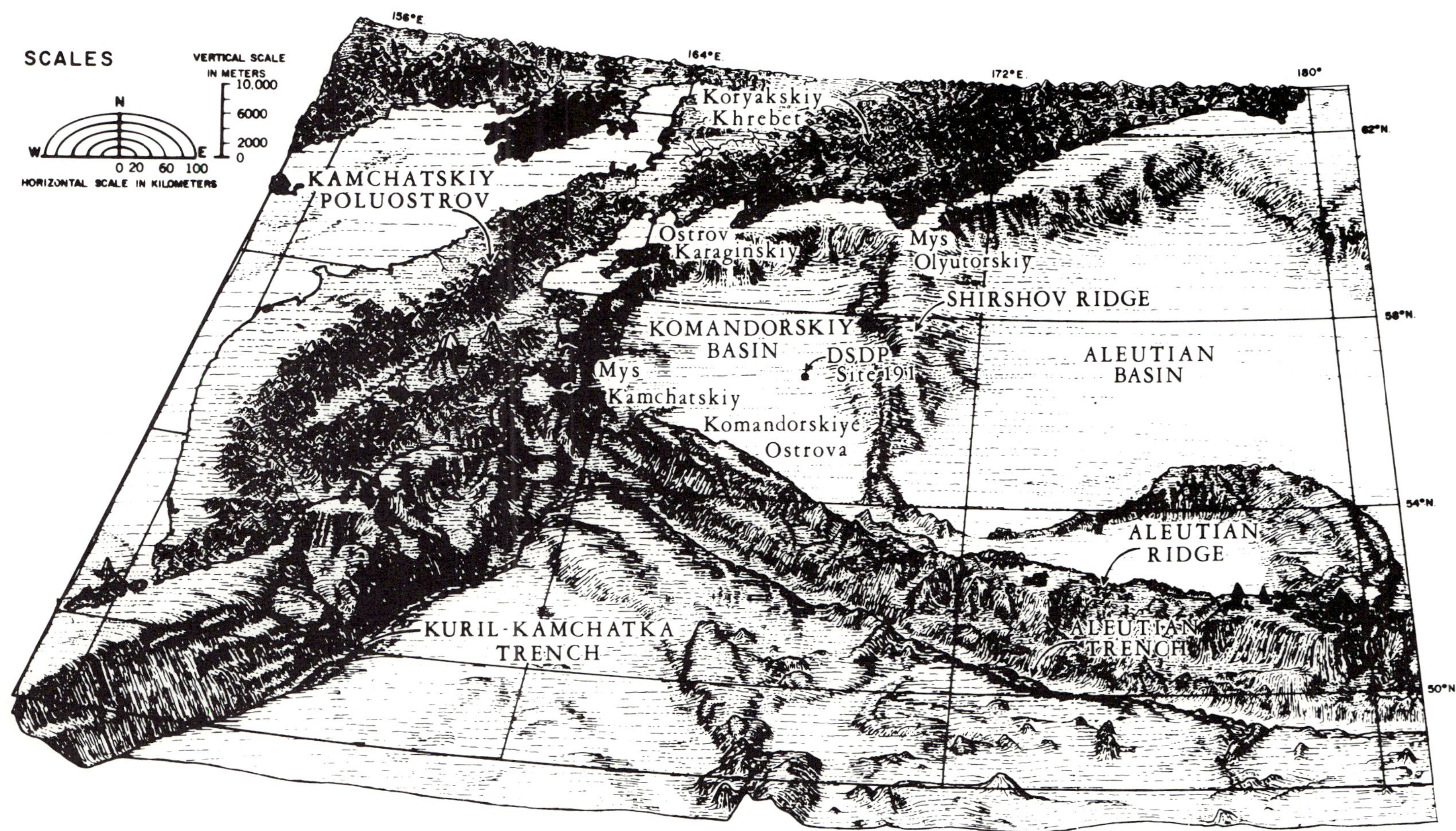

FIG. 1—Index map showing location of Deep Sea Drilling Project Site 191, Komandorskiy basin, western Bering Sea. Figure adapted from Alpha (1974). Vertical exaggeration ×10.

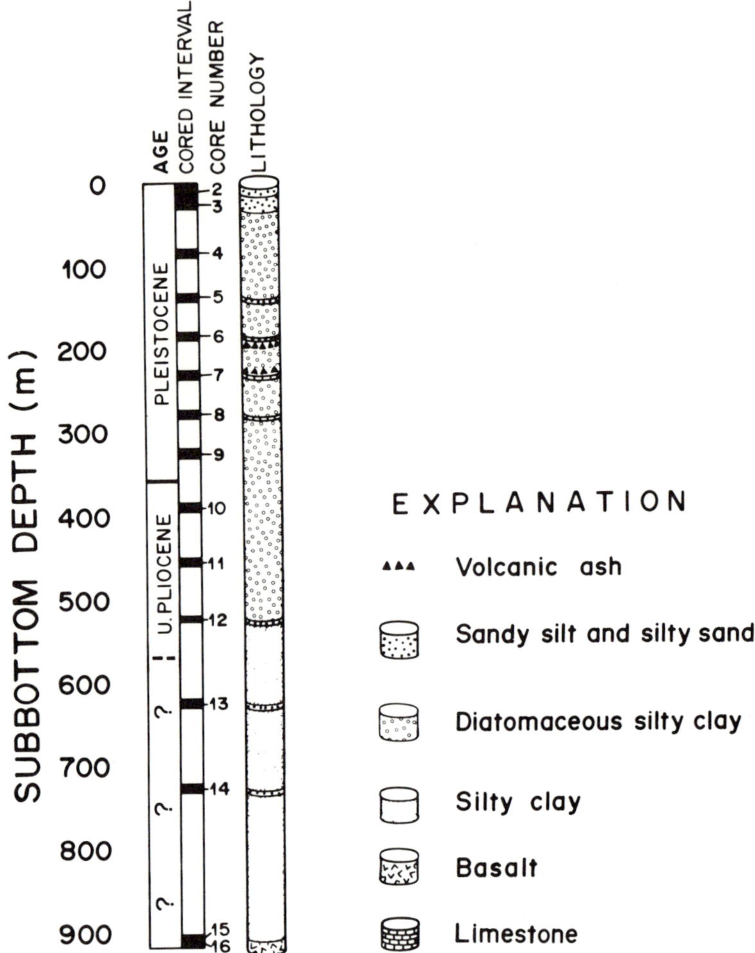

FIG. 2—Diagrammatic stratigraphic section at Deep Sea Drilling Project Site 191, Komandorskiy basin, western Bering Sea. Except for core cuttings, basal 200 m of sediment was not recovered.

canic rocks ranging in age from late Proterozoic(?) through Neogene, and includes a few ultrabasic and granitoid rocks (Nalivkin, 1960; Lisitsyn, 1966; Bogdanov, 1970). Outcrops of Mesozoic and early Cenozoic rocks, including ultrabasic rocks, are known also from the Kamchatskiy Poluostrov and Karaginskiy Ostrov (Karaginsky Island, Fig. 1; Gnibidenko et al, 1972). Shirshov Ridge is a buried submarine ridge blocking flow of sediment between the Aleutian and Komandorskiy basins.

Deep Sea Drilling Project Site 191 is on the floor of Komandorskiy basin just west of Shirshov Ridge (Fig. 1; see Creager et al, 1973, for complete site description). The 900-m sedimentary sequence consists of 520 m of upper Pliocene to upper Pleistocene silty clay, turbidite sand, and scattered ash layers, that overlie 380 m of upper Miocene(?) to Pliocene indurated clay and sand (Fig. 2).[3] Drilling at this site terminated in basalt of uncertain, but possibly late Oligocene(?), age (Stewart et al, 1973).

Most of the material recovered at DSDP Site 191 is diatomaceous silty clay (Creager et al, 1973). Diatoms typically constitute 40 to 50% of the silty clay, and some sediment sections contain 10 to 15% carbonate material. Approximately 8%

[3]Reference to specific samples in this report lists core number, section number, and depth in centimeters within the section. Complete core descriptions were reported in Creager et al, 1973.

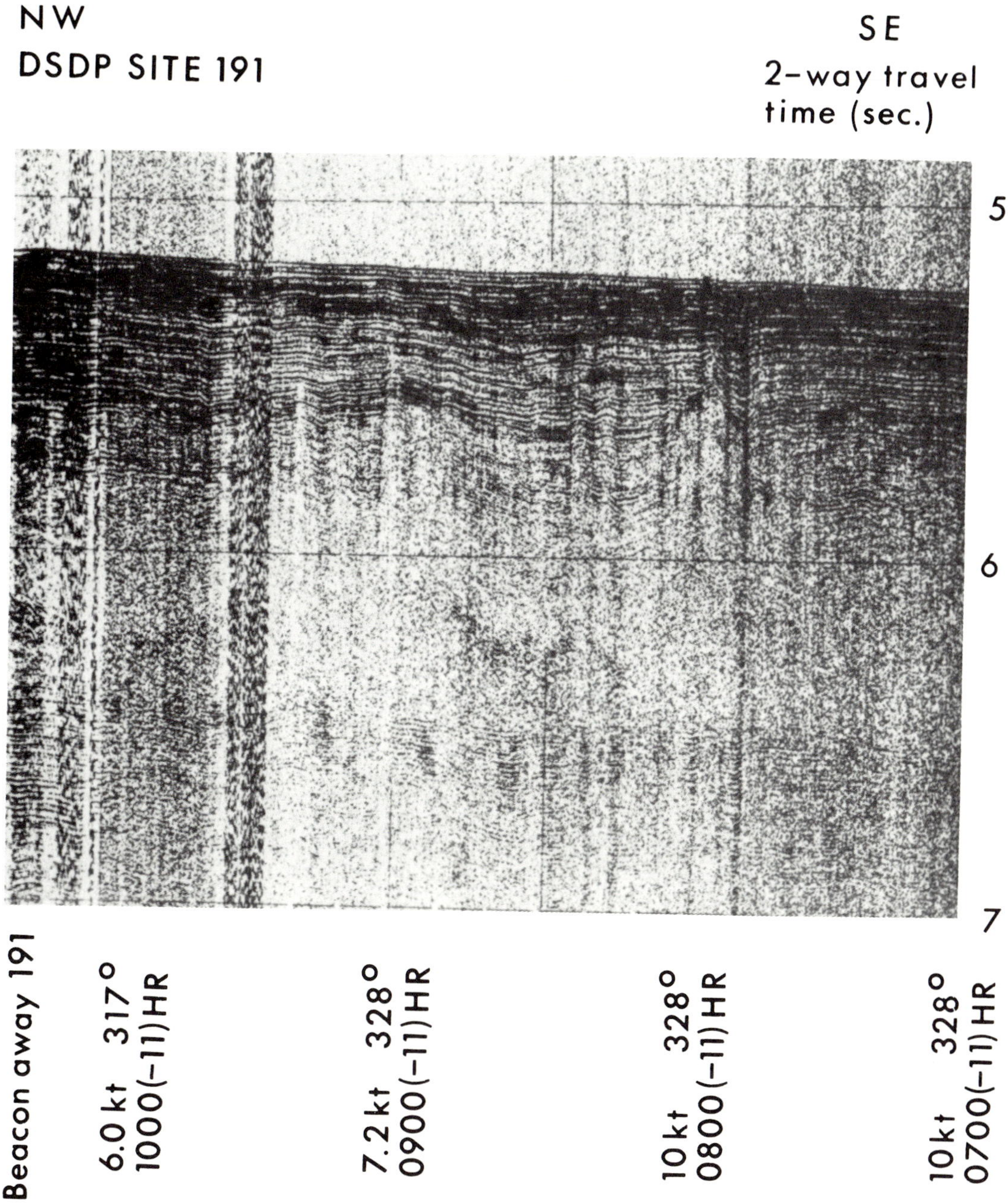

FIG. 3—*Glomar Challenger* seismic turofile in Komandorskiy basin on approach to DSDP Site 191 from southeast. Upper parallel reflectors apparently represent turbidite sand layers that are abundant in upper Pliocene and Pleistocene beds. Acoustically transparent deposits below turbidite layers are indurated mudstone and graywacke. Adapted from Creager et al (1973).

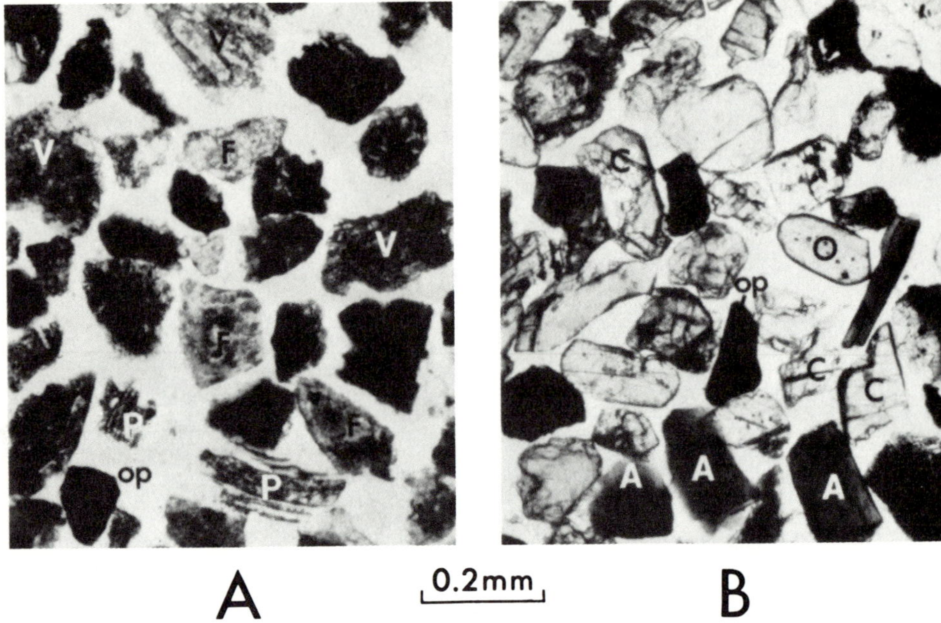

FIG. 4—Representative grain-mount thin sections of sand-sized grains from volcanic-sand layers. **A**, light-mineral separate; *F*, feldspar; *P*, pumice shard; *V*, volcanic lithic fragment; *op*, opaque. **B**, heavy-mineral separate; *C*, clinopyroxene; *O*, orthopyroxene; *A*, amphibole; *op*, opaque. Both samples are from core catcher, core 4.

(4 m) of the sediment recovered at Site 191 is sand. In Pleistocene deposits clay-rich diatom ooze is dominant.

TURBIDITE SAND AND SILT LAYERS

Mineralogy

The upper 300 m of sediment recovered in Komandorskiy basin is a "classic" turbidite sequence (Fig. 3). Size-graded sand layers are interbedded in diatomaceous silty clay and diatom ooze. The sand layers contain reworked fossils and displaced shallow-water and freshwater fossils (Creager et al, 1973). The sand layers are up to 1 m thick and have sharp bottom contacts and indistinct tops (Creager et al, 1973).

Light and heavy-mineral separates of the >62-μ size fraction were prepared by centrifugation in tetrabromoethane (Sp. G. 2.90), mounted in epoxy, and ground to standard thin-section thickness. All light-mineral mounts were stained for potassium feldspar and plagioclase following the methods of Laniz et al (1964). A minimum of 200 nonopaque, nonmicaceous grains was identified in each mount following the procedures of Dickinson (1970).

Typical sand compositions are 45% rock fragments and glass shards, about 30% feldspars, and about 25% chert and quartz (Table 1; Figs. 4, 5).

The ratio of chert to total quartzose debris (C/Q) averages about 0.67 and thus two-thirds of the quartzose debris typically is chert and only 8% of average grain populations is monocrystalline quartz. The ratio of plagioclase to total feldspar (P/F) is quite high, averaging 0.90, although potassium feldspar forms as much as 5% of the grains in some samples. Plutonic and sedimentary-rock fragments form a small but varied part of all samples, and the high average ratio of volcanic to total lithic debris (V/L) indicates that two-thirds of most lithic fragments is volcanic in origin.

Volcanic-rock fragments are mostly microlitic and porphyritic andesite although about 25% of the volcanic-rock fragments are lathy basaltic varieties, and 25% have textures typical of dacitic and rhyolitic rocks. Volcanic-ash grains are mostly elongate pumice shards and subordinate colorless bubble-wall shards.

Heavy minerals from the turbidite sand layers are dominantly clinopyroxene (Table 1, Fig. 6). Orthopyroxenes, epidotes, and amphiboles combined, on the average, comprise about 40% of the heavy crystals. Amphiboles are almost exclusively common hornblende, with traces of basaltic hornblende and tremolite-actinolite. Olivine ranges from 0 to 3% of the heavy fraction. Other miner-

Table 1. Modal Analyses of Sand Layers, DSDP Site 191, Komandorskiy Basin, Western Bering Sea

Core, section, depth in section (cm)	Quartz	Chert	Potassium Feldspar	Plagioclase	Plutonic Fr.	Sedimentary Fr.	Volcanic Rock Fr.	Glass shards	Quartz	Feldspar	Rock Fragments	C/Q	P/F	V/L	Amphiboles	Epidotes	Clinopyroxene	Orthopyroxene
2-2(110-140)	13	21	5	28	1	8	23	1	34	33	33	.63	.85	.72	6	12	72	10
2-2(126-128)	11	30	5	21	5	5	18	5	41	26	33	.74	.82	.71	15	18	56	11
2-3(0-70)	13	22	4	25	3	5	26	2	35	29	36	.63	.87	.78	8	10	74	8
2-3(18-20)	11	20	2	18	2	10	22	15	31	20	49	.63	.90	.75	12	27	48	13
2-3(74-76)	10	21	2	18	2	12	21	14	31	20	49	.69	.90	.70	5	11	78	6
2-3(134-136)	13	23	4	19	3	12	14	12	37	24	39	.64	.85	.63	9	8	73	10
4-2(76-78)	5	15	1	16	2	6	12	43	20	17	63	.75	.94	.88	13	8	62	17
4-4(58-60)	2	12	1	16	1	18	20	30	15	17	68	.84	.91	.70	10	18	59	13
4-5(40-42)	5	8	1	15	3	6	14	48	13	16	71	.64	.94	.88	12	16	60	12
4-5(58-60)	13	28	2	20	4	10	21	2	41	22	37	.32	.92	.61	5	11	80	4
4-6(46-48)	5	10	1	26	1	13	8	36	15	27	58	.67	.98	.76	22	15	49	14
4CC	5	14	2	21	4	5	38	11	20	23	57	.74	.94	.86	13	5	62	20
5-6(43-45)	9	17	2	25	2	10	31	4	25	27	48	.65	.93	.66	17	14	52	17
5-6(54-56)	14	17	5	27	3	6	15	13	31	31	38	.55	.83	.74	21	19	53	7
6-1(116-118)	5	13	2	38	4	3	25	10	18	40	42	.72	.95	.84	34	5	43	18
6-1(130-132)	6	11	2	37	1	1	27	15	17	40	43	.65	.96	.96	47	13	30	10
6-1(138-140)	3	13	4	34	3	0	33	10	16	39	45	.82	.90	.94	36	8	38	18
6-2(16-18)	2	4	1	46	3	5	29	10	6	47	47	.67	.99	.85	40	8	39	13
6-2(18-20)	3	9	4	44	4	3	26	7	13	47	40	.73	.92	.82	32	4	48	16
6-2(31-33)	6	10	1	46	2	7	24	4	16	47	37	.62	.98	.37	36	2	44	18
8-2(81-83)	13	26	4	24	3	8	17	5	40	28	32	.67	.86	.68	6	11	77	6
8-2(145-147)	14	24	5	16	6	10	20	5	38	20	42	.63	.78	.60	6	30	57	7
$\overline{x}$	8.2	12	2.7	26	2.8	7.4	22	14	25	29	46	.67	.90	.74	18	12	57	12
s	4.3	7.2	1.6	10	1.3	4.2	7.2	13	11	10	11	.10	.06	.13	13	7	14	5

The "Quartz", "Feldspar", "Rock Fragments" columns fall under the spanning header **End Member Percent**.

Note: C=chert, Q=total quartzose debris, P=plagioclase, F=total feldspar, V=volcanic lithic fragments, L=total lithic fragments. Chert is summed with quartz for total quartzose debris. Heavy mineral frequencies determined separate from light minerals, and calculated amphiboles + epidotes + pyroxenes = 100.

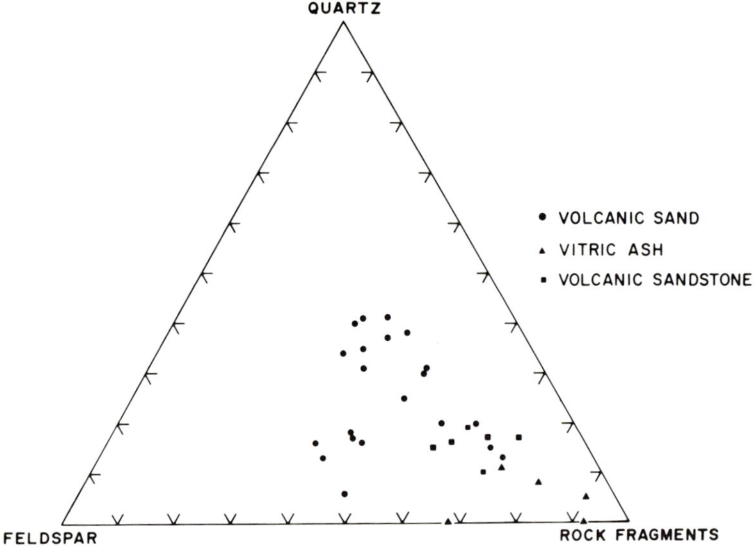

FIG. 5—Triangular diagram showing frequencies of major grain types in sand-sized light-mineral fractions of volcanic sand, vitric ash, and thin sections of volcanic sandstone from DSDP Site 191.

als present in trace amounts include apatite, garnet, kyanite, pumpellyite, sillimanite, sphene, and zircon.

Because most sand-sized grains are angular to subrounded, the sand and silt layers in the upper 300 m are considered texturally immature. However, the available grain-size data (Fig. 7) indicate that clay content is mostly less than 10%, and thus these layers would be "arenites" rather than "wackes" in the most commonly used sandstone classifications (Williams et al, 1954; Pettijohn, 1957; Dott, 1964). The layers are quite varied in composition, ranging from beds rich in volcanic lithic debris and volcanic-ash grains, to those rich in chert. Because there are many rock fragments, most samples probably would be classed as lithic arenite (Williams et al, 1954), feldspathic litharenite (Folk, 1968), or subgraywacke (Pettijohn, 1957). However, if some fine silt is totaled with clay as "matrix," then most of these layers would be feldspatholithic graywackes (Crook, 1960).

Below 300 m most of the recovered sediment is semi-indurated or indurated silty clay. Turbidity-current deposits are present only as indurated sandstone layers and fragments of layers recovered in the interval between 520 and 732 m. One of the sand layers is size graded and others display sharp top and bottom contacts and rip-up(?) shale clasts in the basal part (Creager et al, 1973).

Thin sections from these sandstone layers were stained for potassium feldspar and plagioclase using the techniques of Laniz et al (1964). Modal analyses of the sandstones (Table 2, Fig. 5) suggest these rocks are significantly different from the unlithified sand in the upper 300 m. The main differences include a three-fold average increase in sedimentary-rock fragments, a two-fold increase in volcanic-rock fragments, and an almost total lack of recognizable glass shards in the sedimentary rocks (Tables 1, 2; Fig. 5).

Chert content is considerably lower in the graywacke layers, averaging only about 9%. The ratio of chert to total quartzose debris (C/Q) is roughly similar to the younger unlithified sands, indicating a synchronous increase in both quartz and chert in the younger sands. Feldspars are dominantly plagioclase, and the ratio of plagioclase to total feldspars (P/F) suggests only 5% of all feldspars are potassium varieties. The ratio of volcanic to total lithic debris (V/L) is significantly different from the younger sand layers, apparently reflecting the higher content of sedimentary fragments and lower content of glass shards in the graywackes.

Provenance

In describing bottom sediments of the Bering Sea, Lisitsyn (1959, 1966) delineated several major areas of varied sand compositions (Fig. 8). On

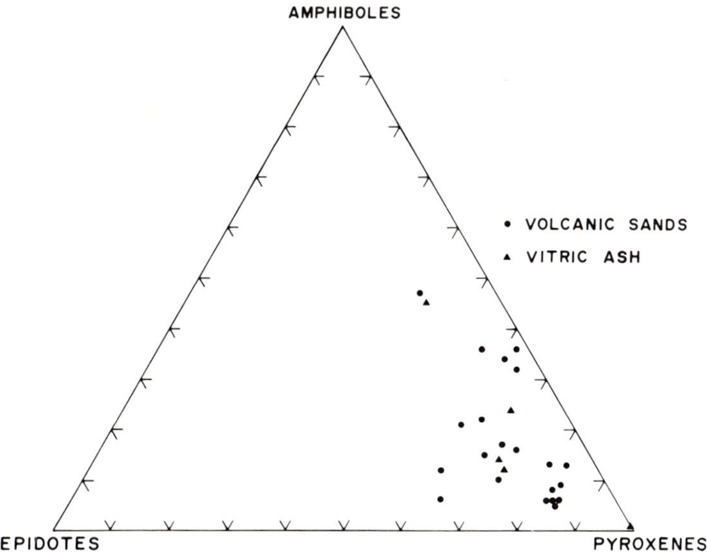

FIG. 6—Triangular diagram showing frequencies of major grain categories in sand-sized, heavy-mineral fractions of volcanic sand and volcanic ash from DSDP Site 191.

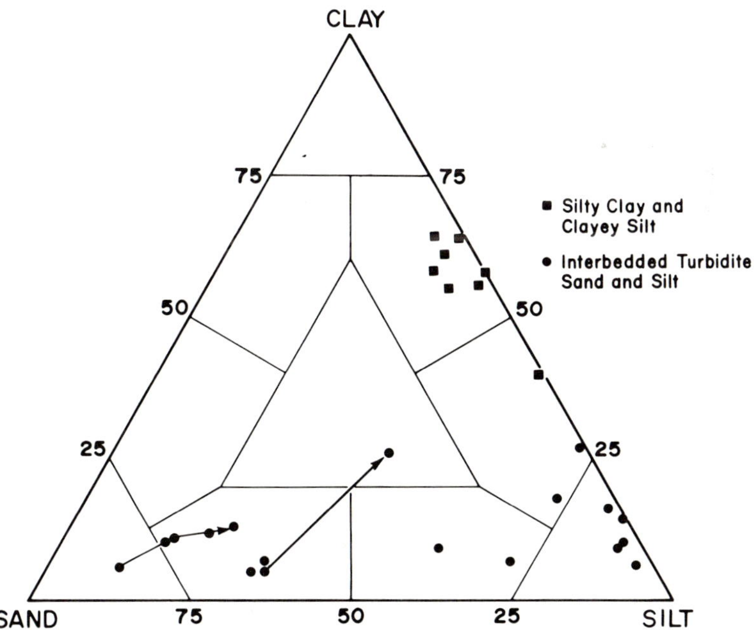

FIG. 7—Grain-size analyses of hemipelagic silt and clay and interbedded turbidite silt and sand at DSDP Site 191. Arrows connect samples from individual graded beds, and demonstrate upward decrease in grain size in turbidites. Data from Bode (1973).

Core, section, depth in section (cm)	Quartz	Chert	Potassium Feldspar	Plagioclase	Plutonic Fr.	Sedimentary Fr.	Volcanic Fr.	End Member Percent Quartz	Feldspar	Rock Fragments	C/Q	P/F	V/L	Matrix percent
12-1(14-18)	8	9	1	26	3	20	36	15	27	59	.59	.98	.61	25
12-2(122-128)	8	9	tr	16	4	24	37	17	16	66	.52	1.00	.57	25
12-2(141-142)	8	9	2	9	4	26	42	17	11	72	.56	.83	.58	28
13-1(4-6)	5	5	1	20	5	13	51	10	21	69	.53	.96	.74	23
13-1(14-18)	7	10	1	21	1	21	38	16	23	61	.59	.93	.62	21
14-2(78-80)	5	14	1	19	3	19	41	19	19	62	.73	.96	.66	21
$\bar{x}$	6.8	9.3	1.0	19	3.3	21	41	16	20	65	.59	.94	.63	24
s	1.5	2.9	0.6	5.7	1.4	4.5	5.5	3.1	5.6	5.0	.08	.06	.06	3

Note: Conventions as on Table 1.

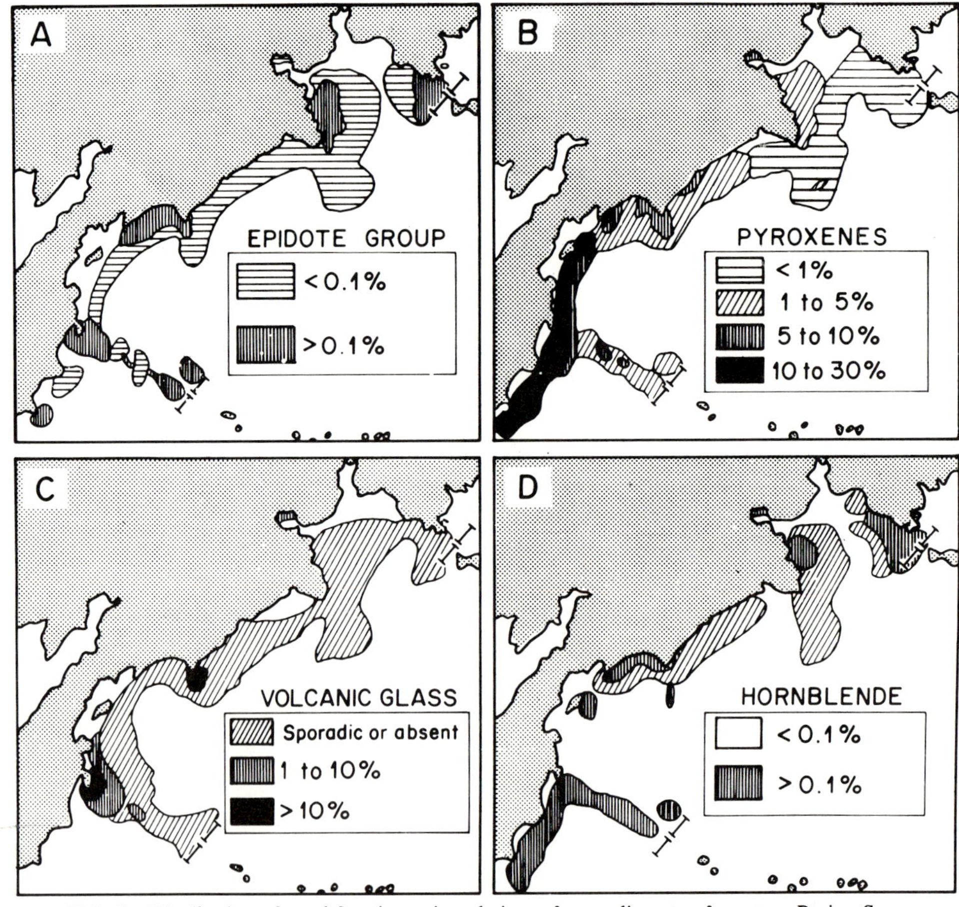

FIG. 8—Distribution of sand-fraction minerals in surface sediments of western Bering Sea. Modified from Lisitsyn (1966).

the basis of his data the likely source for sand rich in chert and quartz is the continental shelf of the westernmost Bering Sea, particularly the shelf between Mys Olyutorskiy (Cape Olyutorsky, Fig. 1) and Karaginskiy Ostrov (Fig. 1). Here the adjacent Koryakskiy Khrebet contain extensive pre-Tertiary eugeosynclinal rocks, including minor intrusive granitic rocks and cherts (Nalivkin, 1960; Lisitsyn, 1966; Bogdanov, 1967). The few crystals of glaucophane in core 8 possibly could have come from the metamorphic rocks recently discovered in the Koryakskiy Khrebet (see Pinus et al, 1970, for a summary).

For those beds relatively high in volcanic-rock-fragment and volcanic-glass content and relatively low in chert the most likely source is the southwestern Bering Sea, probably the continental shelf from Karaginskiy Ostrov south to Mys Kamchatskiy (Cape Kamchatka, Fig. 1). Here the coastline consists mostly of "effusive" volcanic rocks (Lisitsyn, 1966; Gnibidenko et al, 1972) and basaltic and andesitic rock fragments are a major part of the sand fraction in shelf sediments. The Komandorskiye Ostrova provenance is a less likely source, as the recent sand from this region contains relatively abundant authigenic glauconite and biogenic calcite and does not contain appreciable quantities of volcanic glass (Lisitsyn, 1966). Also, the southward regional slope of the basin floor does not suggest derivation from the south, at least in the uppermost part of the core.

If sand compositions are plotted against depth, significant stratigraphic variations in sand composition are apparent (Fig. 9). The change in sand composition between the Pleistocene and late Miocene or early Pliocene time is very significant but difficult to explain. The change possibly is due to diagenetic alteration of volcanic glass in the sandstones to "matrix" and to turbid areas perhaps confused with sedimentary or volcanic-rock fragments. However, few relict shard textures in the sandstones were recognized and the differences between unlithified and lithified sediment more likely are due to change in the nature or location of the source terrane.

The lack of volcanic-ash debris in the sandstones, and the abundance of glass shards in the Pleistocene sediment may reflect increased infalls of tephra in late Pliocene and Pleistocene time. The dramatic increase in explosive magmatic activity in the North Pacific beginning approximately 3 m.y. ago was noted by Scholl and Creager (1973). This event must have provided a significant input of glass shards both to waters overlying the adjacent continental shelves, and to subaerial streams draining lands on the basin rim. Alternately, the differences may be due to erosional destruction of the extensive Pliocene-Pleis-

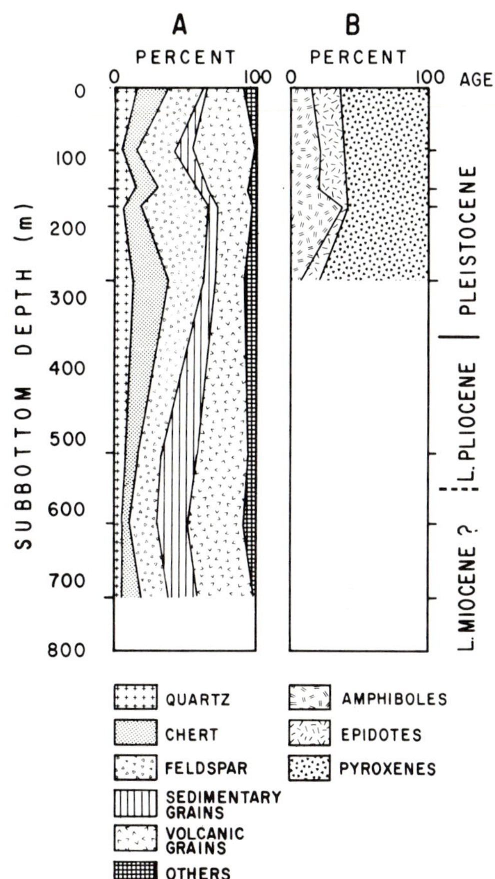

FIG. 9—Stratigraphic variation in abundance of major grain species in sand layers from DSDP Site 191, Komandorskiy basin. **A**, light-mineral fractions; **B**, heavy-mineral fractions.

tocene volcanic deposits present on the Kamchatskiy Poluosrov (Nalivkin, 1960; Avdeiko, 1971; Gnibidenko et al, 1972).

However, the sizable differences in content of sedimentary and volcanic-rock fragments suggest a different area may have been an active sediment source prior to Pleistocene time. It is intriguing to speculate that perhaps the Aleutian arc was an important source of sand during the late Miocene and Pliocene. Scholl et al (1976) documented evidence suggesting that a middle Miocene episode of volcanism may have extended the full length of the Aleutian Ridge, including the Komandorskiye Ostrova. Exposures of Neogene sedimentary and volcanic rocks in the Komandorskiye Ostrova (Zhegalov, 1964; Shmidt, 1973) are not well dated, but Scholl et al (1976) argued persuasively for significant exposures of middle

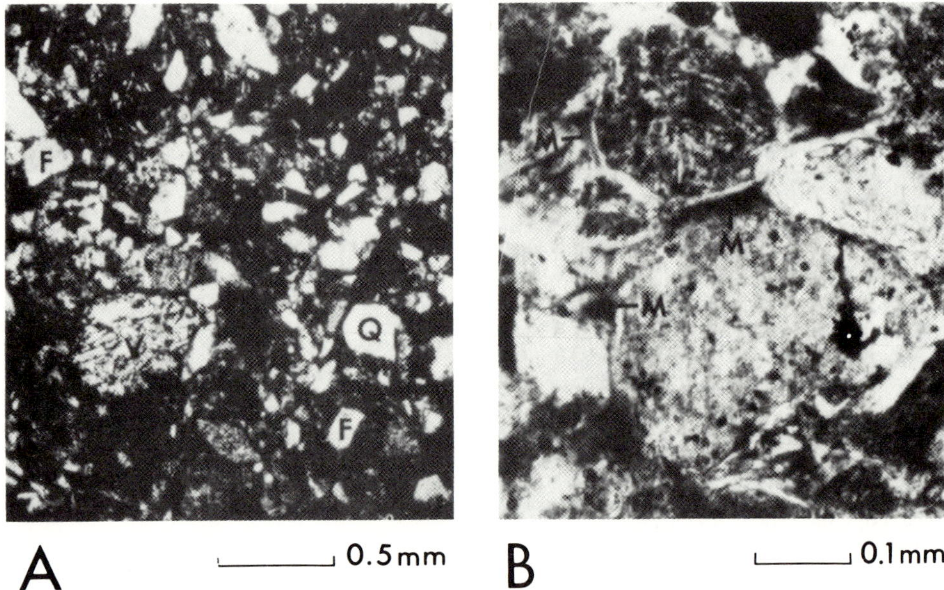

A   ⊢————⊣ 0.5mm   B   ⊢————⊣ 0.1mm

FIG. 10—Representative textures of lithified volcanic sandstone or graywacke layer in strongly lithified mudstone sequence, DSDP Site 191; Q, quartz; F, feldspar; V, volcanic lithic grain; M, clay matrix; **A**, from core 13-1, 4 to 6 cm; **B**, from core 12-1, 14 to 18 cm.

Miocene sedimentary and volcanic rocks. If a significant volcanic edifice was constructed in the westernmost Aleutians in middle Miocene time, then erosional destruction of this assemblage and associated diatomaceous sediments in the late Miocene and Pliocene could have supplied substantial quantities of lithic debris to the adjacent Komandorskiy basin. Crucial to this question are data on possible changes in the gradient of the floor of Komandorskiy basin, data which presently are not available.

DIAGENETIC ALTERATION OF SEDIMENTS

Sediments in Komandorskiy basin are altered diagenetically and below 520 m the silty clay is mudstone and the sand layers are graywacke (Creager et al, 1973). Hein et al (1976) delineated a sequence of diagenetic changes in fine-grained diatomaceous sediment at many DSDP sites in the North Pacific Ocean and Bering Sea. Hein et al (1976) noted that corrosion and dissolution of diatoms are major factors altering sediment originally rich in diatomaceous debris to mudstone with a diatom and clay matrix. Below about 550 m Hein et al (1976) indicated that pyrite, zeolites, clays, and other diagenetic minerals form in abundance.

In the coarser grained sand layers a similar spectrum of alteration products is present, including compaction and induration, sparse carbonate cementation, and significant development of authigenic clay minerals in the lower 400 m. With the exception of slight and possibly insignificant amounts of intrastratal solution, the sand layers in the upper 300 m essentially are unaltered. Pyroxene grains with coxcomb etched and serrate textures typical of grains altered by intrastratal solution are present in small amounts but most of the grains appear unaltered (Fig. 5).

However, sand layers below 520 m are altered and compacted strongly. Alteration of volcanic and other clastic debris to clay minerals and possibly other authigenic species produced firmly consolidated sandstone. Porosity in the sand-sized deposits is destroyed. Any remaining original pore spaces now are filled with clay minerals, often growing radially outward from grain boundaries (Fig. 10). Growth of clay minerals has obliterated many primary clastic textures, and original grain boundaries commonly are difficult to define. Clay minerals extensively replace clinopyroxene grains, leaving pseudomorphs of stubby pyroxenes floating in a clay-mineral matrix (Fig. 11). Volcanic glass gradually is devitrified to clay minerals. Very little glass is devitrified at 520 m, but roughly half of the glass is replaced at 620 m sub-bottom, and all of the glass is altered in samples buried beneath 724 m of sediment. At the deepest levels, phyllosilicates are oriented subparallel with original bedding surfaces, and the sand-

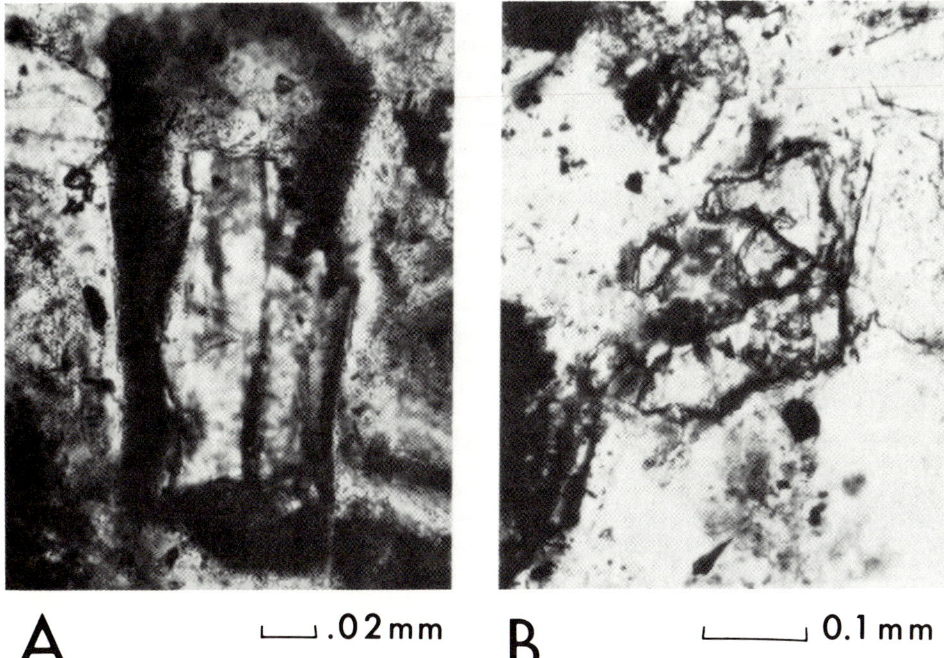

FIG. 11—Pyroxenes partly replaced by clay minerals. Clay minerals are color banded, somewhat turbid, in contrast to clear, high-relief pyroxene. Both **A** and **B** are from core 14-2, 110 to 112 cm.

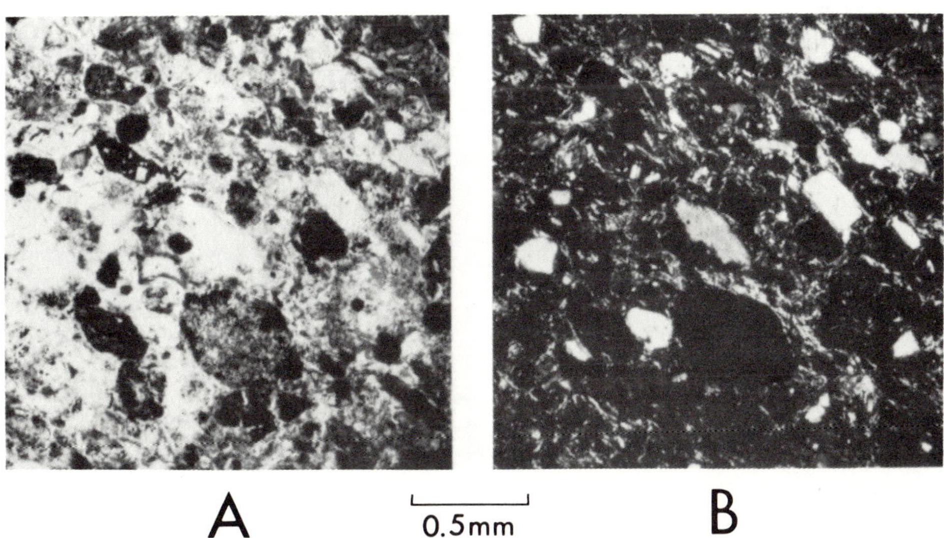

FIG. 12—Representative textures in sandstone, DSDP Site 191. Aggregates of clay platelets display moderate preferred orientation parallel with bedding (running diagonally from lower right to upper left), and are approaching semischist texture. **A**, plane light; **B**, crossed nicols. Sample is from core 14-2, 76 to 78 cm.

stone textures are transitional to semischist (Fig. 12).

X-ray-diffraction analyses indicate that the dominant diagenetic product in sand beds in the lower 400 m is smectite. Examination of the "matrix" component from sandstone samples shows a broad 14Å reflection which sharpens and expands to 17Å on glycolation and collapses completely on heating to 300°C. Zemmels and Cook (1973) also reported a sample of volcanic ash from core 14 with exceedingly strong montmorillonite reflections, and silty clay from core 13 which contains significant amounts of clinoptilolite.

RATE OF ACCUMULATION

The oldest well-dated sediment at Site 191 is late Pliocene in age at 527 m subbottom depth (Creager et al, 1973). Below that, only a sparse arenaceous fauna of possible late Miocene age is present together with a few poorly preserved diatoms.

Because of the uncertainty in the age of the lower mudstone sequence, accumulation rates prior to late Pliocene time are difficult to evaluate. Lee (1973) demonstrated that although diatom ooze is essentially incompressible, silty clay compacts normally. In the sequence below 732 m, the uncorrected rate is about 30 m/m.y.$^{-1}$. Assuming an estimated 50% compaction for the mudstone sequence, the true rate could approach 60 to 70 m/m.y.$^{-1}$.

Accumulation rates for the semilithified silty clays between 200 m and 500 m subbottom were considerably higher than the mudstone-graywacke sequence, and averaged about 170 m/m.y.$^{-1}$ (uncorrected). When corrected for compaction effects this estimate increases significantly, and could exceed 300 m/m.y.$^{-1}$. Accumulation rates apparently were about the same or slightly lower in the diatom-rich section above 200 m, averaging 250 m/m.y.$^{-1}$ (uncorrected). If the diatom ooze is essentially incompressible as suggested by Lee (1973), then this rate is a close approximation to the true value.

These high rates of accumulation were controlled by two major factors, turbidite deposition and diatom production. Turbidites started flooding the basin in great volumes, apparently starting in late Pliocene time, and the marked variation in deposition rates may reflect the increased influx of turbidite sands in the Pleistocene section. Presumably this increased influx resulted from late Cenozoic sea-level transits and increased erosion across the adjacent continental shelves, as suggested by Scholl and Marlow (1974) for Pacific island-arc settings.

The cause for the enormous input of diatoms into the basin in late Pliocene and Pleistocene time is more difficult to specify. Scholl and Creager (1973) suggested that a cooling trend at the end of the middle Miocene may have intensified oceanic circulation and increased diatom productivity, presumably because of increased upwelling of nutrient-rich waters. The slight decrease in accumulation rates in the late Pleistocene could reflect a decline in diatom productivity (Scholl and Creager, 1973).

DISCUSSION AND CONCLUSIONS

Neogene sedimentation in Komandorskiy basin was dominated by particle-by-particle settling of hemipelagic detritus and pelagic diatomaceous debris. Accumulation rates were quite high, probably exceeding 300 m/m.y.$^{-1}$ in late Pliocene and early Pleistocene time, because of extensive shelf erosion during late Cenozoic sea-level transits and because of oceanographic conditions favoring extremely high diatom production. Periodic incursions of sand- and silt-sized debris in turbidite currents deposited thin beds of volcanic and feldspatholithic compositions.

Although the sediment in the upper 300 m virtually is unaltered, all of the sediment below about 500 m is compacted and altered diagenetically (Creager et al, 1973; Hein et al, 1976). The main new mineral phase is smectite which fills voids in sand and silt layers, replaces glass in the vitric-ash layers and in volcanic lithic debris, and is forming in the compacted mudstone (see also Hein et al, 1976). Clinoptilolite also is forming diagenetically in the mudstone (Zemmels and Cook, 1973; Hein et al, 1976), together with pyrite, cristobalite, and other zeolites (Hein et al, 1976). At about 720 m the sediments are altered so extensively that volcanic glass is devitrified totally, and pyroxenes and other reactive minerals are replaced extensively by smectite. Sandstone layers have distinct new mineral fabrics defined by phyllosilicates growing subparallel with bedding, and the rocks approach the textures of semischist. The exact cause of this extensive alteration at relatively shallow burial depths is difficult to specify, but may be related to a combination of diatom dissolution and corrosion (Hein et al, 1976), coupled with high heat flow at Site 191 (Foster, 1962; Fornari et al, 1973).

Sand beds in Komandorskiy basin contain significant quantities of volcanic lithic and vitric-volcanic debris and their proximal counterparts in basins on the adjacent continental margin probably would not be considered possible reservoir rocks. However, many of the beds contain an abundance of chert grains and plagioclase feldspar crystals, and conceivably could be adequate reservoir rocks if present in sufficient thickness and relatively unaltered. Sediment of this composition now is accumulating on the continental shelf between Mys Olyutorskiy and Karaginskiy Ostrov (Lisitsyn, 1966). If the abyssal-plain deposits of Komandorskiy basin are distal equivalents of sediments on the adjacent continental margin, then the poorly known and little explored proximal deposits in this region should be investigated more intensively for hydrocarbon resources.

REFERENCES CITED

Alpha, T. R., 1974, Orthographic drawing, Bering Sea: U.S. Geol. Survey Open File Rept.

Avdeiko, G. P., 1971, Evolution of geosynclines on Kamchatka: Pacific Geology, v. 3, p. 1-13.

Bode, G. W., 1973, Grain size, *in* Initial reports of the Deep Sea Drilling Project, v. 19: Washington, D.C., U.S. Govt. Printing Office, p. 661-662.

Bogdanov, N. A., 1970, Some tectonic features of the eastern part of the Koryak Mountains: Akad. Nauk SSSR Doklady, v. 192, p. 607-610.

Chase, T. E., H. W. Menard, and J. Mammerickx, 1971, Topography of the North Pacific: California Univ. Inst. Marine Research Tech. Rept. TR-17, 1 sheet.

Creager, J. S., et al, 1973, Initial reports of the Deep Sea Drilling Project, v. 19: Washington, D.C., U.S. Govt. Printing Office, 913 p.

Crook, K. A., 1960, Classification of arenites: Am. Jour. Sci., v. 258, p. 419-428.

Dickinson, W. R., 1970, Interpreting detrital modes of graywacke and arkose: Jour. Sed. Petrology, v. 40, p. 695-707.

Dott, R. H., Jr., 1964, Wacke, graywacke and matrix—what approach to immature sandstone classification?: Jour. Sed. Petrology, v. 34, p. 625-632.

Eremenko, N. A., et al, 1973, Geologic structure and oil and gas prospects of USSR continental shelf: AAPG Bull., v. 57, p. 235-243.

Folk, R. L., 1968, Petrology of sedimentary rocks: Austin, Texas, Hemphill's, 170 p.

Fornari, D. J., R. J. Iuliucci, and G. G. Shor, Jr., 1973, Preliminary site surveys in the Bering Sea for the Deep Sea Drilling Project, Leg 19, *in* Initial reports of the Deep Sea Drilling Project, v. 19: Washington, D. C., U.S. Govt. Printing Office, p. 569-614.

Foster, T. D., 1962, Heat-flow measurements in the northeast Pacific and in the Bering Sea: Jour. Geophys. Research, v. 67, p. 2991-2993.

Gnibidenko, G. S., and A. S. Svarichevskiy, 1974, Structure and hydrocarbon potential of the Bering Sea: Sov. Geol., no. 1, p. 89-96; English translation (1975): Internat. Geol. Rev., v. 17, p. 431-438.

———— et al, 1972, Geology and deep structure of Kamchatka Peninsula: Pacific Geology, v. 7, p. 1-32.

Gorshkov, G. S., 1970, Volcanism and the upper mantle: investigations in the Kuril Island arc: New York, Plenum, 385 p.

Hein, J. R., D. W. Scholl, and C. E. Gutmacher, 1976, Diagenesis of Neogene diatomaceous sediment from the far northwest Pacific and southern Bering Sea (abs.): Geol. Soc. America Abs. with Programs, v. 8, p. 379-380.

Laniz, R. V., R. E. Stevens, and M. B. Norman, 1964, Staining of plagioclase feldspar and other minerals with F.D. and C. red no. 2: U.S. Geol. Survey Prof. Paper 501-B, p. B152-B153.

Lee, H. J., 1973, Measurements and estimates of engineering and other physical properties, Leg 19, *in* Initial reports of the Deep Sea Drilling Project, v. 19: Washington, D.C., U.S. Govt. Printing Office, p. 701-719.

Lisitsyn, A. P., 1959, Bottom sediments of the Bering Sea, *in* P. L. Bezrukov, ed., Geographical description of the Bering Sea: Akad. Nauk SSSR Inst. Okean. Trudy, v. 29; English translation (1964): Washington, D.C., U.S. Dept. Commerce, Israel Prog. Sci. Translation, IPST 1151, p. 65-188.

———— 1966, Recent sedimentation in the Bering Sea: Moscow, Izd. Nauka; English translation (1969): Washington, D.C., U.S. Dept. Commerce, Israel Prog. Sci. Translation, IPST 5057, 614 p.

Nalivkin, D. V., 1960, Geology of the U.S.S.R.: New York, Pergamon, 170 p.

Pettijohn, F. J., 1957, Sedimentary rocks: New York, Harper and Bros., 718 p.

Pinus, G. V., L. V. Agafonou, and V. V. Velinsky, 1970, Eclogite-like rocks of the Anadyr-Koryak folded system: Pacific Geology, v. 2, p. 81-91.

Scholl, D. W., and J. S. Creager, 1973, Geologic synthesis of Leg 19 (DSDP) results: far north Pacific, and Aleutian Ridge, and Bering Sea, *in* Initial reports of the Deep Sea Drilling Project, v. 19: Washington, D.C., U.S. Govt. Printing Office, p. 897-913.

———— and M. S. Marlow, 1974, Sedimentary sequence in modern Pacific trenches and the deformed Circum-Pacific eugeosyncline, *in* Modern and ancient geosynclinal sedimentation: SEPM Spec. Pub. 19, p. 193-211.

———— et al, 1974, Base map of the Aleutian–Bering Sea region: U.S. Geol. Survey Misc. Geol. Inv. Map I-879, scale 1:2,500,000.

———— et al, 1976, Episodic Aleutian Ridge igneous activity: implications of Miocene and younger submarine volcanism west of Buldir Island: Geol. Soc. America Bull., v. 87, p. 547-554.

Shmidt, O. A., 1973, New data on the tectonics of the Komandor Islands: Akad. Nauk SSSR Doklady, v. 210, p. 918-920; English translation (1973): Doklady Earth Sci. Sec., v. 210, p. 82-85.

Stewart, R. J., J. H. Natland, and W. R. Glassley, 1973, Petrology of volcanic rocks recovered on DSDP Leg 19 from the North Pacific Ocean and the Bering Sea, *in* Initial reports of the Deep Sea Drilling Project, v. 19: Washington, D.C., U.S. Govt. Printing Office, p. 615-627.

Udintsev, G. B., I. G. Boichenko, and V. R. Kanaeu, 1959, Bottom relief of the Bering Sea, *in* P. L. Bezrukov, ed., Geographical description of the Bering Sea: Akad. Nauk SSSR Inst. Okean. Trudy, v. 29; English translation (1964): Washington, D.C., U.S. Dept.

Commerce, Israel Prog. Sci. Translation, IPST 1151, p. 14-64.

Williams, H., F. J. Turner, and C. M. Gilbert, 1954, Petrography: San Francisco, W. H. Freeman, 406 p.

Zemmels, I., and H. E. Cook, 1973, X-ray mineralogy of sediments from the northern Pacific and the Bering Sea—Leg 19, *in* Initial reports of the Deep Sea Drilling Project, v. 19: Washington, D.C., U.S. Govt. Printing Office, p. 667-698.

Zhegalov, Yu. V., 1964, Komandorskiye Ostrova, *in* G. M. Valasov, ed., Geologiya SSR, v. 31, Kamchatka, Kurilskiye i Komandorskiye ostrova, pt. 1, Geologicheskoye opisaniye: Moscow, Izd. Nedra; English translation, Komandorskiye Islands, *in* Geology of the USSR, v. 31, Kamchatka, Kuril and Komandorskiye Islands, pt. 1, Geological description: Springfield, Virginia, U.S. Dept. Commerce National Tech. Inf. Service Doc. N68-15812, p. 695-729.

TARZANA FAN, DEEP SUBMARINE FAN OF LATE MIOCENE AGE LOS ANGELES COUNTY, CALIFORNIA[1]

HAROLD H. SULLWOLD, JR.[2]
North Hollywood, California

ABSTRACT

The sandstone beds in the Modelo formation (upper Miocene) exposed on the north flank of the Santa Monica Mountains were for the most part deposited from turbidity currents. Evidence of rapid deposition is provided by poor sorting, high clay-silt content, grain angularity, high feldspar content, and load deformation. Evidence for great depth of water (about 3,000 feet) is provided by abundant foraminifers and fish remains in the interbedded shales. That the mechanism for transporting this poorly sorted sand-silt-clay mixture into waters of this depth was that of turbidity currents is borne out by the multitude of syngenetic textures and structures which these rocks have in common with accepted turbidites and with structures which have been produced in laboratory studies.

Some of these structures are oriented with respect to the direction of travel of the current and therefore provide a means of determining the direction of bottom slope along which the currents moved. When plotted on a map these directions reveal a pattern which is unmistakably that of a fan with its apex toward the north. The area studied constitutes a 16-mile-long section through this outcropping north-tilted fan. The point source was most logically the mouth of a submarine canyon. The canyon itself is thus far unrevealed, being concealed beneath the alluviated San Fernando Valley, but the eroding source area is tentatively identified on the basis of mineralogical studies in the San Gabriel Mountains about 22 miles northeast of the fan's apex.

Further studies of this sort can not fail to provide valuable data on earth history and thus aid petroleum exploration in similar rock types elsewhere in California.

INTRODUCTION

Turbidity currents have been recognized for many years under such names as density currents and suspension currents, but it was not until 1950, when Kuenen and Migliorini demonstrated that graded bedding is a natural consequence of turbidity currents, that their importance as a mechanism for sediment transport in the sea came to be recognized. Since then many studies have been made of marine sedimentary rocks, modern sediments on the sea floor, and deposits in experimental tanks which marshal evidence that, far from being freaks, these currents are common now and were common in the past. Indeed, they constitute a very important sedimentation agent. Under certain conditions transport of material by these currents is probably the dominant process in operation. This paper deals with such a situation.

The deposit produced by a turbidity current, named turbidite by Kuenen (1957), may be less than one inch to several tens of feet thick, and may range in area from less than an acre to hundreds, perhaps thousands, of square miles. It may consist of clay, silt, sand, or gravel in various proportions depending upon the nature of the original slumped material and on the effectiveness of sorting during transport.

Because these sediments are deposited from an active bottom current rather than as a gentle rain from above they have certain textural and structural characteristics which can be recognized in the field. Many of these characteristics are difficult or impossible to explain by any other mechanism of deposition. Numerous recent papers on the subject have dealt with the recognition of these features in formations of many ages throughout the world (see starred references at end of this paper.) The deposits generally are poorly sorted, graded, and become thinner and finer-grained away from the source. Fossils in growth position are lacking. Benthonic forms are swept away by the current, and pelagic forms do not fall fast enough to be incorporated in this

[1] Manuscript received, March 19, 1959.

[2] Consulting geologist.
This paper is a condensation of a dissertation submitted in partial satisfaction of the requirements for the Ph.D. degree at the University of California, Los Angeles. The writer is indebted to Professors J. C. Crowell, Cordell Durrell, and E. L. Winterer for stimulation, assistance, and criticism; to John de Grosse for preparation of petrographic thin sections; to W. T. Rothwell, R. L. Pierce, and R. L. Brooks of the Richfield Oil Corporation for micropaleontological control; and to the many graduate students of the University of California, Los Angeles, whose theses provided the geologic maps on which this study is partly based. These former students include F. W. Bergen, T. J. Brady, George Brown, W. W. Duarte, J. G. Elam, J. L. Elliott, T. P. Harding, G. C. Hazenbush, J. T. McGill, J. N. Terpenning, J. D. Traxler, and J. C. West. The critical reading of this manuscript by Professors Crowell and Winterer is greatly appreciated.

229

relatively instantaneous deposit. Any fossils in the deposit will be transported and may show evidence of this by separation of valves, abrasion, presence in cross-bedding, and exotic position of older or shallower types in the turbidity current deposit sandwiched between normal pelagic beds with normal faunas. These transported faunas may play havoc with detailed paleontologic correlation at the stage level or with lesser units. Internal textures include graded bedding, cross-bedding, convolute bedding, lineation of long grains or charcoal fragments, and shale flakes and slabs. External structures include ripple marks at the top of beds and load deformation, groove casts, and flow markings at the bottom.

Many of the textures and structures are oriented with respect to the direction of the current at the time of deposition. Whereas many workers have recently recognized the presence of these structures and thus established the mechanism of deposition for certain formations, only a few have used the oriented structures to show the direction of currents and thus to demonstrate direction toward shore and source. Published directional studies show remarkable constancy of direction of currents and encourage much further effort along these lines throughout the world to work out more detailed paleogeography (Crowell, 1955; Kopstein, 1954; Kuenen and Sanders, 1956; Schneeberger, 1955; Van Houten, 1954; Wilson, Watson, and Sutton, 1953; and Winterer, 1954).

LOCATION

The area investigated is the outcrop of strata of Mohnian age (roughly equivalent to the lower member of the Modelo formation) along the north flank of the Santa Monica Mountains, Los Angeles County, California (Fig. 1). The area is about 20 miles long and 2 miles wide. The lowlands on the north constitute the San Fernando Valley. The mountain range itself resembles an east-west wedge emerging from the sea along the rugged coast between Santa Monica and Oxnard and pointing eastward toward the heart of Los

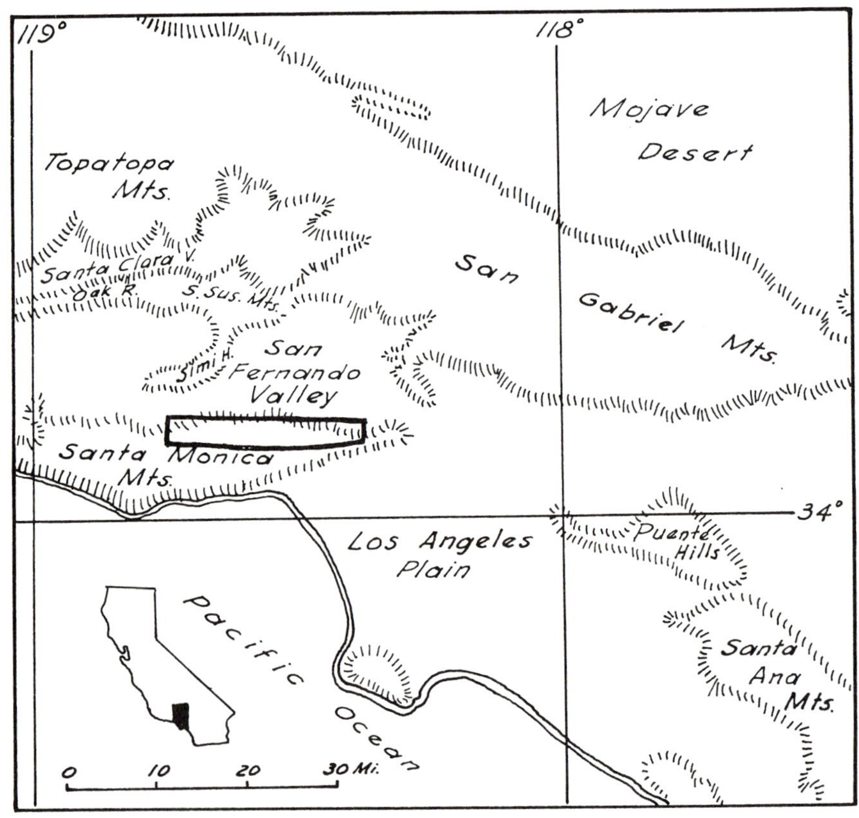

Fig. 1.—Index map.

Angeles. The eastern portion separates the residential districts of Malibu Beach, Santa Monica, Bel Air, Westwood, Beverly Hills, and Hollywood on the south from the communities of Calabasas, Woodland Hills, Tarzana, Encino, Sherman Oaks, and Studio City on the north. The accelerated rate of residential development in the mountains in recent years has resulted in many fresh, large, and easily accessible exposures.

The choice of an area and formation to study was merely a matter of convenience. It was not chosen for its outstanding display of oriented sedimentary structures. Indeed, the pre-eminent work on the geology of the Santa Monica Mountains (Hoots, 1931) devotes 15 two-column pages to a description of the Modelo formation and does not describe the graded bedding, current bedding, load deformation, convolute bedding, and other features which are ubiquitous, but not highly conspicuous, sedimentary features. In other words, the Modelo formation is an ordinary run-of-the-mill California marine Tertiary deposit of sandstone and shale.

METHODS

The areal geology of the region is well known. Hoots (1931) described the part east of Topanga Canyon, Soper (1938) that west of Topanga Canyon, and Durrell (1954) the entire range. Students at the University of California, Los Angeles, have described the geology of most of the area in unpublished Master of Arts theses. The writer thus confined his field activities to actual observations of syngenetic sedimentary structures, especially those related to current direction. About 110 days were spent in the field and perhaps half that number in the laboratory. Artificial exposures are plentiful, and the rocks are relatively soft. Indispensable items of equipment were a whisk broom and broad putty knife for facilitating study of details at the outcrop. By far the best time to view the small structures is after a rain when color contrasts are greatest.

GENERAL GEOLOGY

The geology of the Santa Monica Mountains is complex, both stratigraphically and structurally. Nearly 50,000 feet of Cretaceous and Tertiary sedimentary rocks overlie a basement of (Triassic?) metamorphosed shale and graywacke which was intruded by large bodies of quartz diorite of presumed Jurassic age. This maximum thickness is not represented in any one place, as there have been several periods of deformation, uplift, and erosion, resulting in unconformities with varying amounts of angularity, some as great as 90°. Rocks involved include the basement complex and sedimentary rocks of Cretaceous, Paleocene, Eocene, Oligocene, Miocene, Pliocene, and Pleistocene age.

The dominating structural feature is an east-west-trending anticline, whose axis lies south of the area investigated. This anticline has evidently been growing throughout most of the time represented by the rocks involved, for the "bald-headed" nature of the axial region at various times is represented by overlap of younger rocks over older rocks. The lower Modelo formation described herein lies at various places in outcrop on the Topanga formation (middle Miocene), Chico and Trabuco formations (Cretaceous), plutonic rocks (Jurassic?), and the metamorphosed Santa Monica formation (Triassic?). Faulting has also been fairly continuous; many faults terminate upward against unconformities, and others have had rejuvenated movements.

The Modelo formation is the youngest major stratigraphic unit represented in the range and is therefore the least deformed. There is evidence for only one period of post-Modelo folding, though physiographic evidence indicates several periods of recent uplift. Dips range from horizontal to vertical, but in most places average 20°. The Modelo strikes mostly within 20° of east-west except in the vicinity of the north-plunging Topanga anticline and its adjacent synclines. Many minor faults are present in the Modelo formation, but the only one in the area which seriously interrupts the continuity of the beds is the Hayvenhurst fault which has a strike separation of nearly 2 miles.

MODELO FORMATION

The Modelo formation reaches a thickness of more than 5,000 feet including two members, lower and upper, each constituting roughly half of the total thickness. Its age is late Miocene, including both Mohnian and Delmontian stages. The Mohnian-Delmontian contact is within the upper member. The study was confined to the Mohnian stage (lower upper Miocene) and thus includes the lower member of the Modelo formation and a few observations in the upper member. The type section of the Mohnian stage is along

Topanga Canyon Boulevard in the western part of the area (Kleinpell, 1938, p. 127).

The formation consists mostly of shale but contains many large lenticular bodies of sandstone. The upper member is uniformly light-weight, white, punky, diatomaceous shale, and the lower member contains various sorts of shale described as cherty, porcelaneous, platy, hard, soft, clayey, and silty. The contact between the two members is rarely sharp and is actually erratic toward the west where it crosses other correlatable strata. The important identifying characteristic of both members is the high silica content which qualifies them as belonging to the Monterey formation which Bramlette has thoroughly discussed (1946). Both members contain laminae and lenticular bodies of sandstone, but these are more common in the lower member. The color of the lower member in outcrop is generally light yellowish tan or buff, but the individual laminae may be white, black, brown, gray, tan, or buff. In a 400-foot tunnel south of Calabasas the color gives way, presumably below the zone of weathering, to light gray for sandstone, and dark gray to black for shale.

ENVIRONMENT OF DEPOSITION

Fossils indicate that the rocks are all marine. The basal sandstone was deposited in shallow water, but the remainder of the section was deposited in bathyal depths, probably on the order of 3,000 feet.

Shallow-water conditions at the beginning of Modelo deposition is evidenced by the abundance of rock-dwelling mollusks in the basal sandstone at several localities. These are discussed by Woodring (in Hoots, 1931, p, 110). This sandstone rarely exceeds 50 feet in thickness, and its local derivation is indicated by its habit of containing clasts of the rock on which it lies. Foraminifera and fish remains indicate that the rest of the Modelo formation was deposited in deep water. Evidently the area subsided at least 1,000 feet during the time required to deposit less than 100 feet of sediment. That this did not correspond with a general rise in sea-level is indicated by Kleinpell's statement (1938, p, 128) that Mohnian foraminifers of California are generally neritic.

Paleontologists studying foraminiferal and fish remains in two sections within the mapped area assign bathyal depths to the Mohnian waters (Bergen, 1955; Pierce, 1956). Mohnian depths of 1,300–2,500 feet have been determined in nearby parts of the same seaway by Durham (1954), Natland (1957), and Natland and Rothwell (1954). The depth chart (Fig. 2) shows the present depth ranges offshore from southern California of the surviving benthonic species from Pierce's Mohnian check list. A depth of 2,500–2,900 feet is common to all but one species. Data by D. I. Axelrod (personal communication), Emiliani (1954), and MacNeil (1957) suggest that the Pacific Ocean was warmer in Miocene time than now. It seems reasonable, therefore, to assume a water depth of 3,000 feet for the Tarzana Fan.

Significant down-slope movement of foraminiferal tests has been reported by Crouch (1952), Phleger (1951), Resig (1956), and Uchio (1957), in Recent sediments. Transportation has also been effective in the Tarzana Fan, though an exhaustive study to test its extent was not made. One petrographic thin section is perhaps worth describing in this regard. It contained three fine-grained beds. The lowest was clay shale with scattered silt grains and small foraminiferal tests. The middle bed, 0.2 inch thick, graded from siltstone at the bottom to claystone at the top, and contained Foraminifera in a graded arrangement —they constituted 40 per cent of the rock in the bottom part and diminished to rarity at the top. These tests were notably larger than those in the underlying clay. It seems clear that the tests in this graded siltstone were transported and deposited by the same process that graded the mineral grains. The upper bed is even more striking; it is a siltstone slightly coarser than the basal part of the middle bed, and its pronounced cross-bedded structure can be easily seen with the unaided eye. Foraminiferal tests are confined to the biotite layers in the fore-set laminae of the cross-bedding and have obviously been emplaced by the current which produced the cross-bedding. The fauna in this bed is mixed, containing abundant bathyal forms and rare neritic forms. It is clear that both the middle and top beds were deposited rapidly, each by a single current, and that the Foraminifera did not have time to live out their life spans at their present site of burial; they were clearly swept in by currents.

SANDSTONES

The sandstone in the Modelo formation should be classified as arkosic wacke in the terminology of Gilbert (in Williams, Turner, and Gilbert,

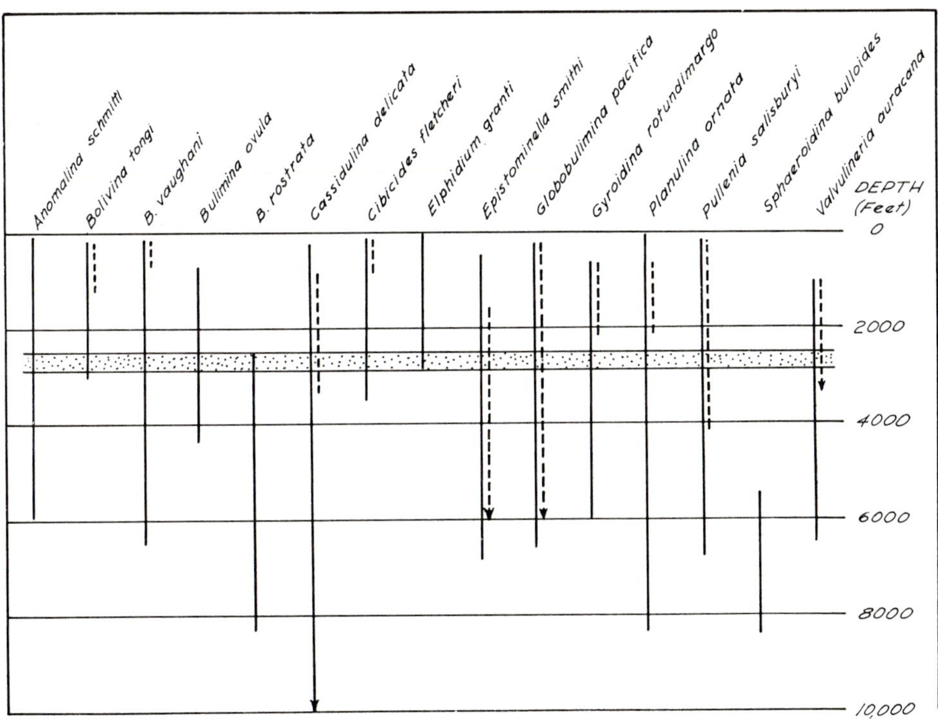

Fig. 2.—Water depths of benthonic Mohnian foraminifera of Pierce's Benedict Canyon section now found in present sediments off California coast. Depths compiled from data of Bandy, Butcher, Crouch, Natland, Phleger, Resig, and Uchio. Solid lines represent total depth ranges of tests reported by all workers; broken lines represent depth ranges of live foraminifera reported by Resig and Uchio only. Stippled band is depth range (2,500 to 2,900 feet) common to all but one species.

1954, p. 290) or as labile graywacke in Packham's classification (1954). It is characterized by high feldspar content, high silt and clay content, angularity of grains, absence of effective cement, poor sorting, graded bedding, abundance of syngenetic sedimentary structures, and virtual absence of megafossils. Individual sedimentation units range from a fraction of an inch to more than 10 feet in thickness and are remarkably consistent within the limits of a single outcrop; their distribution and abundance ranges from occasional thin beds in shale sequences through more abundant, commonly rhythmic, repetitions in shale to sequences of sandstone beds with little or no accompanying shale. The map (Fig. 6) delineates sandstone and shale units, but they are not mutually exclusive. Mapped sandstone units are either alternations of shale and sandstone beds with sandstone predominating in bulk, or a sequence of sandstone beds with little or no intervening shale. These mapped units are markedly lenticular, as shown on the map and section. Because of the paucity,

vagueness, and contradictory nature of the published data on texture and mineralogy of the Modelo sandstones, the writer felt it advisable to undertake some analytical work himself; the results are summarized herewith.

MECHANICAL ANALYSIS

Thirty-eight channel samples were taken in the Modelo sandstone for mechanical analysis. For most of them top and bottom half of beds were analyzed separately in order to study vertical gradation, and multiple samples were taken on some beds in hope of detecting lateral gradation. Nine different beds with a thickness range from 2 to 28 inches were analyzed. All samples were disaggregated and screened dry in a Ro-tap machine, and the silt and clay fraction further subdivided by the hydrometer method.

Grain size and grading.—Median diameters of the analyzed samples were in the fine and very fine sand sizes. Not all the sands were significantly graded, but wherever the difference in median diameter of the top and bottom halves was large the bottom half was coarser. The histograms (Fig. 3) illustrate graded bedding. Granules were present in fifteen bottoms and only six tops. Silt ranged from 11.0 to 24.7 per cent and clay (less than .004 mm.) from 4.2 to 11.2 per cent. Total silt plus clay ranged from 13.5 to 34.5 per cent. There is thus a wide distribution of grain sizes from granules down through clay-size particles, but with no sharp break which would correspond with a clast-matrix system. The writer feels that, though it defies description and measurement, matrix is probably genetically important to turbidite deposition and is worthy of additional thought and study.

No consistent lateral gradation could be detected in the five beds from which several samples were taken. This is not to say that no lateral gradation exists, but that if it does exist the

change in grain size in any bed over a distance of 100–400 feet is less than the error of measurement in determining grain size.

Correlation of bed thickness with grain size is suggested but not fully substantiated. It was noted that the degree of sorting (or difference in median grain size between bottom and top half) in normally graded beds is greater for thick beds than for thin beds as would be expected.

Sorting.—Sorting coefficients, based on the formula

$$So = \sqrt{\frac{\text{coarse quartile}}{\text{fine quartile}}},$$

ranged from 1.66 to 2.27 and averaged 1.86 for top halves and 2.03 for bottom halves. These are in Trask's well sorted category (Trask, 1932, pp. 71–72), but his classification has been questioned (Pettijohn, 1949, p. 24). Figure 4 provides a direct comparison of sorting of an actual Modelo sandstone (whose So most nearly approximates the mean So of all sands analyzed) with a miscellaneous assortment of well known sands and sandstones. The Modelo slope is obviously much gentler than the others usually regarded as well sorted. The long "tails" on the Modelo cumulative curve are characteristic of wackes. Application of the sorting classification of Payne (1942) shows all of the Modelo sandstones studied to be poorly sorted, as they all spread over seven or eight classes and require more than five classes to constitute 90 per cent (Fig. 3).

Roundness.—In studying thin sections and immersion mounts the writer was impressed by the angularity of grains. Many of them have very sharp acute points. Very few display advanced stages of rounding, and these are confined to very coarse sand and small pebbles. No actual measurements or counts were made, but visual comparison with published charts (Krumbein and Sloss, p. 81; Pettijohn, 1949, p. 52; AGI Data Chart 7) suggests that the mean roundness would be in the upper part of the angular group, probably near 0.2 on Wadell's formula (1935). The angularity fits very nicely with the high feldspar content in indicating a first-cycle sediment. No effort was made to measure grain shape or sphericity, as it was felt that this property had little significance in the present study.

MINERALOGICAL COMPOSITION

The Modelo sandstones are remarkably high in feldspar content. No attempt was made to iden-

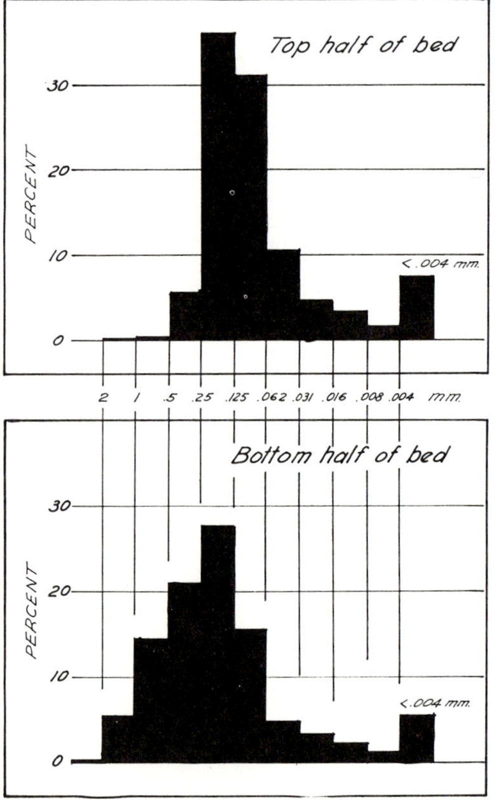

Fig. 3.—Histograms showing grading. Two channel samples were cut on same bed, and two top halves and two bottom halves were analyzed separately and averaged for diagram. Although modes for top and bottom are in the same size class, it is clear that bottom half of bed contains more coarse grains.

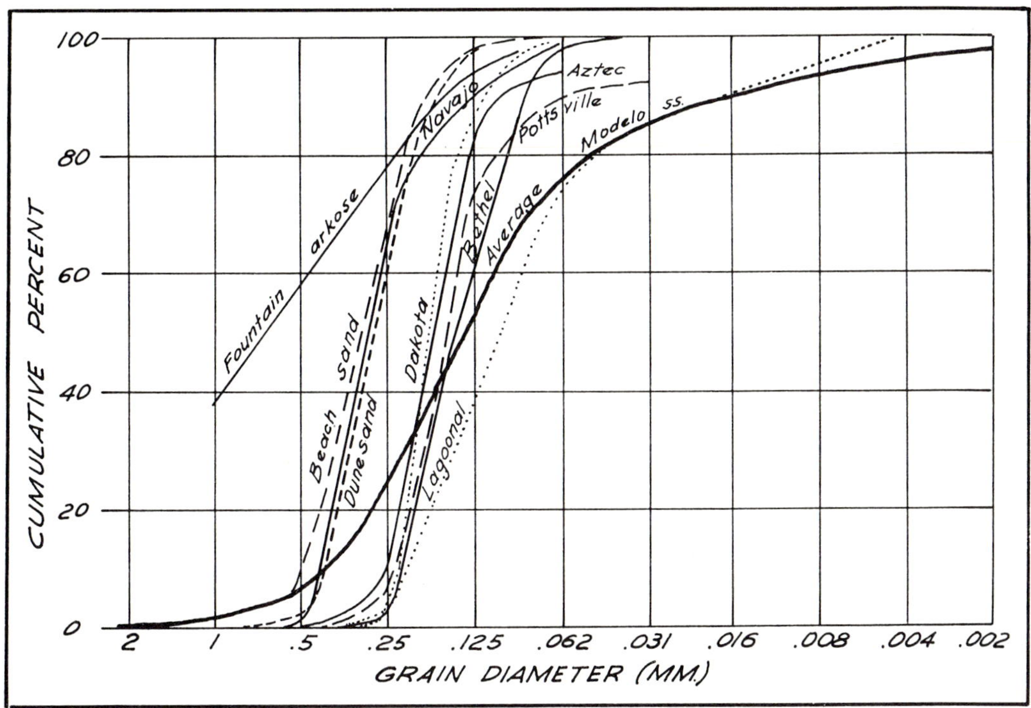

Fig. 4.—Cumulative grain-size curve of Modelo sandstone compared with cumulative curves of other sandstones. Fountain arkose and Dakota sandstone computed from data in L. W. LeRoy, "Size Analysis," pp. 184–92 in "Subsurface Geologic Methods," edited by L. W. LeRoy and published by Colorado School of Mines, 1950. Beach sand, dune sand, and lagoonal sand computed from histograms in Krumbein and Sloss, *Stratigraphy and Sedimentation*, W. H. Freeman and Company, 1951. Pottsville and Bethel sandstones from W. H. Twenhofel, *Principles of Sedimentation*, McGraw-Hill Book Company, Inc., 1950. Navajo, Aztec, and Modelo sandstones drawn from samples collected and analyzed by writer. Modelo curve shown is that of sample which most nearly coincides with mean sorting coefficient of all Modelo samples analyzed.

tify clay minerals, and the heavy minerals were examined only casually, but an effort was made to determine sand-size quartz-feldspar ratios on the assumption that the data would lead to some additional understanding of the environment of deposition and source areas. Ten samples from eight sandstone beds were studied by the index of refraction method in immersion oils. Each sample was split and 200 grains were counted in each of three oils whose indices coincided with the boundary between potash feldspar and plagioclase and the upper and lower boundaries of quartz. The results are summarized in Table I. The modal sand size of these sandstones had the remarkably high feldspar content of 85.6 per cent.

Although the mineralogical analysis is far from complete, as only one size grade was studied and clay minerals were not identified, it is obvious that the samples studied are arkoses. The consistency of the data on these few samples suggests that probably all the Modelo sandstones in the area are arkoses. Four of the analyses represented top and bottom samples from the same bed, and they showed as much variation as that between separate beds.

The mineral percentages of these sandstones are closely comparable with that of quartz diorite (Tyrell, 1950, p. 111), and do not differ greatly from granodiorite or quartz monzonite. The latter is a closer fit with respect to the kind of plagioclase (oligoclase), and the former agrees with respect to the alkali feldspar content.

PEBBLE MINERALOGY

As a check on the rather unusual mineralogy of the sand grains, the writer also made a rough analysis of the pebbles in two fine conglomerate beds, one in Tarzana and the other near Sepulveda Boulevard. Channel samples were cut in both beds, pebbles separated on a coarse screen, and sorted into groups of visually similar types. Twenty-one specimens were selected for thin-

TABLE I. MINERALOGICAL COMPOSITION OF MODAL GRAIN SIZE OF TEN SAMPLES FROM EIGHT MODELO SANDSTONE BEDS DETERMINED BY INDEX OF REFRACTION METHOD ON SPLIT SAMPLES IN THREE IMMERSION OILS

(200 Grains Counted in Each Oil)

| | Per Cent | | |
| | | Range of 10 samples | |
	Mean	Minimum	Maximum
Potash feldspar	22.3	12	31
Plagioclase (An<27)	38.4	32	42
Plagioclase (An 13–45)	21.0*	21	22
Plagioclase (An>30)	3.9	1	10
Total plagioclase	63.3	60	67
Total feldspar	85.6	75	94
Quartz	14.4*	6	25
Quartz-feldspar ratio	.17	.06	.33

* These are estimates for they have the same indices. Maximum possible for either one ranges from 27 to 46 and averages 35.4 per cent if the other is zero. Thus, the average sample contains at least 64.6 per cent feldspar, probably 85.6 per cent, and possibly close to 100 per cent (some quartz was positively identified by other optical properties).

sectioning roughly in proportion to their abundance.

The pebbles averaged about $\frac{1}{4}$ inch in diameter and were, therefore, too small to permit accurate petrographic determinations in all cases, especially were the grains were large. All the pebbles were leucocratic and appeared to be acid igneous plutonic rocks ranging from fine- to coarse-grained. No strong evidence for metamorphism such as foliation or metamorphic minerals was observed, but strain was pronounced in several of them, and it is possible that some are granular metamorphic rocks, perhaps gneisses. Several pebbles were definitely classified as aplite. All the pebbles are types to be expected from granitic rocks of the so-called eastern, or Sierra Nevadan, basement complex, they are clearly not from Franciscan metamorphic terranes. No anorthosite was recognized.

Perhaps more significant than the rock types is the mineralogical analysis. Mineral percentage estimates were made for each pebble. The average mineral content of all the pebbles follows.

Per Cent
43	Potash feldspar
37	Plagioclase
18	Quartz
2	Accessories (muscovite, biotite, opaques, epidote, and garnet)

Perthite is common, and in some pebbles constitutes the bulk of the rock. While these figures are by no means identical with similar data already given for sand grains, they seem sufficiently consistent in view of the relatively small number of samples studied and strongly suggest that the source area for the sands was rock of the type contained in the pebbles. The quartz-feldspar ratios are nearly identical as are the low percentage of heavy minerals and the kind of heavy minerals, especially the abundance of muscovite and biotite and lack of typical Franciscan minerals.

EVIDENCE OF TURBIDITY CURRENTS

Turbidity-current features are so abundant that the writer feels that virtually all of the sandstone beds in the area studied were deposited from these currents. Nearly all the typical characteristics listed by Kuenen and Carozzi (1953) are present. These features are discussed in two groups: those which do not exhibit orientation and those which do.

NON-ORIENTED FEATURES

Graded bedding is obvious in most of the thicker sandstone beds; granules and very coarse sand grains are nearly everywhere more abundant near the base of a given sandstone bed than near the top. However, as pointed out earlier, some beds are not graded, and some may show a slight reverse grading. Although others have recognized a connection between graded bedding and turbidity currents, Kuenen and Migliorini (1950) were the first to explain and expand the idea of turbid flow as the cause of graded bedding. Quantitative data on grading are presented in the section on mechanical analysis.

Poor sorting is characteristic and has already been discussed. Dirtiness is the chief characteristic of a sand which has been dumped rather than spread, winnowed, and reworked. Pettijohn (1950) has suggested that graywackes are the result of turbidity-current deposition, and Packham (1954) goes a step further, regarding deep-water deposition by turbidity currents as part of the definition of graywacke (graywacke in the same sense as wacke of Gilbert.)

Regular bedding, or the even continuity of individual sandstone sedimentation units through the extent of large outcrops, is characteristic. On a larger scale the beds or groups of beds are lenticular, as shown on the map, and on a smaller scale ripple marks, load casts, and scour cause minor irregularities. One would expect to find many beds pinching out, but these are rare. The precise nature of the pinch-out was clearly seen on only one bed and that one thinned from 4 feet to zero in a distance of 30 feet, clearly passing between the converging enclosing shale laminae without any diminution in grain size (Pl. 1a).

Absence of shallow-water features such as clean well sorted sand, oscillation ripple marks, coarse random cross-bedding large-scale channel scour, layers of mollusk shells, and desiccation phenomena is noteworthy. Prior to the acceptance of turbidity currents as an important medium of transportation, the mere presence of sands and gravels would have suggested shallow water, but Kuenen (1948a) has shown that angular boulders weighing 30 tons can be transported by turbidity currents under realistic conditions.

Interbedding of sandstone and shale, either rhythmically or with random thicknesses is typical. The tendency is for sandstone beds to be grouped in sequences in which the separating shale beds are thin or absent and for the shale bodies between sandstone sequences to contain few or no sandstone beds. This may be interpreted as a reflection of relatively long periods of instability in the source area or in nearshore waters causing increased erosion or submarine slumping separated by periods of quiescence when pelagic sediments predominate.

The basal contact of sandstone beds is sharp. The beds lie abruptly and without gradation on shale or on a previously deposited sandstone. The upper contact at some places grades imperceptibly to shale, but commonly it, too, is abrupt. The gradation is to be expected as a consequence of turbidity-current deposition, whereas an abrupt upper contact is believed to be the result of erosion by a subsequent turbidity current.

Absence of fossils in the sandstones is remarkable. Abundant shallow-water megafossils are found in the basal few feet, but only one worn oyster shell was seen in sands higher in the section. Sandstones were not analyzed for foraminiferal content, but foraminifera in one siltstone layer and one silty cross-bedded sandstone were seen in thin section to be transported.

Slump structures were seen in several places. Although not a result of turbidity currents, they are commonly associated as a natural consequence of an appreciable sea-bottom slope in the slump area and perhaps also in the area of deposition. Slump in the general area is indirectly suggested by the presence here and there in sandstone beds of large angular blocks of shale which are more easily understood as having originated in a slump than by scour of shale beds by a current.

Convolute bedding is one of the most common structures and may be seen at nearly any Modelo sandstone outcrop in the area (Pl. 1e). Contortions decrease in intensity downward into undisturbed laminae. In some places the intensity also dies out upward, but more commonly the tops are bevelled off as though by a subsequent current.

One would suspect from the usual cross-sectional view that the axes of the convolutions were mutually parallel and probably at right angles to the direction of current. This may be true locally, but one only sees the plan view after much digging and scraping, and the few excavations made by the writer revealed merely random convolutions much like the section view.

Shale inclusions ranging in size from small flakes to huge slabs are very common in the sandstone beds (Pl. 1b). Fragments may be either angular or rounded, commonly with clearly preserved bedding and of typical Modelo shale lithology, generally monolithologic in any bed as though derived from a localized source. Fragments may occur completely isolated, strung out along the bedding in thin planar groups, or in rather thick masses. These inclusions potentially offer some promise as indicators of current direction through either imbrication or identification of source, but no such indication was found. Further study may be fruitful in this regard.

That the inclusions may have been locally derived by scour of the turbidity currents themselves is strongly suggested by the two field sketches (Fig. 5).

Character individuality of sandstone beds is shown by the fact that a bed generally displays the same feature or features throughout its length of outcrop. That is to say, if a bed contains load casts at one spot it probably displays them at every place where its base can be seen, whereas an adjacent bed may have a sharp plane base throughout its exposure but contain evenly distributed shale slabs. The controlling conditions were evidently constant in some areas for a long enough time to permit deposition of a series of turbidity-current deposits with similar characteristics. In some outcrops nearly every bed displays

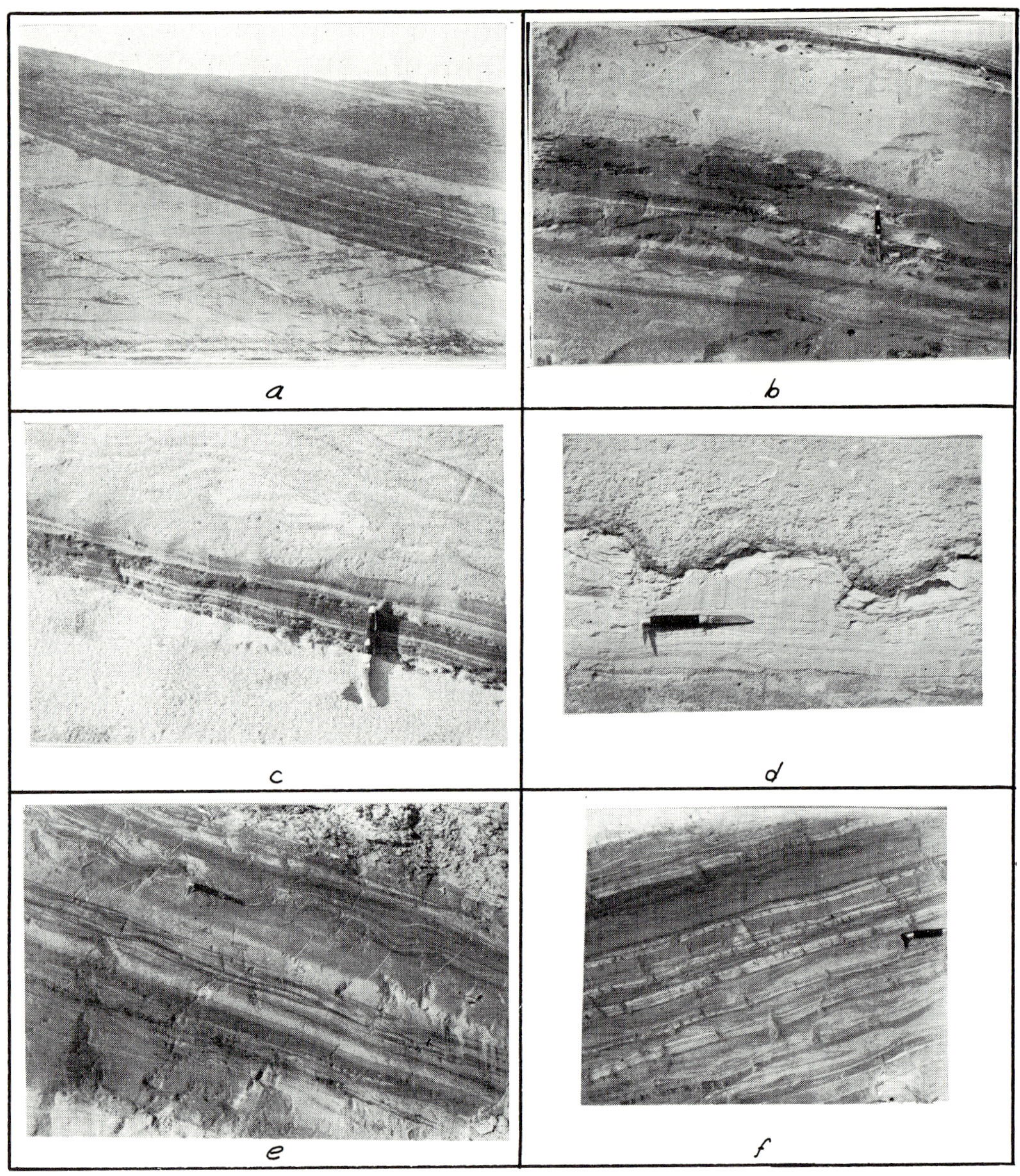

PLATE 1.—Turbidity-current features in Modelo formation. All exposures except *a* were scraped smooth with a garden spade and swept clean with broom before photographed. Knife, including blade, is 6 inches long.

a. A mappable sandstone unit overlain by shale unit. Prominent sandstone pinching out in shale is single graded bed containing shale flakes. At its feather-edge it passes between enclosing shale laminae without facies change. Shale slabs may be seen in main sandstone body near shovel handle. Horizontal streaks are man-made.

b. Graded bedding (thick light-colored bed), rounded clay-shale clasts (dark) in sandstone, and convolute bedding (second bed from bottom). Clay-shale bed at knife handle contains charcoal fragments.

c. Load waves above knife showing rather extreme wispiness and uniformity of inclination.

d. Load pockets and load waves at base of graded sandstone bed and load folds in underlying shale laminae.

e. Convolute bedding (at level of knife), shale clasts in sandstone (below knife), and cross-bedding (in bottom half of photograph).

f. Cross-bedding and ripple marks. Component of current is to right.

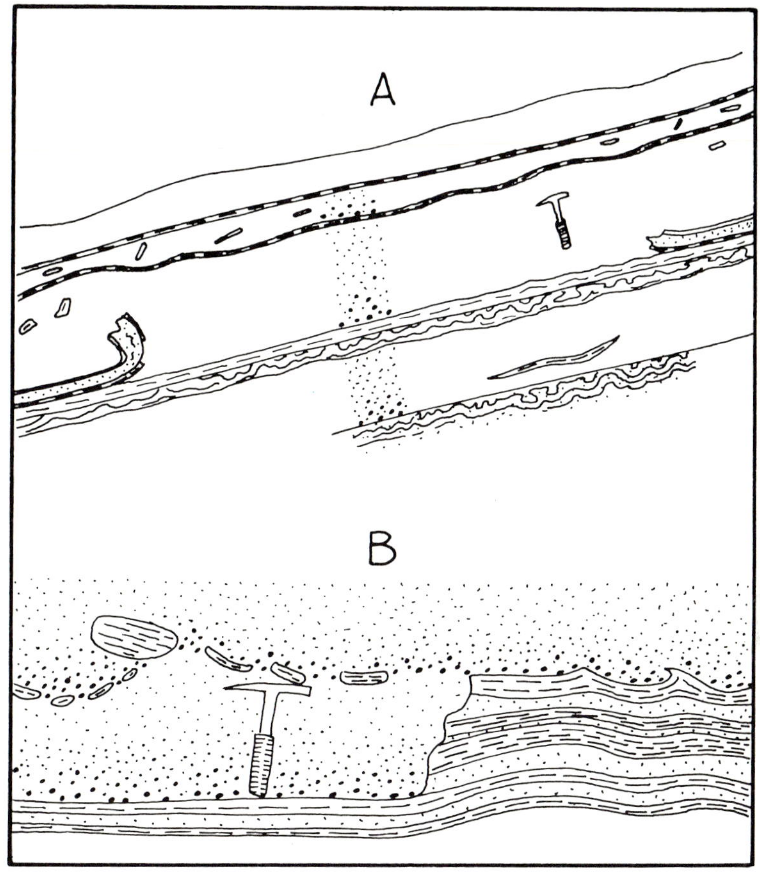

FIG. 5.—Field sketches illustrating scour by turbidity currents in Modelo formation and presumed origin of shale clasts and breccias. In A it appears that current which carried 4-inch layer of sand and shale away (at least 100 square feet was removed) must also have deposited its load nearly simultaneously in order to maintain support for the flap. In B it is interesting to contemplate whether erosion was by scour or slump. Since it is a site for deposition one would suspect insufficient slope for slump. There is also basis here for speculation about rate of consolidation in terms of rate of turbidite deposition. Shales were still soft when the second sand was dumped, as indicated by load cast on top shale bed. Whether this particular submarine erosional feature supplied any cohesive clasts is not known. It was not clear at outcrop whether layer of shale slabs was at top of lower bed or base of upper bed. The exposure of A is in excavation east of Beverly Glen Boulevard and north of Mulholland Drive, and B is in back yard of home east of Sepulveda Boulevard and north of Mulholland Drive.

convolute bedding, and in others nearly all show cross-bedding.

ORIENTED FEATURES

Several different types of structures display orientation with varying degrees of precision and abundance. The current directions are clearly recorded in the rocks and graphically displayed on the map (Fig. 6) by means of arrows for individual beds and rosette diagrams for various segments of the map area.

Because the strata are not horizontal, most current directions were corrected for tilt on the assumption that only one tectonic axis of rotation has been effective and that this corresponds with the present strike of the strata. An additional correction was applied on the flanks of the plunging folds at the western end of the area, where two axes of rotation were involved. Calculations were reduced to a minimum because the rocks are soft enough to permit breaking-out of small slabs on which the structural attitude had been marked in place.

Cross-bedding turned out to be the most useful directional phenomenon in the rocks investigated. It is commonly found in thin sandstone

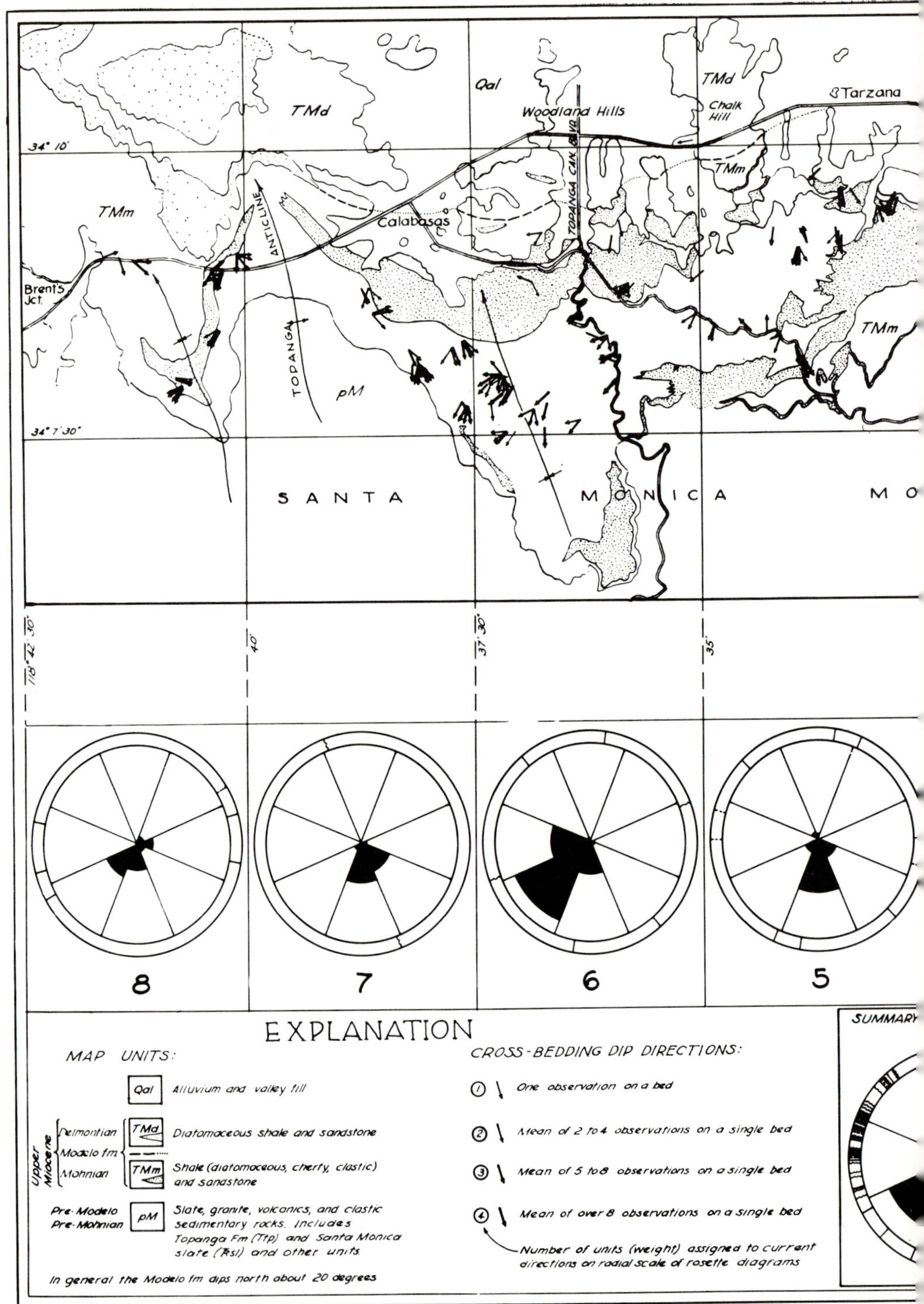

FIG. 6.—Geologic map an

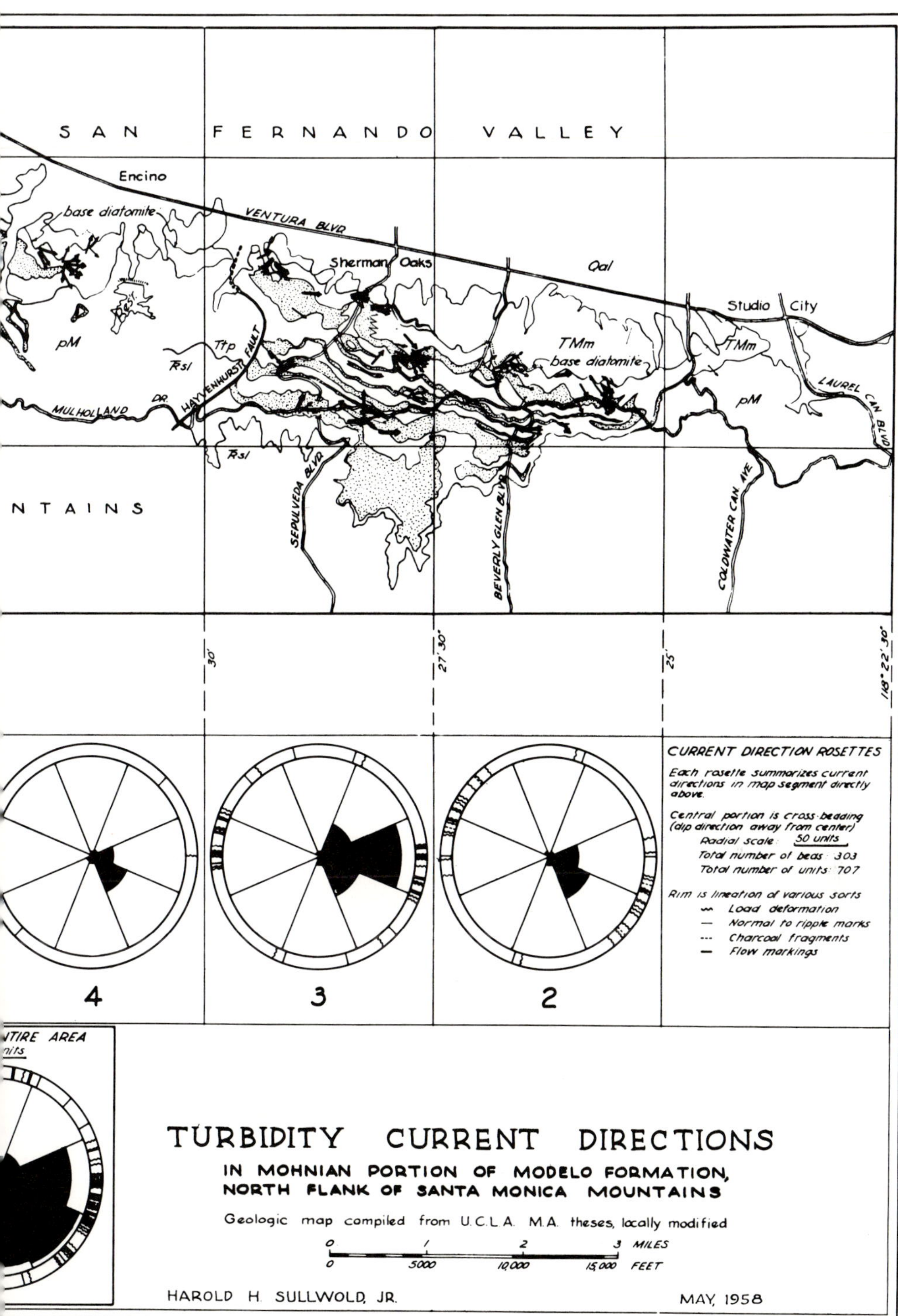

TURBIDITY CURRENT DIRECTIONS

IN MOHNIAN PORTION OF MODELO FORMATION, NORTH FLANK OF SANTA MONICA MOUNTAINS

Geologic map compiled from U.C.L.A. M.A. theses, locally modified

HAROLD H. SULLWOLD, JR.   MAY, 1958

urbidity-current directions.

beds 1–3 inches thick and rarely in the top part of thicker beds. An unusually good exposure is shown in Plate 1f which also shows the ripple marks with which the cross-bedding is commonly associated. Cross-bedded sandstone does not generally display graded bedding, load deformation, convolute bedding, shale slabs, or other features common in thicker beds. Cross-laminae show as color bands rather than as obvious changes in grain size. Thin sections show abundance of biotite along the fore-set layers as well as grain-size variation.

Cross-bedding in conjunction with turbidity currents is generally ascribed to traction transportation during the dying stages of the current after most of its load has been dumped. Cross-bedded sandstones are finer-grained than the average turbidite. They may overlie typical turbidites but are generally seemingly disconnected from them as though the traction currents had either bypassed their main deposits or deposited from their tails before reaching the site of coarse deposition.

Cross-bedding dip directions were determined on more than 300 beds. At each bed 1–19 measurements were made. A true statistical approach to the problem was not made for several reasons. On some beds the cross-bedding is vague and difficult to measure, requiring many readings to get a reliable average; on others it is clear and obviously consistent, requiring only one reading. Some beds are bounded above and below by wavy ripple-marked surfaces which cause difficulty in determining the original horizon, while other beds have parallel plane surfaces at top and bottom. Some beds are so unconsolidated that few beds can be extracted and marked; others come from the outcrops in firm slabs. Some outcrops are on private residence embankments, whereas others are in active excavations where geological digging is less conspicuous. Equal areal distribution of observations was likewise impractical, because outcrops are far from universal and not all exhibit cross-bedding. The writer adopted a normal geological approach rather than a statistical one, because it is difficult to evaluate numerical data which are based in part on judgment.

Nevertheless there is the question of the number of readings required to ascertain reasonably accurate current direction of a particular bed. The writer analyzed the data for the beds on which many readings were taken and concluded

that 4 or 5 readings on a single bed generally give a current direction within about 12° of that determined by three times as many readings. This is well within the accuracy required for studies of this sort.

For purposes of graphic presentation the mean current direction for each observed bed is plotted on the map. Because the values at some beds are more accurate and (or) reliable than at others, an arbitrary factor based on the number of readings was assigned to each mean to give its reliability a quasi-quantitative value. On the map reliability is expressed by conspicuousness of the directional arrows. On the rosettes reliability is expressed in the radial scale as follows: one reading on a bed is assigned one unit on the radial scale; the mean of 2–4 readings, 2 units; the mean of 5–8 readings, 3 units; and more than 8 readings, 4 units. As a step in the direction of evaluating reliability the writer doubled the unit value for single readings which were obviously reliable.

It is obvious from inspection of closely spaced groups of current arrows on the map that for small areas the currents were remarkably constant. For larger areas, as shown by rosettes, constancy decreases, but even for the area as a whole nearly 90 per cent of the current directions lie within a 180° arc toward the southeast. Remarkable as this gross steadiness of current direction may seem, it masks an even more remarkable pattern. That is, the width of spread of current directions for the whole area is not caused by relatively random distribution of current directions within the southeast half-circle but rather by a consistent change of direction from one end of the area to the other, as shown by the family of rosettes representing segments of the map. To minimize chance of coincidence a second set of rosettes (not shown) was drawn with the grid system shifted a half space; the results were in close agreement.

The pattern clearly indicates a point-source for the currents. This, plus the other evidence for rapid dumping in deep water, is convincing evidence, in the writer's opinion, that the lower Modelo formation in this area was deposited as a fan at the mouth of a submarine canyon.

Ripple marks commonly occur at the top surfaces of cross-bedded sandstone layers (Pl. 1f) and less commonly at the tops of some of the more massive sandstone beds. As it is necessary to dig extensively to obtain an accurate strike only a few were plotted. Their wave lengths ranged from 2 to 12 inches and amplitude from 0.25 to 1 inch. As expected, their strike was close to that of the cross-bedding with which they were associated. Were they readily accessible, they would be better than cross-bedding, provided the latter were present to indicate sense of current flow.

Flow markings were found at only one place, on a parting plane at the base of a set of cross-laminae near the top of a sandstone bed. No groove casts were seen in the area. The paucity of observed lineations on the bottoms of sandstone beds is believed to be largely a result of paucity of observed bed bottoms. They are not exposed naturally and are difficult to expose mechanically, because the sandstones and shales do not split along the contact but mutually adhere, thus preventing a view of these fine-grained structures.

Load deformation, generally referred to as load casts or flow casts, is common at the base of Modelo sandstone beds. These features range in size from almost microscopic to several feet across. Two types are illustrated in Plate 1c and 1d. They are normally seen in cross section, for the base of a sandstone bed is not ordinarily exposed. By digging in the relatively soft rocks, however, one may see that some load casts are non-linear or dimpled and others definitely linear. Some of the linear load casts are parallel with and others at right angles to the current as determined from adjacent cross-bedding. Insufficient investigation was made to differentiate the two types, but they offer promise as current indicators. Other workers have noted load casts with lineations parallel with the current (Crowell, 1955), normal to the current (Prentice, 1956), and without lineation (Kuenen, 1957). Evidently in the rocks studied here all three types are present.

The writer is not satisfied with the nomenclature of load deformation and has submitted his views on the subject (Sullwold, 1959 and 1960).

Charcoal fragments are common as isolated particles in the sandstone beds and as flood layers in chocolate-brown silty shale between sandstone beds. These fragments are generally angular and markedly linear in shape, but unfortunately few of them display a preferred orientation. Five localities that did show a preference were split as equally as possible; three seemed to be oriented with the current and two across the current (current direction being surmised from the nearest crossbedding).

INTERPRETATION OF RESULTS AND DISCUSSION

The writer believes that he has offered ample evidence to suggest strongly that the Modelo sandstones in the Santa Monica Mountains are:

1. Wackes (dirty immature sandstones) containing an abundance of interstitial silt and clay.
2. Arkoses whose model size particles have an extremely high feldspar content.
3. Turbidites containing abundant textural and structural characteristics considered by Kuenen and others to be a result of deposition from turbidity currents.
4. Deep-water deposits (almost certainly bathyal; perhaps near 3,000 feet).
5. Deposits from currents traveling from a point source somewhere not far northwest of Tarzana.

These items are direct conclusions of this study and may be used as a basis for some speculation about environment and geography of the area in late Miocene time.

TARZANA FAN

The case for the radial pattern of current directions determined by cross-bedding dips is amply demonstrated on the map (Fig. 6). Since the sands are turbidity-current deposits which normally travel downslope, the conclusion is drawn that the sea bottom in the area investigated lay astride the south flank of a low cone. Projection backward of current directions of arrows or rosettes in map segments 4, 5, and 6 suggests that the point source may be $\frac{1}{2}$ mile north of Ventura Boulevard between Tarzana and Chalk Hill. Map segments 2, 3, 7, and 8 do not fit perfectly, but are not far out of line. Segments 2 and 3 lie east of the Hayvenhurst fault, which has had considerable strike slip, while segments 7 and 8 lie in a more complex structural area where the currents may have been diverted by warping of the sea floor as a prelude to the folding which is now so prominent. As the current directions here are not correlative precisely with the present fold axes, it is possible that the axes have migrated or that a second more westerly fan, whose main part is now eroded away, may have existed and interfingered with the Tarzana Fan.

Further evidence supporting the fan concept may be seen on the partly restored stratigraphic section (Fig. 7), where the marked thinning of the lower Modelo (Mohnian) part of the section both east and west of the Calabasas-Woodland Hills-

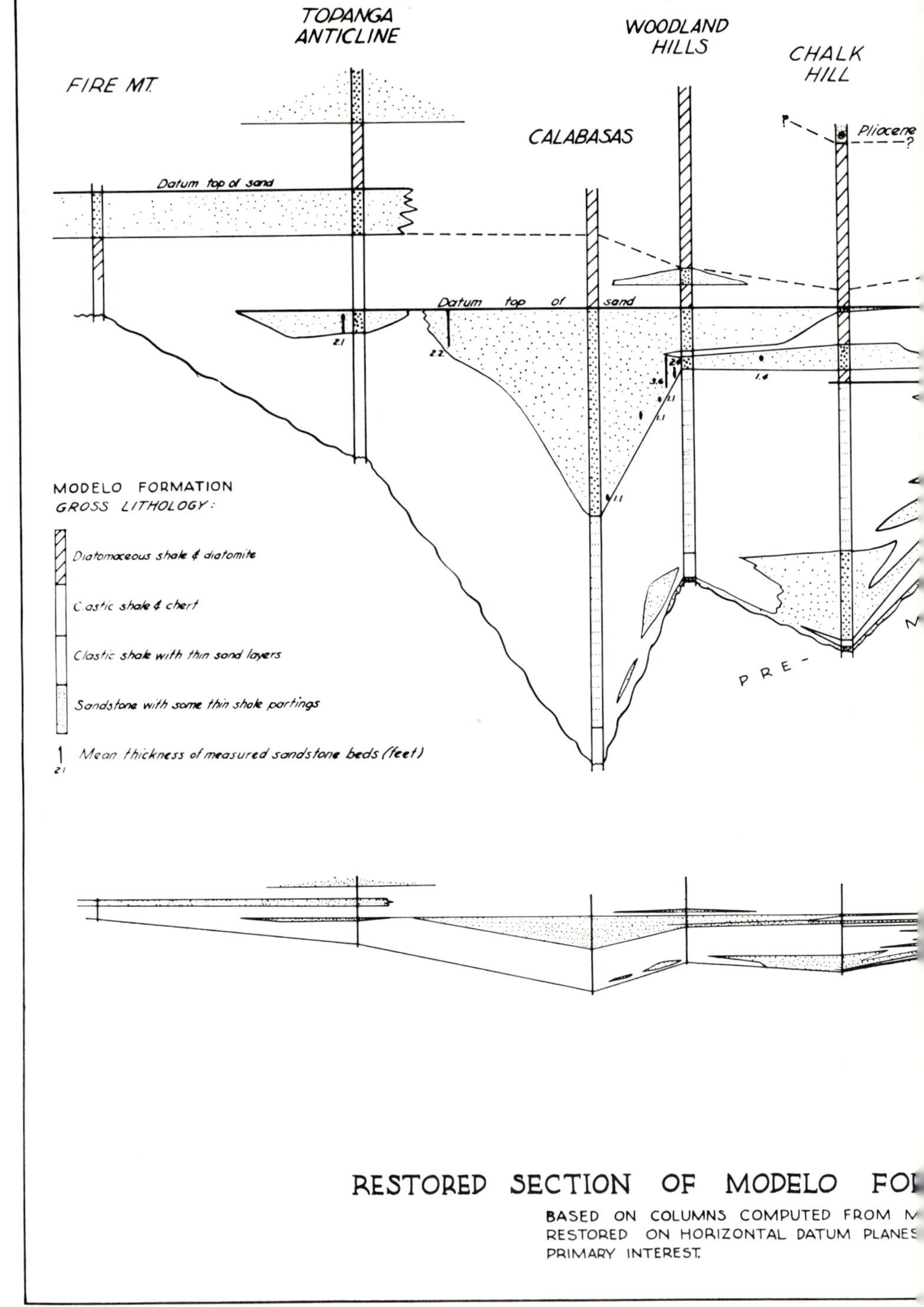

FIG. 7,—Restored section of Modelo formation, Santa Monica Mountains. Essentially a cross section

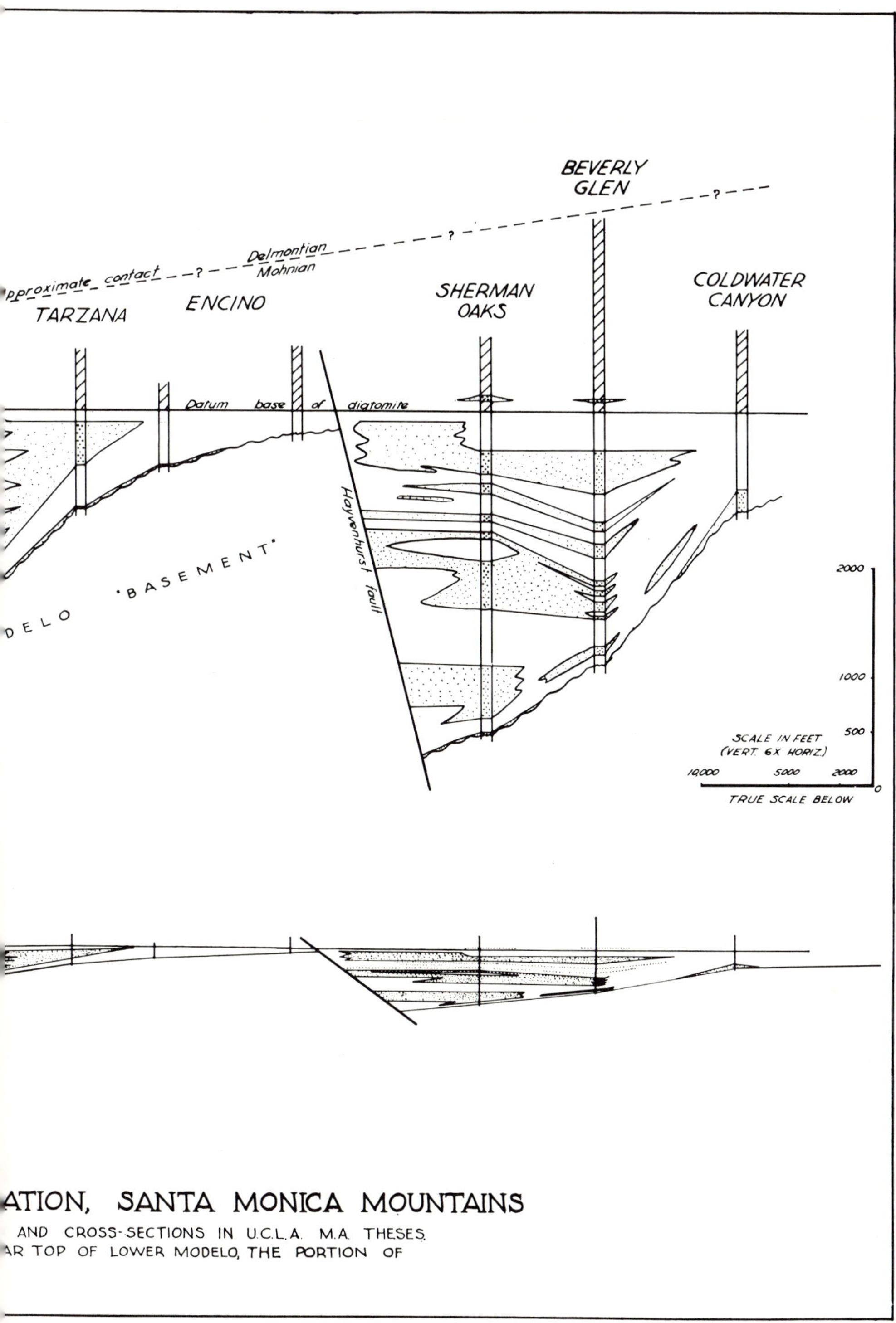

of Tarzana Fan with its top surface horizontal and its eastern end offset by Hayvenhurst fault.

Chalk Hill columns is obvious. The columns were taken from cross sections or plotted from maps in Master of Arts theses at the University of California, Los Angeles, and from the published section by Hoots (1931) locally modified by the writer. The section has the usual shortcomings resulting from conversion of outcrop measurements in gently dipping strata to a pseudo-vertical stratigraphic section. Folding has been eliminated, but the Hayvenhurst fault is shown to emphasize the pronounced change in thickness on its two sides. As the only horizon that could be traced entirely through the area was the base of the Modelo, and, since this is a known erosional surface and probably had considerable relief, an attempt was made to establish a datum near the top of the rocks of interest to portray conditions near the end of Mohnian time. The datum planes are drawn horizontally rather than cone-shaped because the shape and steepness of the cone and location of its axis are not precisely known.

Gross shape of the Mohnian rocks in cross section is in agreement with the point-source concept in that the greatest thickness is near the point source indicated by current directions, and the thickness diminishes away therefrom. The coincidence of thickest section with current point-source is not perfect, but is close. The discrepancy can perhaps be explained by the 2-dimensional aspect of the outcrop area, the possible skewed position of the area on the fan, the "tilt" of the plane of the section, the possible asymmetric shape of the fan, and the irregularities of the basement surface upon which the fan was deposited. The section also reveals the greater abundance of sands in the areas of greater total thickness; that is, the excess thickness of the thicker columns is largely due to increased sand content.

An anomaly exists east of the Hayvenhurst fault where the thickness increases abruptly. This area could be a part of the same fan (or a different fan) brought to its present position by strike-slip movement along the Hayvenhurst fault as shown in Figures 8 and 9. Here the fan is shown elongate toward the east to account for the uniform easterly direction of currents in that area. The fault could have occurred either before or after tilting, but slip was nearly parallel with the bedding planes of the upper Modelo strata as shown by the relatively small separation at the base of the diatomite (Fig. 6).

Figure 8 also attempts to account for the lack of coincidence between current apex and maximum thickness by showing the unusually thick Calabasas sand lens in position west of the apex where it approximately doubles the total thickness. Also shown is the previously mentioned possibility of interference of current directions at the west end by a second fan postulated on the west, now largely eroded away.

The details of the relation of the large mapped sandstone lenses or bundles to the enclosing shales were not worked out in the field and would be a useful project for the future. Several partial sections were measured primarily to obtain individual bed thicknesses, and it is apparent that there is no consistent thinning of individual sand beds toward the edges of the bundles. It may be tentatively concluded that there are *more* sandstone beds in the thicker parts of the bundles. These data are shown in Figure 7.

The dirtiness and grain angularity of the Modelo sandstones are believed to be in accord with the fan concept. Rapid dumping and lack of winnowing would be expected in a fan or delta given the proper source material. Graded sands in modern deltas at the mouths of submarine canyons have been reported by Ericson *et al.* (1951) and Gorsline and Emery (1959). Cores averaging 20 feet in length in the delta of the Hudson submarine canyon show the sediment to be composed of 30 per cent sand, ranging in thickness from thin films to 25 feet, well sorted, graded, with sharp bases and sharp or gradational tops, containing both shallow-water and deep-water foraminifera. These sand bodies are not correlatable between cores and consist mostly of quartz with common feldspar and ferromagnesian grains, both angular and rounded, and vegetal matter. The parameters are not entirely in accord with the Modelo sandstones, especially as to sorting, but the Hudson delta is 400 miles from shore and at a depth of 15,000 feet! Fans offshore from Los Angeles have been recently identified at the mouths of four submarine canyons in the San Pedro and Santa Monica basins at depths of 3,000 feet. The fans and channels are clearly identifiable by their shape, topographic expression, and high sand-shale ratios. The thickest sand layer found at the top of a fan is 4 feet thick and grades from gravel at the base to very fine sand at the top. Sands farther down the flanks of the delta are thinner and finer-grained,

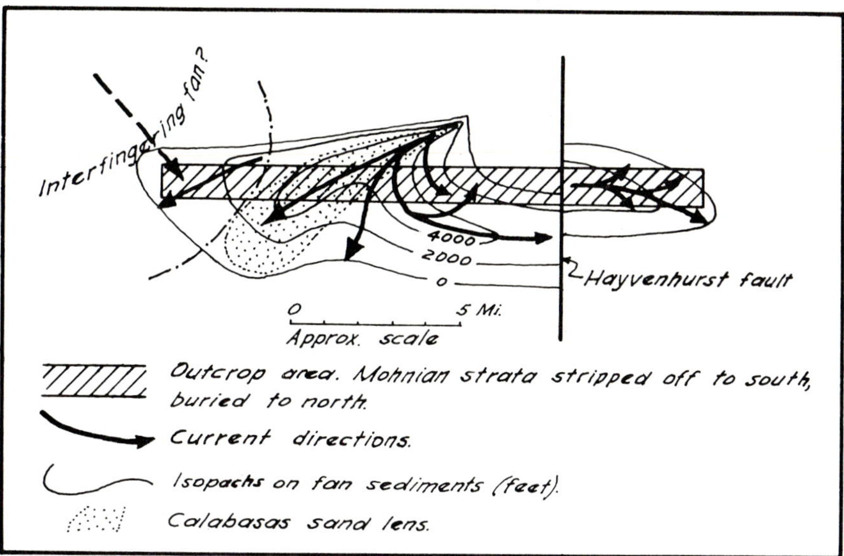

FIG. 8.—Diagrammatic map of Tarzana Fan showing proposed slip along Hayvenhurst fault to account for great change in thickness on its two sides. Presence of unusually thick Calabasas sand lens may account for non-coincidence of thickest part of fan with center as indicated by current directions. Also shown (in broken lines) is part of possible second fan on the west which may account for erratic current directions in that area. Only data are from shaded area of outcrop; hence, over-all configuration is highly conjectural.

and those in the basin floors are $\frac{1}{2}$–1 inch thick with median size of about 70 microns. All sands on the fans are graded, and the material is similar to that moving along the coast in shallow water. These two basins form a part of a group of thirteen offshore basins containing various amounts of sedimentary fill. Their onshore counterparts (Los Angeles Plain, Ventura Plain, and, presumably, the San Fernando Valley) have been completely filled because of proximity to eroding mountains.

SOURCE AREA

It seems reasonably clear that the sandstone beds in the Modelo formation are first-cycle deposits, as borne out by the high feldspar content and grain angularity. The task is to locate an area of crystalline rocks that could (or preferably

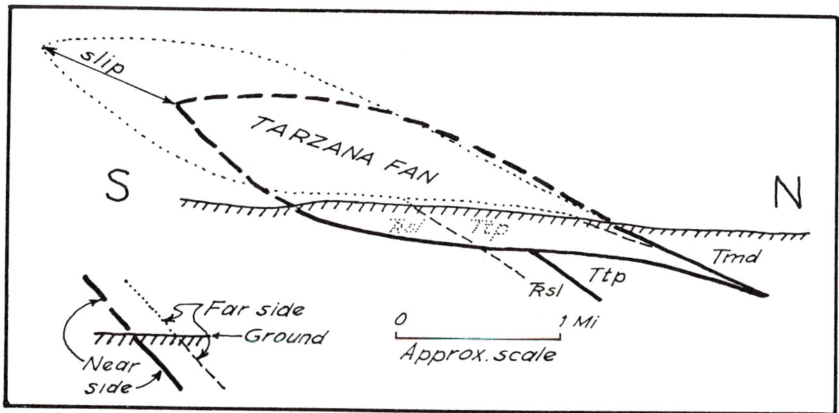

FIG. 9.—Sketch section along Hayvenhurst fault showing both walls and illustrating proposed oblique slip of Tarzana Fan to account for great change in thickness across fault. View westward. Actual cross-sectional shape of fan is unknown, for much of it is eroded away, and there is no well information on subsurface part.

must) have been the eroding source. This calls for a crystalline body which is not overlain by pre-Modelo strata. The regional map (Fig. 10) depicts possible Mohnian eroding areas.

A nearby source under the San Fernando Valley is possible but unlikely, as no evidence for this has been found in the exploratory holes drilled in the search for oil. Most wells have been in the northern and western parts of the valley where, although not uniform in thickness, the Mohnian rocks are everywhere present and overlie older sedimentary rocks.

Similarly, north and west of the San Fernando Valley (west of 118° 30′) sedimentary rocks older than Mohnian are present above the basement. Furthermore, the Mohnian rocks which lie between the nearest possible northwestern crystalline source (in the vicinity of Ridge basin) and the Santa Monica Mountains are locally devoid of sand and contain other faunal and lithologic evidence suggesting the presence in this intervening region of several east-west-trending troughs which would constitute sand traps and prevent turbidity currents from reaching the Santa Monica Mountains (Winterer, 1954, and personal communication; Kew, 1924, cross sections). More specifically, the present Oak Ridge-Simi Valley trend seems to have been a submarine high in Mohnian time as evidenced by its relatively thin and sand-free strata.

At the north-central edge of the valley in the San Fernando-Tujunga area the Modelo formation has been mapped lying on crystalline rocks (Oakeshott, 1954, 1958), where it contains cobble and boulder conglomerate suggesting proximity to shore. A group of megafossil localities of late Miocene age contained 22 genera of mollusca, all of which could have lived in shallow water and 9 of which do not now live in depths greater than 180 feet. The fossils were reported by Beatie (1958), and the depth data were provided by Professor C. A. Hall of the University of California, Los Angeles. These localities are from a single horizon whose position with respect to the entire Mohnian section is uncertain because of faulting. Foraminifera in this area also suggest shallower water than in the Santa Monica Mountains, but the evidence is less conclusive. The writer gathered microfaunal data from several sources[3] and

prepared a depth chart which showed a common depth of 600–1,250 feet for all of the 8 species still living. This depth, though based on less complete faunas, corresponds with the 2,500–2,900-foot depth for the Tarzana Fan (Fig. 2). Thus the evidence suggests shallower Mohnian water on the north side of the San Fernando Valley and proximity to eroding land. This places a possible source area in the San Gabriel Mountains which are composed almost entirely of crystalline rocks. These rocks have not been everywhere mapped or studied in detail, but the distribution of certain major types is fairly well known, and only parts of the range appear to be possible source terrane for the Modelo sandstones.

The modal sand size of the Modelo sandstones has been shown to average about 15 per cent quartz, 22 per cent potash feldspar, and 63 per cent plagioclase, 96 per cent of which is albite or oligoclase. The average mineral content of the few small pebbles studied in thin section was 18 per cent quartz, 43 per cent potash feldspar, 37 per cent plagioclase, and 2 per cent accessories. Perthite was very common. Quartz diorite thus appears to be the ideal source rock, feldspar-rich gneiss might fit the picture, and granodiorite or quartz monzonite would not seriously alter it if present in subordinate amounts.

West of Longitude 118° 05′ in the San Gabriel Mountains about half the exposed crystalline rocks consist of anorthosite, gabbro, and diorite, in which feldspar is too calcic to have contributed heavily to the Modelo sandstones. However, east of this meridian there are large quantities of granodiorite and diorite gneiss which could have been the original source area. These are the rocks labelled "gr" and "gn" on the map (Fig. 10) along the 118° meridian in the vicinity of Pacifico Mountain and Little Rock Creek north of Mt. Wilson. Detailed studies, especially of volumes of various rock types, are lacking in this area, but J. R. Walker, graduate student at University of California, Los Angeles, is currently studying them for the purpose of matching basement rock types across the San Andreas fault. He has looked at more thin sections in this area than anyone else to the writer's knowledge, and, after viewing the Modelo pebble thin sections, stated that they could have come from the San Gabriel Mountains east of the anorthosite body. Several characteristics common to the pebbles and the suspected source area are noteworthy: (1) microperthite is common and of similar aspect in both groups;

---

[3] Harold Lian, Union Oil Company, outcrop samples; W. T. Rothwell, Richfield Oil Corporation, outcrop samples; B. C. Jones, Union Oil Company, several wells; Howard Reynolds, consultant, several wells.

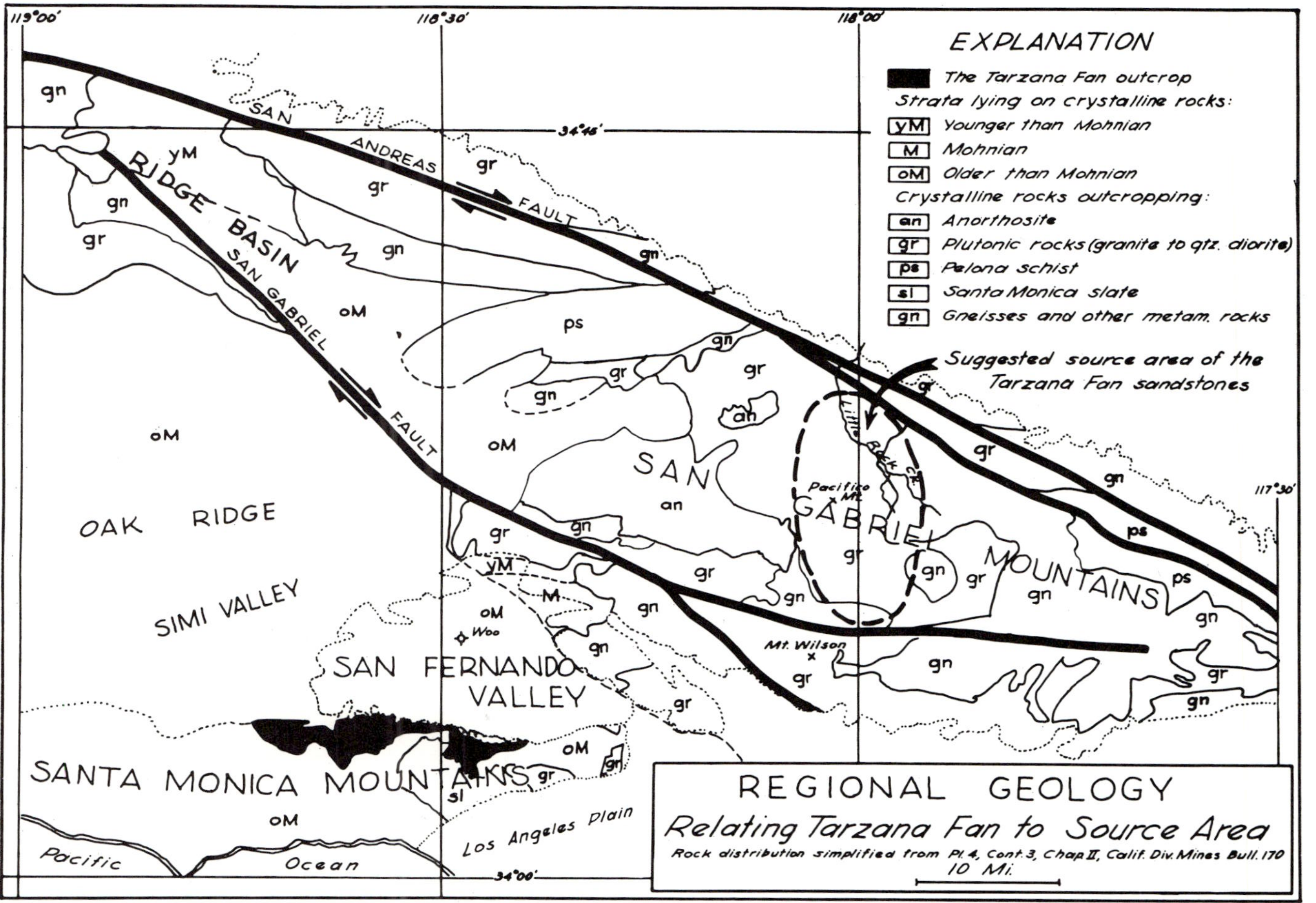

Fig. 10.—Regional geology. Physiographic features indicated by dotted lines, geology by solid and broken lines.

(2) strained quartz and strained feldspar twinning is common in both groups; (3) late quartz is common in both groups; (4) plagioclase in the crystalline rocks is usually about An 25–30 which fits reasonably well with the Modelo sand grains and very well with the pebbles. Farther east of the Pacifico Mountain-Little Rock Creek area the quartz percentage, An content of plagioclase, and degree of metamorphism increase in most of the crystalline rock greatly reducing the likelihood of this area as the source for the Tarzana Fan sediments (Hsu, 1954).

A possible source area is thus identified about 30 miles northeast of the apex of the fan. If the near-shore Mohnian facies here discussed were along the route of travel, the most logical geographic position for the source area would be the region now occupied by anorthosite and gabbro. Since these rocks are evidently not represented in the Tarzana Fan, we may consider that the stream supplying the proper debris by-passed the anorthosite area on the southeast or that during Mohnian time the present site of the anorthosite was occupied by proper lithologic types. The mechanism for this latter notion has already been provided by Crowell (1952) who showed that since Miocene time the San Gabriel fault has had about 20 miles of right strike-slip movement. "Unfaulting" would place the presently favored rock types into position north of the San Fernando-Sunland area and would not require special drainage conditions. It would also result in shorter subaerial transport, thus favoring the high feldspar content and grain angularity of the Tarzana Fan sandstones. Incomplete but highly significant is the cobble inspection of the Cenozoic conglomerates in the San Fernando-Sunland area by Beatie (1958) and Merifield (1958). They report anorthosite clasts rare or absent in the Modelo formation (upper Miocene) and common in the Saugus formation (Plio-Pleistocene).

It is also possible but far less likely that the source area may have been more distant, perhaps across the San Andreas fault. In this case the most abundant source rock would probably be biotite quartz monzonite which, in the Barstow Quadrangle (Bowen, 1954), has the mineral composition of quartz 25–35 per cent, microcline perthite 30–35 per cent, plagioclase (An 28) 25–30 per cent, biotite 2–4 per cent, and muscovite 0–2 per cent. This source area would necessitate a longer stream perhaps involving enrichment of

quartz through destruction of feldspars and possibly more grain-rounding than appears desirable.

In either case the low quartz content of the Modelo sandstones is difficult to explain. In view of the greater stability of quartz relative to feldspar, we should seek a source area with considerably less than 15 per cent quartz, but rocks with the combination of low quartz, high potash feldspar, and lack of calcic plagioclase are rare. Perhaps there is some mechanism by which some of the quartz did not reach the site of deposition (through being coarser as a result of being harder), or does not appear in its true proportions in the median size analyzed. The mineral analysis itself was admittedly inexact, and knowledge of the crystalline rocks is even more so. Hence the writer tentatively concludes that there were possible source areas present in the right direction to supply detritus for the Tarzana Fan, but that its precise identity awaits further study.

Submarine Canyon

A submarine canyon seems required to provide a point source for the sands of the Tarzana Fan in the depth of water indicated. The precise location of this canyon must await further factual knowledge, especially the drilling of wells in the San Fernando Valley. Directional current data presented herein locate its mouth just northwest of the community of Tarzana, and a possible Miocene eroding source area has been tentatively located about 22 miles northeast at the present site of the San Gabriel anorthosite complex. If the Modelo on the north side of the Valley was deposited very near shore, as seems likely from rock distribution, coarseness of clasts, and faunal content, the length of the submarine canyon was about 14 miles with an average slope of about 4 per cent. This slope would increase if the canyon headed seaward from shore and would decrease if the water was relatively deep near shore. However, it appears to have a reasonable order of magnitude (B. C. Heezen, personal communication).

Additional supporting evidence for the canyon is the presence in a well (Standard Oil Company of California's Woo No. 1, Sec. 19, T. 2 N., R. 15 W., S.B.B.M.), roughly half-way between the fan and source area, of the highest Mohnian sand-shale ratio (about 1:1) of any of the central valley bore holes for which data were available to the writer. Much cobble conglomerate was present in this well. This was the easternmost of a series of exploratory holes drilled in the west-central valley, and its high sand and conglomerate content suggests the possibility that it may have penetrated the coarse fill of a depression, which might be a filled submarine canyon. Wells west of the Woo well have Mohnian sand-shale ratios of 1:4 to 1:12, but incompleteness of well data prevents accuracy in these determinations. At any rate the relative coarseness of the Mohnian in the Woo well corroborates the concept that the Tarzana Fan provenance was on the north.

SIGNIFICANCE OF DIRECTIONAL CURRENT STUDIES

Since the Modelo formation is not known to be unique in California, but rather representative of the type of clastic rocks that were deposited in the restricted troughs of the west coast throughout the Cenozoic era, it is a virtual certainty that similar studies in other west coast areas and in rocks of other ages will produce useful results. Oriented syngenetic structures have been observed by the writer in the Stevens sand (core from upper Miocene of San Joaquin Valley), Repetto and Pico formations (Pliocene of Ventura County), Puente formation (upper Miocene outcrops in the Puente Hills), Topanga formation (middle Miocene of Santa Monica Mountains), and Cretaceous outcrops in Sespe Creek and Simi Hills. They undoubtedly are very common, though unobtrusive, and must be sought to be found. Lack of mention of them in the literature does not prove their absence.

Use of oriented syngenetic structures in petroleum exploration in California (and probably elsewhere) is of great significance. The trend of linear shaped sand bodies could be predicted were current directions known. Channels could be predicted crossing anticlines where traps would result without structural closure. Masses of lenticular sand and shale bodies formerly thought to be randomly oriented may be found to have a predictable pattern. Magnitude of net slip on strike-slip faults can be determined by matching offset sedimentational patterns.

Establishment of the current patterns depends largely on well data, especially cores. In areas of known structure, cores whose top and bottom are known could be oriented with respect to the direction of dip, and their oriented features recorded. Where structure is not well known or dips are flat, cores oriented by other means would be required. After the current patterns are established, the reversal of these procedures could produce an oriented structural dip in a core containing a recognizable oriented syngenetic feature. In the case of cores, where a maximum amount of information must be squeezed from a small but expensive piece of rock, grain orientation might prove most useful. This technique has been successful in the Pliocene rocks of the Ventura Avenue oil field. Subsurface investigations would have the advantage of the third dimension of sand body shapes in contrast to the present study which was limited to the outcropping two-dimensional upturned cross section of the submarine fan. Because of the abundance of turbidites in the oil- and gas-bearing part of the geologic column in California (as yet largely undocumented), the writer feels that a whole new frontier of geologic knowledge is awaiting explorers, both academic and commercial, and that many oil pools and many answers to earth history problems will be found by detailed application of turbidity-current studies.

REFERENCES

AMERICAN GEOLOGICAL INSTITUTE, 1958, "Roundness of Sedimentary Particles," *Data Sheet 7.*

BANDY, O. L., 1953, "Ecology and Paleoecology of Some California Foraminifera; Part I: The Frequency Distribution of Recent Foraminifera off California," *Jour. Paleon.,* Vol. 27, pp. 161–203.

BEATIE, R. L., 1958, "The Geology of the Sunland-Tujunga Area, Los Angeles County, California," unpublished Master's thesis, University of California, Los Angeles. 102 pp.

BERGEN, F. W., 1955, "A Restudy of the Upper Mohnian-Lower Delmontian Boundary near Calabasas, California," *ibid.*

*BOKMAN, JOHN, 1953, "Lithology and Petrology of the Stanley and Jackfork Formations," *Jour. Geology,* Vol. 61, pp. 152–70.

BOWEN, O. E., JR., 1954, "Geology and Mineral Deposits of Barstow Quadrangle, San Bernardino County, California," *California Div. Mines Bull. 165.*

BRADY, T. J., 1957, "Geology of Part of the Central Santa Monica Mountains East of Topanga Canyon, Los Angeles County, California," unpublished Master's thesis, University of California, Los Angeles.

BRAMLETTE, M. N., 1946, "Monterey Formation of California and Origin of Its Siliceous Rocks," *U. S. Geol. Survey Prof. Paper 212.* 57 pp.

BROWN, G. E., 1957, "Geology of Parts of the Calabasas and Thousand Oaks Quadrangles, Los Angeles and Ventura Counties, California," unpublished Master's thesis, University of California, Los Angeles. 75 pp.

* Papers dealing with recognition of turbidite structures.

BUTCHER, W. S., 1951, "Foraminifera, Coronado Bank and Vicinity, California," *Univ. California, Scripps Inst. Oceanography, Submarine Geol. Rept. 19,* pp. 1–9.

CROUCH, R. W., 1952, "Significance of Temperatures on Foraminifera from Deep Basins off Southern California," *Bull. Amer. Assoc. Petrol. Geol.,* Vol. 36, pp. 807–43.

CROWELL, J. C., 1952, "Probable Large Lateral Displacement on San Gabriel Fault, Southern California," *ibid.,* Vol. 36, pp. 2026–35.

*———, 1955, "Directional Current Structures from the Prealpine Flysch, Switzerland," *Bull. Geol. Soc. America,* Vol. 66, pp. 1351–84.

*DOREEN, J. M., 1951, "Rubble Bedding and Graded Bedding in Talara Formation of Northwestern Peru," *Bull. Amer. Assoc. Petrol. Geol.,* Vol. 35, pp. 1829–49.

DUARTE, W. W., 1954, "Sherman Oaks Area, Santa Monica Mountains," unpublished manuscript map for Master's thesis, University of California, Los Angeles.

DURHAM, J. W., 1954, "The Marine Cenozoic of Southern California," in "Geology of Southern California," *California Div. of Mines Bull. 170,* pp. 23–31.

DURRELL, C., 1954, "Geology of the Santa Monica Mountains, Los Angeles and Ventura Counties," *ibid.,* Map Sheet 8.

*DZULYNSKI, S., AND RADOMSKI, A., 1955, "Origin of Groove Casts in the Light of Turbidity Current Hypothesis," *Acta Geol. Polonica,* Vol. 5, pp. 47–66. (Ten-page English summary.)

ELAM, J. G., 1948, "Geology of Seminole Quadrangle, Los Angeles County, California," unpublished Master's thesis, University of California, Los Angeles.

ELLIOTT, J. L., 1951, "Geology of Eastern Santa Monica Mountains between Laurel Canyon and Beverly Glen Boulevards, Los Angeles County, California," *ibid.*

EMILIANI, C., 1954, "Temperatures of Pacific Bottom Waters and Polar Superficial Waters during the Tertiary, *Science,* Vol. 119, pp. 853–55.

*ERICSON, D. B., EWING, MAURICE, AND HEEZEN, B. C., 1951, "Deep-Sea Sands and Submarine Canyons," *Bull. Geol. Soc. America,* Vol. 52, pp. 961–66.

*———, AND WOLLIN, G., 1955, "Sediment Deposition in Deep Atlantic," in "Crust of the Earth," *Geol. Soc. America Spec. Paper 62,* pp. 205–19.

*FAIRBRIDGE, R. W., 1946, "Submarine Slumping and Location of Oil Bodies," *Bull. Amer. Assoc. Petrol. Geol.,* Vol. 30, pp. 84–92.

*GILBERT, C. M., 1955, " 'Flow' Cast on Sandstone Beds" (abst.), Geol. Soc. America Cordilleran Section meeting, Berkeley, California.

GORSLINE, DONN S., AND EMERY, K. O., 1959, "Turbidity-Current Deposits in San Pedro and Santa Monica Basins off Southern California," *Bull. Geol. Soc. America,* Vol. 70, pp. 279–90.

HARDING, TOD P., 1952, "Geology of the Eastern Santa Monica Mountains between Dry Canyon and Franklin Canyon," unpublished Master's thesis, University of California, Los Angeles.

HAZENBUSH, G. C., 1950, "Geology of the Eastern Parts of the Dry Canyon and Las Flores Quadrangles, Los Angeles County, California," *ibid.*

*HILLS, E. S., AND THOMAS, D. E., 1954, "Turbidity Currents and the Graptolite Facies in Victoria," *Jour. Geol. Soc. Australia,* Vol. 1, pp. 119–33.

HOOTS, H. W., 1931, "Geology of the Eastern Part of the Santa Monica Mountains, Los Angeles County, California," *U. S. Geol. Survey Prof. Paper 165C.*

HSU, K. J., 1954, "Petrology of the Cucamonga Canyon-San Antonio Canyon Area, Southeastern San Gabriel Mountains, California," unpublished Ph. D. thesis, University of California, Los Angeles.

KEW, W. S. W., 1924, "Geology and Oil Resources of a Part of Los Angeles and Ventura Counties, California," *U. S. Geol. Survey Bull. 753.* 202 pp.

KLEINPELL, R. M., 1938, *Miocene Stratigraphy of California,* Amer. Assoc. Petrol. Geol. 450 pp.

*KOPSTEIN, F. P. H. W., 1954, "Graded Bedding of the Harlech Dome," *Rijksuniversiteit te Groningen, Geol. Inst. Pub. 81.* 97 pp. Groningen-Nederland.

*KRUMBEIN, W. C., AND SLOSS, L. L., 1951, *Stratigraphy and Sedimentation,*" pp. 94–101 (Sedimentary structures). W. H. Freeman and Company, San Francisco.

*KSIAZKIEWICZ, M., 1954, "Graded and Laminated Bedding in the Carpathian Flysch," *Soc. Geol. Pologue Annales,* Vol. 22, Fasc. 4, pp. 442–49. English summary of paper in Polish (Polskiego Towarzystwa Geologicznego, SW., Anny 6, Krakow, Poland).

*KUENEN, P. H., 1948, "Slumping in the Carboniferous Rocks of Pembrokeshire," *Quar. Jour. Geol. Soc. London,* Vol. 104, pp. 365–85.

———, 1948a, "Turbidity Currents of High Density," *18th Internat. Geol. Cong.,* Pt. VIII, Proc. of Sec. G (1950), pp. 44–52.

*———, 1952, "Paleogeographic Significance of Graded Bedding and Associated Features," *Proc. Koninkl. Nederl. Akad. Van Wetenschappen-Amsterdam,* Ser. B, 55, No. 1, pp. 28–36.

*———, 1953, "Significant Features of Graded Bedding," *Bull. Amer. Assoc. Petrol. Geol.,* Vol. 37, pp. 1044–66.

*———, 1954, "Recent Deposits and the Interpretation of Sedimentary Rocks" (summary), *Adv. Sci.,* Vol. 11, pp. 65–66.

*———, 1957, "Sole Markings of Graded Graywacke Beds," *Jour. Geology,* Vol. 65, pp. 231–58.

*———, AND CAROZZI, A., 1953, "Turbidity Currents and Sliding in Geosynclinal Basins of the Alps," *ibid.,* Vol. 61, pp. 363–73.

*———, AND MIGLIORINI, C. I., 1950, "Turbidity Currents as a Cause of Graded Bedding," *ibid.,* Vol. 58, pp. 91–126.

*———, AND SANDERS, J. E., 1956, "Sedimentation Phenomena in Kulm and Flozleeres Graywackes, Sauerland and Oberharz, Germany," *Amer. Jour. Sci.,* Vol. 254, pp. 649–71.

*KUGLER, H. C., 1953, "Jurassic to Recent Sedimentary Environments in Trinidad, " *Bull. Ass. Suisse des Geol. et Ing. du Petrole,* Vol. 20, pp. 27–60.

MACNEIL, F. S., 1957, "Cenozoic Megafossils of Northern Alaska," *U. S. Geol. Survey Prof. Paper 294-C,* pp. 99–126.

McGILL, J. T., 1948, "Geology of a Portion of the Las Flores and Dry Canyon Quadrangles, Los Angeles County, California," unpublished Master's thesis, University of California, Los Angeles.

*MENARD, H. W., 1955, "Deep-Sea Channels, Topography, and Sedimentation," *Bull. Amer. Assoc. Petrol. Geol.,* Vol. 39, pp. 236–55.

MERIFIELD, PAUL, 1958, "Geology of a Portion of the Southwestern San Gabriel Mountains, San Fernando and Oat Mountain Quadrangles, Los Angeles County, California," unpublished Master's thesis, University of California, Los Angeles.

*MIGLIORINI, C. I., 1950, "Dati a conferma della risedimentazione delle arenarie del macigno," *Atti. Soc. Toscana, Sci. Nat. Mem.*, Vol. 57, Ser. A, pp. 3–15. Quoted by Natland and Kuenen (1951).

NATLAND, M. L., 1933, "Temperature and Depth Distribution of Some Recent and Fossil Foraminifera in the Southern California Region," *Bull. Scripps Inst. Oceanography*, Tech. Ser., Vol. 3, pp. 225–30.

———, 1957, "Paleoecology of West Coast Tertiary Sediments," Chap. 19, pp. 543–572, in "Treatise on Marine Ecology and Paleoecology, Vol. 2, Paleoecology," edited by H. S. LADD, *Geol. Soc. America Memoir 67*.

*———, AND KUENEN, P. H., 1951, "Sedimentary History of the Ventura Basin, California, and the Action of Turbidity Currents," *Soc. Econ. Paleon. and Min. Spec. Paper 2*, pp. 76–107.

———, AND ROTHWELL, W. T., JR., 1954, "Fossil Foraminifera of the Los Angeles and Ventura Regions, California," in "Geology of Southern California," *California Div. Mines Bull. 170*, pp. 33–42.

OAKESHOTT, G. B., 1954, "Geology of the Western San Gabriel Mountains, Los Angeles County," *ibid.*, Map Sheet No. 9.

———, 1959, "Geology and Mineral Deposits of San Fernando Quadrangle, Los Angeles County, California," *California Div. Mines Bull. 172*.

*PACKHAM, G. H., 1954, "Sedimentary Structures as an Important Factor in the Classification of Sandstones," *Amer. Jour. Sci.*, Vol. 252, pp. 466–76.

*PASSEGA, R., 1954, "Turbidity Currents and Petroleum Exploration," *Bull. Amer. Assoc. Petrol. Geol.*, Vol. 38, pp. 1871–87.

PAYNE, T. G., 1942, "Stratigraphical Analysis and Environmental Reconstruction," *ibid.*, Vol. 26, pp. 1697–1770.

PETTIJOHN, F. J., 1949, *Sedimentary Rocks*. Harper and Brothers, New York.

*———, 1950, "Turbidity Currents and Graywackes; a Discussion," *Jour. Geology*, Vol. 58, pp. 169–71.

PHLEGER, F. B, 1951, "Displaced Foraminifera Faunas," *Soc. Econ. Paleon. and Min. Spec. Pub. 2*, pp. 66–75.

PIERCE, R. L., 1956, "Upper Miocene Foraminifera and Fish from the Los Angeles Area, California," *Jour. Paleon.*, Vol. 30, pp. 1288–1314.

*PRENTICE, J. E., 1956, "The Interpretation of Flow Markings and Load Casts," *Geol. Mag.*, Vol. 93, pp. 393–400.

REICHE, PARRY, 1938, "An Analysis of Cross-Lamination: the Coconino Sandstone," *Jour. Geology*, Vol. 46, pp. 905–32.

RESIG, J. M., 1956, "Ecology of Foraminifera of Santa Cruz Basin, California," Master's thesis, University of Southern California. Later (1958) published under same title in *Micropaleontology*, Vol. 4, pp. 287–308.

*RICH, J. L., 1950, "Flow Markings, Groovings, and Intra-stratal Crumplings as Criteria for Recognition of Slope Deposits, with Illustrations from Silurian Rocks of Wales," *Bull. Amer. Assoc. Petrol. Geol.*, Vol. 34, pp. 717–41.

*———, 1951, "Three Critical Environments of Deposition and Criteria for Recognition of Rocks in Each of Them," *Bull. Geol. Soc. America*, Vol. 62, pp. 1–20.

*RIVEROLL, D. D., AND JONES, B. C., 1954, "Varves and Foraminifera of a Portion of the Upper Puente Formation (Upper Miocene), Puente, California," *Jour. Paleon.*, Vol. 28, pp. 121–31.

*SCHNEEBERGER, W. F., 1955, "Turbidity Currents, a New Concept in Sedimentation and Its Application to Oil Exploration," *Mines Magazine*, Vol. 45, No. 10, pp. 42–62.

*SHROCK, R. R., 1948, *Sequence in Layered Rocks*. McGraw-Hill Book Company, Inc., New York.

SOPER, E. K., 1938, "Geology of the Central Santa Monica Mountains, Los Angeles County, California," *California Jour. Mines and Geology*, Vol. 34, pp. 131–80.

STOKES, W. L., 1947, "Primary Lineation in Fluvial Sandstones, a Criterion of Current Direction," *Jour. Geology*, Vol. 55, pp. 52–53.

*SUJKOWSKI, ZB. L., 1957, "Flysch Sedimentation," *Bull. Geol. Soc. America*, Vol. 68, pp. 543–54.

*SULLWOLD, HAROLD H., JR., 1959, "Nomenclature of Load Deformation in Turbidites," *ibid.*, Vol. 70, pp. 1247–48.

*———, 1960, "Load Cast Terminology and Origin of Convolute Bedding: Further Comments," *ibid.*, Vol. 71, in press.

TERPENING, J. N., 1951, "Geology of Part of the Eastern Santa Monica Mountains," unpublished Master's thesis, University of California, Los Angeles.

TRASK, P. D., 1932, *Origin and Environment of Source Sediments of Petroleum*. Gulf Publishing Company, Houston, Texas.

TRAXLER, J. D., 1948, "Geology of East-Central Santa Monica Mountains," unpublished Master's thesis, University of California, Los Angeles.

TYRELL, G. W., 1950, *The Principles of Petrology*, 11th ed. Methuen and Co., London.

UCHIO, T., 1957, "Ecology of Living Benthonic Foraminifera from the San Diego, California, Area," unpublished Ph.D. thesis, University of California, Los Angeles (Scripps Institution of Oceanography).

*VAN HOUTEN, F. B., 1954, "Sedimentary Features of Martinsburg Slate, Northwestern New Jersey," *Bull. Geol. Soc. America*, Vol. 65, pp. 813–17.

WADELL, H., 1935, "Volume, Shape, and Roundness of Quartz Particles," *Jour. Geology*, Vol. 43, pp. 250–80.

WEST, J. C., 1947, "Geology of a Small Portion of the Central Santa Monica Mountains, California," Senior thesis in University of California, Los Angeles, library, unpublished.

WILLIAMS, H., TURNER, F. J., AND GILBERT, C. M., 1955, *Petrography*. W. H. Freeman and Company, San Francisco.

*WILSON, G., WATSON, J., AND SUTTON, J., 1953, "Current-Bedding in the Moine Series of Northwestern Scotland," *Geol. Mag.*, Vol. 90, pp. 377–87.

*WINTERER, E. L., 1954, "Geology of Southeastern Ventura Basin, Los Angeles County, California," unpublished Ph.D. dissertation, University of California, Los Angeles.

Depositional Environments of a Reservoir Sandstone in West-Central Texas[1]

RICHARD R. BLOOMER[2]

Abstract Hundreds of late Paleozoic stratigraphic traps in fluvial, deltaic, and marine sandstones are present on the Eastern shelf and slope of the Midland basin in west-central Texas. Widespread fluvial sandstones on the shelf are well known, but there appears to be little recognition of the depositional environments of sandstones in ancient deltas on the shelf and submarine channels and fans on the slope.

The Cook sandstone is a depositional model of late Paleozoic fluvial, deltaic, and slope sandstone facies in the Midland basin. Each of two Cook fluvial and deltaic systems mapped is about 100 mi (160 km) long from outcrop to shelf edge. The "Blackwell" field, in one system, is an excellent example of a Cook point-bar stratigraphic trap.

Basinward from the basal Permian Eastern shelf edge, sand and mud supplied by the two Cook fluvial systems prograded into a shallow sea and became two high-constructive lobate deltas. Delta-plain aggradational facies, delta-front sheet and distributary-channel sandstones, and prodelta facies are present in the deltas. The Group 4000 "Pennsylvanian Cisco" sandstone field produces from Cook deltaic sandstone.

The deltas are the proximal parts of a clastic wedge of sediments on the Eastern slope. The wedge has a maximum thickness of 1,300 ft (396 m) at midslope and thins basinward. The Northeast Bloodworth "Pennsylvanian Canyon" field produces from midslope Cook sandstone which probably was deposited in a submarine channel. The Jameson "Pennsylvanian Strawn" sandstone field produces from Cook lower-slope sandstone and appears to be a submarine fan.

INTRODUCTION

Approximately 300 Lower Permian sandstone reservoirs are present on the Eastern shelf and slope of the Midland basin in west-central Texas (Fig. 1). Most of the reservoirs were discovered between the 1920s and 1950s in shallow fluvial sandstones. First recognition of the fluvial origin of the sandstone in one of these fields was not until 1960 (Lawless, 1960; Shankle, 1960).

The widespread presence of fluvial-sandstone reservoirs on the Eastern shelf now is well known, but there is still little recognition of stratigraphic traps in deltaic deposits on the shelf and submarine-canyon and fan deposits on the slope.

Cook sandstone is a depositional model of late Paleozoic fluvial, deltaic, and slope sandstone facies in the Midland basin. This paper is a regional and local subsurface study of the Cook sandstone which was deposited in two paleodrainage systems on the Eastern shelf and a submarine canyon and fan on the Eastern slope. All of the producing areas in the Cook depositional systems are essentially stratigraphic traps. Exploration for

these traps depends greatly on the interpretation of local environments of deposition. The subsurface methods used to reveal the Cook sandstone depositional environments and associated types of traps are applicable in the exploration for similar sandstone reservoirs in other areas.

GEOLOGIC SETTING

Lower Permian sedimentary rocks crop out along the eastern margin of the Eastern shelf as shown in Figure 1. At the base of the Permian, the Eastern shelf has a regional west-northwest dip of about $0.5°$ (50 ft/mi or 9 m/km) to the shelf edge (Fig. 1). West of the shelf edge, the Eastern slope reaches a maximum regional dip of about $3°$ (276 ft/mi or 53 m/km) at midslope and decreases farther downslope.

Fluvial sandstones (Fig. 2) or their deltaic and slope-stratigraphic equivalents produce in the fields shown in Figure 1. Each sandstone unit is in a cyclic depositional sequence or format (Forgotson, 1957) bounded by thin persistent limestone or coal beds. These beds are excellent surface and subsurface mapping units over a wide area of the Eastern shelf. The widespread and persistent characteristic of these thin limestone and coal beds indicates that they were deposited during eustatic changes in sea level. The limestone beds, with suites of open-marine fossils, probably were deposited during periods of transgression. Regressions of the sea exposed alluvial coastal plains which were incised by meandering streams. Swamps and marshes on the plains became coal beds after burial. There are widespread paleosol zones on the shelf (Harrison, 1973).

Cyclic deposition on the shelf probably was a response to changes in sea level during alternating glacial and interglacial periods in the Southern Hemisphere (Van Siclen, 1958; Shankle, 1960; Jackson, 1964). Brown (1969a, b) and Gal-

[1]Manuscript received, May 24, 1976; accepted, August 16, 1976. Presented before the Southwestern Section of the Association at Fort Worth, Texas, March 16, 1973.

[2]Independent geologist, 132 Devonian Bldg., Abilene, Texas 79603.

The writer is indebted to Daniel A. Busch and Robert A. Berg for their stimulating discussions of the study and critical reading of the manuscript.

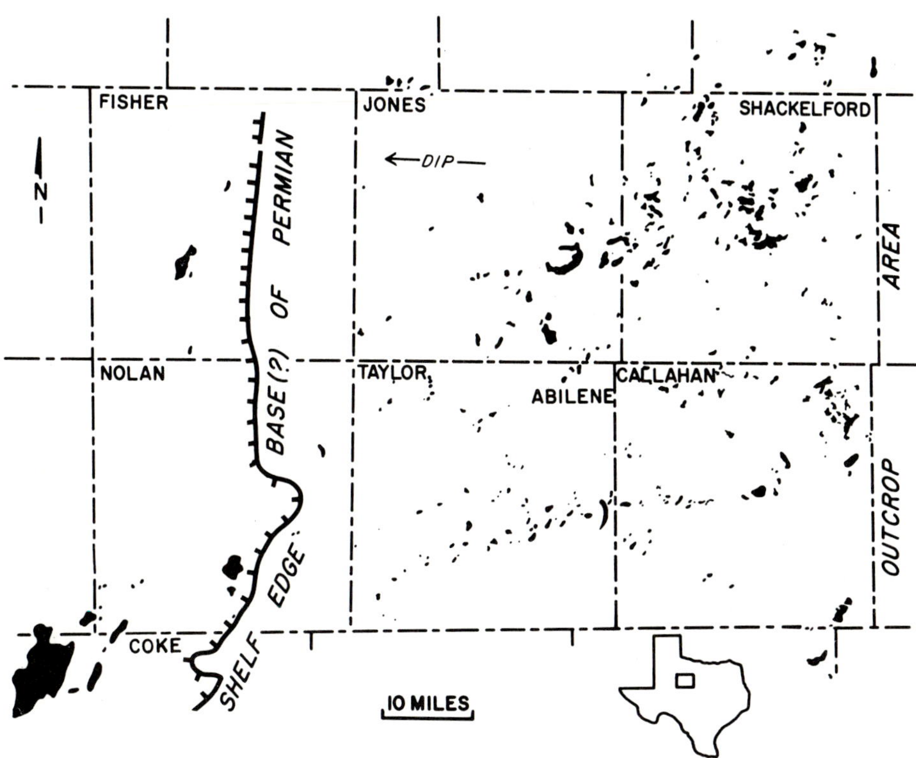

FIG. 1—Map showing basal Permian shelf edge and oil and gas fields producing from Lower Permian sandstones in west-central Texas.

loway and Brown (1972, 1973), however, did not recognize widespread thin limestone, coal, and paleosol beds on the shelf, and, therefore, did not attribute the apparent cyclicity of shelf deposits to eustatic sea-level fluctuations but to intrabasinal variations in uplift and subsidence or sedimentary controls.

The top of the Crystal Falls Limestone Member at the base of the sections shown on Figure 2 is a convenient lithologic marker for the controversial Pennsylvanian-Permian boundary. Overlying the Crystal Falls Limestone Member is the Waldrip Shale Member which contains three thin limestone units named in ascending order, Waldrip limestones 1, 2, and 3 (Fig. 2). The uppermost of the three limestones commonly is referred to as the Flippen limestone in the subsurface. These limestone beds are only a few feet thick in the outcrop (Stafford, 1960) but thicken basinward in the subsurface.

Any sandstone present between the Crystal Falls and Flippen limestones is termed informally "Cook sandstone." Two Cook sandstone paleodrainage systems (post-Waldrip limestone 2 and pre-Flippen limestone in age) are considered in this study. In the fluvial environment, maximum incision of Cook channels was through Waldrip shale and limestone beds to the Crystal Falls limestone (about 100 ft or 30 m). The channels, the extensive coal beds, and the widespread paleosols noted by Harrison (1973) are evidence that a subaerial environment was widespread during deposition of this stratigraphic interval.

COOK PALEODRAINAGE SYSTEMS

Two Cook fluvial and deltaic systems have been mapped (Fig. 3) on the Eastern shelf. About 100 mi (160 km) of each system is preserved in the surface and subsurface. The headwaters of these paleodrainage systems probably were in the ancient Ouachita Mountains, now a buried structural belt (Flawn et al, 1961) about 100 mi (160 km) east.

The Cook sandstone outcrop area shown on Figure 3 is modified after surface maps by Stafford (1960) and Barnes (1972). Every available well log (several thousand) was used to map the paleodrainage systems in the subsurface. Electric-log control is about 95%. Well density ranges

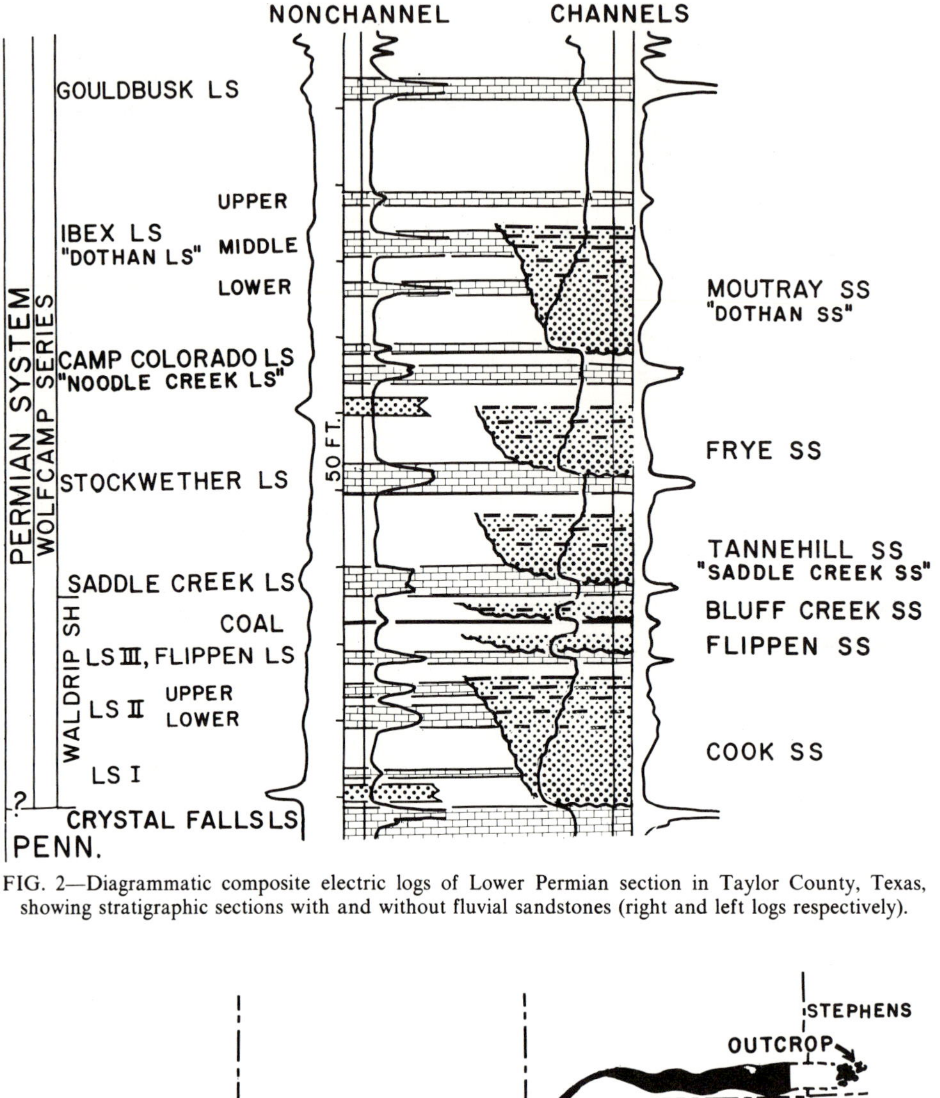

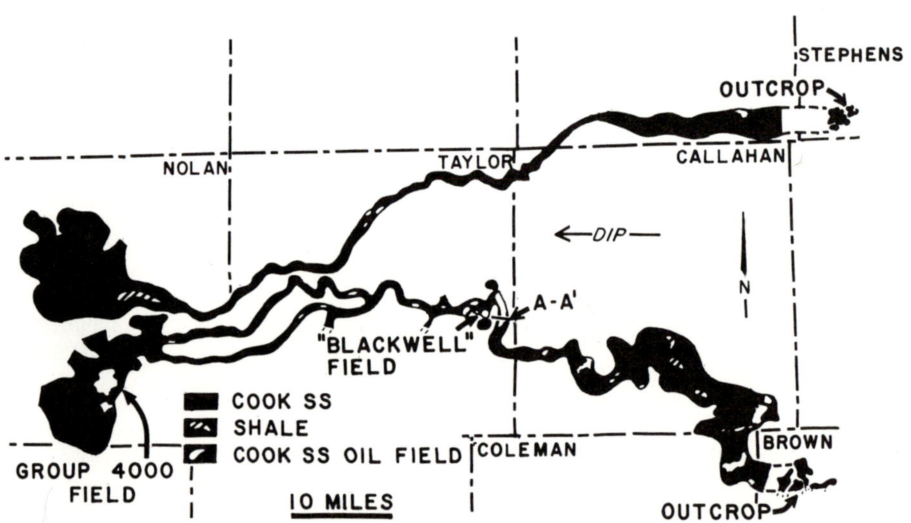

FIG. 2—Diagrammatic composite electric logs of Lower Permian section in Taylor County, Texas, showing stratigraphic sections with and without fluvial sandstones (right and left logs respectively).

FIG. 3—Two Cook paleodrainage systems on Eastern shelf, showing meander belts from outcrop to deltas. *AA'* is line of cross section shown in Figure 5.

FIG. 4—Outcrop in southwest Stephens County, Texas, showing sharp erosional contact (arrow) of Cook point-bar deposit with subjacent Waldrip shale.

from several to over 50 wells/sq mi (20 wells/sq km).

The Cook fluvial systems are considered to be meander belts. The meandering streams within the belts flowed on a low-gradient alluvial paleoslope. Abrupt contacts of the meander-belt deposits with Waldrip limestone beds clearly outline the boundaries of the belts (Fig. 5).

Cook sandstones in the outcrop of the north meander belt (Fig. 3) are characteristic of a fluvial section (Fig. 4). The arrow in Figure 4 points to the sharp basal contact of Cook channel fill with the subjacent Waldrip shale. The basal 2 ft (0.6 m) of channel fill is conglomerate, consisting of pebbles of chert, quartz, and limestone, and probably is channel-lag material. The conglomerate decreases in size upward to medium-grained sandstone at the top of Figure 4. Cross-bedding is common and indicates that the paleocurrent flowed westward. The conglomerate and sandstone in Figure 4 probably are part of a point-bar deposit.

Stratigraphic and structure cross sections of the south meander belt (Fig. 5) show the sharp contact of shale and sandstone with the laterally adjacent Waldrip limestone beds marking the boundaries of the meander belt. The convex-downward base of the meander belt is clear. Two periods of Cook channel cut and fill are indicated.

Brown (1969a, b) and Galloway and Brown (1972, 1973) described downdip elongated Cook and upper Cook/Flippen sandstone units on the Eastern shelf as fluvial-deltaic lobes of progradation and the "interlobe" limestone, shale, and sandstone beds as interdeltaic embayment deposits. Their interpretations were based mainly on

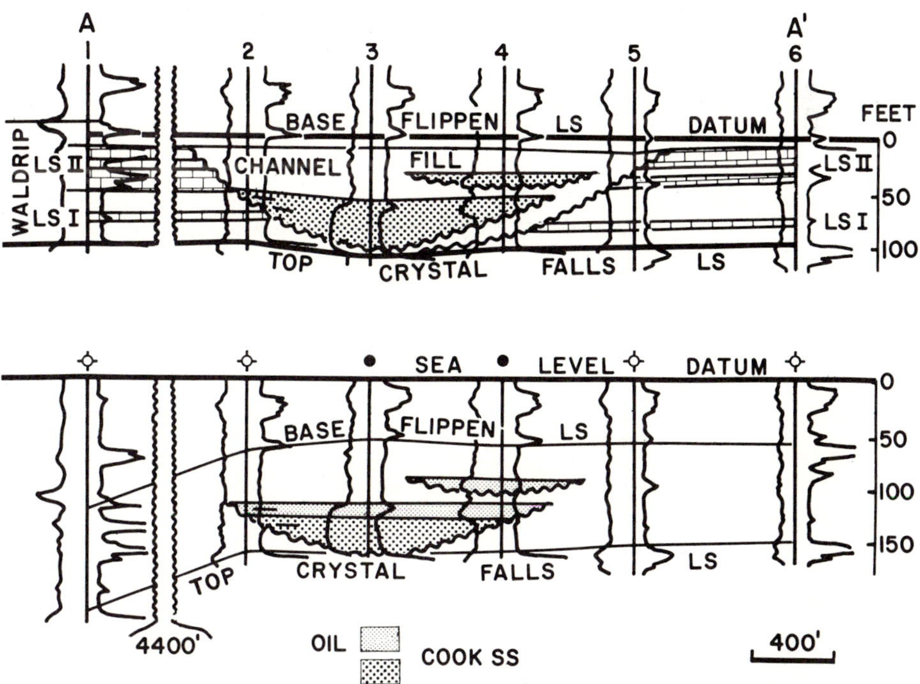

FIG. 5—Stratigraphic and structure cross sections of Cook meander belt. See Figure 3 for location of cross section.

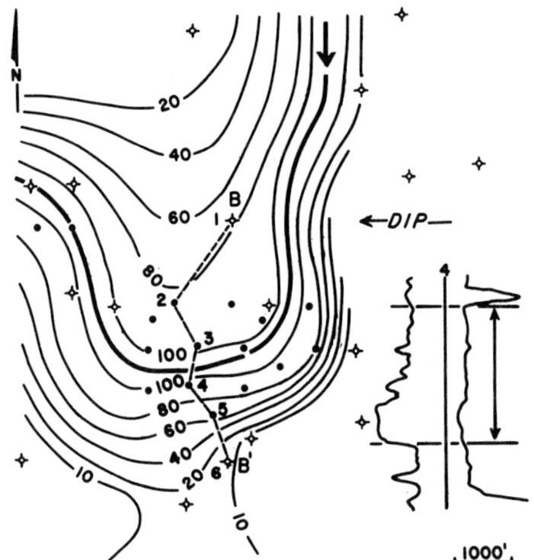

FIG. 6—Isopach map of interval from base of Flippen limestone to base of upper Cook sandstone in Blackwell field, Taylor County, Texas. Contour intervals 10 and 20 ft (3 and 6 m). *BB'* is line of cross section shown in Figure 7.

regional sandstone isopach maps with selected control points at an average 5-mi (8 km) spacing. With such wide well spacing there is little subsurface control for tracing dip-elongated 1 mi (1.6 km) wide belts of Cook sandstone. However, the well spacing can be sufficient for regional isopach mapping of strike-oriented sheet sandstones that thicken basinward.

As Brown and Galloway's upper Cook/Flippen sandstone unit name implied, this isopach interval apparently includes two genetic units or depositional systems (Flippen and Cook sandstones, Fig. 2), making it difficult to interpret the depositional environment of each sandstone unit.

Emphasis in the present study has been to interpret depositional environments from various maps and cross sections of one genetic unit with closely spaced well control.

COOK POINT-BAR STRATIGRAPHIC TRAP

The "Blackwell" (Taylor County Regular) field in east Taylor County (Fig. 3) is a typical fluvial-sandstone reservoir in west-central Texas. Detailed analysis of the field reveals the depositional environment of the sandstone and the type of stratigraphic trap.

An isopach map of the interval from the base of the Flippen limestone, an excellent time-lithologic marker, to the base of the upper Cook fluvial sandstone in the Blackwell field (Fig. 6) illustrates the Cook meander belt. The map can be interpreted as a paleotopographic map of the base of a meander belt or as a "cast" of the meander-belt fill. The heavy line between the 100-ft (30 m) contours is the axial trend or paleovalley axis of the meander belt. The map and cross sections (Fig. 7) show that the meander belt has a convex-downward base.

A subsurface isopach map based on a genetic unit (genetic increment of strata—GIS; Busch, 1971) between a time-lithologic marker bed and the base of a channel-fill deposit (Fig. 6) is essential for mapping channel deposits (Andresen, 1962; Busch, 1971).

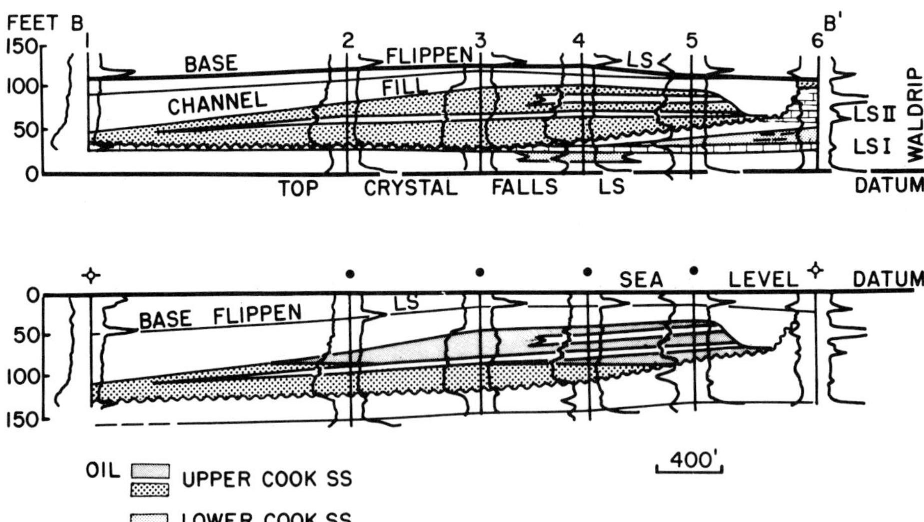

FIG. 7—Stratigraphic and structure cross sections *BB'* of Cook point bar. See Figures 6, 8, 9, 10 for location of section.

Figure 8 is an isopach map of the upper Cook sandstone in the Blackwell field. Inside the bend of the meander belt the sandstone has a maximum thickness of more than 60 ft (18 m). The stippled area between the 0 and 20-ft (6 m) contours is interpreted as the last position of the channel before its abandonment and filling with marine mud. The presence of this updip impermeable barrier to the oil in the point bar is supported by the fact that the upper Cook sandstone in well 6 did not have a show of oil, even though it is higher structurally than the original oil-water contact in well 5 (Fig. 7).

As the channel eroded the south bank of the meander, sandstone deposition on the inside of the meander prograded southward forming a point bar. The *BB'* stratigraphic cross section is, therefore, a profile of the point bar with the vertical scale greatly exaggerated.

The "Christmas tree" or "bell" shape of the Cook sandstone self-potential (SP) curves of the electric logs (Fig. 7) is indicative of a point-bar sandstone deposit. The SP curves characteristically show (1) a sharp basal contact, (2) a decrease in SP amplitude upward, and (3) a more serrate curve upward. Erosion of the subjacent formation made the sharp basal contact of the Cook channel fill (Fig. 4). A decrease in grain size and increase in interstitial clay upward cause the upward decrease in SP amplitude. The change in grain size is very evident in sample cuttings. The sandstone becomes more interbedded with shale

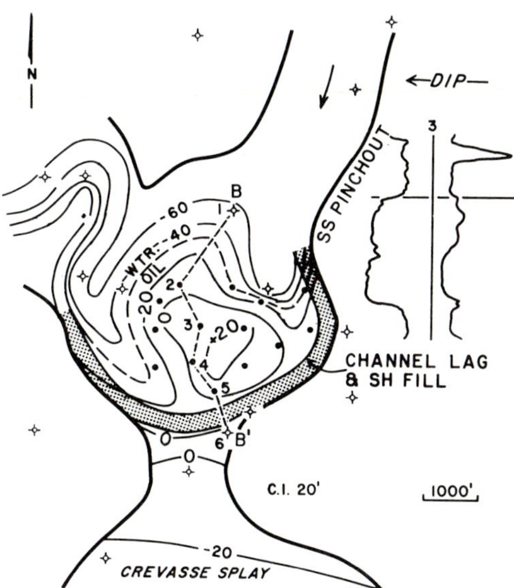

FIG. 9—Structure map of top of upper Cook sandstone in Blackwell field, showing oil trapped in point bar. Contour interval 20 ft (6 m). *BB'* is line of section shown in Figure 7.

upward, making a more serrate SP curve upward until the channel deposit merges into marine siltstone and shale.

About 10 ft (3 m) of shaly sandstone is present in the zone between the Flippen and Waldrip 2 limestone beds in well 6 of Figure 7. South of the meander belt shown in Figure 8 this shaly sandstone is a fan-shaped deposit which is interpreted as an overbank deposit. It pinches out about 5 mi (8 km) south of the area in Figure 8. Probably at flood stage, water overflowed a natural levee bordering the meander belt or cut a crevasse in the levee, and mud and sand were deposited in a fan shape. In all probability it represents a crevasse splay.

Overbank deposits generally border meander belts, deltaic channels, and submarine canyons and fan valleys (Jacka et al, 1968; LeBlanc, 1972; Stanley and Unrug, 1972; Shelton, 1973). When these deposits are recognized in the subsurface, adjoining channel sandstone deposits can be predicted.

In the upper Cook sandstone in the Blackwell field (Fig. 9), oil is trapped by a combination of structural closure and updip pinchout against the clay plug in the channel.

A structure map of the base of the Flippen limestone in the Blackwell field (Fig. 10) demonstrates that the structural closure is a result of differential compaction between the underlying point-bar sandstone and surrounding shale.

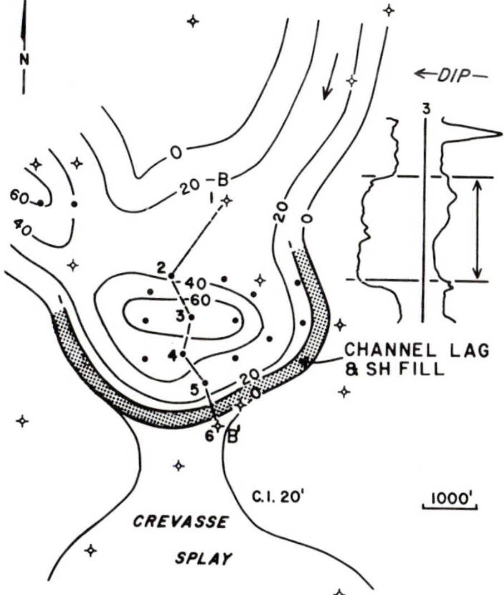

FIG. 8—Isopach map of upper Cook sandstone in Blackwell field, showing point bar. Contour interval 20 ft (6 m). *BB'* is line of section shown in Figure 7.

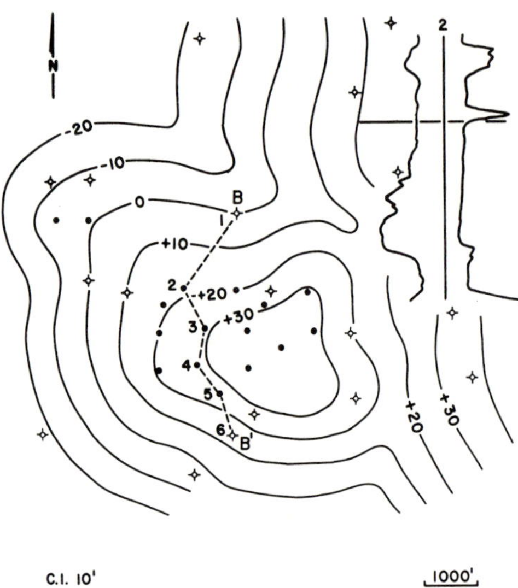

FIG. 10—Structure map of base of Flippen limestone in Blackwell field. Contour interval 10 ft (3 m). *BB'* is line of section in Figure 7.

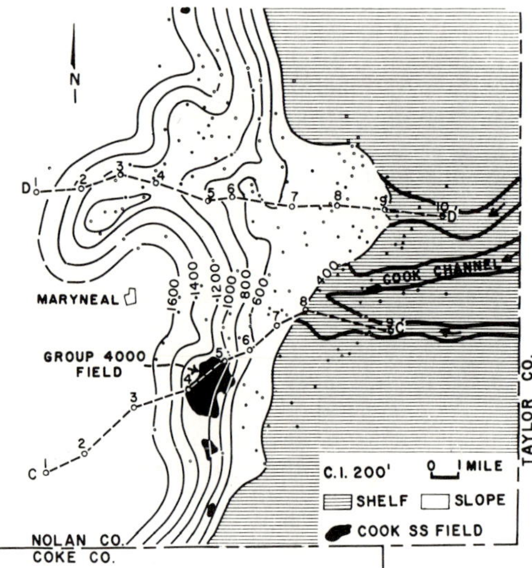

FIG. 11—Isopach map of interval from top of Noodle Creek limestone to top of Waldrip limestone 2, Nolan County, Texas. Contours show paleoslope of top of Waldrip limestone 2. *CC'* is line of section of Figure 12, *DD'* of Figure 14.

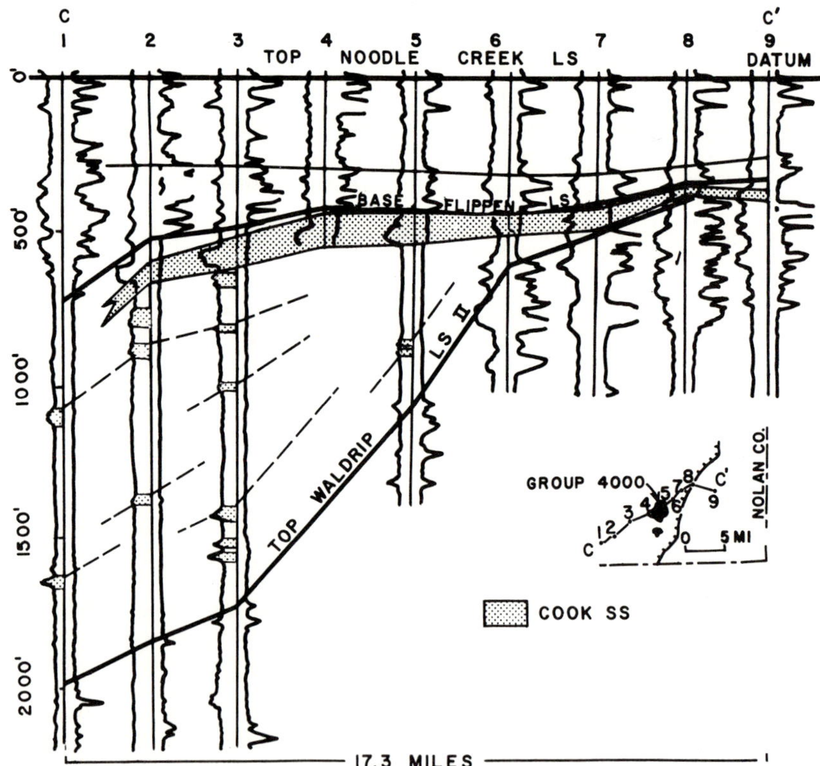

FIG. 12—Stratigraphic cross section *CC'* showing fluvial, deltaic, and slope facies of Cook sandstone. There is no horizontal scale between wells. Location of cross section is shown on Figures 11 and 13.

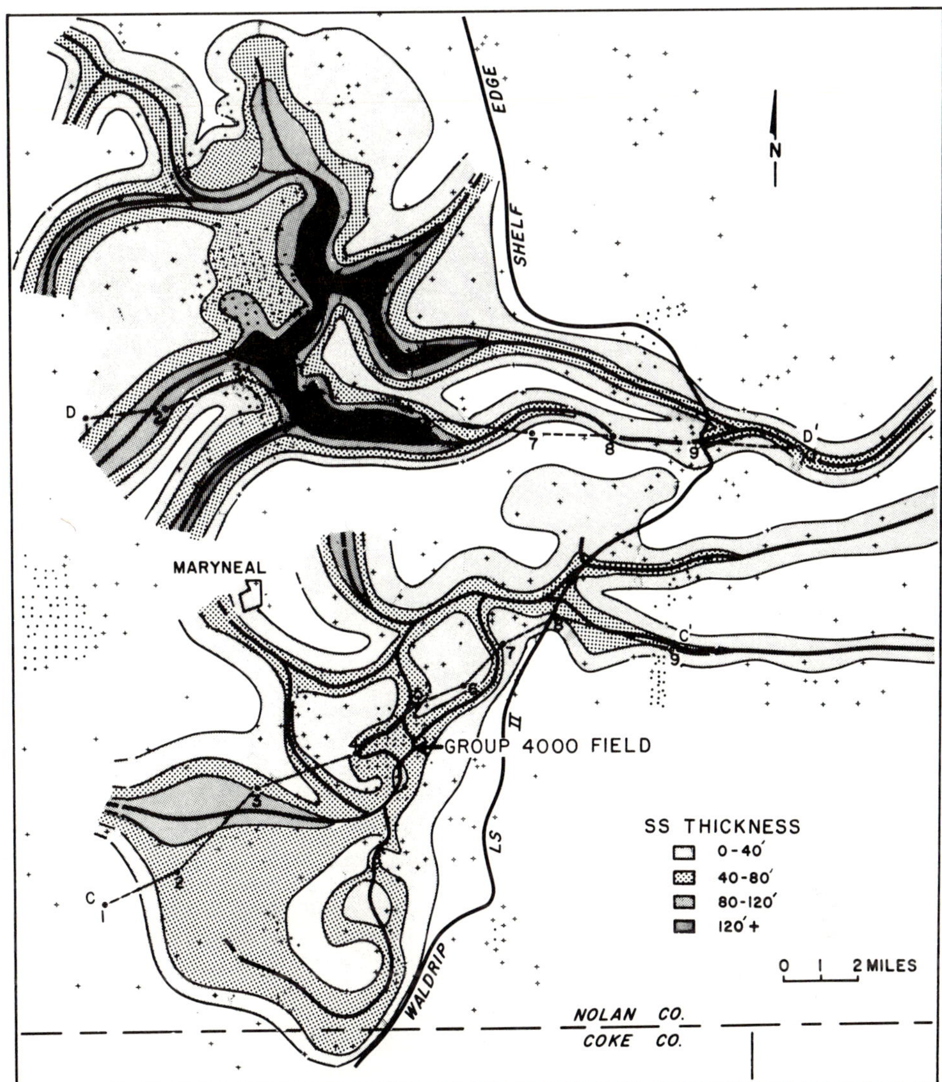

FIG. 13—Isopach map of Cook sandstone in meander belts and deltas in Nolan County, Texas. *CC'* is line of section shown on Figure 12, *DD'* on Figure 14.

At an average depth of 2,200 ft (671 m) the Cook sandstone in the Blackwell field has produced about 100,000 bbl of oil per well.

BASAL PERMIAN SHELF EDGE AND PROXIMAL SLOPE

A paleotopographic map of the top of Waldrip limestone 2 (Fig. 11) in Nolan County, Texas (Figs. 1, 3), shows a steepened slope between wells 3 and 6 (maximum slope angle about 3°) with decreasing slope basinward west of well 3.

Late Paleozoic depositional shelves and slopes along the east margin of the Midland basin first were described by Rall and Rall (1958) and Van Siclen (1958). Before depositional topography was recognized in the Midland basin, most geologists explained the lateral change in lithology, such as the shale-limestone change between wells 5 and 6 of Figure 12, as a change in facies.

The base of the Flippen limestone between wells 2 and 9 in Figure 12 is on the Eastern shelf. The Flippen limestone of well 2 was deposited on the proximal part of the Eastern slope. The clastic wedge of sediments between Waldrip limestone 2 and Flippen limestone (Figs. 12, 14) is equivalent stratigraphically to the Cook sandstone and Wal-

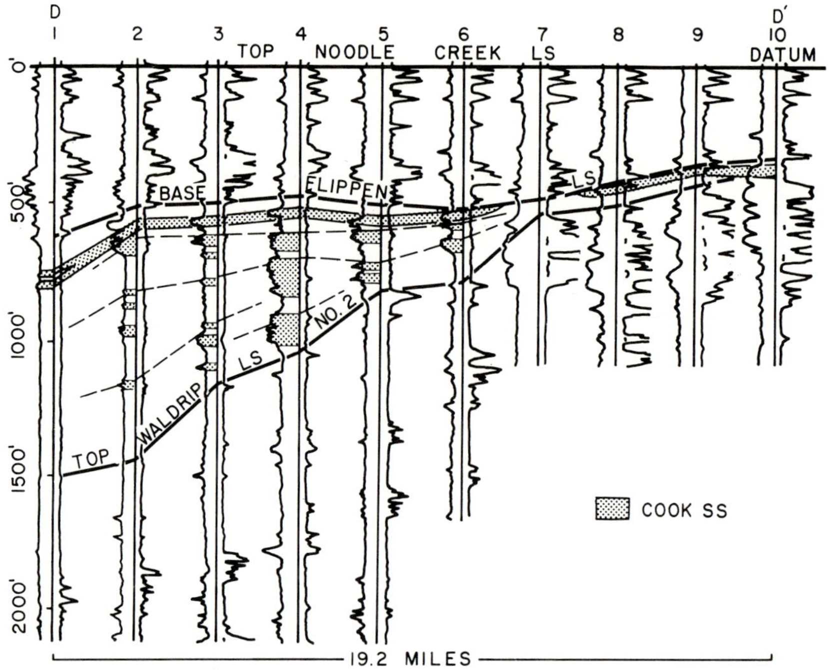

FIG. 14—Stratigraphic cross section *DD'* showing Cook fluvial, deltaic, and slope sandstone facies. There is no horizontal scale between wells. See Figures 11 and 13 for location of cross section.

drip shale between these limestone beds on the shelf.

COOK DELTAS

The shelf edge of Waldrip limestone 2 in Figure 11 coincides approximately with the most westward regressive position of the east shoreline of the embayment. West of the shelf edge, mud and sand supplied by Cook fluvial systems prograded into the shallow sea and became two lobate deltas (Fig. 13). The south delta is approximately 12 mi (19 km) long and 7 mi (11 km) wide and the north delta 16 mi (26 km) long and 10 mi (16 km) wide.

Approximately the eastern third of each delta consists of mostly upper delta-plain facies (wells 6-8, *CC'*, Figs. 12, 13; wells 6-9, *DD'*, Figs. 13, 14). The lithology consists of about 50 ft (15 m) of thin, fine-grained, lenticular sandstone beds interbedded with dark gray to black carbonaceous shale and lenses of coal. These deposits are probably the aggradational facies of swamps and floodplains in a delta plain.

South of wells 7 and 8 of the *DD'* line (Fig. 13) is a double fan-shaped sandstone deposit covering about 3 sq mi (8 sq km). It consists of about

10 ft (3 m) of shaly sandstone. This deposit is probably two coalescing crevasse splays.

Between wells 2 and 5 in Figure 12, an upper unit of Cook sandstone is present. Characteristically the sandstone in each well has an average thickness of 40 ft (12 m) and the electric log SP curves are cylinder- or funnel-shaped with a high amplitude. Nearly every well west of the delta-plain area in the south delta has this sandstone unit in the same stratigraphic position. It apparently consists of delta-front sheet sandstone and distributary-channel sandstone facies. The sandstone units generally are overlain by delta-plain facies (Figs. 12, 14, 16).

The thick wedge of shale and thin lenticular sandstone beds between the upper sandstone facies and Waldrip limestone 2 (Fig. 12) represent prodelta facies.

Figure 14 is a stratigraphic cross section of the north delta in Figures 11 and 13. Cook fluvial sandstone is present in well 10. Wells 8 and 9 have delta-plain facies of Cook sandstone interbedded with carbonaceous shale. In wells 2 through 6, an upper unit of Cook sandstone, with an average thickness of 20 ft (6 m) has the same SP-curve characteristics as the upper unit of sandstone in the south delta and, likewise, is probably

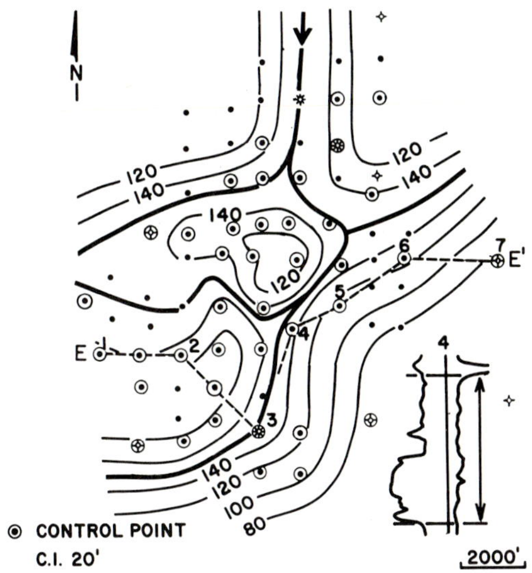

FIG. 15—Isopach map of interval from base of Flippen limestone to base of Cook sandstone in Group 4000 field. *EE'* is line of section shown in Figure 16. Heavy lines are axial trends of thickest deposition. Contour interval 20 ft (3 m).

delta-front sheet and distributary-channel sandstone facies.

In well 4 of Figure 14, more than 300 ft (91 m) of sandstone is present between the delta-front sheet sandstone and Waldrip limestone 2. This thick sandstone unit probably was deposited as a distributary-mouth bar as deposition prograded along the thick sandstone trend shown in Figure 13.

DELTAIC STRATIGRAPHIC TRAP

The Group 4000 "Cisco" sandstone field near the center of the south delta in Figures 11 and 13 produces from Cook sandstone. An isopach map (Fig. 17) and stratigraphic section (Fig. 16) show that the Cook sandstone has a convex-downward base indicative of a channel cut-and-fill deposit. The isopach of the Cook sandstone (Fig. 17) and of the interval between the base of the Flippen limestone and base of the Cook sandstone (Fig. 15) show that the axial trends of the thickest interval and thickest sandstone coincide. These axial trends are probably the thalwegs of distributary channels.

The lower Cook sandstone in wells 1, 2, 5, and 6 of Figure 16 is interpreted to be a delta-front sheet sandstone. It has an average thickness of 40 ft (12 m) and a characteristic sheet sandstone SP curve. The thick upper Cook sandstone in wells 3 and 4 is probably a distributary channel-fill deposit. The base of the channel is characteristically convex downward. As shown on Figure 17, the thick upper Cook sandstone in well 3 is part of an axial trend of thick sandstone.

The thin lenticular beds of upper Cook sandstone in Figure 16 are interbedded with gray to black carbonaceous shale and some coal lentils. These beds probably are aggradational facies de-

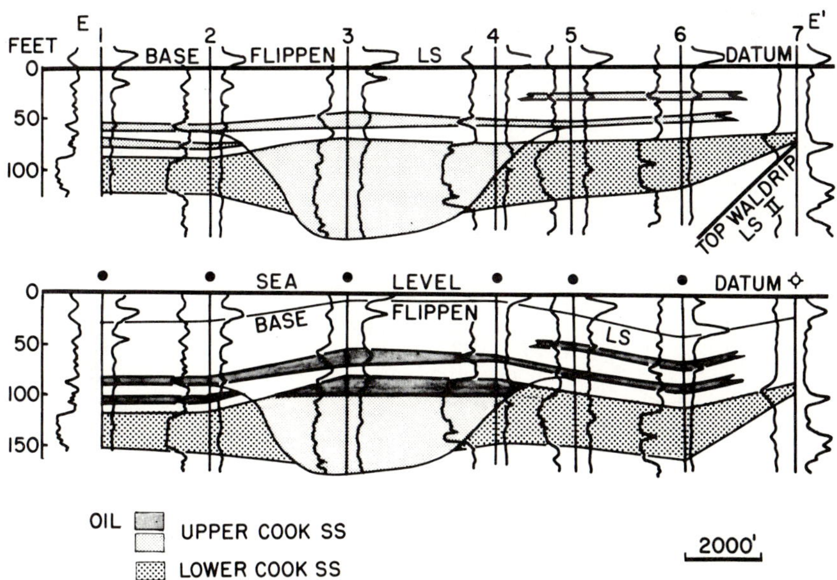

FIG. 16—Stratigraphic and structure cross sections *EE'* of Cook sandstone in Group 4000 field, showing delta-front sheet sandstone incised by distributary channel and filled with channel sandstone (wells 3, 4). See Figures 15 and 17 for location of cross sections.

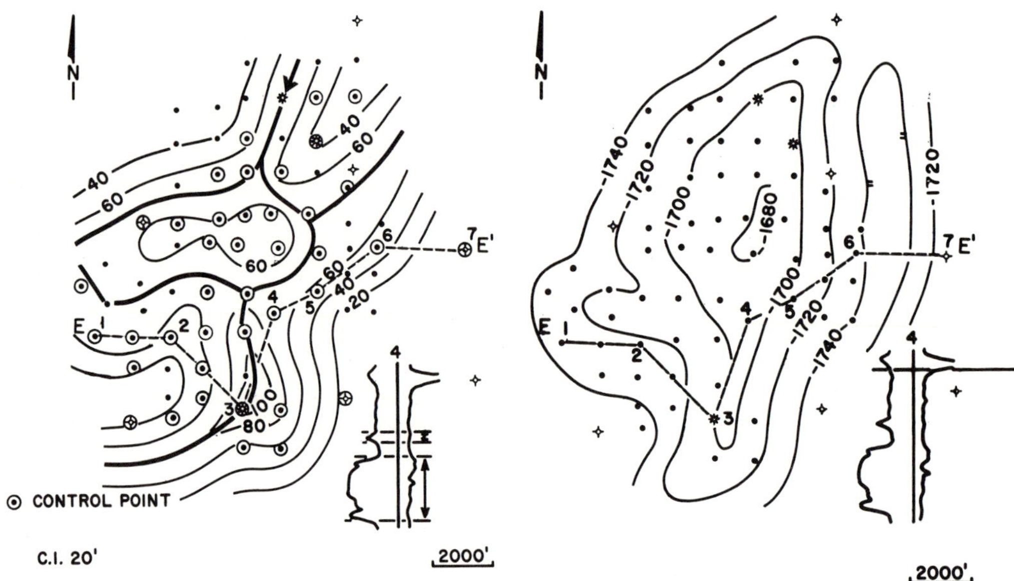

FIG. 17—Isopach map of Cook sandstone in Group 4000 field showing distributary pattern of channel sandstone. Contour interval 20 ft (6 m). *EE'* is line of section of Figure 16. Heavy lines are axial trends of thickest sandstone deposition.

FIG. 18—Structure map of base of Flippen limestone in Group 4000 field. *EE'* is line of section of Figure 16. Contour interval 20 ft (3 m).

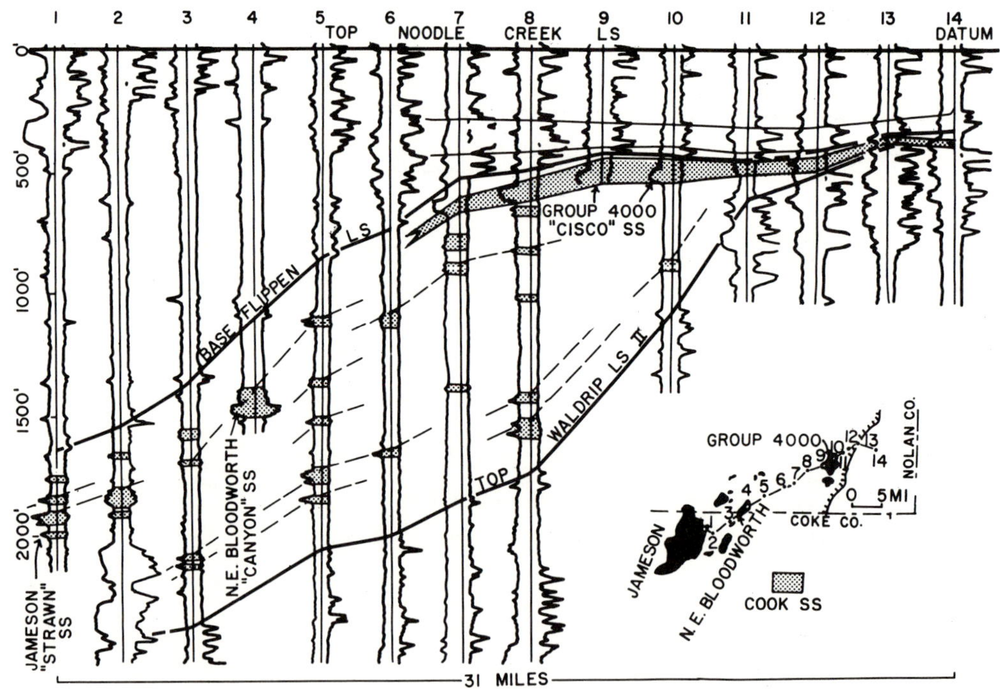

FIG. 19—Stratigraphic cross section of basal Permian clastic wedge of sedimentary rocks on Eastern slope. Wells 6-14 are wells 1-9 in Figure 12 (no horizontal scale between wells).

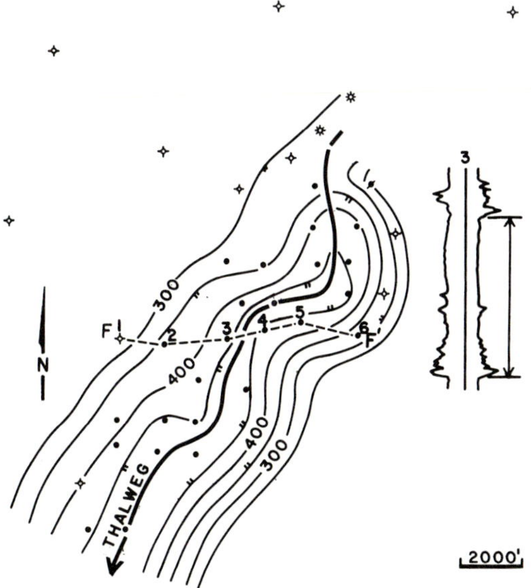

FIG. 20—Isopach map of interval from base of Flippen limestone to base of Cook sandstone in Northeast Bloodworth field, showing convex-downward base of Cook sandstone and axial trend (heavy line). *FF'* is line of cross section shown in Figure 21. Contour interval 50 ft (15 m).

posited in a marsh and flood-plain environment of the delta plain.

Figure 18 is a structure map of the base of the Flippen limestone in the Group 4000 field. The structural closure is due to differential compaction of the underlying Cook sandstone and surrounding shale.

Most of the 2,250,000 bbl of oil that the field has produced has come from the thin sandstone lenses in the delta plain. Reserves of these sandstones are about 200 bbl/acre-ft. The distributary channel sandstones have a reserve of about 400 bbl/acre-ft. Both pay zones have a high gas-oil ratio.

BASAL PERMIAN EASTERN SLOPE

Figure 19 is a stratigraphic cross section of Lower Permian sedimentary rocks from the Eastern shelf westward through the Group 4000 field to the Jameson "Strawn" sandstone field in northwest Coke County, Texas. The top of Waldrip limestone 2 correlates down paleoslope to a Pennsylvanian reef in well 2. Flippen limestone is on the shelf between wells 7 and 14 and on the slope between wells 1 and 7. The shelf edge of the Flippen limestone (well 7) is 7 mi (11 km) west of the shelf edge of Waldrip limestone 2 (well 13). Sun Oil Company geologists (personal commun.)

described Lower Permian Wolfcamp fusulinids in the Flippen limestone of the Jameson field (wells 1, 2).

The clastic wedge of sediments on the slope between Waldrip limestone 2 and the Flippen limestone is considered to be stratigraphically equivalent to the Cook sandstone and Waldrip shale on the shelf. The clastic wedge reaches a maximum thickness of about 1,300 ft (396 m) at midslope (well 7, Fig. 19) and thins basinward. The limestone beds thin basinward, generally change in facies to a pelagic black shale, and finally pinch out in the Midland basin (Van Siclen, 1958). The lens shape of this clastic wedge indicates that the Midland basin was sinking faster than sediments could fill it. Progressively younger clastic wedges prograded basinward and filled the basin.

COOK SUBMARINE-CANYON STRATIGRAPHIC TRAP

The Northeast Bloodworth "Canyon" sandstone field is in southwest Nolan and northwest Coke Counties, Texas (Fig. 19, well 4). The producing sandstone, as the cross section (Fig. 19) demonstrates, is Lower Permian Cook sandstone deposited in a lower slope environment, not a sandstone of Pennsylvanian Canyon age as described by field operators. Isopachs and cross sections of the field (Figs. 19-22) show that, indicative of a channel-fill deposit, the base of the Cook sandstone is asymmetrically convex downward. Maximum thickness of the Cook (Fig. 22) is more than 100 ft (30 m) with a northeast-southwest axial trend coincident with the axial trend of thickest base of Flippen limestone-base of Cook sandstone interval (Fig. 20). This axial trend is probably the paleovalley axis of a channel-fill or canyon-fill deposit.

Inasmuch as the Cook sandstone in the Northeast Bloodworth field is a channel-fill deposit in a lower slope depositional environment (Fig. 19), it is probably a submarine-canyon fill. Additional evidence in support of this conclusion is: (1) the highest amplitudes of the electric-log SP curves in Figure 21 are at the base of the sandstone and have a very sharp contact with the subjacent shale, indicating an erosional surface; (2) an upward decrease in grain size and increase of intergranular clay in the sandstone are inferred by the general upward decrease in SP amplitude (Fig. 21); (3) the lower sandstone is a "bundle" or repetition of thin beds separated by thinner shale beds, a characteristic of turbidite or bottom-current deposits in modern and ancient submarine-canyon deposits (Sullwold, 1961; Stanley and Unrug, 1972) and indicative of relatively short periods of sand deposition and long periods of

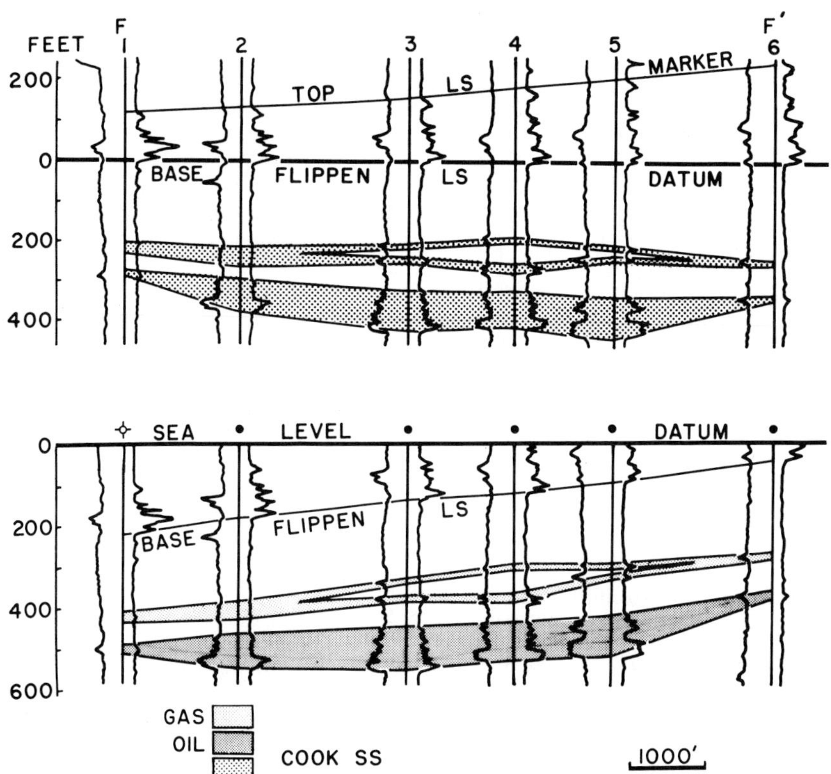

FIG. 21—Stratigraphic and structural cross sections *FF'* of Northeast Bloodworth field showing sharp asymmetric convex-downward base of Cook sandstone. See Figures 20 and 22 for location of cross section.

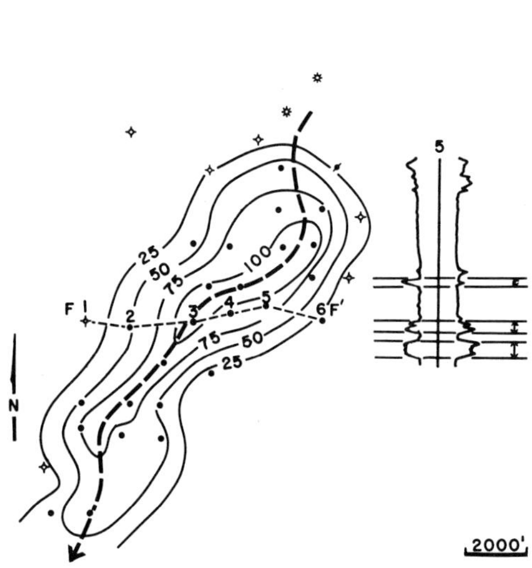

FIG. 22—Isopach map of Cook sandstone in Northeast Bloodworth field. Heavy dashed line is axial trend. *FF'* is line of section shown in Figure 21. Contour interval 25 ft (8 m).

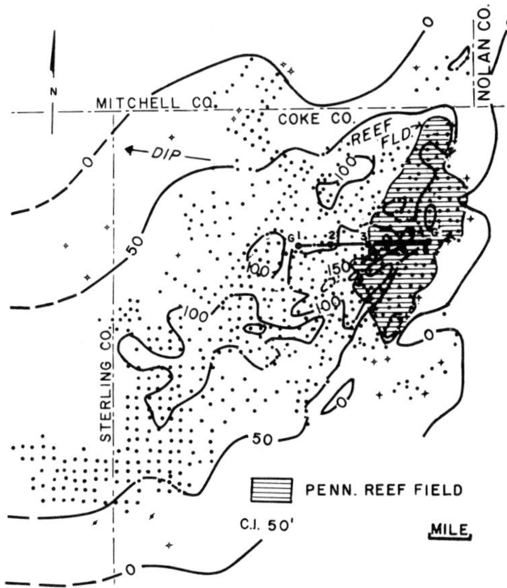

FIG. 23—Isopach map of Cook sandstone in Jameson field, Texas. *GG'* is line of section shown in Figure 24. Contour interval 50 ft (15 m).

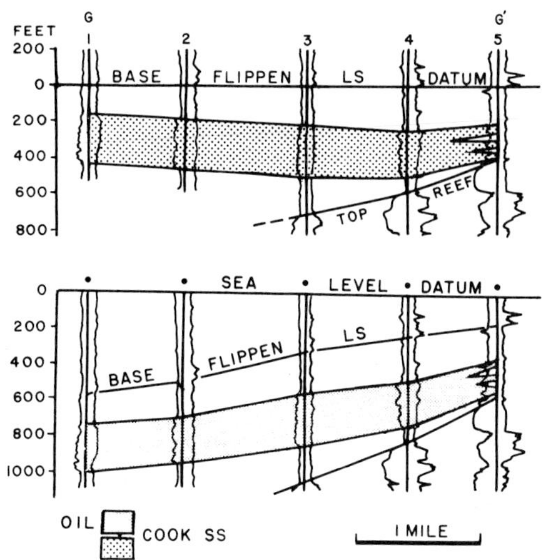

FIG. 24—Stratigraphic and structure cross sections of Cook sandstone in Jameson field. See location of cross sections in Figure 23.

pelagic-mud deposition; (4) individual lower sandstone beds do not correlate between wells (Sullwold, 1961; Stanley and Unrug, 1972); and (5) the top of the uppermost sandstone bed is generally planar.

The Northeast Bloodworth field is a stratigraphic trap with a gas-solution drive and high gas-oil ratio. Since the discovery in 1966, the field has produced approximately 120,000 bbl of oil and 340 million cu ft of gas per well.

COOK SUBMARINE-FAN STRATIGRAPHIC TRAP

In the Jameson "Strawn" sandstone field in northwest Coke County, Texas (Fig. 19, wells 1, 2), the Cook producing sandstone was deposited fan-shaped (Fig. 23) in a Lower Permian lower slope depositional environment (Fig. 19) and is not a sandstone of Pennsylvanian Strawn age as originally described (Riddle, 1954).

In the northeast corner of the mapped area (Fig. 23), the sandstone is about 2 mi (3 km) wide. This is the southwest part of the North Jameson "Strawn" sandstone field (about 6 mi or 10 km long and 2 mi or 3 km wide). The sandstone has a convex-downward base indicative of a submarine-canyon deposit. South of the submarine canyon, sand and mud apparently supplied by the canyon were deposited in a fan shape over an area of about 176 sq mi (458 sq km). The Pennsylvanian reef (Figs. 23, 24) underlying the sandstone near the mouth of the submarine canyon probably was a topographic barrier to sandstone deposition along the west flank of the reef. The

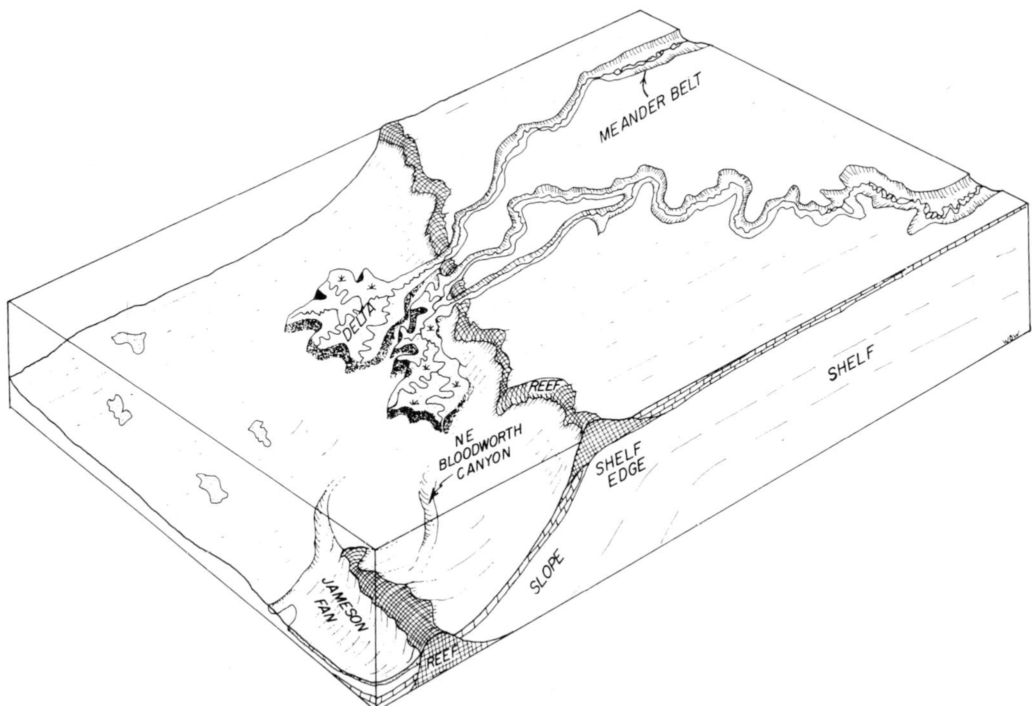

FIG. 25—Block diagram of Eastern shelf and slope at approximate base of Permian. Diagram is hypothetical and illustrates depositional environments of Cook sandstone.

fan-shaped geometry of the large sandstone body and its presence in a lower slope depositional environment suggest that the sandstone is a submarine-fan deposit with the apex on the northeast at the mouth of a submarine canyon. The producing sandstone consists of multiple, thin lenses; individual sandstone units cannot be correlated between wells (Fig. 24). The top and bottom contacts of the Cook producing sandstone can be correlated, inferring that the deposit is one genetic unit. The unit has an average thickness of 200 ft (61 m).

As suggested by the low-SP amplitudes (Fig. 24), the sandstone is fine-grained and shaly in cores. As is characteristic of turbidites or bottom-current deposits (Sullwold, 1961; Stanley and Unrug, 1972), many thin individual sandstone beds are graded. Grain size decreases upward from fine to very fine grained. In scattered zones are clasts of charcoal and red and gray shale and flattened pebbles of red clay. These terrigenous clasts probably were carried by turbidity currents from the Cook deltas that were 20 mi (32 km) northeast.

Since the discovery in 1952 the Jameson stratigraphic trap has produced about 35 million bbl of oil and an unknown quantity of gas. Average porosity is 12%.

SUMMARY

Cook sandstones form essentially continuous depositional systems from fluvial channels, through deltas and submarine canyons, to deep-marine sands (Fig. 25). The total transport route was more than 130 mi (208 km) long from the present outcrop to the Jameson field; originally the distance probably exceeded 230 mi (368 km). Along this route, oil accumulated in stratigraphic traps in fluvial and deltaic sandstones on the shelf and in submarine canyons and fans on the slope. Oil accumulation was controlled largely by lateral facies change to low-permeability siltstone and shale. Consequently, all of these oil fields are the result solely of depositional configurations of the sandstones, and illustrate an unusually complete sequence of environments and stratigraphic traps.

REFERENCES CITED

Andresen, M. J., 1962, Paleodrainage patterns: their mapping from subsurface data, and their paleogeographic value: AAPG Bull., v. 46, p. 398-405.

Barnes, V. E., ed., 1972, Geologic atlas of Texas, Abilene sheet: Texas Univ. Bur. Econ. Geology, scale 1:250,000.

Brown, L. F., Jr., 1969a, Geometry and distribution of fluvial and deltaic sandstones (Pennsylvanian and Permian), north-central Texas, *in* Geology of the American Mediterranean: Gulf Coast Assoc. Geol. Socs. Trans., v. 19, p. 23-47; reprinted, Texas Univ. Bur. Econ. Geology Circ. 69-4.

———— 1969b, Virgil and lower Wolfcamp repetitive environments and the depositional model, north-central Texas, *in* Cyclic sedimentation in the Permian basin—Symposium, Midland, Texas, 1967: West Texas Geol. Soc. Pub. 69-56, p. 115-134; reprinted, Texas Univ. Bur. Econ. Geology Circ. 69-3.

Busch, D. A., 1971, Genetic units in delta prospecting: AAPG Bull., v. 55, p. 1137-1154; reprinted, 1973, AAPG Reprint Ser. 7, p. 137-154.

Fisher, W. L., et al, 1969, Delta systems in the exploration for oil and gas: Texas Univ. Bur. Econ. Geology Syllabus for Delta Colloquium, 102 p.

Fisk, H. N., 1961, Bar-finger sands of Mississippi delta, *in* J. A. Peterson and J. C. Osmond, eds., Geometry of sandstone bodies: AAPG, p. 29-52.

———— et al, 1954, Sedimentary framework of the modern Mississippi delta: Sed. Petrology, v. 24, p. 76-99.

Flawn, P. T., et al, 1961, The Ouachita system: Texas Univ. Bur. Econ. Geology Pub. 6120, 401 p.

Forgotson, J. M., Jr., 1957, Nature, usage and definition of marker-defined vertically segregated rock units: AAPG Bull., v. 41, p. 2108-2113.

Galloway, W. E., and L. F. Brown, Jr., 1972, Depositional systems and shelf-slope relationships in upper Pennsylvanian rocks, north-central Texas: Texas Univ. Bur. Econ. Geology Rept. Inv. 75, 62 p.

———— ———— 1973, Depositional systems and shelf-slope relations on cratonic basin margin, uppermost Pennsylvanian of north-central Texas: AAPG Bull., v. 57, p. 1185-1218.

Harrison, E. P., 1973, Depositional history of Cisco-Wolfcamp strata, Bend arch, north-central Texas: PhD thesis, Texas Tech Univ.

Jacka, A. D., et al, 1968, Permian deep-sea fans of the Delaware Mountain Group (Guadalupian), Delaware basin, *in* Guadalupian facies, Apache Mountain area, West Texas: SEPM Permian Basin Sec. Pub. 68-11, p. 49-90.

Jackson, W. E., 1964, Depositional topography and cyclic deposition in west-central Texas: AAPG Bull., v. 48, p. 317-328.

Lawless, J. E., 1960, Manly-Neas area, Jones County, Texas, *in* Geological contributions, 1960: Abilene Geol. Soc., p. 77-86.

LeBlanc, R. J., 1972, Geometry of sandstone reservoir bodies, *in* Underground waste management and environmental implications: AAPG Mem. 18, p. 133-190; reprinted 1973, AAPG Reprint Ser. 7, p. 155-212.

Rall, R. W., and E. P. Rall, 1958, Pennsylvanian subsurface geology of Sutton and Schleicher Counties, Texas: AAPG Bull., v. 42, p. 839-870.

Riddle, D., 1954, Jameson Strawn sand field, Coke County, Texas, *in* Geological contributions: Abilene Geol. Soc., p. 77-86.

Shankle, J. D., 1960, The "Flippen" sandstone of parts of Taylor and Callahan counties, Texas, *in* Geological contributions, 1960: Abilene Geol. Soc., p. 168-201.

Shelton, J. W., 1973, Models of sand and sandstone deposits; a methodology for determining sand genesis and trend: Oklahoma Geol. Survey Bull. 118, 122 p.

Stafford, P. T., 1960, Geology of the Cross Plains quadrangle, Brown, Callahan, Coleman, and Eastland Counties, Texas: U.S. Geol. Survey Bull. 1096-B, p. 39-72.

Stanley, D. J., and R. Unrug, 1972, Submarine channel deposits, fluxoturbidites and other indicators of slope and base-of-slope environments in modern and ancient marine basins, *in* Recognition of ancient sedimentary environments: SEPM Spec. Pub. 16, p. 287-340.

Sullwold, H. H., Jr., 1961, Turbidites in oil exploration, *in* J. A. Peterson and J. C. Osmond, eds., Geometry of sandstone bodies: AAPG, p. 63-81.

Van Siclen, D. C., 1958, Depositional topography—examples and theory: AAPG Bull., v. 42, p. 1897-1913.

Introduction — Section 2

Deep-water deposits reach their final resting place by a variety of deep-water processes which are sourced in several different ways. The interaction of rivers with the ocean is discussed by Moore (1969). Commonly, relatively coarse-grained clastics reach deep water as a result of downslope flow or slumpage near the front of prograding deltas. Examples of this type of setting are given by Drummond et al (1976), Edmondson (1980), Bloomer (1977), and Burke (1972).

In some basins, sediments are fed from deltas or longshore drift into submarine canyons which carry them into basins. These canyons form "point sources" along the basin margin. Deep-water deposits which have cut down through slope or other sediments are discussed in papers by Heezen et al (1964), Shepard and Emery (1973), and Martin and Emery (1967).

Once the sediments reach deep water, their distribution is dependent on a number of factors including the degree of fan development, the slope of the depositional "basin," the local topography of the depositional basin, the syntectonic activity in the depositional area, and the size of the depositional basin. Most of these variables are treated in the papers on deep water channels and basin controlled deposits.

Roderick W. Tillman
Syed A. Ali

The American Association of Petroleum Geologists Bulletin
V. 53, No. 12 (December, 1969), P. 2421-2430, 3 Figs., 2 Tables

Interaction of Rivers and Oceans—Pleistocene Petroleum Potential[1]

GEORGE T. MOORE[2]

La Habra, California 90631

Abstract The common association of a river (with or without a delta), submarine canyon, and abyssal fan implies a genetic relation. The formation of these features must be considered as one dynamic system and not as independent events. A river entering the ocean may deposit most of its load near its mouth. Factors such as fluctuating discharge and saltwater encroachment, resuspension of material by current action, and other processes may cause much of this material to be conveyed to the deeper parts of the ocean basins. These sediments, commonly transported through submarine canyons or sea valleys, ultimately will be deposited on the abyssal plains, either in the form of a submarine cone or redistributed by deep currents to form a continental rise.

The Los Angeles basin and some fields in the Gulf Coast contain prolific petroleum reserves in deep-water turbidites. The accumulations are in reservoirs of different Cenozoic ages. Oil and gas fields have been discovered in sediments of Pleistocene age; however, the known accumulations are only modest. As exploration moves seaward, greater volumes of late Cenozoic sedimentary strata and petroleum accumulations can be anticipated.

During the Pleistocene, the regimens of rivers were altered significantly. For long periods of time, lowered sea level, increased river gradients, and voluminous glacial meltwater in flood-swollen rivers prevented the formation of large deltas. The net result was transportation of great sediment volumes across the continental shelf and slope with attendant canyon cutting and the deposition of the material at the base of the slope as submarine fans. The associated reservoir rocks of these fans may contain petroleum reserves as great as those of their shoreline equivalents, the deltas.

INTRODUCTION

Ocean surveys repeatedly have shown that submarine canyons cut into the continental shelf and slope, and many submarine fans on

[1] Manuscript received, May 5, 1969; accepted, June 23, 1969.

[2] Chevron Oil Field Research Company. Formerly Associate Professor of Geology, Northeastern Illinois State College, Chicago, Illinois.

Review of the manuscript and helpful suggestions and criticisms by H. L. Ellinwood, M. G. Frey, and J. W. Low are gratefully acknowledged. This paper represents ideas that I developed while employed by Chevron Oil Company. I thank Chevron Oil Company and Standard Oil Company of California for permission to publish this paper. The approval should not be construed as endorsement by the corporations or individuals of the ideas or concepts set forth. I accept the full responsibility for the data, reasoning, and conclusions expressed herein.

abyssal plains are aligned geomorphic features. These features are worldwide in distribution, and are found associated with the mouths of major delta- and nondelta-type rivers, wherever detailed surveys have been made. Probably this relation of river, submarine canyon, and fan is genetic. The processes which explain these associations are controversial, speculative, and largely unknown.

Because of the worldwide association of petroleum provinces with major sites of deposition, such associations must be better understood. Already the results of marine-geology studies and the limited successes offshore indicate that we could apply the concept of "uniformitarianism" more liberally to the yet largely unknown regions beyond the 100-fm line.

DELTAIC SKELETAL FRAMEWORK

Bar-finger sands were recognized and defined by Fisk *et al.* (1954, p. 89) as important elements of the Mississippi delta. They are elongate deposits of sand that occupy the channel and distributary system of the lower delta region. Fisk (1961, p. 29) postulated that they originate as distributary-mouth bars at the delta front and subsequently are incorporated into the main channel as the delta progrades seaward. Infilling by finer sediments between these skeletal elements is accomplished chiefly by the distributive action of tidal and wind-driven waves.

Moore (in press) postulated that another process aids in the continued development and preservation of these sands, particularily in the lower reaches of a river such as the Mississippi, which transports a rather large sediment load in comparison with its discharge. Most rivers, whether of the delta type (Mississippi) or the nondelta type (Congo), are affected to some degree by the presence of a saltwater wedge during the late summer to early spring. The upstream extent and persistence of this wedge are determined by the annual discharge pattern of the river. When the wedge is present in the channel, the decreased current will allow the coarser sediments to settle out of suspension.

There will be virtually no bed-load transport. Some of the finer bed load actually may be transported upstream with the flooding saltwater wedge. Therefore, a net increase of the coarser fractions should occur on the river bed in the lower part of the delta.

Dynamic System

The dynamic system includes the river, its delta, the related submarine canyon, and the submarine fan. As the system is an entity, each part must be considered in its relation to the whole. If the discharge is sufficient to prevent saltwater encroachment, currents near the river bottom will be able to transport most or all of the sediment load to the sea. The interaction between the tides (salt water) and the river discharge (fresh water) will create a zone of turbulence and maximum sediment concentration either in the estuary (Meade, 1968, p. 103) or on the sea floor beyond the mouth of the river. Should this occur on a narrow and relatively steep shelf, the elements would be present for the generation of a turbidity current. Such a current, continually generated in the same area, may be sufficient to cut a channel in soft materials and transport large quantities of sediment into deep water. Shepard and Dill (1966) reviewed the various hypotheses concerning the origin and maintenance of submarine canyons. Excluding sea valleys and canyons of special origin, these authors (p. 340–341) concluded that the most likely processes are mass movement, drowning of subaerially eroded valleys, turbidity currents, and other currents. The relative importance of each of these processes is not known, but their significance may differ from region to region. The sediments would be deposited at the base of the continental slope as a submarine cone. In places the formation of a subaerial delta could be retarded because of excessive sediment loss. The activity of such a proposed mechanism would be related definitely to the volume and regularity of river discharge and tidal action. The cone of sediments might be modified further by currents which flow parallel with the bathymetric contour and redistribute the sediment along the base of the continental slope.

The shaping of the continental rise along the eastern side of the North American continent has been attributed by Heezen *et al.* (1966, p. 507–508) to the action of a deep, south-flowing current, the Western Boundary Undercurrent. Several sparker profiles published by Uchupi and Emery (1967), made off the Atlantic Coast of North America, support the belief that the continental rise is composed of sediments which bury the base of the original continental slope.

The heads of some submarine canyons currently receive large quantities of sediment but are not being filled. The sediment is being transported down the canyons. Pilots have observed that along the coast of Greenland where meltwater streams enter the ocean, the discharge does not always spread over the water of the fjords. Rather, the cold, sediment-laden freshwater streams, plunge below and flow beneath the clear marine water (Gilluly *et al.*, 1968, p. 81). Schubel (1968, p. 1015) studied the sediment concentration in northern Chesapeake Bay. He found that the maximum concentration of suspended sediment (silt and clay) is neither at the mouth of the Susquehanna River, the primary source of discharge into that part of the estuary, nor in the outer part of the estuary. An intermediate area was found to have the highest sediment or turbidity concentration because of tidal scour and wind-caused waves that resuspend the previously deposited sediment. Another factor found to add to the concentration is a "sediment trap" caused by a net nontidal circulation. Meade (1968, p. 107) observed in studies of the Atlantic estuaries that the maximum concentration of sediment appears to be related to the net nontidal circulation. The sediment has few visible mineral grains (10± percent) but much organic matter (21–50 percent; Meade, 1968, p. 98). The zone of net nontidal circulation is where the bottom, landward-moving salt water mixes with the seaward-flowing fresh water. The area of convergence marks the upstream limit of "sea salt" and a zone of maximum suspended sediment. The position of this zone should fluctuate with the seasonal tides as well as with river discharge. Measurements in the Savannah River indicate that the concentration of suspended sediment is greater near the "sea-salt" zone in the estuary than in the river itself (Meade, 1968, p. 102). Moore (in press, Tables I, IV) shows that the particle size of the suspended load (sand, silt, and clay) of both the Mississippi and Columbia Rivers changes with seasonal discharge fluctuations. Thus, there are both direct and indirect methods by which river-borne sediment, once introduced to the marine environment, can be moved by submarine processes into the deep ocean basins.

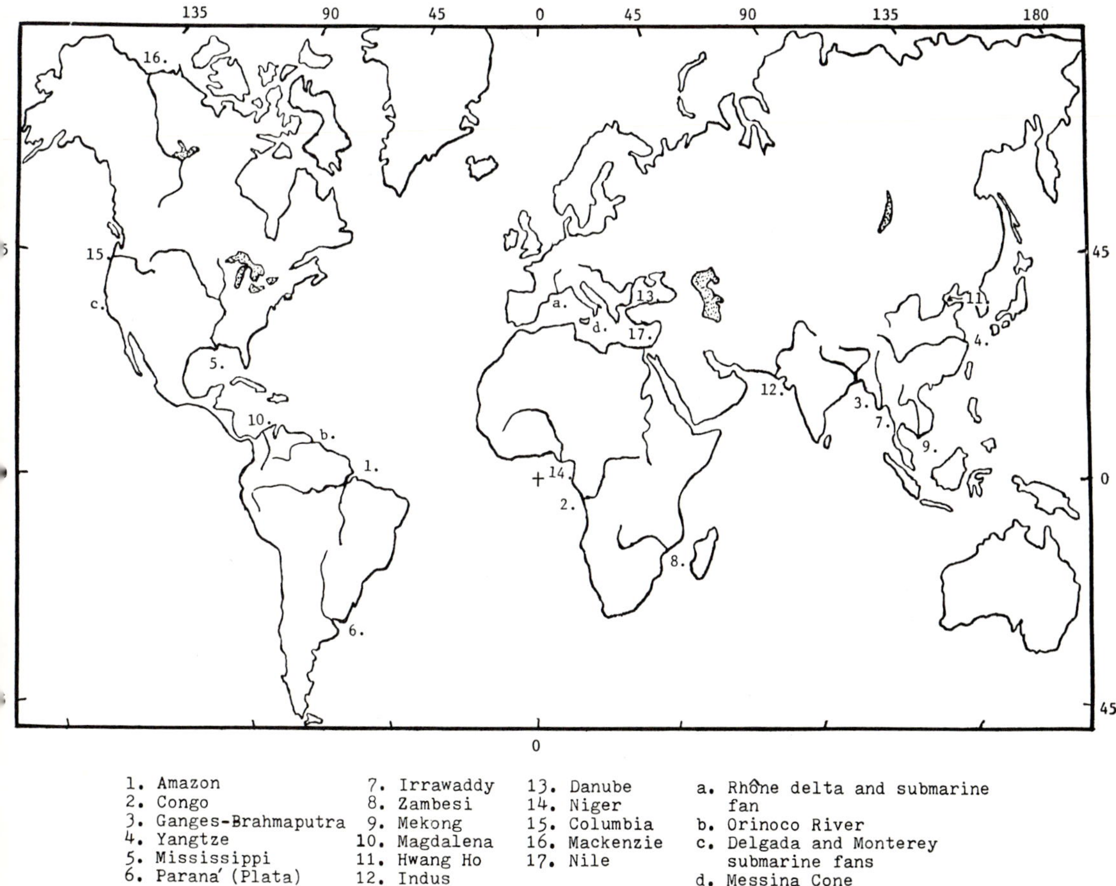

1. Amazon	7. Irrawaddy	13. Danube	a. Rhône delta and submarine
2. Congo	8. Zambesi	14. Niger	fan
3. Ganges-Brahmaputra	9. Mekong	15. Columbia	b. Orinoco River
4. Yangtze	10. Magdalena	16. Mackenzie	c. Delgada and Monterey
5. Mississippi	11. Hwang Ho	17. Nile	submarine fans
6. Parana' (Plata)	12. Indus		d. Messina Cone

Fig. 1.—Index map showing locations of rivers in Tables 1 (1–17) and other features in Table 2 (a–d).

Application of System to Rivers

Figure 1 shows the location of the 17 rivers listed in Table 1,[3] which is a collection of statistical data on these rivers. The list in Table 1 is neither exhaustive nor arbitrary; however, these rivers, with a total discharge of 11,258 km³/year, represent more than 32 percent of the total estimated annual discharge (35,200 km³/year) of the world's rivers (Nace, 1964, p. 13). The choice in Table 1 is based primarily on world rank, particularly in annual volume of discharge, and includes the 11 rivers having the largest average discharges (Ackerman, 1965, p. 88). The selection also was based on the availability of pertinent and reliable data. I am attempting currently to obtain

[3] The numbering sequence of rivers is the same in Figure 1 and Table 1.

additional facts on most of the important rivers of the world. The Nile is listed herein because it is the archetype of delta-forming rivers.

What effect, if any, does the presence or absence of a seasonal saltwater wedge have on the regimen of a major river? Theoretically and logically, a river flowing continually at or near peak discharge rates will prevent a wedge of salt water from transgressing upstream into the channel. Basic to this postulation is an abundant and well-proportioned annual rainfall in the drainage basin of the river, or circumstances regulating the annual flow of the river.

For a river such as the Congo (Table 1), the net change between maximum and minimum discharge rates is relatively small. During periods of minimum flow, salt water does fill the head of the submarine canyon (approximately 35 km upstream from the mouth; Hee-

Table 1. Major Rivers of the World in Order of Annual Discharge

River	Annual Fluid Discharge (km^3)	Annual Sediment Discharge (10^6 Tons)	Maximum Water Discharge (m^3/sec)	Minimum Water Discharge (m^3/sec)
1. Amazon	3,900[1]	900	203,000	—
2. Congo	1,400	70	65,000	27,000
3. Ganges-Brahmaputra	1,210[2]	700[2]	60,000	1,740
4. Yangtze	700	970	80,000	5,270
5. Mississippi	600	600	76,500	3,500
6. Paraná (Plata)	600	90	—	—
7. Irrawaddy	520	290	64,000	1,300
8. Zambesi	500	100	—	—
9. Mekong	400	80	60,000	1,700
10. Magdalena[3]	235	250	12,000	2,000
11. Hwang Ho	200	1,890	25,000	245
12. Indus	200	400	26,000	490
13. Danube	200	80	10,000	—
14. Niger	180	40	30,000	1,200
15. Columbia[4]	174	42	35,200	—
16. Mackenzie[5]	154	15	8,800	1,700
17. Nile	85	60	—	—

From: NEDECO, 1959, Table 1.1.1–1, p. 170 (reproduced with permission).

[1] Oltman *et al.* (1964, p. 1).
[2] Smith (1966, p. 245).
[3] Púlido (1969, p. 52–53).
[4] U. S. Army Corps of Engineers, Portland, Oregon (data received upon request).
[5] Dept. of Energy, Mines and Resources, Ottawa, Canada (data received upon request).

zen *et al.,* 1964, p. 1143, 1147). Nondelta rivers such as the Congo or Columbia (Table 1), which are capable of transporting the vast amount of their sediment into deep water, also have the energy required for cutting and/or maintaining a submarine canyon from the lower part of the river channel across the continental shelf, slope, and rise (Heezen *et al.,* 1964, p. 1143). In contrast, most other rivers are building subaerial deltas. Depending on the volume of fluid and sediment discharge, some also may be building submarine fans. Rivers that have extensive subaerial deltas (*e.g.,* in descending order of delta size, the Hwang Ho, Yangtze, Ganges-Brahmaputra, Mekong, Irrawaddy, Mississippi, Niger, Indus, and Nile) by their nature engender the processes controlling generation, entrapment, and preservation of prodigious oil reserves.

The required elements for the creation of such reserves include abundant nutrients and organic material, a large volume of sediments which locally may dilute the organic-rich material at the prograding delta front, rapid deposition, and preservation of the organic material (Emery, 1963, p. 489). The total amount of organic material annually introduced by rivers into the oceans is 7×10^8 tons; however, this constitutes less than 1 percent of the total amount of organic material in the oceans (Bordovskiy, 1965, p. 8). Rivers are important in that they provide the nutrients required for biologic processes to operate at maximum rates (Bordovskiy, 1965, p. 10). Thus, river mouths are important to the generation of organic matter in the marine environment. The abundance of organic matter in sediments is related to numerous factors. The great regions of high biologic productivity of the oceans result chiefly from upwelling. The concentration of organic material generally increases with decreasing grain size from sand to silt to silty clay. This concentration normally reaches a maximum in the silty clay and shows a decline in the clay-size fraction. Distribution of organic material is controlled partly by the geometric configuration of a basin or continental margin (Bordovskiy, 1965, p. 37–38). Although most of the organic matter develops in the neritic environment of the continental shelf, this material is carried seaward as it settles out and therefore is not concentrated in the shelf sediments. Studies of the distribution of organic carbon in the North Pacific Ocean show a definite concentration on the lower part of the continental slope and the adjacent abyssal plain (Bordovskiy, 1965, Fig. 4). Although the greatest accumulation of organic matter occurs in regions of rapid sedimentation, the organic concentration is low because of dilution by clastic material (Bordovskiy, 1965, p. 44). In mediterraneans, marginal seas, and bodies of water on the continents, the greatest concentration of organic material should be in the central part of the basin where dilution by clastic material is minimal. Thus, the coincidence of high biologic productivity and the deposition of great volumes of clastic sediments forms the nucleus of an important petroleum province. Finally, penecontemporaneous or postdepositional diapiric or slump structures must develop to localize the petroleum as it is generated.

Whether submarine fans that are forming today contain prolific hydrocarbon reserves remains to be established. However, recent offshore drilling by *Glomar Challenger II* has indicated that at least some abyssal-plain structures can contain accumulations of petroleum. One of the Sigsbee Knolls was drilled and cored. Miocene caprock, partly saturated with hydrocarbons, was recovered (Taylor, 1968, p. 39). The Nowlin Knolls, on the continental rise south of the abyssal plain, are probably a southward extension of the Sigsbee Knolls. Two of the cores (depths 1,420 and 1,644 fm) recovered from the knolls contain solid, asphaltic hydrocarbons mixed with sediments. One hydrocarbon zone is 58 cm thick (Bouma and Rezak, 1969, p. 74–75).

FAVORABLE ENVIRONMENT FOR PETROLEUM GENERATION

A delta, although it may contain oil in various channel and bar sands, probably is not an environment that would be classified as ideally suited for the development of prolific hydrocarbon accumulations. Oil and gas accumulations in sediments deposited in the deeper neritic and upper bathyal zones are well documented. Numerous examples of Cenozoic submarine fans and channels are described from California (Gorsline and Emery, 1959; Sullwold, 1961, p. 72, 74) and the Gulf Coast (Hoyt, 1959; Paine, 1966). Large oil and/or gas fields are associated with most of these examples. The principal trapping mechanisms differ greatly. In places, channels form the trap; elsewhere they constitute the reservoir; in other places accumulation is related to tectonic or expansion faulting and/or folding.

Immense reserves of oil and gas can be found in these sedimentary beds. Almost all of the 5 billion bbl of oil produced before 1960 from the Los Angeles basin came from turbidites (Sullwold, 1961, p. 63). More than half of this production was from lower Pliocene sandstone beds that were deposited in water depths of 2,000–2,500 m (Sullwold, 1961, p. 75). The source rocks of this oil province are associated deep-water shale beds (Barbat, 1958, p. 71). Emery (1963, p. 488) reasoned that the continental rises may contain relatively high concentrations of organic material and therefore could contain abundant petroleum reserves.

Cenozoic rocks of different ages contain immense petroleum accumulations, but their relative importance varies among and within geologic provinces. Pleistocene strata now are not particularly rewarding economically, probably because of insufficient exploration (Meyerhoff *et al.*, 1968, p. 554). Petroleum is not unusual in sedimentary strata of Pleistocene age. Numerous fields now produce, or are capable of producing, oil and gas. Meyerhoff *et al.* (1968, p. 553) noted a total of 50 such fields in South Louisiana, 19 of which are producing from known salt domes. Most (38) of these Pleistocene gas and oil reservoirs are in sedimentary beds deposited in a marine environment. Meyerhoff *et al.* (1968, p. 553) also pointed out

that four incompletely explored stratigraphic units—Cretaceous, Wilcox, Pliocene, and Pleistocene—underlie 50 percent of offshore and onshore Louisiana. Obviously most of the offshore area would be underlain by Pliocene and Pleistocene strata. Because these stratigraphic units extend across the continental shelf and beyond, into progressively deeper water, exploration and exploitation will become more expensive. The search, however, is moving seaward at a progressive and irreversible pace.

LATE CENOZOIC SEDIMENTATION RATE

Meyerhoff *et al.* (1968, p. 400, Table I) estimated that the rate of sedimentation in the Gulf Coast geosyncline has been increasing generally since the Cretaceous. They showed a rate in the Pleistocene of 240 cm/1,000 years in Louisiana and 60 cm/1,000 years in Texas. These values are approximately four times (Louisiana) and 1.5 times (Texas) greater than the next highest values in the respective regions. The calculations are based on estimated thicknesses of the Pleistocene sedimentary section (8,000 ft for Louisiana and 2,000 ft for Texas) which have been shown by recent offshore drilling to be conservative. The Gulf Coast geosyncline probably has been explored and studied more than any other offshore geologic province in the world, but only in recent years have the significance and positions of the great accumulations of Pleistocene sediments been recognized. It is logical to assume that the greatest volumes of these sediments are offshore and marginal to drainage from glaciated continental blocks.

The Mississippi cone was deposited on the floor of the Gulf of Mexico. By use of the preliminary map, "Bathymetry of the Gulf of Mexico" (Uchupi and Grady, 1967), a cone with the apex at the 200-m bathymetric contour near South Pass and the base at the 3,200-m contour can be outlined. It has an areal extent of 10.1×10^4 km², and a volume of 10.2×10^4 km³. Calculation is based on certain assumptions:

1. The area occupied by the cone was assumed to be free of salt diapirs. Diapirs are known in the cone (Uchupi and Emery, 1968, Figs. 15, 19), but their number and three-dimensional character are not determined. Therefore, to simplify calculations they have been disregarded, although their presence would certainly reduce the overall volume of sediments in the cone.

2. The base of the cone corresponds roughly to the original abyssal plain. Use of the 3,200-m contour rather than a deeper one is justified because it results

in a simpler geometric figure and because the source and mode of origin of the material below 3,200 m are debatable.

3. Isostatic adjustments caused by the cone have been disregarded.

4. The cone is a depositional feature related directly to the Mississippi River and constructed by the river from its transported load in the recent geologic past. Errors resulting from the first three assumptions can be compensating to some degree.

With an annual sediment load of 600×10^6 tons (Table 1) or 0.3 km³/year (assuming a specific gravity of 2.0), the Mississippi River could have constructed this bathymetric feature in 340,000 years if all the sediment transported by the river were deposited on this cone. Obviously, because the river is actively constructing a delta, such deposition did not occur. If one assumes that, because of mass movement, current action, and sedimentation, 60 percent of the annual sediment volume ultimately was deposited on the cone, the Mississippi River could have constructed the cone in 6×10^5 years. Sedimentation rate is assumed to have been constant during that period. That sedimentation rates during the Pleistocene were greater than those in previous periods has been established (Meyerhoff *et al.*, 1968, p. 400, Table I). All possibilities considered, one must conclude that the Mississippi cone may be rather young. Obviously, because of the many assumptions involved in these calculations, one could not propose an exact date for its origin. Geologically, however, the cone may be no older than late Pliocene.

The Rhone River and related submarine currents (Fig. 1) have constructed a submarine fan on the Balearic abyssal plain, which is approximately 2,800 m deep (Menard *et al.*, 1965, p. 282). Menard *et al.* (1965, p. 283), using present sedimentation rates of the Rhone River, concluded (on the basis of admittedly rather imprecise values) that the Rhone submarine fan and adjoining Balearic abyssal-plain deposits could have been deposited in one third of post-Eocene time (1.2×10^7 years). In this computation a constant depositional rate during the late Cenozoic is assumed. Menard *et al.* also considered the possibility that these deposits are Pleistocene and calculated that the rate of denudation of the source area would have to be 125 cm/10^3 years, a great increase over the 10 cm/10^3 years calculated for an average rate of denudation since post-Eocene time.

ENIGMATIC PLEISTOCENE

Other things being equal, it is reasonable to presume that during "normal" periods in the Cenozoic Era the regimen of rivers was similar to that of today. Orogenesis and epeirogenesis strongly influence the volume of flow and sediment load of rivers. The same kind of seasonal changes that occur in today's rivers, *e.g.*, flooding during periods of heavy seasonal rainfall or melting snows, certainly affected rivers of the earlier Cenozoic as well. Thus, barring any fundamental unbalancing of the factors regulating the erosion, transportation, and deposition of sediment in the oceans, events would progress naturally into the Holocene.

However, this orderly progression was interrupted by extensive continental glaciation. Because glacial meltwater was added to the normal annual precipitation, the discharge of proglacial streams was greater than that of streams today, at least during deglaciation (Flint, 1957, p. 174). Menard *et al.* (1965, p. 283) speculated that during the Pleistocene erosion and sedimentation were more rapid. Meyerhoff *et al.* (1968, p. 400) supported this in a study of the Gulf Coast sedimentation rates. The volume of discharge from those rivers was great through all seasons, compared with the rates of today. Inasmuch as even an advancing glacier yields meltwater, the additional volume of water probably would have kept a proglacial river flowing at capacity during much, if not most, of each glacial stage. The net loss of water from the oceans involved in continental glaciers caused a 60–90 fm eustatic lowering of sea level during periods of maximum glaciation (Shepard, 1963b, p. 264). The volume of water in the oceans varied inversely with the amount of glacial ice present on the continents. Continental glaciation fundamentally altered the regimens of most rivers; however, only the larger rivers emptying directly into an ocean (Table 1) are considered here. On the basis of this setting, the following hypotheses can be established for the Pleistocene.

1. Because of the tremendous supply of meltwater being contributed to the rivers by both the advancing and, to a greater degree, the static or retreating ice fronts, rivers flowing to the sea were in flood stage for millennia.

2. The lowering of sea level significantly increased river gradients, particularly near the mouths, which permitted the rivers to carry larger volumes and coarser material to the river mouths (on developing continental shelf).

3. The presence of a saltwater wedge in a river channel causes large amounts of coarser material to be deposited on the river bed (Moore, 1970). The presence of a saltwater

wedge in a river is greatly dependent on the seasonal variation of discharge. A river like the Congo with a low maximum/minimum ratio (Table 1, Fig. 2) is not actively building a delta, but, rather, is actively cutting and maintaining a canyon across the continental shelf, slope, and rise (Heezen *et al.*, 1964, p. 1127–1131). Inasmuch as the majority of rivers were at flood stage for prolonged periods during the Pleistocene, the saltwater-wedge phenomenon in the channel was probably rare. Thus the Pleistocene rivers flowed into the oceans near the shelf edge and continued eroding and cutting submarine canyons into the lower shelf and continental slope with their heavy loads of coarse sediment. The only force tending to decrease the currents during this period was bottom friction—an effect considered insignificant in contrast to the energy of the system. Under such circumstances great canyons were, or could have been, cut in dense, layered, and crystalline rock. It is possible, even reasonable, that the present submarine canyons cut in igneous rock or well-consolidated sediments, *e.g.*, southern tip of Baja California (Bouma, 1965, p. 292), Carmel Canyon, and other canyons off California (Shepard, 1963a, p. 487), were excavated mainly at that time and are now just kept clean of sediments by submarine current action. A limited test of these ideas might be obtained by a study of the rivers and sea valleys of southwestern Greenland, one of the few areas in the world where conditions are similar to those of the Pleistocene.

4. The corollary follows that a great amount of Pleistocene sediment (coarse to fine) was carried by rivers across the present continental shelf where the rivers began cutting canyons. The material was carried down the canyons cut in the slope and, with the volumes of material eroded from the continental edge (in canyon cutting), was deposited on the abyssal plain in the form of submarine fans, or perhaps was secondarily redistributed by submarine currents as part of the continental rise. It is axiomatic, therefore, that the Pleistocene rivers did not form large subaerial deltas at the shoreline, except in a few closed or semirestricted basins. A greater amount of Pleistocene sedimentation logically would have been beyond the present shelf edge—on the continental rise, submarine fans, and abyssal plains.

5. The role of wind and tidal currents is considered important to the topographic configuration of a delta margin. It is believed, however, that these currents are effective only after delta building has begun. The dominant factor determining whether, or to what extent, a river will form a subaerial delta or cut a submarine canyon is the dynamic energy of the system. If the river has a low maximum/minimum discharge ratio and is the dominant element in this system, it ultimately will cut a submarine canyon and deposit most of its load in the abyssal depths by a direct or indirect (deposition and resuspension) method. However, if the river can only overcome the physical tendency of the ocean to reestablish

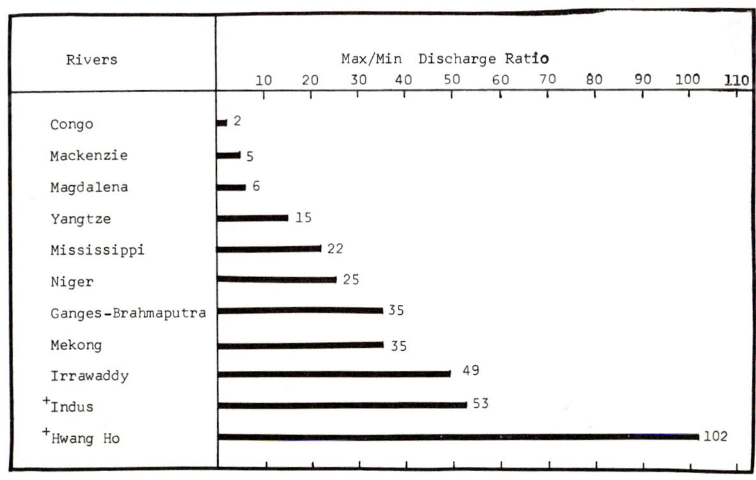

FIG. 2.—Major rivers in order of increasing maximum/minimum discharge ratio. General relation is postulated between this ratio and annual variation of flow.

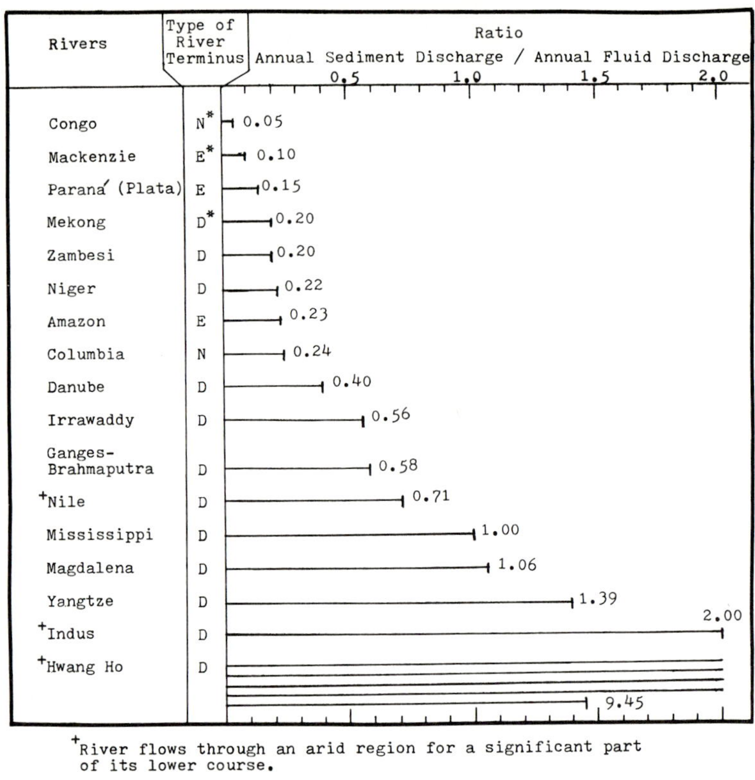

Fig. 3.—Rivers shown in Table 1 arranged in order of increasing ratio of annual sediment discharge/annual fluid discharge.

equilibrium during a part of the year, the net effect will be a buildup of river-transported sediment at the mouth, *i.e.*, a subaerial delta. These two examples represent the extreme end members of an intricate system.

Figure 3 lists the 17 rivers (Table 1) in order of an increase in the ratio of annual sediment discharge to annual fluid discharge, expressed in tons per cubic kilometer. Part of the courses of the Nile, Indus, and Hwang Ho Rivers are through arid regions and the consequent loss of water by evaporation and percolation gives these three rivers high ratios. The type of river-ocean relation is indicated in a separate column. The basis for division into three groups is defined at the bottom of the chart. Rivers that have a low ratio can be of any type and the ratio depends on the particular environmental conditions throughout the river's course. Rivers in Figure 3 which have a ratio greater than 0.24 have deposited extensive, seaward-projecting deltas.

6. Most important of all is the apparent genetic relation between nondelta, submarine-canyon–cutting rivers and a relatively even annual discharge. Information admittedly is sparse, and commonly conflicting or questionable. Rivers with a high maximum/minimum ratio of annual discharge form large subaerial deltas (Fig. 2). At present, one of the "high-grade" sites in which to prospect for vast petroleum reserves is the mouths of river systems that are now, and presumably in the recent past have been, constructing deltas (Table 2, Class I). Recently wildcat activities have been reported in the Mackenzie River delta and estuary (*Oil and Gas Jour.*, 1968, p. 54) and the Irrawaddy River delta (*Oil and Gas Jour.*, 1969, p. 56). Many of the rivers with deltas also have related deep-water fans (Class II). Some of the major

Table 2. Major River Termini Classified
by Location of Petroleum Reserves
(Proved or Potential)

Class I Major Subaerial Delta	Class II Submarine Fan (with Associated Subaerial Delta)	Class III Submarine Fan Only (with Associated Submarine Canyon)
Amazon	Amazon	Congo
Mississippi	Mississippi	Columbia
Paraná	Paraná ?	Delgada[1] (lacks
Orinoco[1]	—	major river)
Irrawaddy	Irrawaddy	Monterey[1] (lacks
Mekong	—	major river)
Ganges-	Ganges-	Messina Cone[1,2]
Brahmaputra	Brahmaputra	
Indus	Indus	
Rhône[1]	Rhône	
Niger	Niger	
Nile	Nile	
Yangtze[3]	Yangtze-	
Hwang Ho[3]	Hwang Ho[3]	

The relation and association of potential petroleum reserves with these localities are the responsibility of the writer. The table is not considered to be a complete classification of these physiographic features.

[1] See Figure 1 for location.

[2] Formed by, and related to, submarine processes.

[3] For political reasons these are inaccessible and are grouped separately.

submarine fans either are associated with rivers that lack a delta—*e.g.*, the Congo and Columbia—or apparently are unrelated to a significant present-day river, such as the Delgada and Monterey abyssal fans (Class III). Exploration of submarine fans for, and the development by submarine drilling of, Pleistocene petroleum is beyond our present technology. However, technology will accomplish what economics dictates.

That many of the foregoing concepts will need future refinement and revision is unquestioned. That others need additional supporting information is being investigated. However, the economic potential of these regions is so vast that consideration and critical evaluation of the various aspects of this geologic framework are of international import.

REFERENCES CITED

Ackerman, E. A., 1965, Rivers, p. 85–88, *in* The American peoples encyclopedia, v. 16: New York, Grolier, Inc., 512 p.

Barbat, W. F., 1958, The Los Angeles basin area, California, p. 62–77, *in* L. G. Weeks, ed., Habitat of oil: Am. Assoc. Petroleum Geologists, 1384 p.

Bordovskiy, O. K., 1965, Accumulation and transformation of organic substances in marine sediments: Marine Geology, v. 3, nos. 1–2, p. 3–114.

Bouma, A. H., 1965, Sedimentary characteristics of samples collected from some submarine canyons: Marine Geology, v. 3, p. 291–320.

———— and R. Rezak, 1969, Oil found on knolls on Gulf's continental rise: Ocean Industry, v. 4, no. 5, p. 73–77.

Emery, K. O., 1963, Oceanographic factors in accumulation of petroleum: 6th World Petroleum Cong. Proc., Sec. I, paper 42-PD 2, p. 483–491.

Fisk, H. N., 1961, Bar finger sands of Mississippi River, p. 29–52, *in* J. A. Peterson and J. C. Osmond, eds., Geometry of sandstone bodies: Am. Assoc. Petroleum Geologists, 240 p.

———— E. McFarlan, Jr., C. R. Kolb, and L. J. Wilbert, Jr., 1954, Sedimentary framework of the modern Mississippi delta: Jour. Sed. Petrology, v. 24, p. 76–99.

Flint, R. F., 1957, Glacial and Pleistocene geology: New York, John Wiley and Sons, 553 p.

Gilluly, James, A. C. Waters, and A. O. Woodford, 1968, Principles of geology, 3d ed.: San Francisco, W. H. Freeman and Company, 678 p.

Gorsline, D. S., and K. O. Emery, 1959, Turbidity-current deposits in San Pedro and Santa Monica basins off southern California: Geol. Soc. America Bull., v. 70, p. 279–290.

Heezen, B. C., R. J. Menzies, E. D. Schneider, W. M. Ewing, and C. L. Granelli, 1964, Congo submarine canyon: Am. Assoc. Petroleum Geologists Bull., v. 48, p. 1126–1149.

———— C. D. Hollister, and W. F. Ruddiman, 1966, Shaping of the continental rise by deep, geostrophic contour currents: Science, v. 152, p. 502–508.

Hoyt, N. A., 1959, Erosional channel in the middle Wilcox near Yoakum, Lavaca County, Texas: Gulf Coast Assoc. Geol. Socs. Trans., v. 9, p. 41–50.

Meade, R. H., 1968, Relations between suspended matter and salinity in estuaries of the Atlantic seaboard, USA: Internat. Assoc. Sci. Hydrology, Pub. 78, v. 4, p. 96–109.

Menard, H. W., S. M. Smith, and R. M. Pratt, 1965, The Rhone deep-sea fan, p. 271–284, *in* W. F. Whittard and R. Bradshaw, eds., Submarine geology and geophysics, Colston Papers, v. 17: London, Butterworths, 464 p.

Meyerhoff, A. A., *et al.*, 1968, Geology of natural gas in south Louisiana, p. 376–581, *in* B. W. Beebe, ed., Natural gases of North America, v. 1: Am. Assoc. Petroleum Geologists Mem. 9, 1226 p.

Moore, G. T., 1970, Role of salt wedge in bar finger sand and delta development: Am. Assoc. Petroleum Geologists Bull., v. 54, no. 2. In press.

Nace, R. L., 1964, Water of the world: Natural History, v. 73, p. 10–19.

NEDECO, 1959, River studies and recommendations on improvement of Niger and Benue: The Hague, Netherlands Engineering Consultants, 1000 p.

Oil and Gas Journal, 1968, Mackenzie delta: new Arctic Coast hot spot: v. 66, no. 34, August 19, p. 54.

———— 1969, Burma ends slump, starts comeback: v. 67, no. 7, February 17, p. 56.

Oltman, R. E., H. O'R. Sternberg, F. C. Ames, and L. C. Davis, Jr., 1964, Amazon River investigations reconnaissance measurements of July 1963: U.S. Geol. Survey Circ. 486, p. 1–15.

Paine, W. R., 1966, Stratigraphy and sedimentation of subsurface Hackberry wedge and associated beds of southwestern Louisiana: Gulf Coast Assoc. Geol. Socs. Trans., v. 16, p. 261–274.

Púlido, E. R., 1969, Las obras de bocas de ceniza: Bogotá, Puertos de Colombia, 100 p.

Schubel, J. R., 1968, Turbidity maximum of the northern Chesapeake Bay: Science, v. 161, p. 1013–1015.

Shepard, F. P., 1963a, Submarine canyons, p. 480–506, *in* M. N. Hill, ed., The oceans, v. 3: New York, John Wiley and Sons, 963 p.
———— 1963b, Submarine geology: New York, Harper and Row, 557 p.
———— and R. F. Dill, 1966, Submarine canyons and other sea valleys: Chicago, Rand McNally, 381 p.
Smith, A. E., Jr., 1966, Modern deltas: comparison maps, p. 233–251, *in* M. L. Shirley, ed., Deltas in their geologic framework: Houston Geol. Survey, 251 p.
Sullwold, H. H., 1961, Turbidites in oil exploration, p. 63–81, *in* J. A. Peterson and J. C. Osmond, eds., Geometry of sandstone bodies: Am. Assoc. Petroleum Geologists, 240 p.

Taylor, D. M., 1968, The adventure of the "Challenger" begins: Ocean Industry, v. 3, no. 10, p. 35–50.
Uchupi, Elazar, and K. O. Emery, 1967, Structure of continental margin off Atlantic coast of United States: Am. Assoc. Petroleum Geologists Bull., v. 51, p. 223–234.
———— and ———— 1968, Structure of continental margin off Gulf Coast of United States: Am. Assoc. Petroleum Geologists Bull., v. 52, p. 1162–1193.
———— and J. R. Grady, 1967, Bathymetry of the Gulf of Mexico: Galveston, Texas, Dept. Interior, Bur. Commercial Fisheries, prelim. map, scale 1 in. = 50 km.

Upper Cretaceous Lithofacies Model, Sacramento Valley, California[1]

K. F. Drummond[2], E. W. Christensen[3], and K. D. Berry[3]

Abstract The Upper Cretaceous Starkey sandstones, Delta shale, and Winters sandstones of the southern Sacramento Valley can be described by a time-transgressive lithofacies model consisting of an eastern shelf sand sequence, a silt and shale foreslope, and a western basinal sand sequence.

Petrographic, paleontologic, and modern seismic evidence show the paleobathymetry, the depositional history, and the time-transgressive nature of the lithofacies model.

Regional stratigraphic cross sections restored on the top of the Starkey Formation show the depositional geometry of these distinct lithofacies and indicate the regressive nature of the Upper Cretaceous following a broad transgression in the upper part of the F Zone (late Campanian) of Goudkoff (1945).

Recent gas discoveries in primary stratigraphic traps of the Winters basinal sandstones underscore the value of studying the detailed depositional history of basinal sandstones.

Introduction

Three distinct depositional facies developed in the Upper Cretaceous of the southern Sacramento Valley (Fig. 1) primarily in response to energy differences related to water depth. The Starkey sandstones, Delta shale, and Winters sandstones were deposited as separate lithofacies across a basinward migrating shelf and their regressive depositional history is the primary subject of this study.

The age of this regressive sequence is late Campanian through Maestrichtian and corresponds to the E through C foraminiferal zones of Goudkoff (1945).

A more regional stratigraphic and structural history of the Sacramento Valley can be found in the California Division of Mines and Geology Bulletin 181 (1962). The generalized formational and age relationships of the southern Sacramento Valley are illustrated in Figure 2. A number of stratigraphic and migrated seismic sections transverse to the basinal axis (Fig. 3) were restored to a top Starkey datum to illustrate the geometry of these lithofacies assemblages. Implications of the lithofacies model developed in this paper can aid both the geologist and geophysicist in reconstructing the stratigraphic history and position of primary stratigraphic traps which may contain hydrocarbons.

Lithofacies Model

The stratigraphic relationship between the Upper Cretaceous Winters sandstones, the Delta shale, and the Starkey sandstones are illustrated by two east-west sections transverse to the depositional strike of the basin and restored to a top Starkey datum. Stratigraphic section 1-1' (Fig. 4) runs from south of the city of Sacramento through the Clarksburg gas field and across the valley to the west. On section 1-1', evidence of foresetting can be demonstrated from electric log traverses on several bentonite horizons in the Delta shale lithofacies. Correlations also show a progressive shingling of the Winters sandstones from the older sands on the east to the younger and stratigraphically higher westernmost sands. It is apparent that the shingling is time-equivalent to regressive western pinchouts of correlative Starkey sands.

In the vicinity of the Thornton Arch, a second regional east-west stratigraphic section (2-2') along the eastern margin of the basin has been restored to a top Starkey datum (Fig. 5). Here again, a well-defined transition exists along a single geologic time marker from an eastern Starkey sandstone into the Delta shale and finally into a time-equivalent Winters sandstone. The restored seismic sections A-A' and B-B' (Fig. 6) show the true scale of this foresetting. Average maximum dips on the foreset slope are about 7 to 8°, implying as much as 600 to 1000 ft (183 to 305 m) of bathymetric relief at the time of deposition.

A seismic time section (Fig. 7) from the east side of the basin is parallel to a portion of sections 1-1' and A-A' and shows the foresetting as it appears on modern 2,400% common depth point records.

The diagrammatic illustration of the relationship between the Starkey sandstones, Delta shale, and Winters sandstones and the depositional history of the E through C zone Cretaceous in the southern Sacramento Valley is shown in Figure 8. The divergence of the time-stratigraphic boundaries from the lithofacies boundaries is shown as the basin edge progresses from east to west. A paleogeographic map (Fig. 9) illustrates the same model and shows conceptual depositional patterns during the Upper

[1]Reprinted in revised form from 1976 AAPG Pacific Section Miscellaneous Publication No. 24, p. 76-88.
[2]Hamilton Brothers Oil Company, Denver, Colorado.
[3]Chevron USA, San Francisco, California

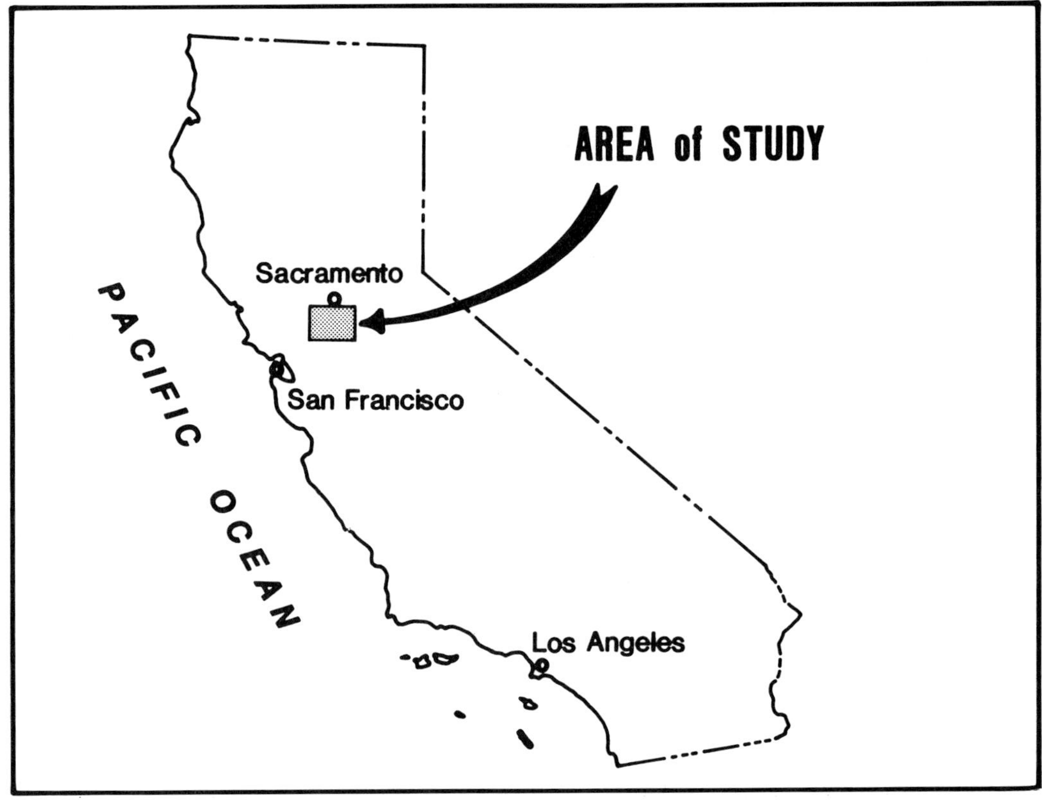

Figure 1 — Area of study in southern Sacramento Valley.

Cretaceous.

Paleobathymetry

Paleoenvironments in the Cretaceous and older rocks are extremely difficult to interpret by analogy with the late Cenozoic and Recent faunas, since all the significant species and many genera are extinct. Therefore, a different technique, which we have called the "oscillation chart" method, was used to determine the relationships between faunal suites and environments. This method consists of actually charting each species present in each of the different faunal suites progressing basinward from an inferred paleoshoreline along presumed geologic time lines. Assuming no structural complications, the faunal suite closest to the inferred shoreline is considered to represent the shallowest water. It is designated as intertidal to inner neritic. Each succeeding faunal suite is then critically analyzed. The differences noted, both in a seaward direction and along strike, are then assumed to be different because of either bathymetric or biofacies change. By studying the relationships of the faunal suites over a broad area, it is usually possible to determine whether the observed changes in the total fauna are due to an increased pal-

eobathymetric range or are due to an alteration in environmental conditions within the same depth range. Biofacies faunas can be separated within many of the depth ranges, and are presumably the result of an interrelation of environmental factors combining with depth, such as salinity, temperature, light, oxygen, turbidity, and food supply. It is not possible, of course, to determine how much influence any one environmental factor has in the total picture, or how to isolate any individual factor.

Two key species highly indicative of paleobathymetric range are shown in Figure 10 for each of the Goudkoff zones under discussion. However, since very few species are restricted to a specific depth zone, and because the bathymetric indicators are not necessarily present in every sample, the entire faunal suite available should be used for these determinations. These paleobathymetric indicators are often different than the main biostratigraphic markers for the foraminiferal zones, since a biostratigraphic indicator should cover several paleobathymetric ranges of the greatest use. Planktonic foraminiferal indicators are also of great value in determining that the benthonic foraminifera are correctly correlated along approximate time lines.

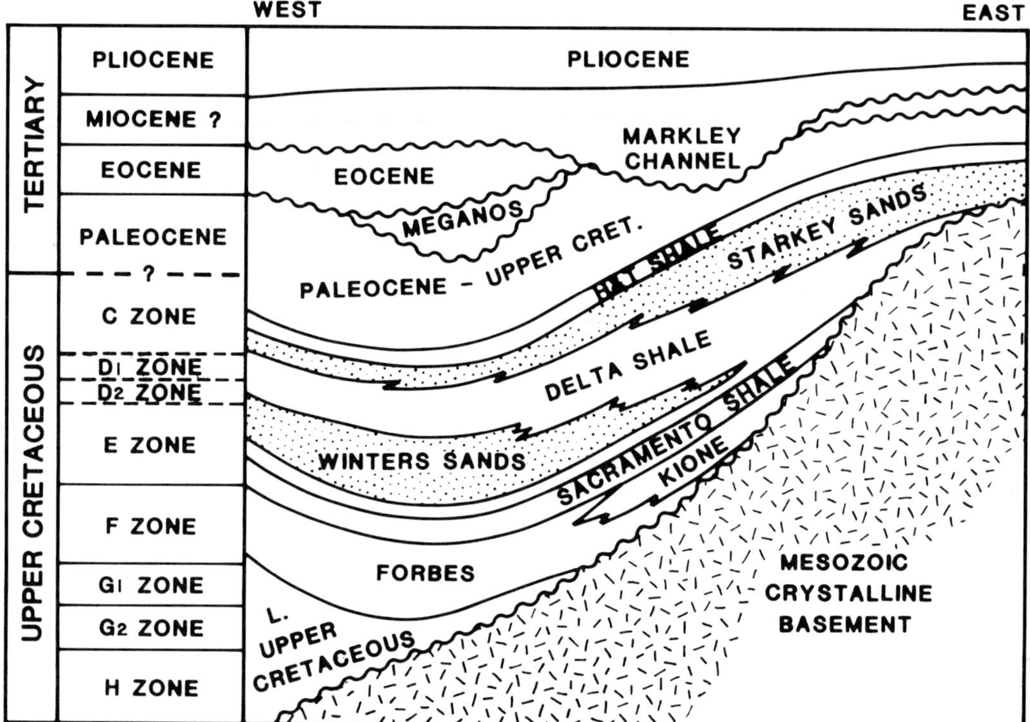

Figure 2 — Diagrammatic cross section showing stratigraphic relationships in the southern Sacramento Valley.

Depth ranges have been assigned somewhat arbitrarily, but are comparable to the ranges found in modern seas. Therefore, these ranges should not be considered as absolute or definite, but only as approximations. Four depth ranges have been established for the portion of the Upper Cretaceous covered in this paper (these are approximate depth ranges): (1) intertidal to inner neritic, 0 to 50 ft (0 to 15 m); (2) middle neritic, 50 to 300 ft (15 to 91 m); (3) outer neritic, 300 to 600 ft (91 to 183 m); and (4) inner bathyal, 600 to 1,500 ft (183 to 457 m). Greater depth ranges than inner bathyal have been found by the "oscillation chart" method in other areas within the Sacramento Valley, but discussion of this aspect is not within the scope of this paper.

Starkey Sandstones

The porous and well-sorted Starkey sandstones were deposited in a high-energy shelf environment where longshore currents and turbulence above wave base may have been responsible for winnowing of the finer clastics. The westward progression of the Starkey shelf and shoreline facies is depicted diagrammatically on Figures 8 and 9 in both the early and late stages of basinal infill. Occasional transgressions during the deposition of the Starkey appear to have occurred rapidly, separating individual regressive

sandstone units of the Starkey with marine shales. The more easily correlatable major units within the Starkey are informally termed H and T sandstone, Upper Peterson sandstone, and Lower Peterson sandstone by Chevron USA. Local difficulties can arise in trying to distinguish between individual units in the Starkey, however, especially in the northern part of the study area where many of the shale units separating individual Starkey sands thin or disappear.

Thirty-eight samples of the Starkey sandstones from eight wells were analyzed in thin section and for heavy minerals. In general, there is a remarkable uniformity of texture and petrography within the Starkey. Most of the samples are well-sorted, fine- to medium-grained sands, although fine to coarse and more poorly sorted samples are also present. All the Starkey sands are angular to subangular first cycle sands with little or no reworking. The sands are feldspathic, grading to lithic-feldspathic with the higher percentages of rock fragments commonly found in the more coarse-grained samples. In order of abundance, the primary constituents are plagioclase, quartz, potash feldspar and rock fragments. Rock fragments are mainly low-grade metamorphic rocks including phyllites and schists. Volcanic rock grains of intermediate composition and devitrified tuffs resembling chert are relatively rare. Minor constituents

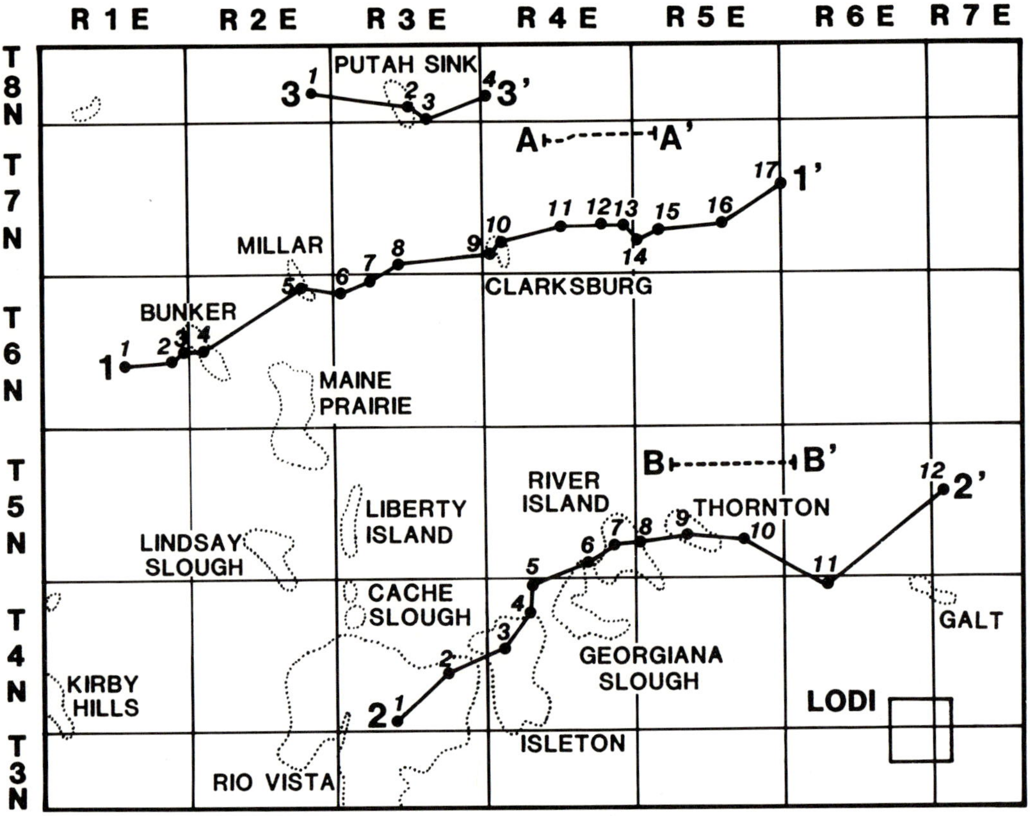

Figure 3 — Index map of the southern Sacramento Valley showing location of restored stratigraphic sections (1-1′, 2-2′, 3-3′) and migrated seismic sections (A-A′, B-B′).

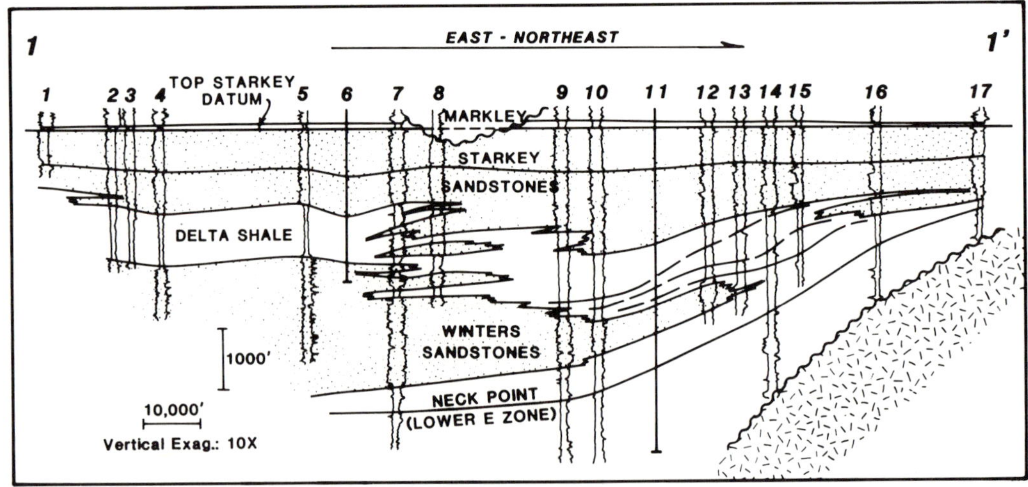

Figure 4 — Regional stratigraphic section 1-1′ restored to a top Starkey datum.

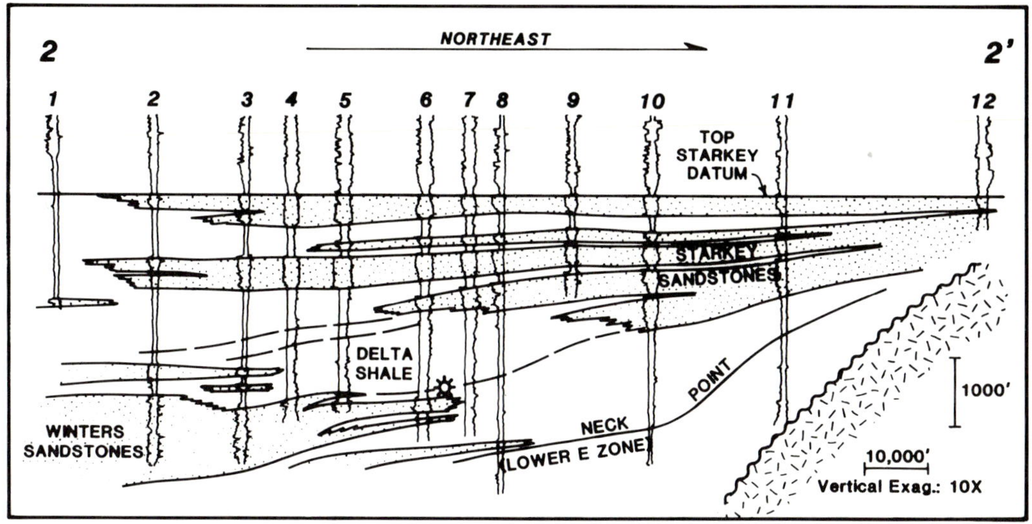

Figure 5 — Regional stratigraphic section 2-2' restored to a top Starkey datum.

observed in thin section include biotite, chlorite, hornblende, epidote, carbonaceous material, iron-oxide, pyrite and "clay."

The Starkey ranges in age from E zone in the lower and easternmost sandstones to C zone in the upper-most unit (H and T sandstone). The associated faunas are typically inner neritic to littoral and numerous barren samples suggest nonmarine or marginal marine deposition.

Delta Shale

The Delta shale lithofacies was deposited on the foreset portion of the paleoslope and contains inner bathyal and outer-neritic to inner-neretic E through C zone faunas. Several well-developed bentonites in the shale unit allow definition of the original geometry and slope characteristics from electric log correlations in certain areas. An interesting aspect of the Delta shale facies is that the fine clastics were apparently deposited more rapidly than the associated shelf and basinal sandstones, as indicated by the divergence of distinct bentonite beds in the upper parts of the foreset slope and their subsequent reconvergence near the base of the paleoslope. This is illustrated dia-grammatically in Figure 11, and the implications are that much of the coarser material on the shelf was being funneled through discrete tributaries into the basin (Fig. 9) and that very little of the sand was "spilled" across the edge of the shelf, where win-nowing above wavebase was the main factor in con-trolling sediment size.

Winters Sandstones

Winters sandstones and correlative sandstones in the southern most Sacramento Valley are basinal de-posits associated with outer-neritic to inner-bathyal faunas. Their mode of depositional emplacement can best be characterized as "sediment gravity flow," a general term used by Middleton and Hampton (1973) to describe any sediment or sediment-fluid mixture which is deposited under the action of gravity. Al-though minimal core data are available to study the Winters sandstones, the similarity of composition and texture between the samples of Starkey and Winters that were examined suggest a grain flow mechanism of depositional emplacement wherein the sands car-ried by longshore currents were funneled into the basin through discrete distributary channels. The ab-sence of significant matrix and high degree of sorting, the abrupt tops and bottoms of the sands, and the lack of internal sedimentary structure (at least in thin section) of the Winters sandstones all support this conclusion. Genetic classification of these sandstones as tractionites, turbidites, contourites, or fluxo-turbidites should be deferred, however, until more petrographic and other data are available.

Only six cores of Winters sandstones from two wells were available for analysis. They provided an insufficient base for adequate petrographic com-parisons between Starkey and Winters sandstones but showed their similar lithology.

The Winters sandstones are fairly well to poorly sorted, very fine to medium-grained and generally angular. There is some interstitial mica, but porosity is high in most samples. Heavy mineral separates differ from the Starkey samples only in the lesser percentage of hornblende. Rock grains, mainly low-grade meta-morphics, are common in the samples examined and the sands can be classified as lithic-feldspathic types. Volcanic grains, including chert-like devitrified tuffs,

285

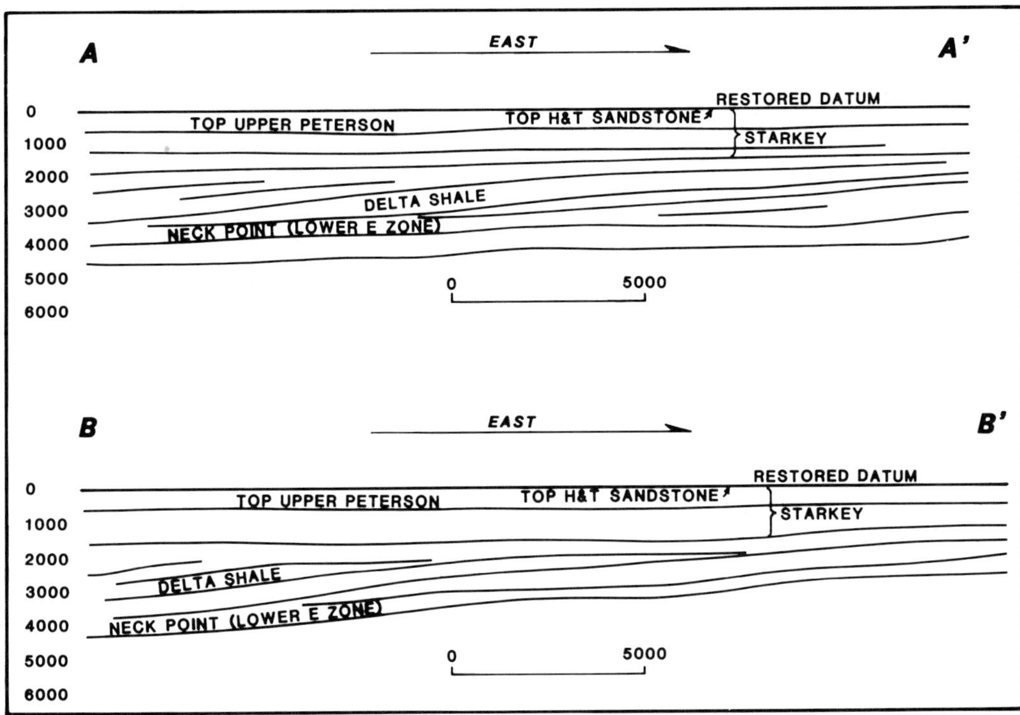

Figure 6 — Migration of traversed seismic sections A-A′ and B-B′ shows true scale foresetting in the Delta Shale lithofacies.

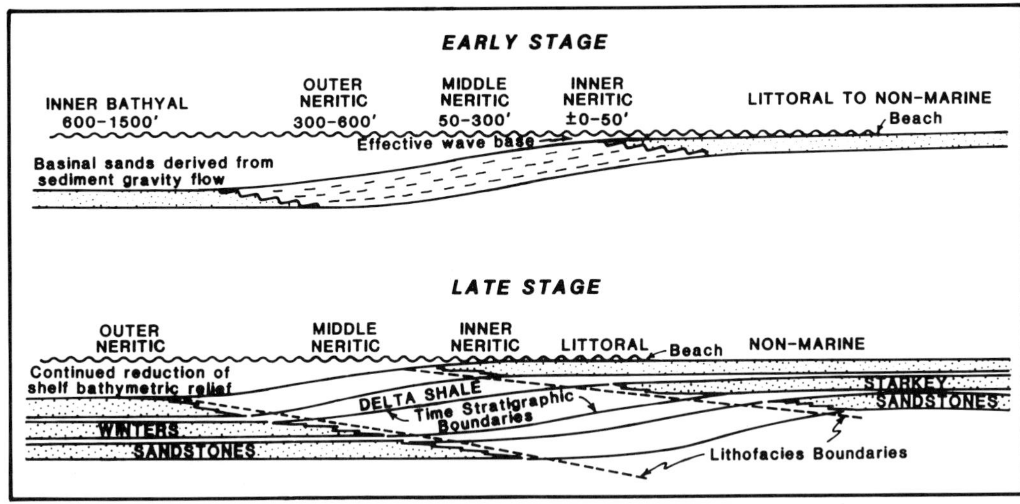

Figure 7 — Modern 2400% CDP time section showing internal geometry of the Delta shale.

	ZONE	PALEOBATHYMETRIC INDICATORS				
		INNER BATHYAL	OUTER NERITIC	MIDDLE NERITIC	INNER NERITIC	NON-MARINE
CRETACEOUS	C	NOT FOUND	"SIPHO." WHITEI "PLANULINA" NACATOCHENSIS	"BUL." PROLIXA (ABUNDANT) "BUL." JOAQUINENSIS	DENTALINA SP. (NEW SPECIES) "ANOMALINA" SP. (NEW SPECIES)	BARREN OF FORAMS
	D-1	NOT FOUND	"SIPHO." CLARKI COSTIFERA "PLANULINA" NACATOCHENSIS	EPON. INGRAMENSIS "BUL." TRIANGULARIS	"BUL." TRIHEDRA ROTALIA BANDYI	
	D-2	EPONIDES SPINEA GYRO. QUADRATA	EPONIDES GOUDKOFFI GAUD. CRASSAFORMIS	"ANOM." CF. RUBIGINOSA "BUL." TRIANGULARIS	"ANOMALINA" SP. (NEW SPECIES) DENTALINA AFF. MARCKI	CHAROPHYTES – VR
	E	EPONIDES SPINEA GYRO. QUADRATA	EPONIDES GOUDKOFFI "PLANULINA" MASCULA	"ANOM." CF. RUBIGINOSA COLOMIA CALIFORNICA	MARG. AFF. CURVISEPTA MARG. AFF. ELONGATA	

Figure 8 — Generalized depositional model for the Upper Cretaceous of the southern Sacramento Valley. Early and later stages of the upper Cretaceous regression are shown.

are present but fairly rare.

The electric-log character of a typical Winters sandstone is shown in Figure 12. The abrupt tops and bottoms of the sandstones imply a different mode of deposition from the time-equivalent Starkey sandstones. The initial updip stratigraphic pinchouts of these sandstones into the Delta shale, their clean and porous nature, and the multiple "shingling" of the sandstones make them potentially attractive hydrocarbon reservoirs. Section 3-3' (Fig. 13) illustrates the stratigraphic setting in the Putah Sink gas field, discovered by Shell Oil Company in 1973.

Implications

Implications of this model are important in dealing with subsurface structural and stratigraphic interpretations. Because of the internal geometry of the Delta shale described earlier, the isopach thicks between two bentonite markers in the Delta shale will occur near the edge of the westward migrating slope break (Fig. 11). This also implies that in any well drilled through the Starkey, Delta shale, and Winters, an isopach interval near the top of the Delta shale would be in a position of westward thickening, whereas an isopach interval near the base of the Delta shale in the same well would be in a position of eastward thickening (Fig. 12). Isopach maps made in the Delta shale should take these factors into consideration. Because of the shingle-like nature of both the Starkey and the Winters sandstones, the uppermost edge of the Win-

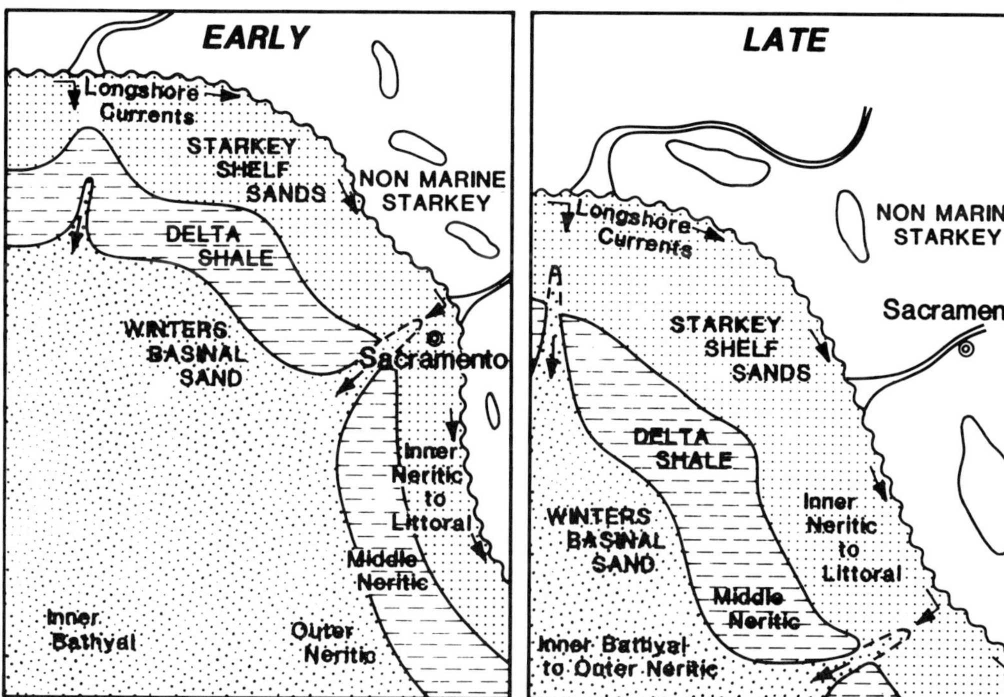

Figure 9 — Generalized paleogeography of the Upper Cretaceous in the southern Sacramento Valley. Direction of sediment source for Winters basinal sandstones is conjectural and is diagrammatically inferred to be dependent on the width of the littoral-inner neritic shelf, longshore currents, and amount of sediment influx.

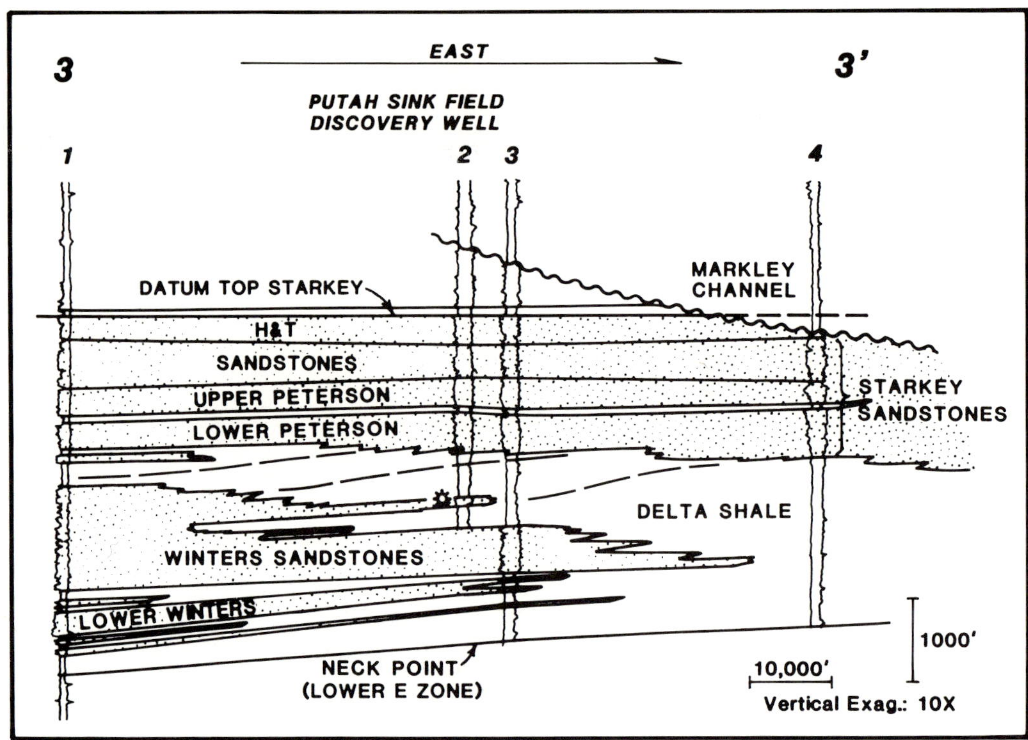

Figure 10 — Two key species most indicative of paleobathymetry are shown for each of the Goudkoff foraminiferal zones discussed.

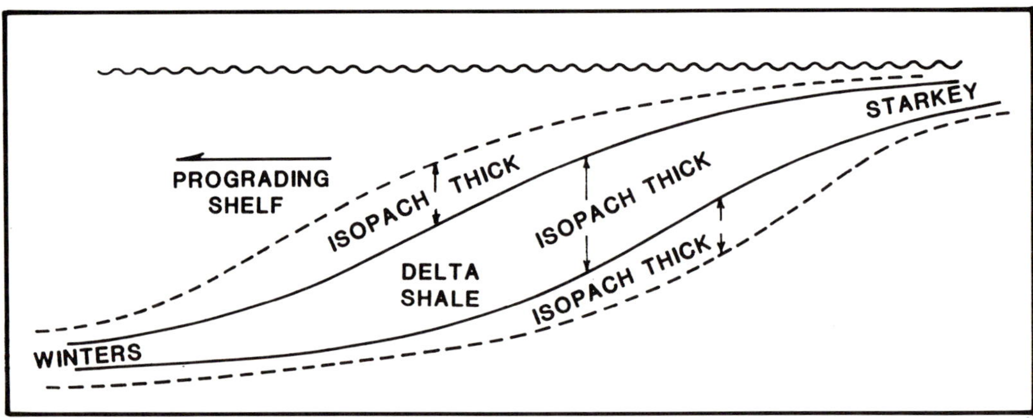

Figure 11 — Migration of the depocenter basinward with corresponding changes in the axis of isopach thickening.

ters lithofacies and the lowermost edge of the Starkey lithofacies are progressively younger toward the west. Electric log correlations in some areas should be made cautiously near these lithofacies boundaries so as not to confuse different sands with similar electric log character.

Applications of this model can aid the geologist in stratigraphic reconstructions. For instance, directions of paleoslope can be mapped from any three wells by mapping the divergence of the plane described by a correlative bentonite marker in the Delta shale from a plane described by an upper Starkey sand where the assumption is made that the topset portion approximates the original sea level (Fig. 14).

In California, similar lithofacies sequences can be found in the upper Miocene Santa Margarita-Stevens

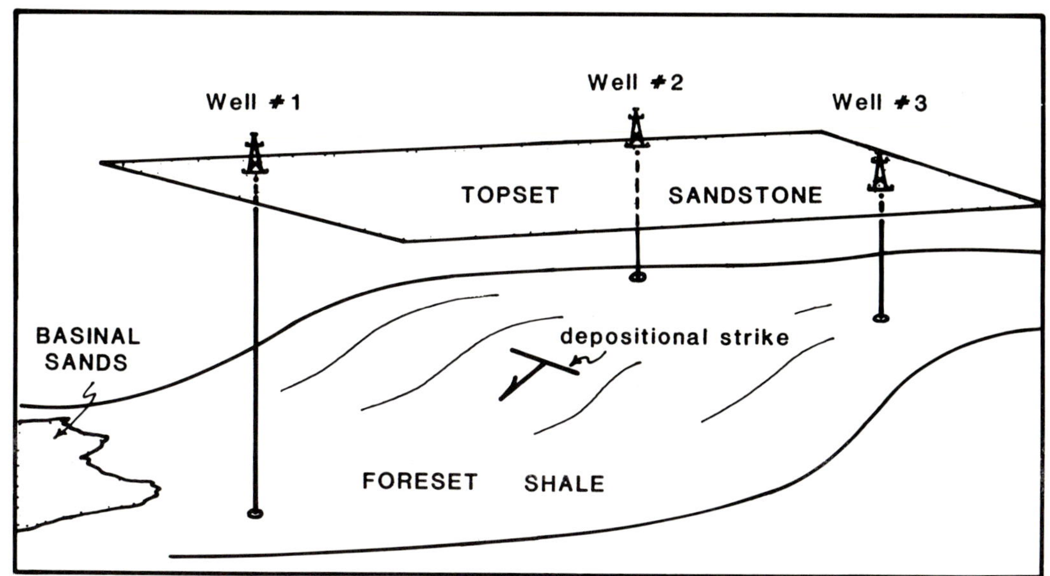

Figure 12 — Discovery well at Putah Sink gas field. Electric log character of Starkey and Winters sandstones and diagrammatic internal geometry of the Delta Shale are shown.

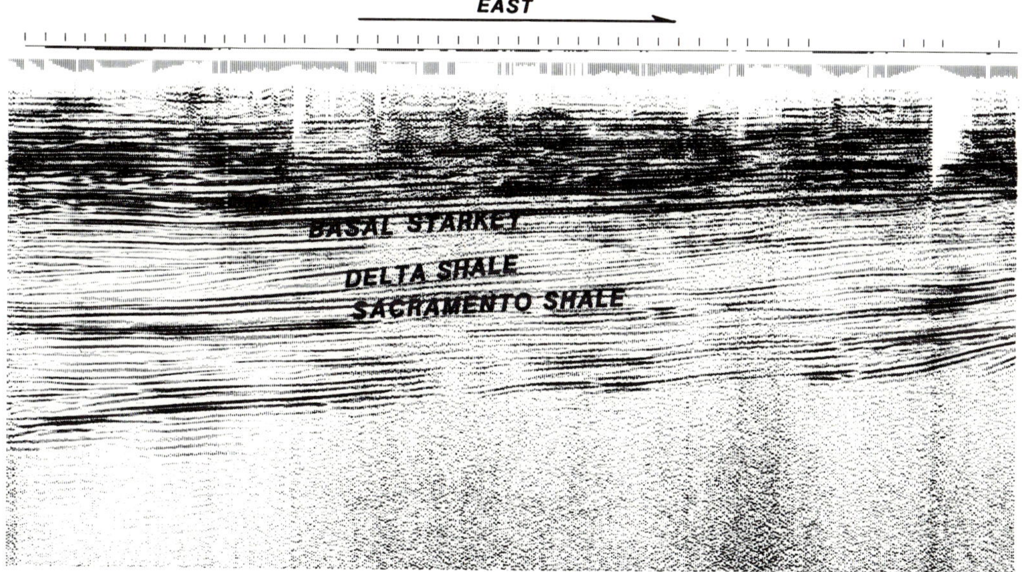

Figure 14 — Direction and magnitude of paleoslope can be inferred from divergence of topset and foreset beds in any three wells.

transition in the southern San Joaquin Valley and in parts of the F zone Cretaceous in the northern Sacramento Valley (Kione-Forbes). A modern example of a similar depositional situation may exist in Puget Sound south of Vancouver, B.C., where the Fraser River delta has a maximum present-day foreslope of approximately 12° (Luternauer and Murray, 1973).

Economic Importance

While both the Starkey and Winters sandstones are excellent reservoir rocks, most of the gas accumulations in the regressive Starkey sandstones of the Sacramento Valley are structurally trapped. Primary stratigraphic traps in the Winters sandstones appear to be related to their basinal character and the updip

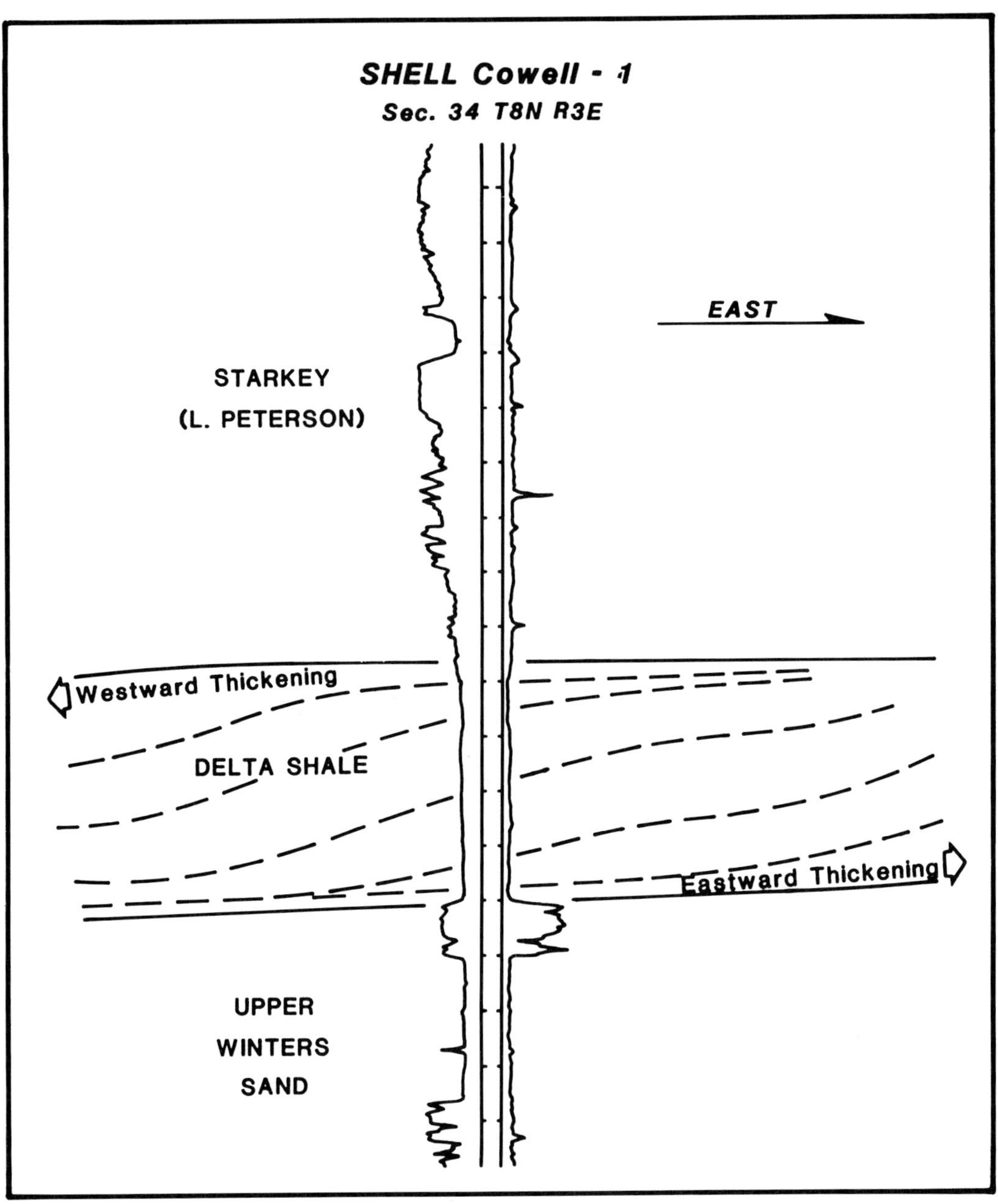

Figure 13 — Stratigraphic section 3-3′ restored to a top Starkey datum and showing the location of the Putah Sink gas field.

equivalence of the Delta shale lithofacies whose existence as potential source and cap rock provides favorable conditions for hydrocarbon entrapment.

The Putah Sink gas field discovered by Shell Oil Company in 1973 (Fig. 13) is an excellent example of the type of primary stratigraphic trap resulting from the model described. Similar production from the Winters sandstones where they pinch out across the Thornton Arch at the West Thornton Field (Silcox, 1962) can be seen on section 2-2′ (Fig. 5). Immediately south of the study area, significant reserves have been discovered at the Union Island gas field by Union in the Winters sandstones, suggesting that additional traps of the type described remain to be discovered.

References Cited

California Division of Mines and Geology, 1962, Geologic guide to the gas and oil fields of northern California: Calif. Div. Mines and Geology, Bull. 181, 412 p.

Goudkoff, P. P., 1945, Stratigraphic relations of Upper Cretaceous in Great Valley, California: AAPG Bull., v. 29, p. 956-1007.

Luternauer, J. L. and J. W. Murray, 1973, Sedimentation on the Western Delta Front of the Fraser River, British Columbia: Canadian Jour. Earth Sci., v. 10, p. 1642-1663.

Middleton, G. V., and M. A. Hampton, 1973, Sediment gravity flows; mechanics of flow and deposition, *in* Turbidites and deep water sedimentation: Short Course Notes, Pac. Sec., SEPM, p. 1-38.

Silcox, J. H., 1962, Geology of West Thornton and Walnut Grove gas fields, California: Calif. Div. Mines and Geology, Bull. 181, p. 140-148.

Member Sands of the Winters Formation[1]

William F. Edmondson[2]

Abstract   Six member sands are defined for the Winters Formation in the subsurface of the Sacramento Valley of California. Relationships between these member sands and the Starkey S-5 sand provide the basis for reconstruction of geologic events and basin geometry.

Two sub-basins are recognized. Deposition of the Winters sands was concentrated in the northern sub-basin though the largest field producing from a Winters sand is at Union Island in the southern sub-basin.

The lower sands (Staten Island and Walnut Grove members) are inner bathyal deposits and part of an oblique tangential depositional pattern. These lower members are overlain unconformably by the Starkey S-5 sand.

Partial filling of the basin brought about a changed pattern of deposition and subsequent sands (Putah Sink, Mound and Unit members) are recognized as the deeper water facies of the Starkey S-5 sand with an intervening area of shale being deposited over a gentle prograding slope. For the uppermost part of the formation (the McCune member) the filling of the basin had progressed to a condition that allowed no intervening shale between the Winters sand facies and the S-5 Starkey sand, and the prograding slope can no longer be identified; the dividing line between these two sands is poorly defined.

Slope distributary channels served as conduits feeding sand into the central portion of the basin. For the lower members, stratigraphic trends associated with these distributary channels are favored areas for gas accumulations. For the higher members, gas accumulations are controlled primarily by faulting.

For both sub-basins initial recoverable reserves in the Winters sands are estimated at 650 million mcf. Future discoveries are anticipated in traps similar to those associated with existing discoveries.

Introduction

The Winters Formation is present in the subsurface of the Sacramento Valley of California and it extends into the northernmost portion of the San Joaquin Valley. Six member sands are defined and described in this paper. The four youngest are contemporaneous with the Starkey S-5 sand and a facies relationship exists between them. This relationship was shown in a

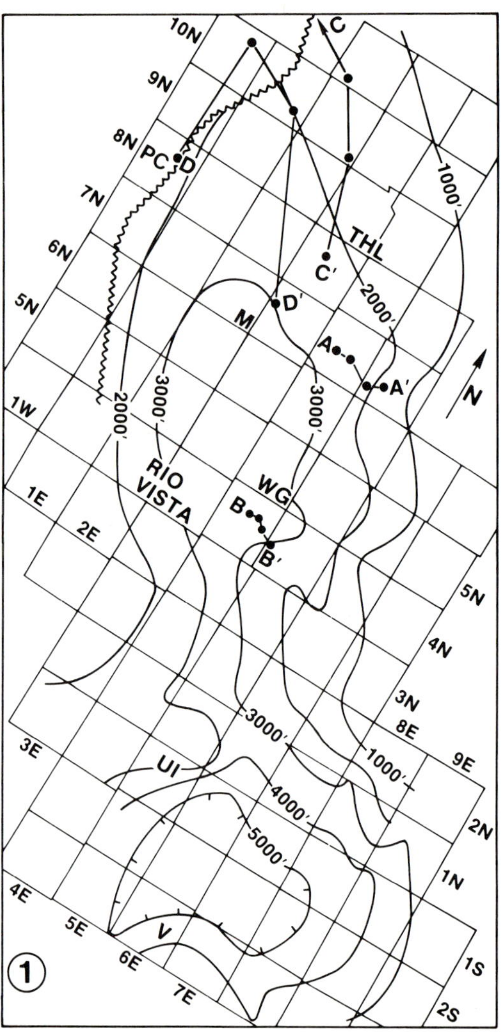

Figure 1 — Isopach: Top Starkey S-5 sand (or equivalent horizon) to "neck."

paper given at the 1976 Pacific Section AAPG meeting (Drummond et al, 1976). Figure 1 gives the

---

[1]Reprinted with revisions from 1980 Program Preprint, Annual Meeting of Pacific Sections, AAPG-SEG, 22 p.
[2]Mariposa Petroleum Company, Bakersfield, California.

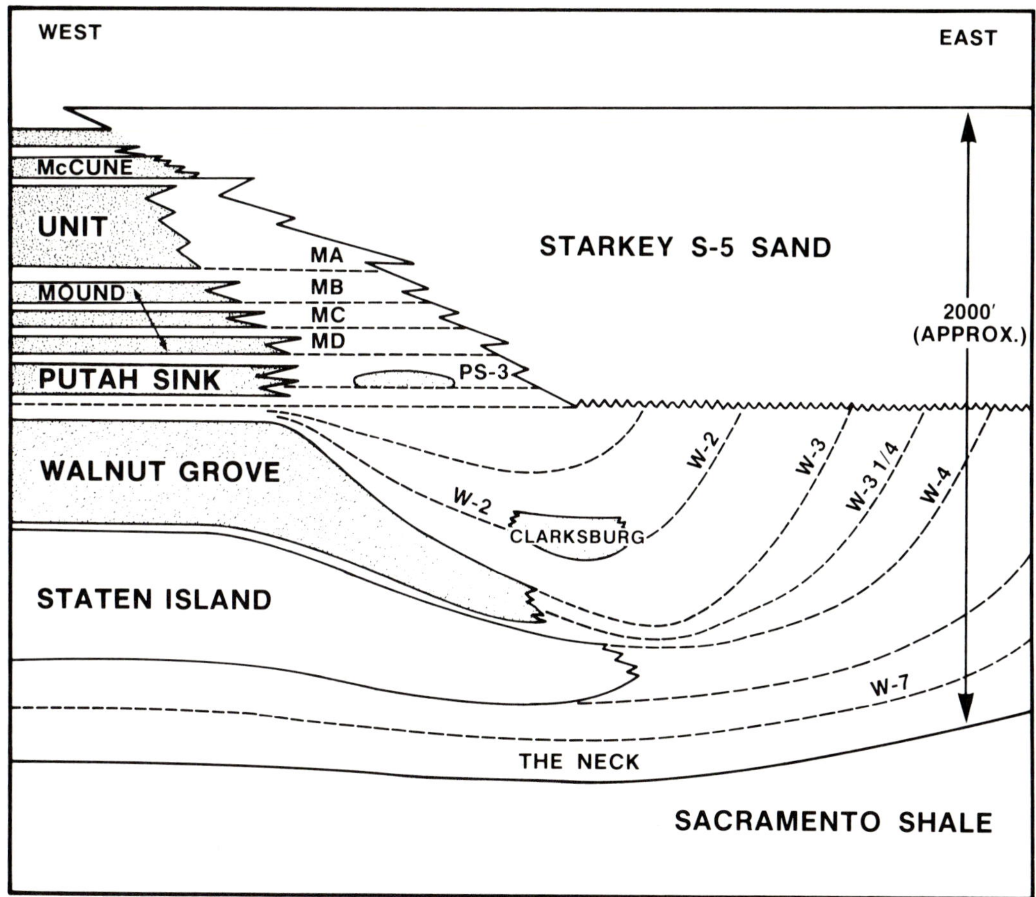

Figure 2 — Idealized schematic section.

thickness of the interval between the top of the Starkey S-5 sand (or its equivalent horizon in areas where the Starkey S-5 sand was not deposited) and the "neck," an electric log marker point at the base of the Winters Formation. This interval includes the Winters Formation, the Starkey S-5 sand and the Lathrop sands. In Figure 1 and in other figures the approximate location for some gas fields are shown by abbreviations: THL for Todhunters Lake; PC for Pleasant Creek; M for Millar; WG for Walnut Grove; UI for Union Island; and LS for Lindsey Slough.

The Winters Formation overlies the Sacramento Shale which was deposited over an extensive area in both the Sacramento and San Joaquin valleys. The Sacramento Shale was limited to the east by the continental margin and its original extent to the north is uncertain because of subsequent erosion; it extends as far north as the Sutter Buttes and into Township 17 north. Its western limit is also marked by subsequent erosion on the west side of the valley. The Alcalde

Shale at Coalinga contains a faunal assemblage correlative to that of the Sacramento Shale (Almgren, 1972) which may indicate continuous deposition from the Sacramento Valley south to the Coalinga area. The Sacramento Shale was deposited in water depths below 1,500 ft (457 m), outer bathyal.

A major regression over the entire basin marks the boundary between the Sacramento Shale and the Winters Formation and the change to shallower deposition is marked on electric logs by the characteristic "neck" marker. At this time the basin can be considered to have started forming two sub-basins; one essentially north of Rio Vista and the other south of Rio Vista expanding into the northern San Joaquin Valley. The division between these two sub-basins is rather arbitrary and shifts back and forth. Although they are designated sub-basins there is no evidence of emergence above sea level at any time of the area between the two.

In most publications, the shale section above the

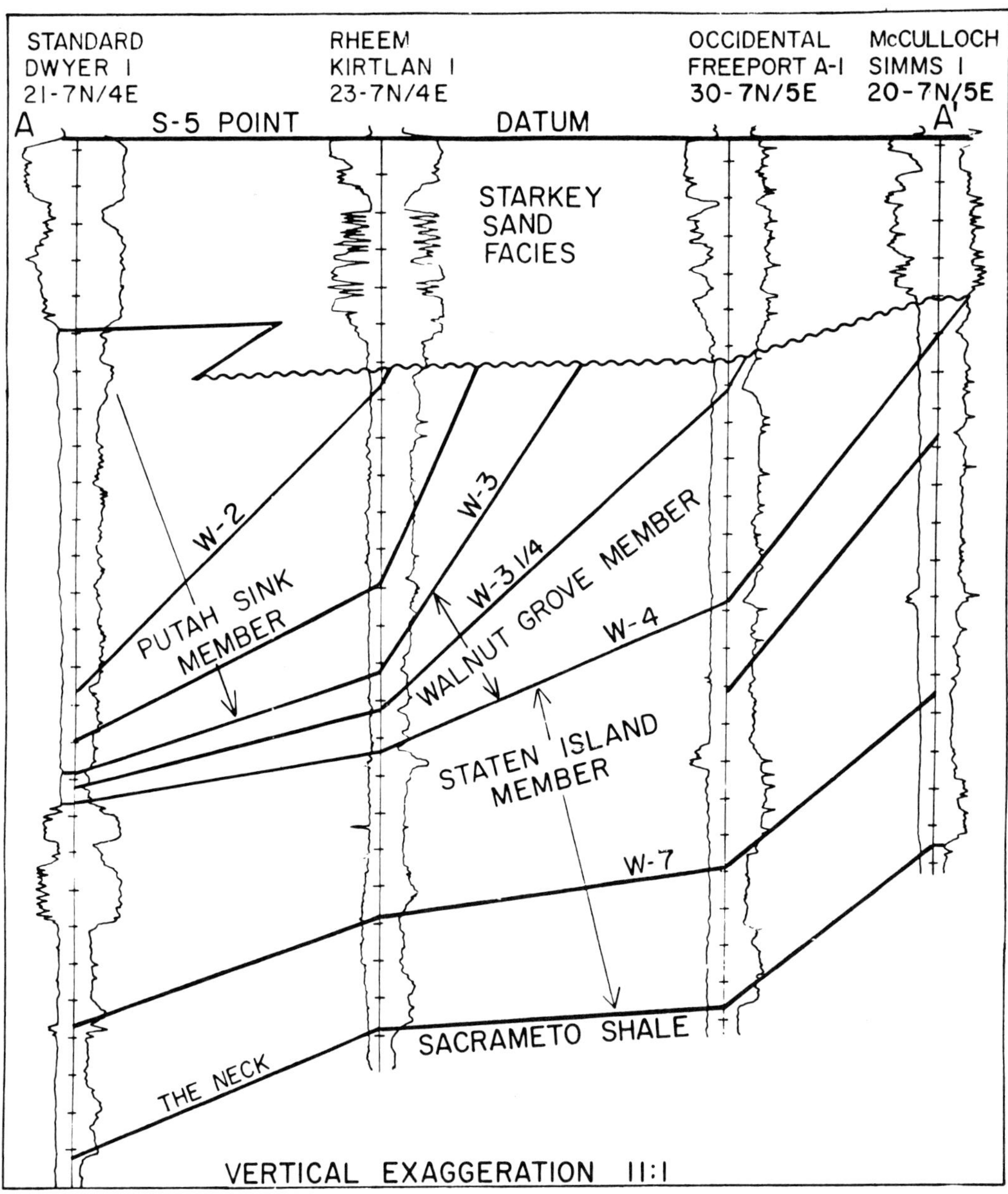

Figure 3 — Correlation section; Clarksburg to Freeport.

"neck" and below the Winters Sands is called Lower Delta Shale and the shale section above the Winters Sands and beneath the Starkey Sands is named Upper Delta Shale. In this paper, the isopach intervals for the separate member sands will include portions of these shales which correspond to the intervals of each member sand.

In the southern sub-basin a thick deltaic unit derived from a westerly source area was deposited in the interval between the "neck" and the Winters Sands. This is the Lathrop Sand which is older than any of the member sands of the Winters Formation. The equivalent section thins rapidly north of the area of Lathrop Sand deposition.

Figure 2 is a highly idealized schematic section which shows the basic stratigraphic relationships between the member sands of the Winters Formation and the Starkey S-5 sand. This section runs from east (on

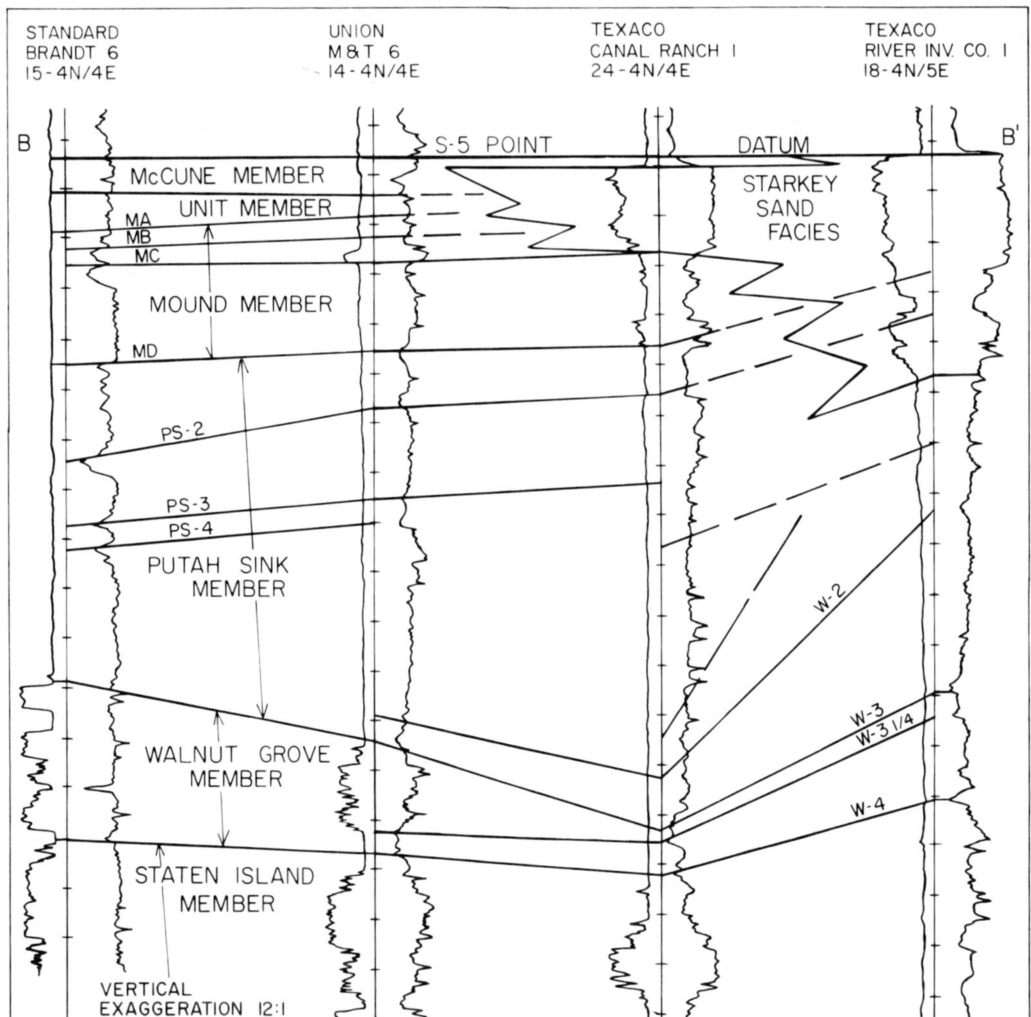

Figure 4 — Correlation section at River Island.

the right) to west and involves about 2,000 to 3,000 ft (610 to 914 m) of section. Each member sand includes several individual sands and the members were defined by grouping together those sands which tend to act as a group in their stratigraphic relationships.

The lower Winters Sands are overlain by the Starkey S-5 sand and a facies relationship does not exist between these two units. The unconformity between them shown in Figure 2 is largely a hiatus representing non-deposition with only limited erosion of the underlying section. This pattern of deposition has been designated oblique tangential (Mitchum et al, 1977). There are some wells south of Freeport which have some regressive sands which may be older than the Starkey S-5 sand and these may represent a near shore facies of the lower portion of the Winters Formation.

There are several good marker beds in the shale section of the lower portion of the Winters Formation; some of these are thin bentonites. These markers are labeled W-2, W-3, W-3¼, W-4 and W-7 on Figures 2, 3 and 4.

During deposition of the Putah Sink member there was a rather abrupt change in formation relationships. Following this change the Winters Sands still represented a deeper water deposit but they were now being deposited concurrently with the Starkey S-5 sand. Between these members and the Starkey S-5 sand there are intervening shales which were deposited on a gentler prograding slope. This change was a result of a shallowing of the basin and there was no longer sufficient relief to allow deposition of such steep (7°) foreset beds. There were now topset beds as

295

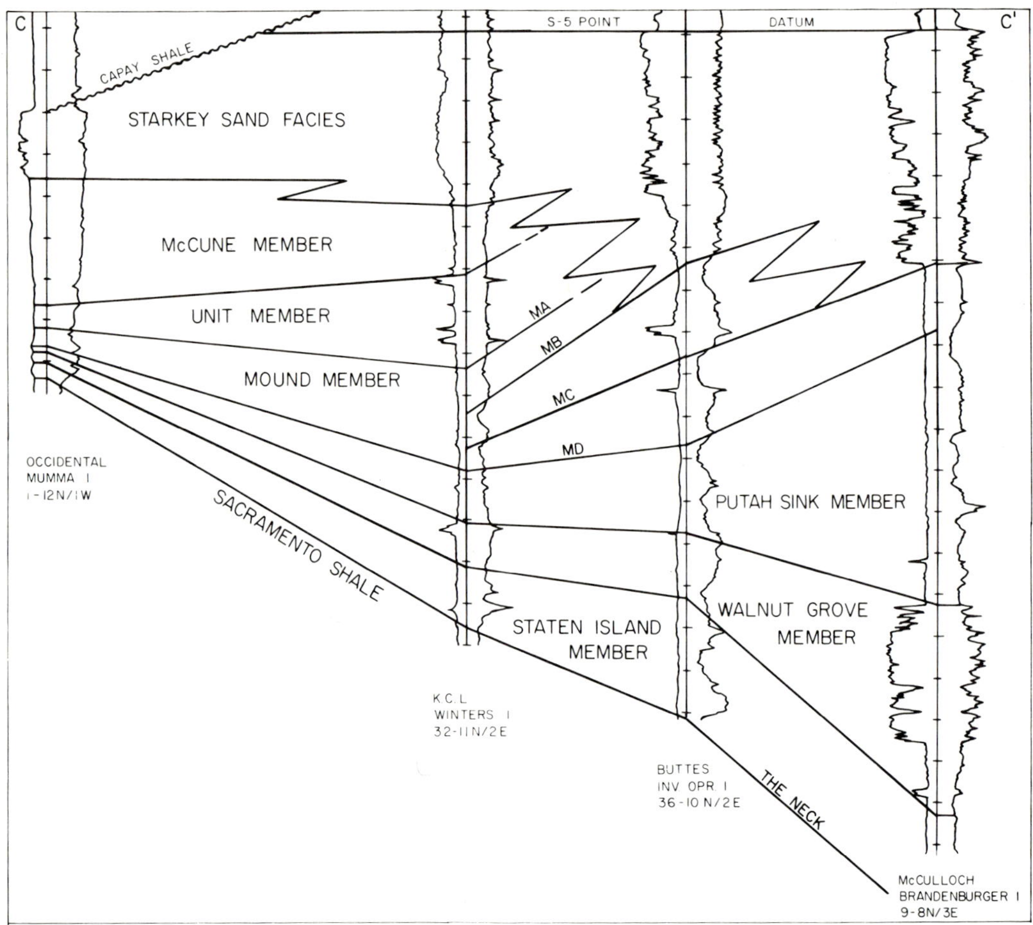

Figure 5 — Correlation section; South Buchege to Todhunters Lake.

well as foreset beds, though the angles involved were gentler. The details of these relationships are not shown in Figure 2 but a very thorough exposition is given by Drummond. This change occurred during deposition of the Putah Sink member but it did not occur at the same time throughout the basin and no single horizon can be designated to correspond to this change.

For the youngest member, the McCune sand, there is no intervening shale separating it from the Starkey S-5 sand and the line of separation between the two is arbitrary. This relationship represents a further filling and maturing of the basin. Succeeding cycles of the regressive Starkey Sands above the Starkey S-5 sand move progressively farther out into the basin, which by this time is a shallower basin than before. For these younger Starkey Sands there are no significant coeval sands in the northern sub-basin. In the southern sub-basin, the Tracy and Blewett sands are contemporaneous with the younger Starkey Sands.

Correlation Sections

Section A-A' (Fig. 3) extends from Clarksburg to Freeport. Isopach maps made between some of the labeled marker beds (W-2, W-3, W-3¼, W-4, and W-7) clearly prove that there is an unconformity at the base of the Starkey S-5 sand on the east side of the basin which truncates the lower portion of the Winters Formation. This hiatus represents mainly nondeposition, however, and very little erosion.

The maximum slope of the foreset beds was 7° and water depth in the central portion of the basin exceeded 1000 ft (305 m) (inner bathyal). Normally these foreset beds should contain considerable sand in the area of maximum slope yet almost no sand is present. This indicates that sand is entering the basin through distributary channels and is confirmed by the pattern of sand distribution in Figures 7 and 8. The base of the prograding slope moved progressively farther out into the basin as new section was deposi-

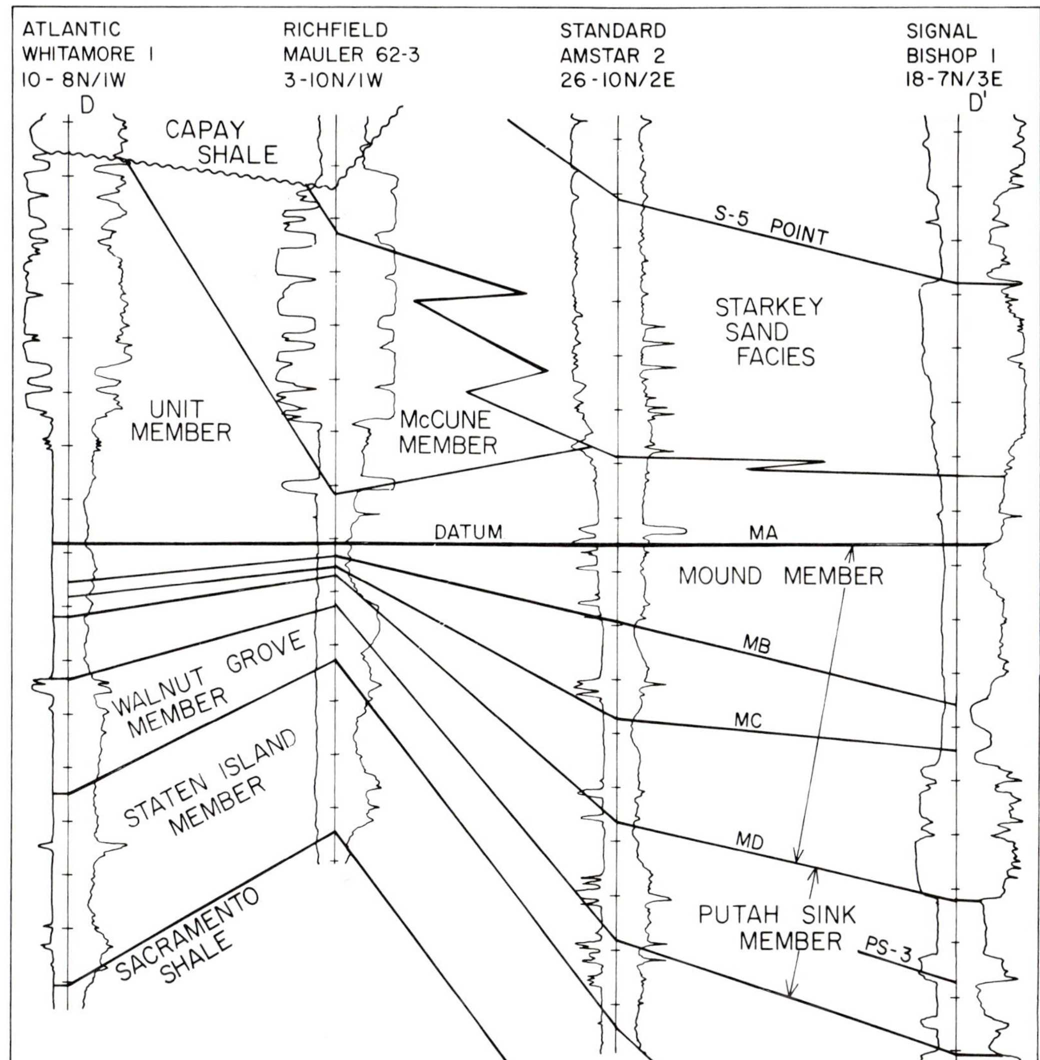

Figure 6 — Correlation section; Pleasant Creek to Saxon.

ted. For the Staten Island member, the base of the prograding slope is near the Rheem "Kirtlan No. 1," and for the Walnut Grove member the base of the prograding slope is near the Standard "Dwyer No. 1," for the Putah Sink member the base would be several miles west of the Standard "Dwyer No. 1."

Section B-B' (Fig. 4) is at River Island and shows a more complete view of relationships involving the younger members. For the McCune and Unit members this section is south of the area of sand deposition and each of these members has a correlative section of less than 100 ft (31 m). The interval for the Mound member is about 300 ft (91 m) thick and though this section shows no sand there is about 100 ft (31 m) of sand in the lower portion of the Mound member just west of this line of section. River Island is one area where it appears possible to trace electrical markers from the shale section into the Starkey S-5 sand.

In Figure 4, the view of the prograding slope is largely restricted to the Putah Sink member. The thickness of the interval of the Putah Sink member is not as dependent on sand content in the central portion of the basin as is the underlying Walnut Grove member whose thickness in the central portion is more directly related to sand content.

Section C-C' (Fig. 5) runs along stratigraphic strike in the northern half of the northern sub-basin. All members except the McCune thin to the north. There is no evidence for any unconformities to explain this thinning. The greater thickness for the McCune mem-

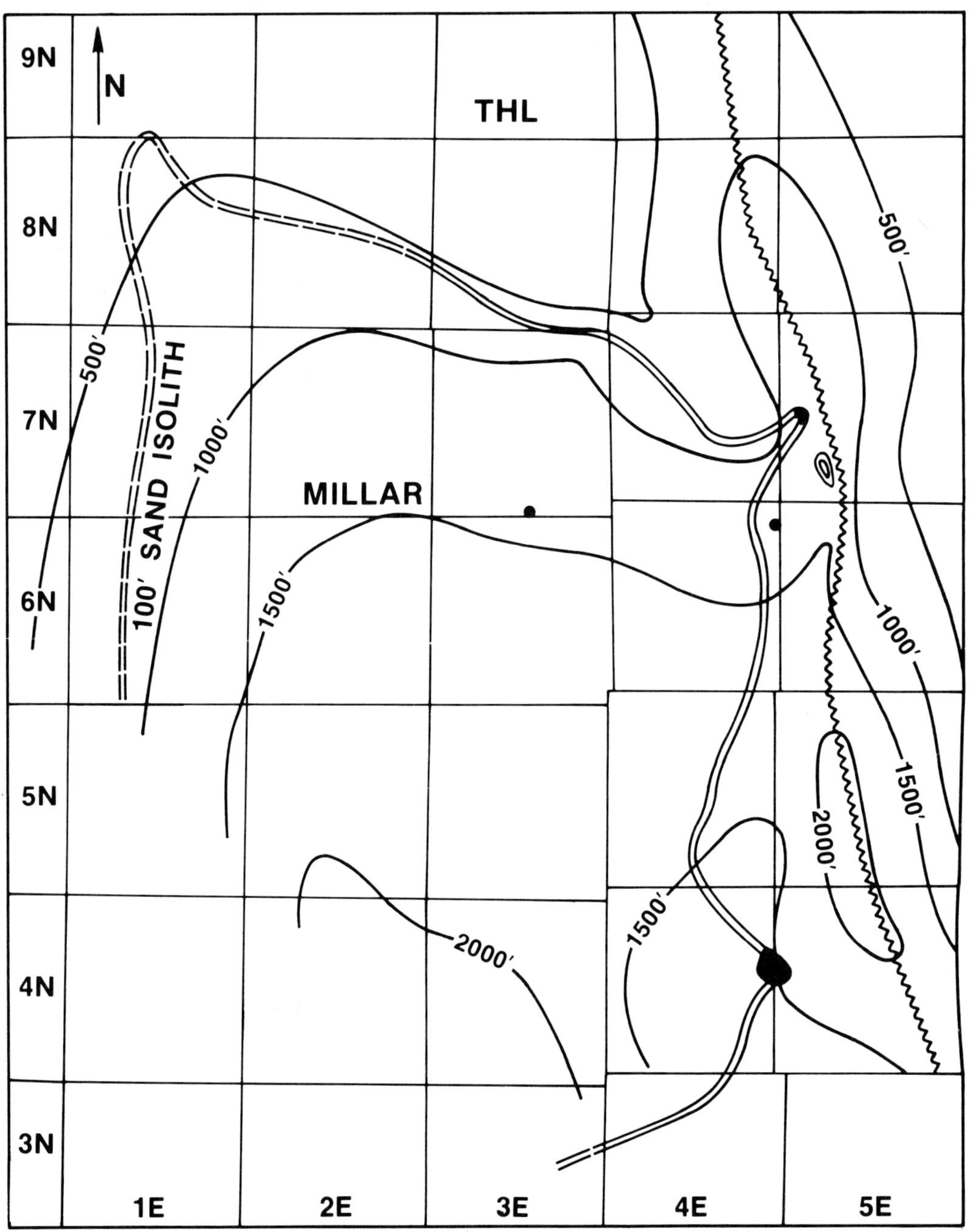

Figure 7 — Isopach for interval of the Staten Island member.

ber shows the northerly shift in area of maximum deposition. Along this line of section, and in most areas of the basin, it is generally impossible to carry the correlative boundary horizons of the individual members into and through the Starkey S-5 sand facies.

Section D-D' (Fig. 6) shows the member break-down of the Winters Formation on both the west and more familiar east sides of the basin. Direct correlation from east to west across the central portion of the basin is not possible with existing well control but these correlations can be made by going north beyond the area's dominant sand deposition and westerly

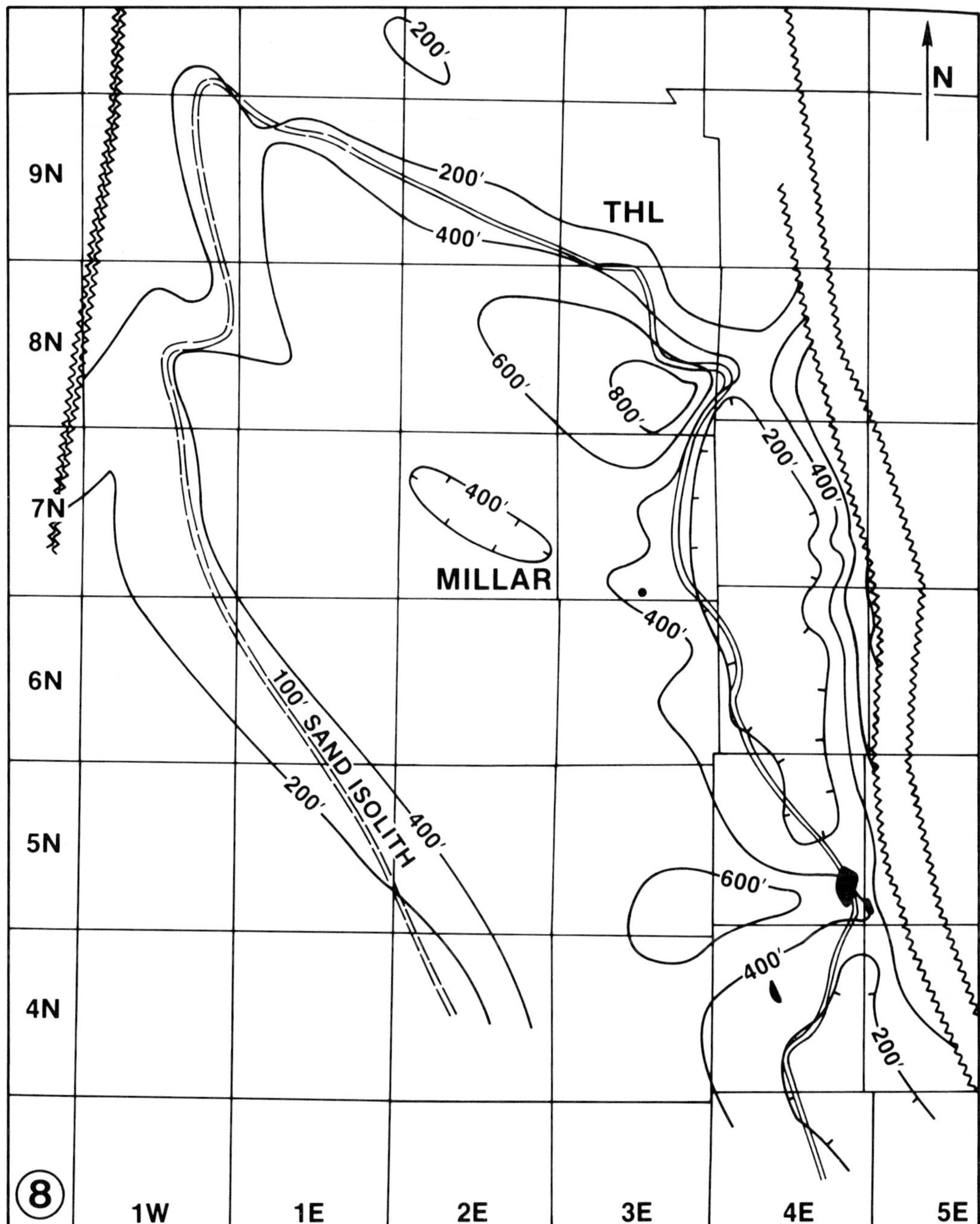

Figure 8 — Isopach for interval of the Walnut Grove member.

through the Dunnigan Hills area and then southerly down the west side of the basin. The shale markers associated with the Mound member (MA, MB, MC and MD) can be correlated with assurance and other correlations are reasonably close.

For the northern sub-basin nothing is seen in the subsurface to indicate the presence of a westerly land area though it can't be said that none was present. This is in contrast to the southern sub-basin where there was a western land mass which was a source area for the Lathrop, Tracey and Blewett sands; and it is reasonable to assume that it contributed some of the sand for the Winters Formation in the southern sub-basin.

Staten Island Sand Member

The Staten Island member is, by definition, the sand in the interval between the W-4 marker bed (or its correlative horizon) and the "neck" at the base of the Winters Formation. In places this interval is more than 2000 ft (610 m) thick though the maximum amount of sand is less than 500 ft (152 m) and usually concentrated near the top of the interval. The 100 ft (31 m) sand isolith is represented by a double line on Figure 7. There are also varying amounts of random thin sands in the lower portion of this interval that are more common to the south and finally develop into the Lathrop Sands; none of these are included as part of the Staten Island sand. The Staten Island sand is younger than the Lathrop Sands but the interval covered by the isopachs in Figure 7 includes section corresponding to the Lathrop Sands. The minimum isopach values on the east side of Figure 7 correspond to the position of the base of the prograding slope. The unconformity line on the east side of Figure 7 represents the intersection of the top of the Staten Island interval with the unconformity at the base of the Starkey S-5 sand (see Figs. 2 and 3).

Two distributary channels can be identified which served to feed sand into the basin; other channels may be present. One is at Freeport (7N/5E) and the other at Staten Island (4N/4E). These distributary channels are favored areas for gas accumulation and have local basal conglomeratic intervals.

Well completions and initial recoverable gas reserves for the Staten Island sand are as follows (see solid areas in Figure 7): River Island (Staten Island Pool) — 6 wells, 36 million mcf; Freeport (7N/5E) — 1 well, 3 million mcf; Elliott Ranch (6N/4E) — 1 well, 1 million mcf; and Saxon (7N/3E) — 1 well, 1 million mcf. These individual figures total to 9 well completions and 41 million mcf in initial recoverable gas reserves for the Staten Island sand.

Walnut Grove Sand Member

The Walnut Grove member is defined as the sand in the interval between the W-3 and W-4 marker beds or equivalent horizons. Areas of maximum interval thickness tend to coincide with areas of lesser interval thickness for the underlying Staten Island member which shows the effect of basin topography on deposition. The minimum isopach values on the east side of Figure 8 show the position of the base of the prograding slope. The two unconformity lines on the east side of Figure 8 represent truncation at the top and bottom of the Walnut Grove interval by the unconformity at the base of the Starkey S-5 sand.

The thickness of the overall interval is closely related to sand content with the double line of Figure 8 representing the 100 ft (31 m) sand isolith. The area of maximum sand content is near Putah Sink (8N/3E) where as much as 600 ft (183 m) of sand is present. This area of maximum sand content is fed by one of the two major distributary channels; the other is at Walnut Grove (5N/4E). Lesser distributary channels may exist but are not as well documented.

Well completions and initial recoverable gas reserve estimates for the Walnut Grove sand are as follows (see solid areas in Fig. 8): Walnut Grove (5N/4E) — 9 wells, 38 million mcf; River Island (4N/4E) — 1 well, 3 million mcf; Saxon (7N/3E) — 1 well, 1 million mcf. These individual figures total to 11 well completions and 42 million mcf in initial recoverable gas reserves for the Walnut Grove sand.

Clarksburg Sand

The Clarksburg sand is a lower sub-member of the Putah Sink sand member. It was not designated as a separate member because it is only a local sand. For the lower Winters Formation this is the only significant sand deposited on the prograding slope and it is more than coincidental that there is no nearby distributary channel for this portion of the section. If there had been a distributary channel the sand would have been funneled out into the basin.

Figure 9 has both sand isoliths for the Clarksburg sand (50 ft; 15 m intervals) and isopaths between the W-2 and W-3 points (100 ft; 31 m intervals). Since the Clarksburg sand extends in section above the W-2 marker bed the sand count exceeds the interval isopachs in portions of Figure 9.

Well completions have been disappointing with initial recoverable gas reserves estimated at only 5 million mcf from 5 wells at Clarksburg (7N/4E), one well completed in the Locke sand at Walnut Grove (5N/4E), and one well at Merritt Island (6N/4E). These areas of production are the solid areas in Figure 9.

Putah Sink Sand Member

Figure 10 shows the thickness of the Putah Sink interval in the basin west of the prograding slope (heavy lines) and gives a sand count for this member (light lines). The facies relationship to the Starkey S-5 sand is shown in Figures 2 and 4. For this and the overlying Mound member, the deposition could best be defined as following a sigmoidal pattern (Sangree and Widmier, 1977). Geometric considerations place water depths in the central portion of the basin at an estimated 500 to 600 ft (152 to 183 m).

Clearly defined distributary channels no longer exist and sand entered the basin over broader, less clearly defined areas. There are many random isolated bar sands east of the main sand body which are

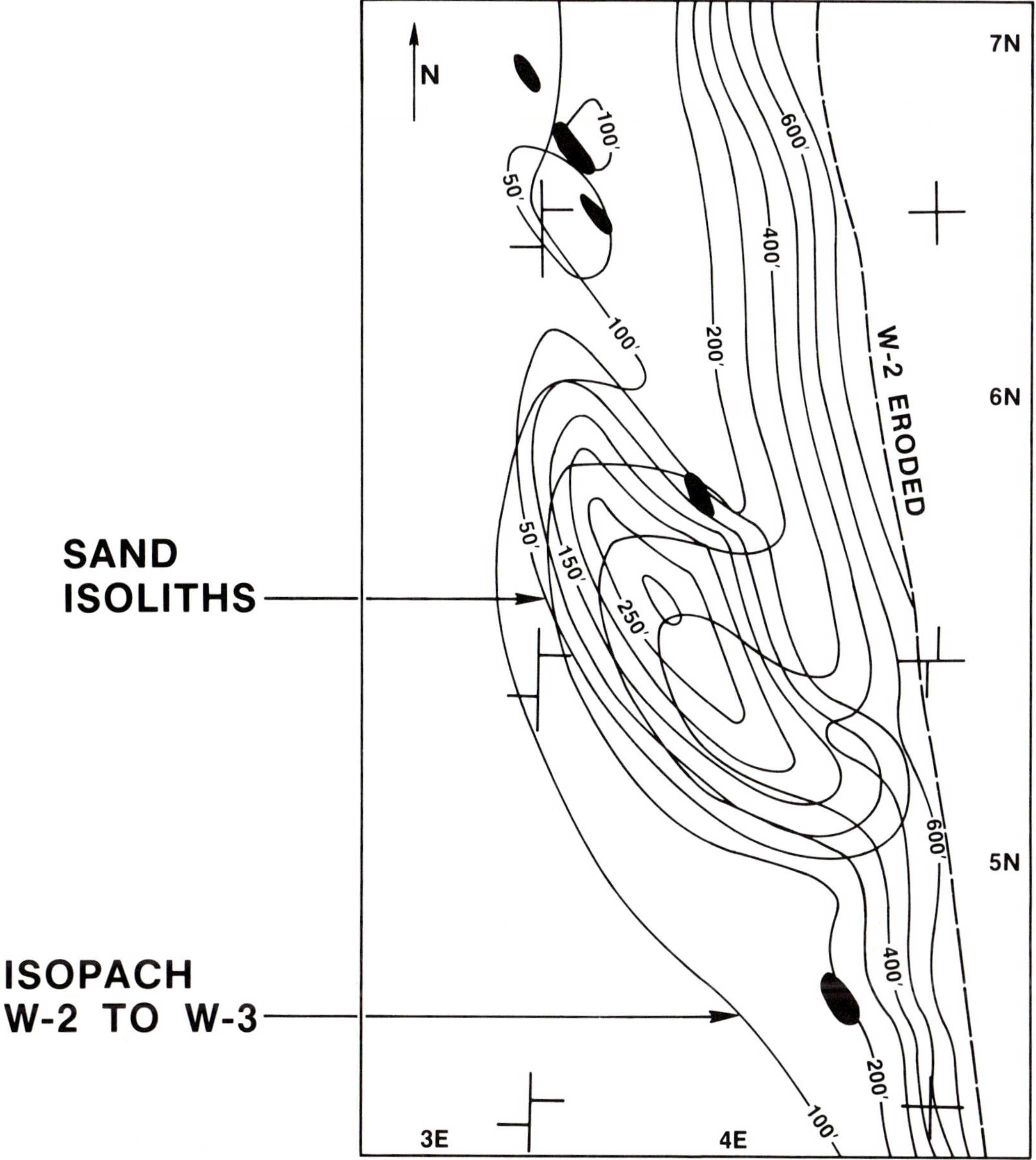

Figure 9 — Isopach: W-2 to W-3 (intervals of 100 ft; 31 m) and sand count for the Clarksburg sand sub-member (intervals of 50 ft; 15 m).

significant because of gas accumulations in these sands at Putah Sink and River Island (4N/4E), and four scattered, less significant, wells in the main sand body have between them initial recoverable gas reserves estimated at 48 million mcf.

Union Island Sand

All of the Winters Sand at Union Island (1S/5E) is considered to be contemporaneous with the Putah Sink member. This is in the southern sub-basin and extends from Township 2 south, Range 5 east, to Township 3 north, Range 4 east. Both eastern and western source areas are postulated for this sand. This sand does not connect with the Putah Sink sand; it is a separate sand body. The geometry of sand distribution suggests that the Union Island field is in a distributary channel and one or more other distributary channels may exist.

This is the only Winters sand whose deposition was restricted to the southern sub-basin. Only one of the

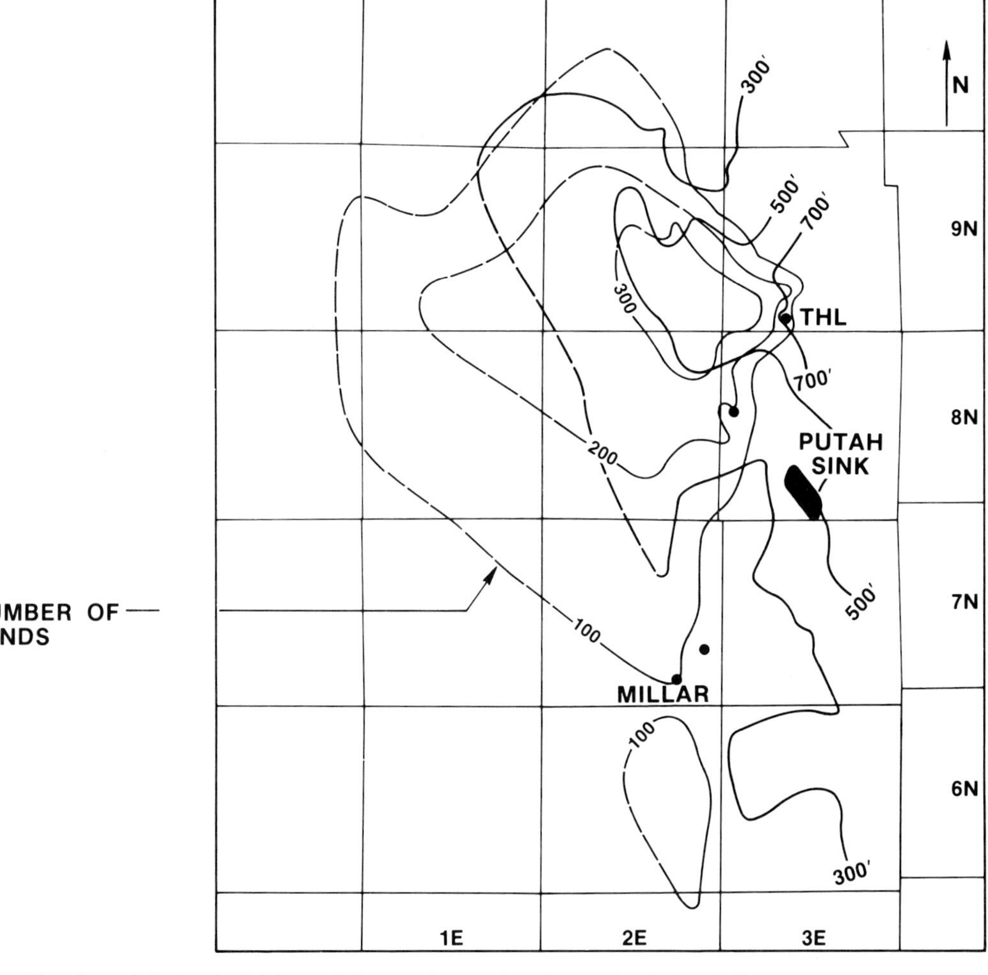

Figure 10 — Isopach for Putah Sink interval (larger values) and sand count for the Putah Sink sand.

member sands from the northern sub-basin extends into the southern sub-basin, the Walnut Grove sand, which extends south of McDonald Island into Township 1 north, Range 4 east.

Initial recoverable gas reserves for the Union Island field are estimated by the California Division of Oil and Gas at 250 million mcf. Ultimate recovery could be greater, possibly exceeding 300 million mcf; which would be almost as much as that for all Winters sand production in the northern sub-basin as listed herein.

Mound Sand Member

The Mound sand was deposited over a greater area than the underlying Putah Sink sand; this probably reflects less relief in the depositional basin. In many areas there are distinctive shale marker beds (MA, MB, MC and MD) which separate the upper, middle and lower sub-members of the Mound sand (see Figs.

2, 4, 5 and 6). Usually these marker beds can be readily correlated but east to west correlation across the central portion of the basin is not possible with any degree of accuracy.

For all older member sands stratigraphy is the dominant factor in providing gas accumulation; for the Mound sand, faulting is the dominant factor. Most of the production for this sand has been found on the east side of the basin but the new discovery at Lindsey Slough (5N/2E) may lead to more extensive exploration to the west.

The solid areas on Figure 11 are the more significant producing areas for the Mound sand and less significant individual wells are shown as solid circles. Seventy wells have been completed in the Mound sand with initial recoverable gas reserves estimated at 140 million mcf.

302

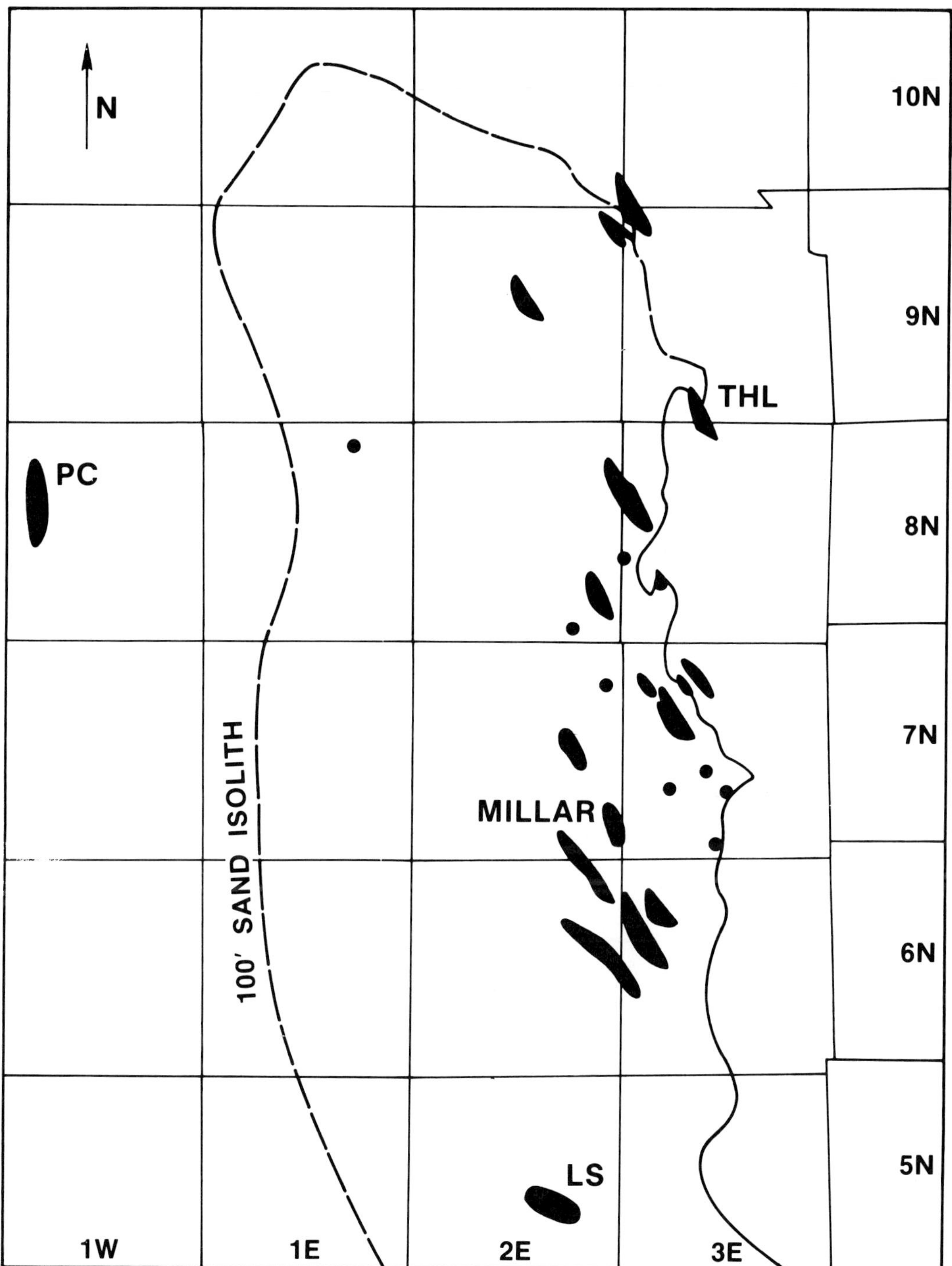

Figure 11 — Productive areas for the Mound sand member.

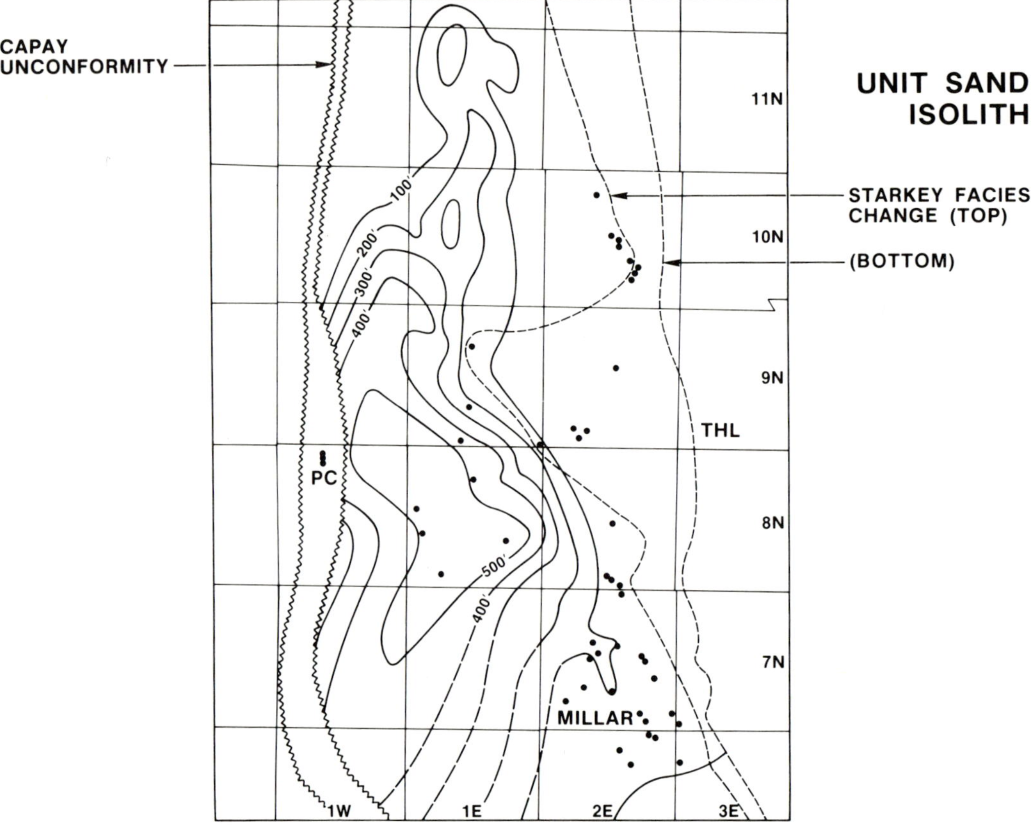

Figure 12 — Sand count for the Unit sand member.

Unit Sand Member

For the Unit sand, the area of deposition shifted considerably to the northwest. Some of the larger sand count values shown in Figure 12 are dependent on proper correlation of the base of this sand which is difficult in the central area of deposition and subject to some error. The unconformity lines on the west side of Figure 12 represent truncation of the top and bottom of this interval by the Capay Shale, the dashed lines on the east side showing the approximate position of the facies change to the Starkey regressive facies for the top and bottom of this interval (see Fig. 5). The pattern of sand deposition suggests that this sand was derived from a northerly source area; all older members show an easterly source area.

Forty-six wells have been completed in the Unit sand (solid circles in Fig. 12) but the combined initial recoverable gas reserves are estimated at a disappointing figure of only 25 million mcf; with Millar (7N/2E) and Crossroads containing the only significant accumulations. However, such a large number of accumulations in widely separated areas and under different conditions of entrapment suggests that larger fields remain to be discovered.

McCune Sand Member

It is hard to separate the McCune sand from the Starkey S-5 sand in areas where they merge. At Winters (8N/1E), the McCune sand is easily recognized but a few miles to the northeast in Township 9 north, Range 1 east, segregation becomes impossible.

This member derives its name from the primary producing sand in the Winters gas field which produced 35 million mcf and is now nearly depleted. The producing sands in the Dunnigan Hills field are part of the McCune sand member and they produced 11 million mcf before the field was abandoned. Other production from the McCune sand member has not been significant.

Summary

The member sands of the Winters Formation are grouped as Upper Winters (McCune sand), Middle Winters (Unit sand; Mound sand; and Putah Sink sand in central basin area), and Lower Winters (Clarksburg sand sub-member, prograding slope area; Walnut Grove sand; and Staten Island sand). Using this breakdown, which applies for the northern sub-basin only, the following summaries can be made:

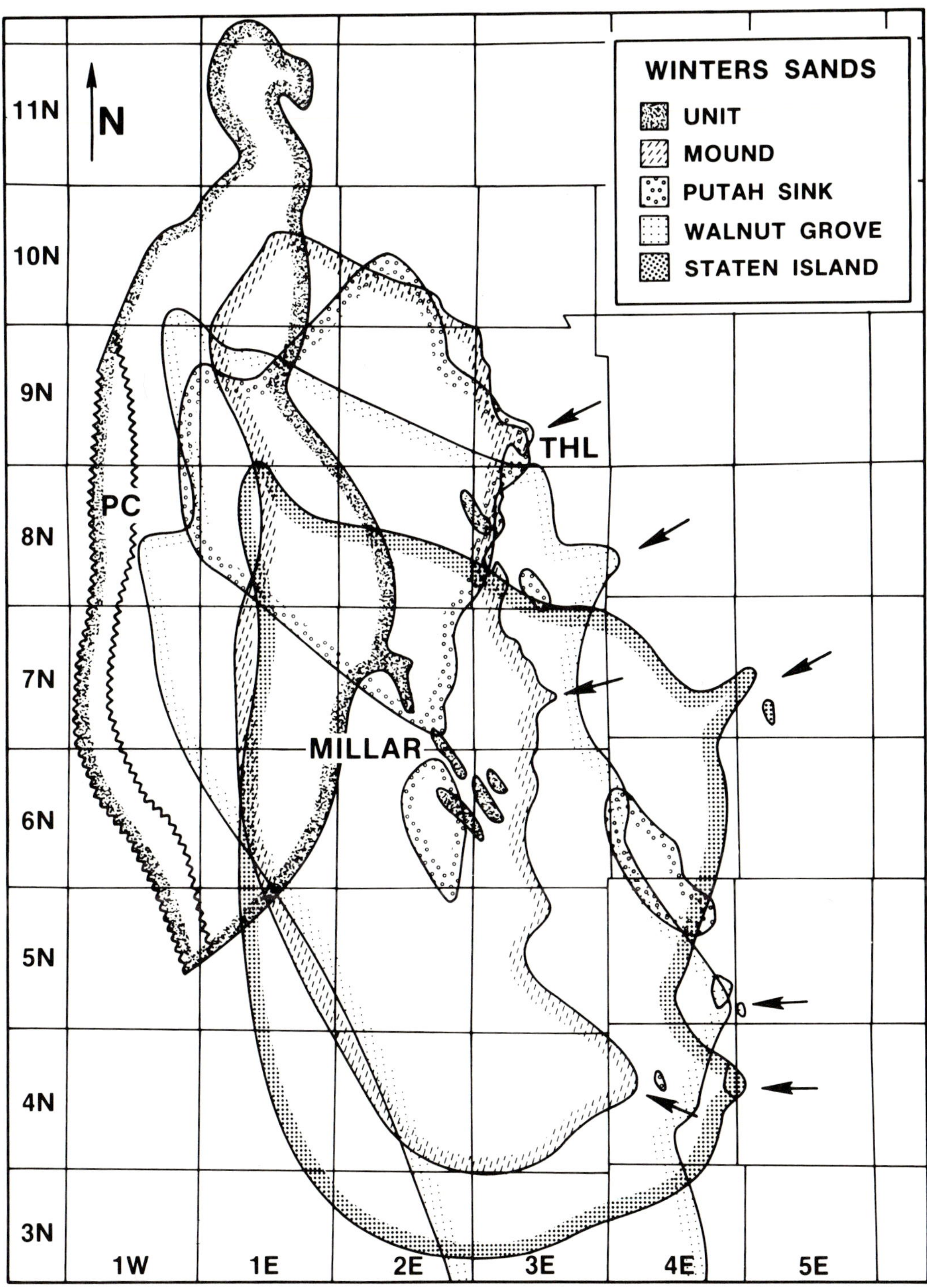

Figure 13 — One hundred foot (31 m) sand limit lines for member sands of the Winter Formation.

Pattern of prograding slope — In the Upper Winters, essentially no pattern is present. In the Middle Winters, the prograding slope pattern is sigmoidal. In the Lower Winters, a dominantly oblique tangential pattern is present.

Angle of prograding slope — In the Upper Winters, since essentially no pattern of slope is present, the question of angle is not applicable. In the Middle Winters, the angle is assumed to be near 3°. In the Lower Winters, the angle is up to 7°.

Water depth — In the Upper Winters, depth is near 300 ft (91 m) or less. In the Middle Winters, depth is 500 to 600 ft (152 to 183 m). In the Lower Winters, depth is below 1000 ft (305 m).

Relationship to Starkey S-5 sand — The Upper Winters shows direct facies relationship with no intervening shale. The Middle Winters shows deeper water facies with intervening shale. The Lower Winters is overridden by Starkey S-5 sand with a separating unconformity.

Distributary channels — Channels are not applicable to the Upper Winters, and are at best poorly defined in the Middle Winters. In the Lower Winters, however, distributary channels are dominant in controlling flow of sand into the basin.

Gas Production — In the Upper Winters, gas production is dominantly stratigraphic on the west side of the basin. In the Middle Winters, production comes from fault and stratigraphic traps for the Unit sand, dominantly fault traps for the Mound sand, and is dominantly stratigraphic in isolated bar sands for the Putah Sink sand. Lower Winters gas production is also dominantly stratigraphic in distributary channels.

Figure 13 shows the 100 ft (31 m) sand count limit lines for each of the member sands except the McCune sand member. The positions for these limit lines shows the northwestward migration of the central portion of the northern sub-basin. Also shown are the more significant productive areas (solid color). Arrows indicate the position of major distributary channels; other lesser distributary channels are present but not identified.

References Cited

Almgren, A.A., 1972, Age and correlation of the Alcalde Shale Formation - an alternative interpretation, *in* Cretaceous of the Coalinga Area: Fall Field Trip Guidebook, Pac. Sec., SEPM, p. 59-65.

Drummond, K. F., E. W. Christensen, and K. D. Berry, 1976, Upper Cretaceous lithofacies model, California: Ann. Mtg., Pac. Sec., AAPG, p. 76-88.

Mitchum, R.M., Jr., P.R. Vail, and J.B. Sangree, 1977, Stratigraphic interpretation of seismic reflection patterns in depositional sequences, *in* Seismic stratigraphy - applications to hydrocarbon exploration: AAPG Mem. 26, p. 117-133.

Sangree, J.B., and J.M. Widmier, 1977, Seismic interpretation of clastic depositional facies, *in* Seismic stratigraphy - applications to hydrocarbon exploration: AAPG Mem. 26, p. 165-184.

The American Association of Petroleum Geologists Bulletin
V. 61, No. 3 (March 1977), P. 344-359, 25 Figs.

Depositional Environments of a Reservoir Sandstone in West-Central Texas[1]

RICHARD R. BLOOMER[2]

Abstract Hundreds of late Paleozoic stratigraphic traps in fluvial, deltaic, and marine sandstones are present on the Eastern shelf and slope of the Midland basin in west-central Texas. Widespread fluvial sandstones on the shelf are well known, but there appears to be little recognition of the depositional environments of sandstones in ancient deltas on the shelf and submarine channels and fans on the slope.

The Cook sandstone is a depositional model of late Paleozoic fluvial, deltaic, and slope sandstone facies in the Midland basin. Each of two Cook fluvial and deltaic systems mapped is about 100 mi (160 km) long from outcrop to shelf edge. The "Blackwell" field, in one system, is an excellent example of a Cook point-bar stratigraphic trap.

Basinward from the basal Permian Eastern shelf edge, sand and mud supplied by the two Cook fluvial systems prograded into a shallow sea and became two high-constructive lobate deltas. Delta-plain aggradational facies, delta-front sheet and distributary-channel sandstones, and prodelta facies are present in the deltas. The Group 4000 "Pennsylvanian Cisco" sandstone field produces from Cook deltaic sandstone.

The deltas are the proximal parts of a clastic wedge of sediments on the Eastern slope. The wedge has a maximum thickness of 1,300 ft (396 m) at midslope and thins basinward. The Northeast Bloodworth "Pennsylvanian Canyon" field produces from midslope Cook sandstone which probably was deposited in a submarine channel. The Jameson "Pennsylvanian Strawn" sandstone field produces from Cook lower-slope sandstone and appears to be a submarine fan.

INTRODUCTION

Approximately 300 Lower Permian sandstone reservoirs are present on the Eastern shelf and slope of the Midland basin in west-central Texas (Fig. 1). Most of the reservoirs were discovered between the 1920s and 1950s in shallow fluvial sandstones. First recognition of the fluvial origin of the sandstone in one of these fields was not until 1960 (Lawless, 1960; Shankle, 1960).

The widespread presence of fluvial-sandstone reservoirs on the Eastern shelf now is well known, but there is still little recognition of stratigraphic traps in deltaic deposits on the shelf and submarine-canyon and fan deposits on the slope.

Cook sandstone is a depositional model of late Paleozoic fluvial, deltaic, and slope sandstone facies in the Midland basin. This paper is a regional and local subsurface study of the Cook sandstone which was deposited in two paleodrainage systems on the Eastern shelf and a submarine canyon and fan on the Eastern slope. All of the producing areas in the Cook depositional systems are essentially stratigraphic traps. Exploration for

these traps depends greatly on the interpretation of local environments of deposition. The subsurface methods used to reveal the Cook sandstone depositional environments and associated types of traps are applicable in the exploration for similar sandstone reservoirs in other areas.

GEOLOGIC SETTING

Lower Permian sedimentary rocks crop out along the eastern margin of the Eastern shelf as shown in Figure 1. At the base of the Permian, the Eastern shelf has a regional west-northwest dip of about 0.5° (50 ft/mi or 9 m/km) to the shelf edge (Fig. 1). West of the shelf edge, the Eastern slope reaches a maximum regional dip of about 3° (276 ft/mi or 53 m/km) at midslope and decreases farther downslope.

Fluvial sandstones (Fig. 2) or their deltaic and slope-stratigraphic equivalents produce in the fields shown in Figure 1. Each sandstone unit is in a cyclic depositional sequence or format (Forgotson, 1957) bounded by thin persistent limestone or coal beds. These beds are excellent surface and subsurface mapping units over a wide area of the Eastern shelf. The widespread and persistent characteristic of these thin limestone and coal beds indicates that they were deposited during eustatic changes in sea level. The limestone beds, with suites of open-marine fossils, probably were deposited during periods of transgression. Regressions of the sea exposed alluvial coastal plains which were incised by meandering streams. Swamps and marshes on the plains became coal beds after burial. There are widespread paleosol zones on the shelf (Harrison, 1973).

Cyclic deposition on the shelf probably was a response to changes in sea level during alternating glacial and interglacial periods in the Southern Hemisphere (Van Siclen, 1958; Shankle, 1960; Jackson, 1964). Brown (1969a, b) and Gal-

[1]Manuscript received, May 24, 1976; accepted, August 16, 1976. Presented before the Southwestern Section of the Association at Fort Worth, Texas, March 16, 1973.

[2]Independent geologist, 132 Devonian Bldg., Abilene, Texas 79603.

The writer is indebted to Daniel A. Busch and Robert A. Berg for their stimulating discussions of the study and critical reading of the manuscript.

344

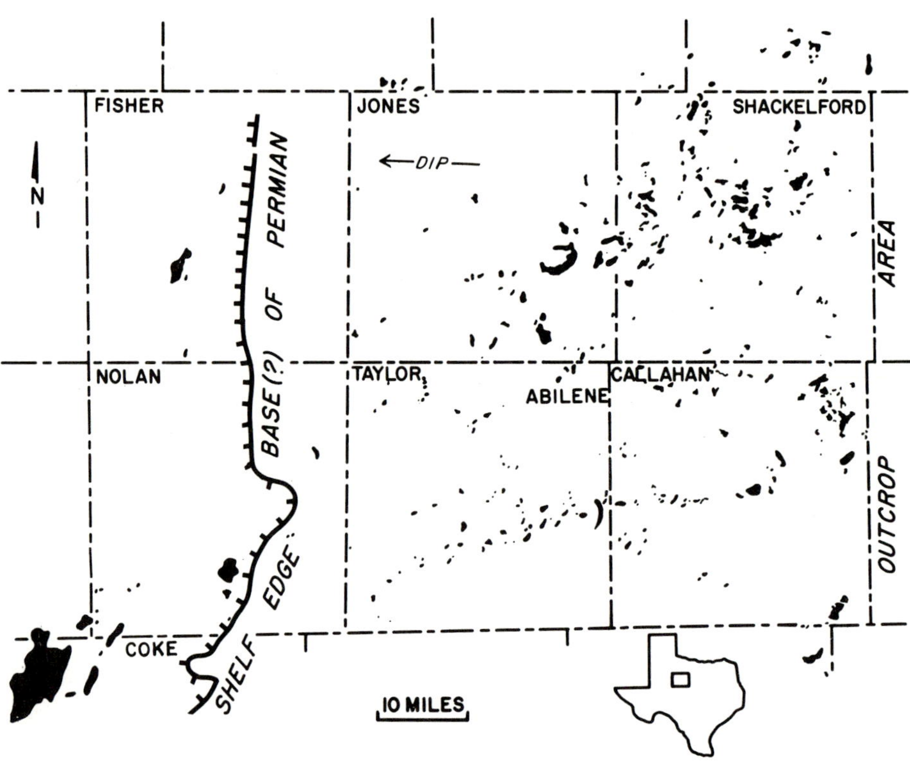

FIG. 1—Map showing basal Permian shelf edge and oil and gas fields producing from Lower Permian sandstones in west-central Texas.

loway and Brown (1972, 1973), however, did not recognize widespread thin limestone, coal, and paleosol beds on the shelf, and, therefore, did not attribute the apparent cyclicity of shelf deposits to eustatic sea-level fluctuations but to intrabasinal variations in uplift and subsidence or sedimentary controls.

The top of the Crystal Falls Limestone Member at the base of the sections shown on Figure 2 is a convenient lithologic marker for the controversial Pennsylvanian-Permian boundary. Overlying the Crystal Falls Limestone Member is the Waldrip Shale Member which contains three thin limestone units named in ascending order, Waldrip limestones 1, 2, and 3 (Fig. 2). The uppermost of the three limestones commonly is referred to as the Flippen limestone in the subsurface. These limestone beds are only a few feet thick in the outcrop (Stafford, 1960) but thicken basinward in the subsurface.

Any sandstone present between the Crystal Falls and Flippen limestones is termed informally "Cook sandstone." Two Cook sandstone paleodrainage systems (post-Waldrip limestone 2 and pre-Flippen limestone in age) are considered in this study. In the fluvial environment, maximum incision of Cook channels was through Waldrip shale and limestone beds to the Crystal Falls limestone (about 100 ft or 30 m). The channels, the extensive coal beds, and the widespread paleosols noted by Harrison (1973) are evidence that a subaerial environment was widespread during deposition of this stratigraphic interval.

COOK PALEODRAINAGE SYSTEMS

Two Cook fluvial and deltaic systems have been mapped (Fig. 3) on the Eastern shelf. About 100 mi (160 km) of each system is preserved in the surface and subsurface. The headwaters of these paleodrainage systems probably were in the ancient Ouachita Mountains, now a buried structural belt (Flawn et al, 1961) about 100 mi (160 km) east.

The Cook sandstone outcrop area shown on Figure 3 is modified after surface maps by Stafford (1960) and Barnes (1972). Every available well log (several thousand) was used to map the paleodrainage systems in the subsurface. Electric-log control is about 95%. Well density ranges

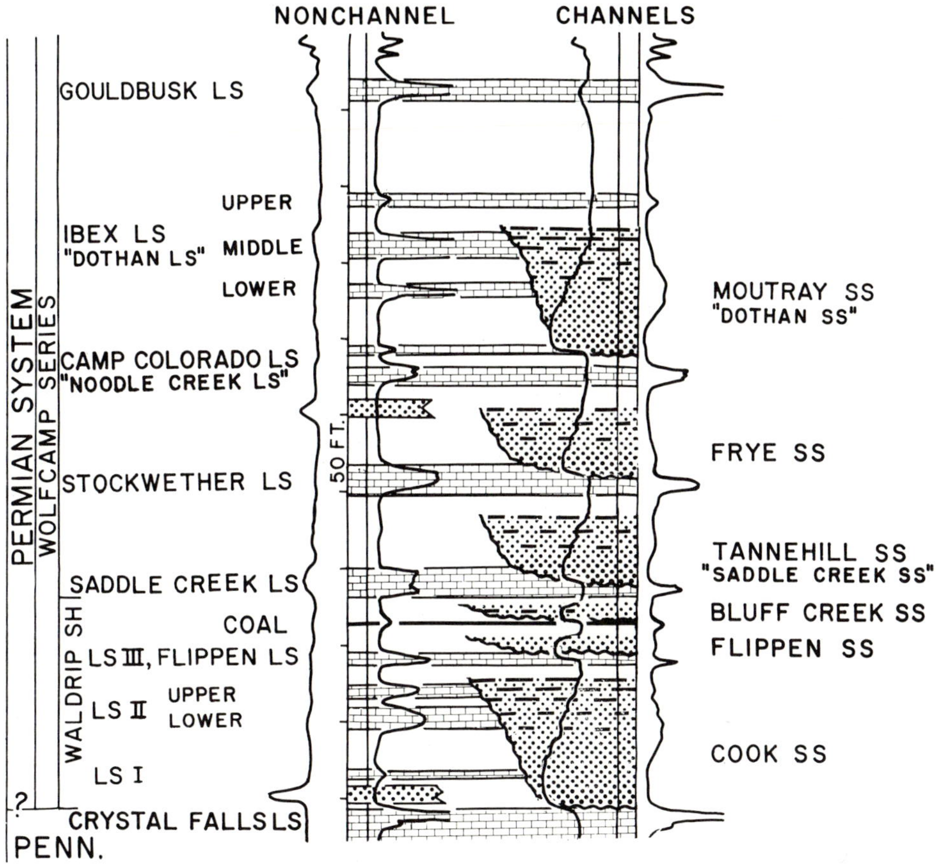

FIG. 2—Diagrammatic composite electric logs of Lower Permian section in Taylor County, Texas, showing stratigraphic sections with and without fluvial sandstones (right and left logs respectively).

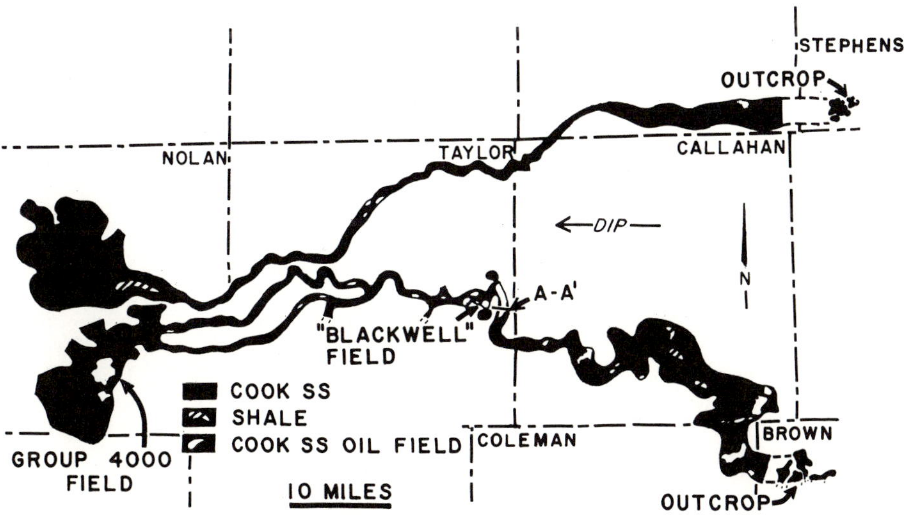

FIG. 3—Two Cook paleodrainage systems on Eastern shelf, showing meander belts from outcrop to deltas. *AA'* is line of cross section shown in Figure 5.

FIG. 4—Outcrop in southwest Stephens County, Texas, showing sharp erosional contact (arrow) of Cook point-bar deposit with subjacent Waldrip shale.

Cook sandstones in the outcrop of the north meander belt (Fig. 3) are characteristic of a fluvial section (Fig. 4). The arrow in Figure 4 points to the sharp basal contact of Cook channel fill with the subjacent Waldrip shale. The basal 2 ft (0.6 m) of channel fill is conglomerate, consisting of pebbles of chert, quartz, and limestone, and probably is channel-lag material. The conglomerate decreases in size upward to medium-grained sandstone at the top of Figure 4. Cross-bedding is common and indicates that the paleocurrent flowed westward. The conglomerate and sandstone in Figure 4 probably are part of a point-bar deposit.

Stratigraphic and structure cross sections of the south meander belt (Fig. 5) show the sharp contact of shale and sandstone with the laterally adjacent Waldrip limestone beds marking the boundaries of the meander belt. The convex-downward base of the meander belt is clear. Two periods of Cook channel cut and fill are indicated.

Brown (1969a, b) and Galloway and Brown (1972, 1973) described downdip elongated Cook and upper Cook/Flippen sandstone units on the Eastern shelf as fluvial-deltaic lobes of progradation and the "interlobe" limestone, shale, and sandstone beds as interdeltaic embayment deposits. Their interpretations were based mainly on

from several to over 50 wells/sq mi (20 wells/sq km).

The Cook fluvial systems are considered to be meander belts. The meandering streams within the belts flowed on a low-gradient alluvial paleoslope. Abrupt contacts of the meander-belt deposits with Waldrip limestone beds clearly outline the boundaries of the belts (Fig. 5).

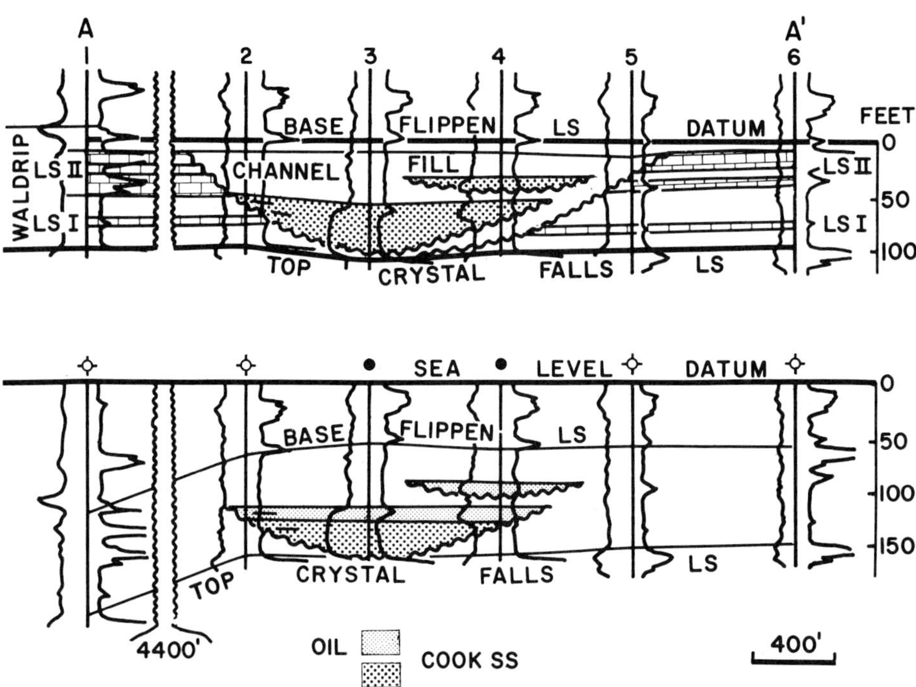

FIG. 5—Stratigraphic and structure cross sections of Cook meander belt. See Figure 3 for location of cross section.

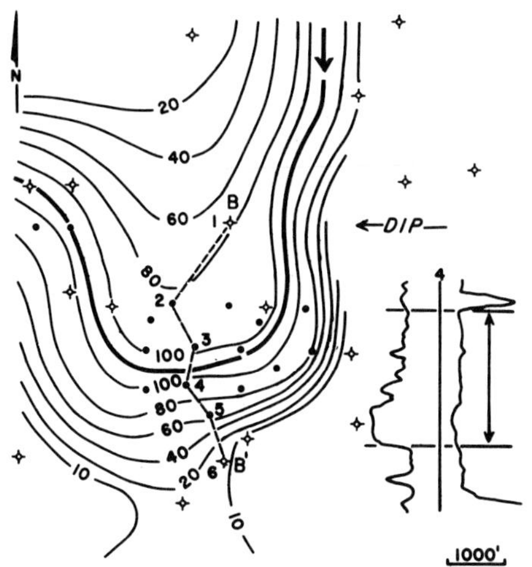

FIG. 6—Isopach map of interval from base of Flippen limestone to base of upper Cook sandstone in Blackwell field, Taylor County, Texas. Contour intervals 10 and 20 ft (3 and 6 m). *BB'* is line of cross section shown in Figure 7.

regional sandstone isopach maps with selected control points at an average 5-mi (8 km) spacing. With such wide well spacing there is little subsurface control for tracing dip-elongated 1 mi (1.6 km) wide belts of Cook sandstone. However, the well spacing can be sufficient for regional isopach mapping of strike-oriented sheet sandstones that thicken basinward.

As Brown and Galloway's upper Cook/Flippen sandstone unit name implied, this isopach interval apparently includes two genetic units or depositional systems (Flippen and Cook sandstones, Fig. 2), making it difficult to interpret the depositional environment of each sandstone unit.

Emphasis in the present study has been to interpret depositional environments from various maps and cross sections of one genetic unit with closely spaced well control.

COOK POINT-BAR STRATIGRAPHIC TRAP

The "Blackwell" (Taylor County Regular) field in east Taylor County (Fig. 3) is a typical fluvial-sandstone reservoir in west-central Texas. Detailed analysis of the field reveals the depositional environment of the sandstone and the type of stratigraphic trap.

An isopach map of the interval from the base of the Flippen limestone, an excellent time-lithologic marker, to the base of the upper Cook fluvial sandstone in the Blackwell field (Fig. 6) illustrates the Cook meander belt. The map can be interpreted as a paleotopographic map of the base of a meander belt or as a "cast" of the meander-belt fill. The heavy line between the 100-ft (30 m) contours is the axial trend or paleovalley axis of the meander belt. The map and cross sections (Fig. 7) show that the meander belt has a convex-downward base.

A subsurface isopach map based on a genetic unit (genetic increment of strata—GIS; Busch, 1971) between a time-lithologic marker bed and the base of a channel-fill deposit (Fig. 6) is essential for mapping channel deposits (Andresen, 1962; Busch, 1971).

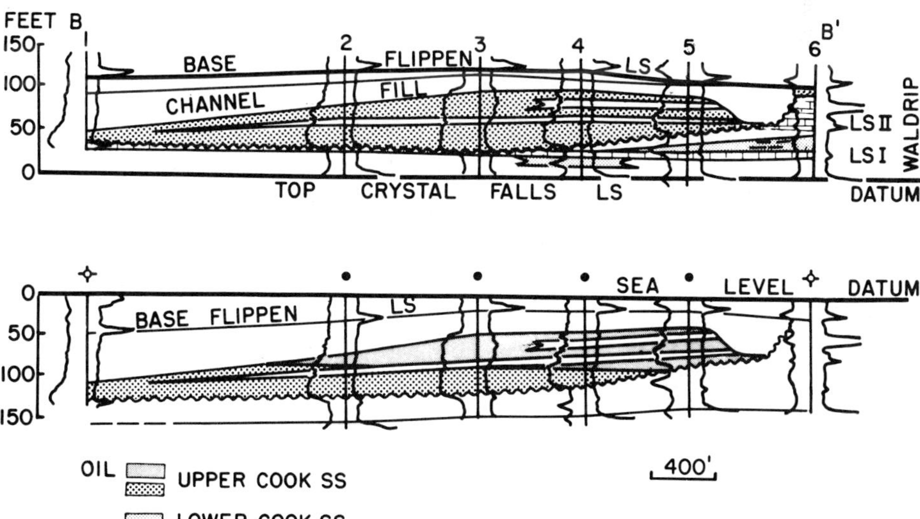

FIG. 7—Stratigraphic and structure cross sections *BB'* of Cook point bar. See Figures 6, 8, 9, 10 for location of section.

Figure 8 is an isopach map of the upper Cook sandstone in the Blackwell field. Inside the bend of the meander belt the sandstone has a maximum thickness of more than 60 ft (18 m). The stippled area between the 0 and 20-ft (6 m) contours is interpreted as the last position of the channel before its abandonment and filling with marine mud. The presence of this updip impermeable barrier to the oil in the point bar is supported by the fact that the upper Cook sandstone in well 6 did not have a show of oil, even though it is higher structurally than the original oil-water contact in well 5 (Fig. 7).

As the channel eroded the south bank of the meander, sandstone deposition on the inside of the meander prograded southward forming a point bar. The *BB'* stratigraphic cross section is, therefore, a profile of the point bar with the vertical scale greatly exaggerated.

The "Christmas tree" or "bell" shape of the Cook sandstone self-potential (SP) curves of the electric logs (Fig. 7) is indicative of a point-bar sandstone deposit. The SP curves characteristically show (1) a sharp basal contact, (2) a decrease in SP amplitude upward, and (3) a more serrate curve upward. Erosion of the subjacent formation made the sharp basal contact of the Cook channel fill (Fig. 4). A decrease in grain size and increase in interstitial clay upward cause the upward decrease in SP amplitude. The change in grain size is very evident in sample cuttings. The sandstone becomes more interbedded with shale

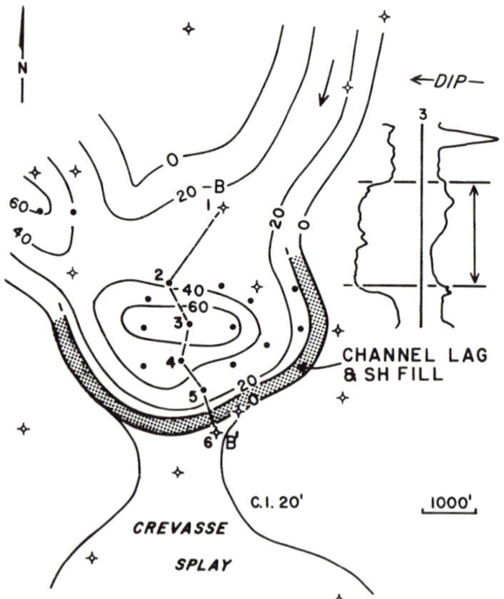

FIG. 8—Isopach map of upper Cook sandstone in Blackwell field, showing point bar. Contour interval 20 ft (6 m). *BB'* is line of section shown in Figure 7.

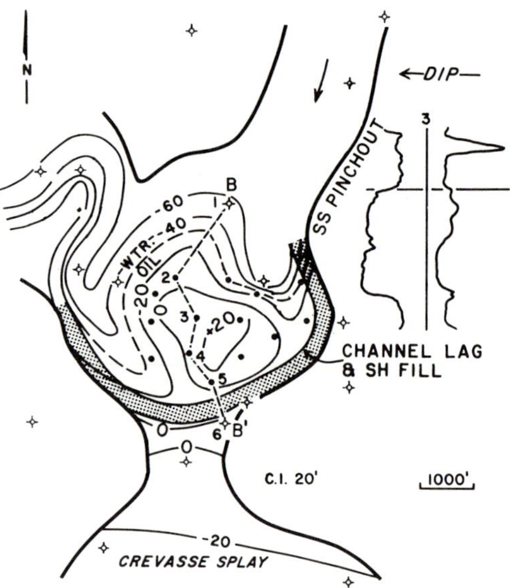

FIG. 9—Structure map of top of upper Cook sandstone in Blackwell field, showing oil trapped in point bar. Contour interval 20 ft (6 m). *BB'* is line of section shown in Figure 7.

upward, making a more serrate SP curve upward until the channel deposit merges into marine siltstone and shale.

About 10 ft (3 m) of shaly sandstone is present in the zone between the Flippen and Waldrip 2 limestone beds in well 6 of Figure 7. South of the meander belt shown in Figure 8 this shaly sandstone is a fan-shaped deposit which is interpreted as an overbank deposit. It pinches out about 5 mi (8 km) south of the area in Figure 8. Probably at flood stage, water overflowed a natural levee bordering the meander belt or cut a crevasse in the levee, and mud and sand were deposited in a fan shape. In all probability it represents a crevasse splay.

Overbank deposits generally border meander belts, deltaic channels, and submarine canyons and fan valleys (Jacka et al, 1968; LeBlanc, 1972; Stanley and Unrug, 1972; Shelton, 1973). When these deposits are recognized in the subsurface, adjoining channel sandstone deposits can be predicted.

In the upper Cook sandstone in the Blackwell field (Fig. 9), oil is trapped by a combination of structural closure and updip pinchout against the clay plug in the channel.

A structure map of the base of the Flippen limestone in the Blackwell field (Fig. 10) demonstrates that the structural closure is a result of differential compaction between the underlying point-bar sandstone and surrounding shale.

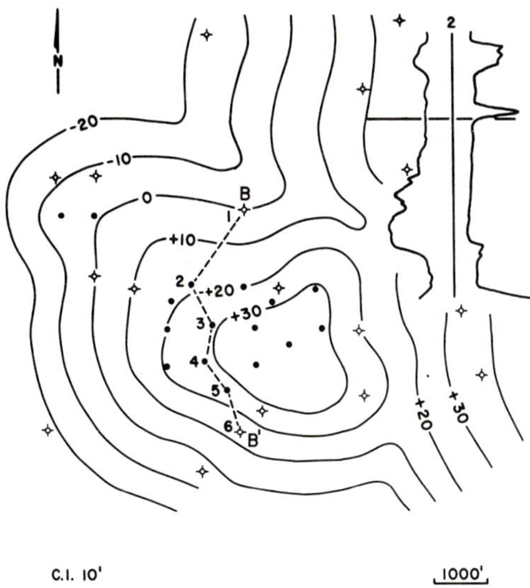

FIG. 10—Structure map of base of Flippen limestone in Blackwell field. Contour interval 10 ft (3 m). *BB'* is line of section in Figure 7.

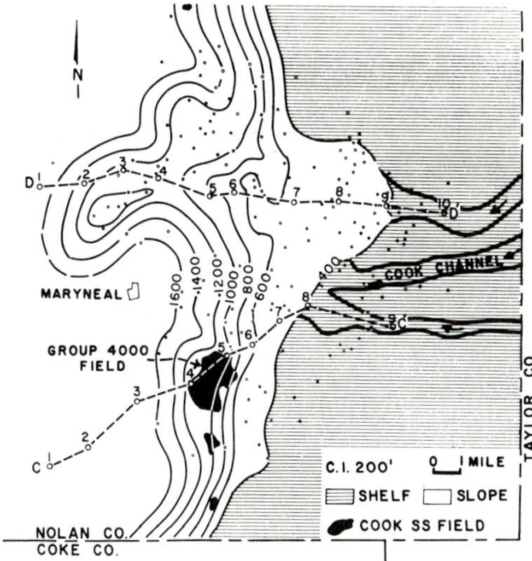

FIG. 11—Isopach map of interval from top of Noodle Creek limestone to top of Waldrip limestone 2, Nolan County, Texas. Contours show paleoslope of top of Waldrip limestone 2. *CC'* is line of section of Figure 12, *DD'* of Figure 14.

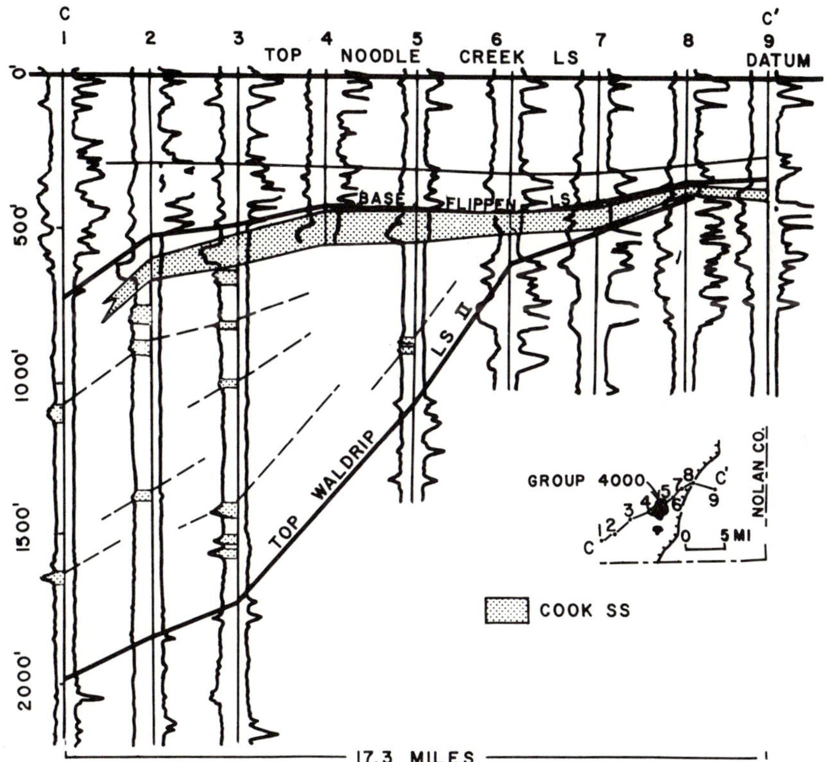

FIG. 12—Stratigraphic cross section *CC'* showing fluvial, deltaic, and slope facies of Cook sandstone. There is no horizontal scale between wells. Location of cross section is shown on Figures 11 and 13.

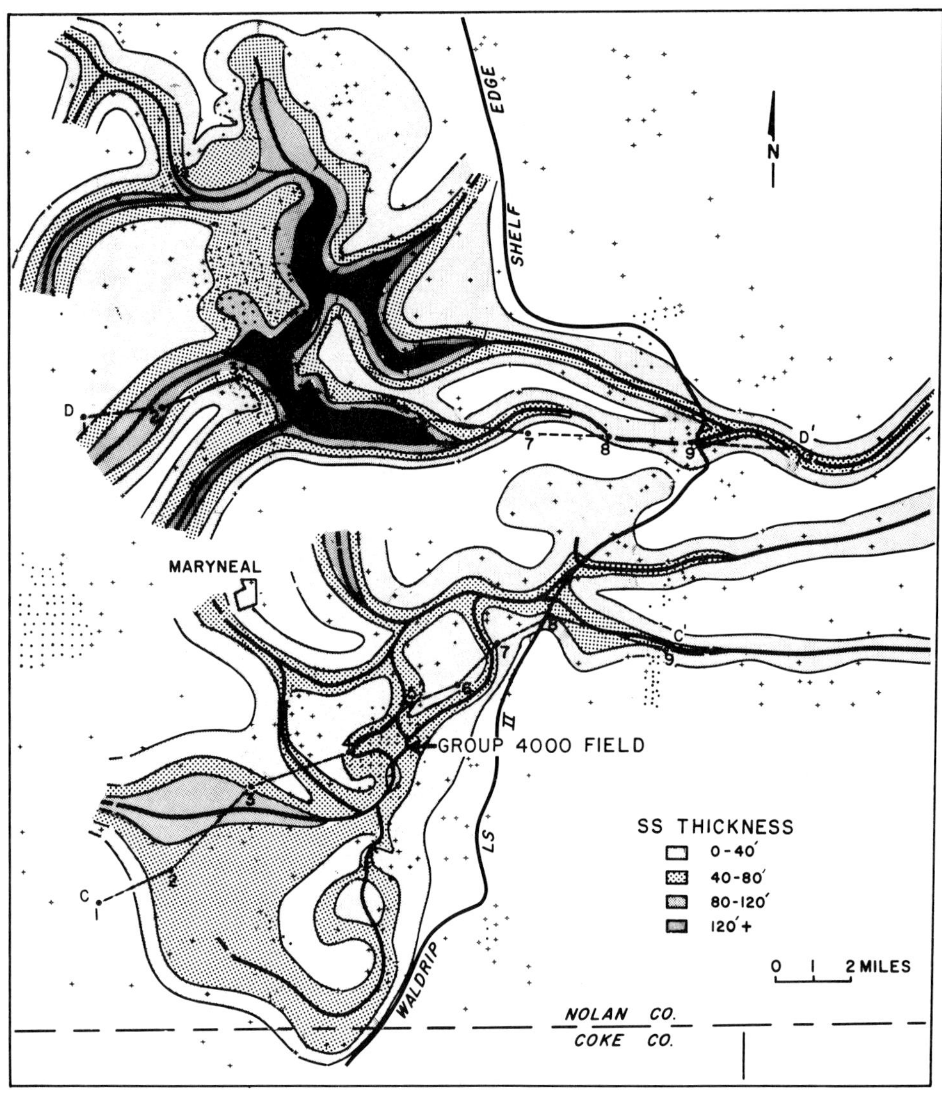

FIG. 13—Isopach map of Cook sandstone in meander belts and deltas in Nolan County, Texas. *CC'* is line of section shown on Figure 12, *DD'* on Figure 14.

At an average depth of 2,200 ft (671 m) the Cook sandstone in the Blackwell field has produced about 100,000 bbl of oil per well.

BASAL PERMIAN SHELF EDGE AND PROXIMAL SLOPE

A paleotopographic map of the top of Waldrip limestone 2 (Fig. 11) in Nolan County, Texas (Figs. 1, 3), shows a steepened slope between wells 3 and 6 (maximum slope angle about 3°) with decreasing slope basinward west of well 3.

Late Paleozoic depositional shelves and slopes along the east margin of the Midland basin first were described by Rall and Rall (1958) and Van Siclen (1958). Before depositional topography was recognized in the Midland basin, most geologists explained the lateral change in lithology, such as the shale-limestone change between wells 5 and 6 of Figure 12, as a change in facies.

The base of the Flippen limestone between wells 2 and 9 in Figure 12 is on the Eastern shelf. The Flippen limestone of well 2 was deposited on the proximal part of the Eastern slope. The clastic wedge of sediments between Waldrip limestone 2 and Flippen limestone (Figs. 12, 14) is equivalent stratigraphically to the Cook sandstone and Wal-

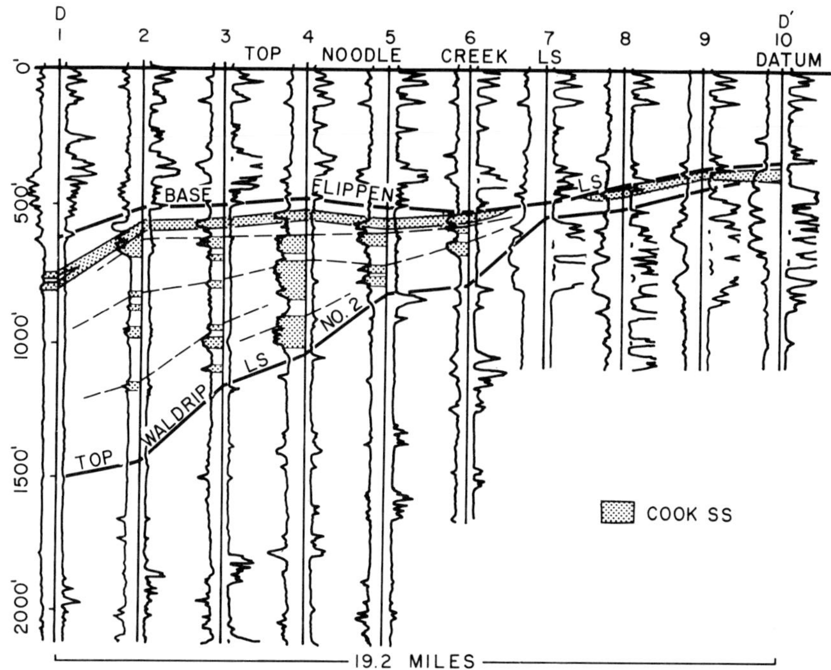

FIG. 14—Stratigraphic cross section *DD'* showing Cook fluvial, deltaic, and slope sandstone facies. There is no horizontal scale between wells. See Figures 11 and 13 for location of cross section.

drip shale between these limestone beds on the shelf.

COOK DELTAS

The shelf edge of Waldrip limestone 2 in Figure 11 coincides approximately with the most westward regressive position of the east shoreline of the embayment. West of the shelf edge, mud and sand supplied by Cook fluvial systems prograded into the shallow sea and became two lobate deltas (Fig. 13). The south delta is approximately 12 mi (19 km) long and 7 mi (11 km) wide and the north delta 16 mi (26 km) long and 10 mi (16 km) wide.

Approximately the eastern third of each delta consists of mostly upper delta-plain facies (wells 6-8, *CC'*, Figs. 12, 13; wells 6-9, *DD'*, Figs. 13, 14). The lithology consists of about 50 ft (15 m) of thin, fine-grained, lenticular sandstone beds interbedded with dark gray to black carbonaceous shale and lenses of coal. These deposits are probably the aggradational facies of swamps and floodplains in a delta plain.

South of wells 7 and 8 of the *DD'* line (Fig. 13) is a double fan-shaped sandstone deposit covering about 3 sq mi (8 sq km). It consists of about

10 ft (3 m) of shaly sandstone. This deposit is probably two coalescing crevasse splays.

Between wells 2 and 5 in Figure 12, an upper unit of Cook sandstone is present. Characteristically the sandstone in each well has an average thickness of 40 ft (12 m) and the electric log SP curves are cylinder- or funnel-shaped with a high amplitude. Nearly every well west of the delta-plain area in the south delta has this sandstone unit in the same stratigraphic position. It apparently consists of delta-front sheet sandstone and distributary-channel sandstone facies. The sandstone units generally are overlain by delta-plain facies (Figs. 12, 14, 16).

The thick wedge of shale and thin lenticular sandstone beds between the upper sandstone facies and Waldrip limestone 2 (Fig. 12) represent prodelta facies.

Figure 14 is a stratigraphic cross section of the north delta in Figures 11 and 13. Cook fluvial sandstone is present in well 10. Wells 8 and 9 have delta-plain facies of Cook sandstone interbedded with carbonaceous shale. In wells 2 through 6, an upper unit of Cook sandstone, with an average thickness of 20 ft (6 m) has the same SP-curve characteristics as the upper unit of sandstone in the south delta and, likewise, is probably

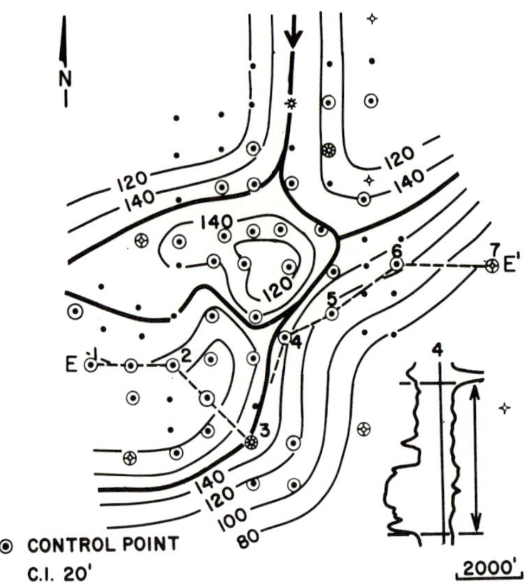

FIG. 15—Isopach map of interval from base of Flippen limestone to base of Cook sandstone in Group 4000 field. *EE'* is line of section shown in Figure 16. Heavy lines are axial trends of thickest deposition. Contour interval 20 ft (3 m).

delta-front sheet and distributary-channel sandstone facies.

In well 4 of Figure 14, more than 300 ft (91 m) of sandstone is present between the delta-front sheet sandstone and Waldrip limestone 2. This thick sandstone unit probably was deposited as a distributary-mouth bar as deposition prograded along the thick sandstone trend shown in Figure 13.

DELTAIC STRATIGRAPHIC TRAP

The Group 4000 "Cisco" sandstone field near the center of the south delta in Figures 11 and 13 produces from Cook sandstone. An isopach map (Fig. 17) and stratigraphic section (Fig. 16) show that the Cook sandstone has a convex-downward base indicative of a channel cut-and-fill deposit. The isopach of the Cook sandstone (Fig. 17) and of the interval between the base of the Flippen limestone and base of the Cook sandstone (Fig. 15) show that the axial trends of the thickest interval and thickest sandstone coincide. These axial trends are probably the thalwegs of distributary channels.

The lower Cook sandstone in wells 1, 2, 5, and 6 of Figure 16 is interpreted to be a delta-front sheet sandstone. It has an average thickness of 40 ft (12 m) and a characteristic sheet sandstone SP curve. The thick upper Cook sandstone in wells 3 and 4 is probably a distributary channel-fill deposit. The base of the channel is characteristically convex downward. As shown on Figure 17, the thick upper Cook sandstone in well 3 is part of an axial trend of thick sandstone.

The thin lenticular beds of upper Cook sandstone in Figure 16 are interbedded with gray to black carbonaceous shale and some coal lentils. These beds probably are aggradational facies de-

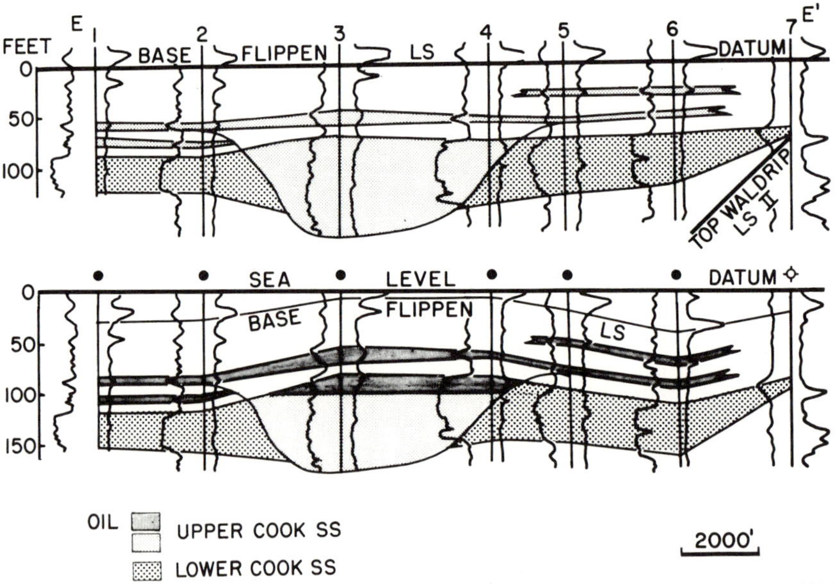

FIG. 16—Stratigraphic and structure cross sections *EE'* of Cook sandstone in Group 4000 field, showing delta-front sheet sandstone incised by distributary channel and filled with channel sandstone (wells 3, 4). See Figures 15 and 17 for location of cross sections.

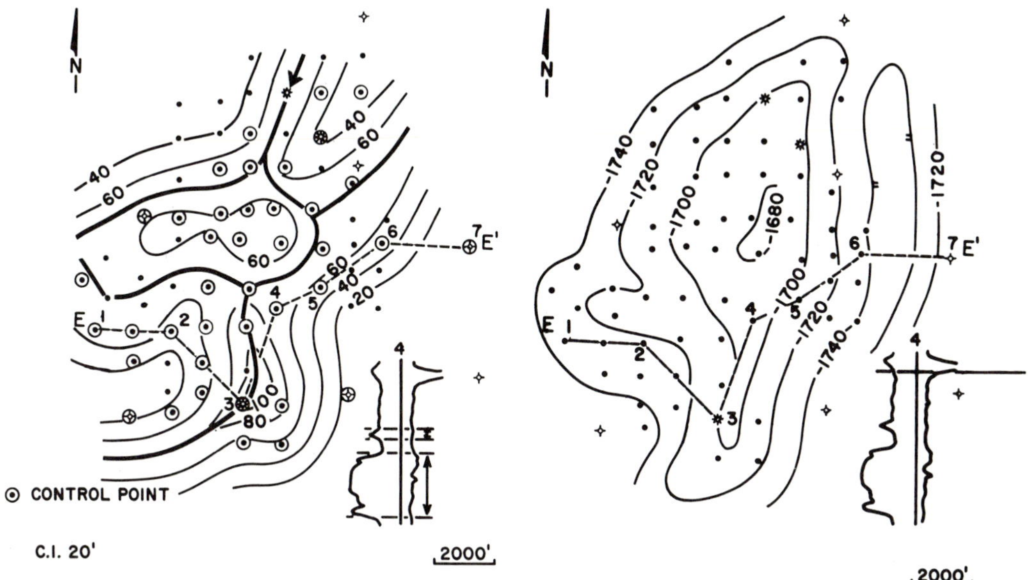

FIG. 17—Isopach map of Cook sandstone in Group 4000 field showing distributary pattern of channel sandstone. Contour interval 20 ft (6 m). *EE'* is line of section of Figure 16. Heavy lines are axial trends of thickest sandstone deposition.

FIG. 18—Structure map of base of Flippen limestone in Group 4000 field. *EE'* is line of section of Figure 16. Contour interval 20 ft (3 m).

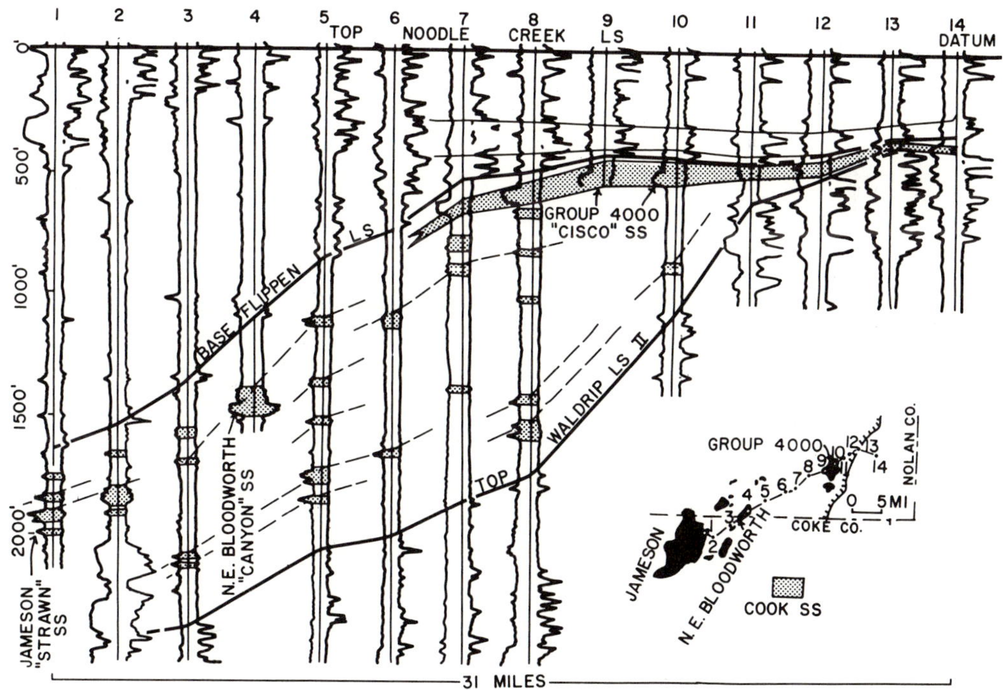

FIG. 19—Stratigraphic cross section of basal Permian clastic wedge of sedimentary rocks on Eastern slope. Wells 6-14 are wells 1-9 in Figure 12 (no horizontal scale between wells).

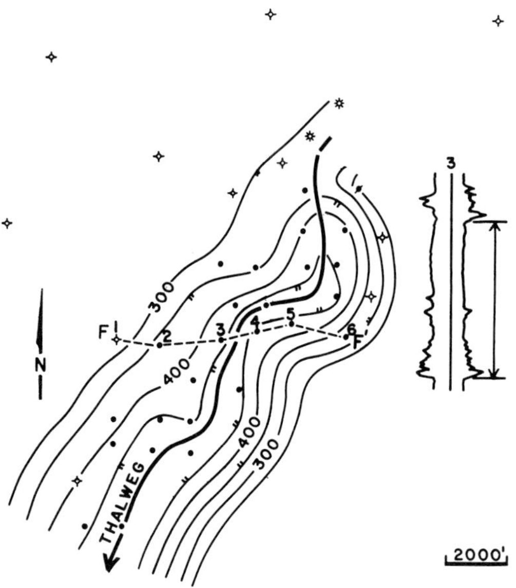

FIG. 20—Isopach map of interval from base of Flippen limestone to base of Cook sandstone in Northeast Bloodworth field, showing convex-downward base of Cook sandstone and axial trend (heavy line). *FF'* is line of cross section shown in Figure 21. Contour interval 50 ft (15 m).

posited in a marsh and flood-plain environment of the delta plain.

Figure 18 is a structure map of the base of the Flippen limestone in the Group 4000 field. The structural closure is due to differential compaction of the underlying Cook sandstone and surrounding shale.

Most of the 2,250,000 bbl of oil that the field has produced has come from the thin sandstone lenses in the delta plain. Reserves of these sandstones are about 200 bbl/acre-ft. The distributary channel sandstones have a reserve of about 400 bbl/acre-ft. Both pay zones have a high gas-oil ratio.

BASAL PERMIAN EASTERN SLOPE

Figure 19 is a stratigraphic cross section of Lower Permian sedimentary rocks from the Eastern shelf westward through the Group 4000 field to the Jameson "Strawn" sandstone field in northwest Coke County, Texas. The top of Waldrip limestone 2 correlates down paleoslope to a Pennsylvanian reef in well 2. Flippen limestone is on the shelf between wells 7 and 14 and on the slope between wells 1 and 7. The shelf edge of the Flippen limestone (well 7) is 7 mi (11 km) west of the shelf edge of Waldrip limestone 2 (well 13). Sun Oil Company geologists (personal commun.)

described Lower Permian Wolfcamp fusulinids in the Flippen limestone of the Jameson field (wells 1, 2).

The clastic wedge of sediments on the slope between Waldrip limestone 2 and the Flippen limestone is considered to be stratigraphically equivalent to the Cook sandstone and Waldrip shale on the shelf. The clastic wedge reaches a maximum thickness of about 1,300 ft (396 m) at midslope (well 7, Fig. 19) and thins basinward. The limestone beds thin basinward, generally change in facies to a pelagic black shale, and finally pinch out in the Midland basin (Van Siclen, 1958). The lens shape of this clastic wedge indicates that the Midland basin was sinking faster than sediments could fill it. Progressively younger clastic wedges prograded basinward and filled the basin.

COOK SUBMARINE-CANYON STRATIGRAPHIC TRAP

The Northeast Bloodworth "Canyon" sandstone field is in southwest Nolan and northwest Coke Counties, Texas (Fig. 19, well 4). The producing sandstone, as the cross section (Fig. 19) demonstrates, is Lower Permian Cook sandstone deposited in a lower slope environment, not a sandstone of Pennsylvanian Canyon age as described by field operators. Isopachs and cross sections of the field (Figs. 19-22) show that, indicative of a channel-fill deposit, the base of the Cook sandstone is asymmetrically convex downward. Maximum thickness of the Cook (Fig. 22) is more than 100 ft (30 m) with a northeast-southwest axial trend coincident with the axial trend of thickest base of Flippen limestone-base of Cook sandstone interval (Fig. 20). This axial trend is probably the paleovalley axis of a channel-fill or canyon-fill deposit.

Inasmuch as the Cook sandstone in the Northeast Bloodworth field is a channel-fill deposit in a lower slope depositional environment (Fig. 19), it is probably a submarine-canyon fill. Additional evidence in support of this conclusion is: (1) the highest amplitudes of the electric-log SP curves in Figure 21 are at the base of the sandstone and have a very sharp contact with the subjacent shale, indicating an erosional surface; (2) an upward decrease in grain size and increase of intergranular clay in the sandstone are inferred by the general upward decrease in SP amplitude (Fig. 21); (3) the lower sandstone is a "bundle" or repetition of thin beds separated by thinner shale beds, a characteristic of turbidite or bottom-current deposits in modern and ancient submarine-canyon deposits (Sullwold, 1961; Stanley and Unrug, 1972) and indicative of relatively short periods of sand deposition and long periods of

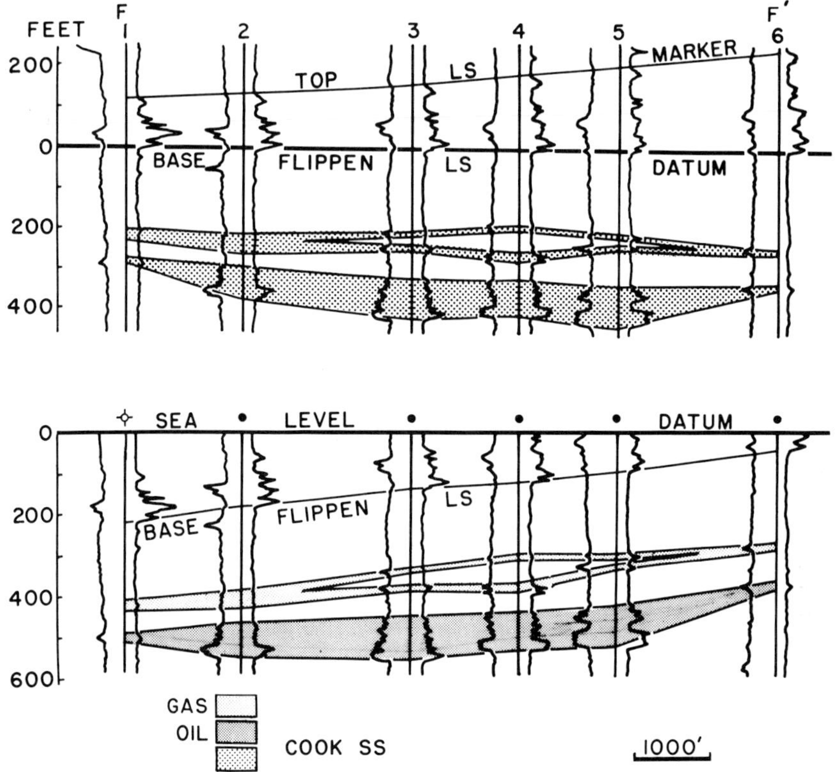

FIG. 21—Stratigraphic and structural cross sections *FF'* of Northeast Bloodworth field showing sharp asymmetric convex-downward base of Cook sandstone. See Figures 20 and 22 for location of cross section.

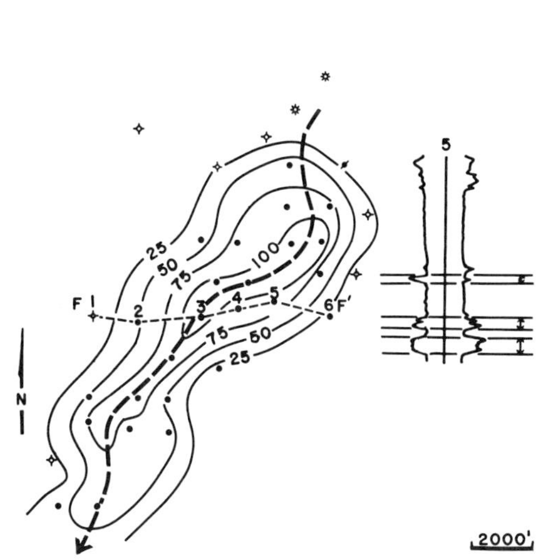

FIG. 22—Isopach map of Cook sandstone in Northeast Bloodworth field. Heavy dashed line is axial trend. *FF'* is line of section shown in Figure 21. Contour interval 25 ft (8 m).

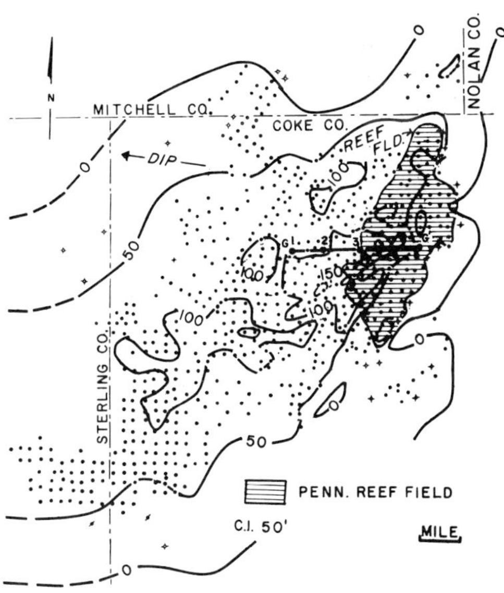

FIG. 23—Isopach map of Cook sandstone in Jameson field, Texas. *GG'* is line of section shown in Figure 24. Contour interval 50 ft (15 m).

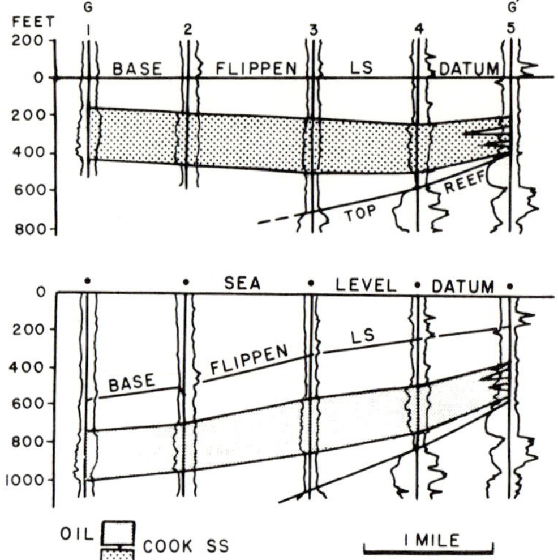

FIG. 24—Stratigraphic and structure cross sections of Cook sandstone in Jameson field. See location of cross sections in Figure 23.

pelagic-mud deposition; (4) individual lower sandstone beds do not correlate between wells (Sullwold, 1961; Stanley and Unrug, 1972); and (5) the top of the uppermost sandstone bed is generally planar.

The Northeast Bloodworth field is a stratigraphic trap with a gas-solution drive and high gas-oil ratio. Since the discovery in 1966, the field has produced approximately 120,000 bbl of oil and 340 million cu ft of gas per well.

COOK SUBMARINE-FAN STRATIGRAPHIC TRAP

In the Jameson "Strawn" sandstone field in northwest Coke County, Texas (Fig. 19, wells 1, 2), the Cook producing sandstone was deposited fan-shaped (Fig. 23) in a Lower Permian lower slope depositional environment (Fig. 19) and is not a sandstone of Pennsylvanian Strawn age as originally described (Riddle, 1954).

In the northeast corner of the mapped area (Fig. 23), the sandstone is about 2 mi (3 km) wide. This is the southwest part of the North Jameson "Strawn" sandstone field (about 6 mi or 10 km long and 2 mi or 3 km wide). The sandstone has a convex-downward base indicative of a submarine-canyon deposit. South of the submarine canyon, sand and mud apparently supplied by the canyon were deposited in a fan shape over an area of about 176 sq mi (458 sq km). The Pennsylvanian reef (Figs. 23, 24) underlying the sandstone near the mouth of the submarine canyon probably was a topographic barrier to sandstone deposition along the west flank of the reef. The

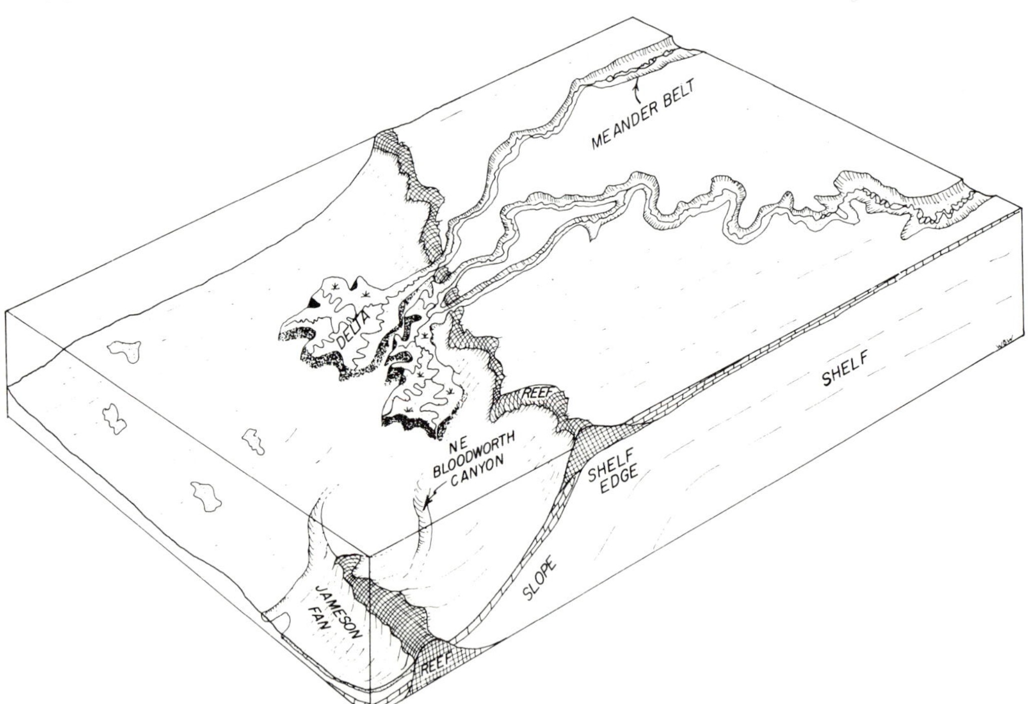

FIG. 25—Block diagram of Eastern shelf and slope at approximate base of Permian. Diagram is hypothetical and illustrates depositional environments of Cook sandstone.

fan-shaped geometry of the large sandstone body and its presence in a lower slope depositional environment suggest that the sandstone is a submarine-fan deposit with the apex on the northeast at the mouth of a submarine canyon. The producing sandstone consists of multiple, thin lenses; individual sandstone units cannot be correlated between wells (Fig. 24). The top and bottom contacts of the Cook producing sandstone can be correlated, inferring that the deposit is one genetic unit. The unit has an average thickness of 200 ft (61 m).

As suggested by the low-SP amplitudes (Fig. 24), the sandstone is fine-grained and shaly in cores. As is characteristic of turbidites or bottom-current deposits (Sullwold, 1961; Stanley and Unrug, 1972), many thin individual sandstone beds are graded. Grain size decreases upward from fine to very fine grained. In scattered zones are clasts of charcoal and red and gray shale and flattened pebbles of red clay. These terrigenous clasts probably were carried by turbidity currents from the Cook deltas that were 20 mi (32 km) northeast.

Since the discovery in 1952 the Jameson stratigraphic trap has produced about 35 million bbl of oil and an unknown quantity of gas. Average porosity is 12%.

SUMMARY

Cook sandstones form essentially continuous depositional systems from fluvial channels, through deltas and submarine canyons, to deep-marine sands (Fig. 25). The total transport route was more than 130 mi (208 km) long from the present outcrop to the Jameson field; originally the distance probably exceeded 230 mi (368 km). Along this route, oil accumulated in stratigraphic traps in fluvial and deltaic sandstones on the shelf and in submarine canyons and fans on the slope. Oil accumulation was controlled largely by lateral facies change to low-permeability siltstone and shale. Consequently, all of these oil fields are the result solely of depositional configurations of the sandstones, and illustrate an unusually complete sequence of environments and stratigraphic traps.

REFERENCES CITED

Andresen, M. J., 1962, Paleodrainage patterns: their mapping from subsurface data, and their paleogeographic value: AAPG Bull., v. 46, p. 398-405.

Barnes, V. E., ed., 1972, Geologic atlas of Texas, Abilene sheet: Texas Univ. Bur. Econ. Geology, scale 1:250,000.

Brown, L. F., Jr., 1969a, Geometry and distribution of fluvial and deltaic sandstones (Pennsylvanian and Permian), north-central Texas, *in* Geology of the American Mediterranean: Gulf Coast Assoc. Geol. Socs. Trans., v. 19, p. 23-47; reprinted, Texas Univ. Bur. Econ. Geology Circ. 69-4.

———— 1969b, Virgil and lower Wolfcamp repetitive environments and the depositional model, north-central Texas, *in* Cyclic sedimentation in the Permian basin—Symposium, Midland, Texas, 1967: West Texas Geol. Soc. Pub. 69-56, p. 115-134; reprinted, Texas Univ. Bur. Econ. Geology Circ. 69-3.

Busch, D. A., 1971, Genetic units in delta prospecting: AAPG Bull., v. 55, p. 1137-1154; reprinted, 1973, AAPG Reprint Ser. 7, p. 137-154.

Fisher, W. L., et al, 1969, Delta systems in the exploration for oil and gas: Texas Univ. Bur. Econ. Geology Syllabus for Delta Colloquium, 102 p.

Fisk, H. N., 1961, Bar-finger sands of Mississippi delta, *in* J. A. Peterson and J. C. Osmond, eds., Geometry of sandstone bodies: AAPG, p. 29-52.

———— et al, 1954, Sedimentary framework of the modern Mississippi delta: Sed. Petrology, v. 24, p. 76-99.

Flawn, P. T., et al, 1961, The Ouachita system: Texas Univ. Bur. Econ. Geology Pub. 6120, 401 p.

Forgotson, J. M., Jr., 1957, Nature, usage and definition of marker-defined vertically segregated rock units: AAPG Bull., v. 41, p. 2108-2113.

Galloway, W. E., and L. F. Brown, Jr., 1972, Depositional systems and shelf-slope relationships in upper Pennsylvanian rocks, north-central Texas: Texas Univ. Bur. Econ. Geology Rept. Inv. 75, 62 p.

———— ———— 1973, Depositional systems and shelf-slope relations on cratonic basin margin, uppermost Pennsylvanian of north-central Texas: AAPG Bull., v. 57, p. 1185-1218.

Harrison, E. P., 1973, Depositional history of Cisco-Wolfcamp strata, Bend arch, north-central Texas: PhD thesis, Texas Tech Univ.

Jacka, A. D., et al, 1968, Permian deep-sea fans of the Delaware Mountain Group (Guadalupian), Delaware basin, *in* Guadalupian facies, Apache Mountain area, West Texas: SEPM Permian Basin Sec. Pub. 68-11, p. 49-90.

Jackson, W. E., 1964, Depositional topography and cyclic deposition in west-central Texas: AAPG Bull., v. 48, p. 317-328.

Lawless, J. E., 1960, Manly-Neas area, Jones County, Texas, *in* Geological contributions, 1960: Abilene Geol. Soc., p. 77-86.

LeBlanc, R. J., 1972, Geometry of sandstone reservoir bodies, *in* Underground waste management and environmental implications: AAPG Mem. 18, p. 133-190; reprinted 1973, AAPG Reprint Ser. 7, p. 155-212.

Rall, R. W., and E. P. Rall, 1958, Pennsylvanian subsurface geology of Sutton and Schleicher Counties, Texas: AAPG Bull., v. 42, p. 839-870.

Riddle, D., 1954, Jameson Strawn sand field, Coke County, Texas, *in* Geological contributions: Abilene Geol. Soc., p. 77-86.

Shankle, J. D., 1960, The "Flippen" sandstone of parts of Taylor and Callahan counties, Texas, *in* Geological contributions, 1960: Abilene Geol. Soc., p. 168-201.

Shelton, J. W., 1973, Models of sand and sandstone deposits; a methodology for determining sand genesis

The American Association of Petroleum Geologists Bulletin
V. 56, No. 10 (October 1972), P. 1975-1983, 10 Figs.

Longshore Drift, Submarine Canyons, and Submarine Fans in Development of Niger Delta[1]

KEVIN BURKE[2]

Toronto, Ontario, Canada

Abstract The southwesterly prevailing wind of the Gulf of Guinea strikes symmetrically on the nose of the Niger delta, causing divergent longshore drifts which meet opposing drifts near Lagos and Fernando Poo. Submarine canyons channel about 1 million cu m of sand a year from each pair of opposing drifts to feed submarine fans on either side of the delta foot. In times of low sea level axial distributaries of the Niger feed a third, now inactive, submarine fan in front of the delta.

At the time of the last low sea level numerous submarine canyons formed on the front of the Niger delta, and their heads cut back into the Benin Formation continental sands. As the sea rose, these canyon heads formed wide estuaries which have since been filled. All Quaternary canyons except three currently scoured by resedimented longshore drift material have been filled.

Because the Niger delta has prograded toward the southwest throughout the Tertiary and because the prevailing wind has blown persistently from the southwest, the longshore drift pattern long has been as it is now, and the two corners of the delta have been areas of high concentrations of submarine canyon formation. There may have been a third area of high concentration of submarine canyons between the Cross River and Niger deltas when these were separate.

Recognition of the submarine fan environment leads to a new symmetrical five-layer interpretation of the structure of the Niger delta: (5) top continental sand unit (Benin Formation); (4) transitional sand/ shale unit (Agbada Formation); (3) marine shale unit (Akata Formation); (2) transitional shale-sand unit (newly distinguished); (1) bottom submarine fan sand unit (newly distinguished). Other deltas feeding into waters of oceanic depths may have a comparable structure.

Introduction

The Cretaceous history of the Niger delta and the closely related Benue trough has been described by Burke *et al.* (1971a) and by Murat (in press). From the Campanian onward deltas prograded down the Benue trough, interrupted by shortlived Maestrichtian and Paleocene marine transgressions. The Paleocene transgression affected only the lower part of the trough.

By late Eocene time delta fronts had pushed down the continental slope and had spread onto the ocean floor. The Eocene to Holocene history and structure of the delta have been described in many papers (Allen, 1964, 1965, 1970; Frankl and Cordry, 1967; Hospers, 1971; Knaap, 1971; Merki, in press; Short and Stäuble, 1967; Weber, 1971). Three major environments—continental,

mixed, and marine—have been distinguished. The deposits of the three environments are the sandy Benin Formation, the alternating sandy and shaly Agbada Formation, and the shaly Akata Formation. Because the delta has been built out into water 4 km deep, and because the coarse, permeable sands of the Benin Formation have a greater bulk density than the fine clays of the underlying Akata Formation, the Akata is squeezed seaward in shale diapirs. As the Benin and Agbada Formations build out on top of the delta these diapirs intrude the Agbada. Within the Agbada a growth fault and rollover anticline environment develops which dominates the oil-producing areas in the delta (Merki, in press).

Longshore Drift and Active Submarine Canyons of Niger Delta

The prevailing wind blows from the southwest onto the nose of the Niger delta except for periods of a few weeks in December and January, when the dust laden *harmattan* blows from the northeast out of the Sahara. Although during Pleistocene glacial maxima the *harmattan* blew for much longer periods (Burke *et al.*, 1971b) it has had no great effect on sedimentation. The southwesterly prevailing wind causes divergent northward and eastward longshore drifts from the nose of the delta. Longshore drift also brings sediments eastward along the coast from the Volta delta (Fig. 1; Pugh, 1953, 1954; Allen 1964, 1965).

Harbor works at Lagos have obstructed the eastward drift from the Volta through most of the present century, and NEDECO (1961) used the volume retained by the Lagos mole to estimate the drift at Lagos as 500,000 cu m/year of sand. Sandy beach stops abruptly 100 km east of Lagos, giving way to soft mud on the Mahin beach at 4°30'E. Some older beach ridges extend farther east, but not for more than 50 km (Fig. 2). The entire volume of the sandy inshore rise (Allen 1964, p. 316) between Lagos and Mahin could

---

[1] Manuscript received, October 27, 1971; accepted, December 2, 1971.

[2] Department of Geology, University of Toronto.

have accumulated from the present beach drift in less than 1,000 years, and it seems that much of the sand passing Lagos does not continue along the coast.

NEDECO (1961, Table VII 5-3) also calculated the longshore sand drift northward up the west face of the Niger delta. This drift increases from the delta nose as successive distributaries contribute and reaches the value of 500,000 cu m/year between the Ramos and Forcados distributary mouths (Fig. 2). A sand beach extends up the west face of the delta as far as the Benin River, north of which the shore is muddy. Although about 1 million cu m/year of sand is carried toward Mahin (half from the west and half from the south) the Mahin beach is made of mud, a fact which requires that the sand be carried away and deposited elsewhere. The Avon and Mahin canyons (Fig. 1; Allen, 1964) are in appropriate places to channel the sand down to the Avon fan at the toe of the delta.

Longshore drift eastward from the delta nose carries sand which is added to at each distributary mouth as far as the Kwa Ibo and Cross Rivers (Fig. 4), where the annual sand drift again reaches about 500,000 cu m (NEDECO, 1961). Northward drift along the Cameroun coast extends at least as far as Souelaba point (Pugh,

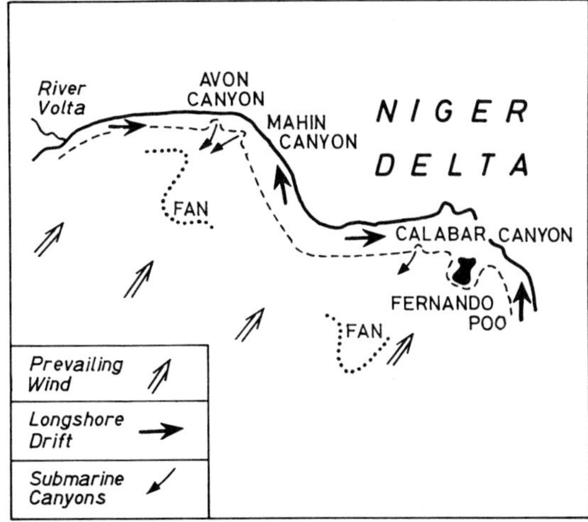

Fig. 1—Sketch showing that prevailing wind strikes nose of Niger delta and causes divergent longshore drifts which meet opposing drifts near Lagos and Fernando Poo. Sediment from meeting drifts is channeled to two submarine fans through three canyons.

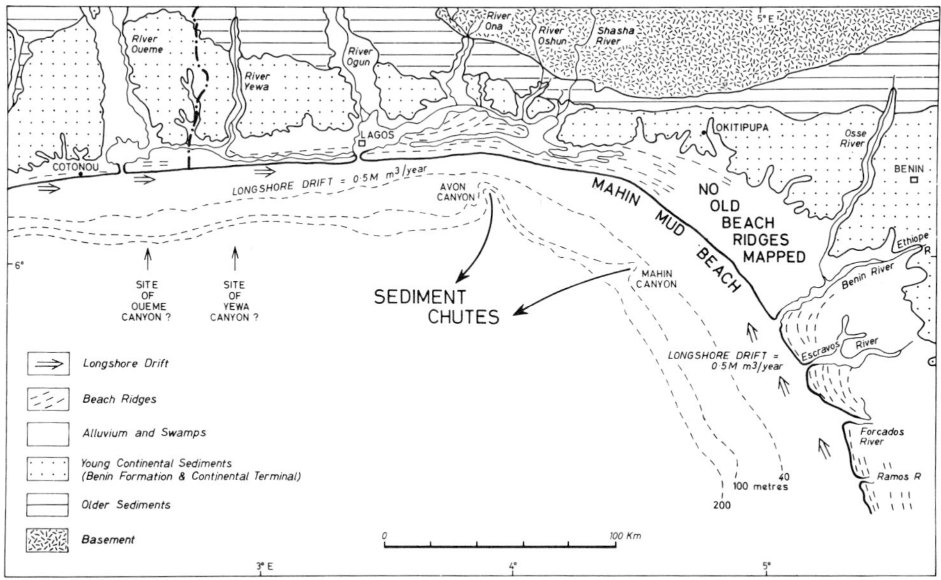

Fig. 2—Longshore drift carries about 1 million cu m of sand toward Mahin beach from west and south every year. Absence of old beach ridges behind Mahin beach indicates that little sand has accumulated in area in last 7,000 years. Sand is believed to pass down sediment chutes of Avon and Mahin canyons. Note narrow outcrop of older sediments north of Okitipupa. Sediments thin across Okitipupa ridge separating Niger delta and Dahomey basins. Slow sediment accumulation in this area in past may have been related to existence of submarine canyons which acted as sediment chutes. (Based mainly on Nigeria Geol. Survey unpub. maps and Allen, 1965, Fig. 27).

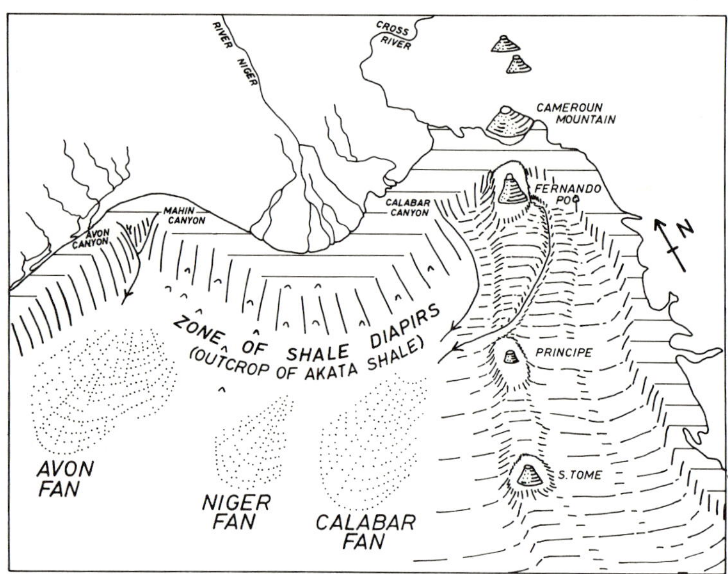

Fig. 3—Physiographic sketch of Niger delta looking northeast and showing Avon and Calabar fans fed by submarine canyons at corners of delta. Niger fan on delta axis is fed from Niger trench at times of low sea level but probably is inactive at present. Two shale diapirs that crop out on continental shelf are plotted from Allen (1964, Fig. 1c); others are represented schematically. Sketch is based on published bathymetric maps, especially Allen (1964, Fig. 2b for water less than 1,000 m deep); International Tectonic Map of Africa (Assoc. African Geol. Surveys, 1968), and Arens *et al.* (1971, Fig. 2).

1954) on the south side of the Mouth of the Camerouns, but no quantitative estimate of the amount of this drift is available. The situation southeast of the delta is thus similar to that in the northwest: about 1 million cu m/year of sand is carried to near Mount Cameroun and disappears. The Calabar canyon (Allen, 1964) is in an appropriate position to channel this sand to the Calabar fan.

Deep-Sea Fans in Front of Niger Delta

Published bathymetric maps indicate the shape of submarine fans in front of the Niger delta; Figure 3 is a physiographic sketch based on this information. The Avon and Calabar fans are fed by the Avon and Calabar canyons (the Mahin canyon is a tributary of the Avon canyon). The Niger fan in front of the delta nose is not now being fed from a submarine canyon, but has presumably received sediment passing through a canyon from a major axial distributary at times of low sea level.

The submarine fans are not large compared with the Mississippi and Nile cones, and are very much smaller than the Indus and Ganges fans. This smaller size is partly because, although the Niger-Benue system brings in 0.02 cu km/year of sediment, much of it is deposited on top of the

delta, which progrades by squeezing out the Akata shale. The zone of shale diapirs that marks the submarine outcrop of the Akata Formation is indicated upslope of the fans on the physiographic diagram. The submarine fans are small because their proximal parts have been buried by the rapidly advancing delta.

Submarine Canyons in Late Quaternary

Allen (1964, 1965) described the late Quaternary history of the Niger delta in detail, but did not discuss the development of submarine canyons in relation to individual rivers. At the time of the late Wurm-Wisconsin low sea level (roughly 20,000 years B.P.) Benin Formation deposits extended as far as the edge of the continental shelf. A crude reconstruction of the courses of rivers crossing the shelf at that time has been attempted in Figure 4 by the projection of present river courses. The Cross and Kwa Ibo Rivers appear to have fed the Calabar canyon; the Ethiope and Osse Rivers fed the Mahin canyon; and the Ona, Oshun, and Shasha Rivers (Fig. 5) emptied into the Avon canyon. The Niger flowed in a deep channel, "the Niger trench," whose exact position is unknown, but which Allen (1964, p. 327) suggested may have been in the area of Cape Formoso, where there is a bulge in the continental slope similar in form to a bulge which marks

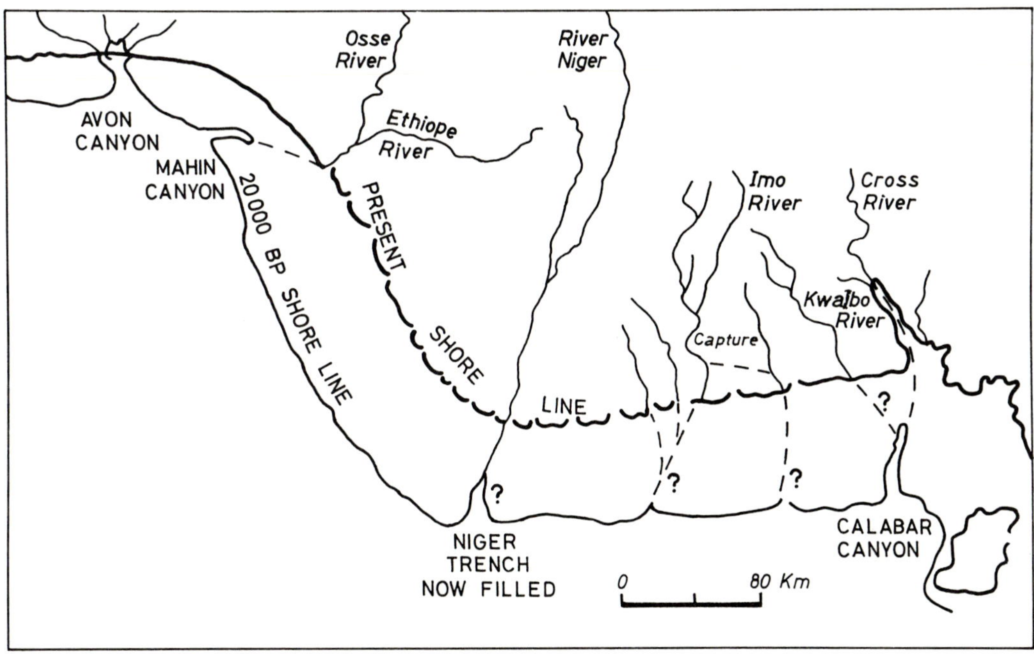

Fig. 4—Possible courses of rivers across Nigerian continental shelf at low sea level of 20,000-years B.P. Only canyons kept open by scouring action of resedimented longshore drifted sand have not yet been filled.

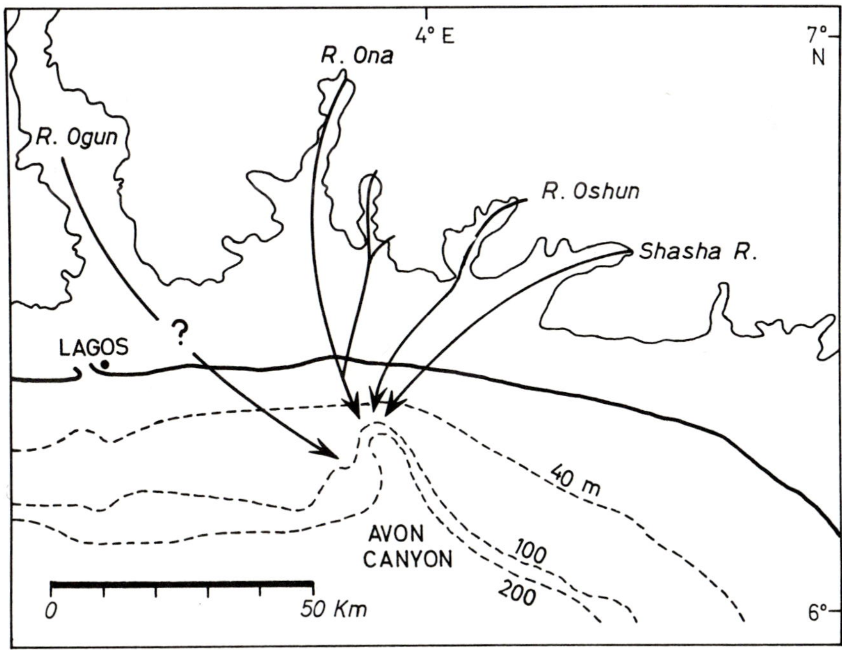

Fig. 5—Courses of rivers heading toward Avon canyon at low sea level. River Ogun may have fed its own canyon. Rising sea level drowned these valleys to produce estuarine coast.

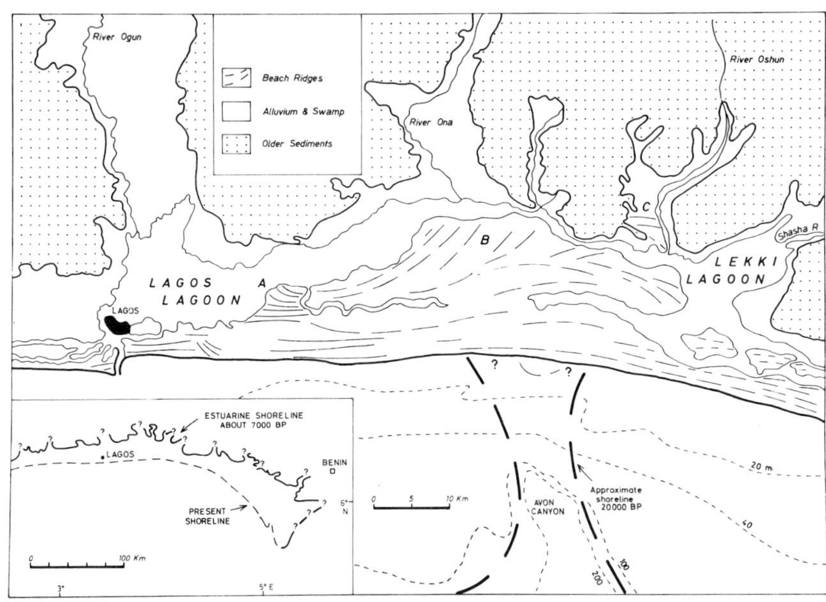

FIG. 6—Beach ridges east of Lagos (based mainly on Allen, 1965, and Nigeria Geol. Survey maps). Shape of oldest beach ridges east of Lagos, especially at *A*, *B*, and *C*, indicates that after maximum extent of last transgression (about 7,000 years B.P.) longshore drift constructed beaches which first filled irregularities in coastline; major estuaries are at Ogun mouth *A*, Ona mouth *B*, and Oshun mouth *C*. Later beach ridges nearer present coast have smooth outline because, when they formed, estuaries had been largely filled. Inset shows approximate shoreline with numerous estuaries about 7,000 years B.P.

the position of the buried Mississippi trench. The River Ogun (Fig. 5) may have fed the Avon canyon or, like the Yewa River 100 km farther west (Fig. 2), may have fed a canyon of its own.

The post-Wisconsin transgression deposited a marine sand unit, the "older sands" of Allen (1965), over the Benin Formation. At the time of the maximum transgression (which is estimated from Allen's, 1964, radiocarbon dates to have taken place between 10,000 and 4,000 years B.P., probably about 7,000 years ago), the sea extended inshore from the present coast. An attempt has been made (inset, Fig. 6) to represent this shoreline in the western part of the delta by using the edge of alluvial fill in incised valleys and the trends of the oldest beach ridges as guides.

The highly curved oldest beach ridges farthest inland indicate in outline the shape of estuaries formed by drowning the incised valleys at the heads of the late Quaternary canyons. Curved ridges at *A* in Figure 6 indicate the site of the Ogun estuary, those at *B* the site of the Ona estuary, and those at *C* the site of the Oshun estuary. Later beach ridges that formed as the estuaries were filled have successively smoother outlines. Ridges striking inland for 25 km south of the Benin River (Fig. 2) indicate the great extent of an estuary formed by the post-Wisconsin transgression in the Mahin beach area.

Canyons and valleys that were open 20,000 years ago have been filled completely except for the Avon, Mahin, and Calabar canyons, which are scoured by resedimentation of longshore-drifted sand, and even these are largely filled. A cycle of cutting and filling of submarine canyons on the front of the Niger delta has been completed in a few tens of thousands of years in response to the most recent ice-controlled fall and rise of sea level.

FOSSIL SUBMARINE CANYONS OF NIGER DELTA

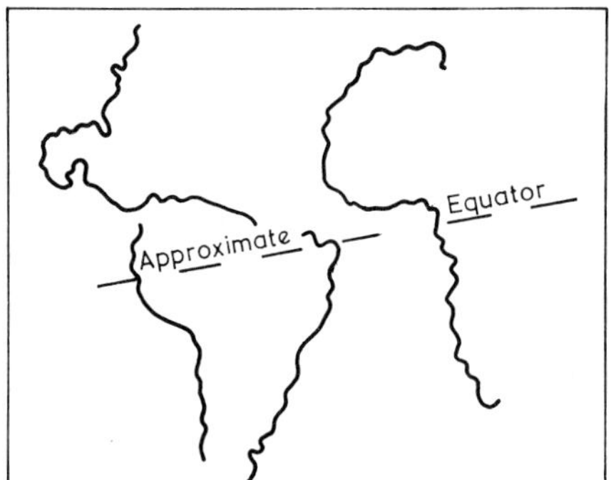

FIG. 7—Southern Atlantic in Paleocene time (from Le Pichon, 1968, Fig. 9, anomaly 31).

Figure 7 is a sketch map of the South Atlantic

as it was in Paleocene time from Le Pichon's (1968, Fig. 9) restoration of magnetic anomalies. The sketch indicates that the ocean, although smaller, was of the same general shape as today. Paleomagnetic studies have shown that the earth's rotation axis was not more than a few degrees away from its present position at that time (Runcorn, 1969). These facts make it probable that the geostrophic winds have changed their orientation little during the last 70 million years and that the prevailing wind in the Gulf of Guinea has blown from the southwest throughout that time.

The Niger-Benue delta system has prograded toward the southwest throughout the Tertiary (Short and Stäuble, 1967), and longshore drift caused by the southwesterly wind has redistributed sand toward the corners of the delta, very much as it does now. Although submarine canyons and small gullies must have developed at many places on the delta front, especially at times of low sea level, major and persistent canyon developments would have been concentrated in the northwest and southeast corners of the delta by the constant longshore drift. Late Quaternary history indicates that the life of individual canyons may have been short but, if the controlling conditions were persistent, canyons would have been cut and filled in the same general areas repeatedly throughout the Tertiary.

A factor which complicates this over-simplified picture is that, during much of the Tertiary, the Cross River (Fig. 8) had a separate delta from the Niger on the eastern side of the Ihuo embayment (Short and Stäuble, 1967). The sketches in Figure 8 illustrate the relations between a southwesterly prevailing wind and the successive Tertiary shorelines mapped by Short and Stäuble (1967, Fig. 6) on the basis of subsurface data. The sketches indicate that when the Cross River delta was separate from that of the Niger delta there was longshore drift from the southwest and southeast into the Ihuo embayment. Material carried into the embayment in this way would soon have filled it and, as it persisted for over 20 million years, a system of submarine canyons probably existed to evacuate sediment brought into the embayment roughly along a line separating the Cross River and Niger deltas. In addition to this system there probably were canyon systems in the northwest and southwest corners of the delta like those of today.

The area of intense canyon development in the northwestern corner of the delta (*1* in Fig. 8) moved westward tens of kilometers during the Tertiary, but has always been close to the Okitipupa ridge (Adegoke, 1968), a structural high

separating the Niger delta and Dahomey basins on which Tertiary sediment accumulation has been notably slow (Fig. 2). If submarine canyons acting as chutes have persistently removed sediment from the ridge, this would help to slow accumulation rates. Subsidence in this area during the last 20,000 years has been less than anywhere else along the Nigerian coast (Allen, 1964, Fig. 4). A rough calculation (after Bloom, 1967) indicates that all the subsidence (80 m in about 15,000 years) can be accounted for by eustatic sea level rise and isostatic adjustment to water load. Sediment is now removed from the area of the Okitipupa ridge down the Avon and Mahin canyons.

Afam Clay and Other Fossil Canyon Fills

The Afam clay (Short and Stäuble, 1967; Figs. 8, 9) is a member of the Benin Formation although it presumably passes down dip into the Agbada and Akata Formations. Its age was not stated in its original definition, but it appears (from Short and Stäuble, 1967, Figs. 2, 6) to be late Miocene. It has the form of a major canyon fill 30 km wide, 100 km long, and 300 m thick at the type locality. Its location downdip of the Ihuo embayment indicates that it may be an upper member of the Ihuo evacuation system. Lubara Creek-1, in which 800 m of Afam clay was found, is assumed to be close to the Afam canyon axis.

There are no other published accounts of canyon fills in the Benin Formation and only the largest canyons could be expected to cut back into the continental sands as the Avon canyon did 20,000 years ago. Thick shale members within the Agbada which may also represent canyon fills have been described from Mobil's MD-1 well and Shell-BP's Opobo South-1 as the Kwa Ibo and Opobo members of the Agbada Formation (Frankl and Cordry, 1967).

It has been suggested (Weber, 1971, p. 560) that the distribution of these and other clay valley fills may be related to basement-controlled epeirogenic movements. The writer prefers the hypothesis that they are submarine canyon fills whose distribution is related to longshore drift.

In analyzing sedimentologic aspects of some Niger delta oil fields, Weber (1971) did not consider the role of submarine canyons, which apparently played no part in the formation of the fields he discussed. However, in some areas, particularly in the southeastern and northwestern corners of the delta, canyon fills within the Agbada Formation can be expected to be of some importance.

If the cross section of the fills generally is as shown for the Afam clay in Figure 9, they may

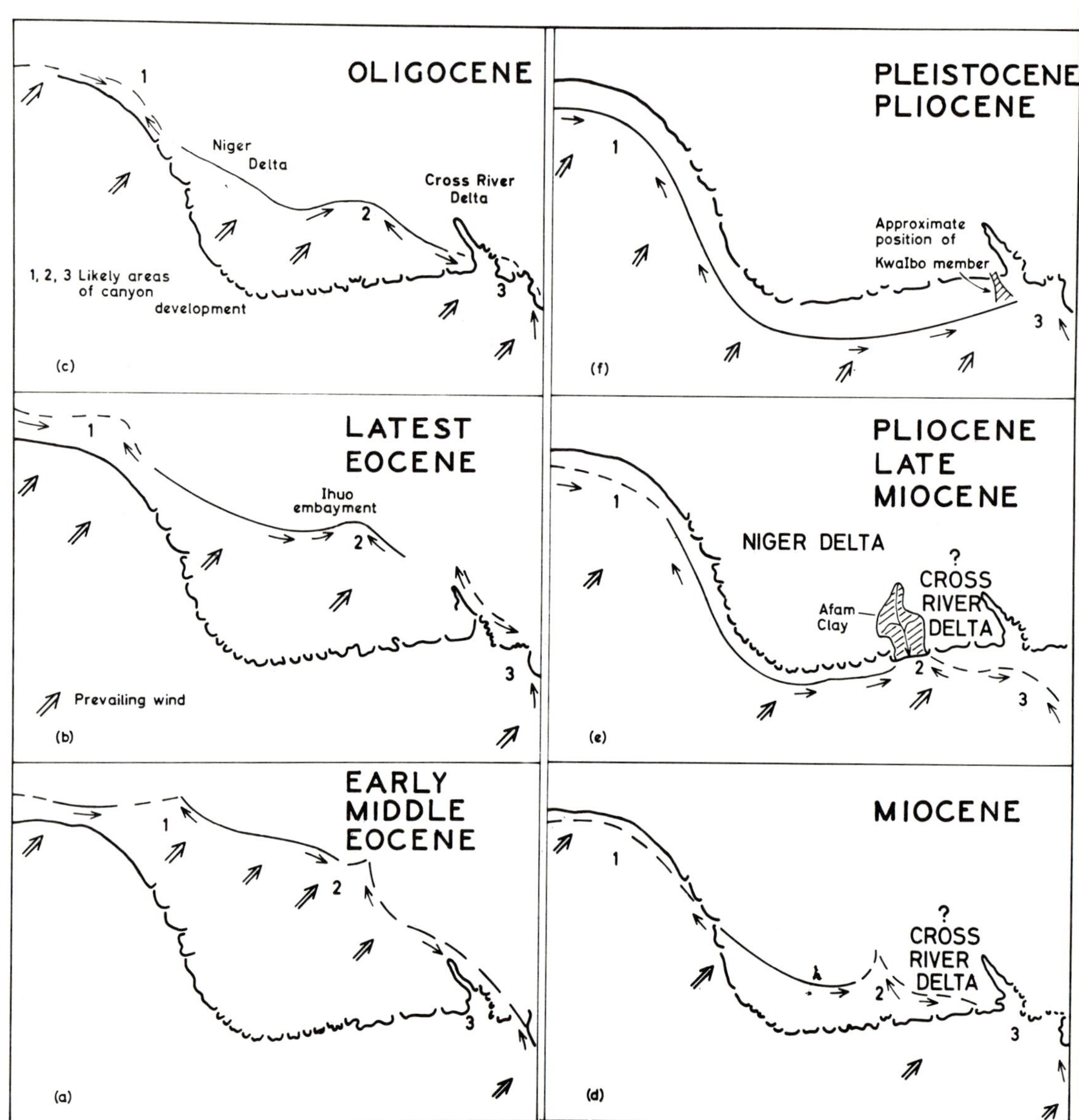

FIG. 8—Sketches showing possible longshore drift directions in Niger delta at various times since middle Eocene. Direction of prevailing wind is assumed to have been constant, and shapes of shorelines have been taken from Short and Stäuble (1967, Fig. 6). Present shoreline is included for reference in all sketches. Areas in northwest *(1)* and southeast *(2)* corners of delta probably always were places where longshore drifts met and canyon development was concentrated. Ihuo embayment *(b)* also may have received drifts from opposite directions and may, therefore, have been area of canyon concentration.

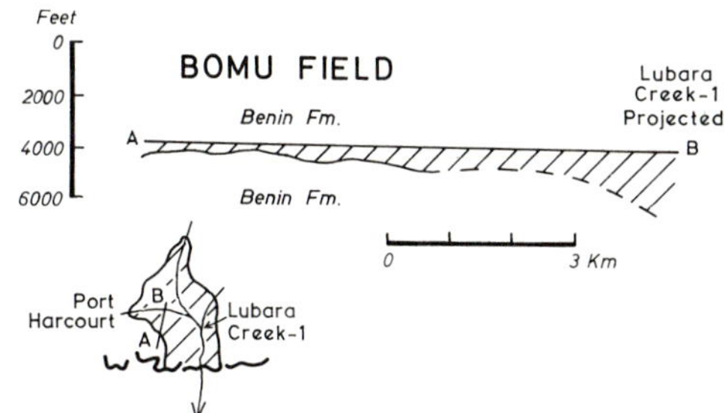

Fig. 9—Sketch cross section across part of Afam clay (based on Frankl and Cordry, 1967, Fig. 6). Inset (based on Short and Stäuble, 1967, Fig. 2) shows approximate position of section. Shape of subcrop indicates that Afam canyon has three branches.

form thick unconformable tops to some structures and will be more effective caprocks than the thin clays of the normal Niger delta offlap sequence (Weber, 1971). On the other hand, wells drilled close to canyon axes are likely to intersect thick sections of canyon-fill clay.

CONCLUSIONS

1. Submarine fans fed largely through submarine canyons are a feature of the Niger delta as they are of other large deltas growing into waters of oceanic depth.

2. Because of the symmetrical way the prevailing southwesterly wind impinges on the Niger delta and the Guinea and Cameroun coasts, converging longshore drifts meeting at the delta corners are main sources of sediment fed to two submarine fans at the foot of the delta.

3. At times of low sea level a third fan at the center of the delta foot is fed by a canyon (or canyons) related to a distributary (or distributaries) close to the delta axis.

4. Persistent longshore drift throughout the Tertiary has contributed to repeated episodes of canyon formation at the delta corners.

5. In the earlier Tertiary, when the Niger and Cross Rivers had separate deltas, a third area of canyon formation may have existed between the deltas in the Ihuo embayment.

6. Interpretations of the structure of the Niger delta in terms of three layers—a continental sand top layer (the Benin Formation), a transitional sand and shale layer (the Agbada Formation), and a bottom marine shale layer (the Akata Formation)—are incomplete. A five-layer structure symmetrical about a middle marine shale (Fig. 10) includes a submarine fan sand layer and an overlying transitional layer. Thin pelagic sediments overlying oceanic crust also are indicated in the section.

Other deltas growing into waters of oceanic

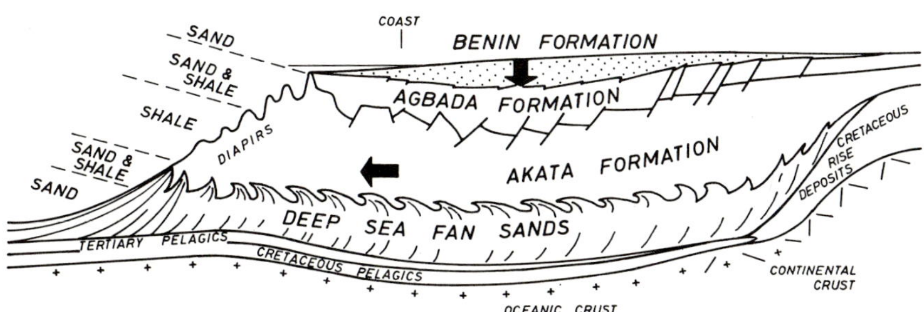

Fig. 10—Schematic structural cross section of Niger delta, modified from Merki (in press) and Weber (1971, Fig. 2) to include submarine fan element. Akata shale overlies submarine fan sand unit at bottom of delta which lies on pelagic sediments over oceanic crust. Transitional shale-sand unit between Akata shale and bottom sand is sketched as dragged forward by motion of Akata.

depths also may have a five-layer sandwich structure like that of the Niger delta.

REFERENCES CITED

Adegoke, O. S., 1968, Eocene stratigraphy of southern Nigeria, *in* Colloque sur l'éocène: Bur. Recherches Géol. et Miniéres Mem., no. 69, p. 23-47.

Allen, J. R. L., 1964, The Nigerian continental margin: Marine Geology, v. 1, p. 289-332.

———— 1965, Late Quaternary Niger delta, and adjacent areas: sedimentary environments and lithofacies: Am. Assoc. Petroleum Geologists Bull., v. 49, no. 5, p. 547-600.

———— 1970, Sediments of the modern Niger delta, *in* Deltaic sedimentation; modern and ancient, p. 138-151: Soc. Econ. Paleontologists and Mineralogists Spec. Pub. 15, 312 p.

Arens, G., J. R. Delteil, P. Valéry, B. Damotte, L. Montadert, and P. Patriat, 1971, The continental margin off the Ivory Coast and Ghana, *in* The geology of the East Atlantic continental margin; 4, Africa: Great Britain Inst. Geol. Sci., Rept. 70/16, p. 61-78.

Association of African Geological Surveys—UNESCO, 1968, International tectonic map of Africa: Paris, scale 1:5,000,000.

Bloom, A. L., 1967, Pleistocene shorelines—a new test of isostasy: Geol. Soc. America Bull., v. 78, p. 1477-1494.

Burke, K., T. F. J. Dessauvagie, and A. J. Whiteman, 1971a, Opening of the Gulf of Guinea and geological history of the Benue depression and Niger delta: Nature Phys. Sci., v. 233, no. 38, p. 51-55.

———— A. B. Durotoye, and A. J. Whiteman, 1971b, A dry phase south of the Sahara 20,000 years ago: West African Jour. Archaeology, v. 1, p. 1-8.

Frankl, E. J., and E. A. Cordry, 1967, The Niger delta oil province: 7th World Petroleum Cong. Proc., v. 2, p. 195-209.

Hospers, J., 1971; The geology of the Niger delta area, *in* The geology of the East Atlantic continental margin; 4, Africa: Great Britain Inst. Geol. Sci., Rept. 70/16, p. 121-142.

Knaap, W. A., 1971, A montane pollen species from the upper Tertiary of the Niger delta: Jour. Geol. Nigerian Mining Geology Metall. Soc., v. 6, p. 23-30.

Le Pichon, X., 1968, Sea-floor spreading and continental drift: Jour. Geophys. Research, v. 73, p. 3661-3697.

Merki, P. J., in press, Structural geology of the Cenozoic Niger delta: African Geology, Ibadan, 1970.

Murat, R. C., in press, Stratigraphy and paleogeography of the Cretaceous and lower Tertiary in southern Nigeria: African Geology, Ibadan, 1970.

NEDECO, 1961, The waters of the Niger delta: The Hague, North Holland Publishing Co., 317 p.

Pugh, J. C., 1953, The Port Novo-Badagri sand· ridge complex: Ibadan Univ. Dept. Geography Research Notes, v. 3, p. 3-15.

———— 1954, Sand movement in relation to wind direction as exemplified on the Nigeria coastline: Ibadan Univ. Dept. Geography Research Notes, v. 5, p. 1-14.

Runcorn, S. K., 1969, The paleomagnetic vector field, *in* The earth's crust and upper mantle: Am. Geophys. Union Geophys. Mon. Ser. 13, p. 447-457.

Short, K. C., and A. J. Stäuble, 1967, Outline of geology of Niger delta: Am. Assoc. Petroleum Geologists Bull., v. 51, no. 5, p. 761-779.

Weber, K. J., 1971, Sedimentological aspects of oilfields in the Niger delta: Geologie en Mijnbouw, v. 50, p. 559-576.

BULLETIN OF THE AMERICAN ASSOCIATION OF PETROLEUM GEOLOGISTS
VOL. 48, NO. 7 (JULY, 1964), PP. 1126-1149, 18 FIGS., 7 TABLES

CONGO SUBMARINE CANYON[1]

B. C. HEEZEN,[2] R. J. MENZIES,[3] E. D. SCHNEIDER,[2] W. M. EWING,[2] AND N. C. L. GRANELLI[4]

ABSTRACT

In May, 1957, a brief survey of the Congo Submarine Canyon was conducted from the Research Vessel VEMA. The canyon was found on the continental slope near the seaward limit of earlier surveys and traced to the west for 150 miles. Where the survey was discontinued, in depths of 2,200 fathoms, the leveed canyon was still a prominent feature. At the 550-fathom contour the canyon is about 5 miles wide at its rim and 500 fathoms deep. The canyon is V-shaped and echoes from the thalweg are usually recorded after the echoes from the steep walls. The outer parts of the canyon, in depths exceeding about 1,800 fathoms, are bounded by huge natural levees. In 1963 a 30-mile-long section of the canyon was surveyed 150 miles west of the 1957 study.

Four biological trawls and ten sediment cores were collected from the canyon region. One trawl in 2,140 fathoms contained abundant tree leaves and a rich fauna. The canyon cores contain silt, sand, and organic debris.

The canyon was discovered in 1886 by a cable route survey. Between 1887 and 1937 the Luanda-Saõ Thomé cable broke 30 times in the canyon. Breaks occurred most frequently during months of maximum river discharge and consequently at the times of greatest bed-load discharge. Cable breaks were limited to periods of years when the river channel was undergoing major changes in position and depth. The cable breaks which occurred up to 120 miles seaward of the river mouth are attributed to turbidity currents generated at the river mouth at times of maximum bed-load transport. These turbidity currents flowed down the continental slope, eroding the deep slope canyon and building the natural levees of the continental rise, and eventually spread out on the Angola Abyssal Plain. In contrast to other great rivers, the Congo is not building a subaerial delta; virtually its entire bed load is being carried by turbidity currents via the Congo Submarine Canyon to the great Congo Cone on the floor of the Angola Basin.

INTRODUCTION

The Congo Submarine Canyon is situated at the mouth of the Congo River of the west coast of Africa (Fig. 1). It is one of the largest submarine canyons in the world. The maximum width is 5 miles and the maximum depth below the rim is 600 fm (Fig. 2b). The Congo Submarine Canyon is now known to be at least 500 miles long (exclusive of meanders). The heads of many submarine canyons are filled with sediment (Fisk and Mc-Farlan, 1955). In contrast, the Congo Canyon extends 20 miles into the mouth of the Congo River with steep sides and a V-shaped profile. A delta is not being formed at the river mouth; instead the bed load of the river is ultimately transported through the canyon and deposited on the continental rise and the Angola Abyssal Plain on the west. Here an immense submarine distributary system was discovered. The transport of great quantities of silt and sand through the canyon at intervals must have eroded the deep canyon and formed an abyssal delta (Congo Cone).

PREVIOUS INVESTIGATIONS OF CONGO SUBMARINE CANYON

The canyon was first discovered in 1886 by J. Y. Buchanan (1888) aboard the British cable-route-exploring ship BUCCANEER (Fig. 2b). The objective of that survey was to select a route for the Saõ Thomé-Luanda (Angola) submarine cable. Buchanan (1887, 1888) discovered three facts about the canyon: (a) its penetration into the Congo River mouth; (b) its great width and depth; (c) the fact that it extended at least to the 1,000-fathom contour. Buchanan correctly presumed that the canyon extends at least to the 2,000-fathom contour. Near shore the canyon has been extensively surveyed by the Congo and Angola governments principally because the axis of the canyon and river forms the international boundary, and changes in the position of the can-

[1] Lamont Geological Observatory Contribution No. 717. Manuscript received, November 4, 1963.

[2] Department of Geology and Lamont Geological Observatory, Columbia University, Palisades, New York.

[3] Department of Zoology and Duke Marine Laboratory, Duke University, Beaufort, North Carolina.

[4] Servicio de Hidrografia Naval, Buenos Aires, Argentina.

This study was supported by the United States Navy, Office of Naval Research, the National Science Foundation, the Bell Telephone Laboratories, and the Argentine Hydrographic Office. Reproduction in whole or in part permitted for any purposes of the United States Government. The assistance of Marion Jacobs, G. L. Johnson, Marie Tharp and Captain V. R. Sinclair, USN (Ret.), is gratefully acknowledged. Discussions with Walter Bucher and Charles Drake were helpful. E. J. Devroey, J. Cl. De Bremaecker, and V. Van Straelen provided valuable information on the Congo River. Engineer-in-Chief of Cable & Wireless, Limited, C. J. V. Lawson, generously made available original cable repair reports.

yon axis have political significance (Devroey, 1946). Unfortunately, these surveys were never extended more than a few miles from the coast. Hull (1900, 1912, in Spencer, 1903) contoured the soundings available at that time. Wire soundings were made in the Congo Canyon by various vessels, including S.M.S. SPERBER [1910], S.M.S. PANTHER [1910], C.S. DUPLEX, C.S. PENDER, and other cable ships. In 1911 S.M.S. MÖWE made 12 sounding lines across the canyon between the river mouth and the shelf edge. Schott and Schulz (1914) present an excellent contour chart of the canyon based on the MÖWE soundings. Both Veatch and Smith (1939) and Shepard (1948) contoured the soundings available before World War II. In 1951 GALATHEA trawled and dredged in the canyon (Bruun, 1957) but the accompanying soundings have not been published.

1957 Reconnaissance Survey

In May, 1957, Menzies, aboard the Research Vessel VEMA, had the opportunity to study the Congo Submarine Canyon, employing three techniques of exploration: (1) precision depth recording; (2) piston coring; and (3) biological trawling (Fig. 2). Additionally standard serial hydrographic work was done.

At the start of this exploration it was expected (1) that the canyon extends well past the 2,000-fathom contour; (2) that graded turbidite sands would be encountered in the floor and at the mouth of the canyon, evidencing submarine turbidity-current activity; (3) and that a relationship between turbidity-current activity and the standing crop of abyssal organisms would be seen. These expectations were all met in the course of the exploration. In addition, a study of the breakage of a submarine cable crossing the submarine canyon provides evidence of contemporary turbidity currents and thus a major clue to the origin of submarine canyons. In 1963, Ewing (VEMA Cruise 19) made a traverse across the western margin of the Angola Abyssal Plain (Figs. 2a, 7), employing the Precision Depth Recorder, magnetometer, seismic reflection profiler, sea-gravimeter, as well as obtaining photographs from the canyon floor, and several cores from the region.

Topography

On VEMA Cruise 12 the Congo region was entered from the south (Fig. 2b) close to the 500-fathom contour. By chance the first crossing was approximately along one of the lines surveyed in 1886 by Buchanan. Once the canyon was found, attempts were made to keep as close to its axis as possible, following it always westward. The axis of the Congo Canyon trends 275° from the river mouth to 11° E; then gradually the canyon arches to the north assuming a trend of 295° T. At about $10\frac{1}{3}°$ E the canyon abruptly bends toward the south, adopting a trend of 225° T. Near the 800-fathom contour, the deep V-shaped profile (Fig. 3) extends 600 fathoms below the adjacent sea floor. The bottom is very narrow, as indicated by the crossing of wall echoes above the floor echo; here the canyon is 5 miles in width. The surface water at this point was discolored, indicating the influence of the Congo River waters 90 miles distant from the mouth of the river. Cable ships have reported river water 120 miles seaward of the river mouth in the vicinity of the canyon. The surface salinity was recorded to be much less than that of the equatorial counter-current water. Here freshwater organisms from the Congo River became mixed with the marine organisms of the northward-flowing Bengela current. It is probable that at least part of the fresh-water diatoms recorded from the deposits on the Mid-Atlantic Ridge have their origin from this source (Kolbe, 1957).

In the vicinity of the 2,000-fathom contour the floor of the canyon lies near the general level of the adjacent continental rise. Extending outward 5 to 20 miles on either side of the various distributaries are well developed natural levees (Figs. 2, 3). The main distributary channel of the canyon shows a characteristic sharp V-shape; other channels are rounded and less distinct. In this region depositional features appear to exceed the erosional characteristics. The canyon was traced to 2,200 fathoms and was still quite evident; dropping 70 fathoms below its natural levees.

In 1963, Ewing (VEMA Cruise 19) made seven traverses of a flat-floored, 3-mile-wide, natural-leveed canyon between 05°34′E and 05°06′E, a distance of about 30 miles (Fig. 2a). The canyon floor dropped continuously westward at about 1:600 from 2,643 fathoms to 2,692 fathoms. The canyon walls are steep and are 10–20 fathoms high. This canyon appears to be a continuation of the southern or main branch of the Congo Canyon System. The survey thus extends the length of the

(*Text continued on page 1132*)

5°
10°
15°
5°
0°
Massif
Cameroon Mt
Sanaga
Fernando Po
Diah
Gowe
Congo
Principe
São Thomé
Annabon
3100
4600
4760
1975

Fig. 1.—Congo Submarine Canyon and Angola Abyssal Plain. This is a part of The Physiographic Diagram of the South Atlantic, published by The Geological Society of America. (Copyright, 1961, by Bruce C. Heezen and Marie Tharp. Reproduced by permission.) Scale: one degree latitude equals 60 nautical miles.

CONGO SUBMARINE CANYON
0 - 2000 CONTOUR INTERVAL 100 FATHOMS
2000-2350 CONTOUR INTERVAL 50 FATHOMS
> 2600 CONTOUR INTERVAL 10 FATHOMS
TRACK OF VEMA
BUCCANEER SOUNDINGS
OTHER SPOT SOUNDINGS
SAÕ THOMÉ LUANDA CABLE
CORE
TRAWL
2250
2350
2600
2610
2620
2650
2670
2680
2630
280
277
278
275
5°
6°
7°
8°

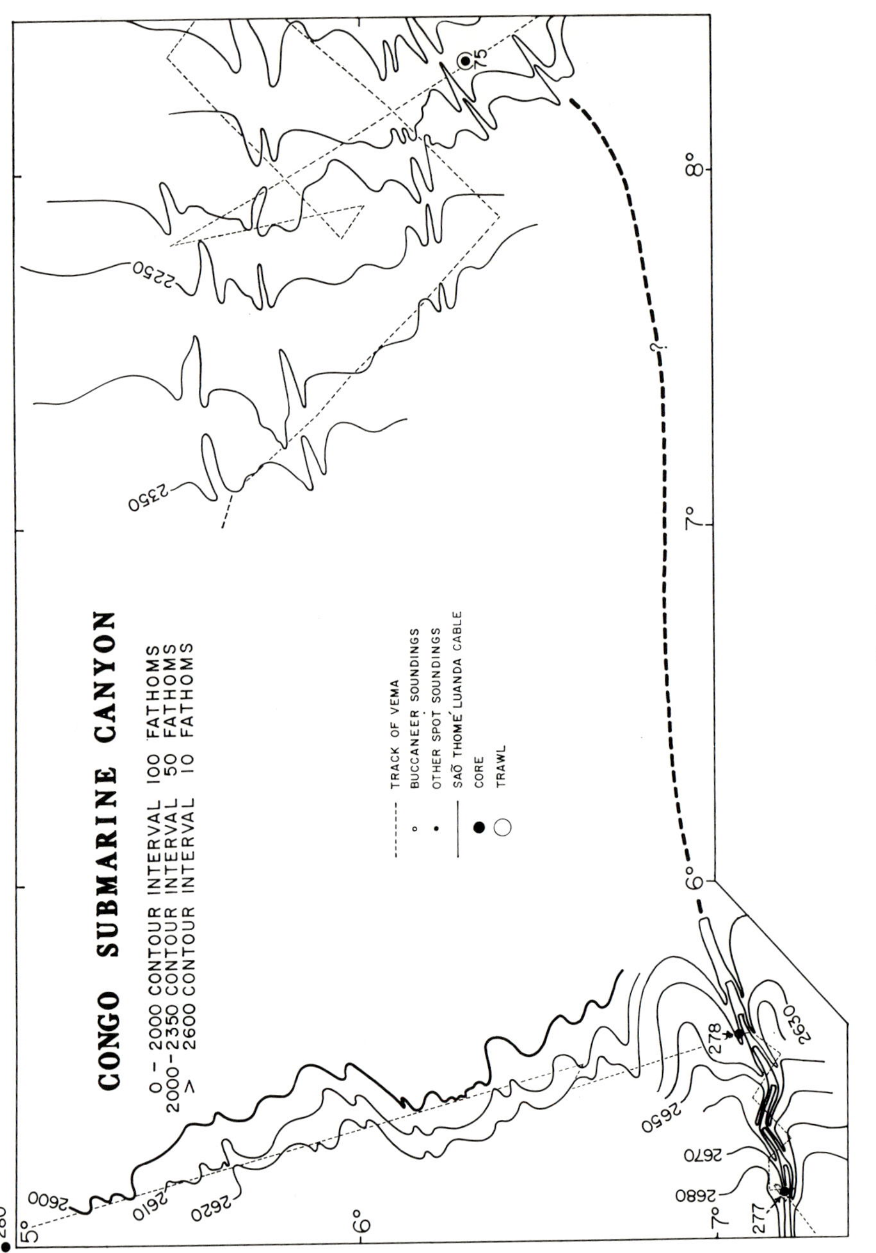

FIG. 2a

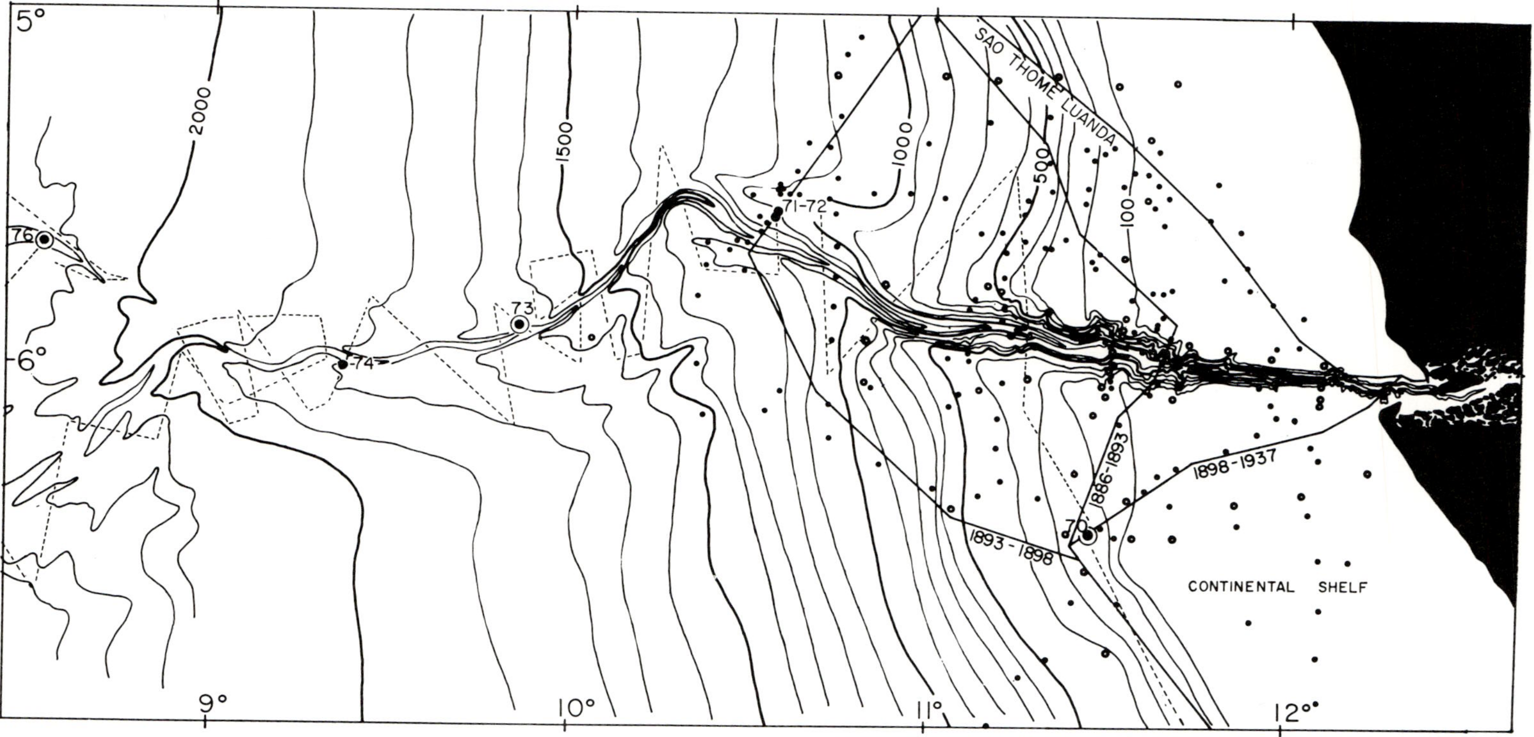

Fig. 2b

Fig. 2. (a, b)—Bathymetric chart of Congo Submarine Canyon. Soundings in fathoms at 800 fm/sec. Routes of Saō Thomé-Luanda submarine cable shown by solid line. One degree latitude equals 60 nautical miles.

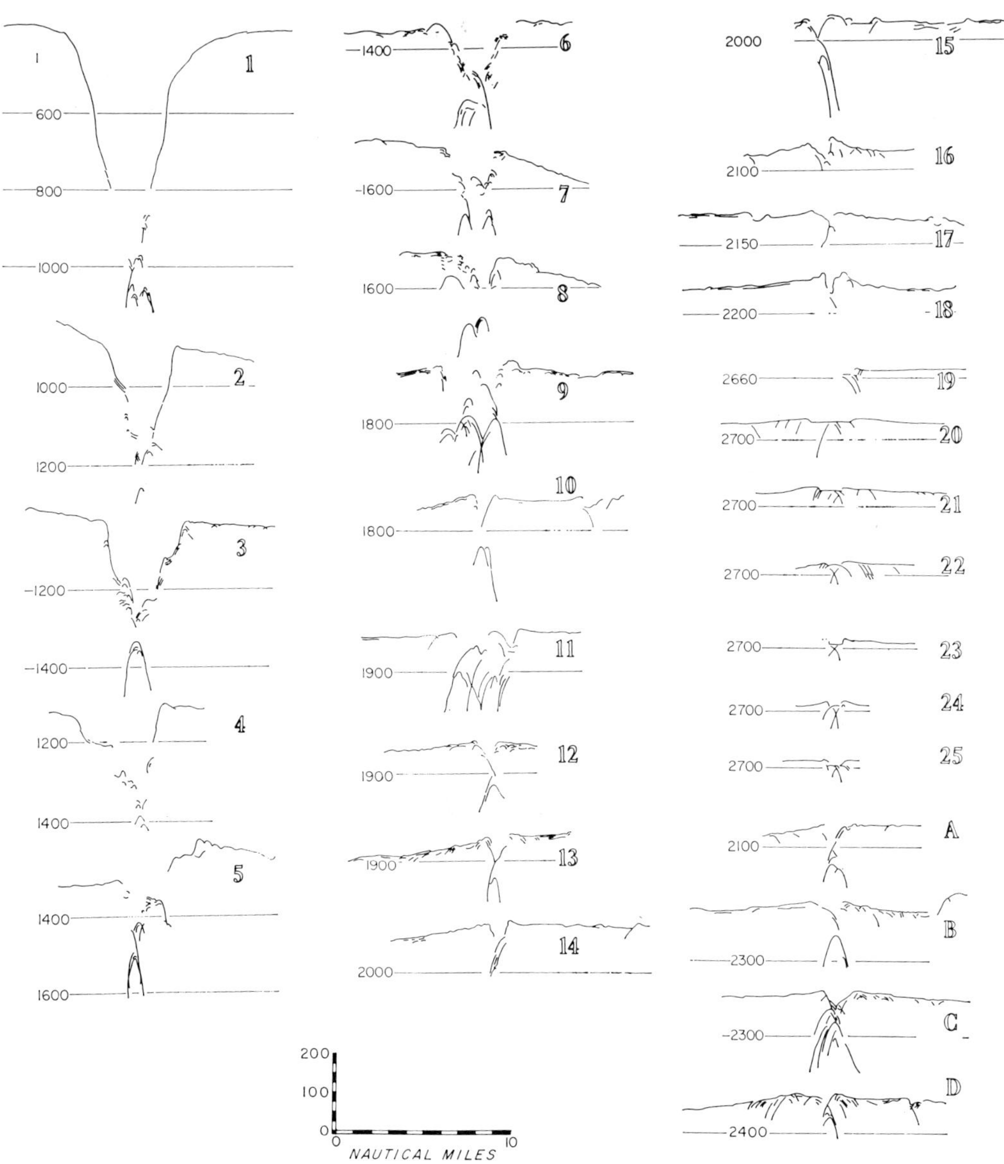

FIG. 3.—Tracings of Precision Depth Recorder echogram of Congo Submarine Canyon.
Depths in fathoms at 800 fms/sec.

canyon by 150 miles, making the entire proven length of the Congo Submarine Canyon 500 miles. Other branches of the canyon continue toward the west but they are much more poorly developed than the main channel.

SEISMIC REFLECTION PROFILE

Difficulty in recording sub-bottom echoes beneath deep-sea cones and on the landward side of abyssal plains is a common experience. This characteristic is generally attributed to numerous thin and highly reflective layers intercalated in the sedimentary column by turbidity-current deposition. In general, good penetration is obtained on gentle rises such as Bermuda Rise as well as at the seaward edges of the abyssal plains where in many places even the 12-kilocycle echo-sounding pulse returns from up to 20 fathoms below the bottom.

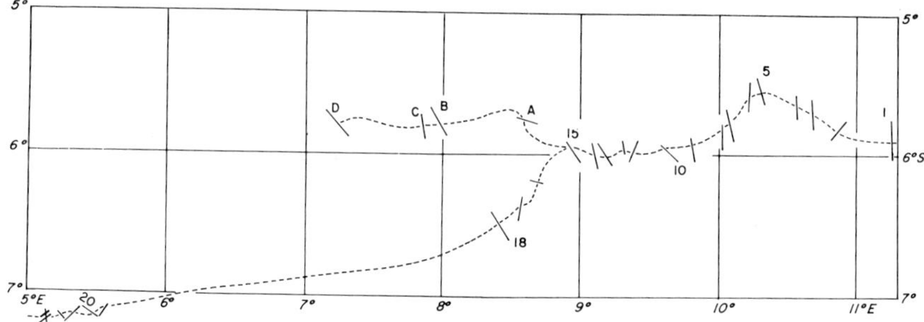

Fɪɢ. 4.—Location of profiles illustrated in Figure 3.

The VEMA-19 seismic reflection profile (Figs. 5, 6, 7) reveals relatively thick sediments up to 1.8 kilometers in thickness overlying the Walfisch Ridge and the Guinea Rise. The Guinea Rise is a broad, smooth arch which exhibits rather irregular mountainous sub-bottom topography beneath a 1-kilometer-thick mantle of sediment. On the Angola Abyssal Plain south of the Congo Canyon a remarkably transoral layer of sediment blankets the underlying relief. However, along a band about 150 miles on either side of the axis of the main channel of the Congo Canyon no sub-bottom penetration was obtained. This is not evidence that the sediment is thin in this area but simply an indication of the difficulty of obtaining reflections from beneath abyssal-cone type sediments. The whole central area of the Congo Cone is highly reflective. However, on the north side of the cone and in a slight depression that lies between the Congo Cone and the Guinea Rise prolonged confused echoes give a hint of closely spaced sub-bottom reflectors. There is a suggestion from the seismic reflection results that turbidity-current sedimentation in the northern part of the cone ceased some time ago, allowing the accumulation of a thin layer of transoral sediments above the lower and more reflective turbidites. This interpretation is supported by the sediment cores obtained in this area.

Sᴇᴅɪᴍᴇɴᴛs

Six piston cores were taken during VEMA Cruise 12 on the continental slope and continental rise in the vicinity of the Congo Canyon (Fig. 2). Cores were taken from the continental slope (V-12-70), the canyon wall (V-12-71, 72, 73), from the continental slope on the canyon rim (V-12-74), on the natural levees (V-12-75), from the floor of a distributary channel on the continental rise (V-12-76).

Cores close to the axis of the canyon contain silts and sand beds (V-12-70, V-12-76); cores farther from the canyon axis (V-12-71, 72, 74) contain mainly hemipelagic silty clays with a few thin layers of silt. Megascopically, four sediment facies can be distinguished in the dried sediment. The first facies, comprising more than half of the sediment, is homogeneous silty lutite (Fig. 8). The silt, largely derived from the suspended sediment discharge of the Congo River, is mixed with the normal pelagic particle-by-particle deposition. A second lutite facies is described as crumbly, silty lutite. In the dried cores these lutites have a hackley broken surface which is distinct from the smooth surface of the dried homogeneous silty lutite. The third facies consists of graded silts and sands (V-12-73, 75, 76) which appear to be turbidites. These beds grade upward from fine sands to silts (in some places laminated). Three sand beds (one in V-12-75, and beds A and C in Core V-12-76; see Figs. 8, 9) contain leaves and twigs of dicotyledon land plants (identified by H. Becker, New York Botanical Gardens). A sample from the top of bed A in Core V-12-76 contains 68 per cent (by weight) leaves and twigs. Leaves and twigs have been found in turbidites originating from the Magdalena River (Heezen, 1956), from the Amazon (Locher, 1954) and in turbidites from the Hatteras Abyssal Plain. Bruun (1957) dredged coconuts and other vegetable matter from the Philippine and other trenches, material which may have floated to the site of deposition but which most probably was deposited by turbidity currents. The fourth facies is laminated silts (V-12-73, 74, 76). These beds consist of alternating lutite-rich and lutite-free laminae.

339

1134

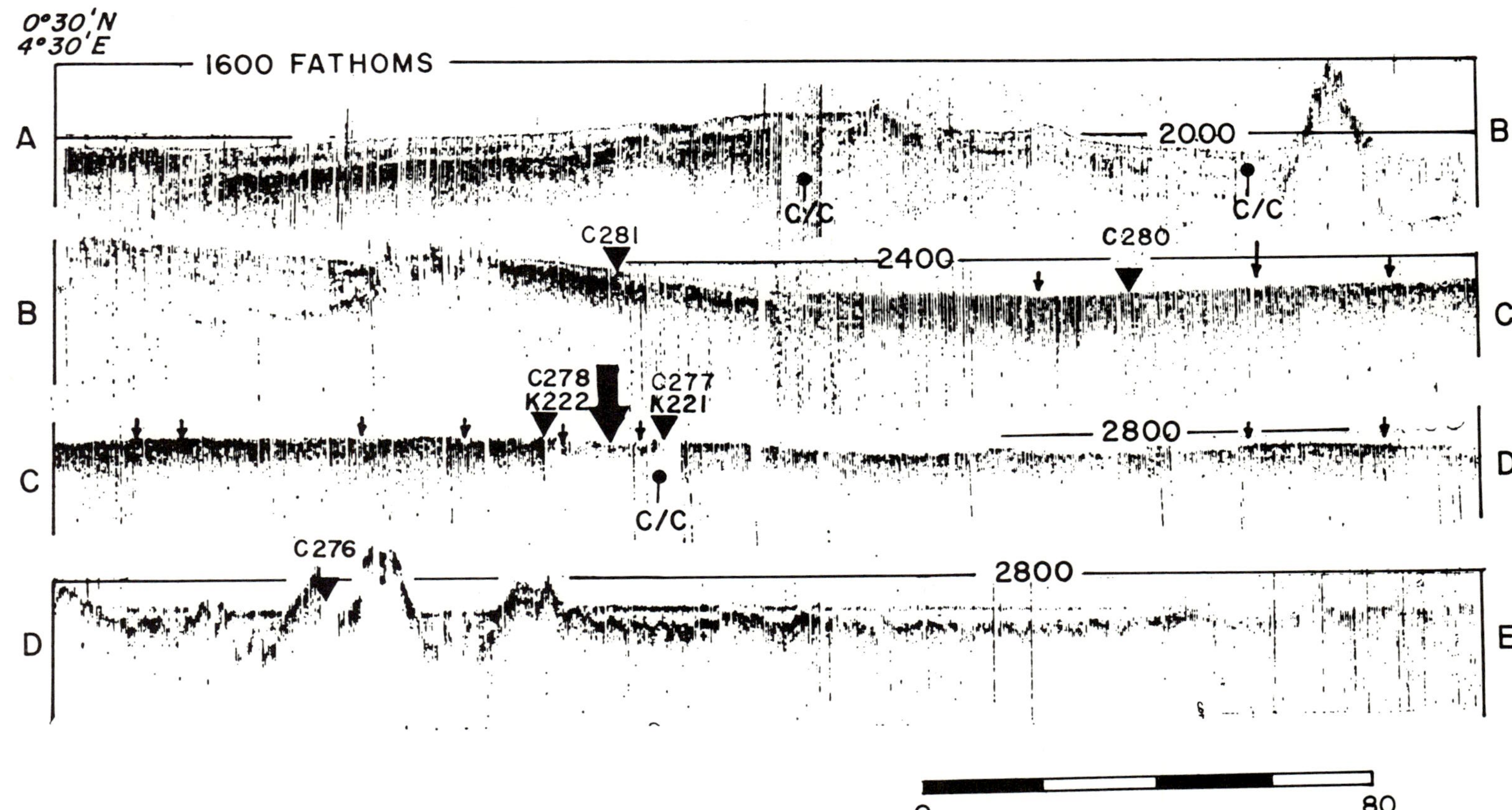

Fig. 5.—Seismic reflection profiler record from Guinea Rise across Congo Cone. Location indicated on Figure 7. Small arrows indicate location of small indistinct channels noted on PDR record. Large arrow indicates location of main branch of Congo Canyon system. Camera stations K221 and K222 and cores C276-C278, C280, C281 obtained on VEMA Cruise 19 indicated by dots. C/C indicates change of ship's course.

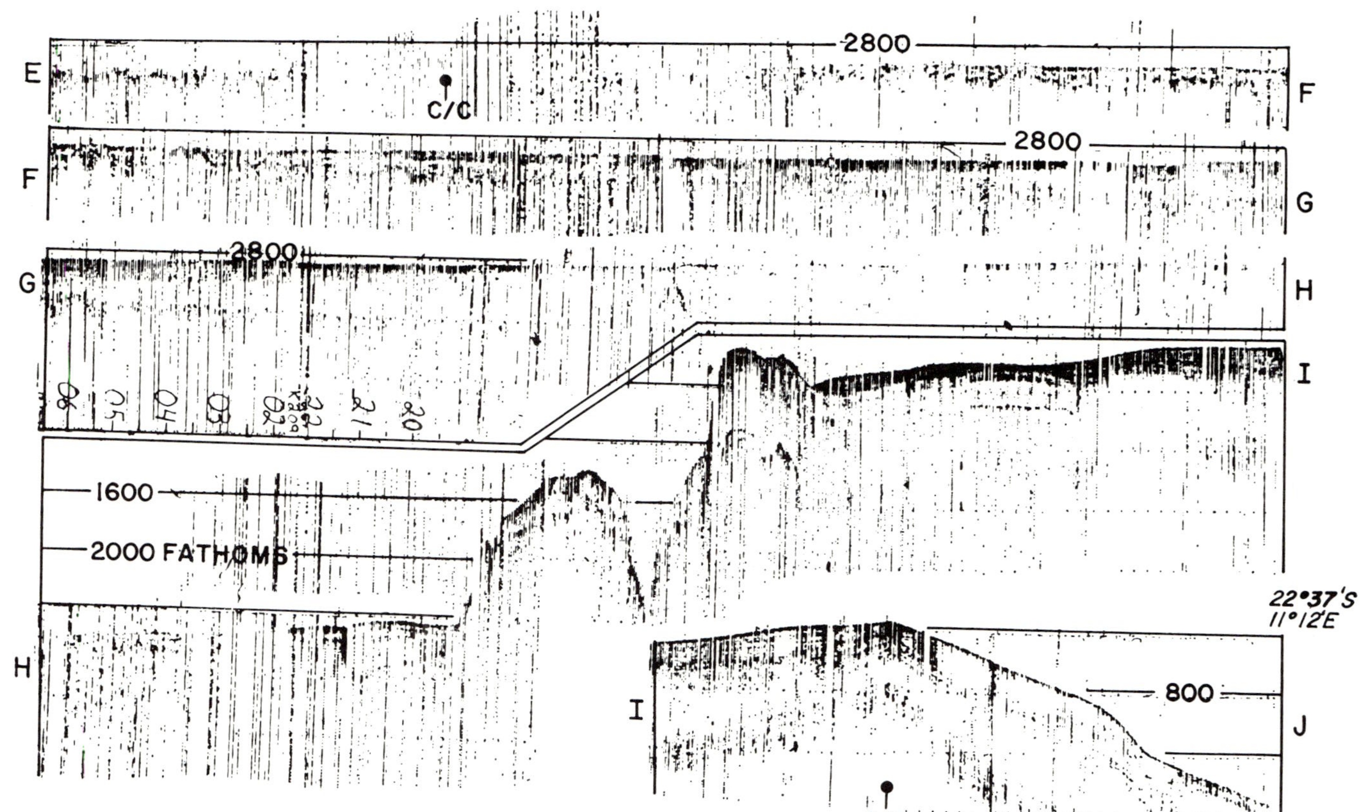

FIG. 6.—Seismic reflection profiler record from abyssal hills across Angola Abyssal Plain to Walfisch Ridge. Location of profile indicated on Figure 7.

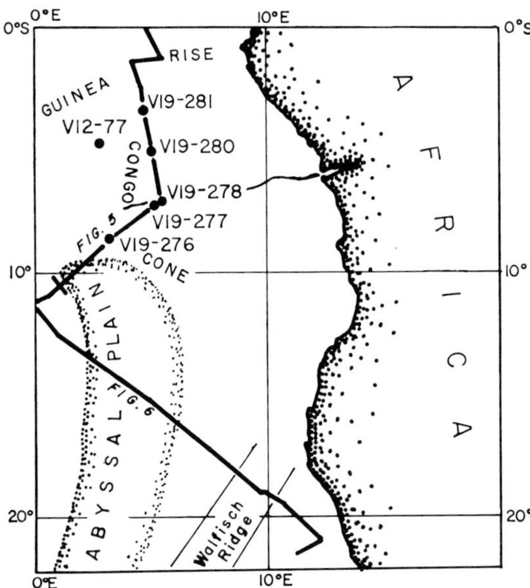

Fig. 7.—Location of VEMA Cruise 19 seismic reflection profile and cores in Angola Basin.

Samples for mechanical and mineralogical analyses were taken at the tops and the bottoms of two of the turbidite sequences (beds A and C of Core V-12-76; Fig. 9). Because of the short penetration of the coring apparatus it is suspected that the deepest sand in Core V-12-76 was considerably thicker than is indicated by the recovery. A single sample was taken from the top turbidite sequence (bed D) in the core because this bed was somewhat deformed. Beds A and C showed normal grading in grain size and the size-distribution curves are more peaked at the tops than at the bottoms of the beds (Table II; Fig. 10).

Heavy-mineral separations were made from six samples of Core V-12-76 (Table III). The percentage of heavy minerals seems to be inversely proportional to the mean grain size (Tables II, III).

Two mineral suites can be tentatively identified. Suite I is characterized by an abundance of hematite and hematite aggregates in the heavy-mineral separation and a preponderance of hematite-coated quartz grains in the light fraction. This suite is also associated with concentrations of leaf and twig fragments. Suite II is characterized by rounded black opaque heavy minerals and a dominance of rounded clean and frosted quartz grains. Residual traces of a former hematite coating are found only on a few grains.

In comparing the two mineral suites, it is noted that the mineral species of Suite I contains many loose aggregates of hematite and somewhat higher concentrates of other easily altered minerals, such as chlorite. This mineral suite is found in beds A and C (V-12-76) and A′ (V-12-73); beds that contain plant debris.

Since a large proportion of the sediment load of the upper Congo River is deposited in the interior Congo Basin above Leopoldville, much of the sediment found in the braided streams of the lower Congo must originate below Stanley Pool. Red sands are common in the alluvial terraces, erosion surfaces (King, 1962), marine terraces and bedrock (Cahen, 1954; Furon, 1963) of the lower Congo region. Red sands are in fact ubiquitous in the Congo drainage basin. The 50 miles of the Congo River cut across raised marine terraces of Pleistocene age which for the most part consist of semi-consolidated red ferruginous sandstones (Mouta and O'Donnell, 1933; Cahen, 1954).

H.E.P. Cust (Anon., 1922), Commander of H.M.S. RAMBLER during an 1899 investigation, found that:

"The nature of the bottom in the Congo River is invariably sand, or in places hard clay, until the deep gully is reached, when there is found everywhere a deep deposit of soft mud and decayed vegetable matter, another proof of the tranquility of the bottom water.

TABLE I. CORE LOCATIONS IN CONGO CANYON AREA

	Lat.	Long.	Depth (Fathoms)	Length (Cm.)
V-12 70	06°29′S	11°26′E	240	610
V-12-71	05°38′S	10°41′E	1,210	1,072
V-12-72	05°38′S	10°40′E	1,131	490
V-12-73	05°54′S	09°53′E	1,635	979
V-12-74	06°00′S	09°20′E	1,845	910
V-12-75	06°19′S	08°19′E	2,145	683
V-12-76	05°43′S	08°30′E	2,137	330
V-12-77	04°48′S	02°45′E	2,747	1,078
V-19-276	08°58′S	02°52′E	2,798	348
V-19-277	07°10′S	05°08′E	2,700	1,344
V-19-278	07°04′S	05°34′E	2,637	996
V-19-280	04°56′S	05°00′E	2,610	1,594
V-19-281	03°19′S	04°39′E	2,430	1,787

CAMERA STATIONS IN CONGO CANYON AREA

Camera Station			No. of Frames
V-19-221	7°10′S	5°08′E	2,679	20
V-19-222	7°04′S	5°34′E	2,721	22

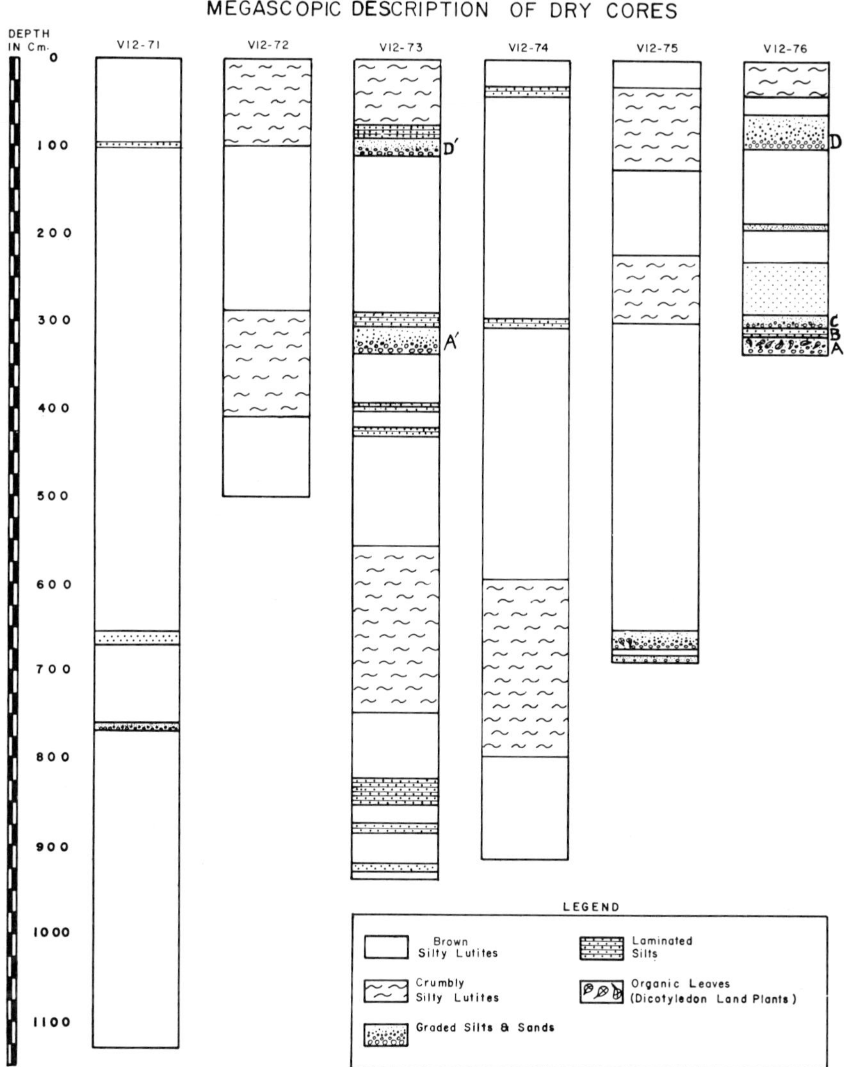

Fig. 8.—Graphic logs of piston cores from region of Congo Submarine Canyon obtained on VEMA Cruise 12.

"Whilst running a sectional line of soundings outside Banana creek, the bottom was invariably mud, but on one single occasion, at a depth of 56 fathoms, although mud was brought up, there were signs on the lead that it had struck a hard substance, the arming was dented and the lead slightly marked.

"The banks of the Congo, the numerous shoals in the river, and the bottom itself, were found during the survey to be invariably sand or hard clay.

"Mud is only met with in the small creeks in the upper portion of the river; and on the banks in the lower portion as far as the mangroves extend, which is from about Kissanga downwards.

"It appears that a very large proportion of the mud found in the lower part of the river in the deeper water is from the washings of the immediate neighbourhood, which is a vast expanse of mangrove swamp, with an innumerable network of creeks drained at every tide.

"The water of the river itself is found to be

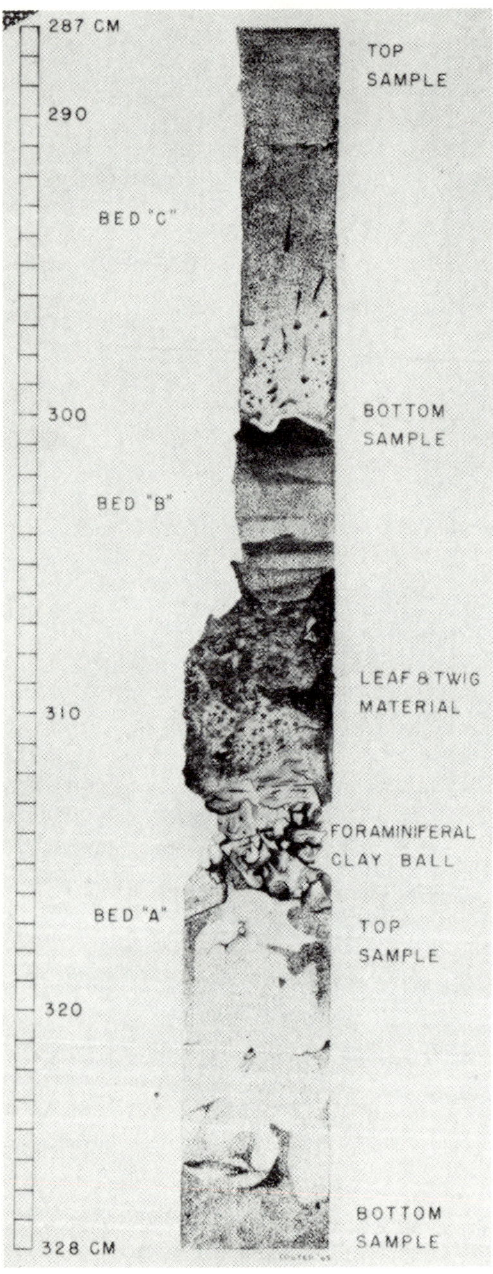

FIG. 9.—Turbidites in Core V-12-76.

deep water. It is stated that on grounding the anchor should never be let go; sooner or later the current will wash bank and ship together down the river until deep water is reached."

Due to the effects of vigorous wave action, clean sands characterize the Atlantic beaches of Angola, Congo, and Cabinda, despite the proximity of the red cliffs of marine terrace deposits which lie behind the beaches. The predominant northward long-shore drift in the region has produced Padrao Point, a sandspit which extends directly to the edge of the Congo Canyon. Two or more subsidiary canyons branch from the main canyon toward this point. Veatch and Smith (1939) note that this strong northerly long-shore current loosened and carried water pipes north from the Luanda. Trask (1955), Shepard (1951), and others have noted that long-shore drift often transports beach sands into the heads of submarine canyons.

The turbidites of mineral Suite I which contain leaf material and a preponderance of hematite and hematite-coated grains, probably are Congo River sediments. The minerals of Suite II which show much evidence of abrasion and transportation (beds B, D, D′; Figs. 8, 9) probably are long-shore drift sands which were swept into the canyon and later carried into deeper water by turbidity currents.

Both heavy-mineral Suites I and II contain appreciable percentages of pyrite. In many places the pyrite is a coating on another mineral and in some places it occurs inside diatom frustules. Gas bubbled profusely from Core V-12-73 when it was extruded on VEMA. Core V-12-70, taken on the continental slope south of the canyon, contains very dark silty clay with some mollusk layers. This core and the cores in the vicinity of the canyon show evidence of reducing conditions through their dark sediment color, pyrite mineralization, and the preservation of organic material which would be destroyed in an oxidizing environment.

Five cores were taken during VEMA Cruise 19 on or near the Congo Cone (Figs. 2a, 7) 150 miles west of the VEMA-12 suite of cores. These sediments represent two distinct facies of pelagic deposition.

The more northern facies consists of alternating layers of carbonate-rich gray lutites and black lutites (V-19-281, V-19-280) (Fig. 11) and the other facies consists of reddish brown abyssal lutite and is the dominant pelagic facies in the

heavily charged with sand. Vessels grounding on banks where the current is strong have had the sand piled up against one side of them nearly to the surface in a few hours, and then a sudden swirl of current has washed it away and left them in

TABLE II. MECHANICAL ANALYSES OF TURBIDITES OF CORE V-12-76: STATISTICAL PARAMETERS
(after Folk 1961)

		Mean Size	Mean Size mm	Sorting	Skewness	Kurtosis
Bed A	Top	3.480	.092	+.823	−.027	1.201
	Bottom	3.180	.291	+.305	−.047	1.262
Bed B		3.683	.085	+.651	−.001	1.667
Bed C	Top	3.470	.091	+.514	−.043	1.204
	Bottom	3.276	.101	+.474	−.003	.978
Bed D		2.780	.258	+.572	−.089	1.038

three southern cores (V-19-278, V-19-277, V-19-276).

The most northern of the cores studied (Core V-19-281) taken on the Guinea Rise, consists of approximately 18 meters of alternating bands of low-carbonate black lutite and gray lutite with carbonate percentages up to 80 per cent. Contacts between these distinct sediment types are in some places sharp and in some places gradational. In places the core is highly worm-burrowed. Radiolarians are found in parts of the core. Core V-12-77 taken somewhat west of V-19-281 contains a large percentage of siliceous material.

Core V-19-280 was taken about 100 miles south of Core 281 on the Congo Cone. The first 12-meter segment of this 16-meter core consists of alternating layers of light gray calcilutites and dark black lutites of the northern facies. The organic carbon content of the black lutite was determined to be 2.9 per cent. Professor William M. Sackett (personal communication) determined the C^{13}/C^{12} ratio of the organic ooze. The isotope ratio (-21.5 $^{o}/_{oo}$ vs. PDB) suggests that the carbon is marine-derived, indicating high organic productivity in this area. Below 12 meters in this core is a distinctly different facies of reddish brown lutite containing little carbonate. The contact between these two facies is gradational over 20 centimeters. The brown lutite is intercalated by numerous sand and silt beds many of which are graded. Cross-bedding and parallel laminations due to lutite films are found in some of the coarser sand beds. In the vicinity of Core V-19-280 the Precision Depth Recorder echogram showed a sub-bottom reflection approximately 10–12 meters below the sediment water interface (.01–.015 sec.). This sub-bottom echo is probably from the contact between the two distinct sediment types found in Core V-19-280. The seismic reflection

profile record previously noted also showed shallow penetration (Fig. 5). Only one turbidite bed is found in the upper 12 meters of this core, whereas the lower 4 meters contain 68 turbidite beds. It can thus be concluded that turbidity current deposition at this site virtually ceased coincidentally with the beginning of deposition at this site of the northern gray-black lutite facies.

Cores V-19-288 and V-19-287 were taken from the northern bank of the canyon. These cores consist of thin turbidites intercalated in reddish brown lutite and resemble the sequence found below 12 meters in Core V-19-280. Both cores contain many turbidites consisting of fine sand to fine silts. Many of these beds are but laminae of fine silt, a centimeter to a few millimeters in thickness. In most places the beds are recognized only

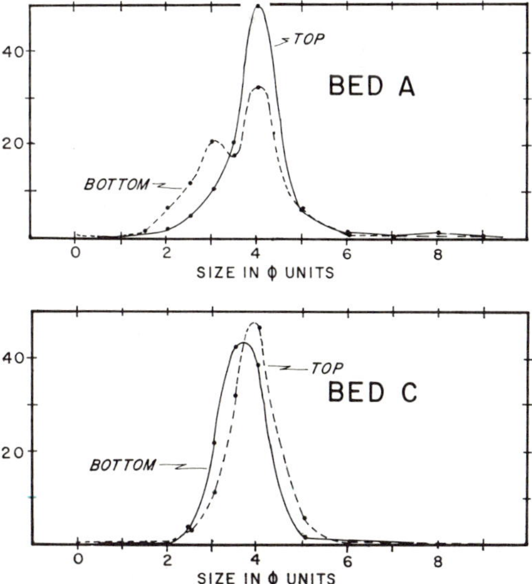

FIG. 10.—Mechanical analyses of turbidites in Core V-12-76.

TABLE III. MINERALOGY OF TURBIDITES IN CORE V-12-76

| Mineral Suite | Bed | | Leaves & Twigs | >90% Qtz. Hematite Coated | >50% Qtz. Frosted | Heavy Minerals* | | | | | | | | | | | % of Heavy Minerals in >62μ Fraction |
| | | | | | | % Opaque ⊗ | | | | % Non-opaque ⊗ | | | | | | | |
						Hematite	Black Opaque	Leucoxene	Pyrite	Chlorite	Hornblende	Tourmaline	Staurolite	Epidote	Sillimanite	Others (<10% Each)	
I	A	Top	+	+	−	61	9	7	23								0.45
		Bott.	+	+	−	57	20	6	15	14	34	8	6	3	8	27*	0.38
	C	Top	+	+	−	65	22	4	8								0.38
		Bott.	+	+	−	62	4	6	27								0.41
II	B		−	−	+	39	46	2	12								0.73
	D		−	−	+	11	69	0	19	5	23	17	10	10	10	25†	.022

* Kyanite, garnet, zircon, rutile, orthopyroxene, biotite, zoisite, augite, and fiberolite.
† Same minerals as above but lacking zoisite and including topaz and sphene.
⊗ % of mineral grains counted >62 μ.

by slight changes of color, or from unequal shrinkage of the thin beds on drying. Core V-19-278 contained 440 such turbidite laminae in 10 meters. Some of the thicker silt and sand beds show grading and some contain microstructures such as cross-bedding and parallel laminations. At 798 cm. in Core V-19-279, a bed of light brown crystalline carbonate sand was found. This material consisted of small rhombohedrals of calcite and showed no evidence of organic deposition. X-ray diffraction indicated the material to be calcite with less than 5 per cent aragonite.

Core V-19-276 was taken in the abyssal hills southwest of the Congo Canyon. This core consisted of reddish brown lutite with a zone of hard indurated yellow lutite between 30 and 90 cm. The core shows no evidence of turbidity-current deposition.

All five cores contain a zone high in foraminifera in the top 14–25 cm. Below this zone, foraminifera are found mainly in the gray lutite facies and in some worm burrows. All the cores except V-19-276 showed evidence of a highly reducing environment; (1) each smelled of H₂S when opened on board ship; (2) each was stained by hydrotroilite; and (3) the lutites of each contained many small micro-crystals of pyrite.

The sediment pattern in the western part of the Angola Basin may be summarized as follows: (1) organic pelagic lutites in the northern equatorial region (either siliceous or carbonate), and (2) reddish brown abyssal lutite in the central and southern regions. Superimposed on this pelagic sedimentation pattern is the effect of the Congo Canyon which has introduced great amounts of silt and sand onto the Congo Cone. Turbidity current deposition apparently ceased in the northern part of the Congo Cone in mid-Pleistocene. A thin blanket of pelagic sediment has softened the topographic forms and obliterated the smaller channels in that area.

Bottom photographs were taken near the submarine canyon on VEMA Cruise 19 at the same positions as Core V-19-277 and Core V-19-278. The photographs (Fig. 12) indicate a generally muddy bottom with fairly abundant tracks and trails, and indicate a tranquil bottom unaffected by ocean currents.

BIOLOGICAL TRAWL RESULTS

Biological trawls were taken simultaneously with several cores (Fig. 2). The same techniques were used with each trawl; hence, the results are comparable among themselves. The standing crop of animals from the upper continental slope (trawl No. 21 at core location V-12-70) exceeded that of deeper trawls by a factor of 100 to 1000 as would be expected, but trawls taken at greater depths showed some interesting and unusual results.

Heezen, Ewing, and Menzies (1955) postulated that submarine turbidity currents might affect abyssal life in two ways: (1) to obliterate life by smothering it; (2) to feed life by providing avalanches of detrital nourishment from shallow water.

Trawl No. 23 was taken from a natural levee and although the water depth was more than 500 fathoms greater than that of trawl No. 22, there was no marked decrease in diversity and abundance of animal life as compared with trawl No. 22 which was taken from a similar situation farther up the continental rise and 90 miles closer to shore. Significantly lower was the population of trawl No. 24 from a sandy distributary floor, although the depth was the same as that of trawl No. 23.

A study was made of the ratio of detritus feeders to filter feeders. Filter feeders, it was reasoned, would be harmed while detritus feeders might be benefited by frequent turbidity currents. The proportion of detritus feeders appears to be sig-

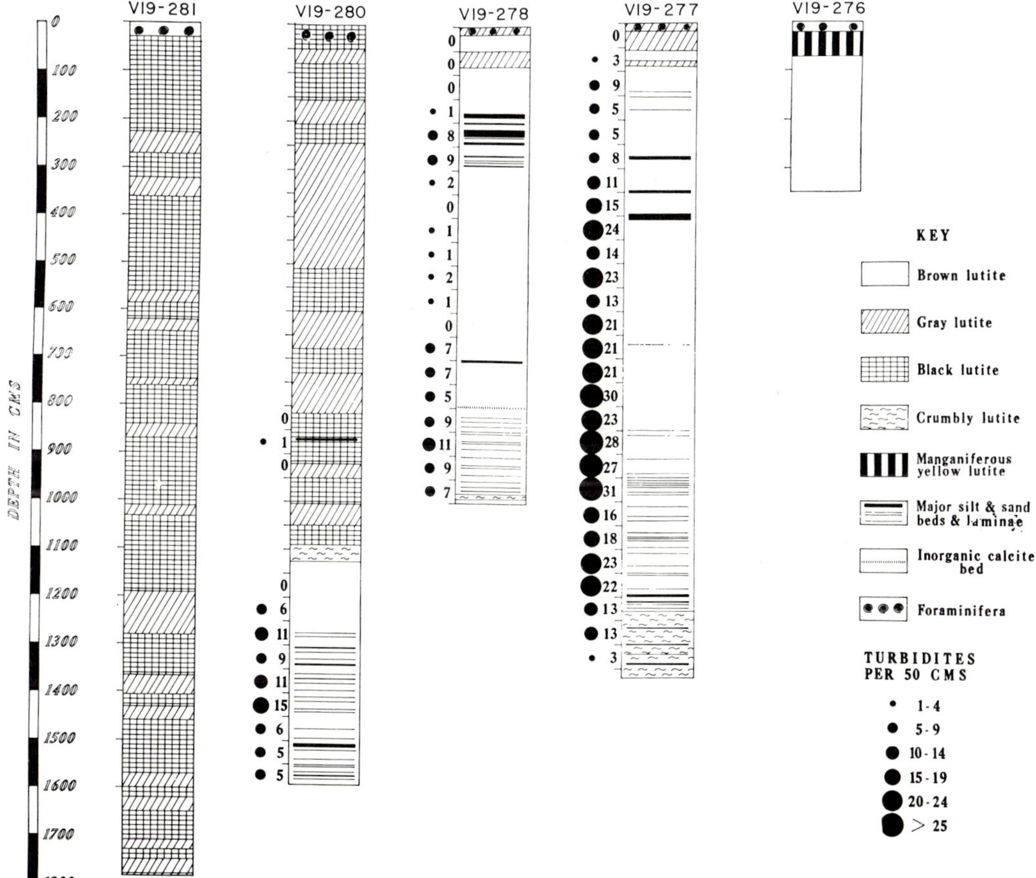

FIG. 11.—Graphic logs of piston cores obtained in vicinity of Congo Cone on VEMA Cruise 19. Locations shown on Figure 7. Positions and depths listed in Table I.

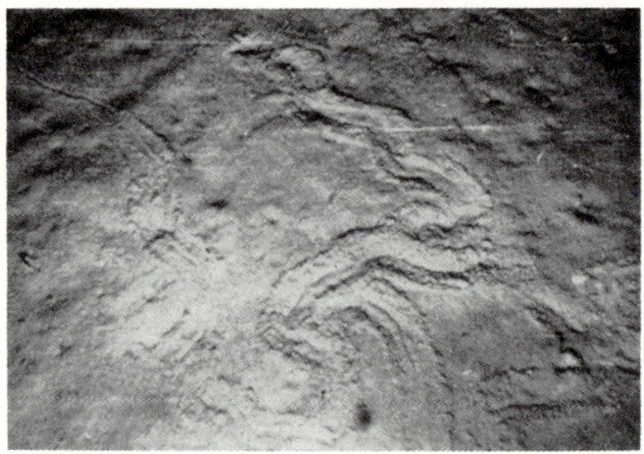

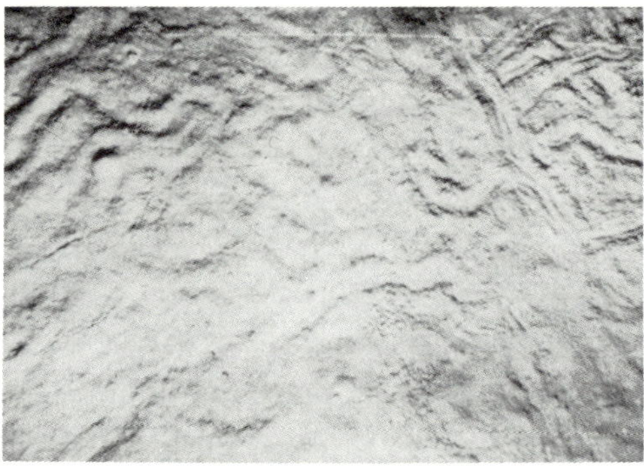

FIG. 12.—Sea-floor photographs of tranquil bottom obtained on natural levees of the Congo Canyon. (Upper photo) Station V-19-221 (18) 2679 fm. (Lower photo) Station V-19-222 (14) 2721 fm. Positions listed in Table I and Figure 5. Dimensions of photographed area approximately 6×9 feet. Note abundant evidence of bottom life and complete absence of current evidence.

nificantly higher in the Congo Canyon distributary system than in adjacent areas. Much more information is needed but the data are at least suggestive.

Submarine Cable Failures

The Saõ Thomé-Luanda (Angola) Cable (Fig. 2), originally laid in 1886 and abandoned in 1937, failed 30 times near the axis of the Congo Canyon.

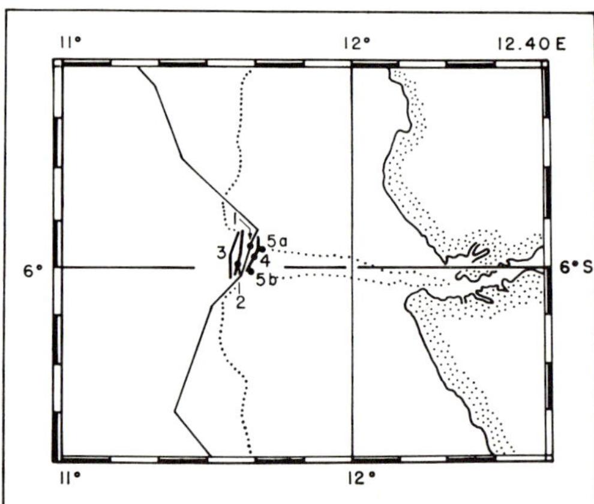

FIG. 13.—Cable failures in Congo Canyon; Cable Route 1, 1886–1894. Dots indicate location of cable breaks. Black line indicates cable replaced during repair.

The first route selected from Buchanan's survey was used from 1886 to 1893, during which time the cable failed five times (Fig. 13). The cable failed each time under tension and in most places was buried under sediment in the floor of the canyon. In 1893 it was decided to divert the cable into deeper water where failures are in general less common. The second route crossed the canyon in much deeper water 70 miles farther from the Congo River, but in this location the cable suffered eight breaks between 1893 and 1897 (Fig. 14). Each of the eight failures was a tension break and in all cases the section of cable near the axis of the canyon was buried and had to be abandoned. The cable engineers recognized that the cable failures were related to the transport of sediment down the canyon from the Congo River. They realized that to avoid further failures they would have to loop the cable hundreds of miles out to sea. Instead they diverted the cable into the river mouth where in relatively shallow depths cable repairs could be more easily effected. Along

TABLE IV. BIOLOGICAL TRAWLS

Trawl No.	At Core Sta.	Surface Sediment	Orders or Classes of Animals	Number of Specimens / Number of Species/Trawl	Approx. No. Species/Trawl	Depth (Fms.)
22	V12-73	Diatomite	12	8.3	25	1635
23	V12-75	Red lutite	8	9.1	14	2145
24	V12-76	Sand	5	3.8	9	2137

this third route the cable failed 15 times between 1897 and 1937 (Fig. 15). Three failures occurred in a tributary canyon leading northwest from the sandspit terminating at Padrao Point. Two failures occurred in another tributary a short distance west. These failures were also all tension breaks and cable recovery was prevented by burial in most instances. The cable ships attempted to lay the·cable across the canyon at about a 45° angle in an effort to decrease the drag on the cable. Repairs of the cable at the river mouth may have taken less time than the deepwater repairs but they were not without difficulties. The high currents at the Congo mouth carried away mark buoys, made grappling runs more difficult, and swamped the boats used in tending buoys. In 1914 it was decided to insert a "T" piece linking Banana with Saō Thomé and Luanda so when the cable failed in the canyon Saō Thomé would be at least linked to Banana. The breaks were not distributed uniformly through the years but occurred at definite seasons during two periods: 1893–1904 and 1924–1928. These cable breaks are associated with some highly significant characteristics of the Congo River itself.

Congo River

The Congo River drains the great inland Congo Basin, reaching Stanley Pool at a low gradient. Below Stanley Pool the Congo drops 300 meters through a series of cataracts to its estuary. The Congo drainage basin encompasses 3,700,000 km.[2] and is second in size only to the Amazon.

In terms of water discharge the Congo also ranks second among the world's rivers (Table VII). In terms of sediment discharge, however, the Congo ranks eleventh. Much of the fine sediment settles above Stanley Pool with the result that the water descending the cataracts carries relatively little sediment despite the high-bed velocities. The sediment of the Congo River below Matadi thus consists largely of material the river has been able to erode in flowing over the Precambrian rocks and terrace deposits in the cataract region. The Congo exhibits a variation in discharge smaller than any other major river measured; minimum discharge being only a little less than half maximum discharge. More than 40 per cent of the total discharge of the braided stream area of the lower Congo is accounted for by discharge through the narrow main channel.

The velocity of the river in the main channel of

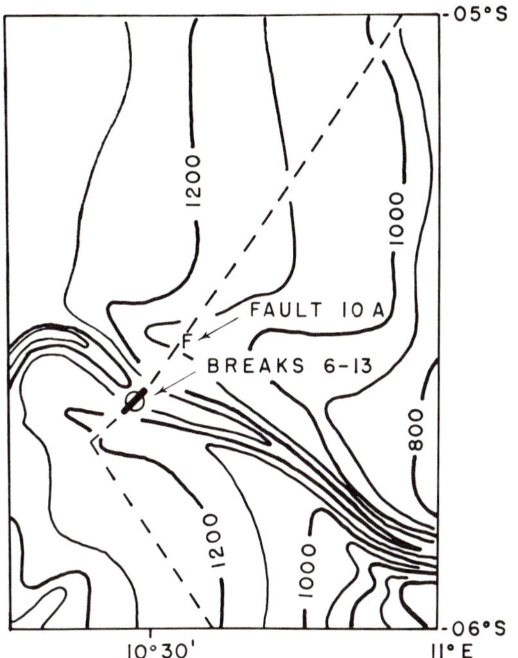

FIG. 14.—Cable failures in Congo Submarine Canyon; Cable Route 2 (deep water) 1894–1897. Circle indicates location of 8 breaks. Letter F indicates location of fault.

the braided stream is of the order of 100 cm/sec. during low water and 150 cm/sec. during high water. However, velocities of more than 250 cm/sec. have been observed during high water at falling tide (Devroey and Vanderlinden, 1951). The velocity at Boma ranges between 150 cm/sec. at low water and 250–300 cm/sec. at high water. Near Matadi velocities of 500 cm/sec. have been recorded at high water. Although bed velocities are of course lower, the velocities observed are competent to erode and transport all sediment sizes through gravel.

The lowest velocities in the Congo channel are considerably higher than most rivers and thus very little clay or fine silt-size material remains in the area of the lower Congo.

On reaching the ocean, the Congo River waters override the relatively motionless heavier sea water. In this area the well known dead-water phenomenon is observed which causes difficulties in navigation. The first investigation of this pheneomenon in the Congo was made by H. E. Cust (Anon., 1922). His observations made in September and October, 1899, from H.M.S. RAMBLER "appeared to show that the fresh water of the

TABLE V. CABLE FAILURES IN CONGO CANYON

Failure No.	Date of Failure	Date of Repair	Reported Cause
		ROUTE 1 (SEE FIG. 8)	
1	(Jan. 1892)	16 Feb. 1892	Break—Fault just off northern bank of gully running out from mouth of Congo.
2	(Dec. 1892)	10 Jan. 1893	Break—Unable to recover either end of break both ends seeming buried in about the same depth of water, think cause of break due to landslip caused by rush of water from Congo.
3	Feb. 1893	5 Mar. 1893	Break—not recovered.
4	Apr. 1893	26 Apr. 1893	Break—heavier type B cable inserted.
5	30 May 1893	18 July 1893	Breaks. The end (a fresh break) was caused by cable being buried and parting under heavy strain—it is evident there were two breaks in the gully, neither end of the southern break being recovered and only the northern end of the other one. Cable diverted to deeper water.
		ROUTE 2 (SEE FIG. 9)	
6	0853 5 Feb. 1894	4 Mar. 1894	Break. 05°37′S-10°29′E., approximately 1300 fms, caused by tension, wire being evenly broken at ends and tape torn off. About $\frac{1}{4}$ inch of copper exposed.
7	19 Mar. 1894	29 Oct. 1894	Faults. Three faults due to wires of sheathing crushing core.
8	1400 21 Jan. 1895	11 Feb. 1895	Break—(ends not recovered).
9	0630 14 Jan. 1896	26 Jan. 1896	Break—total break due to suspension—only northern end of break recovered . . . cable was apparently buried from here to break . . . end of break came in showing cable had parted owing to tension, the wires having broken short off. There was no indication of chafing and the taping and compound were intact up to point where wires had parted.
10	2 May 1896	21 May 1896	Break.
10A	21 May 1896	16 June 1896	Fault—on northern bank of canyon.
11	1530 17 Nov. 1897	24 Nov. 1897	Break—Cable very deeply buried and two miles not recovered.
12	1200 3 Dec. 1897	11 Dec. 1897	Break—Cable parted on drum at a strain of 85 cwts having become very deeply buried during the eleven days that have elapsed since it was laid . . . due probably to the great depth of compressed mud under which the end was undoubtedly buried. Throughout this year very little surface current was observed and the water was nearly clear, presenting a great contrast to its muddy condition during the first repair. Whilst slipping the final splice, however, of the current repair, a strongly marked current was observed settling down towards the ship from the east, bringing thick discolored water with it.
13	0900 12 Dec. 1897	8 Jan. 1898	Break—Cable diverted close to river mouth using type B cable.
		ROUTE 3 (SEE FIG. 10)	
15	1100 4 May 1898	13 May 1898	Break due to sudden strain. Break situated on south side of gully off Congo River. Cable deeply buried. The end presented the appearance of having been broken by a sudden strain, sheathing wires flush and conductor being withdrawn into the gutta percha.
16	1400 2 June 1898	1 July 1898	Break at kink. Cable bright and strained for 63 fathoms from break.
17	0900 11 Dec. 1899	15 Dec. 1899	Break in gully (ends not recovered) deeply buried—cable type E broke on drum while in good mechanical condition—near northern bank.

TABLE V—(*continued*)

Failure No.	Date of Failure	Date of Repair	Reported Cause
18	0900 25 Jan. 1900	30 Jan. 1900	Break—near northern bank of submarine gully.
19	0730 19 May 1900	9 June 1900	Break—break due to cable being buried in southern gully at mouth of Congo.
20	1745 25 Nov. 1901	12 Dec. 1901	Break—ends flush, and corroded.
21	1200 25 Dec. 1901	3 Feb. 1902	Break—total break at kink.
22	0300 30 Aug. 1924	20 Sept. 1924	Break.
23	27 Nov. 1924	5 Dec. 1924	Break—cable showed signs of strain.
24	20 July 1925	31 July 1925	Break—not recovered.
25	0800 21 Oct. 1925	29 Oct. 1925	Break—conductor broken inside of gutta percha.
26	0700 8 Dec. 1925	23 Dec. 1925	Break—caused by strain
27	1445 7 Oct. 1926	18 Oct. 1926	Break—not picked up as cable parted under heavy strain.
28	3 Jan. 1927	13 Jan. 1927	Break—ends not recovered.
29	March 1927	7 April 1928	Fault—not recovered, deeply buried.
30	1000 19 March 1932	2 May 1932	Break—not recovered.

TABLE VI. CABLE FAILURES ON SHELF NORTH OF CANYON

Failure	Date of Failure	Date of Repair	Reported Cause
A	0700 14 Mar. 1902	28 Mar. 1902	Break—total break probably due to ship's anchor.
B	20 May 1907	3 June 1907	Fault—spewed out core—probably caused by ship's anchor.
C		14 Mar. 1916	Fault—corrosion.
D	16 Sept. 1916	19 Oct. 1916	Fault—corrosion.
E	16 Nov. 1918	23 Nov. 1918	Break—corrosion—replace T piece.
F	26 July 1928	1 March 1929	Fault—not recovered due to very strong currents.
G	1440 18 Sept. 1936	18 Oct. 1936	Break—due to chafe and corrosion at T Box.
H	July 1937	2 Aug. 1937	Break—corrosion.

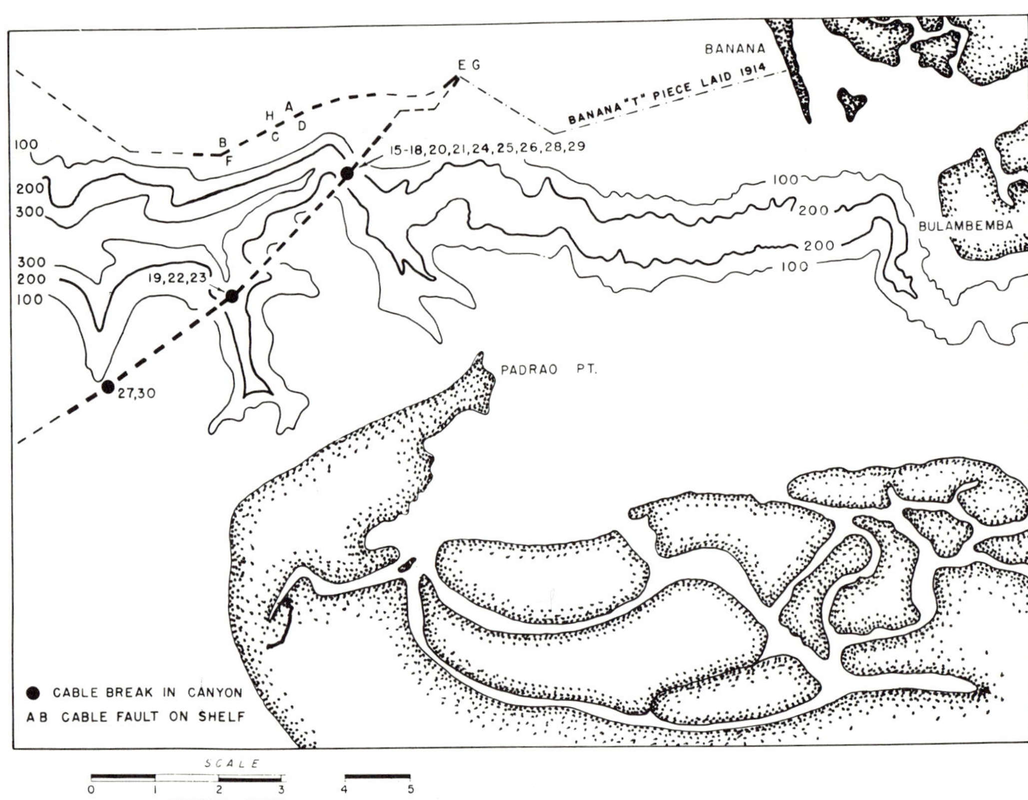

FIG. 15.—Cable failures in Congo Submarine Canyon; Cable Route 3 (shallow water) 1897–1937. Black dots in canyon indicate mean positions of numbered cable breaks. Letters A-H on continental shelf north of canyon indicate cable faults. Heavy dashed line indicates cable replaced during repair. Light dashed line indicates location of undamaged cable.

TABLE VII. MAJOR RIVERS OF WORLD IN ORDER OF ANNUAL WATER DISCHARGE
(After NEDECO 1959)

	Annual Water Discharge $10^9 m^3$	Annual Sediment Discharge 10^6 Tons	(Rank)	Maximum Water Discharge m^3/sec	Minimum Water Discharge m^3/sec
Amazon	3,000	900	(3)	203,000	—
Congo	1,400	70	(11)	65,000	27,000
Yangtze	700	970	(2)	80,000	5,270
Mississippi	600	600	(4)	76,500	3,500
Parana (Plata)	600	90	(8)	—	—
Irrawaddy	520	290	(6)	64,000	1,300
Zambesi	500	100	(7)	—	—
Mekong	400	80	(9)	60,000	1,700
Brahmaputra	380	Probably high	(?)	—	—
St. Lawrence	300	3	—	—	—
Volga	250	25	(14)	—	—
Hwang Ho	200	1890	(1)	25,000	245
Indus	200	400	(5)	26,000	490
Danube	200	80	(10)	10,000	—
Niger	180	40	(13)	30,000	1,200
Nile	85	60	(12)	—	—

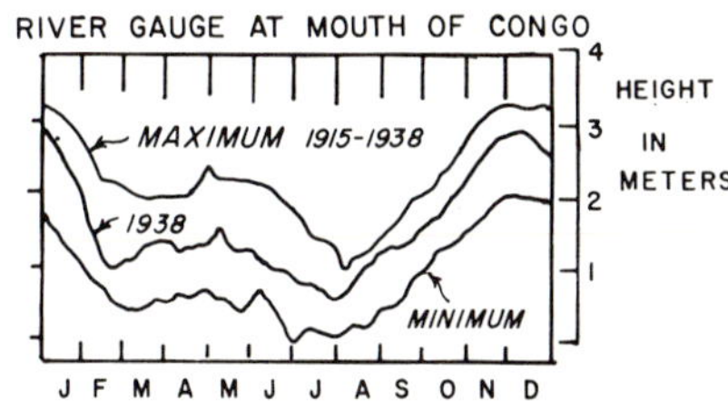

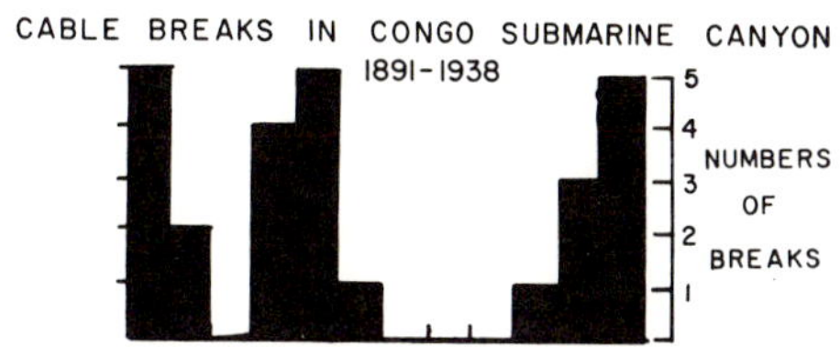

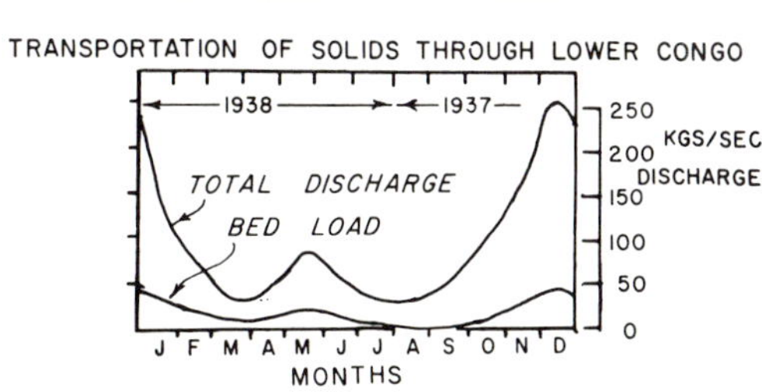

FIG. 16.—Distribution by month of cable failures in the Congo Submarine Canyon, suspended load and bed-load discharge of the Congo River and river stage at Boma. Cable failure data (1889–1937) from the files of Cable & Wireless Limited. Suspended load and bed-load discharge for 1937–38 from Spronck (1941) and average river stage at Boma for the years 1915–1936 from Devroey and Vanderlinden (1951) and Devroey (1956).

Congo extends from the surface to the bottom until the head of the deep Congo reaches just below Kissanga (Lat. 6°01′ S., Long. 12°39′ E.), when it encounters a body of salt water filling this deep gully. It then runs over this denser water with decreased depth and increased velocity, the layer of fresh water being deeper with the ebb tide and shallower with the flood, both decreasing the broader the river becomes, until, from being 3 to 5 fathoms deep just below Bull Island, it is only a few feet deep after passing Bulambemba point. The deep body of salt water is either perfectly still, or has a very slight tidal flow (two-tenths to half a knot), up river with the flood, and down with the ebb tide."

Seasonal variations in river discharge.—Day-to-day and month-to-month variations in the discharge of a river can be easily monitored by the continuous measurement of river height (stage).

Occasional measurements of the river's cross section combined with measurements of current velocity provide calibrations by which the measurements of stage can be converted into absolute discharge. The river height as determined by river gauge data between 1915 and 1938 shows a maximum high water between October and February and a sub-maximum between April and May (Devroey and Vanderlinden, 1951). The minima occur in March and between July, August, and September. The correlation between high river discharge and cable breaks is remarkable* (Fig. 16). The discharge of suspended solids and the bed-load transport fluctuate in response to variations in total discharge (Spronck, 1941). Since some cable breaks occurred in the canyon 150 miles from the river mouth, it seems unlikely that the water-discharge fluctuations had a direct effect on the cable. It seems more likely that the variations in discharge of solids are related to the cable breaks. It would seem likely that it was the bed-load transportation rather than the suspended-load transportation which had the principal effect. We may infer that the 30 cable breaks between 1887 and 1938 are consequently related to the maximum river bed-load transport.

River channel migrations.—The main channel of the Congo periodically changes its position between Boma and Malela, in the broad braided stream region where channels are easily excavated by the river (Devroey and Vanderlinden, 1951). Changes in the main channel are shown in Figure 17. In 1897 a new main channel was established. For a few years prior to 1897, while the river was excavating this new channel, the bed load between Boma and Malela must have been very high. Between 1892 and 1901, 20 submarine cable breaks occurred in the canyon (Fig. 18). In the mid-1920s a new channel was excavated which involved a much smaller change than the previous shift. Dredging operations in some places initiated and in some places modified changes principally effected by the stream itself (Devroey and Vanderlinden, 1951). Between 1924 and 1932, 9 cable breaks occurred. Between 1932 and 1937 the channel made only insignificant alterations without accompanying cable breaks. The cable was abandoned in 1937.

* Cable failure off the Magdalena River (Heezen, 1956) reflects a similar relationship (Elmendorf and Heezen, 1957).

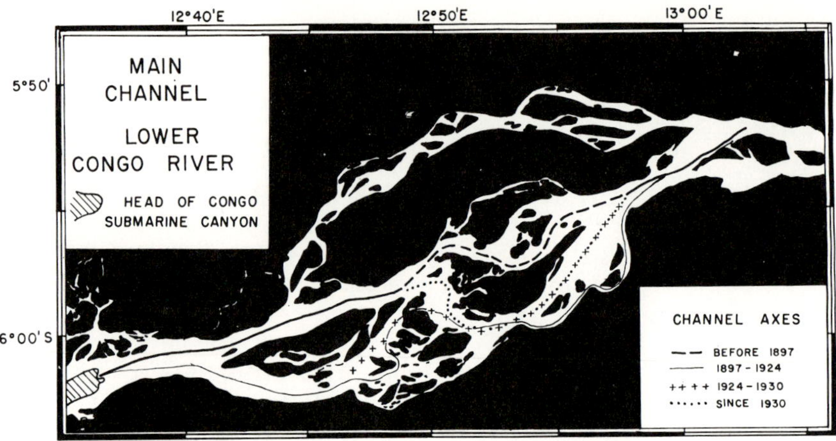

Fig. 17.—Locations of the principal channel in the Lower Congo between 1897 and 1940 (after Devroey and Vanderlinden, 1951).

It is concluded that each major turbidity current between 1886 and 1937 was recorded by a cable break regardless of the depth at which the cable crossed the canyon. The only advantage gained by relocating the cable was that in position 3 near the river mouth repairs could be made in much shallower water ($<$300 fathoms). The near absence of breaks between 1901 and 1924 was due to the remarkable stability of the river channel, not to a safer position of the cable nor to the fact that heavier B cable was used at that position, whereas lighter D deep-sea type was used in deep water.

Because of the good correlation between submarine cable breaks and river bed-load discharge, it is concluded that turbidity currents triggered by the river are responsible for the cable breaks. Seismic triggering is entirely lacking from the area. That the river bed load is carried down the

Congo Submarine Canyon is suggested by the fact that the canyon has not been filled by river deposits near its head. Additional evidence strongly supporting this contention comes from the submarine canyon; the presence of submarine natural levee systems on the continental rise; and the absence of a subaerial delta at the mouth of the Congo River.

The Congo Submarine Canyon today is experiencing about 50 turbidity currents per century, a high frequency comparable with that probably experienced by the canyons off many Pleistocene rivers. The turbidity currents are not only responsible for the breakage of the cables but for the erosion of the V-shaped canyon as well.

Discussion

The exact mechanism by which the river generates turbidity currents at peak-discharge months during periods of river channel excavation is not known. We may only speculate on possible relationships. It might be imagined that at periods of peak discharge enormous quantities of sand and silt are carried in the stream past Malela to the head of the submarine canyon and that appreciable quantities of sand- and silt-size material are thus carried over the underlying salt water. At periods of peak discharge conceivably enough sand would in this manner settle into the underlying sea water to generate turbidity flows.

As another possibility, the bed load carried by traction and saltation along the bottom at times of peak discharge might separate at the head of

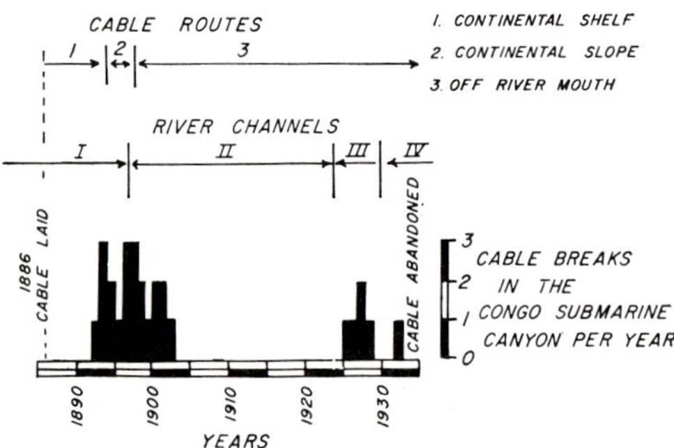

Fig. 18.—Distribution of cable failures in the Congo Submarine Canyon by year (1886–1937).

the canyon and simply continue as a turbidity flow beneath the salt-water wedge which fills the canyon head within the river.

As a third possibility it might be imagined that in addition to one of the above mechanisms, sand bars carried away during peak discharge near the end of the braided stream and near Padrao Point create avalanches of sand which transform into turbidity currents.

A buried "submarine canyon" has been discovered beneath the Mississippi delta (Osterhoudt, 1946). This buried canyon leads to the head of the Mississippi cone distributary system. There has been considerable discussion of the rates of subsidence and rates of compaction, in relation to eustatic events in the Mississippi delta region (Fisk and McFarlan, 1955). Of critical importance to these discussions is the precise relation to sea-level of the braided-stream deposits and the eroded floor of the Pleistocene Mississippi River "trench".

If the submarine canyon of the Pleistocene Mississippi cut into the mouth of the Mississippi and formed a feature similar to the Congo Submarine Canyon, then the floor of the lower Mississippi trench could have been excavated hundreds of feet below sea-level. Buried canyons similar to the modern Congo Canyon may exist below the deltas of other great rivers (Heezen, 1963).

BIBLIOGRAPHY

Anonymous, 1922, The West Coast of Africa: British Admiralty Africa Pilot Pt. II, 276 p.

Bruun, Anton F., 1957, General introduction to the reports and list of deep-sea stations, p. 7–12, *in* GALATHEA Report v. 1, Scientific results of the Danish Deep-Sea Expedition round the world 1950–52: 260 p., Copenhagen.

Buchanan, J. Y., 1887, On the land slopes separating continents and ocean basins, especially those on the West Coast of Africa: Scot. Geograph. Mag., v. III, p. 217–238.

——— 1888, The exploration of the Gulf of Guinea: Scot. Geograph. Mag., v. IV, p. 177–200, 233–251.

Cahen, L., 1954, Geologie due Congobelge: 577 p., Liege, Hvaillant-Carmanne.

Devroey, E. J., 1946, La Vallee Sous-Marine de Fleure Congo Inst. Royal Colonial Belge: Bull. de Seances XVII, p. 1040–1074.

——— 1956, Annuaire Hydrologique du Congo Belge et du Ruanda-Urundi: Ministere des Colonies, Belgique, Pub. 11, 467 p. (for many years such a volume appeared each year).

——— and Vanderlinden, R., 1951, Le Bas-Congo, Artere Vital de Notre Colonie: 350 p., Bruxelles, Goemaere.

Elmendorf, C. H., and Heezen, B. C., 1957, Oceanographic information for engineering submarine cable systems: Bell System Tech. Jour., v. XXXVI, p. 1047–1093.

Fisk, H. N., and McFarlan, Jr., 1955, Late Quaternary deltaic deposits of the Mississippi River: Geol. Soc. America Special Paper 62, p. 279–302.

Folk, R. L., 1961, Petrology of sedimentary rocks: 154 p. Austin, Texas, Hemphill's Bookstore.

Furon, R., 1963, Geology of Africa: 377 p., Edinburgh, Oliver & Boyd.

Heezen, B. C., 1956, Corrientes de Turbidez del Rio Magdalena: Bol. Soc. Geog. Colombia, v. 51 and 52, p. 135–143.

——— 1963, Turbidity currents: Chapter 27, p. 742–775 *in* Hill, N. M., *ed.*, The sea: 963 p., New York, John Wiley.

——— Ewing, M., and Menzies, R. J., 1955, The influence of submarine turbidity currents on abyssal productivity: Oikos, v. 6, p. 170–182.

Hull, E., 1900, The suboceanic river valleys of the West African Continent: Trans. Victoria Inst., v. XXXI.

——— 1912, Monograph on the sub-oceanic physiography of the North Atlantic Ocean: 41 p., London, Edward Stanford.

King, Lester, 1962, Morphology of the earth: 699 p., Edinburgh, Oliver & Boyd.

Kolbe, R. W., 1957, Fresh-water diatoms from Atlantic deep-sea sediments: Science, v. 126, no. 3282, p. 1053–1056.

Locher, F. W., 1954, Ein Beitrag zum Problem der Tiefseesande im westlichen Teil des äquatorialen Atlantiks: Heidelberger Beiträge zu Mineralogie u. Petrographie, v. 4, p. 135–150.

Mouta, F., and O'Donnell, H., 1933, Carte Geologique de l'Angola: Ministerio dans Colonis Republica Portugesa, 87 p.

NEDECO, 1959, River Studies, Niger and Beume: 100 p., Amsterdam, North Holland Publishing Company.

Osterhoudt, W. J., 1946, The seismograph discovery of an ancient Mississippi River channel (abs.): Geophys., v. 9, p. 417.

Schott, G., and Schulz, B., 1914, Die Forschugsreise S.M.S. MÖWE im Jahre 1911: Archiv der Deutschen Seewarte, v. XXVII, no. 1, 104 p., Hamburg.

Shepard, F. P., 1948, Submarine geology: 348 p., New York, Harper Bros.

——— 1951, Mass movements in submarine canyon heads: Trans. Am. Geophys. Union, v. 32, p. 405–418.

Spencer, J. W., 1903, Submarine valleys off the American coast and in the North Atlantic: Geol. Soc. America Bull., v. 14, p. 207–226.

Spronck, R., 1941, Mesures Hydrographiques Effectuees dans la Region Divagante du Bief Maritime du Fleure Congo: Mem. Inst. Roy. Colonial Belge, 156 p.

Trask, P. D., 1955, Movement of sand around Southern California promontories: Beach Erosion Board Tech. Memo 76, 66 p.

Veatch, A. C., and Smith, P. A., 1939, Atlantic submarine valleys of the United States and the Congo submarine valley: Geol. Soc. America Special Paper 7, 101 p.

The American Association of Petroleum Geologists Bulletin
V. 57, No. 9 (September 1973), P. 1679-1691, 10 Figs.

Congo Submarine Canyon and Fan Valley[1]

F. P. SHEPARD[2] and K. O. EMERY[3]
La Jolla, California 92037, and Woods Hole, Massachusetts 02543

Abstract Seventeen transverse profiles of the inner 460 km of the Congo Canyon and the Congo Fan Valley were made during a 4-day study in June 1972. These profiles show that the canyon is V-shaped with side slopes 400–1,400 m high between the coast and a point 240 km seaward, where the axial depth is about 2,700 m. Farther seaward, the continuation as a fan valley narrows and is bordered by levees a few tens of meters high, and distributaries are found. The latter continue beyond the 4,600-m limits of this study for at least an additional 320 km to depths of 4,900 m. Seismic profiles show that the canyon has been cut through a belt of diapirs—probably derived from evaporites of Early Cretaceous age, and still rising through the several kilometers of subsequent sediments. Highest side slopes and a major bend in the canyon are present within the belt of diapirs or in the thick sediments that are dammed by the belt. The absence of a broad delta at the mouth of the Congo River, the presence of a possible temporary fill at the head, a steep axial slope near the head, a submerged fan bordering the seaward side of the diapir belt, and levees at depth support the concept of origin of the canyon–fan-valley system largely through erosion and deposition by turbidity currents. Tidal-current scour probably helps to limit the amount of fill in the canyon head.

INTRODUCTION

Congo Submarine Canyon is unique among the valleys of the ocean floor in that it penetrates as a deep estuary into the continent at the mouth of an enormous river; in fact, the river has the largest water discharge in Africa, although much of its sediment is deposited in an interior delta above Kinshasa (Veatch and Smith, 1939). Submarine canyons are present off other large rivers, including the Columbia, the Eel of northern California, the Hudson, the São Francisco of Brazil, the Tagus, the Rhone, and the Var of southern France. Other deep trough-shaped valleys are off the deltas of the Ganges, the Indus, and the Mississippi Rivers. Trincomalee Canyon in Ceylon, Tokyo Canyon, and a group of canyons along the west coast of Corsica head into relatively deep estuaries of small rivers. However, only the Congo Canyon penetrates the land at the mouth of a large river. The significance of the deep mouth of the Congo is that it exists despite the enormous amount of sediment that is introduced by the river.

The Congo Canyon was first reported by Buchanan (1887) in a survey prior to the laying of a submarine telegraph cable. Considerable interest was aroused because this and other cables laid across the canyon had to undergo frequent repairs (Milne, 1897). Heezen *et al.* (1964) discussed the history of these cable breaks and made a topographic survey of the canyon and its seaward continuation as a fan valley,[4] and obtained cores having characteristics that suggest the movement of turbidity currents along the outer fan valley.

In June 1972, an opportunity for further study of Congo Canyon came during Leg VII of the Eastern Atlantic Continental Margin Program, International Decade of Ocean Exploration (IDOE), when R/V *Atlantis II* was operating in the area. Four days were devoted to seismic profiling of the canyon and fan valley between axial depths of 400 and 4,500 m. In addition, echo profiles were run along the margin of the valley, and a Savonius current meter obtained a 4-day record in the canyon head.

TOPOGRAPHY

Many submarine canyons can be traced seaward to where they merge into leveed valleys

[1] Manuscript received, January 25, 1973; accepted, February 25, 1973. Contribution No. 3097 of the Woods Hole Oceanographic Institution.

[2] Geological Research Division, Scripps Institution of Oceanography, University of California.

[3] Woods Hole Oceanographic Institution.

This study was made possible by funding for the Eastern Atlantic Continental Margin, a program of the International Decade of Ocean Exploration, through National Science Foundation Grant No. 28193. Appreciation is expressed to the many shipboard participants for their help, especially D. E. Koelsch and E. M. Young (chief technicians) and R. C. Groman and D. E. Woods (computer center). We also acknowledge the support of National Science Foundation contract GA-19492 and Office of Naval Research contract Nonr-2216(23). Processing of the soundings and current meter data was conducted by Neil Marshall, Gary Sullivan, and Pat McLoughlin.

[4] Heezen *et al.* referred to the entire valley as a canyon although their profiles and contours show that the floor of the outer valley is only a few meters below its surroundings.

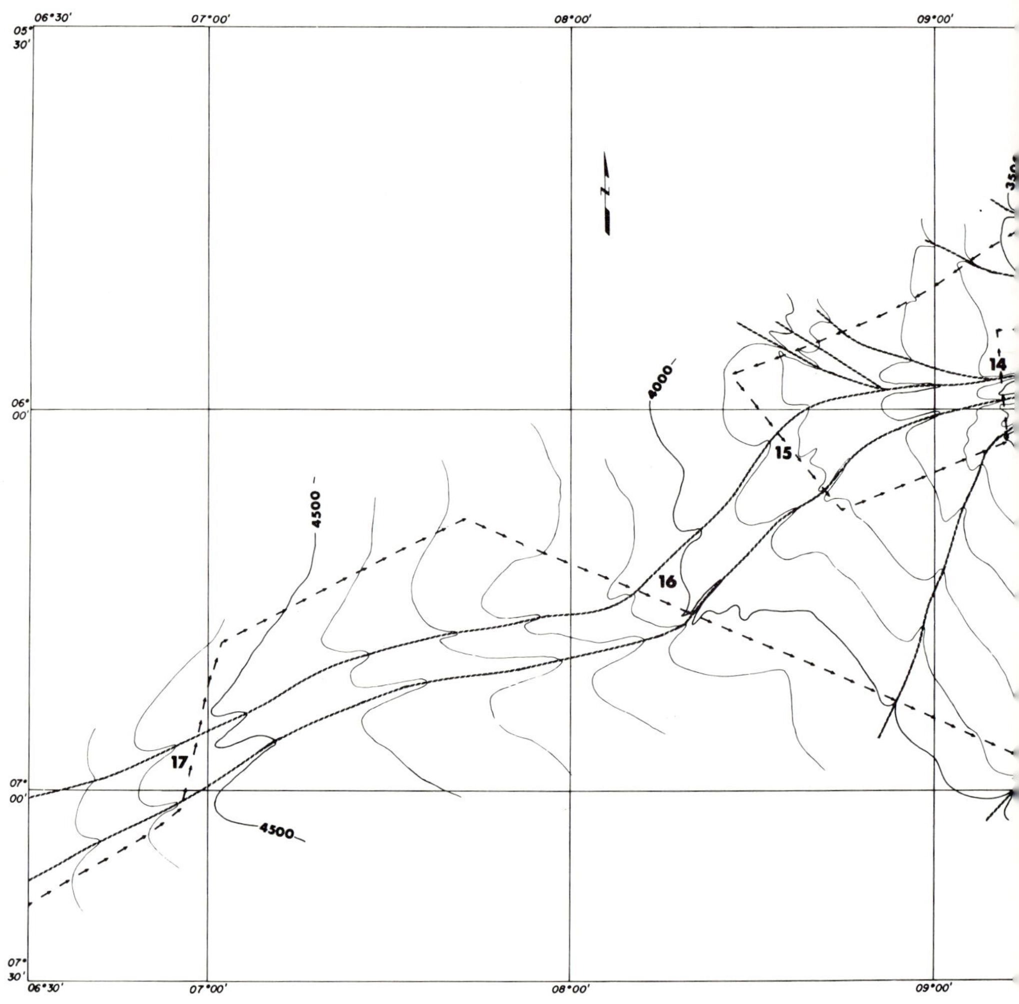

FIG. 1—General relief of Congo Canyon and Fan Valley based on 1972 soundings from *Atlantis II*. Approximate ax
4, and 7 are indi

that wander across the sediment fans at the base of the continental slope. This change in character is well illustrated by the valley off the Congo River, where about 200 km of the V-shaped, deeply incised canyon merges seaward into 500 km or more of leveed valley having distributaries and low side slopes (Fig. 1).

According to available soundings from a 1933 Portuguese chart and other surveys, probably all made by wire rather than by echo sounding, the canyon head has a V-shaped gorge with steep sides and a twisting axis that can be traced at least 25 km up the estuary (Fig. 2). Our operation did not extend landward of Ponto do Padrão, but our profile (Fig.

2 inset) showed a flat floor at 370 m with a slight depression or channel reaching 399 m. This is definitely shoaler than the 500–550-m soundings indicated by the 1933 Portuguese chart and by all other sources. However, only about 2 km west of this point, depths were 580 m, almost as deep as those of the Portuguese chart. It would appear that the head of the canyon may have had an extensive recent fill. Beyond a section of the axis having a seaward slope of about 10 percent, the V shape returns and characterizes all but one profile seaward of the steep area (Fig. 3).

Most of our profiles were made nearly at right angles to the axial trend of the outer val-

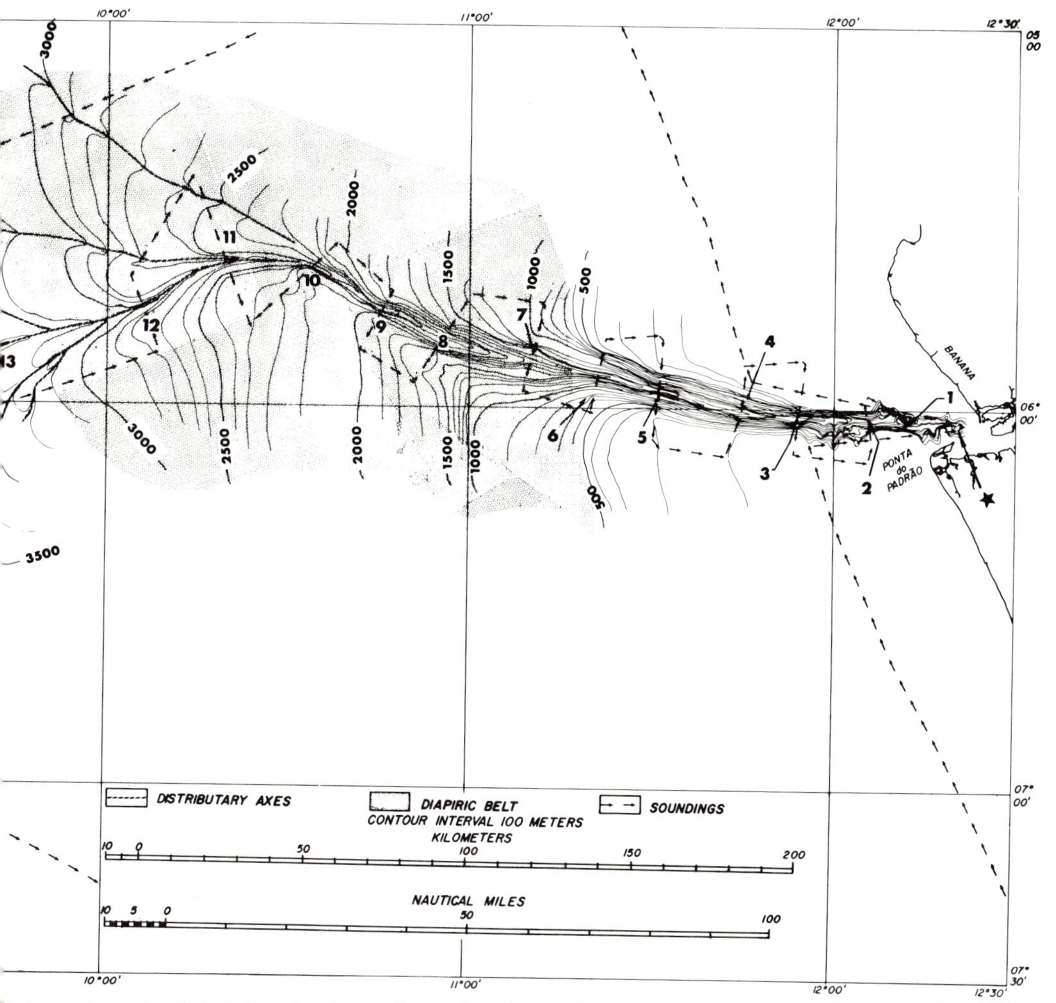

...butaries shown by dashed lines. Locations of soundings (arrows), current-meter station (star), and profiles of Figures 3, ...iric belt is shaded.

ley (Fig. 1), but a few are at a more acute angle, as can be seen from the distinct difference in depth at the top of each side slope. These transverse profiles differ in each of three main structural regions that are crossed by the canyon–fan valley (Figs. 1, 3). Profiles 1–5, in sediment fill landward of a belt of diapirs, have progressively greater axial depths (776–1,582 m) and higher side slopes (400–1,400 m); most of the side slopes also are relatively smooth. Profiles 6–12, within the belt of diapirs, reveal progressive deepening of the axis (1,787–3,045 m) and decrease in height of side slopes (1,300–300 m). All of the side slopes are irregular, partly because of the presence of tributaries (profile 7) or possible distributaries (profiles 11, 12), but probably mostly because of slump or disturbances caused by local diapiric uplifts. Profiles 13–17, seaward of the diapiric belt, exhibit a low axial gradient (3,380–4,587 in 345 km) and low side slopes that diminish seaward (260–40 m) and consist of natural levees that border the fan valley.

The longitudinal profile (Fig. 4) of the axis of the canyon-fan valley was obtained by connecting the greatest depth along each of our 17 transverse profiles (Fig. 3); it also includes the soundings near the current-meter station, 20 km landward of profile 1. The gradient decreases in steepness seaward from perhaps 10

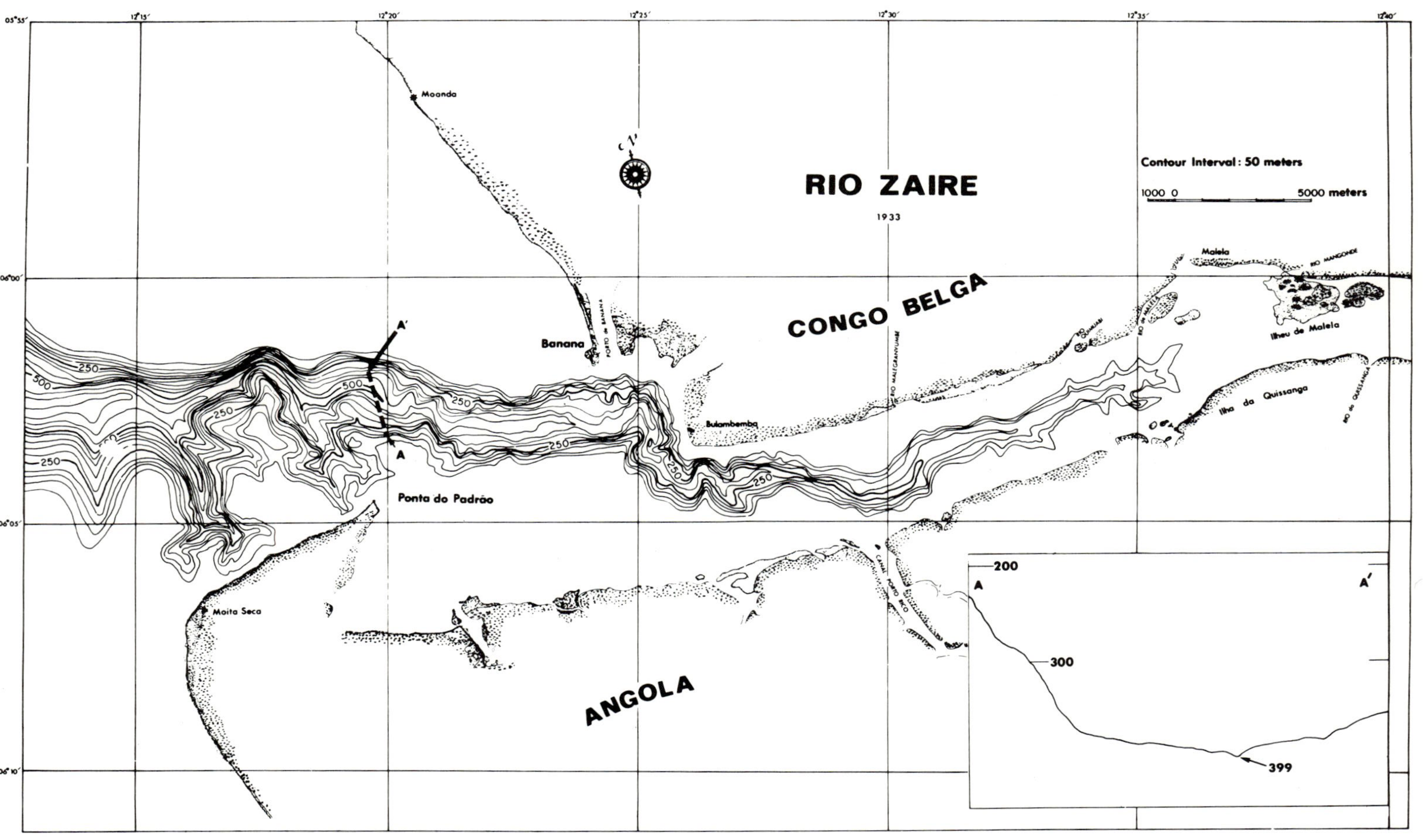

Fig. 2—Contours of head of Congo Canyon based principally on 1933 Portuguese chart. *A-A'* indicates approximate location of 1972 profile with its indications of fill in canyon head.

percent near the current-meter station to an average of 0.8 percent out to profile 12. This is an unusually gentle slope for a submarine canyon. Seaward of profile 12, the slope is still further reduced to 0.45 percent, more or less typical for the gradient of a fan valley.

The earlier surveys depict definite tributaries entering the main canyon along its south side. We were able to run a few profiles along both the north and south sides of the canyon head (Fig. 5). These show that tributaries exist on both sides where the canyon crosses the continental shelf. Evidence for tributaries along the outer part of the canyon is incomplete, but a few short lines suggest that tributaries are rare, if present at all, where the canyon crosses the continental slope.

Seaward of profile 11, the valley has at least eight distributaries (Fig. 1). From the profiles, we can say tentatively that the change from a canyon to a typical fan valley (Shepard, 1965) occurs near profiles 10 and 11, and certainly by profile 12. There is some gradation in this change, because profiles 8–10 show possible levees on their south side. The change between canyon and fan valley is clearly exhibited in the profiles made by Heezen *et al.* (1964). Their profiles extend farther seaward than ours (to long. 5°E), and so far as one can judge from the contours, their outermost crossing shows only a few meters of side slope, approximately equal to the height of the levees.

Seismic Profiles

Our principal addition to previous knowledge relative to the valley off the Congo comes from our seismic profiles (Figs. 6, 7). These profiles support previous studies (Baumgartner and van Andel, 1971; Leyden *et al.*, 1972; Emery, 1972) in showing a diapiric belt that extends along the continental slope off Angola, Zaire, Congo, Gabon, and probably Cameroun. The diapirs are believed to consist of salt of Early Cretaceous age, the same as for the diapirs beneath the adjacent land. Upward movement of the salt at the southern end of the belt continues at present, as shown by the many unconformities in overlying sediments, the protrusion of the diapirs well above their surroundings, and the absence of thick horizontally bedded sediments in ponds between the diapirs. Diapirs off the Congo River are similar except that few of them project above their immediate surroundings; this is interpreted as meaning that sediment is contributed by the Congo and deposited atop the diapir belt at a rate faster

than the rate of diapir protrusion. Correspondingly, a scarp (similar to the Sigsbee escarpment in the Gulf of Mexico) borders the diapir belt except off the Congo River, where its place is taken by an apron of sediments. Moreover, the total thickness of sediments above acoustic basement (probably oceanic basement) is much greater off the Congo River than farther south. Reflection times exceed 4 seconds, meaning that the Congo Canyon is underlain by as much as 6 km of sediments, into the top of which it has been incised.

In interpreting details of the seismic profiles, one must realize that horizontal layers truncated by a canyon appear as a downbend next to the canyon side slope. The profiles clearly indicate that the canyon is cut into sedimentary formations, which are stratified except at the tops of the diapiric intrusions, as in profiles 7 and 8 and possibly profile 5 of Figure 7. Profile 8 has a U shape, in contrast to the V shapes of the other profiles. Apparently, erosion at this point started in the softer sediments and the valley became incised into the top of the diapir near one side. Clear evidence of slumping along the canyon side slopes is shown in the seismic profiles (Fig. 7), notably in profiles 3–7 and 9–12. The terraces for the most part seem to be slump blocks. Only in profile 10 does the layer forming the block appear to continue beneath the adjacent intercanyon surface. In profile 8, where the canyon is cut into a diapir, there are no terraces or other indications of slumping.

The seismic profiles contain many indications of folding in the strata. Some of this folding is related to the diapiric intrusions, as in profile 11, but other folds could be the result of deposition in old valleys and the compaction of the fine-grained sediments. Also, slumping no doubt has contributed to deforming the beds, particularly along the sides of the canyon.

The seismic profiles that cross the fan valley (Fig. 8) indicate that, although erosion may have played a part, deposition on the flanks has been the predominant process of valley formation. Lenses are apparent in profiles 13 and 15, where the fan valley was built by deposition of the levees along the sides.

Canyon-Floor Currents

Emplacement of an Isaacs–Schick-type Savonius rotor current meter (Isaacs *et al.*, 1966) in the head of Congo Canyon provided a record of the currents on the floor for approximately 4 days. The current meter was dropped

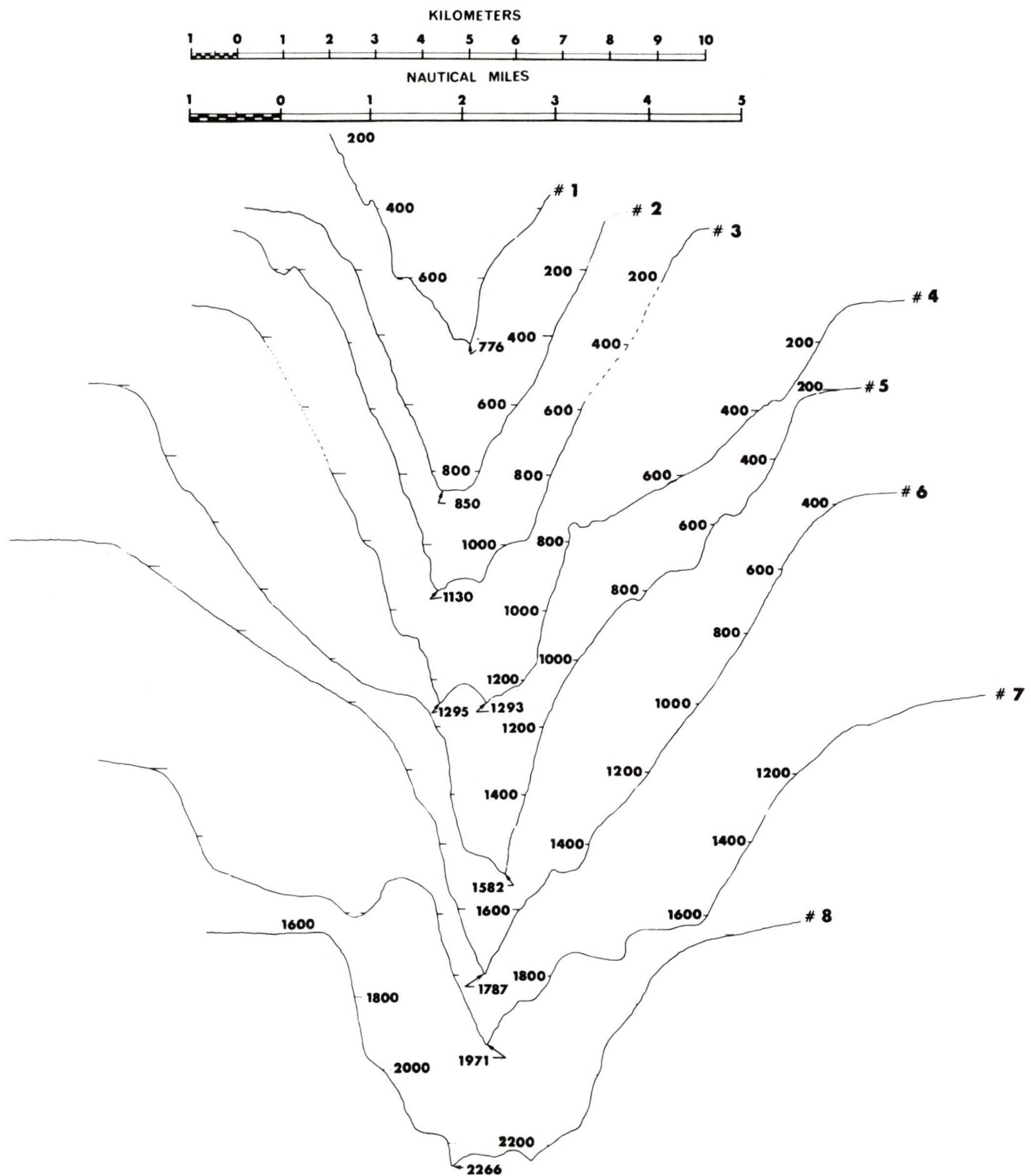

Fig. 3—Transverse profiles of Congo Canyon and Fan Valley.

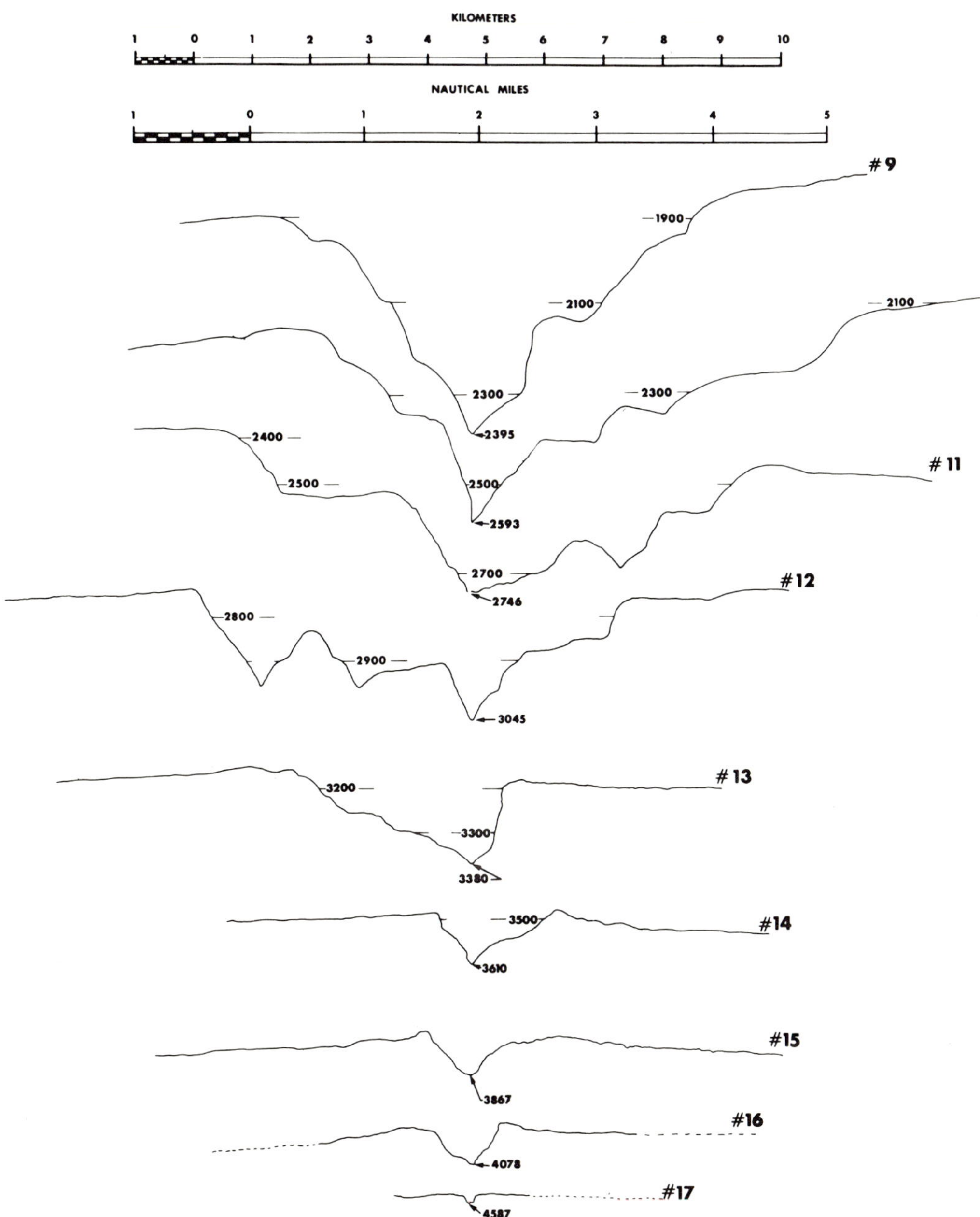

Profiles oriented to face downcanyon. For locations, see Figure 1.

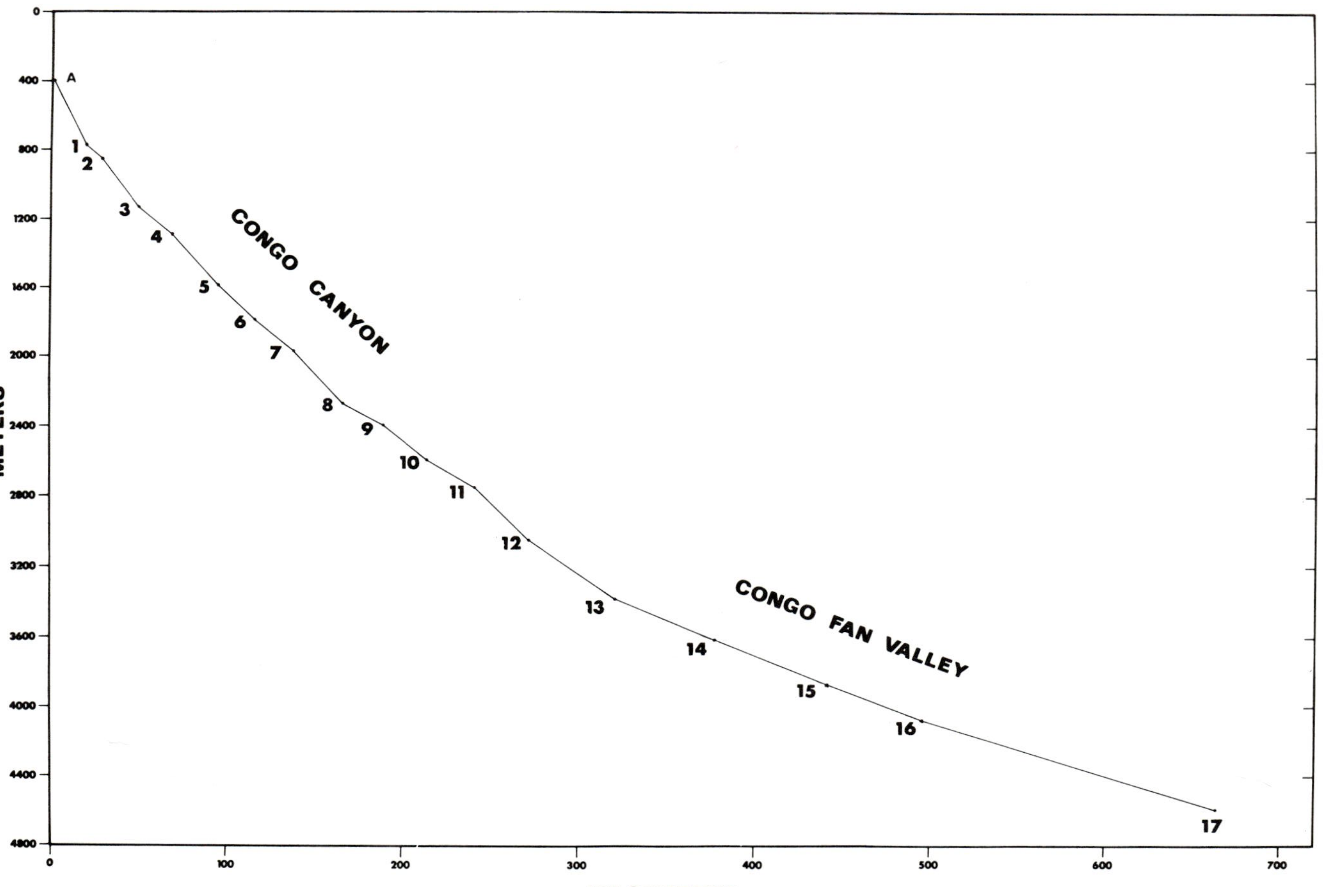

Fig. 4—Longitudinal profile down axis of Congo Canyon and Fan Valley. Note decrease in gradient at approximately point 12 where canyon changes into fan valley.

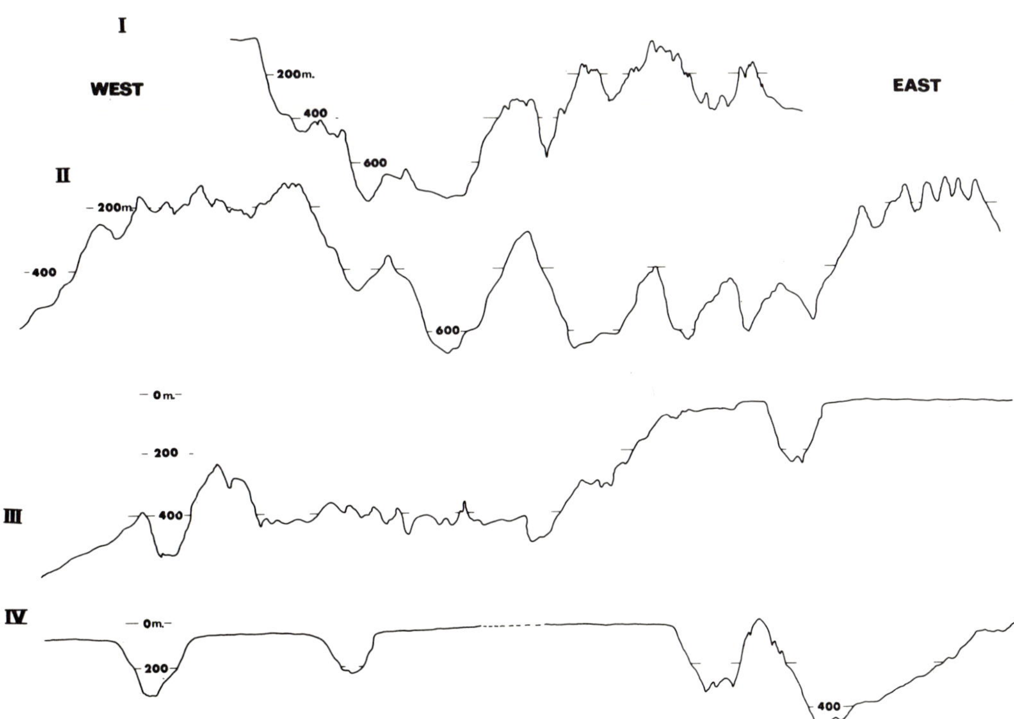

FIG. 5—Longitudinal profiles at head of Congo Canyon. Profiles I and II are on north side and III and IV on south; II and III are not located accurately.

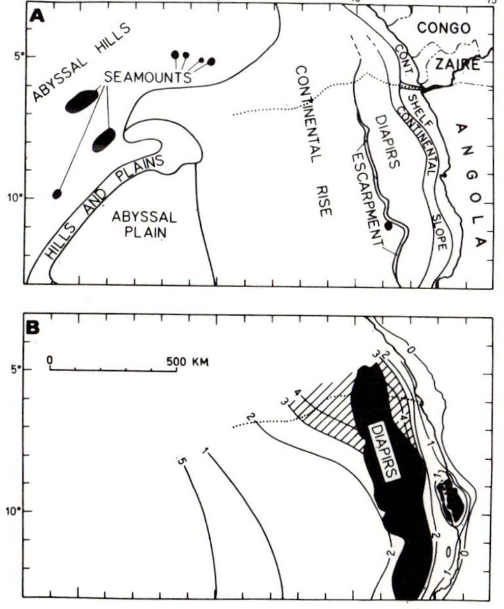

FIG. 6—**A,** Physiographic units of region showing Congo Canyon and Fan Valley (dots) crossing continental shelf, continental slope, diapiric belt, and continental rise. **B,** Thickness of all sediments above basement (in seconds of reflection time). Black denotes areas of salt diapirs of early mid-Cretaceous (Albian-Aptian) age. Diagonal hatching accents positions occupied by sediments thicker than 3 seconds reflection time.

F. P. Shepard and K. O. Emery

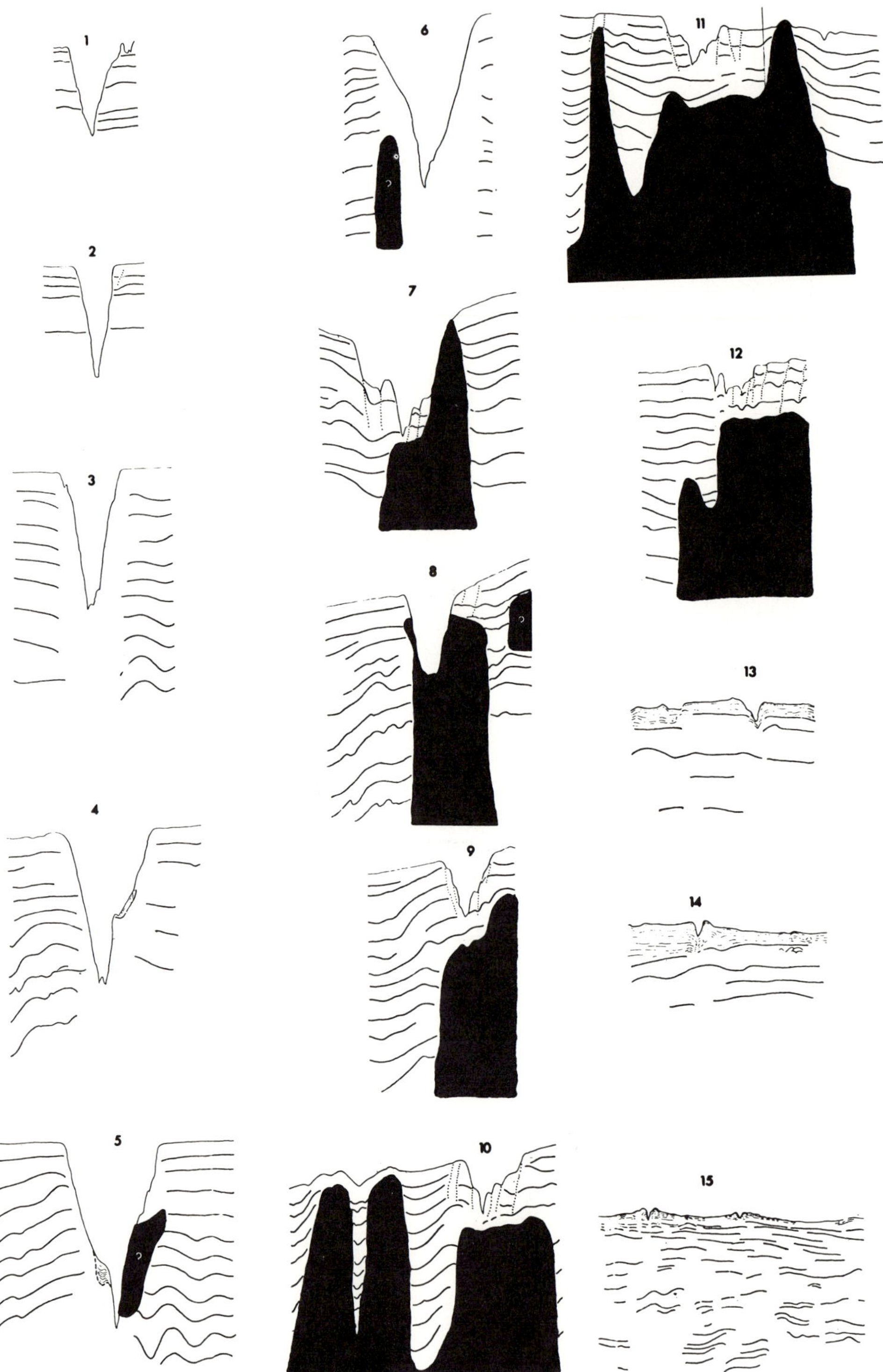

FIG. 7—Diagram from seismic profiles across axis of Congo Canyon and Fan Valley along same lines as in Figure 3. See also Figure 8. Solid zones represent diapirs.

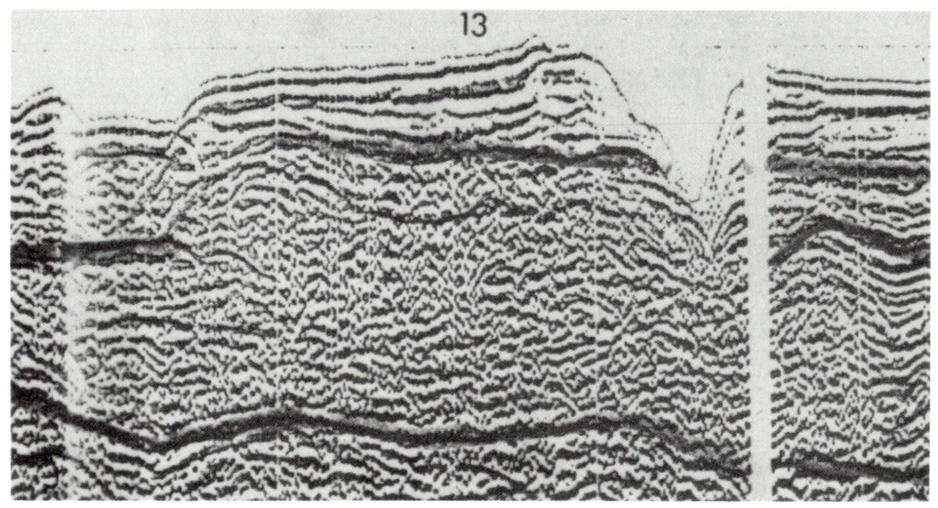

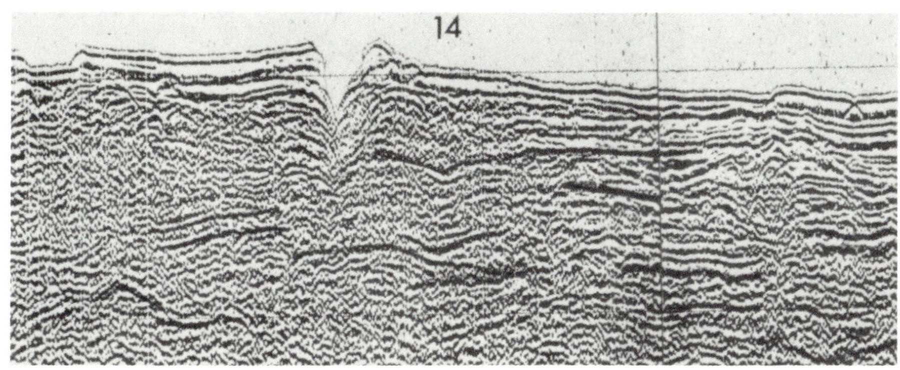

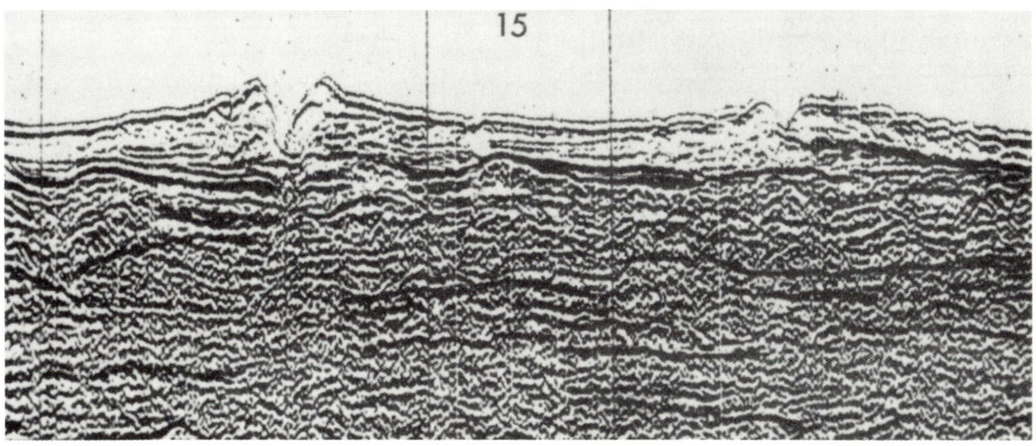

FIG. 8—Photographs of parts of seismic profiles from sections 13–15 showing contrast between recent sediments of levees and underlying beds.

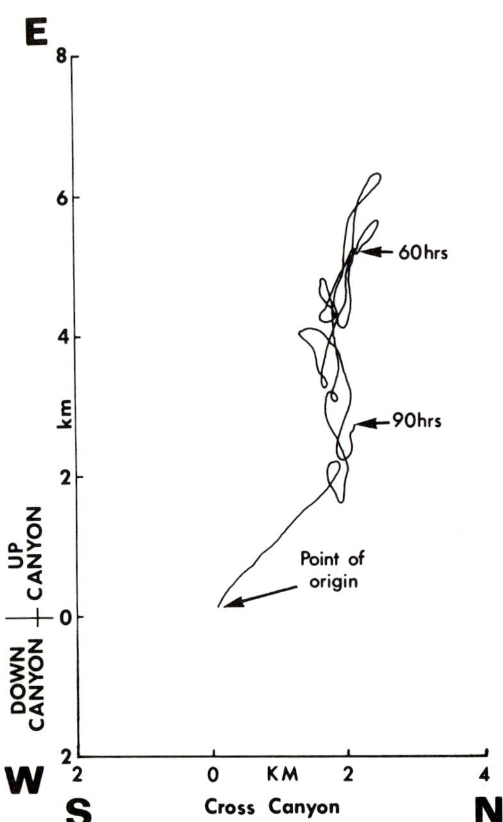

Fig. 9—Trajectory of movement of water particles along axis of Congo Canyon, showing net movement during 4 days of records.

into a 399-m channel that was bordered by a fairly flat floor of about 370 m in a section of the canyon where the axis trends nearly east-west (Fig. 2 inset). Principal results of the analysis of the currents in the record are indicated in Figures 9 and 10. The initial current was southeast (Fig. 9), which was across the general trend and at 180° from the predominantly northwest surface current. However, after the first 5 hours, the current was dominantly upcanyon and downcanyon with almost no net change during the last 84 hours (Fig. 10). Typical canyon currents illustrated by similar measurements at Scripps Institution (Shepard and Marshall, 1973) are downcanyon; apparently, the Congo does not follow this general rule and may well have predominant upcanyon motion. This is not surprising in view of the great seaward flow of water at the surface, presumably overlying a salt wedge (Buchanan, 1887), like that at the mouth of the Mississippi and of other large rivers that carry seawater upvalley.

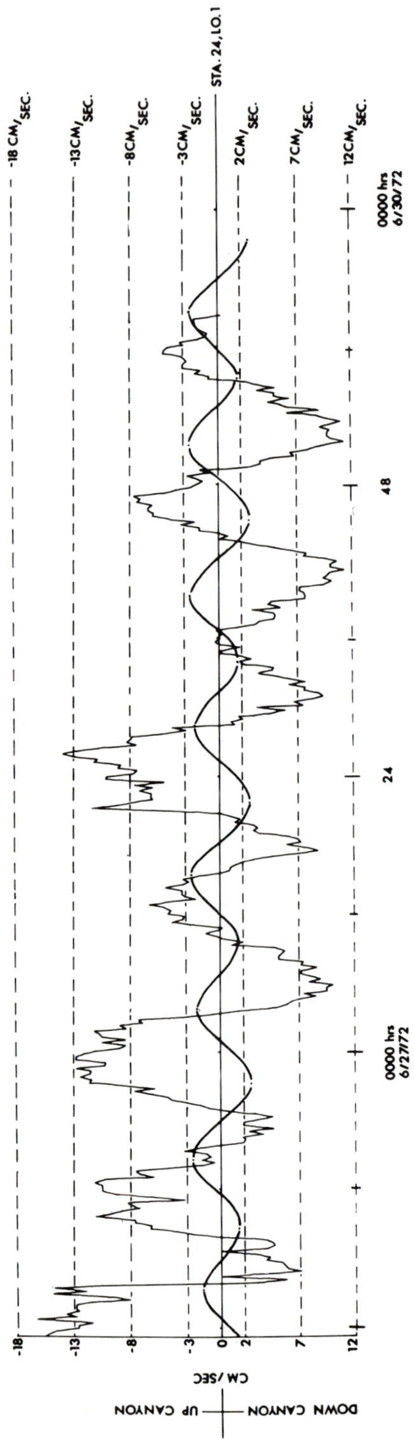

Fig. 10—Showing vectors of upcanyon and downcanyon currents along floor of Congo Canyon at depth of about 400 m. Note general upcanyon currents during rising tides and downcanyon during ebbing tides. Tides are for nearby Ponto do Padrao.

The most interesting feature of the current-meter record is that it seems to follow closely the tidal pattern (Fig. 10). Such a relation was rare in the currents measured off California and Baja California. The exact times of the tides at the mouth of the Congo are known only approximately, but apparently upcanyon flow accompanies the rising tide, and downcanyon flow the ebbing tide.

DISCUSSION

Our recent work on the Congo Canyon provides some new information but fails to explain completely the enigma of a canyon at the mouth of a large river where one should expect a delta. We know from previous information of cable breaking that active slides or strong turbidity currents occur along this canyon, mostly during the high stages of the Congo River (Heezen *et al.,* 1964). Our findings at the canyon head of a flat floor and of much shoaler depths than in 1933 are indicative of filling, or possibly that the previous soundings were made with a lead line dropped from a rapidly drifting boat. Our results here were obtained during the period of relatively low runoff. We need to make further soundings toward the end of a high runoff period to look for seasonal flushing of sediment accumulated in the canyon head.

The canyon appears to have been cut through a belt of diapiric salt masses that has risen as the canyon was eroded. The canyon cuts through only a few of the diapirs, but has cut deeply into the bedded sediments between the diapirs and into a wide thick prism of sediments that have been dammed by the belt. The highest side slopes of 1,400 m occur near the landward side of the diapiric belt. These side slopes probably are locally as steep as 45° and contain many slump blocks near their bases, testifying to instability of the sedimentary

strata. If erosion had been restricted to a remote period, one would expect the slumping to have reduced them to gentle slopes and the canyon floor would have been flattened by the fill. Therefore, it seems likely that the canyon headward of profile 12 is actively eroding at present.

REFERENCES CITED

Baumgartner, T. R., and Tj. H. van Andel, 1971, Diapirs of the continental margin of Angola, Africa: Geol. Soc. America Bull., v. 82, p. 793–802.

Buchanan, J. Y., 1887, On the land slopes separating continents and ocean basins, especially those of the west coast of Africa: Scottish Geog. Mag., v. 3, p. 217–228.

Emery, K. O., 1972, Eastern Atlantic Continental Margin Program of the International Decade of Ocean Exploration (GX-28193) report for period March 1971 to July 1972: Woods Hole Oceanog. Inst. 72-53, 13 p.

Heezen, B. C., R. J. Menzies, E. D. Schneider, W. M. Ewing, and N. C. L. Granelli, 1964, Congo Submarine Canyon: Am. Assoc. Petroleum Geologists Bull., v. 48, no. 7, p. 1126–1149.

Isaacs, J. D., J. L. Reid, G. B. Schick, and R. A. Schwartzlose, 1966, Near-bottom currents measured in 4-kilometers depth off Baja California coast: Jour. Geophys. Research, v. 71, no. 18, p. 4297–4303.

Leyden, R., G. Bryan, and M. Ewing, 1972, Geophysical reconnaissance on African shelf: 2. Margin sediments from Gulf of Guinea to Walvis Ridge: Am. Assoc. Petroleum Geologists Bull., v. 56, no. 4, p. 682–693.

Milne, J., 1897, Sub-oceanic changes: Geog. Jour., v. 10, no. 2, p. 129–146.

Reyre, D., ed., 1966, Bassins sedimentaires du littoral africain, symp., pt. 2, Littoral Atlantiquê (New Delhi, 1964): Paris, Union Internat. Sci. Geol., Assoc. Services Geol. Africains, 304 p.

Shepard, F. P., 1965, Types of submarine valleys: Am. Assoc. Petroleum Geologists Bull., v. 49, no. 3, p. 304–310.

—— and N. F. Marshall, 1973, Currents along the floors of submarine canyons: Am. Assoc. Petroleum Geologists Bull., v. 57, no. 2, p. 244–264.

Veatch, A. C., and P. A. Smith, 1939, Atlantic submarine valleys of the United States and the Congo Submarine Valley: Geol. Soc. America Spec. Paper 7, 101 p.

The American Association of Petroleum Geologists Bulletin
V. 53, No. 2 (February, 1969), P. 390-420, 30 Figs., 1 Table

Physiography and Sedimentary Processes of La Jolla Submarine Fan and Fan-Valley, California[1]

F. P. SHEPARD,[2] R. F. DILL,[3] and ULRICH VON RAD[4]

La Jolla, California 92037, San Diego, California 92132, and Munich, Germany

Abstract The depositional environments of La Jolla canyon, fan-valley, and fan are well known from closely spaced sounding lines, deep-diving vehicle observations, numerous undisturbed box cores, and continuous reflection profiles. The narrow rock-walled canyon changes seaward at 300 fm (549 m) to a wider valley cut into the compacted clayey sediments of a fan, and bordered by discontinuous leveelike embankments. The fan-valley merges gradually into the relatively flat floor of San Diego trough. Numerous dives into the fan-valley have shown precipitous walls along the outside of the bends of the winding channel. Slumping is taking place actively from these walls and large slump blocks of clay are common on the floor. Small scour depressions around isolated erratics suggest the erosive effect of relatively weak currents in some places but, for the most part, the muddy floor seems to have been little disturbed in recent years. Diagonal tension cracks cut the floor locally.

Box cores show that most of the sediment deposited on the valley floor in the past few thousand years consists of poorly sorted clayey silt, underlain by discontinuous layers of well-sorted fine-grained sand with a few coarse sand grains, gravel, and mud balls. Sand layers occur in 94 percent of the valley axis cores, of which 26 percent are graded; 59 percent have parallel laminations; and 41 percent have current-ripple cross-laminations. Sand layers are less common in the cores from levees and from the small discontinuous terraces along the sides of the fan-valley. Cores from the open fan have less and finer grained sand. In all these environments the sand shows no consistent or systematic grain-size variation with increasing water depths. Some of the coarsest sediments, including gravel and mud balls, are found in sand farthest from shore and at the greatest depths.

The character of the sand and the finding of shallow-water Foraminifera indicate the probability that sand has been carried from the coastal area along the valley axes and spilled over the levees onto the open fan. However, there is little evidence of recent high-velocity, high-density turbidity currents, because, in general, the covering mud layer is distinctly separated from the underlying sand deposits, and therefore does not suggest deposition at the terminus of a turbidity current. Also, the discontinuous character of the sands and series of laminae with heavy mineral concentrations indicate introduction by a traction type of pulsating current, such as has been seen during vehicle dives, and also has been measured in the few available current-meter records. The locally precipitous fan-valley walls and outcrops of gravel, and the sand layers on the levees and open fan, may be the product of stronger currents that moved down the valley during earlier more pluvial periods, when greater quantities of sediment entered the canyon heads. Possible confirmation of this idea comes from the available C-14 dates in plant layers, which suggest that deposition in the past few thousand years may have been considerably slower than that indicated for the Pleistocene. The finer sediments may be largely the result of slow downslope movement of slightly higher density muddy waters coming from the coastal areas.

Continuous reflection profiles have shown that the inner La Jolla fan has only a thin cover of unconsolidated sediments overlying the folded and faulted Miocene-Pliocene rocks. The outer fan and adjacent San Diego trough contain a thick section (more than 1,000 m) of Quaternary sediments with probable buried older channels and possible thick lenses of sand sediments.

INTRODUCTION

Because of the convenient location directly outside Scripps Institution, the La Jolla submarine canyon, fan-valley and fan have been investigated more than any other comparable submarine environment in the world. The topography is known from closely spaced, well-located sounding lines (Shepard and Buffington, 1968). At depths too great for scuba dives, the general appearance of the canyon and fan-valley was observed from deep-diving vehicles. Sampling of the area has been very extensive and has led to several reports (summarized in Shepard and Einsele, 1962). Prior to 1963, coring along the canyon and fan-valley was rather unsatisfactory because of the

[1] Manuscript received, December 4, 1967; accepted, February 23, 1968.

[2] University of California, Scripps Institution of Oceanography.

[3] Naval Undersea Research Center.

[4] Institut für Geologie, Technische Hochschule München (Munich), Germany.

The writers acknowledge the help of the crews of various Scripps and Navy Electronics Laboratory ships on which the work was conducted. Also, numerous visiting scientists were helpful in the field work. Andre Rosfelder designed some of the instruments and accompanied us on several trips. Hugo Genser provided descriptions, photographs, and X-radiographs of several box cores. Neil F. Marshall was of great help in field work and in laboratory analyses. We are grateful for critical comments by D. G. Moore and M. L. Natland.

Von Rad gratefully acknowledges a NATO postdoctoral research fellowship granted by the Deutscher Akademischer Austauschdienst (Bad Godesberg, Germany) that enabled him to work for two years at Scripps Institution of Oceanography. The writers' work was supported in part by the Office of Naval Research under Contract Nonr-2216(23) and by the National Science Foundation Grant G.P.-3815.

Photos by Dill courtesy of U. S. Naval Undersea Research Center, San Diego, California 92132.

inadequacy of piston and gravity coring in penetrating compacted sands.

After the initiation of box coring (Reineck, 1963a), much better cores (20.3 × 30.5 cm wide, up to 60 cm long) were obtained, with well-preserved top layers and undisturbed sedimentary fabric. Many were oriented by a locked compass. Ninety-five successful box cores have been taken in this general area. All cores have been examined by the writers. Sediment characteristics were studied especially by von Rad; Foraminifera of many cores, by M. L. Natland; the magnetic fabric of 24 box cores, by A. I. Rees and von Rad (Rees *et al.*, in press); and size analyses were made largely by N. F. Marshall. Kelp layers found in a few box cores have been dated by the C-14 method, giving an estimate of the age of the layers and the relative rate of sediment accumulation. Continuous reflection profiles run by Moore (1966) and the writers along the fan and across the fan-valley were used to interpret the thickness of the fan sediments and the subbottom structure.

BATHYMETRY

The bathymetry of the area is discussed in more detail by Shepard and Buffington (1968). The regional chart with sample locations (Fig. 1) has been contoured mostly from their work.

La Jolla canyon and its concordant tributary, Scripps canyon, continue seaward beyond their 150-fm (274-m) juncture to a depth of about 280 fm (512 m), cutting through rock formations of Cretaceous and Eocene ages. Farther seaward, the channel continues as a fan-valley, cutting semiconsolidated Quaternary fan deposits that have filled San Diego trough. Where the rock walls terminate, the seaward slope of the rim and the valley axis decrease and the wall heights are lower. Also, the valley rim shows the first indication of marginal embankments, generally referred to as "natural levees" (Fig. 2). Between axial depths of 300 and 400 fm (549 and 732 m) these levees are virtually continuous along the north side of the fan-valley, but very intermittent on the south side. In this zone, the valley has relatively straight, steep walls about 40 fm (73 m) high. The valley is about 1 km wide but is cut by a narrow winding inner channel bordered by a low terrace (Fig. 3, secs. 6–8).

About 6 mi[5] (11 km) from its beginning, the fan-valley turns somewhat abruptly from its east-west course, and for about 6.5 mi (12 km) continues southeasterly in a relatively straight course. The walls form a distinct V-shaped profile with a relief of about 30 fm (55 m), and internal terraces of extremely variable heights are present, some near the rim and others near the floor (Fig. 3). The levees are better developed on the northwest wall than on the southeast.

In the next 4 mi (7 km), the V-shaped fan-valley continues to trend southwest in a more

[5] Nautical miles (6,080 ft) are used in the text.

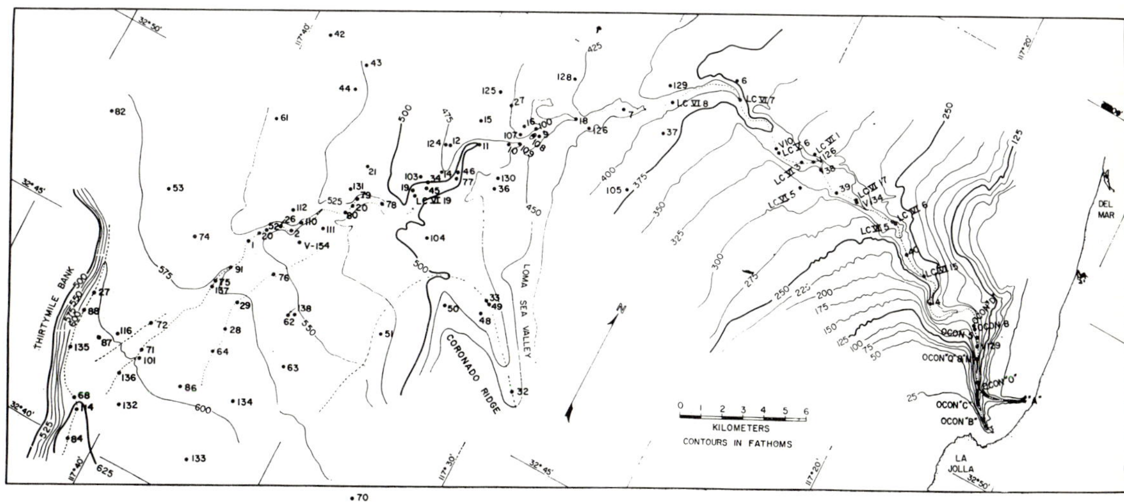

FIG. 1.—Submarine relief and locations of core samples along La Jolla fan and fan-valley. Contours largely from Shepard and Buffington (1968).

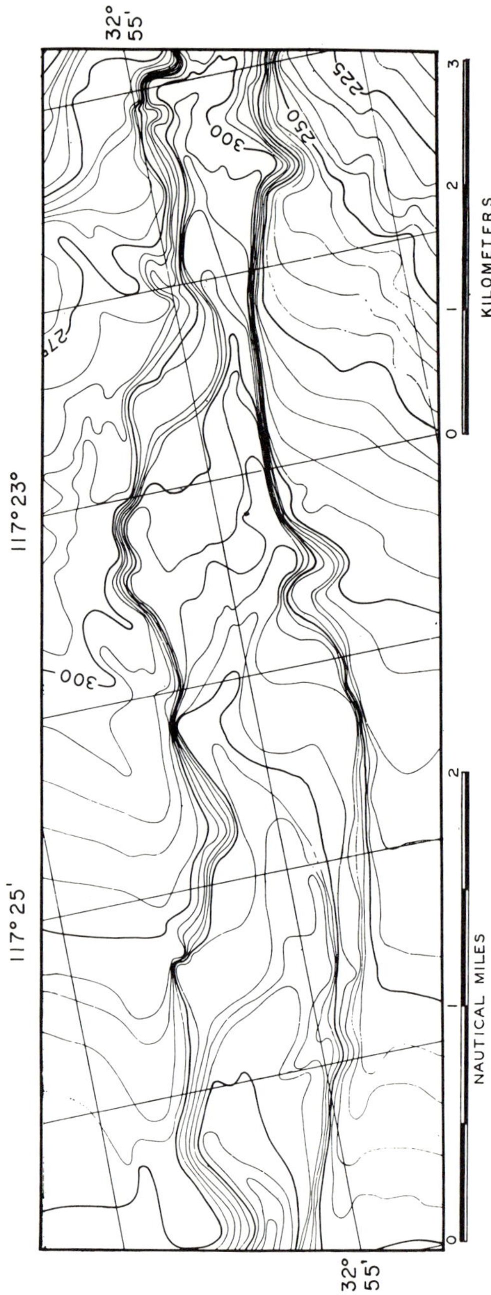

FIG. 2.—Inner La Jolla fan-valley showing winding inner channel and straight valley walls. CI = 5 fm. Original plotting scale 1:24,000. From Shepard and Buffington (1968).

twisting course, the wall relief decreases to about 25 fm (46 m), and the levees become better developed on the southeast side. In the outermost 7 mi (13 km), the fan-valley has a relatively straight course but gradually turns south (Fig. 4). Wall heights decrease until the valley merges into the floor of San Diego trough at 610 fm (1,116 m). In this section, continuous low levees are present on both sides. Small inactive distributary channels on the fan are essentially parallel with the main fan-valley. They are not at present connected with the main valley, and they have shallower depths. Another valley with low relief enters San Diego trough as a continuation of Loma sea valley, which borders the east side of Coronado ridge (Emery *et al.*, 1952, Fig. 1).

At the base of the eastern escarpment of Thirtymile Bank, there is a north-south-trending valley, referred to as a "rift valley" because of its location along a fault scarp. According to Moore (1966), this valley was formed by turbidity currents coming down from Newport and other submarine canyons north of San Diego. However, Dill's observations during *Deepstar-4000* dives in the Newport valley system suggest that the latter is not at present an active distributary system, even though it heads in shallow water.

Direct Observations from Deep-Diving Vehicles

A total of 15 dives (mostly with Dill as observer) have been made into La Jolla fan-valley with the research submersibles *Trieste I* and *II* and *Deepstar-4000*. These have made it possible to gather detailed information on the topography and sediment cover of the valley floors and walls, not otherwise obtainable.

Perhaps the most surprising feature of the fan-valley observed in the deep dives was the extremely steep cliffs and vertical walls (as much as 50 m high) on the outside of some of the bends (Buffington, 1964; Dill, 1961; Moore, 1965). These walls have outcrops of horizontally bedded, semiconsolidated Quaternary sediment. Outcrops of pre-Pleistocene rocks are restricted to depths shallower than 300 fm (549 m), where they are overlain by unconsolidated or semiconsolidated fan deposits (Fig. 5).

Isolated cobbles and boulders (up to 1 m in

FIG. 3.—Profiles taken from fathograms of transverse crossings of La Jolla fan-valley. Includes 5 profiles from outer La Jolla canyon. Indicates character of terraces and levees along fan-valley. From Shepard and Buffington (1968).

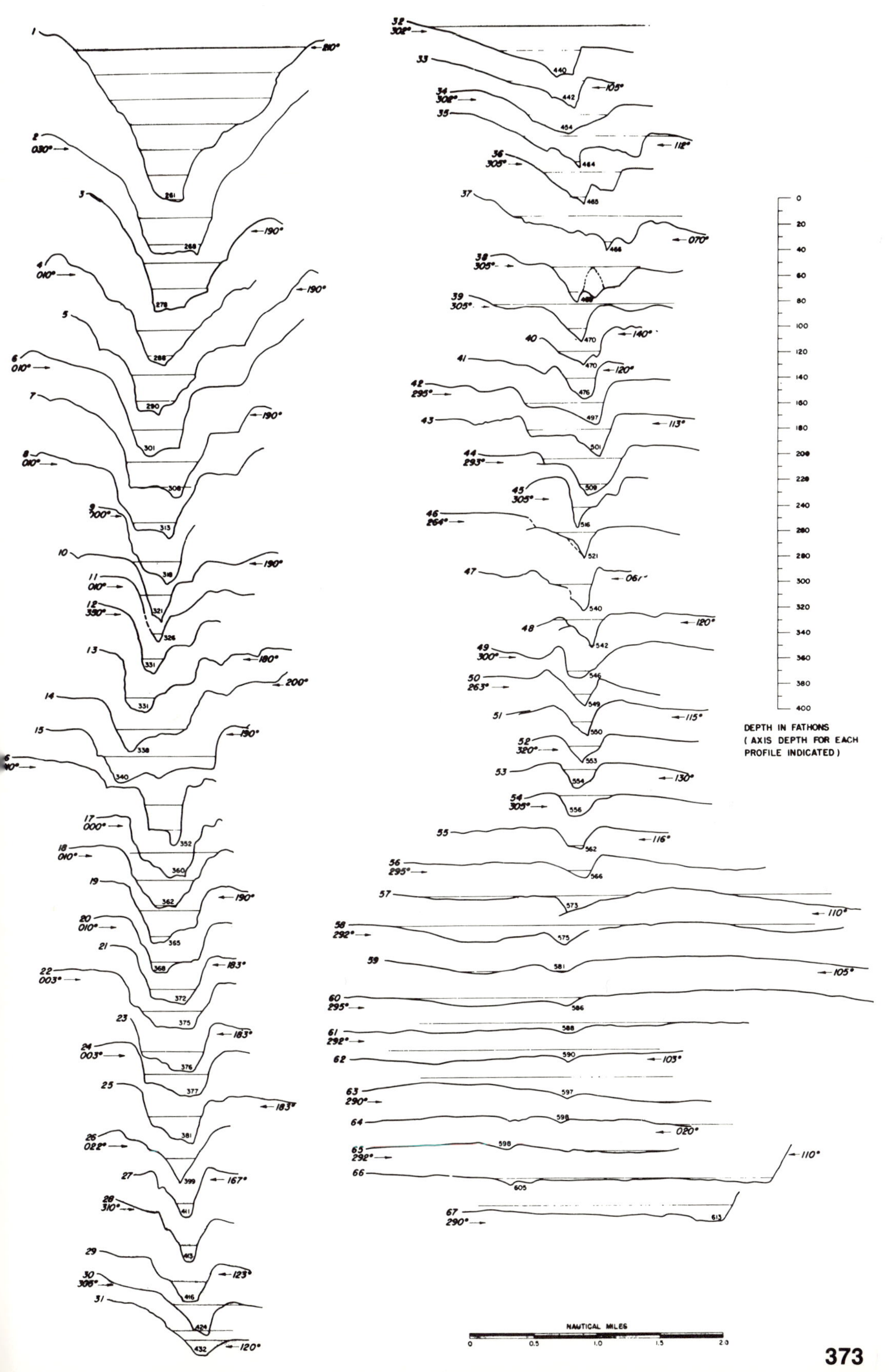

DEPTH IN FATHOMS
(AXIS DEPTH FOR EACH
PROFILE INDICATED)
NAUTICAL MILES

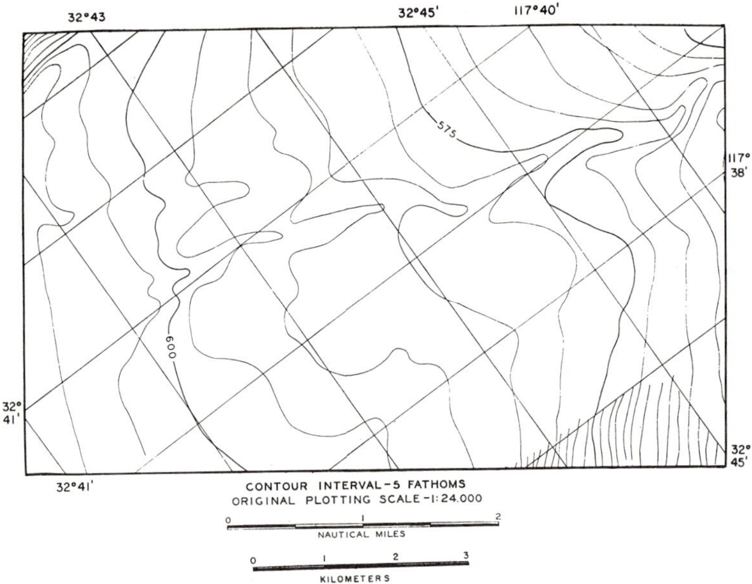

Fig. 4.—Outermost La Jolla fan-valley showing how it merges into San Diego trough and indicating two other fan-valleys and the "rift valley" at the base of Thirtymile Bank. From Shepard and Buffington (1968).

diameter) were first found by Dill in a Bathyscaph-*Trieste* dive near the landward end of the fan-valley at a depth of 325 fm (595 m). Subsequent dives in the *Deepstar-4000* showed that cobbles extend for a considerable distance along the floor of the valley (Fig. 6a). They also crop out in a terrace on the north side about 1–2 m above the valley floor at a constriction in the central axial channel at a depth of 320 fm (585 m) (Fig. 6b).

Rubble blocks of semiconsolidated clay, apparently broken off the steep walls, were found at many depths in the fan-valley. In the outer fan-valley, all stages of this process were observed (Fig. 7), from cracking clay on the walls to broken rubble at the base of steep slopes. Even the sides of the relatively low-walled outer fan-valley show exposures of horizontally bedded clay, locally slumping into the axis. Small vertical escarpments (80–150 cm high) can be seen (Fig. 8) where blocks have slumped away. Blocks of comparatively hard clay apparently are sliding gradually downslope toward the floor, and are seen standing 30 cm or more above the general surface (Fig. 9). This mechanism of slumping is important, because it produces concentrations of clay fragments along the valley axis that later can be picked up, transported, and redistributed, either

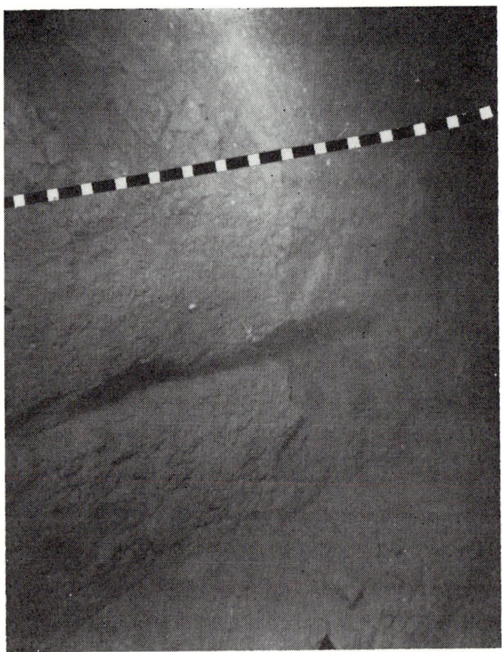

Fig. 5.—Sharp, well-defined contact between Eocene shale and poorly consolidated fan material (dotted line). Channel cuts through both sediment types, proving that downward erosion of canyon occurred following deposition of fan sediments. Photo by Dill.

by strong tractive currents or high-density suspension flows. Also, local spheroidal weathering was observed that produced rounded clay cobbles broken from the sides of the fan-valley (Fig. 10), instead of the more common angular blocks.

Dives also showed that the mud covering the channel floor is generally very thin (1–5 cm). Underlying sand was exposed when the mechanical arm of the submersible pushed aside the thin mud cover. Sand was also seen in

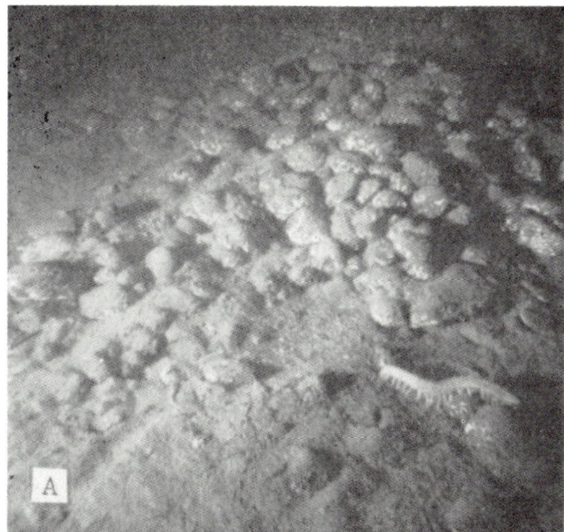

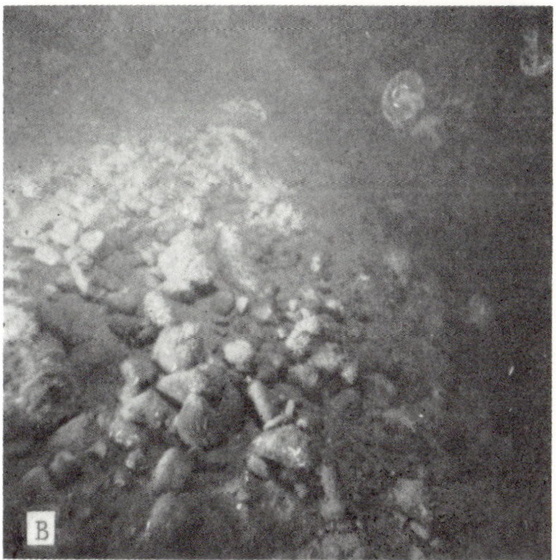

FIG. 6.—**A.** Rounded hard rock cobbles in central channel of fan-valley at depth of 328 fm (600 m). Average diameter is about 15 cm but large boulders were found with diameters up to 1 m. Photo by Dill. **B.** Cobble bed forms an internal wall terrace at base of outermost canyon at about 280 fm (512 m) depth. Wall slopes in area were between 50° and 60°. Photo by Dill.

FIG. 7.—**A.** Large blocks of angular and semi-consolidated fan material from walls of fan-valley form vertical cliffs about 1 m high at about 580 fm (1,061 m) depth. Photo by Dill. **B.** Blocks forming rubble pile at base of vertical walls. Blocks in photo are about 50 cm across. Photo by Dill.

mounds built up where crabs or other benthonic organisms had burrowed into the floor (Fig. 11). In traverses of the internal terraces,

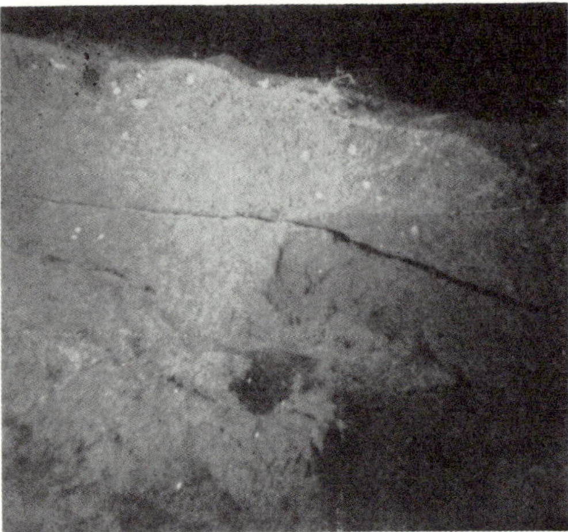

Fig. 8.—Freshly exposed surfaces of semiconsolidated clay forming channel wall at 580 fm (1,061 m) depth. Note cone-shaped mounds formed by burrowing organisms at upper surface of side terrace. Vertical walls were caused by slumping of material into central channel of fan-valley. Photo by Dill.

however, no indications of underlying sand were encountered.

During a dive at 450 fm (823 m), an extraordinary number of small cracks and narrow ridges were seen in the channel fill near the south slope (Shepard and Dill, 1966, Fig. 30). Some of the cracks and ridges appear to extend

Fig. 9.—Relatively angular blocks of semiconsolidated wall sediment forming isolated group of erratics in predominantly sandy fill of fan-valley at 585 fm (1,070 m) depth. This group lacked scour depression. Large central block is about 60 cm across. Photo by Dill.

Fig. 10.—Spheroidal weathering of clay blocks on north wall of La Jolla fan-valley at 445 fm (820 m) depth just above the channel floor. Photo by Dill.

diagonally up-valley, and resemble tension crevasses on the sides of glaciers. Dill has observed similar cracks in the thin blanket of fine sediment that covers the fill in other parts of the valley at depths down to 610 fm (1,116 m). These cracks suggest tension on the floor, due perhaps to a slow downslope creeping motion of the entire channel fill. This is somewhat surprising because the regional slope of the channel axis in this area is slightly less than 1°.

In the outer fan-valley, many deep-sea holothurians (*Scotoplanes* sp.), gastropods (with tracks), and ophiuroids (brittle stars) covered the sea floor (10–20 brittle stars and 5–10 holothurians per m²) (Figs. 11, 12). Locally, these organisms were bowled over by the pressure wave, or water motion, created by the *Deepstar*. The surface layer of flocculated fine-grained mud also was readily set into suspension by the gentle current of the backwash from the craft's propellers. Both these observations suggest that at present the valley floor is not an area of high current velocities. This is indicated also by the lack of scour depressions around corroded and encrusted tin cans that lie on the channel floor (Fig. 13). If strong currents had swept this area recently, the upright can (Fig. 13b) would have been knocked over,

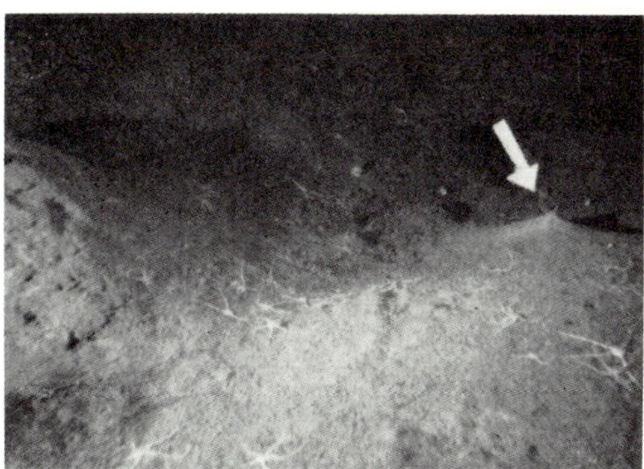

Fig. 11.—Mounds 15 cm high (see arrow), built by burrowing organisms, forming sand cones on fine-grained sediment in central channel at 590 fm (1,079 m) depth. Rounded erratic on left is semiconsolidated clay. Photo by Dill.

as well as the siliceous sponge that is seen in a vertical position (Fig. 13a) which indicates several years of undisturbed growth. However, many of the large clay blocks lying in the sandy fill had scour depressions surrounding them, suggesting the occurrence of local intermittent currents. It is notable that not all of the blocks are surrounded by scour depressions, and that the occurrence of depressions has no relation to depth or location within the valley.

Large sand waves with wave lengths of 15–30 m and heights of up to 1 m were observed in the outer fan-valley in depths of

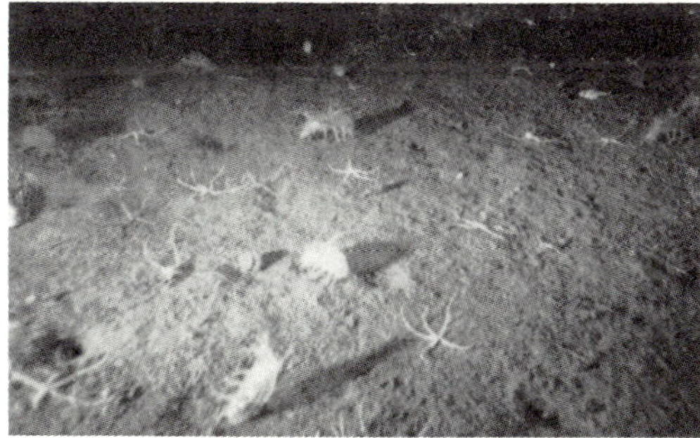

Fig. 12.—Deep-sea holothurians (*Scotoplanes* sp.) at 582 fm (1,065 m) depth, bowled over by pressure wave of slowly moving *Deepstar*. Susceptibility to weak bottom current indicates that present bottom fauna has not developed in high-energy regime. Abundant bottom fauna along channel axis is actively reworking fine-grained flocculent layer that blankets walls and predominantly sandy fill of central channel of fan-valley. Photo by Dill.

Fig. 13.—Encrusted tin cans lying on sediment surface of central channel, indicate that area has not been subjected to strong bottom currents for considerable period of time. **A.** One-quart (0.8-liter) fruit can, at about 580 fm (1,061 m) depth, has siliceous sponge growing from upper rim. **B.** At nearly same depth, highly corroded and encrusted can stands upright surrounded by semiliquid flocculated rust deposit. Note lack of scour depression around base of can. Photo by Dill.

about 600 fm (1,097 m), where the predominantly sand-filled channel merges with the relatively flat floor of San Diego trough. The valley walls in this area are less than 5 fm (9 m) high. The sand waves are associated with large erratic blocks of semiconsolidated clay, which here are generally surrounded by scour depressions. Driftwood and shallow-water plant debris, consisting of waterlogged kelp, eelgrass, *etc.* (Fig. 14), were common in the scour depressions on the upslope side of large clay blocks that had fallen from the steep channel walls.

Fɪɢ. 14.—Gastropods with typical tracks concentrated in vicinity of relatively fresh-appearing clump of shallow-water debris, mostly kelp, 590 fm (1,079 m) depth. Note that partially buried block of semiconsolidated wall sediment is surrounded by shallow scour depression. Photo by Dill.

Surface Sediments

General Remarks

The numerous box cores obtained since 1962 have contributed much improved information, particularly because many of these cores are oriented (Rosfelder and Marshall, 1966), undisturbed, and of sufficient width to study the sedimentary fabric in great detail. The techniques that have been used for the fabric study of the box cores include core slicing, X-radiography of thin sediment slices (Bouma, 1964, 1965), obtaining small-scale lacquer peels, impregnation of small (5 × 5 cm) sediment samples with plastic resin (Reineck, 1963b), preparing "thin sections" (50–80 μ) for fabric study under the petrographic microscope, and measuring the anisotropy of magnetic susceptibility from oriented cylindrical specimens with a torque magnetometer (Rees, 1965).

In discussing the recent sediments of La Jolla fan, it is important to group the box cores according to their natural depositional environments (Table I). More than half the successful box cores (54 out of 95) were taken along the axes of the valleys. An additional 13 were obtained from or near the crest of natural levees, 7 from terraces within the main fan-valley, 16 from the open fan, and 5 from other environments. Table I gives average data for the 95 box cores and 16 gravity cores.[6]

[6] A much more complete table has been deposited as Document No. 9889 with the ADI Auxiliary Publications Project, Photoduplication Service, Library of Congress, Washington, D.C. 20540. A copy may be secured by citing the document number and by remitting $1.25 for photoprints, or $1.25 for 35 mm

Sedimentary Structures Along Canyon and Main Fan-Valley Axis

The sedimentary structures of 40 box and 5 gravity cores, and the longitudinal profile of the canyon and main fan-valley are schematically shown in Figure 15. Of the cores, 42 (93 percent) contain one or several thick layers of fine- to medium-grained terrigenous arkosic sand showing well-preserved sedimentary structures. Of the three mud cores of this group, two consist of still semiconsolidated clayey silt, and probably represent outcrops of the Pleistocene fan deposits. This exposure of older Pleistocene deposits indicates that the recent sediment has been eroded from, or was never deposited at, these locations.

A thin mud layer, rarely more than 5 cm thick, covers the sand layers in most of the cores. It is never greater than 20 cm, except in core 10 with 50 cm of mud and no sand layer. Nine cores of this group contain only sand, three of them are shorter than 10 cm, hence not representative. Predominantly sand cores are much more common than predominantly mud.

The most typical internal sedimentary structures observed in flysch beds and described as the "turbidity sequence" by Bouma (1962) are found in some of the sand layers. The a-b-c from Bouma's idealized sequence (a-e) for a turbidite has the following vertical succession, from top to bottom, for which some subdivisions have been added (see also Table I):

c_4 convolute laminated division
c_3 wavy laminated division
c_2 deformed cross-laminated division
c_1 current-ripple cross-laminated division
b_1 inclined parallel-laminated division
b parallel horizontal-laminated division
a massive graded division

Eighteen sandy cores of this group contain sand layers that are not laminated and show no size grading (a_1, Table I). This "massive ungraded division" is especially typical of the coarse-grained canyon axis cores, but occurs also at intervals all along the fan-valley.

Graded massive bedding was found in 29 percent of the sandy canyon-fan-valley cores,

microfilm. Advance payment is required. Make checks or money orders payable to: Chief, Photoduplication Service, Library of Congress. Samples are listed separately for each of the major depositional environments and given in order of increasing water depth; included are core lengths, the ratio of sand to mud layers, and a few important fabric and size parameters for each core.

Table I. Data from Cores, La Jolla Submarine Fan and Fan-Valley Area, California

	Depositional Environment of Group								
	Canyon Axes	*Inner Fan-Valley*	*Middle Fan-Valley*	*Outer Fan-Valley*	*Total Canyon Fan-Valley Axes*	*Other Valley Axes*	*Terraces*	*Natural Levees*	*Open Fan*
Depth ranges of groups	8–287 fm 15–525 m	310–400 fm 567–732 m	416–545 fm 761–997 m	546–607 fm 999–1,111 m	8–607 fm 15–1,111 m	454–645 fm 831–1,180 m	310–587 fm 567–1,074 m	316–610 fm 578–1,116 m	313–610 fm 571–1,116 m
Number of samples in group	10	7	17	11	45	15	10	15	21
Average ratio of sand layers to mud layers	70	50.7	52.1	63.5	64	44	15	9	12
Percent lacking all sand layers	0	0	17	0	6.6	26	50	26	57
Percent with massive ungraded sand (a_1)	70	28.6	17	45	37	33	0	40	9.5
Percent with massive graded sand (a)	10	0	41	35	26	13	20	7	4.8
Percent with horizontal parallel laminations (b)	80	57	53	54	58	20	30	20	19
Percent with inclined parallel laminations (b_1)	0	14	6	9	7	0	10	7	0
Percent with current-ripple cross-laminations (c_1)	30	43	47	45	42	33	20	47	9.5
Percent with deformed cross-laminations (c_2)	0	0	17	9	9	6.6	10	7	0
Percent with wavy laminations (c_3)	20	28	17	18	20	13	0	13	0
Percent with convolute laminations (c_4)	0	0	0	9	2	13	0	0	0
Percent with micro-erosion channels	10	0	17	18	13	0	0	27	0
Percent with appreciable coarse sand and/or gravel	30	0	6	18	13	33	10	0	4.8
Percent with mud balls in sand	20	14	12	45	22	33	10	0	4.8
Percent with plant-debris layers	30	14	17	18	20	0	0	0	4.8
Average median grain size ϕ of sand or coarse silt layer ϕ (No. of analyses)	2.2 (7)	1.5 (2)	2.5 (11)	1.9 (8)	2.2 (28)	2.5 (8)	2.8 (1)	2.9 (6)	3.17 (8)
Average maximum one- percentile ϕ (No. of analyses)	0.1 (6)	—	1.6 (11)	1.09 (6)	0.9 (23)	1.75 (8)	—	2.2 (2)	2.2 (3)
Standard deviation ϕ (No. of analyses)	.45 (2)	—	0.7 (14)	.08 (8)	0.7 (24)	1.18 (8)	0.7 (1)	0.48 (5)	1.18 (5)

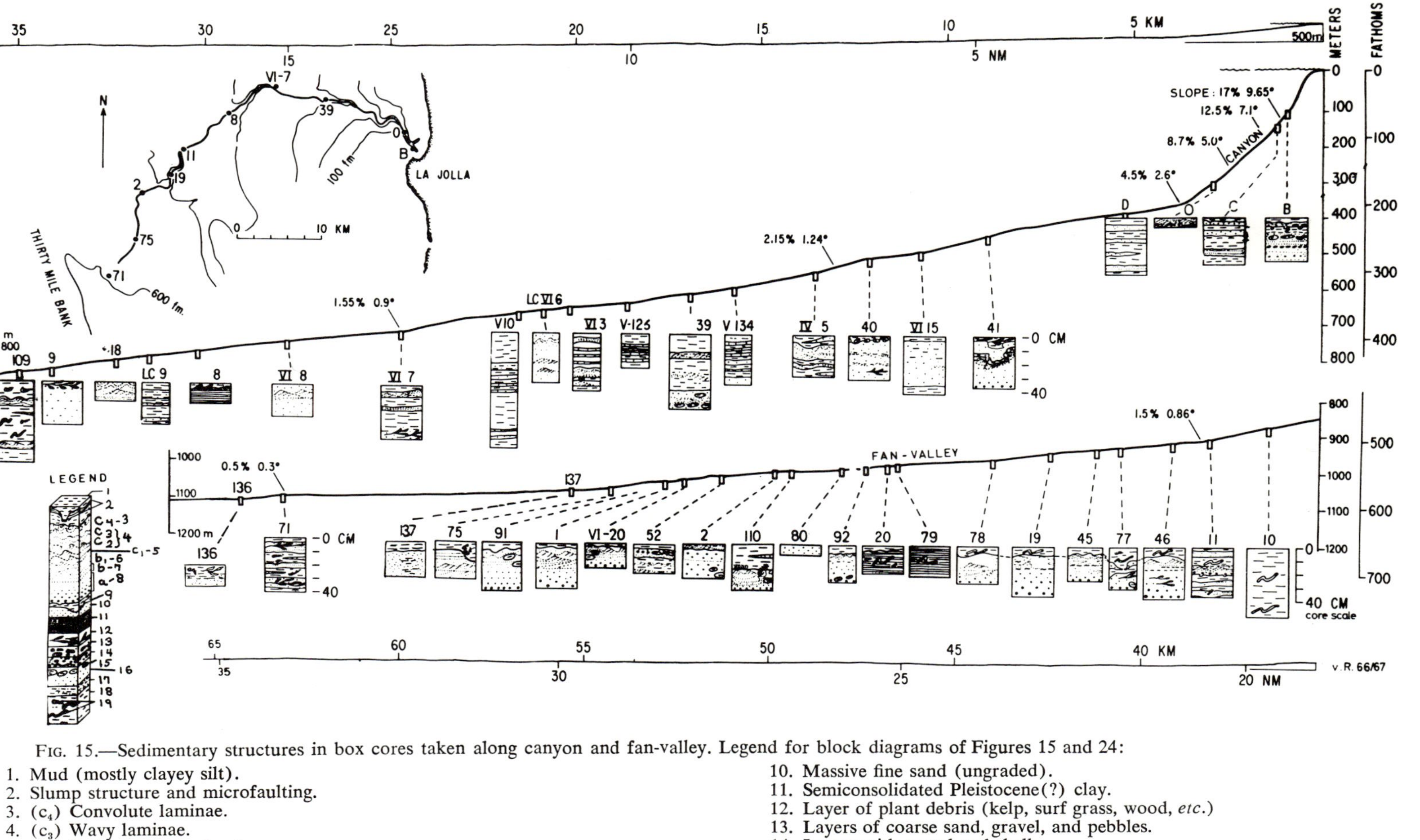

Fig. 15.—Sedimentary structures in box cores taken along canyon and fan-valley. Legend for block diagrams of Figures 15 and 24:

1. Mud (mostly clayey silt).
2. Slump structure and microfaulting.
3. (c₄) Convolute laminae.
4. (c₃) Wavy laminae.
5. (c₂) Deformed cross laminae.
6. (c₁) Current-ripple cross laminae.
7. (b₁) Inclined laminae.
8. (b) Parallel laminae.
9. (a) Massive graded bedding.
10. Erosive channel filled with clayey silt.

10. Massive fine sand (ungraded).
11. Semiconsolidated Pleistocene(?) clay.
12. Layer of plant debris (kelp, surf grass, wood, *etc.*)
13. Layers of coarse sand, gravel, and pebbles.
14. Layers with gravel and shells.
15. Mud balls (rounded fragments of semiconsolidated clay).
16. Fine to medium sand.
17. Sandy silt.
18. Clayey silt.
19. Bioturbation by burrowing organisms (worms, mollusks, echinoids, *etc.*).

380

generally at the base of fully cored sand layers. Grading appears to be conspicuously absent in the canyon and inner fan-valley with the exception of core 41 at 254 fm (455 m). The intermediate and outer parts of the fan-valley yielded a series of cores with graded sand layers: six closely spaced cores (11, 46, 77, 45, 19, 78) between 498 and 531 fm (911 and 971 m), and three cores (LC-VI 20, 1, 9) between 562 and 575 fm (1,028 and 1,051 m). Most sand layers are well to moderately sorted at the base and generally better sorted toward the top. Some grade into overlying, very poorly sorted clayey and sandy silt. Inverse grading occurs rarely.

Parallel lamination (b) is the most abundant structure, and is found in about 60 percent (27 of 45) of all sandy canyon-fan-valley axis cores. It is produced in most cases by changes in composition, such as heavy mineral concentrations, biotite or fecal-pellet layers (Fig. 16b).

Small-scale (mostly less than 1 cm) current-ripple cross-lamination (c_1) (*e.g.*, in core 1, Fig. 17) is the next most important structure, found in 43 percent of the sandy cores. This grouped cross-lamination with scoop-shaped sets of laminae clearly was caused by the migration of small current ripples. The stoss sides of the ripples generally are eroded, and wave lengths and ripple amplitudes are variable, with ripple indices (wave length/amplitude) ranging from 4 to 12 (average 8.5).

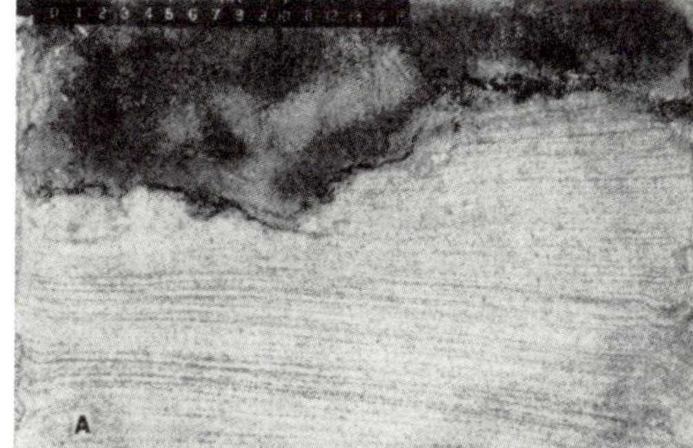

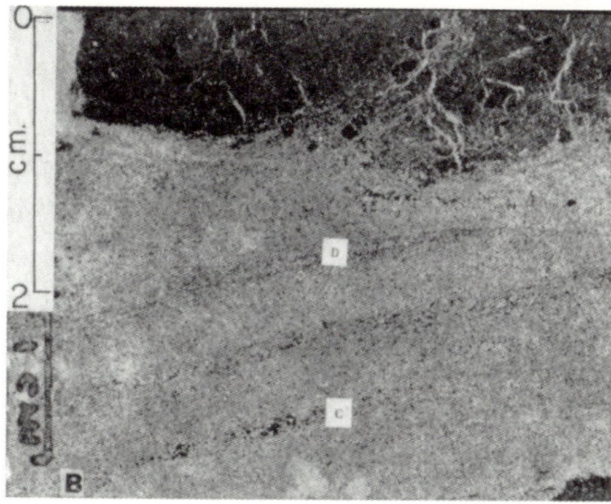

FIG. 16.—**A.** Parallel lamination in sand layer along fan-valley in box core 75 at 578 fm (1,057 m). Dark layers are concentrations of heavy minerals. Note erosional unconformity and overlying mud bed. **B.** Bands of dark minerals shown in X-radiograph of box core 36 along natural levee at depth of 472 fm (864 m).

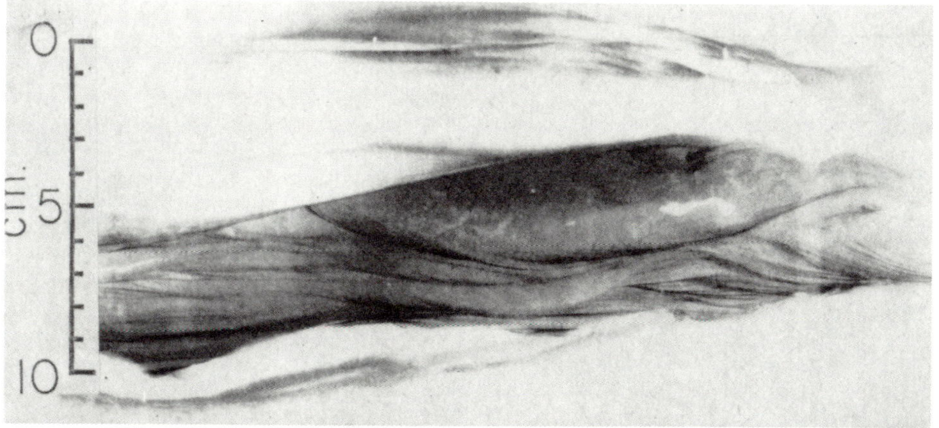

FIG. 17.—X-radiograph of box core 74 taken from natural levee along fan-valley at depth of 556 fm (1,017 m), showing cross-lamination in sand layer.

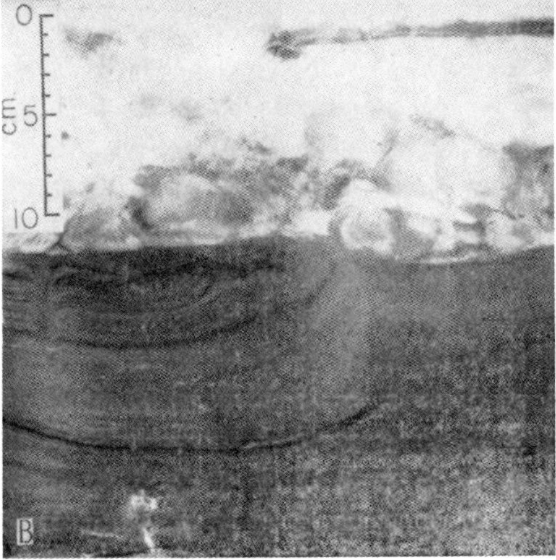

Fig. 18.—Convolute lamination and drag folds. **A.** From X-radiograph of box core 110 taken in fan-valley at depth of 552 fm (1,010 m). **B.** From Pliocene Pico Formation in Hall Canyon, Ventura basin. Laminated fine sands are overlain by coarse-sand formation. Photo by von Rad.

Heavy mineral concentrations (amphibole, epidote, magnetite, *etc.*) in distinct laminae may indicate that reworking by bottom currents is very common in most parts of graded and ungraded beds (Fig. 16). However, these features are rarely found at the base of sand layers. On the other hand, lamination, including alternation of dark and light layers, can be produced without current pulsation, as shown by the tank experiments of Kuenen (1966); but his experiments did not demonstrate that heavy mineral concentrations, like those shown in Figure 16b, can be produced without a distinct increase of current above that which de-

posited the intervening layers. Thin sections reveal micro-unconformities and erosional flutes filled by trapped coarse sand grains or cross-laminated foresets (scour-and-fill structures). The presence of these structures suggests that the sand layers probably have been deposited slowly, grain-by-grain, with intermittent times of nondeposition, erosion, and reworking.

Slightly deformed cross-lamination (c_2) has been produced by syn- or post-depositional deformation and oversteepening of ripple-drift cross-lamination foresets (Fig. 18, left). Wavy parallel laminations (c_3) in some places produced by a bending of laminae around mud pebbles, are relatively common. All transitions exist between the c_1-c_2-c_3 divisions of Table I and the convolute laminated division (c_4), characterized by sharp crested "anticlines" and broadly folded "synclines" (Fig. 18a, right).

Most thin sand layers are very fine grained and distinctly laminated throughout their thickness ("laminites," Lombard, 1963). This b-c sequence was found in about 20 percent of the canyon-fan-valley cores. Fully developed graded parallel and cross-laminated cycles (a-b-c) were present in about 28 percent of the sandy cores of this group (Fig. 19a). The thickness of fully cored sand layers along the canyon-fan-valley axis ranges from a few centimeters to more than 10 cm. Several sand layers more than 40 cm thick were not penetrated completely by the box corer. Hence, it is possible that in some cores, showing only the b-c sequence, the graded division (a) was not penetrated.

Small-scale slump structures with very fine sand layers (*e.g.*, Figs. 15, 20) showing down-slumped and micro-faulted cross-lamination were noted along (or across) the length of some box cores. Channels 5–15 cm wide, cut into parallel-laminated sand and later filled with homogeneous clayey silt (Fig. 16), indicate the activity of strong erosive bottom currents that did not subsequently deposit sandy material.

Semiconsolidated clayey pebbles (called "mud balls") are commonly incorporated in a matrix containing considerable medium- and coarse-grained sand, especially in the outermost fan-valley between 544 and 575 fm (995 and 1,051 m) (Fig. 21a). These inclusions have a diameter range of 6–13 cm. They are similar in texture and foraminiferal content to the scattered blocks of clay rubble, which apparently came from the erosion of the steep channel walls and terraces, and were transported

Fig. 19.—Unlaminated graded sand covered by laminated finer sand and, in turn, by cross-laminated sand. Typical of turbidite beds, according to Bouma (1962). **A.** From box core 1 at 568 fm (1,039 m) along fan-valley. **B.** From Pico Formation of Hall Canyon, Ventura basin. Photo by von Rad.

down-valley as float in a coarse-grained matrix.

Seven of the axis cores, especially those from the outermost part of the fan-valley, have distinct layers of plant debris. These layers consist of waterlogged kelp stipes and holdfasts, eelgrass, land-derived wood, and broken bits of Gorgonian coral fronds, interbedded with or overlying fine-grained laminated sand. Core 71 at 597 fm (1,094 m) (Fig. 29) contains four distinct plant layers, two of them 4–6 cm thick and two 0.5–1.5 cm thick. The plant layers have their closest possible source in the heads of the canyons, where similar mats are being deposited and moved downslope by slow glacierlike gravity creep (Dill, 1964a, p. 158).

Bioturbation and mottling, shown in X-radiographs, considerably affect the sedimentary fabric of the clayey sediments on the sea floor (Fig. 22). Most of the clayey-silt layers have been almost completely homogenized by the action of burrowing organisms. Only exceptionally faint parallel- or cross-lamination can be observed in X-radiographs of sandy silts. The term "mottling" describes a combination of burrow structures (Bouma, 1964, p. 307), such as lumps of silt caused by burrowing or homogeneously mottled "cloudy" structures. To ascribe typical burrow structures to definite types of burrowing animals is difficult. However, annelids and other worms certainly produce most of the small tubelike burrows. Echinoids (*e.g.,* the heart urchin *Echinocardium* sp.), pe-

lecypods and gastropods have been observed reworking the canyon fill and must be responsible for some of the larger burrow structures.

Grain size and thickness of sand layers.—Figure 23 and Table I show that the grain size and thickness of sand layers do not have a consistent decrease along the fan-valley in the

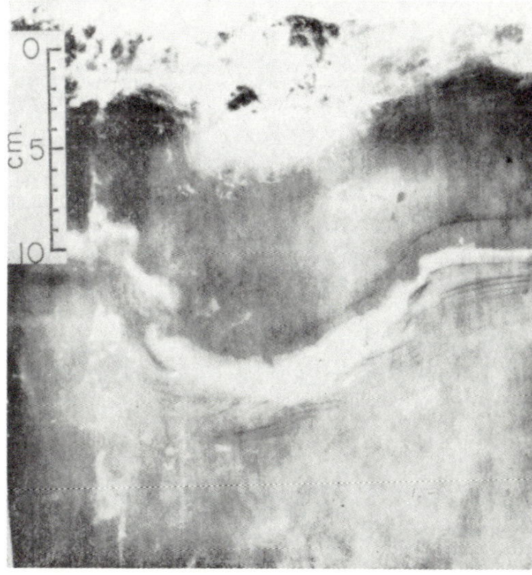

Fig. 20.—X-radiograph of core 41, taken at 250 fm (475 m), showing slumping effects, perhaps result of bioturbation.

FIG. 21.—Mud balls in sand formations. **A.** From box core 2 along fan-valley at 553 fm (1,012 m). **B.** From box core 72 taken in disconnected short fan-valley at depth of 593 fm (1,085 m). **C.** From Miocene at La Jolla, southern California. Scale center in cm. Photo by Neil Marshall.

downslope and offshore direction. This apparently is caused by irregularities in bottom topography and by the irregular and discontinuous processes involved in the initial deposition and reworking of sediment material. It is noteworthy that very coarse sand and gravel associated with reworked Pleistocene(?) mud balls are present in the intermediate and outermost parts of the fan-valley. The distribution of textural parameters along the longitudinal profile of La Jolla canyon and fan-valley is shown in Figure 23. The coarsest representative sand of the upper 20 cm of most cores has been analyzed. As the sand layers occur at different intervals in each core, the parameters shown in this graph cannot be correlated directly from core to core. Along the total length of the canyon-fan-valley axis, at least six samples contain medium- to coarse-grained sand (sometimes with gravel). The coarse sediments also have high medians and maximum one-percentile values, general poor sorting, and bimodality.

A thin layer of clayey silt lies on top of most fine-grained sand. In this top layer, the very poorly sorted mud, influenced by biological mixing, does not show much lower maximum grain sizes than does the underlying well-sorted sand. One exception to the general rule of complete randomness of sediment distribution along the fan-valley is the series of closely spaced cores (11–19) taken within a horizontal distance of about 2.5 mi (4 km) between depths of 498 and 521 fm (911–953 m). The thick sand layers in these cores can be compared, because they are compositionally similar and have several textural parameters that change in a consistent pattern; with one exception, median, maximum one-percentile, and mode-values decrease in the downvalley direction. The sediment from this stretch of the fan-valley may have been transported and deposited by a common mechanism of sedimentation, perhaps a turbidity current.

In two zones along the axis of the fan-valley, the recent sediment cover seems to be missing or very thin. One of these zones with hard semiconsolidated clay (core 8 at 432 fm (790 m)) is in an area where the continuous reflec-

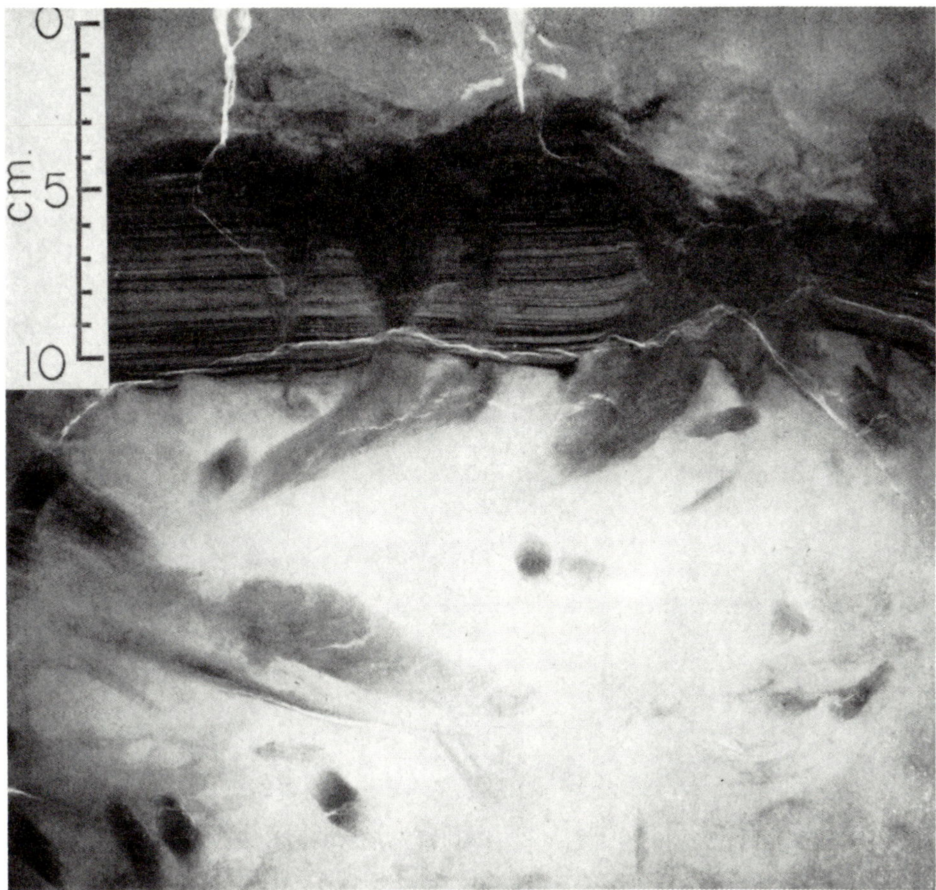

Fig. 22.—Laminated sand layer partly disrupted by bioturbation and considerably disturbed underlying mud formations. From X-radiograph of box core 125 at 453 fm (829 m) on natural levee.

tion profiles show bedrock close to the surface. The other zone (cores 79 and 20) is at a depth of 543 fm (994 m), where the Sparker records show no evidence of near-surface bedrock. The hard clay is apparently part of the fan deposit into which the valley has been cut. F. L. Parker's micropaleontologic examination of cores 8 and 20 suggests that these compact clays are of Pleistocene age (67 and 61 percent, respectively, left-coiling specimens of *Globigerina pachyderma,* including many small and large arctic-type specimens). Surprisingly, the other two samples of semiconsolidated clay (cores 76 and 79, the latter being very close to core 20) showed no indication of "cold Pleistocene" (having 93 percent right-coiling *Globigerina pachyderma*). Possibly the latter clays were deposited during an interglacial or interstadial period, when warmer temperatures similar to

those of the present day (89–98 percent right-coiling *Globigerina pachyderma*) prevailed (F. L. Parker, personal commun.).

Axes of Other Valleys

Examination of Figures 1 and 4 shows four small valleys that are independent of the main La Jolla fan-valley but in close proximity to its outer section. The slightly incised valley forming the continuation of Loma sea valley and extending around the outer end of Coronado ridge was cored in three localities. The cores consist of mud, except for core 49, which has a thin layer of silty sand. Core 76, at 556 fm (1,017 m), from the valley east of and parallel with the outer main fan-valley, contains semiconsolidated Pleistocene(?) clay overlain by soft clayey-silt, similar to that in nearby core

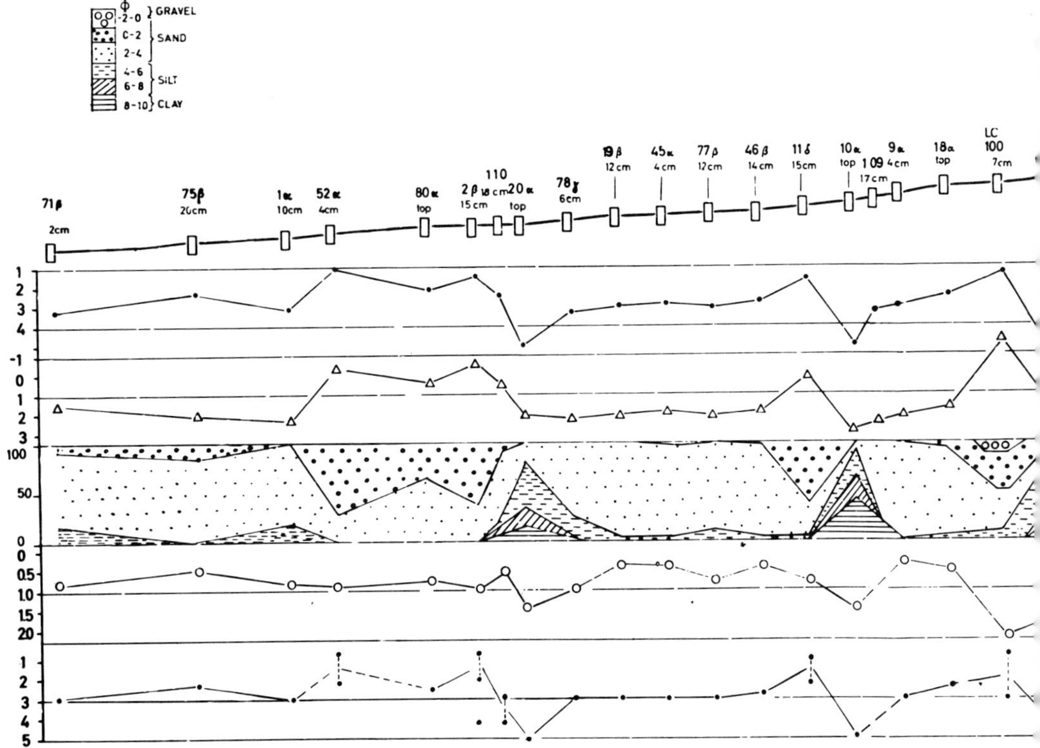

Fig. 23.—Grain-size distribution of top sand layers in cores along canyon and fan-valle

20 in the main fan-valley. Three other cores (29, 28, 86) along this valley have layers of fine to very fine sand. Core 72, taken in the short valley west of the outermost main fan-valley at 595 fm (1,088 m), has a thick lower section of coarse sand and many large (up to 15 cm) rounded and partly imbricated mud balls (Fig. 21b). The matrix of medium- to coarse-grained sand grades upward into a cross-laminated fine-grained sand.

Four of the box cores (27, 135, 114, 84) taken from the rift valley at the east foot of Thirtymile Bank closely resemble core 72, having numerous large, rounded mud balls. Three of these cores contain moderately sorted, medium- to coarse-grained sand, and one has a maximum one-percentile of 2.8 mm ($-1.4\ \phi$). The virtual absence of glauconite and Foraminifera in the coarse-grained sand of the rift valley precludes the possibility that the mud lumps are due to slumping action from the nearby steep slope of Thirtymile Bank.

Natural Levees

The natural levees were sampled in 15 dif-

ferent locations (13 box cores and 2 gravity cores; see Fig. 24). The main difference between the fan-valley axis cores and the levee cores is in the ratio of sand to mud layers. Eleven of 15 levee cores contain sand layers, usually only 1–5 cm thick. Most of the sand is very fine grained and laminated throughout the layer. An exception is core 103 at 496 fm (907 m), which contained a 15-cm thick graded-sand layer with a complete a-b-c_1-c_2-c_3 cycle. The outermost levee sample, core 132 at 608 fm (1,112 m), has the thickest (22 cm) sand layer. The levees, like the channel, appear to have thicker sand layers along the outer part of the fan-valley. All the levee cores lacking sand layers are found at depths of less than 495 fm (905 m), where wall heights are greater than 30 fm (55 m). Eight of 12 levee cores with sand layers are current-rippled and cross-laminated, three are also parallel-laminated, and only three are massive ungraded. One of the levee cores shows an unusually high degree of bioturbation within a sand layer (core 125; see Fig. 22). Similar active reworking of the sea floor along the levees by burrowing organisms was observed by Dill during his deep dives.

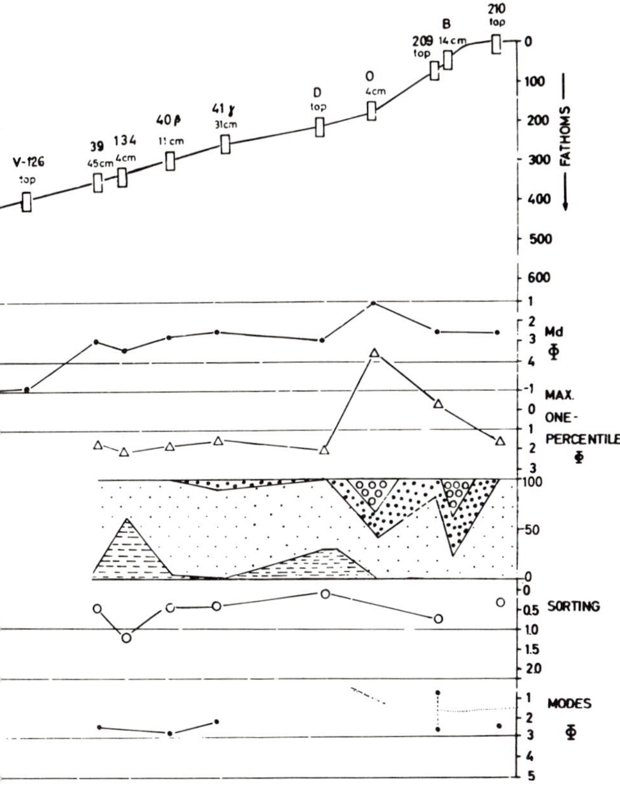

hows that grain size varies irregularly, but with some increase in outer part.

Fan-Valley Terraces

Seven box cores and three gravity cores were obtained from the terraces along the sides of the fan-valley (Fig. 24). The terraces are probably slump blocks except in the inner fan-valley, where they may be due to rejuvenation after a period of deposition. Core 108 at 441 fm (807 m) is from the flat top of a small isolated hill in the fan-valley, and thus is comparable with the terrace samples. Five of the 10 samples contain sand as layers or lenses, generally well below the surface. Most of the samples lacking sand layers were from the upper part of the fan-valley at depths of less than 390 fm (713 m). Most sand layers are only 1–4 cm (maximum, 8 cm) thick, and consist of parallel- and cross-laminated, moderately well-sorted very fine- to fine-grained sand. Core 101 at 587 fm (1,074 m) contains a 7-cm thick sand layer with a complete a-b-b_1-c_1 sequence.

Open La Jolla Fan and Basin Slope

Twenty-one samples (16 box cores, a few of them undisturbed, 4 gravity cores, and 1 piston core) were taken from the open fan away from the fan-valleys and levees (see Fig. 24). Twelve consist of mud only, and most of the remaining 9 have only thin layers of very fine-grained terrigenous sand. However, core 134 at 595 fm (1,088 m) is primarily sand with a thin cover of mud and with large mud balls in the lower part. Core 116, at 598 fm (1,094 m), consists of relatively coarse-grained sand covered with a kelp layer. The proximity of this core to cores 71 (Fig. 30) and 136 in the outermost fan-valley is of interest because they all contain kelp layers. Again, in the open fan environments the deeper cores show an increase of thickness and grain size of sand layers.

Other Environments

Sample 50 from the northernmost tip of Coronado ridge has a poorly sorted, muddy, medium-grained sand layer overlying a very poorly sorted silty-sandy-gravel (maximum one-percentile: 6 mm) with angular rock fragments. Although this sample comes from near the edge of the fan, it does not belong in the open-fan group, because the rock fragments and

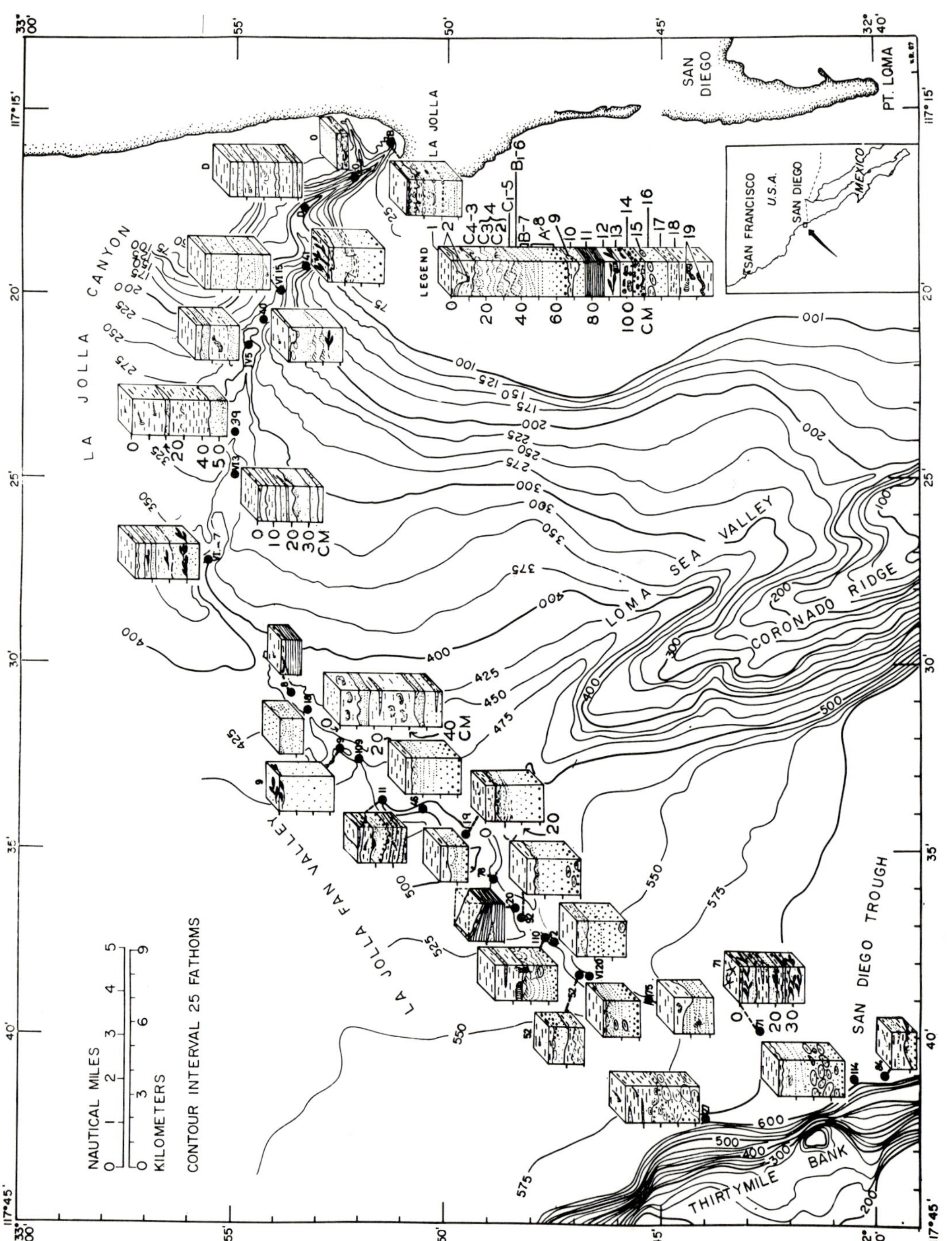

FIG. 24.—Block diagrams of box cores from environments outside of axis of fan-valley. For interpretation of symbols see Figure 15.

sand grains consist of authigenic minerals (collophanite, glauconite, pyrite) and of volcanic, sedimentary, and metamorphic rock fragments, typical of the facies on top of Coronado bank (Emery *et al.*, 1952).

The short box cores from the top of Thirty-mile Bank consist of sand coarser than most of the fan and fan-valley sands, and include considerable glauconite and an abundance of Foraminifera. The two samples taken on the fan-valley slopes are structureless mud with very little sand.

THICKNESS OF FAN DEPOSITS AND UNDERLYING STRUCTURE

Continuous reflection profiling has been used to determine the thickness of the fan sediments and the structure of the rock masses underlying La Jolla fan. The locations of a series of profiles taken with the high-energy, low-frequency Sparker system of the USNS *Davis* (see Curray and Moore, 1963) are shown in Figure 25 and the profiles in Figure 26. Six of the profiles (A-F) run essentially down the slope of the fan (northeast-southwest) and the other 11 (G-Q) cross the fan-valley (northwest-southeast). The profiles show a boundary between what Moore (1966) has called the deformed "preorogenic" bedrock and the "postorogenic" sediments.

The isopach map (Fig. 27), showing the approximate thickness of the postorogenic sediments, was constructed from the writers' seismic profiles and two from Moore (1966, Fig. 16). The approximate thickness of the postorogenic basin fill is calculated from the two-way travel time of sound, according to the values computed by Moore (1966, Fig. 19). These estimates take into consideration the velocity increase with increasing sediment thickness resulting from an increase of average sediment density by diagenetic compaction.

The inner part of the fan consists of a thin veneer (10–100 m) of younger unconsolidated sediments overlying an irregularly sloping bedrock surface. The records show that La Jolla fan is not a "wedge-shaped" cone of clastic sediments, generally considered to be typical of deep-sea fans (Gorsline and Emery, 1959; Menard, 1960; Walker, 1966). Because the Sparker method does not resolve sediment layers thinner than about 4 m, areas with the designation "no sediment cover" in Figure 27 may have a thin cover of sediments, *e.g.*, along the northeast side of profile A (Fig. 26). Profiles A and G also indicate that part of the canyon and

fan-valley may be located along vertical faults, as was suggested by Buffington (1964) in his interpretation of the inner fan topography.

The outer part of La Jolla fan (Fig. 26, profiles L-Q) is underlain by a thick sequence of semiconsolidated[7] Pleistocene(?) sediments (200–1,000 m). A 200–500-m thick sediment wedge comes from the southeast where the inactive Loma sea valley interfingers with the ponded sediments of La Jolla fan. The buried continuation of Coronado ridge, with its partly eroded northwest-plunging anticline, is indicated in Figure 26, and an older, much deeper Loma sea valley is covered by fan deposits. Traces of buried older channels indicate that the fan-valley has migrated considerably during Pleistocene and Holocene times.

The outermost part of La Jolla fan and the central part of San Diego trough have very thick sediment fills, up to 1,000 m, with a small bowl-shaped central "basin" containing up to 1,450 m of mainly Pleistocene and Holocene sediments[8] (Fig. 26, profiles A, B, D, and L-Q; Fig. 27). In some places the bedrock of the trough is not defined clearly by the reflection profiles. Near the eastern part of the trough, the sediment thickness decreases abruptly to less than a few hundred meters, and in the west, the pre-Pleistocene rocks crop out on Thirtymile Bank. Also, rock is found at the surface along much of Coronado bank and ridge. Near the contact of San Diego trough and Thirtymile Bank, the thick basin fill shows abnormal acoustical impedance characteristics, indicated by strong wavy reflection horizons near the escarpment (Fig. 26, profiles A, C, P, O). This may signify a thick sand body shown by weak subparallel and horizontal reflectors. This suggestion conforms with the distribution of surface sediments along the rift valley, as seen in the box cores.

The contact between the trough sediments and the escarpment of Thirtymile Bank (Miocene volcanic rocks) is probably a fault, but profiles A and C (Fig. 26) show no clear evidence of this fault, possibly because of the inability of the Sparker to interpret steep slopes. Coronado Bank and the basin slope on the east consist of folded and faulted Eocene through Pliocene(?) strata that are only partly covered by thin prograding fan deposits (Fig. 26, pro-

---

[7] Geophysicists commonly call sediments "unconsolidated" that have even higher velocities than these, but "semiconsolidated" seems more in keeping with geologic usage.

[8] Age based on probable rates of deposition.

F. P. Shepard, R. F. Dill and Ulrich von Rad

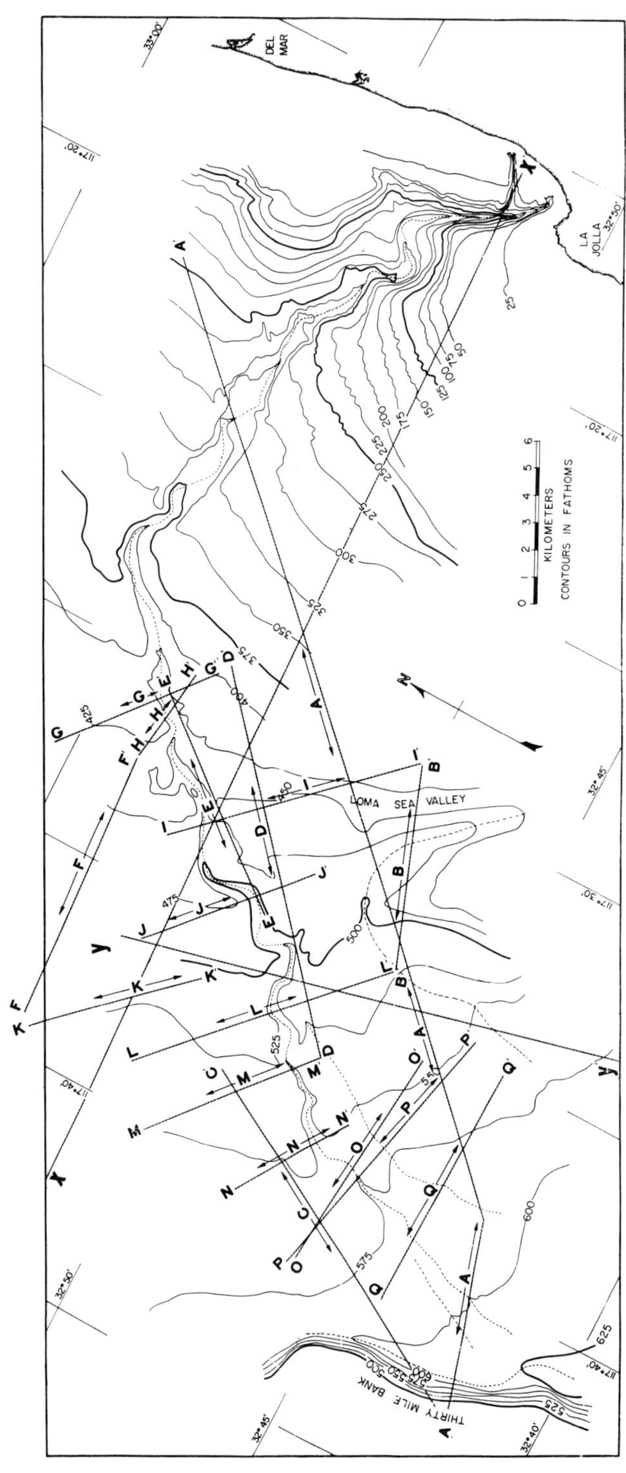

Fig. 25.—Location of continuous reflection profiles by writers (A-Q) and D. G. Moore (X-Y). Sediment thickness data from profiles were used to construct isopach map of sediment thickness for San Diego trough.

file A). Erosion of Miocene rocks is indicated from Foraminifera found in cores taken in the area (M. L. Natland, personal commun.).

SOURCES OF SANDY SEDIMENTS

The main source for sand on La Jolla fan is the sediment eroded from the batholithic and metamorphic rocks of the Peninsular Ranges, transported to the sea by rivers, redistributed along the coast of southern California by marine longshore currents, and funneled downslope by the canyon system. The mineralogic composition and texture of the sand deposits, the large amounts of plant debris, and the occurrence of displaced shallow-water Foraminifera (M. L. Natland, personal commun.) clearly indicate that most of the fan and fan-valley sand has been derived from the near-shore area.

A second, very minor source is from the active submarine erosion of the canyon walls (Dill, 1964b; Shepard and Dill, 1966, p. 38, 42, 58). All formations cropping out along the canyon walls have contributed some sediment material as talus. These include the conglomerate of the Rose Canyon Formation (Eocene), the siltstone and sandstone of the Rosario Formation (Late Cretaceous), and also the Pleistocene gravel beds. Possibly the coarse sediments come from the erosion of the fan-valley walls, although the deep-dive observers saw only clay walls, and a gravel bed in the low terrace near the inner end of the fan-valley (Fig. 6).

Shore-cliff erosion apparently contributes little to the total sand budget (Emery, 1960, p. 11–21). No significant similarities could be found between the size and composition of beach sands and of the sandstone beds exposed in the nearby cliffs (Rose Canyon Formation, La Jolla).

Evidently the submarine ridges and banks do not supply any appreciable amount of sand to the adjacent San Diego trough. Neither the heavy mineral assemblages nor the composition of the rock fragments in the coarse sands is indicative of the Thirtymile Bank sediment cover.

The localized patches of coarse-grained sand and fine gravel cored in the outer fan-valley can be explained as lag deposits or as the result of variation in the type of sediment entering the canyon at its head. This variation is mostly dependent on the intensity of winter storms, which move sediment from the beaches into the heads of the canyons. Coarse sand and cobbles up to 15 cm in diameter are common in the micaceous sands now moving down the

head of Scripps canyon. Reworking of this mixed sediment by currents of variable velocity, like the ones observed and measured indirectly, would tend to concentrate the coarser fractions into patches of sediment with similar grain sizes.

Another possible source of coarse sediment is the submarine erosion of alluvium from the head of La Jolla canyon. The canyon is cutting shoreward locally at a rate as high as 0.6 m/year, through a sediment that was deposited in the canyon head during Pleistocene fluctuations of sea level. The erosion of La Jolla canyon head is most intense during winter storms, when sand from the beach spills over the headwall. On such occasions, coarse gravel and cobbles from the alluvium become part of the sediment fill that is moving downslope. The periodic incorporation of coarse material may result in the observed patchy distribution found in the outer fan-valley.

MECHANISMS OF SAND TRANSPORT

All investigations of the fans at the base of submarine canyons have led to the conclusion that sand and other coarse-grained fan sediments have been transported down along the canyons and their fan-valley continuations. The prevalent opinion has been that these sediments move seaward in turbidity currents. However, in recent years most investigators of these environments have acquired information that suggests a more complex explanation. The complications are discussed more fully elsewhere (see Shepard and Dill, 1966, p. 320–335), but will be considered here briefly.

Turbidity Currents

There are many indications that powerful currents or mass movements of some type transport all sizes of sediment along submarine canyons and prevent canyon filling. The precipitous, freshly eroded walls along the canyons and fan-valleys suggest the presence of powerful currents. Cobbles and even boulders found along many canyon axes imply current strength. Sand on the marginal levees 50 or 100 m above the channel floor shows that currents have had considerable thickness as well as power. All these facts appear to favor the existence of at least periodic high velocity currents.

However, many factors, especially in the area under discussion, are somewhat incompatible with the turbidity-current explanation. The failure of explosives set off in the canyon-head fill to start such currents confirms the work of

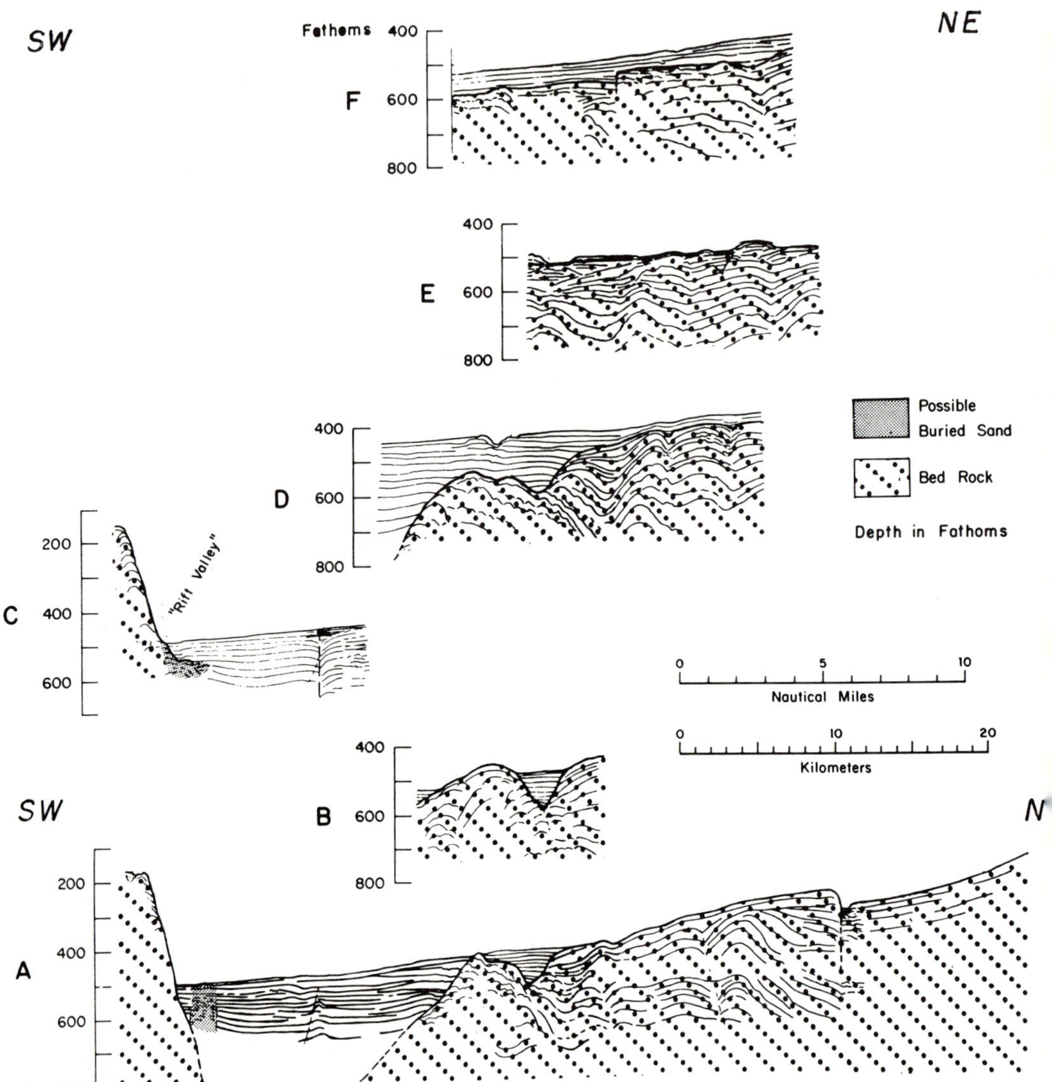

FIG. 26.—Continuous reflection profiles A-E roughly parallel La Jolla fan-valley axis (NE-SW), and F-Q are transverse to axis (NW-SE).

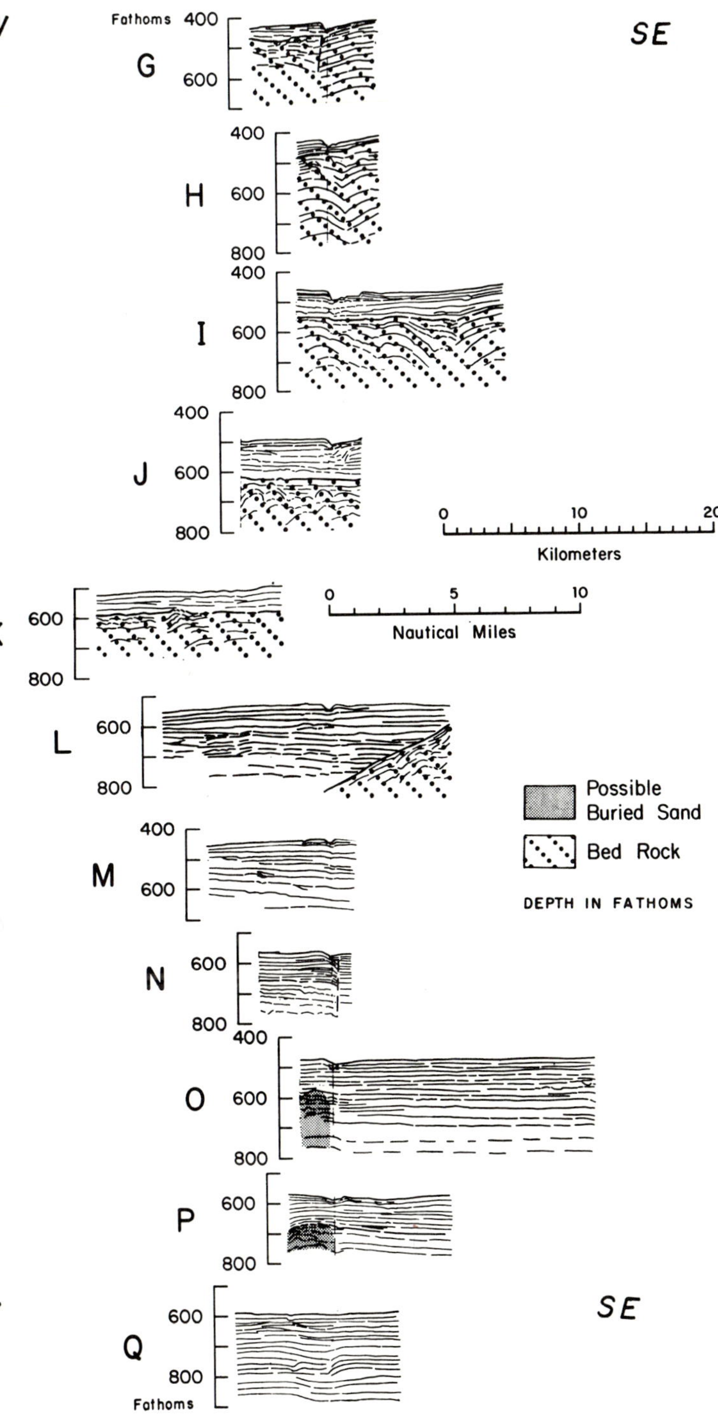
N
Fathoms 400
SE
G
600
400
H
600
800
400
I
600
800
400
J
600
800
0   10   20
Kilometers
0   5   10
Nautical Miles
K
600
800
L
600
800
Possible
Buried Sand
Bed Rock
DEPTH IN FATHOMS
M
400
600
N
600
800
400
O
600
800
P
600
800
SE
Q
600
800
Fathoms

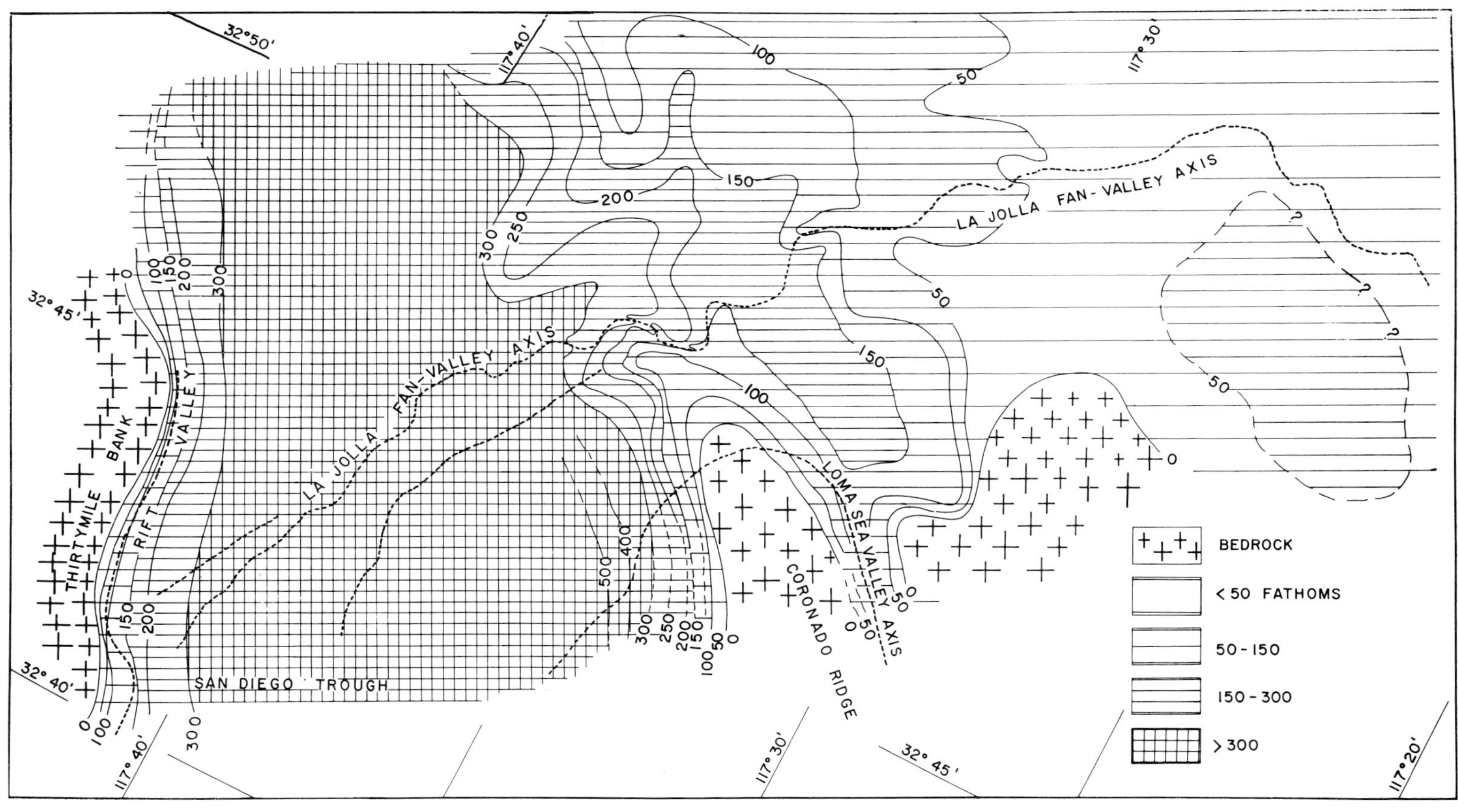

Fig. 27.—Isopach map of La Jolla fan and vicinity, showing approximate thickness of unconsolidated and semiconsolidated postorogenic fan deposits above bedrock boundary. CI = 50 fm.

Dill (*in* Shepard and Dill, 1966, p. 308), which indicated a lack of metastable conditions in the canyon floor fill. The nature of the sand layers along the valley axes also fails to confirm a turbidite origin. Graded sands are relatively rare, and the a-b-c sequence, said to be characteristic of turbidites, was found only in two small sections along the fan-valley axis. If unconformities like those which cut the sand laminae in core 75 (Fig. 16) were the result of a turbidity current, one would expect the tail of the eroding current to deposit a graded sand to mud bed above the eroded surface. Instead, there is only mud. Similarly, the mud layers covering most of the sand on the axis floor show little or no grading into the underlying sand. This fact suggests that the sands were deposited by other types of currents, presumably lacking the cloud of suspended sediment characteristic of turbidity currents. The cleanness of the sand contrasts with the poor sorting of most turbidites. The occurrence of laminae of heavy mineral concentrates (Fig. 16b) indicates alternate winnowing action during the deposition of the sand. Thus, the sand deposits appear to have been what Natland (1967) calls "tractionites" rather than "turbidites."

Observed and Measured Ordinary Bottom Currents

As the result of vehicle dives down to 667 fm (1,219 m) in the canyons and fan-valleys, and of many remote-controlled bottom photographs to depths of 2,000 fm (3,700 m), it is known that tractive bottom currents are capable of transporting sand at all depths. Velocities up to 0.5 knots were recorded during several deep dives. At times, currents move up canyon.

Current-meter measurements confirming the existence of ordinary bottom currents were obtained from 320 fm (585 m) in La Jolla fan-valley by R. S. Schwartzlose and J. D. Isaacs (Fig. 28). Their meter (Isaacs *et al.*, 1966) makes current measurements at 30-minute intervals for a period up to four days. The records show pulsating currents with velocities up to what may be as much as 30 cm/sec (0.6 knots), although the calibration at the time of observation was not well established. Currents move both up and down the valley floor with a rather clear relation to the tidal cycle at Scripps pier. In general, they move up valley during rising tide and down valley during the ebb, although there is a lag from the coastal tide record. Mostly, velocities are lower when the tidal fluctuation is least. However, some of

the highest velocities were observed during a period of small tidal fluctuation (midnight November 6–7, 1967).

Despite the fact that these actual observations show little difference between up- and down-valley currents, many dives have shown depressions scooped out on the up-valley side of obstacles, such as boulders, and dunelike deposits down valley. These features suggest that some type of current, probably neither tidal nor turbidity, is transporting material down the valley floors.

Gravity Creep, Slumps, and Sand Flows

Dill's scuba dives into the canyon heads have shown the importance of a glacial-like creep in moving the sediment fill down the steep floor of the inner gorges. It is possible that this process is also important as a transporting agent along the much reduced gradients of the outer canyon and fan valleys. Cracks and slump scars in the mud fill, noted in some places along La Jolla fan-valley, suggest such action. However, the sedimentary structure of the box cores reveals no distortional movement of this sort, and one could scarcely expect creep to be a major factor with gradients less than 1°.

Sand flows have been observed repeatedly in the 30° sand slopes along the walls of the south side of San Lucas canyon, Baja California, and rarely at the steep head of Scripps canyon. It seems unlikely that these flows would continue where the gradient is much reduced. Some other type of flow takes place in canyon heads causing a periodic deepening of many meters. Because there is little muddy material in these canyon heads to produce a turbidity current, the deepenings probably represent slumps. Presumably they do not travel far beyond the steep heads, but there is no positive information.

Mechanisms of Fine-Grained Sediment Transport

The mud dominating the open fan cores and the relatively thin mud layers along the valley axes appear to have been transported from the land by several processes. In most submersible dives, fine-grained flocculent blobs of clay-sized material have been observed drifting as a slow gravity flow down the valleys, particularly during winter months. This material is stirred up by the waves along shore or introduced into the ocean by runoff from the occasional rains. The presence of suspended matter adds slightly to

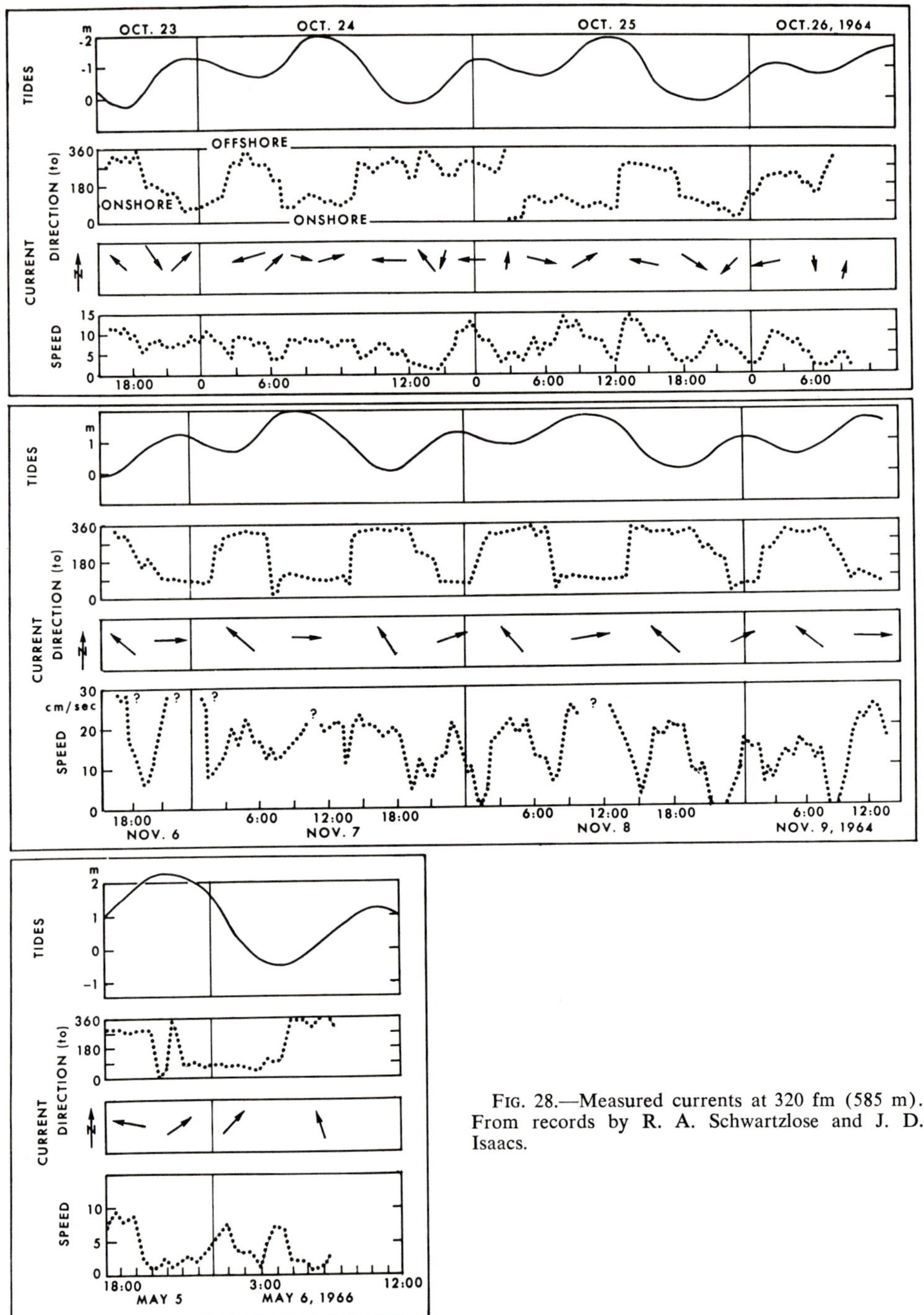

Fig. 28.—Measured currents at 320 fm (585 m). From records by R. A. Schwartzlose and J. D. Isaacs.

the density of the water, producing a slow gravity-controlled movement along the bottom, across the shelf, and down the slopes, until the mud clouds enter the valleys and are moved along them.

In study of the mud on the La Jolla fan, D. J. W. Piper, a student at Scripps Institution, found that the mud contains more silt near the fan-valley. This fact suggests that at least a significant amount of the fine sediment has come down the valleys and spilled over, depositing the coarser material near the valley margin. This is, of course, also implied by the existence of the levees. It is not known whether this process is the result of high-speed turbidity currents moving down the valleys or of some sort of broad density current of low velocity. The fine-grained sand layers found on the levees and less pronounced on the open fan suggest that at least locally the currents that spilled over onto the fan attained velocities of the order of 10 to 25 cm/sec, and probably much faster on the valley floor.

RATE OF SEDIMENT DEPOSITION

The only means of estimating the rate of deposition on the La Jolla fan and in the fan-valley comes from C-14 dates obtained from layers of matted kelp in three fan-valley cores and in one from the outer fan (Fig. 29). Dates from core 71, that has three kelp layers, provide the best information for the fan-valley, although they are inconsistent. The averages of dates at 7 and 30 cm by Isotopes Inc. and Scripps Institution laboratories, yield dates of 1,760 and 4,640 years. This difference of 2,880 gives an average rate for the 23 cm of 8 cm/1,000 years (15.2 mg/cm² for wet weight), which is equivalent to about 4 cm/1,000 years (8.5 mg/cm²) for compacted solid grains with zero porosity. Core 116 from the open fan had a kelp layer at a depth of 34-46 cm, with a date of 2,690 years. This gives an average rate of deposition of the overlying 34 cm of 12.6 cm/1,000 years, or for compacted material with zero porosity about 5.8 cm/1,000 years. These results suggest that deposition is faster on the open fan. However, it seems more likely that deposition has been very discontinuous along the fan-valley and the 2,690-year period probably includes long intervals of nondeposition and/or erosion.

Comparison of the rate of 12.6 cm/1,000 years on the open fan with a date from San Diego trough, obtained by Emery and Bray (1962), shows that their rate of 18 cm/1,000 years for the past 24,500 years is somewhat higher. Also, Inman and Goldberg (1963) found a much higher rate from the MOHOLE experimental drilling slightly north of the fan-valley in 510 fm (933 m). At a depth of 70 m below the surface, they obtained a sample that was dated as 34,000 years old. However, dates as old as this are now considered very suspect. If the rates were actually higher in periods extending back into the last glacial stage, an explanation may be that much more sediment was being introduced during the pluvial period that preceded the present. Possibly during such a pluvial period more powerful currents came down the valleys, building the natural levees and introducing the sand layers into the levees and the fans.

RELATION TO PALEO-ENVIRONMENTS

The sediments of the fan and fan-valley off La Jolla strongly resemble the Pliocene of equally deep-water origin found in the Ventura and Santa Paula Creek areas of southern California (Crowell et al., 1966; Natland, 1933; Natland and Kuenen, 1951; Winterer and Durham, 1962). The writers, with M. L. Natland and others, also have examined outcrops north of Doheny State Park, where an old Miocene or Pliocene channel filled with deep-water sediments is exposed in the sea cliff. Here, also, many features resembling the box cores were noted.

The resemblance between the deep-water Pliocene and the La Jolla fan deposits is illustrated by comparative photographs. Convolute lamination found in some cores in the present study resembles that found in Hall Canyon (Ventura area) Pico Formation (Fig. 18B). The sequence of massive, graded sand covered with parallel-laminated sand and, in turn, with current-rippled sand (the a-b-c sequence of Bouma for typical turbidites) is represented in core 46 and in a lacquer peel from Hall Canyon Pliocene (Fig. 19B). Mud balls commonly present in the outer La Jolla fan are equally common in the Miocene (Fig. 21C). Graded sand covered by a bed with abundant plant debris can be compared in a photograph of core 41 and a lacquer peel from the Ventura basin (Fig. 30). The Pliocene deep-water formations appear to have more graded beds than the recent near-surface fan deposits off La Jolla, but are otherwise very similar.

Longer cores, especially drill holes several hundred meters long, are needed to determine the three-dimensional shape of the sand bodies

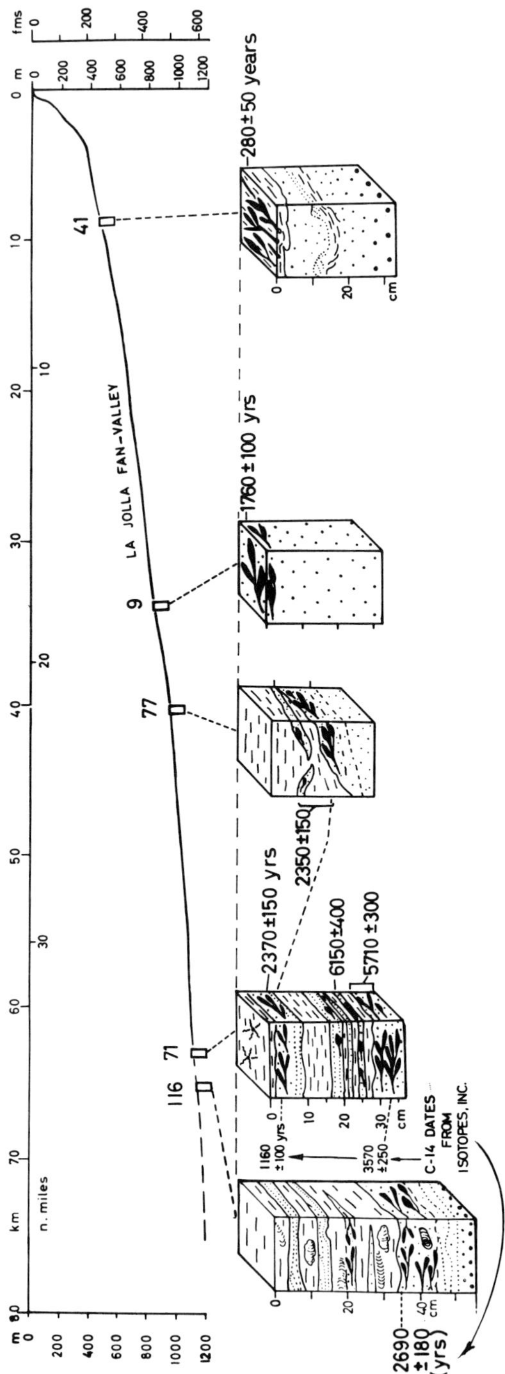

Fig. 29.—C-14 dated layers of matted kelp in cores along La Jolla fan-valley and outer fan (116). Where not otherwise labeled, dates are from Scripps Carbon-14 Lab.

Fig. 30.—Graded sand covered with bed high in plant material. (a) From box core 41 in fan-valley axis at 598 fm (1,094 m). (b) From Ventura basin Pliocene. Photo by von Rad.

in deep, present-day fans, and to find out if the sand layers that are so abundant in short cores have sufficient thickness to develop good stratigraphic traps for oil. The percentage of total organic matter is relatively high (1.5–3.4 percent organic carbon by weight) in the silt and very fine sand of La Jolla fan, so there would be a plentiful organic source for the formation of hydrocarbons during the diagenetic compaction of the basin sediments (Emery, 1960, p. 273–295). The discontinuous "shoestring sands" in direct contact with clay high in organic matter along the narrow fan-valley could become significant reservoirs for hydrocarbons. Sparker profiles indicate the presence of old buried channels, which very likely contain sands covered by muddy sediments.

The asymmetric filling of a basin, in which the coarser material is moved seaward and upward (in the column) with time, will form a wedge of clean sand within the basin fill. This sand will be surrounded by organic-rich clay. The Los Angeles basin, 100 mi north of San Diego trough, is a filled basin that probably received much of its sediment via ancient submarine valleys. The productive Wilmington oil field is on the seaward side of the basin, and production comes from sandstone intervals similar to those cored on the seaward side of San Diego trough.

A particularly intriguing possibility exists along the base of Thirtymile Bank, where the crinkly reflections from the unconsolidated sediments (Fig. 16, profiles A, C, O, and P) suggest the presence of a sand body of considerable thickness. If this proves to be true, it will be in agreement with the surficial sediment distribution patterns where sand becomes much more abundant and coarser as it approaches Thirtymile Bank, both in the channels and on the open fan. Here is an example of sediment becoming coarser seaward and away from the source of sediments. Imagine the difficulties of future geologists investigating the emerged La Jolla fan trying to find the old shoreline by the principle of increasing grain size, or by the use of ripple marks as a criterion for deposition in shallow water. The inclusion of shallow-water Foraminifera and land plants would add to the confusion.

REFERENCES CITED

Bouma, A. H., 1962, Sedimentology of some flysch deposits; a graphic approach to facies interpretation: Amsterdam-New York, Elsevier, 168 p.

——— 1964, Notes on X-ray interpretation of marine sediments: Marine Geology, v. 2, no. 4, p. 278–309.

——— 1965, Sedimentary characteristics of samples collected from some submarine canyons: Marine Geology, v. 3, no. 4, p. 291–320.

Buffington, E. C., 1964, Structural control and precision bathymetry of La Jolla submarine canyon, California: Marine Geology, v. 1, no. 1, p. 44–58.

Crowell, J. C., et al., 1966, Deep-water sedimentary structures, Pliocene Pico Formation, Santa Paula

Creek, Ventura basin, California: California Div. Mines and Geology Spec. Rept. 89, 40 p.

Curray, J. R., and D. G. Moore, 1963, Facies delineation by acoustic-reflections—northern Gulf of Mexico: Sedimentology, v. 2, p. 130–148.

Dill, R. F., 1961, Geological features of La Jolla Canyon as revealed by dive no. 83, of the Bathyscaph *Trieste:* San Diego, California, U.S. Navy Electronics Lab. Tech. Mem. No. TM-516, 27 p.

———— 1964a, Contemporary submarine erosion in Scripps submarine canyon: Ph.D. thesis, Calif. Univ., Scripps Inst. Oceanography, privately printed, 269 p.

———— 1964b, Sedimentation and erosion in Scripps submarine canyon head, *in* Papers in marine geology —Shepard Commemorative Volume: New York, Macmillan, p. 23–41.

Emery, K. O., 1960, The sea off southern California: New York, John Wiley & Sons, 366 p.

———— and E. E. Bray, 1962, Radiocarbon dating of California basin sediments: Am. Assoc. Petroleum Geologists Bull., v. 46, no. 10, p. 1839–1856.

———— *et al.,* 1952, Submarine geology off San Diego, California: Jour. Geology, v. 60, no. 6, p. 511–548.

Gorsline, D. S., and K. O. Emery, 1959, Turbidity-current deposits in San Pedro and Santa Monica basins off southern California: Geol. Soc. America Bull., v. 70, p. 279–290.

Inman, D. L., and E. D. Goldberg, 1963, Petrogenesis and depositional rates of sediment from the experimental Mohole drilling off La Jolla, California (abs.): Am. Geophys. Union Trans., v. 44, no. 1, p. 68.

Isaacs, J. D., *et al.,* 1966, Near-bottom currents measured in 4 kilometers depth off the Baja California coast: Jour. Geophys. Research, v. 71, no. 18, p. 4297–4303.

Kuenen, Ph. H., 1966, Experimental turbidite lamination in a circular flume: Jour. Geology, v. 74, no. 5, pt. 1, p. 523–545.

Lombard, A., 1963, Laminites: a structure of flysch-type sediments: Jour. Sed. Petrology, v. 33, p. 14–22.

Menard, H. W., Jr., 1960, Possible pre-Pleistocene deep-sea fans off central California: Geol. Soc. America Bull., v. 71, p. 1271–1278.

Moore, D. G., 1965, Erosional channel wall in La Jolla sea-fan valley seen from Bathyscaph *Trieste II:*

Geol. Soc. America Bull., v. 76, p. 385–392.

———— 1966, Structure, lith-orogenic units and post-orogenic basin fill by reflection profiling: California continental borderland: Ph.D. thesis, Rijksuniversiteit, Groningen, 151 p.

Natland, M. L., 1933, The temperature and depth distribution of some recent fossil Foraminifera in the southern California region: Scripps Inst. Oceanography Tech. Ser. Bull., v. 3, no. 10, p. 225–230.

———— 1967, New classification of water-laid clastic sediments: Am. Assoc. Petroleum Geologists Bull., v. 51, no. 3, pt. 1, p. 476.

———— and Ph. H. Kuenen, 1951, Sedimentary history of the Ventura basin, California, and the action of turbidity currents: Soc. Econ. Paleontologists and Mineralogists Spec. Pub., no. 2, p. 76–107.

Rees, A. I., 1965, The use of anisotropy of magnetic susceptibility in the estimation of sedimentary fabric: Sedimentology, v. 4, no. 4, p. 257–271.

———— U. von Rad, and F. P. Shepard, 1968, Magnetic fabric of sediments from the La Jolla submarine canyon and fan, California: Marine Geology, v. 6, p. 145–168.

Reineck, H. E., 1963a, Der Kastengreifer: Natur Museum, v. 93, no. 2, p. 65–68.

———— 1963b, Nasshaertung von ungestoerten Bodenproben im Format 5 × 5 cm für projizierbare Dickschliffe: Senckenbergiana Lethaea, v. 44, no. 4, p. 357–362.

Rosfelder, A. W., and N. F. Marshall, 1966, Oriented marine cores: a description of new locking compasses and triggering mechanisms: Jour. Marine Research, v. 24, no. 3, p. 353–364.

Shepard, F. P., and E. C. Buffington, 1968, La Jolla submarine fan-valley: Marine Geology, v. 6, p. 107–143.

———— and R. F. Dill, 1966, Submarine canyons and other sea valleys: Chicago, Rand McNally, 381 p.

———— and G. Einsele, 1962, Sedimentation in San Diego trough and contributing submarine canyons: Sedimentology, v. 1, no. 2, p. 81–133.

Walker, R. G., 1966, Deep channels in turbidite-bearing formations: Am. Assoc. Petroleum Geologists Bull., v. 50, no. 9, p. 1899–1917.

Winterer, E. L., and D. L. Durham, 1962, Geology of the southeastern Ventura basin, Los Angeles County, California: U.S. Geol. Survey Prof. Paper 334 H, p. 275–366.

THE AMERICAN ASSOCIATION OF PETROLEUM GEOLOGISTS BULLETIN
V. 51, NO. 11 (NOVEMBER, 1967), P. 2281-2304, 9 FIGS.

GEOLOGY OF MONTEREY CANYON, CALIFORNIA[1]

BRUCE D. MARTIN[2] AND K. O. EMERY[3]
Annapolis, Maryland 21401, and Woods Hole, Massachusetts 02543

ABSTRACT

New data from marine dredgings off Monterey, California, correlated with wells and outcrops on land indicate that Monterey, Carmel, and Soquel submarine canyons are related closely to the geological history of the continent.

Upper Pliocene siltstone and sandstone bodies in Monterey Canyon correlate with the Pomponio Member of the Purisima Formation in the Santa Cruz Mountains. Granodiorites exposed along the south slope of Monterey Canyon and the east slope of Carmel Canyon correlate with lower Upper Cretaceous Santa Lucia granodiorite on land. Siliceous siltstone, highly-indurated micritic limestone, and metamorphic rocks comprise the west slope of Carmel Canyon. These siltstone and limestone beds contain silicoflagellates, radiolarians, and middle Miocene diatoms, but no Foraminifera. Quartzite from Carmel Canyon probably correlates with the pre-Cretaceous Sur series.

A seaward extension of Sur (Nacimiento) and Palo Colorado faults separates sedimentary and metamorphic rocks from igneous ones in Carmel Canyon; it probably continues northward to connect with the San Gregorio fault in the Santa Cruz Mountains. Monterey fault in Monterey Canyon separates upper Pliocene rocks on one side from granodiorites on the other. The sedimentary rocks in Monterey Canyon contain bathyal foraminifers stratigraphically 400 ft higher than shallow-water Pliocene rocks in the lower Salinas Valley. These relations suggest vertical movement of more than 1,000 ft on a northward extension of the Reliz fault which separates the two areas. A seaward extension of the Carmel Valley fault controls the nearshore east-west trend of Carmel Canyon and the Carmel Canyon fault (seaward extension of the Sur and Palo Colorado faults) controls the northwest trend of Carmel Canyon. Canyon erosion occurred in zones of poor induration and/or weakness along faults and contacts between sedimentary and igneous rocks having diverse induration. Erosion was effected by turbidity currents, by mass movement of sediment down canyon axes, and perhaps was aided by alteration of granodiorite to clay.

The middle Miocene Salinian deformation resulted in more than 15,000 ft of subsidence in Monterey graben, which underlies the shelf at the north side of Monterey Bay. Additional structural features such as Monterey, Gabilan, Carmel Canyon, and Carmel Valley faults either formed at this time or were reactivated. Geomorphic features termed the "Elkhorn erosion surface" and "Pajaro gorge" were formed also at this time.

Late Miocene, Pliocene, and early Pleistocene drainage from the Great Valley of California debouched at Monterey Bay via Elkhorn Slough at the head of the present Monterey Canyon. All canyon heads were cut or modified subaerially to an approximate depth of 300 ft below present sea level. Monterey and Soquel Canyons were cut during late Pliocene or early Pleistocene time. Carmel Canyon and Ascension Canyon initially may have been eroded earlier; they may have funneled sediments seaward before Pleistocene time.

Monterey Canyon and Elkhorn Slough lie directly above the buried middle Miocene Pajaro gorge. The older canyon is not ancestral to Monterey Canyon, but differential compaction in the gorge fill may have directed rivers to reach the sea above the gorge. A sediment trap may have been eroded at the mouth of the slough whereby both littoral-drifted and river-transported sediments were deflected seaward to aid in the erosion of Monterey Canyon.

INTRODUCTION

Monterey Canyon is one of the largest submarine canyons of the world with respect to width, depth, and length. Located about 75 mi south of San Francisco, California, the canyon extends from the shore of Monterey Bay beyond the base of the continental slope (Fig. 1). Its large size and deep notching of the continental shelf made it one of the first canyons identified in the

[1] Manuscript received, April 22, 1966; accepted, December 27, 1966.

[2] University of Southern California; now at the Maryland Department of Water Resources.

[3] University of Southern California; now at Woods Hole Oceanographic Institution.

This study was supported by National Science Foundation Grant No. G-15747. Appreciation is due the officers and crew of *R/V Velero IV* and to many graduate students of the Department of Geology, University of Southern California, who participated in the field work. Special mention is made of the advice and assistance of Elazar Uchupi and Jobst

Hülsemann, now at the Woods Hole Oceanographic Institution. Among the many members of the petroleum industry who supplied data and suggestions are the following: E. H. Stinemeyer (Shell Oil Co.); W. A. Stokesbary (Shell Oil Co., retired); A. J. Macmillan and D. E. Six (Texaco Inc.); T.A. Baldwin and O. D. Hart (Humble Oil Co.); R. W. Rex, F. F. Sabins, Jr., and G. W. Starke (Chevron Research Co.); and W. H. Corey (consultant, formerly with Continental Oil Co.). Other materials were given by E. A. Gribi, Jr. (consultant), R. R. Thorup (consultant), G. D. Hanna (Calif. Acad. Sciences), O. L. Bowen and

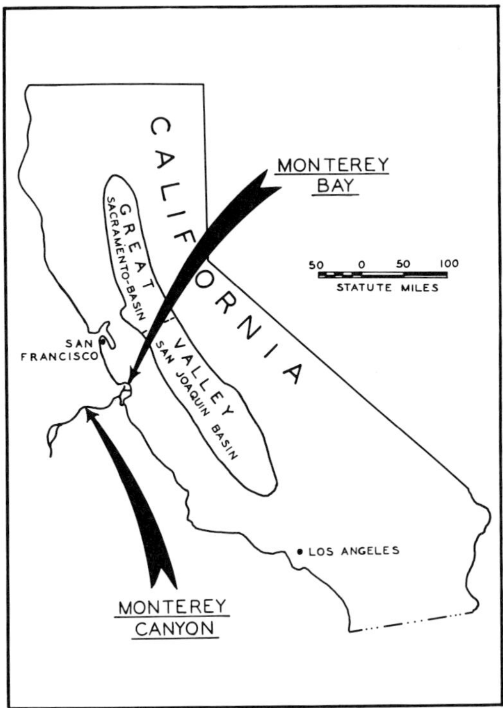

Fig. 1.—General location of Monterey Bay and Monterey Canyon, California.

United States on the basis of lead-line soundings (Davidson, 1887, 1897). Later, echo soundings made by the U.S. Coast and Geodetic Survey in 1932, 1933, and 1934 served as the basis for contour maps (Shepard and Emery, 1941) that clarified canyon relations to the continental margin and to smaller nearby canyons.

Attempts to establish the origin of the canyon were made by supplementing topographic information with studies of geological structure and river drainage systems on land (Beard, 1941) and

C. W. Jennings (Calif. Div. Mines and Geology), J. C. Clark (Stanford Univ.), R. S. Dietz (Environmental Sci. Service Admin.), R. F. Dill (U.S. Navy Electronics Lab., San Diego, Calif.), and F. P. Shepard (Scripps Inst. Oceanography, La Jolla, Calif.). Special thanks are due O. L. Bandy (Univ. Southern Calif.) for verification of foraminiferal identifications, G. D. Hanna (Calif. Acad. Sciences) for identification of diatoms and silicoflagellates, and to E. H. Stinemeyer (Shell Oil Co.) for data on diatoms, silicoflagellates, radiolarians, and sponge spicules in sample C-5 and for additional information on living foraminifers in Monterey Bay. Samples are stored at the Smithsonian Institution, Washington, D.C.

by dredging rock from canyon side slopes at sea (Shepard and Emery, 1941). Neither study proved definitive, chiefly because of insufficient samples and inadequate techniques. Other published investigations of submarine-canyon geology have been confined to topography of the upper part of the canyon (Woodford, 1951), to its seaward extension as a deep-sea channel (Dill et al., 1954), to a study of Monterey and Delgada submarine fans (Menard, 1960), and to sediments of the continental shelf (Galliher, 1935; Cohee, 1938; Pratt, 1962).

Early rock dredging had shown granodiorite to be present along one side of Monterey Canyon and both sides of the head of one tributary (Carmel Canyon), giving rise to speculation that Monterey Canyon had been eroded in granodiorite along much of its length. Investigation of this hypothesis could be made only by dredging additional samples from canyon walls, studying them, and correlating the results with onshore surface and subsurface data. This report summarizes the results obtained from these two avenues of investigation.

Two cruises aboard the R/V Velero IV (research vessel of the University of Southern California) to Monterey Canyon during 1961 and 1962 were made to investigate the geology of the canyon. Eighteen ship-days were spent in dredging canyon slopes, resurveying topography by modern continuous acoustical methods, and obtaining grab samples and cores. Concentration of effort was the greatest in the area east of Long. 122° 15′ W (Fig. 2), in order to relate the geology of the submarine area to that of adjacent land.

TOPOGRAPHY

Monterey Canyon (Fig. 2) has a V-shaped profile and a meandering plan. Two major tributaries join it: Carmel Canyon and Soquel Canyon. The larger tributary, Carmel, is straight along most of its course except for an abrupt bend approximately 3 mi from shore. The meandering plan of Monterey Canyon and the straight course of Carmel Canyon were attributed by Shepard (Shepard and Emery, 1941, p. 72–73) to control of erosion by masses of hard rock. Other control in addition to lithology is suggested by the present study.

Precision soundings made aboard R/V Velero

FIG. 2.—Topography of inner part of Monterey Canyon. Contour interval is 100 fm, with 50-fm contour added. Straight lines show positions of new precision soundings made aboard *R/V Velero IV*, supplementing older spot echo soundings of U. S. Coast and Geodetic Survey.

IV provide far better information about canyon side slopes than did the earlier spot-echo soundings. Many irregularities are shown by profiles that cross the canyons (Fig. 2). Side slopes of Monterey Canyon range from 5° to 34°, those of Carmel Canyon from 8° to 40°, and those of Soquel Canyon from 1° to 25°. Where such irregular and steep slopes are found during marine geological investigations, they generally are associated with rock outcrops. Profiles made at bends

of the canyon axes usually exhibit much steeper slopes on the outside than on the inside of bends, as though the steeper side slopes resulted from undercutting, similar to cutbanks of continental rivers.

Measurements between profiles across the canyon axes show that the axial gradient of Monterey Canyon flattens from 8.5° at its head to 0.5° at the seaward limit of Figure 2, although this decrease is not regular. Both tributaries steepen near their intersection with Monterey Canyon.

Within the area of Figure 2 is shown the topography of the inshore 30 mi of the axial length of Monterey Canyon. This area includes the point of maximum relief, about 1,000 fm near the intersection of Monterey and Carmel Canyons and it contains the maximum width of 13 mi between opposing rims. Beginning at a depth of about 1,450 fm (3 mi west of the limit of Fig. 2), the canyon floor is so flat and broad that the cross section is trough-like. This section, known as Monterey trough (Shepard and Emery, 1941, Pl. III), extends seaward for 32 mi with diminishing relief. Beyond Monterey trough is a long region of low relief atop the continental rise where the axis continues seaward in the form of deeply incised channels that divide and rejoin. It is suggested that the incision of the deep-sea channels into the fan may reflect a cycle of rejuvenation. The channels have been traced for a distance of about 150 mi, giving a total length of at least 215 mi for Monterey Canyon, Monterey trough, and Monterey deep-sea channels. Many sounding profiles were made throughout the length of the canyon and its seaward extensions, but, as rock outcrops have been found only in the inner 30 mi of Monterey Canyon, the present report is limited to the area plus the adjacent land.

Rock Samples from Sea Floor

methods

During 1937 Shepard (Shepard and Emery, 1941, p. 89, Pl. 18) dredged rock from four places on the side slopes of Monterey and Carmel Canyons. When the study was renewed in the 1961 and 1962 cruises of *R/V Velero IV*, 52 more dredgings were attempted. Of these, 21 recovered rock, of which 18 are considered representative of the local outcrop.

Rocks were judged to be in place on the basis of criteria cited by Emery and Shepard (1945): fresh fractures, large size of individual rocks, abundance of a single lithologic type, general angularity, fragile rocks, and firm catching of the dredge on bottom. The abundance of a single lithologic type in specific areas of dredging and the distance between dredgings seem to assure that the dredged rocks are truly representative of the basement and younger strata.

Rocks were obtained with two types of dredges: a chain-mesh biological dredge and a pipe dredge constructed by welding steel rods across one end of a pipe 18 in. in diameter and 3 ft long. Both dredges successfully obtained rocks, but the pipe dredge was used more extensively, as it was less expensive to replace if lost. In water depths greater than 150 fm, the best success occurred when the weight of the dredge exceeded 500 lb which was accomplished by adding additional weights. Dredgings were made both parallel with and at right angles to contours, with the length of cable paid out about 50 per cent greater than water depths. In areas of average to low gradient (approximately 20° or less) covered with a small thickness of pelagic mud and/or where the rock is well-indurated (west wall of Carmel Canyon), dredging parallel with contours appeared to produce better results than dredging at right angles to them.

Positions of all sea-floor samples, both old and new, are shown on Figure 3. Comparison with Figure 4, which shows only samples of rock judged to be from outcrops, indicates that outcrops are restricted generally to the steeper side slopes of canyons. In most other places, bedrock is buried by sediments sufficiently thick to be impenetrable by dredge. Coring might have recovered an inch or so of bedrock in marginal zones covered by less than 10 ft of sediment, but the narrowness of these zones and the small size of resulting rock samples give such samples a low value in comparison with the effort needed to obtain them.

types, ages, and distribution

Sedimentary rocks.—Sedimentary rocks were dredged from outcrops in three areas: (1) the north side slope and nearshore south slope of Monterey Canyon with shoreward limits approximately 6 mi from the canyon head, (2) both slopes of Soquel Canyon, and (3) the west slope of Carmel Canyon. Fifteen dredgings contained sedimentary rocks presumably from outcrops.

The most common rock is brown to gray siltstone cropping out mainly on the north slope of Monterey Canyon, but the rock also is present in a more restricted area on the south slope of the canyon (Fig. 4). This siltstone is similar lithologically to rocks of the upper Pliocene Purisima Formation exposed along the beach near the city of Santa Cruz. Bedding or laminations are rare.

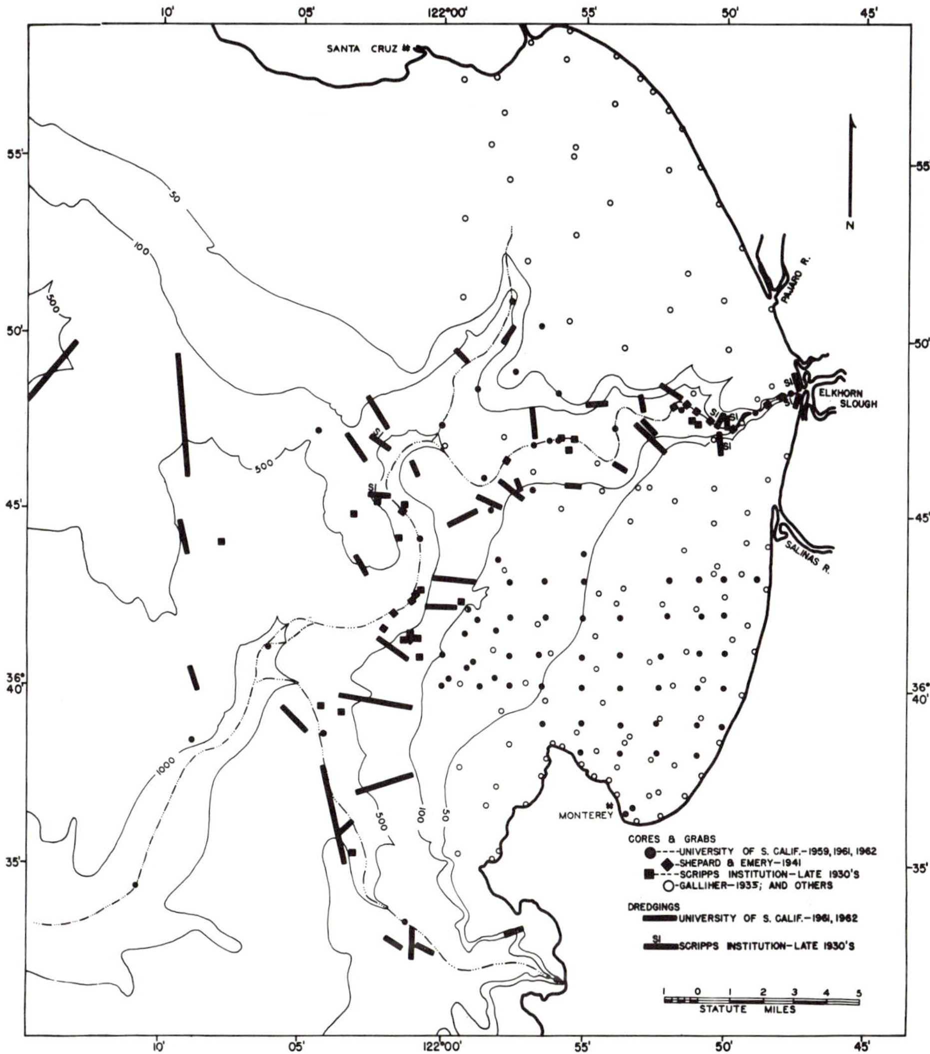

FIG. 3.—Positions of bottom samples.

Worm borings (identified by Olga Hartman, Allan Hancock Foundation, University of Southern California) are common in some samples and almost absent in others. Many borings are filled with sediment, but unfilled holes contained no animals. Density commonly is low (approximately that of diatomite). Lawson (1893) noted a low density in rocks of the upper Miocene Monterey Formation in the Carmel area and attributed this phenomenon to removal by solution of calcareous material, including microfossils. The siltstone consists primarily of quartz, biotite, and feldspar, orthoclase being more abundant than plagioclase.

In thin section, siltstone on the north slope of Monterey Canyon generally is so fine grained that most individual minerals cannot be determined optically; however, quartz grains as large as 75 microns were observed. These larger grains constitute about 5 per cent of the total and occur in a greenish-brown matrix having a grain diameter of approximately 15 microns.

Foraminifera date the siltstone beds as late Pliocene. These rocks are equivalent to all or part of the upper Pliocene Purisima Formation in the Santa Cruz Mountains. The lack of early Pliocene or Miocene assemblages was noted in making the age determination. The following composite list includes Foraminifera from samples M-8, M-14, M-19, and S-1 (Fig. 4). Fauna from each sample is listed by Martin (1964). Identifiable species are:

Angulogerina hughesi (Galloway and Wissler) (worn)

Bolivina decurtata Cushman (Mio-Pliocene)
B. seminuda Cushman (var.)
B. subadvena Cushman var. *spissa* Cushman
B. vaughani Natland
Buliminella california Cushman (distorted)
B. elegantissima (d'Orbigny)
Carpenteria monticularis Carter
Cassidella sp.
Chilostomella ovoidea Reuss
Cibicides fletcheri Galloway and Wissler

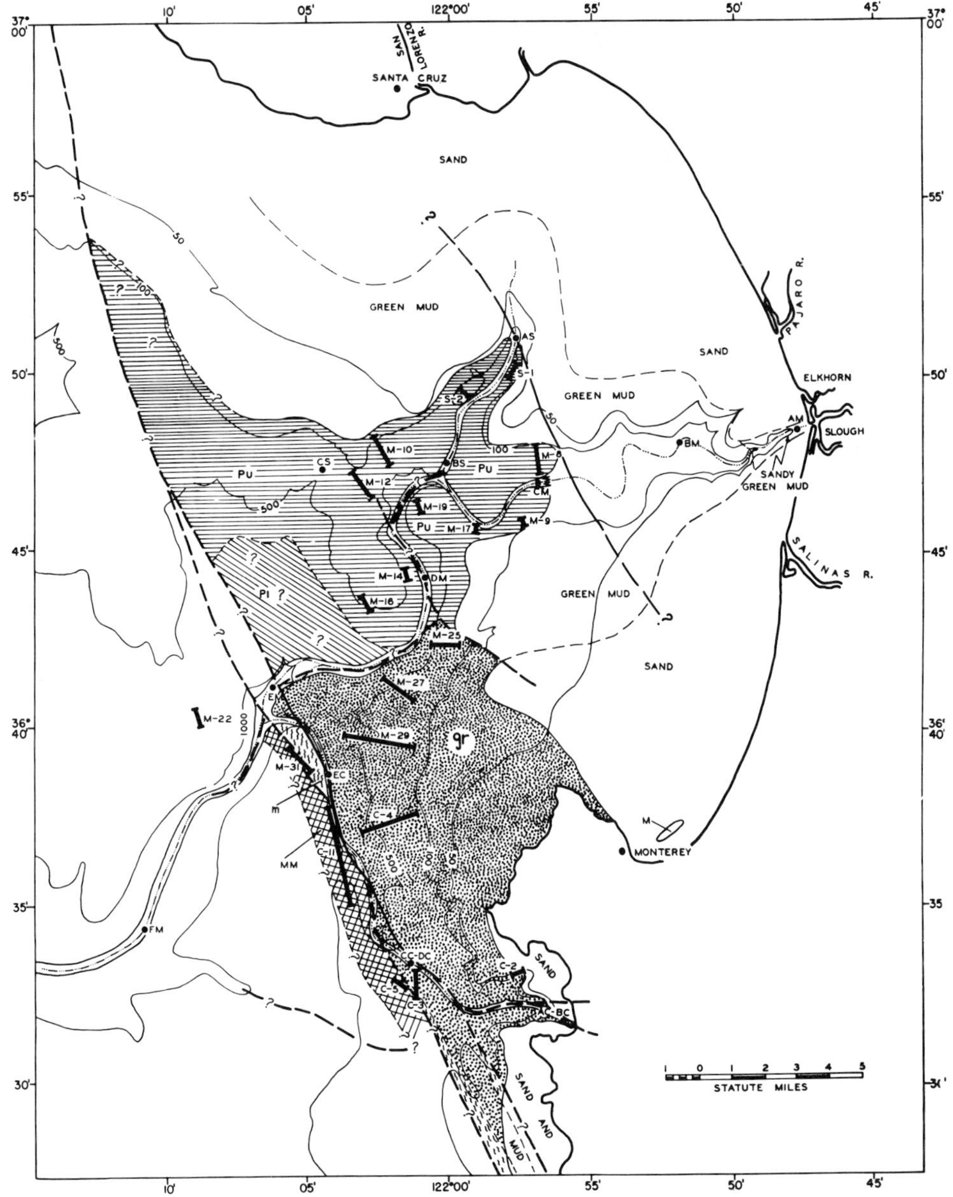

FIG. 4.—Geologic map based on samples from sea floor and extrapolation from outcrops on shore. Symbols are same as Figure 5.

C. lobatus (Montagu) (worn)
C. tuberculatus Natland
Discorbis sp.
Epistominella pacifica (Cushman)
Globigerina concinna Reuss
G. pachyderma (Ehrenberg)
G. uvula (Ehrenberg)
Hanzawaia concentrica (Cushman)
Loxostoma instabile (Cushman)
Nonionella miocenica Cushman
Planulina baggi? Kleinpell
Raphanulina gibba (d'Orbigny)
Uvigerina juncea Cushman and Todd
U. peregrina Cushman
U. subperegrina Cushman and Kleinpell
Valvulineria araucana (d'Orbigny) (fossil?)

Along the west slope of Carmel Canyon (Figs. 2, 4), dredgings C-11 and M-31 retrieved siltstone distinct from that in Monterey Canyon. Distinguishing characteristics are brown color, high density, and high degree of induration caused by siliceous cement. In the extremely fine-grained brown silicic matrix are clear anhedral quartz grains ranging in diameter from 250 to 500 microns; no dark mafic minerals are evident. Foraminifera are absent, but dredging M-31 contains diatoms (approximately 400 per gram of sediment) as well as silicoflagellates and radiolarians. The following list of diatoms and silicoflagellates is from samples M-31 and C-5:

Stephanogonia polygona
Coscinodiscus asteromphalus
G. marginatus
C. lineatus
Actinoptychus undulatus
A. splendens?
Raphonela amphiceros
Diploneis sp.
Xanthiophyxis sp.
Dictyocha fibula
Distephanus sp.

Stephanogonia polygona dates the rocks as middle Miocene. Radiolarians are poorly preserved, but one cubosphaerid type was recognized (W. R. Riedel, written commun., 1963). Little of M-31 and C-11 was dredged, perhaps because of the induration which prevented the dredge from digging into the siltstone exposures.

Sandstone on the south slope of Monterey Canyon is represented by samples M-17, M-19, and perhaps M-9. Dredging M-9 contained a varied group of rocks, including sandstone, siltstone, and granodiorite. Part of the sample probably is material from the canyon axis, but the sandstone appears to be in place. The sandstone samples are generally gray, but some surfaces are oxidized brown. Quartz and feldspar (plagioclase and

orthoclase) are the most important minerals; biotite in small quantities is disseminated throughout. The sandstone ranges from fine grained and well sorted to poorly sorted, the grains ranging from 0.2 to 2.0 mm. Siltstone similar to that from the north slope of Monterey Canyon is associated with the sandstone.

Sample M-19 (sandstone) contained the following Foraminifera:

Bolivina spissa Cushman
Cibicides wuellerstorfi (Schwager)
Epistominella pacifica Cushman
Fissurina sp. cf. *lucida* Williams

These Foraminifera and those described in the next paragraph indicate a late Pliocene age, equivalent to the Purisima Formation.

Another arkosic sandstone crops out on the west slope of Soquel Canyon (sample S-2). Surfaces in contact with the water weather to yellow-brown, but a freshly broken sample exhibits a blue-white appearance. The sandstone is friable and consists principally of quartz, orthoclase, and minor amounts of plagioclase poorly cemented by clay. This matrix appears to be authigenic and derived from alteration of the orthoclase, because individual clay grains within a specific area (where an orthoclase grain had once existed) extinguish simultaneously under crossed nicols. Foraminifera include:

Cibicides sp., probably *lobatus*
C. tuberculata (d'Orbigny) (worn)
C. wuellerstorfi (Schwager)
Epistominella pacifica Cushman
Discorbis sp.
Miliolinella obliquinodus (Riccio)

Limestone dredged from the west slope of Carmel Canyon (sample C-5) is gray, extremely well indurated, and microcrystalline. Calcium carbonate content is 94 per cent. According to the classification of Folk (1962, p. 62), the limestone is a micrite; however, because of the included diatoms, radiolarians, and silicoflagellates, the rock could also be termed a micritic-skeletal limestone (Leighton and Pendexter, 1962, Fig. 1). Contained flora and fauna of middle Miocene age are similar to those from dredging M-31 from the west slope of Carmel Canyon. The strong pull on the cable during dredging and the fresh-appearing fractures suggest that the sample was in place, Dredgings M-9 and M-29 also contained limestone, but in these samples clasts are rounded and pitted on the surface suggesting transport from other regions. As limestone is a common rock type onshore in middle Miocene beds (Relizian Stage of Kleinpell, 1938, p. 119) as well as in the pre-Cretaceous Sur series, the occurrence of this lithologic type is to be expected, both in outcrop and in transported material.

Green-black siltstone, the least common sedimentary rock type, was obtained from dredging M-22. After a mass of green mud was washed away, small pieces (to 1.5 mm in diameter) of moderately indurated siltstone remained on the screen. Associated with the siltstone were rounded and polished pieces of black chert and metamorphic(?) rock. In thin section, the siltstone is extremely fine grained, ranging to 10 microns in diameter with a few grains (less than 1 per cent) having diameters as great as 75 microns. Circular areas in some siltstone fragments suggest the former presence of diatoms.

Igneous rocks.—Igneous rocks of granodioritic composition are present along part of the south slope of Monterey Canyon, the east slope of Carmel Canyon, and on both sides of the nearshore part of Carmel Canyon (Fig. 4). This proximity to land exposures of Santa Lucia granodiorite suggests that igneous rocks in the marine environment are part of the same pluton. Submarine granodioritic rocks, however, lack the well-developed phenocrysts of orthoclase noted in nearby land outcrops (Lawson, 1893), but this textural difference is not considered to be unusual, for on land the porphyritic type grades southward from the Monterey region into a nonporphyritic and orthoclase-deficient quartz diorite (Trask, 1926). Leo (1961) also pointed out a wide compositional variation in intrusive rocks of the nearby Santa Cruz Mountains.

Thin sections of igneous rock from the marine environment present differences in mineral percentages, but a typical sample consists of:

Mineral	Percent	Remarks
Quartz	35	
Orthoclase	30	
Plagioclase	20	(An-29, oligoclase)
Biotite	10	
Chlorite	5	

This rock actually would be classed as a quartz monzonite, but, because of local usage of the term Santa Lucia granodiorite and the variation in mineral percentages of the igneous rocks, the term "granodiorite" is used as an all-inclusive one for the intrusive igneous rocks.

Orthoclase exhibits Carlsbad twinning and is cloudy in some grains and clear in others. Albite twinning is common in plagioclase grains permitting the composition of this mineral to be determined by measurement of extinction angles. Edges of grains interlock somewhat, and boundaries between minerals are sharp.

Many granodiorite samples contain evidence of alteration of the surface in contact with ocean waters; some fragments can be crumbled by hand. Some larger pieces exhibit a progressive change from highly altered material at the surface to fresh at a depth of 4 in. Martin and Rex (1965) and Rex and Martin (1966) investigated this phenomenon in detail by means of X-ray diffractometry and electron microscopy, and noted that potassium feldspars are most severely altered. The authigenic clays formed by submarine weathering of the granodiorite are kaolinite, montmorillonite, K-mica, and halloysite.

Granodiorites in dredging C-3 have slickensides and fault gouge, some of which are cemented by tar (CCl_4 test). The hydrocarbons may have originated in the sedimentary province west of the canyon and later migrated into the granodiorite or they may represent a local seep.

The granodiorites on land adjacent to Monterey Bay (Gabilan Mesa quartz diorite and Santa Lucia granodiorite) were dated by Curtis *et al.* (1958), using the potassium-argon method, as 83.9 and 81.6 m.y., respectively. Thus, granodiorites in the marine environment probably are of early Late Cretaceous age.

Metamorphic rocks.—Dredging M-31 on the west slope of Carmel Canyon near the intersection of Monterey Canyon recovered a small amount of metamorphic rock in addition to sedimentary types previously described. This rock is a white quartzite, having biotite along planes of jointing or fracturing; the largest piece is approximately 1.5 in. in diameter. Strong induration probably prevented more sample from being obtained, as the dredge could not dig into the outcrop.

Thin sections show the rock to consist principally of quartz (85 per cent), plagioclase (10 per cent), and biotite (5 per cent). Feldspar was not noted in hand specimens. An extinction angle of 40° on albite twins indicates that the mineral is labradorite (An-60). Quartz grains are anhedral and range in diameter from 75 to 400 microns;

plagioclase grains exhibit euhedral to subhedral shapes and range in diameter from 25 to 100 microns. Many plagioclase grains are partly or completely altered to a mineral resembling chlorite, and are elongated somewhat by secondary slippage along twinning planes.

GEOLOGICAL MAP

Rocks dredged from the sea floor permit the construction of a geological map of Monterey Canyon and vicinity (Figs. 4, 5). Granodiorite of probable Late Cretaceous age is most widespread, extending northward from outcrops onshore to Monterey Canyon. A sliver of pre-Cretaceous metamorphic rock is along the northwestern border of the granodiorite. West of both granodiorite and metamorphic rock is an area of middle Miocene siltstone.

North of the granodiorite, siltstone of the late Pliocene Purisima Formation crops out. Most of this area lies north of Monterey Canyon, in contrast to the granodiorite, metamorphic rock, siltstone, sandstone, and limestone that only crop out south of the canyon. Because of the large number of tributaries (Fig. 2) whose existence suggests that the underlying strata are easily eroded, an area labeled Pl(?) on the north slope of Monterey Canyon (Fig. 4) is inferred to be sediment although samples could not be obtained. In addition, the area of inferred sediments has a gentler slope than granodiorites on the opposite side of the canyon.

Rock outcrops appear to be absent nearshore and in deep water because of the presence of a thick cover of unconsolidated sediments. These nearshore sediments may be the source of clasts larger than 2 mm found in Monterey Canyon (Fig. 4, axial grab samples DM and EM), because Shepard (1948) determined that material larger than sand size was not entering the head of the canyon. Furthermore, the material greater than 2 mm is not being funneled down Soquel Canyon, because grab samples AS and BS (Fig. 4) indicate that this canyon has silt- and clay-size material in the axis.

In summary, much of the south side of Monterey Canyon and all of the east side of Carmel Canyon consist of granodiorite. Upper Pliocene sedimentary strata lie opposite granodiorites in Monterey Canyon; middle Miocene highly indurated siltstone and limestone crop out opposite granodiorites in Carmel Canyon (except for headward regions where granodiorite occurs on both walls). Also in Carmel Canyon, a sliver of pre-Cretaceous(?) metamorphic rock appears to lie between the Miocene siliceous siltstones and Cretaceous intrusives. Physiographic boundaries separating areas of greatly different lithologic units are strongly suggestive of structural control of both outcrop distribution and canyon trends.

DEPTH OF DEPOSITION OF SEDIMENTARY ROCKS

In the Santa Cruz Mountains (Cummings *et al.*, 1962, p. 210), the Purisima Formation consists of 2,150 ft of shallow-water beds overlain by the deep-water Pomponio Member. The Pomponio is overlain by 1,200 ft of shallow-water beds indicating that this Pliocene section was deposited during a general advance and retreat of the sea. In Monterey and Soquel Canyons, various Foraminifera indicative of both bathyal and open-ocean conditions occur in samples of the Purisima Formation. Dredgings M-8, M-14, M-19, S-1, and S-2 (Fig. 4) include specimens of *Epistominella pacifica* (Cushman), *Uvigerina juncea* Cushman and Todd, *U. peregrina* Cushman, *U. subperegrina* Cushman and Kleinpell, and *Globigerina* sp. The Monterey Canyon Pliocene section represents either deposition in ocean depths not affected by an advance and retreat of the sea (and thus equivalent to the total Pliocene section on land) or the canyon section is equivalent to only a part of the Pliocene section on land.

The presence of the deep-water Foraminifera in the submarine Pliocene section suggests a correlation with the deep-water Pomponio Member of the Purisima Formation of the Santa Cruz Mountains area. *Uvigerina juncea* was described by Goodwin and Thomson (1954) from the Lobitos Mudstone Member of the Purisima Formation in the Santa Cruz Mountains, but all other Foraminifera in this member indicate water depths shallower than 50 fm. Bandy (1953, 1960) discussed the difficulties in correlating foraminiferal ecology between areas of shallow-water and deep-water deposits. The Monterey Canyon Pliocene section correlates only with the Pomponio Member of the Purisima Formation on land, but the submarine section also could be equivalent to other members.

Little can be inferred about the depth of water in which the middle Miocene strata in Carmel

Canyon were deposited, because foraminifers are absent and only diatoms were available for dating purposes.

Structure

Inferences from sea floor

Structural interpretations in the marine environment were based primarily on physiographic and lithologic data from the dredgings and acoustical records. In addition, structural patterns on land were extended into the marine environment; geophysical data from confidential (and thus not quotable) exploration sources were inspected and incorporated into the structural picture; and electric and lithologic logs of wells on land aided greatly in the structural interpretations.

Geologic evidence in the nearshore reaches of Carmel Canyon and on land suggests that the head and perhaps all of the east-west trend of Carmel Canyon are fault controlled. O. L. Bowen, California Division of Mines and Geology (personal commun., 1963), stated that on land, parallel with Carmel Valley, is an east-west fault which he named the Carmel Valley fault (Fig. 5). Documentation of the fault is provided by a seismic low mapped by H. C. Sieck (Stanford University) on the south side of the valley, and by differences in Miocene foraminiferal ecology investigated by Bowen in strata on opposite sides of the valley. In equivalent beds, foraminifers in rocks on the northeast side of Carmel Valley indicate deposition in shallower water than do those on the southwest side, suggesting that the northeast block was up during the Miocene, but still below sea level. Scuba dives into the head of Carmel Canyon revealed the presence of a steep north side composed of granodiorite and a south side completely covered by sand deposits sloping approximately 35° northward; this implies that the north side may be the scarp of an upthrown block. The writers suggest that the fault in the nearshore regions of Carmel Canyon is an extension of the Carmel Valley fault, and that canyon erosion appears to have been facilitated by a zone of weakness along the submarine part of the fault.

The Carmel Canyon fault (Fig. 5) probably is a northerly extension of the Sur fault. Geophysical information suggests that the Carmel Canyon fault returns to land still farther north to become the San Gregorio fault in the Santa Cruz Mountains. In addition, Upper Cretaceous sedimentary rocks lie west of the Sur fault (Nomland and Schenck, 1932) as well as west of the San Gregorio fault (Cummings *et al.*, 1962, p. 183), and granodiorites and metamorphic rocks crop out east of both faults. Considering the similarity of rocks on the same sides of both faults and the fact that the two faults (San Gregorio and Carmel Canyon-Sur) trend toward each other, a connection between them seems probable.

The trend of Monterey Canyon also may be controlled by faults, but evidence is scant except where upper Pliocene rocks and the granodiorites are present on opposite sides (Fig. 4). The writers refer to the possible fault (Figs 4, 6) along the canyon trend as the Monterey fault. If the axial fault is present, the original trend might have been displaced by subsequent movements on transverse faults. Buffington (1964, p. 54) showed that lateral displacement of the La Jolla Canyon trend projected to mapped faults on land. He suggested that the canyon displacements are related to lateral movements on the land faults. One such transverse fault in Monterey Canyon is postulated between Lats. 36°40′N and 36°47′ N in the vicinity of Long. 122°00′ W (Fig. 4). Left-lateral offset is suggested by the local change in axial trend of Monterey Canyon from southwest to southeast; then, approximately 3 mi down canyon the strike again changes toward the southwest. This transverse fault (?) in Monterey Canyon is along the trend of the Tularcitos fault in the Santa Lucia Mountains (Fig. 6), but according to Fiedler (1944, p. 237) movement on the latter fault is principally vertical. Sinuosity of the axial trend of Monterey Canyon east of the transverse fault implies that additional transverse faults may have offset the general southwest trend by varying directions and amounts.

Another fault, possibly of major significance, may occur just east of upper Pliocene outcrops in Monterey Canyon approximately 6 mi from shore (Figs. 5, 7). The top of these Pliocene submarine outcrops is about 400 ft above Pliocene beds penetrated by wells in the lower Salinas Valley (Fig. 7). More important is the fact that canyon outcrops contain bathyal foraminifers, whereas Pliocene beds on land have only shallow-water foraminifers. A downdrop of approximately 1,000 ft is suggested for the Pliocene beds on land relative

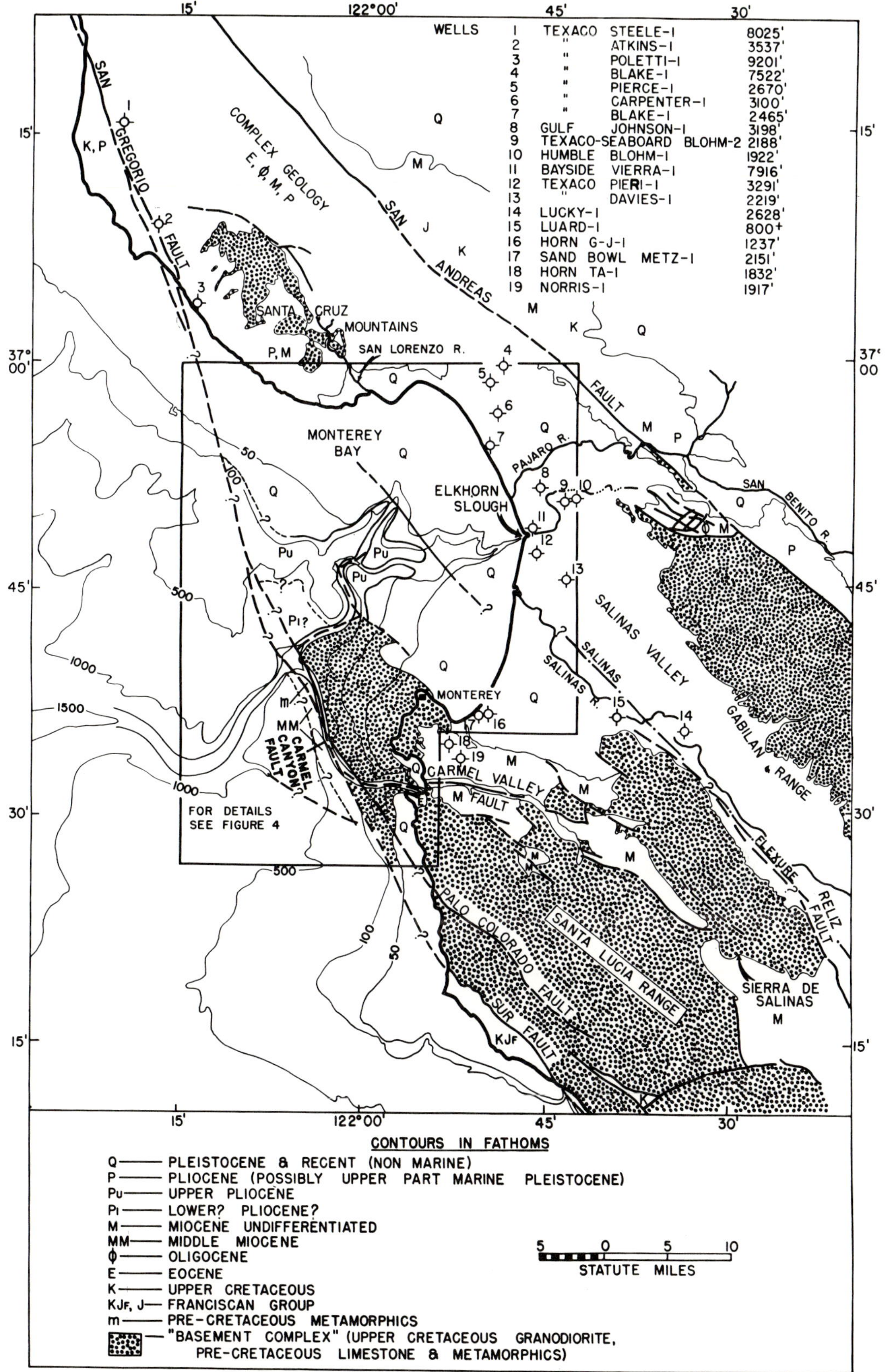

FIG. 5.—Relationship of geology of Monterey Canyon and vicinity in region; mainly after Jennings and Strand (1958) and Jennings and Burnett (1961).

411

to those in the canyon. Because marine Pleistocene and Recent sediments in the lower Salinas Valley contain shallow-water megafossils (Beard, 1941; Burch, 1946) having depth ranges similar to those of the foraminifers in the underlying Pliocene deposits, and because these younger deposits intercalate with nonmarine Pleistocene and Recent sediments, a broad shallow-water shelf must have existed east of the proposed fault. The rate of downwarping of the shelf appears to have kept pace with deposition rates.

Some uplift of upper Pliocene outcrops in Monterey Canyon is postulated (Fig. 7), on the basis of the absence of exposed marine Pleistocene sediments. This fault in Monterey Canyon may be a northward extension of the Reliz fault in the Salinas Valley (Figs. 5, 6), but until the connection can be proved, the writers use the term "Gabilan fault."

INFERENCES LARGELY FROM LAND

Studies of outcrops in California led Reed (1933, p. 28) to classify basement rocks into

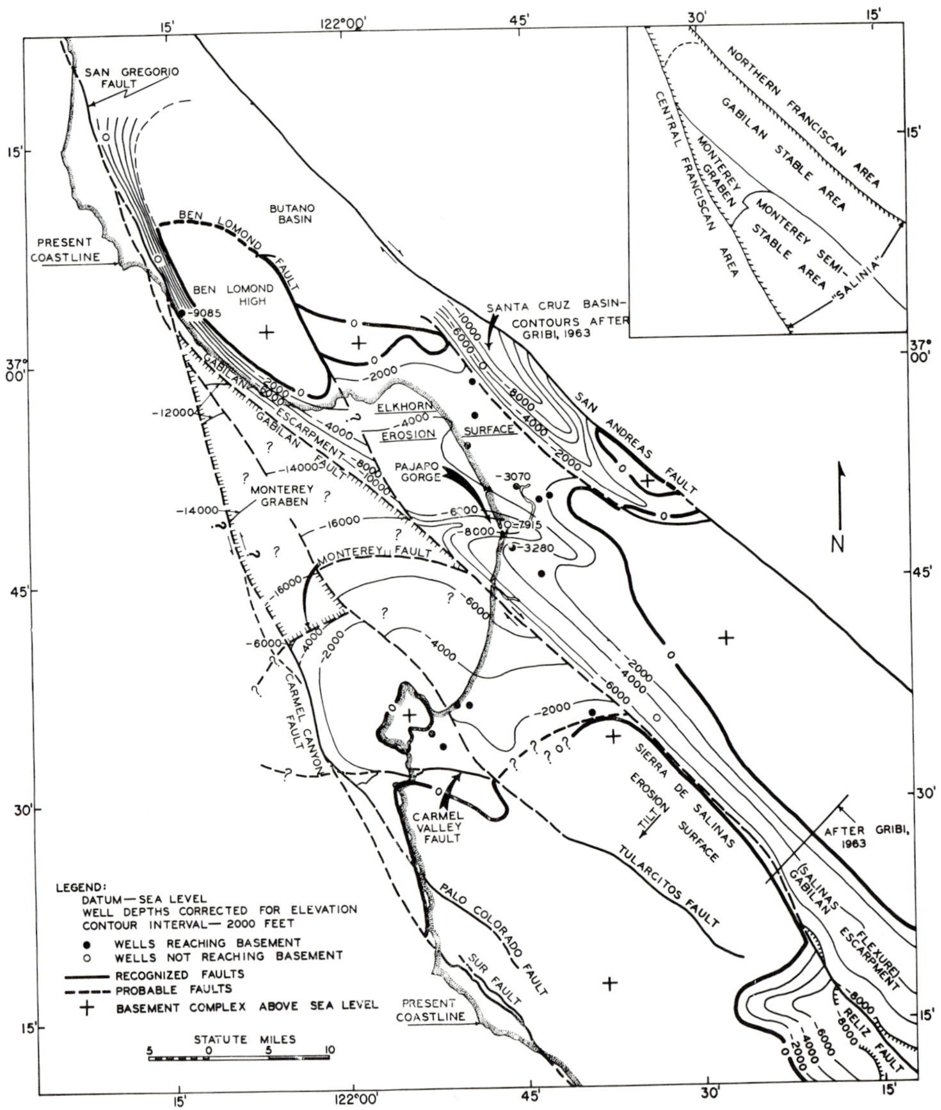

Fig. 6.—Subsea structural contours with datum on basement complex (Cretaceous granodiorites and older rocks). Well data are mainly from files of several oil companies.

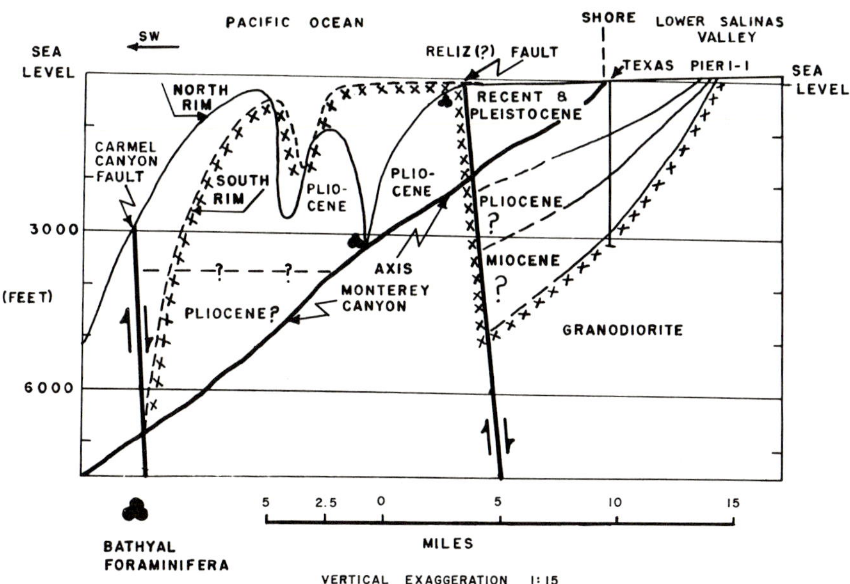

FIG. 7.—Cross section from lower reaches of Monterey Canyon (along axis) to lower Salinas Valley showing structural and stratigraphic relations across northward extension of Reliz(?) fault. East of Reliz(?) fault, stratigraphy is postulated from that known in Texas No. 1 Pieri and from outcrops which reached total depth in granodiorite. Pliocene rocks on north slope of Monterey Canyon west of fault were obtained in dredgings. Exposures of granodiorite (shown by X's on rim of south slope) are known from present study, and from work by Galliher (1935) and Shepard and Emery (1941).

areas of Upper Jurassic to Upper Cretaceous Franciscan types (unmetamorphosed to metamorphosed rocks containing serpentine) separated by major faults from areas of mostly Upper Cretaceous acid intrusive rocks. In the Monterey region (Fig. 6, top inset), areas of Franciscan basement rocks lie northeast of the San Andreas fault and southwest of the San Gregorio-Sur fault. Between these two fault systems (and between the two areas of Franciscan basement rocks) are basement rocks in a long triangular area named "Salinia" by Reed. These rocks consist chiefly of granodiorite, but also include some roof pendants of metasediments (Trask, 1926) quite unlike the serpentine of the Franciscan Group. As shown by the inset of Figure 6 (present structure), Salinia near Monterey consists of three parts: the Gabilan stable area, Monterey semistable area or Monterey high, and Monterey graben—names derived from effects noted in the development of the geologic history. Considerable information about the first two areas has been derived from wells and outcrops, permitting inferences about the third area still submerged below the northern shelf of Monterey Bay.

An attempt was made to construct a structural map, using the top of the basement complex (granodiorites) as datum, from data of present structural conditions. However, the effects of a middle Miocene deformation and a Pleistocene orogeny made it impossible to arrive at a reasonable reconstruction; therefore, a series of paleogeologic maps (Fig. 8) were constructed. Maps I to IV were drawn from many isolated pieces of information, but especially important to their construction was knowledge of present outcrops, well data in the lower Salinas Valley, information supplied by G. W. Starke (Chevron Research Co.) on the Pajaro gorge (but not necessarily his concepts of development), and 3 years of study of the geology of the region. From these maps, structural contours using as datum the top of the basement complex were drawn (Fig. 6). In addition, a fence diagram (Fig. 9) shows a three-dimensional concept of the structural features and stratigraphy. The following discussion of structural geology is from a more detailed study by Martin (1964).

During the construction of the structural map (Fig. 6) and the paleogeologic maps (Fig. 8), structural and geomorphic features related to structural movements were inferred. Structures

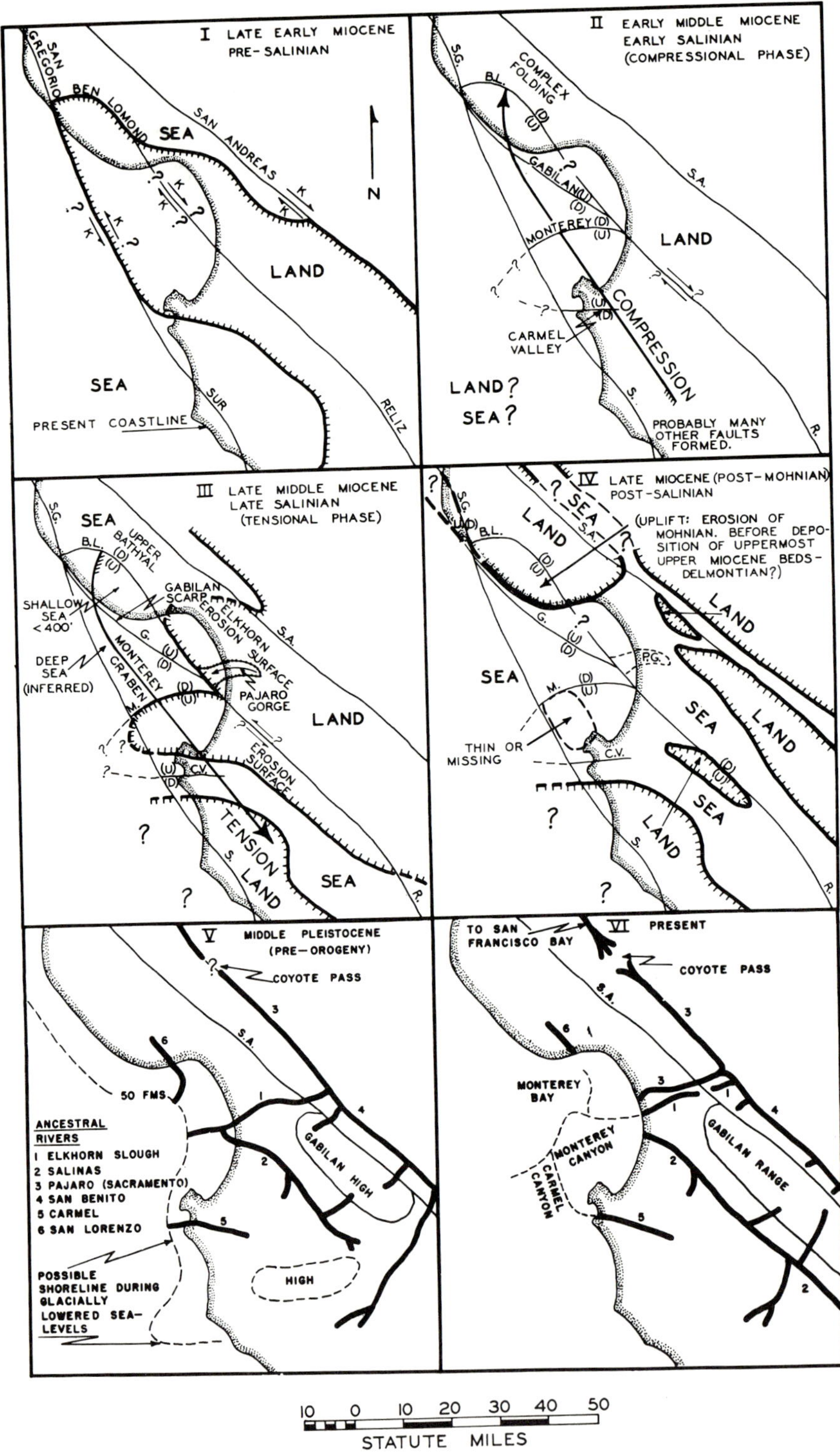

FIG. 8.—Shorelines and main drainage routes of Miocene and Pleistocene compared with present routes.

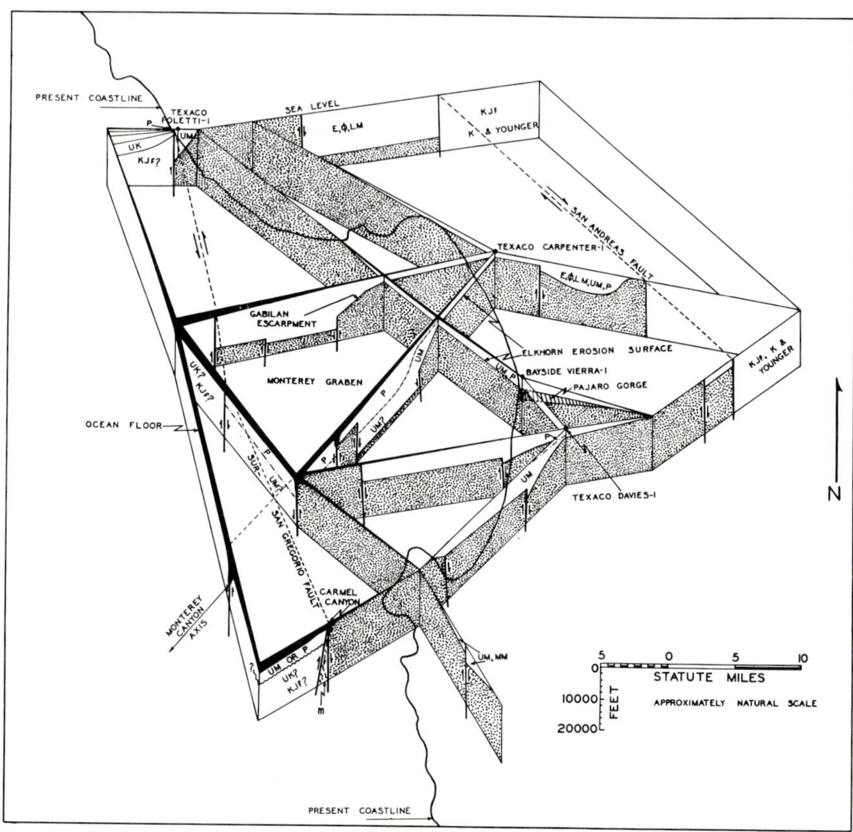

FIG. 9.—Fence diagram showing geology of Monterey Canyon and vicinity. Symbols are same as Figure 5 plus black for water.

are the Gabilan fault, the Gabilan escarpment (in part recognized by other workers), the Monterey fault (parallel with the trend of the canyon), and the unnamed connecting fault between the Carmel Canyon and San Gregorio faults. The previously mentioned Monterey graben is in the category of inferred structure. Geomorphic features include the Elkhorn erosion surface and the Pajaro gorge. E. A. Gribi, Jr., and R. R. Thorup (personal commun., 1966) informed the writers that the Ben Lomond fault (Figs. 6, 8) definitely connects with the fault southwest of the Santa Cruz basin. In order to explain contemporary differential movements of the land, some zone of slippage or boundary fault must exist between the areas of the Elkhorn erosion surface and the Ben Lomond high (Fig. 6). According to Gribi in a subsequent personal communication, the existence of such a boundary fault is reasonable.

Monterey graben is the name proposed for a depression under the north shelf of Monterey Bay. Here Pliocene and earlier (?) sediments, largely derived from rivers and littoral drift, accumulated in a water depth greater than 100 fm. Sediment thickness possibly exceeds 15,000 ft (Fig. 4), if the 6,000 ft of upper Pliocene and Pliocene (?) present on the north side slope of Monterey Canyon is combined with more than 9,000 ft of upper Miocene siltstone recorded in the Texaco No. 1 Poletti well north of Monterey Bay (Fig. 5). The presence of a thick prism of sediment under the north shelf of Monterey Bay also is suggested by the failure of seismic energies to reach the granodiorite basement.

Monterey graben is bounded on the south by the Monterey fault, on the northeast by the Gabilan fault, and on the west by the connecting fault between the San Gregorio and Carmel Canyon faults. Monterey fault is postulated largely on the basis of the diverse lithologic types on opposing canyon side slopes. Gabilan fault is suggested as a connecting link between the Reliz fault along the west side of the Salinas Valley and the San Gregorio fault north of Monterey Bay. This presumed fault must exist to account for discontinuity between local structural provinces on either side. The connecting fault between the Carmel Canyon and San Gregorio faults is indicated both by seismic evidence and by structural and stratigraphic studies of outcrops north and south of Monterey Bay.

The Gabilan escarpment is a new name proposed for buried escarpments known to be present both north and south of Monterey Bay. South of the bay (Figs. 5, 6), the escarpment is termed either the "Salinas flexure" (Baldwin, 1963) or the "King City hinge line" (E. A. Gribi, Jr., personal commun., 1963); north of Monterey Bay, it has been named locally the "Davenport escarpment." The writers consider the two separate escarpments to be parts of a single feature which extends approximately 125 mi from the Santa Cruz Mountains to the central Salinas Valley near King City (southeast of mapped area, Fig. 5). Relief of the buried feature ranges from more than 9,000 ft at the north (as shown by the depth to basement in Texaco No. 1 Poletti compared with surface outcrops a few miles east) to 2,000 ft at the south. The scarp faces in a general west-southwest direction. Under the waters of Monterey Bay, the presence of the Gabilan escarpment is only inferred; uplift occurred along the trends of the Reliz, Gabilan(?), and San Gregorio faults (Fig. 6).

The Elkhorn erosion surface (another proposed name) is recognized in various wells (Figs. 5, 6) drilled to the basement complex east of Monterey Bay in the lower Salinas Valley. The erosional surface on top of the basement complex slopes gently toward a low region directly below the head of present-day Monterey Canyon. This buried surface of low relief probably represents a region reduced by subaerial processes of erosion during pre-middle(?) or early middle(?) Miocene. It may have been contiguous with the erosion surface recognized by Fiedler (1944) in the Sierra de Salinas (Fig. 5) approximately 3,000 ft above sea level.

Indented into the Elkhorn erosion surface is a steep-sided depression that opens westward (Starke, 1956; Starke, personal commun., 1963);

with his concurrence, the feature is named the Pajaro gorge. Bayside No. 1 Vierra was drilled into gorge sediments. Total depth of erosion is not known, as the Vierra No. 1 well did not reach the basement complex at 7,900 ft, but erosion was at least 5,000 ft below the level of the Elkhorn surface. Woodford (1951, p. 834–837, Fig. 20) also recognized the abnormally thick sequence of beds in the Vierra No. 1 well compared with surrounding shallow water wells, and thought that the thick beds represented either deposits on the north edge of a late Pleistocene valley or those related to movement on a fault termed the "Moro Cojo." Later information from wells drilled to the basement (Fig. 5) and seismic information (Starke, personal commun., 1963) delineated the gorge and suggested that the stratigraphic section in Vierra No. 1 represents fill in the gorge. The lithologic log from Vierra No. 1 shows that the fill is in part composed of reddish sandstone and siltstone (especially below the level of the Elkhorn erosion surface) containing a few fresh-water gastropods (*Amnicola* sp.). These data suggested to F. P. Shepard and G. W. Starke (personal commun., 1963) that deposition of fill occurred under subaerial conditions and that gorge erosion probably was effected under similar conditions. This concept of environmental conditions must be re-examined, as Oligocene volcanic rocks in the Santa Cruz Mountains also weather to red colors (Cummings *et al.*, 1962) so detritus from the volcanic rock and the gastropods could have been introduced from land into marine deposits.

GEOLOGIC HISTORY

Two main periods of intense deformation are recognized: the well-known mid-Pleistocene orogeny and a more localized one during the middle Miocene. The effects of the earlier orogenic period, termed herein the "Salinian deformation," are masked by the more widespread mid-Pleistocene orogeny. The Salinian deformation apparently was confined within the region bounded by the Sur-San Gregorio and San Andreas faults. Deformation was more intense in the northwest part of the block (in part Reed's Salinia) than in the southeast.

During the Late Cretaceous and early Tertiary, major faults such as the San Andreas, Sur, San Gregorio, Reliz, and Ben Lomond (Fig. 8-I) developed; seas advanced and retreated in response to the mild diastrophism. In comparison with middle Miocene and mid-Pleistocene conditions, the land appears to have been stable. These conditions are postulated from many works including those of Lawson (1893), Kerr and Schenck (1925), Trask (1926), Fiedler (1944), Allen (1946), Cummings *et al.* (1962), and Gribi (1963a, b).

Beginning in the early part of the middle Miocene (Fig. 8-II), the area of Reed's Salinia (Fig. 6, inset) experienced initial pulses of the Salinian deformation, which eventually became most intense in the area between Ben Lomond-Reliz and Sur-San Gregorio faults. At this time, the Monterey and Carmel Valley faults probably were formed, the Gabilan fault may have been formed or rejuvenated, and the San Gregorio-Carmel Canyon-Sur fault system apparently was reactivated. Compressional stresses appear to have caused block movement northwestward between the bounding Sur-San Gregorio and San Andreas faults; the greatest compression occurred between the Ben Lomond-Reliz and the Sur-San Gregorio faults, with compressional effects becoming increasingly more intense from southeast to northwest. The San Andreas and Sur-San Gregorio faults converge (but do not intersect) toward the northwest.

When compression decreased during the later middle Miocene (Fig. 8-III), the structural and geomorphic history became very complex; many events occurred simultaneously, but in separate physiographic and/or structural provinces. Monterey graben began to subside, possibly as much as 15,000 ft relative to surrounding areas. The site of the Santa Cruz Mountains was extremely unstable, as indicated by two unconformities in middle Miocene and one in upper Miocene rocks. Sediments of Miocene age probably were deposited in Monterey graben, although none are known from that part of the sea floor within the graben; they are present on land and on the west slope of Carmel Canyon. During this time also, east of Ben Lomond-Reliz faults, land surfaces appear to have been high to permit subaerial erosion of the Elkhorn erosion surface and possible initial downcutting of Pajaro gorge by streams (later cutting and filling may have occurred in the marine environment). The area subsided near the end of the middle Miocene and marine upper(?) Miocene sediments were deposited on the eroded surface of the basement complex. Brabb *et al.* (1962) showed lower and/or middle Miocene deposits resting on the basement complex, but the foraminifers are not diagnostic of an age older than late Miocene.

During the early part of the late Miocene (Fig. 8-IV), the intensity of the Salinian deformation decreased and a period of regional downwarping followed. Transgression of the seas was widespread, but controlled by the pre-existing topography. The Elkhorn erosion surface became a slowly subsiding shelf area on which shallow-water marine and nonmarine sediments accumulated. It is postulated also that as the erosion surface subsided below sea level, Pajaro gorge became a submarine canyon through which sediments were transported into Monterey graben.

Among other indications of diverse movements in separate structural provinces is the absence of earliest late Miocene sediments in the Santa Cruz Mountains whereas they are present in the Santa Lucia Mountains (Fig. 8-IV). This suggests that the site of the Santa Cruz Mountains was high and above sea level at this time. However, T. A. Baldwin (personal commun., 1962) found that fragments of earliest late Miocene rocks are present in younger beds so it is evident that the area was uplifted, eroded, and downdropped during a comparatively short interval during the late Miocene. Deposition was continuous in the Santa Lucia Mountains. During the Pliocene, the site of the Santa Lucia Mountains was elevated and remained above sea level when general subsidence occurred elsewhere, including the site of the Santa Cruz Mountains. Fiedler (1944) found no evidence of marine Pliocene or younger marine sediments in the Santa Lucia Mountains.

Monterey graben continued to subside during the Pliocene; foraminiferal ecology indicates that subsidence kept pace with deposition, for bathyal foraminifers occur in Pliocene sediments in Monterey Canyon from the deepest dredging (500 fm) to the shallowest (100 fm).

The land during the Pleistocene was above sea level except in the lower Salinas Valley just east of Monterey Bay. Records of many shallow, water wells in this area (Beard, 1941) indicate that Pleistocene and Recent sediments are shallow-water marine accumulations intercalated with nonmarine sediments (Woodford, 1951, p. 837).

The need for a fault between the Pleistocene-Recent beds on land and the Pliocene Purisima Formation exposed in Monterey Canyon is evident.

Additional Pleistocene and Recent history is discussed with the origin of the canyon and its tributaries.

MONTEREY GRABEN

The existence of Monterey graben could not be discussed adequately until aspects of the structure, stratigraphy, and geologic history had been presented. Aspects most important include (1) the low base level required for the erosion of Pajaro gorge, (2) the need for a depression capable of trapping Miocene and Pliocene sediments to prevent their transportation into deeper parts of the ocean, and (3) the existence of the thick sedimentary prism on the north wall of Monterey Canyon.

Referring to base-level considerations, downcutting of Pajaro gorge could not have been effected unless a deep area was on the west and the channel was attempting to adjust to grade. The presence of an enclosed basin or graben is strongly suggested at the mouth of Pajaro gorge by the stratigraphic section on land west of the Sur-San Gregorio faults (Fig. 9). North of Monterey Bay, Upper Cretaceous, Oligocene or lowermost lower Miocene, and Pliocene rocks are exposed and separated by unconformities (Cummings, *et al.*, 1962, p. 183); south of the bay, Upper Cretaceous rocks occur west of the fault (Nomland and Schenck, 1932) with remnants of upper Miocene in fault blocks. The absence of most of the Miocene section west of the Sur-San Gregorio Faults suggests that this block is structurally high; north of the bay, it may have been below sea level only during Cretaceous, Oligocene to early Miocene, and Pliocene times. That part of the block now below the water of Monterey Bay may have been submerged only recently and reduced in relief by marine erosion, although it could have been high and a barrier to seaward transport of sediments during the Miocene and part of the Pliocene.

The thick prism of sediments on the north slope of Monterey Canyon has been discussed, but the fact remains that upper Pliocene rocks are present on the side opposite igneous rocks. A basin must have been present to permit the accumulation of deposits possibly thicker than 6,000 ft (Fig. 9). It is unlikely that the depression is part of an extremely irregular surface on top of the granodiorites (basement complex); a fault almost certainly must exist between granodiorites and upper Pliocene rocks in Monterey Canyon.

ORIGIN OF MONTEREY AND TRIBUTARY CANYONS

TIMES OF INITIAL EROSION

Erosion of Monterey Canyon, at least to an axial depth of 3,000 ft and within a distance of 15 mi from shore, began no earlier than late Pliocene or early Pleistocene time, because the youngest rocks exposed on the north slope of the canyon are late Pliocene (Fig. 4). Dating of initial canyon erosion by this method has been challenged by Shepard (1952, p. 86, Fig. 1) on the grounds that deposition (on the shelf) and erosion (in the canyon) could occur simultaneously. Thus, the youngest exposures of rock may not be an adequate criterion to determine the time of initial erosion of a submarine canyon. With respect to Monterey Canyon, the hypothesis of Shepard would still place the initial time of cutting within the Pliocene, because no earlier sedimentary strata were dredged in this part of the canyon. Also, no evidence of cut-and-fill is noted in Pliocene sediments east of Monterey Bay in the lower Salinas Valley; such should be noted if the canyon were open and being eroded during the Pliocene.

Soquel Canyon also is considered to have originated by erosion during late Pliocene or early Pleistocene, because the walls consist of indurated upper Pliocene rocks.

On the west slope of Carmel Canyon, middle Miocene rocks are the youngest known and their presence limits initial erosion to post-middle Miocene time for the northwesterly trending part of the canyon. There is even less information about the time when erosion started in the canyon east of Long. 122°00′ W, because the age of canyon wall rock is of early Late Cretaceous age.

The northwesterly trend of Carmel Canyon may have developed during the mid-Pleistocene orogeny when the uplift of the fault block west of Carmel Canyon (movement along the Carmel Canyon fault) diverted an original east-west trend of the canyon. Post-middle Miocene rocks were not dredged on the west slope, suggesting that the block was above sea level until recently.

This lack of post-middle Miocene strata is even more striking, considering that a thick sequence of upper Miocene marine rocks is present on land adjacent to the head of Carmel Canyon. The east-west tributary (Fig. 2) which intersects Monterey Canyon at approximately 1,300 fm and which heads near the 500-fm contour at Long. 122°03′ W possibly is the old beheaded channel of Carmel Canyon.

FACTORS INFLUENCING CANYON POSITIONS

Six major factors influenced the localization or positioning of the canyon complex in the Monterey Bay area:

1. Distribution of late Miocene and post-Miocene seas east of the present coast of California;

2. River systems on the continent;

3. Fluctuations of sea level, especially during glacial and interglacial stages of the Pleistocene Epoch;

4. Structural patterns developed during Late Cretaceous(?), middle Miocene, and Pleistocene;

5. Transportation of sediments and distribution of indurated rocks; and

6. Location of Pajaro gorge.

Distribution of seas and river systems, and fluctuations of sea level.—During late Miocene and Pliocene times, as in earlier Tertiary time, various coastal regions and the southern part of the Great Valley (Fig. 1) were covered by seas (Hoots *et al.*, 1954). Very little post-Eocene marine deposition occured in the northern Great Valley (Sacramento basin) except on its western side (Huey, 1948; Weaver, 1949; Taliaferro, 1951); marine upper Miocene and Pliocene deposits were restricted to an area east of present-day San Francisco Bay and the west side of the Sacramento basin (Howard, 1951; Taliaferro, 1951; Ham, 1952; Hall, 1958). One of the routes for the inland transgression of the seas during Miocene was in the Monterey Bay region. During early Pliocene, ocean water entered the southern Great Valley (San Joaquin basin), probably via both the southeasterly trending San Benito trough from the Monterey Bay region and from the Santa Maria area south of Monterey Bay (Durham and Addicott, 1965, p. A-18 and A-19); later Pliocene seas may have entered the basin via the San Benito trough only. The area of Monterey Bay probably was the only avenue for marine transgressions into the region east of present-day San Francisco Bay.

Before the close of Pliocene (possibly middle Pliocene), mild diastrophism uplifted the land west of the southern San Joaquin basin and the seaway between the basin to the open ocean was disrupted and was never re-established. Shallow, brackish, marine seas in the basin changed quite abruptly to fresh water, creating Lake Tulare. Cores from wells contain mega- and microfossils which document this change in salinity; exposures of rock in the anticlinal Kettleman Hills in the northern part of the basin also show the change (Woodring *et al.*, 1940). The seaway extending from Monterey Bay northward into the area east of San Francisco Bay apparently retreated gradually, as a result of both the uplift of the land and encroachment of alluvial deposits into the seaway. Studies in the area are hampered by a cover of Pleistocene alluvial debris (Livermore and Santa Clara Gravels).

The outflow from Lake Tulare probably followed the course of the earlier marine seaway parallel with the San Andreas fault and eventually reached the ocean in the vicinity of Monterey Bay. Baldwin (1963) showed that the ancestral San Benito River (pre-middle Pleistocene) was much longer and had a greater drainage area than it does today (Fig. 8-V); this river probably was the outflow from Lake Tulare. Similarly, in the north as the northern seaway retreated, the ancestral Sacramento River extended to reach the seaway, for San Francisco Bay was not in existence (Howard, 1951) and drainage from the northern Great Valley (Sacramento basin) had to reach the ocean at some point. The head of the Miocene and Pliocene seaway retreating toward Monterey Bay seems most logical as the point of debouchment.

Work by Allen (1946), Baldwin (1963), and the present writers suggests that the ancestral Sacramento River and the ancestral San Benito River (Fig. 8-V) joined east of the San Andreas fault and flowed to the ocean as a single river. East of Monterey Bay its course was the oversized channel east of present-day Elkhorn Slough. Allen (1946) noted that during the past a much larger flow of water than at present must have coursed down the channel to account for the large cross section. Thus, a greater flow of water down Elkhorn Slough during the Miocene and Pliocene could have transported large volumes of sediment into the marine environment. This sediment contributed to the fill in Monterey graben.

Baldwin (1963) showed that the drainage area of the ancestral San Benito River was reduced as a result of the mid-Pleistocene orogeny; conversely, the drainage area of the Salinas River was increased (Fig. 8-VI). This orogeny probably greatly reduced sediment transport to Monterey Bay by rivers.

Sea-level fluctuations during the Pleistocene are suggested as one of the reasons that Monterey and Soquel Canyons occupy their present positions (Carmel Canyon probably had a somewhat different history). As sea level lowered during a glacial stage, the rivers downcut to meet the lowered base level; as sea level rose during an interglacial stage, the previously eroded channels acted as sediment traps to divert littoral drift into deeper parts of the ocean. Movement of sediment either as turbidity currents or as gravity flows in the canyon axes may have produced the extensive erosion of Monterey and tributary canyons at great depths. The importance of canyon erosion by alteration and removal of the "weathered" granodiorites is not known.

Some of the Pleistocene erosion of the upper regions of Monterey and Soquel Canyons may have been subaerial, as a result of both erosion of exposed upper walls during times of glacially lowered sea levels and uplift of the block west of the Gabilan (Reliz [?]) fault. It is unlikely that uplift was of the magnitude to erode the canyons entirely in the nonmarine environment.

The history of Carmel Canyon erosion is not clear, because of its apparent greater age than Monterey and Soquel Canyons and structural features developed during the middle Miocene and middle Pleistocene. Sediment movements in the canyon and sea-level fluctuations certainly had some effect on eroding the canyon, but the stream (San Jose Creek) debouching into the head of Carmel Canyon cannot be related to conditions prevailing in Elkhorn Slough at the head of Monterey Canyon.

Structural trends.—Before downcutting reached indurated rocks having structural trends, Monterey and its tributary canyons could have had entirely different trends than now. The entire course of Carmel Canyon is structurally controlled: the westerly trend by the Carmel Valley fault and the northwesterly trend by the seaward extension of the Sur-Palo Colorado fault. Monterey Canyon probably was not directly controlled by structure until the upper Pliocene and Upper

Cretaceous rocks (Fig. 4) became exposed either by downcutting of the canyon and/or uplift of the block west of the Gabilan fault (probably during the late Pliocene or Pleistocene); underlying structural features east of Pliocene outcrops in the canyon below the unconsolidated sedimentary cover may account for the sinuosity of the channel. The presence of granodiorite on one side and of less well-indurated Pliocene strata on the other means that structural control is dominant in that part of Monterey Canyon containing rock outcrops.

Sediments and indurated rocks.—Calculations based on volume of sediment (Menard, 1960) in the Monterey and Delgada deep-sea fans suggest that Monterey fan may contain approximately 7,500 cu mi of sediment. Menard (personal commun., 1963) stated that this amount of material could not have accumulated as a result of Pleistocene and post-Pleistocene channeling of sediments down Monterey Canyon (the writers confirm this); therefore, he considers Monterey Canyon to have undergone erosion earlier than Pleistocene, possibly during middle Tertiary or earlier time. Conversely, the present study strongly suggests that Monterey Canyon was not eroded before late Pliocene, yet the great volume of sediment in Monterey deep sea fan had to be channeled in some manner into the oceanic deeps.

This enigma may be resolved if Monterey Canyon is considered to be a tributary to Carmel Canyon; thus, Carmel Canyon is the major canyon and was channeling sediment into the oceanic deeps long before Monterey Canyon was eroded. Monterey trough and its western extensions originally were extensions of Carmel Canyon. Present topographic relations of Monterey Canyon and Carmel Canyon resulted from channel readjustments of Carmel Canyon after the mid-Pleistocene orogeny. Carmel Canyon is south of the barrier to seaward transport of sediment imposed by the Monterey graben. Ascension Canyon north of Monterey Canyon (Shepard and Emery, 1941; Martin, 1964) also may have channeled sediment to oceanic deeps to form a dissected sedimentary apron at its mouth long before the erosion of Monterey Canyon.

Monterey Canyon (Fig. 2) is much wider, deeper, and longer than either Carmel Canyon or Ascension Canyon. These characteristics led earlier workers to consider Monterey Canyon as the major or primary canyon; however, Carmel Canyon is eroded in highly indurated rocks (siliceous siltstone and limestone on one side and granodiorite on the other) whereas the north side of Monterey Canyon consists of poorly indurated siltstone and sandstone. It seems likely that Monterey Canyon could be widened and deepened much faster than could Carmel Canyon.

Monterey Canyon east of the intersection of Soquel Canyon (Fig. 2) may also be younger than Carmel Canyon. Minor submarine tributaries leading into Monterey, Carmel, and Soquel Canyons are concentrated on the north slope of Monterey Canyon west of Soquel Canyon with only two minor tributaries leading into Monterey Canyon east of Soquel Canyon. If the minor tributaries were eroded by seaward movement of sediment from the narrowed shelf during glacially lowered sea levels, Soquel Canyon may have intercepted and prevented sediments from being transported farther east. Soquel Canyon thus effectively prevented erosion of minor tributaries and the head of Monterey Canyon until sea level rose and permitted sediment to bypass the head of Soquel Canyon.

Location of Pajaro gorge.—The position of the head and nearshore reaches of Monterey Canyon directly above the buried Pajaro gorge (Figs. 5, 6) suggests that Monterey Canyon is a present-day expression of the Pajaro gorge; *i.e.*, a previously filled submarine canyon is in the process of being exhumed. Well records from the lower Salinas Valley indicate, however, that deposition was continuous above Pajaro gorge during the late Miocene, Pliocene, and early Pleistocene; thus, Pajaro gorge was filled and covered before the start of erosion of Monterey Canyon.

Conversely, the location of the head and nearshore reaches of Monterey Canyon above Pajaro gorge is not accidental. Differential compaction of the sediment fill of Pajaro gorge probably created a topographic low on the land surface directly above the gorge. Rivers were directed to this low region and sediment being transported by the rivers was most effective in eroding at this point of discharge into the sea.

UNRESOLVED PROBLEMS

Terraces described by Bradley (1956, 1957) indicate that relative uplift above sea level occurred in the Santa Cruz Mountains during the Pleistocene and Recent. The lowest terrace is approximately 100 ft above present sea level, yet

the depositional record just south in the lower Salinas Valley indicates downwarping during the same period. No major fault is recognized between the two areas, but E. A. Gribi, Jr. (personal commun., 1966), considers the existence of such a fault to be reasonable. Prior to obtaining the logs of the wells drilled to basement (Figs. 5, 6), the Moro Cojo fault described by Woodford (1951, p. 834–837; also Fig. 20) might reasonably have been considered to be a zone of slippage to explain the relative movement noted in the terraces in the Santa Cruz Mountains. However, the depths to basement on either side of the Moro Cojo fault(?) are similar (Fig. 5) suggesting that no fault exists (at least not one with vertical movement). The Moro Cojo boundary that Woodford (1951) described as a change in groundwater characteristics is not disputed.

The head of Monterey Canyon begins directly at the shoreline at the mouth of Elkhorn Slough (Fig. 2). Present observations are not considered to have been made fortuitously at the time that sea level rose or fell to find this nickpoint (the gradient of the Salinas River in the lower Salinas Valley is approximately 0.5°; the gradient of the head of Monterey Canyon just seaward of the nickpoint is approximately 8.5°). A delicate balance in the cycle of deposition-erosion may exist between the average slope of the lower Salinas Valley and the slope of the shelf or bottom of Monterey Bay to keep the head of the canyon adjusted to the shifting shoreline.

No direct evidence can be demonstrated that the ancestral Sacramento River (Fig. 8-V) discharged into Monterey Bay, for marine upper Miocene and younger rocks are absent in the vicinity of Coyote Pass (Ortaldo, 1950; Frames, 1955), but the river had to discharge into the sea at some place. Jurassic-Cretaceous Franciscan rocks are present less than 100 ft below the alluvial cover in the pass (Beard, 1941); thus, post-Pliocene uplift and erosion probably have removed evidence of the connection of the seaway (and river) between Monterey Bay and the area east of San Francisco Bay.

CONCLUSIONS

1. Monterey and Carmel Canyons are eroded into rocks of diverse lithologic types. Upper Pliocene sandstone and siltstone bodies and possibly older rocks comprise the north slope of Monterey

Canyon; similar upper Pliocene strata and Upper Cretaceous granodiorites occur on the south slope. Upper Cretaceous granodiorites crop out nearshore (as well as on land) on both side slopes of Carmel Canyon. Along its northwesterly trend, Upper Cretaceous granodiorite is on the east side and middle Miocene siliceous siltstone, middle highly indurated limestone, and pre-Cretaceous(?) metamorphic rocks are exposed on the west. Soquel Canyon is eroded entirely into upper Pliocene sandstone and siltstone.

2. Foraminifera date the Monterey Canyon siltstone and sandstone beds as the late Pliocene Purisima Formation, possibly equivalent to the Pomponio Member of the Purisima in the Santa Cruz Mountains. Depth of deposition was greater than 600 ft, as indicated by the Foraminifera *Epistominella pacifica,* and several species of *Uvigerina;* open-ocean conditions are indicated by the presence of various species of *Globigerina.*

3. Samples of sedimentary strata bordering Carmel Canyon contain only diatoms, silicoflagellates, and radiolarians. The diatom *Stephanogonia polygona* dates the siliceous siltstone as middle Miocene.

4. Igneous rocks of the region are granodioritic with plagioclase determined as An-29, oligoclase. Metamorphic rocks contain plagioclase determined as An-60, labradorite. Granodiorites lack the well-developed feldspar phenocrysts of the granodiorites on land. Those dredged in the marine environment are "weathered" to a depth of approximately 4 in.

5. Rock outcrops in Monterey and Soquel Canyons seem to be absent nearshore and in deep water, presumably because of burial under young unconsolidated sediments. Unconsolidated nearshore Pleistocene sediments may be contributing clasts larger than 2 mm to axial sediments in Monterey Canyon.

6. Structural features are recognized and inferred. A recognized fault is the extension of the Palo Colorado and the Sur (Nacimiento) fault in the Santa Lucia Mountains into Carmel Canyon. Because diverse lithologic types are present on opposite sides of the canyon, this seaward extension is drawn along the northwest part of Carmel Canyon. Geophysical work suggests that this fault extends even farther north to connect with the San Gregorio fault in the Santa Cruz Mountains. An east-west fault parallel with Carmel

Valley on land is suggested to form the north side of Carmel Canyon near its head. Through the difference in lithologies on opposite sides of this canyon, Monterey fault is recognized to be parallel with Monterey Canyon. Numerous transverse faults of different magnitudes appear to offset the Monterey fault. A northern extension of the Reliz fault along the west side of Salinas Valley apparently crosses Monterey Canyon approximately 6 mi from shore. This extension is strongly suggested, because upper Pliocene rocks on the west block contain bathyal Foraminifera whereas Pliocene rocks east of the fault in the lower Salinas Valley contain only shallow-water mollusks and Foraminifera, even though these beds are approximately 1,000 ft stratigraphically lower than those in Monterey Canyon. Uplift also is suggested by the failure to recover in dredgings marine Pleistocene rocks from the west block, although Pleistocene and Recent shallow-water marine and nonmarine sediments are present on the east block.

7. The major inferred structural feature is the Monterey graben underlying the north slope and shelf of Monterey Canyon. The floor of this graben may have subsided 15,000 ft during the Miocene and Pliocene. Additional inferred structural features are the Gabilan fault and the Gabilan escarpment in the vicinity of Monterey Bay (the Gabilan escarpment is documented north and south of the bay). These structural features are necessary to explain differential movements that occurred during the brief, but intense, Salinian deformation of the middle Miocene.

8. Well records from the lower Salinas Valley permitted the identification of the Elkhorn erosion surface at the top of the granodiorite basement and a re-evaluation of the previously known canyon incised into the Elkhorn erosion surface. The canyon, named the Pajaro gorge, may have been eroded during the middle Miocene and filled prior to the Pliocene. The gorge is not an ancestral Monterey Canyon, but differential compaction of the fill appears to have created a topographic low on the land surface toward which surface drainage gravitated. Nearshore reaches and the head of Monterey Canyon lie directly above Pajaro gorge.

9. The trends of Monterey and Carmel Canyons are determined by structural patterns developed during the middle Miocene Salinian de-

formation and by deformation during the mid-Pleistocene orogeny. Carmel Canyon probably was the primary canyon until the middle Pleistocene, when it became a tributary of Monterey Canyon.

10. Monterey Canyon and Soquel Canyon began to be eroded during late Pliocene or early Pleistocene, as indicated by the youngest exposures of upper Pliocene rocks forming the side slopes. The youngest rock in Carmel Canyon is middle Miocene, the northwesterly extension of the canyon being eroded after this time.

This study of Monterey Canyon and its tributaries shows that the origin and geologic history of the canyons are extremely complex. Many factors contributed to the present physiography and no one of them can be considered more important than others. Clearly, the history of submarine canyons cannot be developed without detailed geological study of the land adjacent to the canyons.

BIBLIOGRAPHY CITED

Allen, J. E., 1946, Geology of the San Juan Bautista quadrangle, California: Calif. Div. Mines and Geology Bull. 133, p. 9–75.

Baldwin, T. A., 1963, Land forms of the Salinas Valley, California, *in* Geology of the Salinas Valley and the San Andreas fault: Pacific Sec., Am. Assoc. Petroleum Geologists—Soc. Econ. Paleontologists and Mineralogists Guidebook, 168 p.

Bandy, O. L., 1953, Ecology and paleoecology of some California Foraminifera—Pt 1, The frequency distribution of Recent Foraminifera off California; Pt 2, Foraminiferal evidence of subsidence rates in Ventura basin: Jour. Paleontology, v. 27, p. 161–182, 200–203.

———1960, General correlation of foraminiferal structure with environment: 21st Internat. Geol. Cong., Copenhagen, 1960, Rept., pt. 22, p. 7–19.

Beard, C. N., 1941, Drainage development in the vicinity of Monterey Bay, California: unpubl. Ph.D. dissert., Univ. Illinois, 122 p.

Brabb, E. E., O. L. Bowen, Jr., and E. W. Hart, 1962, Field trip 2—San Francisco to Monterey via California highways 1, 5, 17, and connecting routes, *in* Geologic guide to the gas and oil fields of northern California: Calif. Div. Mines and Geology Bull. 181, p. 381–390.

Bradley, W. C., 1956, Carbon-14 date for a marine terrace at Santa Cruz, California: Geol. Soc. America Bull., v. 67, p. 675–677.

———1957, Origin of marine-terrace deposits in the Santa Cruz area, California: Geol. Soc. America Bull., v. 68, p. 421–444.

Buffington, E. C., 1964, Structural control and precision bathymetry of La Jolla submarine canyon, California: Marine Geology, v. 1, p. 44–58.

Burch, J. Q., 1946, Distributional list of the west American marine mollusks from San Diego, California, to the polar sea, Pts. 1, 2: Los Angeles, privately printed, complexly paged.

Cohee, G. V., 1938, Sediments of the submarine canyons off the California coast: Jour. Sed. Petrology, v. 8, p. 19–13.

Cummings, J. C., R. M. Touring, and E. E. Brabb, 1962, Geology of the northern Santa Cruz Mountains, California, *in* Geologic guide to the gas and oil fields of northern California: Calif. Div. Mines and Geology Bull. 181, p. 179–220.

Curtis, G. H., J. F. Evernden, and J. Lipson, 1958, Age determination of some granitic rocks in California by the potassium-argon method: Calif. Div. Mines and Geology Spec. Rept. 54, 16 p.

Davidson, G., 1887, Submarine valleys on the Pacific coast of the United States: Calif. Acad. Sci. Bull. 2, p. 265–268.

——— 1897, The submerged valleys of the coast of California, U.S.A., and of Lower California, Mexico: Calif. Acad. Sci. Proc., 3rd ser. (geol.), v. 1, p. 73–103.

Dill, R. F., R. S. Dietz, and H. Stewart, 1954, Deep-sea channels and delta of the Monterey submarine canyon: Geol. Soc. America Bull., v. 65, p. 191–194.

Durham, David L., and Warren O. Addicott, 1965, Pancho Rico Formation, Salinas Valley, California: U.S. Geol. Survey Prof. Paper 524-A, 22 p.

Emery, K. O., and F. P. Shepard, 1945, Lithology of the sea floor off southern California: Geol. Soc. America Bull., v. 56, p. 431–478.

Fiedler, W. M., 1944, Geology of the Jamesburg quadrangle, Monterey County, California: Calif. Div. Mines and Geology Jour., v. 40, no. 2, p. 177–250.

Folk, R. L., 1962, Spectral subdivision of limestone types, *in* Classification of carbonate rocks—a symposium: Am. Assoc. Petroleum Geologists Mem. 1, p. 62–84.

Frames, D. W., 1955, Stratigraphy and structure of the lower Coyote Creek area, Santa Clara County, California: unpubl. M. S. thesis, Univ. California, Berkeley, 67 p.

Galliher, E. W., 1935, Geology of glauconite: Am. Assoc. Petroleum Geologists Bull., v. 19, p. 1569–1601.

Goodwin, J. C., and J. N. Thomson, 1954, Purisima Pliocene Foraminifera of the Halfmoon Bay area, San Mateo County, California: Cushman Found. Foram. Research Contr., v. 5, pt. 4, p. 170–178.

Gribi, E. A., Jr., 1963a, Hollister trough oil province, *in* Sacramento Geol. Soc. guidebook: Ann. Field Trip, Univ. Calif. Dept. Geol. Sci., Davis, p. 77–80.

——— 1963b, The Salinas basin oil province, *in* Geology of Salinas Valley and the San Andreas fault: Pacific Secs., Am. Assoc. Petroleum Geologists—Soc. Econ. Paleontologists and Mineralogists Guidebook, p. 16–27.

Hall, C. A., Jr., 1958, Geology and paleontology of the Pleasanton area, Alameda and Contra Costa Counties, California: Univ. Calif. Pubs. Geol. Sci., v. 34, p. 1–89.

Ham, C. K., 1952, Geology of Las Trampas Ridge, Berkeley Hills, California: Calif. Div. Mines and Geology Spec. Rept. 22, 26 p.

Hoots, H. W., T. L. Bear, and W. D. Kleinpell, 1954, Geological summary of the San Joaquin Valley, California, *in* Geology of southern California: Calif. Div. Mines and Geology, Bull. 170, Chap. 2, p. 113–129.

Howard, A. D., 1951, Development of the landscape of the San Francisco Bay counties, *in* Geologic guidebook of the San Francisco Bay counties: Calif. Div. Mines and Geology Bull. 154, p. 95–106.

Huey, A. S., 1948, Geology of the Tesla quadrangle, California: Calif. Div. Mines and Geology Bull. 140, 75 p.

Jennings, C. W., and J. L. Burnett, 1961, Geologic map of California, San Francisco sheet: Calif. Div. Mines, scale, 1:250,000.

——— and R. G. Strand, 1958, Geologic map of California, Santa Cruz sheet: Calif. Div. Mines, scale, 1:250,000.

Kerr, P. F., and H. G. Schenck, 1925, Active thrust faults in San Benito County, California: Geol. Soc. America Bull., v. 36, p. 465–494.

Kleinpell, R. M., 1938, Miocene stratigraphy of California: Am. Assoc. Petroleum Geologists, 450 p.

Lawson, A. C., 1893, The geology of Carmelo Bay: Univ. Calif. Dept. Geology Bull. 1, p. 1–59.

Leighton, M. W., and C. Pendexter, 1962, Carbonate rock types, *in* Classification of carbonate rocks—a symposium: Am. Assoc. Petroleum Geologists Mem. 1, p. 33–61.

Leo, G. W., 1961, The plutonic and metamorphic rocks of Ben Lomond Mountain, Santa Cruz County, California: unpubl. Ph.D. dissert., Stanford Univ., 194 p.

Martin, B. D., 1964, Monterey submarine canyon, California: genesis and relationship to continental geology: unpubl. Ph.D. dissert., Univ. Southern Calif., 249 p.

——— and R. W. Rex, 1965, Clay minerals formed by submarine weathering of granodiorite from the walls of Monterey and Carmel submarine canyons, California (abs.): 3d Mtg., Clay Minerals Society, Berkeley, California.

Menard, H. W., 1960, Possible pre-Pleistocene deep-sea fans off central California: Geol. Soc. America Bull., v. 71, p. 1271–1278.

Nomland, J. O., and H. G. Schenck, 1932, Cretaceous beds at Slate's Hot Springs, California: Univ. Calif. Dept. Geol. Sci. Bull., v. 21, p. 37–49.

Ortaldo, R. A., 1950, Geology of the northern part of the Morgan Hill quadrangle, California: unpubl. M. S. Thesis, Univ. Calif., Berkeley, 55 p.

Pratt, W. L., 1962, Origin and distribution of glauconite from the sea floor of southern California: unpubl. Ph.D. dissert., Univ. Southern Calif., 242 p.

Reed, R. D., 1933, Geology of California: Am. Assoc. Petroleum Geologists, 355 p.

Rex, R. W., and B. D. Martin, 1966, Clay mineral formation in sea water by submarine weathering of K-feldspar, *in* Clays and clay minerals, v. 14: New York, Pergamon Press, p. 235–240.

Shepard, F. P., 1948, Investigation of head of Monterey submarine canyon: Hydrographic Off.—Off. Naval Research, Contr. N6Ori-111, Scripps Inst. Oceanogr., Submarine Geol. Rept. no. 1, 15 p.

——— 1952, Composite origin of submarine canyons: Jour. Geology, v. 60, p. 84–96.

——— and K. O. Emery, 1941, Submarine topography off the California coast—canyons and tectonic interpretations: Geol. Soc. America Spec. Paper 31, 171 p.

Starke, G. W., 1956, Genesis and geologic antiquity of Monterey Submarine Canyon (abs.): Geol. Soc. America Bull., v. 67, p. 1783.

Taliaferro, N. L., 1951, Geology of the San Francisco Bay counties, California, *in* Geologic guidebook of the San Francisco Bay counties: Calif. Div. Mines and Geology Bull. 154, p. 117–150.

Trask, P. D., 1926, Geology of the Point Sur quadrangle, California: Univ. Calif. Dept. Geol. Sci. Bull., v. 16, p. 119–186.

Weaver, C. E., 1949, Geology of the Coast Ranges immediately north of the San Francmsco Bay region, California: Calif. Div. Mines and Geology Bull. 149, 135 p.

Woodford, A. O., 1951, Stream gradients and Monterey sea valley: Geol. Soc. America Bull, v. 62, p. 799–852.

Woodring, W. P., R. B. Stewart, and R. W. Richards, 1940, Geology of the Kettleman Hills oil field, California, stratigraphy, paleontology, and structure: U.S. Geol. Survey Prof. Paper 195, 170 p.

The American Association of Petroleum Geologists Bulletin
V. 60, No. 4 (April 1976), P. 517-527, 11 Figs., 3 Tables

Basinal Sandstone Facies, Delaware Basin, West Texas and Southeast New Mexico[1]

MYRON W. PAYNE[2]
Fairbanks, Alaska 99701

Abstract The Delaware Mountain Group in the Delaware basin of West Texas and southeast New Mexico consists of a thick sequence of Upper Permian clastic rocks deposited in a deep-water, marine setting. The basinal facies and depositional environments of these sediments may be interpreted from the uppermost part of this sequence, and pre-Lamar Bell Canyon sandstones. Three basic lithofacies are present: lutite, laminite, and arenite. These are defined by texture and composition as well as by sedimentary structures.

The lutite lithofacies is a pelagic, organic, fissile shale. It is rich in finely divided organic matter, relatively clay poor, and contains varied amounts of silt and mica. Closely related to the lutite is the laminite lithofacies. It is a laminated, coarse-grained siltstone (mean size 0.05 mm) with alternate laminae having concentrations of organic material, suggesting deposition from suspension. Bedding is marked by *Nereites*-type trace fossils which indicate a bathyal environment. A finer organic-poor laminite also is present in limited amounts and is inferred to have been deposited from turbidity currents because of its turbidite features.

The arenite lithofacies is present in linear submarine-channel areas and is a feldspar-rich quartz sandstone with very fine texture (mean size of 0.065 to 0.95 mm) and low matrix content. Sedimentary structures within this unit indicate deposition from turbidity currents as well as from other fluid-density currents. The channel patterns suggest a major southwest depositional trend but no submarine canyons. Hydrocarbon traps have developed along the channels because of regional tilting eastward and an updip facies change from arenite to laminite and biolaminite lithofacies. The remaining nonchannel sediments are interchannel deposits and wedged lobate deposits along marginal areas, especially in the northern half of the basin.

INTRODUCTION

In recent years, there has been a concentrated effort to understand the processes responsible for filling both recent and ancient deep-marine basins. This study deals with the Delaware basin of West Texas and southeast New Mexico. It is the western segment of the larger Permian basin complex and has a shallower eastern counterpart in the Midland basin (Fig. 1).

Throughout the late Paleozoic, the Delaware basin received dominantly terrigenous clastic sediments, and during Guadalupian time widespread sandstones were deposited in a deep-water environment. The uppermost Guadalupian sandstones, here referred to as the pre-Lamar Bell Canyon section, were deposited in relatively thick units.

The Bell Canyon Formation is the upper part of the Delaware Mountain Group which comprises over 5,000 ft (1,500 m) of terrigenous sedimentary rocks and covers an area approximately 75 mi (110 km) wide and 150 mi (220 km) long. The pre-Lamar Bell Canyon section was examined in detail in cores and exposures to determine petrography and lithofacies distributions. The results can be used to define compositional and textural characteristics that are unique to the lithofacies defined by sedimentary structures.

GEOLOGIC SETTING

The Delaware basin is an asymmetric block-faulted basin containing approximately 25,000 ft (7,600 m) of predominantly Paleozoic sedimentary rocks. Its axis parallels the east side of the basin along the Central Basin platform, and although the basin cannot be separated readily from the Midland basin at the surface, it is defined clearly on subsurface structure (Fig. 2). The ages of the structures range from post-Cretaceous along the western edge of the basin to Paleozoic along the Central Basin platform. Structurally, the subsurface has normal faulting and regional tilt toward the east.

The Upper Permian stratigraphy is reviewed best in terms of three major time-transgressive, sedimentary-topographic divisions: (1) the shelf, (2) the shelf margin, and (3) the basin (King, 1948; Meissner, 1972). The major Middle and Upper Permian rock-stratigraphic units within these divisions are presented in Figure 3. The shelf sections near the shelf margin consist of alternating, thinly bedded dolomites and sandstones with a basal dark limestone (Meinzer et al, 1926; De

[1]Manuscript received, April 28, 1974; accepted, October 10, 1975.

[2]This paper was prepared from a doctoral dissertation at Texas A&M University, College Station, Texas. The writer gratefully acknowledges the suggestions of R. R. Berg, Texas A&M University. Research was conducted under a grant from Gulf Oil Corporation, and cores and samples were obtained from the Tenneco Oil Company and Shell Oil Company. The writer's present address is Geology Department, University of Alaska.

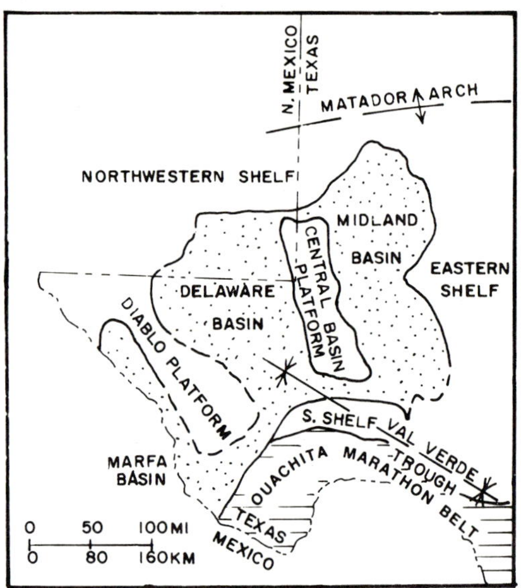

FIG. 1—Tectonic setting in West Texas and southeast New Mexico (modified from Oriel et al, 1967).

Ford and Riggs, 1941; Needham and Bates, 1943; Mear and Yarbrough, 1961). These grade laterally into the shelf margin consisting of massive dolomites and dolomitic limestones (Richardson, 1904; Lloyd, 1929; King, 1948; Newell et al, 1953). The origin of these carbonate rocks as reefs and banks was summarized by Dunham (1972).

The basin sedimentary deposits are strikingly different from the carbonate rocks of the shelf and shelf margin. They comprise the greater volume, consist almost entirely of alternating sandstone and siltstone, and pinch out upslope into the shelf-margin carbonate rocks (Cartwright, 1930; King, 1948; Newell et al, 1953). Correlation of the basin sedimentary rocks with those of the shelf generally is based on fusulinid zones. Of primary interest is the correlation of the Lamar Member of Bell Canyon Formation with either the Tansill or Yates Formations on the shelf. A correlation of the base of the Lamar with the base of the Tansill has been accepted by most recent workers (Tyrell, 1969; Meissner, 1972) with the exception of Kelley (1972) who placed the top of the Lamar at the base of the Yates.

The youngest Permian deposits consist of Ochoan evaporites and red beds. These directly overlie the Bell Canyon Formation and have a maximum thickness of approximately 1,800 ft (550 m). They represent the end of sedimentation in the basin and lie directly on top of the Bell Canyon Formation.

In the subsurface, the pre-Lamar Bell Canyon clastic rocks have been subdivided into three informal members: (1) the upper Ramsay sandstone, (2) the middle Ford shale, and (3) the lower Olds sandstone (Fig. 4). They lie directly below the "Delaware lime" which is an informal name applied to the basinal equivalent of the Lamar Member. The Ramsey, Ford, and Olds were designated by Nottingham (1960) as subsurface units of wide areal extent, and the Ramsey and Olds are oil reservoirs in the basin. Figure 4 illustrates the subsurface units as they are defined in gamma-ray and sonic logs.

PREVIOUS WORK

Studies in the Delaware basin area have been seeking to evaluate possible petroleum prospects since the turn of the century. Early studies were of reconnaissance nature and have been summarized by King (1934). It was not until King (1942, 1948) published his studies of the region that modern concepts of sedimentation processes were given serious consideration.

The basinal facies, Delaware Mountain Group, was studied in detail where it crops out along the Guadalupe, Delaware, and Apache Mountains. The alternating character of clastic and carbonate

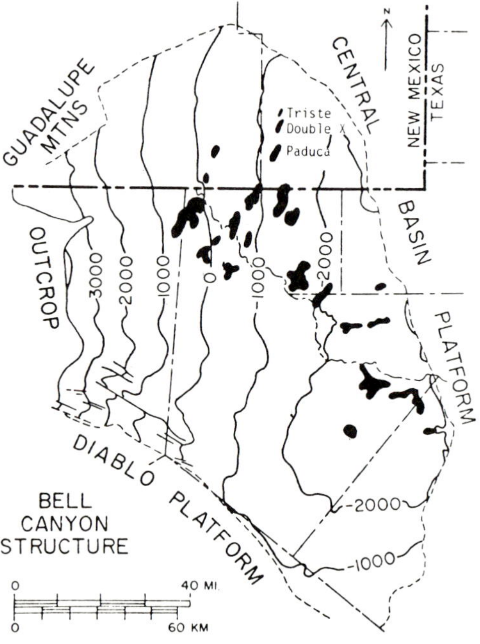

FIG. 2—Regional Bell Canyon structure and location of major Bell Canyon oil fields. Contour interval is 1,000 ft (303 m).

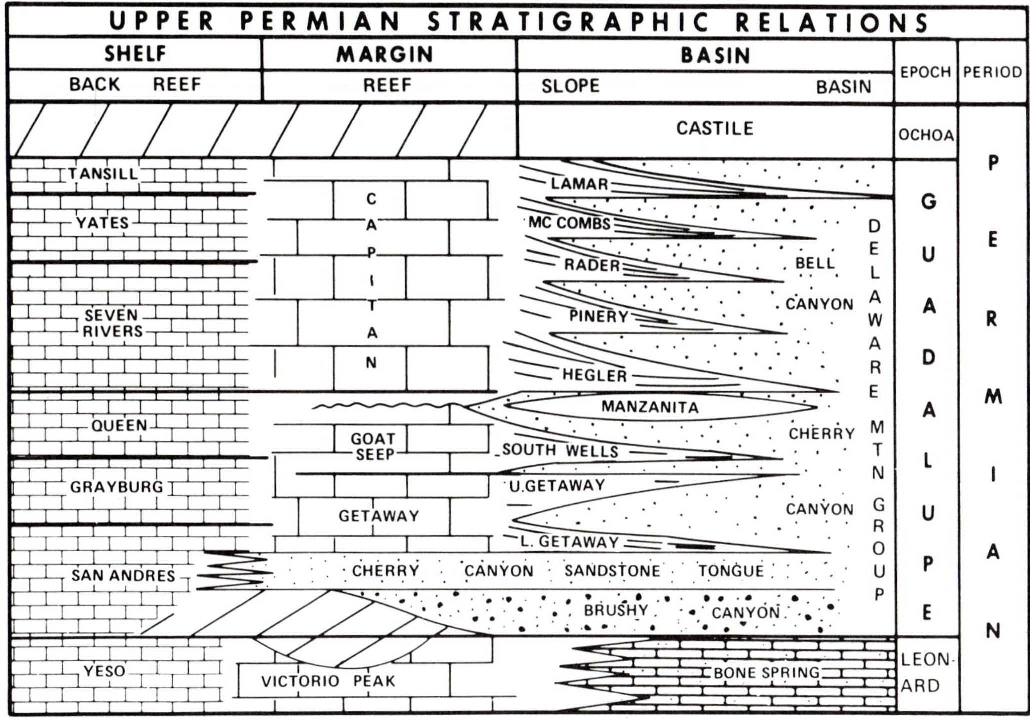

Fig. 3—Upper Permian rock-stratigraphic relations in Guadalupe Mountains (modified from King, 1948).

rocks along the shelf-margin areas eventually led to the concept of cyclic clastic deposition related to eustatic sea-level changes. Silver and Todd (1969) described the major late Guadalupian cycles. These clastic sediments do not seem to have an obvious local source area (Waterschoot van der Gracht, 1931; Adams, 1936; Lang, 1937; King, 1948; McKee, 1951). The processes by which such large volumes of clastic materials were delivered across the shelf and into the basin remain a major sedimentologic problem.

The depositional mechanics of some of the marginal clastic rocks have been studied by Harrison (1966), St. Germaine (1966), and Beck (1967), and summarized by Jacka et al (1968). These authors concluded that the Delaware Mountain Group consists of deep-sea fan and fringe deposits similar to those of the modern continental-borderland basins of the California coast. It was their contention that submarine canyons were almost the exclusive routes for introducing clastic material into the deep basin.

Hull (1957) described the petrology and cyclic nature of the Delaware Mountain Group and developed a model of deposition in which laminated and massive sandstones represent distinct deposits laid down in response to changing subsidence rates.

Meissner's (1972) study extended into the central part of the basin. He recognized two major lithofacies: (1) a massive, silty, very fine-grained sandstone that reached a maximum thickness of 80 ft (25 m) at the toe of the reef, and (2) a southward-thinning tongue of laminated siltstone. He suggested that turbidite fans coalesced to form aprons of massive sandstone near the reef and thin, laminated-siltstone, "sheetwash" deposits basinward.

Harms (1968) noted similar lithofacies for the equivalent deposits along outcrops in the Delaware Mountains. He suggested high-density, non-turbid currents transported sediment into the basin, noting that the sedimentary structures are unique and unlike those associated with turbidites or fluvial deposits.

LITHOFACIES OF BELL CANYON SEDIMENTS

To avoid the problem of genetic implications, the clastic rocks within the pre-Lamar Bell Canyon section have been divided into three purely descriptive lithofacies: lutite, laminite, and arenite. Within each lithofacies, subdivisions are

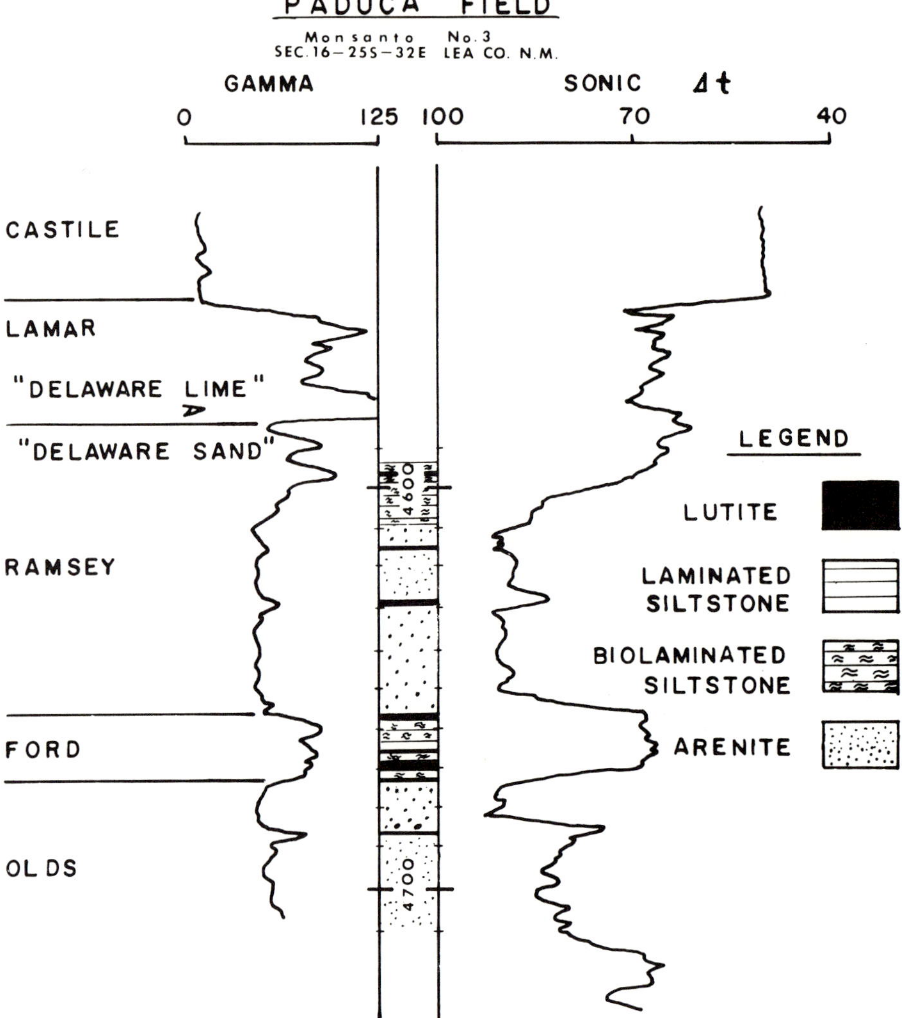

FIG. 4—Subsurface nomenclature for Bell Canyon sedimentary rocks in Delaware basin. Top of Bell Canyon Formation is picked at base of "Delaware lime."

based on the bedding type. A summary of this classification and the lithofacies origin is shown in Table 1.

Lutite

The lutite facies consists of a dark organic fissile shale which may or may not show evidence of bioturbation. It can be subdivided into two parts based on the presence or absence of biogenic structures. Compositionally, the lutite is dominated by organic material. Clay minerals are rare; quartz silt and mica account for most of the terrigenous material present. The origin of the organic material is attributed to planktonic organisms. This view is reinforced by the total absence of macroscopic plant remains.

Laminite

The laminite is a siltstone characterized by parallel laminations which may or may not have alternating organic laminae. The facies is divided into three categories based on the presence or absence of carbonaceous laminae and biogenic structures. Where bioturbation is present, the unit is referred to as "biolaminite" and where it is undisturbed the term "laminite" is used.

TABLE 1. LITHOFACIES CLASSIFICATION

Lithofacies	Bedding Type	Interpreted Origin
Lutite	Fissile	Pelagic sedimentation
	Bioturbated	Bioturbation of pelagic sediments
Laminite	Organic Laminated	Alternating deposition of organic and clastic suspension sediments
	Laminated	Deposition from turbidity current
	Biolaminated	Bioturbation of organic laminite
Arenite	Cross-laminated	Deposition within lower flow regime
	Rippled	Sediment reworking by lower flow regime currents
	Erosional ripple-drift cross-laminated	Net input of sediment during low flow regime ripple formation
	Nonerosional ripple-drift cross-laminated	Deposition of sediment in suspension during waning current in lower flow regime
	Laminated	Possibly deposited from waning currents within upper flow regime
	Graded	Rapid deposition from waning current with large suspension load. Segregation of sizes by hydraulic weight
	Massive	Initial deposition from current with high suspension load. Lack of sedimentary structures may be indicative of very high flow regime

The laminites are of two types: those with alternating organic and silt laminae, and those without significant organic material. The organic-poor laminites are thin units (less than 1 ft or 0.3 m) with extremely fine laminae and an overall upward decrease in texture and laminae thickness. Gradational contacts into the lutite facies are very common. The base of some units grades into an arenite, the origin of which seems related to a waning current and sedimentation from suspension.

Trace fossils along the bedding planes of the biolaminites suggest a fauna characteristic of the *Nereites* community which has been used to define a bathyal environment for marine sediments (Crimes and Harper, 1970).

Compositionally, the laminites and biolaminites average about 30 percent matrix. A plot of the mean quartz-grain size versus the quartz content shows a large degree of scatter (Fig. 5). Quartz ranges from 35 to 68 percent and averages about 50 percent. Size ranges from 0.09 to 0.03 mm and shows a tendency to increase with higher quartz content. The organic-poor laminites show anomalous patterns; they maintain a high quartz content even though they generally are the finest grained.

Arenite

The arenite facies consists of very well-sorted sandstones with varied amounts of interstitial silt and organic-rich matrix. Seven subdivisions are recognized: cross-laminated, rippled, erosional ripple-drift cross-laminated, nonerosional ripple-drift cross-laminated, laminated, graded, and massive.

A plot of mean quartz size versus quartz content for the arenites shows considerable scatter, but in general, higher quartz content is associated with larger sizes (Fig. 5). The separation between laminites, biolaminites, and arenites is not distinct when plotted in this fashion because the matrix content is too variable within the lithofacies.

The bulk compositions for all of the lithofacies are summarized by bedding types in Table 2. The only compositional elements which show a consistent relation are the quartz content and the matrix. Together they account for approximately 75 percent of the bulk composition.

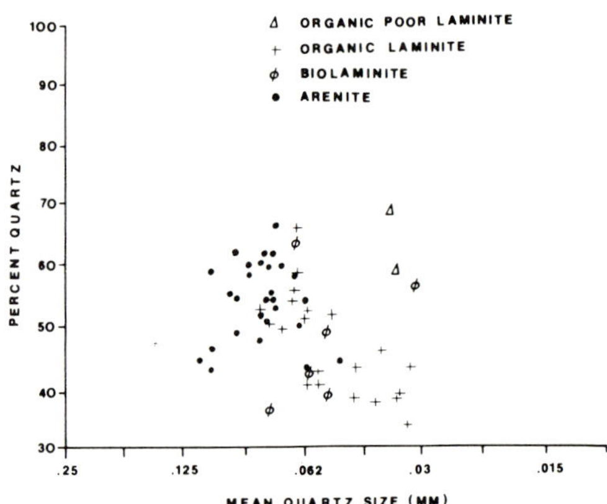

FIG. 5—Plot of quartz content versus mean quartz size for different lithofacies.

As the sedimentary structures indicate progressively lower flow regimes, the matrix content approaches that of the biolaminite. The lowest matrix content is in the massive to graded units.

PETROGRAPHY

The Bell Canyon sandstones are composed primarily of quartz and feldspar grains with an organic-rich matrix. An unusual aspect of these sandstones is that clay is a very minor matrix component. Rock fragments, mica plates, and heavy minerals are of secondary importance. The detrital particles include monocrystalline quartz, feldspar, rock fragments, matrix, and fossils or intraclasts. The cements can be classified as either chert, quartz, or calcite, but only calcite is quantitatively important.

Evidence of diagenesis within the sandstones includes cementation, and development of sericite in clay minerals and some feldspars. Compaction also has caused deformation of shale clasts, mica plates, and some possible chert fragments. Most cementation is present as quartz overgrowths or sparry calcite. Chert is uncommon and only present as stringers within calcareous beds. Overgrowths generally are poorly developed and in many samples form a veneer of small crystals around a single grain, giving the surface a frosted appearance (Hull, 1957). Pressure solution of quartz-grain contacts is evidenced by sutured and concave-convex contacts. Straight-grain contacts are uncommon, giving additional evidence that large overgrowths are poorly developed in most grains.

LITHOFACIES PATTERNS

Though each lithofacies is scattered throughout the entire basin, the degree of development varies

with location. Basin-margin areas show a varied composition of lithofacies whereas the deep-basinal areas have a more predictable distribution. This is demonstrated in a generalized net-sand isolith of the Ramsey member which shows two patterns: (1) lobate wedges thinning basinward from the marginal areas, and (2) linear-thick features trending northeast (Fig. 6). The lobate deposits thin from an average of 60 ft (18 m) at the basin margins to zero at the central part of the basin and indicate that major deposition occurred along the northern margin of the basin with minor wedges developed in the southern areas. The basinal linear patterns suggest a major submarine-channel system trending southwest. Both lobate wedges and linear-thick features have characteristic bed sequences.

CHANNEL SEQUENCE

The lithofacies within the linear sandstone "thicks" of the Ramsey show a very distinctive pattern of bedding types in a dominantly arenite package. Where fully developed, the sequence of lithofacies, unique to these channel areas, is referred to as a mature-channel sequence. Figure 7 shows a fully developed sequence in a core taken more than 30 mi (48 km) from the basin margin.

The alternating nature of the sedimentary structures demonstrates that a highly variable flow system deposited sediment from both traction and suspension loads. The dominance of ag-

TABLE 2.　MEAN COMPOSITION OF THE VARIOUS LITHOFACIES

Bedding Type	Percent of Detrital Grains					
	Qz	Feld	Mica	Rock	Mtx	Other
Laminite	45.9	14.6	5.14	2.4	32.2	1.1
Biolaminite	48.8	19.5	2.8	2.2	26.8	1.0
Arenite						
Cross-laminated	52.5	23.9	1.6	2.4	18.1	1.4
Rippled	49.4	17.2	2.8	3.9	24.4	2.2
Ripple-drift cross-laminated	56.4	17.7	2.5	5.1	17.6	0.4
Laminated	53.4	23.3	2.5	2.4	17.0	0.9
Graded	53.4	23.3	2.5	2.4	17.0	0.9
Massive	64.4	10.8	1.2	7.2	12.0	4.2

Legend:

Qz	–	monocrystalline quartz
Feld	–	feldspar
Mica	–	detrital mica only
Rock	–	includes chert, and polycrystalline quartz
Mtx	–	organic and inorganic matrix material

gradational climbing bedding features, which require sediment influx, suggests a net deposition within the channel.

The lowest interval of shale-chip conglomerate is indicative of a very high-flow regime with erosion of the substrate followed by rapid deposition of the eroded material. The initial deposits therefore represent deposition from a graded suspension under conditions of a waning current. The total thickness of upper-flow-regime deposits is small, seldom over two to three ft (0.6 to 0.9 m).

All of the channel arenites are encased in laminite and biolaminite. Lutite also can be traced across many channels and in many places is overlain by additional channel arenite. The entire system seems to be composed of channel, channel levee, and intrachannel deposits which have a meandering or braided pattern. The channel-transport patterns of the Ramsey unit are shown in Figure 8.

These channels correspond to known oil fields and constitute the reservoir facies for known stratigraphic traps. They are not always apparent on log cross sections unless porosity logs are used as indicators for the arenite lithofacies.

The strong perferred orientation of the channels suggests transport from the northeast across a carbonate platform.

Nonchannel Sequence

Most of the upper Bell Canyon sedimentary rocks were deposited outside of the channel system but no quantitative estimate of their volume has been attempted. They consist of sediment derived from suspension including pelagic and hemipelagic deposition as well as some low-density turbidity currents.

Figure 9 is a composite-facies sequence for nonchannel sediments based on core interpretation. This sequence shows all of the possible facies but does not depict any fixed vertical order; instead, it represents a pattern which most often is repeated.

The basal contact always is sharp and commonly is marked by flame structures ranging from 0.1 to 1 in. (0.2 to 2.5 cm) high. It is overlain by a typical turbidite facies succession of Bouma (1962) which usually shows the CDE pattern, but occasionally develops a complete ABCDE pattern (Fig. 10). The total thickness of an individual turbidite unit seldom exceeds 1 ft (0.3 m). Overlying the turbidite is a random pattern of lutite, biolutite, laminite, and biolaminite lithofacies which generally account for more than 60 percent of any given nonchannel sequence. The only other feature present is local flaser bedding produced by bottom currents.

Reservoir Characteristics

Significant oil reservoirs have been discovered in the Ramsey and Olds intervals of the Bell Canyon sandstones (Fig. 2). With the exception of some structurally controlled fields near the Central Basin platform, all of the fields are stratigraphic traps aided by regional eastward tilting. Paduca, Double X, and Triste Draw oil fields are typical of the stratigraphic traps. Reservoir data for the respective fields are presented in Table 3.

Paduca field (Fig. 11) illustrates the stratigraphic control of hydrocarbon trapping. It has 120 ft (37 m) of oil column and is trapped by a facies change from channel arenites to submarine-fan laminites and biolaminites. The reservoir is a channel arenite with a meander configuration, is tilted 100 ft/mi (19 m/km) toward the east, and has reservoir characteristics averaging 24-percent porosity and 25-md permeability. The updip-trapping facies is a laminite with an estimated 18-percent porosity and 4-md permeability. Berg (1975) demonstrated that the change in relative grain size from the arenite reservoir to laminite-trap facies is sufficient to trap 34 ft of the oil column by capillary trapping. An additional 100 ft of column can be calculated as the result of a tilted oil-water contact (Berg, 1975). The close agreement of observed and calculated oil columns is convincing evidence for the strong in-

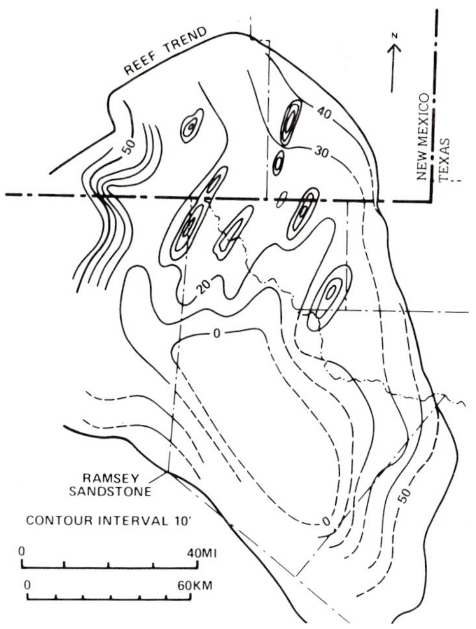

Fig. 6—Generalized net-sandstone isopach of Ramsey interval.

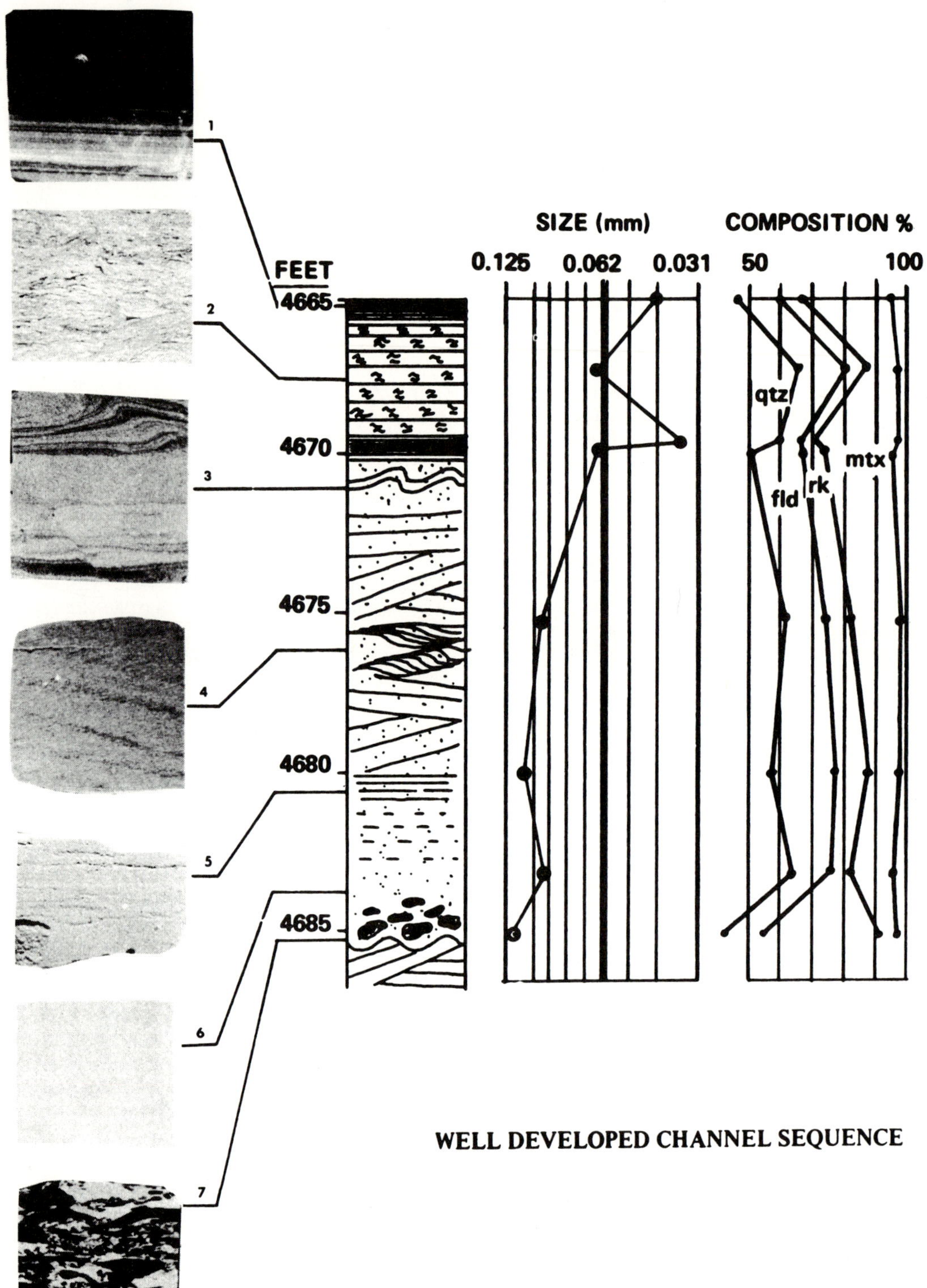

FIG. 7—Complete channel sequence as seen in Paduca field, Lea County, New Mexico. Explanation: *1*, lutite; *2*, biolaminite; *3*, contorted fine-grained arenite and thin lutite lenses; *4*, nonerosional ripple-drift cross-laminations; *5*, relatively coarse-laminated arenite; *6*, massive homogeneous arenite; and *7*, shale-clast conglomerate and coarse arenite mixed. Legend: *Qtz*, quartz; *fld*, feldspar; *rk*, rock fragment; *mtx*, matrix (from Monsanto 3, Sec. 16, T25S, R32E).

432

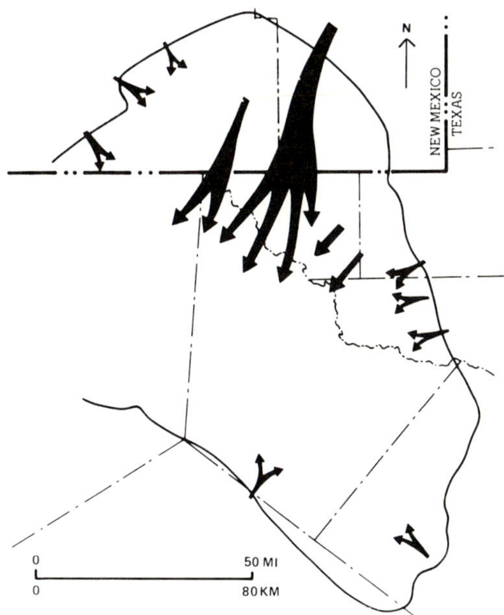

Fig. 8—Proposed major transport directions for upper Bell Canyon sandstones.

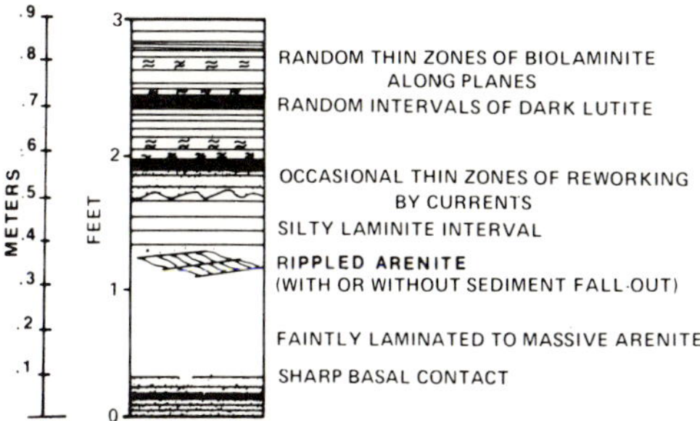

Fig. 9—Composite facies sequence of nonchannel sedimentary rocks.

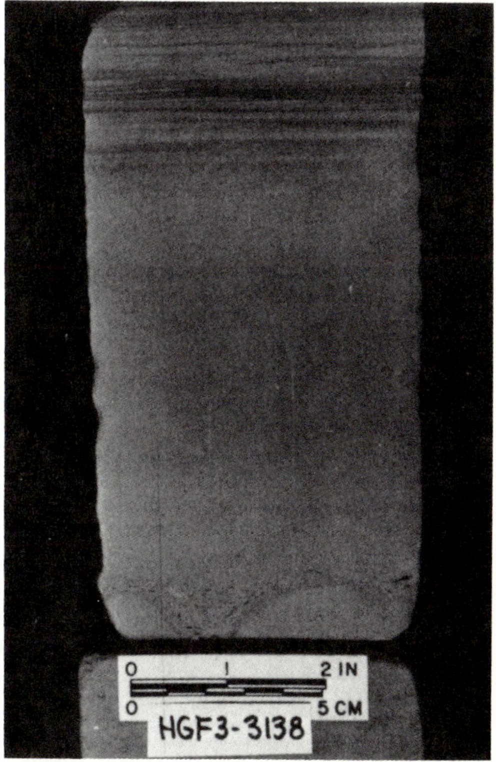

Fig. 10—"Bouma turbidite" sequence of sedimentary structures. These are most common in nonchannel deposits (from Hankamer-Gulf Fed. Beady 3, Sec. 14, T26S, R29E, Eddy County, New Mexico, depth 3,318 ft).

fluence of hydrodynamic and hydrostatic forces in the trapping. It should be remembered that the reservoir facies is subject to minor vertical and lateral variations in porosity and permeability because of lensing within the channel systems themselves.

Conclusions

The Bell Canyon Formation is a Guadalupian (Permian) clastic unit which was deposited in deep water within the Delaware basin. Terrigenous clastics were transported across the carbonate shelf and into the basin where they were deposited as extensive units of sandstone, siltstone, and organic shale. The Ramsey sandstone of the uppermost Bell Canyon displays characteristics which are analogous to clastic units deposited in deep water. They were deposited as wide thin wedges which are thickest in the northern half of the basin and thin to zero at the basin center. Within these wedges are well-developed submarine-channel sandstones along northeast-southwest axes, indicating transport from the northeast.

The sedimentary deposits can be separated into three major lithofacies: the lutite, the laminite, and the arenite. These can be subdivided on the basis of bedding type, organic content, and degree of bioturbation.

The arenite facies are present within laminated sediments in marginal-wedge deposits and in basin channels. The marginal arenites are cyclic with each cycle base marked by contorted and convolute laminae or flame structures over a thin lutite. Individual cycles range from 1 to 20 ft (0.3 to 6 m) in thickness and are virtually featureless

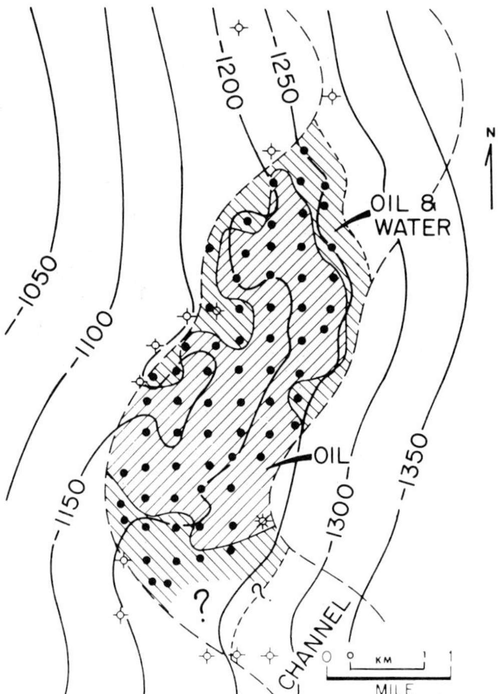

FIG. 11—Paduca field in Lea County, New Mexico. Structural contours are of top of Ramsey sandstone (from Berg, 1975).

TABLE 3.   RESERVOIR DATA

	Triste Draw[1]	Double X[2]	Paduca[3]
Porosity	25	24	24
Permeability	80	10(4)	25
Oil			
API	40°	41°	41°
Sp.Gr.	.82	.82	.82
Water			
Sp.Gr.	1.16	1.16	1.2
Temp °F	106	100	100
Oil Column (ft)	51	55	120
Dip (ft/mile)	100	100	100

(1)   Harrington, G. E., 1966

(2)   Foltz, G. A., 1966

(3)   Scott, R. J., 1966

(4)   Total Ramsey Average

except for the parallel partings and minor turbidite sequences (less than 1 ft or 0.3 m thick).

Most channel arenites were deposited as tractive or suspended loads by fluctuating fluid-density currents. The currents had a high density, relatively high velocity, and may have been overridden occasionally by a low-density turbidity current. Both currents were closely related and originated along the basin-margin slope areas. Hypersaline density currents probably were very important in the process and may have developed in the shelf area where evaporation rates were high. A major depositional trend in the Ramsey sandstone shows channel deposition from the northeast, possibly from the then shelf-like Midland basin. This interpretation is compatible with the known presence of structural sags and poor reef development on that part of the shelf.

It is concluded that during Bell Canyon deposition, fluid-density currents and sediment-suspension clouds were the major sedimentation processes but classic turbidity currents were subordinate. The fluid-density currents transported sediment from the basin margin into the deeper parts of the basin while marginal areas received sediment influxes of very dense sediment-fluid flows which also generated sediment-suspension clouds that separated from the main flow. Direct evidence of well-developed major submarine canyons within the Bell Canyon Formation is lacking. Instead shallow-surge channels probably introduced sediment to the basin, and leveed channels carried the sediment into the distal regions. These channels later become sites for hydrocarbon reservoirs as tilted stratigraphic traps.

REFERENCES CITED

Adams, J. E., 1936, Oil pool of open reservoir type: AAPG Bull., v. 20, p. 780-796.

Beck, R. H., 1967, Depositional mechanics of Cherry Canyon Formation, Texas: Master's thesis, Texas Tech. Univ., 107 p.

Berg, R. R., 1975, Capillary pressures in stratigraphic traps: AAPG Bull., v. 59, p. 939-956.

Bouma, A. H., 1962, Sedimentology of some flysch deposits: a graphic approach to facies interpretation: Amsterdam, Elsevier Pub. Co., 168 p.

Cartwright, L. D., Jr., 1930, Transverse section of Permian basin, West Texas and southeast New Mexico: AAPG Bull., v. 14, p. 969-981.

Crimes, T. P., and J. C. Harper, eds., 1970, Trace fossils, in Liverpool Geol. Soc. and Internat. Union Geol. Sci. Internat. Conf., Liverpool, England, Proc.: Geol. Jour., spec. issue 3, 547 p.

DeFord, R. K., and G. D. Riggs, 1941, Tansill Formation, West Texas and southeastern New Mexico: AAPG Bull., v. 25, p. 1713-1728.

Dunham, R. J., 1972, Capitan reef, New Mexico and Texas: facts and questions to aid interpretation and group discussion: Soc. Econ. Paleontologists and

Mineralogists Permian Basin Sec. Pub. 72-14, Variously paged.

Foltz, G. A., 1966, Double X oil field, *in* The oil and gas fields of southeastern New Mexico: Roswell Geol. Soc., p. 100-101.

Harms, J. C., 1968, Permian deep-water sedimentation by nonturbid currents, Guadalupe Mountains, Texas (abs.): Geol. Soc. America Spec. Paper 121, p. 127.

Harrington, G. E., 1966, Triste Draw oil field, *in* The oil and gas fields of southeastern New Mexico: Roswell Geol. Soc., p. 176-177.

Harrison, S. C., 1966, Depositional mechanics of Cherry Canyon sandstone tongue: Master's thesis, Texas Tech. Univ., 114 p.

Hull, J. P. D., Jr., 1957, Petrogenesis of Permian Delaware Mountain sandstone, Texas and New Mexico: AAPG Bull., v. 41, p. 278-307.

Jacka, A. D., R. H. Beck, L. St. Germain, and S. C. Harrison, 1968, Permian deep-sea fans of the Delaware Mountain Group (Guadalupian), Delaware basin, *in* Guadalupian facies, Apache Mountain area, West Texas: Soc. Econ. Paleontologists and Mineralogists Permian Basin Sec. Pub. 68-11, p. 49-90.

Kelley, V. C., 1972, Geometry and correlation along Permian Capitan escarpment, New Mexico and Texas: AAPG Bull., v. 56, p. 2192-2211.

King, P. B., 1934, Permian stratigraphy of trans-Pecos Texas: Geol. Soc. America Bull., v. 45, p. 697-798.

———— 1942, Permian of West Texas and southeastern New Mexico: AAPG Bull., v. 26, p. 535-763.

———— 1948, Geology of the southern Guadalupe Mountains, Texas: U.S. Geol. Survey Prof. Paper 215, 183 p.

Lang, W. B., 1937, The Permian formations of the Pecos Valley of New Mexico and Texas: AAPG Bull., v. 21, p. 833-898.

Lloyd, E. R., 1929, Capitan limestone and associated formations of New Mexico and Texas: AAPG Bull., v. 13, p. 645-658.

McKee, E. D., 1951, Sedimentary basins of Arizona and adjoining areas: Geol. Soc. America Bull., v. 62, p. 481-506.

Mear, C. E., and D. V. Yarbrough, 1961, Yates Formation in southern Permian basin of West Texas: AAPG Bull., v. 45, p. 1545-1556.

Meinzer, O. E., B. C. Renick, and K. Bryan, 1926, Geology of No. 3 reservoir site of the Carlsbad Irrigation Project, New Mexico, with respect to water-tightness: U.S. Geol. Survey Water-Supply Paper 580, 39 p.

Meissner, F. F., 1972, Cyclic sedimentation in Middle Permian strata of the Permian basin, West Texas and New Mexico, *in* Cyclic sedimentation in the Permian basin, 2d ed.: West Texas Geol. Soc., p. 203-232.

Needham, C. E., and R. L. Bates, 1943, Permian type sections in central New Mexico: Geol. Soc. America Bull., v. 54, p. 1653-1667.

Newell, N. D., J. K. Rigby, A. G. Fisher, A. J. Whiteman, J. E. Hickox, and J. S. Bradly, 1953, The Permian reef complex of the Guadalupe Mountains region, Texas and New Mexico: San Francisco, Freeman and Co., 236 p.

Nottingham, M. W., 1960, Recent Bell Canyon exploration in the North Delaware basin (New Mexico-Texas), *in* Natural gas in the Southwest: Southwestern Federation Geol. Socs. Trans., v. 1, p. 139-153.

Oriel, S. S., D. A. Myers, and E. J. Crosby, 1967, West Texas Permian basin region, *in* Paleotectonic investigations of the Permian System in the United States: U.S. Geol. Survey Prof. Paper 515, p. 21-64.

Richardson, G. B., 1904, Report of a reconnaissance in trans-Pecos Texas north of the Texas and Pacific Railway: Univ. Texas Bull. 23, p. 38-45.

Rigby, J. K., 1958, Mass movement in Permian rocks of trans-Pecos Texas: Jour. Sed. Petrology, v. 28, p. 298-315.

Scott, R. J., 1966, Paduca oil field, *in* The oil and gas fields of southeastern New Mexico: Roswell Geol. Soc., p. 144-145.

Silver, B. A., and R. G. Todd, 1969, Permian cyclic strata, northern Midland and Delaware basins, West Texas and southeastern New Mexico: AAPG Bull., v. 53, p. 2223-2251.

St. Germain, L. C., 1966, Depositional dynamics of the Brushy Canyon Formation, Delaware basin: Master's thesis, Texas Tech. Univ., 119 p.

Tyrell, W. W., 1969, Criteria useful in interpreting environments of unlike but time-equivalent carbonate units (Tansill-Capitan-Lamar), Capitan reef complex, West Texas and New Mexico, *in* Depositional environments in carbonate rocks: Soc. Econ. Paleontologists and Mineralogists Spec. Pub. 14, p. 80-97.

Waterschoot van der Gracht, W. A. J. M., 1931, Permo-Carboniferous orogeny in the south-central United States: AAPG Bull., v. 15, p. 991-1057.

Reprinted for private circulation from
THE AMERICAN ASSOCIATION OF PETROLEUM GEOLOGISTS BULLETIN
Vol. 60, No. 4, April, 1976

ROSEDALE CHANNEL—EVIDENCE FOR LATE MIOCENE SUBMARINE EROSION IN GREAT VALLEY OF CALIFORNIA[1]

BRUCE D. MARTIN[2]
Los Angeles, California

ABSTRACT

West of Bakersfield in the Great Valley of California, eight wells drilled below the upper Miocene-middle Miocene boundary penetrated an anomalous sequence of middle Late Miocene (Middle Mohnian) sediments, principally sandstones. These coarse sediments, within the widespread lower Fruitvale Shale of early Late Miocene age (Early Mohnian), are interpreted to be fill within an early?-middle Late Miocene submarine canyon eroded and filled during a time interval of about 700,000 years. The names, Rosedale Channel and Rosedale Sandstone, are proposed, respectively, for the canyon and the fill.

Electric log correlations, microfossil data, and sedimentary characteristics are used for interpretation. Only a remnant of the originally more extensive canyon is described, because of difficulties in the recognition of the headward and seaward extensions. Seismic data are inadequate for recognition.

Microfossils show that filling occurred entirely in the marine environment in a depth of water probably greater than 1,300 feet. *Uvigerina subperegrina* and *Cyclammina* sp. in the channel fill attest to the depth during the time of deposition of the Rosedale Sandstone.

Erosion was effected also, entirely in the marine environment, at depths similar to that of the Rosedale Sandstone. The ecology of the foraminifera in the Rosedale Sandstone and of the foraminifer *Epistominella* "*Pulvinulinella*" *gyroidinaformis* in the eroded lower Fruitvale Shale shows no uplift of the ocean floor to permit subaerial cutting of the Rosedale Channel.

Cores from the Rosedale Sandstone show many characteristics analogous with turbidites. Turbidity currents or gravity flows cause the erosion. Downcutting was facilitated by the poor induration of the lower Fruitvale shale.

INTRODUCTION

Subsurface geologic investigations in the Late Miocene marine sediments near Bakersfield (Fig. 1) in the Great Valley of California disclosed an anomalous sequence of strata, principally sandstones and pebble conglomerates of a thickness of at least 1,200 feet, within the widespread lower Fruitvale Shale (Fig. 3). Some interbeds of siltstone occur, but because of the preponderance of coarser material, the total unit is called a sandstone. The name Rosedale Sandstone is proposed, because of the proximity to an oil field of the same name (Fig. 2). Conclusions based on electric log

correlations, microfossil ecology, stratigraphic position, sedimentary characteristics, and the geometry of the strata suggest that the Rosedale Sandstone can be (1) compared with similar de-

[1] Manuscript received, May 3, 1962.

[2] Research associate, Geology Department, University of Southern California.
This paper could not have been written without the complete cooperation of the various oil companies whose electric logs and other data are used. The personnel who were helpful are M. G. McAndrews and Lyle Smith, Shell Oil Company; Fred G. Knight, The Ohio Oil Company; C. L. Doyle, Mobil Oil Company; M. C. Barnard, Jr., Richfield Oil Corporation; W. F. Barbat, Standard Oil Company of California; J. E. Kilkenny and Bradford C. Jones, Union Oil Company of California; and R. L. Johnston, Gulf Oil Corporation of California. The last three individuals also kindly supplied additional information which permitted a more complete discussion of the subject matter. The writer is also indebted to Louis O. Heintz, consultant, who permitted the use of more than 100 electric logs from his personal files. K. O. Emery and O. L. Bandy of the University of Southern California critically read the paper and their help and advice are appreciated.

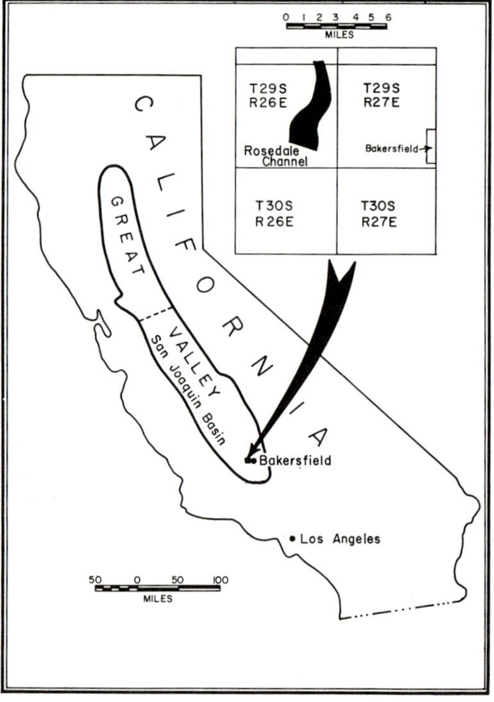

FIG. 1.—Map showing location of Rosedale Channel.

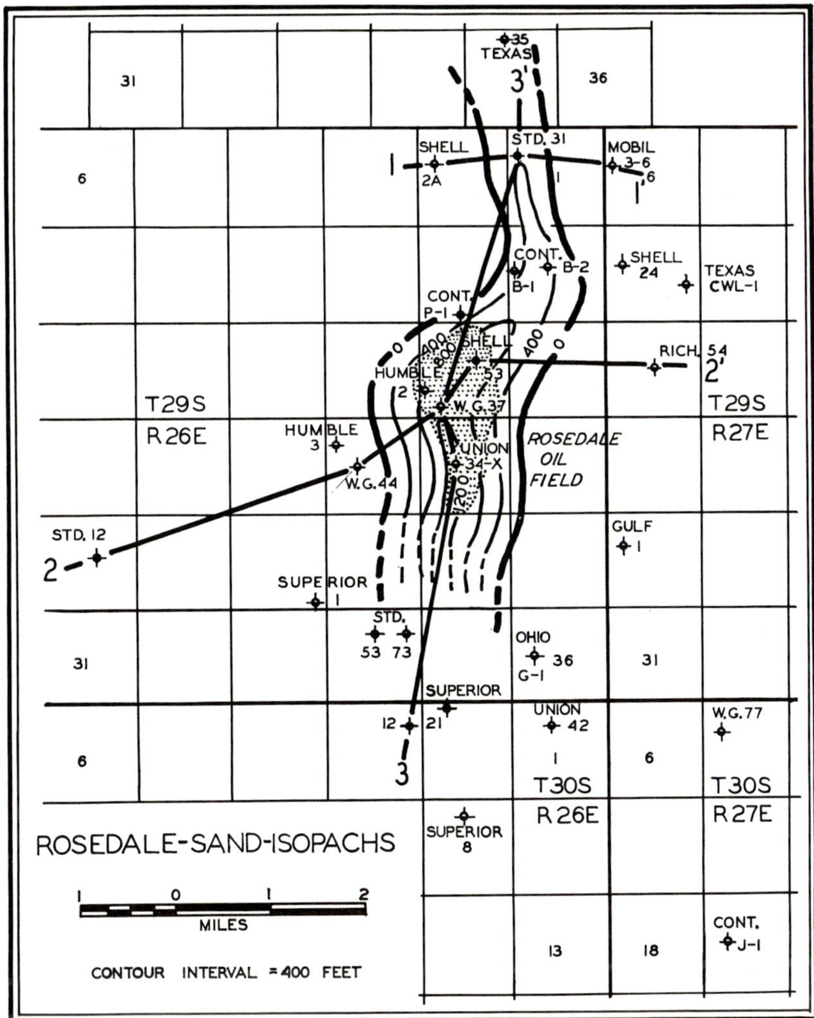

Fig. 2.—Isopachs of Rosedale Sandstone or contours of Rosedale Channel. Only a remnant of originally more extensive channel shown. Not all wells investigated are depicted (see Table I), but only those wells needed to define limits of channel. Figures 5, 6, 7 are cross sections indicated by lines 1-1', 2-2', and 3-3'.

posits in present-day oceans, (2) shown to occur within a certain physiographic feature on the floor of the ocean, and (3) related to submarine processes of erosion. The Rosedale Sandstone is not the principal topic of discussion, as is indicated by the title of the paper, but without the evaluation of the characteristics of the unit, the last two phases of the study can not be adequately discussed.

ROSEDALE SANDSTONE

LOCATION AND CONTROL

The sandstone is approximately 7 miles west of Bakersfield, California, T. 29 S., R. 26 E., Mount Diablo Base and Meridian, trending generally north-south through the eastern half of the Township (Figs. 1, 2). A generalized electric log from wells in the Rosedale oil field (Fig. 3) shows the stratigraphy of the area. A more detailed description of the whole geologic section both in the area of the report and for the entire San Joaquin Basin was given by Hoots, Bear, and Kleinpell (1954). The geology of several oil fields in the vicinity of Bakersfield was described by various authors (Beach; Ferguson; and others, 1947, p. 89–127).

By January, 1962, eight wells penetrated

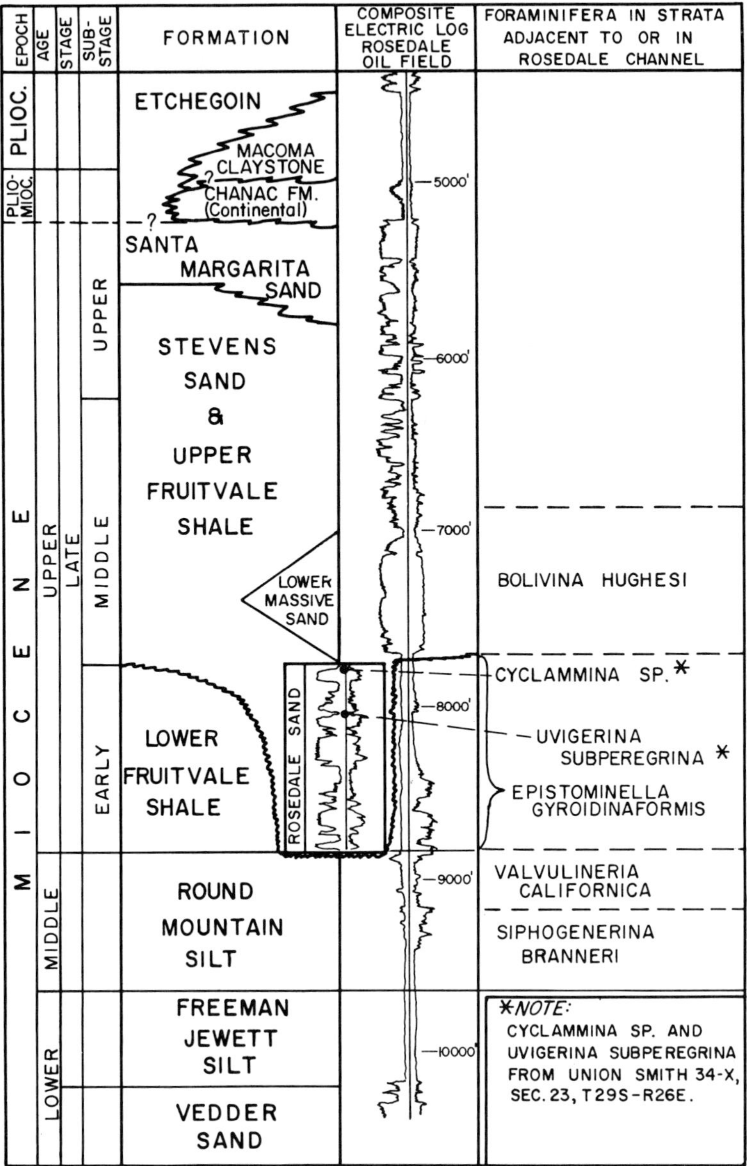

FIG. 3.—Geologic section in vicinity of Rosedale Channel. Rosedale Sandstone shown in stratigraphic relation to lower Fruitvale Shale by insert at left of normal stratigraphic sequence depicted by generalized electric log.

through the sandstone interval according to present correlations; however, other wells especially south of Rosedale oil field may have penetrated the sandstone, but the occurrence of the unit in these wells is questionable. Only four wells of the eight containing the sandstone are used to relate the stratigraphy of the Rosedale Sandstone to other stratigraphic intervals (Figs. 5, 6, 7), as the unit in the other four wells has been extensively faulted and the relationships of the sandstone both to the lower Fruitvale Shale and to the Rosedale Sandstone in other unfaulted wells is not clear. Table I lists the principal wells used. Most of them extend below the base of the upper Miocene sediments; however, many other wells which penetrated well into the upper Mio-

TABLE I. COMPILATION OF SELECT GROUP OF WELLS
USED TO DEFINE ROSEDALE CHANNEL

Well Number	Township Range	Section
Texas KCL 35	28S, 26E	35
*Standard KCL 31-12	29S, 26E	1
Shell KCL 2A-23-2	29S, 26E	2
*Continental KCL P-1	29S, 26E	11
*Continental KCL B-1	29S, 26E	12
*Continental KCL B-2	29S, 26E	12
*Gulf KCL-A 37-14	29S, 26E	14
*Humble KCL 11-14	29S, 26E	14
*Shell KCL 53-14	29S, 26E	14
Humble KCL 3-22	29S, 26E	22
Gulf KCL-A 44-22	29S, 26E	22
*Union Smith 34-X	29S, 26E	23
Moriqui Bros. 17-26	29S, 26E	26
Superior Brandt 1	29S, 26E	28
Standard KCL 12-6	29S, 26E	30
**Standard Karpe 2-53	29S, 26E	34
**Standard Karpe 2-73	29S, 26E	34
**Franco-Western Papoff 85-34	29S, 26E	34
**Superior Berchtold 25	29S, 26E	35
**Standard Borland-Olney 67	29S, 26E	35
**Oceanic Valmar Comm. 35	29S, 26E	35
**Socony Mobil Pioneer 82-36	29S, 26E	36
**Ohio KCL G-1	29S, 26E	36
**Union KCL 42-1	30S, 26E	1
**Superior Quinn et al 21-2	30S, 26E	2
**Superior KCL 12	30S, 26E	3
Tidewater KCL-E 25-7	30S, 26E	7
**Superior-Union KCL 8	30S, 26E	11
**Continental KCL D-2	30S, 26E	14
**Shell KCL-A 53-30	30S, 26E	30
Socony Mobil Kernway 3-6	29S, 27E	6
Shell Shell-Camp-Moncure et al 24-7	29S, 27E	7
Texas Camp-West-Lowe 1	29S, 27E	7
Richfield KCL-E 54-18	29S, 27E	18
Gulf KCL B-45	29S, 27E	22
Gulf Shellabarger-Selden 1	29S, 27E	30
Gulf KCL 77-5	30S, 27E	5
Humble KCL D-2	30S, 27E	15
**Continental KCL J-1	30S, 27E	17
Humble KCL D-1	30S, 27E	22

* Wells penetrating Rosedale Sandstone.
** Wells south of Rosedale oil field which may contain lenses of Rosedale Sandstone and in which Rosedale Channel may be present.

cene section, but not to its base, were examined for additional information. Seismic data are useless for control, as the sandstone body does not appear on any line.

GEOMETRY OF SANDSTONE

Microfossil data, electric log correlations, and sediment characteristics of cores and side-wall samples delimit the vertical and lateral boundaries of the Rosedale Sandstone. Exact areal extent and volume of the sandstone are difficult to calculate, as the values obtained depend in part on the position of the zero isopach which can be correctly drawn only in proportion to the density of well control. Based on the isopach map (Fig. 2), the areal extent is approximately 5.4 square miles and the volume about 1.1 cubic miles. The width ranges from 0.7 mile at the north to 1.3 miles at the south, but these values also are subject to the limitations of correctly positioning the zero isopachs. Thickness ranges from 400 feet to 1,200 feet in Standard KCL 31-12, sec. 1, T. 29 S., R. 26 E., and in Gulf KCL 37-14, sec. 14, T. 29 S., R. 26 E., respectively. The top of the unit occurs between 6,893 feet and 8,002 feet below sea-level with a mean of minus 7,447 feet.

MICROFOSSIL RELATIONSHIPS

Bradford C. Jones, chief paleontologist for the Union Oil Company of California, gave microfossil data from the Rosedale Sandstone in the interval between 7,500 feet and 9,500 feet (well depth) in Union Smith 34-X, sec. 23, T. 29 S., R. 26 E. (Fig. 7). *Uvigerina subperegrina* was identified from a core at 8,363 feet, approximately 800 feet above the base of the Rosedale Sandstone, and *Cyclammina* sp. occurred at 8,010 feet along with species of *Bolivina* and *Haplophragmoides*. At 8,470 feet were specimens of *Nonionella miocenica*, *Bolivina marginata*, *Buliminella elegantissima*, and *Elphidium* sp.

Natland (1933, 1957) discussed the ecology of these foraminifera. *Uvigerina subperegrina* is similar to modern species which live in a depth of water no shallower than 1,300 feet (1957, Pl. 6; Bandy, 1953, 1960). Uvigerinas are considered to be an important group for determining ecological conditions. The other foraminifera in the Union well, except *Cyclammina* sp., indicate shallower-water conditions.

Cyclammina cancellata H. B. Brady collected by Natland off the island of Santa Catalina, California, lives at a depth of 1,300 feet (Natland, 1933, Pl. 1). O. L. Bandy (1960, p. 9–10) stated that Cyclamminas are identified with the bathyal zone throughout present-day oceans. In any case, this genus of foraminifer indicates water depths ranging from about middle bathyal to abyssal.

The lower Fruitvale Shale (Fig. 3) contains the foraminifer *Epistominella "Pulvinulinella" gyroidinaformis*. Natland (1957, Pl. 6) gave a minimum depth limit of 1,500 feet for this species.

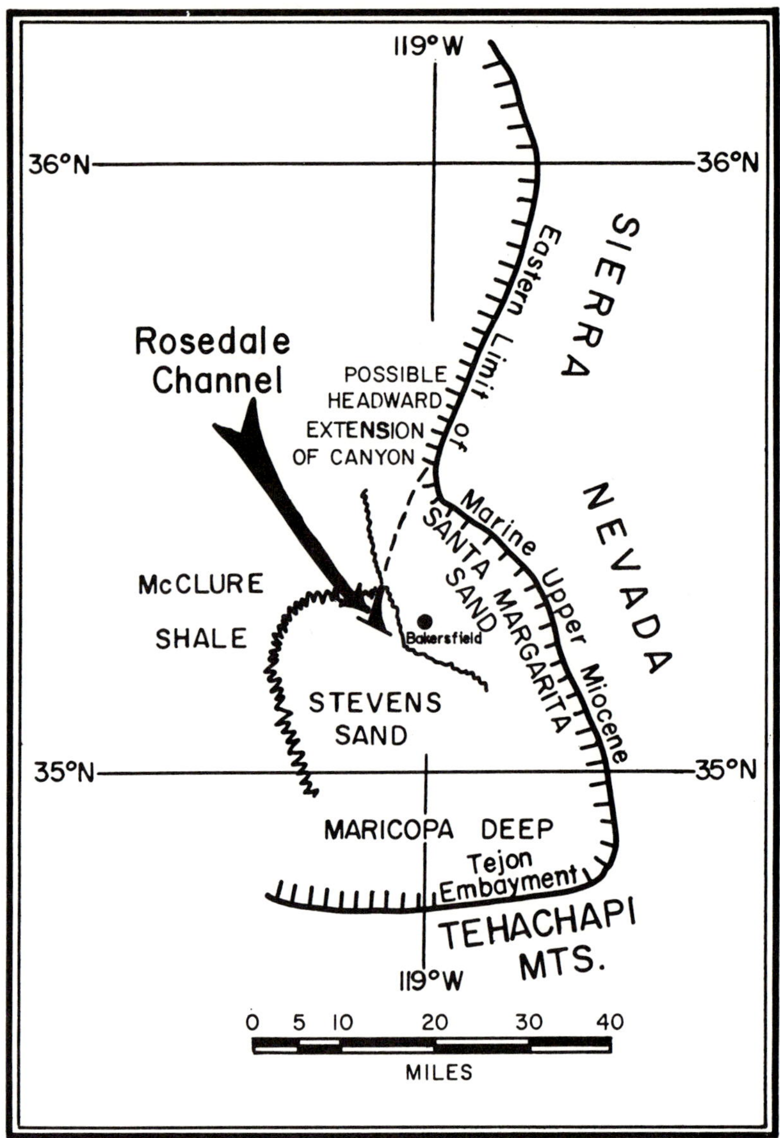

Fig. 4.—Generalized geology and paleogeology of southeastern part of San Joaquin Basin. Line showing eastern limit of marine upper Miocene sediments modified after Hoots, Bear, and Kleinpell (1954).

Microfossil assemblages in the Rosedale Sandstone are dated as middle Late Miocene (Middle Mohnian) and the *Epistominella* fauna in the lower Fruitvale Shale shows the shale to be of early Late Miocene age (Early Mohnian).

Microfossils indicate that: (1) the lower Fruitvale Shale accumulated in water depths greater than 1,500 feet; (2) the Rosedale Sandstone was deposited in a depth of water greater than 1,300 feet; (3) nearshore faunas were displaced into the deeper regions of the marine basin as shown by the occurrence of the shallow-water forms at the same horizon with deep-water species; and (4) faunas in the Rosedale Sandstone are younger than those in the lower Fruitvale Shale.

ELECTRIC LOG RELATIONSHIPS

Electric log correlations show the Rosedale Sandstone in the stratigraphic interval of the lower Fruitvale Shale (Figs. 5,6,7). Without the control afforded by the microfossils, this development of coarse-grained sediments within a normal sequence of fine-grained sediments probably would be considered to be that of an abrupt facies change; but because the Rosedale Sandstone contains foraminifera younger than those in the lower Fruitvale Shale, the facies concept becomes untenable. It is also inadequate to explain the stratigraphic relationships between the sand and shale as a graben filled with the coarser material, for the electric log correlations (Figs. 5, 6, 7) show no downward displacement of pre-upper Miocene strata. Finally, the relationships can not represent contemporaneous deposition of sand and silt from different source areas as indicated again by the age difference in contained microfossils. The best explanation seems to be (1) by erosion which removed all or part of the lower Fruitvale Shale after its deposition and (2) by later sedimentation which deposited the Rosedale Sandstone in the previously eroded channel.

Lateral limits of the Rosedale Sandstone in the region of the cross sections (Figs. 5, 6, 7) are well defined by the relationships to the lower Fruitvale Shale; the base of the sandstone rests either on earlier lower Fruitvale Shale or on the Round Mountain Siltstone; and the upper boundary of the unit is limited by the lower Massive Sandstone member of the Stevens Sandstone. The electric log trace of this last unit aids greatly in delimiting the vertical extent of the Rosedale Sandstone and where the trace is not well developed, the Rosedale Sandstone either can not be recognized or the recognition is highly questionable. Accordingly, some wells may have penetrated this unit, but without the characteristic trace of the lower Massive Sandstone, the correlations are not clear and at present, it is deemed best not to extend the isopachs of the Rosedale Sandstone (Fig. 2) to include such questionable wells.

By correlation, certain wells south of the Union Smith 34-X well (Fig. 2) contain sandstones in the stratigraphic interval of the lower Massive Sandstone, but without the characteristics of the electric log trace. In these wells, no Rosedale Sandstone seems to underlie the lower Massive Sandstone, except for the possiblity of a few thin sandstone lenses in Union KCL 42-1, sec. 1, T. 30 S., R. 26 E.; Superior KCL 12, sec. 3, T. 30 S., R. 26 E.; and Ohio KCL G-1, sec. 36, T. 29 S., R. 26 E. The wells considered to show little or no Rosedale Sandstone are indicated in Table I by a double asterisk.

The trace of the electric log from Gulf KCL-A 44-22, sec. 22, T. 29 S., R. 26 E. (Fig. 6), suggests that the lower Massive Sandstone overlies a somewhat sandy interval, possibly the Rosedale Sandstone. Through the courtesy of R. L. Johnston of the Gulf Oil Corporation of California, Stanley Beck gave microfossil data from this well. The top of the lower Fruitvale Shale occurs at 8,594 feet (well depth). This contact is above the sandy interval and below the lower Massive Sandstone; therefore in this well, the Rosedale Sandstone does not appear and the zero isopach of the sandstone is farther east.

SEDIMENTS

Five side-wall samples were described from the Rosedale Sandstone in Gulf KCL 37-14, T. 29 S., R. 26 E., in the interval between 8,545 feet and 8,870 feet (Fig. 7).

Sand, light gray, medium-grained, fairly well sorted, firm, friable, biotitic, kaolinitic. Some quartz granules up to $\frac{3}{16}$ inch.

One core $1\frac{1}{4}$ feet in length was taken from the Rosedale Sandstone at 7,242 feet in Standard KCL 31-12, sec. 1, T. 29 S., R. 26 E. (Fig. 7).

Sand, light gray, predominantly fine-grained, quartzitic and feldspathic, with abundant interstitial kaolin and clay material.

Seven cores were described from the Rosedale Sandstone in Continental KCL B-2, sec. 12, T. 29 S., R. 26 E.

At 7,010 feet—recovered 10 feet
    1 foot   Dark gray siltstone.
    3 feet   Fine, firm to soft muddy dark gray sand.
    6 feet   Fine sandy to silty dark gray shale.
At 7,020 feet—recovered 6 feet
    Firm to hard dark gray siltstone, scattered sea-shell fragments. Rare streaks dark fractured and slicken-sided shale. Rare pockets of carbonaceous material.
At 7,030 feet—recovered 10 feet
    8 feet   Dark gray silty to sandy shale and siltstone.
            Scattered sea-shell fragments.
   $1\frac{1}{2}$ feet   Muddy fine dark gray sand.
   $\frac{1}{2}$ foot   Muddy conglomeratic sand.
            Pebbles up to $1\frac{1}{4}$ inch.
            Coarse sand streaks.
At 7,040 feet—recovered $7\frac{1}{2}$ feet
    Soft coarse pebbly feldspathic very light gray sand. Few small clay pebbles. 6 inches at top dark gray siltstone. Streak hard shells or boulders at $7,041\frac{1}{2}$, $7,044\frac{1}{2}$, and 6-inch sandy shale streak at 7,045.
At 7,050 feet—recovered 8 feet
    Coarse soft pebbly feldspar light gray sand.

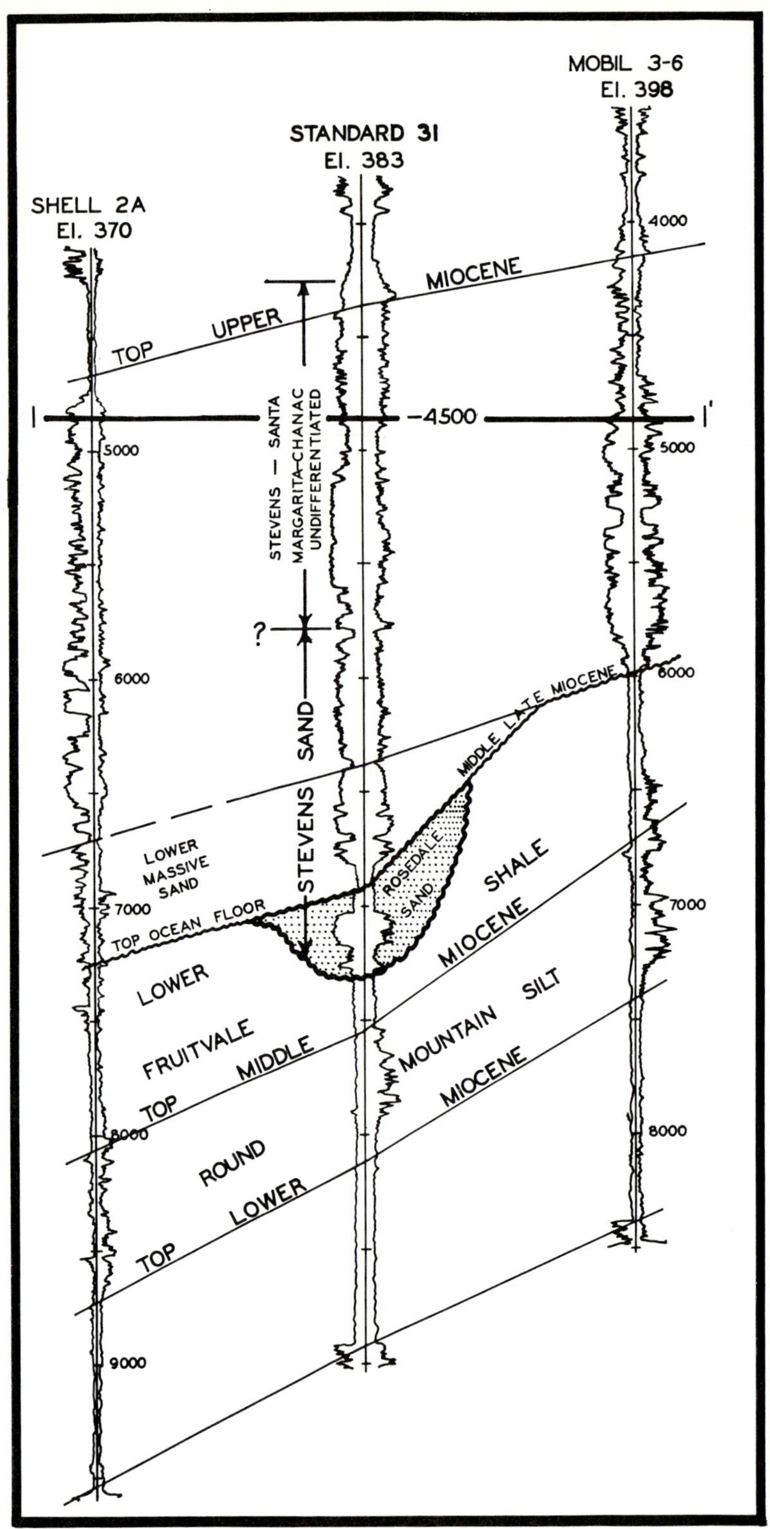

Fig. 5.—West-east cross section (Fig. 2 for location) at north end of Rosedale Channel normal to axial trend. Depth of erosion below Late Miocene ocean floor approximately 400 feet.

At 7,060 feet—recovered 5 feet
  Coarse soft pebbly feldspar light gray sand.
At 7,070 feet—recovered 7 feet
  Coarse soft light and dark gray pebbly feldspar sand with 8 inches of shaly and silty material in center.

Descriptions of the sedimentary characteristics of the Rosedale Sandstone suggest strongly that the unit is composed of a sequence of turbidites. Kuenen and Carozzi (1953, p. 363–373) listed the characteristics of a turbidite, citing such features as graded bedding, poor sorting, regularity of bedding, sharp basal contacts, absence of shallow-water features, convolute bedding, interbeds of sandstone and shale, slump structures, and shale inclusions. Work by Dill and Shumaker (1961) in the bathyscaph Trieste which dived to 320 fathoms in La Jolla Canyon, California, may revise some of this classical thinking concerning turbidites.

The Rosedale Sandstone does not display all the typical sedimentary features of turbidites cited by Kuenen and Carozzi (1953); however, graded beds are suggested in the lowest 2 feet between 7,030 feet and 7,040 feet in Continental B-2 well. All descriptions from the three wells list poor sorting, interbeds of sandstone and shale, and fragments of probable shallow-water megafossils. These data and the occurrence of shallow-water microfossils mixed with the deep-water forms seem conclusive evidence that the Rosedale Sandstone is a turbidite and thus the sediments were transported and deposited under conditions different from the lower Fruitvale Shale.

ROSEDALE CHANNEL

GENERAL REMARKS

The Rosedale Sandstone appears to represent the sedimentary fill within a Late Miocene submarine canyon. The name Rosedale Channel is proposed for the erosional feature for the same reason that the sandstone was earlier named. Filled erosional features have been described in the literature (Bruce, 1959, p. 23–26; Bornhauser, 1948; Hoyt, 1959; Emery, 1960, p. 43; and Neev, 1960, p. 1–20). Other similar physiographic features are known, but not described, as the data usually are in confidential files of oil companies. The Rosedale Channel therefore is not unique, but because microfossil evidence shows that erosion and filling of the channel occurred entirely in the marine environment, the data and the conclusions

obtained may be significant in the discussion of the genesis of present-day submarine canyons.

Because of lack of well control in headward and seaward portions, only a remnant (Fig. 2) of the probably more extensive Late Miocene submarine canyon is recognized; but this fact does not deter the present discussion, for future work may disclose more of the limits of the channel. Data are presented to suggest possible channel trends.

GEOMETRY OF CHANNEL

Many data used in the discussion of the geometry of the Rosedale Sandstone can also be similarly used to show the geometry of the Rosedale Channel. The areal extent is approximately 5.4 square miles and the volume of material removed during the erosion becomes approximately 1.1 cubic miles. Depth of erosion below the Late Miocene ocean floor (Figs. 5, 6, 7) ranges from 400 feet to 1,200 feet in the same wells which show the thickness of the fill (Rosedale Sandstone); however, the Union Smith 34-X well may not be directly above the axis of the channel and the amount of maximum erosion (as well as the thickness of the Rosedale Sandstone) may be greater. The trend of the Rosedale Channel is linear with the length approximately 5 times that of the width. Distances between opposing canyon walls range from 0.7 mile at the north to 1.3 miles at the south (Fig. 2).

The axis of the canyon sloped south in the Late Miocene as shown by the thickening of the channel fill from north to south. The direction of the declivity is the same today. Post-Miocene diastrophism considerably faulted the area around Bakersfield, but little folding accompanied the movement so that the present regional dip is still less than 10° southward. By adjusting the top of the channel fill to a common datum, the gradient of the axis during the Late Miocene is calculated to have been about 4 per cent.

The gradients of the walls can be only approximated, as these values are dependent on the correct positioning of the zero isopach (Fig. 2). Based on this line, the slope of the walls ranges between 19 per cent and 40 per cent. It is not known if the transverse profiles are V-shaped, as no well penetrated the fill high on the wall to give an intermediate point of control between the axial wells and those outside the limits of the canyon.

COMPARISON WITH PRESENT-DAY SUBMARINE CANYONS

It is obvious that the filled and buried Rosedale Channel can not be compared feature for feature with relatively unfilled Recent submarine canyons. Also, the Rosedale Channel represents only a remnant of what is considered to have been a much more extensive canyon during the Late Miocene; however, certain characteristics of the channel still suggest comparative relationships.

Kuenen (1950, p. 487) discussed a typical submarine canyon in present-day oceans. Gradients of profiles from the canyon heads to the basin deeps range from 15 per cent to less than 3 per cent. Walls are steep, some of them exceeding a slope of 100 per cent and the transverse profiles are V-shaped. Canyons may be eroded hundreds to thousands of feet below the adjoining sea floor. Axial trends may be sinuous, but lacking in sharp turns. Nick points and hanging tributaries are not recognized.

The Rosedale Channel is comparable with certain features of Recent canyons, such as axial gradients of the more seaward reaches, slopes of walls, linearity, and depth of erosion. Other physiographic comparisons either are questionable or not recognized. Perhaps in the future when more wells are drilled within the limits of the Rosedale Channel, other points of comparison will be noted, especially those pertaining to the shape of the transverse profiles.

The work of Natland (1957, p. 552–554) emphasized the importance of displaced faunas in the recognition of submarine canyons and turbidites. Because of the displaced shallow-water faunas in the Rosedale Sandstone, this fact is considered as further proof in favor of the Rosedale Channel being a submarine canyon.

Bandy and Arnal (1960, p. 1921–1931) showed several possible channels in the general area in their paleobathymetric map for the middle Miocene.

Paleogeographic and paleogeologic investigations indicate that the shoreline of the Late Miocene sea was approximately 15 miles north and 20 miles east of the channel remnant (Fig. 4). The sandy nature of the sediments in the Rosedale Channel along with fragments of megafossils and nearshore shallow-water microfossils suggest that the head of the canyon extended farther north and possibly approached closer to shore than is now shown by the areal limits (Figs. 2, 4). Emery (1960, p. 217–218) discussed the sediments in present-day canyons. Canyons with heads approaching close to shore contain sandy sediments comparable with the material on adjacent beaches and conversely those with heads far from shore have sandy green muds on the floors. Work by Dill and Shumaker (1961), however, indicated that perhaps other factors besides the distance from shore to the heads of canyons can cause a variation in sediment characteristics in a seaward direction along canyon trends.

Physical recognition of the headward extension of the Rosedale Channel north of T. 29 S. is not possible at present, because of (1) limited well control, (2) sands within the stratigraphic interval of the channel fill similar to nearshore sands, and (3) the absence of deep-water microfossils.

South of Union Smith 34-X (Fig. 2), the more seaward extension of the Rosedale Channel can not be positively recognized. Numerous wells (indicated by a double asterisk in Table I) penetrated the stratigraphic interval of the Rosedale Sandstone, but the relations between the sandstone and shale are not as clear as in the area shown by the cross section (Figs. 5, 6, 7). The enigma of the "missing" southern extension can not be resolved completely at present, but certain evidence suggests possible areas of study. The trend of the channel either (1) has been displaced toward the southeast by post-Miocene lateral faulting or (2) continues without interruption south of the Rosedale oil field (Fig. 2) through the eastern half of T. 30 S., R. 26 E. Either possibility seems reasonable.

Faulting is considered because the thickness of Stevens Sandstone in Continental KCL J-1, sec. 17, T. 30 S., R. 27 E., southeast of the Rosedale oil field, is comparable with the thickness of correlative intervals in Union Smith 34-X, sec. 23, T. 29 S., R. 26 E., and in Gulf KCL 37-14, sec. 14, T. 29 S., R. 26 E. On the other hand, the Stevens Sandstone section is thinner directly south of the Union and Gulf wells in Superior KCL 12, sec. 3, T. 30 S., R. 26 E. (Fig. 7). Investigations of other wells (Table I) also confirm this observation. By this hypothesis, the Stevens Sandstone in Continental KCL J-1 was deposited during the Late Miocene directly south of the Union and Gulf wells and was subsequently dis-

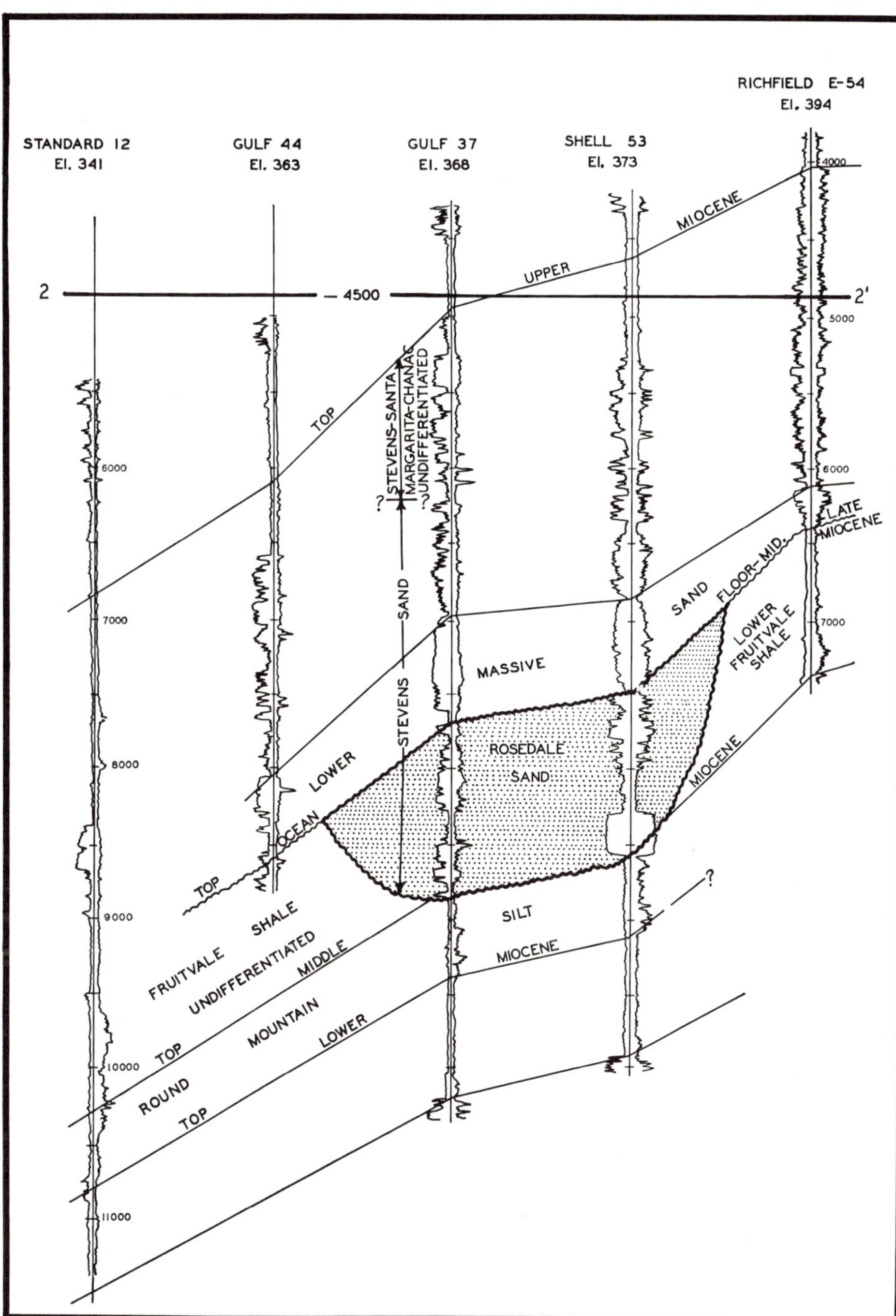

FIG. 6.—West-east cross section (Fig. 2 for location) at south end of Rosedale Channel normal to axial trend. Depth of erosion below Late Miocene ocean floor approximately 1,200 feet.

placed by post-Miocene lateral faulting to the southeast. The Rosedale Sandstone is represented by the lowermost sandstones in Continental KCL J-1. Microfossil ages might confirm this postulation, as was done in the vicinity of the Rosedale oil field (Fig. 2) where the channel is better developed.

Continuation of the channel trend through the eastern half of T. 30 S., R. 26 E., at first does not seem feasible, because of the apparent lack of Rosedale Sandstone and because of no concrete evidence for channelling in the lower Fruitvale Shale. Wells investigated are indicated by a double asterisk in Table I. This hypothesis suggests that the Rosedale Sandstone must grade southward into fine-grained sediments indistinguishable from the sediments of the lower Fruitvale Shale. A few lenses of Rosedale Sandstone possibly are present such as in Union KCL 42-1, sec. 1, T. 30 S., R. 26 E. Three lines of evidence suggest the facies change. (1) correlation between Gulf KCL 37-14, sec. 14, T. 29 S., R. 26 E., and Union Smith 34-X, sec. 23, T. 29 S., R. 26 E. (Fig. 7), shows about 300 feet of Rosedale Sandstone at the top of the section in the former well grading to shale southward. If this rate of change from sand to shale proceeds arithmetically, the sand disappears close to the southern limit of sec. 26, T. 29 S., R. 26 E. (2) Dill and Shumaker (1961) observed the lack of sand in La Jolla Canyon, California, at a depth of 320 fathoms, yet great quantities of sand are trapped and moved basinward from the head of this canyon. By analogy, the grain size of sediments in the Rosedale Channel could grade southward or basinward into increasingly finer sizes and the distinguishing features on electric logs of sandstones abutting against shale no longer are apparent even though the channel may be present. (3) Charles Cary of the Union Oil Company of California kindly supplied microfaunal information from Superior-Union KCL 8, sec. 11, T. 30 S., R. 26 E. In this well, a core from 9,425 feet to 9,431 feet (well depth) contained the foraminifera *Uvigerina subperegrina* and *Bolivina marginata* var. *multicostata*. The electric log shows the cored section to be shale and to occur 300 feet below the base of the Stevens Sandstone. Faunas characteristic of the lower Fruitvale Shale do not appear at less than 9,731 feet in depth. Fine-grained sediments below the base of the Stevens Sandstone contain Middle

Mohnian faunas similar to those in the Rosedale Sandstone, suggesting the presence of the channel, but without the fill of coarse material.

TIME OF EROSION AND FILLING

Erosion and filling of the Rosedale Channel occurred during the early?-middle Late Miocene substage (Fig. 3). This age determination is based not only on the microfossil data, but also on the depositional sequence of the Late Miocene sediments.

Lower Fruitvale Shales (Fig. 3) except within the limits of the channel, range in thickness from 750 feet to 1,300 feet. Because the Rosedale Channel could not have been cut until a greater part of these fine-grained sediments accumulated, the erosion occurred at some time near the close of the early Late Miocene or possibly at the beginning of the middle Late Miocene. The time of erosion could not have been any later, as both the channel fill and the overlying Lower Massive Sandstone (Fig. 3) contain middle Late Miocene microfossils.

Certain data suggest that the channel was cut and filled during an interval of about 700,000 years. Kulp (1961, Table 3) gave figures of 25,000,000 years BP, 17,000,000 years BP, and 13,000,000 years BP for the ages of the Miocene-Oligocene boundary, the middle of the Miocene, and the Pliocene-Miocene boundary, respectively. Each of these figures is accurate to $\pm 1,000,000$ years and in a more detailed study of age dates, this variable should be considered; but in this study where the time of cutting and filling of the Rosedale Channel is in itself only an approximation, the variable does not aid in making the final approximation any more accurate. From the data of Kulp, the length of the interval from the middle of the Miocene to the end is about 4,000,000 years. Dividing this number by 4 results in a figure of about 1,000,000 years for each of the four substages (Fig. 3) of the late Middle and Late Miocene stages. Because sediments overlying the fill in the Rosedale Channel contain middle Late Miocene microfossils, the channel had to be filled before the end of this time, possibly during the first half of the substage. This results in a figure of about 500,000 years during the middle Late Miocene when the channel was open. Another figure of approximately 200,000 years is added to the 500,000 years, making a total of about 700,000 years that the channel was cut and

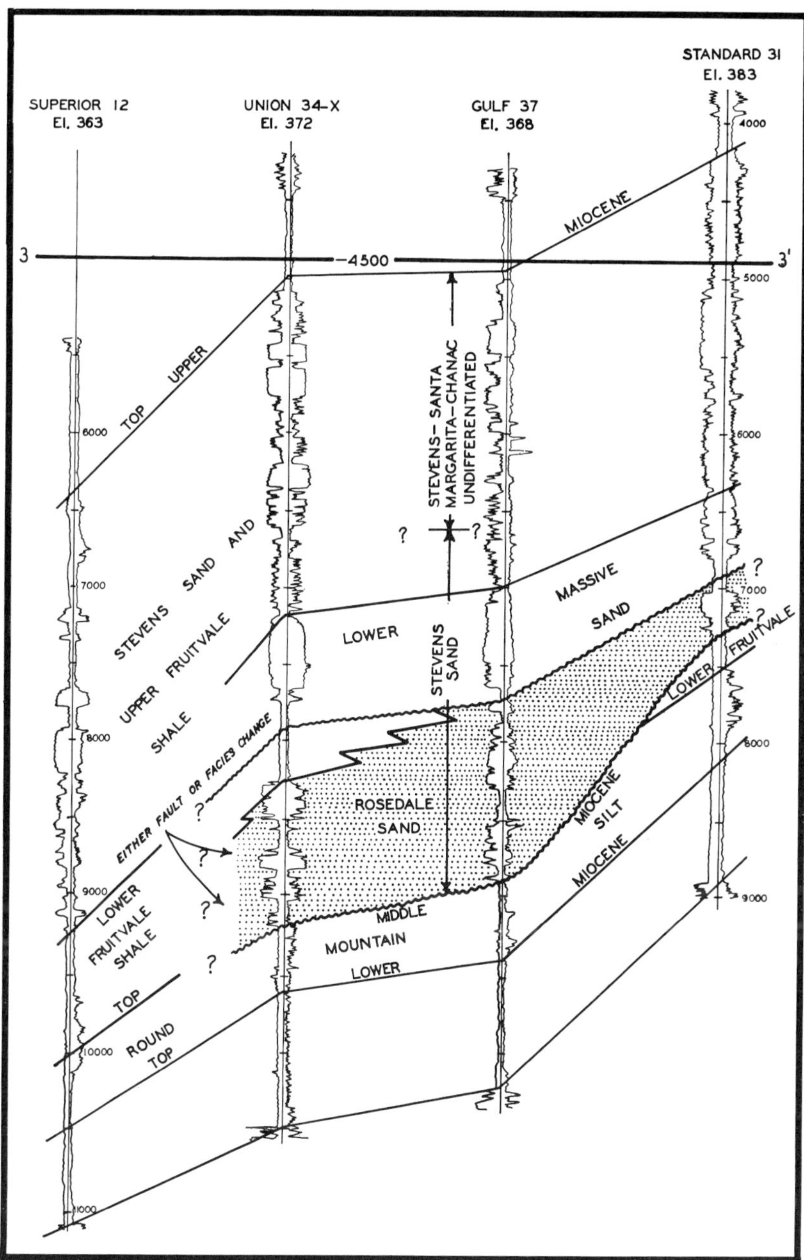

Fig. 7.—South-north cross section (Fig. 2 for location) approximately parallel with axial trend of Rosedale Channel. Relations between Rosedale Sandstone and lower Fruitvale Shale between Superior 12 and Union 34-X suggest complications or facies changes.

filled, because it is possible that the Rosedale Channel began to be eroded during later phases of the early Late Miocene.

ECONOMIC VALUE

In the wells north of Union Smith 34-X (Fig. 2), no oil has been discovered in the sediments in the Rosedale Channel, but the overlying Lower Massive Sandstone (Fig. 3) produces a small quantity of oil from the upper few hundred feet of section. If wells south of the Union Smith well are considered to have penetrated the Rosedale Sandstone interval, then certain wells produce oil or have oil shows in the sands overlying the lower Fruitvale Shale. Two such wells are Standard Karpe 2-53, sec. 34, T. 29 S., R. 26 E., and Superior Berchtold 25, sec. 35, T. 29 S., R. 26 E.

Albright (1954, p. 32) gave data on the oil in the lower Massive Sandstone and in 1957, Shell Oil Company drilled three wells to the oil zone in sec. 14, T. 29 S., R. 26 E. (East Rosedale Extension). Two of the three wells still produce a small quantity of petroleum. This sandstone is considered to be in some way genetically related to the sediments in the Rosedale Channel and for this reason, the discussion of economic value must include petroleum trapped in the lower Massive Sandstone; but the relationships between the Rosedale Sandstone and the lower Massive Sandstone have not been completely studied and at present are not clear.

EFFECT ON NEARSHORE DEPOSITION

Although the principal topic is the discussion of the Rosedale Channel, later Late Miocene deposition seems to have been influenced by the presence of the channel. During the time when the Rosedale Channel was open (along with possibly other presently unrecognized canyons), little sand seems to have been deposited in the nearshore areas. The shallow-water Santa Margarita Sandstone (Fig. 3) began to accumulate only after the Rosedale Channel (and other canyons?) filled with sediments. Sullwold (1961, p. 63–81) discussed more completely the stratigraphic relationships of the nearshore Santa Margarita Sandstone to the deep-water Stevens turbidites. The time difference between the beginning of the deposition of the two units suggests that the Rosedale Channel (and others?) trapped the littoral drift of sands and channelled them into the basin deeps, resulting in little sand available to

form nearshore deposits. Certain present-day submarine canyons, whose heads come close to shore, similarly trap littoral sands and funnel them into the deeper parts of the oceanic basins (Gorcieux, 1922, p. 401–411; Veatch and Smith, 1939, p. 28–29; Shepard and Emery, 1941, p. 93; Medvedev, 1959, p. 10; Emery, 1960, p. 27); therefore, this process probably was active also in the past.

Because Stevens turbidites continued to accumulate (Figs. 5, 6, 7) after the Rosedale Channel was filled (by the probable combination of transgression of the turbidites up the channel because of the filling of the entire basin and deposition within the channel proper directly from turbidity currents or sediment flows coursing down the channel), it is known that other source areas contributed sediments to the basin deeps and it is likely that other submarine canyons channelled these sediments basinward as did the Rosedale Channel. Rose-diagrams of the Stevens turbidite grain orientation indicate that one of the major source areas for the sediments was southeast and south of the Tejon Embayment (Fig. 4). The southern source areas suggest that other submarine canyons probably exist in the Maricopa Deep (Fig. 4) of the San Joaquin Basin, but these channels will be difficult to locate in Miocene and older rocks, because of the deep burial by later marine and non-marine sediments, and the resulting higher costs of deep exploratory drilling. The possibility of locating excellent petroleum traps, however, might overcome objections to the risks involved.

SUBMARINE EROSION OF ROSEDALE CHANNEL
PROCESSES

All evidence in the present study leads to the conclusion that the Rosedale Channel was cut and filled entirely in the marine environment. Accordingly, either an erosion process must be considered which at present may or does account for the physiographic features in present-day oceans, or a new theory must be postulated. No new process of submarine erosion is proposed, for it would be fallacious to propose a process which actively cut canyons during the Late Miocene, but is not active today. One or more of the presently accepted theories of canyon genesis must therefore account for the occurrence of the Rosedale Channel.

Many theories have been proposed, such as

the effects of artesian spring sapping, tsunamis, hydraulic and tidal currents, subaerial erosion during glacially lowered sea-levels, landslides and mudflows, and turbidity currents and flows; however, many of these ideas have been discredited since their promulgation. Shepard and Emery (1941, p. 109–159) and Kuenen (1950 p. 489–526) discussed in detail the merits and failings of these theories.

Ever since Daly (1936, p. 401–420) proposed that submarine canyons are eroded by the action of turbidity currents, the geologic profession has been sharply divided about the erosive power of the currents. The schism becomes even wider when the discussion deals with hard-rock erosion by the flows. R. F. Dill (Naval Electronics Laboratory, San Diego, California) has been studying the problem of submarine erosion in conjunction with his writing a doctoral dissertation. This reference to the work of Dill does not imply that other workers are not intensively studying submarine erosion, but the availability to Dill of the use of the bathyscaph Trieste (Dill and Shumaker, 1961) opens to visual observation regions of the ocean not generally seen by others. Certainly, it is not intended to anticipate the results of the work of Dill, but it seems from personal communication with him that certain presently held concepts of submarine erosion will be revised.

From the investigations of the characteristics of the Rosedale Channel and the Rosedale Sandstone such as the sediments, microfaunal ages, depth of water, and displaced faunas, the evidence strongly suggests that erosion and filling occurred entirely within the marine environment. Thus, a submarine process of erosion cut the channel and whether one wishes to label the process as a turbidity current, gravity flow, or any other name, the evidence seems good for showing that submarine processes of erosion were active during the Late Miocene and there seems to be no reason to doubt that the same or similar processes are eroding present-day submarine canyons. The use of the term "turbidity current" in this report does not signify a rate of flow, amount of material transported, or grain size of sediments.

It is not known from this study whether the currents can erode hard rocks, for it appears likely that the lower Fruitvale Shale into which the Rosedale Channel eroded (Fig. 3) probably was not indurated to any great degree during the Late Miocene and downcutting probably was facilitated. If, however, the currents can cut slightly indurated material, the removal of more indurated sediments may be just a matter of differential erosion rates.

EROSION RATE PROBLEM

It would be gratifying to be able to use the data from this study in order to obtain, or even to approximate, the rate of erosion of the Rosedale Channel. This rate would seem of value as a comparison with rates of erosion in present-day submarine canyons. On first inspection, the data in this study suggest that such a rate can be calculated, for a depth of erosion is given, an approximate time of cutting and filling has been calculated, and depositional and erosional rates in various modern environments had been presented (Emery, 1960, p. 247).

The basic problem is in using rates from modern environments to calculate a Late Miocene erosion rate and then applying the latter rate to a study of the rapidity of downcutting of present-day submarine canyons. Other problems inherent in this phase of the study are: (1) nothing is known of rates of erosion and deposition in Late Miocene submarine canyons; (2) the time between the initial cutting and final filling of the Rosedale Channel in itself is only an approximation; (3) the erosive force of Late Miocene turbidity currents may have been different from the force in modern oceans; and (4) the induration of the lower Fruitvale Shale (Fig. 3) during the Late Miocene is unknown. Other factors not even suspected may have affected also the Late Miocene erosion rates.

What appears at first to be a fairly simple matter of calculation becomes in the final analysis, a rather complex one which can not be resolved at this time.

SUMMARY

In the Great Valley of California, a sequence of middle Late Miocene strata, principally sandstones, occurs within the lower Fruitvale Shale of early Late Miocene age. These strata are interpreted to be fill within a submarine canyon. The names, Rosedale Channel and Rosedale Sandstone, are proposed for the canyon and fill, respectively. The conclusion regarding the strati-

449

graphic relations was reached by the following data.

1. Microfossils (Middle Mohnian) contained in the channel fill are younger than the microfossils (Early Mohnian) in the lower Fruitvale Shale.

2. Electric log correlations show that the Rosedale Sandstone within the lower Fruitvale Shale is anomalous within the regional stratigraphy of the area.

3. The lower Fruitvale Shale does not occur where the Rosedale Sandstone is well developed. Microfossil ages prove that the sand-shale relation is not a local facies change, but that the shale was eroded before the deposition of the sands. The sands were not deposited within a graben, as electric log correlations do not show a local downfaulting of the pre-upper Miocene strata.

4. Cores taken within the interval of channel fill show many characteristics of turbidites such as graded bedding, interbedded sandstones and shales, poor sorting, and displaced faunas.

5. Foraminiferal ecology indicates that the cut-and-fill occurred in water depths greater than 1,300 feet.

6. Certain physiographic characteristics of the Rosedale Channel are comparable with present-day submarine canyons such as axial and wall gradients, linearity, depth of erosion below the ocean floor, and depth of water in which erosion took place. Other features at present are not comparable, because of lack of well control in key areas and the fact that the channel represents only a remnant of a probably more extensive Late Miocene feature. A V-shaped cross-profile can not be recognized because of lack of control wells normal to the axial trend.

Erosion and filling to a depth of approximately 1,200 feet occurred within a period of about 700,000 years during the early?-middle Late Miocene. These data suggest that the rate of erosion of the channel could be calculated, but too little is known of certain variables affecting the accuracy of the rate calculations to make any such study significant. Variables include depositional and erosional rates during the Late Miocene, induration of the eroded shale body, the force of turbidity currents in the past, and a more accurate time calculation for the interval between the initial cutting of the channel and final filling.

The Rosedale Channel during the time that it was open probably funnelled nearshore sediments into the basin deeps resulting in a time lag between the start of Stevens turbidite sedimentation and the beginning of the nearshore Santa Margarita deposition.

All data lead to the conclusion that submarine processes eroded the Rosedale Channel. Turbidity currents or gravity flows of sediment are considered to have affected the downcutting; however, it seems likely that the lower Fruitvale Shale into which the canyon cut probably was not indurated to any great degree during this time and because of the lack of much induration, erosion probably was facilitated.

Little or no petroleum has been discovered in the sediments of the Rosedale Channel, but this does not preclude the possibility that future discoveries will be made. Filled and buried submarine canyons should have an excellent potential for any exploratory efforts although from the nature of the sandstone bodies, location and recognition may be difficult.

REFERENCES

AAPG-SEG-SEPM, 1947, Field trip guidebook with discussion of several oil fields: Am. Assoc. Petroleum Geologists—Soc. Explor. Geophysicists—Soc. Econ. Paleontologists and Mineralogists, Joint Ann. Meeting, Los Angeles, Calif., 132 p.

Albright, M. B., Jr., 1954, Rosedale oil field: Summary of Operations, Calif. Oil Fields, Div. Oil and Gas, v. 40, p. 31–39.

Bandy, O. L., 1953, Ecology and paleoecology of some new California Foraminifera. Part I. The frequency distribution of Recent Foraminifera off California. Part II. Foraminiferal evidence of subsidence rates in the Ventura Basin: Jour. Paleontology, v. 27, p. 161–182; 200–203.

———— 1960, General correlation of foraminiferal structure with environment: Report Internat. Geol. Cong. Rept., XXI Sess., Norden, 1960, pt. XXII, Internat. Paleont. Union, Copenhagen, p. 7–19.

———— and Arnal, R. E., 1960, Concepts of foraminiferal paleoecology: Am. Assoc. Petroleum Geologists Bull., v. 44, no. 12, p. 1921–1931.

Beach, J. H., 1947, Geology of basement complex, Edison oil field, Kern County, California: Guidebook Am. Assoc. Petroleum Geologists—Soc. Explor. Geophysicists—Soc. Econ. Paleontologists and Mineralogists, Joint Ann. Meeting, Los Angeles, Calif. 132 p.

Bornhauser, Max, 1948, A possible ancient submarine canyon in southwestern Louisiana: Am. Assoc. Petroleum Geologists Bull., v. 32, p. 2287–2294.

Bruce, Donald D., 1959, Princeton gas field: Summary of Operations, Calif. Oil Fields, Div. Oil and Gas, v. 45, p. 23–26.

Daly, R. A., 1936, Origin of submarine "Canyons": Am. Jour. Science, 5th ser., v. 31, p. 401–420.

Dill, R. F., and Shumaker, L. A., 1961, Geologic features of La Jolla Canyon as revealed by Dive No. 83 of the Bathyscaph Trieste: Unpub. Memo. TM-516, U. S. Naval Electronics Lab., San Diego, Calif.

Emery, K. O., 1960, The sea off southern California:

N. Y. and London, John Wiley & Sons, Inc., 366 p.

Ferguson, Glenn C., 1947, Correlation of oil field formations on east side of San Joaquin Valley: Guidebook Am. Assoc. Petroleum Geologists—Soc. Explor. Geophysicists—Soc. Econ. Paleontologists and Mineralogists, Joint Ann. Meeting, Los Angeles, Calif., 132 p.

Gorcieux, Charles, 1922, Le Gouf de Cap Breton: Revue de Geog., v. 37, p. 401–411.

Hoots, Harold W., Bear, Ted L., and Kleinpell, William D., 1954, Geologic summary of the San Joaquin Valley, California: Calif. Div. Mines Bull. 170, chap. II, p. 113–129.

Hoyt, W. V., 1959, Erosional channel in the middle Wilcox near Yoakum, Lavaca County, Texas: Gulf Coast Assoc. Geol. Soc. Trans., v. 9, p. 41–50.

Kuenen, Ph. H., 1950, Marine geology: New York, John Wiley & Sons, Inc.; London, Chapman & Hall, Limited, 568 p.

——— and Carozzi, A., 1953, Turbidity currents and sliding in geosynclinal basins in the Alps: Jour. Geology, v. 61, p. 363–373.

Kulp, J. L., 1961, Geologic time scale: Science, v. 133, no. 3459, p. 1105–1114.

Medvedev, V. C., 1959, Certain problems of the analysis of dynamics and morphology of the western shore of Sakhalin in connection with the construction of small fishing ports: Trans. Oceanogr. Comm., Presidium Acad. Science, USSR, v. 4, p. 1–13 (in Russian).

Natland, M. L., 1933, The temperature and depth distribution of some recent and fossil foraminifera in the southern California region: Bull. Scripps Inst. Oceanography, Univ. Calif., v. 3, p. 225–230.

——— 1957, Paleoecology of west coast Tertiary sediments: Geol. Soc. America Memoir 67, v. 2, p. 543–572.

Neev, D., 1960, A Pre-Neogene erosion channel in the southern coastal plain of Israel: Israel Ministry Devel., Geol. Survey Bull. 25, Oil Div. Paper 7, p. 1–20.

Shepard, F. P., and Emery, K. O., 1941, Submarine topography off the California coast; Canyon and tectonic interpretation: Geol. Soc. America Special Paper 31, 171 p.

Sullwold, Harold H., Jr., 1961, Turbidites in oil exploration, *in* Geometry of sandstone bodies: Am. Assoc. Petroleum Geologists, p. 63–81.

Veatch, A. C., and Smith, P. A., 1939, Atlantic submarine valleys off the United States and Congo submarine valley: Geol. Soc. America Special Paper 7, 101 p.

THE AMERICAN ASSOCIATION OF PETROLEUM GEOLOGISTS BULLETIN
V. 51, NO. 6 (JUNE, 1967), P. 873-882, 6 FIGS.

UPPER PALEOCENE BURIED CHANNEL IN SACRAMENTO VALLEY, CALIFORNIA[1]

A. B. DICKAS[2] AND J. L. PAYNE[2]
Superior, Wisconsin, and Oildale, California

ABSTRACT

The Meganos channel, a fossil submarine channel, is in the central part of the Sacramento Valley of California. This channel was cut and filled in a marine environment during a relatively short interval of late Paleocene time. It has been traced in the subsurface more than 50 miles and has a maximum width of 6 miles. The greatest known thickness of sediments filling the channel is 2,015 feet. Formations truncated by the channel range in age from latest Paleocene through Late Cretaceous.

The rocks eroded by the channel are principally arenaceous; however, sediments filling the channel are composed of more than 95 per cent shale. Paleontologic studies indicate that the channel deposits accumulated in water depths ranging from neritic to upper bathyal.

A major factor contributing to the formation of the Meganos channel is thought to be regional faulting. Before the channel existed, the Midland fault, a major north-south-striking feature, began to form. It is the writers' opinion that the declivity created by this down-to-the-west normal fault set up conditions favorable for extensive slumping and turbidity currents which caused erosion of the sea floor and development of the channel. Subsequently, fine-grained terrigenous clastics filled this erosional feature and, at the same time, were deposited in a thin layer in the areas outside of the channel.

The channel shale contained an original large volume of interstitial water. Burial and overburden pressures compacted the shale to 35–60 per cent of its original volume.

Comparison among ancient subsurface and existing submarine canyons shows that features comparable with the Meganos channel have been formed in the past and are being eroded today, all under differing geologic settings.

Truncation of underlying formations by the channel shale combined with the local structure formed commercial accumulations of hydrocarbons at the Brentwood oil and gas field and the Dutch Slough, River Break, and West Thornton gas fields.

INTRODUCTION

An anomalous shale section, ranging in thickness from less than 100 feet to more than 2,000 feet, has been traced 50 miles in the central part of the Sacramento Valley of California. Subsurface data and isopachous studies indicate that this shale sequence occupies an ancient marine channel which was eroded and then filled with fine-grained clastics (Fig. 1).

Paleontologic and regional lithologic studies indicate that this channel was eroded in a marine environment into a predominantly arenaceous section. After erosion, sedimentary infilling took place in a very short period of geologic time during late Paleocene time. The sediments which filled the eroded channel have been assigned to the upper Paleocene Meganos C formation of Clark and Woodford (1927); this assignment is based on the Foraminifera found in the channel sediments.

[1] Manuscript received, March 10, 1966; accepted, July 22, 1966. Presented to the San Joaquin Geological Society by J. L. Payne on June 14, 1966.

[2] Geologists, Standard Oil Company of California, Western Operations, Inc., Northern Division, Oildale, California. A. B. Dickas' present address: Wisconsin State University. J. L. Payne's present address: Chevron Oil Company, Houston, Texas.

Sediments ranging in age from late Paleocene through Late Cretaceous were truncated during the creation of this feature. Where the regional dip is opposite to that of the slope of the channel walls, favorable local structure created hydrocarbon traps in the truncated sandstone beds abutting channel shale. To date, gas and oil production has been developed in four fields along the trend of this channel.

EARLY DEVELOPMENT OF CHANNEL CONCEPT

Although the anomalously thick shale section comprising the Meganos channel has been known for many years, a better understanding of the morphology of this shale section was not possible until detailed information became available through the discovery and development of the Brentwood field by the Shell Oil Company in 1962. In this field, the channel shale provides the updip closure on a northwest-dipping homocline. Lateral closure is created by a northwest-trending system of parallel faults (Sullivan, 1963).

As the Brentwood field was developed, isopachous studies of the Meganos shale made it possible to identify these sediments as a subsurface-filled erosion channel. However, in this area, only the northern flank of the channel remains intact;

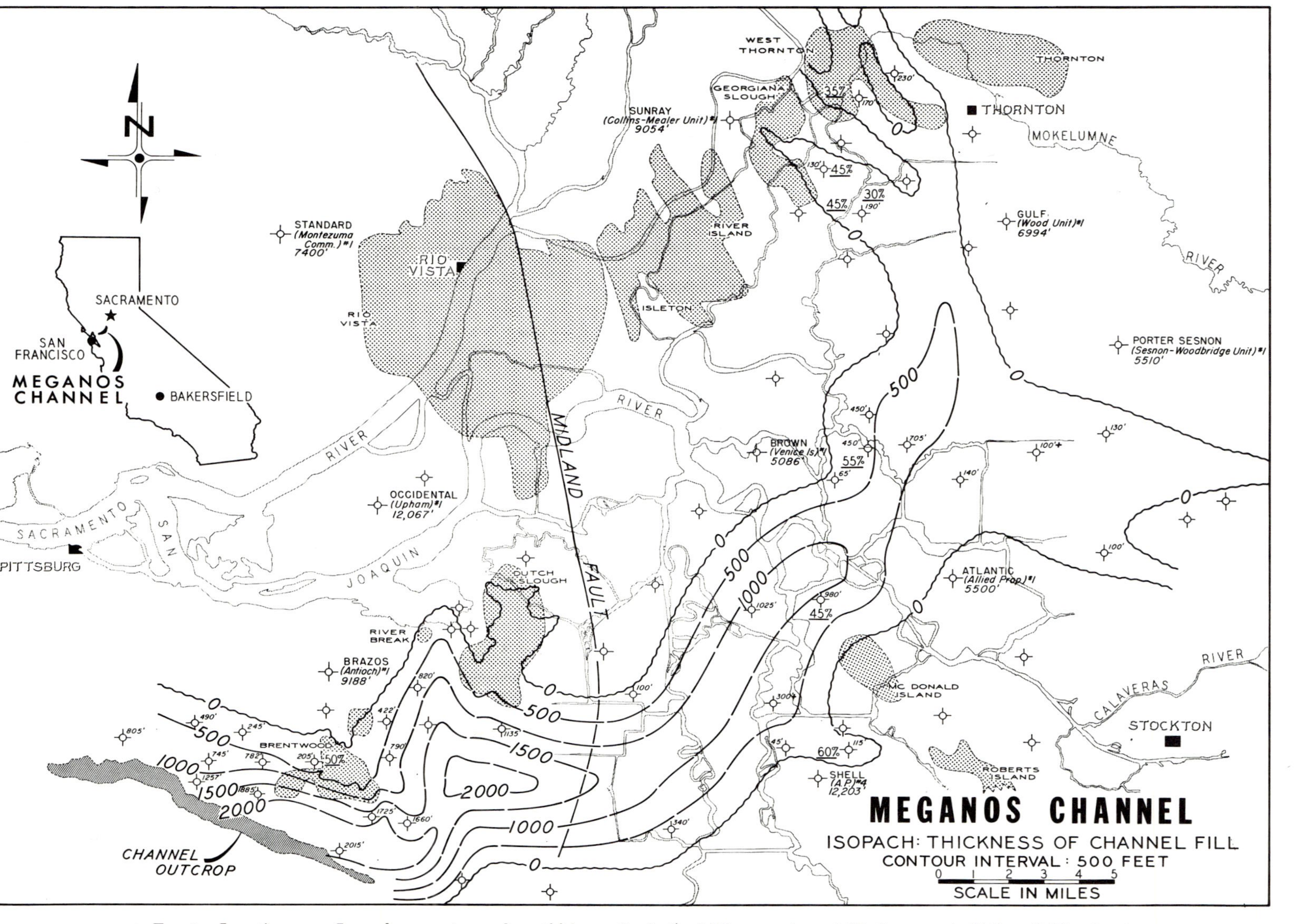

Fig. 1.—Location map. Isopachous contours show thickness (in feet) of Meganos channel fill, Sacramento Valley, California.

the original extent and morphology of the southern flank have been destroyed by post-channel uplift and erosion along the Mount Diablo mountain range. Remnants of this southern flank are present at the surface on the north side of the Mount Diablo uplift (see Figs. 1, 5).

Once the existence of the buried Meganos channel was recognized and understood in the Brentwood area, other "anomalous shale" wells, some of which had been considered evidence for local channels (Silcox, 1962; Safonov, 1962), were for the first time fully comprehended. The Meganos channel was extended eastward from Brentwood to the position of the McDonald Island field and northward to what is now recognized as the "headward" area of the channel in the vicinity of the West Thornton field. The present extent of the Meganos channel was first defined publicly by W. F. Edmondson at a meeting of the San Joaquin Geological Society on October 13, 1964, and later in the selected papers presented to the San Joaquin Geological Society (Edmondson, 1965).

Regional Morphology of Channel

From the headward area of the channel north of the West Thornton gas field, the course of the Meganos channel has been traced 50 miles south and west to the Brentwood field as shown in Figure 1. At present, the downslope termination is not known because of the paucity of drilling in the area west of Brentwood and because of the extensive unconformities found in the general San Francisco Bay area. However, it is the writers' hypothesis that the effluence of the channel was situated in the Pacific Ocean west of San Francisco. If this is correct, the large volume of arenaceous sediments which was eroded as the channel was formed should be present in the offshore regions of the Pacific Ocean. To date, however, it is not known where the large volume of eroded sediments was redeposited.

The channel width is variable, ranging from a minimum of 2 miles to a maximum of 6 miles. This variation is probably a result of sea-floor topography at the time of channel erosion. Except in the headward region, tributary channels flow into the main channel course with an orientation of approximately 90° to the main axial trend and thus form a modified trellis drainage pattern.

In the headward area in the vicinity of the West Thornton gas field, a complex drainage pattern developed. From south to north in this region, the channel course divides into several distinct tributary channels. One very interesting feature associated with these tributary channels is the complete erosional isolation of a large island of the older sandstone sediments (Figs. 1, 3). Some geologists (Edmondson, 1965) have considered this sandstone island to be part of the channel fill, with channel sandstones resting directly on older Martinez sandstones. However, the writers would correlate these sandstones with the older Martinez sandstones in the manner shown in Figure 3. In view of the documented tributary channels in this region, it would not be unreasonable for an isolated island of older sandstone to develop as shown in Figure 1.

Isopachous studies, based on electric logs and paleontologic data, show that the deepest development of this channel is in the Brentwood area. Here, a maximum thickness of 2,015 feet of channel shale is mapped. Because of the characteristic electric-log pattern of the gorge shale and age dating by Foraminifera, there is no difficulty in establishing the base of the sediment fill even though in this area some shale-on-shale contacts are found.

Although the average plunge of the channel axis is southwesterly at approximately 2°, local areas exist where the axis plunges opposite the direction of the channel current flow. This is probably a result of later structural warping. Numerous measurements along the entire course of the gorge indicate that the slope of the walls ranges from 5° to 15°. This variation is random.

The regional morphology of the Meganos channel is outlined in Figure 1. The shale isopachous contours show that the channel depth increases in a downstream direction. The width also increases gradually in this direction, thus producing an overall gorge morphology very similar to present-day offshore channels. It was not possible to make a detailed interpretation of the channel configuration in the vicinity of the Midland fault because of the sparse well control in this region. However, as more subsurface information becomes available, it would be expected that some effect of the Midland fault on the gorge morphology will be noted.

Three electric-log cross sections (Figs. 3–5) have been constructed across the Meganos chan-

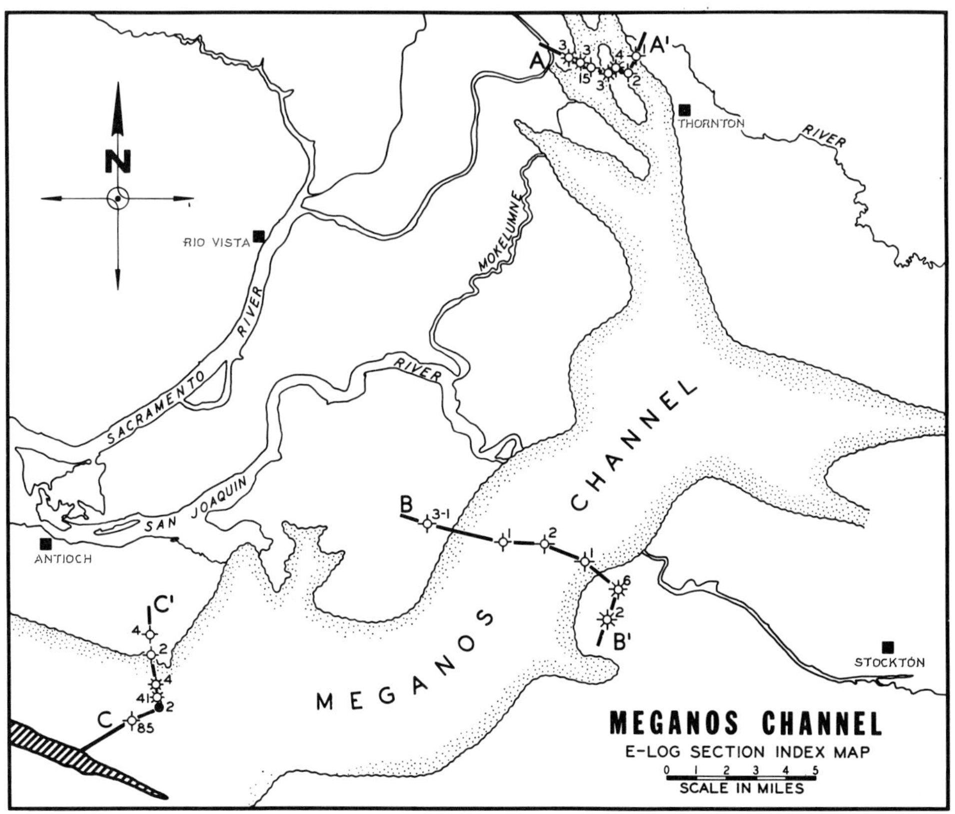

FIG. 2.—Location of electric-log sections and wells used, Meganos channel, Sacramento Valley, California. Cross section A-A′ = Figure 3; cross section B-B′ = Figure 4; and cross section C-C′ = Figure 5.

nel and their locations are shown in Figure 2. The erosional origin and later infilling of shale sediments is well demonstrated by these sections, especially Figures 4 and 5.

LITHOLOGY OF CHANNEL FILL

Silty shale and shaly siltstone comprise more than 95 per cent of the channel-fill sediments. In some places glauconitic units are found. Though most of the siltstone is confined to the basal sedimentary units, the overall fill sequence is remarkably uniform. This dominant fine-grained lithologic character is readily recognized by the nondescript spontaneous potential curve on electric logs.

In some areas, especially near the "headwaters," sandstone units of measurable thickness are recorded. These sandstone units appear to be restricted in their geographic development and, like the siltstone, are found generally in the older part of the channel section. To date no production has been discovered in these sandstones.

The principal shale lithologic features contrast sharply with the underlying section into which the channel was eroded. Except for a minor sandstone and shale sequence west of the Midland fault, the channel has been eroded into massive sandstone units. This abutment of channel shale with regional sandstone makes recognition of the channel floor possible from electric logs alone. However, in the areas of deepest channel erosion, the channel shale is in juxtaposition with Upper Cretaceous shale. Here the base of the gorge can be recognized best by paleontologic studies.

GEOLOGIC HISTORY OF CHANNEL FORMATION

During the middle of the Paleocene Epoch, tectonic stresses caused initial movements along the Midland fault, a major structural feature of

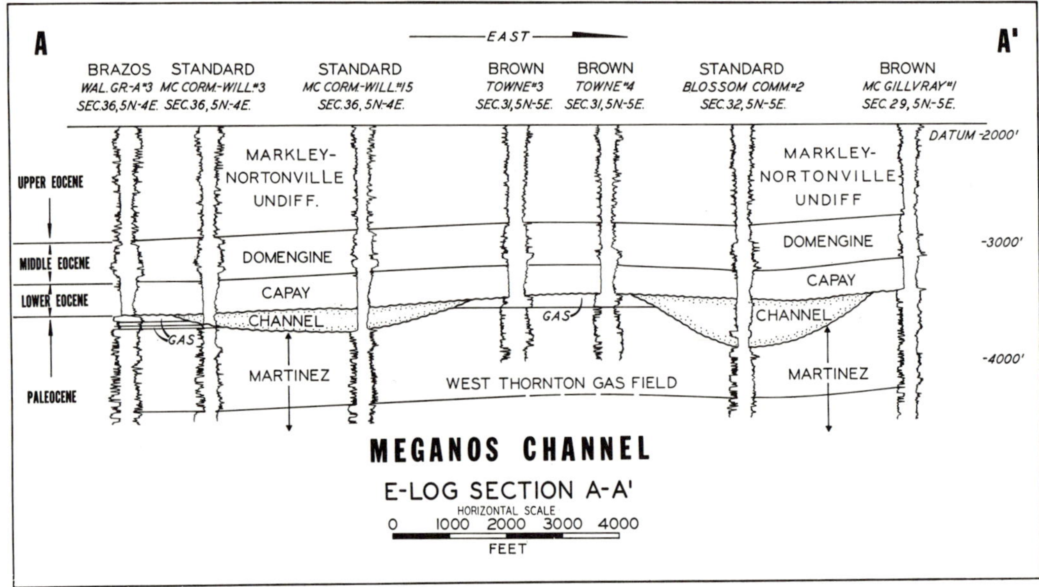

FIG. 3.—West-east cross section A-A′, showing erosional isolation of island of older sandstone at West Thornton gas field. Location of section shown on Figure 2.

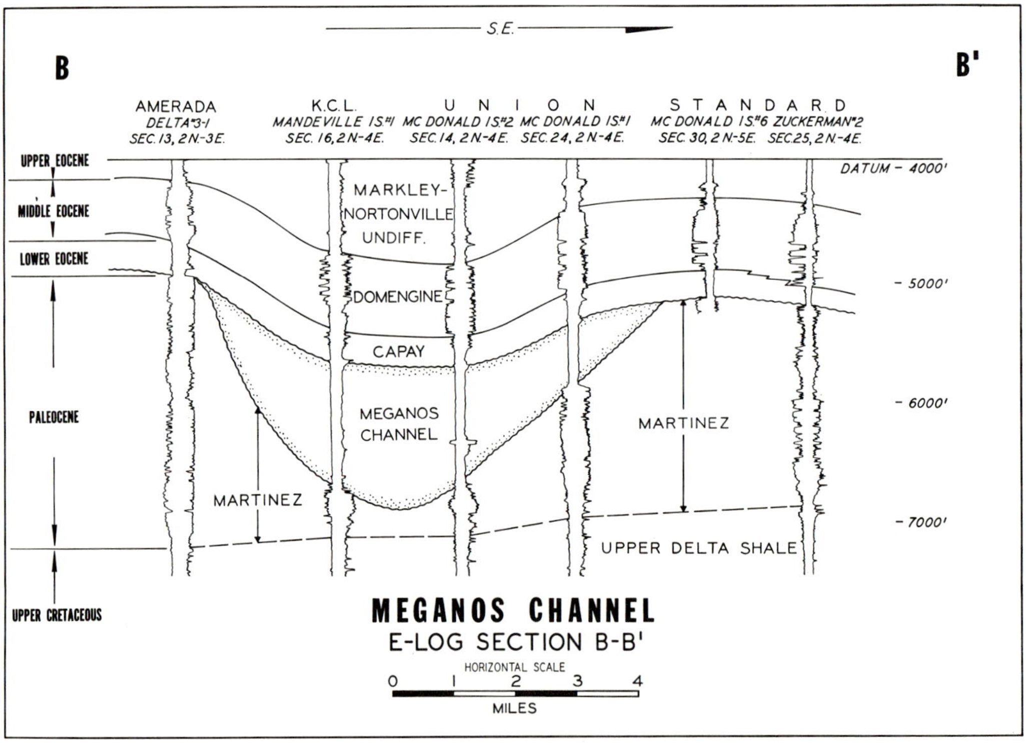

FIG. 4.—West-northwest—south-southeast cross section B-B′. Section indicates magnitude of differential compaction of channel shale beds. Location of section shown on Figure 2.

A. B. DICKAS AND J. L. PAYNE

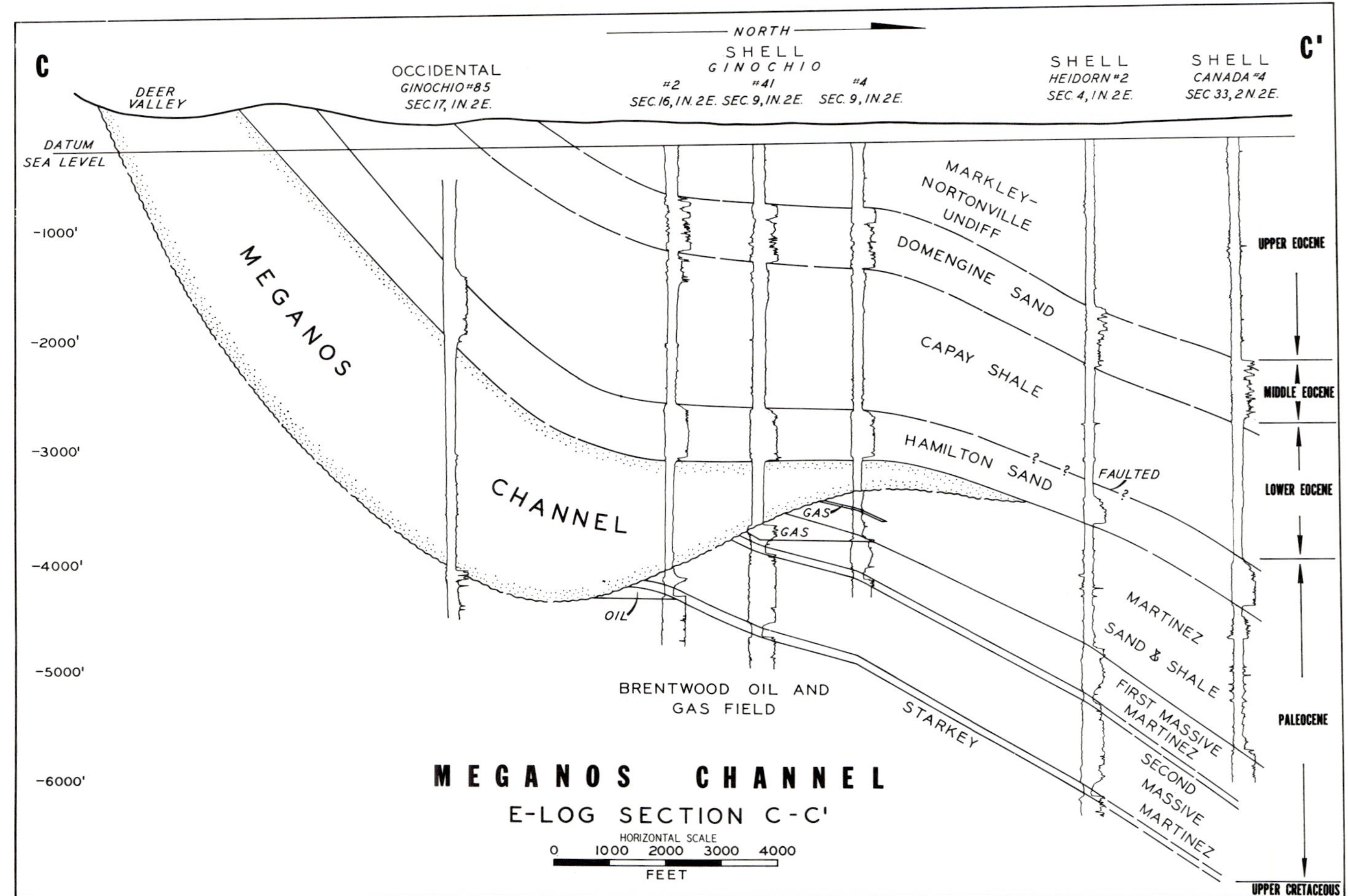

Fig. 5.—South-north cross section C-C'. Section shows updip entrapment of oil and gas by channel shale fill at Brentwood oil and gas field. Location of section shown on Figure 2.

457

the central Sacramento Valley (Fig. 1). This north-south-striking, down-to-the-west, normal fault was intermittently active through the Eocene Epoch.

Restored regional cross-section studies indicate that the first vertical movements along this fault took place just before the formation of the channel. Because of the relationship in the timing of these two events, it is believed that the faulting directly contributed to conditions favorable for channel formation.

The initial declivity caused by rupture along the Midland fault is believed to have increased local slope to the extent that turbidity currents were triggered. These turbidity currents probably were localized by ocean-bottom topography and a major onshore river system, such as an ancestral Sacramento River. The evidence of any river system has been destroyed by basin-edge erosion. The turbidity currents flowed westward down the footwall of the fault and initially eroded massively bedded sandstone. The relatively high velocity of these currents, and the abrasive nature of their suspended bed load quickly eroded a trough across the fault scarp. This trough was the beginning of the Meganos channel.

In a short period of time, lateral extension took place both landward toward the east and north, and seaward toward the west, by continued sporadic turbidity-current activity. As this development proceeded, the channel became a focal point for other turbid currents developing in the area. These added currents hastened vertical erosion. Tributary channels soon formed and flowed into the trunk channel, thus greatly altering its overall outline.

Eventually, a balance was established between discharge, on one hand, and gradient and current velocity, on the other. The resulting gorge profile established a state of erosive equilibrium and the further extension of this feature was retarded. A period of transgression of the area followed the erosional cycle and initiated a period of rapid infilling of the channel and the deposition of a thin shale section outside of the channel. Paleobathymetric analysis by M. Polugar[3] of the foraminiferal suite found within the channel fill suggests that the fill was deposited in a marine envi-

[3] Standard Oil Co. of California, Western Operations, Inc.

ronment ranging in depth from neritic to upper bathyal.

The shales of Meganos channel age which were deposited outside of the limits of the channel generally are 100 feet thick, a figure which is relatively constant. As a result, the channel edge was established arbitrarily at the 100-foot isopachous contour of the Meganos shale. Because of post-channel erosion, this blanket of Meganos shale is present only on the hanging-wall side of the Midland fault.

The presence of a thin layer of Meganos shale outside of the channel confines indicates that deposition was taking place throughout the area at the same time that the eroded channel was filling. The thinness of the shale section outside of the channel is explained by the very process of channel infilling. It is postulated that offshore currents were continually sweeping the sediments from the ocean floor and infilling the channel cavity. Therefore, only a very thin layer of sediments could build up outside of the channel as the channel itself was infilled from the sides. This method of infilling submarine channels might actually be the rule rather than the exception, because the phenomenon of a thin section of channel-age sediments outside of the channel confines has been noted elsewhere (Hoyt, 1959). After infilling of the erosional channel, normal sedimentation continued in the area. The overburden of the post-channel sediments caused differential compaction of the channel shale.

DIFFERENTIAL COMPACTION OF CHANNEL SHALE

The marine environment of the channel shale and the high initial porosity inherent to shale caused the fine-grained clastics which originally filled the eroded channel to have a very high volume of water. Weller (1960, p. 297) indicates that the initial porosity values for fine-grained marine clay, which constitutes most of the channel fill, can exceed 80 per cent, whereas the initial porosity of sandstone is estimated by Weller (1960, p. 292) to be approximately 37 per cent. Gradual elimination from the shale of the water and the closer packing of the sedimentary particles as a result of the pressure exerted by the accumulating overburden resulted in greater compaction of the channel shale relative to the arenaceous section. This is well displayed in Figure 4.

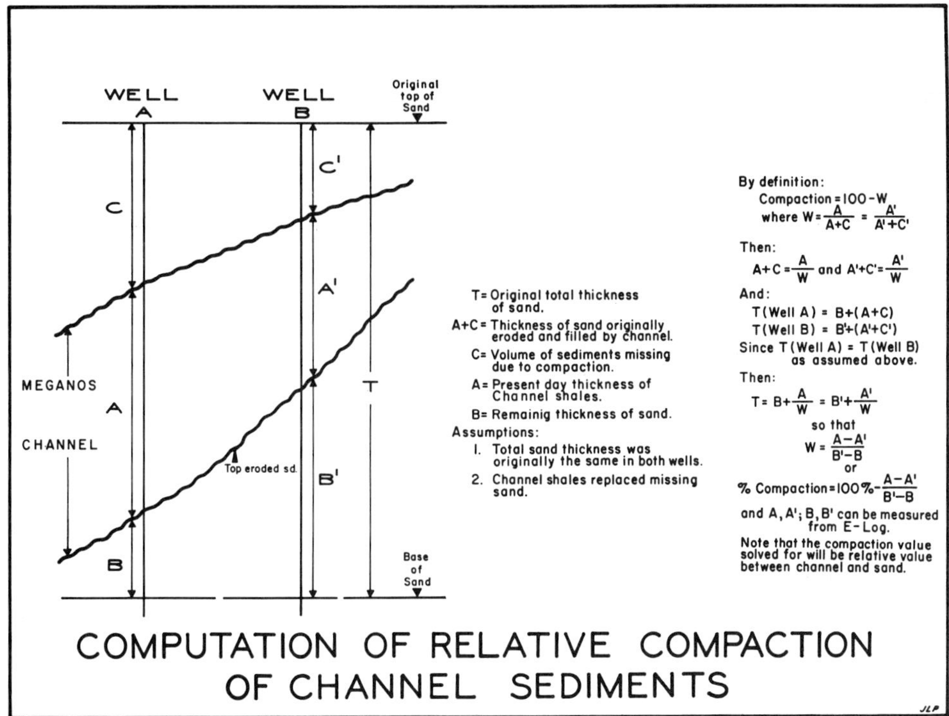

Fig. 6.—Method for computing relative compaction of channel shale beds.

The magnitude of the differential compaction of the channel sediments was computed along the length of the channel, and representative values are shown in Figure 1. In general, the differential-compaction values for the channel shale range from 60 to 35 per cent.

The simplest way to determine the differential compaction of the channel shale would be to compare a channel well with a non-channel well where the two are located close to each other and where both penetrated adequate section below the channel base to have reached good pre-channel correlation markers. Unfortunately, these conditions were not found in many places, so that it was necessary to devise the method outlined in Figure 6. The accuracy of this method should be as good as that of comparing a channel with a non-channel well.

COMPARISON OF MEGANOS CHANNEL WITH
OTHER SUBSURFACE AND SUBMARINE
CHANNELS

Much has been written on the origin and morphology of both ancient and modern submarine channels. It is not the writers' intent to present a detailed comparison between these many documented features and the Meganos channel. However, in reviewing the literature, certain aspects of the Meganos channel in relation to documented subsurface and modern channels should be noted.

One of the most outstanding examples found, and certainly the most interesting from the viewpoint of comparison with the Meganos channel, is the Yoakum channel (Hoyt, 1959). This mid-Eocene feature is in the central Gulf Coast of Texas. From the viewpoint of geometry, age of gorge erosion, thickness, lithologic character of the sedimentary fill, and the presence of a regional thin blanket of shale which is correlatable with the channel shale, the Meganos and Yoakum channels are unusually similar. In fact, the only apparent difference between these two features is the well-documented tributary system associated with the Meganos channel and the striking absence of such associated with the Yoakum channel.

In comparing the Meganos channel with existing submarine valleys, one must picture the Meganos channel before deposition of the shale fill. Just after complete erosion of the Meganos channel and before sedimentary infill, this gorge resembled many of the well-documented, present-day submarine canyons. One of the largest and best known of these is the Congo Canyon off the mouth of the Congo River on the western coast of Africa. With a main-branch length of approximately 200 miles, a maximum width of 5 miles, and a maximum depth below the canyon rim of 3,000 feet, the physical parameters of the Congo Canyon are similar to those of the Meganos channel. The erosion mechanism of the Congo Canyon is thought by Heezen *et al.* (1964) to be turbidity currents which were triggered in the confines of the mouth of the Congo River during months of maximum bed-load transport of that river.

ECONOMIC IMPORTANCE OF MEGANOS CHANNEL

Truncation of reservoir rock by channel shale has aided in the entrapment of economic quantities of gas and oil in four fields. These are the Brentwood oil and gas field and the Dutch Slough, River Break, and West Thornton gas fields (Fig. 1). Economically the Brentwood and Dutch Slough fields are classed as major discoveries in California. The Brentwood field, discovered in July, 1962, contains the only economic accumulation of oil discovered in the Sacramento Valley. The Dutch Slough field, a 1963 discovery, is presently ranked as the third largest gas field in California, with total recoverable gas reserves calculated at 300 billion cubic feet (Reedy, 1966).

To date, the thin but porous sandstone beds associated with the channel fill have not produced hydrocarbons.

CONCLUSIONS

There is sufficient subsurface evidence to prove conclusively the existence of the Meganos channel described here. The evidence indicates an erosional origin which is similar to that of the many submarine canyons which are known to exist on the present-day continental shelves. It is postulated that the method of channel infilling which is indicated for the Meganos channel occurred in the past in other areas and could be expected to take place again in the future.

The Meganos channel is not unique in that other fossil channels have been recorded previously in the literature. In fact, it is probable that submarine channeling is a much more common occurrence than generally is realized. This might be especially true if the possibility is considered that, should the depositional fill of a channel resemble the sediments of the formation eroded (as might be expected to occur in many areas), it would be difficult to detect the presence of a channel except through paleontological information.

It is hoped that this paper has added additional information to the study of submarine channels or canyons and possibly has raised some questions regarding the mechanisms of erosion and infilling of the channels. Certainly, the part of the Meganos channel discussed in this paper provides a good study of fossil channels because of the degree of documentation possible from the many wells drilled in the area.

SELECTED BIBLIOGRAPHY

Almgren, A. A., and W. N. Schlax, 1957, Post-Eocene age of "Markley Gorge" fill, Sacramento Valley, California: Am. Assoc. Petroleum Geologists Bull., v. 41, p. 326–330.

Bornhauser, Max, 1948, A possible ancient submarine canyon in southwestern Louisiana: Am. Assoc. Petroleum Geologists Bull., v. 32, p. 2287–2294.

Clark, B. L., and A. O. Woodford, 1927, The geology and paleontology of the type section of the Meganos Formation of California: Univ. Calif. Pub. Dept., Geol. Sci., Bull. 17, p. 63–142.

Edmondson, E. F., 1965, The Meganos gorge: San Joaquin Geol. Soc., Selected Papers, v. 3, p. 36–51.

Frick, J. D., T. P. Harding, and A. W. Marianos, 1950, Eocene gorge in northern Sacramento Valley (abs.): Am. Assoc. Petroleum Geologists Bull., v. 43, p. 255.

Goudkoff, P. P., 1945, Stratigraphic relations of Upper Cretaceous in Great Valley, California: Am. Assoc. Petroleum Geologists Bull., v. 29, p. 956–1007.

Heezen, B. C., R. J. Menzies, E. D. Schneider, W. M. Ewing, and N. C. L. Gionelli, 1964, Congo submarine canyon: Am. Assoc. Petroleum Geologists Bull., v. 48, p. 1126–1149.

Hoyt, N. W., 1959, Erosional channel in the middle Wilcox near Yoakum, Lavaca County, Texas: Gulf Coast Assoc. Geol. Soc. Trans., v. 9, p. 40–50.

Hunter, W. J., 1964, Dutch Slough gas field, *in* Summary of operations—California oil fields: Calif. Div. Oil & Gas, v. 50, no. 2, p. 63–69.

Kuenen, P. H., Marine geology: New York, John Wiley & Sons, 568 p.

Martin, B. D., 1963, Rosedale channel—evidence for late Miocene submarine erosion in Great Valley of California: Am. Assoc. Petroleum Geologists Bull., v. 47, p. 441–456.

Reedy, R. D., 1966, Dutch Slough field typifies Sacramento Valley gas finds: World Oil, January, p. 85–88.

Safonov, Anatole, 1962, The challenge of the Sacramento Valley, California: Calif. Div. Mines and Geology Bull. 181, p. 89–97.

Shepard, F. P., and K. O. Emery, 1941, Submarine topography off the California coast—canyons and tectonic interpretation: Geol. Soc. America Spec. Paper 31, 171 p.

Silcox, J. H., 1962, West Thornton and Walnut Grove gas fields, California: Calif. Div. Mines and Geology Bull. 181, p. 140–148.

Sullivan, J. C., 1963, Brentwood oil field, *in* Summary of operations—California oil fields: Calif. Div. Oil & Gas, v. 49, no. 2, p. 5–15.

Veatch, A. C., and P. A. Smith, 1939, Atlantic submarine valleys of the United States and the Congo submarine valley: Geol. Soc. America Spec. Paper 7, 101 p.

Weller, J. Marvin, 1960, Stratigraphic principles and practice: New York, Harper and Brothers, 725 p.

The American Association of Petroleum Geologists Bulletin
V. 61, No. 2 (February 1977), P. 207-226, 18 Figs.

Origin and Distribution of Tertiary Conglomerates, Veracruz Basin, México[1]

PABLO CRUZ HELU,[2] RODOLFO VERDUGO V.,[2] and RODOLFO BARCENAS P.[2]

Abstract Commercial quantities of gas have been produced from several Miocene conglomerates in the southwestern part of the Veracruz basin. These conglomerates, with their associated sandstones and shales, make up a zone 1,000 m thick, all of which was deposited in a bathyal environment.

The conglomerates consist of recycled Upper Cretaceous limestone clastic rocks containing minor amounts of igneous and metamorphic rock fragments. The clastic material was derived from the west, where a very thick section of carbonate and terrigenous rocks, Jurassic to Paleocene in age, was uplifted, folded, and thrust faulted during the Laramide orogeny. Erosion of the deformed source materials is indicated by two significant unconformities. One unconformity truncates sedimentary rocks from the Paleocene to the Upper Cretaceous, whereas the other is at the base of the Miocene. Most of the conglomerate beds are in the lower Miocene. A reconstruction of the pre-lower Miocene erosional surface serves to locate paleodrainage patterns and the areas of debouchment of deep-water fanglomerates.

Sediments derived from the source area west of Veracruz basin were transported in a general northeasterly direction by fluvial currents which fed into deep submarine canyons. Through these canyons most of the coarse clastic sediments were carried down to the basin floor and deposited as a series of prograding submarine fans.

INTRODUCTION

The Veracruz basin in eastern México (Fig. 1) is a Tertiary basin with an area of approximately 30,000 sq km in which more than 8,500 m of predominantly deep-water marine clastic sediments were deposited during Paleocene through recent times. The basin is limited on the north by the Santa Ana uplift, on the west and southwest by the folded mountains of the Sierra Madre Oriental, on the south by the Sierra de Chiapas, and on the southeast by the Isthmus Salt basin and the Tuxtlas uplift. To the northeast it extends into the Gulf of México.

The Veracruz basin is in an early stage of exploration so there is only a limited amount of subsurface well data for interpretation. However, seismic information is fairly complete. The general outline of the basin is shown on Figure 2. The eastern edge of the Cretaceous subcrop passes through the central part of the study area.

In the basin, as in other Gulf Coast areas, Tertiary sedimentary deposits consist of a monotonous sequence of alternating shales and sandstones which locally are interrupted by conglomerate zones. Thus, it is very difficult to

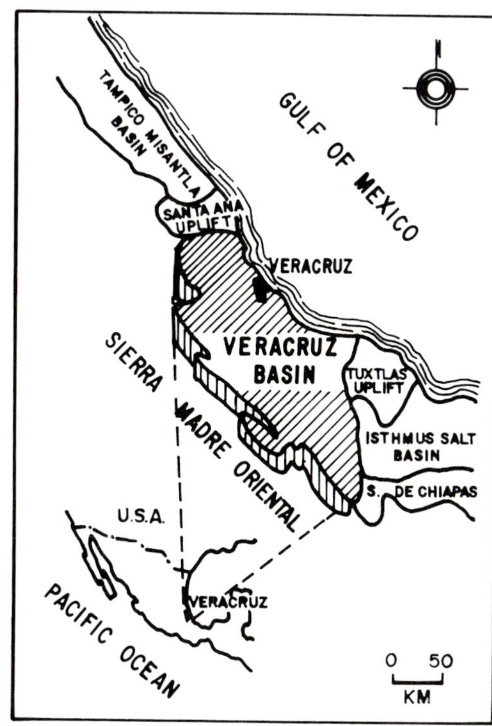

FIG. 1—Index map of geologic provinces in eastern México showing location of Veracruz basin.

define formational boundaries on the basis of lithologic changes and detailed micropaleontologic studies were made to establish the biostratigraphy, biozonation, and paleobathymetry of the sediments. Such information is essential for a correct interpretation of the stratigraphy and structure of the area and is also very helpful in establishing log correlations and unconformities.

[1]Manuscript received, December 5, 1975; accepted, July 21, 1976.

[2]Petroleos Mexicanos, Exploración, Córdoba, Veracruz, México.

The writers express their sincere appreciation to Petroleos Mexicanos, Gerencia de Exploración, for permission to publish this paper. We especially acknowledge Daniel A. Busch for his cooperation, helpful suggestions, and critical review of the manuscript.

207

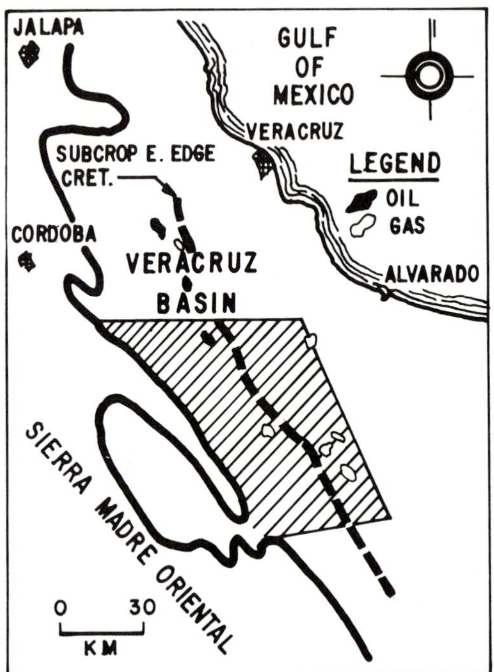

FIG. 2—Map of northern part of Veracruz basin. Studied area is shaded with diagonal lines. Dashed line indicates eastern limit of Cretaceous sedimentary rocks in subsurface.

In areas of sparse well control recognition of different directions of dip and variations in thickness of certain stratigraphic intervals was based almost entirely on seismic-profile data.

Investigation was restricted predominantly to the western part of the basin where commercial gas production from Tertiary sedimentary rocks has been established in several pools. The gas is produced from several thick Miocene conglomerates. The conglomerates, with their associated sandstones and shales, make up a zone about 1,000 m thick. The main purpose of the study was to determine the origin and distribution of the gas-productive conglomerates.

STRATIGRAPHY AND STRUCTURE

The stratigraphic subdivisions of Tertiary sedimentary rocks in the Veracruz basin are summarized in Figure 3. The manner in which the various Tertiary formations thin westward as they onlap the underlying Cretaceous also is shown grammatically in the figure. In the deepest part of the basin, on the east, they have an estimated combined thickness of 8,900 m. Sixty-five percent of the total thickness consists of Oligocene-Miocene sedimentary rocks.

Several significant unconformities are present along the western margin of the basin. The lowermost is between middle and upper Eocene rocks and loses its identity as it rises to the west and merges with a pre-early Miocene unconformity (Fig. 3). The unconformity at the base of the Miocene is the most significant in the basin; Miocene strata may overlie directly Oligocene, Eocene, Paleocene, or even Cretaceous strata. In addition, several minor unconformities are present in the area.

Structural cross section *1-1'* (Fig. 4) shows the general stratigraphic and structural characteristics across the central Veracruz basin. The three easternmost wells were drilled offshore in the Gulf of México. The manner in which the Ter-

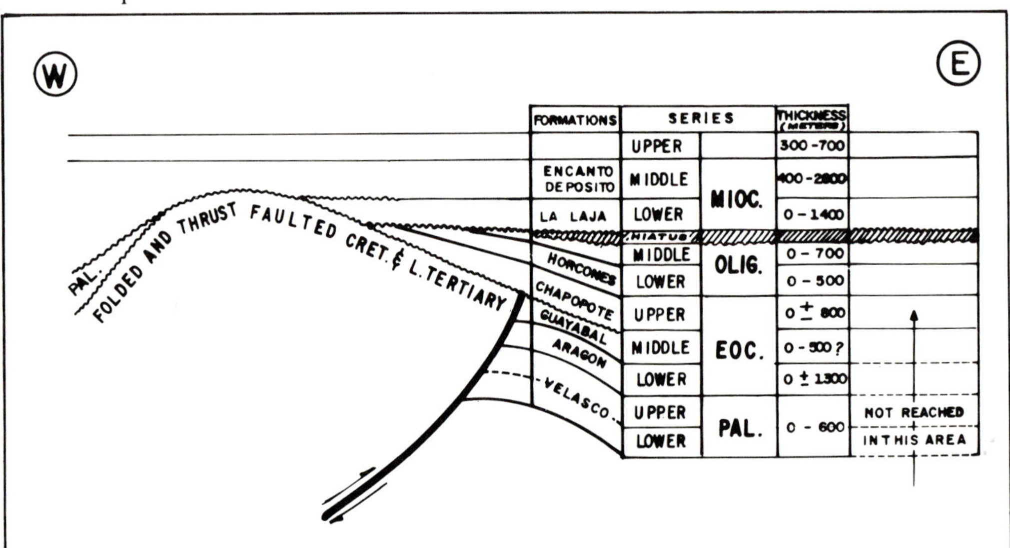

FORMATIONS	SERIES		THICKNESS (METERS)	
	UPPER		300 - 700	
ENCANTO DEPOSITO	MIDDLE	MIOC.	400 - 2800	
LA LAJA	LOWER		0 - 1400	
	HIATUS			
HORCONES	MIDDLE	OLIG.	0 - 700	
CHAPOPOTE	LOWER		0 - 500	
GUAYABAL	UPPER		0 ± 800	
ARAGON	MIDDLE	EOC.	0 - 500 ?	
	LOWER		0 ± 1300	
	UPPER	PAL.	0 - 600	NOT REACHED
	LOWER			IN THIS AREA

FIG. 3—Stratigraphic subdivisions of Tertiary sedimentary rocks of Veracruz basin.

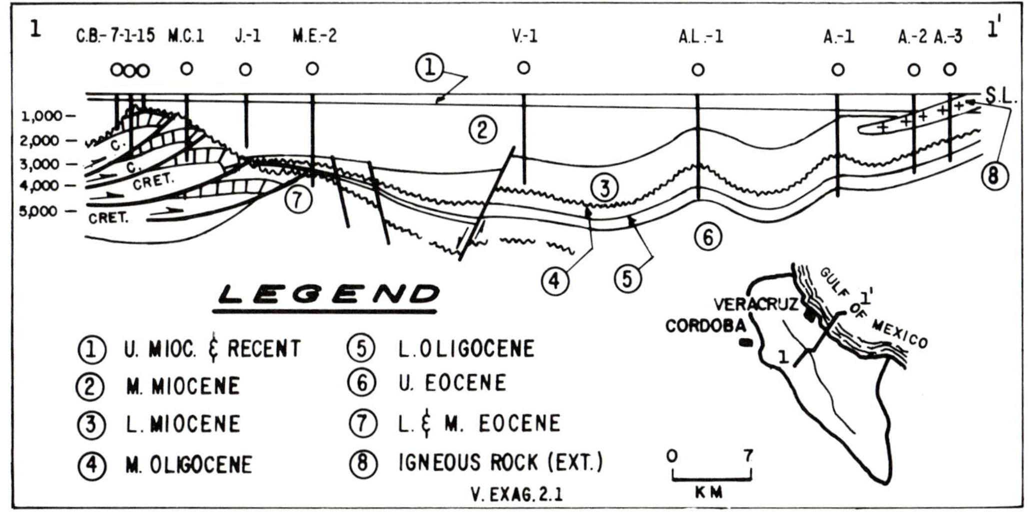

FIG. 4—Structural cross section *1-1'* across central Veracruz basin. Vertical exaggeration 2.1, vertical scale in meters.

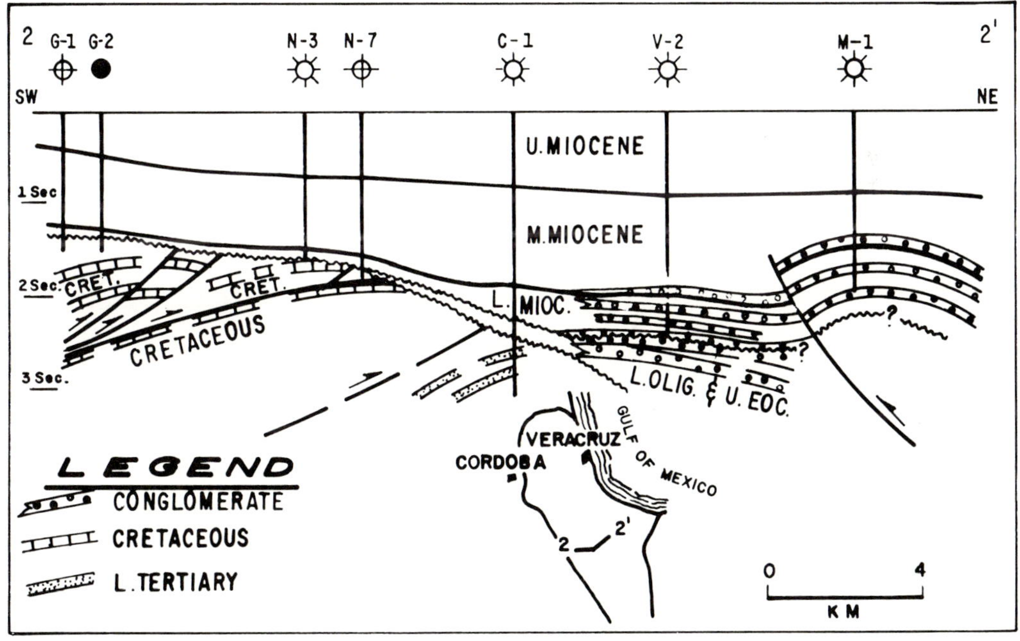

FIG. 5—Southwest-northeast structural cross section *2-2'* across western half of Veracruz basin.

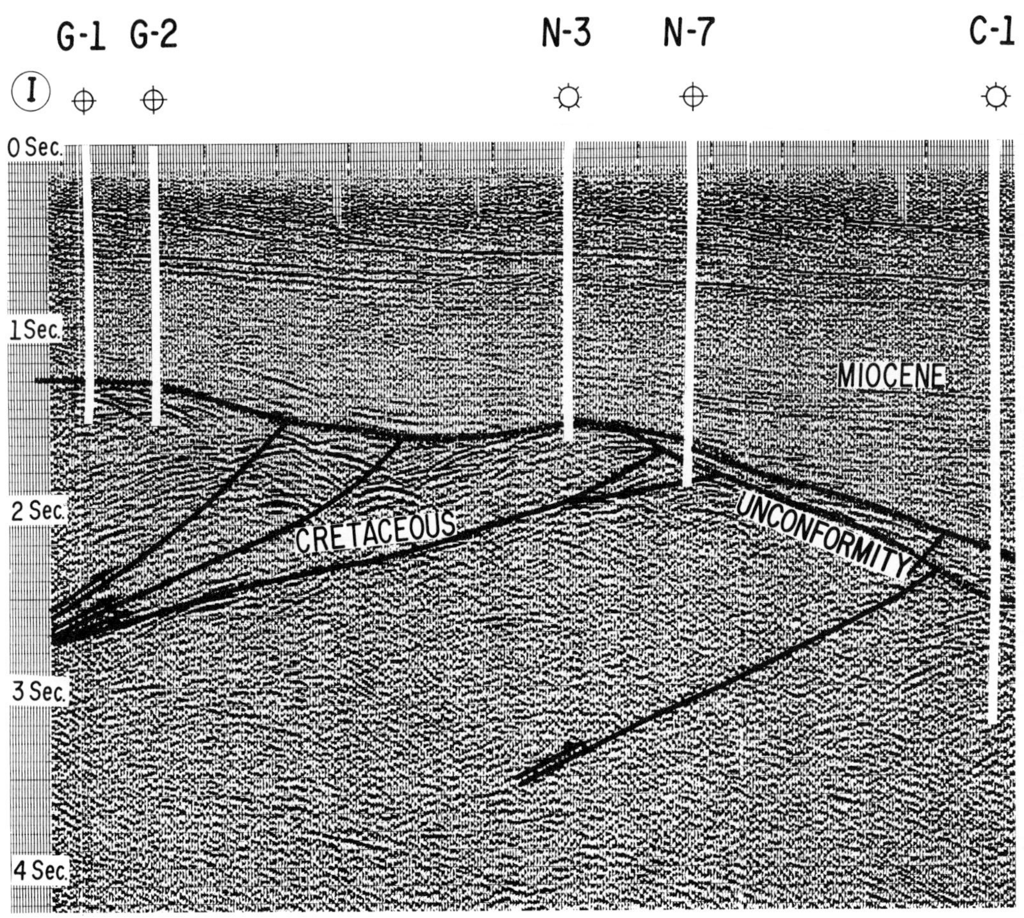

FIG. 6—Seismic profile *I-I'* along line of cross section *2-2'* (Fig. 5).

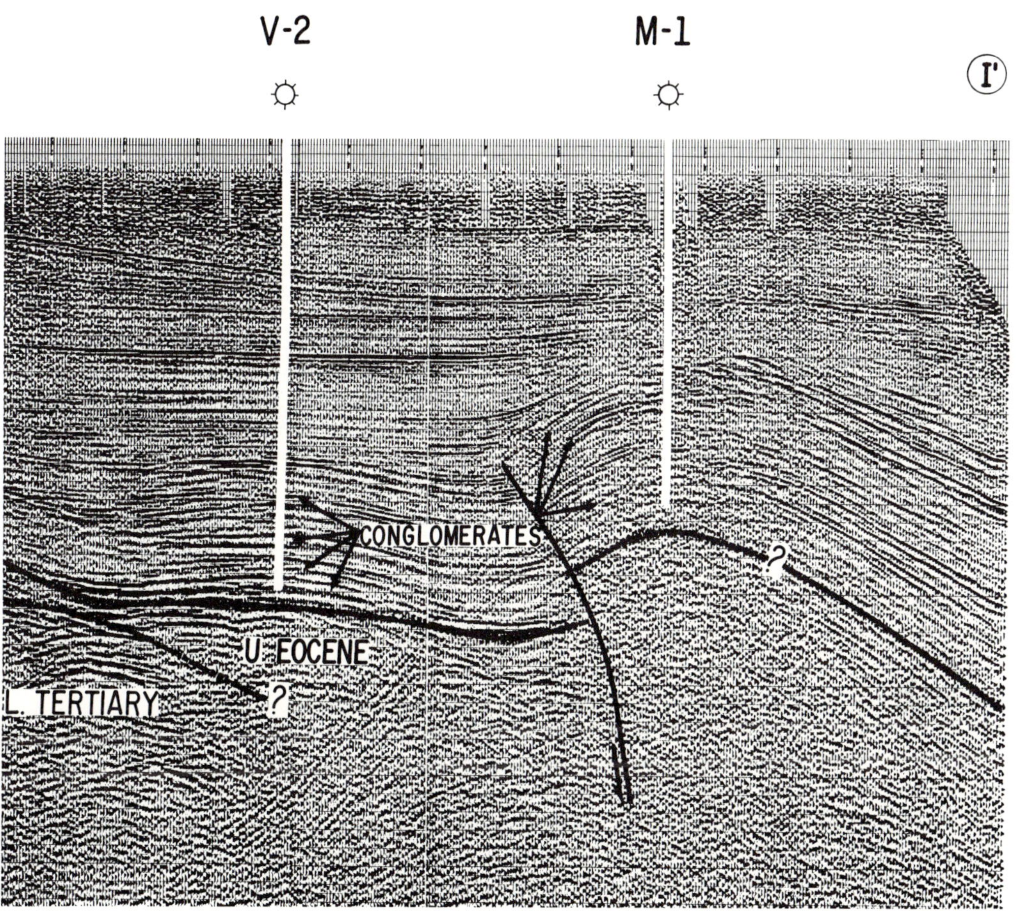

V-2
M-1
I'
CONGLOMERATES
U. EOCENE
L. TERTIARY
?
?

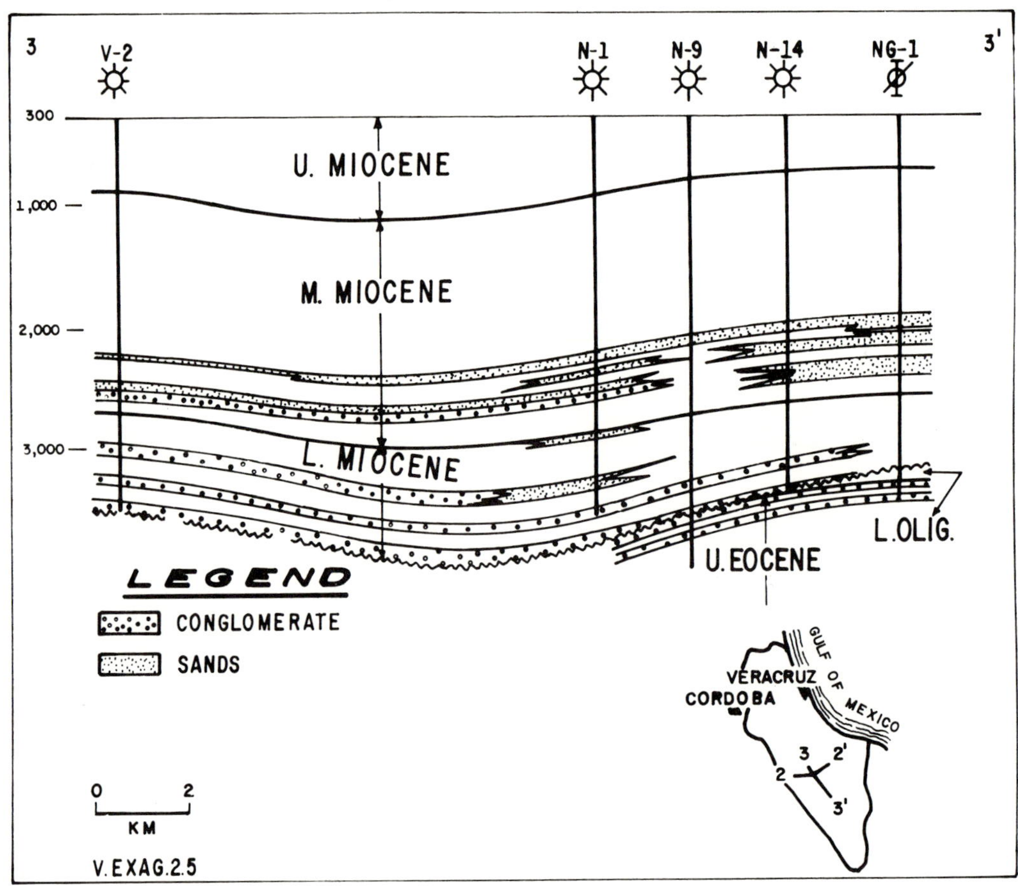

FIG. 7—Northwest-southeast structural cross section *3-3'* parallel with structural strike. Vertical exaggeration 2.5, vertical scale in meters.

tiary sedimentary deposits onlap the Cretaceous to the southwest is illustrated clearly. An unconformity at the base of the upper Eocene truncates Cretaceous, Paleocene, and lower and middle Eocene sedimentary rocks. Below this unconformity Cretaceous and lower Tertiary sedimentary rocks are folded and thrust faulted. The thrust faults and the structural axes of the Cretaceous folds are oriented northwest-southeast. A second unconformity at the base of the Miocene merges to the west with the older unconformity.

Several prominent structures are present in the eastern half of the cross section (Fig. 4). The middle Miocene sedimentary rocks are much thicker in the synclines than over the structural ridges because of compensatory deposition in submarine paleotopographic lows, indicating that the deformation occurred in post-early Miocene time. The dating of the folding is confirmed by the es-

sentially uniform thickness of the Oligocene and lower Miocene strata. Normal faulting is of the same age as the folding. The depositional environment of the sediments ranged from bathyal to outer neritic.

The structural and stratigraphic characteristics of the Tertiary shown on the profile of Figure 4 are fairly uniform along the southwestern margin of the Veracruz basin. Cross section *2-2'* (Fig. 5), about 50 km south of the profile of Figure 4, shows a similar stratigraphic and structural relation.

TERTIARY CONGLOMERATES

The previously mentioned Tertiary unconformities are shown in structural cross section *2-2'* (Fig. 5). Middle Miocene sedimentary rocks thin southwestward by depositional convergence. Lower and middle Miocene conglomerates are

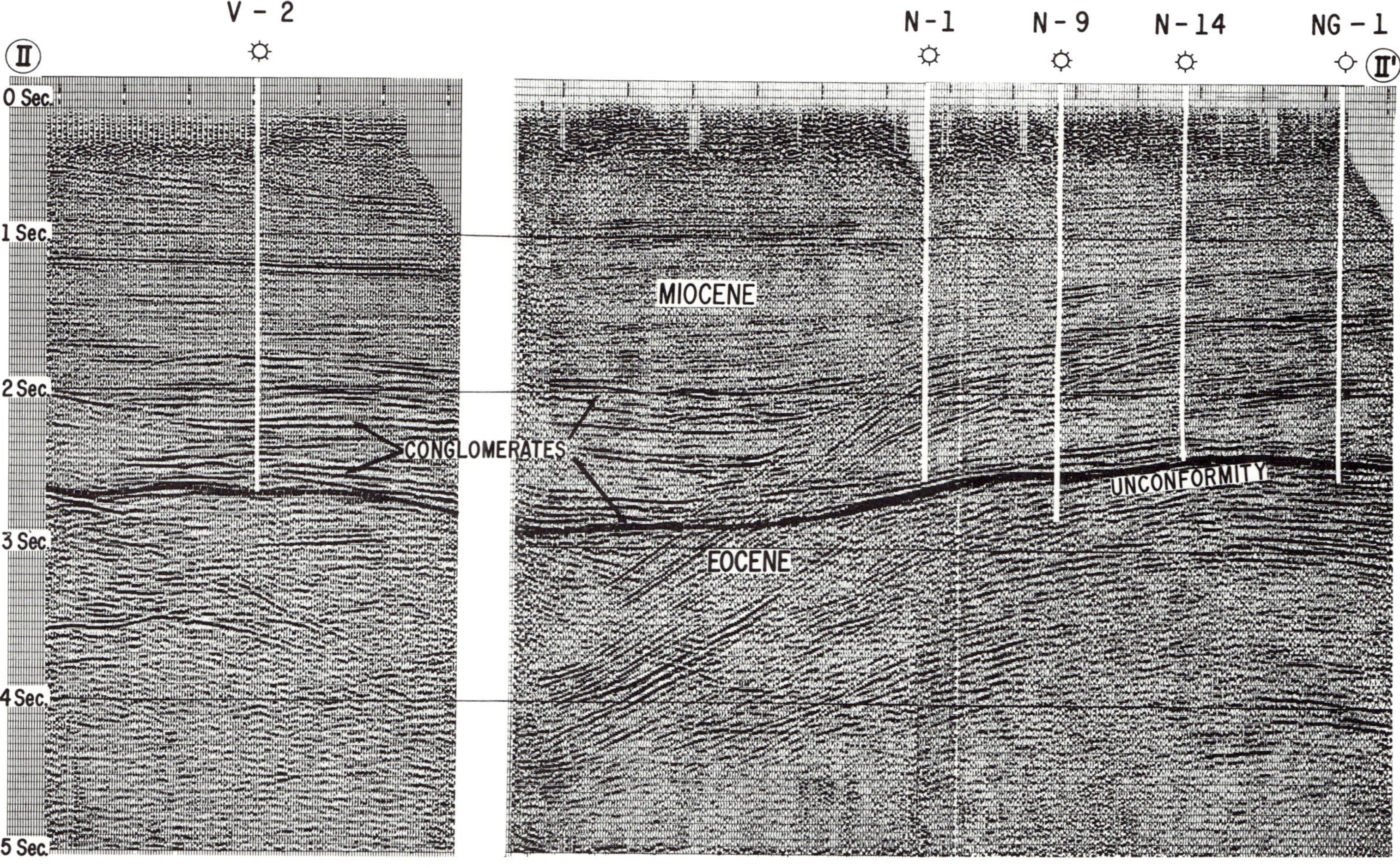

FIG. 8—Seismic profile *II-II'* along line of cross section *3-3'* (Fig. 7).

FIG. 9—Core 1, Novillero 9, depth 3,358 to 3,367 m, showing conglomerates and associated sedimentary rocks. Graded bedding from coarse conglomerate on bottom to silty sand at top is most conspicuous sedimentary structure. This graded sequence is not always complete. At bottom of core is about 1.4 m of pebbly mudstone (see Fig. 11 for details). Conglomerates are poorly sorted; pebbles are subrounded to well rounded.

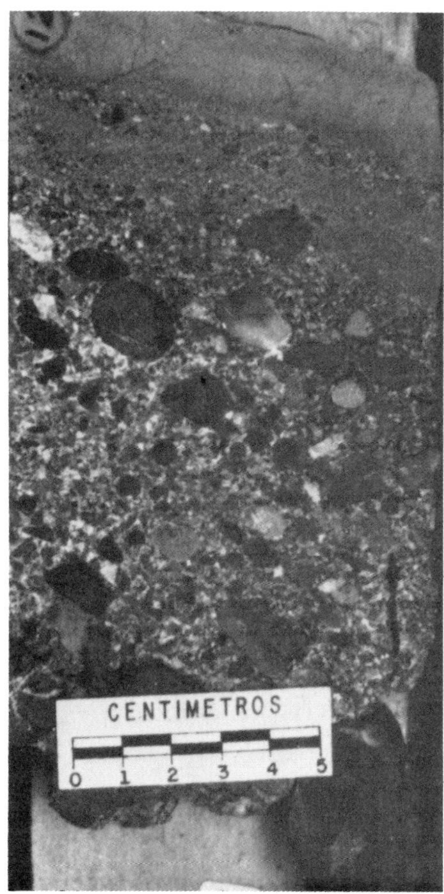

FIG. 10—Close view of graded bed (from core of Fig. 9). Fine grains are distributed throughout bed.

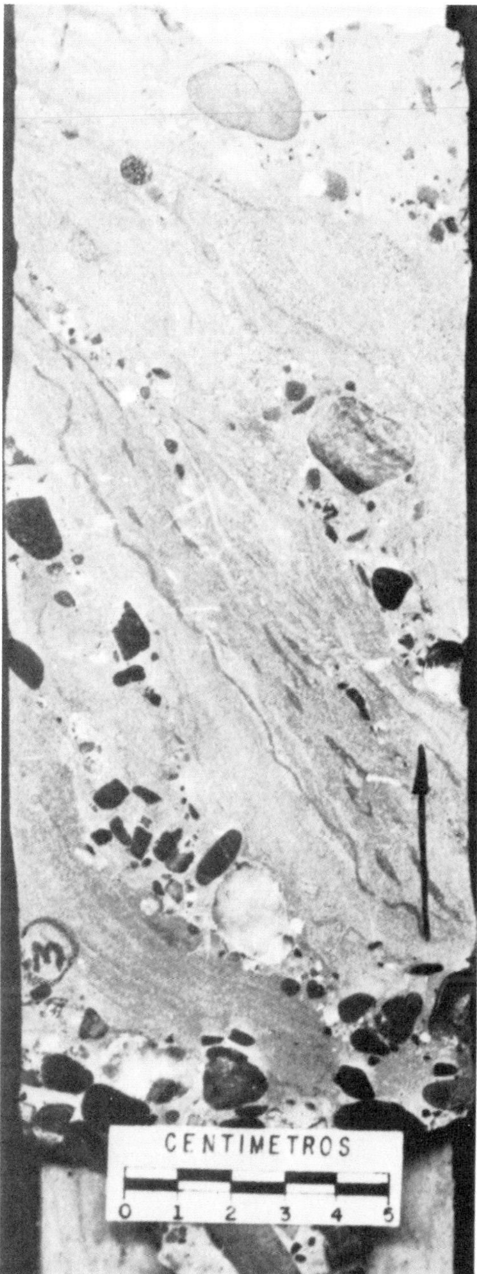

FIG. 11—Pebbly mudstone (Core 1, Novillero 9) sub-rounded to well-rounded pebbles in contorted silty-shale matrix.

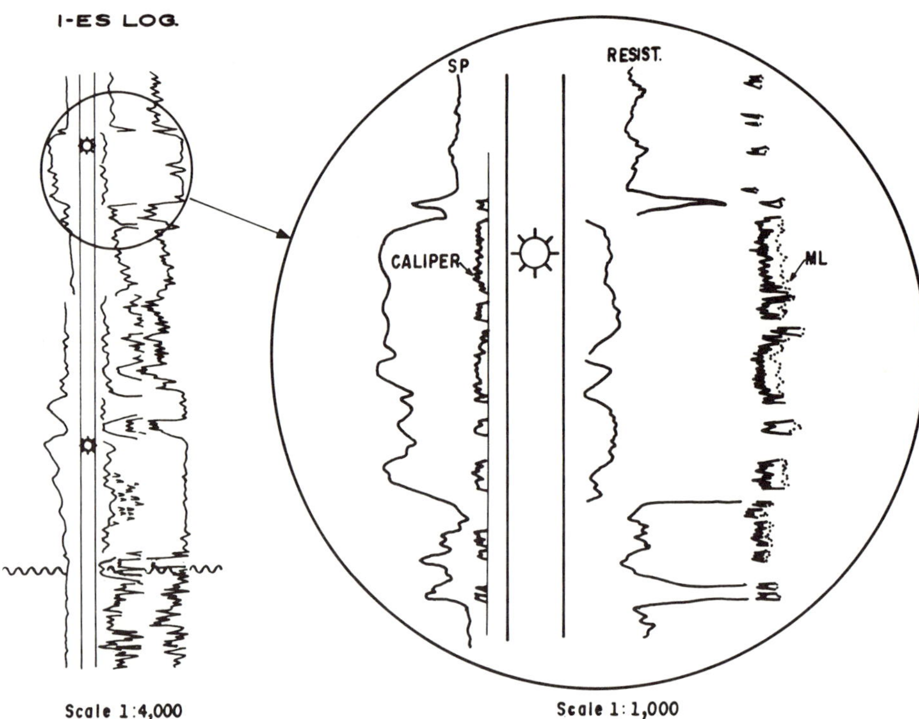

FIG. 12—Part of electric log from well 9 showing characteristics of producing intervals associated with conglomerates. Conglomeratic sections are indicated by high-resistivity intervals. In detail picture (right) microlog (ML) and caliper curves indicate porous and permeable intervals.

well developed along the profile. Individual conglomerate bodies range in thickness up to a maximum of 100 m and are interbedded with shales and sandstones; associated shales contain deepwater microfauna. *Osangularia culter* (Parker and Jones), *Usbekistania charoides* (Jones and Parker), *Bulimina alazanensis* (Cushman), *Planulina wüellerstorfii* (Schwager), and *Cibicidoides robertsoniana* (Brady) are common. For this reason the conglomerates are assigned to a bathyal depositional environment.

Seismic profile *I-I'* (Fig. 6) along the line of the section shown in Figure 5 shows the nature of the folding and faulting. The strong low-frequency seismic reflections at the position of the V-2 and M-1 wells correspond in time with the known depths of conglomerates in the wells. These strong conglomerate reflections disappear toward the west before reaching well C-1, in which no conglomerates were present.

Structural cross section *3-3'* (Fig. 7), oriented northwest-southeast, is essentially parallel with the paleodepositional and structural strike of the sedimentary rocks. Thus, the formations maintain

fairly uniform thickness and structures are subdued greatly.

The lower Miocene lies unconformably on upper Eocene and lower Oligocene sedimentary deposits. Conglomerates are well developed in this area; gas production is obtained from the two lower Miocene conglomerates.

Seismic profile *II-II'* (Fig. 8) is in the same position as the cross section of Figure 7. Here again, there is a close correlation between the known depths of conglomerates in the wells and the strong low-frequency seismic reflections. This suggests that in areas of sparse well control it is possible to interpret the presence of conglomerates from the characteristics of the seismic reflections.

Upper Eocene, Oligocene, and lower Miocene conglomerates were deposited along the western margin of the Veracruz basin and are related closely to the pre-early Eocene and pre-early Miocene unconformities. These conglomerates are lithologically very similar regardless of age. They consist mainly of middle and Upper Cretaceous subangular to well-rounded limestone frag-

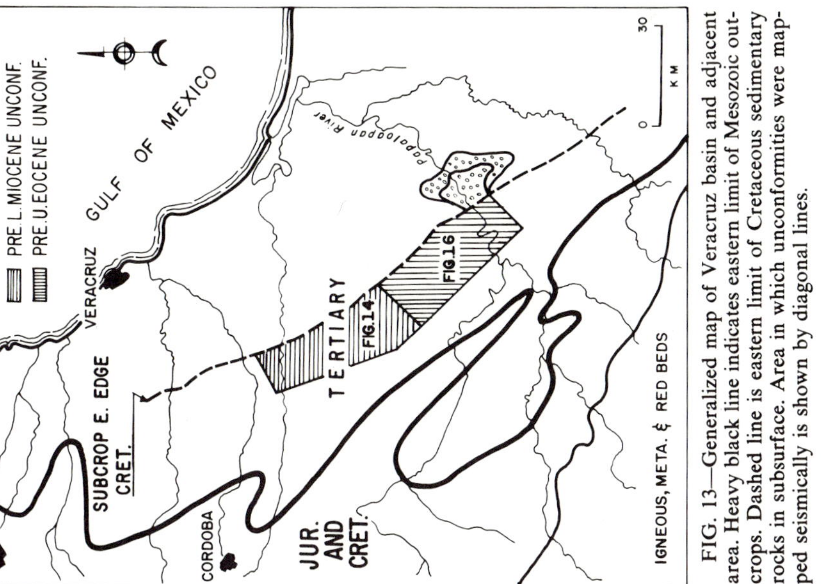

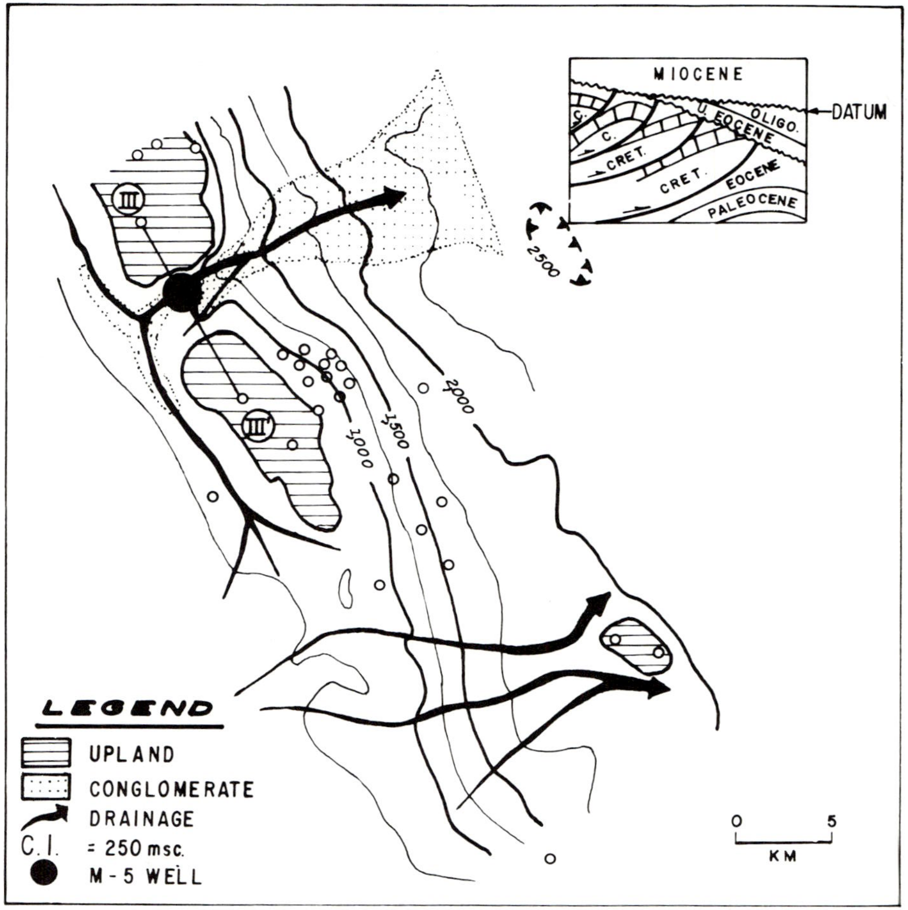

FIG. 14—Seismic-structure map of pre-early Miocene unconformity in part of western margin of Veracruz basin.

FIG. 13—Generalized map of Veracruz basin and adjacent area. Heavy black line indicates eastern limit of Mesozoic outcrops. Dashed line is eastern limit of Cretaceous sedimentary rocks in subsurface. Area in which unconformities were mapped seismically is shown by diagonal lines.

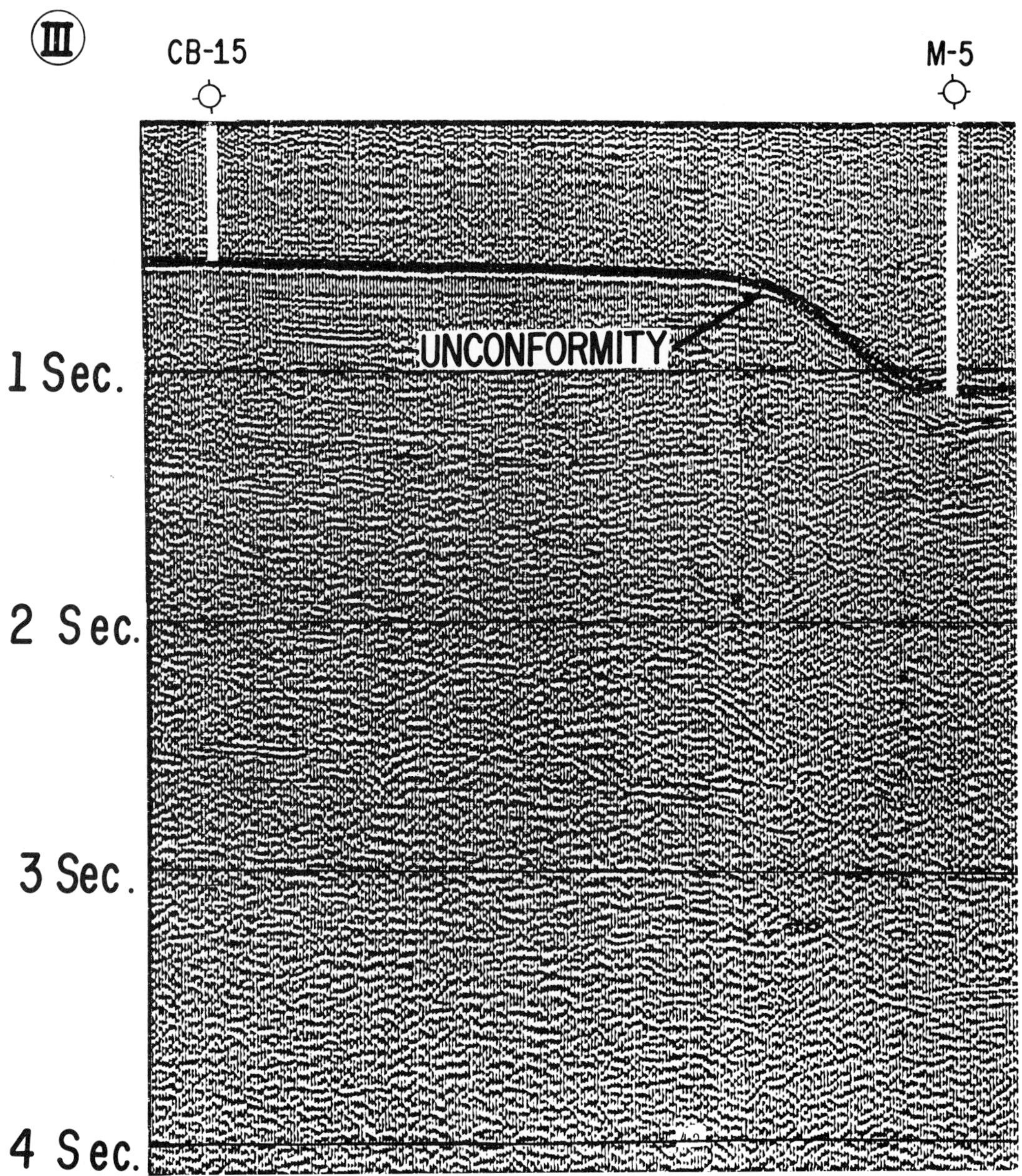

FIG. 15—Seismic profile *III-III'*, oriented northwest-southeast (see Fig. 14 for location) parallel with structural strike of Cretaceous.

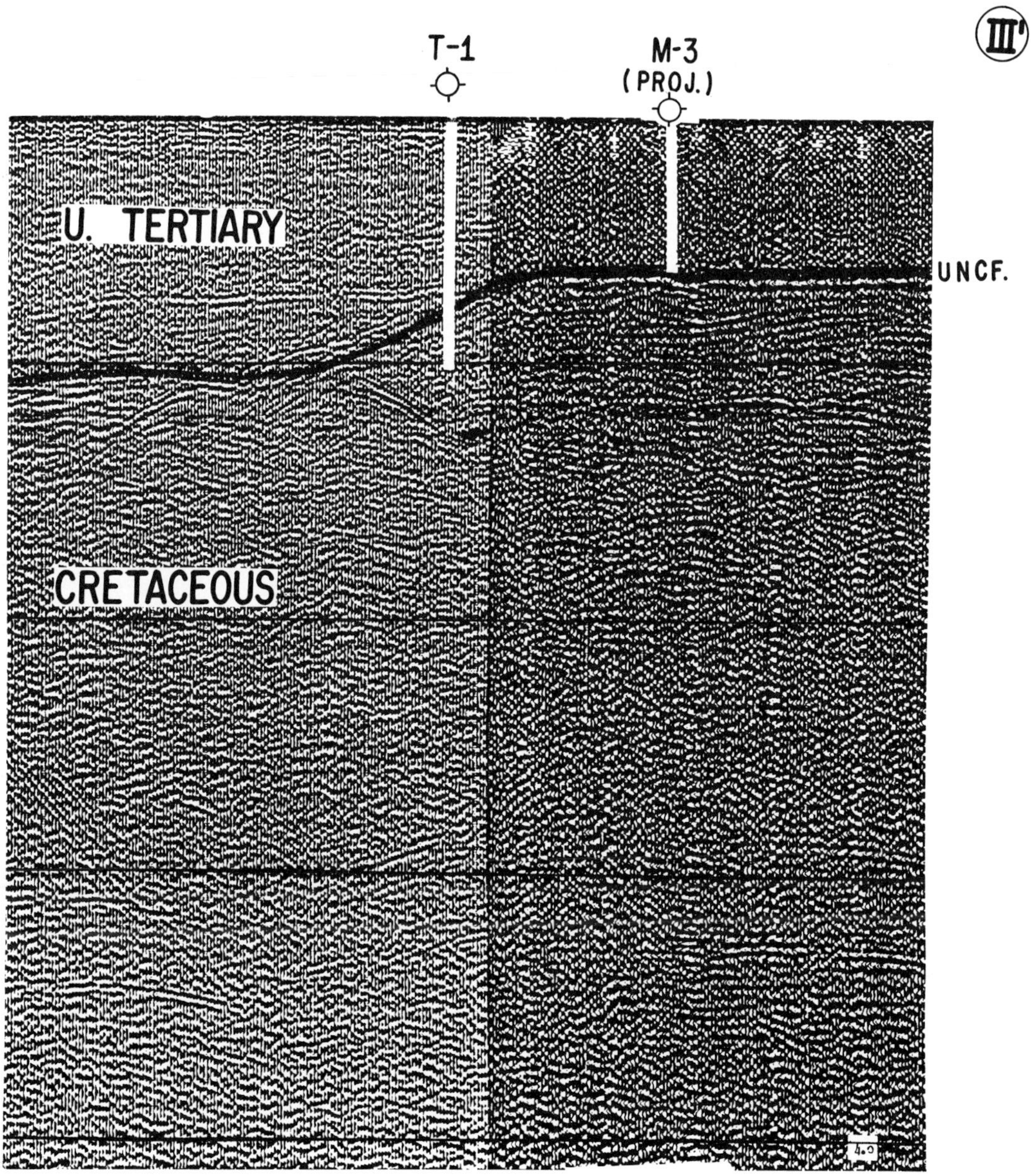
T-1
M-3
(PROJ.)
III'
U. TERTIARY
UNCF.
CRETACEOUS

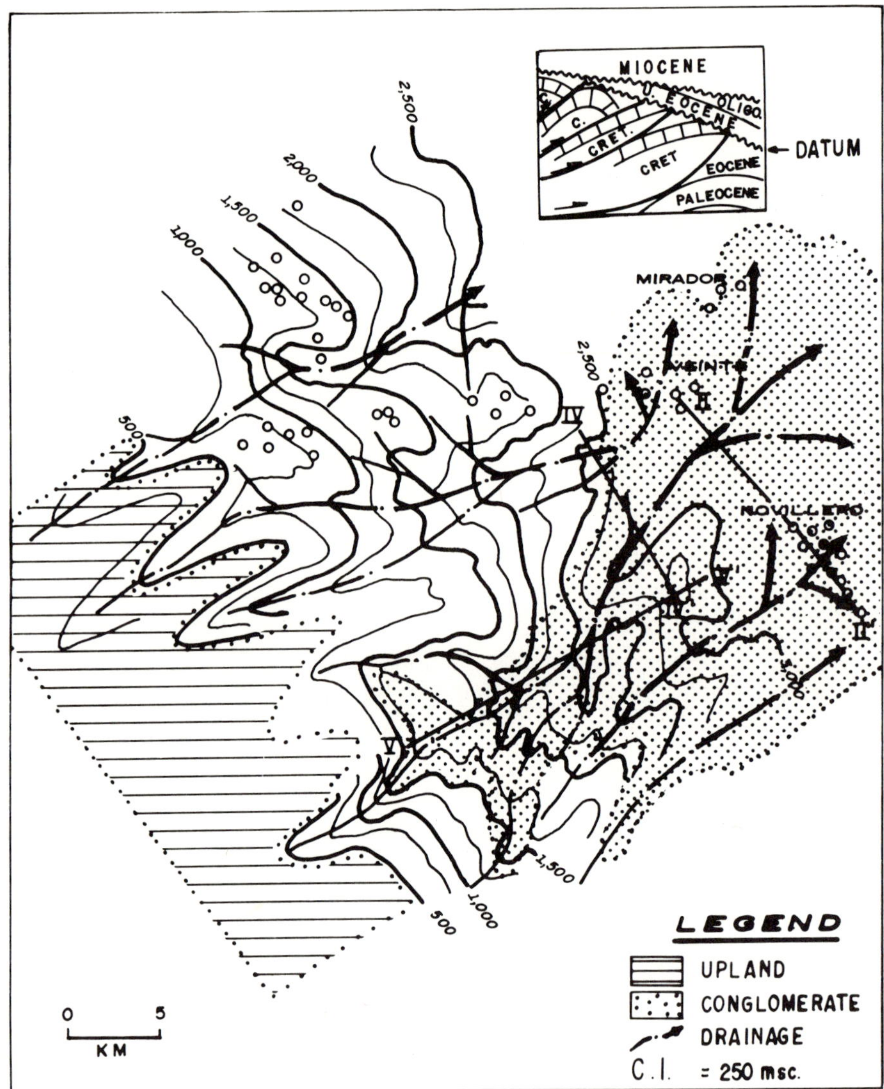

FIG. 16—Seismic-structure map of pre-late Eocene unconformity in part of the Veracruz basin, showing paleodrainage courses and possible conglomerate distribution.

ments, dolomite, and variable amounts of metamorphic and igneous rock fragments. There are admixtures of chert and sedimentary-rock fragments. Sorting is poor and the size of the pebbles ranges up to 10 cm. The matrix generally is sandy or shaly (Fig. 9). Locally the conglomerates are tightly cemented by calcite. Graded bedding is a common sedimentary structure. Pebbly mudstone locally is associated with the conglomerates (Figs. 10, 11).

Data obtained from electric logs and core analysis indicate that the average porosity of the producing conglomerates ranges from 6 to 8%. Figure 12 reproduces part of an induction log showing the characteristics of an interval associated with gas-productive conglomerates.

The conglomerates were derived primarily from a source area west and southwest of the basin. Probably the conglomerate zones merge vertically in the direction of the source. Southwestward from the area of vertical convergence they become linear and are confined to submarine canyons. The updip wedge edges of canyon conglomerates should be ideal stratigraphic traps for hydrocarbon accumulation, especially in view of the fact that the conglomerates are overlain directly by lower and middle Miocene shales which serve as seal and possible source rock.

Figure 13 is a generalized map which indicates the eastern limit of the outcrops of Mesozoic rocks of the Sierra Madre Oriental—a thick sequence of Jurassic and Cretaceous limestones and terrigenous sedimentary deposits. Further southwest is an extensive area of metamorphic (schists and gneisses) and igneous rocks together with a thick sequence of red beds (sandstones and shales) of pre-Late Jurassic age. This upland area southwest of the Veracruz basin served as a source area for the Tertiary conglomerate pebbles. Figure 13 also indicates the known eastern edge of the Cretaceous sedimentary rocks in the subsurface where they are overlain by late Tertiary rocks above the two big unconformities, previously discussed. Also shown is the area in which the Tertiary conglomerates are well developed. All of the area west of the dashed line was uplifted, folded, and thrust faulted during the Laramide orogeny.

Present drainage systems are superimposed on Figure 13. The big fluvial systems, like that of the Papaloapan River, probably were initiated early in the Tertiary in response to the tectonic movements of the Laramide orogeny and were preserved throughout the Tertiary. The eroded sediments from the deformed source area were transported in a general northeasterly direction by fluvial currents which flowed into deep submarine canyons. Through these canyons density currents transported the coarse clastic sediments and deposited them as a series of prograding submarine fans on the basin floor.

A reconstruction of the two main unconformity surfaces locates and delineates paleodrainage patterns and areas of accumulation of deep-water fanglomerates. From subsurface well control and seismic and paleontologic data, it was possible to reconstruct the paleotopography of the pre-late Eocene unconformity on the south and the pre-Miocene unconformity on the north.

Figure 14 shows the topographic expression of the pre-early Miocene unconformity. The southwestern two-thirds of the contoured area shows the eroded topography of the top of the Cretaceous. In the northeastern one-third, Oligocene and upper Eocene sedimentary rocks underlie this unconformity. A major northeast-trending paleodrainage course in the northern half of the mapped area has two major tributaries; one flowed northwest parallel with the Cretaceous structural strike; the other flowed southeast in the same strike valley.

Well M-5 (Fig. 14) was drilled into the channel fill where a conglomeratic (middle Miocene) section of about 150 m is unconformable on Upper Cretaceous sedimentary rocks. The distribution of canyon conglomerates and fanglomerates in this northern area also is indicated on Figure 14.

Seismic profile *III-III'* (Fig. 15), in the northwestern part of the area, trends parallel with the structural strike of the Cretaceous and illustrates very clearly the topographic expression of the unconformity surface. A channel crossing 2 km wide and 800 m deep is present at the top of the Cretaceous.

Figure 16 illustrates the contouring of the pre-late Eocene unconformity surface in the southern area. Paleocene and lower and middle Eocene sedimentary rocks underlie the eastern one-third of the mapped area, whereas the western two-thirds represent the eroded topography of the top of the Cretaceous. Several northeasterly trending paleodrainage systems are interpreted in this area. For the most part, these are modified dendritic-drainage systems with the positions of some of the tributaries being controlled by rocks of lesser resistance to erosion, between successive overthrust blocks of the underlying Cretaceous.

Doubtless the Eocene, Oligocene, and Miocene conglomerates that are so common along the western edge of the Veracruz basin were derived from the Cretaceous and older upland area on the southwest, and it is probable that the avenues of transport of the conglomerates were these conspicuous drainage courses. The gradients of the channels are steep as are the canyon walls. The possible conglomerate distribution coincides with the Papaloapan drainage system. In the Novillero-Veinte-Mirador area (Fig. 16), a composite deep-water fanglomerate is present. In this area a pronounced decrease in paleogradient caused the fanglomerates to spread out radially on the seafloor. The extension of these conglomerates upstream in the ancestral Papaloapan River canyon is based on seismic-reflection data. Wherever conglomerate is shown in the canyons, the evidence consists of sharp seismic reflections directly above the unconformity. These reflections are not parallel with either the unconformity or the strata directly beneath it. They are essentially horizontal when traced at right angles to a canyon and are tilted downstream when traced parallel with a canyon axis.

Seismic profile *IV-IV'* (Fig. 17) is at right angles to the axis of the canyon and parallel with the structural strike of the Cretaceous. It shows the erosional surface, the width of the canyon, and sharp seismic reflections interpreted as possible conglomerates. Seismic profile *V-V'* (Fig. 18)

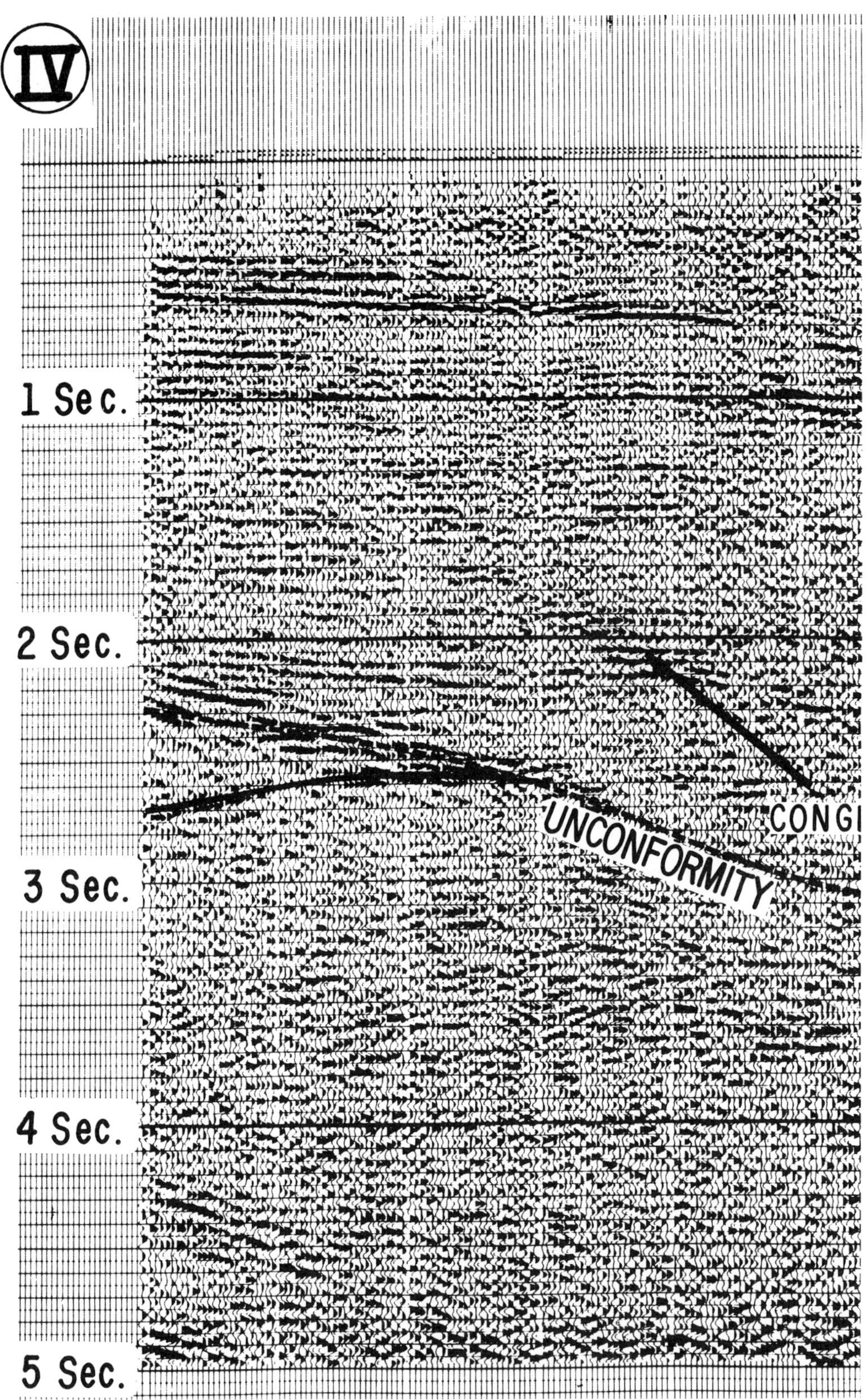

FIG. 17—Seismic profile *IV-IV'* oriented southwest-northeast, parallel with main axis of ancestral Papaloapan canyon (see Fig. 16 for location).

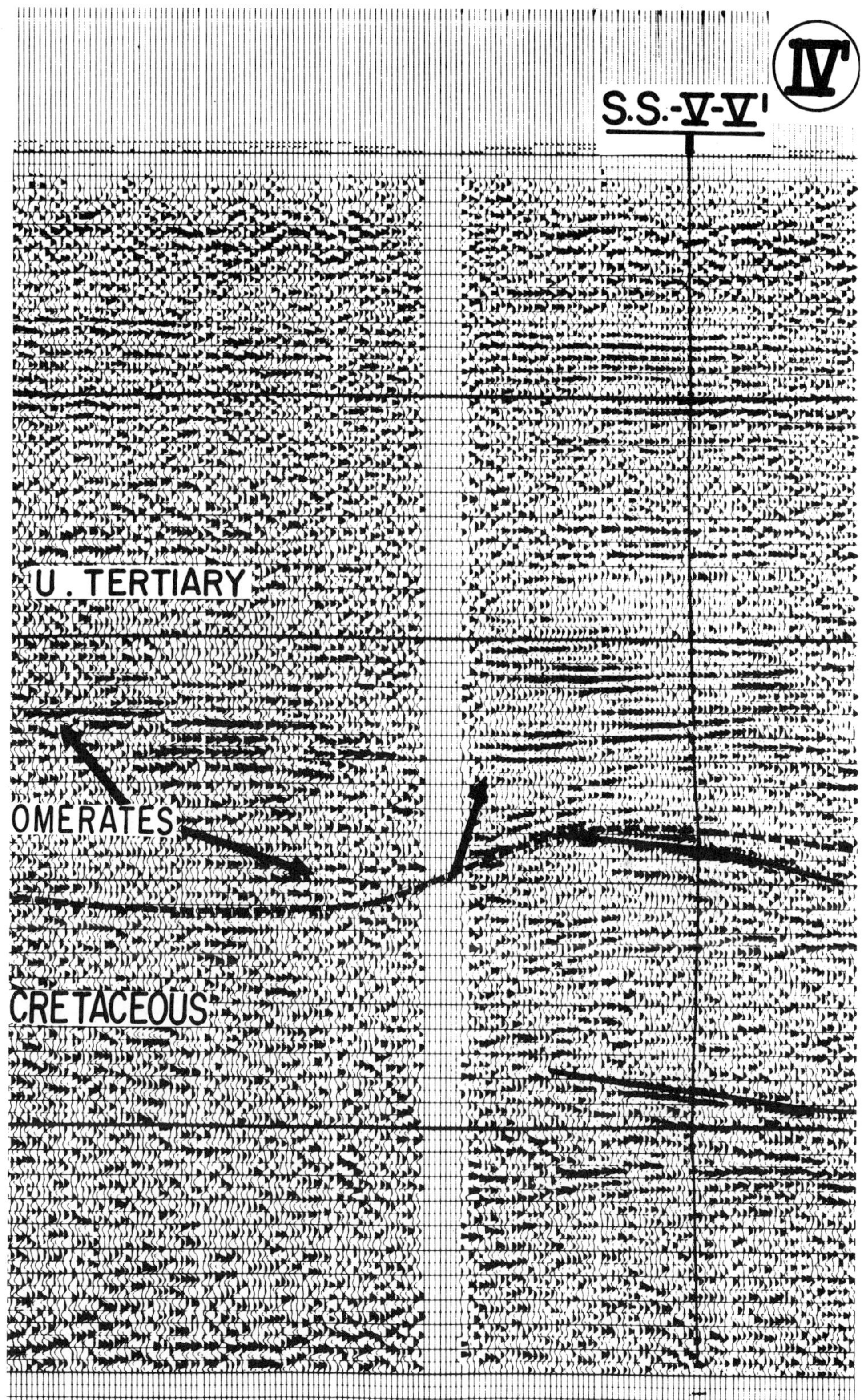

IV
S.S.-V-V'
U. TERTIARY
OMERATES
CRETACEOUS

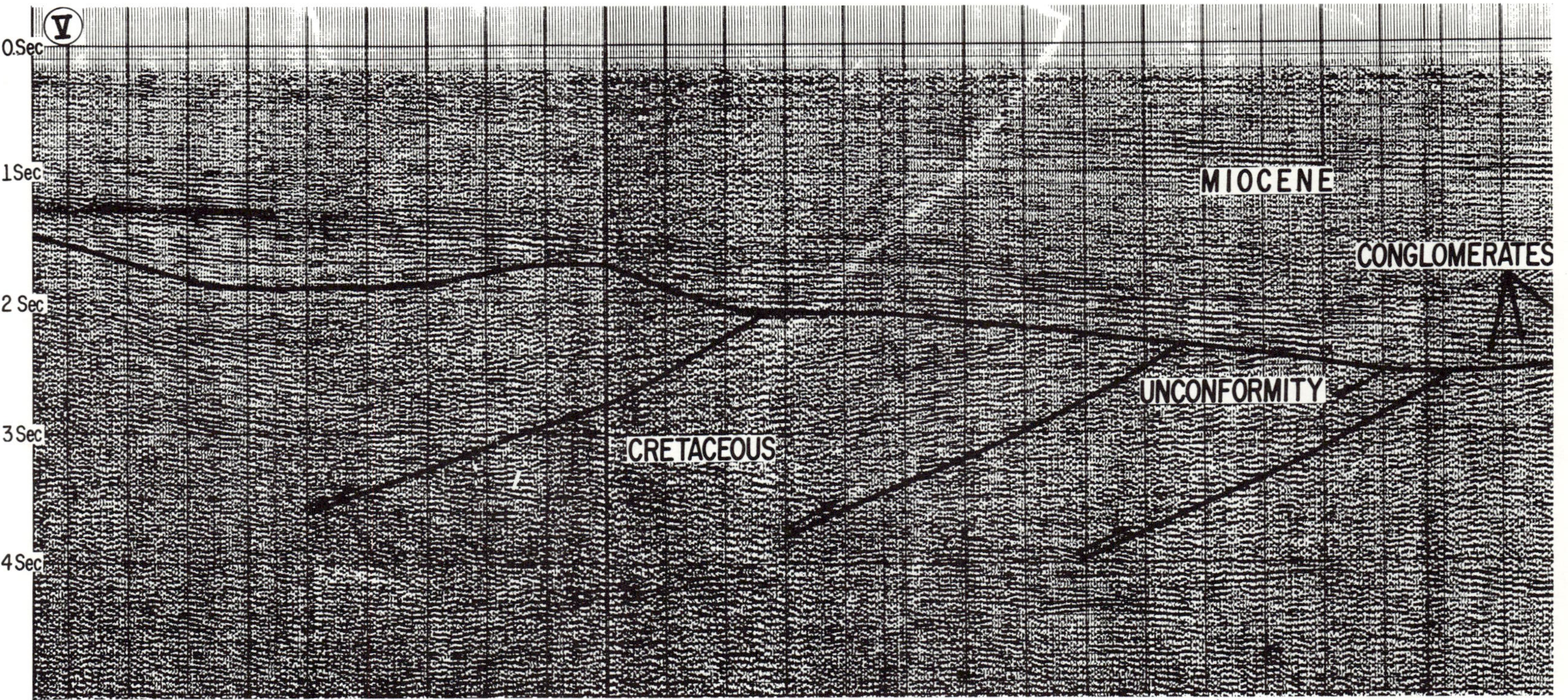

0.Sec
1 Sec
2 Sec
3 Sec
4 Sec
MIOCENE
CONGLOMERATES
UNCONFORMITY
CRETACEOUS

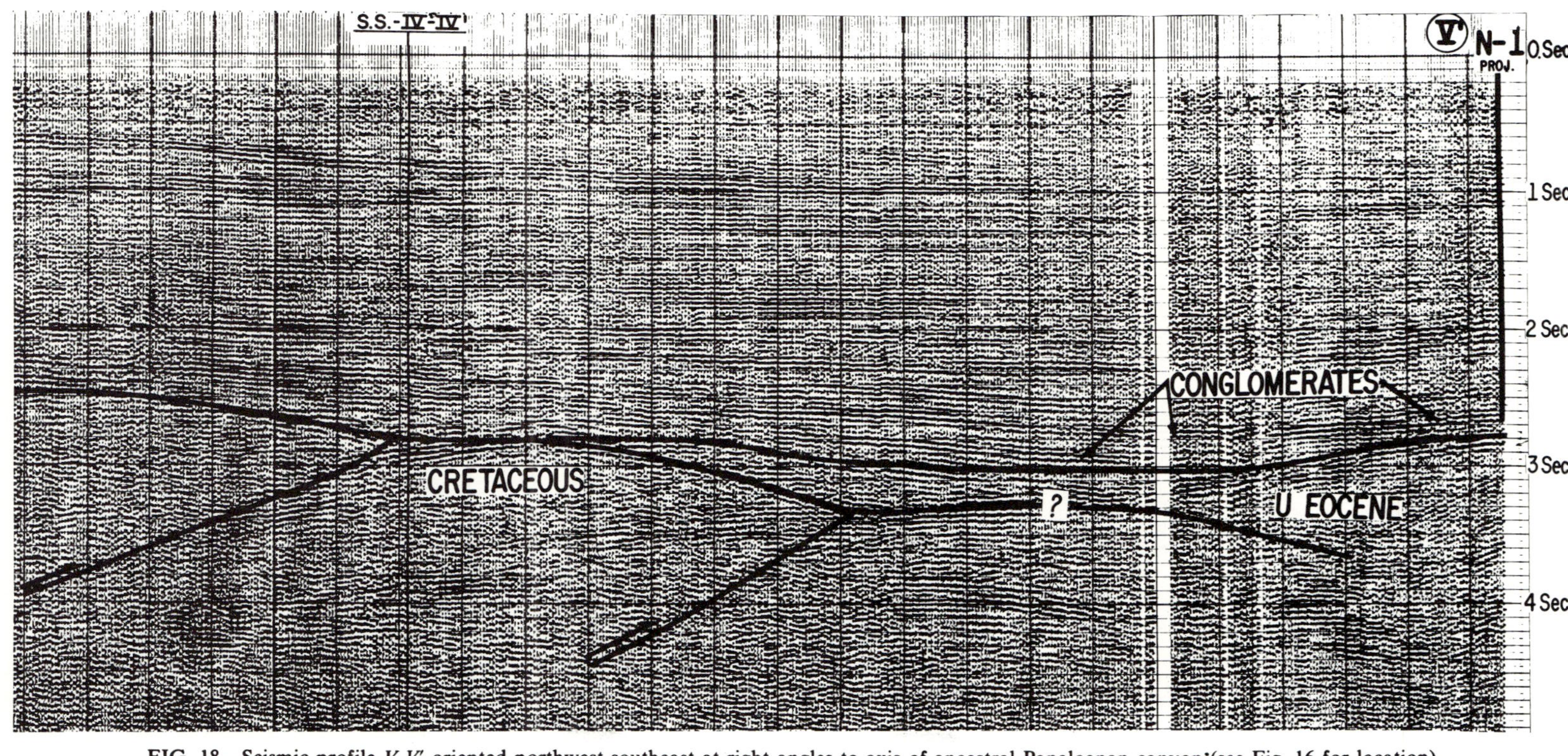

FIG. 18—Seismic profile V-V' oriented northwest-southeast at right angles to axis of ancestral Papaloapan canyon (see Fig. 16 for location). Section shown in two continuous parts.

along the general axis of the ancestral Papaloapan canyon shows the sharp seismic reflections tilted in a downstream direction, the unconformity, and the structural deformation of the Cretaceous.

EXPLORATION IMPLICATIONS

Seismic reflections of lower Tertiary strata have been traced eastward into the Novillero pool (Fig. 16) where their depths coincide with conglomerate reservoirs. It remains to be determined by drilling in the ancestral Papaloapan canyon whether this lateral extension of reflections from the Novillero pool is a valid basis for predicting conglomerates.

The distribution of conglomerates postulated here would be an excellent stratigraphic trap for hydrocarbons with more than 1,000 m of Miocene shales overlying the possible reservoirs. Conditions appear to be ideal for a perfect updip seal of hydrocarbons in the conglomerates.

CONCLUSIONS

The conglomerates along the central part of the western border of the Veracruz basin were derived from weathering and erosion of the Meso-zoic overthrust belt on the west. They were transported by density currents down submarine canyons and then spread out on the basin floor as submarine fans. A bathyal environment for these sediments is indicated by the associated fauna.

The possibility of predicting the presence of conglomerates in the subsurface by the analysis of seismic information in areas of very sparse well control is considered to be of great economic importance for petroleum exploration in the Veracruz basin. However, the validity of this hypothesis still remains to be proved by drilling.

SELECTED REFERENCES

Gorsline, D. S., and K. O. Emery, 1959, Turbidity-current deposits in San Pedro and Santa Monica basins off southern California: Geol. Soc. America Bull., v. 70, p. 279-290.

Normark, W. R., and D. J. W. Piper, 1969, Deep-sea fan-valleys, past and present: Geol. Soc. America Bull., v. 80, p. 1859-1866.

Stanley, D. J., 1969, Submarine channel deposits and their fossil analogs ("fluxoturbidites"), in The new concepts of continental margin sedimentation—AGI Short course lecture notes, Philadelphia: Am. Geol. Inst., p. DJ59-1–DJ59-17.

Studies of Ventura Field, California, I: Facies Geometry and Genesis of Lower Pliocene Turbidites[1]

KENNETH J. HSÜ[2]

Abstract The Pliocene sediments of Ventura field, more than 10,000 ft (3,048 m) thick, were deposited in a deep-sea basin, and subsequently were folded into an anticline. Three approaches were used to investigate sand trends within the basin: (1) facies relations using isopach maps, fence diagrams, and textural analyses of core samples; (2) grain orientation by visual inspection, thin-section analysis, and dielectric-anisotropy measurements; (3) sedimentary structures, mainly cross-bedding.

Facies relations suggest that the sands were deposited as elongate lenticular bodies by laterally restricted westerly flowing currents in deep parts of a deep-sea basin. The sand trends are oriented predominantly east-northeast, approximately parallel with the axis of the present Ventura anticline. Studies of grain orientation and cross-bedding support evidence for approximately east-northeast lower Pliocene sand trends in the Ventura field.

Maximum sand development in the central parts of ancient deep-sea basins in tectonically active areas is in direct contrast to the sand-development patterns in shallower water paralic and platform types of depositional basins. This study suggests that the search for sand reservoirs in tectonically active regions might be directed toward the central parts of ancient depositional basins if paleoecologic and sedimentologic data suggest conditions of deep-sea sedimentation. Parallelism of structural axis and trend of sand bodies in folded deep-sea basins suggests an influence on structural trends by sediment-distribution pattern; anticlinal traps might be expected to coincide with maximum sand development, and stratigraphic traps might be expected on the flanks of such folds.

From detailed isopach maps of sand trends offset by faults it is possible to determine fault displacements accurately. Such evidence indicates that the thrust faults in Ventura field have had right-lateral movements. The Padre Juan fault northwest of the Ventura field is perhaps also a right-lateral fault, offsetting an originally continuous structure into the en-echelon Rincon and Ventura anticlines.

INTRODUCTION

The Ventura field is on the Ventura anticline in the northwestern onshore part of the modern Ventura basin, about 2 mi (3 km) north of the city of Ventura, California (Fig. 1).

The Pliocene series in the Ventura field is composed of the Pico (upper Pliocene) and Repetto (lower Pliocene) Formations and consists of a se-

[1]Manuscript received, November 24, 1975; accepted, July 13, 1976.

[2]Geological Institute, Swiss Federal Institute of Technology, CH-8006 Zurich, Switzerland.

Publication No. 85 of the Laboratory of Experimental Geology, ETH, Zurich, Switzerland.

The work was carried out as part of the Shell Development Company project, "Studies of Sandstone Reservoirs," a research project headed by Robert H. Nanz. I am indebted to Shell Development for the release of this material, and am happy to acknowledge my gratitude to many of my former colleagues in the Shell organization who made this research possible. Robert Nanz suggested the theme. John Shelton, Dave Pontius, the late Lloyd Yerkel, the late Hugh Bernard, the late Bernard Wilson, the late Jim Taylor, and many of the former "Clastics" group of Shell inspired me with discussions and advice. Rufus LeBlanc, the late Gordon Rittenhouse, Noyes Smith, H. Gershinowitz, Bert Mull, Jack Fouts, Leo Newfarmer, and others formerly in administrative positions of Shell Development or of Shell Oil Company (Pacific Coast Area) provided opportunity for me to get free access to all company files; they also kept a keen interest in all stages of the research and I was most grateful for their continued encouragement. Many of the laboratory experiments and data-reduction chores were carried out by an excellent team of technicians, headed by Bill Begin and Arly Walters, and by summer student-assistants. Jaime LaVerde and the late Rickie Blackwood drafted the illustrations. Thanks also are due to P. H. Kuenen, who reviewed the Shell report as a consultant to the Royal Dutch Shell Group; his "stamp of approval" facilitated the propagation of the main theme of this work—the longitudinal transport of turbidites in oblong basins.

I cannot help but recall with nostalgia the golden era of Shell research, and my many good friends in Houston, Los Angeles, and Ventura. This work is dedicated, however, to the memory of my late wife, Ruth, with whom I spent many happy days at Ventura.

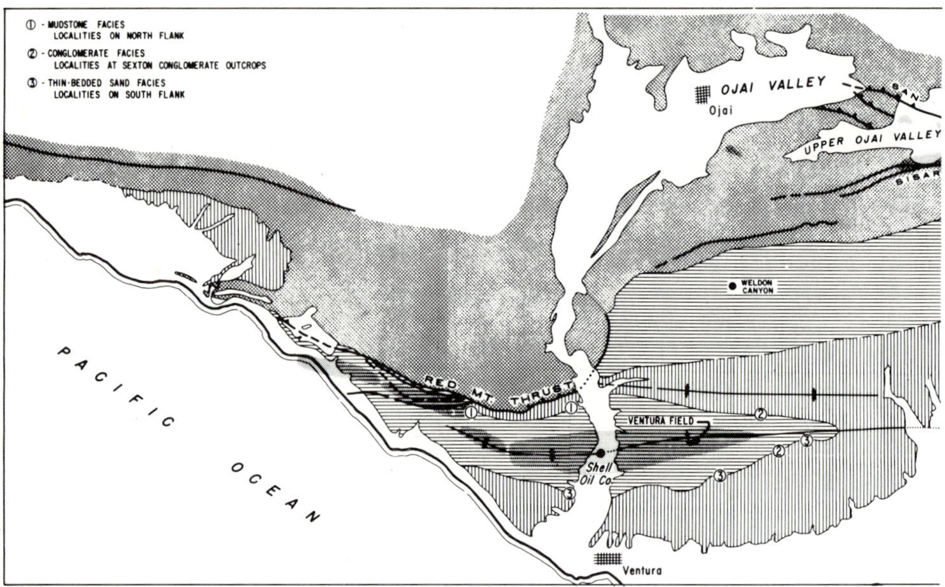

FIG. 1—Generalized geologic map of western Ventura basin.

quence of sands, silts, and shales more than 10,000 ft (3,048 m) thick. The coarser Pliocene sediments have been considered to have been emplaced in deep water by turbidity currents in tectonically active regions (e.g., Natland and Kuenen, 1951; Winterer and Durham, 1962; Crowell et al, 1966). This report concerns the Repetto Formation only and is designed to define the following features of these deep-water deposits—configuration, facies relation, source areas, path of transportation, and modes of deposition. It is demonstrated further that a knowledge of the sand trend is applicable to solving some fault problems in the field.

The work was carried out during the spring and summer of 1955 at a time when the idea of turbidity-current deposition was at best a working hypothesis. There was a general impression, reinforced perhaps by the report of the broad-front advance of the turbidity current generated by the Grand Banks earthquake (Heezen and Ewing, 1952), that turbidites are sheet sands or blanket-type deposits. The conclusion based on isopach studies that the Repetto oil reservoirs are shoestring sands did not meet immediate acceptance. However, further drilling did confirm the concept of preferential accumulation of turbidites in central depressions of basins, and the postulate of lateral pinchout along basin margin has proved useful for efficient exploitation of petroleum in the Ventura basin. Furthermore, studies of ancient turbidites on outcrops (e.g., Kuenen, 1957) and recent turbidites in Holocene environment

(e.g., von Rad, 1968) have all but confirmed the idea of longitudinal transport of turbidity currents after their arrival on the flat bottom of oblong basins.

This work was published as an internal report of the Shell Development Company in 1956. Since then the sedimentology of turbidity currents has been the theme of many outstanding investigations of outcrops (see review by Walker, 1970). Nevertheless, subsurface studies based on closely spaced data are few, and the documentation of the three-dimensional geometry of oil-producing turbidite sands is still inadequate (see Sullwold, 1961). I was, therefore, invited by the Editor of the *Bulletin* to present the information from the Ventura field. Thanks to the courtesy of the Shell Development Company, the report was released and is presented here as it was written except for editorial changes to include some new and relevant literature citations.

SEDIMENTARY FACIES IN REPETTO FORMATION

The Repetto sediments of the Ventura field consist of a conformable sequence of sands, silts, and shales, more than 5,000 ft (1,524 m) thick (Fig. 2). The formation has been subdivided into microfaunal zones, and electric-log correlations are recognized within the Repetto. These two types of correlation tools have allowed the unraveling of the extremely complicated structures in the field.

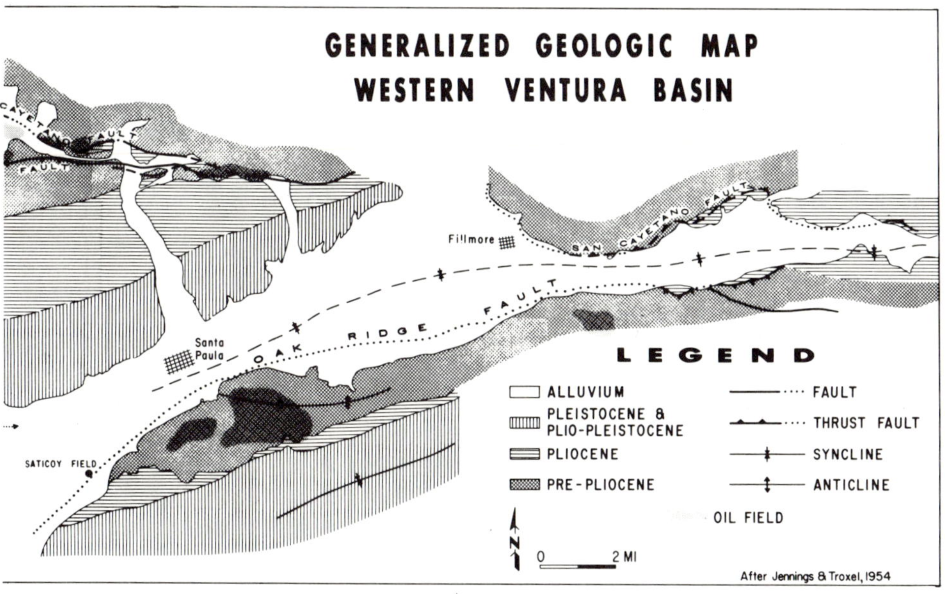

The type and diversity of the Foraminifera indicate that the sands were deposited in a deep-marine basin. Sand-percentage maps and columnar sections indicate that sand percentages are higher in the central part of the basin than on the flanks. Electric-log correlations (Figs. 3, 4) also reveal facies changes within the field. The present work describes the detailed facies relations and the approximate configurations of several genetic units.

Lithofacies in Upper Producing Zones

Correlation section AA' in Figure 4 illustrates the facies relation of sands, silts, and shale in the interval between Ventura field electric log markers AM to AQ_1 in zone 3 and zone 4 (Fig. 2). The interval is approximately 350 ft (107 m) thick. Lithology was interpreted from electric logs, supplemented by examination of cores. Because the wells are at or near the crest of the anticline, the thickness of the units is approximately equal to the apparent thickness as shown by the electric logs. The upper interval AM-AO and the lower interval AP-AQ_1 consist predominantly of sands with thin shale intercalations. Individual sand units within these intervals cannot be correlated with certainty. On the other hand, a sand unit of varied thickness between the electric-log markers AO and AP can be correlated within the field because of prominent shale breaks above and below (Fig. 4). This sand unit is designated as the AO_1 sand. Thickness data for the AO_1 sand were obtained from electric logs. Thickness was determined by well penetrations converted to true (stratigraphic) thicknesses by a graph method. Inclinations of dips of the folded strata and of well deviations were taken into account. The calculated thickness data should have an error of less than 5%, except those from Tidewater Associated wells which are less accurate because well-survey data were not available and well deviations can be estimated only by comparing nearby electric logs.

The isopach map indicates that the sand is an elongate lenticular deposit with an east-west trend approximately coincident with the crest of the Ventura anticline (Fig. 5). Within the limits of the field, the sand body is approximately 7 mi (11 km) long and more than 1 mi (1.6 km) wide. It has a maximum thickness of about 80 ft (24 m) near the eastern margin of the field at Tidewater Associated Ventura Land & Water Co. lease, and thins gradually westward along the sand axis to a thickness of about 20 ft (6 m) at the southwestern corner of Shell Taylor lease. The sand pinches out completely to the north and is very thin in some south-flank wells, suggesting that the sand also may be bordered on the south by shales and silts. Whereas the AO_1 sand thins westward, the sands just below, between the electric-log markers AP and AQ_1, apparently thicken westward (Fig. 4). This pattern of compensating facies changes is very typical of the Pliocene sediments of the Ventura field. Whereas the thickness of any unit may be reduced by half or even by three-quarters, the total thickness and sand percentages of the Plio-

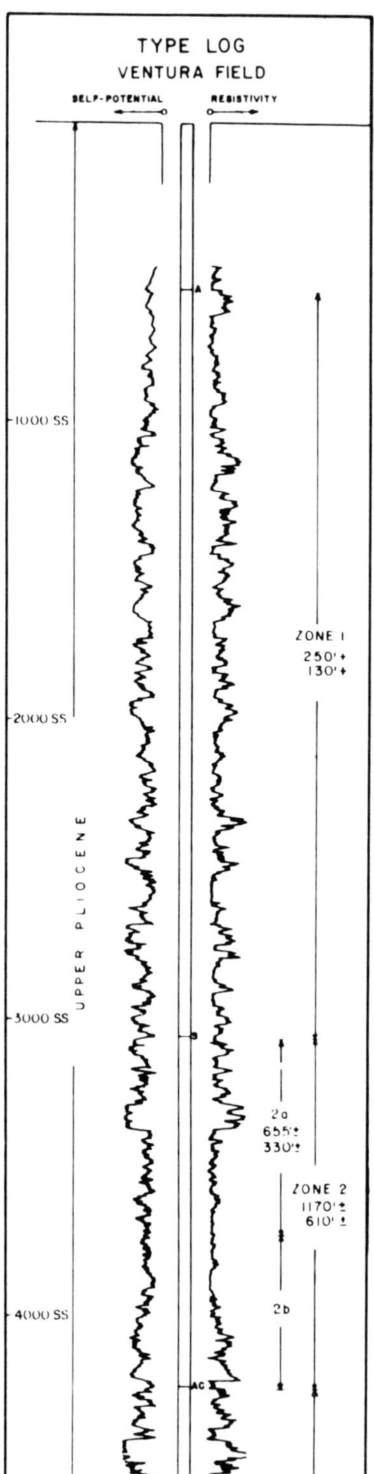

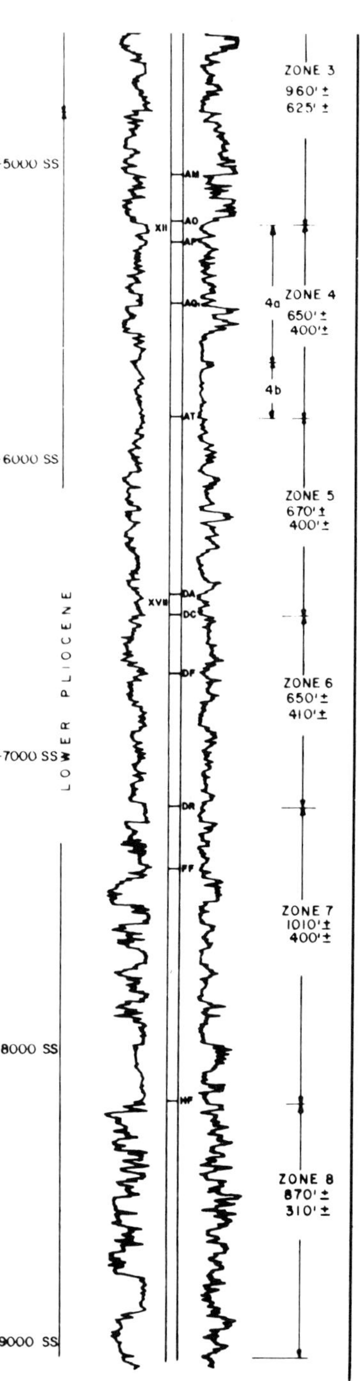

FIG. 2—Type electric log, Ventura field; composite of logs from several wells. Zones 1, 2, 3, 4, ... refer to oil-producing zones of field. Thicknesses are in feet.

cene section are not grossly different from the east to the west end of the Ventura Avenue field.

Lithofacies in Middle Producing Zones

The facies relation of sands, silts, and clay in the intervals between electric-log markers DA and DF in zone 5 and zone 6 (Fig. 2) is shown in Figure 4 (correlation section *BB'*). The interval is approximately 300 ft (91 m) thick; it consists predominantly of sands with thin shale intercalations.

A sand unit, a few tens of feet thick, is present between electric-log markers DA and DB in the western part of the Ventura field where the sand is separated by correlatable shale breaks from sands above and below. The underlying interval, DB-DC, is shaly and silty in the west (in Shell Taylor and Edison leases, Fig. 3), but it becomes increasingly sandy farther east. Near Tidewater Associated Lloyd well 168 (Fig. 4), a well-defined sand unit is developed in the interval DB-DC. Farther east, the DB shale marker which serves to separate the two sand units is no longer recognizable; there a massive sand unit about 75 ft (23 m) thick is present between electric-log markers DA and DC. The approximate southeastern limit of the DB shale is indicated in Figure 6. Farther east in Tidewater Associated Lloyd (well 165) and Ventura Land and Water leases (well 84), the sand between markers DA and DC combines with the topmost unit of the DC-DF sequence to form a more massive unit about 110 ft (34 m) thick.

Whereas the DA-DC interval becomes more sandy eastward, the DC-DF interval below shows an apparent decrease in thickness and in sand percentage as suggested by correlation along the anticlinal crest (Fig. 4). The eastward thinning may be real, or it may be only apparent if sand trend for the DC-DF interval crosses the structural axis obliquely. Nevertheless, the correlation illustrates again the pattern of compensating facies changes.

An isopach map (Fig. 6) was prepared for the DA sand, which is defined as the interval between the DA and DB markers where it is separated by correlatable shale breaks. The map indicates that the DA sand, like the upper Repetto AO_1 sand, is an elongate lenticular deposit with an approximately east-southeast trend (Fig. 6). The sand is almost 40 ft (12 m) thick near the southeast limit of the DB shale where it merges with the underlying sand to form a massive sandstone sequence about 75 ft (23 m) thick. The DA sand thins gradually westward and silts out between the Shell Taylor wells 404 and 505 (Fig. 4). The sand also

silts out completely toward the north and southwest. The southeastern limit of the sand body has not been reached.

Sand units recognizable by electric-log correlation in the Ventura field consist actually of many smaller amalgamated genetic units as revealed by a careful examination of samples from continuously cored intervals. Such genetic units vary in texture and graded bedding is very common. Examination of the cores of many wells of the Shell Taylor lease indicated that the "massive sand" between the electric-log markers DC and DE consists of 20 or more graded beds which range in thickness from a few feet to more than 10 ft (3 m). The basic genetic unit in the Ventura field is a graded bed.

Individual graded beds rarely exceed 15 ft (5 m) and are commonly only a few feet thick in the Ventura field. Few of the beds can be correlated from well to well. Very few beds have been cored continuously in more than one well. Only the continuous coring of zone 6 from three wells in Shell Taylor lease (310, 316, 372) affords an opportunity to study the lateral textural variation of a graded bed.

The correlation of this graded bed, about 10 ft (3 m) thick, was established through a careful analysis of electric logs and cores; the graded bed is just above the prominent shale maker DE. In the Shell Taylor 310 near the northeastern corner of the Shell Taylor lease, the graded bed is 13.5 ft (4 m) thick and is a coarse sand at the base (Fig. 7). The bottom of the graded sand is poorly sorted ($\delta\phi = 2.7$) with a median diameter of 0.92 mm and a maximum diameter, P_5, of 4.35 mm; P_5 is that diameter that has 5% of the distribution coarser than that size and 95% finer. Immediately above the base the sand is poorly sorted with a maximum diameter of 4.5 mm. However, the median diameter of this sand is slightly greater than the bottommost sample because it is skewed strongly to the coarse side. The sands decrease in maximum and median grain size and become better sorted upward; the upper sands also are less skewed in general. The topmost sample is a well-sorted, very fine sand with a median diameter of 0.120 mm.

In Shell Taylor 372 and 316 on the west and southwest, the same graded unit is thinner (12 to 10 ft or 4 to 3 m thick respectively) and is poorly sorted medium sand at the base. The westward and southwestward decrease in thickness and in median grain size is inferred to indicate a decrease in competence of the transporting current in these directions. A similar lateral variation of grain size is present in the graded sediments deposited by the turbidity currents of Lake Mead

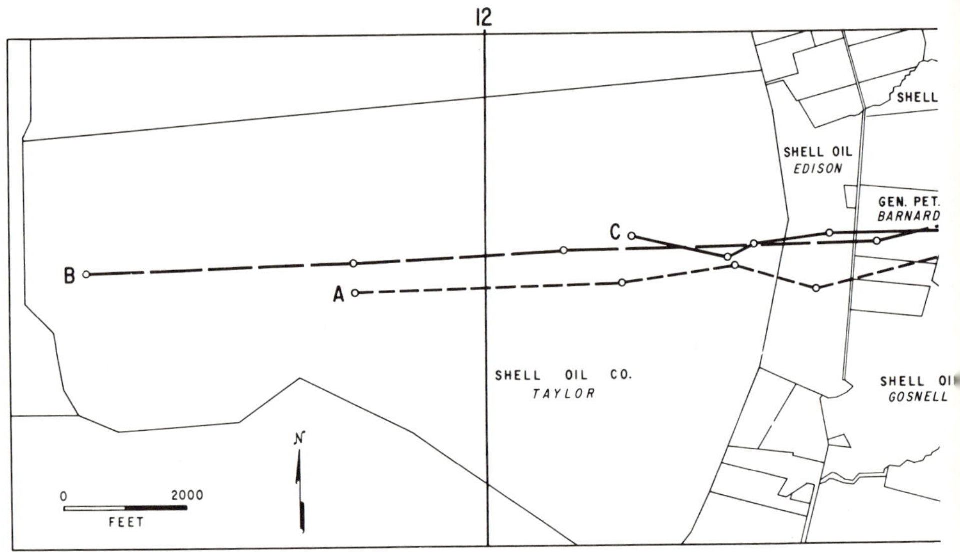

FIG. 3—Index map showing location of wells chosen for well-log correlation (Fig. 4) and structural cross section 12 (Fig. 14).

(Gould, 1951) and in ancient sediments (e.g., Enos, 1969).

A west-southwesterly direction of transport is suggested by the facies relations and by the lateral textural variation of the bottom graded bed in the DC-DE sequence. The sand axis crosses the axis of the Ventura anticline at a slight angle (Fig. 7).

Lithofacies in Lower Producing Zones

The facies relation of sands, silts, and clay in the interval between electric-log markers DR and FF_1 in zone 7 is illustrated by correlation section CC' in Figure 4. Four major sand units are recognized within the 250-ft (76 m) interval.

The correlatability of these sands varies considerably. The DR sand is developed best in the eastern part of the field in the Tidewater leases; it thins and silts out abruptly to the north and more gradually to the west. The best developed FA sand is in wells in the south-central part of the field and the FA sand also thins and silts out northward. The FC sand is developed best in the northeast part of the field, and thins toward the south and west. The FF sand (above FF_1 marker), like the DS sand, is developed best in the Tidewater leases and becomes very thin in the Shell Taylor lease. In general, better sand development within the interval DR-FF_1 is in the eastern part of the field in the Tidewater Associated Ventura Land and Water lease. The pattern is similar to that for the sediments of the upper and middle producing zones.

GRAIN ORIENTATION

Thirty-five lower Pliocene and Miocene samples from the Ventura field were analyzed for grain orientation. Cores initially were oriented using the dip directions of the folded strata, and later were checked by a magnetic method. The preferred grain orientation of all but a few samples can be detected by dielectric-anisotropy measurements (Nanz, 1960). As indicated by Figures 8 and 10, the directions of preferred grain orientation of most samples are easterly or northeasterly.

Carbonized wood fragments are very abundant in finely laminated silts and very fine sands near the top of a graded sequence. Silts and sands, when split along laminations, commonly show an alignment of wood fragments in the bedding plane (Fig. 9). Some coarser grained sandstones also show an alignment of carbonized wood fragments accompanied by a parallel alignment of large quartz grains. The direction of alignment of such carbonized wood fragments in 20 samples can be estimated visually, and the orientations so determined are generally parallel or subparallel with those determined by the dielectric-anisotropy method (Fig. 10). The thin-section method of determining grain orientation was used to analyze only one sample from zone 6 from a well in the Shell Frazer lease (Fig. 3). More than 500 grains were measured, and the preferred grain orientation of the sample is N 75° E (Fig. 10).

A close correlation among the three different methods used to determine grain orientation is

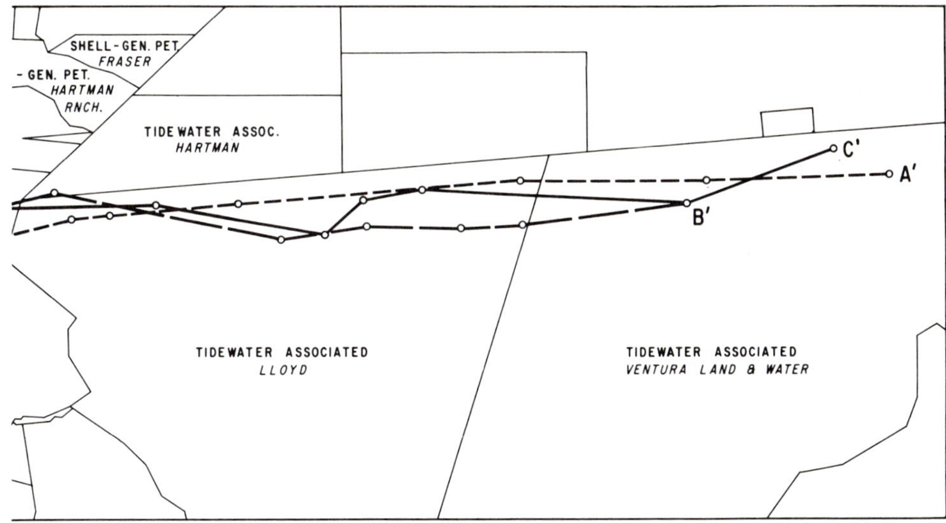

illustrated in Figure 10. If we assume that the preferred grain orientation is, in most cases, parallel with the sand trend (Nanz, 1960), the average or the predominant lower Pliocene sand trend is east-northeast. This then would support the conclusion based on sedimentary-facies studies.

SEDIMENTARY STRUCTURES

Cross-laminations are common in fine sands and silts near the top of several of the sequences. Many of the cross-laminae also are graded. The dip of the lamellae range is from 15 to 20° (see Fig. 11). Cross-laminations and other directional-current structures are observed easily in the field, and they have been mapped to indicate directions of sediment transport in the eastern part of the Ventura basin (Winterer and Durham, 1962). Mapping of transport direction based on subsurface sedimentary structures in the Ventura field is impracticable because of the relatively few core samples showing such structures. Only three core samples out of the several hundred examined had cross-laminations. The fact that all three observed cross-laminations dip to the west (Fig. 8) supports the westerly transport direction of the Ventura field sediments.

GENETIC INTERPRETATIONS

Paleoecologic studies have shown that the Ventura field was a site of deep-sea sand accumulations during early Pliocene time. Sand deposition was interrupted now and then by accumulation of muds (under normal deep-marine conditions) which form intercalations of shale in a sequence composed predominantly of sand. Individual sand layers range from a few inches to a few tens of feet thick. Graded bedding is common. Amalgamation of many graded beds forms apparently massive sand sequences more than 100 ft (30 m) thick, such as the AM-AO sand sequence in the upper Repetto and the DC-DF sequence in the middle Repetto. Correlation of individual graded beds, which are usually less than 30 ft (9 m) thick, is difficult. Thin sand units such as the AO_1 and the DA sands perhaps consist of only a few graded beds; isopach maps of those thinner sand units may approximate the configuration of an individual genetic unit.

The geometry of both the AO_1 and the DA sands suggests that (1) individual sand units were deposited by laterally restricted westerly flowing currents; (2) the currents decreased in competence westward and deposited thinner and finer sands; (3) muds and silts accumulated in areas not frequented by the westerly flowing currents.

The facies changes of sediments between the isopached units also were studied. The more rapid northward and southward thinning of many sands in contrast to the relatively less rapid westward thinning supports the idea of an approximately east-west sand trend. Gradual westward thinning and silting of some sands agree with the hypothesis of westerly flowing currents. Eastward thinning of some sands along the anticlinal crest such as $AP-AO_1$ and DC-DF may be only apparent; possibly trends of those units cross the struc-

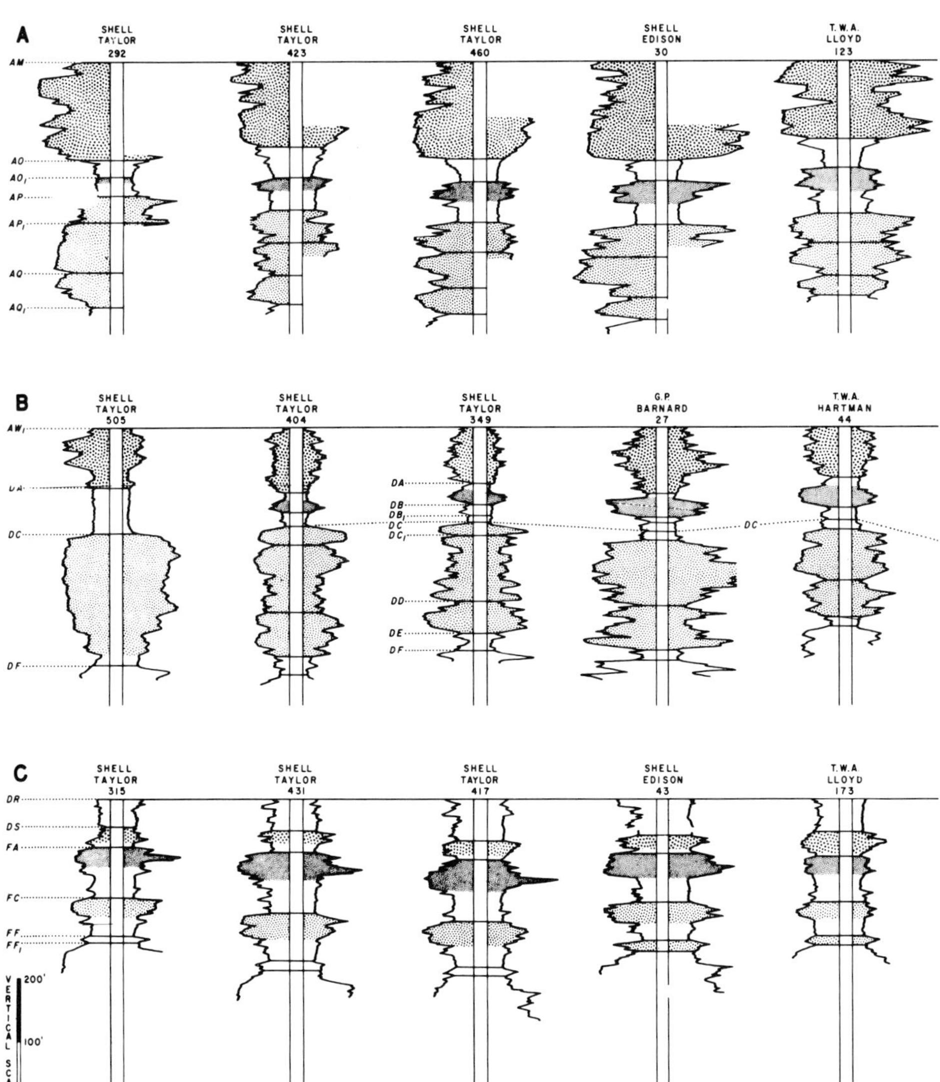

FIG. 4—Electric-log correlation of Repetto sediments along crest of Ventura anticline. Thicknesses are in feet. For locations see Fig. 3.

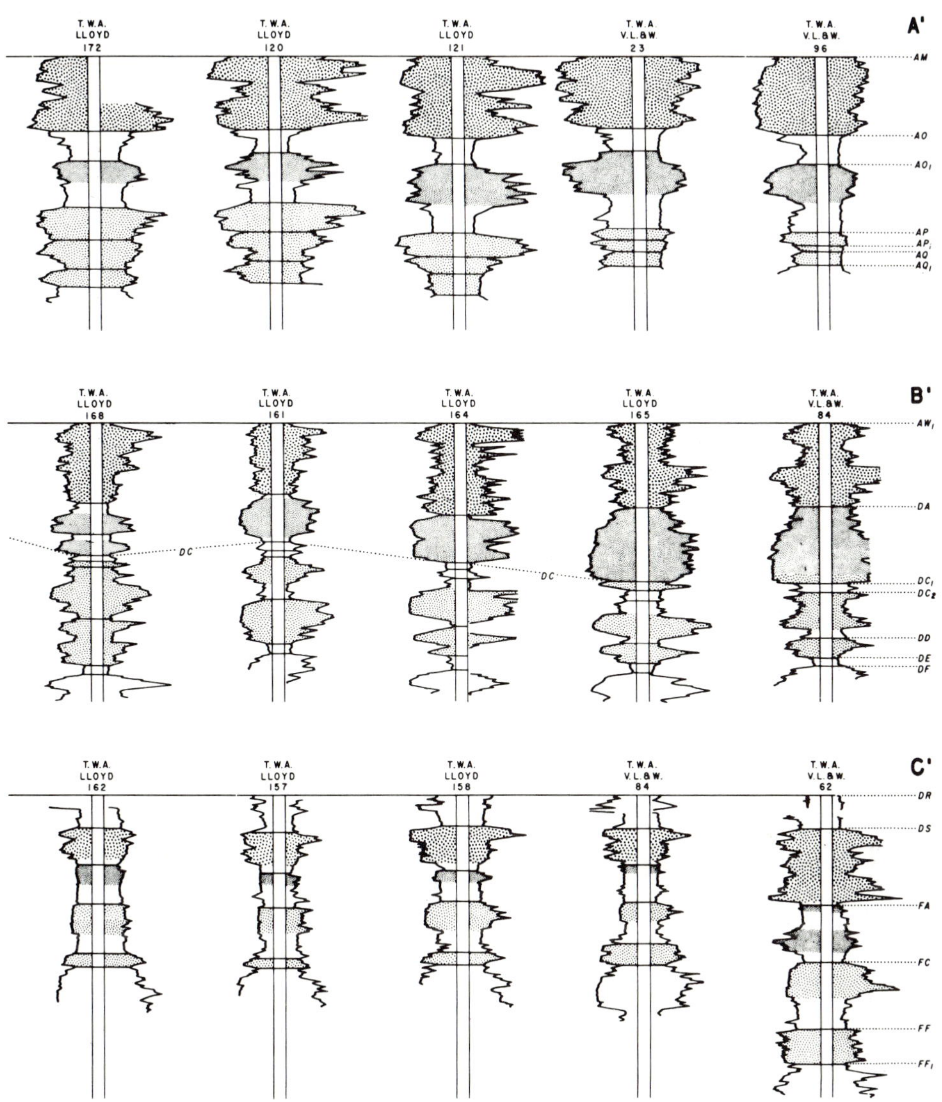

T.W.A. LLOYD 172
T.W.A. LLOYD 120
T.W.A. LLOYD 121
T.W.A. V.L.&W. 23
T.W.A. V.L.&W. 96
A'
AM
AO
AO₁
AP
AP₁
AQ
AQ₁
T.W.A. LLOYD 168
T.W.A. LLOYD 161
T.W.A. LLOYD 164
T.W.A. LLOYD 165
T.W.A. V.L.&W. 84
B'
AW₁
DA
DC
DC₁
DC₂
DD
DE
DF
T.W.A. LLOYD 162
T.W.A. LLOYD 157
T.W.A. LLOYD 158
T.W.A. V.L.&W. 84
T.W.A. V.L.&W. 62
C'
DR
DS
FA
FC
FF
FF₁

Kenneth J. Hsü

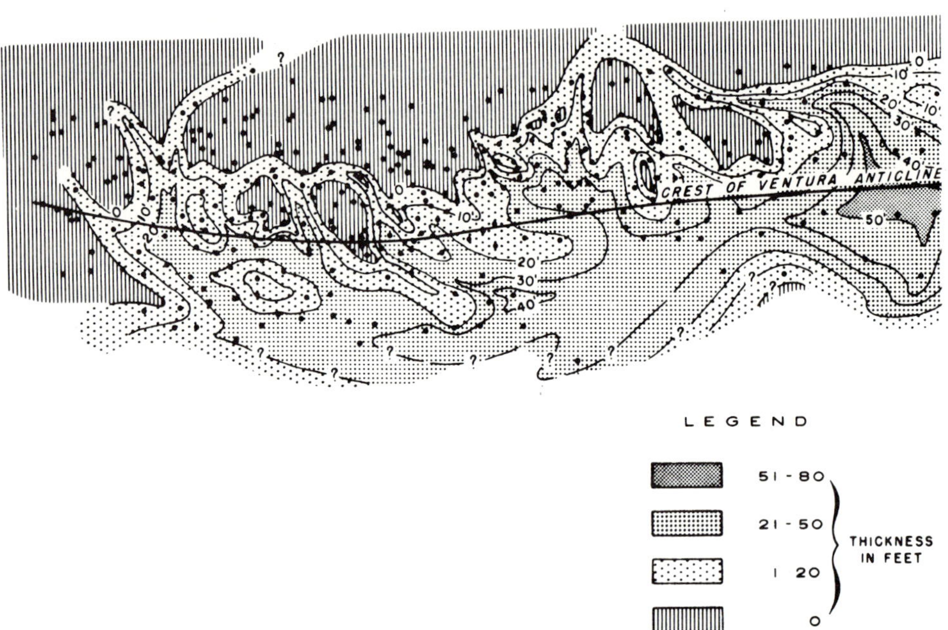

FIG. 5—Isopach map of AO₁ sand in upper producing zone.

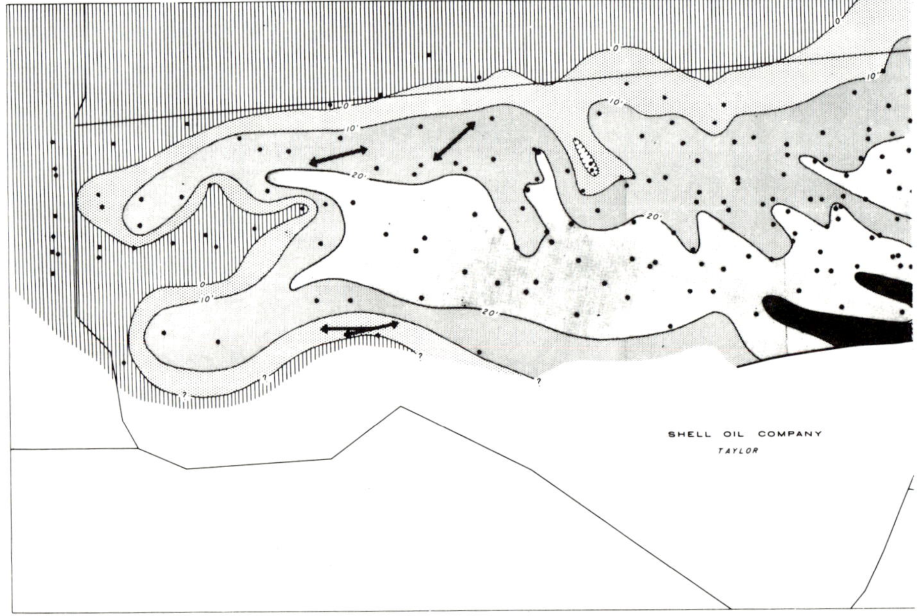

FIG. 6—Isopach map of DA sand in middle producing zone.

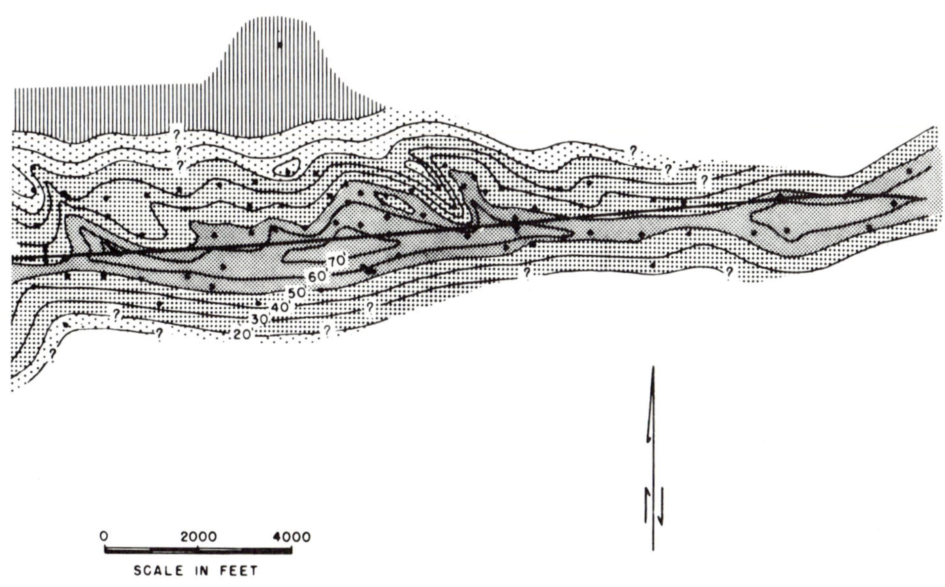
70
60'
50
40'
30
20
0 2000 4000
SCALE IN FEET
N

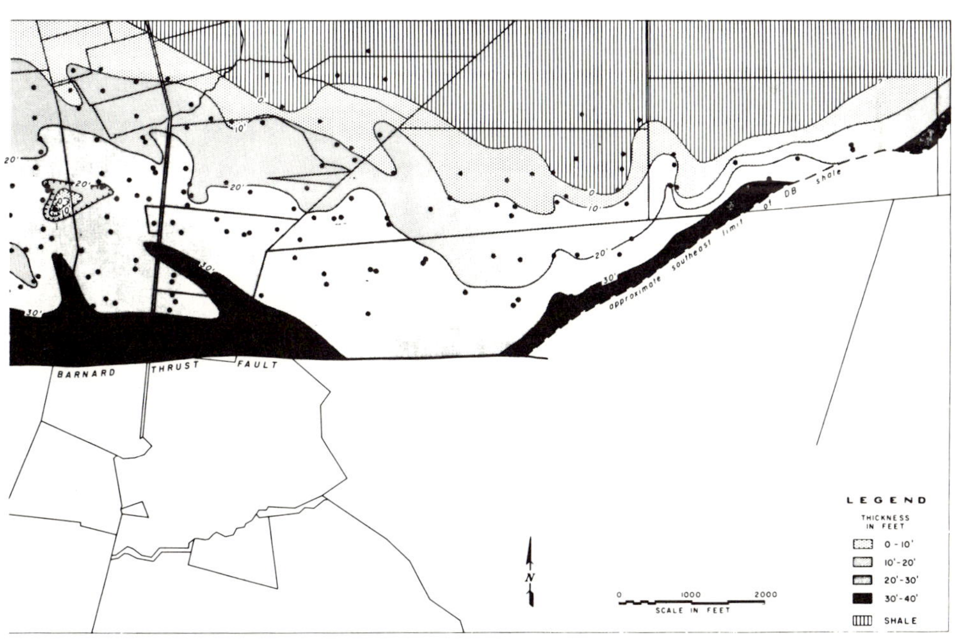
20'
20'
20'
10'
0
10'
0
10'
20'
30'
approximate southeast limit of DB shale
BARNARD THRUST FAULT
N
LEGEND
THICKNESS
IN FEET
0 - 10'
10' - 20'
20' - 30'
30' - 40'
SHALE
0 1000 2000
SCALE IN FEET

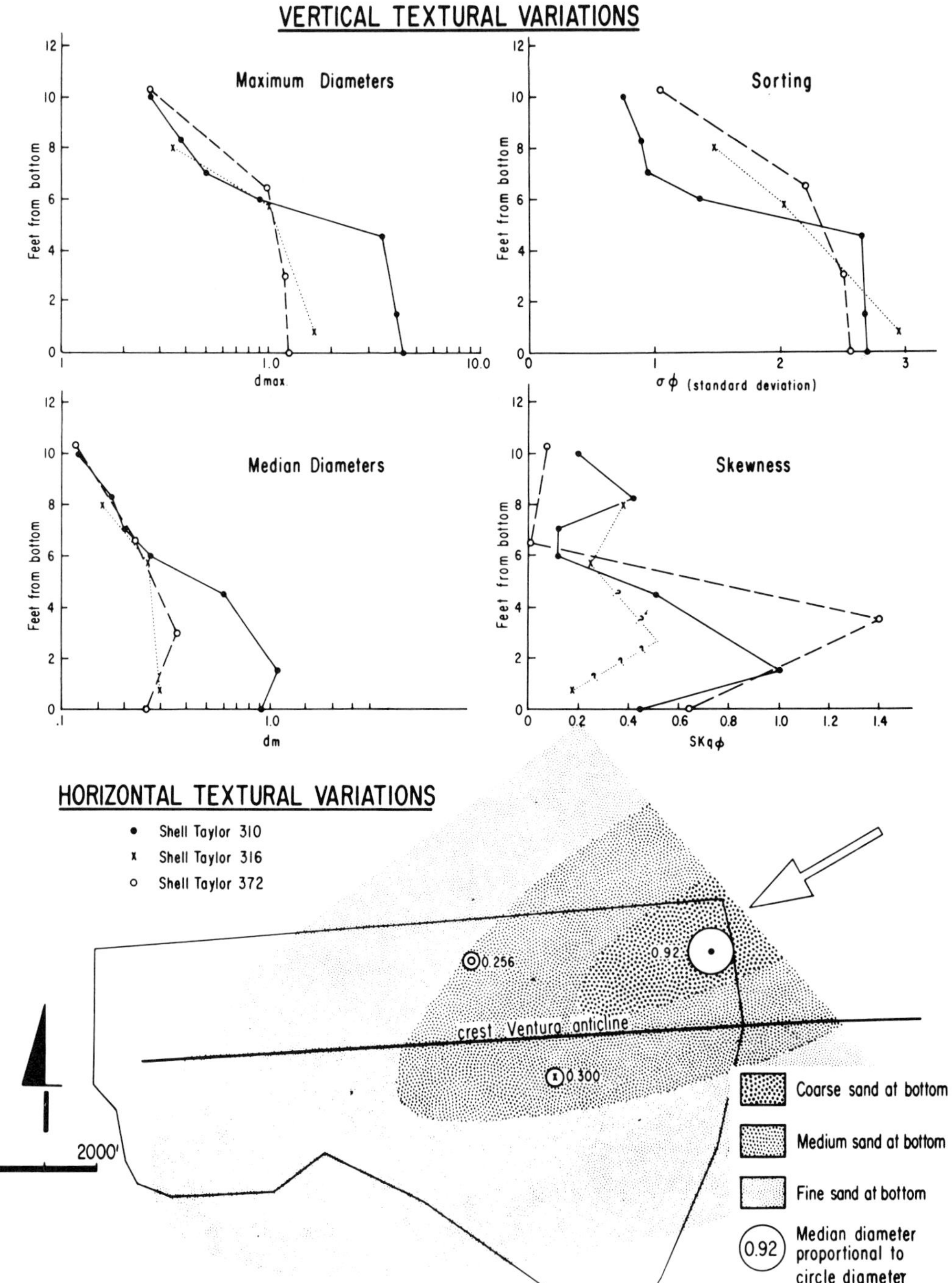

FIG. 7—Vertical and lateral textural variation of graded bed in middle producing zone.

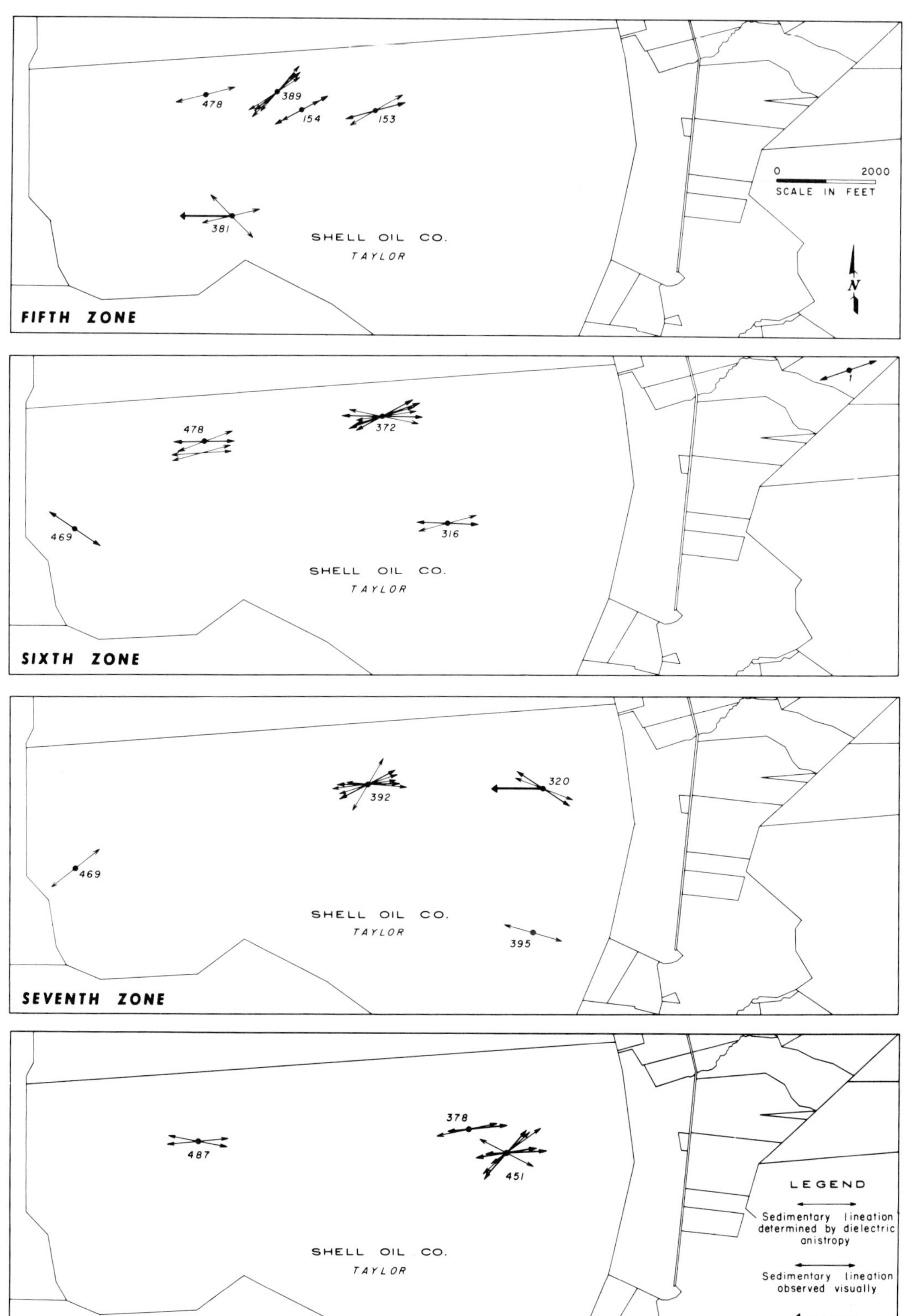

FIG. 8—Preferred grain orientation and cross-bedding direction of Repetto sediments, core samples from zones 5, 6, 7, and 8. See Figure 2 for stratigraphic position of zones.

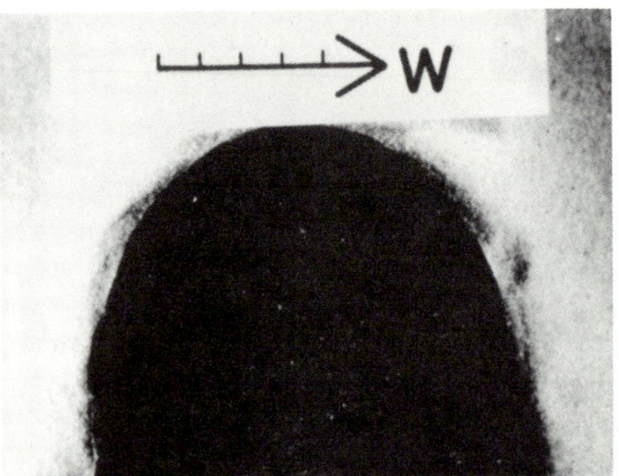

FIG. 9—Alignment of carbonized wood fragments, viewing down bedding planes. Cores oriented by their steep dips.

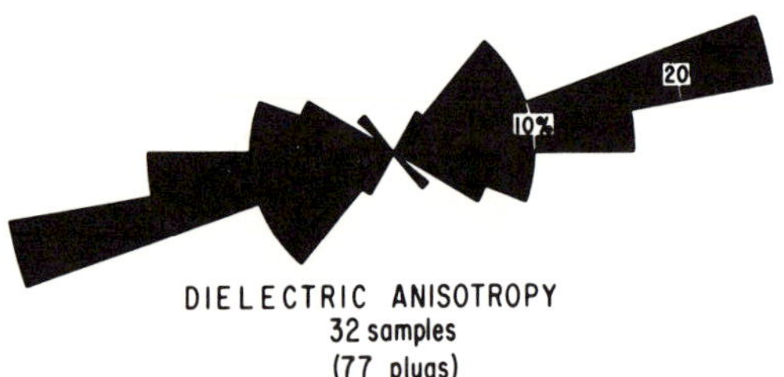

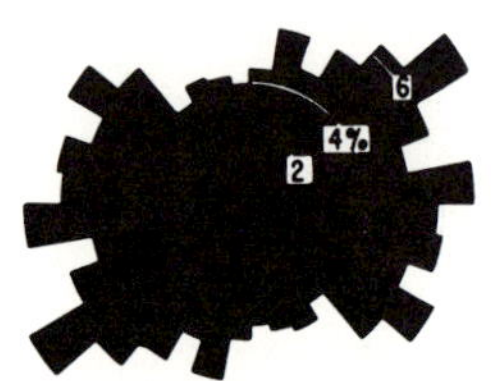

FIG. 10—Comparison of results from three different methods of evaluation of preferred grain orientation.

tural axis at an angle. However, the fact remains that there is little change of the thickness of Pliocene sand reservoirs from east to west of the field; some sand beds are thicker on the east, others on the west. Apparently much of the detritus was carried down by currents that flowed down the submarine canyons on the sides of the basin before they were deflected into a westerly transport direction; not all sands were derived by longitudinal transport from the eastern end.

The concept of linear sand-transporting currents flowing westward into deeper parts of the Pliocene basin may help to explain the following observations:

1. Sand percentages are higher in the central part of ancient elongate deep-water basins of California (such as the Los Angeles and Ventura basins).

2. Many sands in the Ventura field are not present in outcrop sections of equivalent age north of the field.

3. Lower Pliocene sediments of the Rincon field on the west are less sandy than sediments of equivalent age at Ventura.

4. Abrupt facies changes occur locally in the Ventura basin.

The laterally restricted sand-depositing currents were thought to have been turbidity currents. The hypothesis had been put forward by Natland and Kuenen (1951) and subsequently was reinforced by several other sedimentologic studies (e.g., Winterer and Durham, 1962; Crowell et al, 1966). The Ventura basin has become the classic "type locality," and a mecca of turbiditophiles. It is no longer necessary now, 20 years after the conclusion of this work, to belabor the evidence for turbidite origin of the sands from the Ventura field.

The sedimentary facies of the Ventura field sediments suggests a control of sand deposition by the topographic relief of the basin floor. As shown in Figure 4, the site of best sand development during AP-AQ interval was also the site where the AO_1 sand was not deposited; and the site of best sand development during the DC-DE interval was also the site where DA sand was not deposited. Such facies relation is attributed to deposition by turbidity currents. Sands carried down by turbidity currents were guided by gravity down the steep slopes of submarine canyons to fill topographic depressions of a deep-sea basin. Sedimentation, in general, tends to mask the irregularities of the submarine relief, but deposition by laterally restricted turbidity currents on a level basin floor should form levees, leaving depressions between levees as the site for subsequent deposition by density currents. Thus, areas of thick sand accumulation at one time should be succeeded by silt and clay accumulations.

The detailed three-dimensional configuration of sand deposits in Holocene deep-sea basins is not well known. An example of the studies is the one by Ludwick (1950) who investigated the deep-water sands off San Diego, California. He noted that sands in the La Jolla Submarine Canyon are concentrated in the axial part and at the mouth, where a submarine fan was formed. A reconstructed map based on Ludwick's data (Fig. 12) shows the sand-shale distribution. The map suggests that a lenticular sand deposit now is accumulating in the central and deep parts of the San Diego Trough with a sand trend parallel with the elongate axis of the trough. Sands notably are absent along the base of the Coronado Bank on the flank of the San Diego Trough. A more detailed study by von Rad (1968) confirmed Ludwick's reconnaissance. This pattern of Holocene sand distribution in a deep-sea trough is remarkably similar to the early Pliocene pattern in the Ventura basin.

The conclusions based on sedimentary facies relation are in accord with grain-orientation measurements and the meager sedimentary-structure data. Sand trends and preferred grain orientation are in general parallel, both trending east or east-northeast.

Compared with the Holocene sand-distribution pattern in San Diego Trough, the maximum sand development in Ventura field is not surprising, because the site of the field was the central or deeper part of a Pliocene depositional basin as suggested by paleoecology studies (Winterer and Durham, 1962). It was noted further that the coarse clastic deposits of the Ventura field are mainly sands. Few conglomerates or pebbly mudstones were found in the thousands of feet of cores examined by the writer. Cobbles and gravels obviously were left behind in the submarine canyons and were deposited on submarine fans. Only sands were transported by turbidity currents to the basin center, where they formed potential oil reservoirs.

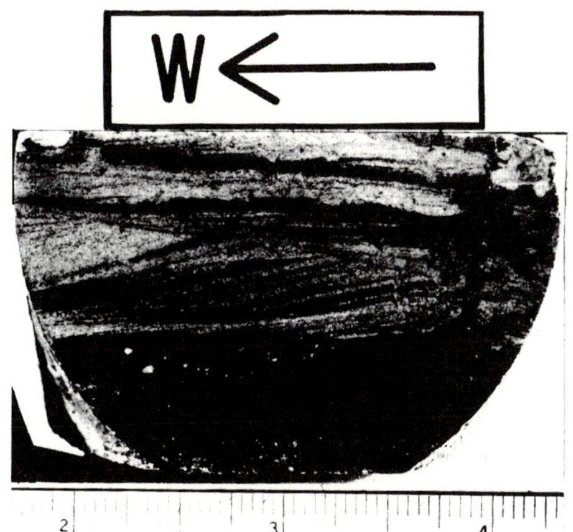

FIG. 11—Cross-lamination of sandy silt.

The axis of maximum sand development is approximately coincident with the axis of the Ventura anticline. The sand geometry provides an explanation. It is difficult for connate water to escape a lenticular body of sand deposited in the central part of a deep-sea basin and flanked on both sides by shales. It, therefore, was not surprising that deeply buried and highly compacted sands of the Ventura field are subjected to high pore pressure; the pressure is almost lithostatic in deeper zones (Watts, 1948). It now is well established that zones of high pore pressure are also zones of very low strength (Hubbert and Rubey, 1959). When the Ventura basin was being compressed, the sands with high pore pressure in the central part of the basin should be the logical loci for failure and for fracture. The placement of a faulted anticline developed at the site of maximum sand development contributes thus to the high productivity of the Ventura field.

APPLICATION TO OIL EXPLORATION

The submarine topography of the late Tertiary seas of the Pacific Coast region was probably similar to that off the southern California coast today. The "continental borderland" was dissected by fault systems forming a basin-and-range structure. The down-faulted blocks formed elongate deep-sea basins. The upthrown blocks formed banks and islands. Shelf deposits accumulated along narrow shelves or banks nearshore. Most of the sediments, however, mainly were carried down submarine canyons by turbidity currents, or gradually filled slowly subsiding basins. "Escarpments" or "slopes" separated deep-sea basins from shelves (or banks). Turbidity currents

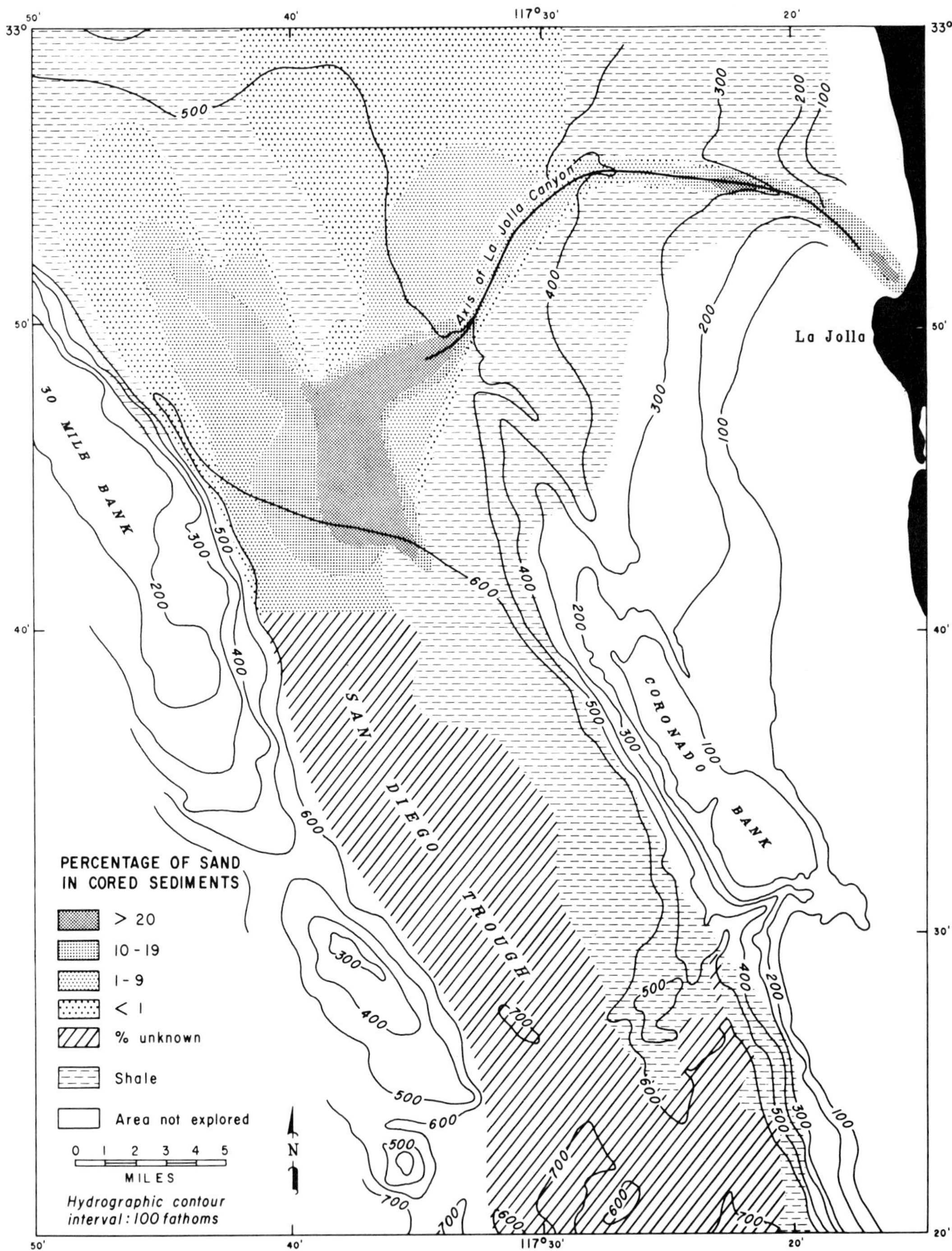

FIG. 12—Distribution of Holocene sand in San Diego Trough, offshore California.

probably were channelized by submarine canyons. Between canyons and on the slopes bypassed by turbidity currents, hemipelagic sediments settled. This pattern of sedimentation would lead to a preferential accumulation of potential reservoir sands in the central parts of deep-water basins. Paleoecologic studies of producing intervals of oil fields from the Ventura, Los Angeles, and San Joaquin basins have shown that many reservoir beds are indeed deep-water turbidites.

The sand distribution pattern in a deep-sea basin is exemplified by that of the Pliocene sediments in Ventura basin. Sands carried down the submarine canyons filled in topographic depressions and accumulated to greatest thickness in areas of maximum subsidence. Optimum sand development in central parts of ancient deep-sea basins thus can be expected, and exploration for sand reservoirs might be directed thereto if ecologic and sedimentologic data should suggest deep-sea sedimentation. However, it must be pointed out that the nature of the sedimentary load carried by a turbidity current depends on the source materials and the velocity of the current; sediments deposited by turbidity currents in many areas are not sandy. Similar situations may be present in some offshore deep-sea basins where sand deposition is limited to the submarine-fan areas near the mouth of submarine canyons, while clays and sills were deposited by turbidity currents in the central parts of these basins.

The approximate coincidence of maximum sand development with the axis of the Ventura anticline may be explained by a hypothesis that the folding was controlled by sediment distribution. It has been shown that sediment distribution in a deep-sea basin is determined by the configuration of the basin which perhaps is delimited by faulting prior to and contemporaneous with sedimentation. The Ventura basin during the Pliocene epoch was perhaps an elongate deep-sea basin oriented east-northeast. Subsidence of the basin perhaps took place along east-northeast-trending faults. Sediments which accumulated in the basin were lenticular bodies trending parallel with the axis of the trough and flanked on both sides by shales. When the region was subjected to compression during the middle Pleistocene, an anticline was developed along the sand axis which was apparently a zone of weakness. Oil producing structures in the Los Angeles basin may have undergone similar genesis.

The parallelism of sedimentary trend and regional tectonic trend suggests the possibility of stratigraphic types of oil traps on the flank of a fold where lenticular sand bodies pinchout updip. Examples of stratigraphic pinchouts in shallow-water sediments are well known in San Joaquin basin, such as the Pliocene pool of the Sunset-Midway field (Hoots et al, 1954). The present investigation suggests that similar stratigraphic traps might be expected in sediments deposited in a deep-water environment.

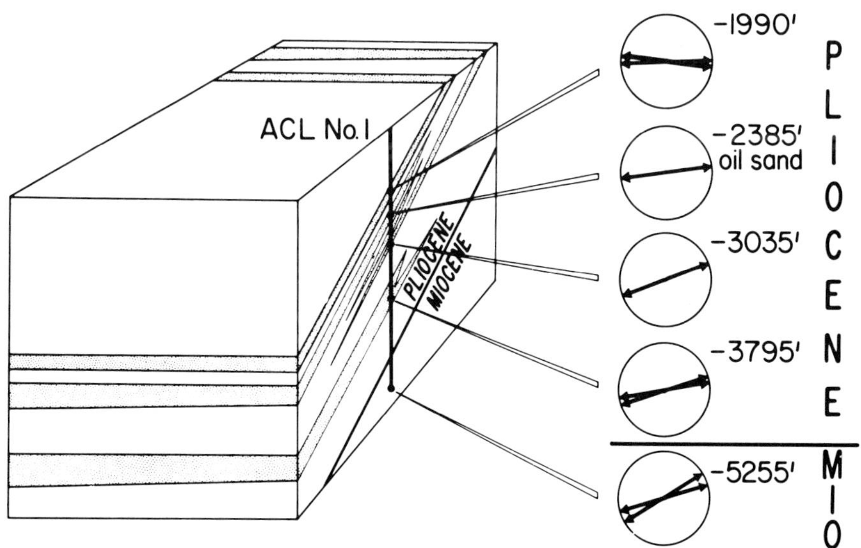

FIG. 13—Generalized stratigraphic relations and preferred grain orientations of Pliocene and Miocene sediments in Weldon Canyon area. Depths are in feet below sea level.

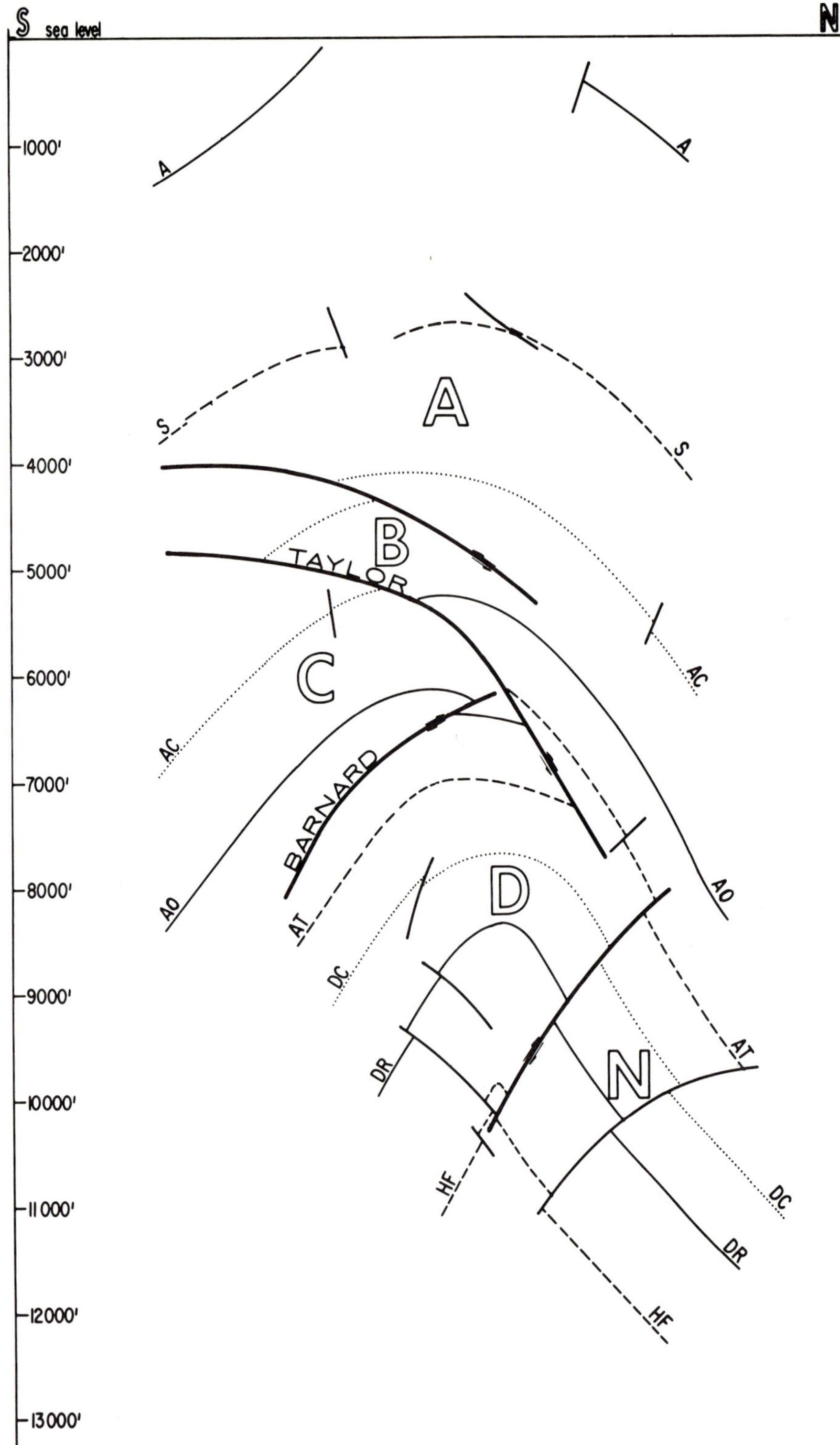

FIG. 14—Structural cross section 12 of Ventura field; *A, B, C, D,* and *N* refer to producing blocks. For location see Figure 3. Horizontal and vertical scales (in feet) are same.

The deep-sea basin sands form elongate, lenticular bodies, trending parallel with the axis of the basin. They are restricted laterally in extent. Small sand bodies pinchout laterally within the basin. Large sand bodies, on the other hand, may extend laterally to butt against a steep slope or escarpment on the border of a basin. Thus, many stratigraphic pinchouts are to be expected in the narrow transition zone where the density-current deposits of the basin facies grade into the pelagic sediments of the slope facies. Such pinchouts are ideal stratigraphic traps after the slope facies have been faulted or folded over the basin facies. The late Pliocene pool of the Saticoy field of Ventura basin is such a faulted stratigraphic trap. The possibility that an escarpment or steep slope may have existed along the Oak Ridge fault during much of Pliocene time suggests that more stratigraphic traps are to be found below the present producing zones. Small oil fields also have been discovered north of Ventura on the north flank of the Ventura syncline, but trapping factors for these oil accumulations never have been ascertained (Waterfall, 1943). An updip stratigraphic pinchout is present also in the Weldon Canyon area (Figs. 1, 13).

APPLICATIONS TO PRODUCTION GEOLOGY
Development of Field

The discovery in Weldon Canyon area (Fig. 1) is thought to be an updip pinchout of a lenticular body (Fig. 13). Ten core samples from the discovery well, Bell & Burton ACL 1, were examined for grain orientation after orientation by the southerly dips of the folded strata. The grain-orientation results suggest that the lower Pliocene sands in Weldon Canyon also have an approximately east-northeast trend parallel with the sand trend of Ventura field (Fig. 16). The successful stepout wells, from analysis of facies and directional features, were located along strike.

Planning of Secondary Recovery

A knowledge of lithofacies relation commonly is necessary when secondary recovery is being planned for a field. Although the present work was not undertaken in direct connection with a secondary recovery program, the results of the Repetto lithofacies studies should be of value in secondary-recovery efforts at Ventura field.

Determination of Fault Displacement by Isopach

The Pliocene sediments of the Ventura field were folded into an anticline during the middle Pleistocene orogeny. The anticline has been sliced into many blocks by curved thrust faults. Three major fault blocks separated by two major thrust faults are recognized: the B block is thrust over the C block along the north-dipping Taylor thrust, and the D block is thrust under the C block along the south-dipping Barnard fault (Fig. 14). Neville (1952) summarized the geologic structure of the Ventura field and noted that the Taylor thrust shows a maximum vertical stratigraphic repetition of 1,000 ft (305 m) and the Barnard fault shows a maximum of 1,300 ft (396 m) of vertical repetition. Dip-slip components of these faults could be estimated and the strike-slip component of displacement could not be determined.

The direction and net slip of the movement along the two major faults of Ventura field can be estimated by comparing isopach patterns in separate fault blocks. Isopach maps of the AO_1 sand in B, C, and D blocks are shown by Figures 15-17. The AO_1 sand shows a braided pattern. The braided sand-depositing distributaries have been offset by faulting. A matching of offset distributaries gives a measure of relative displacement between the three major fault blocks. The distribution pattern of the AO_1 sand before faulting is portrayed by Figure 18. The map shows that the B block has been overthrust southeastward along the north-dipping Taylor thrust and the D block has been underthrust eastward along the south-dipping Barnard fault. In other words, both the Taylor and Barnard faults are thrust faults with considerable right-lateral movement.

The horizontal projections of the net-slip directions and displacements along the Taylor and Barnard faults are indicated on Figure 19. The direction of the overthrust movement along the Taylor thrust is southeasterly, and the displacement increases gradually westward to a maximum in the west-central part of the Shell Taylor lease. The direction of the underthrust movement along the Barnard fault is easterly to southeasterly and the displacement increases gradually eastward. The axis of the Ventura anticline also has been offset by the Taylor and Barnard faults. Restored to their positions before faulting, as indicated in Figure 18, the anticline was originally continuous.

As both the Taylor and the Barnard faults strike approximately east-west, the throws and heaves and apparent displacement can be estimated from north-south cross sections. In the western part of the Shell Taylor lease, the section of the Taylor thrust which offsets the AP marker has a throw (de) of 250 ft (76 m), a heave (cd) of 1,400 ft (427 m), and a stratigraphic separation (ab) of about 950 ft (290 m) in Shell Taylor 136 (Fig. 20). As the strike of the strata approximately parallels the strike of the fault, the dip slip is approximately equal to the 1,450 ft (442 m) of apparent displacement (ce) measured from the cross

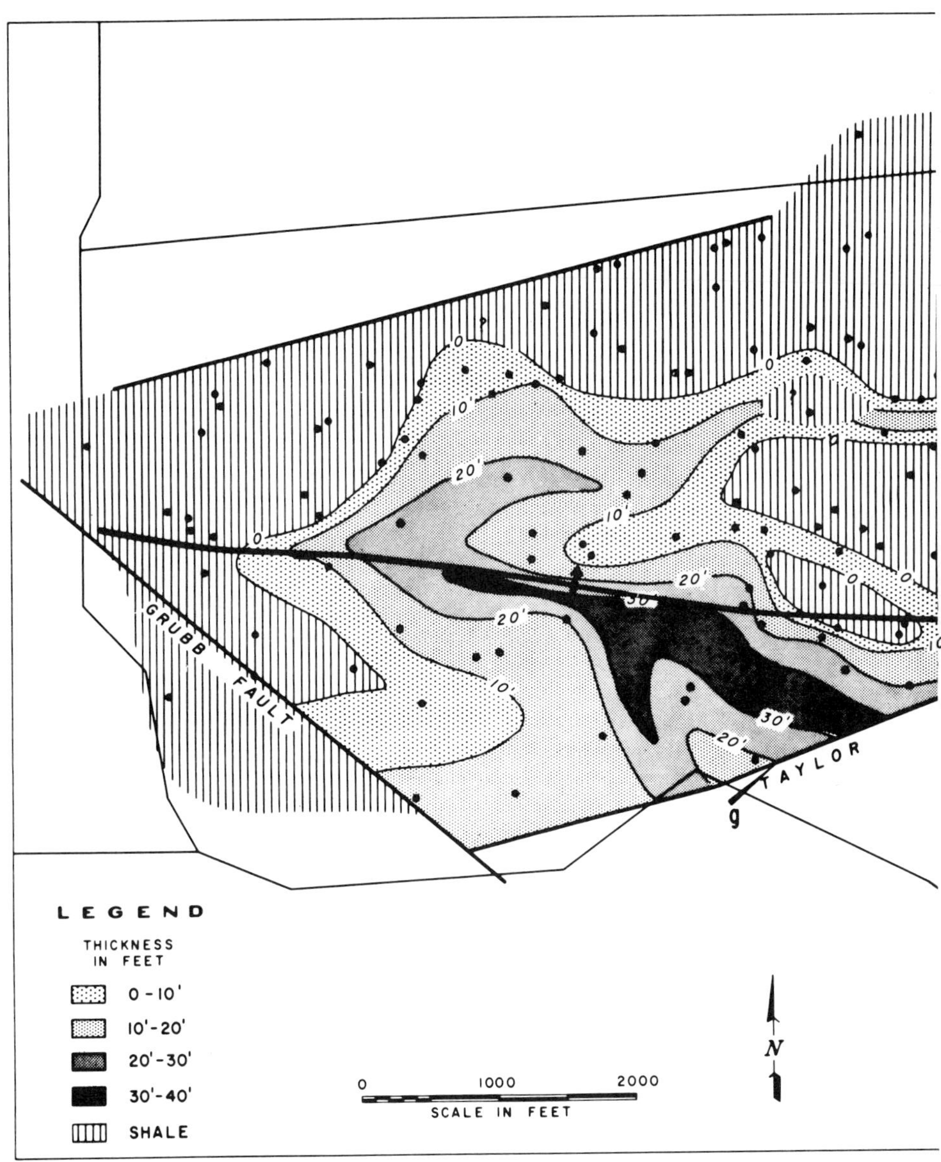

FIG. 15—Isopach map of AO_1 sand in Shell Taylor lease, B block.

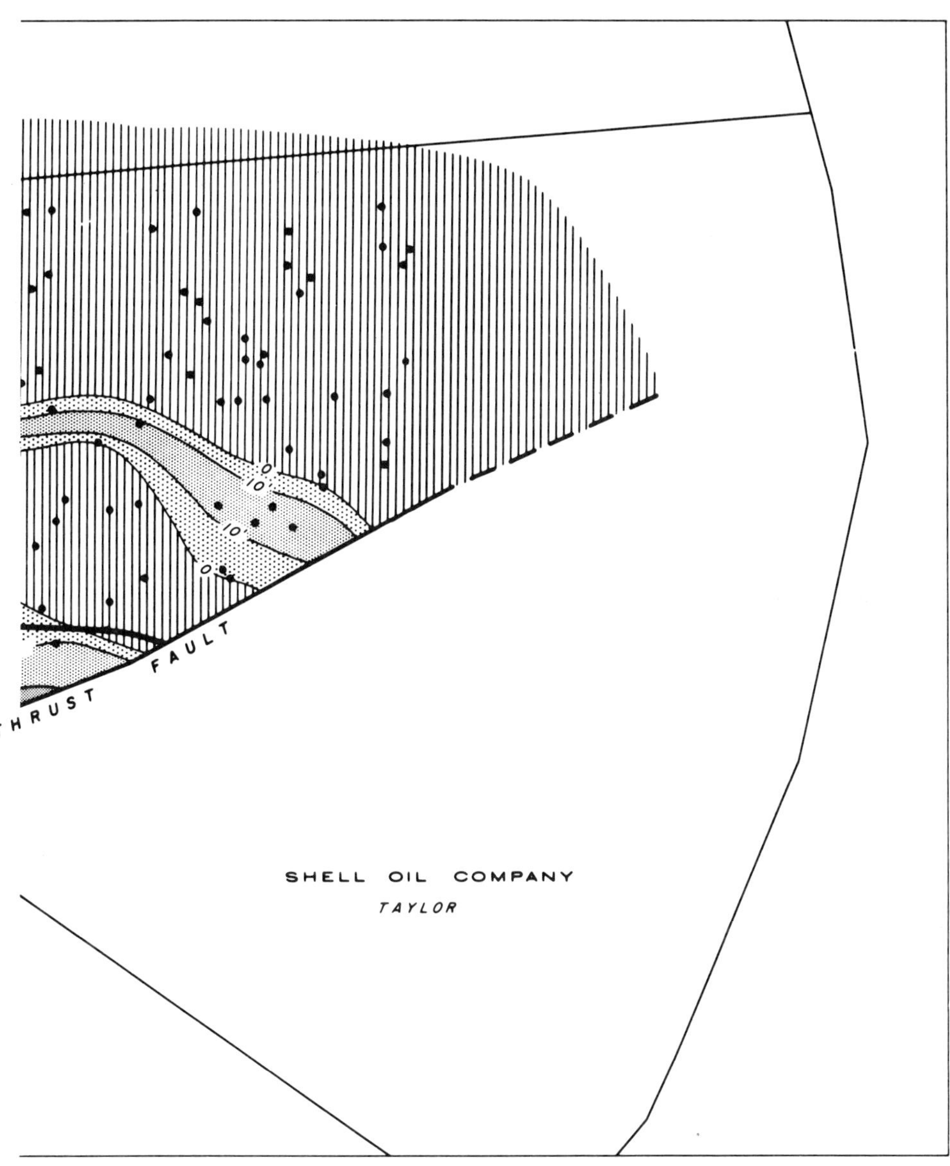
THRUST FAULT
SHELL OIL COMPANY
TAYLOR

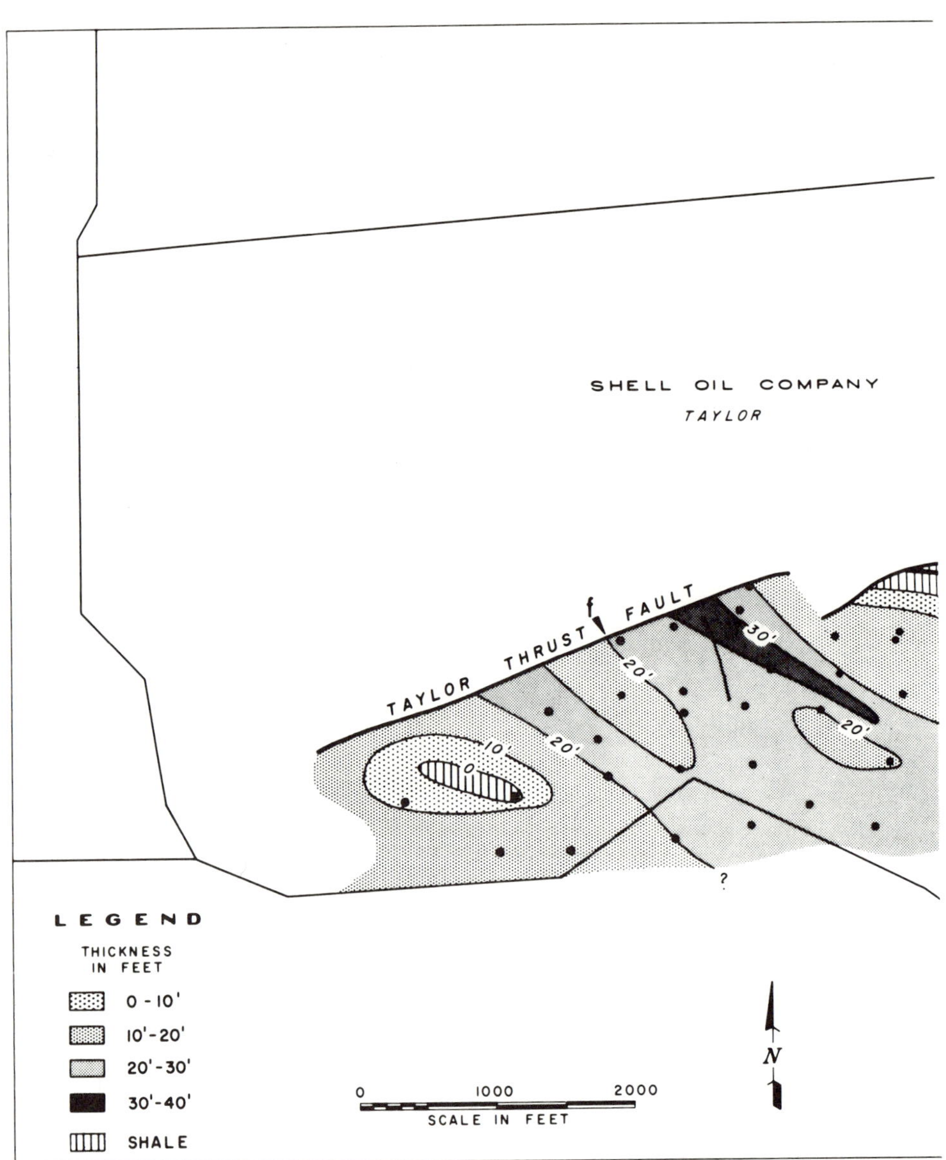

FIG. 16—Isopach map of AO_1 sand in Shell Taylor lease, C block.

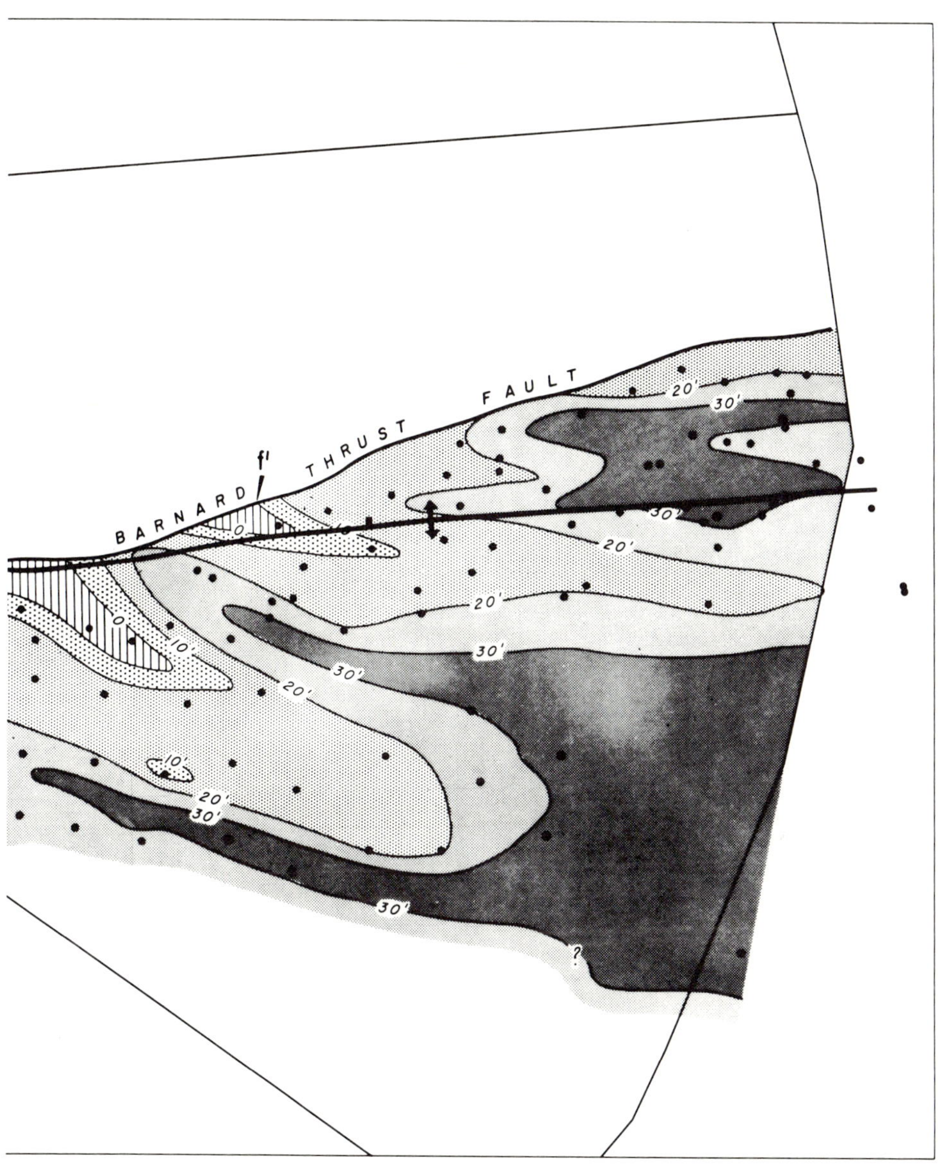

BARNARD THRUST FAULT
f'
0
0
10'
20'
20'
30'
30'
30'
20'
20'
30'
10'
20'
30'
30'
30'
?

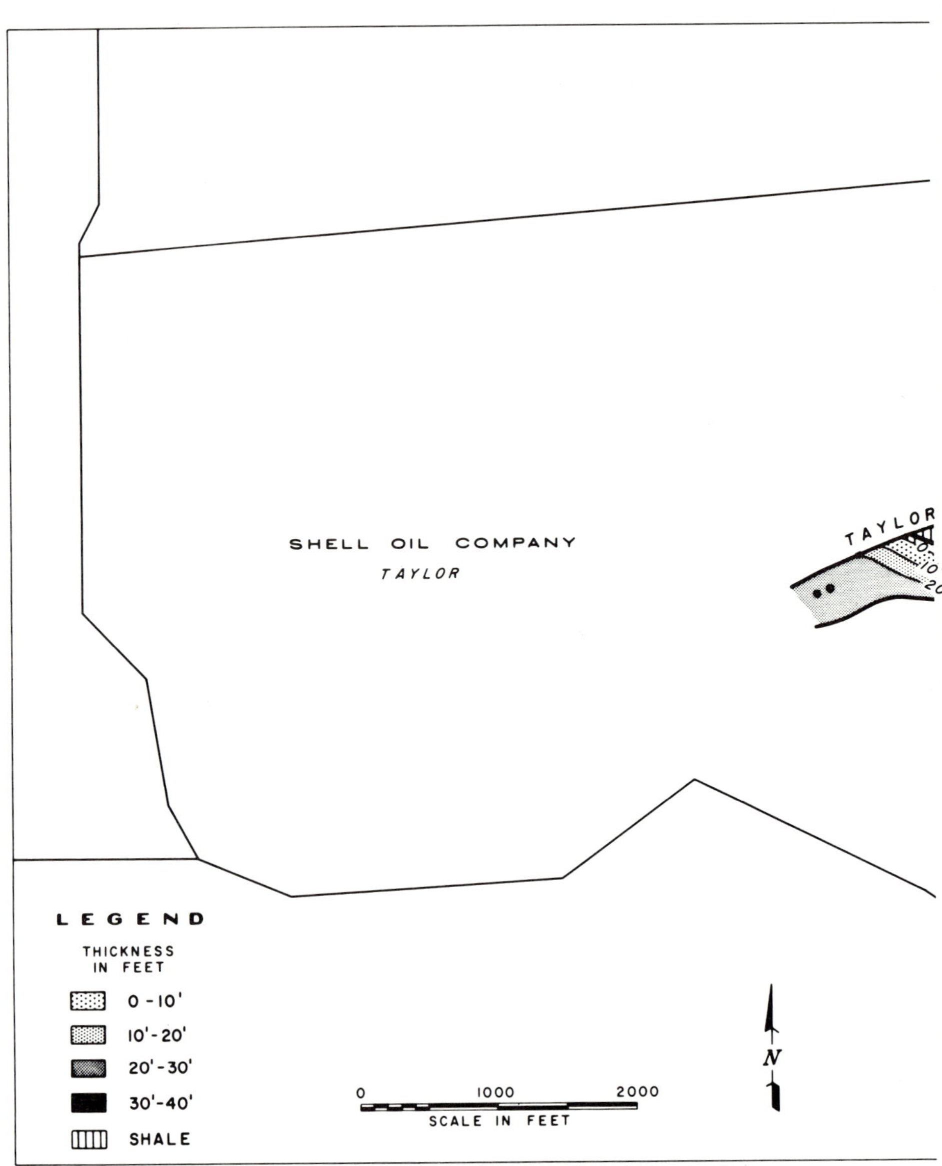

FIG. 17—Isopach map of AO_1 sand in Shell Taylor lease, D block.

507

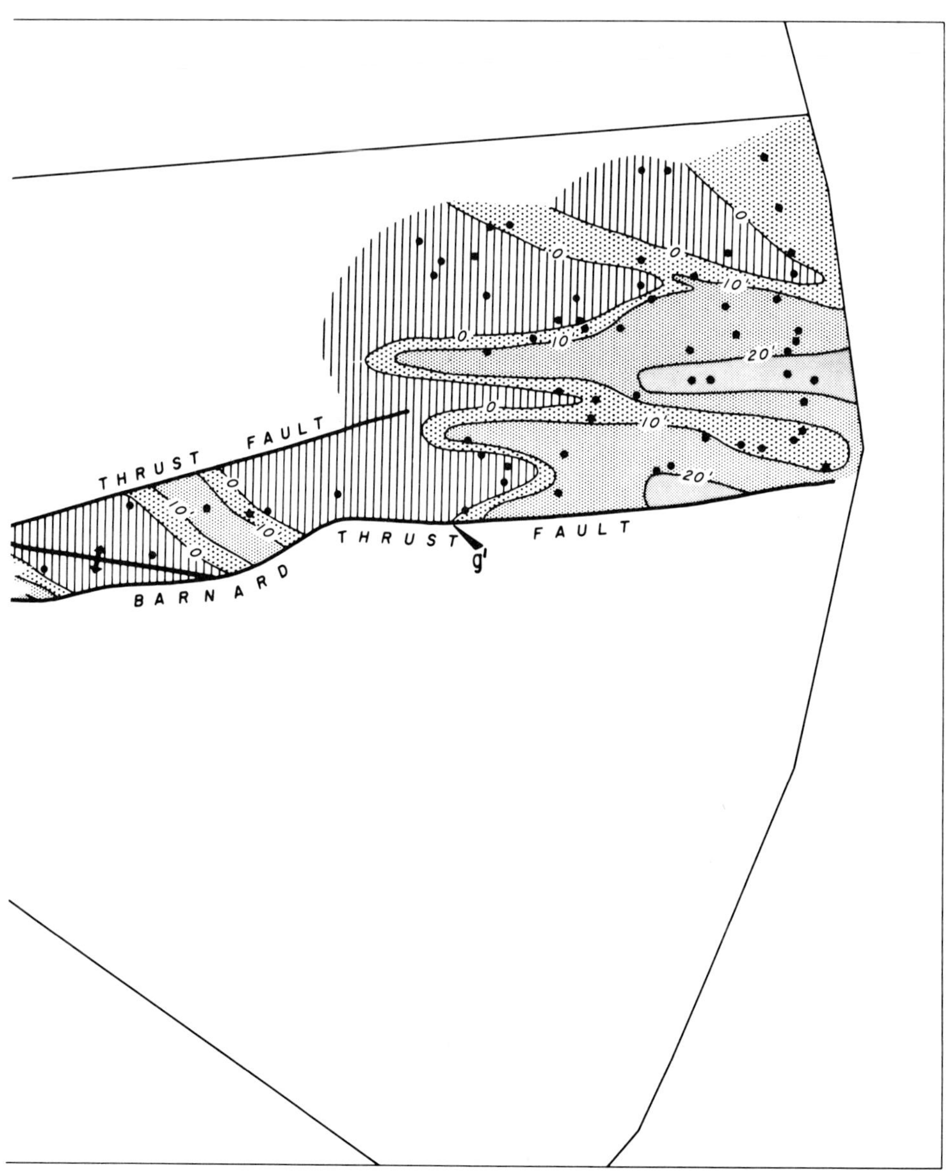

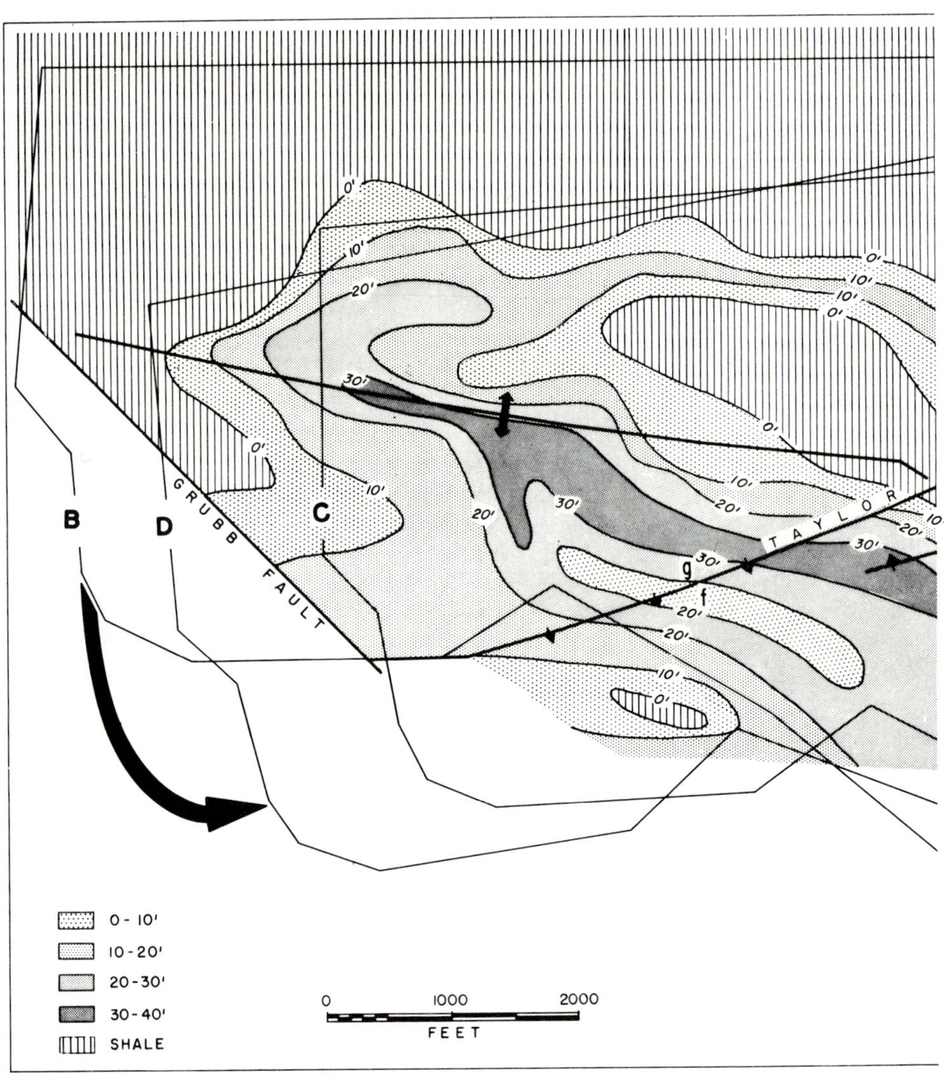

FIG. 18—Isopach map depicting configuration of AO$_1$ sand in Shell Taylor lease before faulting.

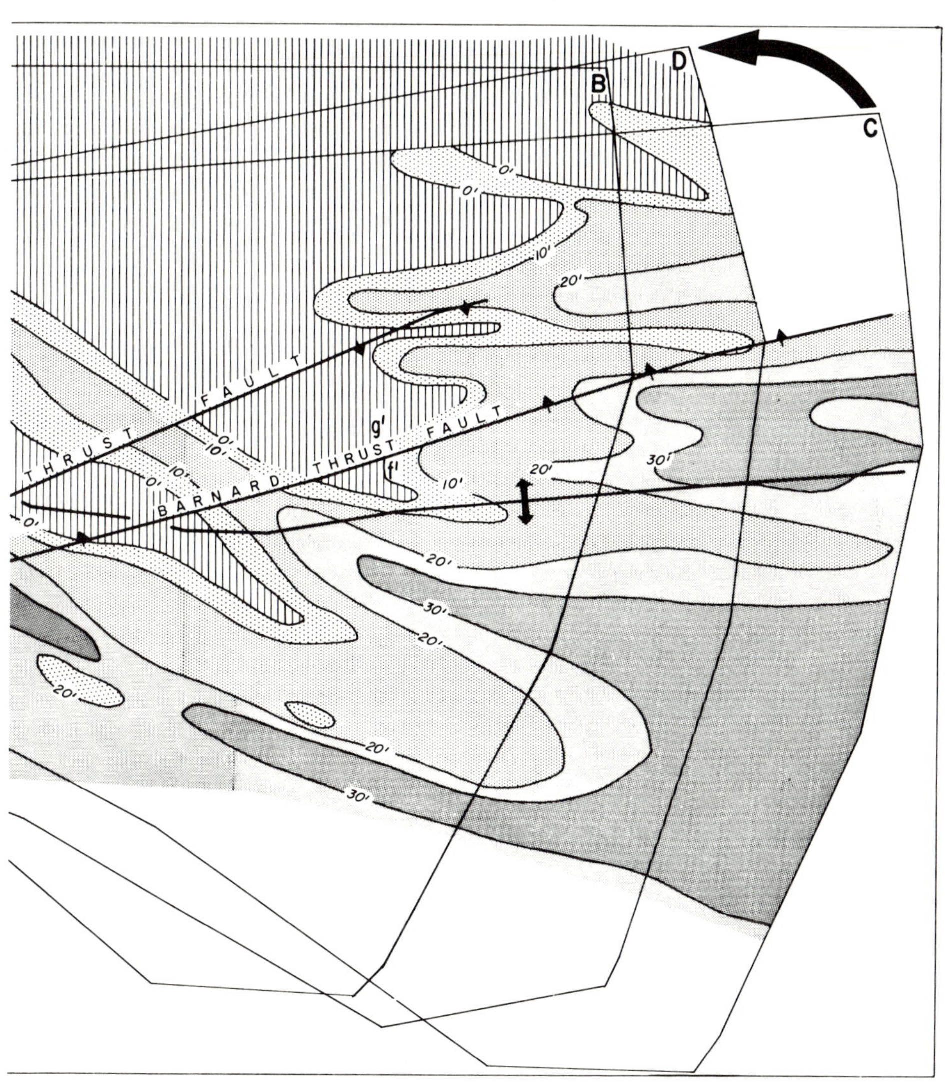
D
B
C
0'
0'
10'
20'
THRUST FAULT
THRUST
g'
f'
BARNARD THRUST FAULT
0'
10'
10'
0'
0'
10'
20'
30'
20'
20'
30'
20'
20'
30'
20'

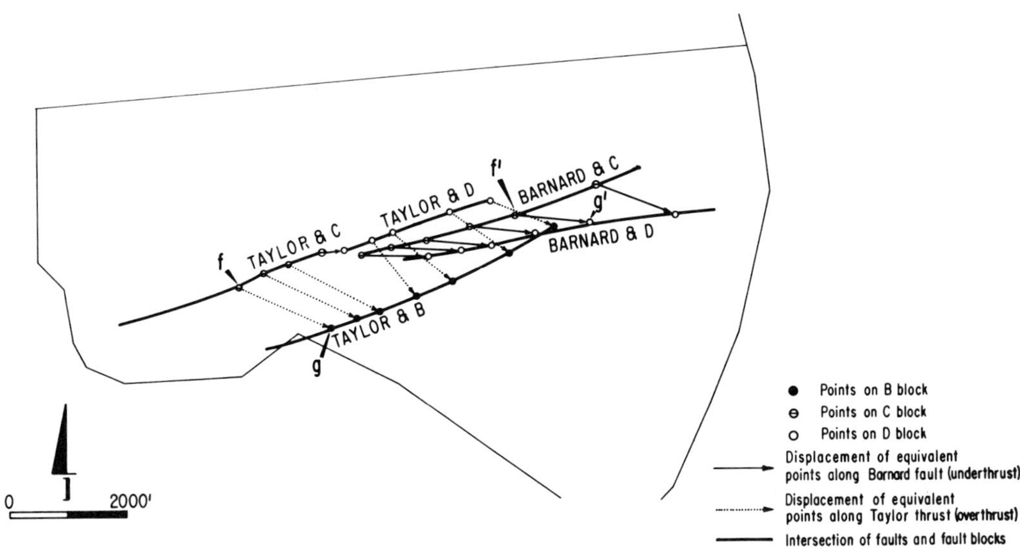

FIG. 19—Horizontal projection of net-slip movements along Taylor thrust and Barnard fault.

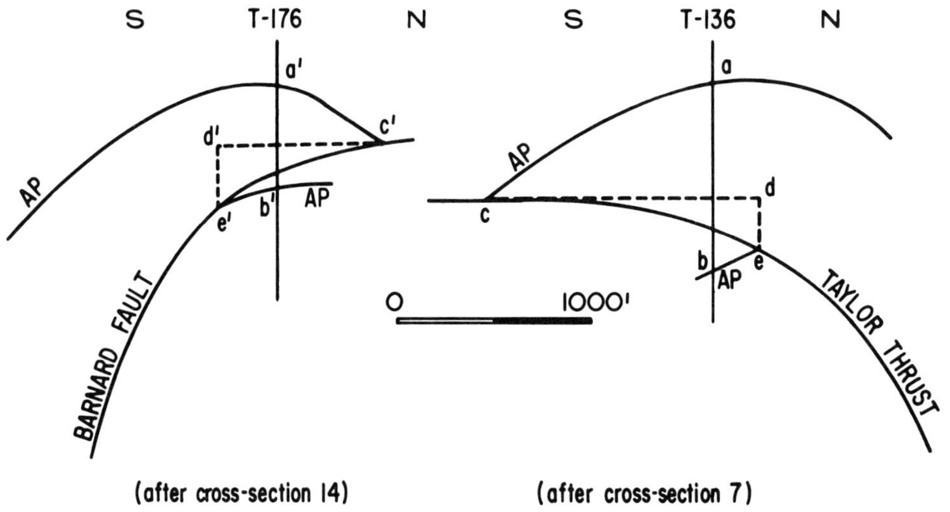

FIG. 20—Displacements along Taylor thrust and Barnard fault.

section. The horizontal projection of net slip (fg), as measured from Figure 18, is approximately 1,800 ft (549 m). In the east-central part of the Shell Taylor lease, the section of the Barnard fault which offsets the AP marker has a throw (d'e') of 350 ft (107 m), a heave (c'd') of 900 ft (274 m), and a stratigraphic separation of about 500 ft (152 m) in Shell Taylor 176 (Fig. 20). As the strike of the beds almost parallels the strike of the fault, the dip slip is approximately equal to the 950 ft (290 m) of apparent displacement (c'e') measured from the cross section. The horizontal projection of the net slip (f'g'), as measured from Figure 19, is approximately 300 ft (91 m). The strike-slip movement of the Barnard fault in that part of the field is calculated to be about 1,200 ft (366 m).

Speculations on Nature of Ventura Anticline and Padre Juan Fault Problem

The present en-echelon arrangement of the Ventura and Rincon anticlines (Fig. 20) leaves open two alternative interpretations for the origin of these anticlines. Either (1) the two anticlines were originally a simple long continuous structure whose trend was broken only by the Padre Juan fault, or (2) the two anticlines were not a continuous unbroken structure, but rather they were originally two anticlines en echelon in arrangement before faulting. If the two anticlines were originally en echelon, the continuation of the Ventura anticline might be west of the San Miguelito field. On the other hand, if the Rincon anticline has been offset from the Ventura anticline by faulting, the underthrust part of the anticline under the Padre Juan fault should be in the Shell Taylor lease, and such a faulted anticline should be an ideal trap for oil accumulation.

If the Rincon and the Ventura anticlines originally were one continuous structure, the Padre Juan fault must have been a complex fault which involved a large horizontal component of motion as well as a northerly thrusting. It is estimated that the dip-slip component of such a fault would have to be approximately 7,000 ft (2,134 m), the strike-slip component 9,000 ft (2,743 m), and the horizontal projection of the net slip 9,500 ft (2,896 m) with a direction of underthrust movement due east. As there was no evidence to indicate any strike-slip component for the Padre Juan fault, previous workers were inclined to favor the idea that the Rincon anticline and the Ventura Avenue anticline had been originally two anticlines en echelon in arrangement before thrusting movement took place along the Padre Juan fault.

Fabric studies of intensely deformed rocks have indicated that movements along a parallel set of shear planes are commonly of the same direction, although they may differ in magnitude. As the Padre Juan fault and the Barnard fault long have been recognized as belonging in the same fault system, the fact that the Barnard fault has a considerable strike-slip component suggests strongly right-lateral movement along the Padre Juan fault. The direction and displacements along the Barnard fault have been discussed previously. The direction of the movement of the underthrust block is practically due east, and the ratio of the strike-slip component to the dip-slip component is 1,200/950 or approximately 4/3. Now the dip-slip component of the Padre Juan fault is estimated to be 7,000 ft (2,134 m). If the movement along the Padre Juan fault is parallel with that of the Barnard fault, then the strike-slip component of the Padre Juan fault, maintaining the same ratio to dip-slip component as in Barnard fault, is computed to be 9,000 ft (2,743 m) and the direction of the underthrust is due east with a horizontal projection of net-slip movement of about 9,000 ft. These displacements are exactly what is required for support of the hypothesis that the Ventura anticline and the Rincon anticline originally were one continuous structure and that the Rincon anticline represents the underthrust continuation of the Ventura anticline.

The distribution of the upper Repetto AO_1 sand in the San Miguelito and Padre Canyon fields has been studied to estimate the displacement along the Padre Juan fault. A sand in most places less than 20 ft (6 m) thick is present between the AO and AP markers in the southern part of the San Miguelito field, but the sand, with a few possible exceptions, is not present in the northern part of the field where the interval between the AO and AP markers is entirely shaly (Fig. 21). In the eastern part of the Padre Canyon field only one well has penetrated to the AP electric-log marker. There a thin sand layer about 10 ft (3 m) thick is present between the AO and AP markers. In the western part of the Padre Canyon field and in the Rincon field, no sand is present between the electric-log markers AO and AP.

The postulated distribution of the AO_1 sand in the San Miguelito and Padre Canyon fields is shown in Figure 21. Even though the control in the Padre Canyon field is very poor, the sand-distribution pattern suggests also a right-lateral movement along the Padre Juan fault (see Figs. 21, 22).

SUMMARY

The Pliocene sediments of Ventura field consist of more than 10,000 ft (3,048 m) of sand and thin silts and shale intercalations. Sand units a few

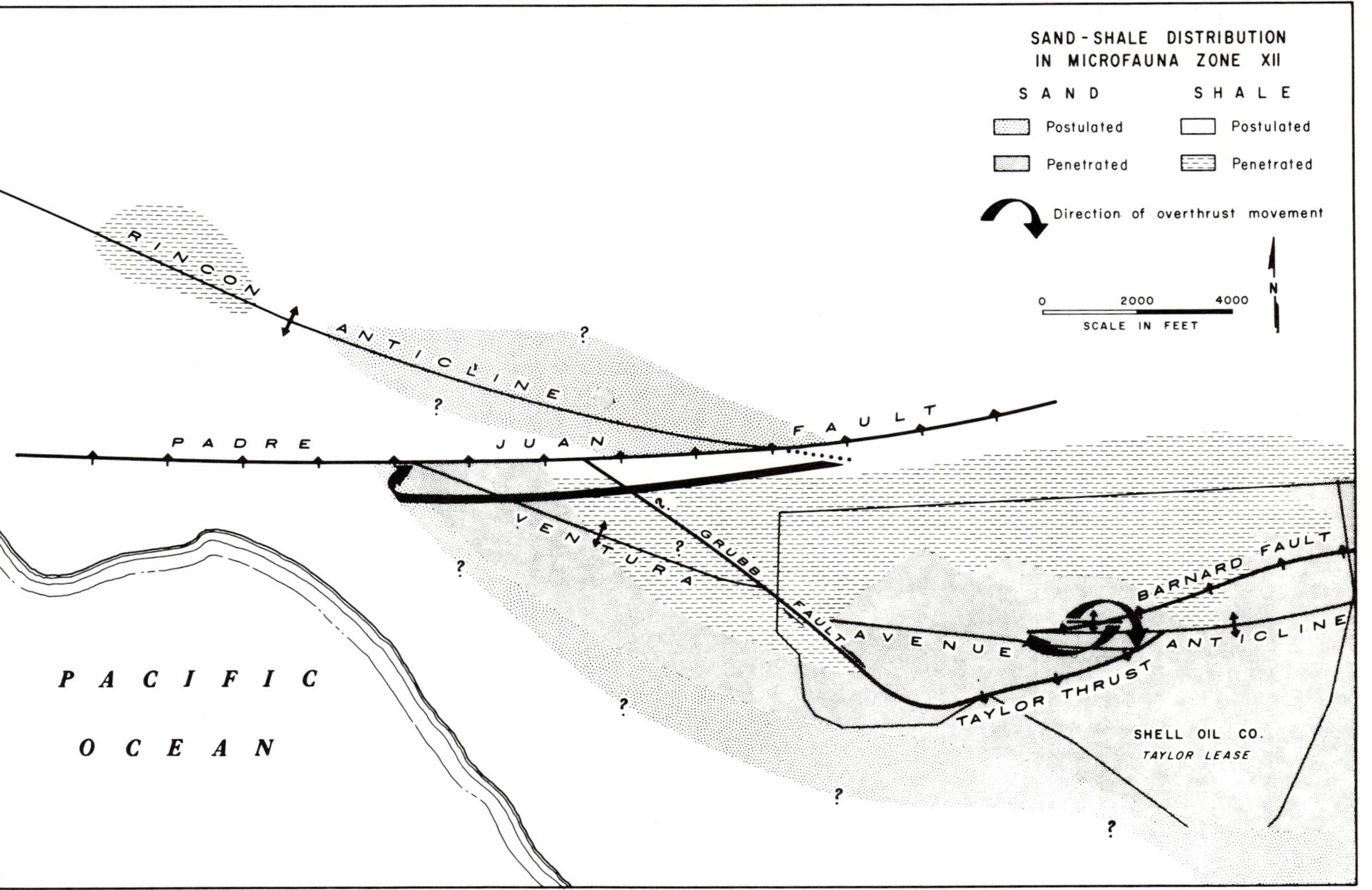

FIG. 21—Distribution of AO_1 sand in western Ventura basin and its bearing on Padre Juan fault problem.

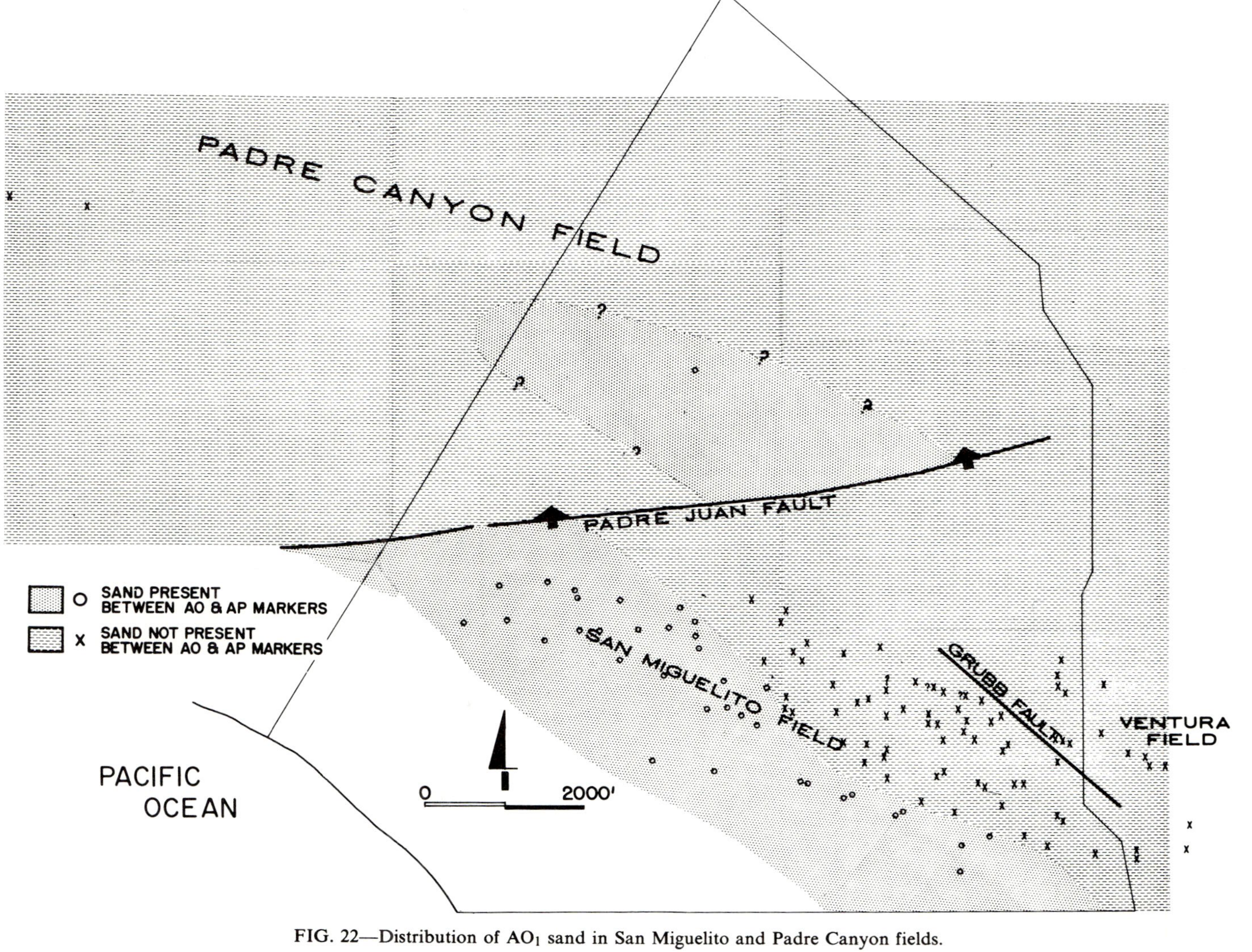

FIG. 22—Distribution of AO$_1$ sand in San Miguelito and Padre Canyon fields.

tens to a few hundred feet thick separated by prominent shale intervals are recognized by electric-log correlation. This study indicates:

1. Such sand units commonly are elongate lenticular bodies with a predominant east-northeast trend and were deposited by laterally restricted westerly flowing currents in deep-sea basins.

2. Such sand units consist of a few to many individual beds which are a few inches to more than 10 ft (3 m) thick and which commonly are graded. A gradual decrease in grain size in a graded bed is accompanied by an increase in degree of sorting.

3. Grain-orientation studies by visual inspection, thin-section analysis, and dielectric-anisotropy measurements all show a predominantly east-northeast preferred grain orientation.

4. Preferred grain orientations of Ventura Pliocene sediments are, in general, parallel with sand trends revealed by facies studies. Grain-orientation analysis by any of the three methods might be a useful tool in determining sand trend.

5. Some lower Pliocene core samples show recognizable cross-bedding which indicates westerly direction of transport.

6. Maximum sand development might be expected in central parts of a deep-sea basin in tectonically active regions.

7. Coincidence of maximum sand development with anticlinal axes might be expected in folded deep-sea basins in tectonically active regions.

8. Stratigraphic pinchouts might be expected on the flanks of such folded deep-sea basins.

9. Results of sand-trend studies can be applied to the development of a newly discovered field, or to the planning of secondary recovery of a depleted field.

10. Grain-orientation studies indicate that the Pliocene sand trends in the Weldon Canyon area north of Ventura field also are predominantly east-northeast.

11. Isopach maps may be used to determine fault displacement in a field where considerable structure and stratigraphic data have been accumulated.

12. Thrust faults of the Ventura field have considerable right-lateral displacement; strike-slip components of movement are somewhat greater than dip-slip components.

13. The Padre Juan fault perhaps is also a right-lateral fault.

14. Ventura anticline and Rincon anticline originally may have been one continuous structure before they were offset en echelon by the right-lateral Padre Juan fault.

REFERENCES CITED

Crowell, J. C., et al, 1966, Deep-water sedimentary structures, Pliocene Pico Formation, Santa Paula Creek, Ventura basin, California: California Div. Mines Spec. Rept. 89, 40 p.

Enos, P., 1969, Anatomy of a flysch: Jour. Sed. Petrology, v. 39, p. 680-723.

Gould, H. R., 1951, Some quantitative aspects of Lake Mead turbidity currents, in Turbidity currents and the transportation of coarse sediments to deep water: SEPM Spec. Pub. 2, p. 34-52.

Heezen, B. C., and M. Ewing, 1952, Turbidity currents and submarine slumps, and the 1929 Grand Banks (Newfoundland) earthquake: Am. Jour. Sci., v. 250, p. 849-873.

Hoots, H. W., T. R. Bear, and W. D. Kleinpell, 1954, Stratigraphic traps for oil and gas in the San Joaquin Valley, in Geology of southern California: California Div. Mines Bull. 170, p. 29-32.

Hubbert, M. K., and W. W. Rubey, 1959, Role of fluid pressure in mechanics of overthrust faulting: Geol. Soc. America Bull., v. 70, p. 115-166.

Kuenen, P. H., 1957, Longitudinal filling of oblong sedimentary basins: Koninkl. Nederlandsch Geol.-Mijnb. Gen., Verhandl., Geol. Ser., v. 18, p. 189-195.

Ludwick, J. C., Jr., 1950, Deep water sand layers off San Diego, California: PhD thesis, Univ. California, Los Angeles.

Nanz, R. H., 1960, Exploration of earth formations associated with petroleum deposits: U.S. Patent 2,963, p. 641.

Natland, M. L., and P. H. Kuenen, 1951, Sedimentary history of the Ventura basin, California, and the action of turbidity currents, in Turbidity currents and the transportation of coarse sediments to deep water: SEPM Spec. Pub. 2, p. 76-107.

Neville, B., 1952, A geological evaluation of the land portion of the Ventura anticline: Shell Oil Co., Pacific Coast Area, Coastal Div., unpub. rept., 51 p.

Sullwold, H. H., Jr., 1961, Turbidites in oil exploration, in Geometry of sandstone bodies—a symposium, 45th annual meeting, Atlantic City, N.J., 1960: AAPG, p. 63-81.

Von Rad, U., 1968, Comparison of sedimentation in the Bavarian flysch (Cretaceous) and recent San Diego trough (California): Jour. Sed. Petrology, v. 38, p. 1120-1154.

Walker, R. G., 1970, Review of the geometry and facies organization of turbidites and turbidite-bearing basins, in Flysch sedimentary in North America: Geol. Assoc. Canada Spec. Paper 7, p. 219-251.

Waterfall, L. N., 1943, Santa Paula oil field: California Div. Mines Bull. 118, p. 394.

Watts, E. V., 1948, Some aspects of high pressures in the D-7 zone of the Ventura Avenue field: AIME Trans., v. 174, p. 191-200; discussion, p. 200-205.

Winterer, E. L., and D. L. Durham, 1962, Geology of the southeastern Ventura basin, Los Angeles County, California: U.S. Geol. Survey Prof. Paper 334-H, p. 275-366.

BULLETIN OF THE AMERICAN ASSOCIATION OF PETROLEUM GEOLOGISTS
VOL. 49, NO. 5 (MAY, 1965), PP. 526-546, 14 FIGS., 1 TABLE

PLIOCENE SEAKNOLL AT SOUTH MOUNTAIN, VENTURA BASIN, CALIFORNIA[1]

ROBERT S. YEATS[2]
Ventura, California

ABSTRACT

Extensive drilling in the southern Ventura basin in the last decade has provided useful data about Pliocene basin floor topography. The pre-basinal Miocene Modelo Formation, mainly siliceous shale, limestone, and organic shale with a bathyal microfauna, underwent considerable submarine tilting and faulting, which produced a surface of high relief upon which the deep-sea Plio-Pleistocene Pico Formation was deposited. Among the land forms on this surface was a seaknoll at South Mountain.

The seaknoll is characterized by relatively uneroded fault scarps with slopes exceeding 40 degrees, a veneer of glauconite sandstone locally containing a talus of Modelo limestone fragments from the scarps, and, near the seaknoll summit, a small biostrome in a shelly glauconite sandstone matrix. The steep fault scarps at South Mountain, near the center of the basin, contrast with the bevelled fault scarps at nearby Berylwood anticline and the Oxnard Plain, near the southern edge of the Pliocene basin and on the site of a pre-Pico submarine slope eroded in the Modelo Formation.

The Pico Formation has onlapped and buried the seaknoll and the submarine slope. The Pico contains graded sandstones interbedded with siltstones containing indigenous bathyal and displaced neritic microfaunas. The sandstone beds lens out and the enclosing siltstones thin in the direction of the seaknoll, suggesting that the seaknoll underwent less subsidence than the surrounding sea floor during its burial by Pico sediments.

Because the Berylwood-Oxnard submarine slope was in the path of turbidity currents carrying sediment from subaerially eroded highlands on the south and east, it underwent submarine erosion. The seaknoll, because it was separated from the subaerial highlands by deep-sea channels, was relatively unaffected by turbidity current scour. Both submarine slope and seaknoll remained below sea level until buried.

Erosional submarine unconformities of the Berylwood and Oxnard Plain submarine slope type may be relatively common in diastrophic basins where rapid sedimentation and strong deformation are closely related. Above these unconformities, the immediately overlying sediments are deep-sea rather than shallow-water. Because certain deep-sea currents can erode the sea floor if given enough time, erosional unconformities cannot be used as evidence for subaerial erosion without additional supporting data, nor can onlapping depositional surfaces be equated with former sea levels.

INTRODUCTION

The Plio-Pleistocene Ventura basin is an ideal area for comparing ancient and modern deep-sea sediments (Natland, 1933; Shepard and Emery, 1941; Natland and Kuenen, 1951), because of its proximity to modern, structurally similar, and genetically related submarine basins in which bathyal sedimentation is now taking place (Gorsline and Emery, 1959). Subsequent to Natland and Kuenen's study in the Ventura basin, many wells have been drilled for oil in its southern part, particularly in and near the South Mountain,

Oxnard, and West Montalvo oil fields (index map, Fig. 1). These wells penetrate the Pliocene section, reaching Miocene and older beds, and thereby providing the data necessary for subsurface study of the Pliocene basin floor. This study indicates that the Pliocene deep-sea sediments accumulated on a sea-floor surface of high relief including, in the area later to become South Mountain, a fault-block hill which persisted as a seaknoll until it was buried by Pliocene sediments.

GENERAL GEOLOGIC SETTING, SOUTH MOUNTAIN AREA

PRE-BASINAL DEPOSITS

The geology of South Mountain has been described by Baddley (1954) and Lung (1958). The producing Sespe Formation (Eocene-Miocene) is overlain by Miocene sandstone, which is itself intruded by an andesite sill, locally 90 meters thick (Fig. 2) (Yeats, 1964).

The Miocene sandstone is overlain conformably by the Modelo Formation (Kew, 1924) of Miocene (Luisian[3] to at least late Mohnian[3]) age,

[1] Manuscript received, October 22, 1964. Published by permission of Shell Oil Company.

[2] Senior production geologist, Shell Oil Company, Ventura, California. The writer is indebted to J. L. Cowell, J. C. Crowell, and S. H. Hamann for valuable discussions related to the paper. The microfossils and their paleoecology were studied by U. S. Armstrong and J. J. Conniff of the Shell Oil Company, Ventura, California. V. S. Mallory of the University of Washington examined the megafossils. The K-Ar age determinations of ash beds were done by George Edwards, Shell Development Company, Houston, Texas. Illustrations were prepared by L. A. Watts. The paper was critically reviewed by J. H. Mackin, University of Texas, and D. C. Pontius of Bataafse Internationale Petroleum, Mij.

[3] Miocene microfaunal stages of Kleinpell (1938).

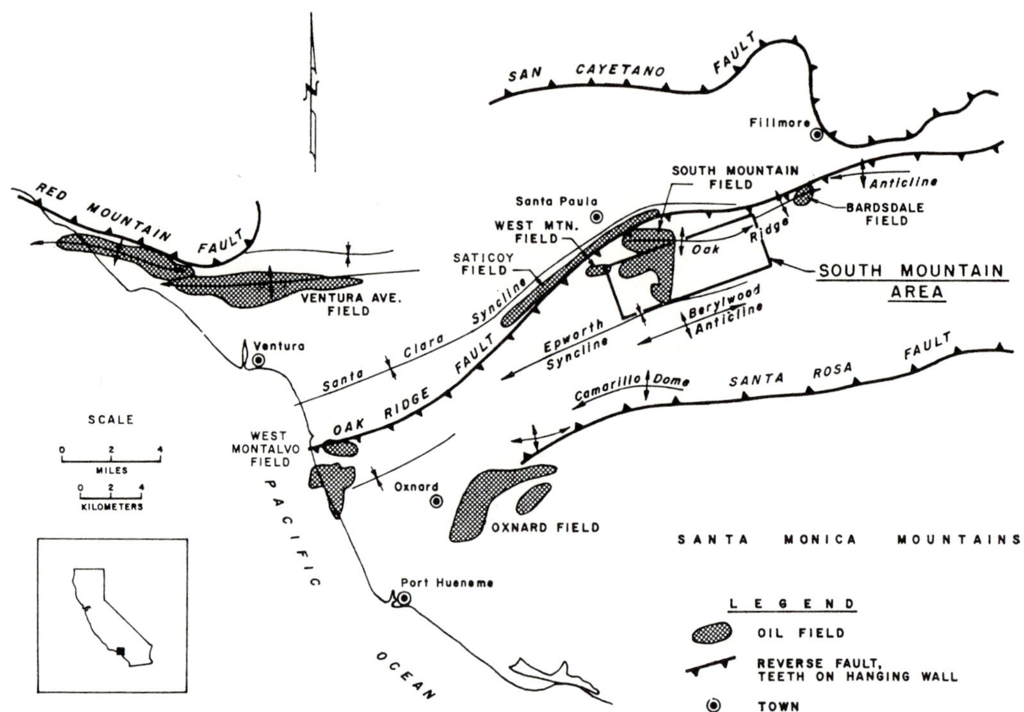

Fig. 1.—Part of Ventura basin, showing major tectonic units, oil fields, and South Mountain area.

based on fossil Foraminifera (Lung, 1958; U. S. Armstrong and J. J. Conniff, personal communication). The Modelo is composed of diatomite, punky brown massive or laminated shale, siliceous shale, and lesser amounts of chert, limestone, sandstone, and volcanic ash.

The uppermost Modelo strata in the South Mountain area have not yielded age-diagnostic Foraminifera, although these strata conformably overlie lithologically similar rocks of late Mohnian age. However, these overlying shales are rich in diatoms, radiolarians, and siliceous sponge spicules. They unconformably underlie beds of early Pliocene age at Balcom Canyon (Fig. 3), and a Delmontian[3] age has been suggested (Brown, 1959). A silver-gray ash bed near the top of the Modelo in the SE$\frac{1}{4}$ of NE$\frac{1}{4}$, sec. 19, T. 3 N., R. 20 W., at South Mountain field, gave a glass-concentrate K-Ar age of 11.5 million years ($\pm$10%) (George Edwards, personal communication).

Lung (1958) suggested that the basal Miocene sandstones were laid down in shallow water. His evidence showed that the Modelo Formation was deposited in a basin which was subsiding more rapidly than it was being filled; a bathyal marine sedimentary environment persisted to the end of Modelo deposition. Microfossils from Balcom Canyon and from the McCulloch-Standard-Terry No. 1 well (sec. 31, T. 3 N., R. 20 W.) at South Mountain (Fig. 4) and the Shell Berylwood No. 1 and 2 wells (sec. 32, T. 3 N., R. 20 W.) at Berylwood anticline have been studied by U. S. Armstrong and J. J. Conniff. They observed a gradual transition within the Modelo from a neritic microfauna (water depth shallower than 90 meters) upsection into a bathyal microfauna (water depth 180–600 meters), confirming Lung's observations for this area and Bandy's observations (1953a) for the western and central Ventura basin in general.

PICO FORMATION

The Modelo is unconformably overlain by the Pico Formation (Kew, 1924; Lung, 1958).[4] On the

[4] The upper part of the Pico at South Mountain has been correlated by means of faunal evidence with the Santa Barbara Formation by Bailey (1943) and others. In the writer's opinion, Bailey's Santa Barbara cannot be mapped in this area as a separate formation; hence, the return to the older usage of Kew.

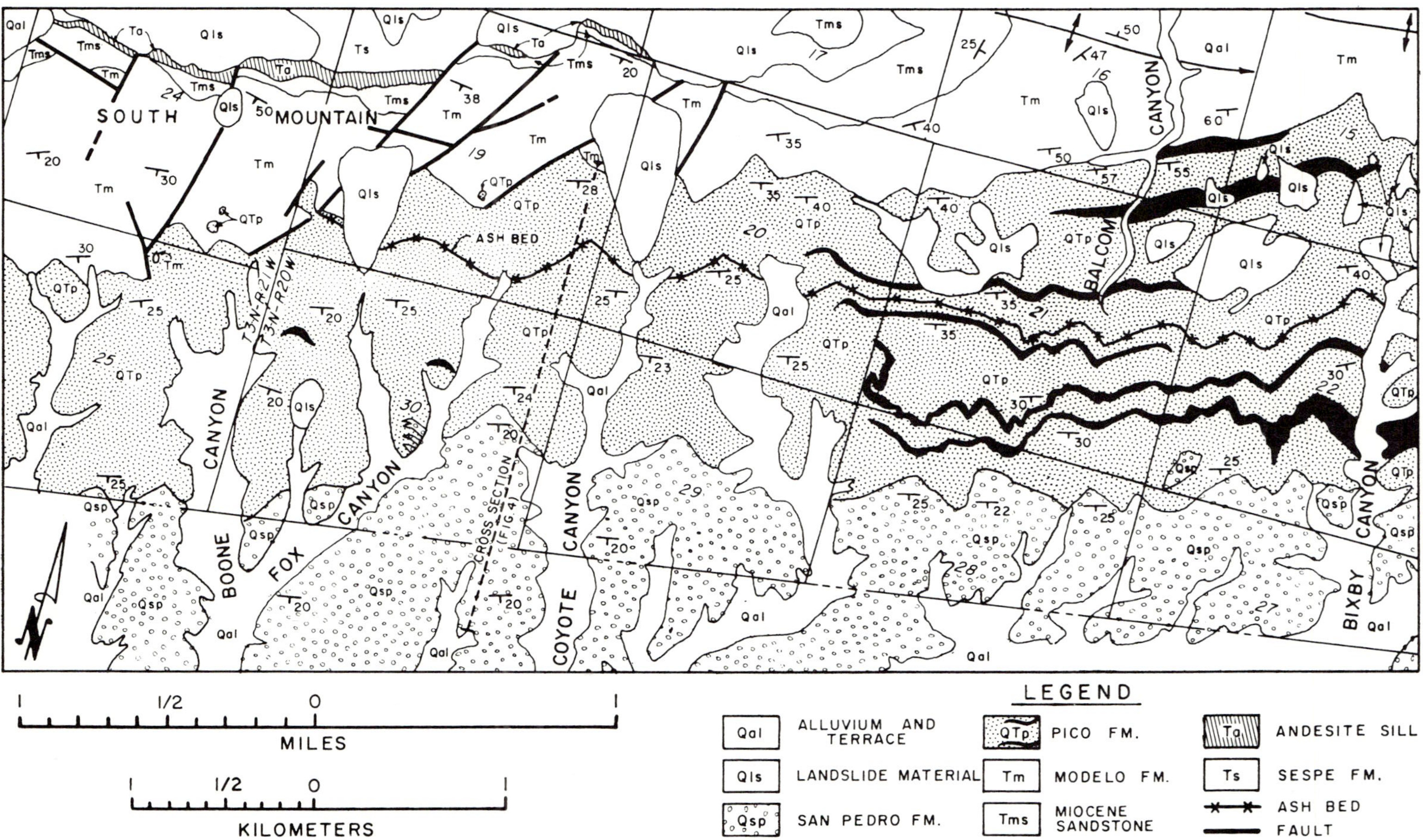

Fig. 2.—Geologic map of South Mountain-Oak Ridge area (see Fig. 1 for location). Within the Pico Formation, sandstones are shown in black, and fine-grained sediments in fine stipple pattern. Intra-Pico ash bed continues unfaulted over tops of Modelo fault scarps. Sandstones are abundant at Bixby and Balcom Canyons, but lens out westward toward South Mountain. (Base map Moorpark and Santa Paula 7½-minute quadrangles, U. S. Geological Survey, 1951.)

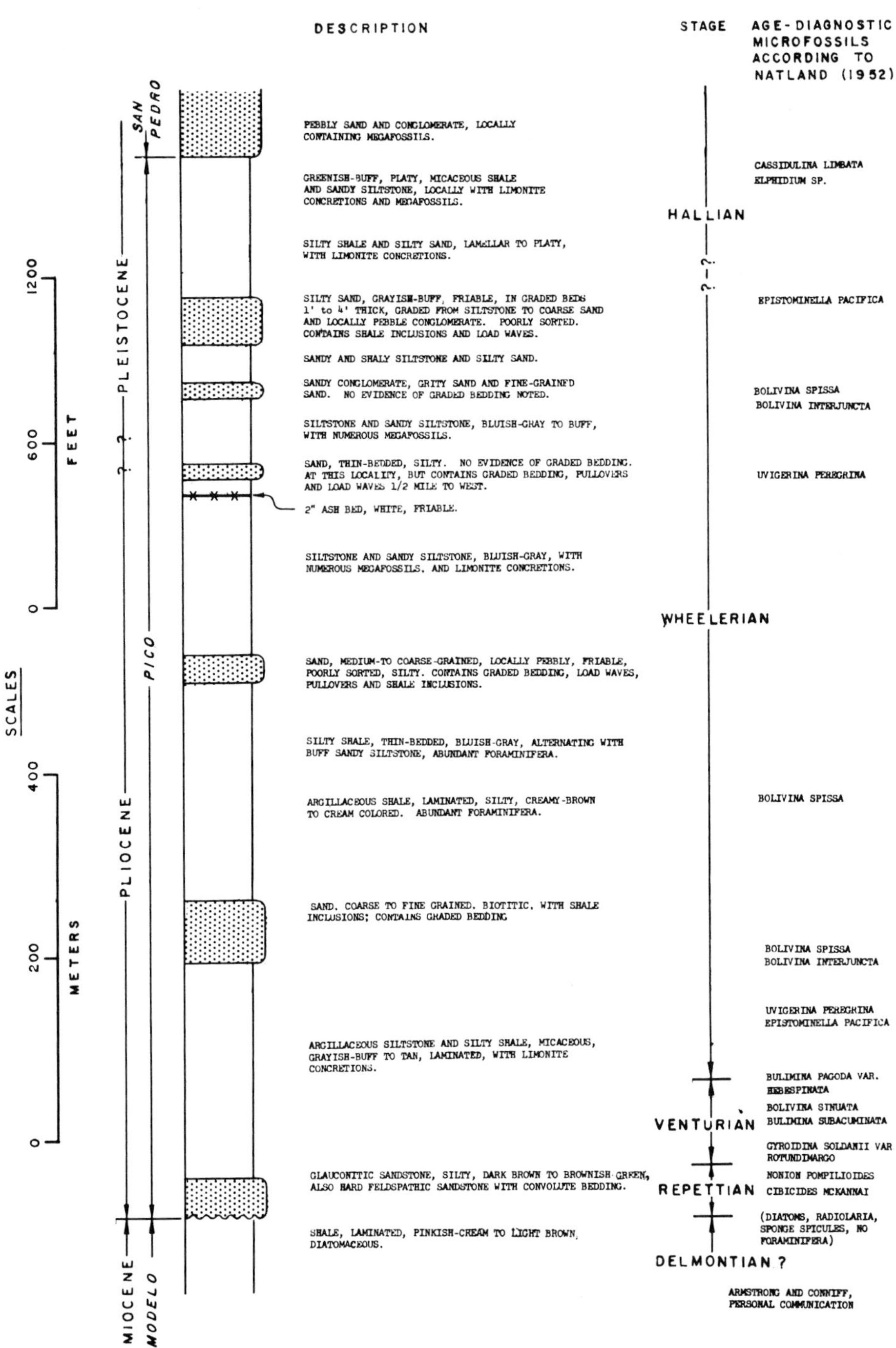

FIG. 3.—Pliocene columnar section, Balcom Canyon (map, Fig. 2, for location), revised from Lung (1958). Microfaunal data based on 200 outcrop samples of shale and siltstone at 10–20-foot intervals.

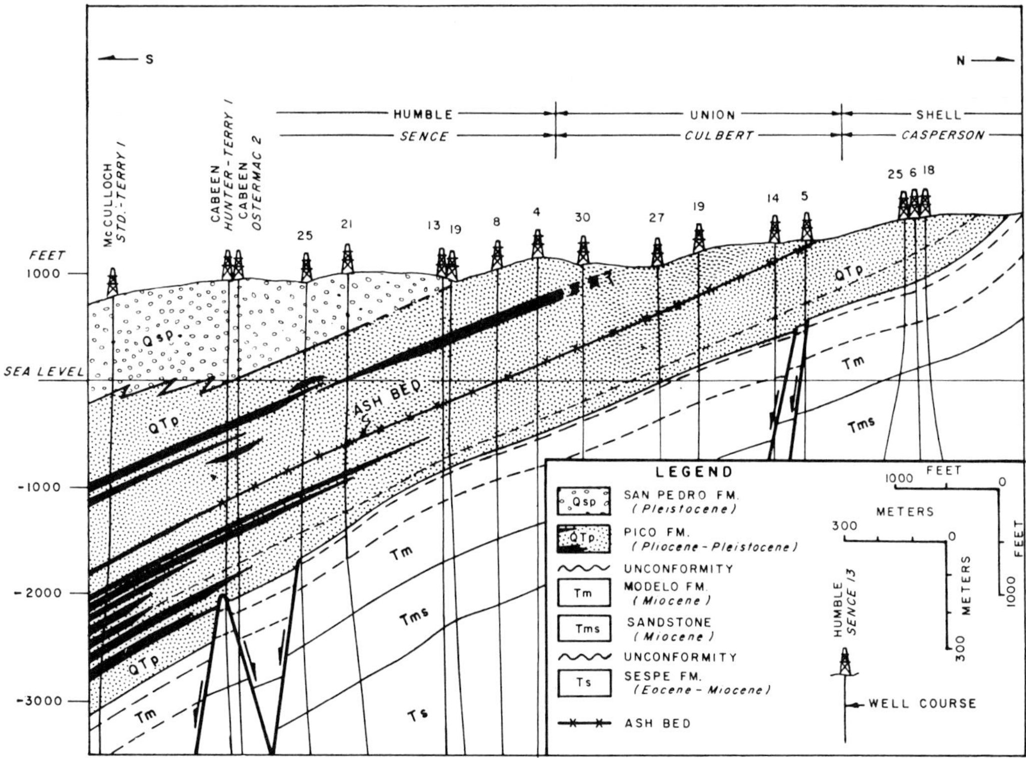

Fig. 4.—South–north cross section (Fig. 2 for location) through part of South Mountain oil field. Within Pico Formation (QTp), sandstones are in black, siltstones and shales in fine-stipple pattern. Sandstones in north limb of Epworth syncline (left) lens out northward toward South Mountain.

south side of South Mountain and in the area to the east, the Pico is predominantly sandy mudstone, siltstone, and claystone, locally laminated, and locally with thin beds of fine-grained silty sandstone. Intercalated in these fine, grained sediments at Balcom Canyon (Fig. 2) are six sandstone members, all with sedimentary features ascribed to turbidity current deposition, including repeated graded bedding, flame structure, convolute bedding, and imbricated shale clasts. North of South Mountain, in the Saticoy field, the Pico contains many sandstone members which are characterized by repeated graded bedding and displaced shallow-water faunas; these are similar to outcropping turbidity current deposits north of the Saticoy field, described by Natland and Kuenen (1951).

The Pico ranges in age from early Pliocene to early Pleistocene, which is late Repettian to Hallian, according to the microfaunal stages of Natland (1952). The most complete section in the South Mountain area is at Balcom Canyon (Fig.

3), where an upper Repettian foraminiferal assemblage occurs in beds lying on the Modelo. These Repettian beds are overlain conformably by Pliocene and Pleistocene sediments with Venturian, Wheelerian, and Hallian microfaunas. The uppermost Pico (Hallian) is considered to be Pleistocene on the basis of a horse tooth in the upper Pico at Grimes Canyon, south of Bardsdale oil field (Bailey, 1943). Furthermore, a glass concentrate from an ash bed nearly 400 meters below the top of the Pico at Balcom Canyon yielded a K-Ar age of 1.1 million years ($\pm 10\%$) (George Edwards, personal communication), close to the Plio-Pleistocene time boundary according to the revised geologic time scale of Holmes (1960).[5]

[5] Evernden and others (1964) have cautioned against the use of volcanic glass for K-Ar dating because devitrification of glass may cause the calculated age to be too young. Because of this, and considering the possible significance of this ash bed in Plio-Pleistocene chronology, Edward's comments on the ash bed are here included:

"A concentrate of glass from volcanic ash was sepa-

TABLE I. DEEP-WATER LIVING FORAMINIFERA CONSPECIFIC WITH FOSSILS IN BALCOM CANYON SECTION

Species	Recent Locality Studied	Present Bathymetric Range (Living Specimens Only) (Fathoms)	Reference	Location in Balcom Canyon Section Above Top Miocene
Bolivina spissa Cushman	Todos Santos Bay	150–600+. Most abundant at 350–400	Walton, 1955	Abundant 80± to 870± meters
Bolivina spissa Cushman	Santa Cruz Basin	200–650	Resig, 1958	
Bolivina spissa Cushman	Offshore San Diego	100–625. Maximum 350–525	Uchio, 1960	
Cassidulina delicata Cushman	Santa Cruz Basin	250–550	Resig, 1958	Most abundant 120± to 340± meters
Cassidulina delicata Cushman	Offshore San Diego	160–650+	Uchio, 1960	Less abundant 340± to top of Pico
Epistominella pacifica (Cushman)	Santa Monica Bay	300 and deeper	Zalesny, 1959	Most abundant 110± to 930± meters
Planulina ornata d'Orbigny	Santa Cruz Basin	100–350	Resig, 1958	Abundant 240–260 meters. Rare to common 0–210± meters, 310± to 340± meters
Gyroidina altiformis (Stewart and Stewart)	Santa Cruz Basin	220–1,000+	Resig, 1958	Abundant 620–640 meters. Rare to absent 0–140 meters, and 640 meters to top of Pico.

The Pliocene and lower Pleistocene benthonic microfaunas from Balcom Canyon, Saticoy oil field, South Mountain oil field, and adjacent areas resemble the microfaunas of Natland's (1952) stages in that, when compared with Recent assemblages, they exhibit a gradual shallowing up-section from bathyal to neritic water depths. U. S. Armstrong and J. J. Conniff (personal communication) found that the Pico Formation in these areas, except for its youngest part, was deposited in depths of water greater than 200 meters. At Balcom Canyon, they observed abundant *Bolivina spissa*, *Cassidulina delicata*, *Epistominella pacifica*, *Planulina ornata*, and *Uvigerina peregrina* and, less commonly, *Angulogerina angulosa* and *Bulimina subacuminata*, a microfauna characteristic of the bathyal zone (200–2,000 meters) off California today (Bandy, 1953b). The youngest Pico at Balcom Canyon contains *Elphidium* sp., *Cassidulina limbata*, and *Angulogerina angulosa*, a neritic microfauna according to Bandy (1953b).

The presence in the Pico Formation of fossil Foraminifera which are conspecific with living

species of benthonic Foraminifera in bathyal waters off the southern California coast is strong evidence for deep-sea deposition of the Pico. Table I lists several species found at Balcom Canyon for which the depth range of living representatives (identified as living because of the presence of organic tissue, which is preferentially stained by rose Bengal) has been determined by detailed bottom sample traverses across the shelf, slope, and basin floor.

SAN PEDRO FORMATION

The folded marine and non-marine sandstones and conglomerates, locally with siltstone and clay interbeds and conformably overlying the Pico Formation, were considered part of the Saugus Formation of late Pliocene and early Pleistocene age by Kew (1924). Bailey (1943) correlated these beds with the San Pedro Formation, based on (a) the priority of the term as a reference to the southern California coarse clastic marine Pleistocene, (b) the faunal similarity of the lower part to Arnold's (1903) lower San Pedro of the Los Angeles basin, and (c) the difficulty of correlating the Oak Ridge beds with the apparently non-marine, unfossiliferous type Saugus Formation. The retention of the term "San Pedro" in this area is tentative, pending a review of the Saugus-San Pedro problem in both the Ventura and Los Angeles basins.

In the South Mountain area, the formation is Pleistocene on the basis of vertebrate remains (Pressler, 1929). Natland (1952) considered the lower, marine part of the San Pedro Formation to

---

rated by the use of heavy liquids. That material having a specific gravity of less than 2.50 appears to be about 95 per cent glass. The remaining 5 per cent is composed of approximately equal amounts of quartz and sanidine fragments. The glass has an index of refraction between 1.50 and 1.51 and does not appear to be devitrified. Crystallites, such as globulites, longulites, and spiculites, are abundant. Many crystallites exhibit preferred orientation along possible flow lines. A few spherulites are present. Many glass fragments show either a reddish brown or a yellow stain." In the analysis, %K =4.25, Ar_{40}=1.72×10^{-7} radiogenic ml/g.

be upper Hallian; the upper part is non-marine, but still Pleistocene. However, the Pico-San Pedro contact occurs at older horizons eastward; in the eastern Ventura basin, the coarse clastic deposits (Saugus) are in part Pliocene (Winterer and Durham, 1958, 1962).

POST-SAN PEDRO DEFORMATION
(PASADENAN OROGENY)

During the medial Pleistocene, the area was strongly folded and faulted (Pasadenan orogeny). The principal structural feature of the Pasadenan orogeny in the South Mountain area is the Oak Ridge high-angle reverse fault, which brings Eocene and Oligocene sedimentary rocks of the Oak Ridge anticline against Pliocene and Pleistocene sediments of the Santa Clara syncline. The Oak Ridge anticline, also formed during the Pleistocene, is composed of a series of *en echelon* folds generally paralleling the Oak Ridge fault on the south, or hanging wall, side (Fig. 1). The Santa Clara syncline occurs on the site of the main Plio-Pleistocene depositional trough of the Ventura basin. The strongly negative character of this trough during the Pliocene and early Pleistocene resulted in the accumulation of a much greater thickness of Pico there than south of the Oak Ridge fault. The Pico is so thick that the underlying Miocene deposits have yet to be reached by exploratory drilling.

South of the Oak Ridge anticline, the Pleistocene Epworth syncline is a broad sag reflecting a minor Plio-Pleistocene trough. This syncline is on strike with another southwest-trending syncline in the Oxnard Plain (Fig. 1). Farther south, the Berylwood anticline and the Camarillo dome comprise broadly upwarped Plio-Pleistocene sediments unconformably overlying a northwest- to north-dipping homocline of Miocene and older beds.

MIO-PLIOCENE UNCONFORMITY
DURATION OF HIATUS

The unconformity between the Modelo and Pico Formations is of great regional importance in the Ventura basin. At South Mountain, the youngest beds below the unconformity are at least as young as latest Mohnian, and may be in part Delmontian. The oldest beds above the unconformity are late Repettian. Beds correlative with the type lower Repettian and most of the type middle Repettian of the Los Angeles basin have not been found in the central and western Ventura basin (Natland, 1952), although beds of that age are probably present in the eastern Ventura basin (Winterer and Durham, 1958). Based on the difference in K-Ar ages between an ash bed near the top of the Modelo and an intra-Pico ash bed which locally intersects the Modelo at South Mountain, the unconformity at South Mountain encompasses a time gap of about 10 million years.

South of South Mountain, the time interval represented by the hiatus increases. Near the Santa Rosa fault, the Miocene marine beds were eroded completely prior to Plio-Pleistocene deposition. The Pico onlaps the unconformity, so that locally the overlying San Pedro coarse clastic sediments rest with angular unconformity directly on the Sespe Formation.

In the Santa Susana Mountains of the eastern Ventura basin, the unconformity apparently disappears (Winterer and Durham, 1958, 1962). The Mohnian Modelo Formation is overlain by the Delmontian and early Pliocene Towsley Formation, and the Towsley is overlain by the type Pico Formation of early to late Pliocene age. The entire sequence is conformable, with interfingering and gradation at formation contacts.

PRE-PICO FAULTING AND FAULT SCARPS

South Mountain oil field is cut by many diversely oriented normal faults which break up the producing Sespe Formation into small, fault-bounded oil reservoirs. Some of these faults extend upward into Miocene beds, offsetting an andesite sill which forms the brow of South Mountain (Fig. 2). Several surface faults, the east side of each being downthrown relative to the west side, cut the sill and extend to the top of the Modelo where they coincide with apparent offsets of the Modelo-Pico contact.

However, intra-Pico marker beds, even those very close to the Modelo-Pico contact, are not faulted. Figures 5 and 6 are structural contour maps. The horizons on which they are drawn are Miocene and Pliocene electric log markers. The base of the Modelo (Fig. 5) is cut by many normal faults whereas the intra-Pico marker bed (Fig. 6) is not faulted. The Pico ash bed (Fig. 2), which crops out at the south end of the field, is unbroken by faults, although the Modelo adjacent to the ash bed is cut by several north-trending faults.

Hence the faults, despite the apparent offset of the Modelo-Pico contact, are older than the overlying Pico Formation.

The apparently contradictory evidence of unfaulted Pico beds with a basal contact which appears to be faulted can be resolved by interpreting the top of the Modelo as having been virtually unmodified by erosion since faulting. The contact, thus interpreted, includes a series of post-Modelo fault scarps subsequently buried by Pico sediments. From Balcom Canyon west, the Pico wedges out against the Modelo and several buried, east-facing fault scarps which cut the Modelo. This gives the Pico-Modelo contact a zig-zag appearance on air photos and on the map (Fig. 2). An isopachous map of the Pico between the ash bed and the unconformity at South Mountain field (with over 70 points of well control) emphasizes the scarps, because the Pico acts as a mold with respect to the underlying fault topography. The thickness of Pico sediments below the ash bed decreases from 780 meters at Balcom Canyon to 180 meters north of Coyote Canyon, and to 3 meters on the east side of Boone Canyon. West of Boone Canyon, the ash bed is absent; beds containing a microfauna characteristic of sediments younger than the ash bed directly overlie the Modelo Formation (U. S. Armstrong and J. J. Conniff, personal communication). The paleo-relief on the Modelo surface exceeds 150 meters within the South Mountain oil field and

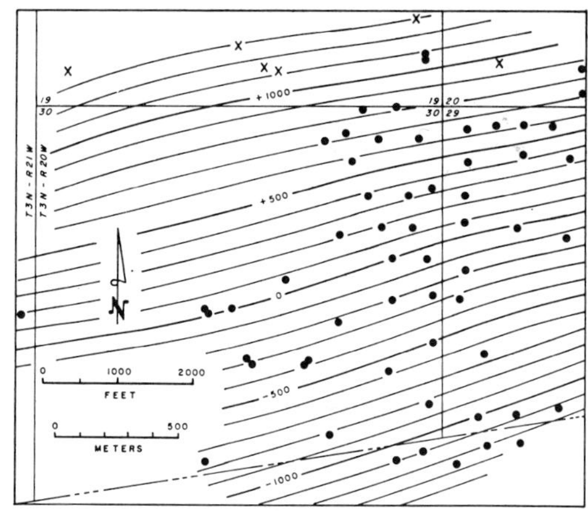

Fig. 6.—Structure contour map. Datum is ash bed, Pico Formation, South Mountain field. Dots represent well control; X represents outcrop control. Contour interval, 100 feet.

may exceed 800 meters in the area as a whole, including Balcom Canyon.

Electric log study of 82 South Mountain field wells containing correlative Modelo and Pico marker beds shows that the Modelo beds, between a given upper Modelo electric log marker and the unconformity (Fig. 7), change in thickness by less than 70 meters. In most wells, the thickness change is much less, in the order of a few meters. Generally, Modelo bedding parallels both the unconformity and Pico bedding, as evidenced in both the subsurface (Fig. 4) and outcrop. An angular unconformity has been observed at only one locality, at a cut-bank adjacent to the Shell-Casperson No. 36 well in the $S\frac{1}{2}$ of sec. 19, T. 3 N., R. 20 W., where the Pico is deposited across the truncated edges of Modelo shales (Fig. 8). The parallelism between Miocene bedding and the unconformity (except for the scarps) and the small variation in thickness between the unconformity and a given Modelo marker strongly suggest that the present configuration of the Modelo-Pico contact at South Mountain is structural in nature and that it is relatively unaffected by post-faulting, pre-Pico erosion.

Estimates of scarp slopes in Pliocene time are based on observed slopes of minor scarps in road cuts (Fig. 9), geometry of the zig-zag Mio-Pliocene contact, and subsurface control. Well spacing (120–150 meters) is not sufficient to contour scarp slopes, but is close enough to permit subsur-

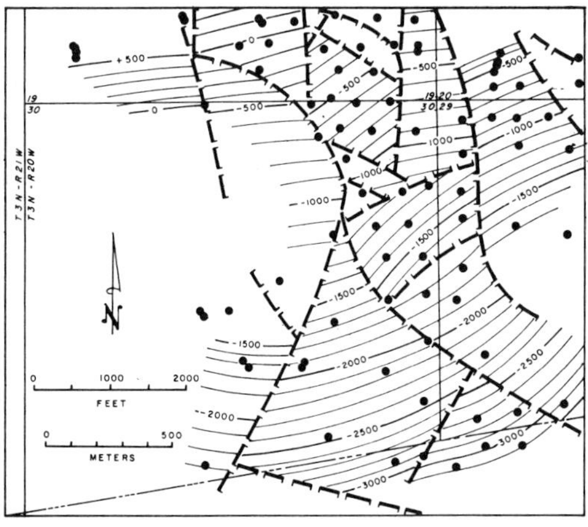

Fig. 5.—Structural contour map. Datum is base of the Modelo Formation, South Mountain field. Dashed lines are normal faults, with hachures in direction of downthrow. Dots represent well control. Contour interval, 100 feet.

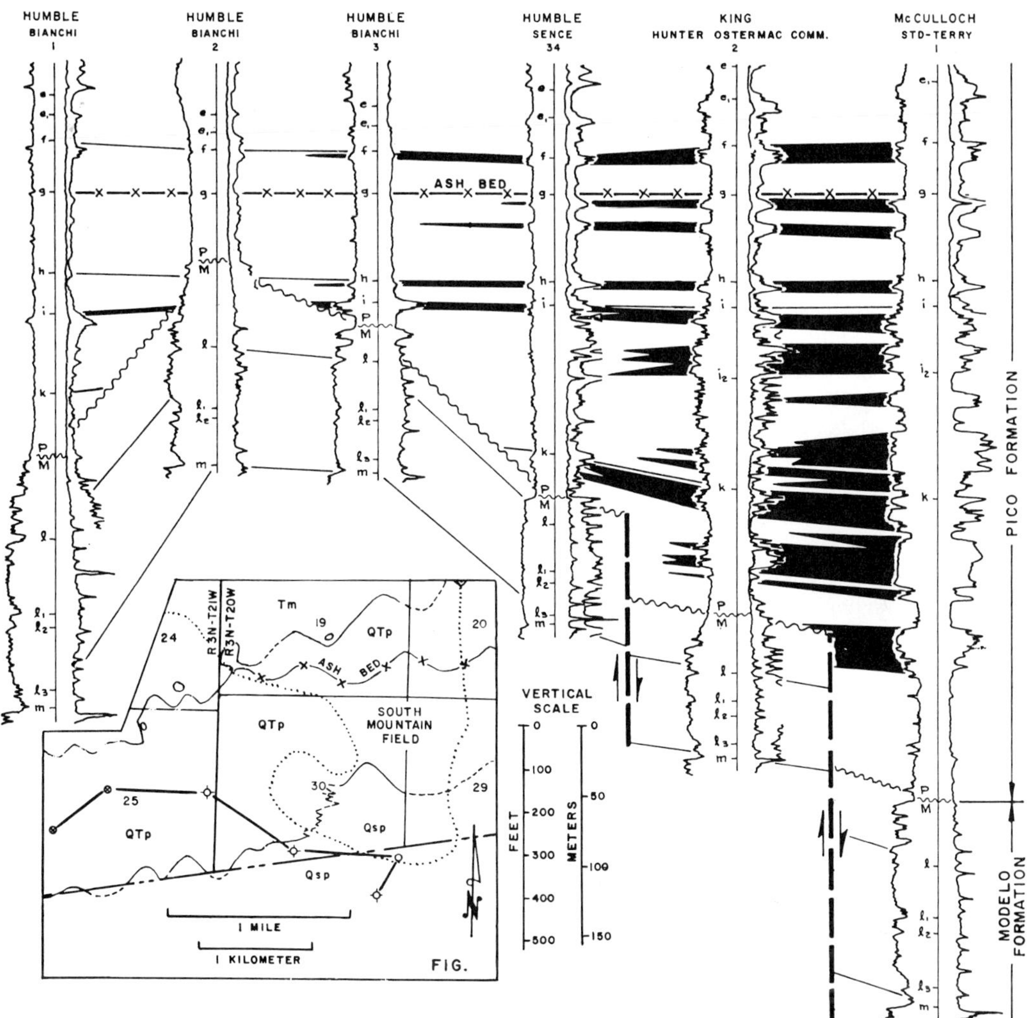

FIG. 7.—Electric log correlation section showing beds adjoining Mio-Pliocene unconformity at South Mountain. Pico sandstones are shown in black. On well location map, Tm = Modelo Fm., QTp = Pico Fm., and Qsp = San Pedro Fm.

face contouring of some of the faults that originally formed the scarps. These faults dip 50 to 60 degrees with respect to Pico bedding (Fig. 4). The limited data indicate that most scarp slopes exceeded 40 degrees, but no evidence suggests that any of the scarps were vertical.

Scarp heights before burial were mostly less than 50 meters, but one east-facing scarp in the subsurface in the SW¼ of sec. 19 and the NW¼ of sec. 30, T. 3 N., R. 20 W., may have exceeded 150 meters in height. The south-facing scarp at the south end of South Mountain field (left side of Figure 4) was at least 100 meters high.

SEDIMENTARY FEATURES ASSOCIATED WITH UNCONFORMITY

Lung (1958) noted a veneer of glauconite sandstone at the Modelo-Pico contact. The glauconite sandstone apparently transgresses time; at Balcom Canyon, it is conformable beneath beds with Repettian microfossils, whereas west of Boone Canyon, it conformably underlies Wheelerian sediments. The sandstone grades upward into glauconite-free mudstone, and the contact with the Modelo Formation is sharp. The uppermost Modelo is extensively bored by marine organisms; these borings are filled with glauconite. The sandstone, where present, ranges up to six meters in thickness. It is lemon-yellow, yellow-green, orange, or greenish brown (in contrast to the dull brown-weathering mudstones above and the white-weathering diatomaceous shales below), and is soft, friable, and clayey. The sandstone has the appearance of having been churned and reworked by marine animals. Microscopic examination reveals that the glauconite occurs as subrounded to rounded coarse to fine grains with bluish-green, green, and yellow mottling and a waxy luster. In addition to glauconite, siliceous sponge spicules and angular to subangular detrital grains, principally of quartz, are present.

In secs. 24 and 25, T. 3 N., R. 21 W., and sec. 19, T. 3 N., R. 20 W., the glauconite sandstone contains angular to subrounded fragments of limestone ranging from a few millimeters to greater than 60 centimeters in diameter. The limestone fragments, gray to buff when fresh, are mottled buff, yellow, green, brown, and ocher, apparently by weathering. The weathered surface, 2–10 millimeters thick on individual fragments, is penetrated by animal borings of two types (Fig. 10): (1) shallow depressions, slightly ovoid, up to 10 millimeters long, and (2) narrow, curved holes with circular or ovoid cross-sections 1–4 millimeters in diameter; some are normal to, others parallel with, the rock surface. Many of these narrow holes turn back on themselves, forming a "U"; others are only tiny pits in the rock surface. Many borings are lined or filled with glauconite sandstone and (or) brown limonitic material.

The glauconite sandstone and the limestone fragments post-date the formation of the fault

Fig. 9.—Steep scarp at Pico-Modelo contact, SE$\frac{1}{4}$, sec. 24, T. 3 N., R. 21 W. Modelo (M) is light colored in contrast to Pico (P). Entire contact is mantled by Pico glauconite sandstone with angular boulders of limestone containing marine borings. Scarp is about 4 meters high, and view is northward.

scarps. In the SW$\frac{1}{4}$ of sec. 19, T. 3 N., R. 20 W., glauconite sandstone containing many limestone fragments with animal borings abuts against a scarp of Modelo Formation with large, irregularly shaped concretionary masses of limestone. This limestone, except for the absence of borings and a weathering rind, is lithologically similar to the limestone comprising the fragments in the glauconite sandstone (Fig. 9). The limestone rubble is most abundant adjacent to the scarp; 20 meters east, away from the scarp, the fragments are smaller, and still farther east, glauconite sandstone without fragments is found. This suggests that the fragments originated as a subaqueous rockfall (*cf.* Dott, 1963) from the steep Modelo scarp and accumulated as talus at the foot of the scarp. Borings on all sides of most limestone fragments are proof that at least some of the

Fig. 8.—Angular unconformity (dashed white line) between Pico glauconite sandstone (above) and Modelo shale (below) in S$\frac{1}{2}$, sec. 19, T. 3 N., R. 20 W. View is northward.

Fig. 10.—Angular limestone fragment from basal Pico glauconite sandstone, sec. 24, T. 3 N., R. 21 W., showing animal borings. Rock is 22 centimeters across.

boring activity occurred after the fragments had fallen from the scarp. Smaller fragments are also somewhat rounded, suggesting that they were rolled about on the sea floor by bottom currents. The absence of a weathering rind in the Modelo limestone is evidence that the weathering of the fragments occurred prior to their burial by Pico sediments. The presence of the rind on all sides of the fragments indicates that at least part of the weathering occurred after the fragments had fallen from the scarps.

Accompanying the limestone rubble in some localities are well-rounded pebbles, cobbles, and boulders of pre-Tertiary granite, gneiss, diorite, argillite, and volcanic rock. Pebbles of pre-Tertiary rock, together with some carbonaceous fragments, are found in glauconite sandstone on the east, in secs. 19 and 20, T. 3 N., R. 20 W.; there the limestone fragments are rare to absent. The significance of these rounded stones is discussed in a later section. In the SE$\frac{1}{4}$ of sec. 24, T. 3 N., R. 21 W., the glauconite sandstone encloses a small biostrome composed of juvenile and adult forms of the large clam *Thyasira disjuncta* (Gabb). The clams are generally found with both valves in the closed position, suggesting little or no transport after death. *T. disjuncta* lives today off the coasts of Oregon, Washington, and the Alaska Peninsula in waters 4–140 fathoms deep (V. S. Mallory, personal communication).

Limestone pebbles and cobbles riddled with borings were reported from the basal San Pedro Formation at the now-destroyed Deadman Island in Los Angeles Harbor (Woodring and others, 1946). The San Pedro overlay Timms Point Silt containing *T. disjuncta*. Some of the bored cobbles had masses of the worm-like gastropod *Aletes squamigerus* attached to them.

Pico Sedimentation Relative to South Mountain

South of the Oak Ridge fault, South Mountain can be shown by direct evidence of Pico onlap to be the site of a former hill underlain by the Modelo Formation and slowly buried by Pico sediments. In the outcrop belt on Oak Ridge, the Pico onlaps the Modelo from Balcom Canyon west toward South Mountain; paleo-relief here perhaps was greater than 800 meters (Fig. 2). South of South Mountain, the Pico below the ash bed decreases in thickness northward from 360 meters in the McCulloch-Standard-Terry well

(sec. 31, T. 3 N., R. 20 W.) in the Epworth syncline to 180 meters at the outcrop, partly by onlap (Fig. 4). West of Boone Canyon, the pre-ash bed sediments are present in Humble-Bianchi Core Holes 1 and 2 (sec. 25, T. 3 N., R. 21 W.), but are absent at the outcrop on the north because of onlap (Fig. 7). South and southwest of the West Mountain oil field, the Pico onlaps the Modelo toward the northeast, in the direction of South Mountain. To the north, across the Oak Ridge fault, the Modelo Formation is too deep to have been reached by wells, but pre-ash bed Pico sediments are present there at least from West Montalvo to Bardsdale and Shiells Canyon oil fields. The fact that post-ash bed sediments at South Mountain locally onlap the Modelo shows that South Mountain was the site of a hill with respect to Pico sediments on the north, southwest, south, and east. This hill was separated from positive areas east and south of it by structural depressions which received Pico sediments.

In addition to onlap, the South Mountain hill influenced Pico deposition in another way. Pico deep-sea sandstones lens out in the direction of South Mountain. The six turbidite sandstone members of the Pico at Balcom Canyon diminish in thickness and disappear one mile west of Balcom Canyon (Fig. 2). In wells at the south end of South Mountain field and southwest of West Mountain field, several Pico sandstones thin and lens out northward in the direction of Pico wedge-out (Fig. 4).

Across the Oak Ridge fault, Pico turbidite sandstones pinch out southward in the direction of South Mountain (Schultz, 1960); this southward pinch-out provides the primary closure for some of the oil accumulations in Saticoy field. The Saticoy sandstone pinch-outs to the south are dependent upon a Pliocene positive feature other than the Oak Ridge fault. The zero thickness sandstone isopachous contours at the Saticoy field trend more westerly than the Oak Ridge fault, they intersect the fault at the east end of Saticoy field (S. H. Hamann and J. C. Taylor, personal communication). At the West Montalvo field (Hardoin, 1961) near the coast, more than 20 kilometers southwest of South Mountain but still close to the Oak Ridge fault, Pico sandstones, except for several Repettian sandstones near the base of the formation, show no change in facies to mudstone near the fault.

The reason for the thinning and disappearance

of sandstone in the direction of the South Mountain hill is not obvious. The hill was not the source of the sandstones; they contain detritus indicative of a granitic source east of the area, whereas the entire surface of the hill is composed of Modelo Formation. Not only do the sandstones thin, but individual mudstone beds also thin from all sides in the direction of South Mountain (Fig. 7). The thinning is not caused by increasing distance from the source, because the easterly-derived turbidites are found in abundance not only east, but also west and southwest of South Mountain. In fact, a subsurface section from the Montalvo area northeast toward South Mountain shows thinning *toward* the source.

Why, then, do the sandstones thin? The most likely explanation is that the South Mountain hill subsided slightly less than the area surrounding it while the Pico was being deposited. This differential subsidence persisted during burial of the hill, with the result that, at a given time, the South Mountain region, including the already-deposited Pico sediments around the yet-unburied hill, maintained a low slope toward the surrounding sea floor. Although the main body of the sandstones undoubtedly was deposited by turbidity currents traveling downhill, that part of the current depositing the feather edges of the sandstones around South Mountain may have been traveling uphill, a suggestion made to the writer by M. L. Natland based on experimental studies of laboratory-produced turbidites.

In summary, the wedge-out of the Pico in the direction of South Mountain from all sides indicates that South Mountain in Pico time was on the site of a topographic prominence gradually buried by Pico sediments. The lensing-out of turbidite sandstones and the general thinning of associated mudstone beds in the direction of South Mountain suggest that the South Mountain hill underwent less subsidence than its surroundings while it was being buried by the Pico. Relatively uneroded fault scarps, bathyal Pliocene microfaunas, and a mantle of glauconite sandstone at the Pico-Modelo unconformity overlain by fine-grained clastic sediments relatively free of South Mountain-derived debris are evidence that the South Mountain hill was a seaknoll, below sea level in its entirety. This evidence, the absence of shallow-water sedimentary features at the unconformity, and the absence of

any shallowing, indicated by paleobathymetric study of Foraminifera from adjacent areas where the Mio-Pliocene hiatus is greatly reduced or absent (Natland, 1952), indicate that the unconformity probably was submarine during its entire existence, a period of about 10 million years.

PRE-PICO EROSION IN ADJACENT AREAS

Although no shallow-water features are present at the Mio-Pliocene unconformity in nearby areas, the unconformity is obviously erosional in areas south and west of South Mountain.

South of the Epworth syncline is the Berylwood anticline area, where the Pico Formation overlies the Modelo Formation with angular unconformity (Fig. 11). The Pico onlaps the Modelo southward, but, in contrast to the South Mountain area, is coarse-grained, with the sandstone percentage increasing in the direction of onlap. On the crest of the Berylwood anticline, the Modelo underwent considerable pre-Pico erosion; in the Shell-McBean No. 1 well, pre-Pico erosion penetrated within 45 meters of the base of the Modelo Formation. The Las Posas fault underwent at least 90 meters of vertical movement after Modelo deposition; the upthrown block was bevelled by erosion prior to Pico deposition. Yet the microfauna is characteristic of bathyal water depths and includes such bathyal Foraminifera as *Bolivina spissa*, *Bulimina subacuminata*, *Cassidulina delicata*, *Gyroidina altiformis*, and *Planulina ornata*. The presence of a bathyal Pico microfauna and the absence of any evidence for shallowing at the unconformity make it likely that the erosion and subsequent deposition at Berylwood took place in deep water.

The same type of evidence suggests that the Mio-Pliocene erosional unconformity is submarine in the Oxnard Plain. Here, 300 meters of Modelo Formation have been eroded in a southeasterly direction (Kaplow, 1947). At the southeastern edge of the Oxnard oil field, Pico sandstones directly overlie pre-Modelo volcanic rocks. Microfaunas in the Pico indicate bathyal conditions, as at Berylwood anticline, and indicate shallower-water conditions upsection. Much of the Oxnard area apparently was a submarine slope titled about 15 degrees; this slope subsequently was eroded and buried by Pico sediments, all at bathyal water depths.

Studies off the California coast by marine geologists show that some of the offshore slopes are

Recent analogs of the Oxnard and Berylwood erosion surface.

Sub-bottom profiles obtained by Moore (1960) of submarine slopes off southern California show that Recent slope deposits are thin and locally are absent, despite the long exposure of the slope to marine sedimentation processes. The base of these Recent deposits is a sharply defined horizon on sub-bottom profile records. In some profiles, the underlying beds dip more steeply basinward than do the Recent beds, suggesting angular unconformity. The Recent deposits directly overlie basinward-dipping strata as old as Eocene. The Recent deposits thus are not gradational with the underlying deposits, but rest unconformably on them.

One explanation for the lack of Recent deposits is provided by Menard (1960), who suggests that continental shelf sediments off California are moved parallel with the shore by longshore drift until they are trapped in the heads of submarine canyons. Later, they are transported down the canyons by deep-sea processes and redeposited on the deep-sea floor. The California shelf, according to Menard, is a realm of sedimentary transport, not of deposition.

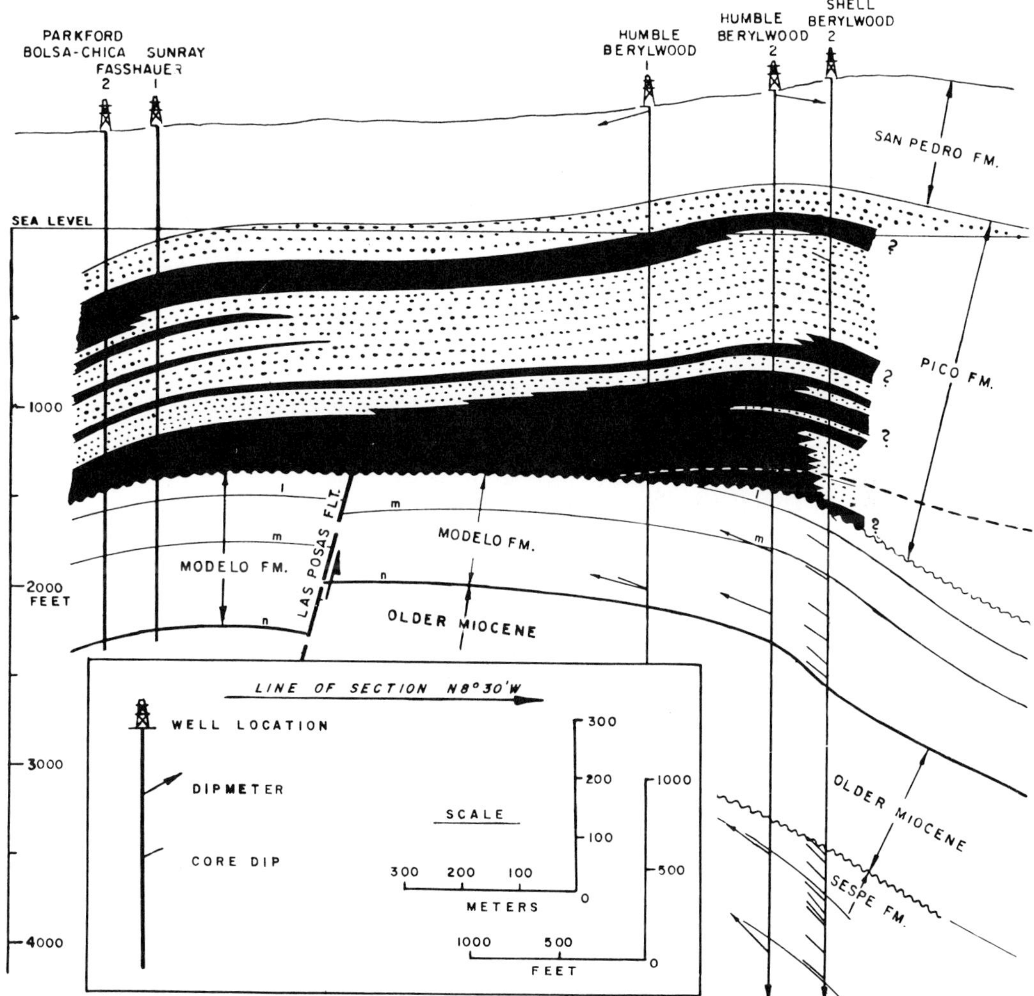

FIG. 11.—Cross section through Berylwood anticline. Within Pico Formation, sandstones are in black, siltstones and shales are in stippled pattern; no control north of Shell Berylwood No. 2 well. Dashed line represents horizon approximately contemporaneous with ash bed at South Mountain.

The concept of the continental shelf and slope as realms of erosion has been presented by Dietz (1952, 1963). Although the general applicability of this concept has been questioned (Moore and Curray, 1963), it is an attractive explanation for certain submarine slopes off the coast of California, and adequately explains the erosion surface at Berylwood anticline and the Oxnard Plain. The slope is eroded back by turbidity currents which deposit their sedimentary load in the adjacent basin. The basin fills with sediment, gradually burying the slope; such burial could occur in deep water.

An adequate explanation of the submarine nature of the unconformity must include the reason why fault scarps are eroded in some areas but not in others. It is also necessary to consider how the scarps could remain unaffected by erosion and current action for a long time. It is of interest to compare features described at South Mountain with similar features found on the present sea floor.

Comparison with Present-Day Offshore Features

SEAKNOLLS

Certain features of the Pliocene South Mountain seaknoll are similar to those of present-day seaknolls and banks in the continental borderland off southern California. Holzman (1952), in a study of Cortes and Tanner Banks, noted that the bank tops contain boulders principally of Miocene rock types, most of which are encrusted with organisms. These boulders possibly are analogous with the South Mountain Miocene limestone fragments with borings (Fig. 10). Holzman considered the presence of encrusted Miocene boulders as an indication that sediment accumulation on the banks was insufficient to cover the rocks. Holzman further observed that very coarse sand lies on the tops of the banks; the grain size of this sand decreases downslope to a very fine size. On the banks, the sand is mainly coarse calcareous debris composed of tests of benthonic organisms, but also contains echinoid spines, bryozoans and sponge spicules, glauconite, and collophane. Except for the collophane, echinoid spines, and bryozoans, the sand is very similar to the glauconite sandstone near the summit of the South Mountain seaknoll where the Modelo Formation is directly overlain by coarse-grained, glauconitic Pico sediments younger than the ash bed. The South Mountain seaknoll is overlain at the crest by a small biostrome of *Thyasira*, suggesting that in the shoalest part of the seaknoll, the Pliocene beds directly overlying the Modelo were deposited in neritic rather than in bathyal water depths.

STEEP SCARPS ON SEA FLOOR

Evidence for steep slopes on the ocean bottom is of two types: (1) direct observation by divers, from submerged vessels, or by ocean-floor photographs, and (2) echo soundings.

Direct observation provides evidence that, on a very small scale at least, very steep topography exists on the sea floor off southern California (Emery, 1960, p. 44). Recent dives off La Jolla, California, in the Costeau diving saucer have shown the existence of vertical and overhanging walls (Shepard and others, 1964).

Echo soundings, particularly those made by the precision depth recorder, give evidence of larger scarps, but are unable to measure slopes steeper than 45 degrees. Slightly less steep slopes have been measured; examples are slopes of 37 degrees on the flanks of Horizon Guyot in the Mid-Pacific Mountains (Hamilton, 1956) and 43 degrees on the side slopes of Anegada Passage east of Puerto Rico (Frassetto and Northrop, 1957). Slopes steeper than this generally appear as an abrupt change of depth on the echogram which cannot be accurately measured because of vertical exaggeration, dispersion of the sound beam with depth, and side slope reflection. Gaskell and others (1953) noticed that soundings taken at one-minute intervals in the Marianas Trench in some places showed an abrupt depth change of 50 to 70 fathoms after traversing a flat-bottomed stretch for several minutes. Koczy (1954) reported vertical[7] scarps on deep-sea plains on the floor of the Indian Ocean southeast of Ceylon and on the eastern Pacific Ocean floor. Dietz and others (1954) wrote that the Marshall scarp in the western Pacific has an essentially vertical[6] cliff more than 300 meters high. Stewart and others (1964) report that the continental slope south of Trincomalee, Ceylon, deepens 4,000 feet in a horizontal distance of 3,800 feet, possibly the steepest continental slope in the world.

It is concluded that scarps of a magnitude and

[6] Scarps reported as "vertical" are perhaps more correctly described as too steep to be measured by the sounding equipment, and could possibly be as gentle as 45 degrees.

steepness comparable with those at South Mountain occur on the present day sea floor, although a survey of the literature suggests that they are not common.

PROBLEM OF LITHIFICATION OF SEAKNOLL SCARPS

It has been concluded that the scarps of Modelo Formation on the South Mountain seaknoll existed for a long time, during which only a few meters of sediment were removed by erosion. If the lack of erosion of the scarps means the absence of any kind of significant erosion after the cessation of Modelo deposition, it would follow that the uppermost Modelo beds beneath the unconformity would be essentially the last sediments deposited before the scarps were formed, and that the unconformity would represent a period of non-deposition. Should this be the fact, the topmost Modelo sediments, never having been buried, would have been non-lithified at the time the scarps were formed. But would it be possible for non-lithified, fine-grained sediments to form scarps capable of remaining standing for millions of years?

The angle of internal friction of non-lithified sediments off the southern California coast has been found by Moore (1960) to be less than the angle of slope on which they lie (24–40 degrees); hence, these sediments are stable under ordinary conditions. But several workers have shown that sediments can slide down slopes much gentler than their angle of internal friction; this may be due to their being jarred by earthquakes. Terzaghi (1957) suggests that sediments ranging in grain size from fine sand to medium silt may be deposited in a metastable condition; jarring action caused by an earthquake would cause a temporary loss of bearing strength, resulting in liquefaction and net downslope movement. The formation of the South Mountain fault scarps must have been attended by many earthquakes; it appears unlikely that non-lithified sediments could have remained standing even long enough to allow the scarps to form, let alone for 5 to 10 million years.

Slump deposits have been reported in southern California diastrophic basins, but are not found in the South Mountain scarp area. Modelo debris is almost completely absent in the Pico, except for the marine-bored rubble associated with the scarps. The rubble itself is not plastically deformed, nor are the scarp beds plastically deformed. It is probably that the topmost Modelo

preserved today was lithified at the time the scarps were formed.

Non-lithified sediments could have been removed by erosion prior to scarp formation, exposing lithified, compacted beds with greater shear strength. The minimum amount of erosion necessary to remove these non-lithified sediments and expose lithified beds can be estimated both by studying Recent abyssal sea-floor deposits which are unaffected by later erosion and by determining the depth below the sea floor at which lithification takes place. Hamilton (1959) subjected abyssal sea-floor sediments to load tests and noted a sharp decrease in porosity with increasing pressure in deep-sea clays underlying the sea floor and a more gradual decrease with greater pressure. This change in the rate of porosity decrease was determined in the laboratory to occur at 35–45 per cent porosity at a depth below the sea floor of 150–300 meters (calculated by converting load pressure in his laboratory tests to overburden pressure). Hamilton considered this to be the depth at which lithification (transformation of clay to shale) takes place. Hedberg (1936) believed that lithification occurs at about 35 per cent porosity, at which point compaction and dewatering have proceeded to the stage that solid grains begin to come into contact with one another. Hamilton (1959) correlated the lithification depth to the base of a low-velocity unconsolidated layer of sea-floor sediments which, in all ocean basins, is about 300 meters or more thick. According to Hamilton, sediments are added to the sea floor, causing an increase in overburden pressure at the lower limit of the non-lithified layer; this results in lithification of the underlying sediments. Thus, the 300-meter-thick unconsolidated layer maintains the same thickness with continued sedimentation. If the scarp materials at South Mountain had to be lithified in order for the scarps to stand without collapsing, assuming Hamilton's conclusions to be applicable, at least 300 meters of Modelo sediments had to be removed by erosion before the scarps were formed (Fig. 12).

Subaerial erosion of the non-lithified Modelo cover would have required that the bathyal sea floor be raised to at least sea level in post-Modelo time, then returned to bathyal depths before Pico sedimentation began. Such an isolated tectonic pulse would have had to be of such a nature that no direct evidence of its presence was

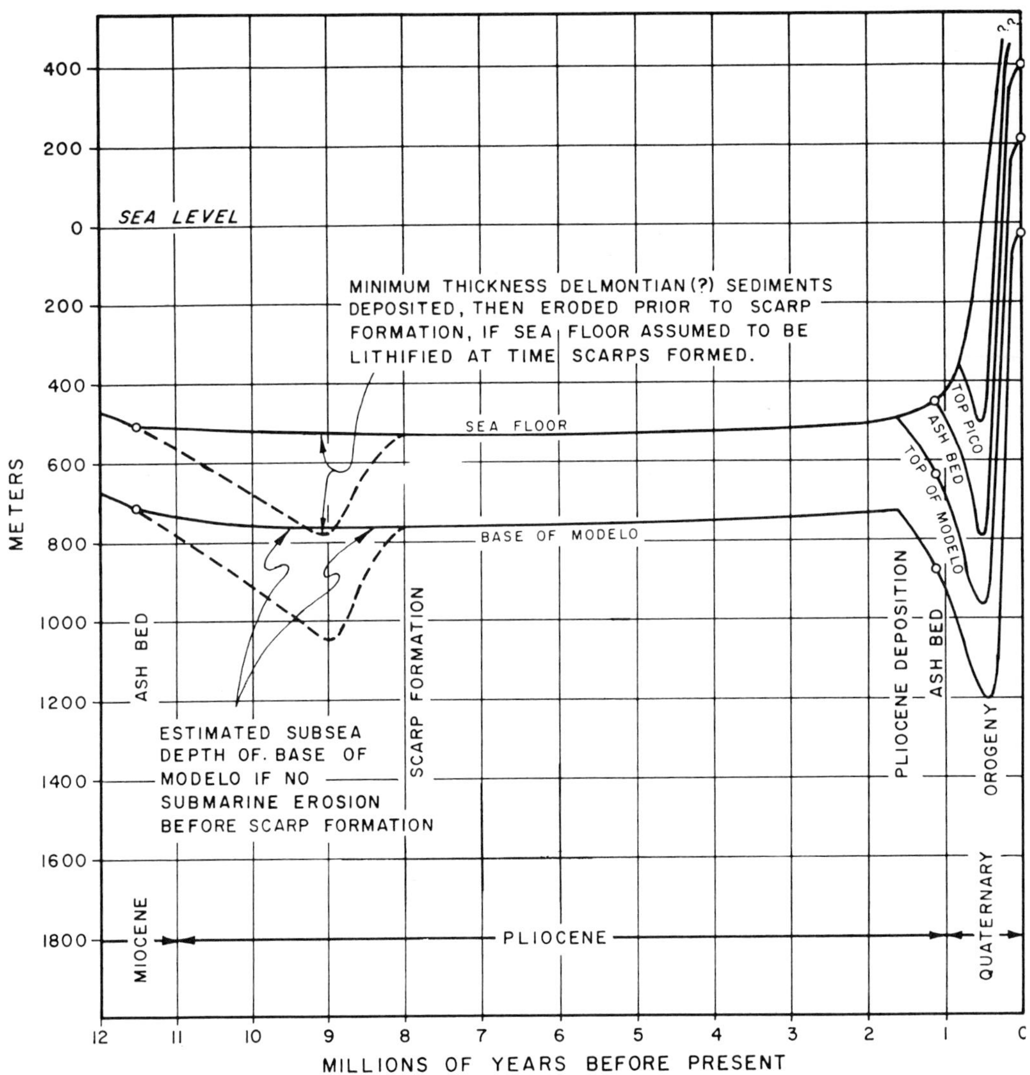

Fig. 12.—Altitude vs. time graph, South Mountain area. Thicknesses and present-day altitudes are those in SE¼, SE¼, sec. 19, T. 3 N., R. 20 W. The absolute ages of the beginning and end of the Pliocene are taken from Holmes (1960). The absolute ages of other events (Pasadenan orogeny, etc.) are geologic estimates not based on radiometry. The altitude-time graph technique was adapted from von Bubnoff (1931).

preserved anywhere in the Ventura basin, either as shallow-water or non-marine deposits, or as debris in adjacent sedimentary basins. In the adjacent Los Angeles basin and in the eastern Ventura basin, where sedimentation was continuous from late Miocene through Pliocene time, no evidence for a shallowing during the late Miocene or early Pliocene has been found. The erosion was probably not subaerial but submarine, similar to the erosion at Berylwood anticline and Oxnard Plain.

PROPOSED HYPOTHESIS

If South Mountain was subjected to submarine erosion which removed 150 to 300 meters of Modelo prior to scarp formation, it is clear that the Mio-Pliocene unconformity encompasses at least two distinct events. First, the sea floor changed from a realm of Modelo deposition to one of erosion, perhaps by regional tilting westward. Other factors remaining the same, erosion rates were rapid at first, as non-lithified material was removed. Then, as the base of the non-lithified layer was approached, erosion was slower. Erosion continued long enough to expose lithified rock on the sea floor, but not long enough for the ponded sediment level in the adjacent basin to encroach upon and cover the erosion surface. The second event was the formation of the scarps accompanying the rise of the seaknoll. When the scarps were formed, erosion essentially ceased, as shown by the near lack of subsequent modification of the scarps.

What sort of erosive mechanism would cease when the area was uplifted? Turbidity currents are thought by many workers to be able to erode the sea bottoms over which they move. Such currents, more dense than the enclosing clear water, tend to seek low places on the sea floor. Turbidity currents are deflected around uplifts on the sea floor. These currents would have been the most likely mechanism for eroding the Modelo at South Mountain because (1) they would be able to act in deep water, and (2) they would be effective only when the area was low.

The rounded pebbles, cobbles, and boulders of basement found in the glauconite sandstone at the unconformity at South Mountain are probably the only remaining evidence of the turbidity currents there; these rounded stones, incidentally, illustrate both events included in the time that the unconformity existed. If South Mountain, after the scarps formed, was isolated from basement provenances by deep channels, how did these pebbles get to its summit? Exotic methods of transport (kelp rafting, ice rafting, transport in the stomach of swimming animals) are considered unlikely because of the large size of some of the stones and their restriction to the glauconitic Pico. It is more likely that these stones were deposited from a turbidity current immediately prior to scarp (and seaknoll) formation. When the seaknoll was formed, subsequent turbidity currents were deflected around its base and the deposits of the earlier flow were stranded on top. Later, bottom currents on the seaknoll top were able to winnow out the finer clastic sediments, but were too weak to move the large stones, which are preserved as the only direct evidence of the earlier strong currents.

The erosional event could have been a regional tilting affecting the entire southern Ventura basin. The second event may have been local, affecting only South Mountain. Possibly related to scarp formation was the emplacement of a sill of hornblende andesite, which may have occurred later than the deposition of the Modelo (Yeats, 1964). The andesite sill roughly coincides in plan with the seaknoll. It may have been emplaced rapidly, under a thin Modelo cover, which was buckled and block-faulted at the same time. Induration of the Modelo by the sill, in addition to compaction by Delmontian sediments, may have caused the scarp-forming material to be too hard to slump.

SUGGESTED RELATIONSHIP OF SEAKNOLL TO
LATE CENOZOIC VENTURA BASIN HISTORY

The Mohnian Modelo Formation was buried by younger sediments, probably Delmontian, but possibly as young as early Pliocene. Regional westward tilting occurred, resulting in a shift from deposition to erosion. This tilting may have accompanied pronounced uplift of the mountains east of the Ventura basin, because the sediments brought to the area were probably much coarser-grained than the Modelo. The coarse-grained sediments were not deposited in the South Mountain area, but were swept past by turbidity currents to a final resting place far to the west. As the turbidity currents continued westward, their erosive power diminished, with the result that the underlying Modelo west of South Mountain was less effectively eroded. At West Montalvo oil field, on the present coastline, over 300 meters of Delmontian sediments overlie the Mohnian Modelo Formation (Hardoin, 1961). Correlative sediments at South Mountain may have been removed by turbidity current scour.

Then, in the middle of the basin, a seaknoll, accompanied by faulting, rose on the sea floor. This episode may have included the emplacement of a sill of andesite. The rise of the seaknoll caused turbidity currents to be deflected around its base. The remnants of the last turbidity flow prior to seaknoll formation were stranded on its summit. Weak bottom currents winnowed out the finer-grained elements, leaving only larger stones as a lag gravel.

On Figures 13 and 14A the geography and geology of the South Mountain area at the beginning of upper Pliocene deposition have been reconstructed. North of the seaknoll was the flat-floored Santa Clara basin, the main Pliocene depositional trough of the Ventura basin, already containing a thick sequence of lower Pliocene sediments. South and east of the South Mountain seaknoll, where the upper Pliocene beds today overlie the Modelo directly, the sea was floored

531

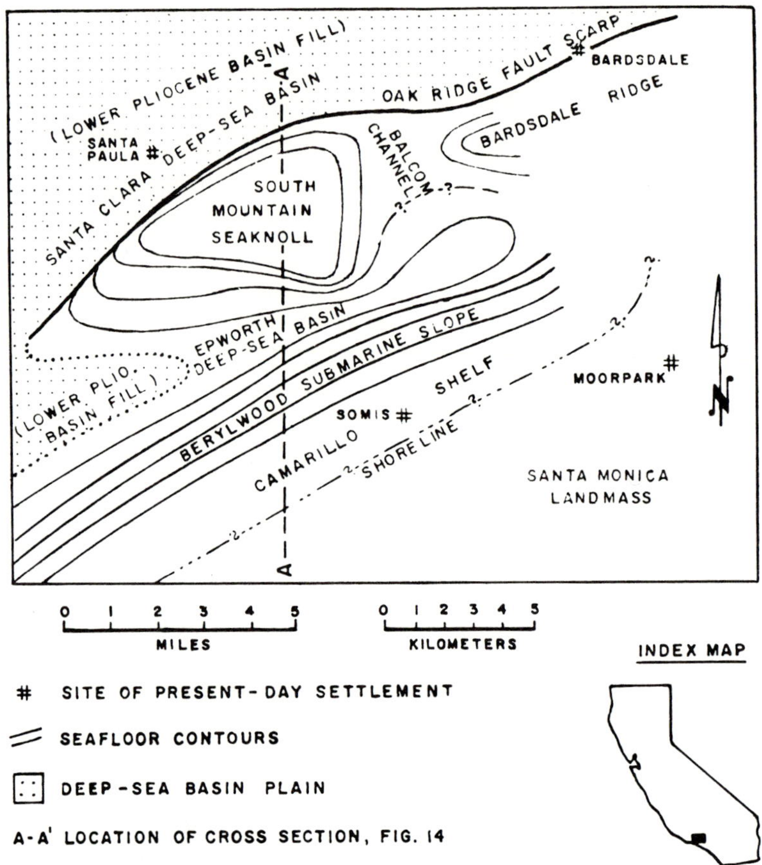

FIG. 13.—Paleogeography of South Mountain-Oak Ridge area at end of lower Pliocene deposition.

entirely by the Modelo Formation. The Epworth basin possibly was a submarine canyon, still receiving turbidity currents. Farther south, a submarine slope merged southward into a mainland shelf where today San Pedro coarse clastic, shallow-water sediments directly overlie Modelo or older beds. The nearest area undergoing subaerial erosion was probably the Santa Monica Mountains. But the main source of sediments then, as now, was the high mountainous area east of the Ventura basin.

The South Mountain seaknoll, not being exposed to subaerial erosion during Pliocene time, shed very little detritus; most of that which formed was rockfall from the Modelo scarps. Because of the lack of sediment from the seaknoll, turbidity currents were not generated on its slopes, and the scarps were not scoured. As a topographic prominence, the seaknoll continued to deflect sediment-laden currents from nearby

land areas into channels around its base. As a result, these currents also were unable to erode the scarps.

In contrast, the submarine slope east and south of the Epworth basin continued to be scoured. Sediments derived from subaerially eroded land areas, originally deposited in shallow water, were transported across the shelf, then carried down the slope, at least partly by turbidity currents. These currents, loaded with abrasive sediments, scoured the slope as they transported the sediments, bevelling fault scarps and locally removing the Modelo Formation entirely. The sediments carried by these currents were deposited on the basin floor, which, through aggradation, encroached upon the submarine slope and seaknoll.

Once, during Wheelerian time, the sea was blanketed by an ash fall which settled through the water column onto the sea floor (Fig. 14B).

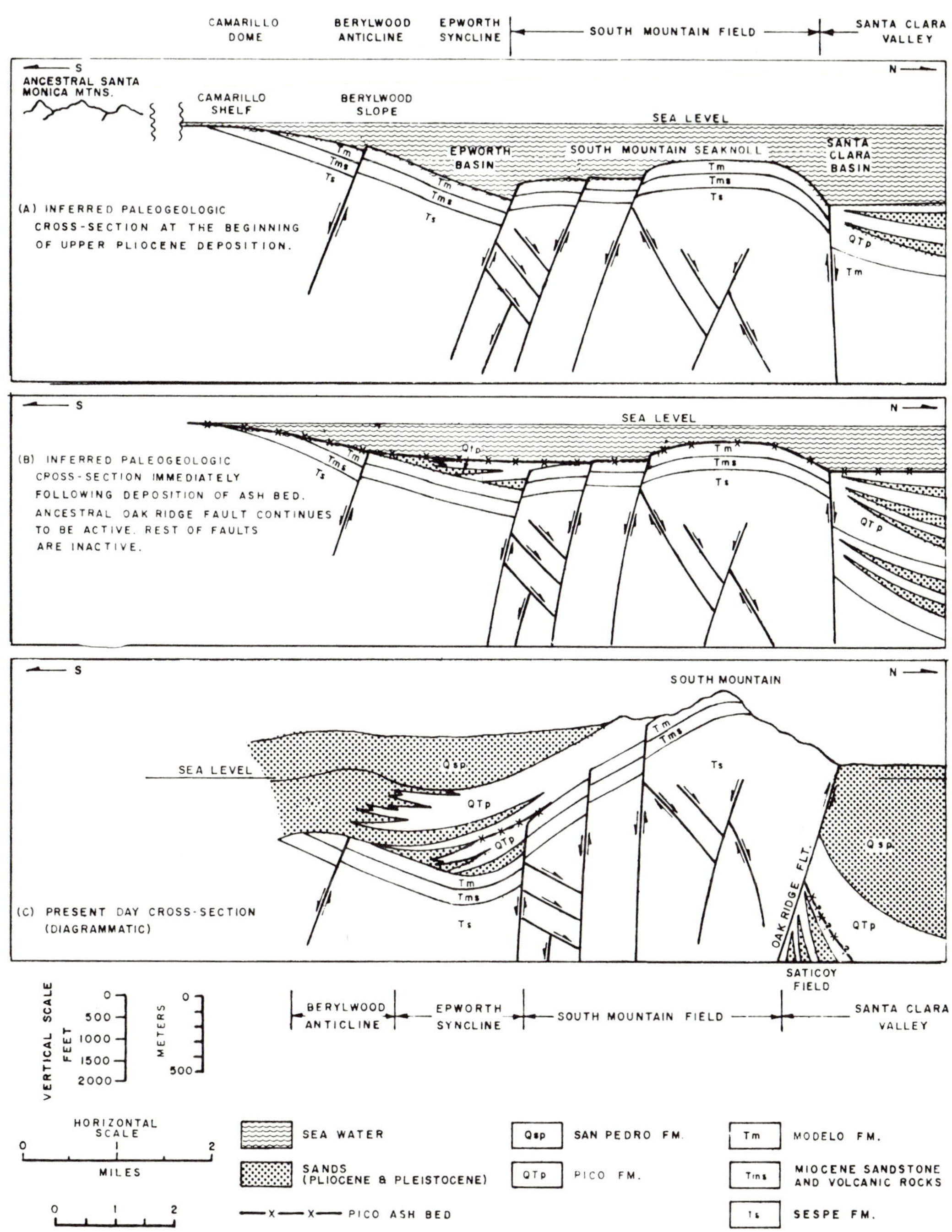

Fig. 14.—Evolution of South Mountain area from beginning of upper Pliocene deposition to present. Location of cross-section is indicated in Figure 13.

By this time, the basin floor sediments occupied the Epworth basin and the Balcom channel. The South Mountain seaknoll, its surface coated with relict sediments and glauconite, rose like a steptoe above the aggrading basin plain. Ash deposited on the seaknoll or on the slope soon was carried downslope by bottom currents in perhaps the same way that pelagic sediments, deposited by slow diffusion through the water column, must have been removed from the seaknoll. Basin plain sedimentation slowly overtook subsidence, so that the later sediments were deposited in water shoaler than 200 meters. The shoalest areas of the South Mountain seaknoll were suitable for clam biostrome formation.

Eventually, the seaknoll and the slope were buried, and the basin plain approached emergence. Easterly-derived coarse sands and gravels now were deposited, perhaps reflecting the early stages of Pasadenan orogeny in the eastern uplands, or climatic change related to Pleistocene glaciation. During the deposition of these coarse clastic deposits, aggradation overtook subsidence, and the basin floor became emergent; the youngest deposits were non-marine. Finally, during the Pleistocene Pasadenan orogeny, the beds were folded and faulted (Fig. 14C), exhuming the South Mountain seaknoll.

GENERAL APPLICABILITY OF SUBMARINE UNCONFORMITY CONCEPT

Rocks and faunas, above and below the Mio-Pliocene unconformable contact in the southern Ventura basin, provide evidence that the unconformity developed and was maintained under submarine rather than subaerial conditions. Perhaps the most striking feature of the unconformity surface is the South Mountain seaknoll, flanked by relatively uneroded fault scarps and coated by a veneer of glauconite sandstone. These non-erosional features were preserved at South Mountain, not only because of the submarine nature of the unconformity but also because of the physiographic isolation of the seaknoll from turbidity currents.

It is probable that topographic prominences or seaknolls in the midst of near-shore, deep-sea basins occupy a relatively small area, compared with that of submarine shelves and slopes which lie directly in the path of density-generated, scouring bottom currents transporting sediment from subaerial source areas. Certainly this is true in the submarine basins off the present-day southern California coast. In the Ventura basin, the South Mountain seaknoll occupies a very small area in contrast to the eroded submarine slope at Berylwood anticline, Camarillo dome, and the Oxnard Plain. If the seaknoll had not been exhumed by later subaerial erosion, and if oil had not been discovered in association with it, the origin of this topographic feature might not have been recognized at all.

It seems likely, then, that the eroded submarine unconformity, with its more subtle evidence of submarine origin, may be much more common than the more spectacular uneroded unconformity. The latter type of unconformity probably would lack a glauconite sandstone veneer, and probably would lack fault-scarp topography. The signature of the eroded submarine unconformity would be deep-sea sediments instead of transgressive shallow-water sediments directly above the unconformity. In the absence of faunal evidence of bathymetry (as would be expected in most pre-Cretaceous sequences), physical criteria for deep-sea sedimentation, such as turbidity current features, would have to suffice.

This study indicates that buried topography or erosional unconformities are inadmissible by themselves as evidence for subaerial erosion, and that onlapping depositional surfaces cannot be equated with former sea levels. In arriving at a reasonable hypothesis of origin, each unconformity must be studied as a separate entity, and both subaerial and submarine origins must be considered.

REFERENCES

Arnold, Ralph, 1903, Paleontology and stratigraphy of the marine Pliocene and Pleistocene of San Pedro, California: Calif. Acad. Sci., Proc., v. 3, p. 1–420.

Baddley, E. R., 1954, Geology of the South Mountain oil field, Ventura County, California: Calif. Div. Mines Bull. 170, Map Sheet 29.

Bailey, T. L., 1943, Late Pleistocene Coast Range orogenesis in southern California: Geol. Soc. America Bull., v. 54, p. 1549–1568.

Baldwin, E. Joan, 1959, Pliocene turbidity current deposits in Ventura basin, California: unpub. M.A. thesis, Univ. Sou. Calif., 66 p.

Bandy, O. L., 1953a, Ecology and paleoecology of some California Foraminifera. Part II. Foraminiferal evidence of subsidence rates in the Ventura basin: Jour. Paleontology, v. 27, p. 200–203.

——— 1953b, Ecology and paleoecology of some California Foraminifera. Part I. The frequency distribution of Recent Foraminifera off California: Jour. Paleontology, v. 27, p. 161–192.

Brown, R. S., 1959, The geology of the Grimes Canyon area, Moorpark and Fillmore quadrangles, Ventura County, California: unpub. M.A. thesis, Univ. Calif. Los Angeles, 121 p.

Dietz, R. S., 1952, Geomorphic evolution of continental terrace: Am. Assoc. Petroleum Geologists Bull., v. 36, p. 1802–1819.

——— 1963, Wave-base, marine profile of equilibrium, and wave-built terraces: a critical appraisal: Geol. Soc. America Bull., v. 74, p. 971–990.

Dietz, R. S., Menard, H. W., and Hamilton, E. L., 1954, Echograms of the Mid-Pacific expedition: Deep-Sea Research, v. 1, p. 258–272.

Dott, R. H., Jr., 1963, Dynamics of subaqueous gravity depositional processes: Am. Assoc. Petroleum Geologists Bull., v. 47, p. 104–128.

Emery, K. O., 1960, The sea off southern California: New York, John Wiley & Sons, Inc., 366 p.

Evernden, J. F., Savage, D. E., Curtis, G. H., and James, G. T., 1964, Potassium-argon dates and the Cenozoic mammalian chronology of North America: Am. Jour. Sci., v. 262, p. 145–198.

Frassetto, R., and Northrop, J., 1957, Virgin Islands bathymetric survey: Deep-Sea Research, v. 4, p. 138–146.

Gaskell, T. F., Swallow, J. C., and Ritchie, G. S., 1953, Further notes on the greatest oceanic sounding and the topography of the Marianas Trench: Deep-Sea Research, v. 1, p. 60–63.

Gorsline, D. S., and Emery, K. O., 1959, Turbidity current deposits in San Pedro and Santa Monica basins off southern California: Geol. Soc. America Bull., v. 70, p. 279–290.

Hamilton, E. L., 1956, Sunken islands of the Mid-Pacific Mountains: Geol. Soc. America Mem. 64, 97 p.

——— 1959, Thickness and consolidation of deep-sea sediments: Geol. Soc. America Bull., v. 70, p. 1399–1424.

Hardoin, J. L., 1961, McGrath area of West Montalvo oil field: Summary of Operations, California Oil Fields, v. 47, p. 43–53.

Hedberg, H. D., 1936, The gravitational compaction of clays and shales: Am. Jour. Sci., v. 231, p. 241–287.

Holmes, Arthur, 1960, A revised geological time-scale: Trans. Edinburgh Geol. Soc., v. 17, p. 183–215.

Holzman, J. E., 1952, Submarine geology of Cortes and Tanner Banks: Jour. Sed. Petrology, v. 22, p. 97–118.

Kaplow, E. J., 1947, Oxnard oil field: Summary of Operations, California Oil Fields, v. 33, p. 16–20.

Kew, W. S. W., 1924, Geology and oil resources of a part of Los Angeles and Ventura Counties, California: U. S. Geol. Survey Bull. 753, 202 p.

Kleinpell, R. M., 1938, Miocene stratigraphy of California: Tulsa, Oklahoma, Am. Assoc. Petroleum Geologists, 450 p.

Koczy, F. F., 1954, A survey on deep-sea features taken during the Swedish Deep-Sea expedition: Deep-Sea Research, v. 1, p. 176–184.

Lung, Richard, 1958, Geology of the South Mountain area, Ventura County, California: unpub. M.A. thesis, Univ. Calif. Los Angeles, 80 p.

Menard, H. W., 1960, Possible pre-Pleistocene deep-sea fans off central California: Geol. Soc. America Bull., v. 71, p. 1271–1278.

Moore, D. G., 1960, Acoustic-reflection studies of the continental shelf and slope off southern California: Geol. Soc. America Bull., v. 71, p. 1121–1136.

——— and Curray, J. R., 1963, Sedimentary framework of continental terrace off Norfolk, Virginia, and Newport, Rhode Island: Am. Assoc. Petroleum Geologists Bull., v. 47, p. 2051–2054.

Natland, M. L., 1933, The temperature- and depth-distribution of some Recent and fossil Foraminifera in the southern Califonia region: Bull. Scripps Inst. Oceanography, Univ. Calif., v. 3, no. 10, p. 225–230.

——— 1952, Pleistocene and Pliocene stratigraphy of southern California: unpub. Ph.D. thesis, Univ. Calif. Los Angeles, 165 p.

——— and Kuenen, P. H., 1951, Sedimentary history of the Ventura basin, California, and the action of turbidity currents: Soc. Econ. Paleontologists and Mineralogists Special Pub. 2, p. 76–107.

Pressler, E. D., 1929, The Fernando group in the Las Posas-South Mountain district, Ventura County, California: Univ. Calif. Pub., Bull. Dept. Geol. Sci., v. 18, No. 13, p. 325–345.

Resig, J. M., 1958, Ecology of Foraminifera of the Santa Cruz basin, California: Micropaleontology, v. 4, p. 287–308.

Schultz, C. H., 1960, Saticoy oil field: Summary of Operations, California Oil Fields, v. 46, p. 58–69.

Shepard, F. P., and Emery, K. O., 1941, Submarine topography off the California coast: canyons and tectonic interpretation: Geol. Soc. America Special Paper 31, 171 p.

Shepard, F. P., Curray, J. R., Inman, D. L., Murray, E. A., Winterer, E. L., and Dill, R. F., 1964, Saucer dives in submarine canyons off La Jolla, California: Program, 1964 Annual Meetings, Geol. Soc. America, p. 182–183.

Stewart, H. B., Shepard, F. P., and Dietz, R. S., 1964, Submarine canyons off eastern Ceylon: Program, 1964 Annual Meetings, Geol. Soc. America, p. 197.

Taylor, D. W., 1948, Fundamentals of soil mechanics: New York, John Wiley & Sons, Inc., 700 p.

Terzaghi, Karl, 1957, Varieties of submarine slope failures: Proc. 8th Texas Conf. Soil Mechanics and Foundation Eng., Sept. 14–15, 40 p.

Uchio, Takayasu, 1960, Ecology of living benthonic Foraminifera from the San Diego, California, area: Cushman Found. Foram. Research Special Pub. 5, 72 p.

von Bubnoff, Serge, 1931, Grundprobleme der Geologie: Berlin, Gebrüder Borntraeger, p. 44–46.

Walton, W. R., 1955, Ecology of living benthonic Foraminifera, Todos Santos Bay, Baja California: Jour. Paleontology, v. 29, p. 952–1018.

Winterer, E. L., and Durham, D. L., 1958, Geologic map of a part of the Ventura basin, Los Angeles County, California: U.S. Geol. Survey Oil and Gas Map OM 196.

——— and ——— 1962, Geology of southeastern Ventura basin, Los Angeles County, California: U. S. Geol. Survey Prof. Paper 334-H, p. 275–366.

Woodring, W. P., Bramlette, M. N., and Kew, W. S. W., 1946, Geology and paleontology of Palos Verdes Hills, California: U. S. Geol. Survey Prof. Paper 207, 145 p.

Yeats, R. S., 1964, Andesite sill at South Mountain oil field, Ventura County, California: Geol. Soc. America Special Paper 76, p. 233.

Zalesny, E. R., 1959, Foraminiferal ecology of Santa Monica Bay, California: Micropaleontology, v. 4, p. 101–126.

GEOGRAPHICAL LISTING

AFRICA

Bandy, O. L., 1960, General correlation of foraminiferal structure with environment: Copenhagen, 21st Internat. Geol. Cong., pt. 22, p. 7-19.

Barr, F. T., and B. R. Walker, 1966, Late Tertiary channel system in northern Libya and its implications on Mediterranean Sea level changes, *in* Washington, D.C., Initial Repts. Deep Sea Drilling Proj., v. 13, p. 1244-1251.

Burke, K., 1972, Longshore drift, submarine canyons, and submarine fans in development of Niger delta: AAPG Bull., v. 56, p. 1975-1983.

Carr, M. D., and E. L. Miller, 1979, Overthrust emplacement of the Numidian flysch complex in the westernmost Mogod Mountains, Tunisia; summary: Geol. Soc. America Bull., pt. 1, v. 90, p. 513-515.

Cossey, S. P. J., 1978, Jurassic mass movement and turbidite sequences associated with growth faulting, Northern Tunisia: South Carolina Univ., Ph.D. dissert., 68 p.

Cossey, S. P., and R. Ehrlich, 1978, Growth fault-controlled submarine carbonate debris flow and turbidite deposits from the Jurassic of northern Tunisia; possible canyon fill sequences, *in* D. J. Stanley and G. Kelling, eds., Sedimentation in submarine canyons, fans, and trenches: Stroudsburg, Pa., Dowden, Hutchinson and Ross, Inc., p. 127-137.

Dietz, R. S., and H. J. Knebel, 1971, Trou sans fond submarine canyon; Ivory Coast, Africa: Deep Sea Research, v. 18, p. 441-447.

———— ———— and L. H. Somers, 1968, Cayar submarine canyon: Geol. Soc. America Bull., v. 79, p. 1821-1828.

Dingle, R. V., 1977, The anatomy of a large submarine slump on a sheared continental margin (S. E. Africa): London, Jour. Geol. Soc., v. 134, p. 293-310.

Heezen, B. C., et al, 1964, Congo submarine canyon: AAPG Bull., v. 48, p. 1126-1149.

Hoyt, J. H., B. L. Oostdam, and D. D. Smith, 1969, Offshore sediments and valleys of the Orange River (south and southwest of Africa): Marine Geol., v. 7, p. 69-84.

Padgett, G., R. Ehrlich, and M. Moody, 1977, Submarine debris flow deposits in an extension setting - Upper Devonian of western Morocco: Jour. Sed. Petrology, v. 47, p. 811-818.

Shepard, F. P., and K. O. Emery, 1973, Congo submarine canyon and fan valley: AAPG Bull., v. 57, p. 1679-1691.

ASIA - AUSTRALIA

General

Damuth, J. E., 1979, Migrating sediment waves created by turbidity currents in the northern South China Basin: Geology, v. 7, p. 520-523.

Moore, G. F., et al, 1980, Sedimentology and paleobathymetry of Neogene trench-slope deposits, Nias Island, Indonesia: Jour. Geology, v. 88, p. 161-180.

Von Der Borch, C. C., 1969, Submarine canyons of southeastern New Guinea; seismic and bathymetric evidence for the modes of origin: Deep-Sea Research, v. 16, p. 323-328.

Whittle, A. P., and G. A. Short, 1978, The petroleum geology of the Tembungo Field, East Malaysia: Offshore S. E. Asia Conf., S. E. Asia Petroleum Explor. Soc., 4 p.

Indian Ocean

Curray, J. R., and D. G. Moore, 1971, Growth of the Bengal deep-sea fan and denudation in the Himalayas: Geol. Soc. American Bull., v. 82, p. 563-572.

Ingersoll, R. V., and C. A. Suczek, 1979, Petrology and provenance of Neogene sand from Nicobar and Bengal fans, DSDP sites 211 and 218: Jour. Sed. Petrology, v. 49, p. 1217-1228.

Jipa, D., and R. B. Kidd, 1974, Sedimentation of coarser grained interbeds in Arabian Sea and sedimentation processes of the Indus Cone *in* R. B. Whitmarsh, et al, eds.: Washington, D.C., Initial Repts. of Deep Sea Drilling Proj., v. 23, p. 471-495.

Mallik, T. K., 1978, Mineralogy of deep-sea sands of the Indian Ocean: Marine Geol., v. 27, p. 161-176.

Moore, D. G., J. R. Curray, and F. J. Emmel, 1976, Large submarine slide (olistostrome) associated with Sunda Arc subduction zone, northeast Indian Ocean: Marine Geol., v. 21, p. 211-226.

Rodolfo, K. S., 1969, Sediments of the Andaman Basin, northeastern Indian Ocean: Marine Geol., v. 7, p. 371-402.

Siebold, E., 1972, Sedimentary regimes at continental margins and in abyssal cones in Indian Ocean: 24th Internat. Geol. Cong., Sec. 8, p. 75-84.

Weser, O. E., 1974, Sedimentological aspects of strata encountered on Leg 23 in the northern Arabian Sea, *in* R. B. Whitmarsh, et al, eds.: Washington, D. C., Initial Repts. of Deep Sea Drilling Proj., v. 23, p. 503-519.

Australia

Cas, R., 1979, Mass-flow arenites from a Paleozoic Interarc Basin, New South Wales, Australia; mode and environment of emplacement: Jour. Sed. Petrology, v. 49, p. 29-44.

Conolly, J. R., 1968, Submarine canyons of the continental margin, East Bass Strait (Australia): Marine Geol., v. 6, p. 449-461.

Corbett, K. D., 1972, Features of thick-bedded sandstones in a proximal flysch sequence, upper Cambrian, southwest Tasmania: Sedimentology, v. 19, p. 99-114.

Korsch, R. J., 1978, Petrographic variations within thick turbidite sequences; an example from the Late Paleozoic of eastern Australia: Sedimentology, v. 25, p. 247-265.

Winterer, E. L., 1970, Submarine valley systems around the Coral Sea Basin, Australia: Marine Geol., v. 8, p. 229-244.

New Zealand

Carter, R. M., 1979, Trench-slope channels from the New Zealand Jurassic; the Otekura Formation, Sandy Bay, South Otago: Sedimentology, v. 26, p. 475-496.

———, et al, 1978, Sedimentation patterns in an ancient arc-trench-ocean basin complex, Carboniferous to Jurassic rangitata orogen, New Zealand, *in* D. J. Stanley and G. Kelling, eds., Sedimentation in submarine canyons, fans and trenches,: Stroudsburg, Pa., Dowden, Hutchinson and Ross, Inc., p. 340-361.

——— and R. J. Norris, 1977, Redeposited conglomerates in Miocene flysch sequence at Blackmount, western Southland, New Zealand: Sed. Geology, v. 18, p. 289-319.

——— and J. K., Linkqvist, 1977, Balleny Group, Chalky Island, southern New Zealand; an inferred Oligocene submarine canyon and fan complex: Pacific Geol., v. 12, p. 1-46.

——— and ———, 1975, Sealers Bay submarine fan complex, Oligocene, southern New Zealand: Sedimentology, v. 22, p. 465-483.

Gregory, M. R., 1969, Sedimentary features and penecontemporaneous slumping in the Waitemata Group, Whangaparoa Peninsula, North Auckland, New Zealand: Jour. Geology Geophysics, v. 12, p. 248-282.

Herzer, R. H., 1979, Submarine slides and submarine canyons on the continental slope off Canterbury, New Zealand: New Zealand Jour. Geology Geophysics, v. 22, p. 391-406.

——— and D. W. Lewis, 1979, Growth and burial of a submarine canyon off Motunau, north Canterbury, New Zealand: Sed. Geology, v. 24, p. 69-83.

Laird, M. G., 1972, Sedimentology of the Greenland group in the Paparoa Range, West Coast, South Island: New Zeal. Jour. Geology and Geophysics, v. 15, p. 372-393.

Lewis, D. W., 1976, Subaqueous debris flows of Early Pleistocene age at Motunau, North Canterbury, New Zealand: New Zeal. Jour. Geology Geophysics, v. 19, p. 535-567.

Lewis, K. B., and B. P. Kohn, 1973, Ashes, turbidites, and rates of sedimentation on the continental slope off Hawkes Bay: New Zeal. Jour. Geology Geophysics, v. 16, p. 439-454.

Japan

Abe M., 1980, The sedimentary history of Minami-Aga Sandstone in Niigata Basin, Japan: Offshore S.E. Asia Conf. Proc.; SEAPEX Session, 7 p.

Hesse, R., 1977, Softex-radiographs of sliced piston cores from the Japan and southern Kurile trench and slope areas, *in* E. Honza, ed., Geological investigation of Japan and southern Kurile trench and slope areas: Japanese Geol. Survey, Cruise Rept., no. 7, p. 86-108.

Hirayama, J., and T. Nakajima, 1977, Analytical study of turbidites, Otadai Formation, Boso Peninsula, Japan: Sedimentology, v. 24, p. 747-779.

O'Brien, N. R., K. Nakazawa, and S. Toluhashi, 1980, Use of clay fabric to distinguish turbiditic and hemipelagic siltstones and silts: Sedimentology, v. 27, p. 47-61.

Sibley, D. F., and K. J. Pentony, 1978, Provenance variation in turbidite sediments, Sea of Japan: Jour. Sed. Petrology, v. 48, p. 1241-1247.

Suzuik, U., 1979, Petroleum geology of the Sea of Japan Northern Honshu: Japanese Assoc. of Petroleum Technol., v. 44, p. 59-75 (in Japanese, figures have English captions).

EUROPE

General

Aiello, E., et al, 1977, Sedimentary features of the Srednogorie zone (Bulgaria); an Upper Cretaceous intra-arc basin: Sed. Geology, v. 19, p. 39-68.

Bouma, A. H., 1964, Ancient and recent turbidites: Geologie en Mijnbouw, v. 43, p. 375-379.

———, 1972, Fossil contourites in lower Niesenflysch, Switzerland: Jour. Sed. Petrology, v. 42, p. 917-921.

Dimitrijevic, M.D., and M.N. Dimitrijevic, 1976, Transgressive overlap of a turbiditic sequence - Priabonian of Ovce Pole, Macedonia: Sed. Geology, v. 16, p. 223-234.

Dzulynski, S., M. Ksiazkiewicz, and P. H. Kuenen, 1959, Turbidites in flysch of the Polish Carpathian Mountains: Geol. Soc. America Bull., v. 70, p. 1089-1118.

———— and A. Slaczka, 1959, Directional structures and sedimentation of the Krosno Beds (Carpathian Flysch): Ann. Soc. Geol. Pologne, v. 28, p. 205-259.

———— and A. J. Smith, 1964, Flysch facies: Ann. Soc. Geol. Pologne, v. 34, p. 245-266.

———— and E. K. Walton, 1965, Sedimentary features of flysch and greywackes: Amsterdam, Elsevier Sci. Pub., 274 p.

Hesse, R., 1961, The flysch area of the Zwiesel west of Bad Tolz, Upper Bavaria: Deut. Geol. Ges., Z., v. 113, p. 293-304.

————, 1972, Lithostratigraphy, petrography, and genesis of the Bavarian (Germany) Flysch; Lower Cretaceous [lithostratigraphie, petrographie und entstehungsbedingungen des bayerischen flysches; unterkreide]: Geol., Bavarica, no. 66, p. 148-222.

————, 1974, Long-distance continuity of turbidites; possible evidence for an Early Cretaceous trench-abyssal plain in the east Alps: Geol. Soc. America Bull., v. 85, p. 859-870.

————, 1975a, Turbiditic and non-turbiditic mudstone of Cretaceous flysch sections of the east Alps and other basins: Sedimentology, v. 22, p. 387-416.

————, 1975b, Turbidite sedimentation systems and the recognition of the new Cretaceous plate boundaries in the Alpine-Carpathian arc: Nice, France, 9th Internat., Sed. Cong. Proc. [Theme 4] v. 2, p. 189-194.

———— and A. Butt, 1976, Paleobathymetry of Cretaceous turbidite basins of the east Alps relative to the calcite compensation level: Jour. Geology, v. 84, p. 505-533.

Ksiazkiewicz, M., 1958, Submarine slumping in the Carpathian Flysch: Ann. Soc. Geol. Pologne, v. 28, p. 123-151.

Marschalko, R., 1964, Sedimentary structures and paleocurrents in the marginal lithofacies of the Central-Carpathian flysch, in A. H. Bouma, and A. Brouwer, eds., Turbidites: Amsterdam, Devel. in sedimentology, 3, Elsevier Sci. Pub., p. 106-126.

Picha, F., 1974, Ancient submarine canyon of the Carpathian miogeosyncline, in R. H. Dott, Jr., and R. H. Shaver, eds., SEPM Spec. Pub. 19, p. 126-127.

————, 1979, Ancient submarine canyon of Tethyan continental margins, Czechoslovakia: AAPG Bull., v. 63, p. 67-86.

Pickering, K. T., 1979, Possible retrogressive flow slide deposits from the Kongsfjord Formation; a Precambrian submarine fan, Finnmark, N. Norway: Sedimentology, v. 26, p. 295-306.

Piper, D. J. W., A. G. Panagos, and G. G. Pe, 1978, Conglomeratic Miocene flysch, western Greece: Jour. Sed. Petrology, v. 48, p. 117-125.

Sestini, G., 1970, Flysch facies and turbidite sedimentology: Sed. Geology, v. 4, p. 559-597.

Stanley, D. J., 1967, Comparing patterns of sedimentation in some modern and ancient submarine canyons: Earth and Planetary Sci. Letters, v. 3, . 371-380.

————, 1969, Submarine channel deposits and their fossil analogs ('Fluxoturbidites'), in New concepts of continental margin sedimentation: AGI Short Course Notes, Lecture 9, p. DJS-9-1 to DJS-9-17.

————, 1974, Dish structures and sand flow in ancient submarine valleys, French Maritime Alps: Bull. Centre Rech. Pau-SNPA, v. 8, p. 351-371.

————, 1975, Submarine canyon and slope sedimentation (Gres D'Annot) in the French Maritime Alps: Nice, France, 9th Internat. Sed. Cong., 131 p.

———— and B. A. Hall, 1978, The Bucegi Conglomerate; A Romanian Carpathian submarine slope deposit: Nature, v. 275, p. 60-64.

———— H. D. Palmer, and R. F. Dill, 1978, Coarse sediment transport by mass flow and turbidity current processes and downslope transformations in Annot Sandstone canyon-fan valley system, in D. J. Stanley and G. Kelling, eds., Sedimentation in submarine canyons, fans, and trenches: Stroudsburg, Pa., Dowden, Hutchinson, Ross, Inc., p. 85-115.

———— and R. Unrug, 1972, Submarine channel deposits, fluxoturbidites and other indicators of slope and base-of-slope environments in modern and ancient marine basins, in J. K. Rigby and W. K. Hamblin, eds., Recognition of ancient sedimentary environments: SEPM Spec. Pub. no. 16, p. 287-340.

Swarbrick R. E., and M. A. Naylor, 1980, The Kathikas melange, S. W. Cyprus; Late Cretaceous submarine debris flows: Sedimentology, v. 27, p. 63-78.

Unrug, R., 1963, Istebna beds - a fluxoturbidite formation in the Carpathian Flysch: Ann. Soc. Geol. Pologne, v. 33, p. 49-92.

Von Rad, U., 1968, Comparison of sedimentation in the Bavarian Flysch (Cretaceous) and Recent San Diego Trough (California): Jour. Sed. Petrology, v. 38, p. 1120-1154.

England-Ireland

Allen, J. R. L., 1960, The Mam Tor Sandstones; a "turbidite" facies of the Namurian deltas of Derbyshire, England: Jour. Sed. Petrology, v. 30, p. 193-208.

———, 1964, Primary current lineation in the Lower Old Red Sandstone (Devonian), Anglo-Welsh basin: Sedimentology, v. 3, p. 89-108.

Anketell, J. M., and J. P. B. Lovell, 1976, Upper Llandoverian Grogal Sandstones and Aberystwyth Grits in the New Quay area, central Wales; a possible upward transition from contourites to turbidites: Geol. Jour., v. 11, p. 101-108.

Collinson, J. D., 1969, The sedimentology of the Grindslow Shales and the Kinderscout Grit; a deltaic complex in the Namurian of northern England: Jour. Sed. Petrology, v. 39, p. 194-221.

D'Olier, B., and R. J. Maddrell, 1970, Buried channels of the Thames Estuary: Nature, v. 226, p. 347-348.

Fowler, C., 1975, Geology of the Montrose Field, *in* A. W. Woodland, ed., Petroleum and continental shelf of northwest Europe: New York, John Wiley and Sons, p. 467-476.

James, D. M. D., 1972, Sedimentation across an intra-basinal slope; the Graneddwen Formation (Ashgillian), west-central Wales: Sed. Geology, v. 7, p. 291-307.

———, 1975, Caradoc turbidites at Poppit sands (Pembrokeshire), Wales: Geol. Mag., v. 112, p. 295-304.

Kelling, G., 1968, Submarine channel and fan deposits, Silurian of central Wales, United Kingdom (abs.), AAPG Bull., v. 52, p. 535-536.

——— and J. Holroyd, 1978, Clast size, shape, and composition in some ancient and modern fan gravels, *in* D. J. Stanley, and G. Kelling, eds., Sedimentation in submarine canyons, fans, and trenches: Stroudsburg, Pa., Dowden, Hutchinson and Ross, Inc., p. 138-159.

——— and M. H. Woollands, 1969, The stratigraphy and sedimentation of the Llandoverian rocks of the Rhayader District, *in* A. Woods, ed., Precambrian and Lower Paleozoic Rocks of Wales: Univ. of Wales Press, p. 255-281.

Kenyon, N. H., R. H. Belderson, and A. H. Stride, 1978, Channels, canyons and slump folds on the continental slope between South West Ireland and Spain: Oceanol. Acta, v. 1, p. 369-380.

McCabe, P. J., 1978, The Kinderscoutian Delta (Carboniferous) of northern England; a slope influenced by density currents, *in* D. J. Stanley and G. Kelling, eds., Sedimentation in submarine canyons, fans and trenches: Stroudsburg, Pa., Dowden, Hutchinson and Ross, Inc., p. 116-126.

Piper, D. J. W., 1970, A Silurian deep sea fan deposit in western Ireland and its bearing on the nature of turbidity currents: Jour. Geology, v. 78, p. 509-522.

———, 1975, A reconnaissance of the sedimentology of lower Silurian mudstones, English Lake District: Sedimentology, v. 22, p. 623-630.

———, 1972, Turbidite origin of some laminated mudstones: Geol. Mag., v. 109, p. 115-126.

Rich, R. L., 1950, Flow markings, groovings and intra-stratal crumplings as criteria for recognition of slope deposits, with illustrations from Silurian rocks of Wales: AAPG Bull., v. 34, p. 717-741.

Walker, R. G., 1966a, Deep channels in turbidite-bearing formations: AAPG Bull., v. 50, p. 1899-1917.

———, 1966b, Shale grit and Grindslow Shales; transition from turbidite to shallow water sediments in the Upper Carboniferous of northern England: Jour. Sed. Petrology, v. 36, p. 90-114.

Woodcock, N.H., 1976, Ludlow series slumps and turbidites and the form of the Montgomery Trough, Powys, Wales: Proc. Geol. Assoc., v. 87, pt. 2, p. 169-182.

Woodland, A. W., ed., 1975, Petroleum and the continental shelf of northwest Europe; 1, Geology: London, Applied Sci. Pubs., 500 p.

Italy

Angelucci, A., et al, 1967, Sedimentological characteristics of some Italian turbidites: Geol. Rom., v. 6, p. 345-420.

Bartolini C., S. Berato, and V. Bortolotti, 1975, Upper Miocene shallow-water turbidites from western Tuscany: Sed. Geology, v. 14, p. 77-122.

Bouma, A. H., 1964, Ancient and recent turbidites: Geologie en Mijnbouw, v. 43, p. 375-379.

Clari, P., and G. Ghibaudo, 1979, Multiple slump scars in the Tortonian type area (Piedmont Basin, northwestern Italy): Sedimentology, v. 26, p. 719-730.

Hesse, R., and A. Butt, 1976, Paleobathymetry of Cretaceous turbidite basins of the east Alps relative to the calcite compensation level: Jour. Geology, v. 84, p. 505-533.

Martini, I. P., and M. Sagri, 1977, Sedimentary fillings of ancient deep-sea channels; two examples from northern Apennines (Italy): Jour. Sed. Petrology, v. 47, p. 1542-1553.

——— ——— and J. H. Doveton, 1978, Lithologic transition and bed thickness — periodicities in turbidite successions of the Antola Formation, northern Apennines, Italy: Sedimentology, v. 25, p. 605-623.

McBride, E. F., 1978, Reservoir aspects of submarine fan facies; example of the Marnoso Arenacea Formation, northern Apennines, Italy: Proc. Offshore Tech. Conf., v. 1, no. 10, p. 587-591.

Mutti, E., and G. Ghibaudo, 1972, Un esempio di torbiditi di conoide sottomarina esterna; le arenarie di San Salvatore (Formazione di Bobio, Miocene) nell'Apennino di Piacenza: Mem. Acc. Sci. Torino, Classe, Sci. Fis., Mat., Nat., Serie 4, no. 16, 40 p.

———— T. H. Nilsen, and F. Ricci-Lucchi, 1978, Outer fan depositional lobes of the Laga Formation (Upper Miocene and Lower Pliocene), East-central Italy, *in* D. J. Stanley and G. Kelling, eds., Sedimentation in submarine canyons, fans, and trenches: Stroudsburg, Pa., Dowden, Hutchinson and Ross, Inc., p. 210-223.

————, et al, 1975, Examples of turbidite facies and facies associations from selected formations of the northern Apennines: Nice, France, 9th Internat. Sed. Cong., Guidebook to field trip A-11, 120 p.

———— and F. Ricci-Lucchi, 1972, Le torbiditi dell'Appennino setten-trionale; introduzione all'analisi di facies: Mem. Soc. Geol. Italiana, v. 11, p. 161-199.

———— and ————, 1975, Turbidite facies and facies associations, *in* E. Mutti, et al, eds., Example of turbidite facies and associations from selected formations of the northern Apennines: Nice, France, 9th Internat. Cong. of Sed., Guidebook to field trip A-11, p. 21-36.

———— and ————, 1978, Turbidites of the northern Apennines; introduction to facies analysis, T. H. Nilsen, translator: Intern. Geol. Review, v. 20, p. 125-166.

Pescatore, T., 1978, The irpinids; a model of tectonically controlled fan and base-of-slope sedimentation in southern Italy, *in* D. J. Stanley and G. Kelling, eds., Sedimentation in submarine canyons, fans, trenches: Stroudsburg, Pa., Dowden, Hutchinson and Ross, Inc., p. 325-339.

Ricci-Lucchi, F., 1969, Channelized deposits in the Middle Miocene flysch of Romagna (Italy): Giorn. Geol., v. 36, p. 203-282.

————, 1975, Sediment dispersal in turbidite basins; examples from the Miocene of northern Apennines: Nice, France, Rapports, 9th Internat. Cong. of Sed., [Theme 5], v. 1, p. 347-352.

————, 1975, Depositional cycles in two turbidite formations of northern Apennines (Italy): Jour. Sed. Petrology, v. 45, p. 3-43.

————, 1975, Miocene paleogeography and basin analysis in the Periadriatic Apennines, *in* C. Squyres, ed., Geology of Italy: Tripoli, Petroleum Explor. Soc., of Libya, p. 5-111.

————, 1978, Turbidite dispersal in a Miocene deep-sea plain; the Marnoso-Arenacea of the northern Apennines: Geologie en Mijnbouw, v. 57, p. 559-576.

Sagri, M., 1979, Upper Cretaceous carbonate turbidites of the Alps and Apennines deposited below the calcite compensation level: Jour. Sed. Petrology, v. 49, p. 23-28.

Scholle, P.A., 1971, Sedimentology of fine-grained deep-water carbonate turbidites, Monte Antola Flysch (Upper Cretaceous), northern Apennines: Geol. Soc. America Bull., v. 82, p. 629-658.

————, 1971, Diagenesis of deep-water carbonate turbidites, Upper Cretaceous Monte Antola Flysch, northern Apennines, Italy: Jour. Sed. Petrology, v. 41, p. 233-250.

North Sea

Bjorlykke, K., and A. Hurst, 1980, The control of clastic mineral composition on Jurassic sandstone diagenesis in the Statfjord Field of the North Sea, (abs.): Bochum, Internat. Assoc. Sedimentol. 1st Europe Mtg., p. 157.

Hendry, H. E., 1973, Sedimentation of deep water conglomerates in Lower Ordovician rocks of Quebec - composite bedding produced by progressive liquefaction of sediment?: Jour. Sed. Petrology, v. 43, p. 125-136.

Heritier, F. E., P. Lossel, and E. Wathne, 1979, Frigg field — large submarine - fan trap in Lower Eocene rocks of North Sea Viking Graben: AAPG Bull., v. 63, p. 1999-2020.

Parker, J. R., 1975, Lower Tertiary sand development in the Central North Sea, *in* A. W. Woodland, ed., Petroleum and the continental shelf of northwest Europe; 1, Geology: London, Applied Sci. Pubs., p. 447-452.

————, 1977, Deep-sea sands, *in* G. D. Hobson, ed., Developments in petroleum geology I: London, Applied Sci. Pubs., p. 225-242.

Sellevoll, M. A., and E. Sundvor, 1974, The origin of the Norwegian Channel - a discussion based on seismic measurements: Canadian Jour. Earth Sci., v. 11, p. 224-231.

Skipper, K., 1977, Offshore petroleum developments in northwest Europe - an update: Geosci. Canada, v. 4, p. 31-40.

Thomas, A. N., P. J. Walmsley, and D. A. L. Jenkins, 1974, Forties Field, North Sea: AAPG Bull., v. 58, p. 396-406.

Walmsley, P. J., 1975, The Forties Field, *in* A. W. Woodland, ed., Petroleum and the continental shelf of northwest Europe: New York, John Wiley and Sons, p. 477-485.

Spain

Crimes, T. P., 1976, Sand fans, turbidites, slumps and the origin of the Bay of Biscay; a facies analysis of the Guipuzcoan flysch: Palaeogeog., Palaeoclim., Palaeoecol., v. 19, p. 1-15.

————, 1977, Trace fossils of an Eocene deep-sea sand fan, northern Spain: Geol. Jour., Spec. Issue no. 9, p. 71-90.

———— A. Marcos, and A. Perez Estaun, 1974,

Upper Ordovician turbidites in western Asturias; a facies analysis with particular reference to vertical and lateral variations: Palaeogeog., Palaeoclim., Palaeoecol., v. 15, p. 169-184.

Kenyon, N. H., R. H. Belderson, and A. H. Stride, 1978, Channels, canyons and slump folds on the continental slope between south west Ireland and Spain: Oceanol. Acta, v. 1, p. 369-380.

Knox, G. J., 1976, The Early Tertiary deep-water sandstones near San Sebastian, Spain, some aspects of diagenesis: Geol. Mag., v. 113, p. 341-348.

Kruit, C. J. Brouwer, and P. Ealey, 1972, A deep-water sand fan in the Eocene Bay of Biscay: Nature Phys. Science, v. 240, p. 59-61.

———— ————, et al, 1975, Eocene deep-water fan deposits of San Sebastian (Spain): Nice, France, 9th Congres International de Sedimentologie, Field Guide Z-23, 75 p.

Mutti, E., 1977, Distinctive thin-bedded turbidite facies and related depositional environments in the Eocene Hecho Group, (south-central Pyrenees, Spain): Sedimentology, v. 24, p. 107-131.

Rupke, N.A., 1976, Large-scale slumping in a flysch basin, southwestern Pyrenees: Jour. Geol. Soc. London, v. 132, p. 121-130.

————, 1976, Sedimentology of very thick calcarenite-marlstone beds in a flysch succession, southwestern Pyrenees: Sedimentology, v. 23, p. 43-65.

————, 1977, Growth of an ancient deep-sea fan: Jour. Geology, v. 85, p. 725-744.

———— and D. J. Stanley, 1974, Distinctive properties of turbiditic and hemipelagic mud layers in the Algero-Balearic Basin, Western Mediterranean Sea: Smithsonian Contr. to Earth Sci., v. 13, 40 p.

————, 1969, Aspects of bed thickness in some Eocene turbidite sequences, Spanish Pyrenees: Jour. Geology, v. 77, p. 482-484.

————, 1972, Geologic studies of an Early and Middle Eocene flysch formation, south-western Pyrenees, Spain: Princeton Univ., Masters thesis, 208 p.

Van De Graaff, W. J. E., 1971, Three Upper Carboniferous, limestone-rich, high-destructive, delta systems with submarine fan deposits, Cantabrian Mountain, Spain: Leidse Geologische Mededelingen, decl 46, blg, p. 157-235.

Van Hoorn, B., 1969, Submarine canyon and fan deposits in the Upper Cretaceous of the south-central Pyrenees, Spain: Geologie en Mijnbouw, v. 48, p. 67-72.

————, 1970, Sedimentology and paleogeography of an Upper Cretaceous turbidite basin in the south-central Pyrenees, Spain: Leidse Geologische Mededelingen, v. 45, p. 73-154.

Van Vliet, A., 1978, Early Tertiary deepwater fans of Guipuzcoa, Northern Spain, in D. J. Stanley and G. Kelling, eds., Sedimentation in submarine canyons, fans, and trenches: Stroudsburg, Pa., Dowden, Hutchinson, and Ross, Inc., p. 190-210.

Mediterranean Sea

Barr, F. T., and B. R. Walker, 1966, Late tertiary channel system in northern Libya and its implications on Mediterranean Sea level changes: Washington, D.C., Initial Repts. of Deep Sea Drilling Proj., v. 13, p. 1244-1251.

Bartolini, C., S. Berato, and V. Bortolotti, 1975, Upper Miocene shallow-water turbidites from western Tuscany: Sed. Geology v. 14, p. 77-122.

———— C. DeGiuli, and G. Gianelli, 1974, Turbidites of the southern Tyrrhenian Sea: Boll. Soc. Geol. Ital., v. 93, p. 3-22.

———— C. Gehin, and D. J. Stanley, 1972, Morphology and recent sediments of the western Alboran Basin in the Mediterranean Sea: Marine Geol., v. 13, p. 159-223.

Byramjee, R. S., J. F. Mugniot, and B. Biju-Duval, 1975, Petroleum potential of deep-water areas of the Mediterranean and Caribbean Seas: Proc., 9th World Petroleum Cong., v. 2, p. 299-312.

Got, H., and D. J. Stanley, 1974, Sedimentation in two Catalonian canyons, northwestern Mediterranean: Marine Geol., v. 16, p. M91-M100.

Heezen, B. C., and M. Ewing, 1955, Orleansville earthquake and turbidity currents: AAPG Bull., v. 39, p. 2505-2514.

Horn, D. R., J. I. Ewing, and M. Ewing, 1972, Graded-bed sequences emplaced by turbidity currents north of 20°N in the Pacific, Atlantic and Mediterranean: Sedimentology, v. 18, p. 247-275.

Maldonado, A., and D. J. Stanley, 1976, The Nile Cone; submarine fan development by cyclic sedimentation: Marine Geol., v. 20, p. 27-40.

———— and ————, 1978 Nile Cone depositional processes and patterns in the Late Quaternary, in D. J. Stanley, and G. Kelling, eds., Sedimentation in submarine canyons, fans, and trenches: Stroudsburg, Pa., Dowden, Hutchinson and Ross, Inc., p. 239-260.

———— and ————, 1979, Depositional patterns and Late Quarternary evolution of two Mediterranean submarine fans; a comparison: Marine Geol., v. 31, p. 215-250.

Menard, H. W., Jr., S. M. Smith, and R. M. Pratt, 1965, The Rhone deep-sea fan, in W. F. Whittard and R. Bradshaw, eds., Submarine geology and geophysics: Butterworths, London, Proc. 17th Symp. Colston Res. Soc., p. 271-285.

Mutti, E., 1974, Examples of ancient deep-sea fan deposits from circum - Mediterranean geosynclines, *in* R. H. Dott, Jr., and R. H. Shaver, eds., Modern and ancient geosynclinal sedimentation: SEPM Special Pub., no. 19, p. 92-105.

Ross, D. A., et al, 1978, Sedimentation and structure of the Nile Cone and Levant Platform area, *in* D. J. Stanley and G. Kelling, eds., Sedimentation in submarine canyons, fans and trenches: Stroudsburg, Pa., Dowden, Hutchinson and Ross, Inc., p. 261-275.

Rupke, N.A., 1975, Deposition of fine-grained sediments in the abyssal environment of the Algero-Balearic Basin, western Mediterranean Sea: Sedimentology, v. 22, p. 95-109.

——— D. J. Stanley, and R. Stuckenrath, 1974, Late Quaternary rates of abyssal mud deposition in the western Mediterranean Sea: Marine Geol., v. 17, p. M9-M16.

Ryan, W. B. F., and B. C. Heezen, 1965, Ionian Sea submarine canyons and the 1908 Messina turbidity current: Geol. Soc. America Bull., v. 76, p. 915-932.

——— K. Venkatarathnam, and F. C. Wezel, 1973, Mineralogical composition of Nile Cone, Mediterranean Ridge, and Strabo Trench sandstones and clays: *in* R. B. Whitmarsh, O. E. Weser, and D. A. Ross, et al, eds., Washington, D. C., Initial Repts. of Deep Sea Drilling Proj., v. 13, p. 731-746.

Sarnthein, M., and C. Bartolini, 1973, Grain size studies on turbidite components from Tyrrhenian deep sea cores: Sedimentology, v. 20, p. 425-436.

Stanley, D. J., 1973, Basin plains in the eastern Mediterranean — significance in interpreting ancient marine deposits; I. basin depth and configuration: Marine Geol., v. 15, p. 295-307.

———, 1974, Modern flysch sedimentation in a Mediterranean Island Arc setting, *in* R. H. Dott, Jr., and R. H. Shaver, eds., Modern and ancient geosynclinal sedimentation: SEPM Spec. Pub. no. 19, p. 240-259.

——— and A. Maldonado, 1977, Nile Cone; Late Quaternary stratigraphy and sediment dispersal: Nature, v. 266, p. 129-135.

Summerhayes, C., D. Ross, and P. Stoffers, 1977, Nile submarine fan; sedimentation, deformation, and oil potential: Proc. 9th Offshore Tech. Conf., v. 1, p. 35-40.

Van Straaten, L. M. J. U., 1964, Turbidite sediments in the southeastern Adriatic Sea, *in* A. H. Bouma and A. Brouwer, eds., Turbidites: Amsterdam, Elsevier Sci. Pub., p. 142-147.

———, 1968, Turbidites, ash layers, and shell beds in the bathyal zone of the southern Adriatic Sea: Revue Geogreaphie Physique Geologie Dynamique, v. 9, p. 219-240.

NORTH AMERICAN

Alaska

Bouma, A. H., and T. H. Nilsen, 1978, Turbidite facies and deep-sea fans - with examples from Kodiak Island, Alaska: Proc. 10th Ann. Offshore Tech. Conf., v. 1, p. 559-567.

Eisbacher, G. H., 1976, Sedimentology of the Dezadeash flysch and its implications for strike-slip faulting along the Denali fault, Yukon Territory and Alaska: Canadian Jour. Earth Sci., v. 13, p. 1495-1513.

Galloway, W. E., 1974, Deposition and diagenetic alteration of sandstone in northeast Pacific arc-related basins; implications for graywacke genesis: Geol. Soc. America Bull., v. 85, p. 379-390.

Hamilton, E. L., 1967, Marine geology and Abyssal Plains in the Gulf of Alaska: Jour. Geophys. Research, v. 72, p. 4189-4213.

Mammerickx, J., 1970, Morphology of the Aleutian Abyssal Plain: Geol. Soc. America Bull., v. 81, p. 3457-3464.

Piper, D. J. W., 1973, Sedimentology of silt turbidites from the Gulf of Alaska, *in* L. D. Kulm and R. van Huene, et al, eds., Washington, D.C., Initial Repts. of Deep Sea Drilling Proj., v. 18, p. 847-867.

——— R. von Huene, and J. R. Duncan, 1973, Late Quaternary sedimentation in the active eastern Aleutian trench: Geology, v. 1, p. 19-22.

Scholl, D. W., et al, 1970, Structure and origin of the large submarine canyons of the Bering Sea: Marine Geol., v. 8, p. 187-210.

Stewart, R. J., 1976, Turbidites of the Aleutian abyssal plain; mineralogy, provenance, and constraints for Cenozoic motion of the Pacific plate: Geol. Soc. America Bull., v. 87, p. 793-808.

———, 1977, Neogene turbidite sedimentation in Komandorskiy basin, western Bering Sea: AAPG Bull., v. 61, p. 192-206.

Winn, R. D., Jr., R. J. Bailes, and K. I. Lu, 1981, Debris flows, turbidites, and lead-zinc sulfides along a Devonian submarine fault scarp, Jason Prospect, Yukon Territory, *in* Deep water clastic sediments, a core workshop: San Francisco, SEPM Core Workshop no. 2, p. 396-416.

Appalachians

Brown, W. R., 1970, Investigations of the sedimentary record in the Piedmont and Blue Ridge of Virginia, *in* G. Fisher, et al, eds., Studies of Appalachian geology, central and southern: New York, Wiley-Interscience, p. 335-349.

Cheema, M. R., 1977, Sedimentation and gas production of the Upper Devonian Benson Sand in north-central West Virginia - a model for exo-geosynclinal mid-fan: West Virginia Univ., Ph.D. dissert., 127 p.

Davies, I. C., and R. G. Walker, 1974, Transport and deposition of resedimented conglomerates; the Cap Enrage Formation, Gaspe, Quebec: Jour. Sed. Petrology, v. 44, p. 1200-1216.

Enos, P., 1969, Anatomy of a flysch: Jour. Sed. Petrology, v. 39, p. 680-723.

————, 1969, Cloridorme Formation, Middle Ordovician flysch, northern Gaspe Peninsula, Quebec: Geol. Soc. America Spec. Paper 117, 66 p.

Fisher, G. W., 1970, The metamorphosed sedimentary rocks along the Potomac River near Washington, D.C., in G. W. Fisher, et al, eds., Studies of Appalachian geology, central and southern, New York: New York, Wiley Interscience, p. 229-315.

Glaeser, J. D., 1979, Catskill delta slope sediments in the central Appalachian basin; source deposits and reservoir deposits, in L. J. Doyle and O. H. Pilkey, eds., Geology of continental slopes: SEPM Spec. Pub. no. 27, p. 343-357.

Graham, S. A., W. R. Dickinson, and R. V. Ingersoll, 1975, Himalayan-Bengal model for flysch dispersal in the Appalachian-Ouachita system: Geol. Soc. America Bull., v. 85, p. 273-286.

Hand, B. M., G. V. Middleton, and K. Skipper, 1972, Antidune cross-stratification in a turbidite sequence, Cloridorme Formation, Gaspe, Quebec: Sedimentology, v. 18, p. 135-138.

Hendry, H. E., 1973, Sedimentation of deep water conglomerates in Lower Ordovician rocks of Quebec - composite bedding produced by progressive liquefaction of sediment?: Jour. Sed. Petrology, v. 43, p. 125-136.

————, 1978, Cap Des Rosiers Formation at Grosses Roches, Quebec-deposits of the mid-fan region on a Ordovician submarine fan: Canadian Jour. Earth Sci., v. 15, p. 1472-1488.

Hiscott, R. N., 1978, Provenance of Ordovician deep-water sandstones, Tourelle Formation, Quebec, and implications for initiation of the Taconic Orogeny: Canadian Jour. Earth Sci., v. 15, p. 1579-1597.

————, 1979, Clastic sills and dikes associated with deep-water sandstones, Tourelle Formation, Ordovician, Quebec: Jour. Sed. Petrology, v. 49, p. 1-9.

———— and G. V. Middleton, 1979, Depositional mechanics of thick-bedded sandstones at the base of a submarine slope, Tourelle Formation (Lower Ordovician), Quebec, Canada, in L. J. Doyle and O. H. Pilkey, eds., Geology of continental slopes: SEPM Spec. Pub. no. 27, p. 307-326.

Hubert, C., J. Lajoie, and M. A. Leonard, 1970, Deep sea sediments in the Lower Paleozoic Quebec Supergroup, in J. Lajoie, ed., Flysch sedimentology in North America: Geol. Assoc. Canada, Spec. Paper 7, p. 103-125.

Johnson, B. A., 1974, Deep-sea fan-valley conglomerate; Cap Enrage Formation, Gaspe, Quebec: Hamilton, Ont., McMaster Univ., Masters thesis, 108 p.

———— and R. G. Walker, 1979, Paleocurrents and depositional environments of deep water conglomerates in the Cambro-Ordovician Cap Enrage Formation, Quebec Appalachians: Canadian Jour. Earth Sci., v. 16, p. 1375-1387.

Lajoie, J., 1969, Dispersal and petrology of the Silurian Val Brillant and Robitaille sandstones, Appalachians, Quebec: Jour. Sed. Petrology, v. 38, p. 643-647.

———— and A. Chagnon, 1973, Origin of red beds in a Cambrian flysch sequence, Canadian Appalachians, Quebec: Sedimentology, v. 20, p. 91-103.

———— Y. Heroux, and B. Mathey, 1974, The Precambrian shield and the Lower Paleozoic shelf; the unstable provenance of the Lower Paleozoic flysch sandstone and conglomerates of the Appalachians between Beaumont and Bic, Quebec: Canadian Jour. Earth Sci., v. 11, p. 951-963.

————, 1979, Origin of megarhythms in flysch sequences of the Quebec Appalachians: Canadian Jour. Earth Sci., v. 16, p. 1518-1523.

McBride, E. F., 1962, Flysch and associated beds of the Martinsburg Formation (Ordovician), central Appalachians: Jour. Sed. Petrology, v. 32, p. 39-91.

McIver, N. L., 1970, Appalachian turbidites, in G. W. Fisher, et al, eds., Studies of Appalachian geology, central and southern: New York, Wiley-Interscience, p. 69-81.

Parkash, B., 1970, Downcurrent changes in sedimentary structure in Ordovician turbidite greywackes: Jour. Sed. Petrology, v. 40, p. 572-590.

———— and G. V. Middleton, 1970, Downcurrent textural changes in Ordovician turbidite greywackes: Sedimentology, v. 14, p. 259-293.

Rocheleau, M., and J. Lajoie, 1974, Sedimentary structures in resedimented conglomerate of the Cambrian flysch L'Islet, Quebec Appalachians: Jour. Sed. Petrology, v. 44, p. 826-836.

Shanmugam, G., and G. L. Benedict, III, 1978, Fine-grained carbonate debris flow, Ordovician basin margin, southern Appalachians: Jour. Sed. Petrology, v. 48, p. 1233-1240.

———— and K. R. Walker, 1978, Tectonic significance of distal turbidites in the Middle Ordovician Blockhouse and lower Sevier Formations in east Tennessee: Am. Jour. Sci., v. 278, p. 551-578.

Skipper, K., and S. B. Battacharjee, 1978, Backset bedding in turbidites; a further example from the

Cloridorme Formation (Middle Ordovician), Gaspe, Quebec: Jour. Sed. Petrology, v. 48, p. 193-201.

——— and G. V. Middleton, 1975, The sedimentary structures and depositional mechanics of certain Ordovician turbidites, Cloridorme Formation, Gaspe Peninsula, Quebec: Canadian Jour. Earth Sci., v. 12, p. 1934-1952.

Stow, D. A. V., and G. Shanmugam, 1980, Sequence of structures in fine-grained turbidites; comparison of recent deep-sea and ancient flysch sediments: Sed. Geology, v. 25, p. 23-42.

Thompson, A. M., 1972, Shallow-water distal turbidites in Ordovician flysch, central Appalachian mountains, U.S.A., 24th Internat. Cong., Sec. 6, p. 89-99.

Walker, R. G., 1971, Nondeltaic depositional environments in the Catskill Clastic Wedge (Devonian) of central Pennsylvania: Geol. Soc. America Bull., v. 82, p. 1305-1326.

California - Nevada

General

Cook, H. E., 1979, Ancient continental slope sequences and their value in understanding modern slope development, *in* L. J. Doyle and O. H. Pilkey, eds., Geology of continental slopes: SEPM Spec. Pub. no. 27, p. 287-305.

——— and W. C. Chamberlain, 1978, Ancient continental slope sequences and their usefulness in understanding modern slope development: Proc., 10th Ann. Offshore Tech. Conf., v. 1, p. 547-553.

Offshore California

Beer, R. M., and D.S., Gorsline, 1971, Distribution, composition and transport of suspended sediment in Redondo Submarine Canyon and vicinity (California): Marine Geol., v. 10, p. 153-175.

Bouma, A. H., 1964, Ancient and recent turbidites: Geologie en Mijnbouw, v. 43, p. 375-379.

Buffington, E. C., 1964, Structural control and precision bathymetry of La Jolla submarine canyon, California: Marine Geol., v. 1, p. 44-58.

Chamberlain, C. K., 1964, Mass transport of sediment in the heads of Scripps Canyon, *in* Papers in marine geology. New York, Macmillan Co., p. 42-64.

Chase, T. E., W. R. Normark, and P. Wilde, 1975, Oceanographic data of the Monterey deep-sea fan, 34°-37°N, 120°-127°W: Univ. California Inst. Marine Resources Tech. Rept. TR-58, 2 p.

Crouch, R. W., 1952, Significance of temperature on foraminifera from deep basins off southern California coast: AAPG Bull., v. 36, p. 807-843.

Crowell, J. C., 1952, Submarine canyons bordering central and southern California: Jour. Geology, v. 60, p. 58-83.

Davies, J. R., 1971, Sedimentation of Pliocene sandstones in Santa Barbara Channel, California (abs.): AAPG Bull., v. 55, p. 335.

Dill, R. F., 1964, Sedimentation and erosion in Scripps submarine canyon head, *in* R. L. Miller, ed., Papers in marine geology: New York, McMillan, p. 23-41.

———, 1969, Earthquake effects on fill of Scripps submarine canyon: Geol. Soc. America Bull., v. 80, p. 321-328.

Emery, K. O., 1960, Basin plains and aprons off southern California: Jour. Geology, v. 68, p. 464-479.

Erickson, J. W., 1975, Sedimentology of the South Point Formation (Eocene), Santa Rosa Island, California: Proc., AAPG-SEPM-SEG Pac. Secs. Mtg. p. 169-190.

Felix, D. W., and D. S., Gorsline, 1971, Newport submarine canyon, California; an example of the effects of shifting loci of sand supply upon canyon position: Marine Geology, v. 10, p. 177-198.

Gorsline, D. S., and K. O. Emery, 1959, Turbidity current deposits in San Pedro and Santa Monica basins off southern California: Geol. Soc. America Bull., v. 70, p. 279-290.

Hand, B. M., and K. O. Emery, 1964, Turbidites and topography of north end of San Diego Trough, California: Jour. Geology, v. 72, p. 526-542.

Haner, B. E., 1971, Morphology and sediments of Redondo submarine fan, southern California: Geol. Soc. America Bull., v. 82, p. 2413-2432.

Hess, G. R., and W. R. Normark, 1976, Holocene sedimentation history of the major fan valleys of Monterey Fan: Marine Geology, v. 22, p. 233-251.

Komar, P. D., 1969, The channelized flow of turbidity currents with application to Monterey deep-sea fan channel: Jour. Geophys. Research, v. 74, p. 4544-4558.

Marshall, N. F., 1978, A large storm-induced slump re-opens an unknown Scripps submarine canyon tributary, *in* D. J. Stanley and G. Kelling, eds., Sedimentation in submarine canyons, fans, and trenches: Stroudsburg, Pa., Dowden, Hutchinson and Ross, Inc., p. 73-84.

Martin, B. D., and K. O. Emery, 1967, Geology of Monterey Canyon, California: AAPG Bull., v. 51, p. 2281-2304.

Menard, H. W., Jr., 1960, Possible pre-Pleistocene deep-sea fans off central California: Geol. Soc. America Bull., v. 71, p. 1271-1278.

Normark, W. R., 1970, Channel piracy on Monterey deep-sea fan: Deep-sea Research, v. 17, p. 837-846.

———, 1970, Growth patterns of deep-sea fans: AAPG Bull., v. 54, p. 2170-2195.

———, 1971, Mini-topography of deep-sea fans;

geometric considerations for facies interpretations in turbidites: SEPM, Pac. Sec., Field Trip Guidebook, p. 22-36.

————, 1974, Submarine canyons and fan valleys; factors affecting growth patterns of deep-sea fans, *in* R. H. Dott, Jr., and R. H. Shaver, eds., Modern and ancient geosynclinal sedimentation: SEPM Spec. Pub. no. 19, p. 56-68.

———— and D. J. W. Piper, 1969, Deep-sea fan-valleys past and present: Geol. Soc. America Bull., v. 80, p. 1859-1866.

———— and ————, 1972, Sediments and growth pattern of Navy deep-sea fan, San Clemente Basin, California borderland: Jour. Geology, v. 80, p. 198-223.

———— ———— and G. R. Hess, 1979, Distributary channels, sand lobes, and mesotopography of Navy submarine fan, California borderland, with applications to ancient fan sediments: Sedimentology, v. 26, p. 749-774.

Palmer, H. D., 1976, Erosion of submarine outcrops, LaJolla submarine canyon, California: Geol. Soc. America Bull., v. 87, p. 427-432.

Piper, D. J. W., 1970, Transport and deposition of Holocene sediment of La Jolla deep sea fan, California: Marine Geol., v. 8, p. 211-227.

———— and W. R. Normark, 1971, Re-examination of a Miocene deep-sea fan and fan-valley, southern California: Geol. Soc. America Bull., v. 82, p. 1823-1830.

Shepard, F. P., 1966, Meander in valley crossing a deep-ocean fan: Science, v. 154, p. 385-386.

———— and E. C. Buffington, 1968, La Jolla submarine fan-valley: Marine Geology, v. 6, 107-143.

———— R. F. Dill, and U. Von Rad, 1969, Physiography and sedimentary processes of La Jolla submarine fan and fan-valley, California: AAPG Bull., v. 53, p. 390-420.

———— and N. F. Marshall, 1969, Currents in La Jolla and Scripps submarine canyons: Science, v. 165, p. 177-178.

Smith, D. L., and W. R. Normark, 1976, Deformation and patterns of sedimentation, south San Clemente basin, California borderland: Marine Geology v. 22, p. 175-188.

Taylor, J. C., 1976, Geologic appraisal of the petroleum potential of offshore southern California; the borderland compared to onshore coastal basins: U.S. Geol. Survey Circ. 730, 43 p.

Van Rad, U., 1968, Comparison of sedimentation in the Bavarian Flysch (Cretaceous) and Recent San Diego Trough (California): Jour. Sed. Petrology, v. 38, p. 1120-1154.

Wilde, P., 1965, Recent sediments of the Monterey deep-sea fan: Berkeley, California Univ., Hydrol. Eng. Lab. Rept. HEL 2-13, 153 p.

———— W. R. Normark, and T. E. Chase, 1976, Petroleum potential of continental rise off central California - summary, *in* M. T. Halbouty, J. C. Maher, and H. M. Lian, eds., Circum-Pacific Energy and Mineral Resources: AAPG Mem. 25, p. 313-317.

———— ———— and ————, 1978 Channel sands and petroleum potential of Monterey deep-sea fan, California: AAPG Bull., v. 62, p. 967-983.

Northern California

Aalto, K. R., 1976, Sedimentology of a melange; Franciscan of Trinidad, California: Jour. Sed. Petrology, v. 46, p. 913-929.

Almgren, A. A., 1978, Timing of Tertiary submarine canyons and marine cycles of deposition in the southern Sacramento Valley, California, *in* D. J. Stanley and G. Kelling, eds., Sedimentation in submarine canyons, fans, and trenches: Stroudsburg, Pa., Dowden, Hutchinson and Ross, Inc., p. 276-291.

Bailey, E. H., W. P. Irwin, and D. L. Jones, 1964, Franciscan and related rocks, and their significance in the geology of western California: Cal. Div. Mines Geology Bull., v. 183, 177 p.

Bartow, J. A., 1966, Deep submarine channel in Upper Miocene, Orange County, California: Jour. Sed. Petrology, v. 36, p. 700-705.

Blake, M. C., Jr., and D. L. Jones, 1974, Origin of Franciscan melanges in northern California, *in* R. H. Dott, Jr., and R. H. Shaver, eds., Modern and ancient geosynclinal sedimentation: SEPM Spec. Pub. no. 19, p. 345-357.

Brown, R. D., Jr., and E. I. Rich, 1960, Early Cretaceous fossils in submarine slump deposits of Late Cretaceous age, northern Sacramento Valley, California: U.S. Geol. Survey Prof. Paper 400-B, p. 318B-320B.

———— and ————, 1967, Implications of two Cretaceous mass transport deposits, Sacramento Valley, California; comment on a paper by G. L. Peterson: Jour. Sed. Petrology, v. 37, p. 240-248.

Chipping, D. H., 1972, Sedimentary structures and environment of some thick sandstone beds of turbidite type: Jour. Sed. Petrology, v. 42, p. 587-595.

Church, V. R., 1979, Display of cores from the Winters Sand (Upper Cretaceous), Sacramento Valley, California: Calif. Well Sample Repository Spec. Pub. no. 2, 56 p.

Colburn, I. P., 1968, Grain fabrics in turbidite sandstone beds and their relationship to sole mark trends on the same beds: Jour. Sed. Petrology, v. 38, p. 146-158.

Crowell, J. C., 1957, Origin of pebbly mudstones: Geol. Soc. America Bull., v. 68, p. 993-1009.

————, 1974, Sedimentation along the San Andreas Fault, California, *in* Modern and ancient geosynclinal sedimentation: SEPM Spec. Pub. no. 19, p. 292-303.

Dickas, A. B., and J. L. Payne, 1967, Upper Paleocene buried channel in Sacramento Valley, California: AAPG Bull., v. 51, p. 873-882.

Dickinson, W. R., R. V. Ingersoll, and S. A. Graham, 1979, Paleogene sediment dispersal and paleotectonics in northern California; summary: Geol. Soc. America Bull., pt. 1, v. 90, p. 897-898.

—— R. W. Ojakangas, and R. J. Stewart, 1969, Burial metamorphism of the Late Mesozoic Great Valley Sequence, Cache Creek, California: Geol. Soc. America Bull., v. 80, p. 519-526.

—— and E. I. Rich, 1972, Petrologic intervals and petrofacies in the Great Valley Sequence, Sacramento Valley, California: Geol. Soc. America Bull., v. 63, p. 3007-3024.

Drummond, K. F., E. W. Christensen, and K. D. Berry, 1976, Upper Cretaceous lithofacies model, Sacramento Valley, California: AAPG Pac. Sec. Misc. Pub. no. 24, p. 76-88.

Edmondson, W. F., 1965, The Meganos Gorge of the southern Sacramento Valley: San Joaquin Geol. Soc., Selected Papers, no. 3, p. 36-51.

Ernst, W. G., 1970, Tectonic contact between the Franciscan melange and the Great Valley Sequence - crustal expression of a Late Mesozoic Benioff zone: Jour. Geophys. Research, v. 75, p. 886-901.

Fischer, P. J., 1971, An ancient (Upper Paleocene) submarine canyon and fan; the Meganos channel, Sacramento Valley, California, (abs.): 67th Annual Geol. Soc. America, Cordilleran Sec. Mtg. Program, v. 3, p. 120.

——, 1979, The evolution of a Late Paleocene submarine canyon and fan system; the Meganos Formation, southern Sacramento Basin, California: Field Guide, Geol. Soc. America, Cordilleran Sec. Mtg, 10 p.

Galloway, A. J., 1966, Field trip Point Reyes Peninsula and San Andreas fault zone, *in* E. H. Bailey, ed., Geology of northern California: Calif. Div. Mines Geol. Bull. 190, p. 429-440.

Garrison, L. E., ed., 1961, East central Sacramento Valley; Marysville (Sutter) Buttes-Chico Creek-Oroville: Field Trip, Geol. Soc. of Sacramento, 51 p.

Gilbert, W. G., and W. R. Dickinson, 1970, Stratigraphic variations in sandstone petrology, Great Valley sequence, central California coast: Geol. Soc. America Bull., v. 81, p. 949-954.

Goudkoff, P. P., 1945, Stratigraphic relations of Upper Cretaceous in Great Valley, California: AAPG Bull., v. 29, p. 956-1007.

Ingersoll, R. V., 1978, Submarine fan facies of the Upper Cretaceous Great Valley Sequence, northern and central California: Sed. Geology, v. 21, p. 205-230.

——, 1979, Evolution of the Late Cretaceous fore-arc basin, northern and central California: Geol. Soc. America Bull., v. 90, pt. 1, p. 813-826.

—— E. I. Rich, and W. R. Dickinson, 1977, Great Valley Sequence, Sacramento Valley: Sacramento, Geol. Soc. America Annual Meeting, Cordilleran Sec., Field Trip Guide, 71 p.

Link, M. H., and T. H. Nilsen, 1980, The Rocks sandstone, an Eocene sand-rich deep-sea fan deposit, northern Santa Lucia Range, California: Jour. Sed. Petrology, v. 50, p. 583-602.

Lowe, D. R., 1972, Submarine canyon and slope channel sedimentation model as inferred from Upper Cretaceous deposits, western California: 24th Internat. Geol. Cong., sec. 6, p. 75-81.

——, 1972, Implications of three submarine mass-movement deposits, Cretaceous, Sacramento Valley, California: Jour. Sed. Petrology, v. 42, p. 89-101.

——, 1973, Conglomerate-filled submarine fan channel in the Cretaceous of western California: Geol. Soc. America Abs. with Programs, v. 5, no. 7, p. 718.

——, 1979, Stratigraphy and sedimentology of the Pigeon Point Formation, San Mateo County, California, *in* T. H. Nilsen, ed., Geology of the Santa Cruz Mountains, California: Field Trip Guide, Geol. Soc. America, Cordilleran Sec., p. 17-29.

Martin, B. D., 1963, Rosedale channel - evidence for Late Miocene submarine erosion in Great Valley California: AAPG Bull., v. 47, p. 441-456.

Nelson, C. H., and T. H. Nilsen, 1974, Depositional trends of modern and ancient deep-sea fans, *in* R. H. Dott, Jr., and R. H. Shavers, eds., Modern and ancient geosynclinal sedimentation, SEPM Spec. Pub. no. 19, p. 69-91.

Nilsen, T. H., 1979, Sedimentology of the Butano Sandstone Santa Cruz Mountains, California, *in* Geology of the Santa Cruz Mountains, California: Field Trip Guide, Geol. Soc. America, Cordilleran Sec., p. 30-39.

—— and E. E. Brabb, 1979, Geology of the Santa Cruz Mountains, California: Field Trip Guide, Cordilleran Sec., Geol. Soc. America, 97 p.

—— and T. W. Dibblee, Jr., 1979, Geology of the Central Diablo Range between Hollister and New Idria, California: Field Trip Guide, Geol. Soc. America, Cordilleran Sec., 106 p.

—— and M. H. Link, 1975, Stratigraphy, sedimentology and offset along the San Andreas fault of Eocene to Lower Miocene strata of the northern Santa Lucia Range and the San Emigdio Mountains, coast ranges, central California, *in* D. S. Weaver, G. R. Hornaday, and A. Tipton, eds., Paleogene Symp., Conf. in Future Energy Horizons of the Pac. Coast, p. 367-400.

———— and T. R. Simoni, 1973, Deep-sea fan paleocurrent patterns of the Eocene Butano sandstone, Santa Cruz Mountains, California: Jour. Research, U.S. Geol. Survey, v. 1, p. 439-452.

Ojakangas, R. W., 1964, Petrology and sedimentation of the Cretaceous Sacramento Valley Sequence, Cache Creek, California: Stanford Univ., Unpub. Ph.D. dissert.: 190 p.

————, 1968, Cretaceous sedimentation, Sacramento Valley, California: Geol. Soc. America Bull., v. 79, p. 973-1008.

Pessagno, E. A., Jr., 1974, Radiolarian zonation and stratigraphy of the Upper Cretaceous portion of the Great Valley Sequence, California coast ranges: Micropaleontology Spec. Paper 1, 233 p.

Peterson, G. L., 1965, Implications of two Cretaceous mass transport deposits, Sacramento Valley, California: Jour. Sed. Petrology, v. 35, p. 401-407.

Piper, D. J. W., W. R. Normark, and J. C. Ingle, Jr., 1976, The Rio Dell Formation; a Plio-Pleistocene basin slope deposit in northern California: Sedimentology, v. 23, p. 309-328.

Poole, F. G., 1974, Flysch deposits of the Antler foreland basin, western United States, in W. R. Dickinson, ed., Tectonics and sedimentation: SEPM Special Pub. no. 22, p. 58-82.

Rich, E. I., and R. V. Ingersoll, 1975, Spring field trip guide and road log - western margin Sacramento Valley-Clear Lake area: Northern Cal. Geol. Soc., (AAPG, Pac. Sec.), 21 p.

Smith, G. W., D. G. Howell, and R. V. Ingersoll, 1979, Late Cretaceous trench-slope basins of central California: Geology, v. 7, p. 303-306.

Stauffer, P. H., 1967, Sedimentologic evidence on Eocene correlations, Santa Ynez Mountains, California: AAPG Bull., v. 51, p. 607-611.

————, 1967, Grain flow deposits and their implications, Santa Ynez Mountains, California: Jour. Sed. Petrology, v. 37, p. 487-508.

Taira, A., and P. A. Scholle, 1979, Deposition of resedimented sandstone beds in the Pico Formation, Ventura Basin, California, as interpreted from magnetic fabric measurements: Geol. Soc. America Bull., pt. 1, v. 90, p. 952-962.

Tillman, et al, 1981, Upper Cretaceous deep-water Winters sandstone, Cities Service Nixon Community No. 1, Solano County, California, in Deep water clastic sediments, a core workshop: San Francisco, SEPM Core Workshop No. 2, p. 45-76.

Weagant, F. E., 1972, Grimes gas field, Sacramento Valley, California, in R. E. King, ed., Stratigraphic oil and gas fields: AAPG Mem. 16, p. 428-439.

Williamson, C. R., and D. R. Hill, 1981, Submarine-fan deposition of the Upper Cretaceous Winters Sandstone, Union Island Field, Sacramento Valley, California, in Deep water clastic sediments, a core workshop: San Francisco, SEPM Core Workshop No. 2, p. 77-115.

Winterer, E. L., J. R. Curray, and M. N. A. Peterson, 1968, Geological history of the Pioneer Fracture Zone with the Delgada deep-sea fan, northeast Pacific: Deep-Sea Research, v. 15, p. 509-520.

Southern California

Barbat, W. F., 1958, The Los Angeles Basin area, California, in L. G. Weeks, ed., Habitat of Oil: AAPG Spec. Pub., p. 62-77.

Bazley, W. J. M., 1977, Late Miocene geology and new oil fields of the southern San Joaquin Valley: Guidebook, Pac. Sec. AAPG, SEG, SEPM, 88 p.

Biddle, K. T., J. C. Maher, and R. D. Carter, 1975, Channel turbidite sandstone in the Elk Hills shale member of the Monterey Shale, in J. C. Maher, et al, eds., Petroleum geology of Naval Petroleum Reserve No. 1, Elk Hills, Kern County, California: U.S. Geol. Survey Prof. Paper 912, p. 79-84.

Blackie, G. W., and R. J. Yeats, 1976, Magnetic-reversal stratigraphy of Pliocene-Pleistocene producing section at Saticoy oil field, Ventura Basin, Calif.: AAPG Bull., v. 60, p. 1985-1992.

Church, V. R., ed., 1978, Display of cores from the Stevens Sand (Upper Miocene), southern San Joaquin Valley, Calif.: Calif. Well Repository Spec. Pub. No. 1, 39 p.

Colburn, I. P., and A. E. Fritsche, eds., 1973, Cretaceous stratigraphy of the Santa Monica Mountains and Simi Hills, southern California: Pac. Sec. SEPM Guidebook (Fall 1973), 93 p.

Cole, M. R., 1977, Eocene paleocurrents and sedimentation, San Nicolas Island, California: AAPG Bull., v. 61, p. 237-247.

Crowell, J. C., 1975, The San Gabriel Fault and Ridge Basin, southern California: Cal. Div. Mines Geology Spec. Rept. no. 18, p. 208-219.

————, 1976, Implications of crustal stretching and shortening of coastal Ventura Basin, Calif.: AAPG Pac. Sec., Misc. Pub. no. 24, p. 365-382.

————, et al, 1966, Deep-water sedimentary structures, Pliocene Pico Formation, Santa Paula Creek, Ventura Basin, California: Cal. Div. Mines Geology, Spec. Rept. no. 89, 40 p.

————, et al, 1973, Sedimentary facies changes in Tertiary rocks - California transverse and southern coast ranges: Annual AAPG-SEPM-SEG Field Trip Guidebook, SEPM Trip no. 2, 60 p.

Dott, R. H., Jr., 1963, Dynamics of subaqueous gravity depositional processes: AAPG Bull., v.

47, p. 104-128.

Edmondson, W. R., 1980, Member sands of the Winters Formation: AAPG Pac. Sec. Program Preprint, 20 p.

Gardett, P. H., 1971, Petroleum potential of Los Angeles Basin, California, *in* I. H. Cram, ed., Future petroleum provinces of the United States - their geology and potential: AAPG Mem. 15, p. 298-308.

Harman, R. A., 1964, Distribution of foraminifera in the Santa Barbara Basin, California: Micropaleontology, v. 10, p. 81-96.

Hill, D. R., 1979, Union Island gas field, San Joaquin County, California: Calif. Well Sample Repository Spec. Pub. no. 2, 8-17.

Howell, D. G., and M. H. Link, 1979, Eocene conglomerate sedimentology and basin analysis, San Diego and the southern California borderland: Jour. Sed. Petrology, v. 49, p. 517-540.

Hsu, K. J., 1977, Studies of Ventura field, California, 1; facies geometry and genesis of Lower Pliocene turbidites: AAPG Bull., v. 61, p. 137-138.

Kennedy, M. P., and G. W. Moore, 1971, Stratigraphy and structure of the area between Oceanside and San Diego, California, field trip no. 8, *in* W. A. Elders, ed., Geological excursions in southern California: Riverside, Calif., Cordilleran Sec., Geol. Soc. America, p. 152-166.

Kohlbush, R. L., 1977, Tule Elk oil field *in* B. Bazeley, ed., Late Miocene geology and new oil fields of the southern San Joaquin Valley: Guidebook, Pac. Secs. AAPG-SEG-SEPM, p. 73-78.

Link, M. H., 1975, Matilija sandstone; a transition from deep-water turbidite to shallow marine deposition in the Eocene of California: Jour. Sed. Petrology, v. 45, p. 63-78.

Lohmar, J. M., and J. E. Warme, 1978, Anatomy of an Eocene submarine canyon-fan system, southern California borderland: Proc., 10th Ann. Offshore Tech. Conf. v. 1, p. 571-577.

MacPherson, B. A., 1978, Sedimentation and trapping mechanism in Upper Miocene Stevens and older turbidite fans of Southeastern San Joaquin Valley: AAPG Bull., v. 62, p. 2243-2274.

Maher, J. C., R. D. Carter, and R. J. Lantz, 1975, Petroleum geology of Naval Petroleum Reserve No. 1, Elk Hills, Kern County, California: U. S. Geol. Survey Prof. Paper 912, 109 p.

Mansfield, C. F., 1972, Petrofacies units and sedimentary facies of the Late Mesozoic strata west of Coalinga, California, *in* Cretaceous of the Coalinga area: Guidebook, SEPM Pac. Sec., p. 19-26.

———, 1979, Upper Mesozoic subsea fan deposits in the southern Diablo Range, California; record of the Sierra Nevada magmatic arc: Geol. Soc. America Bull., pt. 1, v. 90, p. 1025-1046.

Mayuga, M. N., 1970, Geology and development of California's giant - Wilmington oil field, *in* M. T. Halbouty, ed., Geology of giant petroleum fields: AAPG Mem. 14, p. 158-184.

Nagle, H. E., and E. S. Parker, 1971, Future oil and gas potential of onshore Ventura Basin, California, *in* I. H. Cram, ed., Future petroleum provinces of the United States - their geology and potential: AAPG Mem. 15, p. 254-297.

Natland, M. L., 1933, The temperature and depth distribution of some recent and fossil foraminifera in the southern California region: Bull. of Scripps Inst. of Oceanography, Tech. Series, v. 3, p. 225-230.

——— and P. H. Kuenen, 1951, Sedimentary history of the Ventura Basin, California, and the action of turbidity currents, *in* J. L. Hough, ed., Turbidity currents and the transportation of coarse sediment into deep water: SEPM Spec. Pub. no. 2, p. 76-107.

Redwine, L. E., ed., 1952, Cenozoic correlation section paralleling north and south margins, western Ventura Basin, from Point Conception to Ventura and Channel Islands, California: Geology Names and Correlation Comm., Sub. Comm. on Cenozoic, Pac. Sec., AAPG, 2 sheets.

Rennie, E. W., Jr., ed., 1972, Guidebook, geology and oil fields, west side central San Joaquin Valley: Pac. Secs., AAPG-SEG SEPM, 102 p.

Scott, R. M., and R. W. Tillman, 1981, Stevens Sandstone (Miocene), San Joaquin Basin, California, *in* Deep water clastic sediments, a core workshop: San Francisco SEPM Core Workshop No. 2, p. 116-248.

Spotts, J. H., and O. E. Weser, 1964, Directional properties of a Miocene turbidite, California, *in* A. H. Bouma and A. Brouwer, eds., Turbidites: New York, Elsevier Sci. Pub., p. 198-221.

Sullwold, H. H., 1960, Tarzana fan, deep submarine fan of Late Miocene age, Los Angeles County, California: AAPG Bull., v. 44, p. 433-457.

———, 1961, Turbidites in oil exploration, *in* J. A. Peterson and J. C. Osmond, eds., Geometry of sandstone bodies: AAPG Spec. Pub., p. 63-81.

Van De Kamp, P. C., and J. D. Harper, 1969, The Cretaceous and Lower Tertiary of the Wheeler Springs-Ojai area, Ventura Basin, California: Guidebook, Upper Sespe Creek Field Trip, Pac. Sec., SEPM p. 24-26.

——— ———, et al, 1974, Facies relationships in the Eocene-Oligocene in the Santa Ynez Mountains, California: Jour. Geol. Soc. London, v. 130, p. 545-565.

Walker, R. G., 1975, Nested submarine-fan channels in the Capistrano Formation, San Clemente, California: Geol. Soc. America Bull., v. 86, p. 915-924.

———, 1975, Upper Cretaceous resedimented conglomerates at Wheeler Gorge, California; description and field guide: Jour. Sed. Petrology, v. 45, p. 105-112.

Webb, G. W., 1977, Stevens and earlier Miocene turbidite sands, San Joaquin Valley, Calif., *in* W. J. M. Bazeley, ed., Late Miocene, geology and new oil fields of the southern San Joaquin Valley: Guidebook, Pac. Secs. AAPG-SEG-SEPM, p. 37-55.

Winterer, E. L., and D. L. Durham, 1962, Geology of southeastern Ventura Basin, Los Angeles County, Calif.: U. S. Geol. Survey Prof. Paper 334-H, p. 275-366.

Yeats, R. S., 1965, Pliocene seaknoll at South Mountain, Ventura Basin, California: AAPG Bull., v. 49, p. 526-546.

Yerkes, R. F., et al, 1965, Geology of the Los Angeles Basin, California - an introduction: U. S. Geol. Survey Prof. Paper 420A, 57 p.

Canada-Greenland

Campbell, J. S., and D. L. Clark, 1977, Pleistocene turbidites of the Canada abyssal plain of the Arctic Ocean: Jour. Sed. Petrology, v. 47, p. 657-670.

Chough, S. K., 1978, Morphology, sedimentary facies and processes of the Northwest Atlantic Mid-Ocean Channel between 61° and 52° N, Labrador Sea: McGill Univ., Ph.D. dissert., 167 p.

——— and R. Hesse, 1976, Submarine meandering thalweg and turbidity currents flowing for 4000 km in the Northwest Atlantic Mid-Ocean Channel, Labrador Sea: Geology, v. 4, p. 529-533.

——— and ———, 1980, Northwest Atlantic Mid-Ocean Channel of the Labrador Sea; III, head spills versus body spill deposits from turbidity currents on natural levees: Jour. Sed. Petrology, v. 50, p. 227-234.

Cook, H. E., et al, 1972, Allochthonous carbonate debris flows at Devonian bank ("reef") margins, Alberta, Canada: Bull. Canadian Petroleum Geology v. 20, p. 439-497.

Davies, G. R., 1977, Turbidites, debris sheets, and truncation structures in upper Paleozoic deep-water carbonates of the Sverdrup Basin, Arctic Archipelago, *in* H. E. Cook and P. Enos, eds., Deep-water carbonate environments, SEPM Spec. Pub. no. 25, p. 221-239.

Davies, I. C., and R. G. Walker, 1974, Transport and deposition of resedimented conglomerates; the Cap Enrage Formation, Gaspe, Quebec: Jour. Sed. Petrology, v. 44, p. 1200-1216.

Embry, A., and J. E. Klovan, 1976, The Middle-Upper Devonian clastic wedge of the Franklinian geosyncline: Bull. Canadian Petroleum Geology, v. 24, p. 485-639.

Enos, P., 1969, Anatomy of a flysch: Jour. Sed. Petrology, v. 39, p. 680-723.

———, 1969, Cloridorme Formation, Middle Ordovician flysch, northern Gaspe Peninsula, Quebec: Geol. Soc. America Spec. paper 117, 66 p.

Hamblin, A. P., and R. G. Walker, 1979, Storm-dominated shallow marine deposits; the Fernie-Kootenay (Jurassic) transition, southern Rocky Mountains: Canadian Jour. Earth Sci., v. 16, p. 1673-1690.

Hamilton-Smith, T., 1971, A proximal-distal turbidite sequence and a probable submarine canyon in the Siegas Formation (Early Llandovery) of northwestern New Brunswick: Jour. Sed. Petrology, v. 41, p. 752-762.

Hand, B. M., G. V. Middleton, and K. Skipper, 1972, Antidune cross-stratification in a turbidite sequence, Cloridorme Formation, Gaspe, Quebec: Sedimentology, v. 18, p. 135-138.

Hendry, H. E., 1978, Cap Des Rosiers Formation at Grosses Roches, Quebec - deposits of the mid-fan region on a Ordovician submarine fan: Canadian Jour. Earth Sci., v. 15, p. 1472-1488.

Hesse, R., 1976, Some unusual secondary sedimentary structures in the lacustrine Lower Carboniferous of Nova Scotia, Canada, and their significance as earthquake indicators [Einige ungewoehnliche sekundaere Sedimenstrukturen im lakustrinen Unterkarbon Neuschottlands (Kanada) und ihre Deutung als Erdbebenanzeiser]: Eclogae Geol. Helv. (EGHVAG), v. 69, p. 196-201.

Hiscott, R. N., 1978, Provenance of Ordovician deep-water sandstones, Tourelle Formation, Quebec, and implications for initiation of the Taconic Orogeny: Canadian Jour. Earth Sci., v. 15, p. 1579-1597.

———, 1979, Clastic sills and dikes associated with deep-water sandstones, Tourelle Formation, Ordovician, Quebec: Jour. Sed. Petrology, v. 49, p. 1-9.

——— and G. V. Middleton, 1979, Depositional mechanics of thick-bedded sandstones at the base of a submarine slope, Tourelle Formation (Lower Ordovician), Quebec, Canada, *in* L. J. Doyle and O. H. Pilkey, eds., Geology of continental slopes: SEPM Spec. Pub. no. 27, p. 307-326.

Hubert, C., J. Lajoie, and M. A. Leonard, 1970, Deep sea sediments in the Lower Paleozoic Quebec Supergroup, *in* J. Lajoie, ed., Flysch sedimentology in North America: Geol. Assoc. Canada, Spec. Paper 7, p. 103-125.

Johnson, B. A., 1974, Deep-sea fan-valley conglomerate; Cap Enrage Formation, Gaspe, Quebec: Hamilton, Ont., McMaster Univ., Masters thesis, 108 p.

——— and R. G. Walker, 1979, Paleocurrents and depositional environments of deep water conglomerates in the Cambro-Ordovician Cap Enrage Formation, Quebec Appalachians: Canadian Jour.

Earth Sci., v. 16, p. 1375-1387.

Krause, F. F., and A. E. Oldershaw, 1979, Submarine carbonate breccia beds - a depositional model for two layer, sediment gravity flows from the Sekwi Formation (Lower Cambrian), Mackenzie Mountains, Northwest Territories, Canada: Canadian Jour. Earth Sci., v. 16, p. 189-199.

Lajoie, J., 1969, Dispersal and petrology of the Silurian Val Brillant and Robitaille sandstones, Appalachians, Quebec: Jour. Sed. Petrology, v. 38, p. 643-647.

———— and A. Chagnon, 1973, Origin of red beds in a Cambrian flysch sequence, Canadian Appalachians, Quebec: Sedimentology, v. 20, p. 91-103.

———— Y. Heroux, and B. Mathey, 1974, The Precambrian shield and the Lower Paleozoic shelf; the unstable provenance of the Lower Paleozoic flysch sandstones and conglomerates of the Appalachians between Beaumont and Bic, Quebec: Canadian Jour. Earth Sci., v. 11, p. 951-963.

————, 1979, Origin of megarhythms in flysch sequences of the Quebec Appalachians: Canadian Jour. Earth Sci., v. 16, p. 1518-1523.

Lonsdale, P., and C. D. Hollister, 1979, Cut-offs at an abyssal meander south of Iceland: Geology, v. 7, p. 597-601.

McIlreath, I. A., 1977, Accumulation of a Middle Cambrian, deep-water limestone debris apron adjacent to a vertical, submarine carbonate escarpment, southern Rocky Mountains, Canada *in* H. E. Cook and P. Enos, eds., Deep-water carbonate environments: SEPM Spec. Pub. no. 25, p. 113-124.

Michaelis, E. R., and G. Dixon, 1969, Interpretation of depositional processes from sedimentary structures in the Cardium sand: Bull. Canadian Petroleum Geology v. 17, p. 410-443.

Morrow, D. W., 1978, The Prairie Creek embayment and associated slope, shelf and basin deposits: Canadian Geol. Surv. Paper no. 78-1A, p. 361-370.

Parkash, B., 1970, Downcurrent changes in sedimentary structures in Ordovician turbidite greywackes: Jour. Sed. Petrology, v. 40, p. 572-590.

———— and G. V. Middleton, 1970, Downcurrent textural changes in Ordovician turbidite greywackes: Sedimentology, v. 14, p. 259-293.

Parsons, M. G., 1975, The geology of the Laurentian fan and Scotia Rise, *in* C. J. Yorath, E. R. Parker and D. J. Glass, eds., Canada's continental margins and offshore petroleum exploration: Canadian Soc. Petroleum Geology Mem. 4, p. 155-167.

Pettijohn, F. J., 1936, Early Precambrian carved slate in northwestern Ontario: Geol. Soc. America Bull., v. 47, p. 621-628.

Piper, D. J. W., 1975, Late Quaternary deep water sedimentation off Nova Scotia and western Grand Banks, *in* C. J. Yorath, E. R. Parker and D. J. Glass, eds., Canada's continental margins and offshore petroleum exploration: Canadian Soc. Pet. Geol. Mem. 4, p. 195-204.

Rocheleau, M., and J. Lajoie, 1974, Sedimentary structures in resedimented conglomerate of the Cambrian flysch, L'Islet, Quebec Appalachians: Jour. Sed. Petrology, v. 44, p. 826-836.

Skipper, K., and S. B. Battacharjee, 1978, Backset bedding in turbidites; a further example from the Cloridorme Formation (Middle Ordovician), Gaspe, Quebec: Jour. Sed. Petrology, v. 48, p. 193-201.

———— and G. V. Middleton, 1975, The sedimentary structures and depositional mechanics of certain Ordovician turbidites, Cloridorme Formation, Gaspe Peninsula, Quebec: Canadian Jour. Earth Sci., v. 12, p. 1934-1952.

Srivastava, P., C. W. Stearn, and E. W. Mountjoy, 1972, A Devonian megabreccia at the margin of the ancient wall carbonate complex, Alberta: Bull. Canadian Petroleum Geology, v. 20, p. 412-438.

Stanley, D. J., 1967, Comparing patterns of sedimentation in some modern and ancient submarine canyons: Earth and Planetary Sci. Letters, v. 3, p. 371-380.

———— and N. Silverberg, 1969, Recent slumping on the continental slope off Sable Island bank, southeast Canada: Earth and Planetary Sci. Letters, v. 6, p. 123-133.

Stow, D. A. V., 1976, Deep water sands and silts on the Nova Scotian Continental Margin: Maritime Sediments, v. 12, p. 81-90.

————, 1979, Distinguishing between fine-grained turbidites and contourites on the Nova Scotian deep water margin: Sedimentology, v. 26, p. 371-387.

———— and A. J. Bowen, 1978, Origin of lamination in deep sea, fine-grained sediments: Nature, v. 274, no. 5669, p. 324-328.

———— and ————, 1980, A physical model for the transport and sorting of fine-grained sediment by turbidity currents: Sedimentology, v. 27, p. 31-46.

———— and G. Shanmugam, 1980, Sequence of structures in fine-grained turbidites; comparison of recent deep-sea and ancient flysch sediments: Sedimentary Geology, v. 25, p. 23-42.

Surlyk, F., and L. B. Clenimensen, 1975, A Valanginian turbidite sequence and its paleogeographical setting (Kuhn, 0, East Greenland): Bull. Geol. Soc. Denmark, v. 24, p. 61-73.

Gulf of Mexico

Bergantino, R. N., 1971, Submarine regional geomorphology of the Gulf of Mexico: Geol. Soc.

America Bull., v. 82, p. 741-752.

Bouma, A. H., 1972, Recent and ancient turbidites and contourites: Trans., Gulf Coast Assoc. Geol. Socs., v. 22, p. 205-221.

Caughey, C. A., and C. J. Stuart, 1976, Where the potential is in the deep Gulf of Mexico: World Oil, v. 183, p. 67-72.

Davies, D. K., 1972, Mincralogy, petrography, and derivationof sands and silts of continental slope, rise, and abyssal plain of the Gulf of Mexico: Jour. Sed. Petrology, v. 42, p. 59-65.

————, 1972, Deep sea sediments and their sedimentation, Gulf of Mexico: AAPG Bull., v. 56, p. 2212-2239.

Ewing, M., and B. C. Heezen, 1958, Sediments and topography of the Gulf of Mexico: in L. G. Weeks, ed., Habitat of Oil: AAPG Spec. Pub., p. 995-1053.

Handford, C. R., 1981, Deep-water facies of the Spraberry formation (Permian), Regan County, Texas, in Deep water clastic sediments, a core workshop: San Francisco, SEPM Core Workshop no. 2, p. 372-395.

Huang, T. C., and H. G. Goodell, 1970, Sediments and sedimentary processes of eastern Mississippi cone, Gulf of Mexico: AAPG Bull., v. 54, p. 2070-2100.

Lund, J. W., et al, 1978, Pre-platform exploration of High Island blocks A-a560 and A561: Trans., Gulf Coast Assoc. Geol. Socs., v. 28, p. 273-294.

Moore, G. T., and T. J. Fullam, 1973, Deep water channels and their potential value in petroleum localization: Trans., Gulf Coast Assoc. Geol. Socs., v. 23, p. 256-258.

———— and T. J. Fullam, 1975, Submarine channel systems and their potential for petroleum localization, in M. L. Broussard, ed., Deltas - models for exploration: Houston Geol. Soc., p. 165-189.

————, et al, 1978, Mississippi fan, Gulf of Mexico - physiography, stratigraphy, and sedimentational patterns, in A. H. Bouma, G. T. Moore, and J. M. Coleman, eds., Framework, facies and oil trapping characteristics of the Upper Continental Margin: AAPG Stud. in Geol. no. 7, p. 155-191.

———— and H. O. Woodbury, 1978, Mississippi fan - morphology, sedimentational history, petroleum potgential: 10th Ann. Offshore Tech. Conf., v. 1, p. 391-398.

Sabate, R. W., 1968, Pleistocene oil and gas in coastal Louisiana: Trans., Gulf Coast Assoc. Geol. Socs., v. 18, p. 373-386.

Siemers, C. T., 1981, Sedimentological core analysis of deep-water clastic sediments in the down-dip Woodbine-Eagle Ford interval (Upper Cretaceous), Tyler County, Texas, in Deep water clastic sediments, a core workshop: San Francisco, SEPM Core Workshop, no. 2, p. 249-371.

Walker, J. R., and J. V. Massingill, 1970, Slump features on the Mississippi fan, northeastern Gulf of Mexico: Geol. Soc. America Bull., v. 81, p. 3101-3108.

Woodbury, H. O., J. H. Spotts, and W. H. Akers, 1978, Gulf of Mexico continental-slope sediments and sedimentation, in A. H. Bouma, G. T. Moore, and J. M. Coleman, eds., Framework facies, and oil-trapping characteristics of the Upper Continental Martin: AAPG Stud. in Geology, no. 7, p. 117-137.

Marathon Basin

McBride, E. F., 1966, Sedimentary petrology and history of the Haymond Formation (Pennsylvanian), Marathon Basin, Texas: Austin, Univ. of Texas, Bur. Econ. Geol., Rept. Invest. 57, 101 p.

————, 1970, Flysch sedimentation in the Marathon region, Texas, in J. LaJoie, ed., Flysch sedimentology in North America: Geol. Assoc. Canada Spec. Paper no. 7, p. 67-83.

————, 1978, Olistostrome in the Tesnus Formation (Mississippian-Pennsylvanian), Payne Hills, Marathon region, Texas: Geol. Soc. America Bull., v. 89, p. 1550-1558.

———— and R. M. Flores, 1975, Characteristics of the Pennsylvanian lower-middle Haymond delta-front sandstones, Marathon Basin, West Texas [discussion and reply]: Geol. Soc. America Bull., v. 86, p. 264-266.

———— and S. J. Mazzullo, eds., 1978, Tectonics and Paleozoic facies of the Marathon geosyncline, West Texas: Permian Basin Sec., Pub. 78-17, SEPM, 271 p.

Thomson, A. F., 1978, Summary of the geologic history of the Marathon geosyncline, in E. F. McBride and S. J. Mazzullo, eds., Tectonics and Paleozoic facies of the Marathon geosyncline, West Texas: Permian Basin Sec., SEPM p. 79-88.

———— and M. R. Thomasson, 1964, Sedimentology and stratigraphy of the Dimple Limestone, Marathon Region, Texas, in The filling of the Marathon geosyncline: Symp. and Guidebook, Permian Basin Sec., SEPM, p. 22-30.

———— and ————, 1969, Shallow to deep water facies development in the Dimple Limestone (Lower Pennsylvanian), Marathon Region, Texas, in G. M. Friedman, ed., Depositional environments in carbonate rocks: SEPM Spec. Pub. no. 14, p. 57-78.

Mexico

Busch, D. A., and S. A. Govela, 1978, Stratigraphy and structure of Chicontepec turbidites, southeastern Tampico-Misantla basin, Mexico: AAPG Bull., v. 62, p. 235-246.

Helu, P. C., V. R. Verdugo, and P. R. Barcenas, 1977, Origin and distribution of Tertiary conglomerates, Veracruz basin, Mexico: AAPG Bull., v. 61, p. 207-226.

Normark, W. R., 1970, Growth patterns of deep-sea fans: AAPG Bull., v. 54, p. 2170-2195.

————, 1974, Ranger submarine slide, northern Sebastian Vizcaino Bay, Baja California, Mexico: Geol. Soc. America Bull., v. 85, p. 781-784.

———— and D. J. W. Piper, 1969, Deep-sea fan-valleys past and present: Geol. Soc. America Bull., v. 80, p. 1859-1866.

Ouachita Mountains - Arkoma Basin

Briggs, G., 1974, Carboniferous depositional environments in the Ouachita Mountains - Arkoma Basin area of southeastern Oklahoma: Geol. Soc. America Spec. Paper no. 148, p. 225-239.

————, ed., 1974, Carboniferous rocks of the southeastern United States: Geol. Soc. America Spec. Paper no. 148, 361 p.

———— and L. M. Cline, 1967, Paleocurrents and source areas of late Paleozoic sediments of the Ouachita Mountains, southeastern Oklahoma: Jour. Sed. Petrology, v. 37, p. 985-1000.

———— and D. Roeder, 1975, Sedimentation and plate tectonics, Ouachita Mountains and Arkoma Basin: in Guidebook to the sedimentology of Paleozoic flysch and associated deposits, Ouachita Mountains - Arkoma Basin, Oklahoma: Field Trip Guidebook, Dallas Geol. Soc. p. 1-22.

Chamberlain, C. K., 1971, Morphology and ethology of trace fossils from the Ouachita Mountains, southeast Oklahoma: Jour. Paleontology, v. 45, p. 212-246.

————, 1971, Bathymetry and paleoecology of Ouachita geosyncline of southeastern Oklahoma as determined from trace fossils: AAPG Bull., v. 55, p. 34-50.

————, 1975, Biogenetic sedimentary structures - trace fossils of the Ouachitas: in Guidebook to sedimentology of Paleozoic flysch and associated deposits, Ouachita Mountains - Arkoma Basin, Oklahoma: Dallas Geol. Soc. Pub., p. 51-68.

Cline, L. M., 1966, Late Paleozoic rocks of Ouachita Mountains, a flysch facies: in Oklahoma Guidebook, 29th Kansas Geol. Soc. Field Conf., p. 91-111.

————, 1968, Comparison of main geologic features of Arkoma Basin and Ouachita Mountains, southeastern Oklahoma: Okla. City Geol. Soc., AAPG-SEPM Joint Mtg. Field Conf., p. 63-74.

————, 1968, A guidebook to the geology of the western Arkoma Basin and Ouachita Mountains, Oklahoma: Okla. City Geol. Soc., AAPG-SEPM Joint Field Conf., 126 p.

————, 1970, Sedimentary features of Late Paleozoic flysch, Ouachita Mountains, Oklahoma, in J. Lajoie, ed., Flysch sedimentology of North America: Geol. Assoc. Canada Spec. Paper no. 7, p. 85-101.

Graham, S. A., W. R. Dickinson, and R. V. Ingersoll, 1975, Himalayan-Bengal model for flysch dispersal in the Appalachian-Ouachita System: Geol. Soc. America Bull., v. 85, p. 273-286.

————, R. V. Ingersoll, and W. R. Dickinson, 1976, Common provenance for lithic grains in Carboniferous sandstones from Ouachita Mountains and Black Warrior Basin: Jour. Sed. Petrology, v. 46, p. 620-632.

Johnson, K. E., 1966, A depositional interpretation of the Stanley Group of the Ouachita Mountains, Oklahoma, in Flysch facies and structure of the Ouachita Mountains:, 29th Kansas Geol. Soc. Field Conf. Okla. Guidebook, p. 140-163.

Klein, G. D., 1966, Dispersal and petrology of sandstones of Stanley-Jackfork boundary, Ouachita fold belt, Arkansas and Oklahoma: AAPG Bull., v. 50, p. 308-326.

Lajoie, J., Y., and A. Chagnon, 1973, Origin of red beds in a Cambrian flysch sequence, Canadian Appalachians, Quebec: Sedimentology, v. 20, p. 91-103.

McBride, E. F., 1975, The Ouachita trough sequence; Marathon region and Ouachita Mountains, in Guidebook to sedimentology of Paleozoic flysch and associated deposits, Ouachita Mountains - Arkoma Basin, Oklahoma: Field Trip Guidebook, Dallas Geol. Soc., p. 23-41.

Moiola, R. J., and E. F. McBride, 1975, Sedimentology of Ouachita turbidites, southeastern Oklahoma; a summary, in Sedimentology of Paleozoic flysch and associated deposits, Ouachita Mountains - Arkoma Basin, Oklahoma: Field Trip Guidebook, Dallas Geol. Soc., p. 42-50.

Morris, R. C., 1971, Classification and interpretation of disturbed bedding types in Jackfork flysch rocks (Upper Mississippian), Ouachita Mountains, Arkansas: Jour. Sed. Petrology, v. 41, p. 410-424.

————, 1971, Stratigraphy and sedimentology of Jackfork Group, Arkansas: AAPG Bull., v. 55, p. 387-402.

————, 1974, Carboniferous rocks of the Ouachita Mountains, Arkansas - a study of facies patterns along the unstable slope and axis of a flysch trough, in G. Briggs, ed., Symposium on the carboniferous rocks of the southeastern United States: Geol. Soc. America Spec. Paper no. 148, p. 241-279.

————, 1974, Sedimentary and tectonic history of the Ouachita Mountains, in W. R. Dickinson, ed.,

Tectonics and sedimentation: SEPM Spec. Pub. no. 22, p. 120-142.

————, et al, 1975, Stratigraphy and structure of part of frontal Ouachita Mountains, Arkansas: AAPG Bull., v. 59, p. 747-765.

———— K. E. Proctor, and M. R. Koch, 1979, Petrology and diagenesis of deep-water sandstones, Ouachita Mountains, Arkansas and Oklahoma, *in* P. A. Scholle and P. R. Schluger, eds., Aspects of diagenesis: SEPM Spec. Pub. no. 26, p. 263-279.

Niem, A. R., 1976, Patterns of flysch deposition and deep-sea fans in the lower Stanley Group (Mississippian), Ouachita Mountains, Oklahoma and Arkansas: Jour. Sed. Petrology, v. 46, p. 633-646.

Picha, F., and L. M. Cline, 1973, Radiographic investigation of laminated and ripple cross-laminated flysch sandstones, Ouachita Mountains, Oklahoma: Jour. Sed. Petrology, v. 43, p. 466-470.

———— and A. R. Niem, 1974, Distribution and extent of beds in flysch deposits, Ouachita Mountains, Arkansas and Oklahoma: Jour. Sed. Petrology, v. 44, p. 328-335.

Shideler, G. L., 1970, Provenance of Johns Valley boulders in Late Paleozoic Ouachita facies, southeastern Oklahoma and southwestern Arkansas: AAPG Bull., v. 54, p. 789-806.

Stark, P. H., 1966, Stratigraphy and environment of deposition of the Atoka Formation in the Central Ouachita Mountains, Oklahoma: Oklahoma Guidebook, 29th Kansas Geol. Soc. Field Conf.

Stone, C. G., B. R. Haley, and G. W. Viele, 1973, A guidebook to the geology of the Ouachita Mountains, Arkansas: Little Rock, Ark. Geol. Comm., p. 1-113.

Vedros, S. G., and G. S. Visher, 1978, The Red Oak Sandstone; a hydrocarbon producing submarine fan deposit, *in* J. D. Stanley and G. Kelling, eds., Sedimentation in submarine canyons, fans, and trenches: Stroudsburg, Pa., Dowden, Hutchinson, and Ross, Inc., p. 292-310.

Walthall, B. H., 1967, Stratigraphy and structure, part of Athens Plateau, southern Ouachitas, Arkansas: AAPG Bull., v. 51, p. 504-528.

Oregon and Washington

Carson, B., 1973, Acoustic stratigrahy, structure, and history of Quaternary deposition in Cascadia Basin: Deep-Sea Research, v. 20, p. 387-396.

———— J. Yuan, and P. B. Myers, Jr., 1974, Initial deep-sea sediment deformation at the base of the Washington continental slope; a response to subduction: Geology, v. 2, p. 561-564.

Dickinson, W. R., 1979, Mesozoic forearc basin in central Oregon: Geology, v. 7, p. 166-170.

Dott, R. H., Jr., 1963, Dynamics of subaqueous gravity depositional processes: AAPG Bull., v. 47, p. 104-128.

———— and K. J. Bird, 1979, Sand transport through channels across an Eocene shelf and slope in southwestern Oregon, U.S.A., *in* L. J. Doyle and O. H. Pilkey, eds., Geology of continental slopes: SEPM Spec. Pub. no. 27, 327-242.

Duncan, J. R., and L. D. Kulm, 1970, Mineralogy, provenance and dispersal history of Late Quaternary deep-sea sands in Cascadia Basin and Blanco fracture zone off Oregon: Jour. Sed. Petrology, v. 40, p. 874-887.

Griggs, G. B., A. G. Carey, Jr., and L. D. Kulm, 1969, Deep-sea sedimentation and sediment-fauna interaction in Cascadia Channel and on Cascadia Abyssal Plain: Deep-Sea Research, v. 16, p. 157-170.

———— and L. D. Kulm, 1970, Sedimentation in Cascadia Deep-Sea Channel: Geol. Soc. America Bull., v. 81, p. 1361-1384.

———— and ————, 1973, Origin and development of Cascadia Deep-Sea Channel: Jour. Geophys. Research, v. 78, p. 6325-6339.

———— ————, et al, 1970, Deep-sea gravel from Cascadia Channel: Jour. Geology, v. 78, p. 611-619.

Hurley, R. J., 1964, Analysis of flow in Cascadia deep-sea channel, *in* R. L. Miller, ed., Papers in marine geology: New York, Macmillan Co., p. 117-132.

Kulm, L. D., and G. A. Fowler, 1974, Cenozoic sedimentary framework of the Gorda-Juan de Fuca Plate and adjacent continental margin - a review, *in* R. H. Dott, Jr., and R. H. Shaver, eds., Modern and ancient geosynclinal sedimentation: SEPM Spec. Pub. no. 19, p. 212-229.

———— and ————, 1974, Oregon continental margin structure and stratigraphy; a test of the imbricate thrust model, *in* C. A. Burk and C. L. Drake, eds., The geology of the continental margins: New York, Springer-Verlag, p. 261-283.

Lovell, J. P. B., 1969, Tyee Formation; a study of proximality in turbidites: Jour. Sed. Petrology, v. 39, p. 935-953.

McLean, H., 1977, Lithofacies of the Blakeley Formation, Kitsap County, Washington; a submarine fan comlex?: Jour. Sed. Petrology, v. 47, p. 78-88.

Nelson, C. H., 1976, Late Pleistocene and Holocene depositional trends, processes, and history of Astoria deep-sea fan, northeast Pacific: Marine Geol., v. 20, p. 129-173.

————, et al, 1970, Development of the Astoria canyon-fan physiography and comarison with similar systems: Marine Geol., v. 8, p. 259-291.

———— and T. H. Nilsen, 1974, Depositional trends of modern and ancient deep-sea fans, *in* R. H. Dott, Jr., and R. H. Shavers, eds., Modern and Ancient geosynclinal sedimentation: SEPM Spec. Pub. no. 19, p. 69-91.

Poole, F. G., 1974, Flysch deposits of the Antler foreland basin, western United States, *in* W. R. Dickinson, ed., Tectonics and sedimentation: SEPM Spec. Pub. no. 22, p. 58-82.

Snavely, P. D., H. C. Wagner, and N. S. MacLeod, 1964, Rhythmic-bedded eugeosynclinal deposits of the Tyee Formation, Oregon Coast Range, *in* D. F. Merriam, ed., Symposium on cyclic sedimentation: Bull. Kansas Geol. Survey, no. 169, p. 461-480.

Walker, R. G., 1977, Deposition of Upper Mesozoic resedimented conglomerates and associated turbidites in southwestern Oregon: Geol. Soc. America Bull., v. 88, p. 273-285.

Rocky Mountains

Chamberlain, C. K., K. O. Stanley, and J. H. Stewart, 1978, Depositional setting of some eugeosynclinal Ordovician rocks and structurally interleaved Devonian rocks in the Cordilleran mobile belt, *in* Paleozoic paleogeography of the western United States: Symp. Proc., SEPM Pacific Sec., p. 259-274.

Hamblin, A. P., and R. G. Walker, 1979, Storm-dominated shallow marine deposits; the Fernie-Kooatenay (Jurassic) transition, southern Rocky Mountains: Canadian Jour. Earth Sci., v. 16, p. 1673-1690.

Krutak, P. R., 1970, Origin and depositional environment of the Codell sandstone member of the Carlile shale (Upper Cretaceous), southeastern Colorado: Mtn. Geol., v. 7, p. 185-204.

McIlreath, I. A., 1977, Accumulation of a Middle Cambrian, deep-water limestone debris apron adjacent to a vertical, submarine carbonate escarpment, southern Rocky Mountains, Canada *in* H. E. Cook and P. Enos, eds., Deep-water carbonate environments: SEPM Spec. Pub. no. 25, p. 113-124.

Texas

Berg, R. R., 1979, Characteristics of Lower Wilcox reservoirs, Valentine and south Hallettsville fields, Lavaca County, Texas: Trans., Gulf Coast Assoc. Geol. Socs., v. 29, p. 11-23.

———— and R. L. Findley, 1973, Deep-water interpretation of Upper Wilcox sandstones from core study, Katy Field, Texas: Trans., Gulf Coast Assoc. Geol. Socs., v. 23, p. 259-265.

———— and R. R. Powell, 1976, Density-flow origin for Frio reservoir sandstones, Nine Mile Point Field, Aransas County, Texas: Trans., Gulf Coast Assoc. Geol. Socs., v. 26, p. 310-319.

———— and F. J. Tedford, 1977, Characteristics of Wilcox gas reservoirs, northeast Thompsonville Field, Jim Hogg and Webb Counties, Texas: Trans., Gulf Coast Assoc. Geol. Socs., v. 27, p. 6-19.

Bloomer, R. E., 1977, Depositional environments of a reservoir sandstone in west-central Texas: AAPG Bull., v. 61, p. 344-359.

Bornhauser, M., 1960, Depositional and structural history of northwest Hartburg Field, Newton County, Texas: AAPG Bull., v. 44, p. 458-470.

Foss, D. C., 1979, Depositional environment of Woodbine sandstone, Polk County, Texas: Trans., Gulf Coast Assoc. Geol. Socs., v. 29, p. 83-94.

Galloway, W. E., and L. F. Brown, Jr., 1972, Depositional systems and shelf-slope relationships in Upper Pennsylvanian rocks, north-central Texas: Austin, Univ. Texas, Bureau Econ. Geology, Rept. Inv. no. 75, 62 p.

Handford, C. R., and S. P. Dutton, 1980, Pennsylvanian-Early Permian depositional systems and shelf-margin evolution, Palo Duro basin, Texas: AAPG Bull., v. 64, p. 88-106.

Harms, J. C., 1974, Brushy Canyon Formation, Texas; a deep water density current deposit: Geol. Soc. America Bull., v. 85, p. 1763-1784.

———— and R. K. Fahnestock, 1965, Stratification, bed forms and flow phenomena (with example from the Rio Grande), *in* G. V. Middleton, ed., Primary sedimentary structures and their hydrodynamic interpretation: SEPM Spec. Pub. no. 12, p. 84-115.

Hoyt, W. V., 1959, Erosional channel in the middle Wilcox near Yoakum, Lavaca County, Texas: Trans., Gulf Coast Assoc. Geol. Socs., v. 9, p. 41-50,.

Jacka, A. D., et al, 1968, Permian deep-sea fans of the Delaware Mountain group (Guadalupian), Delaware Basin, *in* Guadalupian facies, Apache Mountain area, west Texas: Permian Basin Sec., SEPM Pub. 68-11, p. 49-90.

————, et al, 1967, Guadalupian depositional cycles of Delaware Basin and northwest shelf, *in* J. G. Flam and S. Chuber, eds., Cyclic sedimentation in the Permian Basin: Midland, West Texas Geol. Soc., p. 152-196.

Koss, G. M., 1977, Carbonate mass flow sequences of the Permian Delaware Basin, West Texas, *in* Upper Guadalupian Facies Permian Reef Complex, Guadalupe Mountains: New Mexico and West Texas Field Conf. Guidebook, Permian Basin Sec., SEPM Pub. 77-16, p. 391-431.

Mitchell, M. H., 1975, Depositional environment and facies relationships of the canyon sandstones, Val Verde Basin, Texas: Texas A&M Univ. Masters thesis, 211 p.

Payne, M. W., 1976, Basinal sandstone facies Delaware Basin, west Texas and southeast New Mexico: AAPG Bull., v. 60, p. 517-527.

Siemers, C. T., 1978, Submarine fan deposition of the Woodbine-Eagle Ford interval (Upper Cretaceous), Tyler County, Texas: Trans., Gulf Coast Assoc. Geol. Soc., v. 28, p. 493-533.

Vormelker, R. S., 1979, Mid-Wilcox channel; deep exploration potential: Bull. South Texas Geol. Soc., v. 20, p. 10-40.

————, 1980, Texas Middle Wilcox channel; deep exploration potential: Oil & Gas Jour. (March 10, 1980), p. 136-138, 140, 143, 146, 148, 151-154.

Williamson, C. R., 1977, Deep sea channels of the Bell Canyon Formation (Guadalupian) Delaware Basin, Texas-New Mexico, *in* Upper Guadalupian Facies Permian Reef Complex, Guadalupe Mountains: New Mexico and West Texas Field Conf. Guidebook, Permian Basin Sec., SEPM Pub. 77-16, v. 1, p. 409-431.

Gulf Coast

Benson, P. H., 1971, Geology of the Oligocene Hackberry trend, Gillis English Bayou - Manchester area, Calcasieu Parish, Louisiana: Trans., Gulf Coast Assoc. Geol. Socs., v. 21, p. 1-14.

Bornhauser, M., 1948, Possible ancient submarine canyon in southwestern Louisiana: AAPG Bull., v. 32, p. 2287-2294.

Paine, W. R., 1966, Stratigraphy and sedimentation of subsurface Hackberry wedge and associated beds of southwestern Louisiana: Trans., Gulf Coast Assoc. Geol. Socs., v. 16, p. 261-274.

Sangree, J. B., et al, 1978, Recognition of continental-slope seismic facies, offshore Texas-Louisiana, *in* A. H. Bouma, G. T. Moore, and J. M. Coleman, eds., Framework, facies, and oil-trapping characteristics of the Upper Continental Margin: AAPG Stud. in Geology, no. 7, p. 87-116.

Atlantic Ocean

Ayers, M. W., and W. J. Cleary, 1980, Wilmington fan; mid-Atlantic lower rise development: Jour. Sed. Petrology, v. 50, p. 235-245.

Ball, M. M., 1979, Petroleum potential of passive margin slopes, *in* L. J. Doyle and O. H. Pilkey, eds., Geology of continental slopes: SEPM Spec. Pub., no. 27, p. 43-47.

Bradley, W. H., et al, 1940, Geology and biology of North Atlantic deep-sea cores between Newfoundland and Ireland: U.S. Geol. Survey Prof. Paper 196A, 56 p.

————, et al, 1942, Geology and biology of North Atlantic deepsea cores: U.S. Geol. Survey, Prof. Paper 196, 163 p.

Cacchione, D. A., G. T. Rowe, and A. Malahoff, 1978, Submersible investigation of outer Hudson submarine canyon, *in* D. J. Stanley and G. Kelling, eds., Sedimentation in submarine canyons, fans, and trenches: Stroudsburg, Pa., Dowden, Hutchinson and Ross, Inc., p. 42-50.

Cherkis, N. Z., H. S. Fleming, and R. H. Feden, 1973, Morphology and structure of Maury Channel, northest Atlantic Ocean: Geol. Soc. America Bull., v. 84, p. 1601-1606.

Chough, S. K., 1978, Morphology, sedimentary facies and processes of the Northwest Atlantic Mid-Ocean Channel between 61° and 52° N, Labrador Sea: McGill Univ., Ph.D. dissert., 167 p.

———— and R. Hesse, 1976, Submarine meandering thalweg and turbidity currents flowing for 4000 km in the Northwest Atlantic Mid-Ocean Channel, Labrador Sea: Geology, v. 4, p. 529-533.

———— and ————, 1980, Northwest Atlantic Mid-Ocean Channel of the Labrador Sea; III, head spills versus body spill deposits from turbidity currents on natural levees: Jour. Sed. Petrology, v. 50, p. 227-234.

Cleary, W. J., and J. R. Conolly, 1974, Hatteras deep-sea fan: Jour. Sed. Petrology, v. 44, p. 1140-1154.

———— and ————, 1974, Petrology and origin of deep-sea sands; Hatteras Abyssal Plain: Marine Geology, v. 17, p. 263-279.

———— O. H. Pilkey, and M. Ayers, 1977, Morphology and sediments of three ocean basin entry points, Hatteras Abyssal Plain: Jour. Sed. Petrology, v. 47, p. 1157-1170.

Damuth, J. E., and R. W. Embley, 1979, Upslope flow of turbidity currents on the northwest flank of Ceara Rise; western Equatorial Atlantic: Sedimentology, v. 26, p. 825-834.

Davies, T. A., and A. L. Laughton, 1972, Sedimentary processes in the North Atlantic: Washington, D.C., Initial Repts. of Deep Sea Drilling Proj., v. 12, p. 905-934.

Drake, D. E., P. G. Hatcher, and G. H. Keller, 1978, Suspended particulate matter and mud deposition in Upper Hudson submarine canyon, *in* D. J. Stanley and G. Kelling, eds., Sedimentation in submarine canyons, fans, and trenches: Stroudsburg, Pa., Dowden, Hutchinson and Ross, Inc., p. 33-41.

Elmore, R. D., et al, 1979, Black Shell turbidite, Hatteras Abyssal Plain, western Atlantic Ocean: Geol. Soc. America Bull., pt. 1, v. 90, p. 1165-1176.

Emery, K. O., and E. Uchupi, 1972, Western North Atlantic Ocean; topography, rocks, structure, water, life, and sediments: AAPG Mem. 17, 532 p.

Ericson, D. B., M. Ewing, and B. C. Heezen, 1952, Turbidity currents and sediments in North Atlantic: AAPG Bull., v. 36, p. 489-511.

——— ——— ——— and G. Wollin, 1955, Sediment deposition in deep Atlantic: Geol. Soc. America Spec. Paper 62, p. 205-219.

——— ——— ——— and B. C. Heezen, 1961, Atlantic deep-sea sediment cores: Geol. Soc. America Bull., v. 72, p. 193-285.

Ewing, J., X. LePichon, and M. Ewing, 1963, Upper stratification of Hudson apron region: Jour. Geophys. Research, v. 68, p. 6303-6316.

Fallaw, W. C., 1976, Factors affecting petroleum accumulation beneath the eastern United States continental margin: Southeastern Geology, v. 17, p. 189-205.

Field, M. E., and O. H. Pilkey, 1971, Deposition of deep-sea sands; comparison of two areas of the Carolina continental rise: Jour. Sed. Petrology, v. 41, p. 526-536.

Fritz, S. J., and O. H. Pilkey, 1975, Distinguishing bottom and turbidity current coarse layers on the continental rise: Jour. Sed. Petrology, v. 45, p. 57-62.

Heezen, B. C., 1968, Atlantic ocean floor map: Natl. Geog. Mag., June, p. 794 (insert).

——— and C. L. Drake, 1964, Grand banks slump: AAPG Bull., v. 48, p. 221-233.

——— D. B. Ericson, and M. Ewing, 1954, Further evidence for a turbidity current following the 1929 Grand Banks earthquake: Deep-Sea Research, v. 1, p. 193-202.

——— and M. Ewing, 1952, Turbidity currents and submarine slumps and the 1929 Grand Banks earthquake: Am. Jour. Sci., v. 250, p. 849-873.

——— and C. D. Hollister, 1964, Deep-sea current evidence from abyssal sediments: Marine Geology, v. 1, p. 141-174.

——— M. Tharp, and M. Ewing, 1959, The floors of the ocean, 1 - the North Atlantic: Geol. Soc. America Spec. Paper 65, 122 p.

Horn, D. R., et al, 1971, Turbidites of the Hatteras and Sohm abyssal plains, western North Atlantic: Marine Geology, v. 11, p. 287-323.

——— J. I. Ewing, and M. Ewing, 1972, Graded-bed sequences emplaced by turbidity currents north of 20°N in the Pacific, Atlantic and Mediterranean: Sedimentology, v. 18, p. 247-275.

Hubert, J. F., 1964, Textural evidence for deposition of many western North Atlantic deep-sea sands by ocean-bottom currents rather than turbidity currents: Jour. Geology, v. 72, p. 757-785.

——— and W. J. Neal, 1967, Mineral composition and dispersal patterns of the deep-sea sands in the western North Atlantic petrologic province: Geol. Soc. America Bull., v. 78, p. 749-772.

Keller, G. H., and F. P. Sherpard, 1978, Currents and sedimentary processes in submarine canyons off the northeast United States, in D. J. Stanley and G. Kelling, eds., Sedimentation in submarine canyons, fans and trenches: Stroudsburg, Pa., Dowden, Hutchinson and Ross, Inc., p. 15-32.

Kelling, G., and D. J. Stanley, 1970, Morphology and structure of Wilmington and Baltimore canyons, eastern United States: Jour. Geology, v. 78, p. 637-660.

Knebel, H. J., and D. W. Folger, 1976, Large sand waves on the Atlantic outer continental shelf around Wilmington Canyon, off eastern United States: Marine Geology, v. 22, p. M7-M15.

Krause, D. C., et al, 1970, Turbidity currents and cable breaks in the western New Britain Trench: Geol. Soc. America Bull., v. 81, p. 2153-2160.

Kuenen, P. H., 1952, Estimated size of the Grand Banks turbidite currents: Am. Jour. Sci., v. 250, p. 874-884.

Lonsdale, P., and C. D. Hollister, 1979, Cut-offs at an abyssal meander south of Icealdn: Geology, v. 7, p. 597-601.

Mattick, R. E., et al, 1978, Petroleum potential of U.S. Atlantic slope, rise and abyssal plain: AAPG Bull., v. 62, p. 592-608.

McMaster, R. L., and A. Ashraf, 1973, Drowned and bruied valleys on the southern New England continental shelf: Marine Geology, v. 15, p. 249-268.

——— and ———, 1973, Extent and formation of deeply buried channels on the continental shelf off southern New England: Jour. Geology v. 81, p. 374-379.

Milliman, J. D., C. P. Summerhayes, and H. T. Barretto, 1975, Quaternary sedimentation on the Amazon continental margin; a model: Geol. Soc. America Bull., v. 86, p. 610-614.

Normark, W. R., 1978, Fan valleys, channels and depositional lobes on modern submarine fans; character for recognition of sandy turbidite environments: AAPG Bull., v. 62, p. 912-931.

Parsons, M. G., 1975, The geology of the Laurentian fan and Scotia Rise in C. J. Yorath, E. R. Parker and D. J. Glass, eds., Canada's continental margins and offshore petroleum exploration: Canadian Soc. Petroleum Geols. Mem. 4, p. 155-167.

Piper, D. J. W., 1975, Late Quaternary deep water sedimentation off Nova Scotia and western Grand Banks, in C. J. Yorath, E. R. Parker and D. J. Glass, eds., Canada's continental margins and offshore petroleum exploration: Canadian Soc. Petroleum Geols. Mem. 4, p. 195-204.

Poag, C. W., 1979, Stratigraphy and depositional environments of Baltimore Canyon trough: AAPG Bull., v. 63, p. 1452-1466.

Pratt, R. M., 1968, Atlantic continental shelf and slope of the United States - physiography and sediments of the deep-sea basin: U.S. Geol. Survey Prof. Paper 529-B, 44 p.

Roberson, M. I., 1964, Continuous seismic profiler survey of oceanographer, Gilbert, and Lydonia submarine canyons, Georges Bank: Jour. Geophys. Research, v. 69, p. 4779-4789.

Rona, P. A., 1969, Middle Atlantic continental slope of United States; deposition and erosion: AAPG Bull., v. 53, p. 1453-1969.

———, 1970, Submarine canyon origin on upper continental slope off Cape Hatteras: Jour. Geology, v. 78, p. 141-152.

Ruddiman, W. F., F. A. Bowles, and B. Molnia, 1972, Maury Channel and fan: 24th Internat. Geol. Cong., Sec. 8, p. 100-108.

Schneider, E. D., et al, 1967, Further evidence of contour currents in the western North Atlantic: Earth and Planetary Sci. Letters, v. 2, p. 351-359.

Sheridan, R. E., X. Golovchenko, and J. I. Ewing, 1974, Late Miocene turbidite horizon in Blake-Bahama basin: AAPG Bull., v. 58, p. 1797-1805.

Stanley, D. J., 1967, Comparing patterns of sedimentation in some modern and ancient submarine canyons: Earth and Planetary Sci. Letters, v. 3, p. 371-380.

———, 1969, Submarine channel deposits and their fossil analogs (Fluxoturbidites), in New concepts of continental margin sedimentation: AGI Short Course Notes, Lecture 9, p. DJS-9-1 to DJS-9-17.

———, 1970, Flyschoid sedimentation on the outer Atlantic margin off northeast North America, in J. Lajoie, ed., Flysch sedimentology in North America: Geol. Assoc. Canada Spec. Paper no. 7, p. 179-210.

———, 1974, Pebbly mud transport in the head of Wilmington Canyon: Marine Geology v. 16, p. M1-M8.

——— and G. Kelling, 1968, Sedimentation patterns in the Wilmington submarine canyon area, in Symposium on ocean sciences and engineering of the Atlantic Shelf: Philadelphia, Trans. Delaware Valley Sec., Mar. Tech. Soc., p. 127-142.

——— and ———, 1970, Interpretation of a levee-like ridge and associated features, Wilmington submarine canyon, eastern United States: Geol. Soc. America Bull., v. 81, p. 3747-3752.

——— H. Sheng, and C. P. Pedraza, 1971, Lower continental rise east of the Middle Atlantic states; predominant sediment dispersal perpendicular to isobaths: Geol. Soc. America Bull., v. 82, p. 1831-1840.

——— and R. Unrug, 1972, Submarine channel deposits, fluxoturbidites and other indicators of slope and base-of-slope environments in modern and ancient marine basins, in J. K. Rigby and W. K. Hamblin, eds., Recognition of ancient sedimentary environments: SEPM Spec. Pub. no. 16, p. 287-340.

Stow, D. A. V., 1976, Deep water sands and silts on the Nova Scotian continental margin: Maritime Sediments, v. 12, p. 81-90.

———, 1979, Distinguishing between fine-grained turbidites and contourites on the Nova Scotian deep water margin: Sedimentology, v. 26, p. 371-387.

——— and A. J. Bowen, 1978, Origin of lamination in deep sea, fine-grained sediments: Nature, v. 274, no. 5669, p. 324-328.

——— and ———, 1980, A physical model for the transport and sorting of fine-grained sediment by turbidity currents: Sedimentology, v. 27, p. 31-46.

Uchupi, E., and J. A. Austin, Jr., 1979, The stratigraphy and structure of the Laurentian Cone region: Canadian Jour. Earth Sci., v. 16, p. 1726-1752.

Pacific Ocean

Bandy, O. L., and K. S. Rodolfo, 1964, Distribution of foraminifera and sediments, Peru-Chile Trench area: Deep Sea Research, v. 11, p. 817-837.

Carlson, P. R., and C. H. Nelson, 1969, Sediments and sedimentary structures of the Astoria submarine canyon-fan system, northeast Pacific: Jour. Sed. Petrology, v. 39, p. 1269-1282

Carson, B., 1973, Acoustic stratigraphy, structure, and history of Quaternary deposition in Cascadia Basin: Deep-Sea Research, v. 20, p. 387-396.

Chamberlain, C. K., 1976, Trace fossils in DSDP cores of the Pacific: Jour. Paleontology., v. 49, p. 1074-1096.

Chase, T. E., W. R. Normark, and P. Wilde, 1975, Oceanograhic data of the Monterey deep-sea fan, 34°-37°N, 120°-127°W: Univ. Calif. Inst. Marine Resources Tech. Rept. TR-58, 2 p.

Dickinson, W. R., 1970, Clastic sedimentary sequences deposited in shelf, slope, and trough settings between magmatic arcs and associated trenches: Pacific Geology, v. 3, p. 15-30.

———, 1974, Sedimentation within and beside ancient and modern magmatic arcs, in R. H. Dott, Jr., and R. H. Shaver, eds., Modern and ancient geosynclinal sedimentation: SEPM Spec. Pub. no. 19, p. 230-239.

Duncan, J. R., and L. D. Kulm, 1970, Minerology, provenance and dispersal history of Late Quarternary deep-sea sands in Cascadia Basin and Blanco fracture zone off Oregon: Jour. Sed. Petrology, v. 40, p. 874-887.

Griggs, G. B., A. G. Carey, Jr., and L. D. Kulm, 1969, Deep-sea sedimentation and sediment-fauna interaction in Cascadia Channel and on Cascadia Abyssal Plain: Deep-Sea Research, v. 16, p. 157-170.

——— and L. D. Kulm, 1970, Sedimentation in Cascadia Deep-Sea Channel: Geol. Soc. America,

Bull., v. 81, p. 1361-1384.

—— and ——, 1973, Origin and development of Cascadia Deep-Sea Channel: Jour. Geophys. Research, v. 78, p. 6325-6339.

—— ——, et al, 1970, Deep-sea gravel from Cascadia Channel: Jour. Geology, v. 78, p. 611-619.

Horn, D. R., et al, 1971, Turbidites of the northeast Pacific: Sedimentology, v. 16, p. 55-69.

——J. I. Ewing, and M. Ewing, 1972, Graded-bed sequences emplaced by turbidity currents north of 20°N in the Pacific, Atlantic and Mediterranean: Sedimentology, v. 18, p. 247-275.

——, B. M. Horn, and M. N. Delach, 1970, Sedimentary provinces of the North Pacific: Geol. Soc. America Mem. 126, p. 1-21.

Hurley, R. J., 1964, Analysis of flow in Cascadia Deep-Sea Channel, in R. L. Miller, ed., Papers in marine geology: New York, Macmillan Co., p. 117-132.

Johnson, D. A., and T. C. Johnson, 1970, Sediment redistribution by bottom currents in the Central Pacific: Deep-Sea Research, v. 17, p. 157-169.

Klein, G. D., 1975, Sedimentary tectonics in southwest Pacific marginal basins based on leg 30 Deep Sea Drilling Project cores from the south Fiji, Hebrides, and Coral Sea basins: Geol. Soc. America Bull., v. 86, p. 1012-1018.

Kulm, L. D., and G. A. Fowler, 1974, Cenozoic sedimentary framework of the Gorda-Juan de Fuca Plate and adjacent continental margin - a review, in R. H. Dott, Jr., and R. H. Shaver, eds., Modern and ancient geosynclinal sedimentation: SEPM Spec. Pub. no. 19, p. 212-229.

—— and ——, 1974, Oregon continental margin structure and stratigraphy; a test of the imbricate thrust model, in C. A. Burk and C. L. Drake, eds., The geology of the continental margins: New York, Springer-Verlag, p. 261-283.

Menard, H. W., Jr. 1964, Marine geology of the Pacific: New York, McGraw Hill, 271 p.

Moore, D. G., 1961, Submarine slumps: Jour. Sed. Petrology, v. 31, 343-357.

Mordojovich, C., 1974, Geology of a part of the Pacific margin of Chile, in C. A. Burk and C. L. Drake, eds., The geology of continental margins, New York, Springer-Verlag, p. 591-598.

Nelson, C. H.,1976, Late Pleistocene and Holocene depositional trends, processes, and history of Astoria deep-sea fan, northeast Pacific: Marine Geology, v. 20, p. 129-173.

—— et al, 1970, Development of the Astoria canyon-fan physiography and comparison with similar systems: Marine Geology, v. 8, p. 259-291.

Normark, W. R., 1978, Fan valleys, channels and depositional lobes on modern submarine fans; characters for recognition of sandy turbidite environments: AAPG Bull., v. 62, p. 912-931.

——G. R. Hess, and F. N. Spiess, 1978, Mapping of small scale (outcrop-size) sedimentological features on modern submarine fans: Proc., 10th Ann. Offshore Tech. Conf. v. 1, p. 593-597.

—— and D. J. W. Piper, 1969, Deep-sea fan-valleys past and present: Geol. Soc. America Bull., v. 80, p. 1859-1866.

Scholl, D. W., and M. S. Marlow, 1974, Sedimentary sequence in modern Pacific trenches and the deformed circum-Pacific eugeosyncline, in R. H. Dott, Jr., and R. H. Shaver, eds., Modern and ancient geosynclinal sedimentation: SEPM Spec. Pub. no. 19, p. 193-211.

Schweller, W. J., and L. D. Kulm, 1978, Depositional patterns and channelized sedimentation in active eastern Pacific trenches, in D. J. Stanley and G. Kelling, eds., Sedimentation in submarine canyons, fans, and trenches: Stroudsburg, Pa., Dowden, Hutchinson and Ross, Inc., p. 311-324.

Shepard, F. P., 1973, Sea floor off Magdalena Delta and Santa Marta area, Columbia: Geol. Soc. America Bull., v. 84, p. 1955-1972.

—— and N. F. Marshall, 1975, Dives into outer Coronado Canyon system: Marine Geology, v. 18, p. 313-323.

Sliter, W. V., 1972, Upper Cretaceous planktonic foraminiferal zoogeography and ecology - eastern Pacific margin: Palaeogeog., Palaeoclim., Paleoecol., v. 12, p. 15-31.

Stewart, R. J., 1976, Turbidites of the Aleutian abyssal plain; mineralogy, provenance, and constraints for Cenozoic motion of the Pacific plate: Geol. Soc. America Bull., v. 87, p. 793-808.

Vallier, T. L., P. J. Harold, and W. A. Girdley, 1973, Provenance and dispersal patterns of turbidite sands in Escanaba Trough, northeastern Pacific Ocean: Marine Geology, v. 15, p. 67-87.

Winterer, E. L., J. R. Curray, and M. N. A. Peterson, 1968, Geological history of the Pioneer Fracture Zone with the Delgada Deep-Sea Fan, northeast Pacific: Deep-Sea Research, v. 15, p. 509-520.

New York-Vermont

Carbo, S., 1979, Vertical distribution of trace fossils in a turbidite sequence, Upper Devonian, New York State: Palaeogeog. Paleoclim., Palaeoecol., v. 28, p. 81-101.

Clarke, J. M., 1917, Strand and undertow markings of upper Devonian time as indication of the prevailing climate: New York State Mus. Bull., v. 196, p. 199-210.

Hall, B. A., and D. J. Stanley, 1973, Levee-bounded submarine base-of-slope channels in the Lower Devonian Seboomook Formation, northern

Maine: Geol. Soc. America Bull., v. 84, p. 2101-2110.

Keith, B. D., and G. M. Friedman, 1977, A slope-fan-basin-plain model, Taconic sequence, New York and Vermont: Jour. Sed. Petrology, v. 47, p. 1220-1241.

Middleton, G. V., 1965, Paleocurrents in Normanskill graywackes north of Albany, New York: Geol. Soc. America Bull., v. 76, p. 841-844.

Onions, D., and G. V. Middleton, 1968, Dimensional grain orientation of Ordovician turbidite grey-wackes: Jour. Sed. Petrology, v. 38, p. 164-174.

Walker, R. G., and R. G. Sutton, 1967, Quantitative analysis of turbidites in the Upper Devonian Sonyea Group, New York: Jour. Sed. Petrology, v. 37, p. 1012-1022.

Weber, J. N., and G. V. Middleton, 1961, Geochemistry of the turbidites of the Normanskill and Charny formations - pt. 1, effect of turbidity currents on the chemical differentiation of turbidites; pt. 2, distribution of trace elements: Geochim. et Cosmochim, Acta, v. 22, p. 200-288.

Kentucky-Illinois-Indiana

Kepferle, R. C., 1977, Stratigraphy, petrology and depositional environment of the Kenwood Siltstone Member, Borden Formation (Mississippian), Kentucky and Indiana: U.S. Geol. Survey Prof. Paper 1007, 49 p.

————, 1978, Prodelta turbidite fan apron in Borden Formation (Mississippian), Kentucky and Indiana, in D. J. Stanley and G. Kelling, eds., Sedimentation in submarine canyons, fans, and trenches: Stroudsburg, Pa., Dowden, Hutchinson and Ross, Inc., p. 224-238.

Lineback, J. A., 1968, Turbidites and other sandstone bodies in the Borden Siltstone (Mississippian) in Illinois: Ill. State Geol. Surv., Circ. 425, 29 p.

Sedimentation Seminar, 1969, Bethel Sandstone (Mississippian) of western Kentucky and south-central Indiana, a submarine-channel fill: Kentucky Geol. Surv. Rept. Invest., v. 11, 24 p.

Antarctica

Kurtz, D. D. and J. B. Anderson, 1979, Recognition and sedimentologic description of recent debris flow deposits from the Ross and Weddell Seas, Antarctica: Jour. Sed. Petrology, v. 49, p. 1159-1169.

Piper, D. J. W., and C. D. Brisco, 1975, Deep-water continental-margin sedimentation, DSDP Leg 28, Antarctica, in D. E. Hayes, L. A. Frakes, et al, eds.: Washington, D.C., Initial Repts. of Deep Sea Drilling Proj., v. 28, p. 727-755.

SOUTH AMERICA

Bandy, O. L., and R. E. Arnal, 1957, Distribution of recent foraminifera off the coast of Central America: AAPG Bull., v. 41, p. 2037-2053.

———— and K. S. Rodolfo, 1964, Distribution of foraminifera and sediments, Peru-Chile Trench area: Deep Sea Research, v. 11, p. 817-837.

Brown, L. F., Jr., and W. L. Fisher, 1977, Seismic-stratigraphic interpretation of depositional systems; examples from Brazilian rift and pull-apart basins, in C. E. Payton, ed., Seismic stratigraphy-applications to hydrocarbon exploration: AAPG Mem. 26, p. 213-248.

Carozzi, A. V., 1979, Petroleum geology in the Paleozoic clastics in the middle Amazon Basin, Brazil: Jour. Petroleum Geology, v. 2, p. 55-74.

Damuth, J. E., and M. A. Gorini, 1976, The Equatorial Mid-Ocean Canyon; a relict deep-sea channel on the Brazilian continental margin: Geol. Soc. America Bull., v. 87, p. 340-346.

Dorreen, J. M., 1951, Rubble bedding and graded bedding in Talara Formation of northwestern Peru: AAPG Bull., v. 35, p. 1829-1849.

Gonzalez-Bonorino, G., and G. V. Middleton, 1976, A Devonian submarine fan in western Argentina: Jour. Sed. Petrology, v. 46, p. 56-69.

Milliman, J. D., C. P. Summerhayes, and H. T. Barretto, 1975, Quarternary sedimentation on the Amazon continental margin; a model: Geol. Soc. America Bull., v. 86, p. 610-614.

Mordojovich, C., 1974, Geology of a part of the Pacific margin of Chile, in C. A. Burk and C. L. Drake, eds., The geology of continental margins: New York, Springer-Verlag, p. 591-598.

Mutti, E., et al, 1980, Deep-sea fan turbidite sediments winnowed by bottom currents in the Eocene of the Campos Basin, Brazilian offshore, (abs.): Bochum, Internat. Assoc. Sedimentols. 1st European Mtg., 114 p.

Scott, K. M., 1966, Sedimentology and dispersal pattern of a Cretaceous flysch sequence, Patagonian Andes, southern Chile: AAPG Bull., v. 50, p. 72-107.

Shepard, F. P., 1973, Sea floor off Magdalena Delta and Santa Marta area, Columbia: Geol. Soc. America Bull., v. 84, p. 1955-1972.

Winn, R. D., Jr., 1978, Upper Mesozoic flysch of Tierre del Fuego and south Georgia Island; a sedimentologic approach to lithosphere plate restoration: Geol. Soc. America Bull., v. 89, p. 533-547.

———— and R. H. Dott, Jr., 1977, Large-scale traction-produced structures in deep-water fan-channel conglomerates in southern Chile: Geology, v. 5, p. 41-44.

———— and ————, 1978, Submarine-fan turbidites

and resedimented conglomerates in a Mesozoic arc-rear marginal basin in southern South America, *in* D. J. Stanley and G. Kelling, eds., Sedimentation in submarine canyons, fans and trenches: Stroudsburg, Pa., Dowden, Hutchinson and Ross, Inc., p. 362-376.

———— and ————, 1979, Deep-water fan-channel conglomerates of Late Cretaceous age, southern Chile: Sedimentology, v. 26, p. 203-228.

Caribbean

Appelbaum, B. S., and A. H. Bouma, 1972, Geology of the upper continental slope in Alaminos Canyon region: Trans., Gulf Coast Assoc. Geol. Socs., v. 22, p. 157-164.

————, 1974, Surface microtextures of deep water quartz sands from Colombia and Sigsbee basins: Texas A&M Univ., Unpub. Ph.D. dissert., 220 p.

Bennetts, K. R. W., and O. H. Pilkey, 1976, Characteristics of three turbidites, Hispaniola-Caicos Basin: Geol. Soc. America, Bull., v. 87, p. 1291-1300.

Bouma, A. H., et al, 1972, Deep sea sedimentation and correlation of strata off Magdalena River in Beata Strait: Margarita, Venezuela, 6th Conf. Geol. Del Caribe, p. 430-438.

Byramjee, R. S., J. F. Mugniot, and B. Biju-Duval, 1975, Petroleum potential of deep-water areas of the Mediterranean and Caribbean Seas: Proc., 9th World Petroleum Congress, v. 2, p. 299-312.

Dillion, W. P., and J. G. Vedder, 1973, Structure and development of the continental margin of British Honduras: Geol. Soc. America Bull., v. 84, p. 2713-2732.

Embley, R. W., J. I. Ewing, and M. Ewing, 1970, The Vidal deep-sea channel and its relationship to the Demerara and Barracude abyssal plains: Deep-Sea Research, v. 17, p. 539-552.

Grippi, J., and K. Bruke, 1980, Submarine-canyon complex among Cretaceous island-arc sediments, western Jamaica: Geol. Soc. America Bull., v. 91, pt. 1, p. 179-184.

Moussa, M. T., 1977, Bioclastic sediment gravity flow and submarine sliding in the Juana Diaz Formation, southwestern Puerto Rico: Jour. Sed. Petrology, v. 47, p. 593-599.

Seiglie, G. A., P. N. Froelich, and O. H. Pilkey, 1976, Deep-sea sediments of Navidad Basin; correlation of sand layers: Deep-Sea Research, v. 23, p. 89-101.

Shepard, F. P., 1979, Submarine slopes and canyons on north side, St. Croix Island: Marine Geology, v. 31, p. M69-M76.

SEDIMENTARY PROCESSES AND STRUCTURES

Allen, J. R. L., 1964, Primary current lineation in the Lower Old Red Sandstone (Devonian), Anglo-Welsh basin: Sedimentology, v. 3, p. 89-108.

Andrews, J. E., and R. J. Hurley, 1978, Sedimentary processes in the formation of a submarine canyon: Marine Geology, v. 26, p. M47-M50.

Beer, R. M., and D. S. Gorsline, 1971, Distribution, composition and transport of suspended sediment in Redondo Submarine Canyon and vicinity (California): Marine Geology, v. 10, p. 153-175.

Bell, H. S., 1942, Density currents as agents for transporting sediments: Jour. Geology, v. 50, p. 512-547.

Buller, A. T., and J. McManus, 1973, Modes of turbidite deposition deduced from grain-size analyses: Geol. Mag., v. 109, p. 491-500.

Carter, R. M., 1975, A discussion and classification of subaqueous mass-transport with particular application to grain-flow, slurry-flow and flu-xoturbidites: Earth Sci. Review, v. 11, p. 145-177.

Chipping, D. H., 1972, Sedimentary structures and environment of some thick sandstone beds of turbidite type: Jour. Sed. Petrology, v. 42, p. 587-595.

Chough, S. K., 1978, Morphology, sedimentary facies and processes of the Northwest Atlantic Mid-Ocean Channel between 61° and 52° N, Labrador Sea: McGill Univ., Ph.D. dissert., 167 p.

———— and R. Hesse, 1976, Submarine meandering thalweg and turbidity currents flowing for 4000 km in the Northwest Atlantic Mid-Ocean Channel, Labrador Sea: Geology, v. 4, kp. 529-533.

———— and ————, 1980, Northwest Atlantic Mid-Ocean channel of the Labrador Sea; III. head spills body spill deposits from turbidity currents on natural levees: Jour. Sed. Petrology, v. 50, p. 227-234.

Clarke, J. M., 1917, Strand and undertow markings of upper Devonian time as indication of the revailing climate: New York State Mus. Bull., v. 196, p. 199-210.

Colburn, I. P., 1968, Grain fabrics in tyurbidite sandstone beds and their relationship to sole mark trends on the same beds: Jour. Sed. Petrology, v. 38, p. 146-158.

Cole, M. R., 1977, Eocene paleocurrents and sedimentation, San Nicolas Island, California: AAPG Bull., v. 61, p. 237-247.

Crowell, J. C., et al, 1966, Deep-water sedimentary structures, Pliocene Pico Formation, Santa Paula Creek, Ventura Basin, California: Calif. Div. Mines Geology, Spec. Rept. 89, 40 p.

Curray, J. R., and D. G. Moore, 1971, Growth of the Bengal deep-sea fan and denudation in the Himalayas: Geol. Soc. America Bull., v. 82, p.

563-572.

Damuth, J. E., 1979, Migrating sediment waves created by Turbidity currents in the northern South Cina Basin: Geology, v. 7, p. 520-523.

———— and R. W. Embley, 1979, Upslope flow of turbidity currents on the northwest flank of Ceara Rises; western Equatorial Atlantic: Sedimentology, v. 26, p. 825-834.

Davies, G. R., 1977, Turbidites, debris sheets, and truncation structures in uper Paleozoic deep-water carbonates of the Sverdrup Basin, Artic Archipelago, in H. E. Cook and O. Enos, eds., Deep-water carbonate environments, SEPM Spec. Pub. no. 25, p. 221-239.

Davies, T. A., and A. L. Laughton, 1972, Sedimentary processes in the North Atlantic: Washington, D.C., Initial Repts. Deep Sea Drilling Proj., v. 12, p. 905-934.

Davis, J. R., 1971, Sedimentation of Pliocene sandstones in Santa Barbara Channel, California (abs.): AAPG Bull., v. 55, p. 335.

Dickinson, W. R., 1970, Clastic sedimentary sequences deposited in shelf, slope, and trough settings between magmatic arcs and associated trenches: Pac. Geology, v. 3, p. 15-30.

———— 1974, Sedimentation within and beside ancient and modern magmatic arcs, in R. H. Dott, Jr. and R. H. Shaver, eds., Modern and ancient geosynclinal sedimentation, SEPM Spec. Pub. no. 19, p. 230-239.

———— R. V. Ingersoll, and S. A. Graham, 1979, Paleogeneiment dispersal and paleotectonics in northern California; summary: Geol. Soc. America Bull., pt. I, v. 90, p. 897-898.

Dill, R. F., 1964, Sedimentationand erosion in Scripps submarine canyon head, in R. L. Miller, ed., Papers in marine geology: New York, McMillan, p. 23-41.

Ditty, P. S., et al, 1977, Mixed terrigenous-carbonate sedimentation in the Hispaniola-Caicos turbidite basin: Marine Geology, v. 24, p. 1-20.

Dorreen, J. M., 1951, Rubble bedding and graded bedding in Talara Formation of northwestern Peru AAPG Bull., v. 35, p. 1829-1849.

Dott, R. H., Jr., 1963, Dynamics of subaqueous gravity depositional processes: AAPG Bull., v. 47, p. 104-128.

———— and K. J. Bird, 1979, Sand transport through channels across an Eocene shelf and slope in southwestern Oregon, U.S.A., in L. J. Doyle and O. H. Pilkey, eds., Geology of continental slopes: SEPM Spec. Pub. no. 27, 327-242.

Drake, D. E., P. G. Hatcher, and G. H. Keller, 1978, Suspended particulate matter and mud deposition in Upper Hudson submarine canyon, in D. J. Stanley and G. Kelling, eds., Sedimentation in submarine canyons, fans, and trenches: Stroudsburg, Pa., Dowden, Hutchinson and Ross, Inc., p. 33-41.

Duncan, J. R., and L. D. Kulm, 1970, Mineralogy, provenance and dispersal history of Late Quaternary deep-sea sands in Cascadia Basin and Blanco fracture zone off Oregon: Jour. Sed. Petrology, v. 40, p. 874-887.

Dzulynski, S., and A. Slaczka, 1959, Directional structures and sedimentation of the Krosno Beds (Carpathian Flysch): Ann. Soc. Geol. Pologne, v. 28, p. 205-259.

———— and E. K. Walton, 19654, Sedimentary features of flysch and greywackes: Amsterdam, Elsevier Sci. Pub., 274 p.

Einsele, G., and K. Kelts, 1980, Physical properties of unconsolidated mud turbidites, and their significance for diagenetic processes, (abs.): Bochum, Internat. Assoc. Sedimentols. p. 155-156.

Ellis, B. A., 1980, The lateral variation of structures within basinwide turbidites, (abs.): Bochum, Internat. Assoc. Sedimentols. p. 115.

Embley, R. W., 1976, New evidence for occurrence of debris flow deposits in the deep sea: Geology, v. 4, p. 371-374.

Enos, P., 1977, Flow regimes in debris flow: Sedimentology, v. 24, p. 133-142.

Ericson, D. B., M. Ewing, and B. C. Heezen, 1952, Turbidity currents and sediments in North Atlantic: AAPG Bull., v. 36, p. 489-511.

———— ———— ———— and G. Wollin, 1955, Sediment deposition in deep Atlantic: Geol... Soc. America Spec. Paper 62, p. 205-219.

Ewing, J., X. LePichon, and M. Ewing, 1963, Upper stratification of Hudson apron region: Jour. Geophys. Research, v. 68, p. 6303-6316.

Field, M. E., and O. H. Pilkey, 1971, Deposition of deep-sea sands; comparison of two areas of the Carolina continental rise: Jour. Sed. Petrology, v. 41, p. 526-536.

Fisher, F. V., and J. M. Mattinson, 1968, Wheeler Gorge turbidite-conglomerate series, California; inverse grading: Jour. Sed. Petrology, v. 38, p. 1013-1023.

Fox, P. J., B. C. Heezen, and A. M. Harian, 1968, Abyssal anti-dunes: Nature, v. 220, p. 470-472.

Fritz, SA. J., and O. H. Pilkey, 1975, Distinguishing bottom and turbidity current coarse layers on the continental rise: Jour. Sed. Petrology, v. 45, p. 57-62.

Galloway, W. E., 1974, Deposition and diagenetic alteration of sandstone in northwest Pacific arc-related basins; implications for graywacke genesis: Geol. Soc. America Bull., v. 85, p. 379-390.

Griggs, G. B., A. G. Carey, Jr., and L. D. Kulm, 1969, Deep-sea sedimentation and sediment-fauna interaction in Cascadia channel and on Cascadia abyssal plain: Deep-Sea Research, v. 16, p.

157-170.

Grover, N. C., and C. S. Howard, 1938, The passage of turbid water through Lake Mead: Am. Soc. Civil Engineers, v. 103, p. 720-732.

Hampton, M. A., 1972, The role of subaqueous debris flows in generating turbidity currents: Jour. Sed. Petrology, v. 42, p. 775-793.

Hand, B. M., G. V. Middleton, and K. Skipper, 1972, Antidune cross-stratification in a turbidite sequence, Cloridorme Formation, Gaspe, Quebec: Sedimentology, v. 18, p. 135-138.

Harms, J. C., and R. K. Fahnestock, 1965, Stratification, bed forms and flow phenomena (with example from the Rio Grande), in G. V. Middleton, ed., Primary sedimentary structures and their hydrodynamic interpretation: SEPM Spec. Pub. no. 12, p. 84-115.

———— et al, 1975, Depositional environments as interpreted from primary sedimentary structures and stratification sequences: Dallas, Tex., SEPM Short Course 2, 161 p.

Heezen, B. C., 1959, Dynamic processes of abyssal sedimentation; erosion, transportation, and redeposition on the deep-sea floor: Geophys. Jour., v. 2, p. 142-163.

————, 1963, Turbidity currents, in N. M. Hill, ed., The sea, v. 3; the earth beneath the sea: New York, John Wiley & Sons, p. 742-775.

———— and C. L. Drake, 1964, Grand banks slump: AAPG Bull., v. 48, p. 221-233.

———— D. B. Ericson, and M. Ewing, 1954, Further evidence for a turbidity current following the 1929 Grand Banks earthquake: Deep-Sea Research, v. 1, p. 193-202.

———— and M. Ewing, 1952, Turbidity currents and submarine slumps and the 1929 Grand Banks earthquake: Am. Jour. Sci., v. 250, p. 849-873.

———— and ————, 1955, Orleansville earthquake and turbidity currents: AAPG Bull., v. 39, p. 2505-2514.

———— and C. D. Hollister, 1964, Deep-sea current evidence from abyssal sediments: Marine Geology, v. 1, p. 141-174.

———— and W. F. Ruddiman, 1966, Shaping of the continental rise by deep geostrophic contour currents: Science, v. 152, p. 502-508.

Hesse, R., 1976, Some unusual secondary sedimentary structures in the lacustrine Lower Carboniferous of Nova Scotia, Canada, and their significance as earthquake indicators [Einige ungewoehnliche sekundaere Sedimenstrukturen im lakustrinen Unterkarbon Neuschottslands (Kanada) und ihre Deutung als Erdbebenanzeiser]: Eclogae Geol. Helv. (EGHVAG), v. 69, p. 196-201.

————, 1977, Softex-radiographs of sliced piston cores from the Japan and southern Kurile trench and slope areas, in E. Honza, ed., Geological investigation of Japan and southern Kurile trench and slope areas: Japanese Geol. Surv., Cruise Rept., no. 7, p. 86-108.

Hiscott, R. N., and G. V. Middleton, 1979, Depositional mechanics of thickbedded sandstones at the base of a submarine slope, Tourelle Formation (Lower Ordovician), Quebec, Canada, in L. J. Doyle and O. H. Pilkey, eds., Geology of continental slopes: SEPM Spec. Pub. no. 27, p. 307-326.

Horn, D. R., J. I. Ewing, and M. Ewing, 1972, Graded-bed sequences emplaced by turbidity currents north of 20°N in the Pacific, Atlantic and Mediterranean: Sedimentology, v. 18, p. 247-275.

Hough, J. L., ed., 1951, Turbidity currents and the transportation of coarse sediment into deep water: SEPM Spec. Pub. no. 2, 107 p.

Huang, T. C., and H. G. Goodell, 1970, Sediments and sedimentary processes of eastern Mississippi cone, Gulf of Mexico: AAPG Bull., v. 54, p. 2070-2100.

Hubert, J. F., 1964, Textural evidence for deposition of many western North Atlantic deep-sea sands by ocean-bottom currents rather than turbidity currents: Jour. Geology, v. 72, p. 757-785.

———— and W. J. Neal, 1967, Mineral composition and dispersal patterns of the deep-sea sands in the western North Atlantic petrologic province: Geol. Soc. America Bull., v. 78, p. 749-772.

Hurley, R. J., 1964, Analysis of flow in Cascadia deep-sea channel, in R. L. Miller, ed., Papers in marine geology: New York, Macmillan Co., p. 117-132.

Ingersoll, R. V., and C. A. Suczek, 1979, Petrology and provenance of Neogene sand from Nicobar and Bengal fans, DSDP sites 211 and 218: Jour. Sed. Petrology, v. 49, p. 1217-1228.

Jipa, D., and R. B. Kidd, 1974, Sedimentation of coarser grained interbeds in Arabian Sea and sedimentation processes of the Indus Cone: Washington, D. C., Init. Repts. Deep Sea Drilling Proj., v. 23, p. 471-495.

Johnson, D. A., and T. C. Johnson, 1970, Sediment redistribution by bottom currents in the central Pacific: Deep Sea Research, v. 17, p. 157-169.

Keller, G. H., and F. P. Sherpard, 1978, Currents and sedimentary processes in submarine canyons off the northeast United States, in D. J. Stanley and G. Kelling, eds., Sedimentation in submarine canyons, fans, and trenches: Stroudsburg, Pa., Dowden, Hutchinson and Ross, Inc., p. 15-32.

Kersey, D. G., and K. J. Hsu, 1976, Energy relations and density-current flows; an experimental investigation: Sedimentology, v. 23, p. 761-789.

Komar, P. D., 1969, The channelized flow of turbidity currents with application to Monterey deep-sea fan channel: Jour. Geophys. Research, v. 74, p. 4544-4558.

————, 1970, The competence of turbidity current flow: Geol. Soc. America Bull., v. 81, p. 1555-1562.

————, 1973, Continuity of turbidity current flow and systematic variations in deep-sea channel morphology: Geol. Soc. America Bull., v. 84, p. 3329-3338.

Krause, D. C., et al, 1970, Turbidity currents and cable breaks in the western New Britain trench: Geol. Soc. America Bull., v. 81, p. 2153-2160.

Ksiazkiewicz, M., 1958, Submarine slumping in the Carpathian flysch: Ann. Soc. Geol. Pologne, v. 28, p. 123-151.

Kuenen, P. H., 1951, Properties of turbidity currents of high density: in J. L. Hough, ed., Turbidity currents and the transportation of coarse sediments to deep water: SEPM Spec. Pub. no. 2, p. 14-33.

————, 1952, Estimated size of the Grand Banks turbidite currents: Am. Jour. Sci., v. 250, p. 874-884.

————, 1966, Matrix of turbidites; experimental approach: Sedimentology, v. 7, p. 267-297.

————, 1967, Emplacement of flysch-type sand beds: Sedimentology, v. 9, p. 203-243.

———— and H. W. Menard, 1952, Turbidity currents, graded and nongraded deposits: Jour. Sed. Petrology, v. 22, p. 83-96.

———— and C. I. Migliorini, 1950, Turbidity currents as a cause of graded bedding: Jour. Geology, v. 58, p. 91-127.

Lajoie, J., 1972, Slump fold axis orientations; an indication of paleoslope?: Jour. Sed. Petrology, v. 42, p. 584-586.

Lawson, L. M., 1914, Movement of silt, Elephant Butte reservoir: Reclamation Record, v. 10, 411 p.

Lowe, D. R., 1975, Water escape structure in coarse-grained sediments: Sedimentology, v. 22, p. 157-204.

————, 1976, Subaqueous liquefied and fluidized sediment flows and their deposits: Sedimentology, v. 23, p. 285-308.

————, 1979, Sediment gravity flow; their classification and some problems of application to natural flows and deposits, in L. J. Doyle and O. H. Pilkey, eds., Geology of continental slopes: SEPM Spec. Pub. no. 27, p. 75-82.

———— and R. D. LoPiccolo, 1974, The characteristics and origins of dish and pillar structures: Jour. Sed. Petrology, v. 44, p. 484-501.

Marschalko, R., 1964, Sedimentary structures and paleocurrents in the marginal lithofacies of the central-Carpathian flysch, in A. H. Bouma, and A. Brouwer, eds., Turbidites: Amsterdam, Devel. in sedimentology, 3, Elsevier Sci. Pub., p. 106-126.

Marshall, N. F., 1978, A large storm-induced slump re-opens an unknown Scripps submarine canyon tributary, in D. J. Stanley and G. Kelling, eds., Sedimentation in submarine canyons, fans, and trenches: Stroudsburg, Pa., Dowden, Hutchinson and Ross, Inc., p. 73-84.

Martin, B. D., 1963, Rosedale channel - evidence for Late Miocene submarine erosion in Great Valley California: AAPG Bull., v. 47, p. 441-456.

———— and K. O. Emery, 1967, Geology of Monterey Canyon, California: AAPG Bull., v. 51, p. 2281-2304.

McCabe, P. J., 1978, The Kinderscoutian Delta (Carboniferous) of northern England; a slope influenced by density currents, in D. J. Stanley and G. Kelling, eds., Sedimentation in submarine canyons, fans and trenches: Stroudsburg, Pa., Dowden, Hutchinson and Ross, Inc., p. 116-126.

Michaelis, E. R., and G. Dixon, 1969, Interpretation of depositional processes from sedimentary structures in the Cardium sand: Bull. Canadian Petroleum Geology v. 17, p. 410-443.

Middleton, G. V., 1962, Size and sphericity of quartz grains in two turbidite formations: Jour. Sed. Petrology, v. 32, p. 725-742.

————, 1965, Antidune cross-bedding in a large flume: Jour. Sed. Petrology, v. 35, p. 922-927.

————, 1965, Paleocurrents in Normanskill graywackes north of Albany, New York: Geol. Soc. America Bull., v. 76, p. 841-844.

————, ed., 1965, Primary sedimentary structures and their hydrodynamic interpretation - a symposium: SEPM Special Pub. no. 12, 265 p.

————, 1966, Experiments on density and turbidity currents - pt. 1; motion of the head: Canadian Jour. Earth Sci., v. 3, p. 523-546.

————, 1966, Experiments on density and turbidity currents - pt. 2; uniform flow of density currents: Canadian Jour. Earth Sci., v. 3, p. 627-637.

————, 1966, Small-scale models of turbidity currents and the criterion for auto-suspension: Jour. Sed. Petrology, v. 36, p. 202-208.

————, 1967, Experiments on density and turbidity currents - pt. 3; deposition of sediment: Canadian Jour. Earth Sci., v. 4, p. 475-505.

————, 1967, The orientation of concavo-convex particles deposited from experimental turbidity currents: Jour. Sed. Petrology, v. 37, p. 229-232.

————, 1969, Grain flows and other mass movements down slopes, in The new concepts of continental margin sedimentation: Am. Geophys. Instit. Short Course Lecture Notes no. 11, 14 p.

————, 1969, Turbidity currents, in The new concepts of continental margin sedimentation: Am. Geophys. Instit. Short Course Lecture Notes no. 10, 20 p.

————, 1971, Experimental studies related to problems of flysch sedimentation, *in* J. Lajoie, ed., Flysch sedimentology of North America: Canadian Geol. Assoc. Spec. Paper no. 7, p. 253-272.

———— and M. A. Hampton, 1973, Sediment gravity flows; mechanics of flow and deposition, *in* G. V. Middleton, ed., Turbidites and deep-water sedimentation: SEPM Short Course, Pac. Sec., p. 1-38.

———— and ————, 1976, Subaqueous sediment transport and deposition by sediment gravity flows, *in* D. J. Stanley and D. J. P. Swift, eds., Marine sediment transport and environment management. New York, Wiley Interscience, p. 197-218.

Moore, D. G., 1961, Submarine slumps: Jour. Sed. Petrology, v. 31, 343-357.

Nardin, T. R., et al, 1979, A review of mass movement processes, sediment and acoustic characteristics, and contrasts in slope and base-of-slope systems versus canyon-fan-basin floor systems, *in* L. J. Doyle and O. H. Pilkey, eds., Geology of continental slopes: SEPM Spec. Pub. no. 27, p. 61-73.

Nelson, C. H., 1976, Late Pleistocene and Holocene depositional trends, processes, and history of Astoria deep-sea fan, northeast Pacific: Marine Geology, v. 20, p. 129-173.

Nelson, C. H., E. Mutti, and F. Ricci Lucchi, 1974, Criteria for distinguishing thin-bedded turbidites deposited in proximal overbank and distal basin plain environments: Geol. Soc. America Abs. with Programs, v. 6, no. 7, p. 2887-2888.

———— ———— and ————, 1975, Comparison of proximal and distal thin-bedded turbidites with current winnowed deep sea sands: Nice, France, 9th Internat. Sedimentols. Cong., (Theme 5), v. 2, p. 317-324.

————, et al, 1978, Thin-bedded turbidites in modern submarine canyons and fans, *in* D. J. Stanley and G. Kelling, eds., Sedimentation in submarine canyons, fans, and trenches: Stroudsburg, Pa., Dowden, Hutchinson and Ross, Inc., p. 177-189.

Nilsen, T. H., et al, 1977, New occurrences of dish structures in the stratigraphic record: Jour. Sed. Petrology, v. 47, p. 1299-1304.

Normark, W. R., 1970, Channel piracy on Monterey deep-sea fan: Deep-Sea Research, v. 17, p. 837-846.

————, 1970, Growth patterns of deep-sea fans: AAPG Bull., v. 54, p. 2170-2195.

Palmer, H. D., 1976, Erosion of submarine outcrops, LaJolla submarine canyon, California: Geol. Soc. America Bull., v. 87, p. 427-432.

Parkash, B., 1970, Downcurrent changes in sedimentary structures in Ordovician turbidite greywackes: Jour. Sed. Petrology, v. 40, p. 572-590.

———— and G. V. Middleton, 1970, Downcurrent textural changes in Ordovician turbidite greywackes: Sedimentology, v. 14, p. 259-293.

Picha, F., and L. M. Cline, 1973, Radiographic investigation of laminated and ripple cross-laminated flysch sandstones, Ouachita Mountains, Oklahoma: Jour. Sed. Petrology, v. 43, p. 466-470.

Piper, D. J. W., 1970, Transport and deposition of Holocene sediment of La Jolla deep sea fan, California: Marine Geology, v. 8, p. 211-227.

————, 1978, Turbidite muds and silts on deep sea fans and abyssal plains, *in* D. J. Stanley and G. Kelling, eds., Sedimentation in submarine canyons, fans and trenches: Stroudsburg, Pa., Dowden, Hutchinson and Ross, p. 163-176.

Rich, R. L., 1950, Flow markings, groovings and intra-stratal crumplings as criteria for recognition of slope deposits, with illustrations from Silurian rocks of Wales: AAPG Bull., v. 34, p. 717-741.

Rocheleau, M., and J. Lajoie, 1974, Sedimentary structures in resedimented conglomerate of the Cambrian flysch, L'Islet, Quebec Appalachians: Jour. Sed. Petrology, v. 44, p. 826-836.

Rona, P. A., 1969, Middle Atlantic continental slope of United States; deposition and erosion: AAPG Bull., v. 53, p. 1453-1969.

Sanders, J. E., 1965, Primary sedimentary structures formed by turbidity currents and related sedimentation mechanisms, *in* G. V. Middleton, ed, Primary sedimentary structures and their hydrodynamic interpretation: SEPM Spec. Pub. no. 12, p. 192-219.

Sarnthein, M., and L. Diester-Haass, 1977, Eolian - sand turbidites: Jour. Sed. Petrology, v. 47, p. 868-890.

Schneider, E. D., et al, 1967, Further evidence of contour currents in the western North Atlantic: Earth and Planetary Sci. Letters, v. 2, p. 351-359.

Schweller, W. J., and L. D. Kulm, 1978, Depositional patterns and channelized sedimentation in active eastern Pacific trenches, *in* D. J. Stanley, and G. Kelling, eds., Sedimentation in submarine canyons, fans and trenches: Stroudsburg, Pa., Dowden, Hutchinson and Ross, Inc., p. 311-324.

Shepard, F. P., R. F. Dill, and U. Von Rad, 1969, Physiography and sedimentary processes of La Jolla submarine fan and fan-valley, California: AAPG Bull., v. 53, p. 390-420.

———— and N. F. Marshall, 1969, Currents in La Jolla and Scripps submarine canyons: Science, v. 165, p. 177-178.

———— and ————, 1973, Currents along floors of submarine canyons: AAPG Bull., v. 57, p. 244-264.

———— and ————, 1978, Currents in submarine canyons and sea valleys *in* D. J. Stanley and G. Kelling, eds., Sedimentation in submarine canyons, fans, and trenches: Stroudsburg, Pa., Dowden, Hutchinson and Ross, Inc., p. 3-14.

———— ———— and P. A. McLoughlin, 1974, Currents in submarine canyons: Deep-Sea Research., v. 21, p. 691-706.

———— ———— ———— and G. G. Sullivan, 1979, Currents in submarine canyons and other sea valleys: AAPG Stud. in Geology, no. 8, 173 p.

Skipper, K., and S. B. Battacharjee, 1978, Backset bedding in turbidites; a further example from the Cloridorme Formation (Middle Ordovician), Gaspe, Quebec: Jour. Sed. Petrology, v. 48, p. 193-201.

———— and G. V. Middleton, 1975, The sedimentary structures and depositional mechanics of certain Ordovician turbidites, Cloridorme Formation, Gaspe Peninsula, Quebec: Canadian Jour. Earth Sci., v. 12, p. 1934-1952.

Stanley, D. J., 1974, Dish structures and sand flow in ancient submarine valleys, French Maritime Alps: Bull. Centre Rech. Pau-SNPA, v. 8, p. 351-371.

————, 1974, Pebbly mud transport in the head of Wilmington Canyon: Marine Geology v. 16, p. M1-M8.

———— H. D. Palmer, and R. F. Dill, 1978, Coarse sediment transport by mass flow and turbidity current processes and downslope transformations in Annot Sandstone canyon-fan valley system, *in* D. J. Stanley and G. Kelling, eds., Sedimentation in submarine canyons, fans, and trenches: Stroudsburg, Pa., Dowden, Hutchinson, Ross, Inc., p. 85-115.

———— H. Sheng, and C. P. Pedraza, 1971, Lower continental rise east of the middle Atlantic states; predominant sediment dispersal perpendicular to isobaths: Geol. Soc. America Bull., v. 82, p. 1831-1840.

———— and N. Silverberg, 1969, Recent slumping on the continental slope off Sable Island bank, southeast Canada: Earth and Planetary Sci. Letters, v. 6, p. 123-133.

Stow, D. A. V., and A. J. Bowen, 1978, Origin of lamination in deep sea, fine-grained sediments: Nature, v. 274, no. 5669, p. 324-328.

Summerhayes, C., D. Ross, and P. Stoffers, 1977, Nile submarine fan; sedimentation, deformation, and oil potential: Proc., 9th Offshore Tech. Conf. v. 1, p. 35-40.

Walker, R. G., 1965, The origin and significance of the internal sedimentary structures of turbidites: Proc., Yorkshire Geol. Soc., v. 35, p. 1-32.

————, 1967, Turbidite sedimentary structures and their relationship to proximal and distal depositional environments: Jour. Sed. Petrology, v. 37, p. 25-43.

————, 1969, Geometrical analysis of ripple-drift cross-lamination: Canadian Jour. Earth Sci., v. 6,

p. 383-391.

Walton, E. K., 1967, The sequence of internal structures in turbidites: Scottish Jour. Geology, v. 3, p. 306-317.

Winn, R. D., Jr., and R. H. Dott, Jr., 1977, Large-scale traction-produced structures in deep-water fan-channel conglomerates in southern Chile: Geology, v. 5, p. 41-44.

SUBMARINE CANYONS

Almgren, A. A., 1978, Timing of Tertiary submarine canyons and marine cycles of deposition in the southern Sacramento Valley, California; *in* D. J. Stanley and G. Kelling, eds., Sedimentation in submarine canyons, fans, and trenches: Stroudsburg, Pa., Dowden, Hutchinson and Ross, Inc., p. 276-291.

Andrews, J. E., and R. J. Hurley, 1978, Sedimentary processes in the formation of a submarine canyon: Marine Geology, v. 26, p. M47-M50.

Appelbaum, B. S., and A. H. Bouma, 1972, Geology of the upper continental slope in Alaminos Canyon region: Trans., Gulf Coast Assoc. Geol. Soc., v. 22, p. 157-164.

Beer, R. M., and D. S. Gorsline, 1971, Distribution, composition and transport of suspended sediment in Redondo submarine canyon and vicinity (California): Marine Geology, v. 10, p. 153-175.

Bornhauser, M., 1948, Possible ancient submarine canyon in southwestern Louisiana: AAPG Bull., v. 32, p. 2287-2294.

Bouma, A. H., 1965, Sedimentary characteristics of samples collected from some submarine canyons: Marine Geology, v. 3, p. 291-320.

Buffington, E. C., 1964, Structural control and precision bathymetry of La Jolla submarine canyon, California: Marine Geology, v. 1, p. 44-58.

Burke, K., 1972, Longshore drift, submarine canyons, and submarine fans in development of Niger delta: AAPG Bull., v. 56, p. 1975-1983.

Cacchione, D. A., G. T. Rowe, and A. Malahoff, 1978, Submersible investigation of outer Hudson submarine canyon, *in* D. J. Stanley and G. Kelling, eds., Sedimentation in submarine canyons, fans, and trenches: Stroudsburg, Pa., Dowden, Hutchinson and Ross, Inc., p. 42-50.

Carlson, P. R., and C. H. Nelson, 1969, Sediments and sedimentary structures of the Astoria submarine canyon-fan system, northeast Pacific: Jour. Sed. Petrology, v. 39, p. 1269-1282.

Carter R. M., and J. K. Linkqvist, 1977, Balleny Group, Chalky Island, southern New Zealand; an inferred Oligocene submarine canyon and fan complex: Pacific Geology, v. 12, p. 1-46.

Chamberlain, C. K., 1964, Mass transport of sediment in the heads of Scripps Canyon; *in* Papers in marine geology; New York, Macmillan Co., p. 42-64.

Cohen, Z., 1976, Early Cretaceous buried canyon; influence on accumulation of hydrocarbons in Helez oil field, Israel: AAPG Bull., v. 60, p. 108-114.

Conolly, J. R., 1968, Submarine canyons of the continental margin, East Bass Strait (Australia): Marine Geology, v. 6, p. 499-461.

Cossey, S. P., and R. Ehrlich, 1978, Growth fault-controlled submarine carbonate debris flow and turbidite deposits from the Jurassic of northern Tunisia; possible canyon fill sequences, *in* D. J. Stanley and G. Kelling, eds., Sedimentation in submarine canyons, fans, and trenches: Strouds-burg, Pa., Dowden, Hutchinson and Ross, Inc., p. 127-137.

Crowell, J. C., 1952, Submarine canyons bordering central and southern California: Jour. Geology, v. 60, p. 58-83.

Daly, R. A., 1936, Origin of submarine "canyons": Am. Jour. Sci., v. 31, p. 401-420.

Damuth, J. E., and M. A. Gorini, 1976, The Equato-rial Mid-Ocean Canyon; a relict deep-sea channel on the Brazilian continental margin: Geol. Soc. America Bull., v. 87, p. 340-346.

Dickas, A. B., and J. L. Payne, 1967, Upper Pal-eocene buried channel in Sacramento Valley, Cal-ifornia: AAPG Bull., v. 51, p. 873-882.

Dietz, R. S., and H. J. Knebel, 1971, Trou sans fond submarine canyon; Ivory Coast, Africa: Deep Sea Research, v. 18, p. 441-447.

———— ———— and L. H. Somers, 1968, Cayar submarine canyon: Geol. Soc. America Bull., v. 79, p. 1821-1828.

Dill. R. F., 1964, Sedimentation and erosion in Scripps submarine canyon head, *in* R. L. Miller, ed., Papers in marine geology: New York, Mc-Millan and Co., p. 23-41.

————, 1969, Earthquake effects on fill of Scripps submarine canyon: Geol. Soc. America Bull., v. 80, p. 321-328.

Emery, K. O., 1950, A suggested origin of con-tinental slopes and of submarine canyons: Geol. Mag., v. 87, p. 102-104.

Ericson, D. B., M. Ewing, and B. C. Heezen, 1951, Deep-sea sands and submarine canyons: Geol. Soc. America Bull., v. 62, p. 961-965.

Felix, D. W., and D. S. Gorsline, 1971, Newport submarine canyon, California; an example of the effects of shifting loci of sand supply upon canyon position: Marine Geol., v. 10, p. 177-198.

Fischer, P. J., 1971, An ancient (Upper Paleocene) submarine canyon and fan; the Meganos channel, Sacramento Valley, California, (abs.): Cordilleran Sec. Mtg. Prog., 67th Ann. Geol. Soc. America

Mtg., v. 3, p. 120.

————, 1979, The evolution of a Late Paleocene submarine canyon and fan system; the Meganos Formation, southern Sacramento Basin, Cali-fornia: Cordilleran Sect. Mtg. Field Guide, Geol. Soc. America, 10 p.

Fisher, F. V., and J. M. Mattinson, 1968, Wheeler Gorge turbidite-conglomerate series, California; inverse grading: Jour. Sed. Petrology, v. 38, p. 1013-1023.

Got, H., and D. J. Stanley, 1974, Sedimentation in two Catalonian canyons, northwestern Medi-terranean: Marine Geology, v. 16, p. M91-M100.

Grippi, J., and K. Bruke, 1980, Submarine-canyon complex among Cretaceous island-arc sediments, western Jamaica: Geol. Soc. America Bull., v. 91, pt. 1, p. 179-184.

Hamilton-Smith, T., 1971, A proximal-distal turbidite sequence and a probable submarine canyon in the Siegas Formation (Early Llandovery) of north-western New Brunswick: Jour. Sed. Petrology, v. 41, p. 752-762.

Heezen, B. C., et al, 1964, Congo submarine canyon: AAPG Bull., v. 48, p. 1126-1149.

Herzer, R. H., 1979, Submarine slides and submarine canyons on the continental slope off Canterbury, New Zealand: New Zealand Jour. Geology, Geo-physics v. 22, p. 391-406.

———— and D. W. Lewis, 1979, Growth and burial of a submarine canyon off Motunau, north Can-terbury, New Zealand: Sed. Geology, v. 24, p. 69-83.

Johnson, D., 1938, The origin of submarine canyons: Jour. Geomorphology, v. 1, p. 11-129, 230-243, 324-340.

————, 1939, The origin of submarine canyons: Jour. Geomorphology, v. 2, p. 42-60, 133-158, 213-236.

Keller, G. H., and F. P. Sherpard, 1978, Currents and sedimentary processes in submarine canyons off the northeast United States, *in* D. J. Stanley and G. Kelling, eds., Sedimentation in submarine can-yons, fans, and trenches: Stroudsburg, Pa., Dow-den, Hutchinson and Ross, Inc., p. 15-32.

Kelling, G., and J. Holroyd, 1978, Clast size, shape, and composition in some ancient and modern fan gravels, *in* D. J. Stanley, and G. Kelling, eds., Sedimentation in submarine canyons, fans and trenches: Stroudsburg, Pa., Dowden, Hutchinson and Ross, Inc., p. 138-159.

———— and D. J. Stanley, 1970, Morphology and structure of Wilmington and Baltimore canyons, eastern United States: Jour. Geology, v. 78, p. 637-660.

———— and ————, 1976, Sedimentation in canyon, slope, and base-of-slope environments, *in* D. J. Stanley and D. J. P. Swift, eds., Marine sediment transport and environmental management: New York, John Wiley and Sons, p. 379-435.

———— and ————, 1978, Sedimentation in submarine canyons, fans, and trenches; appraisal and augury, *in* D. J. Stanley and G. Kelling, eds., Sedimentation in submarine canyons, fans, and trenches: Stroudsburg, Pa., Dowden, Hutchinson and Ross, Inc., p. 377-388.

Kenyon, N. H., R. H. Belderson and A. H. Stride, 1978, Channels, canyons and slump folds on the continental slope between south west Ireland and Spain: Oceanol. Acta, v. 1, p. 369-380.

Knebel, H. J., and D. W. Folger, 1976, Large sand waves on the Atlantic outer continental shelf around Wilmington Canyon, off eastern United States: Marine Geology, v. 22, p. M7-M15.

Lohmar, J. M., and J. E. Warme, 1978, Anatomy of an Eocene submarine canyon-fan system, southern California borderland: Proc., 10th Ann. Offshore Tech. Conf. v. 1, p. 571-577.

Lowe, D. R., 1972, Submarine canyon and slope channel sedimentation model as inferred from Upper Cretaceous deposits, western California: 24th Internat. Geol. Cong., Sec. 6, p. 75-81.

Marshall, N. F., 1978, A large storm-induced slump re-opens an unknown Scripps submarine canyon tributary, *in* D. J. Stanley and G. Kelling, eds., Sedimentation in submarine canyons, fans, and trenches: Stroudsburg, Pa., Dowden, Hutchinson and Ross, Inc., p. 73-84.

Moore, G. T., and T. J. Fullam, 1975, Submarine channel systems and their potential for petroleum localization, *in* M. L. Broussard, ed., Deltas-models for exploration: Houston Geol. Soc., p. 165-189.

Nardin, T. R., et al, 1979, A review of mass movement processes, sediment and acoustic characteristics, and contrasts in slope and base-of-slope systems versus canyon-fan-basin floor systems, *in* L. J. Doyle and O. H. Pilkey, eds., Geology of continental slopes: SEPM Spec. Pub. no. 27, p. 61-73.

Nelson, C. H., et al, 1970, Development of the Astoria canyon-fan physiography and comparison with similar systems: Marine Geology, v. 8, p. 259-291.

Palmer, H. D., 1976, Erosion of submarine outcrops, LaJolla submarine canyon, California: Geol. Soc. America Bull., v. 87, p. 427-432.

Picha, F., 1974, Ancient submarine canyons of the Carpathian miogeosyncline, *in* R. H. Dott, Jr., and R. H. Shaver, eds., Modern and ancient geosynclinal sedimentation: SEPM Spec. Pub. no. 19, p. 126-127.

————, 1979, Ancient submarine canyon of Tethyan continental margins, Czechoslovakia: AAPG Bull., v. 63, p. 67-86.

Roberson, M. I., 1964, Continuous seismic profiler survey of oceanographer, Gilbert, and Lydonia submarine canyons, Georges Bank: Jour. Geophys. Research, v. 69, p. 4779-4789.

Rona, P. A., 1970, Submarine canyon origin on upper continental slope off Cape Hatteras: Jour. Geology, v. 78, p. 141-152.

Scholl, D. W., et al, 1970, The structure and origin of the large submarine canyons of the Bering Sea: Marine Geology, v. 8, p. 187-210.

Shepard, F. P., 1937, Daly's submarine canyon hypothesis: Am. Jour. Sci., v. 38, p. 369-379.

————, 1951, Mass movements in submarine canyon heads: Trans., Am. Geophys. Union, v. 32, p. 405-418.

————, 1951, Transportation of sand into deep water, *in* J. L. Hough, ed., Turbidity currents, and the transportation of coarse sediment into deep water: SEPM Spec. Pub. no. 2, p. 53-65.

————, 1960, Deep-sea sands: Copenhagen, 21st Internat. Geol. Cong., pt. 23, p. 26-42.

————, 1963, Submarine canyon, *in* M. N. Hill, ed., The sea; ideas and observations on progress in the study of the seas, v. 3: New York, Interscience, p. 480-506.

————, 1972, Submarine canyons: Earth Sci. Review, v. 8, p. 1-12.

————, 1979, Submarine slopes and canyons on north side, St. Croix Island: Marine Geology, v. 31, p. M69-M76.

———— and E. C. Buffington, 1968, La Jolla submarine fan-valley: Marine Geology, v. 6, p. 107-143.

———— and R. F. Dill, 1966, Submarine canyons and other sea valleys: Chicago, Ill., Rand McNally, 381 p.

———— and U. Von Rad, 1969, Physiography and sedimentary processes of La Jolla submarine fan and fan-valley, California: AAPG Bull., v. 53, p. 390-420.

———— and G. Einsele, 1962, Sedimentation in San Diego trough and contributing submarine canyons: Sedimentology, v. 1, p. 81-133.

———— and K. O. Emery, 1973, Congo submarine canyon and fan valley: AAPG Bull., v. 57, p. 1679-1691.

———— and N. F. Marshall, 1969, Currents in La Jolla and Scripps submarine canyons: Science, v. 165, p. 177-178.

———— and ————, 1973, Currents along floors of submarine canyons: AAPG Bull., v. 57, p. 244-264.

———— and ————, 1975, Dives into outer Coronado Canyon system: Marine Geology, v. 18, p. 313-323.

———— and ————, 1978, Currents in submarine canyons and sea valleys, *in* J. D. Stanley and G.

Kelling, eds., Sedimentation in submarine canyons, fans, and trenches: Stroudsburg, Pa., Dowden, Hutchinson and Ross, Inc., p. 3-14.

—————— —————— and P. A. McLoughlin, 1974, Currents in submarine canyons: Deep-Sea Research, v. 21, p. 691-706.

—————— —————— —————— and G. G. Sullivan, 1979, Currents in submarine canyons and other sea valleys: AAPG Stud. in Geology, no. 8., 173 p.

Stanley, D. J., 1970, Bioturbation and sediment failure in some submarine canyons, in 3rd European symposium on marine biology: Vie et Milieu, Supplement 22, v. 2, p. 541-555.

————, 1975, Submarine canyon and slope sedimentation (Gres D'Annot) in the French maritime Alps: Nice, France, 9th Internat. Sedimentol. Cong., 131 p.

———— and G. Kelling, 1968, Sedimentation patterns in the Wilmington submarine canyon area, in Symposium on ocean sciences and engineering of the Atlantic Shelf: Philadelphia, Pa., Trans., Delaware Valley Sec., Mar. Tech. Soc., p. 127-142.

———— and ————, 1970, Interpretation of a levee-like ridge and associated features, Wilmington submarine canyon, eastern United States: Geol. Soc. America Bull., v. 81, p. 3747-3752.

———— and ————, eds., 1978, Sedimentation in submarine canyons, fans and trenches: Stroudsburg, Pa., Dowden, Hutchinson and Ross, Inc., 395 p.

———— H. D. Palmer, and R. F. Dill, 1978, Coarse sediment transport by mass flow and turbidity current processes and downslope transformations in Annot Sandstone canyon-fan valley system, in D. J. Stanley and G. Kelling, eds., Sedimentation in submarine canyons, fans and trenches: Stroudsburg, Pa., Dowden, Hutchinson, Ross, Inc., p. 85-115.

Trimonis, E. S., and K. M. Shimjus, 1970, Sedimentation at the head of a submarine canyon: Acad. Science U.S.S.R., Oceanology, Engl. Ed., v. 10, p. 74-85.

Van Berckel, F. L., 1976, On the origin of submarine canyons: Geologie en Mijnbouw, v. 55, p. 7-17.

Van Hoorn, B., 1969, Submarine canyon and fan deposits in the Upper Cretaceous of the south-central Pyrenees, Spain: Geologie en Mijnbouw, v. 48, p. 67-72.

Von Der Borch, C. C., 1969, Submarine canyons of southeastern New Guinea; seismic and bathymetric evidence for the modes of origin: Deep-Sea Research, v. 16, p. 323-328.

Warme, J. E., R. A. Slater, and R. A. Cooper, 1978, Bioerosion in submarine canyons, in D. J. Stanley and G. Kelling, eds., Sedimentation in submarine canyons, fans and trenches: Stroudsburg, Pa., Dowden Hutchinson, and Ross, Inc., p. 65-72.

Whitaker, J. H. McD., 1974, Ancient submarine canyons and fan valleys, in R. H. Dott, Jr., and R. H. Shaver, eds., Modern and ancient geosynclinal sedimentation: SEPM Spec. Pub. no. 19, p. 106-125.

TRACE FOSSILS

Carbo, S., 1979, Vertical distribution of trace fossils in a turbidite sequence, Upper Devonian, New York State: Palaeogeog., Paleoclimatol., Palaeoecol., v. 28, p. 81-101.

Chamberlain, C. K., 1971, Morphology and ethology of trace fossils from the Ouachita Mountains, southeast Oklahoma: Jour. Paleontology, v. 45, p. 212-246.

————, 1971, Bathymetry and paleoecology of Ouachita geosyncline of southeastern Oklahoma as determined from trace fossils: AAPG Bull., v. 55, p. 34-50.

————, 1975, Biogenetic sedimentary structures - trace fossils of the Ouachitas, in Guidebook to sedimentology of Paleozoic flysch and associated deposits, Ouachita Mountains - Arkoma Basin, Oklahoma: Dallas Geol. Soc., p. 51-68.

————, 1976, Trace fossils in DSDP cores of the Pacific: Jour. Paleontology, v. 49, p. 1074-1096.

———— and D. L. Clark, 1973, Trace fossils and conodonts as evidence for deep-water deposits in the Oquirrh Basin of central Utah: Jour. Paleontology, v. 47, p. 663-682.

Scott, R. M., 1978, Physical and biogenic characteristics of sediments from Hueneme submarine canyon, California, in D. J. Stanley, and G. Kelling, eds., Sedimentation in submarine canyons, fans, and trenches: Stroudsburg, Pa., Dowden, Hutchinson and Ross, Inc., p. 51-64.

Seilacher, A., 1962, Paleontological studies on turbidite sedimentation and erosion: Jour. Geology, v. 70, p. 227-234.

————, 1964, Biogenic sedimentary structures, in J. Imbrie and N. Newell, eds., Approaches to paleoecology: New York, John Wiley & Sons, p. 296-316.

————, 1967, Bathymetry of trace fossils: Marine Geology, v. 5, p. 413-428.

FORAM-PALEOECOLOGY

Bandy, O. L., 1960, General correlation of foraminiferal structure with environment: Copenhagen, 21st Internat. Geol. Cong., pt. 22, p. 7-19.

———— and R. E. Arnal, 1957, Distribution of recent foraminifera off the coast of Central America:

AAPG Bull., v. 41, p. 2037-2053.
———— and ————, 1960, Concepts of foraminiferal paleoecology: AAPG Bull., v. 44, p. 1921-1932.
———— and K. S. Rodolfo, 1964, Distribution of foraminifera and sediments, Peru-Chile trench area: Deep Sea Research, v. 11, p. 817-837.
Crouch, R. W., 1952, Significance of temperature on foraminifera from deep basins off southern California coast: AAPG Bull., v. 36, p. 807-843.
Douglas, R. G. 1979, Benthic foraminiferal ecology and paleoecology; a review of concepts and methods, *in* Foraminiferal ecology and paleoecology: Houston, SEPM Short Course no. 6, p. 21-53.
Harman, R. A., 1964, Distribution of foraminifera in the Santa Barbara Basin, California: Micropaleontology, v. 10, p. 81-96.
Natland, M. L., 1933, The temperature and depth distribution of some recent and fossil foraminifera in the southern California region: Bull., Scripps Inst. Oceanography, Tech. Ser., v. 3, p. 225-230.
————, 1957, Paleoecology of West Coast Tertiary sediments; *in* H. D. Ladd, ed., Treatise on marine ecology and paleoecology: Geol. Soc. America Mem. 67, v. 2, p. 543-572.
————, 1963, Presidential address; paleoecology and turbidites: Jour. Paleontology, v. 37, p. 946-951.
Phleger, F. B., 1960, Ecology and distribution of recent foraminifera: Baltimore, John Hopkins Univ. Press, 297 p.
———— F. L. Parker, and J. F. Peirson, 1953, North Atlantic foraminifera: Repts., Swedish Deep-Sea Expedition, v. 7, p. 3-122.
Sliter, W. V., 1972, Upper Cretaceous planktonic foraminiferal zoogeography and ecology - eastern Pacific margin: Palaeogeog., Palaeoclimatol., Paleoecol., v. 12, p. 15-31.

LACUSTRINE FAN

Dickson, F. H., and W. R. Normark, 1976, Sublacustrine fan morphology in Lake Superior: AAPG Bull., v. 60, p. 1021-1036.
Gould, H. R., 1951, Some quantitative aspects of Lake Mead turbidity currents *in* J. L. Hough, ed., Turbidity currents and the transportation of coarse sediments to deep water: SEPM Spec. Pub. no. 2, p. 34-52.
————, 1960, Turbidity currents, *in* Comprehensive survey of sedimentation in Lake Mead, 1948-49: U. S. Geol. Survey Prof. Paper 295, p. 201-207.
Grover, N. C., and C. S. Howard, 1938, The passage of turbid water through Lake Mead: Am. Soc. Civil Engrs., v. 103, p. 720-732.
Harrison, S. S., 1975, Turbidite origin of glaciolacustrine sediments, Woodcock Lake, Pennsylvania: Jour. Sed. Petrology, v. 45, p. 738-744.
Normark, W. R., and F. H. Dickson, 1976, Man-made turbidity currents in Lake Superior: Sedimentology, v. 23, p. 815-831.

ALPHABETICAL LISTING

Aalto, K. R., 1976, Sedimentology of a melange; Franciscan of Trinidad, California: Jour. Sed. Petrology, v. 46, p. 913-929.

Abe, M., 1980, The sedimentary history of Minami-Aga Sandstone in Niigata Basin, Japan: Proc., Offshore S. E. Asia Conf., SEAPEX Session, 7 p.

Aiello, E., et al, 1977, Sedimentary features of the Srednogorie zone (Bulgaria); an Upper Cretaceous intra-arc basin: Sed. Geology, v. 19, p. 39-68.

Allen, J. R. L., 1960, The Mam Tor Sandstones; a "turbidite" facies of the Namurian deltas of Derbyshire, England: Jour. Sed. Petrology, v. 30, p. 193-208.

————, 1964, Primary current lineation in the Lower Old Red Sandstone (Devonian), Anglo-Welsh basin: Sedimentology, v. 3, p. 89-108.

Almgren, A. A., 1978, Timing of Tertiary submarine canyons and marine cycles of deposition in the southern Sacramento Valley, California, in D. J. Stanley and G. Kelling, eds., Sedimentation in submarine canyons, fans, and trenches: Stroudsburg, Pa., Dowden, Hutchinson and Ross, Inc., p. 276-291.

Andrews, J. E., and R. J. Hurley, 1978, Sedimentary processes in the formation of a submarine canyon: Marine Geol., v. 26, p. M47-M50.

Angelucci, A., et al, 1967, Sedimentological characteristics of some Italian turbidites: Geol. Rom., v. 6, p. 345-420.

Anketell, J. M., and J. P. B. Lovell, 1976, Upper Llandoverian Grogal Sandstones and Aberystwyth Grits in the New Quay area, central Wales; a possible upward transition from contourites to turbidites: Geol. Jour., v. 11, p. 101-108.

Appelbaum, B. S., and A. H. Bouma, 1972, Geology of the upper continental slope in Alaminos Canyon region: Trans., Gulf Coast Assoc. Geol. Socs., v. 22, p. 157-164.

————, 1974, Surface microtextures of deep water quartz sands from Colombia and Sigsbee basins: College Station, Tex., Texas A&M Univ., Unpub. Ph.D. dissert., 220 p.

Ayers, M. W., and W. J. Cleary, 1980, Wilmington fan; mid-Atlantic lower rise development: Jour. Sed. Petrology, v. 50, p. 235-245.

Bailey, E. B., 1930, New light on sedimentation and tectonics: Geol. Mag., v. 67, p. 77-92.

Bailey, E. H., W. P. Irwin, and D. L. Jones, 1964, Franciscan and related rocks, and their significance in the geology of western California: Cal. Div. Mines Geology Bull., v. 183, 177 p.

Ball, M. M., 1979, Petroleum potential of passive margin slopes, in L. J. Doyle and O. H. Pilkey, eds., Geology of continental slopes: SEPM Spec. Pub. no. 27, p. 43-47.

Bandy, O. L., 1960, General correlation of foraminiferal structure with environment: Copenhagen, 21st Internat. Geol. Cong., pt. 22, p. 7-19.

———— and R. E. Arnal, 1957, Distribution of recent foraminifera off the coast of Central America: AAPG Bull., v. 41, p. 2037-2053.

———— and ————, 1960, Concepts of foraminiferal paleoecology: AAPG Bull., v. 44, p. 1921-1932.

———— and ————, 1969, Middle Tertiary Basin development, San Joaquin Valley, California: Bull. Geol. Soc. America, v. 80, p. 783-820.

———— and K. S. Rodolfo, 1964, Distribution of foraminifera and sediments, Peru-Chile Trench area: Deep Sea Research, v. 11, p. 817-837.

Barbat, W. F., 1958, The Los Angeles Basin area, California, in L. G. Weeks, ed., Habitat of Oil: AAPG Spec. Pub., p. 62-77.

Barr, F. T., and B. R. Walker, 1966, Late Tertiary channel system in northern Libya and its implications on Mediterranean Sea level changes: Deep Sea Drilling, v. 13, p. 1244-1251.

Bartolini, C., S. Berato, and V. Bortolotti, 1975, Upper Miocene shallow-water turbidites from western Tuscany: Sed. Geology, v. 14, p. 77-122.

———— C. DeGiuli, and G. Gianelli, 1974, Turbidites of the southern Tyrrhenian Sea: Boll. Soc. Geol. Ital., v. 93, p. 3-22.

———— C. Gehin, and D. J. Stanley, 1972, Morphology and recent sediments of the western Alboran Basin in the Mediterranean Sea: Marine Geol., v. 13, p. 159-223.

Bartow, J. A., 1966, Deep submarine channel in Upper Miocene, Orange County, California: Jour. Sed. Petrology, v. 36, p. 700-705.

Bates, C. C., 1953, Rational theory of delta formation: AAPG Bull., v. 37, p. 2119-2162.

Bazeley, W., 1972, San Emidio Noze Field, in R. E. King, ed., Stratigraphic oil and gas fields: AAPG Mem. 16, p. 297-312.

————, 1977, Late Miocene geology and new oil fields of the southern San Joaquin Valley: Guidebook, Pac. Sec. AAPG, SEG, SEPM, 88 p.

Beer, R. M., and D. S. Gorsline, 1971, Distribution, composition and transport of suspended sediment in Redondo Submarine Canyon and vicinity (California): Marine Geol., v. 10, p. 153-175.

Bell, H. S., 1942, Density currents as agents for transporting sediments: Jour. Geology, v. 50, p. 512-547.

Bennetts, K. R. W., and O. H. Pilkey, 1976, Characteristics of three turbidites, Hispaniola-Caicos Basin: Geol. Soc. America Bull., v. 87, p. 1291-1300.

Benson, P. H., 1971, Geology of the Oligocene Hackberry trend, Gillis English Bayou - Manchester area, Calcasieu Parish, Louisiana: Trans., Gulf Coast Assoc. Geol. Socs., v. 21, p. 1-14.

Berg, R. R., 1979, Characteristics of Lower Wilcox reservoirs, Valentine and south Hallettsville fields, Lavaca County, Texas: Trans., Gulf Coast Assoc. Geol. Socs., v. 29, p. 11-23.

———— and R. L. Findley, 1973, Deep-water interpretation of Upper Wilcox sandstones from core study, Katy Field, Texas: Trans., Gulf Coast Assoc. Geol. Socs., v. 23, p. 259-265.

———— and R. R. Powell, 1976, Density-flow origin for Frio reservoir sandstones, Nine Mile Point Field, Aransas County, Texas: Trans., Gulf Coast Assoc. Geol. Socs., v. 26, p. 310-319.

———— and F. J. Tedford, 1977, Characteristics of Wilcox gas reservoirs, northeast Thompsonville Field, Jim Hogg and Webb Counties, Texas: Trans., Gulf Coast Assoc. Geol. Socs., v. 27, p. 6-19.

Bergantino, R. N., 1971, Submarine regional geomorphology of the Gulf of Mexico: Bull. Geol. Soc. America, v. 82, p. 741-752.

Biddle, K. T., J. C. Maher, and R. D. Carter, 1975, Channel turbidite sandstone in the Elk Hills shale member of the Monterey Shale, in J. C. Maher et al, eds., Petroleum geology of Naval Petroleum Reserve No. 1, Elk Hills, Kern County, California: U. S. Geol. Survey Prof. Paper 912, p. 79-84.

Bjorlykke, K., and A. Hurst, 1980, The control of clastic mineral composition on Jurassic sandstone diagenesis in the Statfjord Field of the North Sea, (abs): Bochum, 1st Europe Meeting, Internat. Assoc. Sedimentol., p. 157.

Blackie, G. W., and R. J. Yeats, 1976, Magnetic-reversal stratigraphy of Pliocene-Pleistocene producing section at Saticoy oil field, Ventura Basin, Calif.: AAPG Bull., v. 60, p. 1985-1992.

Blake, M. C., Jr., and D. L. Jones, 1974, Origin of Franciscan melanges in northern California, in R. H. Dott, Jr., and R. H. Shaver, eds., Modern and ancient geosynclinal sedimentation: SEPM Special Pub. no. 19, p. 345-357.

Bloomer, R. E., 1977, Depositional environments of a reservoir sandstone in west-central Texas: AAPG Bull., v. 61, p. 344-359.

Bornhauser, M., 1948, Possible ancient submarine canyon in southwestern Louisiana: AAPG Bull., v. 32, p. 2287-2294.

————, 1960, Depositional and structural history of northwest Hartburg Field, Newton County, Texas: AAPG Bull, v. 44, p. 458-470.

Bouma, A. H., 1962, Sedimentology of some flysch deposits: Amsterdam, Elsevier Sci. Pub., 168 p.

————, 1964, Ancient and recent turbidites: Geologie en Mijnbouw, v. 43, p. 375-379.

————, 1965, Sedimentary characteristics of samples collected from some submarine canyons: Marine Geol., v. 3, p. 291-320.

————, 1972, Fossil contourites in lower Niesenflysch, Switzerland: Jour. Sed. Petrology, v. 42, p. 917-921.

————, 1972, Recent and ancient turbidites and contourites: Trans., Gulf Coast Assoc. Geol. Socs., v. 22, p. 205-221.

————, 1973, Leveed-channel deposits, turbidites, and contourites in deeper part of Gulf of Mexico: Trans., Gulf Coast Assoc. Geol. Socs., v. 23, p. 368-376.

————, 1979, Continental slopes, in L. J. Doyle and O. H. Pilkey, eds., Geology of continental slopes: SEPM Spec. Pub. No. 27, p. 1-15.

———— and A. Brouwer, eds., 1964, Turbidites: Amsterdam, Developments in sedimentology, 3, Elsevier Sci. Pub., p. 247-256.

————, et al, 1972, Deep sea sedimentation and correlation of strata off Magdalena River in Beata Strait: Margarita, Venezuela, 6th Conf. Geol. Del Caribe, p. 430-438.

———— and C. D. Hollister, 1973, Deep ocean basin sedimentation, in G. V. Middleton, and A. H. Bouma, eds., Turbidites and deep water sedimentation: Short Course notes, SEPM Pac. Sec., p. 79-118.

———— and T. H. Nilsen, 1978, Turbidite facies and deep-sea fans - with examples from Kodiak Island, Alaska: Proc., 10th Ann. Offshore Tech. Conf. v. 1, p. 559-567.

Bradley, W. H., et al, 1940, Geology and biology of North Atlantic deep-sea cores between Newfoundland and Ireland: U. S. Geol. Survey Prof. Paper 196A, 56 p.

————, et al, 1942, Geology and biology of North Atlantic deep-sea cores: U. S. Geol. Survey Prof. Paper 196, 163 p.

Briggs, G., 1974, Carboniferous depositional environment in the Ouachita Mountains - Arkoma Basin area of southeastern Oklahoma: Geol. Soc. America Spec. Paper No. 148, p. 225-239.

————, ed., 1974, Carboniferous rocks of the southeastern United States: Geol. Soc. America Spec. Paper No. 148, 361 p.

———— and L. M. Cline, 1967, Paleocurrents and source areas of late Paleozoic sediments of the Ouachita Mountains, southeastern Oklahoma: Jour. Sed. Petrology, v. 37, p. 985-1000.

———— and D. Roeder, 1975, Sedimentation and plate tectonics, Ouachita Mountains and Arkoma Basin, in Guidebook to the sedimentology of Paleozoic flysch and associated deposits, Ouachita Mountains - Arkoma Basin, Oklahoma: Dallas Geol. Soc. p. 1-22.

Brown, L. F., Jr., and W. L. Fisher, 1977, Seismic-stratigraphic interpretation of depositional systems; examples from Brazilian rift and pull-apart basins, in C. E. Payton, ed., Seismic stratigraphy - applications to hydrocarbon exploration: AAPG Mem. 26, p. 213-248.

Brown, R. D., Jr., and E. I. Rich, 1960, Early Cretaceous fossils in submarine slump deposits of Late Cretaceous age, northern Sacramento Valley, California: U. S. Geol. Survey Prof. Paper 400-B, p. 318B-320B.

———— and ————, 1967, Implications of two Cretaceous mass transport deposits, Sacramento Valley, California; comment on a paper by G. L. Peterson: Jour. Sed. Petrology, v. 37, p. 240-248.

Brown, W. R., 1970, Investigations of the sedimentary record in the Piedmont and Blue Ridge of Virginia, in G. W. Fisher, et al, eds., Studies of Appalachian geology, central and southern: New York, Wiley-Interscience, p. 335-349.

Buffington, E. C., 1952, Submarine "natural levees": Jour. Geology, v. 60, p. 473-479.

————, 1964, Structural control and precision bathymetry of La Jolla submarine canyon, California: Marine Geol., v. 1, p. 44-58.

Buller, A. T., and J. McManus, 1973, Modes of turbidite deposition deduced from grain-size analyses: Geol. Mag., v. 109, p. 491-500.

Burke, K., 1972, Longshore drift, submarine canyons, and submarine fans in development of Niger delta: AAPG Bull., v. 56, p. 1975-1983.

Busch, D. A., and S. A. Govela, 1978, Stratigraphy and structure of Chicontepec turbidites, southeastern Tampico-Misantla basin, Mexico: AAPG Bull., v. 62, p. 235-246.

Byramjee, R. S., J. F. Mugniot, and B. Biju-Duval, 1975, Petroleum potential of deep-water areas of the Mediterranean and Caribbean Seas: Proc., 9th World Petroleum Cong., v. 2, p. 299-312.

Cacchione, D. A., G. T. Rowe, and A. Malahoff, 1978, Submersible investigation of outer Hudson submarine canyon, in D. J. Stanley and G. Kelling, eds., Sedimentation in submarine canyons, fans, and trenches: Stroudsburg, Pa., Dowden, Hutchinson and Ross, Inc., p. 42-50.

Campbell, J. S., and D. L. Clark, 1977, Pleistocene turbidites of the Canada Abyssal Plain of the Arctic Ocean: Jour. Sed. Petrology, v. 47, p. 657-670.

Carbo, S., 1979, Vertical distribution of trace fossils in a turbidite sequence, Upper Devonian, New York State: Palaeogeog., Paleoclimatol., Palaeoecol., v. 28, p. 81-101.

Carlson, P. R., and C. H. Nelson, 1969, Sediments and sedimentary structures of the Astoria submarine canyon-fan system, northeast Pacific: Jour. Sed. Petrology, v. 39, p. 1269-1282.

Carozzi, A. V., 1979, Petroleum geology in the Paleozoic clastics in the middle Amazon Basin, Brazil: Jour. Petroleum Geology, v. 2, p. 55-74.

Carr, M. D., and E. L. Miller, 1979, Overthrust emplacement of the Numidian flysch complex in the westernmost Mogod Mountains, Tunisia; summary: Geol. Soc. America Bull., pt. 1, v. 90, p. 513-515.

Carson, B., 1973, Acoustic stratigraphy, structure, and history of Quaternary deposition in Cascadia Basin: Deep-Sea Research v. 20, p. 387-396.

———— J. Yuan, and P. B. Myers, Jr., 1974, Initial deep-sea sediment deformation at the base of the Washington continental slope; a response to subduction: Geology, v. 2, p. 561-564.

Carter, R. M., 1975, A discussion and classification of subaqueous mass-transport with particular application to grain-flow, slurry-flow and fluxoturbidites: Earth Sci. Rev., v. 11, p. 145-177.

————, 1979, Trench-slope channels from the New Zealand Jurassic; the Otekura Formation, Sandy Bay, South Otago: Sedimentology, v. 26, p. 475-496.

———— and J. K. Lindqvist, 1975, Sealers Bay submarine fan complex, Oligocene, southern New Zealand: Sedimentology, v. 22, p. 465-483.

———— and ————, 1977, Balleny Group, Chalky Island, southern New Zealand; an inferred Oligocene submarine canyon and fan complex: Pac. Geology, v. 12, p. 1-46.

———— and R. J. Norris, 1977, Redeposited conglomerates in Miocene flysch sequence at Blackmount, western Southland, New Zealand: Sed. Geology, v. 18, p. 289-319.

————, et al, 1978, Sedimentation patterns in an ancient arc-Trench-Ocean Basin complex, Carboniferous to Jurassic Rangitata Orogen, New Zealand, in D. J. Stanley and G. Kelling, eds., Sedimentation in submarine canyons, fans and trenches: Stroudsburg, Pa., Dowden, Hutchinson and Ross, Inc. p. 340-361.

Cas, R., 1979, Mass-flow arenites from a Paleozoic Interarc Basin, New South Wales, Australia; mode and environment of emplacement: Jour. Sed. Petrology, v. 49, p. 29-44.

Caughey, C. A., and C. J. Stuart, 1976, Where the potential is in the deep Gulf of Mexico: World Oil, v. 183, p. 67-72.

Chamberlain, C. K., 1964, Mass transport of sediment in the heads of Scripps Canyon, in Papers in Marine Geology, New York, Macmillan Co., p. 42-64.

————, 1971, Morphology and ethology of trace fossils from the Ouachita Mountains, southeast

Oklahoma: Jour. Paleont., v. 45, p. 212-246.

————, 1971, Bathymetry and paleoecology of Ouachita geosyncline of southeastern Oklahoma as determined from trace fossils: AAPG Bull., v. 55, p. 34-50.

————, 1975, Biogenetic sedimentary structures - trace fossils of the Ouachitas: *in* Guidebook to sedimentology of Paleozoic flysch and associated deposits, Ouachita Mountains - Arkoma Basin, Oklahoma: Dallas Geol. Soc., p. 51-68.

————, 1976, Trace fossils in DSDP cores of the Pacific: Jour. Paleont., v. 49, p. 1074-1096.

———— and D. L. Clark, 1973, Trace fossils and conodonts as evidence for deep-water deposits in the Oquirrh Basin of central Utah: Jour. Paleont., v. 47, p. 663-682.

———— K. O. Stanley, and J. H. Stewart, 1978, Depositional setting of some eugeosynclinal Ordovician rocks and structurally interleaved Devonian rocks in the Cordilleran mobile belt: Paleozoic Paleography of the Western United States, Symp. Proc., SEPM Pac. Sec., p. 259-274.

Chase, T. E., W. R. Normark, and P. Wilde, 1975, Oceanographic data of the Monterey deep-sea fan, 34°-37°N, 120°-127°W: Univ. Calif. Inst. Marine Resources Tech. Rept. TR-58, 2 p.

Cheema, M. R., 1977, Sedimentation and gas production of the Upper Devonian Benson Sand in north-central West Virginia - a model for exogeosynclinal mid-fan: W. Va. Univ., Ph.D. thesis, 127 p.

Cherkis, N. Z., H. S. Fleming, and R. H. Feden, 1973, Morphology and structure of Maury Channel, northeast Atlantic Ocean: Geol. Soc. America Bull., v. 84, p. 1601-1606.

Chipping, D. H., 1972, Sedimentary structures and environment of some thick sandstone beds of turbidite type: Jour. Sed. Petrology, v. 42, p. 587-595.

Chough, S. K., 1978, Morphology, sedimentary facies and processes of the Northwest Atlantic Mid-Ocean Channel between 61° and 52° N, Labrador Sea: McGill Univ., Ph.D. thesis, 167 p.

———— and R. Hesse, 1976, Submarine meandering thalweg and turbidity currents flowing for 4000 km in the Northwest Atlantic Mid-Ocean Channel, Labrador Sea: Geology, v. 4, p. 529-533.

———— and ————, 1980, Northwest Atlantic Mid-Ocean Channel of the Labrador Sea; III. head spills vs. body spill deposits from turbidity currents on natural levees: Jour. Sed. Petrology, v. 50, p. 227-234.

Church, V. R., ed., 1978, Display of cores from the Stevens Sand (Upper Miocene), southern San Joaquin Valley, Calif.: Calif. Well Repository Spec. Pub. no. 1, 39 p.

————, ed., 1979, Display of cores from the Winters Sand (Upper Cretaceous), Sacramento Valley, Calif.: Calif. Well Sample Repository Spec. Pub. no. 2, 56 p.

Clari, P., and G. Ghibaudo, 1979, Multiple slump scars in the Tortonian type area (Piedmont Basin, northwestern Italy): Sedimentology, v. 26, p. 719-730.

Clarke, J. M., 1917, Strand and undertow markings of upper Devonian time as indication of the prevailing climate: New York State Mus. Bull., 196, p. 199-210.

Cleary, W. J., and J. R. Conolly, 1974, Hatteras deep-sea fan: Jour. Sed. Petrology, v. 44, p. 1140-1154.

———— and ————, 1974, Petrology and origin of deep-sea sands; Hatteras Abyssal Plain: Marine Geol., v. 17, p. 263-279.

———— O. H. Pilkey and M. Ayers, 1977, Morphology and sediments of three ocean basin entry points, Hatteras Abyssal Plain: Jour. Sed. Petrology, v. 47, p. 1157-1170.

Cline, L. M., 1966, Late Paleozoic rocks of Ouachita Mountains, flysch facies: Guidebook, 29th Kansas Geol. Soc. Field Conf., p. 91-111.

————, 1968, Comparison of main geologic features of Arkoma Basin and Ouachita Mountains, southeastern Oklahoma: AAPG-SEPM Joint Mtg. Field Conf., Oklahoma City Geol. Soc., p. 63-74.

————, 1968, A guidebook to the geology of the western Arkoma Basin and Ouachita Mountains, Oklahoma: AAPG-SEPM Joint Mtg. Field Conf., Oklahoma City Geol. Soc., 126 p.

————, 1970, Sedimentary features of Late Paleozoic flysch, Ouachita Mountains, Oklahoma, *in* J. Lajoie, ed., Flysch sedimentology of North America: Geol. Assoc. Canada Spec. Paper No. 7, p. 85-101.

Cohen, Z., 1976, Early Cretaceous buried canyon; influence on accumulation of hydrocarbons in Helez oil field, Israel: AAPG Bull., v. 60, p. 108-114.

Colburn, I. P., 1968, Grain fabrics in turbidite sandstone beds and their relationship to sole mark trends on the same beds: Jour. Sed. Petrology, v. 38, p. 146-158.

———— and A. E. Fritsche, eds., 1973, Cretaceous stratigraphy of the Santa Monica Mountains and Simi Hills, southern California: Guidebook, Pac. Sec. SEPM, 93 p.

Cole, M. R., 1977, Eocene paleocurrents and sedimentation, San Nicolas Island, California: AAPG Bull., v. 61, p. 237-247.

Collinson, J. D., 1969, The sedimentology of the Grindslow Shales and the Kinderscout Grit; a deltaic complex in the Namurian of northern England: Jour. Sed. Petrology, v. 39, p. 194-221.

Conolly, J. R., 1968, Submarine canyons of the

continental margin, East Bass Strait (Australia): Marine Geol., v. 6, p. 449-461.

Cook, H. E., 1979, Ancient continental slope sequences and their value in understanding modern slope development, *in* L. J. Doyle and O. H. Pilkey, eds., Geology of continental slopes: SEPM Spec. Pub. no. 27, p. 287-305.

——— and W. C. Chamberlain, 1978, Ancient continental slope sequences and their usefulness in understanding modern slope development: Proc., 10th Ann. Offshore Tech. Conf., v. 1, p. 547-553.

——— and P. Enos, eds., 1977, Deep-water carbonate environments: SEPM Spec. Pub. no. 25, 336 p.

———, et al, 1972, Allochthonous carbonate debris flows at Devonian bank ("reef") margins, Alberta, Canada: Bull. Canadian Petroleum Geology, v. 20, p. 439-497.

Corbett, K. D., 1972, Features of thick-bedded sandstones in a proximal flysch sequence, upper Cambrian, southwest Tasmania: Sedimentology, v. 19, p. 99-114.

Cossey, S. P. J., 1978, Jurassic mass movement and turbidite sequences associated with growth faulting, Northern Tunisia: S.C. Univ., Ph.D. thesis, 68 p.

Cossey, S. P., and R. Ehrlich, 1978, Growth fault-controlled submarine carbonate debris flow and turbidite deposits from the Jurassic of northern Tunisia; possible canyon fill sequences, *in* D. J. Stanley and G. Kelling, eds., Sedimentation in submarine canyons, fans, and trenches: Stroudsburg, Pa., Dowden, Hutchinson and Ross, Inc., p. 127-137.

Crimes, T. P., 1976, Sand fans, turbidites, slumps and the origin of the Bay of Biscay; a facies analysis of the Guipuzcoan flysch: Palaeogeog., Palaeoclimatol., Palaeoecol., v. 19, p. 1-15.

———, 1977, Trace fossils of an Eocene deep-sea sand fan, northern Spain: Geol. Jour., Spec. Issue no. 9, p. 71-90.

——— A. Marcos, and A. Perez Estaun, 1974, Upper Ordovician turbidites in western Asturias; a facies analysis with particular reference to vertical and lateral variations: Palaeogeog., Palaeoclimatol., Palaeoecol., v. 15, p. 169-184.

——— and J. D. Crossley, 1980, Inter-turbidite bottom current orientation from trace fossils with an example from the Silurian flysch of Wales: Jour. Sed. Petrology, v. 50. p. 821-830.

Crouch, R. W., 1952, Significance of temperature on foraminifera from deep basins off southern California coast: AAPG Bull., v. 36, p. 807-843.

Crowell, J. C., 1952, Submarine canyons bordering central and southern California: Jour. Geology, v. 60, p. 58-83.

———, 1957, Origin of pebbly mudstones: Geol. Soc. America Bull., v. 68, p. 993-1009.

———, 1974, Sedimentation along the San Andreas Fault, California: *in* Modern and ancient geosynclinal sedimentation: SEPM Spec. Pub. no. 19, p. 292-303.

———, 1975, The San Gabriel Fault and Ridge Basin, southern California: Cal. Div. Mines Geology Spec. Rept. no. 18, p. 208-219.

———, 1976, Implications of crustal stretching and shortening of coastal Ventura Basin, California: Pac. Sec. AAPG, Misc., Pub. no. 24, p. 365-382.

———, et al, 1966, Deep-water sedimentary structures, Pliocene Pico Formation, Santa Paula Creek, Ventura Basin, California: Cal. Div. Mines Geology, Spec. Rept. 89, 40 p.

———, et al, 1973, Sedimentary facies changes in Tertiary rocks - California transverse and southern coast ranges: Ann. AAPG-SEPM-SEG Field Trip Guidebook SEPM Trip no. 2, 60 p.

Curray, J. R., and D. G. Moore, 1971, Growth of the Bengal deep-sea fan and denudation in the Himalayas: Geol. Soc. America Bull., v. 82, p. 563-572.

Daly, R. A., 1936, Origin of submarine "canyons": Am. Jour. Sci., v. 31, p. 401-420.

———, 1942, The floor of the ocean: Chapel Hill, Univ. of N. C. Press, p. 112-157.

Damuth, J. E., 1979, Migrating sediment waves created by turbidity currents in the northern South China Basin: Geology, v. 7, p. 520-523.

——— and R. W. Embley, 1979, Upslope flow of turbidity currents on the northwest flank of Ceara Rise; western equatorial Atlantic: Sedimentology, v. 26, p. 825-834.

——— and M. A. Gorini, 1976, The equatorial Mid-Ocean Canyon; a relict deep-sea channel on the Brazilian continental margin: Geol. Soc. America Bull., v. 87, p. 340-346.

——— and N. Kumar, 1975, Amazon Cone; morphology, sediments, age, and growth pattern: Geol. Soc. America Bull., v. 86, p. 863-878.

Davies, D. K., 1972, Mineralogy, petrography, and derivation of sands and silts of continental slope, rise and abyssal plain of the Gulf of Mexico: Jour. Sed. Petrology, v. 42, p. 59-65.

———, 1972, Deep sea sediments and their sedimentation, Gulf of Mexico: AAPG Bull., v. 56, p. 2212-2239.

Davies, G. R., 1977, Turbidites, debris sheets, and truncation structures in upper Paleozoic deep-water carbonates of the Sverdrup Basin, Arctic Archipelago, *in* H. E. Cook and P. Enos, eds., Deep-water carbonate environments: SEPM Spec. Pub. no. 25, p. 221-239.

Davies, I. C., and R. G. Walker, 1974, Transport and deposition of resedimented conglomerates; the Cap Enrage Formation, Gaspe, Quebec: Jour. Sed.

Petrology, v. 44, p. 1200-1216.

Davies, T. A., and A. L. Laughton, 1972, Sedimentary processes in the North Atlantic: Washington, D.C., Initial Repts. of Deep Sea Drilling Project, v. 12, p. 905-934.

Davies, J. R., 1971, Sedimentation of Pliocene sandstones in Santa Barbara Channel, California (abs.): AAPG Bull., v. 55, p. 335.

DeRaaf, J. F. M., 1968, Turbidites et associations sedimentaires apparentees; I and II. Koninkl: Proc., Nederl. Akad. Van Wetenschappen, v. 71, p. 1-23.

Dickas, A. B., and J. L. Payne, 1967, Upper Paleocene buried channel in Sacramento Valley, California: AAPG Bull., v. 51, p. 873-882.

Dickinson, W. R., 1970, Clastic sedimentary sequences deposited in shelf, slope, and trough settings between magmatic arcs and associated trenches: Pac. Geology, v. 3, p. 15-30.

———, ed., 1974, Plate tectonics and sedimentation: SEPM Spec. Pub. no. 22, 204 p.

———, 1974, Sedimentation within and beside ancient and modern magmatic arcs, in R. H. Dott, Jr., and R. H. Shaver, eds., Modern and ancient geosynclinal sedimentation: SEPM Spec. Pub. no. 19, p. 230-239.

———, 1974, Plate tectonics and sedimentation, in W. R. Dickinson, ed., Tectonic and sedimentation: SEPM Spec. Pub. no. 22, p. 1-27.

———, 1979, Mesozoic forearc basin in central Oregon: Geology, v. 7, p. 166-170.

——— R. V. Ingersoll, and S. A. Graham, 1979, Paleogene sediment dispersal and paleotectonics in northern California; summary: Geol. Soc. America Bull., pt. 1, v. 90, p. 897-898.

——— R. W. Ojakangas, and R. J. Stewart, 1969, Burial metamorphism of the Late Mesozoic Great Valley Sequence, Cache Creek, California: Geol. Soc. America Bull., v. 80, p. 519-526.

——— and E. I. Rich, 1972, Petrologic intervals and petrofacies in the Great Valley Sequence, Sacramento Valley, California: Geol. Soc. America Bull., v. 63, p. 3007-3024.

——— and C. A. Suczek, 1979, Plate tectonics and sandstone compositions: AAPG Bull., v. 63, p. 2164-2182.

Dickson, F. H., and W. R. Normark, 1976, Sublacustrine fan morphology in Lake Superior: AAPG Bull., v. 60, p. 1021-1036.

Dietz, R. S. and H. J. Knebel, 1971, Trou sans fond submarine canyon; Ivory Coast, Africa: Deep-Sea Research, v. 18, p. 441-447.

——— ——— and L. H. Somers, 1968, Cayer submarine canyon: Geol. Soc. America Bull., v. 79, p. 1821-1828.

Dill, R. F., 1964, Sedimentation and erosion in Scripps submarine canyon head, in R. L. Miller, ed., Papers in marine geology, New York, McMillan & Co., p. 23-41.

———, 1969, Earthquake effects on fill of Scripps submarine canyon: Geol. Soc. America Bull., v. 80, p. 321-328.

Dillion, W. P., and J. G. Vedder, 1973, Structure and development of the continental margin of British Honduras: Geol Soc. America Bull., v. 84, p. 2713-2732.

Dimitrijevic, M. D., and M. N. Dimitrijevic, 1976, Transgressive overlap of a turbiditic sequence - Priabonian of Ovce Pole, Macedonia: Sed. Geol., v. 16, p. 223-234.

Dingle, R. V., 1977, The anatomy of a large submarine slump on a sheared continental margin (S. E. Africa): London, Jour. Geol. Soc., v. 134, p. 293-310.

Ditty, P. S., et al, 1977, Mixed terrigenous-carbonate sedimentation in the Hispaniola-Caicos turbidite basin: Marine Geol., v. 24, p. 1-20.

Dixon, W. H., 1972, Reservoir geology of the Sage Farm, Bradford Oil Field, McKean County, Pennsylvania: Field Trip Guidebook, Eastern Section AAPG, p. 111-1 to 111-22.

D'Olier, B., and R. J. Maddrell, 1970, Buried channels of the Thames Estuary: Nature, v. 226, p. 347-348.

Dorreen, J. M., 1951, Rubble bedding and graded bedding in Talara Formation of northwestern Peru: AAPG Bull., v. 35, p. 1829-1849.

Dott, R. H., Jr., 1963, Dynamics of subaqueous gravity depositional processes: AAPG Bull., v. 47, p. 104-128.

——— and K. J. Bird, 1979, Sand transport through channels across an Eocene shelf and slope in southwestern Oregon, U.S.A., in L. J. Doyle and O. H. Pilkey, eds., Geology of continental slopes: SEPM Spec. Pub. no. 27, 327-242.

——— and R. H. Shaver, eds., 1974, Modern and ancient geosynclinal sedimentation: SEPM Spec. Pub. No. 19, 380 p.

Douglas, R. G. 1979, Benthic foraminiferal ecology and paleoecology; a review of concepts and methods, in Foraminiferal ecology and paleoecology: Houston, SEPM Short Course No. 6, p. 21-53.

Drake, D. E., P. G. Hatcher, and G. H. Keller, 1978, Suspended particulate matter and mud deposition in Upper Hudson submarine canyon, in D. J. Stanley and G. Kelling, eds., Sedimentation in submarine canyons, fans, and trenches: Stroudsburg, Pa., Dowden, Hutchinson and Ross, Inc., p. 33-41.

Drummond, K. F., E. W. Christensen, and K. D. Berry, 1976, Upper Cretaceous lithofacies model, Sacramento Valley, California: Pac. Sec. AAPG Misc. Pub. no. 24, p. 76-88.

Duff, P. M. D., A. Hallam, and E. K. Walton, 1967, Flysch, *in* Cyclic sedimentation: Amsterdam, Devel. in Sedimentology, no. 10, Elsevier Sci. Pub., p. 215-231.

Duncan, J. R., and L. D. Kulm, 1970, Mineralogy, provenance and dispersal history of Late Quaternary deep-sea sands in Cascadia Basin and Blanco fracture zone off Oregon: Jour. Sed. Petrology, v. 40, p. 874-887.

Dzulynski, S., M. Ksiazkiewicz, and P. H. Kuenen, 1959, Turbidites in flysch of the Polish Carpathian Mountains: Bull. Geol. Soc. America, v. 70, p. 1089-1118.

————— and A. Slaczka, 1959, Directional structures and sedimentation of the Krosno Beds (Carpathian Flysch): Ann. Soc. Geol. Pol., v. 28, p. 205-259.

————— and A. J. Smith, 1964, Flysch facies: Ann. Soc. Geol. Pol., v. 34, p. 245-266.

————— and E. K. Walton, 1965, Sedimentary features of flysch and greywackes: Amsterdam, Elsevier Sci. Pub., 274 p.

Edmondson, W. R., 1965, The Meganos Gorge of the southern Sacramento Valley: Selected Papers, San Joaquin Geol. Soc., no. 3, p. 36-51.

—————, 1980, Member sands of the Winters Formation: Pac. Sec. AAPG, Prog. Preprint, 20 p.

Einsele, G., and K. Kelts, 1980, Physical properties of unconsolidated mud turbidites, and their significance for diagenetic processes, (abs.): Bochum, 1st Europe Meeting, Internat. Assoc. Sedimentols., p. 155-156.

Eisbacher, G. H., 1976, Sedimentology of the Dezadeash flysch and its implications for strike-slip faulting along the Denali fault, Yukon Territory and Alaska: Canadian Jour. Earth Sci., v. 13, p. 1495-1513.

Ellis, B. A., 1980, The lateral variation of structures within basinwide turbidites, (abs.): Bochum, 1st Europe Mtg., Internat. Assoc. Sedimentols. p. 115.

Elmore, R. D., et al, 1979, Black Shell turbidite, Hatteras Abyssal Plain, western Atlantic Ocean: Geol. Soc. America Bull., pt. 1, v. 90, 1165-1176.

Embley, R. W., 1976, New evidence for occurrence of debris flow deposits in the deep sea: Geology, v. 4, p. 371-374.

————— J. I. Ewing, and M. Ewing, 1970, The Vidal deep-sea channel and its relationship to the Demerara and Barracude abyssal plains: Deep-Sea Research, v. 17, p. 539-552.

Embry, A., and J. E. Klovan, 1976, The Middle-Upper Devonian clastic wedge of the Franklinian geosyncline: Bull. Canadian Petroleum Geology, v. 24, p. 485-639.

Emery, K. O., 1950, A suggested origin of continental slopes and of submarine canyons: Geol. Mag., v. 87, p. 102-104.

—————, 1960, Basin plains and aprons off southern California: Jour. Geology, v. 68, p. 464-479.

—————, 1980, Continental margins - classification and petroleum prospects: AAPG Bull., v. 64, p. 297-315.

————— and E. Uchupi, 1972, Western North Atlantic Ocean; topography, rocks, structure, water, life, and sediments: AAPG Mem. 17, 532 p.

Enos, P., 1969, Anatomy of a flysch: Jour. Sed. Petrology, v. 39, p. 680-723.

—————, 1969, Cloridorme Formation, Middle Ordovician flysch, northern Gaspe Peninsula, Quebec: Geol. Soc. America Spec. Paper 117, 66 p.

—————, 1977, Flow regimes in debris flow: Sedimentology, v. 24, p. 133-142.

Erickson, J. W., 1975, Sedimentology of the South Point Formation (Eocene), Santa Rosa Island, California: Proc., Pac. Secs. AAPG-SEPM-SEG Mtg., p. 169-190.

Ericson, D. B., M. Ewing, and B. C. Heezen, 1951, Deep-sea sands and submarine canyons: Geol. Soc. America Bull., v. 62, p. 961-965.

————— ————— and —————, 1952, Turbidity currents and sediments in North Atlantic: AAPG Bull., v. 36, p. 489-511.

————— ————— ————— and G. Wollin, 1955, Sediment deposition in deep Atlantic: Geol. Soc. America Spec. Paper 62, p. 205-219.

————— ————— G. Wollin, and B. C. Heezen, 1961, Atlantic deep-sea sediment cores: Geol. Soc. America Bull., v. 72, p. 193-285.

Ernst, W. G., 1970, Tectonic contact between the Franciscan melange and the Great Valley Sequence - crustal expression of a Late Mesozoic Benioff zone: Jour. Geophys. Research, v. 75, p. 886-901.

Ewing, M., D. B. Ericson, and B. C. Heezen, 1958, Sediments and topography of the Gulf of Mexico, *in* L. G. Weeks, ed., Habitat of Oil: AAPG Spec. Pub. p. 995-1053.

Ewing, J., X. LePichon, and M. Ewing, 1963, Upper stratification of Hudson apron region: Jour. Geophys. Research, v. 68, p. 6303-6316.

Fallaw, W. C., 1976, Factors affecting petroleum accumulation beneath the eastern United States continental margin: Southeast Geol., v. 17, p. 189-205.

Felix, D. W., and D. S. Gorsline, 1971, Newport submarine canyon, California; an example of the effects of shifting loci of sand supply upon canyon position: Marine Geol., v. 10, p. 177-198.

Field, M. E., and O. H. Pilkey, 1971, Deposition of deep-sea sands; comparison of two areas of the Carolina continental rise: Jour. Sed. Petrology, v. 41, p. 526-536.

Fischer, P. J., 1971, An ancient (Upper Paleocene) submarine canyon and fan; the Meganos channel,

Sacramento Valley, California, (abs.): Prog., Ann. Geol. Soc. America Cordilleran Sec., v. 3, p. 120.

————, 1979, The evolution of a Late Paleocene submarine canyon and fan system; the meganos Formation, southern Sacramento Basin, California: Field Guide, Geol. Soc. America Cordilleran Sec. Mtg., 10 p.

Fisher, F. V., and J. M. Mattinson, 1968, Wheeler Gorge turbidite-conglomerate series, California; inverse grading: Jour. Sed. Petrology, v. 38, p. 1013-1023.

Fisher, G. W., 1970, The metamorphosed sedimentary rocks along the Potomac River near Washington, D. C., in G. W. Fisher et al, eds., Studies of Appalachian geology, central and southern, New York: Wiley Interscience, p. 299-315.

Forel, F. A., 1885, Les ravins sous-lacustres des fleuves Glaciaires: Acad. Sci. Paris, Ct. Rend., v. 101, p. 725-728.

Foss, D. C., 1979, Depositional environment of Woodbine sandstone, Polk County, Texas: Trans., Gulf Coast Assoc. Geol. Socs., v. 29, p. 83-94.

Fowler, C., 1975, Geology of the Montrose Field, in A. W. Woodland, ed., Petroleum and continental shelf of northwest Europe: New York, John Wiley and Sons, p. 467-476.

Fox, P. J., B. C. Heezen, and A. M. Harian, 1968, Abyssal anti-dunes: Nature, v. 220, p. 470-472.

Fritz, S. J., and O. H. Pilkey, 1975, Distinguishing bottom and turbidity current coarse layers on the continental rise: Jour. Sed. Petrology, v. 45, p. 57-62.

Galloway, A. J., 1966, Field trip Point Reyes Peninsula and San Andreas fault zone, in E. H. Bailey, ed., Geology of northern California: Calif. Div. Mines Geol. Bull. 190, p. 429-440.

Galloway, W. E., 1974, Deposition and diagenetic alteration of sandstone in northeast Pacific arc-related basins; implications for graywacke genesis: Geol. Soc. America Bull., v. 85, p. 379-390.

———— and L. F. Brown, Jr., 1972, Depositional systems and shelf-slope relationships in Upper Pennsylvanian rocks, north-central Texas: Austin, Univ. Texas, Bureau Econ. Geology Rept. Inv. no. 75, 62 p.

Gardett, P. H., 1971, Petroleum potential of Los Angeles Basin, California, in I. H. Cram, ed., Future petroleum provinces of the United States - their geology and potential: AAPG Mem. 15, p. 298-308.

Garrison, L. E., ed., 1961, East Central Sacramento Valley; Marysville (Sutter) Buttes-Chico Creek-Oroville: Field Trip, Geol. Soc. of Sacramento, 51 p.

Ghibaudo, G., 1980, Deep-sea fan deposits in the Macigno Formation (Middle-Upper Oligocene) of the Gordana Valley, northern Apennines, Italy: Jour. Sed. Petrology, v. 50, p. 723-742.

Gilbert, W. G., and W. R. Dickinson, 1970, Stratigraphic variations in sandstone petrology, Great Valley Sequence, central California coast: Geol. Soc. America Bull., v. 81, p. 949-954.

Glaeser, J. D., 1979, Catskill delta slope sediments in the central Appalachian basin; source deposits and reservoir deposits, in L. J. Doyle and O. H. Pilkey, eds., Geology of continental slopes: SEPM Spec. Pub. no. 27, p. 343-357.

Goetz, J. F., W. J. Prins, and J. F. Logar, 1977, Reservoir delineation by wireline techniques: The Log Analyst, v. 18, p. 12-40.

Gonzalez-Bonorino, G., and G. V. Middleton, 1976, A Devonian submarine fan in western Argentina: Jour. Sed. Petrology, v. 46, p. 56-69.

Gorsline, D. S., 1978, Anatomy of margin basins - presidential address: Jour. Sed. Petrology, v. 48, p. 1055-1068.

———— and K. O. Emery, 1959, Turbidity current deposits in San Pedro and Santa Monica basins off southern California: Geol. Soc. America Bull., v. 70, p. 279-290.

Got, H., and D. J. Stanley, 1974, Sedimentation in two Catalonian canyons, northwestern Mediterranean: Marine Geol., v. 16, p. M91-M100.

Goudkoff, P. P., 1945, Stratigraphic relations of Upper Cretaceous in Great Valley, California: AAPG Bull., v. 29, p. 956-1007.

Gould, H. R., 1951, Some quantitative aspects of Lake Mead turbidity currents in J. L. Hough, ed., Turbidity currents and the transportation of coarse sediments to deep water: SEPM Spec. Pub. No. 2, p. 34-52.

————, 1960, Turbidity currents, in Comprehensive survey of sedimentation in Lake Mead, 1948-49: U.S. Geol. Survey Prof. Paper 295, p. 201-207.

Graham, S. A., W. R. Dickinson, and R. V. Ingersoll, 1975, Himalayan-Bengal model for flysch dispersal in the Appalachian-Ouachita System: Geol. Soc. America Bull., v. 85, p. 273-286.

———— R. V. Ingersoll, and W. R. Dickinson, 1976, Common provenance for lithic grains in Carboniferous sandstones from Ouachita Mountains and Black Warrior Basin: Jour. Sed. Petrology, v. 46, p. 620-632.

Gregory, M. R., 1969, Sedimentary features and penecontemporaneous slumping in the Waitemata Group, Whangaparoa Peninsula, North Auckland, New Zealand: Jour. Geology Geophysics, v. 12, p. 248-282.

Griggs, G. B., A. G. Carey, Jr., and L. D. Kulm, 1969, Deep-sea sedimentation and sediment-fauna interaction in Cascadia Channel and on Cascadia Abyssal Plain: Deep-Sea Research, v. 16, p. 157-170.

———— and L. D. Kulm, 1970, Sedimentation in Cascadia Deep-sea Channel: Geol. Soc. America Bull., v. 81, p. 1361-1384.

———— and ———— , 1973, Origin and development of Cascadia Deep-sea Channel: Jour. Geophys. Research, v. 78, p. 6325-6339.

———— ———— A. C. Waters, and G. A. Fowler, 1970, Deep-sea gravel from Cascadia Channel: Jour. Geology, v. 78, p. 611-619.

Grim, P.J., 1969, Seamap Deep-sea Channel: Seattle, Wash., U.S. Dept. Commerce, ESSA Tech. Rept., ERL 93-Pol. 2, p. 1-27.

Grippi, J., and K. Bruke, 1980, Submarine-canyon complex among Cretaceous island-arc sediments, western Jamaica: Geol. Soc. America Bull., v. 91, pt. 1, p. 179-184.

Grover, N. C., and C. S. Howard, 1938, The passage of turbid water through Lake Mead: Am. Soc. Civil Engineers, v. 103, p. 720-732.

Hall, B. A., and D. J. Stanley, 1973, Levee-bounded submarine base-of-slope channels in the Lower Devonian Seboomook Formation, northern Maine: Geol. Soc. America Bull., v. 84, p. 2101-2110.

Hamblin, A. P., and R. G. Walker, 1979, Storm-dominated shallow marine deposits; the Fernie-Kootenay (Jurassic) transition, southern Rocky Mountains: Canadian Jour. Earth Sci., v. 16, p. 1673-1690.

Hamilton, E. L., 1967, Marine geology of Abyssal Plains in the Gulf of Alaska: Jour. Geophys. Research, v. 72, p. 4189-4213.

Hamilton-Smith, T., 1971, A proximal-distal turbidite sequence and a probable submarine canyon in the Siegas Formation (Early Llandovery) of north-western New Brunswick: Jour. Sed. Petrology, v. 41, p. 752-762.

Hampton, M. A., 1972, The role of subaqueous debris flows in generating turbidity currents: Jour. Sed. Petrology, v. 42, p. 775-793.

Hand, B. M., and K. O. Emery, 1964, Turbidites and topography of north end of San Diego Trough, California: Jour. Geology, v. 72, p. 526-542.

———— G. V. Middleton, and K. Skipper, 1972, Antidune cross-stratification in a turbidite sequence, Cloridorme Formation, Gaspe, Quebec: Sedimentology, v. 18, p. 135-138.

Handford, C. R., and S. P. Dutton, 1980, Pennsylvanian-Early Permian depositional systems and shelf-margin evolution, Palo Duro basin, Texas: AAPG Bull., v. 64, p. 88-106.

Haner, B. E., 1971, Morphology and sediments of Redondo submarine fan, southern California: Geol. Soc. America Bull., v. 82, p. 2413-2432.

Harman, R. A., 1964, Distribution of foraminifera in the Santa Barbara Basin, California: Micropaleontology, v. 10, p. 81-96.

Harms, J. C., 1974, Brushy Canyon Formation, Texas; a deep water density current deposit: Geol. Soc. America Bull., v. 85, p. 1763-1784.

———— and R. K. Fahnestock, 1965, Stratification, bed forms and flow phenomena (with example from the Rio Grande), in G. V. Middleton, ed., Primary sedimentary structures and their hydrodynamic interpretation: SEPM Spec. Pub. no. 12, p. 84-115.

———— and P. Tackenberg, 1972, Seismic signatures of sedimentation models: Geophysics, v. 37, p. 45-58.

———— , et al, 1975, Depositional environments as interpreted from primary sedimentary structures and stratification sequences: Dallas, SEPM Short Course No. 2, 161 p.

Harrison, S. S., 1975, Turbidite origin of glacio-lacustrine sediments, Woodcock Lake, Pennsylvania: Jour. Sed. Petrology, v. 45, p. 738-744.

Hedberg, H. D., 1970, Continental margins from viewpoint of the petroleum geologist: AAPG Bull., v. 54, p. 3-43.

Heezen, B. C., 1959, Dynamic processes of abyssal sedimentation; erosion, transportation, and redeposition on the deep-sea floor: Geophys. Jour., v. 2, p. 142-163.

———— , 1963, Turbidity currents, in N. M. Hill, ed., The Sea, v. 3, the Earth beneath the Sea: New York, John Wiley & Sons, p. 742-775.

———— , 1968, Atlantic ocean floor map: Natl. Geog. Mag., June, p. 794 insert.

———— and C. L. Drake, 1964, Grand banks slump: AAPG Bull., v. 48, p. 221-233.

———— D. B. Ericson, and M. Ewing, 1954, Further evidence for a turbidity current following the 1929 Grand Banks earthquake: Deep-Sea Research, v. 1, p. 193-202.

———— and M. Ewing, 1952, Turbidity currents and submarine slumps and the 1929 Grand Banks earthquake: Am. Jour. Sci., v. 250, p. 849-873.

———— and ———— , 1955, Orleansville earthquake and turbidity currents: AAPG Bull., v. 39, p. 2505-2514.

———— and C. D. Hollister, 1964, Deep-sea current evidence from abyssal sediments: Marine Geol., v. 1, p. 141-174.

———— , et al, 1964, Congo submarine canyon: AAPG Bull., v. 48, p. 1126-1149.

———— C. D. Hollister, and W. F. Ruddiman, 1966, Shaping of the continental rise by deep geostrophic contour currents: Science, v. 152, p. 502-508.

———— M. Tharp, and M. Ewing, 1959, The floors of the ocean, 1. The North Atlantic: Geol. Soc. America Spec. Paper 65, 122 p.

Helu, P.C., V.R. Verdugo, and P. R. Barcenas, 1977, Origin and distribution of Tertiary conglomerates, Veracruz basin, Mexico: AAPG Bull., v. 61,

p. 207-226.

Hendry, H. E., 1973, Sedimentation of deep water conglomerates in Lower Ordovician rocks of Quebec - composite bedding produced by progressive liquefaction of sediment?: Jour. Sed. Petrology, v. 43, p. 125-136.

———, 1978, Cap Des Rosiers Formation at Grosses Roches, Quebec – deposits of the mid-fan region on an Ordovician submarine fan: Canadian Jour. Earth Sci., v. 15, p. 1472-1488.

Heritier, F. E., P. Lossel, and E. Wathne, 1979, Frigg field - large submarine - fan trap in Lower Eocene rocks of North Sea Viking Graben: AAPG Bull., v. 63, p. 1999-2020.

Herzer, R. H., 1979, Submarine slides and submarine canyons on the continental slope off Canterbury, New Zealand: New Zeal. Jour. Geology Geophysics, v. 22, p. 391-406.

——— and D. W. Lewis, 1979, Growth and burial of a submarine canyon off Motunau, north Canterbury, New Zealand: Sed. Geology, v. 24, p. 69-83.

Hess, G. R., and W. R. Normark, 1976, Holocene sedimentation history of the major fan valleys of Monterey Fan: Marine Geol., v. 22, p. 233-251.

Hesse, R., 1961, The flysch area of the Zwiesel west of Bad Tolz, Upper Bavaria: Deut. Geol. Ges., Z., v. 113, p. 293-304.

———, 1965, Herkunft und Transport der Sedimente im bayerischen Flyschtrog: (incl. English summary), Deut. Geol. Ges., Z., v. 116, p. 403-426.

———, 1965, Herkunft und Transport der Sedimente im bayerischen Flyschtrog: Verh. Geol. B-A, p. 147-170.

———, 1972, Lithostratigraphy, petrography, and genesis of the Bavarian (Germany) Flysch; Lower Cretaceous [Lithostratigraphie, Petrographie und Entstehungsbedingungen des bayerischen Flysches; Unterkreide]: Geol., Bavarica, no. 66, p. 148-222.

———, 1974, Long-distance continuity of turbidites; possible evidence for an Early Cretaceous trench-abyssal plain in the east Alps: Geol. Soc. America Bull., v. 85, p. 859-870.

———, 1975, Turbiditic and non-turbiditic mudstone of Cretaceous flysch sections of the east Alps and other basins: Sedimentology, v. 22, p. 387-416.

———, 1975, Turbidite sedimentation systems and the recognition of the Cretaceous plate boundaries in the Alpine-Carpathian arc: Proc., 9th Internat. Sedimentology Cong. [Theme 4], v. 2, p. 189-194.

———, 1976, Some unusual secondary sedimentary structures in the lacustrine Lower Carboniferous of Nova Scotia, Canada, and their significance as earthquake indicators [Einige ungewoehnliche sekundaere Sedimenstrukturen im lakustrinen Unterkarbon Neuschottlands (Kanada) und ihre Deutung als Erdbebenanzeiser]: Eclogae Geol. Helv. (EGHVAG), v. 69, p. 196-201.

———, 1977, Softex-radiographs of sliced piston cores from the Japan and southern Kurile Trench and slope areas, in E. Honza, ed., Geological investigation of Japan and southern Kurile Trench and slope areas: Jap. Geol. Surv., Cruise Rept., no. 7, p. 86-108.

——— and A. Butt, 1976, Paleobathymetry of Cretaceous turbidite basins of the east Alps relative to the calcite compensation level: Jour. Geology, v. 84, p. 505-533.

Hill, D. R., 1979, Union Island gas field, San Joaquin County, California: Calif. Well Sample Repository Spec. Pub. no. 2, p. 8-17.

Hirayama, J. and T. Nakajima, 1977, Analytical study of turbidites, Otadai Formation, Boso Peninsula, Japan: Sedimentology, v. 24, p. 747-779.

Hiscott, R. N., 1978, Provenance of Ordovician deep-water sandstones, Tourelle Formation, Quebec, and implications for initiation of the Taconic Orogeny: Canadian Jour. Earth Sci., v. 15, p. 1579-1597.

———, 1979, Clastic sills and dikes associated with deep-water sandstones, Tourelle Formation, Ordovician, Quebec: Jour. Sed. Petrology, v. 49, p. 1-9.

——— and G. V. Middleton, 1979, Depositional mechanics of thick-bedded sandstones at the base of a submarine slope, Tourelle Formation (Lower Ordovician), Quebec, Canada, in L. J. Doyle and O. H. Pilkey, eds., Geology of continental slopes: SEPM Spec. Pub. no. 27, p. 307-326.

——— and ———, 1980, Fabric of coarse deep-water sandstones, Tourelle Formation, Quebec, Canada: Jour. Sed. Petrology, v. 50, p. 703-722.

Horn, D. R., B. M. Horn, and M. N. Delach, 1970, Sedimentary provinces of the North Pacific: Geol. Soc. America Mem. 126, p. 1-21.

——— M. Ewing, M. N. Delach, and B. M. Horn, 1971, Turbidites of the northeast Pacific: Sedimentology, v. 16, p. 55-69.

——— ——— B. M. Horn, and M. N. Delach, 1971, Turbidites of the Hatteras and Sohm abyssal plains, western North Atlantic: Marine Geol., v. 11, p. 287-323.

———J. I. Ewing, and M. Ewing, 1972, Graded-bed sequences emplaced by turbidity currents north of 20°N in the Pacific, Atlantic and Mediterranean: Sedimentology, v. 18, p. 247-275.

Hough, J. L., ed., 1951, Turbidity currents and the transportation of coarse sediment into deep water: SEPM Spec. Pub. no. 2, 107 p.

Howell, D. G., and M. H. Link, 1979, Eocene conglomerate sedimentology and basin analysis, San

Diego and the southern California borderland: Jour. Sed. Petrology, v. 49, p. 517-540.

Hoyt, J. H., B. L. Oostdam, and D. D. Smith, 1969, Offshore sediments and valleys of the Orange River (south and southwest of Africa): Marine Geol., v. 7, p. 69-84.

Hoyt, W. V., 1959, Erosional channel in the middle Wilcox near Yoakum, Lavaca County, Texas: Trans., Gulf Coast Assoc. Geol. Socs., v. 9, p. 41-50.

Hsu, K. J., 1977, Studies of Ventura field, California, 1; Facies geometry and genesis of Lower Pliocene turbidites: AAPG Bull., v. 61, p. 137-168.

Huang, T. C., and H. G. Goodell, 1970, Sediments and sedimentary processes of eastern Mississippi cone, Gulf of Mexico: AAPG Bull., v. 54, p. 2070-2100.

Hubert, C., J. Lajoie, and M. A. Leonard, 1970, Deep sea sediments in the Lower Paleozoic Quebec Supergroup, in J. Lajoie, ed., Flysch sedimentology in North America: Geol. Assoc. Canada, Spec. Paper, 7, p. 103-125.

Hubert, J. F., 1964, Textural evidence for deposition of many western North Atlantic deep-sea sands by ocean-bottom currents rather than turbidity currents: Jour. Geology, v. 72, p. 757-785.

——— and W. J. Neal, 1967, Mineral composition and dispersal patterns of the deep-sea sands in the western North Atlantic petrologic province: Geol. Soc. America Bull., v. 78, p. 749-772.

Hughes, S. M., 1979, Winters Sand: Calif. Well Sample Repository Spec. Pub. no. 2, p. 3-7.

Hurley, R. J., 1964, Analysis of flow in Cascadia deep-sea channel, in R. L. Miller, ed., Papers in marine geology: New York, Macmillan Co., p. 117-132.

Ingersoll, R. V., 1978, Submarine fan facies of the Upper Cretaceous Great Valley sequence, northern and central California: Sed. Geology, v. 21, p. 205-230.

———, 1979, Evolution of the Late Cretaceous fore-arc basin, northern and central California: Geol. Soc. America Bull., v. 90, pt. 1, p. 813-826.

——— E. I. Rich, and W. R. Dickinson, 1977, Great Valley Sequence, Sacramento Valley: Sacramento, Field Trip Guide, Cordilleran Sec., Geol. Soc. America Ann. Mtg., 71 p.

——— and C. A. Suczek, 1979, Petrology and provenance of Neogene sand from Nicobar and Bengal fans, DSDP sites 211 and 218: Jour. Sed. Petrology, v. 49, p. 1217-1228.

Jacka, A. D., et al, 1968, Permian deep-sea fans of the Delaware Mountain group (Guadalupian), Delaware Basin: in Guadalupian facies, Apache Mountain area, west Texas: Permian Basin Sec., SEPM Pub. 68-11, p. 49-90.

———, et al, 1967, Guadalupian depositional cycles of Delaware Basin and northwest shelf, in J. G. Flam and S. Chuber, eds., Cyclic sedimentation in the Permian Basin: Midland, West Texas Geol. Soc. p. 152-196.

James, D. M. D., 1972, Sedimentation across an intra-basinal slope; the Graneddwen Formation (Ashgillian), west-central Wales: Sed. Geology, v. 7, p. 291-307.

———, 1975, Caradoc turbidites at Poppit sands (Pembrokeshire), Wales: Geol. Mag., v. 112, p. 295-304.

Jipa, D., and R. B. Kidd, 1974, Sedimentation of coarser grained interbeds in Arabian Sea and sedimentation processes of the Indus Cone: Washington, D. C., Repts. of Deep Sea Drilling Proj., v. 23, p. 471-495.

Johnson, B. A., 1974, Deep-sea fan-valley conglomerate; Cap Enrage Formation, Gaspe, Quebec: Hamilton, Ont., McMaster Univ., Master's thesis, 108 p.

——— and R. G. Walker, 1979, Paleocurrents and depositional environments of deep water conglomerates in the Cambro-Ordovician Cap Enrage Formation, Quebec Appalachians: Canadian Jour. Earth Sci., v. 16, p. 1375-1387.

Johnson, D., 1938, The origin of submarine canyons: Jour. Geomorphology, v. 1, p. 11-129, 230-243, 324-340.

———, 1939, The origin of submarine canyons: Jour. Geomorphology, v. 2, p. 42-60, 133-158, 213-236.

Johnson, D. A., and T. C. Johnson, 1970, Sediment redistribution by bottom currents in the central Pacific: Deep-Sea Research, v. 17, p. 157-169.

Johnson, K. E., 1966, A depositional interpretation of the Stanley Group of the Ouachita Mountains, Oklahoma, in Flysch facies and structure of the Ouachita Mountains: Guidebook, 29th Field Conf., Kansas Geol. Soc. p. 140-163.

Keith, B. D., and G. M. Friedman, 1977, A slope-fan-basin-plain model, Taconic sequence, New York and Vermont: Jour. Sed. Petrology, v. 47, p. 1220-1241.

Keller, G. H., and F. P. Sherpard, 1978, Currents and sedimentary processes in submarine canyons off the northeast United States, in D. J. Stanley and G. Kelling, eds., Sedimentation in submarine canyons, fans, and trenches: Stroudsburg, Pa., Dowden, Hutchinson and Ross, Inc., p. 15-32.

Kelling, G., 1968, Submarine channel and fan deposits, Silurian of central Wales, United Kingdom (abs.): AAPG Bull., v. 52, p. 535-536.

——— and M. H. Woollands, 1969, The stratigraphy and sedimentation of the Llandoverian rocks of the Rhayader District, in A. Woods, ed., The Precambrian and Lower Paleozoic rocks of Wales: Univ. Wales Press, p. 255-281.

———— and D. J. Stanley, 1970, Morphology and structure of Wilmington and Baltimore Canyons, eastern United States: Jour. Geology, v. 78, p. 637-660.

———— and ————, 1976, Sedimentation in canyon, slope, and base-of-slope environments, in D. J. Stanley and D. J. P. Swift, eds., Marine sediment transport and environmental management: New York, John Wiley and Sons, Inc., p. 379-435.

———— and ————, 1978, Sedimentation in submarine canyons, fans, and trenches; appraisal and augury, in D. J. Stanley and G. Kelling, eds., Sedimentation in submarine canyons, fans, and trenches: Stroudsburg, Pa., Dowden, Hutchinson and Ross, Inc., p. 377-388.

———— and J. Holroyd, 1978, Clast size, shape, and composition in some ancient and modern fan gravels, in D. J. Stanley, and G. Kelling, eds., Sedimentation in submarine canyons, fans, and trenches: Stroudsburg, Pa., Dowden, Hutchinson and Ross, Inc., p. 138-159.

Kennedy, M. P., and G. W. Moore, 1971, Stratigraphy and structure of the area between Oceanside and San Diego, California, Field Trip No. 8, in W. A. Elders, ed., Geological excursions in southern California: Riverside, Calif., Cordilleran Sec., Geol. Soc. of America, p. 152-166.

Kenyon, N. H., R. H. Belderson and A. H. Stride, 1978, Channels, canyons and slump folds on the continental slope between south west Ireland and Spain: Oceanol. Acta, v. 1, p. 369-380.

Kepferle, R. C., 1977, Stratigraphy, petrology and depositional environment of the Kenwood Siltstone Member, Borden Formation (Mississippian), Kentucky and Indiana: U. S. Geol. Survey Prof. Paper 1007, 49 p.

————, 1978, Prodelta turbidite fan apron in Borden Formation (Mississippian), Kentucky and Indiana, in D. J. Stanley and G. Kelling, eds., Sedimentation in submarine canyons, fans, and trenches: Stroudsburg, Pa., Dowden, Hutchinson and Ross, Inc., p. 224-238.

Kersey, D. G., and K. J. Hsu, 1976, Energy relations and density-current flows; an experimental investigation: Sedimentology, v. 23, p. 761-789.

Klein, G. D., 1966, Dispersal and petrology of sandstones of Stanley-Jackfork boundary, Ouachita fold belt, Arkansas and Oklahoma: AAPG Bull., v. 50, p. 308-326.

————, 1975, Sedimentary tectonics in southwest Pacific marginal basins based on leg 30 Deep Sea Drilling Project cores from the south Fiji, Hebrides, and Coral Sea basins: Geol. Soc. America Bull., v. 86, p. 1012-1018.

Knebel, H. J., and D. W. Folger, 1976, Large sand waves on the Atlantic outer continental shelf around Wilmington Canyon, off eastern United States: Marine Geol., v. 22, p. M7-M15.

Knox, G. J., 1976, The Early Tertiary deep-water sandstones near San Sebastian, Spain, some aspects of diagenesis: Geol. Mag., v. 113, p. 341-348.

Kohlbush, R. L., 1977, Tule Elk oil field, in B. Bazeley, ed., Late Miocene geology and new oil fields of the southern San Joaquin Valley: Guidebook, Pac. Sec. AAPG, SEG, SEPM, p. 73-78.

Komar, P. D., 1969, The channelized flow of turbidity currents with application to Monterey Deep-Sea Fan Channel: Jour. Geophys. Research, v. 74, p. 4544-4558.

————, 1970, The competence of turbidity current flow: Bull. Geol. Soc. America, v. 81, p. 1555-1562.

————, 1973, Continuity of turbidity current flow and systematic variations in deep-sea channel morphology: Bull., Geol. Soc. America, v. 84, p. 3329-3338.

Korsch, R. J., 1978, Petrographic variations within thick turbidite sequences; an example from the Late Paleozoic of eastern Australia: Sedimentology, v. 25, p. 247-265.

Koss, G. M., 1977, Carbonate mass flow sequences of the Permian Delaware Basin, west Texas, in Upper Guadalupian facies Permian reef complex Guadalupe Mountains, New Mexico and west Texas: Field Conf. Guidebook, Permian Basin Sec. SEPM Pub. 77-16, p. 391-431.

Krause, D. C., et al, 1970, Turbidity currents and cable breaks in the western New Britain Trench: Bull. Geol. Soc. America, v. 81, p. 2153-2160.

Krause, F. F., and A. E. Oldershaw, 1979, Submarine carbonate breccia beds - a depositional model for two layer, sediment gravity flows from the Sekwi Formation (Lower Cambrian), Mackenzie Mountains, Northwest Territories, Canada: Canadian Jour. Earth Sci., v. 16, p. 189-199.

Kruit, C., J. Brouwer, and P. Ealey, 1972, A deepwater sand fan in the Eocene Bay of Biscay: Nature Phy. Sci., v. 240, p. 59-61.

———— ———— G. Know, W. Schollnberger, and A. Van Vliet, 1975, Eocene deep-water fan deposits of San Sebastian (Spain): Nice, France, Field Guide Z-23, IXe Congres Internat. de Sedimentologie, 75 p.

Krutak, P. R., 1970, Origin and depositional environment of Codell sandstone member of the Carlile shale (Upper Cretaceous), southeastern Colorado: Mtn. Geol., v. 7, p. 185-204.

Ksiazkiewicz, M., 1958, Submarine slumping in the Carpathian Flysch: Ann. Soc. Geol. Pologne, v. 28, p. 123-151.

Kuenen, P. H., 1951, Properties of turbidity currents of high density, in J. L. Hough, ed., Turbidity currents and the transportation of coarse sediments to deep water: SEPM Spec. Pub. no. 2, p. 14-33.

———— , 1952, Estimated size of the Grand Banks turbidite currents: Am. Jour. Sci., v. 250, p. 874-884.

———— , 1964, Deep-sea sands and ancient turbidites, *in* A. H. Bouma and A. Brouwer, eds., Turbidites: Developments in sedimentology No. 3: Elsevier Sci. Pub., p. 3-33.

———— , 1966, Matrix of turbidites; experimental approach: Sedimentology, v. 7, p. 267-297.

———— , 1967, Emplacement of flysch-type sand beds: Sedimentology, v. 9, p. 203-243.

———— and C. I. Migliorini, 1950, Turbidity currents as a cause of graded bedding: Jour. Geology, v. 58, p. 91-127.

———— and H. W. Menard, 1952, Turbidity currents, graded and nongraded deposits: Jour. Sed. Petrology, v. 22, p. 83-96.

Kulm, L. D., and G. A. Fowler, 1974, Cenozoic sedimentary framework of the Gorda-Juan de Fuca Plate and adjacent continental margin - a review, *in* R. H. Dott, Jr., and R. H. Shaver, eds., Modern and ancient geosynclinal sedimentation: SEPM Spec. Pub. no. 19, p. 212-229.

———— and ————, 1974, Oregon continental margin structure and stratigraphy; a test of the imbricate thrust model, *in* C. A. Burk and C. L. Drake, eds., The geology of the continental margins: New York, Springer-Verlag, p. 261-283.

Kurtz, D. D. and J. B. Anderson, 1979, Recognition and sedimentologic description of recent debris flow deposits from the Ross and Weddell Seas, Antarctica: Jour. Sed. Petrology, v. 49, p. 1159-1169.

Laird, M. G., 1972, Sedimentology of the Greenland group in the Paparoa Range, West Coast, South Island: New Zeal. Jour. Geology and Geophysics, v. 15, p. 372-393.

Lajoie, J., 1969, Dispersal and petrology of the Silurian Val Brillant and Robitaille sandstones, Appalachians, Quebec: Jour. Sed. Petrology, v. 38, p. 643-647.

———— , ed., 1970, Flysch sedimentology in North America: Geol. Assoc. Canada Spec. Paper No. 7, 272 p.

———— , 1972, Slump fold axis orientations; an indication of paleoslope?: Jour. Sed. Petrology, v. 42, p. 584-586.

———— , 1979, Origin of megarhythms in flysch sequences of the Quebec Appalachians: Canadian Jour. Earth Sci., v. 16, p. 1518-1523.

———— and A. Chagnon, 1973, Origin of red beds in a Cambrian flysch sequence, Canadian Appalachians, Quebec: Sedimentology, v. 20, p. 91-103.

———— Y. Heroux, and B. Mathey, 1974, The Precambrian shield and the Lower Paleozoic shelf; the unstable provenance of the Lower Paleozoic flysch sandstones and conglomerates of the Appalachians between Beaumont and Bic, Quebec: Canadian Jour. Earth Sci., v. 11, p. 951-963.

Lawson, L. M., 1914, Movement of silt, Elephant Butte Reservoir: Reclamation Record, v. 10, 411 p.

Le Blanc, R. J., 1972, Geometry of sandstone reservoir bodies, *in* T. D. Cook, ed., Symposium on underwater management and environmental implications: AAPG Mem. 18, p. 133-190.

Lewis, D. W., 1976, Subaqueous debris flows of Early Pleistocene age at Motunau, North Canterbury, New Zealand: New Zeal. Jour. Geology Geophys., v. 19, p. 535-567.

Lewis, K. B., and B. P. Kohn, 1973, Ashes, turbidites, and rates of sedimentation on the continental slope off Hawkes Bay: New Zeal., Jour. Geology Geophysics, v. 16, p. 439-454.

Lineback, J. A., 1968, Turbidites and other sandstone bodies in the Borden Siltstone (Mississippian) in Illinois: Ill. State Geol. Surv., Circ. 425, 29 p.

Link, M. H., 1975, Matilija Sandstone; a transition from deep-water turbidite to shallow marine deposition in the Eocene of California: Jour. Sed. Petrology, v. 45, p. 63-78.

———— and T. H. Nilsen, 1980, The Rocks sandstone, an Eocene sand-rich deep-sea fan deposit, northern Santa Lucia Range, California: Jour. Sed. Petrology, v. 50, p. 583-602.

———— and ————, in press, Sedimentology of the rocks, sandstone and Eocene paleogeography of the Northern Santa Lucia Basin, California: Jour. Sed. Petrology.

Lohmar, J. M., and J. E. Warme, 1978, Anatomy of an Eocene submarine canyon-fan system, southern California borderland: Proc., 10th Ann. Offshore Tech. Conf. v. 1, p. 571-577.

Lonsdale, P., and C. D. Hollister, 1979, Cut-offs at an abyssal meander south of Iceland: Geology, v. 7, p. 597-601.

Lovell, J. P. B., 1969, Tyee Formation; a study of proximality in turbidites: Jour. Sed. Petrology, v. 39, p. 935-953.

Lowe, D. R., 1972, Submarine canyon and slope channel sedimentation model as inferred from Upper Cretaceous deposits, western California: 24th Internat. Geol. Cong., Sec. 6, p. 75-81.

———— , 1972, Implications of three submarine mass-movement deposits, Cretaceous, Sacramento Valley, California: Jour. Sed. Petrology, v. 42, p. 89-101.

———— , 1973, Conglomerate-filled submarine fan channel in the Cretaceous of western California: Geol. Soc. America Abs. with Programs, v. 5, p. 718.

———— , 1975, Water escape structure in coarse-grained sediments: Sedimentology, v. 22, p.

157-204.

———, 1976, Subaqueous liquefied and fluidized sediment flows and their deposits: Sedimentology, v. 23, p. 285-308.

———, 1979, Stratigraphy and sedimentology of the Pigeon Point Formation, San Mateo County, California, *in* T. H. Nilsen, ed., Geology of the Santa Cruz Mountains, California: Field Trip Guide, Cordilleran Sec., Geol. Soc. America, p. 17-29.

———, 1979, Sediment gravity flow; their classification and some problems of application to natural flows and deposits, *in* L. J. Doyle and O. H. Pilkey, eds., Geology of continental slopes: SEPM Spec. Pub. no. 27, p. 75-82.

——— and R. D. LoPiccolo, 1974, The characteristics and origins of dish and pillar structures: Jour. Sed. Petrology, v. 44, p. 484-501.

Lund, J. W., et al, 1978, Pre-platform exploration of High Island Blocks A-560 and A561: Trans., Gulf Coast Assoc. Geol. Socs. v. 28, p. 273-294.

MacPherson, B. A., 1978, Sedimentation and trapping mechanism in Upper Miocene Stevens and older turbidite fans of southeastern San Joaquin Valley: AAPG Bull., v. 62, p. 2243-2274.

Maher, J. C., R. D. Carter, and R. J. Lantz, 1975, Petroleum geology of Naval Petroleum Reserve no. 1 Elk Hills, Kern County, California: U.S. Geol. Survey Prof. Paper 912, 109 p.

Maldonado, A., and D. J. Stanley, 1976, The Nile Cone; submarine fan development by cyclic sedimentation: Marine Geol., v. 20, p. 27-40.

——— and ———, 1978, Nile Cone depositional processes and patterns in the Late Quaternary, *in* D. J. Stanley, and G. Kelling, eds., Sedimentation in submarine canyons, fans, and trenches: Stroudsburg, Pa., Dowden, Hutchinson and Ross, Inc., p. 239-260.

——— and ———, 1979, Depositional patterns and Late Quarternary evolution of two Mediterranean submarine fans; a comparison: Marine Geol., v. 31, p. 215-250.

Mallik, T. K., 1978, Mineralogy of deep-sea sands of the Indian Ocean: Marine Geol., v. 27, p. 161-176.

Mammerickx, J., 1970, Morphology of the Aleutian Abyssal Plain: Geol. Soc. America Bull., v. 81, p. 3457-3464.

Mansfield, C. F., 1972, Petrofacies units and sedimentary facies of the Late Mesozoic strata west of Coalinga, California, *in* Cretaceous of the Coalinga area: Guidebook, Pac. Sec., SEPM, p. 19-26.

———, 1979, Upper Mesozoic subsea fan deposits in the southern Diablo Range, California; record of the Sierra Nevada magmatic arc: Geol. Soc. America Bull., pt. 1, v. 90, p. 1025-1046.

Marschalko, R., 1964, Sedimentary structures and paleocurrents in the marginal lithofacies of the Central-Carpathian flysch, *in* A. H. Bouma, and A. Brouwer, eds., Turbidites: Amsterdam, Devel. in sedimentology, 3, Elsevier Sci. Pub., p. 106-126.

Marshall, N. F., 1978, A large storm-induced slump re-opens an unknown Scripps submarine canyon tributary, *in* D. J. Stanley and G. Kelling, eds., Sedimentation in submarine canyons, fans and trenches: Stroudsburg, Pa., Dowden, Hutchinson and Ross, Inc., p. 73-84.

Martin, B. D., 1963, Rosedale channel - evidence for Late Miocene submarine erosion in Great Valley California: AAPG Bull., v. 47, p. 441-456.

——— and K. O. Emery, 1967, Geology of Monterey Canyon, California: AAPG Bull., v. 51, p. 2281-2304.

Martini, I. P., and M. Sagri, 1977, Sedimentary fillings of ancient deep-sea channels; two examples from northern Apennines (Italy): Jour. Sed. Petrology, v. 47, p. 1542-1553.

——— ——— and J. H. Doveton, 1978, Lithologic transition and bed thickness periodicities in turbidite succcessions of the Antola Formation, northern Apennines, Italy: Sedimentology, v. 25, p. 605-623.

Mattick, R. E., et al, 1978, Petroleum potential of U.S. Atlantic slope, rise and abyssal plain: AAPG Bull., v. 62, p. 592-608.

Mayuga, M. N., 1970, Geology and development of California's giant - Wilmington Oil Field, *in* M. T. Halbouty, ed., Geology of giant petroleum fields: AAPG Mem. 14, p. 158-184.

McBride, E. F., 1962, Flysch and associated beds of the Martinsburg Formation (Ordovician), central Appalachians: Jour. Sed. Petrology, v. 32, p. 39-91.

———, 1966, Sedimentary petrology and history of the Haymond Formation (Pennsylvanian), Marathon Basin, Texas: Austin, Univ. Texas, Bur. Econ. Geol., Rept. Invest. 57, 101 p.

———, 1970, Flysch sedimentation in the Marathon region, Texas, *in* J. LaJoie, ed., Flysch sedimentology in North America: Geol. Assoc. Canada Spec. Paper no. 7, p. 67-83.

———, 1975, The Ouachita trough sequence; Marathon region and Ouachita Mountains, *in* Guidebook to sedimentology of Paleozoic flysch and associated deposits, Ouachita Mountains - Arkoma Basin, Oklahoma: Dallas Geol. Soc., p. 23-41.

———, 1978, Olistostrome in the Tesnus Formation (Mississippian-Pennsylvanian), Payne Hills, Marathon region, Texas: Geol. Soc. America Bull., v. 89, p. 1550-1558.

———, 1978, Reservoir aspects of submarine fan facies; example of the Marnoso Arenacea Formation, northern Apennines, Italy: Proc., Offshore Tech. Conf., no. 10, v. 1, p. 587-591.

————, 1978, Tesnus and Haymond Formations; siliciclastic flysch, *in* E. F. McBride and S. J. Mazzullo, eds., Tectonics and Paleozoic facies of the Marathon Geosyncline, west Texas: SEPM Permian Basin Sec., p. 131-147.

———— and R. M. Flores, 1975, Characteristics of the Pennsylvanian lower-middle Haymond Delta-Front sandstones, Marathon Basin, west Texas [discussion and reply]: Geol. Soc. America Bull., v. 86, p. 264-266.

———— and S. J. Mazzullo, eds., 1978, Tectonics and Paleozoic facies of the Marathon geosyncline, west Texas: SEPM Permian Basin Sec. Pub. 78-17, 271 p.

McCabe, P. J., 1978, The Kinderscoutian Delta (Carboniferous) of northern England; a slope influenced by density currents, *in* D. J. Stanley and G. Kelling, eds., Sedimentation in submarine canyons, fans and trenches: Stroudsburg, Pa., Dowden, Hutchinson and Ross, Inc., p. 116-126.

McIlreath, I. A., 1977, Accumulation of a Middle Cambrian, deep-water limestone debris apron adjacent to a vertical, submarine carbonate escarpment, southern Rocky Mountains, Canada, *in* H. E. Cook and P. Enos, eds., Deep-water carbonate environments: SEPM Spec. Pub. no. 25, p. 113-124.

McIver, N. L., 1970, Appalachian turbidites, *in* G. W. Fisher, et al, eds., Studies of Appalachian geology, central and southern: New York, Wiley-Interscience, p. 69-81.

McLean, H., 1977, Lithofacies of the Blakeley Formation, Kitsap County, Washington; a submarine fan complex?: Jour. Sed. Petrology, v. 47, p. 78-88.

McMaster, R. L., and A. Ashraf, 1973, Drowned and buried valleys on the southern New England continental shelf: Marine Geol., v. 15, p. 249-268.

———— and ————, 1973, Extent and formation of deeply buried channels on the continental shelf off southern New England: Jour. Geology, v. 81, p. 374-379.

Menard, H. W., Jr., 1955, Deep-sea channels, topography, and sedimentation: AAPG Bull., v. 39, p. 236-255.

————, 1960, Possible pre-Pleistocene deep-sea fans off central California: Geol. Soc. America Bull., v. 71, p. 1271-1278.

————, 1964, Marine geology of the Pacific: New York, McGraw Hill, 271 p.

———— S. M. Smith, and R. M. Pratt, 1965, The Rhone deep-sea fan, *in* W. F. Whittard and R. Bradshaw, eds., Submarine geology and geophysics: Butterworths, London, Proc. 17th Symp. Colston Res. Soc. p. 271-285.

Michaelis, E. R., and G. Dixon, 1969, Interpretation of depositional processes from sedimentary structures in the Cardium sand: Bull. Canadian Petroleum Geology, v. 17, p. 410-443.

Middleton, G. V., 1962, Size and sphericity of quartz grains in two turbidite formations: Jour. Sed. Petrology, v. 32, p. 725-742.

————, 1965, Antidune cross-bedding in a large flume: Jour. Sed. Petrology, v. 35, p. 922-927.

————, 1965, Paleocurrents in Normanskill graywackes north of Albany, New York: Geol. Soc. America Bull., v. 76, p. 841-844.

————, ed., 1965, Primary sedimentary structures and their hydrodynamic interpretation - a symposium: SEPM Special Pub. no. 12, 265 p.

————, 1966, Experiments on density and turbidity currents - 1, motion of the head: Canadian Jour. Earth Sci., v. 3, p. 523-546.

————, 1966, Experiments on density and turbidity currents - 2, uniform flow of density currents: Canadian Jour. Earth Sci., v. 3, p. 627-637.

————, 1966, Small-scale models of turbidity currents and the criterion for auto-suspension: Jour. Sed. Petrology, v. 36, p. 202-208.

————, 1967, Experiments on density and turbidity currents - 3, deposition of sediment: Canadian Jour. Earth Sci., v. 4, p. 475-505.

————, 1967, The orientation of concave-convex particles deposited from experimental turbidity currents: Jour. Sed. Petrology, v. 37, p. 229-232.

————, 1969, Grain flows and other mass movements down slopes, *in* The new concepts of continental margin sedimentation: Am. Geol. Inst. Short Course Lecture Notes no. 11, 14 p.

————, 1969, Turbidity currents, *in* The new concepts of continental margin sedimentation: Am. Geol. Inst. Short Course Lecture Notes no. 10, 20 p.

————, 1971, Experimental studies related to problems of flysch sedimentation, *in* J. Lajoie, ed., Flysch sedimentology of North America: Canadian Geol. Assoc. Spec. Paper no. 7, p. 253-272.

———— and A. H. Bouma, eds., 1973, Turbidites and deep-water sedimentation: Pac. Sec. SEPM Short Course, 157 p.

———— and M. A. Hampton, 1973, Sediment gravity flows; mechanics of flow and deposition, *in* G. V. Middleton, ed., Turbidites and deepwater sedimentation: Pac. Sec. , SEPM Short Course, p. 1-38.

———— and ————, 1976, Subaqueous sediment transport and deposition by sediment gravity flows, *in* D. J. Stanley and D. J. P. Swift, eds., Marine sediment transport and environmental management: New York, Wiley Interscience, p. 197-218.

Milliman, J.D., C.P. Summerhayes, and H. T. Barretto, 1975, Quaternary sedimentation on the

Amazon continental margin; a model: Geol. Soc. America, Bull. v. 86, p. 610-614.

Mitchell, M. H., 1975, Depositional environment and facies relationships of the canyon sandstones, Val Verde Basin, Texas: College Station, Tex., Texas A&M Univ., M.S. thesis, 211 p.

Moiola, R. J., and E. F. McBride, 1975, Sedimentology of Ouachita turbidites, southeastern Oklahoma; a summary, *in* Sedimentology of Paleozoic flysch and associated deposits, Ouachita Mountains - Arkoma Basin, Oklahoma: Field Trip Guidebook, Dallas Geol. Soc., p. 42-50.

Moore, D. G., 1961, Submarine slumps: Jour. Sed. Petrology, v. 31, 343-357.

———— J. R. Curray, and F. J. Emmel, 1976, Large submarine slide (olistostrome) associated with Sunda Arc subduction zone, northeast Indian Ocean: Marine Geol., v. 21, p. 211-226.

Moore, G. F., et al, 1980, Sedimentology and paleobathymetry of Neogene trench-slope deposits, Nias Island, Indonesia: Jour. Geology, v. 88, p. 161-180.

Moore, G. T., 1969, Interaction of rivers and oceans - Pleistocene petroleum potential: AAPG Bull., v. 53, p. 2421-2430.

———— and T. J. Fullam, 1973, Deep water channels and their potential value in petroleum localization: Trans., Gulf Coast Assoc. Geol. Soc., v. 23, p. 256-258.

———— and T. J. Fullam, 1975, Submarine channel systems and their potential for petroleum localization, *in* M. L. Broussard, ed., Deltas - models for exploration: Houston Geol. Soc., p. 165-189.

———— G. W. Starke, L. C. Bonham, and H. O. Woodbury, 1978, Mississippi fan, Gulf of Mexico - physiography, stratigraphy, and sedimentational patterns, *in* A. H. Bouma, G. T. Moore, and J. M. Coleman, eds., Framework, facies and oil trapping characteristics of the upper continental margin: AAPG Stud. in Geology no. 7, p. 155-191.

———— and H. O. Woodbury, 1978, Mississippi fan - morphology, sedimentational history, petroleum potential: 10th Ann. Offshore Tech. Conf., v. 1, p. 391-398.

Mordojovich, C., 1974, Geology of a part of the Pacific margin of Chile, *in* C. A. Burk and C. L. Drake, eds., The geology of continental margins: New York, Springer-Verlag, p. 591-598.

Morris, R. C., 1971, Classification and interpretation of disturbed bedding types in Jackfork flysch rocks (Upper Mississippian), Ouachita Mountains, Arkansas: Jour. Sed. Petrology, v. 41, p. 410-424.

———— , 1971, Stratigraphy and sedimentology of Jackfork Group, Arkansas: AAPG Bull., v. 55, p. 387-402.

———— , 1974, Carboniferous rocks of the Ouachita Mountains, Arkansas - a study of facies patterns along the unstable slope and axis of a flysch trough, *in* G. Briggs, ed., Symposium on the Carboniferous rocks of the southeastern United States: Geol. Soc. America Spec. Paper 148, p. 241-279.

———— , 1974, Sedimentary and tectonic history of the Ouachita Mountains, *in* W. R. Dickinson, ed., Tectonics and sedimentation: SEPM Spec. Pub. no. 22, p. 120-142.

———— M. R. Burkart, P. W. Palmer, and R. R. Russell, 1975, Stratigraphy and structure of part of frontal Ouachita Mountains, Arkansas: AAPG Bull., v. 59, p. 747-765.

———— K. E. Proctor, and M. R. Koch, 1979, Petrology and diagenesis of deep-water sandstones, Ouchita Mountains, Arkansas and Oklahoma, *in* P. A. Scholle and P. R. Schluger, eds., Aspects of diagenesis: SEPM, Spec. Pub. no. 26, p. 263-279.

Morrow, D. W., 1978, The Prairie Creek embayment and associated slope, shelf and basin deposits: Canadian Geol. Surv. Paper no. 78-1A, p. 361-370.

Moussa, M. T., 1977, Bioclastic sediment gravity flow and submarine sliding in the Juana Diaz Formation, southwestern Puerto Rico: Jour. Sed. Petrology, v. 47, p. 593-599.

Mutti, E. 1969, Sedimentologia delle Arenarie di Messanagros (Oligocene — Aquitaniano) nell' isola di Rodi: Mem. Soc. Geol. Italiana, v. 8, p. 1027-1070.

———— , 1974, Examples of ancient deep-sea fan deposits from circum - Mediterranean geosynclines, *in* R. H. Dott, Jr., and R. H. Shaver, eds., Modern and ancient geosynclinal sedimentation: SEPM Spec. Pub. no. 19, p. 92-105.

———— , 1977, Distinctive thin-bedded turbidite facies and related depositional environments in the Eocene Hecho Group, (south-central Pyrenees, Spain): Sedimentology, v. 24, p. 107-131.

———— M. Barros, S. Possato, and L. Rumenos, 1980, Deep-sea fan turbidite sediments winnowed by bottom currents in the Eocene of the Campos Basin, Brazilian offshore, (abs.): Bochum, 1st European Mtg., Internat. Assoc. Sedimentols. p. 114.

———— and G. Ghibaudo, 1972, Un esempio di torbiditi di conoide sottomarina esterna; le arenarie di San Salvatore (Formazione di Bobio, Miocene) nell'Appennino di Piacenza: Mem. Acc. Sci. Torino, Classe, Sci. Fis., Mat., Nat., Serie 4, n. 16, 40 p.

———— T. H. Nilsen, and F. Ricci Lucchi, 1978, Outer fan depositional lobes of the Laga Formation (Upper Miocene and Lower Pliocene), east-central Italy, *in* D. J. Stanley and G. Kelling, eds., Sedimentation in submarine canyons, fans, and trenches: Stroudsburg, Pa., Dowden, Hutchinson and Ross, Inc., p. 210-223.

————, et al, 1975, Examples of turbidite facies and facies associations from selected formations of the northern Apennines: Nice, France, Field Trip Guidebook A-11, 9th Internat. Sedimentology Cong., 120 p.

———— and F. Ricci Lucchi, 1972, Le torbiditi dell'Appennino settentrionale; introduzione all'analisi di facies: Mem. Soc. Geol. Ital., v. 11, p. 161-199.

———— and ————, 1975, Turbidite facies and facies associations, *in* E. Mutti et al, eds., Example of turbidite facies and associations from selected formations of the northern Apennines: Nice, France, Field Trip Guidebook A-11, 9th Internat. Sedimentology Cong., p. 21-36.

———— and ————, 1978, Turbidites of the northern Apennines; introduction to facies analysis, T. H. Nilsen, translator: Internat. Geol. Rev., v. 20, p. 125-166.

Nagle, H. E., and E. S. Parker, 1971, Future oil and gas potential of onshore Ventura Basin, California, *in* I. H. Cram, ed., Future petroleum provinces of the United States - their geology and potential: AAPG Mem. 15, p. 254-297.

Nardin, T. R., et al, 1979, A review of mass movement processes, sediment and acoustic characteristics, and contrasts in slope and base-of-slope systems versus canyon-fan-basin floor systems, *in* L. J. Doyle and O. H. Pilkey, eds., Geology of continental slopes: SEPM Spec. Pub. no. 27, p. 61-73.

Natland, M. L., 1933, The temperature and depth distribution of some recent and fossil foraminifera in the southern California region: Bull., Scripps Inst. Oceanography, Tech. Ser., v. 3, p. 225-230.

————, 1957, Paleoecology of west coast Tertiary sediments, *in* H. D. Ladd, ed., Treatise on marine ecology and paleoecology: Geol. Soc. America Mem. 67, v. 2, p. 543-572.

————, 1963, Presidential address; paleoecology and turbidites: Jour. Paleont., v. 37, p. 946-951.

———— and P. H. Kuenen, 1951, Sedimentary history of the Ventura Basin, California, and the action of turbidity currents, *in* J. L. Hough, ed., Turbidity currents and the transportation of coarse sediment into deep water: SEPM Spec. Pub. no. 2, p. 76-107.

Nelson, C. H., et al, 1970, Development of the Astoria canyon-fan physiography and comparison with similar systems: Marine Geol., v. 8, p. 259-291.

———— and L. D. Kulm, 1973, Submarine fans and deep-sea channels, *in* G. V. Middleton and A. H. Bouma, eds., Turbidites and deep-water sedimentation: Pac. Sec. SEPM Short Course, p. 39-78.

———— E. Mutti, and F. Ricci Lucchi, 1974, Criteria for distinguishing thin-bedded turbidites deposited in proximal overbank and distal basin plain environments: Geol. Soc. America Abs. with Programs, v. 6, no. 7, p. 2887-2888.

———— ———— and ————, 1975, Comparison of proximal and distal thin-bedded turbidites with current winnowed deep sea sands: Nice, France, 9th Internat. Sedimentology Cong., v. 2, p. 317-324.

———— and T. H. Nilsen, 1974, Depositional trends of modern and ancient deep-sea fans, *in* R. H. Dott, Jr., and R. H. Shavers, eds., Modern and ancient geosynclinal sedimentation: SEPM Spec. Pub. no. 19, p. 69-91.

———— W. R. Normark, A. H. Bouma, and P. R. Carlson, 1978, Thin-bedded turbidites in modern submarine canyons and fans, *in* D. J. Stanley and G. Kelling, eds., Sedimentation in submarine canyons, fans, and trenches: Stroudsburg, Pa., Dowden, Hutchinson and Ross, Inc., p. 177-189.

————, 1976, Late Pleistocene and Holocene depositional trends, processes, and history of Astoria deep-sea fan, northeast Pacific: Marine Geol., v. 20, p. 129-173.

Niem, A. R., 1976, Patterns of flysch deposition and deep-sea fans in the lower Stanley Group (Mississippian), Ouachita Mountains, Oklahoma and Arkansas: Jour. Sed. Petrology, v. 46, p. 633-646.

Nilsen, T. H., 1979, Sedimentology of the Butano Sandstone Santa Cruz Mountains, California, *in* Geology of the Santa Cruz Mountains: Calif. Field Trip Guidebook, Cordilleran Sec., Geol. Soc. America, p. 30-39.

———— J. A. Bartow, E. Stump, and M. H. Link, 1977, New occurrences of dish structures in the stratigraphic record: Jour. Sed. Petrology, v. 47, p. 1299-1304.

———— and E. E. Brabb, 1979, Geology of the Santa Cruz Mountains, California: Field Trip Guidebook, Geol. Soc. America Cordilleran Section, 97 p.

———— and T. W. Dibblee, Jr., 1979, Geology of the Central Diablo Range between Hollister and New Idria, California: Field Trip Guidebook, Cordilleran Sec., Geol. Soc. America, 106 p.

———— and M. H. Link, 1975, Stratigraphy, sedimentology and offset along the San Andreas fault of Eocene to Lower Miocene strata of the northern Santa Lucia Range and the San Emigdio Mountains, Coast Ranges, central California, *in* D. S. Weaver, G. R. Hornaday, and A. Tipton, eds., Paleogene symposium, conference in future energy horizons of the Pacific Coast, p. 367-400.

———— and T. R. Simoni, 1973, Deep-sea fan paleocurrent patterns of the Eocene Butano Sandstone, Santa Cruz Mountains, California: Jour. Research U.S. Geol. Survey, v. 1, p. 439-452.

————— and J. H. Stewart, 1980, The Antler orogeny - Mid-Paleozoic tectonism in western North America, Penrose conference report: Geology, v. 8, p. 298-302.

Normark, W. R., 1970, Channel piracy on Monterey deep-sea fan: Deep-Sea Research, v. 17, p. 837-846.

—————, 1970, Growth patterns of deep-sea fans: AAPG Bull., v. 54, p. 2170-2195.

—————, 1971, Mini-topography of deep-sea fans; geometric considerations for facies interpretations in turbidites: Field Trip Guidebook, Pac. Sec., SEPM, p. 22-36.

—————, 1974, Submarine canyons and fan valleys; factors affecting growth patterns of deep-sea fans, in R. H. Dott, Jr., and R. H. Shaver, eds., Modern and ancient geosynclinal sedimentation: SEPM Spec. Pub. no. 19, p. 56-68.

—————, 1974, Ranger submarine slide, northern Sebastian Vizcaino Bay, Baja California, Mexico: Geol. Soc. America Bull., v. 85, p. 781-784.

—————, 1978, Fan valleys, channels and depositional lobes on modern submarine fans; characters for recognition of sandy turbidite environments: AAPG Bull., v. 62, p. 912-931.

————— and F. H. Dickson, 1976, Man-made turbidity currents in Lake Superior: Sedimentology, v. 23, p. 815-831.

————— G. R. Hess, and F. N. Spiess, 1978, Mapping of small scale (outcrop-size) sedimentological features on modern submarine fans: Proc., 10th Ann. Offshore Tech. Conf., v. 1, p. 593-597.

————— and D. J. W. Piper, 1969, Deep-sea fan-valleys past and present: Geol. Soc. America Bull., v. 80, p. 1859-1866.

————— and —————, 1972, Sediments and growth pattern of Navy deep-sea fan, San Clemente Basin, California Borderland: Jour. Geology, v. 80, p. 198-223.

————— ————— and G. R. Hess, 1979, Distributary channels, sand lobes, and mesotopography of Navy submarine fan, California borderland, with applications to ancient fan sediments: Sedimentology, v. 26, p. 749-774.

O'Brien, N. R., K. Nakazawa, and S. Toluhashi, 1980, Use of clay fabric to distinguish turbiditic and hemipelagic siltstones and silts: Sedimentology, v. 27, p. 47-61.

Ojakangas, R. W., 1964, Petrology and sedimentation of the Cretaceous Sacramento Valley Sequence, Cache Creek, California: Stanford Univ., unpub. Ph.D. dissert., 190 p.

—————, 1968, Cretaceous sedimentation, Sacramento Valley, California: Geol. Soc. America Bull., v. 79, p. 973-1008.

Onions, D., and G. V. Middleton, 1968, Dimensional grain orientation of Ordovician turbidite grey-wackes: Jour. Sed. Petrology, v. 38, p. 164-174.

Padgett, G., R. Ehrlich, and M. Moody, 1977, Submarine debris flow deposits in an extensional setting - Upper Devonian of western Morocco: Jour. Sed. Petrology, v. 47, p. 811-818.

Paine, W. R., 1966, Stratigraphy and sedimentation of subsurface Hackberry wedge and associated beds of southwestern Louisiana: Trans., Gulf Coast Assoc. Geol. Socs., v. 16, p. 261-274.

Palmer, H. D., 1976, Erosion of submarine outcrops, LaJolla submarine canyon, California: Geol. Soc. America Bull., v. 87, p. 427-432.

Parea, G. C., and F. Ricci Lucchi, 1975, Turbidite key beds as indicators of ancient deep-sea plains: Nice, France, 9th Intern. Sedimentology Cong., p. 235-240.

Parkash, B., 1970, Downcurrent changes in sedimentary structures in Ordovician turbidite grey-wackes: Jour. Sed. Petrology, v. 40, p. 572-590.

————— and G. V. Middleton, 1970, Downcurrent textural changes in Ordovician turbidite grey-wackes: Sedimentology, v. 14, p. 259-293.

Parker, J. R., 1975, Lower Tertiary sand development in the central North Sea, in A. W. Woodland, ed., Petroleum and the Continental Shelf of northwest Europe; 1, Geology: London, Applied Sci. Pub., p. 447-452.

—————, 1977, Deep-sea sands, in G. D. Hobson, ed., Developments in petroleum geology I: London, Applied Sci. Pub., p. 225-242.

Parsons, M. G., 1975, The geology of the Laurentian fan and Scotia Rise, in C. J. Yorath, E. R. Parker and D. J. Glass, eds., Canada's continental margins and offshore petroleum exploration: Canadian Soc. Petroleum Geology Mem. no. 4, p. 155-167.

Passega, R., 1954, Turbidity currents and petroleum exploration: AAPG Bull., v. 38, p. 1871-1887.

Payne, M. W., 1976, Basinal sandstone facies Delaware Basin, west Texas and southeast New Mexico: AAPG Bull., v. 60, p. 517-527.

Payre, X., and O. Serra, 1979, A case history; turbidites recognized through dipmeter: London, Trans., 6th European Symp. SPWLA, Paper K.

Pescatore, T., 1978, The Irpinids; a model of tectonically controlled fan and base-of-slope sedimentation in southern Italy, in D. J. Stanley and G. Kelling, eds., Sedimentation in submarine canyons, fans, trenches: Stroudsburg, Pa., Dowden, Hutchinson and Ross, Inc., p. 325-339.

Pessagno, E. A., Jr., 1974, Radiolarian zonation and stratigraphy of the Upper Cretaceous portion of the Great Valley Sequence, California coast ranges: Micropaleontology Spec. Paper 1, 233 p.

Peterson, G. L., 1965, Implications of two Cretaceous mass transport deposits, Sacramento Valley, California: Jour. Sed. Petrology, v. 35, p. 401-407.

Pettijohn, F. J., 1936, Early Precambrian varved slate in northwestern Ontario: Geol. Soc. America Bull., v. 47, p. 621-628.

———, 1950, Turbidity currents and greywackes - a discussion: Jour. Geology, v. 58, p. 169-171.

Phleger, F. B., 1960, Ecology and distribution of recent foraminifera: Baltimore, Md., Johns Hopkins Univ. Press, 297 p.

——— F. L. Parker, and J. F. Peirson, 1953, North Atlantic foraminifera: Swedish Deep-Sea Exped., Repts.: v. 7, p. 3-122.

Picha, F., 1974, Ancient submarine canyons of the Carpathian Miogeosyncline: SEPM Special Pub. no. 19, p. 126-127.

———, 1979, Ancient submarine canyon of Tethyan continental margins, Czechoslovakia: AAPG Bull., v. 63, p. 67-86.

——— and L. M. Cline, 1973, Radiographic investigation of laminated and ripple cross-laminated flysch sandstones, Ouachita Mountains, Oklahoma: Jour. Sed. Petrology, v. 43, p. 466-470.

——— and A. R. Niem, 1974, Distribution and extent of beds in flysch deposits, Ouachita Mountains, Arkansas and Oklahoma: Jour. Sed. Petrology, v. 44, p. 328-335.

Pickering, K. T., 1979, Possible retrogressive flow slide deposits from the Kongsfjord Formation; a Precambrian submarine fan, Finnmark, N. Norway: Sedimentology, v. 26, p. 295-306.

Piper, D. J. W., 1970, Transport and deposition of Holocene sediment of La Jolla deep sea fan, California: Marine Geol., v. 8, p. 211-227.

———, 1970, A Silurian deep sea fan deposit in western Ireland and its bearing on the nature of turbidity currents: Jour. Geology, v. 78, p. 509-522.

———, 1972, Turbidite origin of some laminated mudstones: Geol. Mag., v. 109, p. 115-126.

———, 1973, Sedimentology of silt turbidites from the Gulf of Alaska, in L. D. Kulm et al, eds., Initial reports of the Deep Sea Drilling Project, v. XVIII: Washington, D.C., U.S. Govt. Printing Office, p. 847-867.

———, 1975, Late Quaternary deep water sedimentation off Nova Scotia and western Grand Banks, in C. J. Yorath, E. R. Parker and D. J. Glass, eds., Canada's continental margins and offshore petroleum exploration: Canadian Soc. Petroleum Geology Mem. no. 4, p. 195-204.

———, 1975, A reconnaissance of the sedimentology of lower Silurian mudstones, English Lake District: Sedimentology, v. 22, p. 623-630.

———, 1978, Turbidite muds and silts on deep sea fans and abyssal plains, in D. J. Stanley and G. Kelling, eds., Sedimentation in submarine canyons, fans and trenches: Stroudsburg, Pa., Dowden, Hutchinson and Ross, p. 163-176.

——— and C. D. Brisco, 1975, Deep-water continental-margin sedimentation, DSDP Leg 28, Antarctica, in D. E. Hayes, et al, eds., Initial reports of the Deep Sea Drilling Project: Washington D.C., U.S. Govt. Printing Office, v. 28, p. 727-755.

——— R. von Huene, and J. R. Duncan, 1973, Late Quaternary sedimentation in the active eastern Aleutian trench: Geology, v. 1, p. 19-22.

——— and W. R. Normark, 1971, Re-examination of a Miocene deep-sea fan and fan-valley, southern California: Geol. Soc. America Bull., v. 82, p. 1823-1830.

——— ——— and J. C. Ingle, Jr., 1976, The Rio Dell Formation; a Plio-Pleistocene basin slope deposit in northern California: Sedimentology, v. 23, p. 309-328.

——— A. G. Panagos, and G. G. Pe, 1978, Conglomeratic Miocene flysch, western Greece: Jour. Sed. Petrology, v. 48, p. 117-125.

Poag, C. W., 1979, Stratigraphy and depositional environments of Baltimore Canyon trough: AAPG Bull., v. 63, p. 1452-1466.

Poole, F. G., 1974, Flysch deposits of the Antler foreland basin, western United States, in W. R. Dickinson, ed., Tectonics and sedimentation: SEPM Spec. Pub. no. 22, p. 58-82.

Pratt, R. M., 1968, Atlantic continental shelf and slope of the United States - physiography and sediments of the deep-sea basin: U.S. Geol. Survey Prof. Paper 529-B, 44 p.

Redwine, L. E., ed., 1952, Cenozoic correlation section paralleling north and south margins, western Ventura Basin, from Point Conception to Ventura and Channel Islands, California: Sub-Comm. on Cenozoic, Geology Names and Correlation Comm., Pac. Secs. AAPG, 2 sheets.

Rennie, E. W., Jr., ed., 1972, Guidebook, geology and oil fields, west side central San Joaquin Valley: Pac. Sec; AAPG, SEG, SEPM, 102 p.

Ricci Lucchi, F., 1965, Alcune strutture di re-sedimentazione nella formazione marnosa-arenacea romagnola: Giorn. Geol., v. 33, p. 265-292.

———, 1969, Channelized deposits in the Middle Miocene flysch of Romagna (Italy): Giorn. Geol., v. 36, p. 203-282.

———, 1975, Sediment disperal in turbidite basins; examples from the Miocene of northern Apennines: Nice, France, Rapports, 9th Internat. Sedimentology Cong., v. 1, p. 347-352.

———, 1975, Depositional cycles in two turbidite formations of northern Apennines (Italy): Jour. Sed. Petrology, v. 45, p. 3-43.

———, 1975, Miocene paleogeography and basin analysis in the Periadriatic Apennines, in C.

588

Squyres, ed., Geology of Italy: Tripoli, Petroleum Explor. Soc. of Libya, p. 5-111.

————, 1978, Turbidite dispersal in a Miocene deep-sea plain; the Marnoso-Arenacea of the northern Apennines: Geologie en Mijnbouw, v. 57, p. 559-576.

Rich, E. I., and R. V. Ingersoll, 1975, Spring field trip guide and road log - western margin Sacramento Valley-Clear Lake area: Northern Calif. Geol. Soc., Pac. Sec., AAPG, 21 p.

Rich, R. L., 1950, Flow markings, groovings and intra-stratal crumplings as criteria for recognition of slope deposits, with illustrations from Silurian rocks of Wales: AAPG Bull., v. 34, p. 717-741.

Roberson, M. I., 1964, Continuous seismic profiler survey of oceanographer, Gilbert, and Lydonia submarine canyons, Georges Bank: Jour. Geophysics Research, v. 69, p. 4779-4789.

Rocheleau, M., and J. Lajoie, 1974, Sedimentary structures in resedimented conglomerate of the Cambrian flysch, L'Islet, Quebec Appalachians: Jour. Sed. Petrology, v. 44, p. 826-836.

Rodolfo, K. S., 1969, Sediments of the Andaman Basin, northeastern Indian Ocean: Marine Geol., v. 7, p. 371-402.

Rona, P. A., 1969, Middle Atlantic continental slope of United States; deposition and erosion: AAPG Bull., v. 53, p. 1453-1969.

————, 1970, Submarine canyon origin on upper continental slope off Cape Hatteras: Jour. Geology, v. 78, p. 141-152.

Ross, D. A., E. T. Degens, and J. MacIlvaine, 1970, Black Sea; recent sedimentary history: Science, v. 170, p. 163-165.

————, et al, 1978, Sedimentation and structure of the Nile Cone and Levant Platform area, in D. J. Stanley and G. Kelling, eds., Sedimentation in submarine canyons, fans and trenches: Stroudsburg, Pa., Dowden, Hutchinson and Ross, Inc., p. 261-275.

Ruddiman, W. F., F. A. Bowles, and B. Molnia, 1972, Maury Channel and fan: 24th Internat. Geol. Cong., Sec. 8, p. 100-108.

Rupke, N. A., 1969, Aspects of bed thickness in some Eocene turbidite sequences, Spanish Pyrenees: Jour. Geology, v. 77, p. 482-484.

————, 1972, Geologic studies of an Early and Middle Eocene flysch formation, southwestern Pyrenees, Spain: Princeton Univ. Ph.D. thesis, 208 p.

————, 1975, Deposition of fine-grained sediments in the abyssal environment of the Algero-Balearic Basin, western Mediterranean Sea: Sedimentology, v. 22, p. 95-109.

————, 1976, Large-scale slumping in a flysch basin, southwestern Pyrenees: Jour. Geol. Soc. London, v. 132, p. 121-130.

————, 1976, Sedimentology of very thick calcarenite-marlstone beds in a flysch succession, southwestern Pyrenees: Sedimentology, v. 23, p. 43-65.

————, 1977, Growth of an ancient deep-sea fan: Jour. Geology, v. 85, p. 725-744.

———— and D. J. Stanley, 1974, Distinctive properties of turbiditic and hemipelagic mud layers in the Algero-Balearic Basin, western Mediterranean Sea: Smithsonian Contrib. to Earth Sci., v. 13, 40 p.

———— ———— and R. Stuckenrath, 1974, Late Quaternary rates of abyssal mud deposition in the western Mediterranean Sea: Marine Geol., v. 17, p. M9-M16.

Ryan, W. B. F., and B. C. Heezen, 1965, Ionian Sea submarine canyons and the 1908 Messina turbidity current: Geol. Soc. America Bull., v. 76, p. 915-932.

———— K. Venkatarathnam, and F. C. Wezel, 1973, Mineralogical composition of Nile Cone, Mediterranean Ridge, and Strabo Trench sandstones and clays, in Initial reports of Deep Sea Drilling Project, v. XIII: Washington, D.C., U.S. Govt. Printing Office, p. 731-746.

Sabate, R. W., 1968, Pleistocene oil and gas in coastal Louisiana: Trans., Gulf Coast Assoc. Geol. Socs., v. 18, p. 373-386.

Sagri, M., 1979, Upper Cretaceous carbonate turbidites of the Alps and Apennines deposited below the calcite compensation level: Jour. Sed. Petrology, v. 49, p. 23-28.

Sanders, J. E., 1965, Primary sedimentary structures formed by turbidity currents and related sedimentation mechanisms, in G. V. Middleton, ed., Primary sedimentary structures and their hydrodynamic interpretation: SEPM Spec. Pub. no. 12, p. 192-219.

Sangree, J. B., et al, 1978, Recognition of continental-slope seismic facies, offshore Texas-Louisiana, in A. H. Bouma, G. T. Moore, and J. M. Coleman, eds., Framework, facies, and oil-trapping characteristics of the upper continental margin: AAPG Stud. in Geology No. 7, p. 87-116.

Sarnthein, M., and C. Bartolini, 1973, Grain size studies on turbidite components from Tyrrhenian deep sea cores: Sedimentology, v. 20, p. 425-436.

———— and L. Diester-Haass, 1977, Eolian - sand turbidites: Jour. Sed. Petrology, v. 47, p. 868-890.

Schneider, E. D., et al, 1967, Further evidence of contour currents in the western North Atlantic: Earth and Planetary Sci. Letters, v. 2, p. 351-359.

Scholl, D. W., and M. S. Marlow, 1974, Sedimentary sequence in modern Pacific trenches and the deformed circum-Pacific eugeosyncline, in R. H. Dott, Jr., and R. H. Shaver, eds., Modern and ancient geosynclinal sedimentation: SEPM Spec. Pub. no. 19, p. 193-211.

———— E. C. Buffington, D. M. Hopkins, and T. R. Alpha, 1970, The structure and origin of the large submarine canyons of the Bering Sea: Marine Geol., v. 8, p. 187-210.

Scholle, P. A., 1971, Sedimentology of fine-grained deep-water carbonate turbidites, Monte Antola Flysch (Upper Cretaceous), northern Apennines: Geol. Soc. America Bull., v. 82, p. 629-658.

———— , 1971, Diagenesis of deep-water carbonate turbidites, Upper Cretaceous Monte Antola Flysch, northern Apennines, Italy: Jour. Sed. Petrology, v. 41, p. 233-250.

Schweller, W. J., and L. D. Kulm, 1978, Depositional patterns and channelized sedimentation in active eastern Pacific trenches, in D. J. Stanley and G. Kelling, eds., Sedimentation in submarine canyons, fans, and trenches: Stroudsburg, Pa., Dowden, Hutchinson and Ross, Inc., p. 311-324.

Scott, K. M., 1966, Sedimentology and dispersal pattern of a Cretaceous flysch sequence, Patagonian Andes, southern Chile: AAPG Bull., v. 50, p. 72-107.

Scott, R. M., 1978, Physical and biogenic characteristics of sediments from Hueneme Submarine Canyon, California, in D. J. Stanley and G. Kelling, eds., Sedimentation in submarine canyons, fans, and trenches: Stroudsburg, Pa., Dowden, Hutchinson and Ross, Inc., p. 51-64.

Sedimentation Seminar, 1969, Bethel Sandstone (Mississippian) of western Kentucky and south-central Indiana, a submarine-channel fill: Kentucky Geol. Survey Rept. Invest., v. 11, 24 p.

Seiglie, G. A., P. N. Froelich, and O. H. Pilkey, 1976, Deep-sea sediments of Navidad Basin; correlation of sand layers: Deep-Sea Research, v. 23, p. 89-101.

Seilacher, A., 1962, Paleontological studies on turbidite sedimentation and erosion: Jour. Geology, v. 70, p. 227-234.

———— , 1964, Biogenic sedimentary strucures, in J. Imbrie and N. Newell, eds., Approaches to paleoecology: New York, John Wiley & Sons, p. 296-316.

———— , 1967, Bathymetry of trace fossils: Marine Geol., v. 5, p. 413-428.

Sellevoll, M. A., and E. Sundvor, 1974, The origin of the Norwegian Channel - a discussion based on seismic measurements: Canadian Jour. Earth Sci., v. 11, p. 224-231.

Sestini, G., 1970, Flysch facies and turbidite sedimentology: Sed. Geology, v. 4, p. 559-597.

Shanmugam, G., and G. L. Benedict, III, 1978, Fine-grained carbonate debris flow, Ordovician basin margin, southern Appalachians: Jour. Sed. Petrology, v. 48, p. 1233-1240.

———— and K. R. Walker, 1978, Tectonic significance of distal turbidites in the Middle Ordovician Blockhouse and lower Sevier formations in east Tennessee: Am. Jour. Sci., v. 278, p. 551-578.

Sheldon, P. G., 1928, Some sedimentation conditions in Middle Portage rocks: Am. Jour. Sci., v. 15, p. 243-252.

Shepard, F. P., 1937, Daly's submarine canyon hypothesis: Am. Jour. Sci., v. 38, p. 369-379.

———— , 1951, Mass movements in submarine canyon heads: Trans., Am. Geophys. Union, v. 32, p. 405-418.

———— , 1951, Transportation of sand into deep water, in J. L. Hough, ed., Turbidity currents, and the transportation of coarse sediment into deep water: SEPM Spec. Pub. no. 2, p. 53-65.

———— , 1960, Deep-sea sands: Copenhagen, 21st Internat. Geol. Cong., pt. 23, p. 26-42.

———— , 1963, Submarine canyon, in M. N. Hill, ed., The sea; ideas and observations on progress in the study of the seas, v. 3: New York, Interscience, p. 480-506.

———— , 1963, Submarine geology: New York, Harper and Row Pub. 557 p.

———— , 1965, Importance of submarine valleys in funneling sediments to the deep sea, in M. Sears, ed., Progress in oceanography, v. 3, p. 321-332.

———— , 1966, Meander in valley crossing a deep-ocean fan: Science, v. 154, p. 385-386.

———— , 1972, Submarine canyons: Earth Sci. Review, v. 8, p. 1-12.

———— , 1973, Sea floor off Magdalena Delta and Santa Marta area, Columbia: Geol. Soc. America Bull., v. 84, p. 1955-1972.

———— , 1979, Submarine slopes and canyons on north side, St. Croix Island: Marine Geol., v. 31, p. M69-M76.

———— and E. C. Buffington, 1968, La Jolla submarine fan-valley: Marine Geol., v. 6, p. 107-143.

———— and R. F. Dill, 1966, Submarine canyons and other sea valleys: Chicago, Rand McNally, 381 p.

———— ———— and U. Von Rad, 1969, Physiography and sedimentary processes of La Jolla submarine fan and fan-valley, California: AAPG Bull., v. 53, p. 390-420.

———— and G. Einsele, 1962, Sedimentation in San Diego Trough and contributing submarine canyons: Sedimentology, v. 1, p. 81-133.

———— and K. O. Emery, 1973, Congo submarine canyon and fan valley: AAPG Bull., v. 57, p. 1679-1691.

———— and N. F. Marshall, 1969, Currents in La Jolla and Scripps Submarine Canyons: Science, v. 165, p. 177-178.

———— and ———— , 1973, Currents along floors of submarine canyons: AAPG Bull., v. 57, p. 244-264.

———— and ———— , 1975, Dives into outer Coronado Canyon system: Marine Geol., v. 18,

p. 313-323.

———— and ————, 1978, Currents in submarine canyons and sea valleys, *in* D. J. Stanley and G. Kelling, eds., Sedimentation in submarine canyons, fans, and trenches: Stroudsburg, Pa., Dowden, Hutchinson and Ross, Inc., p. 3-14.

———— ———— and P. A. McLoughlin, 1974, Currents in submarine canyons: Deep-Sea Research, v. 21, p. 691-706.

———— ———— ———— and G. G. Sullivan, 1979, Currents in submarine canyons and other sea valleys: AAPG Stud. in Geology No. 8, 173 p.

Sheridan, R. E., X. Golovchenko, and J. I. Ewing, 1974, Late Miocene turbidite horizon in Blake-Bahama basin: AAPG Bull., v. 58, p. 1797-1805.

Shideler, G. L., 1970, Provenance of Johns Valley boulders in Late Paleozoic Ouachita facies, southeastern Oklahoma and southwestern Arkansas: AAPG Bull., v. 54, p. 789-806.

Sibley, D. F., and K. J. Pentony, 1978, Provenance variation in turbidite sediments, Sea of Japan: Jour. Sed. Petrology, v. 48, p. 1241-1247.

Siebold, E., 1972, Sedimentary regimes at continental margins and in abyssal cones in Indian Ocean: 24th Internat. Geol. Cong., Sec. 8, p. 75-84.

Seimers, C. T., 1978, Submarine fan deposition of the Woodbine-Eagle Ford interval (Upper Cretaceous), Tyler County, Texas: Trans., Gulf Coast Assoc. Geol. Soc., v. 28, p. 493-533.

Signorini, R., 1936, Determinazione del senso di sedimentazione degli strati nelle formazioni arenacee dell'appennino settentronale: Bull. Soc. Geol. Italiana, v. 55, p. 259-265.

Silver, E. A., and E. C. Beutner, 1980, Melanges (Penrose Conf. Rept.): Geology, v. 8, p. 32-34.

Skipper, K., 1977, Offshore petroleum developments in northwest Europe - an update: Geosci. Canada, v. 4, p. 31-40.

———— and S. B. Battacharjee, 1978, Backset bedding in turbidites; a further example from the Cloridorme Formation (Middle Ordovician), Gaspe, Quebec: Jour. Sed. Petrology, v. 48, p. 193-201.

———— and G. V. Middleton, 1975, The sedimentary structures and depositional mechanics of certain Ordovician turbidites, Cloridorme Formation, Gaspe Peninsula, Quebec: Canadian Jour. Earth Sci., v. 12, p. 1934-1952.

Sliter, W. V., 1972, Upper Cretaceous planktonic foraminiferal zoogeography and ecology - eastern Pacific margin: Palaeogeog., Palaeoclimatol., Paleoecol., v. 12, p. 15-31.

Smith, G. W., D. G. Howell, and R. V. Ingersoll, 1979, Late Cretaceous trench-slope basins of central California: Geology, v. 7, p. 303-306.

Smith, D. L., and W. R. Normark, 1976, Deformation and patterns of sedimentation, south San Clemente basin, California borderland: Marine Geol., v. 22, p. 175-188.

Snavely, P. D., H. C. Wagner and N. S. MacLeod, 1964, Rhythmic-bedded eugeosynclinal deposits of the Tyee Formation, Oregon Coast Range, *in* D. F. Merriam, ed., Symposium on cyclic sedimentation: Bull. Kansas Geol. Survey No. 169, p. 461-480.

————, et al, 1980, Makah Formation - a deep-marginal-basin sedimentary sequence of Late Eocene and Oligocene age in the northwestern Olympic Peninsula, Washington: U.S. Geol. Survey Prof. Paper No. 1162-B, 31 p.

Spearing, D. R. (compil.), 1974, Summary sheet of sedimentary deposit-chart 7; turbidity current deposits: Geol. Soc. America.

Spotts, J. H., and O. E. Weser, 1964, Directional properties of a Miocene turbidite, California, *in* A. H. Bouma and A. Brouwer, eds., Turbidites: New York, Elsevier, Sci. Pub., p. 198-221.

Srivastava, P., C. W. Stearn, and E. W. Mountjoy, 1972, A Devonian megabreccia at the margin of the ancient wall carbonate complex, Alberta: Bull. Canadian Petroleum Geology, v. 20, p. 412-438.

Stanley, D. J., 1967, Comparing patterns of sedimentation in some modern and ancient submarine canyons: Earth and Planetary Sci. Letters, v. 3, p. 371-380.

————, 1969, Submarine channel deposits and their fossil analogs ('Fluxoturbidites') *in* New concepts of continental margin sedimentation: Am. Geophy. Inst. Short Course Notes, Lecture 9, p. DJS-9-1 to DJS-9-17.

————, 1970, Flyschoid sedimentation on the outer Atlantic margin off northeast North America, *in* J. Lajoie, ed., Flysch sedimentology in North America: Geol. Assoc. Canada Spec. Paper No. 7, p. 179-210.

————, 1970, Bioturbation and sediment failure in some submarine canyons, *in* 3rd European Symposium on Marine Biology: Vie et Milieu, Supplement 22, v. 2, p. 541-555.

————, 1973, Basin plains in the eastern Mediterranean; significance in interpreting ancient marine deposits, I. basin depth and configuration: Marine Geol., v. 15, p. 295-307.

————, 1974, Dish structures and sand flow in ancient submarine valleys, French Maritime Alps: Bull. Centre Rech. Pau-SNPA, v. 8, p. 351-371.

————, 1974, Modern flysch sedimentation in a Mediterranean Island Arc setting, *in* R. H. Dott, Jr., and R. H. Shaver, eds., Modern and ancient geosynclinal sedimentation: SEPM Spec. Pub. no. 19, p. 240-259.

————, 1974, Pebbly mud transport in the head of Wilmington Canyon: Marine Geol. v. 16,

p. M1-M8.

————, 1975, Submarine canyon and slope sedimentation (Gres D'Annot) in the French maritime Alps: Nice, France, 9th Internat. Sediment. Cong., 131 p.

———— and B. A. Hall, 1978, The Bucegi Conglomerate; A Romanian Carpathian submarine slope deposit: Nature, v. 275, p. 60-64.

———— and G. Kelling, 1968, Sedimentation patterns in the Wilmington submarine canyon area, *in* Symposium on ocean sciences and engineering of the Atlantic shelf: Philadelphia, Trans., Mar. Tech. Soc., Delaware Valley Sec., p 127-142.

———— and ————, 1970, Interpretation of a levee-like ridge and associated features, Wilmington submarine canyon, eastern United States: Geol. Soc. America Bull., v. 81, p. 3747-3752.

———— and ————, eds., 1978, Sedimentation in submarine canyons, fans, and trenches: Stroudsburg, Pa., Dowden, Hutchinson and Ross, Inc., 395 p.

———— and A. Maldonado, 1977, Nile Cone; Late Quaternary stratigraphy and sediment dispersal: Nature, v. 266, p. 129-135.

———— H. D. Palmer, and R. F. Dill, 1978, Coarse sediment transport by mass flow and turbidity current processes and downslope transformations in Annot Sandstone canyon-fan valley system, *in* D. J. Stanley and G. Kelling, eds., Sedimentation in submarine canyons, fans, and trenches: Stroudsburg, Pa., Dowden, Hutchinson, Ross, Inc., p. 85-115.

———— H. Sheng, and C. P. Pedraza, 1971, Lower continental Rise east of the Middle Atlantic States; predominant sediment dispersal perpendicular to isobaths: Geol. Soc. America Bull., v. 82, p. 1831-1840.

———— and N. Silverberg, 1969, Recent slumping on the continental slope off Sable Island bank, southeast Canada: Earth and Planetary Sci. Letters, v. 6, p. 123-133.

———— and R. Unrug, 1972, Submarine channel deposits, fluxoturbidites and other indicators of slope and base-of-slope environments in modern and ancient marine basins, *in* J. K. Rigby and W. K. Hamblin, eds., Recognition of ancient sedimentary environments: SEPM Spec. Pub. no. 16, p. 287-340.

Stark, P. H., 1966, Stratigraphy and environment of deposition of the Atoka Formation in the central Ouachita Mountains, Oklahoma: 29th Field Conf., Kansas Geol. Soc.

Stauffer, P. H., 1967, Sedimentologic evidence on Eocene correlations, Santa Ynez Mountains, California: AAPG Bull., v. 51, p. 607-611.

————, 1967, Grain flow deposits and their implications, Santa Ynez Mountains, California:

Jour. Sed. Petrology, v. 37, p. 487-508.

Stewart, R. J., 1976, Turbidites of the Aleutian abyssal plain; mineralogy, provenance, and constraints for Cenozoic motion of the Pacific plate: Geol. Soc. America Bull., v. 87, p. 793-808.

————, 1977, Neogene turbidite sedimentation in Komandorskiy basin, western Bering Sea: AAPG Bull., v. 61, p. 192-206.

Stone, C. G., B. R. Haley, and G. W. Viele, 1973, A guidebook to the geology of the Ouachita Mountains, Arkansas: Little Rock, Ark. Geol. Comm., p. 1-113.

Stow, D. A. V., 1976, Deep water sands and silts on the Nova Scotian Continental Margin: Maritime Sediments, v. 12, p. 81-90.

————, 1979, Distinguishing between fine-grained turbidites and contourites on the Nova Scotian deep water margin: Sedimentology, v. 26, p. 371-387.

———— and A. J. Bowen, 1978, Origin of lamination in deep sea, fine-grained sediments: Nature, v. 274, p. 324-328.

———— and ————, 1980, A physical model for the transport and sorting of fine-grained sediment by turbidity currents: Sedimentology, v. 27, p. 31-46.

———— and J. P. B. Lovell, 1979, Contourites; their recognition in modern and ancient sediments: Earth Sci. Rev., v. 14, p. 251-291.

———— and G. Shanmugam, 1980, Sequence of structures in fine-grained turbidites; comparison of recent deep-sea and ancient flysch sediments: Sed. Geology, v. 25, p. 23-42.

Sullwold, H. H., 1960, Tarzana fan, deep submarine fan of Late Miocene age, Los Angeles County, California: AAPG Bull., v. 44, p. 433-457.

————, 1961, Turbidites in oil exploration, *in* J. A. Peterson and J. C. Osmond, eds., Geometry of sandstone bodies: AAPG Spec. Pub., p. 63-81.

Summerhayes, C., D. Ross, and P. Stoffers, 1977, Nile submarine fan; sedimentation, deformation, and oil potential: Proc., 9th Offshore Tech. Conf., v. 1, p. 35-40.

Surlyk, F., 1978, Submarine fan sedimentation along fault scarps on tilted fault blocks (Jurassic-Cretaceous boundary, East Greenland): Greenland Geol. Surv. Bull. No. 128, 134 p.

————, 1980, Models of resedimented conglomerates (abs.): Bochum, 1st Europe Meeting, Internat. Assoc. Sedimentols. p. 43.

———— and L. B. Clenimensen, 1975, A Valanginian turbidite sequence and its paleogeographical setting (Kuhn, East Greenland): Bull. Geol. Soc. Denmark, v. 24, p. 61-73.

Suzuik, U., 1979, Petroleum geology of the Sea of Japan Northern Honshu: Japanese Assoc. Petroleum Technols. v. 44, p. 59-75 (in Japanese, figs. have English captions).

Swarbrick, R. E., and M. A. Naylor, 1980, The Kathikas melange, S. W. Cyprus; Late Cretaceous submarine debris flows: Sedimentology, v. 27, p. 63-78.

Taira, A., and P. A. Scholle, 1979, Deposition of resedimented sandstone beds in the Pico Formation, Ventura Basin, California, as interpreted from magnetic fabric measurements: Geol. Soc. America Bull., pt. 1, v. 90, p. 952-962.

Taylor, J. C., 1976, Geologic appraisal of the petroleum potential of offshore southern California; the borderland compared to onshore coastal basins: U.S. Geol. Survey Circ. 730, 43 p.

Thomas, A. N., P. J. Walmsley, and D. A. L. Jenkins, 1974, Forties field, North Sea: AAPG Bull., v. 58, p. 396-406.

Thomson, A. F., 1978, Summary of the geologic history of the Marathon Geosyncline, in E. F. McBride and S. J. Mazzullo, eds., Tectonics and Paleozoic facies of the Marathon Geosyncline, west Texas: Permian Basin Sec., SEPM, p. 79-88.

———— and M. R. Thomasson, 1964, Sedimentology and stratigraphy of the Dimple Limestone, Marathon Region, Texas, in The filling of the Marathon Geosyncline, symposium and guidebook: Permian Basin Sec., SEPM, p. 22-30.

———— and ———— , 1969, Shallow to deep water facies development in the Dimple Limestone (Lower Pennsylvanian) Marathon Region, Texas, in G. M. Friedman, ed., Depositional environments in carbonate rocks: SEPM Spec. Pub. no. 14, p. 57-78.

Thompson, A. M., 1972, Shallow-water distal turbidites in Ordovician flysch, central Appalachian Mountains, U.S.A.: Sec. 6, 24th Internat. Cong., p. 89-99.

Tillman, R. W., R. M. Scott and J. Rennison, 1979, Core description and interpretation, Cities Service Nixon Community No. 1, California: Calif. Well Sample Depository Spec. Pub. no. 2, p. 1-25.

Trimonis, E. S., and K. M. Shimjus, 1970, Sedimentation at the head of a submarine canyon: Acad. Sci. U.S.S.R., Oceanology, Engl. Ed., v. 10, p. 74-85.

Uchupi, E., and J. A. Austin, Jr., 1979, The stratigraphy and structure of the Laurentian Cone region: Canadian Jour. Earth Sci., v. 16, p. 1726-1752.

Unrug, R., 1963, Istebna Beds - a fluxoturbidite formation in the Carpathian Flysch: Ann. Soc. Geol. Pol., v. 33, p. 49-92.

———— , 1964, Turbidites and fluxoturbidites in the Morvia-Silesia Kulm zone: Acad. Pol. Sci. Bull. Ser. Sc. Geol. Geograph., v. 12, p. 187-194.

Vallier, T. L., P. J. Harold, and W. A. Girdley, 1973, Provenance and dispersal patterns of turbidite sands in Escanaba Trough, northeastern Pacific Ocean: Marine Geol., v. 15, p. 67-87.

Van Berckel, F. L., 1976, On the origin of submarine canyons: Geologie en Mijnbouw, v. 55, p. 7-17.

Van De Graaff, W. J. E., 1971, Three Upper Carboniferous, limestone-rich, high-destructive, delta systems with submarine fan deposits, Cantabrian Mountain, Spain: Leidse Geol. Meded., decl 46, blg, p. 157-235.

Van De Kamp, P. C. and J. D. Harper, 1969, The Cretaceous and Lower Tertiary of the Wheeler Springs-Ojai area, Ventura Basin, California, in Guidebook Upper Sespe Creek field trip: Pac. Sec., SEPM, p. 24-26.

————————J. J. Conniff, and D. A. Morris, 1974, Facies relationships in the Eocene-Oligocene in the Santa Ynez Mountains, California: London, Jour. Geol. Soc., v. 130, p. 545-565.

Van Der Lingen, G. J., 1969, The turbidite problem: New Zealand Jour. Geology Geophysics, v. 12, p. 7-50.

Van Hoorn, B., 1969, Submarine canyon and fan deposits in the Upper Cretaceous of the south-central Pyrenees, Spain: Geologie en Mijnbouw, v. 48, p. 67-72.

———— , 1970, Sedimentology and paleogeography of an Upper Cretaceous turbidite basin in the south-central Pyrenees, Spain: Leidse Geol. Meded., v. 45, p. 73-154.

Van Straaten, L. M. J. U., 1964, Turbidite sediments in the southeastern Adriatic Sea, in A. H. Bouma and A. Brouwer, eds., Turbidites: Amsterdam, Elsevier, Sci. Pub., p. 142-147.

———— ,.1968, Turbidites, ash layers, and shell beds in the bathyal zone of the southern Adriatic Sea: Revue Geogreaphie Physique Geologie Dynamique, v. 9, p. 219-240.

Van Vliet, A., 1978, Early Tertiary deepwater fans of Guipuzcoa, Northern Spain, in D. J. Stanley and G. Kelling, eds., Sedimentation in submarine canyons, fans, and trenches: Stroudsburg, Pa., Dowden, Hutchinson, and Ross, Inc., p. 190-210.

Vedros, S. G., and G. S. Visher, 1978, The Red Oak Sandstone; a hydrocarbon producing submarine fan deposit, in J. D. Stanley and G. Kelling, eds., Sedimentation in submarine canyons, fans, and trenches: Stroudsburg, Pa., Dowden, Hutchinson, and Ross, Inc., 292-310.

Von Der Borch, C. C., 1969, Submarine canyons of southeastern New Guinea; seismic and bathymetric evidence for the modes of origin: Deep-Sea Research, v. 16, p. 323-328.

Von Rad, U., 1968, Comparison of sedimentation in the Bavarian Flysch (Cretaceous) and Recent San Diego Trough (California): Jour. Sed. Petrology, v. 38, p. 1120-1154.

Vormelker, R. S., 1979, Mid-Wilcox channel; deep exploration potential: Bull. S. Tex. Geol. Soc., v.

20, p. 10-40.

———— , 1980, Texas Middle Wilcox channel; deep exploration potential: Oil & Gas Jour., Mar. 10, p. 136-138; 140, 143, 146, 148, 151-154.

Walker, J. R., and J. V. Massingill, 1970, Slump features on the Mississippi fan, northeastern Gulf of Mexico: Geol. Soc. America Bull., v. 81, p. 3101-3108.

Walker, R. G., 1965, The origin and significance of the internal sedimentary structures of turbidites: Proc., Yorkshire Geol. Soc., v. 35, p. 1-32.

———— , 1966, Deep channels in turbidite-bearing formations: AAPG Bull., v. 50, p. 1899-1917.

———— , 1966, Shale grit and Grindslow Shales; transition from turbidite to shallow water sediments in the Upper Carboniferous of northern England: Jour. Sed. Petrology, v. 36, p. 90-114.

———— , 1967, Turbidite sedimentary structures and their relationship to proximal and distal depositional environments: Jour. Sed. Petrology, v. 37, p. 25-43.

———— , 1969, Geometrical analysis of ripple-drift cross-lamination: Canadian Jour. Earth Sci., v. 6, p. 383-391.

———— , 1970, Review of the geometry and facies organization of turbidites and turbidite-bearing basins, in J. Lajoie, ed., Flysch sedimentology in North America: Geol Assoc. Canada Spec. Paper No. 7, p. 219-251.

———— , 1971, Nondeltaic depositional environments in the Catskill Clastic Wedge (Devonian) of central Pennsylvania: Geol. Soc. America Bull., v. 82, p. 1305-1326.

———— , 1973, Mopping-up the turbidite mess, in R. N. Ginsburg, ed., Evolving concepts in sedimentology: Baltimore, Md., Johns Hopkins Univ. Press, p. 1-37.

———— , 1975, Generalized facies models for re-sedimented conglomerates of turbidite association: Geol. Soc. America Bull., v. 86, p. 737-748.

———— , 1975, Nested submarine-fan channels in the Capistrano Formation, San Clemente, California: Geol. Soc. America Bull., v. 86, p. 915-924.

———— , 1975, Upper Cretaceous resedimented conglomerates at Wheeler Gorge, California; description and field guide: Jour. Sed. Petrology, v. 45, p. 105-112.

———— , 1976, Facies Models; 1, general introduction: Geosci. Canada, v. 3, p. 21-24.

———— , 1976, Facies Models; 2, turbidites and associated coarse clastic deposits: Geosci. Canada, v. 3, p. 25-36.

———— , 1977, Deposition of Upper Mesozoic resedimented conglomerates and associated turbidites in southwestern Oregon: Geol. Soc. America Bull., v. 88, p. 273-285.

———— , 1978, Deep water sandstone facies and ancient submarine fans; models for exploration for stratigraphic traps: AAPG Bull., v. 62, p. 932-966.

———— , 1978, Deep-water pebbly sandstones and conglomerates - facies and reservoir characteristics: 10th Ann. Offshore Tech. Conf., v. 1, p. 581-586.

———— , ed., 1979, Facies models: Geosci. Canada Repr. Ser. 1, 211 p.

———— , 1979, Turbidites and associated coarse clastic deposits, in R. G. Walker, ed., Facies models: Geosci. Canada Repr. Ser. 1, p. 91-103.

———— and E. Mutti, 1973, Turbidite facies and facies associations, in G. V. Middleton and A. H. Bouma, eds., Turbidites and deep-water sedimentation: Pac. Sec. SEPM, Short Course, p. 119-157.

———— and R. G. Sutton, 1967, Quantitative analysis of turbidites in the Upper Devonian Sonyea Group, New York: Jour. Sed. Petrology, v. 37, p. 1012-1022.

Walmsley, P. J., 1975, The Forties Field, in A. W. Woodland, ed., Petroleum and the continental shelf of northwest Europe: New York, John Wiley and Sons, p. 477-485.

Walthall, B. H., 1967, Stratigraphy and structure, part of Athens Plateau, southern Ouachitas, Arkansas: AAPG Bull., v. 51, p. 504-528.

Walton, E. K., 1967, The sequence of internal structures in turbidites: Scottish Jour. Geology, v. 3, p. 306-317.

Warman, H. R., 1979, Hydrocarbon potential of deep water: Phil. Trans. Royal Soc. London, Ser. A., v. 290, p. 33-42.

Warme, J. E., R. A. Slater, and R. A. Cooper, 1978, Bioerosion in submarine canyons, in D. J. Stanley and G. Kelling, eds., Sedimentation in submarine canyons, fans and trenches: Stroudsburg, Pa., Dowden, Hutchinson, and Ross, Inc., p. 65-72.

Weagant, F. E., 1972, Grimes gas field, Sacramento Valley, California, in R. E. King, ed., Stratigraphic oil and gas fields: AAPG, Mem. 16, p. 428-439.

Webb, G. W., 1977, Stevens and earlier Miocene turbidite sands, San Joaquin Valley, Calif., in W. J. M. Bazeley, ed., Late Miocene, geology and new oil fields of the southern San Joaquin Valley: Guidebook, Pac. Secs. AAPG, SEG, SEPM, p. 37-55.

Weber, J. N., and G. V. Middleton, 1961, Geochemistry of the turbidites of the Normanskill and Charny formations - pt. 1, effect of turbidity currents on the chemical differentiation of turbidites; pt. 2, distribution of trace elements: Geochim. et Cosmochim. Acta, v. 22, p. 200-288.

594

Welsh, W., 1979, A discussion of the criteria for distinguishing proximal from distal turbidites: Sed. Geology, v. 22, p. 121-126.

Weser, O. E., 1974, Sedimentological aspects of strata encountered on Leg 23 in the northern Arabian Sea, *in* R. B. Whitmarsh et al, eds., Initial reports of Deep Sea Drilling Project v. XXIII: Washington, D.C., U.S. Govt. Printing Office, p. 503-519.

———— , 1977, Deep-water sand reservoirs - ancient case histories and modern concepts: AAPG Continuing Educ. Course Notes, 174 p.

———— , 1978, Oil trapping characteristics of turbidites, *in* A. H. Bouma, G. T. Moore, and J. M. Coleman, eds., Framework, facies, and oil trapping characteristics of the upper continental margin: AAPG Stud. in Geology no. 7, p. 227-241.

Whitaker, J. H. McD., 1974, Ancient submarine canyons and fan valleys, *in* R. H. Dott, Jr. and R. H. Shaver, eds., Modern and ancient geosynclinal sedimentation: SEPM Spec. Pub. no. 19, p. 106-125.

Whittle, A. P., and G. A. Short, 1978, The petroleum geology of the Tembungo Field, East Malaysia: Offshore S.E. Asia Conf., Asia Petroleum Explor. Soc., 4 p.

Wilde, P., 1965, Recent sediments of the Monterey deep-sea fan: Berkeley, Univ. California, Hydrol. Eng. Lab. Rept. HEL 2-13, 153 p.

———— W. R. Normark, and T. E. Chase, 1976, Petroleum potential of continental rise off central California - summary, *in* M. T. Halbouty, J. C. Maher, and H. M. Lian, eds., Circum-Pacific energy and mineral resources: AAPG Mem. 25, p. 313-317.

———— ———— and ———— , 1978, Channel sands and petroleum potential of Monterey deep-sea fan, California: AAPG Bull., v. 62, p. 967-983.

Williamson, C. R., 1977, Deep sea channels of the Bell Canyon Formation (Guadalupian) Delaware Basin, Texas-New Mexico: SEPM Permian Basin Sec., Pub. 77-16, v. 1, p. 409-431.

Winn, R. D., Jr., 1978, Upper Mesozoic flysch of Tierre del Fuego and south Georgia Island; a sedimentologic approach to lithosphere plate restoration: Geol. Soc. America Bull., v. 89, p. 533-547.

———— and R. H. Dott, Jr., 1977, Large-scale traction-produced structures in deep-water fan-channel conglomerates in southern Chile: Geology, v. 5, p. 41-44.

———— and ———— , 1978, Submarine-fan turbidites and resedimented conglomerates in a Mesozoic arc-rear marginal basin in southern South America, *in* D. J. Stanley and G. Kelling, eds., Sedimentation in submarine canyons, fans, trenches: Stroudsburg, Pa., Dowden, Hutchinson and Ross, Inc., p. 362-376.

———— and ———— , 1979, Deep-water fan-channel conglomerates of Late Cretaceous age, southern Chile: Sedimentology, v. 26, p. 203-228.

Winterer, E. L., 1970, Submarine valley systems around the Coral Sea Basin, Australia: Marine Geol., v. 8, p. 229-244.

———— J. R. Curray, and M. N. A. Peterson, 1968, Geological history of the Pioneer Fracture Zone with the Delgada deep-sea fan, northeast Pacific: Deep-Sea Research, v. 15, p. 509-520.

———— and D. L. Durham, 1962, Geology of southeastern Ventura Basin, Los Angeles County, Calif.: U.S. Geol. Survey Paper 334-H, p. 275-366.

Woodbury, H. O., J. H. Spotts, and W. H. Akers, 1978, Gulf of Mexico continental-slope sediments and sedimentation, *in* A. H. Bouma, G. T. Moore, and J. M. Coleman, eds., Framework, facies, and oil-trapping characteristics of the upper continental margin: AAPG Stud. in Geology no. 7, p. 117-137.

Woodcock, N. H., 1976, Ludlow series slumps and turbidites and the form of the Montgomery Trough, Powys, Wales: Proc. Geol. Assoc., v. 87, pt. 2, p. 169-182.

Woodland, A. W., ed., 1975, Petroleum and the Continental Shelf of northwest Europe; 1, Geology: London, Applied Sci. Pubs., 500 p.

Yeats, R. S., 1965, Pliocene seaknoll at South Mountain, Ventura Basin, California: AAPG Bull., v. 49, p. 526-546.

Yerkes, R. F., et al, 1965, Geology of the Los Angeles Basin, California - an introduction: U.S. Geol. Survey Prof. Paper 420A, 57 p.

RECOMMENDED READINGS

The following papers are what might be considered the most significant papers on deep water clastic deposits. The papers by Walker, Mutti and Ricci Lucchi, Normark and Seimers and Tillman discuss the specifics of the various models now being utilized to analyze deep water clastic deposits. The papers by Middleton and Hampton give insight into the processes of deposition. The papers by Drummond and others and Scott and Tillman are among the most definitive on facies distribution of California deep water clastics. Cook's paper is one of several on deep water detrital carbonates.

Cook, H. E., 1979, Ancient continental slope sequences and their value in understanding modern slope development, *in* L. J. Doyle and O. H. Pilkey, eds., Geology of continental slopes: SEPM Spec. Pub. no. 27, p. 287-305.

Drummond, K. F., E. W. Christensen, and K. D. Berry, 1976, Upper Cretaceous lithofacies model, Sacramento Valley, California: AAPG Pac. Sec. Misc. Pub. no. 24, p. 76-88.

Hampton, M. A., 1972, The role of subaqueous debris flows in generating turbidity currents: Jour. Sed. Petrology, v. 42, p. 775-793.

Middleton, G. V., and M. A. Hampton, 1976, Subaqueous sediment transport and deposition by sediment gravity flows, *in* D. J. Stanley and D. J. P. Swift, eds., New York, Science, p. 197-218.

Mutti, E., and F. Ricci Lucchi, 1978, Turbidites of the Northern Appenines; introduction to facies analysis, T. H. Nilsen, translator: Internat. Geol Review, v. 20, p. 125-166.

Normark, W. R., 1978, Fan valleys, channels and depositional lobes on modern submarine fans: characters for recognition of sandy turbidite environments: AAPG Bull., v. 62, p. 912-931.

Scott, R. M., and R. W. Tillman, 1981, Stevens Sandstone (Miocene), San Joaquin Basin, California, *in* Deep water clastic sediments, a core workshop: San Francisco, SEPM Core Workshop No. 2, p. 116-248.

Siemers, C. T., and R. W. Tillman, 1981, Deep water clastic sediments; an introduction to the core workshop and review of depositional models, *in* Deep water clastic sediments, a core workshop: San Francisco, SEPM Core Workshop No. 2, p. 1-19.

Walker, R. G., 1975, Generalized facies models for resedimented conglomerates of turbidite association: Geol. Soc. America Bull., v. 86, p. 737-748.

Walker, R. G., 1978, Deep water sandstone facies and ancient submarine fans; models for exploration for stratigraphic traps: AAPG Bull, v. 62, p. 932-966.

Walker, R. G., 1979, Turbidites and associated coarse clastic deposits, *in* R. G. Walker, ed., Facies models: Geoscience Canada Rep. Ser. 1, p. 91-103.

ACKNOWLEDGEMENTS

This bibliography was prepared by Dr. R. W. Tillman and Dr. Syed A. Ali. Typing and editing were provided by Darlene Haynes, Mikki Brown, Darlene Kastl, Mary Walker, Melinda Everley, and Paige Graening.